ESSENTIAL READINGS IN

MAGNESIUM TECHNOLOGY

ESSENTIAL READINGS IN
MAGNESIUM TECHNOLOGY

Edited by

Suveen N. Mathaudhu, Alan A. Luo, Neale R. Neelameggham,
Eric A. Nyberg, Wim H. Sillekens

Editors
Suveen N. Mathaudhu
Alan A. Luo
Neale R. Neelameggham
Eric A. Nyberg
Wim H. Sillekens

ISBN 978-3-319-48588-1 ISBN 978-3-319-48099-2 (eBook)
DOI 10.1007/978-3-319-48099-2

Chemistry and Materials Science: Professional

Published by Springer International Publishers, Switzerland, 2016
Reprint of the original edition published by John Wiley & Sons, Inc., 2014, 978-1-118-85894-3

Printed on acid-free paper

This Springer imprint is published by Springer Nature
The registered company is Springer International Publishing AG
The registered company address is: Gewerbestrasse 11, 6330 Cham, Switzerland

TABLE OF CONTENTS
Essential Readings in Magnesium Technology

Part 1: Magnesium Technology History and Overview

Part 2: Electrolytic and Thermal Primary Production

Part 3: Melting, Refining, Recycling, and Life-Cycle Analysis

Part 4: Casting and Solidification

Part 5: Alloy and Microstructural Design

Part 6: Wrought Processing

Part 7: Modeling and Simulation

Part 8: Joining

Part 9: Corrosion, Surface Treatment, and Coating

PREFACE

"Magnesium will lighten the tasks of man in countless ways as yet undreamed of, except in the minds of far-seeing engineers [...] who are already planning the future."

– 1942 Dow Chemical Advertisement

Due to their extraordinarily low densities, magnesium ("Mg") and its alloys have continued to be the focus of intensive research and development over the past century though widespread application has been restricted by property, cost, and performance limitations with respect to other metallic materials such as Al-alloys, Ti-alloys, and ferrous alloys. However, in light of Mg's availability and global efforts to reduce weight in the transportation sector, there has been a tremendous resurgence in research, development, and applications of these remarkable alloys.

Along with this renaissance came renewed interest from government, academia, and industry, and in the year 2000, the first Magnesium Technology Symposium was held at The Minerals, Metals & Materials Society (TMS) Annual Meeting and Exhibition in Nashville, Tennessee. A proceedings volume with 56 manuscripts was published in parallel with the new symposium, and the tremendous interest that was generated enabled the symposium to be held annually with the *Magnesium Technology* proceedings becoming the *de facto* publication for the magnesium industry.

In your hands (or perhaps on your screen), you have some of the most influential and impactful papers of the 1024 manuscripts published in the proceedings between 2000 and 2012. This single reference chronicles and condenses the major global advances made over the last decade. To select the best papers, a team of editors was selected based on their extensive experience with various aspects of Mg technology, with all editors having served as organizers and editors on *Magnesium Technology* proceedings published during the selected timeframe (Mathaudhu, '11–'12; Luo, '04,'06,'07; Neelameggham, '06–'12; Nyberg, '08–'10; Sillekens, '10–'12).

Manuscripts were comprehensively reviewed for prospective inclusion, and the final papers were selected based on the following three criteria:

1. *General Relevance.* Beginning in 2002, the symposium organizers and session chairs began selecting "best paper" awards for one student and one contributed manuscript. These papers, at the time, represented the latest scientific breakthroughs, and thus are included as "essential reading" (see Table 1 on page xv). Also included based on this criteria are a number of "review" type papers that summarize the state-of-the-art in Mg technology, and thus frame the true impact of recent advances.

2. *Scientific Relevance.* In the academic world, the largest indication of impact is based on citations. The *Magnesium Technology* proceedings are indexed by a number of databases, from which citation data were collected and analyzed and papers with a high number of citations were selected for inclusion.

3. *Industrial Relevance.* In 2007, a "best paper" award in application was added to the student and contributed awards. These manuscripts were selected for inclusion due to their representation of leap-forward advances and novel demonstrations of industrial applications.

Introductions to each part further frame the importance and significance of the selected manuscripts. The papers selected for inclusion were divided along nine topical thematic areas, with the papers arranged by subject area rather than chronologically to support the cohesiveness of subtopics within each theme. The nine themes are as follows:

1. Magnesium Technology History and Overview
2. Electrolytic and Thermal Primary Production
3. Melting, Refining, Recycling, and Life-Cycle Analysis
4. Casting and Solidification
5. Alloy and Microstructural Design
6. Wrought Processing
7. Modeling and Simulation
8. Joining
9. Corrosion, Surface Treatment, and Coating

A significant amount of time and resources have gone into this volume, and many thanks are owed to those who have donated their energy to its success. Primary thanks goes to the many authors whose work is presented in this volume, the many symposium organizers and participants who have continually maintained a high level of scholarship and ensured the continual growth of this discipline. The production team at TMS and Wiley, primarily Matt Baker, are acknowledged for their continual support and motivation to see this project to completion. It is our hope that this volume will provide a key resource for those "*far-seeing engineers*" who have, and will, dedicate their livelihoods to the advancement of this amazing metal.

Suveen N. Mathaudhu, U.S. Army Research Office
Alan A. Luo, The Ohio State University
Neale R. Neelameggham, IND LLC
Eric A. Nyberg, Pacific Northwest National Laboratory
Wim H. Sillekens, European Space Agency

Table 1: Magnesium Technology Best Paper Award Winners

PART	AWARD	AUTHOR(S)	TITLE	PAGE
Part 1: Magnesium Technology History and Overview	2013 Best Application Paper	D'Errico F., Screnci A.	High-capacity hydrogen-based green-energy storage solutions for the grid balancing	77
Part 2: Electrolytic and Thermal Primary Production	2010 Best Student Paper	McLean K., Pettingill J., Davis B.	Cathode wetting studies in magnesium electrolysis	107
Part 3: Melting, Refining, Recycling, and Life-Cycle Analysis	2003 Best Student Paper	Aarstad K., Syvertsen M., Engh T.A.	Solubility of fluorine in molten magnesium	181
Part 4: Casting and Solidification	2004 Best Contributed Paper	St.John D.H., Dahle A.K., Abbott T., Nave M.D., Qian M.	Solidification of cast magnesium alloys	193
	2007 Best Student Paper	Lee S.G., Gokhale A.M.	Phenomena of formation of gas induced shrinkage porosity in pressure die-cast Mg-alloys	211
	2005 Best Student Paper	Moore A.R., Torbet C.J., Shyam A., Jones J.W., Walukas D.M., Decker R.F.	Fatigue behavior of thixomolded® magnesium AZ91D using ultrasonic techniques	227
	2010 Best Application Paper	Duygulu O., Ucuncuoglu S., Oktay G., Temur D.S., Yucel O., Kaya A.A.	Development of 1500mm wide wrought magnesium alloys by twin roll casting technique in Turkey	239
Part 5: Alloy and Microstructural Design	2007 Best Application Paper	Klösch G., McKay B.J., Schumacher P.	Preliminary investigation on the grain refinement behavior of ZrB_2 particles in Mg-Al alloys	255
	2006 Best Student Paper	Blake A.H., Cáceres C.H.	Solid solution effects on the tensile behaviour of concentrated Mg-Zn alloys	263
	2002 Best Contributed Paper	Powell B.R., Rezhets V., Balogh M.P., Waldo R.A.	The relationship between microstructure and creep behavior in AE42 magnesium die casting alloy	275
	2013 Best Fundamental Paper	Sasaki T.T., Ohkubo T., Hono K.	Age hardening behavior of Mg-1.2Sn-1.7Zn alloy containing Al	269
	2006 Best Contributed Paper	Suzuki A., Saddock N.D., Jones J.W., Pollock T.M.	Phase transformation and creep of Mg-Al-Ca based die-cast alloys	291
	2011 Best Student Paper	Lee J.-K., Kim S.K.	Fire-proof evaluation of CaO added Mg-3Al, Mg-6Al, and Mg-9Al Mg cast products	305
	2010 Best Fundamental Paper	Rosalie J.M., Somekawa H., Singh A., Mukai T.	Directionally controlled precipitation on twin-boundaries in Mg-Zn-Y alloys	319
	2012 Best Application Paper	Okamoto K., Sasaki M., Takahashi N., Wang Q., Gao Y., Yin D., Chen C.	Applicability of Mg-Zn-(Y,Gd) alloys for engine pistons	325
Part 6: Wrought Processing	2009 Best Student	Ebeling T., Hartig Ch., Laser T., Nürnberg M.R., Bormann R.	Deformation mechanisms of AZ31 magnesium alloy	333
	2008 Best Student Paper	Slooff F.A., Zhou J., Duszczyk J., Katgerman L.	Constitutive behavior of wrought magnesium alloy AZ61	339
	2011 Best Fundamental Paper	Chun Y.B., Davies C.H.J.	The evolution of in-grain misorientation axes (IGMA) during deformation of wrought magnesium alloy AZ31	345
	2012 Best Student Paper	Stutz L., Bohlen J., Letzig D., Kainer K.U.	Formability of magnesium sheet ZE10 and AZ31 with respect to initial texture	357
	2009 Best Application Paper	Mishra R.K., Gupta A.K., Rao P.R., Sachdev A.K., Kumar A.M., Luo A.A.	Influence of cerium on texture and ductility of magnesium extrusions	363
	2005 Best Contributed Paper	Kurz G.	Heated hydro-mechanical deep drawing of magnesium sheet metal	389
	2013 Best Student Paper	Sagapuram D., Efe M., Moscoso W., Chandrasekar S., Trumble K.P.	Non-basal textures in magnesium alloy strips produced by extrusion-machining	395
Part 7: Modeling and Simulation	2009 Best Fundamental Paper	Janz A., Gröbner J., Schmid-Fetzer R.	The Mg-Al-Zn-Mn-Ca-Sr alloy system: Backbone of understanding phase formation in AXJ alloys and modifications of AZ and AM alloys with Ca or Sr	437
	2008 Best Fundamental Paper	Schmid-Fetzer R., Gröbner J., Mirković D., Janz A., Kozlov, A.	Constitution of magnesium alloys	451
	2011 Best Application Paper	Turski M., Grandfield J.F., Wilks T., Davis B., DeLorme R., Cho K.	Computer modeling of DC casting magnesium alloy WE43 rolling slabs	457
	2007 Best Fundamental Paper	Lévesque J., Inal K., Neale K.W., Mishra R.K., Luo A.A.	Numerical modelling of large strain deformation in magnesium	467
	2012 Best Fundamental Paper	Solanki K.N., Moitra A., Bhatia M.	Effect of substituted aluminum in magnesium tension twin	479
Part 9: Corrosion, Surface Treatment, and Coating	2003 Best Contributed Poster	Tiwari B.L., Bommarito J.J.	A novel technique to evaluate the corrosion behavior of magnesium alloys	573
	2004 Best Student Paper	Mandagie M., Brandt M., Durandet Y., Jahedi M.	Parametric study of laser cladding of AS 21 magnesium alloy with aluminium silicon/tungsten carbide powder	617

LEAD EDITOR

Suveen N. Mathaudhu

Suveen Mathaudhu serves as the Program Manager for Synthesis and Processing of Materials with the U.S. Army Research Office (ARO), Materials Science Division. He received his B.S.E. in Mechanical Engineering from Walla Walla University in 1998, and Ph.D. in Mechanical Engineering from Texas A&M University in 2006. In his current position, he manages programs that focus on the use of innovative approaches for processing high performance structural materials reliably and at lower costs. He also concurrently serves as an Adjunct Assistant Professor in the Department of Materials Science and Engineering at North Carolina State University, where his research interests include ultrafine-grained and nanostructured materials by severe plastic deformation, consolidation of metastable particulate materials and processing-microstructure-property relationships of refractory metals and lightweight metals, and thermally stable nanocrystalline materials. He has co-authored more than 60 technical publications in these areas. He is an active member of ASM, TMS, and MRS, and has served as the Chair of the TMS Magnesium Committee, and as organizer of the 2012 Magnesium Technology symposium.

EDITORIAL TEAM

Alan A. Luo

Alan Luo is a Professor of Materials Science and Engineering and Professor of Integrated Systems Engineering (Manufacturing Group) at The Ohio State University (OSU) in Columbus, Ohio, USA. Prior to joining OSU in July 2013, Dr. Luo was a GM Technical Fellow at General Motors Global Research and Development Center (Warren, Michigan, USA) with 20 years of industrial experience. Dr. Luo won two John M. Campbell Awards for his fundamental research, and three Charles L. McCuen Awards for research applications at GM. He has 16 patents and more than 170 technical publications in advanced materials, manufacturing, and applications. Dr. Luo is an elected Fellow of ASM (American Society of Metals) International. He received the TMS (The Minerals, Metals & Materials Society) Brimacombe Medalist Award and SAE (Society for Automotive Engineers) International Forest R. McFarland Award in 2013, and USCAR (United States Council for Automotive Research) Special Recognition Award in 2009, and ASM Materials Science Research Silver Medal in 2008. Dr. Luo's research is also recognized by several Best Paper awards from TMS, SAE, and AFS (American Foundry Society). He is the Vice Chair of TMS Light Metals Division and SAE Materials Division and the Chair of SAE Non-Ferrous Metals Committee.

Eric A. Nyberg

Eric A. Nyberg is a Chief Engineer at the Pacific Northwest National Laboratory (PNNL) in Washington state. Nyberg has conducted materials processing and characterization research of lightweight metals such as aluminum, magnesium, and titanium for more than 20 years. His research in magnesium has included producing rapidly solidified and extruded magnesium alloys, developing superplastically formed sheet panels, and developing high strain rate deformation models of magnesium for automotive applications. He has been recognized as an invited speaker at multiple international conferences for his work on lightweight automotive applications. He holds three U.S. Patents, an R&D 100 Award, and a Federal Laboratory Consortium Award for titanium injection molding. Both his B.S. and M.S. degrees are in Materials Science and Engineering from Washington State University. He has been actively involved with the TMS Magnesium Committee and TMS Magnesium Technology symposium since their inception in 2000.

Neale R. Neelameggham

Neale R. Neelameggham is "Guru" at IND LLC, involved in technology marketing and consulting in the field of light metals and associated chemicals, [boron, magnesium, titanium, lithium and alkali metals], rare earth elements, battery and energy technologies, etc. He has more than 38 years of expertise in magnesium production technology from the Great Salt Lake brine in Utah. He was involved in the Process Development of its startup company, NL Magnesium, through the presently known US Magnesium LLC, and was its Technical Development Scientist from where he retired. Dr. Neelameggham's expertise includes all aspects of the magnesium process, from solar ponds through the cast house including solvent extraction, spray drying, molten salt chlorination, electrolytic cell and furnace designs, lithium ion battery chemicals and byproduct chemical processing. In addition, he has an in-depth and detailed knowledge of alloy development as well as all competing technologies of magnesium production, both electrolytic and thermal processes worldwide. Dr. Neelameggham holds 13 patents and a pending patent on boron production, and has several technical papers to his credit. As a member of TMS, AIChE, and a former member of the American Ceramics Society, he is well-versed in energy engineering, bio-fuels, rare-earth minerals and metal processing, and related processes. Dr. Neelameggham has served in the TMS Magnesium Committee since its inception in 2000, chaired it in 2005, and has been a co-organizer of the Magnesium Technology Symposium since 2004. In 2007 he was made a permanent co-organizer for the Magnesium Technology Symposium. He has been a member of the Reactive Metals Committee and Recycling Committee, and a Programming Committee representative of the TMS Light Metals Division. In 2008, the TMS Energy Committee was created following the symposium on CO_2 Reduction Metallurgy initiated by him, and he was selected as the inaugural Chair for the committee. He received the Light Metals Division Distinguished Service Award in 2010 and has been a co-editor of the Energy Technology symposium proceedings volume. Dr. Neelameggham holds a doctorate in extractive metallurgy from the University of Utah.

Wim H. Sillekens

Wim H. Sillekens is a project manager in the New Materials & Energy Unit at the research and technology center of the European Space Agency (ESA–ESTEC), where he is currently acting as the coordinator of the European Community research project ExoMet. He obtained his Ph.D. from Eindhoven University of Technology, Netherlands, on a subject relating to metal-forming technology. He has been engaged in aluminum and magnesium research, among others on (hydromechanical) forming, recycling/refining, (hydrostatic) extrusion, forging, magnesium-based biodegradable implants, and as of late on light-metal matrix nanocomposites. His professional career includes positions as a post-doc researcher at his alma mater and as a research scientist / project leader at the Netherlands Organization for Applied Scientific Research (TNO). International working experience covers a placement as a research fellow at MEL (now AIST) in Tsukuba, Japan, and – more recently – shorter stays as a visiting scientist at GKSS (now HZG) in Geesthacht, Germany, and at PNNL in Richland, Washington, USA. He has authored or coauthored book chapters, journal papers, patents, conference papers, oral presentations (keynote/invited/contributed), and so on (about 135 entries to date). Other professional activities include an involvement in association activities (among others, as the lead organizer of the TMS Magnesium Technology Symposium in 2011), international conference committees, and as a peer reviewer of manuscripts for scientific journals and conference proceedings as well as of research proposals. Research interests are in physical and mechanical metallurgy in general and in light-metals technology in particular.

PART 1: MAGNESIUM TECHNOLOGY HISTORY AND OVERVIEW

Alan A. Luo

Magnesium, an alkaline earth metal, is the ninth most abundant element in the universe by mass, the eighth most abundant element in the earth crust, and the third most abundant element dissolved in seawater. According to the U.S. Geological Survey, the leading use (41%) of primary magnesium is as an alloying element in aluminum-based alloys for packaging, transportation, and other applications. Structural uses of magnesium (castings and wrought products) accounted for about 32% of primary metal consumption in 2010. Desulfurization of iron and steel accounted for 13% of U.S. consumption of primary metal, and other uses were 14%.

Brown provided an excellent review on the historical development of magnesium production projects around the world in the 1990s. Many of these projects failed due to technical and commercial reasons, providing good lessons for future development of magnesium industry. Fast forward to 2010, Patzer reviewed the global magnesium production in the new century, with China becoming the major producer, and forecasted increasing use of magnesium in die casting and wrought alloys for automotive and 3C (computer, communication, and consumer products) industries. Dieringa and colleagues summarized the latest developments in manufacturing processes of cast and wrought alloys in Europe, especially Germany.

Luo compared the mechanical properties, structural performance, and mass saving potential of magnesium alloys with several major structural materials, and provided an overview of magnesium applications in automotive subsystems. Cole summarized a Magnesium Vision 2020 study sponsored by the United States Automotive Materials Partnership (USAMP), outlining the strategy and roadmap for increased use of magnesium components in the automotive industry. The next two papers by Luo and colleagues and Allison and colleagues described a first-of-its-kind Canada-China-USA collaboration on magnesium automotive applications focusing on a front end body structure and test bed and related Integrated Computational Materials Engineering (ICME) development, which is followed by a paper by Logan describing a lightweight automobile body concept utilizing a hybrid materials approach with emphasis on a magnesium and aluminum structure to support a high fuel-efficiency vehicle project.

The paper by Wendt and colleagues reviewed the past and current applications of magnesium castings in aircrafts and described an international research project (IDEA) to develop new magnesium alloys for special use aerospace industry. Mathaudhu and Nyberg provided an excellent review of historical, current, and potential future magnesium applications with a focus on scientific, engineering, and social barriers relevant to integration of magnesium alloys. The final paper, by D'Errico and Screnci, explored new applications in high-capacity hydrogen-based green-energy storage solutions for the grid balancing.

MAGNESIUM INDUSTRY GROWTH IN THE 1990 PERIOD

Robert E. Brown

Magnesium Monthly Review
226 Deer Trace
Prattville, Alabama 36067

Abstract

Electrolytic magnesium production has been the mainstay of the world's magnesium industry since magnesium was first discovered by Davy in 1808. Many of the early workers developed small advances until the electrolysis of anhydrous magnesium chloride became the standard method of production. From the very first days, the importance of anhydrous magnesium chloride has been recognized. It remains the major problem area of economic and efficient electrolytic magnesium production.

There has been a dramatically increased usage of magnesium in the past ten years by the automotive industry. This usage is projected to continue a large growth as automakers continue to strive for better fuel economy with reduced emission. The use in die casting alone has been projected to increase at 10-15% per year for the next 10 years.

Cost of magnesium and its alloys is constantly compared to aluminum and its alloys by the automakers on all continents. Magnesium usually loses this battle, in spite of the different densities. Aluminum is 50% heavier than magnesium, hence for the same casting shape a pound of magnesium would make three castings while a pound of aluminum would make only two. Automakers feel that to be fully competitive, magnesium should be priced at 1.5 times the price of aluminum. This only takes into account the densities and not the other advantages offered by magnesium such as damping capacity and strength and rigidity.

In recent years, the interest in magnesium has grown dramtically and there is a great deal of basic research and pilot plant work going on to identify better and more economic ways to produce electrolytic magnesium metal. There is more technical brainpower being applied to magnesium than ever before at anytime in history. The work has no boundries or restrictions and can be found on all the major continents (except maybe Antartica).

ELECTROLYTIC MAGNESIUM HISTORY

Magnesium as an element was discovered by Sir Humphrey Davy in 1808. There is no record showing that Davy ever isolated magnesium as a metal. The first actual production of pure magnesium metal in the metallic form has been credited to the French scientist Bussy, who fused anhydrous magnesium chloride with metallic potassium. The German scientists Liebig and Bunsen and the French scientists St. Claire Deville and Caron all worked on methods of producing magnesium from anhydrous magnesium chloride. Michael Faraday is credited with producing the first magnesium by electrolysis of molten magnesium chloride salt. A major step toward the mass production of magnesium was made by Robert Bunsen, the German scientist, who made a small laboratory cell for the electrolysis of fused magnesium chloride in 1852. Graetzel and Fischer, also German scientists, started investigating the use of carnallite from Stassfurt as the raw material for electrolytic production of magnesium.

The original electrolytic cells that were developed by Chemische Fabrik Griesheim-Elektron used a cell feed of molten carnallite ($MgCl_2 \cdot KCl \cdot 6H_2O$). The dehydration of carnallite is much easier than the dehydration of pure magnesium chloride solution ($MgCl_2 \cdot 6H_2O$). Most of the electrolytic magnesium production cells operating today are derivatives of the original Chemische Fabrik which became part of I. G. Farben in 1927.

Up until 1915, Germany was the only magnesium producer. When WWI started in 1914, there was a shortage of magnesium for pyrotechnics in military ordnance. Magnesium had long been used for flares and for tracer bullets The price of magnesium was $5.00 to $6.00 per pound. Eight companies in North America went into the production of magnesium metal. As soon as the war ended, the number of companies dropped off to two, Dow in Midland, Michigan and American Magnesium Corporation (Alcoa) at Niagara Falls, NY.

Dow Chemical had gotten into the production of magnesium from underground brines at the Midland, Michigan plant. Dow could not solve the anhydrous problem and developed a special process that fed "wet" feed ($MgCl_2 \cdot 1.5H_2O$) to the Dow-designed electrolytic cell. American Magnesium Corporation used the oxide fluoride process developed by Harvey. This process was similar to the aluminum production process in that it was electrolysis of magnesium oxide in fluoride melts. However, the slight solubility of magnesium in the fluoride created many problems, including high operating temperatures, and high specific gravity of the eutectic mixtures (largely barium chloride) which caused the liberated magnesium to immediately rise to the top of the bath and oxidize.

The Dow process was cheaper and gave more pure magnesium, so American Magnesium Corporation ceased production in 1927. AMC costs of production were 43 cents per pound in 1927 while Dow's were 22.5 cents per pound. AMC negotiated a purchase agreement with Dow to purchase all their magnesium requirements from Dow and ceased production. It was an 18 month contract and a new five year contract was signed in 1928. The prices paid by AMC were below the market and decreased even more as the Dow total sales increased.

For most of the 1930 period, Dow Chemical in the US and MEL in England and two companies in France continued to produce and market magnesium metal in small quantities.

It took the start of hostilities in Europe for the US to become aware of the need for strategic materials. To assist in a build up, the US created the Defense Plant Corporation. This group had the job of building plants for production and fabrication of strategic materials. Magnesium was a strategic material. The US increased their magnesium production rapidly by building 6 new electrolytic plants and 7 silicothermic plants

Most of these plants were shut down after the war and eventually Dow found itself in the US as the sole producer. The electrolytic plant in Michigan was closed and all of the magnesium metal production concentrated in Texas.

Magnesium production appears to be quite simple. This is merely a deceptive appearance that has trapped many of the world's foremost metals companies and several junior companies. A normal question would be, "If it is simple why aren't more people doing it?" For some reason that idea has not seemed to cause second thoughts among too many experts, who usually feel that they can solve the problem of profitable magnesium production that few people have ever been able to solve.

It is the "profitable" part that has given the most problems. There have been many plants and many processes that have produced magnesium, but not very many that have produced profits.

Since 1950 , there have been many projects attempted as is shown in Table 1. See Appendix attached.

NEW MAGNESIUM PROJECTS BEING DISCUSSED

There are 18 new magnesium projects under consideration somewhere in the world. They range from new smelters to expansions to feasibility studies. Surprisingly, in the USA, there are no new magnesium projects being discussed. The largest and oldest US producer, Dow Chemical, shutdown their total magnesium operations in 1998, removing 65,000 tons per year of magnesium production from the market. This metal supply has been picked up by many of the other magnesium production sources.

NORTH AMERICA

The leading new project is the **Magnola** project of **Noranda Magnesium** being constructed at Asbestos, Quebec. (see LMA Feb 1998). The C$733 million project includes a total facility to produce anhydrous magnesium chloride cell feed from asbestos tailings (serpentine) using a proprietary process. Alcan electrolytic cells (24) will be used to convert the magnesium chloride to 63,000 metric tons per year of magnesium metal. Design and engineering is approximately 98% complete with construction about 70% complete. The plant is presently on schedule to start producing magnesium metal by mid- 2000 with commercial production by March 2001. Hatch Engineers did the process work on the project and plant design and engineering and construction is by SNC-

Lavalin of Montreal.

Norsk Hydro Canada Inc has plans to expand its primary magnesium plant at Becancour, Quebec. Plans are to take the 45,000 metric ton per year plant to 85,000 metric tons per year in two phases. The decision to start construction has on the first phase has been postponed because of economic problems at the parent company. It had been expected to produce metal by 2001. The plant uses imported magnesite from the Pacific Rim for its feed and converts it to anhydrous magnesium chloride by a proprietary process. Norsk Hydro also has a proprietary electrolytic cell design which operates at over 400 kA.

The plant expansion will make use of much of the infrastructure already in place. The existing dehydration units will be modified to accommodate the first phase. The project includes new electrolytic capacity and technological improvements leading to higher productivity per cell. Other changes, recently identified will lead to reduced energy consumption and increased throughput in various units in the plant.

Gossan Resources continues to work on their high purity dolomite property at Inwood, Manitoba, 50 miles north of the city of Winnepeg. Metallurgical testing was carried out by Hazen Research of Golden, Colorado in mid 1997. Hatch and Associates of Montreal were engaged to carry out a prefeasibility study on the Inwood magnesium project. The report indicated that a 50,000 metric ton per year magnesium metal production plant using off-the-shelf technology of a silicothermic plant would be about US$0.89 per pound and US$1.13 per pound after financing. Gossan is working on a market study for magnesium metal that would be produced from this project.

Minroc Mines Inc. has started shipment of high-grade Chrysotile product from the Cassiar operation in British Columbia, Canada, utilizing the company's own proprietary wet process technology. Minroc has recently announced that it will proceed with investigations and work on a project at the Cassiar Mine for the production of magnesium metal from the existing tailings source at the mine. The high purity of the chrysotile at Cassiar, with its low iron content, they feel is a prime requirement for economic and competitive production. The Cassiar chrysotile feed is 23% magnesium and 5.5% iron. This compares with the Noranda Chrysotile project in Quebec which is announced at 21%-23% magnesium and 4%-8% iron. The Minroc Cassiar Magnesium project is being evaluated at production rates of 30,000 and 60,000 mtpy. The project has been assured of competitive rates for power, which is the key factor in costs and economics of any magnesium project.

Minroc also signed a memorandum of understanding with a division of the Korean automaker, **Hyundai**. The latest plans say that the magnesium plant will produce 90,000 mtpy of magnesium and the Aluminum Company of Korea (a Hyundai company) will be entitled to as much of the product as it may require with the remainder to be made available for sale to the international markets. Aluminum of Korea may acquire a 35% interest in the project in conjunction with an initial $25 million financing under the agreements, and could acquire a 65% interest in the project by providing additional project funding.

A preliminary assessment report by Kilborn/SNC Lavalin suggests that the mine's reclamation pile contains enough magnesium for 100 years of production. SNC Lavalin is assisting the company in securing financing. Detailed tests for producing magnesium metal will take place simultaneously with the reclamation activities.

A great new future for **Newfoundland-Labrador** was discussed at the 11th Annual Mining Conference in Baie Verte, Newfoundland. The Minister of Mines, Chuck Furey said, " A new industry is being investigated for the old asbestos mine site. Geotech Survey Ltd was given permission by the government a few months ago to determine if it's feasible to produce magnesium from the old asbestos ore tailings.

The **Northwest Alloys (Alcoa)** plant continues to operate the modified Magnetherm process at Addy, Washington. The plant is is rated at 41,000 mtpy. Production is of high purity magnesium used for alloying in Alcoa aluminum plants. In recent years, there has been an interest in the production of diecasting alloy at this plant. The plant has been upgraded over its entire life and further work to improve the process has been announced. NWA will work with Mintek of South Africa at developing a modified reduction reactor at Addy. The original plant had nine furnaces rated at 4,000 mtpy each. In later years, the plant has increased its rated capacity while reducing the number of furnaces in service.

Magnesium Corporation of America at Rowley, Utah is investing $46 million to evaluate new electrolytic magnesium cell technology and to install a new magnesium DC caster. The new cells when installed will have the potential to provide manufacturing efficiencies and reduce costs of magnesium production. With improved efficiencies, the production of the plant at Rowley could increase its capacity from the present announced rating of 41,000 mtpy. The new caster enables MagCorp to produce various sizes, shapes and weights of magnesium ingots and billets at lower cost. It can also produce T-Bar ingot which is used for aluminum alloying and offers a void-free, large shape.

AUSTRALIA

Australia has a number of magnesium projects being actively discussed and studies for magnesium metal plants are being conducted in all of the 6 major states. Tasmania and Western Australia have two each. The first large magnesium metal production project and the one that is furthest along is the project in Queensland which was originally developed by Queensland Metals as one of the end uses of its large high-purity magnesite deposit at Kunwarara, north of Rockhampton. It has been incorporated as Australian Magnesium Corporation. (AMC).

Australian Magnesium Corporation is a company that is owned by Queensland Metals (50%) and Normandy Mining (50%) subject to financial support and a 5% interest held by Fluor Daniel for engineering services. Normandy also holds a 36.85% interest in QMC. AMC is the most advanced of the potential new producers in Australia and has a partner (Normandy) which is financially sound. Ford invested A$40 million to assist the project into a pilot plant stage. Ford has also signed a long term off-take agreement for one-half of the planned production of 90,000 metric tons per year. AMC is operating a 1500 tpy demonstration plant to prove the process and gather operational data to complete the feasibility study. The construction of the commercial plant slated for completion by mid 2002 with commissioning and commercial operations by the end of 2002. The plant uses a process patented by CSIRO, an Australian Government Research arm, for production of anhydrous magnesium chloride from magnesite. Alcan electrolytic cells will be used in the commercial plant. The plant is expected to cost A$780 million including working capital.

The demonstration plant consisting of the CSIRO feed process and one full-scale Alcan Multipolar cell has been run since August 1999. The cash cost of producing magnesium is estimated to be A$0.65.

The electrical contract was signed at 20 mils which is somewhat less than the original number that was used in their feasibility calculations. The new plant site at Stanwell is near the power station, near a major gas pipeline and only a short distance from a major ocean port.

Crest Magnesium was one of the leading projects, but recently seems to be struggling to keep all of the partners working to get a plant built. Located on a very good deposit of magnesite in NW Tasmania, the project seemed to be going quite well until late in 1999. Discussions with a potential JV partner, Xstrata of Switzerland, were broken off.

The project as originally planned had **Crest** with the exclusive rights for Australia and New Zealand to use technology developed by the Ukrainian National Research and Design Titanium Institute and VAMI JSC over 20 years. There was talk of doubling the plant capacity in three stages over an 11 year period – taking production to 190,000 mtpy. The large Australian engineering and construction firm, Multiplex and Hatch Associates Limited of Canada were reviewing the technology in conjunction with Hatch or other approved consultants, and would provide a performance guarantee as to the nameplate operation of the plant (95,000 metric tons per year). The Tasmanian government will act as an intermediary for the supply of all energy: gas, electricity and commercial steam at a price that meets the indicative price already supplied by Duke Energy. Electrical costs are estimated to be 40% of the total cash production costs and Crest estimates an electrical cost of US$0.28 to produce each pound of magnesium.

Crest and Multiplex agreed to dissolve their JV partnership in October 1999.

Golden Triangle Resources NL which was orgiinally investigating the possibility of another magnesium project based on another section of the Main Creek Magnesite Deposit (adjacent to the Crest/Multiplex section), six to seven kilometers south of the Savage River Iron Ore Mine. The projects lie southwest of Burnie in northwest Tasmania. Golden Triangle exercised an option to acquire this portion of the Main Creek Magnesite Resource (47 million tonnes) from Savage Resources Limited in September 1998. First stage bench scale hydrometallurgical test work by Oretest Pt'y Ltd in Perth has now been augmented by Lakefield Research Limited of Canada who have begun work on the second phase of laboratory test work that will lead to a pilot plant program. This project has been deferred in favor of the Woodsreef project.

Bass Resources, a Tasmanian mining company has announced that it has identified the site for a new magnesium production plant at Bell Bay. Bass Resources is planning to develop an arrangement with Pasminco which will provide access to a mineral resource based on the Main Creek magnesite deposit. If that proceeds, Bass Resources would be in a position to obtain access to Golden Triangle Resources' exploration results and process technology.

Golden Triangle now intends to make the development of the magnesium project, the "Woodsreef" project in New South Wales, its main focus. Drill testing of the 24 million tonne Woodsreef asbestos tailings dump, located in Northern New South Wales has been completed. This is similar to the resource being developed in Canada by Noranda Magnesium.

In January 1999 **Golden Triangle** announced that it had awarded a contract to carry out "Comparative Magnesium Production Scoping Study" between the Tasmanian and New South Wales magnesium projects to South African engineering group, Bateman Brown and Root. The Bateman Group has some recent magnesium experience, having worked with the Israeli Chemical Industries in the development of the Dead Sea Magnesium Project. Golden Triangle has engaged the services of Mintek, South Africa's national minerals-research organization, which is separately involved with the development of Plasma magnesium processing technology.

Pima Mining N.L. is a mineral exploration company. In September of 1998, Pima's 80% owned subsidiary, South Australian Magnesium Corporation (SAMAG) acquired a 100% interest in a number of magnesite deposits in the Leigh Creek area of Australia, and plans to establish a magnesium metal production plant at Port Augusta, South Australia. Magnesite has been mined intermittently in South Australia's Flinders Ranges since 1919. Currently, **SAMAG** is proceeding towards the development of a proposed magnesite mine in the Willouran Ranges, North West of Leigh Creek. Their estimated mineral resources total 205 million tonnes of magnesite, with 16 million tonnes being in the "measured" category.

SAMAG has recently announced that they purchased Dow magnesium process and plant design. Plus they purchased the research records from Dow and hired several top Dow technical employees from Texas Division. The key component is electrical energy at a competitive price. This power situation has become a problem and recent statements from South Australian power officials indicate that project power costs of about 2 cents per Kwh may not be obtained.

Hatch Associates have completed a pre-feasibility study of the Port Augusta magnesium metal project, based on Dow electrolytic cell technology. SAMAG indicates that the study confirms that there is significant potential to produce magnesium metal at a cash operating cost of less than the US$0.61 per pound originally stated. Pima recently stated that, "The SAMAG project should produce 52,500 tpa of magnesium or magnesium alloys." A detailed study on this project by Hatch has confirmed the low projected costs of production. First commercial production is scheduled for the first half of 2003.

In the Northern Territory, **Mt. Grace Gold Mining** NL acquired a 100% interest in the Batchelor Magnesite Deposit in late 1998. The Company has reported extensive occurences of magnesite in their tenement near Batchelor, some 85 kilometers south of Darwin. The Company has now initiated a metallurgical testing program to demonstrate that Bachelor magnesite is amenable to beneficiation by flotation and is suitable for the production of magnesium metal. The stated aim is to construct a 50,000 tpa magnesium metal smelter with commissioning by July 2002. Energy may be available from a proposed development of Timor Sea natural gas together with the existing natural gas pipeline infrastructure crossing Mt. Grace's reserve.

Mt. Grace had retained DevMin Consultants to do a pre-feasibility study which will lead to a six to 12 month bankable feasibility study to prove the project's viability.

Mt. Grace has signed an agreement with Magnesium Developments International to use the Heggie process for their magnesium production plant. The Heggie process is a thermal process.

Pilbara Magnesium Metal Associates (PMMA) is a joint venture based on Onslow Salt deposits in Western Australia. It was reported that HCC Pty Ltd and Multiplex Construction were part of this project. The plant would use bitterns from existing salt operations for the source of magnesium credits. This would require technology somewhat similar to that used in Israel or at the Great Salt Lake. It has been reported that Uri Ben Noon, the former CEO of Dead Sea Magnesium, is a consultant to this project. An Israeli engineering company is providing a preliminary feasibility study. It is also reported that test work is being conducted in Russia and Israel. The venture proposes a 50,000 metric ton per year magnesium plant.

CRA in conjunction with **Fluor Daniel Australia** and **St. Joe Minerals** conducted testing and pre-feasibility work in this same area in 1985-86 with the intention of using by- product magnesium-rich liquor from the CRA gypsum operations. At that time, the anhydrous magnesium chloride feed production process investigated was the Nalco process. There was an earlier 1970's feasibility study for a magnesium metal production plant done by CRA and Ube Industries Ltd of Japan.

Electrolytic magnesium production plants produce more pounds of chlorine than they do magnesium. Shell and Dow are considering an integrated chemical plant in the same region and could possibly use the chlorine by-product stream for chemical production. With good sound basic technology for magnesium chloride production and an efficient electrolytic cell, the metal cost could be competitive with the other Australian projects.

It has been rumored that PMMA is in discussions with the Solikamsk magnesium production facility in Russia to obtain the latest electrolytic magnesium production technology.

Hazelwood Power is again investigating the possibility of recovering magnesium metal from fly ash. Hazelwood Power is a 1600 MW brown coal fired electricity generator located in the LaTrobe Valley of Victoria. The Victoria state power commission looked at recovering magnesium from fly ash in 1970's. The private company (Hazelwood) is working with HRL Technology Ltd to conduct pre-feasibility studies into the possibility of using a magnesium chloride feed liquor produced from flyash for magnesium metal production. It has been reported that there is sufficient fly ash available to supply a 30,000 metric ton per year smelter for 30/40 years. The big advantage would be transmission-free energy contracts, excellent water resources, and waste disposal potential. The study was based on the Alcan process for the production of anhydrous magnesium chloride and Alcan electrolytic cells.

Anaconda Nickel has announced an A$1 billion magnesium smelter will be built near a magnesite deposit they discovered when looking for nickel. The project development plans have not been clearly established at the present time. Shortly after the announcement of the magnesium project, Anglo American bought 23% of Anaconda Nickel and are said to be very interested in magnesium.

A summary of some of the planned magnesium projects was presented by Chris Laughton of Golden Triangle at a magnesium meeting in Sydney, Australia in June 1999. See Table 2.

REPUBLIC OF CONGO (Brazzaville)

Magnesium Alloy Corporation (MAC) commissioned SNC-Lavalin in Montreal to perform a feasibility study for the Kouilou hydroelectric site conditional upon certain financing arrangements by SNC-Lavalin. Upon completion of the study SNC-Lavalin may assist MAC in the financing and/or construction of the Kouilou hydroelectric site in Congo. SNC-Lavalin with its engineering expertise in hydroelectric facilities as well as its extensive construction experience in Africa, makes an important addition to MAC's technical team. MAC has an option to develop the Kouilou hydroelectric site as a potential low-cost energy source for this extraction plant. The Kouilou River site lies 50 km north of Pointe Noire.

The lead contractor for the feasibility study is Salzgitter Anlagenbau GmbH (SAB), an engineering and general contracting division of Preussag AG of Germany. Kavernen Bau-und Betriebs (KBB), another member of the Preussag Group, has extensive experience in all phases of solution mining including modeling of reserves, solution mining simulation, drilling production wells together with brine extraction and transport.

VAMI, SAB and KBB, the principal contractors along with several sub-contractors have done detailed studies. VAMI and Ukrainian State Titanium Institute have been performing evaluation of advanced and improved modifications to proven magnesium extraction technologies. VAMI and the Titanium Institute took part in the design and implementation of the magnesium extraction technology for the Dead Sea Works Magnesium facility. They also took part in the technical design for the proposed magnesium facility in Iceland in conjunction with Salzgitter. VAMI developed the technology and took part in the design and construction of all the magnesium plants in the former Soviet Union (Berezniki, Solikamsk, Kalush, Zaporozhe, Ust-Kamenogorsk).

Preliminary reviews of the MAC project, subject to low energy costs as currently indicated, indicate very low cost magnesium production. MAC anticipates a first phase annual production rate of 58,000 metric tons with a second phase of 16,000 tons. Production decision due in 1999 with production possible by 2001.

NETHERLANDS

The Dutch development of the magnesium project for the Northern Netherlands is proceeding at this time. The project is part of the Antheus public-private project organization charged with developing the metal business climate in Northern Netherlands. The **Magnesium Development Project Delfzijl (MDPD)** project team is led by Reinder Rentema as chairman. A plan for a magnesium metal production plant of 40-60,000 metric tons per year has been presented to interested magnesium producers and magnesium users and the investment community.

It has an estimated installed cost of US$400 million. A study run by Hatch Associates of Quebec, Canada recently evaluated and compared the "Antheus" option with existing magnesium-producing technologies. That study shows that thanks to the high purity brine and other favorable production factors the planned region can offer a proposition that will feature one of the lowest cost structures of all existing and planned magnesium producing plants worldwide.

The exact technology to be used has not been chosen, but the planned project location has operating magnesium chloride solution mines that are presently being mined at a rate of 200,000 tons of magnesium chloride per year. Hydro Terra of Canada is working on a feasibility study for this project. Ample electrical energy is available and power deregulation in Europe in 2002 will help keep costs competitive. The brine is reported to be very pure. Long term plans call for a combined plant that will use the chlorine by-product of the magnesium operations to combine with ethylene to make ethylene dichloride. One of the partners is **Nedmag,** a former Billiton Company.

NORWAY

The original **Norsk Hydro** magnesium production plant in Norway was built at Porsgrunn during World War II using I.G.Farben technology. This plant has been upgraded and modified to reach the present capacity of 40,000 mtpy. Presently, this plant uses seawater and dolomite to produce its anhydrous magnesium chloride cell feed. The plant has been upgraded to take care of environmental concerns and additional 10,000 tons of recycling capacity has been added in recent years. Further details will be available from a more detailed presentation by Norsk Hydro.

FORMER SOVIET UNION (FSU)

Now the oldest magnesium production operation in the world is the **Solikamsk** facility in Russia. It has been in operation since 1934. Solikamsk has installed a magnesium powder production plant and has a contract with GM for magnesium alloy. Solikamsk produces about 10,000 tpy of primary metal and 10,000 tons of alloy. The new magnesium granule plant is rated at 2,000 tpy with potential to expand to 8,000 tpy. Solikamsk also produces recycled magnesium. There have been plans and discussions to double the size of the primary magnesium plant. It was reported in 1998 that Solikamsk would participate in a project to use asbestos tailings from Uralbest. The plans were to double the Solikamsk production of primary metal and alloy. Estimated project costs were reported at US$300-500 million.

Avisma which is the magnesium plant at Berezniki produces an estimated 15,000 mtpy and no immediate plans were known about expansion.

In Kazakstan, the magnesium plant at Ust-Kamenogorsk produced an estimated 10,000 tons in 1998 with no announced plans of expansion.

In the Ukraine, there are two magnesium plants: **Zaporozhe** which did not operate in 1998 and **Kalush** produced an estimated 10,000 tons of primary magnesium in 1998. No plans for expansion have been seen although there have been several announcements made about start-up plans.

ICELAND

The Iceland Magnesium Project has been around in variou forms since 1971. Promoted by the Sudenes Heatin Corporation, a producer of heat and electricity from geotherma steam, the project has had new life. A consortium of Salzgitte Magniy (VAMI and UTI), and Amalgamet did a feasibilit study for a 50,000 metric ton primary magnesium meta production plant. The proposed plant used an electrolyti process with cell feed produced using VAMI technology. Th study confirmed the technical viability of such a project. Bot seawater and imported magnesite were reviewed. Again, th potential supply of low cost electricity made the productio costs attractive.

In 1998, **Australian Magnesium Investments**, purchased a 40% share in the Icelandic Magnesium Project. No decision ha been made at this time as to when the design and constructio will start. AMI is part of the Australian Magnesium Corporation and the acquisition presumably gives them the access to the Russian-Ukrainian technology or they coulc possibly use the technology that is being developed in Queensland. No immediate decision to proceed is expected until AMC gets the final results of the feasibility study based on demonstration plant operation.

ISRAEL

Dead Sea Magnesium has struggled to get into full production. In 1998, they produced 25,000 mt. Israeli Chemical Limited, the parent of DSM, will put up $50 million more for debottlenecking work. It was reported that this money will be used to improve equipment serving the DSM electrolytic reduction plant. It was said that the present auxiliary equipment can only process 27,000 tpy, but DSM hopes to develop a capacity of 35,000 tons The latest $50 million is in addition to the $460 million already spent.

The board of directors of Israel Chemicals (ICL) and its subsidiary Dead Sea Works Ltd have authorized the deal (on October 18, 1999) in which the magnesium unit which was a subsidiary of Dead Sea Works will be transferred to Israel Chemicals," ICL Joseph Rosen reported recently. Dead Sea Works hold a 65% stake in the unit, while Volkswagen AG holds 35% of the joint venture. After the deal, ICL will hold 65% of the Magnesium unit.

ICL, a chemical holding company, will inject $65 million into the magnesium unit to promote growth and sales. Officials from Volkswagen and ICL have agreed on a joint business plan to invest $100 million into the magnesium unit to promote growth and profitability. According to the plan, Volkswagen will invest $35 million into the unit.

JORDAN

The **Jordan Magnesia Company** has built a US$70 million magnesium oxide plant. The project has a planned production of 50,000 metric tons per year of high quality magnesium oxide and 10,000 tons of specialty products from Dead Sea brine. The plant will be near the potash project. The Jordan Magnesia Company is owned by Arab Potash and JODICA.

The Near East Group is currently involved with the Arab Potash Company and is working to develop a magnesium production project using Dead Sea brine as a raw material. The plant will be a 25,000 ton per year facility. The Arab Potash group had signed an agreement to use Russian and Ukrainian Technology for the magnesium production facility. A major sponsor or partner is being sought by NEEC.

ALBANIA

Albania has a found a magnesium hydrosilicate deposit of Crysotil – Antigorite type, formed by tectonic myllonitization of ultarmafic rocks and their hydrothermal elaboration. The deposit has enormous reserves (over 100 million tones). There is strong interest in building a magnesium metal plant near the deposit.

UNITED ARAB EMIRATES

Construction of a Dh734 million magnesium alloy plant is being planned for Sharjah's Hmriyah Free Zone. The smelter project is being promoted by the Sahari Group of Abu Dhabi and Normans of Albania. "This project is currently owned 50:50 by the two partners. The group is seeking European and Gulf partnership and funding. The ownership profile will be changed then according to spokesmen. The plant will have an initial capacity to produce 20,000 tons per year of magnesium products, to be increased to a 60,000 ton plant upon completion in the next 24 months. For the Sharjah project, raw material will come from Albanian mines which are estimated to have reserves of over 400 million tons. Magnesium products made at the plant will be sold to buyers in Japan, the United States and Europe.

CHINA

China has more installed magnesium production capacity than any country in the world. The exact capacity and how much is actually producing at any one time is unknown, because the bulk of the production is from small and widely scattered silicothermic Pidgeon process plants (i.e. small, horizonal steel retorts charged with briquettes of calcined dolomite and ground 75% ferrosilicon as a reductant). There are 24 magnesium production plants with larger than 3000 tons per year production (three are electrolytic, the others thermal).

Fifty plants have capacities of 1,000 to 2,000 mt. There are announced plans for expansion by several of the larger thermal plants. One of the large electrolytic smelters, Minhe, has also made announcements about expansion, but there has been no confirmation that this move is happening. The continued internal competition to sell magnesium has caused the Chinese to lower the selling price CIF their ports to very low numbers. In 1999, pure magnesium was available in port of Tanjain for US$1650 per metric ton. An internal group has attempted to establish a minimum export price of $2150 per metric ton. These efforts have not been completely successful.

The small plants can be shut down and started up relatively quickly. So as the price goes up, the country's production will increase. It has been speculated that the very small plants need a price of US$1800 to break even, but based on the transaction prices the break even must be be closer to $1400. It is very subjective and varies widely according to plant and location.

PERU

One company in Peru has been reported to have a pilot plant running using Epsomite/Magnesium sulfate heptahydrate (Epsom Salt). A company called TRC Technologies was looking for someone to proved a lump sum, turnkey plant in 1998. There has been no further announcements from this area.

COLUMBIA

Over the past several years, there have been reports that Columbia was investigating magnesium containing ore bodies with the idea of producing magnesium metal. This has not been updated in the last two years.

BRAZIL

Brasmag has operated a modified Ravelli silicothermic process in Minas Gervais for a number of years. On and off there have been announcements for expansions. At this time, nothing has been actually done to increase the 7,000 metric ton capacity.

GENERAL DISCUSSION

During project evaluation and feasibility studies, there are several key items that must be in place. Power costs must be competitive. The source of magnesium feed for any process must be readily available and free of any of the contaminating elements that could adversely affect the production process. There must be a sufficient differential between the production costs and the selling price to produce profit.

Australia and Quebec have power prices that are among the lowest in the world. The prefeasibility studies on new plants in these areas are indicating production costs that are lower than the estimated production costs of the present magnesium producers. The selling prices that have been used in the studies have been adjusted downward to reflect the impact of increased production. It appears from this quick summary that all is rosy.

Unfortunately there are a few areas that are not being addressed or referred to very often. The newest operating magnesium project, the Dead Sea Magnesium plant, had trouble getting production up to the design production levels. Costs for the facility tended to overrun the original budget. Final product quality was reported to be substandard for almost a year. The costs of production are not available, but are known to be higher than originally budgeted. This plant uses the Russian/Ukrainian magnesium production process, which has been proven to successfully produce magnesium using carnallite as a feed stock, but is still fairly complicated.

The Noranda Magnola project has had their design, build, construct costs increase to C$733 million while planned production went from 58,000 mtpy to 63,000 mtpy. Design engineering is complete and construction is about 50% complete. This plant is using proprietary technology to produce anhydrous magnesium chloride from asbestos tailings. Alcan electrolytic cells will be used to produce the magnesium metal. The process was proven in a pilot plant operation, but the design scale-up is very large.

The Australian project that is furthest along is the Australian Magnesium Corporation work in Queensland. A modified Nalco process was developed (AM process) by the Australian Magnesium Research and Development project and Australian patents obtained. AMC has built a 1500 mtpy demonstration plant at Gladstone, in Queensland. This plant uses the AM process to produce anhydrous magnesium chloride from Queensland magnesite from Kunwarara. A single full scale Alcan electrolytic cell is installed to produce magnesium metal and data for the feasibility study. There have been numerous problems in starting up and running the demonstration plant, including major electrolytic cell operating problems caused by initial introduction of iron contaminated feed. The cell has been cleaned, modified and is restarted and has now produced the first magnesium metal. Much of the feasibility study has been completed. The demonstration plant will supply operational data that are needed to verify the initial conclusions. The development work for this project will have taken 13 years and expended almost $100 million by the time sufficient data for a go or no-go decision is accumulated. It is hard to believe that there are many projects or many companies that would be willing to spend this amount of money on development.

The major stumbling block to magnesium production is the lack of a standard proven commercial technology. There is also a very limited experience base. For most processes, the experience lies in the hands of the present producers and most do not want to share or license their technology.

There are basically two processes, electrolytic and thermal. Each of these has many sub-groups or sub divisions. The electrolytic is based on converting magnesium credits to some form of magnesium chloride and breaking the bonds to create magnesium metal and chlorine gas. There are several versions of this technology, but the predominant is reduction of anhydrous magnesium chloride in a version of the I.G. Farben sealed cell which produces magnesium metal and chlorine gas.

The thermal process converts magnesium credits to magnesium oxide and removes the oxide by use of a reducing agent. There are many types of these processes being used. The most widely used is the silicothermic process, which utilizes ferrosilicon as the reducing agent.

The Australian projects are mainly looking at electrolytic reduction for their process. One project is looking at a thermal type plant. **AMC** uses the AM (Australian) process for production of anhydrous magnesium chloride and uses the Alcan cell for reduction. **Crest** has signed to use the VAMI/Ukrainian technology. **SAMAG** has signed to license Dow technology for the cell feed and the reduction areas. **Golden Triangle** has retained Bateman Brown and Root and is looking at some new technology for feed production, with an Alcan cell being mentioned for reduction. **Hazelwood** mentions Alcan technology for their process to convert flyash to magnesium metal. **Batchelor** has signed an agreement with Magnesium Developments International to license the Heggie thermal magnesium process. **PMMA** is reportedly working with some Israeli engineers with no specific mention of the process. It has been mentioned that Solikamsk technology is being discussed. No process was mentioned for **Anaconda Nickel.**

A summary table of projected costs and returns is shown in Table 3 (attached.)

This summary shows that there is profit in magnesium projects. These figures are basically taken from preliminary company announcements. As the projects develop further, the costs will be more accurately presented. However, it can be quickly seen that the rate of return is good if the projects are built and operated for the budget numbers. It can also be seen that there is some advantage in larger scale plants, IF they run according to plan. There are a lot of IF's in all the Australian Projects (perhaps Dreams would be a better choice of word) and, for that matter, in all the new proposed magnesium developments around the World.

No one has ever designed or built a 90,000 metric ton per year plant. A plant that size will require a large amount of special design engineering and special construction. It will require special crew training and a long start up period. Unfortunately, the magnesium cells run in series and when you start one line, you must run them all. The cost of designing a plant where each cell could be electrically isolated by switching would be cost prohibitive. The cells must be charged with molten magnesium chloride, which must come from a special melter or from cells already on line.

The estimated rate of return on the thermal process that is planned for Mt. Graces' Batchelor Project is surprising. Of course, that is the main great attribute of the thermal process. Low construction costs and the fact that small plants can be built and justified. Modular construction can be used to expand the plants in a fairly simple fashion, compared to an electrolytic plant.

Future Needs

For magnesium to be highly successful, especially in the automotive area, the price must be lower than the published prices of today. The ideal price ratio discussed is a magnesium price that equals the price of aluminum x 1.6. As mentioned in other places, this is only to take into account the density differences. There are some additional value-added properties of magnesium other than weight savings. These include machinability, die casting speeds, rigidity, and damping capacity, die wear. It has been mentioned by Ford that 1.8 x the price of aluminum would be acceptable while GM, Fiat, VW, Toyota and others suggest a much lower ratio. Accepting that price is the most critical factor there is also the absolute need for long term price stability, long term alloy development efforts, physical and chemical data on the alloy performances and a marketing effort that would provide assistance in the design and development of magnesium parts for specific vehicle platforms.

NEW TECHNOLOGY

The search for new and better efficient magnesium production is being carried on by major companies and small private research firms. Government sponsored programs can be found in most of the industrialized countries. Any researchers with a sound idea and the ability to present this idea will find funding easier now than ever before . There is great need for a simpler and more efficient electrolytic process and there are a large number of very talented technicians and scientists working on the problems. A major breakthrough could come most any day.

Table 1: MAJOR MAGNESIUM PROJECTS SINCE 1950

Bold Face indicates plant that are still in operation

Startup	Company	Location	Mg Source	Process	Type	Initial Capacity	Comment
1951	**Norsk Hydro**	**Porsgrun**	**Sea Watr**	**Electrolyt**	**I.G. Farb**	**18,000**	**1.**
1959	Alabama Met	Selma, Al	Dolomite	Thermal	Pidgeon	7,000	X 2.
1960	Furakawa	Japan	Dolomite	Thermal	Pidgeon	5,000	X 3.
1964	**Pechiney**	**France**	**Dolomite**	**Thermal**	**Magneth**	**9,000**	**4.**
1964	Ube Kosan	Japan	Dolomite*	Thermal	Pidgeon	5,000	X 5.
1969	**Nat. Lead**	**Utah**	**Brine**	**Electrolyt**	**Modif IG**	**40,000**	**6.**
1970	Am Magnes	Texas	Brine	Electrolyt	Modif IG	25,000	X 7.
1972	Dow Chem	Texas	Sea H2O	Electrolyt	Dow Cell	25,000	X 8.
1975	**NoWst Alloy**	**Addy WA**	**Dolomite**	**Thermal**	**Magneth**	**36,000**	**9.**
1985	MagCan	Canada	Magnesi	Electrolyt	MPLC	12,500	X 10.
1992	**Norsk Hydro**	**Canada**	**Magnesit**	**Electrolyt**	**Norsk**	**45,000**	**11.**
1994	Noranda	Canada	Asbestos	Electrolyt	Alcan	63,000	12.
1997	AusMagCorp	Quenslnd	Magnesite	Electrolyt	Alcan	1500	13.

Comments:

1. **Plant expanded and process improved. Output in 1998 at 43,000 tpy + 10,000 tpy recycling**
2. Pidgeon process plant was put out of business by Dow lowering price from 36 to 30 cents
3. Plant became uneconomic, labor, electricity and process costs became too high
4. **Plant at Marginac has been expanded to 20,000 tpy cap but electrical costs are high**
5. Made dolomite from seawater magnesia and calcined limestone. Process became too costly. Moved plant to China in a joint venture
6. **National Lead sold to Amax who sold to Renco. Plant is looking at expanding to 60,000 tpy**
7. Mag plant cells were not good, bought Russian technology, had environmental problems, fixed much of that, then evaporation ponds flooded, sold out to MPLC
8. Dow built a Mag-Chlor plant to produce magnesium and strong chlorine, it did not run well and was closed and demolished in one year.
9. **Alcoa established a magnesium production plant using Pechiney technology. The plant is located just north of the old WWII magnesium plant that used the same dolomite deposit. Plant efficiency and capacity have been expanded by Alcoa applied research. Also looking at Mintek process.**
10. MagCan was designed and built to process magnesite using a carbo-chlorination process developed by MPLC and piloted in England. The plant used the modified Russian electrolytic cells developed by American Mag. Failed due to combination of technical and partner reasons.
11. **Norsk Hydro built a new plant in Canada to use magnesite. Process was developed in Norway. The plant was late and over budget and the operations took a long time to get running well. The plant was built in North America to access the auto industry and aluminum alloying. Anti dumping charges virtually barred the shipment of magnesium from this plant to the US, causing the plant to run at half capacity for several years and lose money. Plans to expand are on hold.**
12. **Noranda has been working on a process for asbestos for many years. The process was piloted and now a commercial plant is being built in Danville, Quebec. Work on the process development took about ten years before the pilot plant operation. Hatch did process, SNC is doing design and construction.**
13. **Australian Magnesium (Queensland Metals) has been working on the magnesium process for producing magnesium from magnesite for over 10 years. A 1500 tpy demonstration plant is in operation and is producing several tons of magnesium per day to prove the process and provide data for the feasibility study to be completed by end of 1999. The commercial plant is planned to produce 90,000 mtpy and Ford Motor has signed an off-take agreement for ½ of the production for 5 years with an option to renew for another 5. Base price is said to beUS$1.30. Ford also put A$40 million into the project to help build the demonstration plant.**

Table 2. Electrolytic Magnesium Producers – Current and Proposed

Location	Plant/Company	Capacity	Operating Cost* ¢/lb US	Capital Cost US$ Million
USA	Magcorp	45,000	94	N/a
USA **	Dow	60,000	91	N/a
Norway	Norsk Hydro	44,000	85	N/a
Israel	Dead Sea	55,000	81	460
Canada	Norsk Hydro	68,000	74	
Australia	SAMAG	52,000	71 (60-65)	420
Canada	Magnola	58,000	66	515
Australia	QMC	90,000	66	520
Australia	Crest	95,000	65	561
Australia	GTR (Main Creek)	80,000	62	421
Australia	GTR (Woodsreef)	80,000	57	423

* Derived from a variety of sources   **Now closed

Table 3. Proposed Magnesium Projects with Estimated Costs

Company	Mtpy	Est Op cost/lb	Capital cost US	Capital cost/tn	Opcratc cost/ton	8%-15* yr Loan	Total cost/ton	Cost lb $US	%ROI @ 1.50	%ROI @ 1.20
SAMAG recent	52,500 Sep 99	$0.61 $0.59	$420M $375M	8000 7142	1342 1300	917 819	2259 2119	1.024 0.96	12.0 15.12	4.5 6.72
AMC	96,000	$0.66	$520M	5416	1452	570	2022	0.917	21.5	10.45
Crest	95,000	$0.65	$561M	5905	1430	677	2107	0.995	18.45	8.2
GTR M	80,000	$0.62	$421M	5262	1364	603	1967	0.898	22.88	11.47
Congo	60,000	$0.55	$514	5666	1210	648	1858	0.843	23.19	12.60
Magnola	63,000	$0.66	$492M	7809	1452	900	2352	1.067	11.09	3.4
GTRW	80,000	$0.57	$423M	5287	1254	604	1858	0.843	24.8	13.5
Hazelw	34,000	$0.55	$178M	5235	1210	600	1810	0.82	25.9	14.5
Batchelor	10,000	$0.70	$154M	1540	1540	176	1716	0.778	93.7	54.8
PMMA	50,000	N/a	N/a	---	---	---	---	----	----	----
Anaconda	90,000	N/a	N/a		N/a	N/a	----	---	----	----

*Annual cost for a 15 year loan at 8% interest, basis $/ton installed capacity

ROI= 100x2000x [sales price/lb – total cost/lb] / Installed capital cost/ton

From *Magnesium Technology 2010*, S.R. Agnew, N.R. Neelameggham, E.A. Nyberg, and W.H. Sillekens, Editors

THE MAGNESIUM INDUSTRY TODAY...THE GLOBAL PERSPECTIVE

Author: Greg Patzer, Executive Vice President

The International Magnesium Association
1000 N. Rand Road, Suite 214, Wauconda, IL, USA 60084

Keywords: Magnesium, Industry Update

Introduction

World demand for magnesium will show a decline in 2009. The outlook for 2010, which is guardedly optimistic, will be for a resumption of slow growth. The industry has seen marked changes in the sources of supply for primary and alloyed magnesium in recent years. Technological advances in magnesium continue at a strong pace as does interest in the material as a substitute for other light metals. The automotive segment remains the end-use area with the largest growth potential, if for no other reason than the size and quantity of the potential materials substitution applications. However, the shrinkage of that market, particularly in North America will have a definite impact on expectations for magnesium. The 3C market (computers, communications & consumer electronics) will continue to show above average growth. Other niche markets related to medical and construction industries also offer potential.

This document will present the current situation in the global magnesium industry in 2009. The immediate past industry changes will be assessed as to the impact that is being realized today, while also examining those developing factors, which will impact the industry's future in 2010 in particular. Analysis is based on information and data gathered by the International Magnesium Association from numerous sources such as, but not limited to: allied industry associations such as the China Magnesium Association, statistical sources such as Antaike in China, the United States Geological Survey and import/export data from several countries. Worldwide industry publications as well as anecdotal references from producers and manufacturing users of magnesium were also used extensively.

The Changing Face of Global Supply/Capacity

As we know the onset of 2007 saw a dramatic change in the landscape of the magnesium industry with the announcement that Norsk Hydro, arguably the largest producer of magnesium at that time, was departing the magnesium business With the departure of Hydro on the heels of years of having seen the departure of numerous western producers, end users of magnesium obviously voiced concerns relative to the available capacity of our industry.

Has global primary magnesium retracted, remained the same or expanded during recent years as the industry has been in a state of dramatic changes? Using 2001 as a baseline (albeit arbitrarily) we can see (Figure 1) that in year 2001 the total capacity for producers of primary magnesium was approximately 530,000 (MT). Of this total capacity, the capacity at that time of China producers was 209,000 (MT). The capacity of all other producers of primary magnesium worldwide was 321,000 (MT).

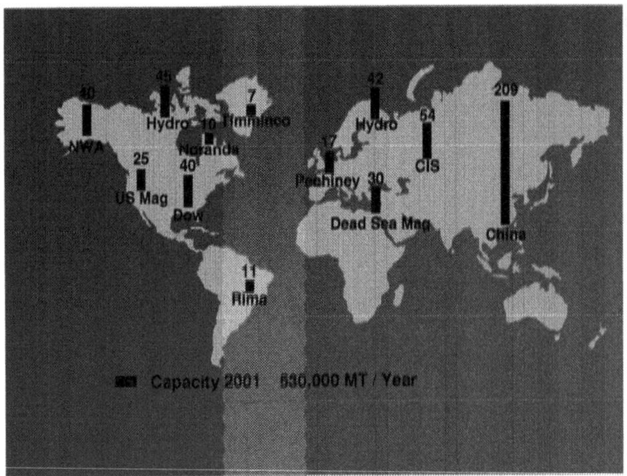

Figure 1. Magnesium Primary Production Capacity - 2001

In 2008 the estimated total global capacity of primary producers is seen to have risen significantly to a total capacity for producers of primary magnesium of 1,313,000 (MT), an increase of 783,000 (MT). Of this total capacity, the capacity of China producers at present is estimated at 1,162,000 (MT) an increase of 953,000 (MT) or 556%. The capacity of all other producers of primary magnesium worldwide is 151,000 (MT), a loss of 170,000(MT) or 53%. (Figure 2) It needs to be noted that in 2009 the aforementioned total Chinese capacity must be tempered by the reality of an industry running at approximately 50% capacity.

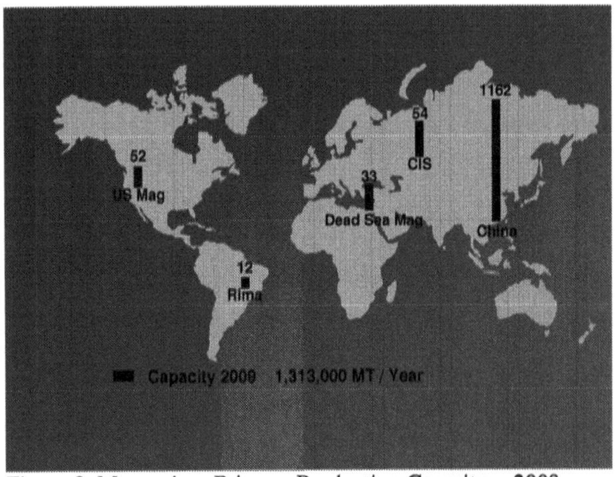

Figure 2. Magnesium Primary Production Capacity - 2008

As can be seen, during the time period between 2001 – 2008 industry capacity has in fact seen a marked increase despite the closure of numerous facilities worldwide. An underlying fact is that during the mid to late 1990's China saw the substantial growth not only in capacity but also in the number of primary producers. Although no firm numbers are available anecdotal references indicate the number of primary producers reached a high saturation point between 700 – 900 during the late1990's. The number of primary producers has remained relatively flat in recent years following a dramatic decrease precipitated by closures, mergers & acquisitions. Total number of primary producers in China in 2001 was estimated at 120 and 100 in 2007. [1,2] We expect the number of producers to remain relatively flat in the short term but with a small number or consolidations continuing as government pressures continue to discourage new plants coming online.

Due to the high price levels we have see we are now seeing for the first time in many years several non-Chinese primary magnesium projects that are in various stages of development. There are numerous magnesium-rich magnesite deposits in Canada, but economic challenges remain before production becomes feasible. In Canada the Gossan Resources Ltd. Project holds promise with a primary production capacity when operational of 25,000 tons/year. The planned process is the Improved Pidgeon process using Dolomite as source material. Planned start was 2011 but the current recession has likely pushed that back. The Mag Metals, Pointe Noire project in the Congo promise production at 90,000 ton/year but is still far from realization, with infrastructure needing first to be put in place. In Europe, SilMag Technology AS, in Norway has planned production of 35,000 tons/year with a start date of 2011. Source material is olivine using an advanced Electrolytic process. The affects of the current global recession on the start-up times for all these projects is yet to be seen.

Production

In 2004 the International Magnesium Association suspended the statistics program, which it had provided to the industry over the years following the dramatic upheaval in the sources of primary magnesium, and the resulting difficulties in attaining accurate production data at that time. In 2006 the International Magnesium Association revisited the methodology previously utilized and published production data which was a combination of statistics from the United States Geological Survey and data from the China Magnesium Association. [1,2,3,4,5] At the same time data from the years from 2000 through 2005 was also redone so as to develop a consistent baseline for comparison. In Figure 3 you will find this data from the period 2002 through 2008.

Years 2002 – 2008
(000's of Metric Tons)

	2002	2003	2004	2005	2006	2007	2008
USA	35	43	43	43	43	43	52
Brazil	7	6	11	6	6	18	18
Canada	86	50	55	54	50	16	0
China	232	354	450	470	490	627	559
France	0	0	0	0	0	0	0
Israel	34	30	33	28	28	25	30
Kazakhstan	10	14	14	20	20	21	20
Norway	10	0	0	0	0	0	0
Russia	52	45	45	45	50	37	35
Ukraine	0	0	0	2	2	3	3
Serbia	2	2	4	2	1	2	2
Total	468	544	655	670	693	792	719
Annual Change	+4%	+16%	+20%	+2%	+3%	+14%	-9%

Figure 3. – Primary Magnesium Production 2002 - 2008

The rapid decrease in overall magnesium production in the latter half of 2008 was in sharp contrast to 2007, which set the high-water mark for magnesium production (792,000mt) and firmly established China as the leader in primary production at approximately 80 percent of world production. The year 2008 ended with a sizable decline of -9 percent for the year, bringing total global production down to 719,000mt. Not surprisingly China experienced the majority of decline in production. Without a doubt the artificial trade barriers of punitive anti-dumping duties against Chinese exported magnesium to the U.S. had no small bearing on the year's final results with the North American market having been virtually eliminated to Chinese magnesium.

Opportunities

Traditionally the largest usage segment for magnesium in the past had been in the alloying of aluminum. Over the last decade the estimated annual growth for aluminum alloy had been between 1.5 and 3 %. However, in approximately 2005 indications, but not hard statistics, indicated that die-casting had minimally equaled or exceeded aluminum alloying as the largest usage segment. Our expectation is that the die-casting segment will for the near future be a driving factor in increased magnesium usage/demand. However it must be noted that in 2009 the two key markets, both die casting as well as aluminum alloying appear to be loss years for the respective segments with projections showing a potential decline in world aluminum production.

As mentioned earlier the 3C market (computers, communications & consumer electronics) has continued to show above average growth and is expected to continue to do so. This segment has seen rapid growth in recent years in such areas as cell phone bodies, laptop computer bodies/enclosures and in personal entertainment devices such as Apple's iPod.

Medical applications are a niche market which has not been hurt as drastically as others during this recessionary period and is expected to show growth in 2010. Construction tools and anodes have been good markets in the past but have been hurt with the current situation in that market. Wrought applications are also attracting increased interest, particularly in extrusions and rolled products and are expected to show growth in 2010.

The automotive market has traditionally been viewed as the segment with the highest potential for having the greatest tonnage increase effect on our industry. The current woes of the automotive industry need not be recounted here. It is apparent that this market itself has decreased substantially for the foreseeable future. The drivers for an increase in use of magnesium as an alternative material in automotive applications are threefold and of course relate to the need for weight reduction: 1. Increased fuel prices are driving consumers to more fuel efficient vehicles; 2. Increased environmental concerns such as global warming and CO2 emissions are becoming a recognized global issue; and 3. Global sustainable mobility goals are gaining significant attention by government and industry. The obvious solution to achieving emissions targets while meeting the needs of consumers and the needs of automotive industry to stay competitive is magnesium. Magnesium offers 33% weight savings versus aluminum and a 50% weight savings versus steel. Unfortunately the aforementioned drivers are being minimized by the global downturn and losses experienced by the industry. Also of note is the incidence of material reversal wherein immediate past periods of high magnesium prices applications were lost to competing materials. Reversal back is typically not a quick endeavor. Current and potential applications are noted. (Figure 4).

Body Structures	**Powertrain**
Instrument Panel Beams	4WD Transfer Cases
Radiator Supports	Transmission cases/housings
Seat Frames	Clutch housings
Center Consoles	Axle Carriers
Steering Column	**Engine Components**
Brackets	Head covers
Steering Wheel	Intake manifolds
Armatures	Oil pans
Front of Dash Panels	Bed plates
Doors	Engine block
Lift gates	Engine mount brackets
Head lamp bezels	Alternator brackets
Roof bows	A/C Compressors
Mirror frames	**Chassis**
Spare tire carriers	Front sub-frame/
Shifter Housings	Engine cradle
	Rear sub-frame

Figure 4. Current & Potential Magnesium Automotive Applications

Challenges to the Industry

As is always the case where there are opportunities there are challenges that go hand in hand, with the converse being true also. There are several areas which will present growing challenges both to our industry in general and for the growth in the expansion of use of metallic magnesium specifically. When looking at the challenges facing the opportunity it cannot be ignored that the concentration of magnesium production within one country, in the current case China, often has or might have an amplifying effect.

In the case of looking at the opportunities for magnesium earlier discussed relative to the United States it cannot be ignored that the largest producing country of both pure and magnesium alloys has been effectively barred from selling in what is arguably one of the world's primary magnesium markets. The harsh reality, without judging in any way the events leading to the situation, is that duties that have been imposed by the US government effectively & economically having ended the exportation from China and importation to the United States of both pure and alloy magnesium. If such duties are rescinded or lessened substantially the affect could be significant.

When one is looking at competitive advantages and disadvantages the most obvious cannot be ignored, price. For magnesium the most critical factor relative to price becomes its comparative price to the material it is most often competing with, aluminum. In the following charts you will see the recent price movements of magnesium and can also note the secondary factor of "where." The following charts are based on the average monthly "Low" and the average monthly "High" for each month 2007 through YTD 2009. Prices are for Primary Magnesium, minimum 99.9 % pure. [6]

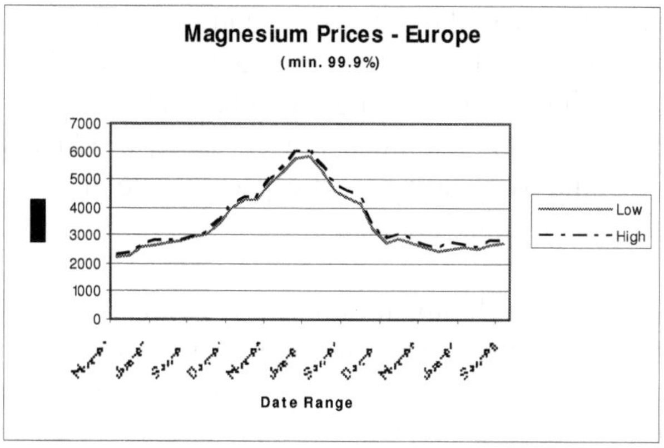

Figure 5. Magnesium Prices - Europe

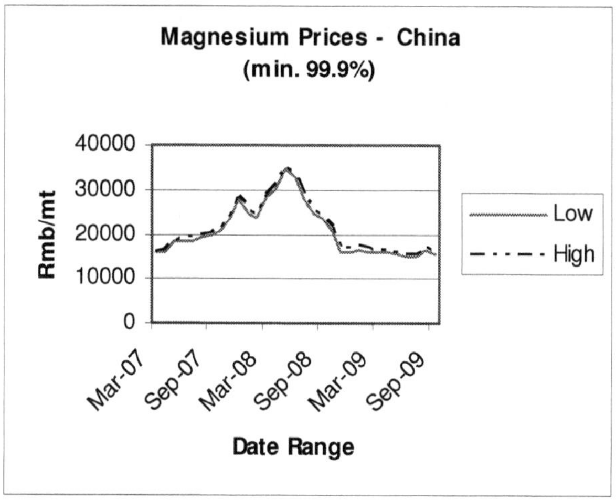

Figure 6. Magnesium Prices - China

Magnesium prices for Europe are charted in Figure 5. Please note that prices peaked in May 2008 with an Average Low of $5761 metric ton and an Average High of $5956 metric ton. In Figure 6 magnesium prices for China are charted. Prices also reached their high in May 2008 with an Average Low of 34,556 RMB metric ton and an Average High of 35,333 RMB metric ton.

Magnesium Prices FOB China is shown in Figure 7. Prices again peaked in May 2008 with an Average Low of $5644 metric ton and an Average High of $5744 metric ton. In the final Figure 8, magnesium prices for USA (Delivered) are indicated. Prices plateaued for a three month period June through August 2008. Average Low was $3.50 per pound and the Average High was $3.70 per pound. As can be seen there is a marked difference between Delivered USA and the rest of the world.

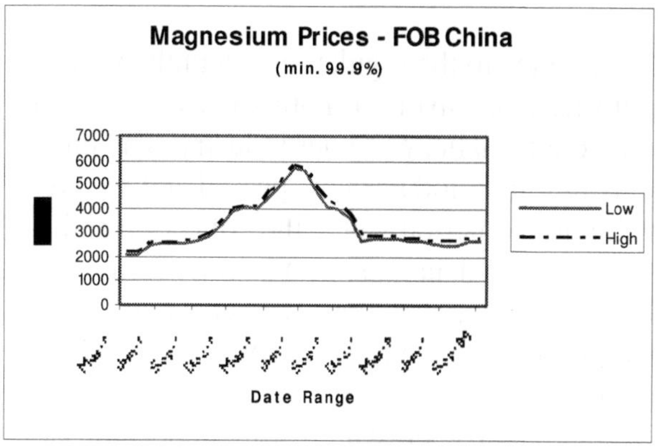

Figure 7. Magnesium Prices – FOB China

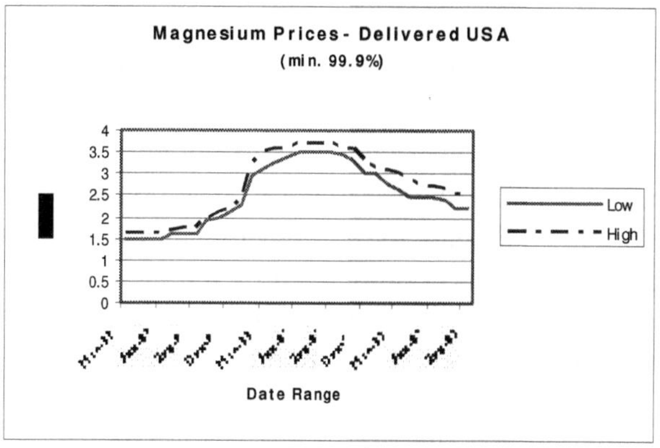

Figure 8. Magnesium Prices – Delivered USA

A dramatic issue for our industry, as well as for all other metals industries, is also looming, largely misunderstood, often denied, and due to the potential scope and impact was thought to make its realization unlikely. Unfortunately this proved untrue and this sweeping legislation and regulation has come to pass. I am, of course speaking of REACH – which stands for **R**egistering, **E**valuating, **A**uthorizing, of **Ch**emicals. Our industry's first obvious reaction was that magnesium is not a chemical, nor for that matter neither is aluminum, copper, zinc or all other metals. Unfortunately for our industry and the other metals industries it was determined by the European Union that for purposes of this far reaching regulation all metals is considered chemicals and thus fall under the regulation which went into effect on June 1, 2007. The regulation itself is over 900 pages of text so a comprehensive summary will be impossible within the context of this paper and presentation.

Compliance to the regulation is a multi-year, multi-task endeavor. Of immediate concern, the immediate challenge to our industry. The pre-registration period has now passed and we are well into the period where the collection and generation of data must be a priority so as to meet the Registration deadlines. The costs of complying with the requirements of REACH for any individual company would and will be substantial. The only mitigating opportunity is that registered participants can use the data collected and submitted to the EU relative to compliance, thus allowing a group, i.e. consortium to spread out the costs of compliance over the members of the consortium. For the magnesium industry the consortium which has been formed is named MaREC. The potentially monumental challenge for our industry relative to Europe again is compounded by the concentration of the source for magnesium and its alloys and how the producing and exporting companies within China react or do not react to a regulation that can cut-off exportation of magnesium if it is not complied to.

Another challenge that will affect our industry and others in general is the growing global environmental concerns. The environmental effects in the primary production of magnesium and its alloys as well as those issues such as cover gases in the manufacturing processes will continue to face increased scrutiny. It is significant that a trend is occurring within China to move away from raw coal as the energy source in using the Pidgeon process of Magnesium production.

Conclusions

We expect that the demand for magnesium will continue to be flat or very minimal for the remainder of the year (2009). In 2010 we foresee the beginning of a turnaround with small growth, particularly in the die-casting and wrought segments, and in particular in automotive and 3C segments. Although there will be continued consolidation at a much slower rate by producers in China we would expect capacity to be in excess of future demands.

China's development into downstream magnesium production (alloys) will continue to expand. Likewise the downstream manufacturing processes in China will continue to grow (die-casting) thus fueling internal magnesium consumption. This trend was well established with an estimated increase in internal consumption in 2006 of 50%. Automotive, 3C applications and niche applications will fuel the growth of magnesium but at a slower growth rate (2-3% in 2010) which will be highly dependent on stability in both pricing and supply from China.

References

1. G. Patzer, International Magnesium Association, Wauconda, IL
2. China Magnesium Association, Beijing, China
3. United States Geological Survey, 2007 Minerals Yearbook-Magnesium
4. United States Geological Survey, Mineral Commodity Summaries, Magnesium-Metal, 2009-2004
5. United States Geological Survey, Mineral Industry Surveys-Magnesium, Quarterly, 2009-2006
6.Metal-Pages, www.metal-pages.com, Historical Pricing Archives

From *Magnesium Technology 2007*, R.S. Beals, A.A. Luo, N.R. Neelameggham, and M.O. Pekguleryuz, Editors

Advances in Manufacturing Processes for Magnesium Alloys

Hajo Dieringa, Jan Bohlen, Norbert Hort, Dietmar Letzig, Karl Ulrich Kainer
GKSS Research Center, MagIC – Magnesium Innovation Center
Max-Planck-Str. 1, 21502 Geesthacht, GERMANY

Keywords: Magnesium, High Pressure Die Casting, Protective Gases, Wrought Magnesium, Research Projects

Abstract

The present paper reviews the present potential of magnesium and its alloys as structural light-weight material in transportation industry. An overview will be given on the state of the art in casting technologies of magnesium alloys in order to broaden possible applications in light weight construction. Further, wrought magnesium will be emphasized as a challenge for developing magnesium towards a competitive constructional material. Future needs in all regards involve further alloy development which has to need the special requirements for each process. An overview on actual research activities in Europe and also especially in Germany will be given.

Introduction

Light weight construction is one of the modern answers to the increasing usage of natural resources as well as increasing emissions in the transportation industries. Magnesium and its alloys offer a remarkable potential in this regard [1] since their low density makes them the lightest available constructional metals.

Applications that can be realized by casting technologies are already well established [2]. The good castability of magnesium and its alloys allows the efficient production of components that require reasonable mechanical properties as well as temperature stability. The development of protective gases for casting processes is an important issue due to newly introduced restrictions in Europe.

An even higher potential which is not readily used today is the application of semi-finished components of magnesium alloys. The body-in-white of an automobile, for example, has a fraction of 20 to 25% of the total mass and therefore offers opportunities for saving weight by using magnesium. Rolled sheets or extruded profiles allow a broader application by using post-processing shape giving technologies, but also require improved formability. Generally, semi-finished components exhibit an improved quality and microstructural homogeneity than their cast counterparts [3]. While the technology for processing magnesium sheets and profiles is available and common alloys for sheets and profiles have existed for more than 60 years [4], there is no large-volume application for such components at present. The worldwide consumption of wrought magnesium alloys is only about 8.000 – 10.000 tons with 25 – 30 % of this being used in structural applications. In this regard requirements for improved alloys for cast and wrought purposes have to be discussed. Various manufacturing processes and alloy compositions result in specific microstructures and properties which determine the potential use of these materials. Today, the spectrum of available alloys covers standard alloys as well as alloys for use at elevated temperatures. The trend will be to further diversify the alloy compositions and to improve the alloy properties.

Since an increased interest in magnesium and its alloys for structural application especially in the automotive industry came up almost a decade ago, an appreciable number of R&D activities were carried out. Driven by a significant demand for magnesium alloys designed for special purposes, research programs for the improvement in the property profile and the production technology were started. The research topics include fundamental and applied research with the aim to extend the potential for using of magnesium components in automotive applications. It seems worthwhile to review these activities.

Developments in cast alloys

Magnesium die casting is the largest application field when magnesium is used as structural metal. The increase in die casting application in the last decade occurs due to the wider range of applications for products. While in 1983 27.000 tons had been used for constructional purposes based on die casting there is in 2004 already more than 150.000 tons of magnesium alloy used for this application.

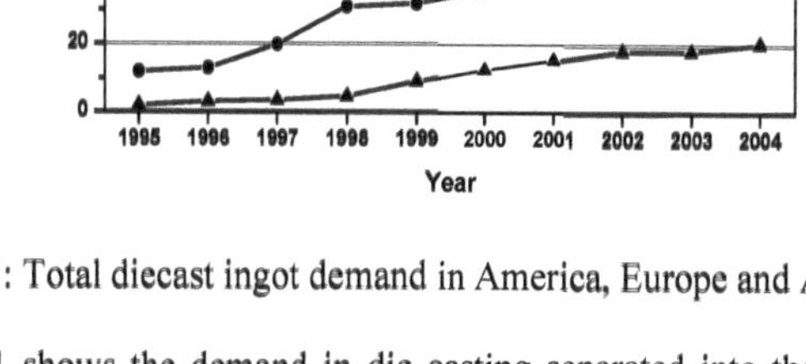

Fig. 1: Total diecast ingot demand in America, Europe and Asia

Fig. 1 shows the demand in die casting separated into the three most important areas America, Europe and Asia over the last ten years. Whereas the demand in Europe and Asia increases continuously there was a significant cutback in America after 1999. Fig 2 and 3 show the different application fields of magnesium die casting over the last five years and forecasting for powertrain and structural application until 2010. Data are based

on Hydros Magnesium probability forecast for automotive application presented in [5].
The main target of alloy development for cast magnesium alloys in the automotive industry is high temperature stability. Silicon and rare earth elements (RE) have already been identified as alloying elements that form stable phases with aluminum. The AS-alloy family (Mg-Al-Si) that has been in use for example in the Volkswagen Beetle and the AE-alloy family (Mg-Al-RE), that shows even improved properties at elevated temperatures are examples for die casting alloys with improved mechanical properties. Alloy development has been continued with development of gravity cast QE-alloys (Mg-Ag-RE) and WE-alloys (Mg-Y-RE). These alloys show a further improvement in high temperature properties but these systems are not suitable for high pressure die casting (HPDC).

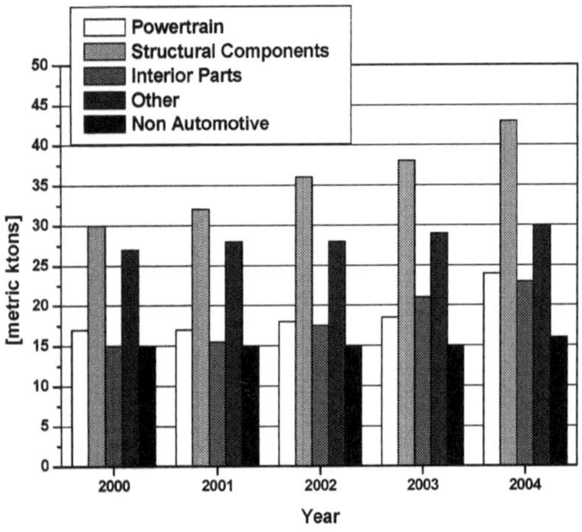

Fig. 2: Applications of diecast magnesium

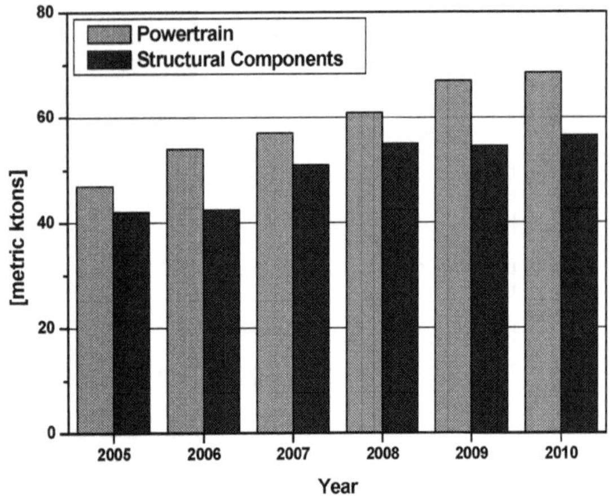

Fig. 3: Forecast of growing magnesium usage in cars

The modification of AZ-family and AM-family (Mg-Al-Mn) started with the use of Si, Sn, Ca, RE and Sr as additional alloying elements [6]. Several patents have been written with these metals as alloying elements but only a few of them have been introduced to a series production. Close to a production are alloy developments produced from the Magnesium Research Institute in cooperation with Volkswagen [7].

The aluminum and silicon containing AS-family shows improved creep properties compared to AZ91 and was already used for years in the Volkswagen Beetle. It shows limited castability in high pressure die casting and also some problems in corrosion behavior. Norsk Hydro made developments in the AS-alloys by using high purity alloying elements and it could also be shown that modern HPDC facilities are able to handle this alloy as well. A modified AS31 alloy was used by Daimler Chrysler [8] for production of an automatic transmission housing [Fig. 4].

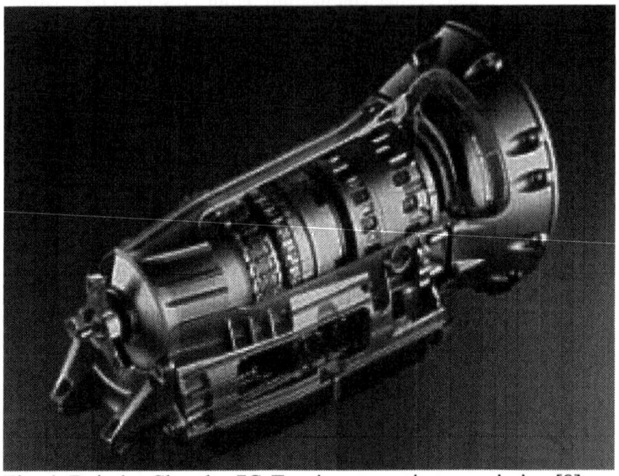

Fig. 4: Daimler Chrysler 7G-Tronic automatic transmission [8]

Australian researchers at Cooperative Research Centre for Cast Metals Manufacturing (CAST) in cooperation with the Australian Magnesium Corporation (AMC) developed the sand cast alloy AMC SC1. The alloy is based on the magnesium-rare earth system and will be used in the fully heat-treated T6 condition. Creep and bolt load retention properties are comparable to the aluminum alloy A319 [9].
Sr is another newly introduced alloying element for magnesium alloys. Noranda developed the AJ-alloy family (Mg-Al-Sr) [10, 11, 12], which is in use for instance in the BMW hybrid crankcase (Fig. 5). This part is composed of an open-deck insert made of AlSi17Cu4Mg alloy, which for better bonding is coated with AlSi12 using an arc spraying process. This engine block will be produced in a series of more than 600.000 pieces.
Another newly developed magnesium alloy is the ACM-family (Mg-Al-Ca-Mn) [6]. It was developed by Toyota but at this time there is no real application that is in service.

Fig. 5: BMW hybrid crankcase

Protective gases for casting

Protective gases are mainly in use during melting operations of magnesium alloys. In the past SF_6 was in use in combination with carrier gases like dry air, nitrogen or argon [13-19]. In accordance with a decrease of the amount of green house gases SF_6 also came into discussion due to its enormous potential as a green house gas with a global warming potential (GWP) of 23.600 compared to CO_2. There is no doubt that the amount of SF_6 used in magnesium foundries will be limited in the future. In the EU a threshold of 800 t per year usage is discussed. If this value is not reached, SF_6 will still be allowed. For higher amounts of magnesium other alternatives need to be introduced.

The suppliers of magnesium alloys and protective gases are aware of this discussion and they already started to develop and to test new gases in combination with appropriate feeding systems. Mainly fluor containing gases (so called F-Gases) which are already used in cooling systems became interesting candidates for the use in magnesium foundries. The gas HFC134a (R134a, CH_2FCF_3) introduced by AMG as a protective gas for magnesium melts already proved its ability to be used as a replacement for SF_6. This gas is having a GWP of 1600 and is already commercially available. By 3M another fluor containing protective gas has been introduced under the name Novec 612 ($C_3F_7C(O)C_2F_5$). It shows a GWP of approximately 1. Also manufactured by 3M is HFE 7100 ($C_4F_9OCH_3$), a fluorinated ether that is used as a cleaning agent. This gas is showing a GWP of 320 and is also an alternative for SF_6. Unfortunately another discussion raises also doubt about future use of fluorinated gases at all. In line with the Council position last year, a ban will apply. The EU and the USA agreed to restrict the use of HFC134a for uses where containment is deemed inappropriate (magnesium die-casting, vehicle tires, windows, footwear, non-refillable containers, certain foams, self-chilling drinking cans, certain aerosols, new fire protection systems, fire extinguishers) starting at 2011 while HFC152 is still allowed. From 2017 all HFC gases should be completely banned in cars [20, 21].

While in most European countries the discussion of the allowed amount of SF_6 used in foundries is still not completed Austria has totally abandoned the use of SF_6 and it has been replaced completely by SO_2. By using SO_2 special care has to be taken because this is a toxic gas and also corrosive but this gas has proved its ability to replace SF_6 and additionally it is showing no GWP at all. Additionally the use of CO_2 snow as a protective gas has been shown successfully [22].

Developments in wrought magnesium

Sheets, extruded profiles and forged parts are in the focus of interest increasingly. Their advantages compared to cast parts are better mechanical properties and the ability to produce thin walled or long structures. There are in present only few applications for magnesium wrought alloy. One reason is the small number of alloys. Commercially established wrought magnesium alloys are those containing aluminum, zinc or manganese as the main alloying elements [23]. Typical examples of magnesium-aluminum alloys are those of the AZ-series that also have a small amount of zinc. Alloys containing zinc but no aluminum often make use of the strong grain refinement that can be achieved with small additions of zirconium, e.g. alloys of the ZK-series. Manganese is used to improve formability (e.g. M1A). It also tends to form harmless intermetallic compounds with impurities such as Fe, Ni, or Cu which originally enhance corrosion. Lithium as an alloying element has also been discussed in terms of its influence on the c/a ratio of the hexagonal lattice structure and a resulting improvement in formability [24, 25]. Rare earth elements have been used to form precipitates that lead to improvements in mechanical properties, e.g. in ZE10 or its modern modification with zirconium, ZEK100 or in alloys of the WE-series which are used for extrusion. Actual alloy development focuses on grain refinement strategies of the feedstock for an improvement in formability at least at elevated temperature during processing. Examples are given in Fig. 6. In aluminum-free alloys zirconium offers a very effective solution in this regard, however, in aluminum containing magnesium alloys there is still a need. A number of grain refining additives are discussed, such as Ca, Sr [26, 27] or even Mn. While Ca shows a tendency to grain refinement, it has recently been shown, that Mn is not suitable in this regard, although a primary formation of β-Mn or intermetallic compounds with Al such as Al_8Mn_5 forms during solidification [28]. Small pinning particles from stable intermetallic compounds could be used as grain stabilizers preventing further grain growth to a certain extent.Sheets are important semi-finished products with regard to their universal applicability for press forming. Together with their improved mechanical properties, they are an important extension to allow increasing usage of magnesium in industrial applications. As with extrusion, the technology to process various alloys to sheets is available today but only few suppliers can be found with small product lines regarding different alloys. One of the most important issues with magnesium sheets is the occurrence of a distinct yield stress asymmetry with quite remarkable variations dependent on the orientation. This is the result of strong textures that are generated during rolling [29].

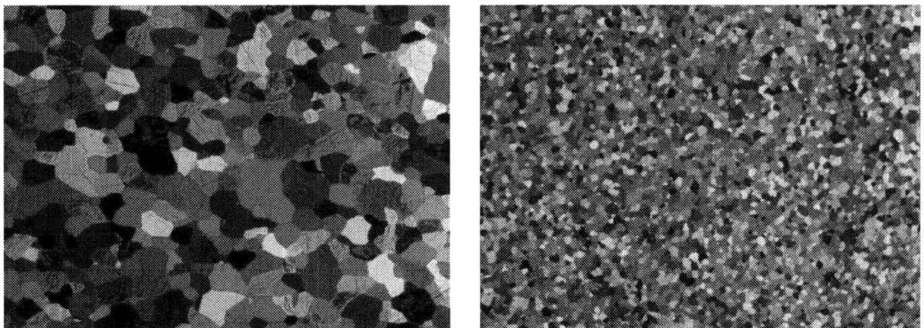

Fig. 6: Micrograph (longitudinal section) taken from a ZR modified cast billet of a) M1 based: 120 – 170 μm based b) ZE10: 50 – 70 μm

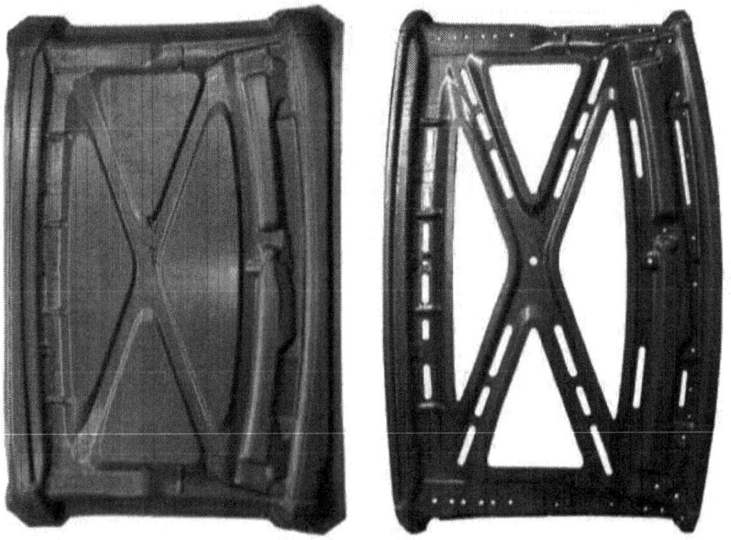

Fig. 7: Drawn component made of magnesium sheets. Left: after drawing; right: after cutting [30].

The optimization of this is one of the major issues for current alloy and process development activities. Furthermore, rolling strategies for magnesium alloys have to be rather different than for example aluminum alloys, because of the generally low formability of magnesium. Also all processing steps have to be carried out at elevated temperatures. In summary, this increases the costs for magnesium sheet production for a product that has to be optimized regarding its mechanical properties.

In Fig. 7 components made of magnesium sheets delivered from Salzgitter Magnesium-Technologie GmbH after drawing are presented. Responsible for drawing and industrial application was DaimlerChrysler. It could be shown that drawing of magnesium sheets needs some technological modifications compared to steel or aluminum sheets but generally it is possible.

Forging of magnesium alloys is a great research challenge and is used with the perspective to add to die-cast products in terms of an improved mechanical behavior. The intention is the substitution of aluminum forgings by magnesium forgings showing the same mechanical properties. Therefore the development of new fine grained magnesium alloys with improved formability at high strain rates and lower temperature is necessary.

The potential of magnesium wrought components is not readily used today. Present R&D activities lead to an increasing understanding of magnesium alloys and the related forming processes. An application will be the result of further integrated development of alloys and related processes. Nevertheless, potential application of magnesium profiles will strongly depend on the question whether established forming processes for aluminum and steel can be easily adapted to magnesium.

Research and development activities in Europe

The main objective of considering magnesium and its alloys as an alternative construction metal is its weight saving potential. Focusing on the European region and also especially Germany as an initial driving focus in magnesium research it has been understood that in this regard a potential is available which is not readily used today. This is also the reason for an initiation of national and international funding for projects that can help to overcome the limitations of the material for its industrial success over the last decade and still ongoing.

Magnesium based projects supported by European Commission are instrumented in terms of the framework programs of the EC. Actual projects that are still operating are part of the 6th Framework Program whereas the 7th framework program is presently under consideration. An overview can be found by using the database at CORDIS (Community Research & Development Information Service) [31]. This includes detailed information about projects, grants and different framework programs. While in the beginning pure research with only a few contractors involved was funded, the projects changed to larger programs with a great number of contractors or even to thematic networks. Actually, the funding is given to investigate the integration of magnesium in combination with different materials and existing production chains. Some examples will be given in the following:

MAGDOOR was a project with the objective to decrease emissions through reduced fuel consumption by reducing the weight of a car body door. This required an industrial application of magnesium at same or lower component cost than existing steel door concepts. These Objectives were approached by replacing steel sheet with magnesium shell castings on a car door. The project started in January 1996 and ended in December 1999. It was coordinated by Volvo.

MAGJOIN was a project dealing with the development of a suitable joining technology for magnesium components and for aluminum-magnesium joints to overcome problems of fragile failure of the joints due to metallurgical and technological aspects. This included the development of a new welding process based on laser melting. Friction welding was also promising in joining of different materials and was investigated in this project. The project started in January 2000 and ended in March 2003. It was coordinated by CRF (Fiat).

MAG-ENGINE was a project that aimed to reduce pollution by introducing lightweight magnesium as construction material in the engine area. The specific objectives were the development of Mg-alloy(s) with improved high-temperature properties and the determination of high-temperature mechanical properties for selected Mg-alloys. Generic design criteria for magnesium based engine blocks were to be described and engine block prototypes from magnesium to be produced. The project started in May 2005 and ended in December 2005. The project coordinator was Risoe National Laboratory.

MAGNEXTRUSCO was a project that aimed the development of a new cost-efficient process for the production of lightweight, cost-effective and safe magnesium structures, based on the hydrostatic extrusion principle. These extrusion profiles and die-cast components, meant for applications in transport means, are targeted to be produced with extrusion speeds of the same order of magnitude as. The project furthermore aimed to generate the know-how for the processing of magnesium alloys tailored for hydrostatic extrusion, which have better extrudability and improved functional performance and to gather the know-how on working and joining techniques for magnesium extrusion profiles.

AEROMAG is a project initiated by EADS for evaluating the usage of magnesium as a break through technology in the aerospace industry if used for cost efficient, low weight components and airframe structures. The technical focus of the university driven project is the development of new Magnesium wrought products such as sheets and extrusions that provide significantly improved static and fatigue strength properties. Stakeholder is the European aeronautical and aerospace industry. Main targets are the corrosion behavior and flammability of magnesium and its alloys. Material adapted design and the evaluation of structural behavior will be investigated to close the process and development chain for aeronautic components. The project started in March 2005 and shall end in February 2008.

MAGFORGE is a project that addresses the issue of weight reduction of structural components by using magnesium in terms of providing tailored and cost-effective technologies for the industrial manufacturing of magnesium forged components. Current constraints are to be resolved in order to prepare the European forging industry for this innovation. Hence, the technological basis for the design and manufacture of magnesium forgings needs to be reinforced and dissemination and training needs to be undertaken. The targeted advancements are to enable high-volume applications of lightweight magnesium forgings. Stakeholders are the forging industry (associations and their SMEs), but also the material producers and end users as well as suppliers of auxiliaries and engineering services. The project leader is TNO Industrial Research in the Netherlands. Details are available on the projects website [32] and [33]. The project is launched in July 2006 and shall end in June 2009.

ACRONYM	PROJECT TITLE	PRO-GRAMME	DURATION	COORDINATOR	REF.
MAGDOOR	Magnesium in Industrial Application by Manufacturing a Complete Car Door with a Magnesium Door Inner Shell	FP 4	01/1996 – 12/1999	AB Volvo Technological Development	[31]
MAGJOIN	New joining techniques for light magnesium components	FP5	01/200 – 03/2003	Centro Ricerche Fiat S.C.P.A.	[31]
MAG-ENGINE	Light weight engine construction through extended and sustainable use of mg-alloys	FP5	05/2005 – 12/2005	Risoe National Laboratory	[31]
MAGNEXTRUSCO	Hydrostatic extrusion process for efficient production of magnesium structural components	FP5	03/2001 - 08/2004	AUDI AG	[34]
AEROMAG	Aeronautical Application of Wrought magnesium	FP 6	03/2005 – 02/2008	EADS Deutschland GMBH	[31]
MAGFORGE	Magnesium Forged Components for Structural Lightweight Transport Applications	FP 6	07/2006 – 06/2009	Netherlands Organisation for Applied Scientific Research (TNO)	[32, 33]

Research and development activities in Germany

National funded projects in Germany experienced a strong focus in the last 3 years. The German government supports joint research projects between research facilities and industry partners. The most important aspect of funding is to support innovations for applied research. Projects funded by the BMBF have to be conducted in Germany. Funding is non-repayable and especially granted to small and medium-sized companies. After a variety of projects present activities are mainly focusing in to area of magnesium sheet production.

An actual initiative in this regard is named "M^3: Mobil mit Magnesium (Mobile with magnesium)". The objective of this project is the development of solutions for weight-based construction based on magnesium sheets in the automotive industry. The project is mainly application driven and focused on a realization of a cost effective processing of semi-finished components. However, it includes an evaluation of the complete process chain from cast feedstock to the design of selected properties. The project started in October 2006 and is planned for 3 years. Coordinator of the project is the Salzgitter Magnesium-Technologie GmbH SZMT in Salzgitter in Germany.

Projects funded by German Research Association "Deutsche Forschungsgemeinschaft" (DFG) are mainly focused on basic research to show the feasibility of processes, properties of the chosen material etc. DFG funded projects are located mainly at research institutes and universities. Normally the funding is given for two years with the possibility of extension for one more year. Details about grants, projects etc. are available on the website of DFG [35]. Funded by this research organization is one major priority program entitled "Extending the Range of Applications of Magnesium Alloys". It combines 16 university based projects with a funding of almost 3.5 million EURO. The aim of this priority program is the development of a continuous concept for the development of new and innovative magnesium alloys with tailored properties to support the increasing requirements of future light weight constructions. Therefore the entire production chain has to be evaluated critically. This starts with alloy development over the influences that processing (casting, deformation, surface treatment) gives to the microstructure. Based on know facts in modeling and simulation and the property-microstructure relationships new tools have to be developed to allow an entire view on the processing route. This is accompanied by modeling and simulation over the entire process chain. Therefore the following topics describe the major working topics of the priority program:

- Systematic investigations on the influences of micro-alloying elements in known and new magnesium alloys including their use ability as grain refining agents during casting
- Determination of the influences of thermo-mechanical treatments on microstructure during solidification and deformation processes
- Investigation of relationships between processing parameters and resulting microstructure with a special regard to materials and process models, including the life time behavior of components
- Modeling of materials properties for a latter use in modeling of components properties and life time behavior.
- Development of innovative joining techniques to improve similar and dissimilar joints (Mg-Al, Mg-Fe, Mg-polymers).
- Systematic investigations to improve corrosion behavior including the influence of micro-alloying elements and processing parameters on the microstructural development during processing
- Development of suitable surface coatings to improve corrosion and wear

References

[1] H. Friedrich, S. Schumann, J. Mat. Processing Technology 117 (2001) 276 – 281.

[2] H. Dieringa, P. Maier, D. Fechner, J. Bohlen, K. U. Kainer, Proceedings of the First Russian International Conference and Exhibition of Magnesium – Broad Horizons, 29.11 – 01.12.2005.

[3] J. Fischer, G. Becker, In: Magnesium – Eigenschaften, Anwendungen, Potenziale, edited by K.U. Kainer, VCH Wiley Weinheim (2000), 99 – 110.

[4] A. Beck, Magnesium und seine Legierungen, Springer, Berlin (1939).

[5] D. Webb: Magnesium Supply and Demand 2004; presented at "IMA Annual Conference" Berlin (D) 2005; http//www.intlmag.org

[6] A.A. Luo: Recent magnesium alloy development for powertrain applications; Mat. Sci. Forum 419-422 (2003) 57-66.

[7] E. Aghion et al.: The environmental impact of new magnesium alloys on the transportation industry; Magnesium Technology 2004 Editor: A.A. Luo, TMS (2004) 167-172.

[8] A. Barth: Mg application in the 7G Tronic gear; Proc. 61st Annual World Magnesium Conference 2004 New Orleans/USA (2004) 81-86.

[9] C.J. Bettles et al.: AMC SC1: A new magnesium alloy suitable for powertrain applications; SAE Technical Paper 2003-01-1365, (2003).

[10] M.O. Pekguleryuz and A.A. Kaya: Magnesium diecasting alloys for high temperature applications; Magnesium Technology 2004 Editor: A.A. Luo, TMS (2004) 281-287.

[11] M.O. Pekguleryuz et al.: Magnesium diecasting alloy AJ62x with superior creep resistance; Magnesium Technology 2003 Ed.: H.I. Kaplan, TMS (2003) 201-206.

[12] E. Baril et al.: AJ (Mg-Al-Sr) alloy system used for new engine block; SAE Technical Paper Series 2004-01-0659.

[13] J. E. Hillis, The international program to identify alternatives to SF_6 for magnesium melt protection, The International Conference on SF_6 and the Environment, 2002, San Diego, California, USA

[14] G. Tranell, T. A. Engh, Alternatives to SF_6 for the magnesium processor – A technical, environmental and econonomic assessment, Proc. 61st Annual World Magnesium Conference 2004, New Orleans, Louisiana, USA, 161-174

[15] S. P. Cashion, et al., The protection of molten magnesium and its alloys during die casting, The Annual IMA Automotive Seminar, 2000, Aalen, Germany

[16] D. Milbrath, Development of 3M Novec 612 Magnesium protection fluid as substitute for SF_6 over molten magnesium, The International Conference on SF_6 and the Environment, 2002, San Diego, California, USA

[17] G. Franke, J. Schlimbach, Use of SO_2 as protection gas for Mg-recycling, Proc. 61st Annual World Magnesium Conference 2004, New Orleans, Louisiana, USA, 185-190

[18] K. Renger, et al., The protection of molten magnesium by SO_2 gas mixtures, Proc. 61st Annual World Magnesium Conference 2004, New Orleans, Louisiana, USA, 191-198

[19] R. Bailey, N. Ricketts, AM-cover: A production proven SF6 replacement for the magnesium die casting industry, Proc. 61st Annual World Magnesium Conference 2004, New Orleans, Louisiana, USA, 199-204

[20] http://www.euractiv.com

[21] http://europa.eu.int

[22] F-W. Bach, A. Karger, C., Pelz, M. Schaper, Use of CO2-Snow for Protecting Molten Magnesium from Oxidation, In: Magnesium Technology 2005, Eds.: Neale R. Neelameggham, Howard I. Kaplan and Bob R. Powell (2005) 33-36.

[23] C.J. Bettles, M.A. Gibson, JOM 5 (2005) 46 – 49.

[24] F.-W. Bach, M. Rodmann, A. Roßberg, Materialprüfung, 47 (2005) 661-665.

[25] A. Ahmadieh, Jack Mitchell, J.E. Dorn, Trans. Met. Soc. AIME (1965) 1130 – 1138.

[26] Y. C. Lee, A. K. Dahle and D. H. StJohn, Metallurgical and Materials Transactions, Nov. 2000, 31A, 11, pp. 2895-2906.

[27] P. Li, B. Tang, E.G. Kandalova, Materials Letters 59 (2005) 671-675.

[28] T. Laser, M. R. Nürnberg, A. Janz, Ch. Hartig, D. Letzig, R. Schmid-Fetzer, R. Bormann, Acta Materialia (2006) accepted.

[29] S.R. Agnew: Plastic Anisotropy of Magnesium Alloy AZ31B Sheet, Proceedings TMS2002, Seattle, pp. 169-174.

[30] P. Bogon: Werkzeugtechnik zur Magnesium-Blechunformung – ein Beispiel; www.UTFscience.de 2 (2005) 1-10.

[31] www.cordis.lu

[32] www.magforge.eu

[33] W.H. Sillekens, D. Letzig, Proceedings DGM Conference Magnesium, Dresden 2006 (2006) in press

[34] J. Bohlen, J. Swiostek, W.H. Sillekens, P.-J. Vet, D. Letzig, K.U. Kainer, In: Neale R. Neelameggham, Howard I. Kaplan, Bob R. Powell Ed. „Magnesium Technology 2005", TMS Annual Meeting San Francisco (2005), 241.

[35] www.dfg.de

MATERIALS COMPARISON AND POTENTIAL APPLICATIONS OF MAGNESIUM IN AUTOMOBILES

Alan A. Luo

Materials & Processes Laboratory
General Motors Research & Development Center
30500 Mound Road
Warren, Michigan 48090-9055

Abstract

In this paper, the material properties, structural performance, mass saving potential, design and manufacturing characteristics of magnesium are compared with various competing materials such as cast iron, steel sheet, aluminum alloys and polymers. The current and potential automotive applications of magnesium are reviewed, and the technical challenges for these applications are also discussed. Recent alloy development for powertrain applications and the creep resistance of several experimental magnesium alloys are reviewed.

Introduction

Weight reduction for automobile fuel economy has spurred the growth of magnesium consumption over the last decade at an annual rate of 15%. This growth is predicted to continue at a rate of 12% per year over the next 10 years [1].

Fig. 1 shows the estimated magnesium consumption by the Big Three automotive companies, and Fig. 2 compares the aluminum and magnesium contents in 1999 model year vehicles built in North America [2, 3]. The data shows that General Motors is the leading user of magnesium in both total consumption and content per vehicle. The new GMT800 full-size pickup trucks (Silverado and Sierra), for example, contain 27 pounds of magnesium (instrument panel, transfer cases, steering wheel and side mirror brackets) per vehicle. Despite the recent strong growth, the use of magnesium still lags aluminum: 271 pounds of Al vs. 8 pounds of Mg per GM vehicle [3].

This paper compares the material and process properties of magnesium against several major automotive materials including cast iron, steel sheet, aluminum and polymers. The current and potential magnesium applications in automobiles are reviewed, and technical challenges are also discussed.

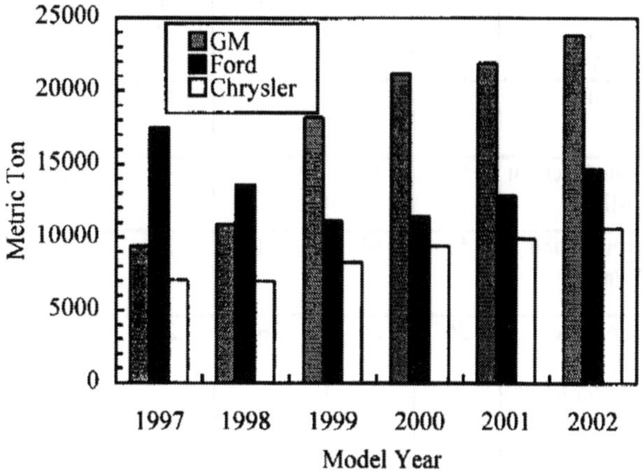

Fig. 1: Estimated magnesium consumption for the Big Three [2].

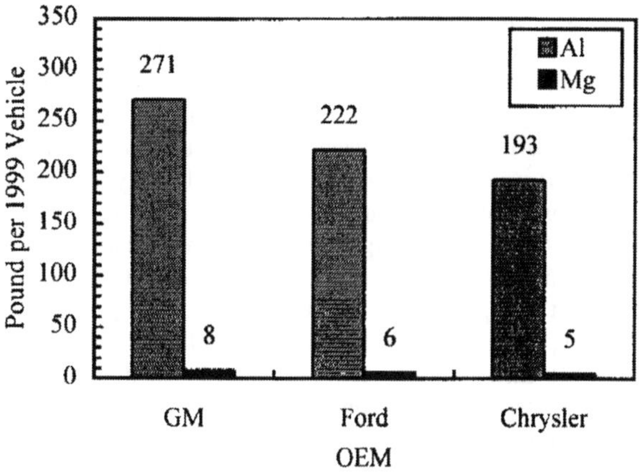

Fig. 2: Average aluminum and magnesium contents in 1999 vehicles of Big Three [2, 3].

Magnesium vs. Competing Materials

Material Properties

Table I summarizes the mechanical and physical properties of typical cast and wrought magnesium alloys in comparison with other competing materials [4-8]. Being the lightest structural metal, magnesium has a density only slightly higher than that of plastics (Fig. 3). Where strength is concerned, the workhorse die cast magnesium alloy AZ91 has the same yield strength and ductility as its aluminum counterpart A380. In spite of its lower fatigue strength, AZ91 has made significant inroads in replacing

Table I Comparison of mechanical and physical properties of various materials [5-8]

Material	Cast Mg		Wrought Mg		Cast Iron	Steel	Cast Al		Wrought Al		Plastics (PC/ABS)
Alloy/Grade	AZ91	AM50	AZ80-T5	AZ31-H24	Class 40	Galvanized	A380	A356-T6	6061-T6	5182-H24	Dow Pulse 2000
Process/Product	die cast	die cast	extrusion	sheet	sand cast	sheet	die cast	P/M* cast	extrusion	sheet	injection molding
Density (d, g/cm^3)	1.81	1.77	1.80	1.77	7.15	7.80	2.68	2.76	2.70	2.70	1.13
Elastic Modulus (E, GPa)	45	45	45	45	100	210	71	72	69	70	2.3
Yield Strength (YS, MPa)	160	125	275	220	N/A	200	159	186	275	235	53
Ultimate Tensile Strength (S_t, MPa)	240	210	380	290	293	320	324	262	310	310	55
Elongation (e_f, %)	3	10	7	15	0	40	3	5	12	8	5 at yield and 125 at break
Fatigue Strength (S_f, MPa)	85	85			128		138	90	95		
Thermal Cond. (l, W/m.K)	51	65	78	77	41	46	96	159	167	123	
Thermal Exp. Coeff. (d, mm/m.K)	26	26	26	26	10.5	11.70	22	21.5	23.6	24.1	74
Melting Temp. (T_m, °C)	598	620	610	630	1175	1515	595	615	652	638	143 (softening temp.)

* P/M: permanent mold

A380 aluminum in non-structural and low-temperature components such as brackets, covers, cases and housings, providing essentially the same functionality with significant weight savings. For structural applications such as instrument panels, steering systems and seating structures where the crashworthiness is important, magnesium die cast alloys, AM50 or AM60, offer unique advantages of high ductility (10-15% elongation) and impact strength over aluminum die cast A380 alloy. To achieve such a high ductility in aluminum alloys, primary A356 alloy has to used in heat-treated conditions and it has to be cast by higher cost processes such as gravity permanent mold casting, low-pressure casting, squeeze casting or semi-solid metal (SSM) casting. The newly developed low-iron ductile aluminum alloys for die casting [9] are not as castable as magnesium alloys.

For extrusion alloys, Mg alloy AZ80 provides comparable tensile strength as Al alloy 6061, but is less ductile. Mg sheet metal, such as AZ31 alloy, offers slightly lower strengths, but a higher ductility, than commonly used 5xxx series Al sheet alloys. Although slightly heavier than plastics, magnesium components are much stiffer due to the fact that its elastic modulus is almost 20 times as high as a plastic material such as PC/ABS (a blend of polycarbonate and acrylonitrile-butadiene-styrene). Plastic materials are generally not suitable for elevated temperature applications due to their low softening temperatures (e.g., 143°C for the PC/ABS material).

Although the general corrosion resistance of high purity magnesium alloys is even better than A380 aluminum alloy in atmospheric or saline environments, galvanic corrosion remains a concern for design and applications engineers in developing magnesium components. The degree of galvanic corrosion attack on magnesium is strongly influenced by the compatibility of the second metal in contact with the magnesium parts. Fig. 4 shows a compatibility series for various metals with magnesium [10], with Al alloy 5052 being the most compatible metal and steels the least compatible. Proper design, combined with the use of non-conductive or compatible aluminum shims or washers, is a very simple and efficient way to prevent galvanic corrosion when a magnesium part is attached to another metal.

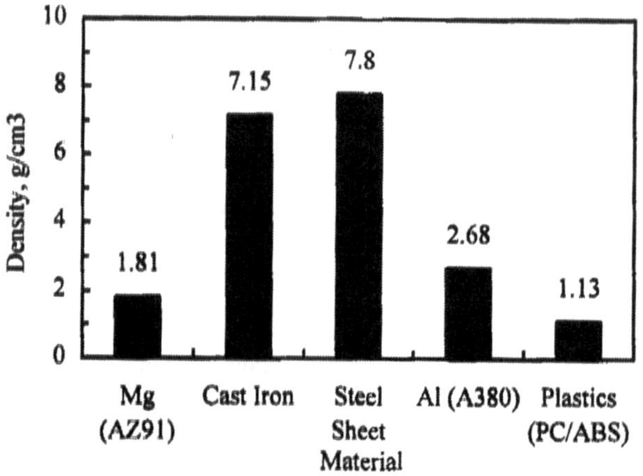

Fig. 3: Materials density comparison.

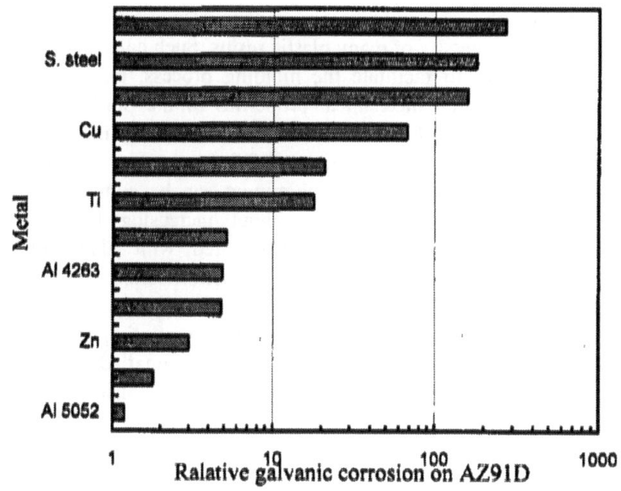

Fig. 4: Relative compatibility series of various metals coupled to magnesium [10].

Another disadvantage of current magnesium alloys is that they have poor creep resistance at temperatures above 125°C [11], which is inadequate for powertrain applications; transmission cases can operate up to 150°C, and engine blocks up to 200°C. Creep resistance is a major requirement for the use of magnesium in automotive powertrains, which are currently made of aluminum or cast iron. The poor creep strength of such components can result in clamping load reduction in bolted joints and poor bearing/housing contact leading to oil leakage and/or increased NVH (noise, vibration and harshness) in powertrains. New magnesium alloys with improved creep resistance are being developed for elevated temperature powertrain applications.

Other properties of magnesium, such as the higher thermal expansion coefficients and lower thermal conductivity compared with aluminum, should also be considered in designing magnesium components.

Structural Performance and Mass Saving Potential

In most applications of magnesium or other lightweight materials, such as an instrument panel beam or a body panel, the mode of loading is primarily bending. Bending stiffness and bending strength are the limiting factors in designing these components. For a flat beam or panel under bending load, the minimum thickness (t) and mass (m) can be calculated using the "materials performance index" concept [4]. Designating steel and magnesium properties with subscripts S and Mg, the thickness ratios and mass ratios of components made of the two materials for an equal stiffness design may be expressed as:

$$t_{Mg}/t_S = \sqrt[3]{E_S/E_{Mg}} \quad (1)$$

$$m_{Mg}/m_S = (d_{Mg}/d_S)\sqrt[3]{E_S/E_{Mg}} \quad (2)$$

where E and d are the elastic modulus and density of the materials, respectively. Using the property data as shown in Table I, the thickness and mass ratios of magnesium (AZ91 alloy) vs. steel beam can be calculated:

$$t_{Mg}/t_S = 1.67 \quad (3)$$

$$m_{Mg}/m_S = 0.39 \quad (4)$$

Therefore, in order to achieve the same bending stiffness, a magnesium beam or panel will be required to have 1.67 times the thickness of a steel one, with a mass (weight) saving of 61%. For bending strength-limited design (same bending strength at minimum mass), such ratios for AZ91 magnesium alloy vs. steel become:

$$t_{Mg}/t_S = \sqrt{YS_S/YS_{Mg}} = 1.12 \quad (5)$$

$$m_{Mg}/m_S = (d_{Mg}/d_S)\sqrt{YS_S/YS_{Mg}} = 0.26 \quad (6)$$

where YS is the yield strength of the materials. Based on the above nomenclature, Table II summarizes the thickness and mass ratios of various lightweight materials compared with steel, for bending stiffness- and strength-limited design, respectively. Figs. 5 and 6 highlight the thickness ratios and the resultant percentage mass savings of the materials when replacing a steel component. These results show that magnesium alloys have a slightly higher weight saving potential when compared to aluminum and plastics in substituting steel structures for equal stiffness or strength. To overcome its much lower elastic modulus, a plastic part will have to be designed 4.5 times the thickness of the steel part for equal bending stiffness, which significantly undermines its weight saving advantage. In most cases, a metal-plastic hybrid structure will have to be used.

Table II Thickness and Mass Ratios of Various Materials Compared with Steel for Equal Bending Stiffness- and Strength-Limited Design

Material	Equal Bending Stiffness		Equal Bending Strength	
	Thicknes s Ratio	Mass Ratio	Thickness Ratio	Mass Ratio
Mg (AZ91)	1.67	0.39	1.12	0.26
Mg (AM50)	1.67	0.38	1.26	0.29
Mg (AZ80-T5)	1.67	0.38	0.85	0.20
Mg (AZ31-H24)	1.67	0.38	0.95	0.22
Al (A380)	1.43	0.49	1.12	0.38
Al (A356-T6)	1.43	0.50	1.04	0.37
Al (6061-T6)	1.45	0.50	0.85	0.30
Al (5182-H24)	1.44	0.50	0.92	0.32
Plastics (Pulse 2000)	4.50	0.65	1.94	0.28

Design and Manufacturing Considerations

Magnesium die castings offer attractive flexibility in design and manufacturing. The excellent die filling characteristics of magnesium allows large, thin-walled and complex castings to be economically produced, replacing steel structures made of numerous stampings and weldments. The world's largest magnesium die casting is the one-piece full instrument panel (IP) for the GM G-van (Savana and Express). In addition to the weight reduction (27 pounds in magnesium vs. 40 pounds in steel), the magnesium IP design provides significant performance improvements (improved crashworthiness and reduced vibration) and cost savings due to the part consolidation (25 parts in Mg design vs. 67 parts in steel) [12]. More efficient magnesium IP designs in recent GM models have achieved even greater weight savings and part consolidation. Fig. 7 shows a 12-pound magnesium IP beam casting for GM H-car (Bonneville and LeSabre), made by Meridian Magnesium Products of Canada.

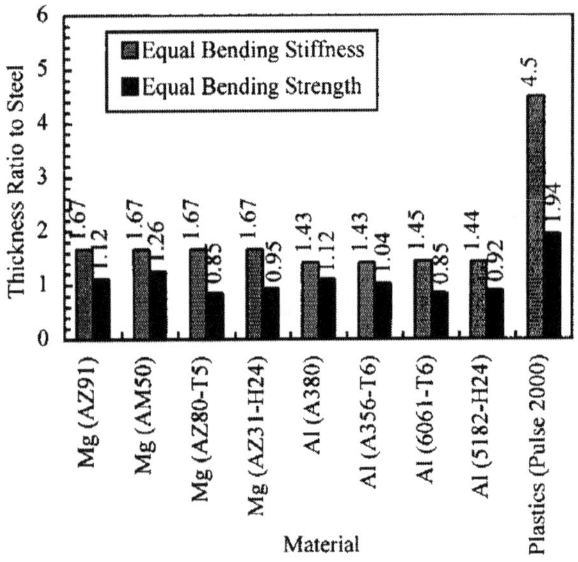

Fig. 5: Thickness ratios of various materials vs. steel for designing equivalent bending stiffness or bending strength.

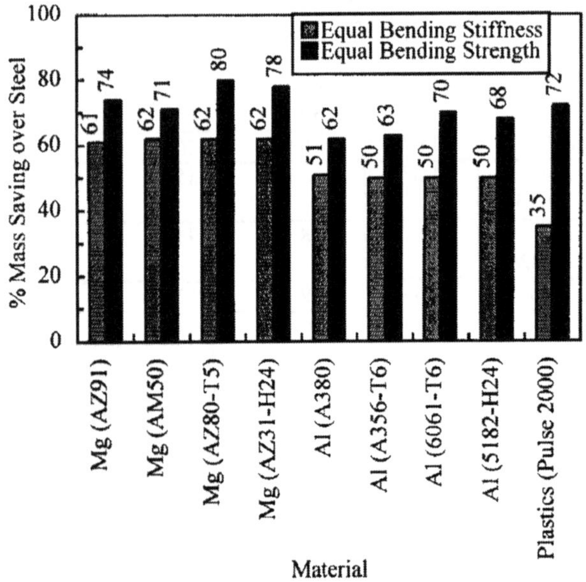

Fig. 6: Percentage mass savings of various materials vs. steel for designing equivalent bending stiffness or bending strength.

Table III lists some design parameters and manufacturing characteristics for magnesium, aluminum and plastic components. Magnesium die castings can be designed with thin walls in areas where strength is not a concern and with thicker walls in areas where strength requirements are higher. However, uniform thickness is a necessity of many plastic resins. Such consistency is needed not only to facilitate the molding process, but also to achieve the required mechanical properties of the plastic parts [13]. Magnesium can be cast with thinner walls (1-1.5 mm) than plastics (2-3 mm) or aluminum (2-2.5 mm). The stiffness disadvantage of magnesium vs. aluminum can be compensated with appropriately located ribs without increasing the wall thickness of a magnesium part. Compared with aluminum, magnesium has a lower latent heat for solidification and less affinity to iron in steel tooling. This leads to considerably shorter casting cycle times and longer die lives in die casting operations. Table III also shows that magnesium offers additional advantages in machinability, recyclability, and good surface finishing over thermoplastics.

Table III: Comparison of Design Parameters and Manufacturing Characteristics for Magnesium, Aluminum and Plastic Materials

Material	Mg	Al	Thermoplastics
Dimensional tolerance (mm/mm)	±0.001	±0.002	±0.002
Draft angle, (°)	0-1.5	2-3	2-3
Minimum wall thickness (mm)	1-1.5	2-2.5	2-3
Casting/molding cycle time (unit)	1.4-1.6	1.1-1.4	1
Typical die life (x1,000 shots)	250-300	100-150	500
Triming cycle time (unit)	1	1	2-4
Machinability	excellent	good	poor
Welding/joining	fair	good	poor
Surface finishing	good	excellent	fair
Recyclability	good	good	poor

Cost Comparison and Case Study

The cost comparison of different materials based on price per unit weight is often misleading because of varying density and performance. In this report, the most recent average prices (in relative values of the steel price) of various materials [14] are compared on both weight and volume basis. Table IV shows that the current price per pound of magnesium is least competitive. On a per unit volume basis, magnesium price becomes more favorable: the forecast price for magnesium in 2003 [15] is essentially the same as the current aluminum price on per unit volume basis. However, even at this price, magnesium compares unfavorably to plastics on a per volume basis.

When a specific application is concerned, however, product engineers are most interested in the cost of a material relative to its mechanical and physical properties, since this information reflects what one is paying for in terms of performance. Therefore, the *performance cost index (PCI)* of a lightweight material can be defined as its cost per unit volume times its thickness ratio relative to steel for equal stiffness or strength as calculated in Table II. PCI, the performance-adjusted cost, reflects the cost of a new material to provide the same stiffness or strength of a steel part. As shown in Table IV, magnesium performs better than plastics in both strength-limited and stiffness-limited designs. The PCI

Fig. 7: GM H car instrumental panel beam (12 pounds). Made by Meridian Magnesium Products.

Table IV Cost Ratios over Steel and Performance Cost Index (PCI) of Various Materials

Material	Steel (galvanized)	Al (A380)	Al (6061)	Plastic (PC/ABS)	Mg (1999)	Mg (2003)
Cost ratio per unit weight	1	3.36	3.59	6.14	7.5	5.91
Cost ratio per unit volume	1	1.17	1.33	0.83	1.67	1.33
PCI for equal stiffness	**1**	**1.67**	**1.93**	**3.74**	**2.79**	**2.22**
PCI for equal strength	**1**	**1.13**	**1.13**	**1.61**	**1.87**	**1.49**

Table V Overall Materials Comparison for Cam/Valve Cover Applications [16]

Performance Criterion	Weight Factor	Magnesium	Aluminum	Thermoset	Thermo-plastic
Manufacturing	4	37.33	39.00	36.00	37.33
Development time	3	34.60	34.60	32.00	31.20
Application performance	5	37.40	31.00	33.80	33.00
Cost	4	26.77	25.30	23.88	31.33
Assembly	3	30.40	28.00	31.20	34.20
Recycling	1	23.66	24.00	18.66	26.33
Material	3	33.00	37.50	33.33	32.50
Service	2	33.00	33.00	33.00	31.33
Normalized score		103.38	100.31	100.35	102.79

disadvantage of magnesium over steel or aluminum is often offset by its higher potential for part consolidation, yielding a lower-cost finished product.

The materials comparison for specific applications requires more factors to be considered. A case study on cam/valve cover applications is given in Table V [16], which shows a higher normalized score for magnesium over aluminum and plastics.

Current And Potential Applications In Automobiles

The global automotive applications of magnesium have been reviewed by a number of authors [1, 17-18], and Table VI is an updated summary of the current major magnesium applications. Table VI shows that magnesium has made significant gains in world-wide interior applications, replacing mostly steel stampings in instrumental panels, steering wheels and steering column components. In the powertrain area, GM and Ford lead the applications of magnesium 4WD (four-wheel-drive) transfer cases in high-volume truck production; while Volkswagen is aggressively expanding the use of magnesium in manual transmission cases produced in both Europe and Asia (China). Only a limited number of body and chassis components are currently made of magnesium. Table VII lists the future potential magnesium applications in four vehicle systems: interior, body, chassis and powertrain. The technical challenges for those applications are also discussed as follows.

Interior

Since corrosion is of less concern in interior applications, this area has seen the most magnesium applications (Table VI), with the biggest growth in the instrument panels and seating structures. The design and die casting of magnesium IP's have advanced dramatically in the last decade. For example, the current IPs normally have a thickness of 2.5 mm (compared to 4 mm for the earlier G-van IP) with more part consolidation and weight savings. Most recently, PC/ABS type of plastics is emerging as a

Table VI Global Magnesium Applications in Automobiles [1, 17-18]

System	Product	GM	Ford	Chrysler	Europe	Asia
Interior	instrument panel	yes		yes	yes	
	knee bolster retainer	yes				
	seat frame	yes		yes	yes	yes
	seat riser	yes	yes	yes		
	seat pan		yes	yes		
	console bracket		yes			
	airbag housing			yes		
	steering wheel	yes	yes	yes	yes	yes
	keylock housing	yes	yes	yes		
	steering column parts	yes	yes	yes	yes	yes
	radio housing		yes			
	glove box door	yes				
	window motor housing		yes		yes	
Powertrain	valve cover/cam cover	yes	yes	yes	yes	yes
	4WD transfer case	yes	yes			
	manual transmission case				yes	yes
	clutch housing & piston	yes				
	intake manifold	yes			yes	
	engine oil pan				yes	
	Alternator/AC bracket			yes		
	transmission stator	yes				
	oil filter adapter	yes				yes
	electric motor housing	yes				
Body	door frame				yes	
	tailgate				yes	
	roof frame	yes			yes	
	sunroof panel	yes	yes		yes	
	mirror bracket	yes	yes	yes	yes	
	fuel filler lid				yes	
	door handle				yes	yes
Chassis	wheel	yes				
	ABS mounting bracket	yes		yes		
	brake padel bracket	yes	yes	yes		yes
	brake/accelerator bracket	yes	yes			
	brake/clutch bracket		yes			
	brake padel arm		yes			

very competitive material to magnesium in IP construction. To maintain and grow its use in IP production, magnesium design and thin-wall casting technology must continue to improve in reducing weight and cost.

The use of magnesium seat structures began in Germany, where Mercedes used magnesium die castings in the integrated seat structure with a three-point safety belt in the SL Roadster. This seat structure consisted of 5 parts (two parts for the seat back frame and three parts for the cushion frame) with a total weight of 8.5 kg and varying wall thickness of 2-20 mm [19]. In the material selection for the seat program, magnesium was chosen over plastic, steel sheet and aluminum gravity casting designs. Magnesium die castings made of the high-ductility alloys of AM50 and AM20 offered the best combination of high strength, extreme rigidity, low weight and cost [19]. Similar to the IP development, magnesium seat design and manufacturing have gone through significant improvements in recent years. The latest example is the two-piece (backrest and cushion) design used in the Alfa Romeo 156. The die castings used today for seating structures are as thin as 2-4 mm, providing even greater weight savings. While other materials such as aluminum and plastics are also being developed for these applications, it is expected that magnesium will make significant inroads into seat components as a lightweight and cost-effective solution.

Table VII Potential Magnesium Applications and Technical Challenges

System	Product	Technical Challenges
Interior	airbag housing Window regulator housing glove box	improved casting process rapid prototyping design for magnesium
Body	Door frame/inner A & B pillar hatchback frame spare tire jack	thin-wall design thin-wall casting joining and welding
Chassis	Wheel spare tire rim control arm rack and pinion housing brackets for rail frames	high-strength alloy development new casting processes (squeeze and semi-solid metal casting) low-cost coatings
Powertrain	automatic transmission case engine block crankcase oil pan starter housing oil/water pump housing intake manifold engine mount	creep-resistant alloy development design for magnesium fastening strategy engine coolant compatibility

Body

European carmakers, especially Volkswagen, are pioneering the use of thin-wall magnesium die castings in body panel applications. The use of magnesium door inners and tailgates in VW's three-liter Lupo contributes to its potential 3.0 L/100 km (78 mpg) fuel economy. The key to make those thin-wall (around 1.5 mm) castings lies in casting design using proper radii and ribs to smooth die filling and to stiffen the parts. These thin-wall die castings, such as door inners, A and B pillars, can offer cost savings over steel sheet metal constructions, due to part consolidation. In body panel applications where bending stiffness is frequently the design limit, magnesium sheet metal can offer as much as 62% weight saving (see Fig. 6).

Chassis

GM has offered cast magnesium wheels for Corvettes since 1998, which is the first original equipment option of magnesium wheels in North America. However, the extremely high cost and the potential corrosion problems of magnesium wheels prevent their use in high volume vehicle production. The production of low-cost and high-integrity magnesium wheels and other chassis components such as control arms depends on the improvement of magnesium casting processes. Various casting processes have been developed for the production of aluminum wheels and chassis parts. These processes include permanent mold casting, low-pressure casting, squeeze casting and SSM casting. The successful adaptation of these processes to magnesium alloys will make magnesium castings more competitive to aluminum in the chassis area. The development of low-cost, corrosion-resistant coatings and new magnesium alloys with improved fatigue and impact strength will also accelerate the further penetration of magnesium in chassis applications.

Fig. 8: GMT-800 transfer cases (6 and 9 pounds, respectively). Made by Meridian Magnesium Products.

Powertrain

At present nearly one million of General Motors' GMT800-based trucks and sport utility vehicles (SUVs) produced annually have two magnesium transfer cases (total weight of 15 pounds) per truck (Fig. 8). At Volkswagen, around 600 manual transmission cases (28 pound each) are produced daily for use in VW Passat and Audi A4/A6 [20], and magnesium transmission cases are also used in the Santana model built in China. The operating temperatures for these applications are below 120°C, and AZ91 is the alloy of choice due to its excellent combination of mechanical properties, corrosion resistance and castability. In fact, VW has made it public that all their new transmissions will be designed to have magnesium cases in the future [20]. However, for automatic transmissions and engine applications (blocks and crankcases) where the operating temperatures are significantly higher, new magnesium alloys with better creep resistance must be developed. This subject is treated in detail in the next section. Other technical challenges for powertrain applications can be tackled by using proper design and engineering strategies.

Alloy Development for Powertrain Applications

The poor creep resistance of existing magnesium alloys has prevented magnesium applications in major powertrain components such as engine blocks and automatic transmission cases. Recent efforts to develop creep-resistant magnesium alloys for such applications have resulted in a number of experimental alloys [21-28]. These newly developed alloys can be classified into three categories:

- RE (rare earth elements)-containing alloys;
- Si-containing alloys; and
- Ca-containing alloys.

This section reviews the development, creep properties (see Table VIII) and potential applications for these new alloys.

RE-Containing Alloys

The beneficial effect of rare earth elements on the strength of magnesium alloys was discovered in the 1930's [29]. In 1972, it was reported [30] that 1% addition of RE in the form of misch metal improved the creep resistance of Mg-Al-based alloys, especially when the aluminum content was low (less than 4%). This led to the development of AE series alloys (AE41, AE42, and AE21) which contained 2-4%Al. Several alloys from this Mg-Al-RE system were extensively tested and studied by Dow Magnesium [21] and General Motors [31]. The GM results suggest that the alloy AE42 is viable for use in powertrain components, especially transmissions which operate at 150°C and below. However, the relatively high cost associated with the RE additions in AE42 alloy excluded it from consideration for volume production.

In an attempt to reduce the cost of AE42 alloy, a group of Nissan and Ube engineers in Japan has introduced calcium additions (0.25-5.5%) to partially replace the expensive rare earth additions [25]. The resultant Mg-Al-RE-Ca alloys are claimed to have improved creep strength over AE42 alloy, while retaining comparable tensile properties. Another new development is the most recent "Elektron MEZ alloy" (Mg-2.5%RE-0.35%Zn-0.3%Mn) patented by Magnesium Elektron Ltd. (MEL) in the UK. The data for this alloy [22] shows that it has superior creep resistance compared to AE42 alloy and good mechanical property retention up to 177°C. The capability to produce automotive castings such as gearboxes, housings and oil pans has also been reported. However, the alloy remains expensive as it has a even higher content of rare earth elements than AE42.

Si-Containing Alloys

The Mg-Si system was studied as a potential high-temperature alloy system due to the high melting point (1085°C), high hardness (460 $HV_{0.3}$), low density (1.9 g/cm^3) and low thermal expansion coefficient (7.5x10^{-6} K^{-1}) of the Mg_2Si intermetallic phase. The advantages of Mg_2Si have been demonstrated in die cast Mg-Al-Si alloys, such as AS41 (Mg-4%Al-1%Si) and AS21 (Mg-2%Al-1%Si), that offer a creep strength slightly higher than AZ91D or AM50 alloy. However, the experience with AS alloys also proves that Si additions below the Mg-Mg_2Si eutectic point (1.34% Si [32) offer a limited improvement in creep resistance, due to the small volume fractions of Mg_2Si particles present. Significant improvement in creep resistance would require a much higher Si content, i.e., hypereutectic alloys.

The Mg-Si phase diagram [32] shows that the liquidus of hypereutectic Mg-Si alloys increases about 40°C for every 1% addition of silicon, which is a serious challenge to die casting operations. Consequently, gravity casting or squeeze casting applications, such as engine pistons, have been targeted for the Mg-Si hypereutectic alloys. Another drawback in this alloy system is that the large dendritic morphology of the Mg_2Si phase in the microstructure impairs both mechanical properties and castability of the alloys. It was found that rare earth elements such as Nd can modify the Mg_2Si phase, leading to a homogeneous distribution of fine Mg_2Si particles [33]. More recently, it was reported that small additions of calcium (0.2-0.4%) to Mg-2%Si and Mg-4%Si alloys can modify the dendritic morphology into finely dispersed round particles [34]. Calcium additions can also improve the castability of Mg-Si based alloys via increased fluidity and reduced oxidation for the alloy melts at temperatures up to 900°C. The excellent high-temperature properties and the wear resistance of Mg-Si-Ca alloys suggest that this maybe a promising system for the development of magnesium pistons.

Ca-Containing Alloys

The addition of calcium to Mg-Al based alloys for improved creep resistance was reported in a British patent [35] as early as 1960. This patent disclosed that calcium additions of 0.5-3% provided creep resistance to Mg alloys containing up to 10% Al, up to 0.5% Mn and a possible Zn content up to 4%. However, the patent also acknowledged that Ca-containing alloys were prone to hot-cracking during die casting.

Volkswagen attempted the use of Mg-Al-Ca alloys in 1970's, and reported an improvement of creep resistance with the addition of about 1% calcium to magnesium alloy AZ81 (Mg-8%Al-1%Zn). However, die-sticking and hot-cracking prevented the application of this alloy to die cast crankcases [36]. In recent years, Mg-Al-Ca alloy system was revisited by researchers at the Institute of

Table VIII Tensile Creep Data for Various Magnesium Alloys

Alloy	Nominal Composition	Test Condition			Total Creep Extension (%)	Ref. #
		Temp. (°C)	Stress (ksi)	Duration (h)		
AZ91D	Mg-9Al-0.7Zn	150	5	200	2.5	24
AM50A	Mg-5Al-0.4Mn	150	5	200	2.15	24
AE42	Mg-4Al-2RE	150	5	200	0.33	24
		177	5	100	0.08	21
		177	7	100	0.10	21
		177	8.5	100	0.14	21
		177	9	100	0.19	21
AC51	Mg-5Al-0.8Ca	150	5	200	0.26	23
ZAC8506	Mg-8Zn-5Al-0.6Ca	150	5	200	0.14	24

Magnesium Technology (ITM) [23]. An alloy, known as AC51 (Mg-5%Al-0.8%Ca), was developed, which provided creep resistance similar to AE42 and corrosion resistance similar to AZ91D. Presently, this alloy is being evaluated by potential users for creep resistance and castability.

Castability issues with Mg-Al-Ca alloys have prompted the development of new alloys. In a recent alloy development program at IMRA [24, 37], it was found that 8% zinc addition effectively restored the die-castability of AC alloys. The new ZAC (Mg-Zn-Al-Ca) alloys offer a good combination of mechanical, physical and corrosion properties at temperatures up to 150°C, with a die-castability rating close to that of the benchmark AZ91D alloy. Recently, the new ZAC alloys were granted a US patent [38], and are receiving a worldwide attention from major automotive companies for potential applications.

Summary

Magnesium offers a good combination of physical and mechanical properties as well as attractive processing characteristics during die casting. In comparison with steel, aluminum, cast iron and polymer materials for structural applications, magnesium castings provide the most weight savings in a stiffness-limited design. Cost comparison of materials should be based on the "performance cost index" concept, which is the material cost per unit volume adjusted for equal stiffness or strength. For a stiffness-limited design such an instrument panel beam, magnesium is significantly less expensive than PC/ABS-type of polymers. However, it is slightly more expensive than PC/ABS in a strength-driven application.

Magnesium offers a lightweight solution in all vehicle sub-systems. While interior applications have seen a considerable growth, body, chassis and powertrain present significant opportunities for future growth, but there are substantial technical challenges. With the availability of high purity magnesium alloys as standard material, galvanic corrosion of magnesium components can be easily overcome by proper part design and assembly. Several creep-resistant magnesium alloys have also been developed for elevated-temperature applications: RE-containing alloys are considered too expensive for volume production, and other low-cost Ca-containing alloys are now available.

Acknowledgments

The author wishes to acknowledge Anil Sachdev, Bob Powell, Dave Barris and other members of GM magnesium technology team for many helpful discussions, and General Motors Corporation for permission of publishing this paper. Lisabeth Riopelle of Hydro Magnesium is acknowledged for sharing the forecast information on magnesium consumption at Big Three.

References

1 D. Magers and J. Willekens, "Global Outlook on the Use of Magnesium Diecastings in Automotive Applications", Magnesium Alloys and Their Applications, B.L. Mordike and K.U. Kainer, eds. (Werkstoff-Informationsgesellschaft, Frankfurt, Germany, 1998), 105-112.

2 L. Riopelle, private communication, Hydro Magnesium, Livonia, MI, January 1999.

3 R.A. Schultz and W.J. Haupricht, "Trends in Aluminum Use for Passenger Cars and Light Trucks in North America", Light Metals Age, February (1999), 108-113.

4 ASM: Metals Handbook, Vol. 20 (ASM International, Materials Park, OH, 1998).

5 ASM: Metals Handbook, Desk Edition (ASM International, Materials Park, OH, 1998).

6 K. Gong: private communication, Alcan International, Kingston, ON, August 1999.

7 Hydro Magnesium, "Data Sheet: Die Cast Magnesium Alloys", Hydro Magnesium, Porsgrunn, Norway, November 1996.

8 Dow Plastics, "Pulse 2000 Engineering Resin", Dow Plastics, Midland, MI, 1999.

9 H. Koch and A.J. Franke: "Ductile Pressure Die Castings for Automotive Application", Automotive Alloys, S.K. Das, ed. (TMS, Warrendale, PA, 1997), 185-194.

10 Hydro Magnesium, "Corrosion and Finishing of Magnesium Alloys", Hydro Magnesium, Porsgrunn, Norway, November 1997.

12 M. Nehan and R. Maloney, "Magnesium AM60B Instrument Panel Structure for Crashworthiness FMVSS 204 and 208 Compliance", SAE Paper No. 960419 (SAE, Warrendale, PA, U.S.A.), 1996.

13 T. Ruden, "Magnesium vs. Plastics: A Comparative Study", (Paper G-T89-113, presented at the 15th International Die Casting Congress & Exposition, St. Louis, MO, 1989).

14 American Metal Market, May 1999.

15 M.M. Jones: "Magnesium: Its Place in the Automotive Industry" (Paper presented at IMA Automotive Seminar, Dearborn, MI, April 7, 1999).

16 M. Dierks and T. Heizer: "A Materials Comparison for Cam Cover and Valve Cover Applications" (Paper presented at IMA Automotive Seminar, Dearborn, MI, April 7, 1999).

17 A. Luo et al., "Magnesium Castings for Automotive Applications", JOM, 47 (7) (1995), 28-31.

18 G.S. Cole, "Use of Magnesium Components in the Automotive Industry", Proc. of International Symposium on Light Metals, C.M. Bickert and R.I.L. Guthrie, Eds. (The Metallurgical Society of CIM, Canada, 1997), 51-67.

19 A. Hector and W. Heiss, "Magnesium Die-Castings as Structural Members in the Integral Seat of the New Mercedes-Benz Roadster", SAE Paper No. 900798 (SAE, Warrendale, PA, U.S.A., 1990).

20 S. Schumann and F. Friedrich, "The Use of Magnesium in Cars - Today and in Future", Magnesium Alloys and Their Applications, B.L. Mordike and K.U. Kainer, ed. (Werkstoff-Informationsgesellschaft, Frankfurt, Germany, 1998), 3-13.

21 W.E. Mercer II, "Magnesium Die Cast Alloys for Elevated Temperature Applications", SAE Paper No. 900788 (SAE, Warrendale, PA, U.S.A., 1990).

22 P. Lyon, J.F. King and K. Nuttall: "A New Magnesium HPDC Alloy for Elevated Temperature Use", Proc. 3rd Intl. Magnesium Conf., G.W. Lorimer, ed. (The Institute of Materials, London, UK, 1996), 99-108.

23 M.O. Pekguleryuz and A. Luo, "Creep Resistant Magnesium Alloy for Die Casting", PCT (Patent Cooperative Treaty) application WO 96/25529, filed on Feb. 14, 1996 and published on Aug. 22, 1996.

24 A. Luo and T. Shinoda: "A New Magnesium Alloy for Automotive Powertrain Applications", SAE Paper No. 980086 (SAE, Warrendale, PA, U.S.A., 1998).

25 K. Makino et al., "Magnesium Alloy", UK Patent Application, GB 2 296256 A, filed on Dec. 22, 1994 and published on June 26, 1996.

26 K. Sakamoto et al, "Heat-Resistant Magnesium Alloy Member", European Patent Application, EP 0 799 901 A1, filed on April 4, 1997 and published on Aug. 10, 1997.

27 T. Baba, K. Honma, M. Ichikawa, "Heat-Resistant Magnesium Alloy", European Patent Application, EP 0 791 662 A1, filed on Feb. 2, 1997 and published on Aug. 27, 1997.

28 S. Park et al., "Magnesium Alloy for a High Pressure Casting and Process for Preparation Thereof", PCT (Patent Cooperative Treaty) application WO 97/40201, filed on April 25, 1997 and published on Oct. 30, 1997.

29 A. Luo and M.O. Pekguleryuz, "Cast Magnesium Alloys for Elevated Temperature Applications", Journal of Materials Science, 29 (1994), 5259-5271.

30 G.S. Foerster, Proc. 7th SDCE International Die Casting Congress (1972), Paper No. 9372.

31 E.G. Sieracki, J.J. Velazquez and K. Kabiri, "Compressive Stress Retention Characteristics of High Pressure Die Cast Magnesium Alloys", SAE Paper No. 960421 (SAE, Warrendale, PA, U.S.A., 1996).

32 T.B. Massalski, ed., "Binary Alloy Phase Diagrams", vol. 2 (ASM, Materials Park, Ohio, 1986), 1545.

33 S. Beer, G. Frommeyer and E. Schmid, "Development of Mg- Mg_2Si Light Weight Alloys", Magnesium Alloys and Their Applications, B.L. Mordike and F. Hehmann, eds. (DGM Informationsgesellschaft, Germany, 1998), 317-324.

34 Y. Carbonneau et al., "On the Ca Modification of Binary Mg-Si Alloys" Proc. of International Symposium on Light Metals, C.M. Bickert and R.I.L. Guthrie, Eds. (The Metallurgical Society of CIM, Canada, 1997), 69-84.

35 H.J. Fuchs, "Magnesium Alloys having a High Resistance to Permanent Creep Deformation at Elevated Temperatures", British Patent No. 847,992, published in 1960.

36 F. Hollrigl-Rosta et al., "Magnesium in Volkswagen", Light Metals Age (August 1980), 22-29.

37 A. Luo and T. Shinoda, "Development of a Creep-Resistant Magnesium Alloy for Die Casting Applications", Magnesium Alloys and Their Applications, B.L. Mordike and K.U. Kainer, ed. (Werkstoff-Informationsgesellschaft, Frankfurt, Germany, 1998), 151-156.

38 A. Luo and T. Shinoda, "Magnesium Alloys Having Superior Elevated-Temperature Properties and Die Castability", United States Patent 5,855,697, January 5, 1999.

SUMMARY OF "MAGNESIUM VISION 2020: A NORTH AMERICAN AUTOMOTIVE STRATEGIC VISION FOR MAGNESIUM"

Gerald S. Cole
LightWeightStrategies, LLC
25800 Romany Way, Franklin, MI 48025, *geraldscole@sbcglobal.net*

Keywords: Magnesium Research and Technology Development, Infrastructure, USCAR, Vision for Magnesium's Future

Abstract

This paper summarizes the monograph, "Magnesium Vision 2020. A North American Automotive Strategic Vision for Magnesium"[1] prepared under the auspices of the United States Automotive Materials Partnership The objective was to understand the infrastructural and technical challenge that can increase the use of magnesium in the automotive industry. One hundred sixty three (163) Research and Technology Development Themes (RTDTs), or RTD projects were developed that addressed issues of corrosion, fastening, and processing-other-than-high pressure die casting to produce automotive magnesium parts. A major problem identified in the study is the limited ability of the current magnesium industrial infrastructure to supply RTD and implementation-ready automotive magnesium components. One solution is to create a magnesium cyber center where globally networked experts would be able to innovate in process and product development, model metalworking and non-HPDC foundry processes, and integrate theoretical predictions/models of metallurgical structure with component function.

Introduction

The monograph, "Magnesium Vision 2020: A North American Automotive Strategic Vision for Magnesium"[1], originated within the Structural Cast Magnesium Development and the Magnesium Powertrain Cast Components project teams of the United States Automotive Materials Partnership (USAMP)-Automotive Metals Division (AMD). The monograph is based on a framework initially developed by the author with suggestions, reviews and amendments by representatives of the U.S. automakers (USOEMs) and other participants in the two AMD projects.

Background

A typical U.S-built 1525 kg vehicle comprises 114 kg of plastics (7.6%), 127 kg of Al (8.3%), and 975 kg of steel/cast iron (64%), while Mg use is only in 5-6 kg (0.3% of a vehicle mass), see Figure 1. But growth has increased by 10-15% pa over the past 15 years based on magnesium's ability to reduce vehicle mass and the potential impact this has on increasing fuel efficiency, reducing environmental emissions (important in Europe), and improving many vehicle attributes.

Mass Reduction With Mg Using Mg in a mass-reduction scenario requires an understanding of how much mass can be replaced by substituting existing components with engineered Mg derivatives. Even though Mg is 75% lighter than steel/cast iron, Mg parts can

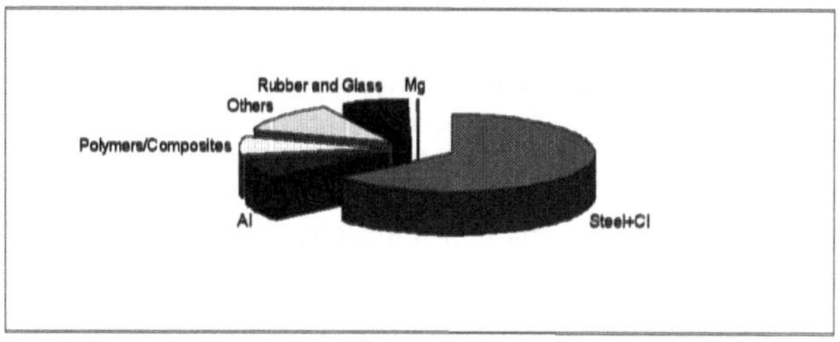

Figure 1. Proportion of materials in a Current Vehicle

replace steel/cast iron typically only on a 1:2 basis (100 kg of Mg substituting for 200 kg of steel/cast iron parts) and typically replace Al on a 30% basis (45 kg of Mg substituting for 65 kg of Al). Additional mass reduction is possible if the chassis, suspension and powertrain can be resized or downweighted. However, the amount of secondary mass reduction possible is highly controversial varying from an equal mass[3], to none, depending on the vehicle (e.g. trailer tow requirements).

Effect of Mass Reduction and Fuel Consumption Unfortunately, there is no easy way to calculate the relationship between mass reduction and improved fuel consumption since it depends on the vehicle's mass and powertrain employed. There is an old "rule-of thumb" in the auto industry that a 1.0% fuel efficiency improvement requires 1.5% mass reduction (22.5 kg mass reduction in our average vehicle). But newer data [1] suggests it could require as much 3.0 % (45 kg). Nevertheless, mass reduction is cost-effective compared to other more expensive fuel efficiency procedures such as diesel/hybrid powertrains, six/seven-speed automatic transmissions and so forth [2]. In the United States, mass reduction primarily is used to achieve a specific Inertial Weight Class (IWC), which the U.S. employs to calculate Corporate Average Fuel Economy (CAFÉ is 27.5 mpg for cars and 22.2 mpg for trucks after 2007). It is also solves axle loading problems if the original vehicle design is too heavy for the allowable Gross Vehicle Weight (GVW).

Other attributes of Mg. Mg parts are predominantly produced via High Pressure Die Casting (HPDC). The HPDC process allows many fabrications to be integrated into a simpler one piece part. This can reduce noise, vibration and harshness, improve overall craftsmanship and allow unique customer feature designs. As well, low mass in the front can improve performance and safety. Mg has improved mechanical and physical properties vs. competitive materials: higher strength, stiffness, thermal stability

To realize the aggressive fuel efficiency and acceleration objectives for the program, several performance

stiffening/NVH foam between the two layers of metal.

and thermal conductivity vs. some plastics, and higher specific strength, ductility and impact resistance vs. conventional HPDC Al. Most importantly, for the first time in 40 years Mg provides a cost advantage….. Mg alloy ingots now are less costly than Al while at production volumes < 50,000 units pa Mg castings are less costly than fabricated steel (because of initial high tooling and gauging costs for steel fabrications).

What Prevents Magnesium from Achieving Its Potential?

During the past 20 years over 160 kg of chassis, powertrain, interior and exterior Mg parts have been designed, cast, assembled and approved to be implementation-ready via a broad testing protocol varying from laboratory testing through test track and fleet trials. So why are there not more than 55,000T of Mg parts (6 kg/vehicle x 9.2 M vehicles) used on USOEM vehicles? The basic intent of this document is to identify issues and concerns that have limited Mg applications and develop strategies that will enhance its permanent place in automotive construction.

Cost. At $2.90/kg (the typical ingot price until recently) Mg was an expensive lightweight option ingot and components were perceived as being higher cost compared to competitive materials. The usual product experience had been that if a vehicle had a mass problem during its development stage (GVW- or CAFÉ-related) a Mg component would have been rushed into production to reduce mass and "save the day". However when a lower cost solution later became available, the Mg design was discarded. ***Now, for the first time in generations, Mg alloys are 20% less costly than Al and cost is less of a concern.***

Infrastructure Limitations. Infrastructural problems permeate all other concern. The U.S. auto industry has evolved considerably since the 1960-1987 time-frame, when Al applications increased from 24 to 113 kg. The industrial infrastructure that designed, produced championed, approved and marketed implementation-ready Al components, does not exist now for Mg as it did for Al. In addition, other commodities have a stronger design, manufacturing, and marketing supply base that can cope more rapidly to changing conditions. Other commodities also have technical societies and trade associations which support RTD and the growth of their interest group; Mg has none.

Quality, Prototyping, Perception. The next set of issues is associated with quality, design flexibility, and time required for prototyping. Almost all Mg parts are cast via HPDC. The quality and design flexibility of this process is perceived to be lower than components produced by other processes. Prototyping and engineering changes also take longer and are more expensive vs steel. Lastly, there is the perception that Mg burns and that there could be costs associated with handling flammability issues.

Engineering/Technical Concerns. A major set of issues revolves around the concerns and perception that Mg parts are not as durable compared to Al and plastics, due to galvanic corrosion, assembly and joining Mg parts to a vehicle where stress relaxation could occur.

Limited manufacturing outside of HPDC. A final set of issues is related to the limited manufacturing capability for Mg parts, including the small number of alloys available: properties cannot be optimized and therefore the development of new Mg parts vs. competitive materials could be compromised. In the Mg 2020 document many non-HPDC casting processes are examined and RDT issues to make them commercial are discussed. Stamping, extrusion and forging processes also are examined and RDT protocols are presented to increase their automotive utilization.

Infrastructural Limitations

Analogue of Aluminum

Al and OEM R&D. In 1960 the average NA car contained 24 kg of Al parts; by 1987 it contained 113 kg (all of them cast components). How did this happen for Al and could it happen in the future for Mg? The Al industry spent $ billions to support its automotive customers and suppliers with designs, as well as developing new processes to produce components with those designs. Al engineers and scientists provided 24/7 managerial support to convert steel and cast iron components into Al and automotive engineers were supported with unique Al products that improved function while reducing cost. At the same time the laboratories of the USOEMs provided reciprocal support with funding, facilities and manpower to further leverage their foray into lightweight automotive components. Who can do this in Mg?

There is no western industry to support RTD in Mg, since the demise of DOW Magnesium, Magnola Magnesium, Australian Magnesium Corporation, Pechiney and most recently Hydro Magnesium. China currently produces ~80% of the world's supply but does not support Western technology, RTD and automotive product development.

Manpower Limitations Compared to the enormous scientific and engineering effort that nurtured Al automotive growth in the 1960s-'80s, the manpower to develop a Mg industry is not available. Fewer students are studying metallurgy and manufacturing in NA. In addition, since the 1970s, NA has gradually deindustrialized and many manufacturing jobs have moved off-shore. For example, in 1952, 33% (>40 M workers) were involved in manufacturing; 50 years later < 10 % of the workforce (13M workers) are so employed (Bureau of Labor Statistics). It appears that there just is not enough manufacturing-metallurgical engineering-scientific expertise to support the development of a US Mg-intensive automotive infrastructure.

Tier One Evolution There has been a considerable shift in the U.S. automotive industry over the past 30 years with USOEMs outsourcing more activities to (fewer and fewer) Tier One assemblers. The Tier Ones have become multibillion-dollar entities that design and manufacture steel, plastic and Al components and subassemblies which they ship to OEM production lines for direct installation into vehicles. Tier Ones are responsible for much of the new product development and technology used on USOEM vehicles. But only one Tier One has the infrastructure to design/develop, produce and assemble implementation ready Mg components. For most Tier Ones, Mg production lines compete with their existing product/alloy mix, and there is a high cost for displacing that production.

With this limited Tier One competency and interest, the large-scale production of Mg applications will require a major change in how Mg RTD is performed, how/where components are developed and manufactured, and how they are approved to be implementation-ready for vehicle service.

Technical Societies/Trade Associations

The ferrous industry has the Iron and Steel Association and the Auto Steel Partnership, the Al industry has the Al Association and Auto Al Alliance, and the plastics industry has the Society of Plastics Engineers with an automotive section. These strong, viable western-based industries spend funds to sponsor automotive-relevant RTD. Even though the International Magnesium Association represents Mg interests, it has limited funds to invest in RTD and promote automotive Mg applications. We ignore here the organizations that provide only a forum for Mg: the Mg Division of TMS, and the American Foundry and North American Die Casting Societies.

The US Department of Energy (DOE) and the US automakers (via USCAR) have been the recent dominant agents promoting Mg RTD in NA, using funds provided by DOE and in-kind resources emanating from USOEMs and other USAMP project participants.

New Paradigm for Mg Research and Technology Development

One hundred and sixty (160) kg of implementation-ready Mg components on a vehicle will not occur by 2020 (as proposed in the 2020 Strategy), with the limited technical/scientific, corporate and academic infrastructure currently available in NA. A new paradigm is needed that could provide Mg with the "right stuff" to compete with Al, steel and plastics for vehicle lightweighting. The following two actions are proposed:

1. Create a virtual, distributed magnesium cyber center that will coordinate the production and acquisition of knowledge and perform the modeling that will obviate the current expensive implementation-readiness approach.
2. Simultaneously develop a system of cooperative and collaborative modalities among global organizations and international Mg scientists, academicians and experts.

1. Magnesium CyberCentre of Excellence

Implementation-Readiness The Magnesium CyberCentre conceptualizes a "virtual" distributed laboratory that will not entail a bricks-and-mortar edifice. It will be a distribution of cyber-nodes, where the multidisciplinary scientific and technology-based global community can network with market-driven industrial participants. The CyberCentre will coordinate the RTD in materials science, engineering and computer modeling that is required to develop/design new Mg components and their associated manufacturing processes so that cost models can be integrated into implementation ready products. The CyberCentre will incubate the product and processing technology required for the Magnesium Vision 2020 to succeed.

Simulation Tools The Mg CyberCentre will incubate automotive-related Mg RTD around the world toward developing a microstructural engineering knowledge base that uses computational materials engineering tools with multi-scale modeling and simulation-specific capabilities. Processing (casting, metalworking and heat treatment) will be integrated with materials microstructural evolution to understand how Mg microstructure can be manipulated to improve formability through identification, quantification and control of critical microstructural parameters. This will allow prediction of properties and component performance. Appropriate laboratory testing will be developed to validate the models and thus obviate expensive and lengthy test track approval for general implementation-readiness capability. The objective is to develop cradle-to-grave-to-cradle simulation tools that cover different alloys and various material scales of analysis (atomistic, electronic, nano, micro and macro-scales) with product models that demonstrate Mg component implementation-ready functionality. An obvious requirement is the financial cost modeling to ensure economic competitiveness.

Information Resource. The CyberCentre will be a repository for automotive-relevant scientific and engineering data and information equivalent to that required by the Al, steel and plastics industries and OEM laboratories. It will include chemical, physical and mechanical properties, and automotive parts standards in NA and off-shore markets. The CyberCentre will also centralize marketing and supplier/community educational support functions; e.g. supply activities to Mg-needy customers, similar to a university agricultural extension program. We foresee that that the CyberCentre would aid US suppliers and USOEMs until the Mg knowledge-base became sufficiently mature, the market sufficiently strong, and the suppliers sufficiently capable to support their own in-house RTD.

2. International Collaboration

For this magnesium strategic vision to succeed, global cooperation and collaboration will be required among industry, academia and state/federal government laboratories. Many countries expend a great deal of money and are acquiring valuable fundamental information about magnesium that could complement NA data. There is a storehouse of R&D knowledge held by researchers around the world that could be applied to promote the 2020 Mg Vision; that information needs to be accessed and made universally available to researchers at universities/technical colleges, state/provincial science and technology facilities, and government/industry laboratories.

Research and Technology Development (RTD) Required to Promote Automotive Magnesium Growth

There is a large body of knowledge that automotive OEMs and their suppliers require to enhance their comfort with Mg. The USCAR monograph examines the major technical issues and automotive concerns, via a series of 163 Research and Technology Development Themes or projects (RTDTs). They deal with perception issues in HPDC processing (cost, quality and prototyping), technical issues (corrosion and fastening) and manufacturing (non-HPDC and wrought processing). Each RTDT identifies a current problem and outlines a task that could eliminate the concern. Successful RTD will show the automotive community that the use of Mg components proffers no different risks (including fire) than any other new material. With current commodity economics (Mg alloy ingots less costly than Al and the high price of oil), RTD to promote more Mg components will be an important automotive industry strategy.

Cost, Quality, Prototyping Current HPDC Magnesium

Cost Issues The Magnesium Vision 2020 monograph outlines 19 RTDTs that address cost reduction strategies, the cost of quality and the cost of prototyping. For example, the data used to design a component so that it will survive its service lifetime often is based on mechanical properties obtained from separately cast test bars.

However, test bars do not mimic production and can produce lower 3 sigma values that are often considerably less than that achievable with good casting practice. Accurate production sampling data allow engineers to design for higher properties; as well, factors of safety can be reduced allowing for increased confidence in thinner wall designs, The result will be reduced Mg mass and lower cost components. In cases where there might be wall thickness changes because of environmental degradation, section thickness can be increased to insure long-term durability, thus enhancing the perception of quality.

This is what was found in the two USAMP projects, Design for Product Optimization (primarily in Al castings)[5] and the Structural Cast Magnesium Development project (for Mg castings).[6] More accurate data allowed castings to be designed with thinner walls, and thus be 5-10% lighter. Less metal was needed and cost was able to be reduced.

Quality. The HPDC process often scares auto engineers. They see cracks and properties that vary from the ingate into the casting and they worry that their components could fail in service. CAE modeling of die filling and solidification have improved die design capability. Coupled with improved HPDC processing algorithms (use of vacuum, overflows, shot controls) defects can be reduced, quality variability reduced and mechanical properties increased so that Mg can be considered for structural components not just for "pots and pans". As long as non-metallic inclusions and impurities are controlled, high elongations are possible in HPDC Mg parts - higher than almost all other casting processes - and certainly higher than HPDC Al. Ten RTDTs were developed in the monograph to address these issues but are not included here because of space constraints.

Prototyping. An important issue that affects automakers is the attainment of prototyped parts that simulate "make-like-production" conditions. Time is an important economic variable in deciding materials and processing strategy. Product cycles are becoming shorter and designs must be completed in a shorter time frame, often < 18 months. In addition, parts are often redesigned during the time between initial prototype vehicle build and final vehicle acceptance, so that reprototyping and revalidation must occur quickly. Materials/components that can be prototyped and validated quickly and at low cost with "make-like" production functionality have an advantage over HPDC slower more costly procedures. HPDC components currently require several weeks to prototype and fabricate. The monograph includes several RTDTs that treat this subject in a more detailed manner.

Corrosion Issues

Introduction The automotive community has serious concerns with Mg corrosion. However, modern HPDC Mg alloys have almost the same corrosion resistance to normal environmental exposure as mild steel, and considerably better than HPDC Al which contains 2 % Fe and 2-4 % Cu. The size and distribution of impurity-rich phases (that cause Mg to corrode) and the $Mg_{17}Al_{12}$ beta phase (the more corrosion-resistant high Al phase) have a large influence on corrosion. One advantage of HPDC is that the rapid freezing in the metal die produces finer microstructure, smaller inclusions, a higher tolerance to impurities and less corrosion than sand castings which freeze slower and have increased precipitate size and shape.

General Corrosion Fourteen (14) RTDTs are presented in the monograph, including the following more explicit 7 themes:

- Develop fundamental understanding of surface chemistry and reaction kinetics for low cost Cr-free protection.
- Develop low cost shiny, hard, wear resistant surfaces for Class A surface applications.
- Examine use of nano-particles in solutions and electrolytes to impart self-healing.
- Develop colored/texture coatings for unique design effects.
- Develop peck-resistant (e.g. urethane) coatings.
- Study effect of cyclic strain which ruptures protective films and exposes new unprotected metal.
- Develop a "stainless" Mg alloy.

Galvanic Corrosion Galvanic corrosion presents one of the most pressing technical issues that limits Mg use in vehicles and adds considerable cost to contain. Of the 21 RTDTs developed to reduce concerns with galvanic corrosion 7 are listed below:

- Model the dimensions/geometry of the electrolytic effects associated with the fluctuating (dry-wet) conditions from salts and road debris which bridge Mg and a ferrous part.
- Develop impermeable coatings for use on the cathode (ferrous part) rather than coating the anode, the Mg part.
- Determine the effects of corrosion on mechanical properties, especially for Mg used in structural applications.
- Corrosion can take many years to appear. Accelerated methods need to be developed that can predict corrosion throughout all vehicle locations and which can be correlated to laboratory and test track tests.
- Address gaps in understanding the thermodynamics of corrosion including passivation and hydride formation.
- Develop standardized testing protocols, and procedures to characterize corrosion in the presence of engine coolants. Understand the effects of engine coolants (and inhibitors) on engine materials. Determine if/how inhibitors reduce corrosion and degrade during engine operation.

Surface Contamination: Pitting may occur from surface contamination with small particles of a dissimilar metal, especially in the absence of a surface-protective oxide film. The possible sources of surface contamination include: die lubricants that contain molybdenum disulphide and carbon, iron particles that can be transferred from dies during casting, forging, extrusion, rolling and stamping, or from heavy metals present in shot blasting media. Pickling solutions can also cause problems if they contain heavy-metal salts.

Fastening and Joining Issues

Mg can be joined by most of the welding, mechanical fastening and chemical bonding methods common to the metalworking industries. Mechanical fastening technologies, (rivets/self-piercing rivets, bolts, screws, crimping, over-molding and self-fastening devices) require unique materials/processing to eliminate galvanic corrosion and/or damage. Depending on the Mg alloy, fastener material, service temperature and applied loads, the Mg part could sufficiently relax that clamp loads would reduce to below a desirable level. Acceptable, durable clamp loads can be maintained by spreading the loads with washers and shims, by using higher temperature capable Mg alloys, and/or by using Al alloy fasteners. Components sometimes experience the greatest fatigue stresses at a fastener and may have to be redesigned to eliminate a potential fatigue condition.

Of the 16 RTDTs issues/concerns with fastening durability presented in the monograph, 11 of the most important are listed below:

- Develop CAE/CAD design guidelines to define shim/washer requirements and critical distances between Mg and the ferrous component.
- Develop design rules to ensure no liquid can seep into joints.
- Develop guidelines for clamp loads, joint design, and rapid/robust test methodologies for use with threaded fasteners.
- Develop creep-resistant plastics or other non-conducting washers that can maintain clamp load in service.
- Develop a process to affix a washer to a part so that it remains in place during transport to the assembly plant.
- Develop criteria for riveting, crimping and thread-forming without cracking Mg parts.
- Ensure no joint damage from body-in-white processing.
- Develop low cost Al and/or coated steel fasteners.
- Develop full database of adhesive bonding for joining castings/stampings of Mg-Mg, Mg-Al, and Mg-steel joints.
- Determine if corrosion occurs during adhesive bonding Mg to Mg, Al and steel and develop appropriate low cost solutions.
- Develop process parameters/CAE predictive tools for spot/continuous sheet welding.

Non-HPDC Casting

Introduction All Mg components are produced via injection into a die, either via HPDC or semisolid casting processing (<1% of current parts). This allows thin (<3 mm) walls, accurate dimensions and high production rates that define low cost, and the high solidification rates that produce useful mechanical properties. The concerns with HPDC were considered earlier: prototyping costs and timing, and variable quality for high strength structural applications. Other processing choices such as sand and semi-permanent mold (SPM) casting processes have been used in the past, but only to produce premium Mg parts (e.g. helicopter rotors and road wheels). Non-HPDC castings also require thermal processing to develop viable mechanical properties; conventional heat treatment requires atmosphere protection, is slow (>40 hours) and therefore is costly. The "bible" on the subject was written by Emley 40 years ago[4].

Is low cost non-HPDC processing viable for automotive? Many of the issues are presented in The Magnesium Vision 2020 monograph where 34 RTD themes focus on the issues that need elucidation. Because of space constraints, only the general concerns are described below; the reader is referred to the Magnesium 2020 monograph for details [1]. As well, the American Foundry Society prepared a Mg road map in 2005 [5] which forms the basis for the recently commissioned USAMP HI-MAC project[6].

Heat Transfer Issues. In HPDC, grains are in the 20-30 μm size range due to the rapid quenching action of the massive mold, which produce high thermal gradients and freezing rates. On the other hand, arts cast in sand solidify slowly, at almost 0 temperature gradient. In sand castings, positive temperature gradients have to be forced on the system, such as by using high thermal conductivity sands (steel granules, zircon, etc.), shaped-chills, or via a recent invention that uses a moving wall of water[7] over the mold,

Grain Refinement. In the absence of grain refinement, large (> 5 mm) grains form that can crack and lead to reduced mechanical properties. Thus, exogenous grain refinement is required to achieve the desirable mechanical properties. Grain refinement in Al has been extensively studied for over 50 years and low cost grain refiners have been developed for ingots and shaped castings. The effort in Mg has been much less and currently there are no commercial inoculants for Mg, especially the important Mg 1-9% Al alloy families. However, there is a world-wide research effort to address this problem. Appropriate inoculant-introduction methods also need development, similar to the gas injection, plunging and in-mold coating and inoculation procedures used with Al and Fe castings. Since molten Mg is so chemically active, it is important to ensure no potentially strength-reducing discontinuities or impurities are thereby introduced.

Mold and Metal Reactions: Molten Mg is very reactive. Core/mold media, additives and coatings are required that will produce a smooth casting surface, will not react, and will collapse after solidification and shake-out without cracking the relatively weak casting (at/near solidification temperatures). Filling a mold without oxidation-induced turbulence is required for high quality structural Al castings as well as for Mg. Mechanical/electrical pumping systems (or tilt-pour arrangement[8]) can produce large thin-walled castings at a controlled/slow fill with no turbulence and thus no inclusion-generation. Obtaining long pump life in molten Mg requires new materials and designs, which will be developed in the HI-MAC project [6].

New Alloys New, low cost, high strength sand castable automotive alloys are required. Properties can also be improved with composite additions containing macro and nano-particulates, fibers and composite performs. The choices are more limited in Mg vs Al since Mg will react with almost all non-metallics to form spinels and other reaction products that can induce corrosion.

Metalworking Technologies

Unlike most metals that are stronger in compression than tension, Mg is ~ 2x weaker in compression than tension. Mg also demonstrates a strong preferred structure-orientation with extensive twinning, and properties changing in the working direction. Both these issues require unique designs to compensate for the fabrication process. Importantly, since the Mg crystal structure is hexagonal closed packed, room temperature deformability is considerably lower than steel and Al; not until the temperature is ~ 225 °C is there enough metalworking ductility. However ultra-fine grained materials (<3 μm) may be formable at lower temperatures[4]. The many scientific/technical issues in developing commercial wrought components are discussed in the monograph [1]; German researchers have been the leaders in this field [10, 11].

Low cost sheet production Ingot-processing is the only method used currently to produce Mg sheet, but at 5 times more expensive than Al it is not a viable automotive material. Twin roll casting could make Mg sheet more cost-competitive and is being developed at Posco Steel Works (Korea), Thyssen Krupp (Germany), Hydro Magnesium (Norway) and CSIRO (Australia)[9]. This potentially could make stamped components economically competitive with Al at an ~ 30% mass reduction, or even with steel due to the ~ 50% mass reduction possible. A great deal of

RTD is needed to develop casting, rolling and heat treatment that can produce cost-effective automotive-formable sheet.

Stamping. Alloy composition, grain structure and thermo-mechanical processing have a significant impact on automotive stamping; texture, spring-back, and stamping temperature are not well characterized for Mg parts. Ultra-fine grain continuous cast sheet might be warm formable (in the 90-150C range), and would lower processing cost. Rules for stamping/drawing, lubricants/coatings and designs for bead dies to hem, join inner and outer that compensate for magnesium's unique attributes require development. More information will be learned during the Magnesium Front End Design and Development Project [12], recently commissioned by USAMP.

Extrusion The fundamental advantage of Mg extrusions lies in their better mechanical properties compared to cast components (developing up to 15 % elongation at fracture[10]). Mg offers mass savings potential in single axis tensile and bending stress applications, where thin walls and component cross-sections can be enlarged. The energy absorption capabilities of Mg are lower under dynamic axial stress with dynamic transverse and diagonal loading, but new designs can reduce the differences. New alloys have an increased potential for higher specific energy absorption Vs Al or steel [11].

RTD Projects The monograph details 59 different RTDTs activities that can improveg our ability to utilize stamping, forging, extrusion and powder metal consolidation to produce Mg components. Nine examples are shown below:

- Determine the influence of microstructure, process and alloy on mechanical properties.
- Develop a property database for wrought automotive structures.
- Develop new alloys and fine-grained metallurgical structure that can allow ductile forming at 90-150 °C.
- Develop a thermomechanical processing strategy to obtain warm forming ductility.
- Develop design rules that take into account magnesium's lower compressive vs. tensile strength.
- Predict static/dynamic behavior vs testing/working direction vs. starting material orientation and temper.
- Develop the microstructure that allows superplastic behavior and use superplasticity to form unique, exciting automotive shapes.
- Develop multi-scale modeling methodology that include history effects and validate models with experiments.

Summary

Lightweighting with magnesium can enhance fuel efficiency and excite customers with innovation and performance. The 163 research and technology development activities identified in the Magnesium Vision 2020 document represent a key for overcoming the cost, infrastructure, and technical issues of corrosion, fastening and manufacturing that could lead to increased magnesium automotive applications. Successful completion of the Magnesium Vision 2020 strategy will give US automaker and suppliers confidence that magnesium components are durable, and add function and value while providing potential cost and mass savings.

Achieving the Magnesium 2020 Vision will require developing a new infrastructure to engineer, design, manufacture and test/approve/certify magnesium components for automotive use. A Magnesium CyberCentre is proposed that would digitally network/integrate global magnesium scientists, engineers and manufacturing/product development specialists. Computational materials engineering models and technical cost models would be integrated with component specifications, process data and financial assumptions to develop implementation-ready magnesium products and processes. This approach could provide for magnesium some of the industrial research and technology development support and marketing infrastructure that sponsored aluminum's growth over the past 50 years.

References

1. Magnesium Vision 2020: A North American Automotive Strategic Vision for Magnesium, U. S. Automotive Materials Partnership, (USAMP) U.S. Council for Automotive Research (USCAR) www.uscar.org, Nov 2006
2. R. H. Bezdek and R. M. Wendling, Fuel Efficiency and the Economy, American Scientist, Vol. 93, p 132-139 (2005)
3. F. R. Field III and J. P.Clark, A Practical Road to Light-weight Cars, Technology Review.com (January 1997)
4. E. F. Emley, Principles of Magnesium Technology, Pergammon Press (1966)
5. Magnesium Casting Industry Technology Roadmap. American Foundry Society, September 2005
6. High Integrity Mg Automotive Components (HI-MAC) Project, USAMP, commissioned April 1, 2006
7. J. Grassi, Alotech Corporation, private communication
8. T-Mag Process, CSIRO, Manufacturing and Infrastructure Technology, Melbourne Australia
9. J. Hunt, Projections for Sheet, Stamping and Extrusion in Magnesium (March 2005).
10. H. Friedrich and S. Schumann, Trans IMM, p. C65-71, May 2002.
11. H.E. Friedrich and B.L. Mordike, Magnesium technology: Metallurgy, Design Applications, Springer Verlag 2006.
12. "Magnesium Front End Design and Development Project", USAMP, commissioned November 2006

Acknowledgments

This material is based upon work supported by the Department of Energy National Energy Technology Laboratory under Award Number DE-FC26-02OR22910. The report, Magnesium Vision 2020: A North American Automotive Strategic Vision for Magnesium was prepared as an account of work sponsored by an agency of the United States Government. Neither the United States Government nor any agency thereof, nor any of their employees, makes any warranty, express or implied, or assumes any legal liability or responsibility for the accuracy, completeness, or usefulness of any information, apparatus, product, or process disclosed, or represents that its use would not infringe privately owned rights. Reference herein to any specific commercial product, process, or service by trade name, trademark, manufacturer, or otherwise does not necessarily constitute or imply its endorsement, recommendation, or favoring by the United States Government or any agency thereof. The views and opinions of authors expressed herein do not necessarily state or reflect those of the United States Government or any agency thereof.

MAGNESIUM FRONT END RESEARCH AND DEVELOPMENT: A CANADA-CHINA-USA COLLABORATION

Alan A. Luo[1], Eric A. Nyberg[2], Kumar Sadayappan[3], Wenfang Shi[4]
[1]General Motors Research & Development Center, Warren, MI, USA
[2]Pacific Northwest National Laboratory, Richland, WA, USA
[3]CANMET - Materials Technology Laboratory, Ottawa, Ontario, Canada
[4]China Magnesium Center, Beijing, China

Keywords: Magnesium, Automotive applications, Casting, Extrusion, Sheet forming, Joining.

Abstract

The Magnesium Front End Research & Development (MFERD) project is an effort jointly sponsored by the United States Department of Energy, the United States Automotive Materials Partnership (USAMP), the Chinese Ministry of Science and Technology and Natural Resources Canada (NRCan) to demonstrate the technical and economic feasibility of a magnesium-intensive automotive front end body structure which offers improved fuel economy and performance benefits in a multi-material automotive structure. The project examines novel magnesium automotive body applications and processes, beyond conventional die castings, including wrought components (sheet or extrusions) and high-integrity body castings. This paper outlines the scope of work and organization for the collaborative (tri-country) task teams. The project has the goals of developing key enabling technologies and knowledge base for increased magnesium automotive body applications. The MFERD project began in early 2007 by initiating R&D in the following areas: crashworthiness, NVH, fatigue and durability, corrosion and surface finishing, extrusion and forming, sheet and forming, high-integrity body casting, as well as joining and assembly. Additionally, the MFERD project is also linked to the Integrated Computational Materials Engineering (ICME) project that will investigate the processing/structure/properties relations for various magnesium alloys and manufacturing processes utilizing advanced computer-aided engineering and modeling tools.

Introduction

Increasing worldwide energy demand due, in part, to emerging economies, coupled with probable peaking of conventional petroleum production in the coming decades, are increasing the need to reduce dependence on petroleum-based fuels for transportation [1]. Vehicle lightweighting is seen as a significant strategy to improve fuel economy of vehicles with conventional gasoline internal combustion engines or alternative energy powertrains. Magnesium, the lightest structural metal, has recently emerged as a promising material for lightweighting and become a focus of research and development in North America, China and Europe [1-3]. Over the last few years, China has become the world's largest magnesium producer, supplying more than 70% of the world market in 2006. Magnesium components are routinely used by major automotive companies including General Motors (GM), Ford, Chrysler, BMW, Volkswagen and Toyota [4-11].

Despite increasing magnesium automotive applications in North America (NA), Europe and China, automotive magnesium applications remain principally at the single component level and dominated by die-casting applications. To evaluate the viability of magnesium as a major automotive structural material for vehicle mass reduction and performance improvement, a large scale magnesium development project was needed. It is also necessary to develop enabling technologies for magnesium castings and wrought products, in order to bring automotive magnesium applications from the single component level to the subsystem level.

Front end mass reduction and a near 50/50 front-to-rear mass ratio are important to vehicle fuel economy, driving and handling performance. In 2005, BMW introduced an aluminum front-end structure in its 5/6 series cars to reduce mass (about 20 kg) and achieve a near 50/50 mass distribution, at considerable cost penalty compared to steel [12]. Based on the better castability of magnesium compared to aluminum (especially in making large thin-wall castings for part consolidation) and the overall reduction in cost of the primary metal, it is anticipated that a magnesium-based, front-end structure can be designed and manufactured at significantly lower cost compared to aluminum and at near cost parity with steel.

Despite both opportunities for the automotive and magnesium industries and potential benefits to society from increased use of magnesium for automotive applications, there are many technical challenges in materials and manufacturing technologies as well as knowledge-base development. These challenges are huge and global, and require close collaboration among industries, governments and academia from many countries.

Through the efforts of many organizations and individuals, a Canada-China-USA collaborative project on Magnesium Front End Research & Development (MFERD) has been started in early 2007, to address the challenges for primary body applications of magnesium. The project is sponsored by Natural Resources Canada, the Chinese Ministry of Science and Technology, the United States Department of Energy and the United States Automotive Materials Partnership (USAMP) which is a partnership of Chrysler, Ford and GM. The goal of this project is to develop the enabling technologies in magnesium extrusion, sheet, high-integrity body casting, joining and assembly as well as the knowledge base in magnesium corrosion protection, crashworthiness, fatigue and durability, noise, vibration and harshness (NVH) performance.

Two attributes of this project are particularly significant:

- The focus on entire "front end" structures as the development object as opposed to individual components, thereby offering critical vehicle-level benefits in terms of mass distribution and performance, while challenging the technical community to devise materials and manufacturing approaches to permit mass production of such structures.

- The establishment of a truly international effort bringing together scientific and engineering expertise in the field of magnesium technology from the United States, Canada and China, in what is believed to be a first-of-its-kind collaboration.

This paper describes the project goals, structure, technical tasks, progress and future prospects of this first-of-its-kind Canada-China-USA collaboration.

Project Goals and Structure

The Magnesium Front End Research and Development project is divided into the following two phases:

- Phase I. Enabling Technology Development (three years); and
- Phase II. Components Manufacturing Validation (three years).

As shown in Fig. 1, the development of enabling technologies in this project will draw product requirements from an accompanying USAMP project "Magnesium Front End Design and Development" (MFEDD) and will select target products from the front end design. In return, this project will provide critical material properties, performance (such as crash and NVH) and manufacturability data to the MFEDD design and technical cost modeling tasks.

Fig. 1. Magnesium front end project structure.

This paper provides the details of the Phase I work for the first three years, with funding approved by all three countries. A decision gate is placed at the end of Phase I for both MFEDD and MFERD projects. Nonetheless, the development of the enabling technologies in MFERD Phase I would be beneficial to expanded use of magnesium in body applications. These technologies will not only enable the magnesium front end development, but also accelerate magnesium applications in automotive body structures and closures as well as other structural applications.

The development target has been set as the magnesium front end structures designed in a related USAMP project, Magnesium Front End Design and Development [13]. Two architectures, rear-wheel-drive (RWD) unibody and body-on-frame (BOF), have been selected for the USAMP magnesium front end design and feasibility study. The "donor" design target vehicle structures have been provided by GM (a RWD unibody architecture represented by Cadillac CTS) and Ford (a light truck BOF architecture represented by Ford F150). The "donor" vehicle teams have provided the baseline steel math data and the vehicle design/performance targets to the USAMP project team for design and technical cost modeling of magnesium front end structures. Fig. 2 shows the unibody front end steel baseline design [13] and the magnesium/steel upper front end design of the BOF architecture [14]. A 50-60% mass reduction goal has been set for the magnesium front end designs with equivalent performance and comparable cost (based on a technical cost modeling study) to the baseline vehicles. The critical material performance and manufacturing requirements for these magnesium structures are used as the development targets for the Canada-China-USA collaborative project on Magnesium Front End Research & Development (MFERD).

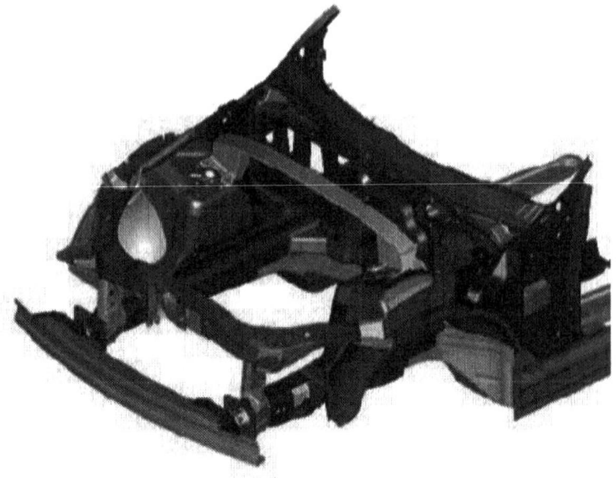

(a) Unibody front end steel baseline [13]

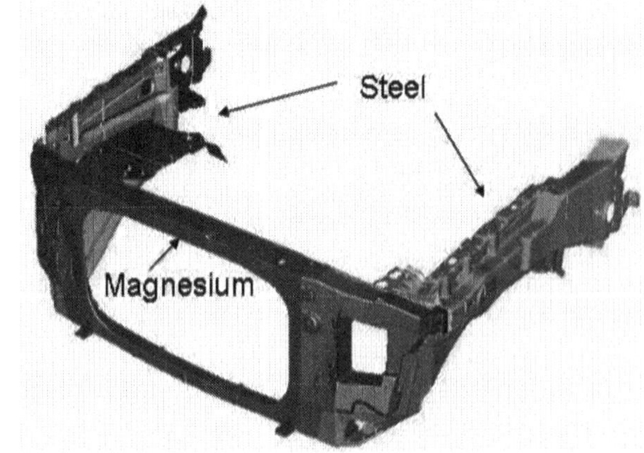

(b) BOF Mg/steel upper front end baseline [14]

Fig. 2. Magnesium front end development target.

Fig. 3 illustrates in "fishbone" format, how the various organizational aspects of the project contribute to the overall objective of a durable, affordable and manufacturable magnesium body structure. The project is divided into two major elements:

- Knowledge base development incorporating features of: crashworthiness (Task 1), NVH (Task 2), fatigue and durability (Task 3), and corrosion and surface treatment (Task 4); and
- Enabling technologies development including: extrusion and forming (Task 5), sheet manufacture and forming (Task 6), body castings (Task 7), and welding and joining (Task 8). Additionally, ICME (Task 9) provides an integrated computational tool to support the development of magnesium body applications.

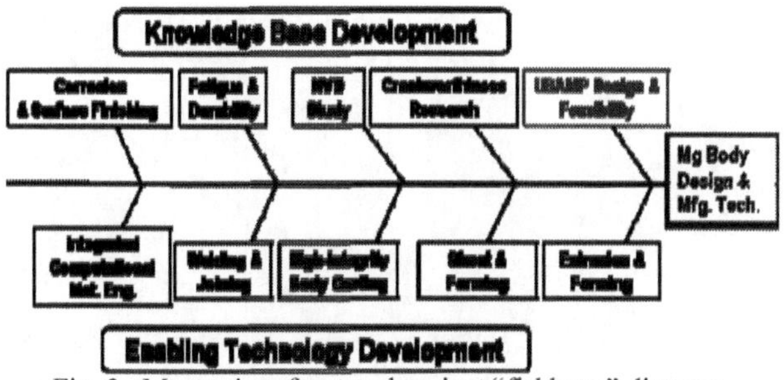

Fig. 3. Magnesium front end project "fishbone" diagram.

Project Tasks

Task 1. Crashworthiness Research

Magnesium castings have been used in many automotive components such as the instrument panel beams and radiator support structures. High-ductility AM50 or AM60 alloys are used in these applications and performed well in crash simulation and tests and many vehicles, with these magnesium components, achieved five-star crash rating. However, an entire front end structure of magnesium presents a significant challenge in vehicle crash and energy management. There is limited material performance data available for component design and crash simulation.

This task will determine the effect of high strain rates (as observed during crash tests) and anisotropic characteristics on the mechanical properties and energy absorption capability of magnesium wrought and cast alloys. The crashworthiness research includes material testing, failure characterization, CAE analyses and selected testing of prototype sections in crush (low loading rate) and crash (high loading rate) component tests. The strain rates for the material testing ranges from quasi-static, 10^{-3} sec^{-1}, through the mid rates of 10 sec^{-1}, up to 10^3 sec^{-1} which is seen in the high deformation areas in a high speed crash. Presently, high strain rate testing is ongoing at four different labs in Canada, China and US on the common materials specified by the three-country team. Fig. 4 shows the Split Hopkinson bar test setup available to the Chinese crashworthiness team [15].

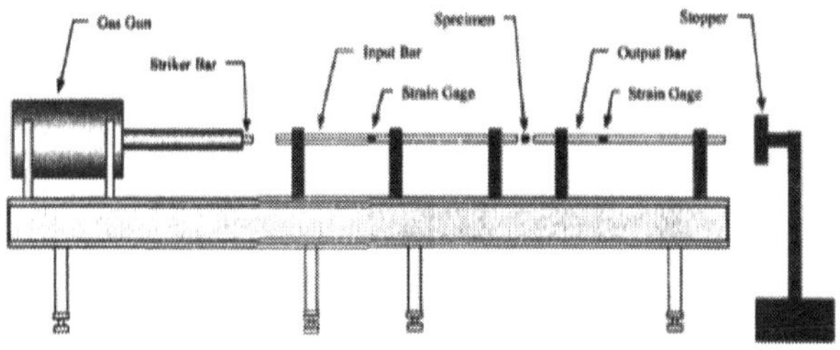

Fig. 4. Split Hopkinson bar test setup in China [15].

Task 2. Noise, Vibration and Harshness (NVH)

It is well known that magnesium has high damping capability, but this can be translated into better NHV performance only for a limited range of the sound frequency, 100-1000 Hz. The low-frequency (<100 Hz) structure-borne noise needs to be controlled by the panel stiffness between the source and receiver of the sound. The lower modulus of magnesium, compared to steel, is often compensated by the thicker gages and/or ribbing designs. For high-frequency (>1000 Hz) airborne noise, a lightweight magnesium panel would transmit significantly more noise into the occupant compartment than a steel or aluminum panel, Fig. 5 [16]. A two layer construction with an expandable low density (0.52 g/cm^3) urethane stiffening/NVH foam between the two layers of metal was used in the underbody prototype of Chrysler's magnesium-intensive body study vehicle [17].

The NVH study in this project is focused on the sound transmission characteristics of magnesium alloys used in dash panels. The task will also include analysis of sound of different frequencies (such as powertrain and airborne and white noise at audible frequencies, 20-20,000 Hz) and the effect of dash panel features on the sound transmission. This task will also compare the tested NVH properties of magnesium alloys with other materials such as aluminum and steel as well as a laminated dash panel from a typical production vehicle. While the fundamental NVH study is carried out in China and Canada, the US team is focused on the acoustic mitigation of the typical dash panel configurations and component NVH testing.

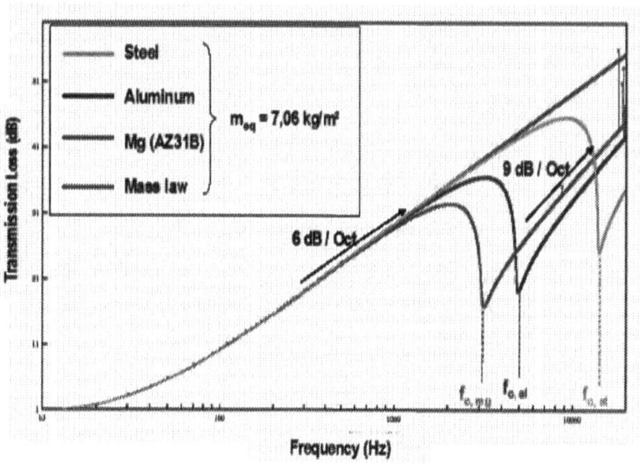

Fig. 5. Transmission loss vs. sound frequency of a magnesium panel compared to steel and aluminum [16].

Task 3. Fatigue and Durability

Fatigue and durability are critical in magnesium structural applications and there are very limited data in the literature, especially on wrought alloys. The Fatigue and Durability team has structured an international arrangement to address the fatigue behavior of magnesium castings, extrusions and stampings which have been joined into an integrated front end body structure. The effect of alloy chemistry, processing and microstructure on the fatigue characteristics of magnesium alloys will be studied. Fig. 6 is a total strain amplitude plot as a function of the number of cycles to failure for an extruded AZ31 section, generated by the fatigue team [18].

This task will study, compare and improve the fatigue predictive capability for analyzing magnesium components and sub-system behavior through better understanding the critical interactions of design, material and processing parameters that influence component microstructure and subsequent properties. Casting, extrusion and sheet products will be characterized sufficiently to establish links between microstructural features and fatigue behavior. Required monotonic and cyclic material properties for both multi-scale and traditional fatigue analysis methods will be defined, measured and made available in a final report. Preferred joining methods will be evaluated and CAE analysis parameters developed that allow accurate simulations of joints. The team will analyze magnesium front end components and sub-systems with both multi-scale and traditional methods, compare results and make recommendations that ensure component designs/sub-assemblies meets vehicle durability requirements. Functional durability of magnesium samples and typical magnesium joints will eventually be tested to validate the fatigue test and simulation results.

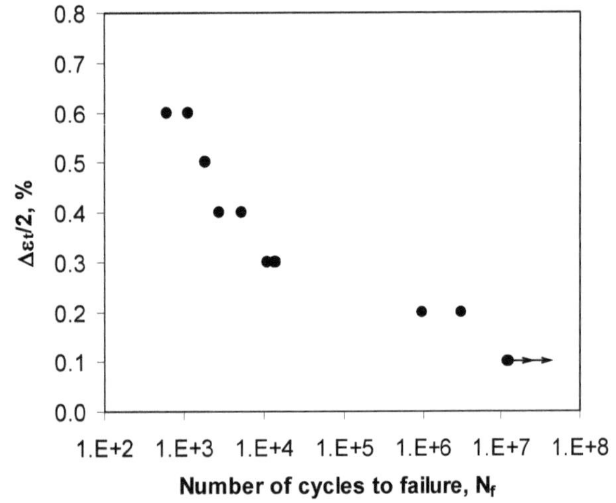

Fig. 6.Total strain amplitude as a function of the number of cycles to failure for the extruded AZ31 magnesium alloy [18].

Task 4. Corrosion and Surface Finishing

A major challenge in this project is to establish the surface finishing processes required for the intended front end application. The challenge is twofold in the sense that surface treatments to magnesium play roles in both manufacturing processes (e.g. adhesive bonding) as well as the use phase of the product life cycle demanding corrosion resistance. Furthermore, the current manufacturing paradigm for steel-intensive body structures employs chemistries in the paint shop that are corrosive to magnesium and are additionally aggravated by galvanic couples, primarily steel fasteners.

The project will explore novel coating and surface treatment technologies that have been developed in the U.S., China and Canada [19, 20]. These include novel pretreatments such as micro-arc anodizing, non-chromated conversion coatings, and "cold" metal spraying of aluminum onto aluminum surfaces. Fig. 7 shows the protective capability for a selectively cold-sprayed region in the vicinity of a steel fastener on a magnesium coupon [20]. Since most studies of corrosion protection and pre-treatment of magnesium have focused on die castings, the behaviors of sheet, extrusion and high-integrity castings will be explored for process compatibility.

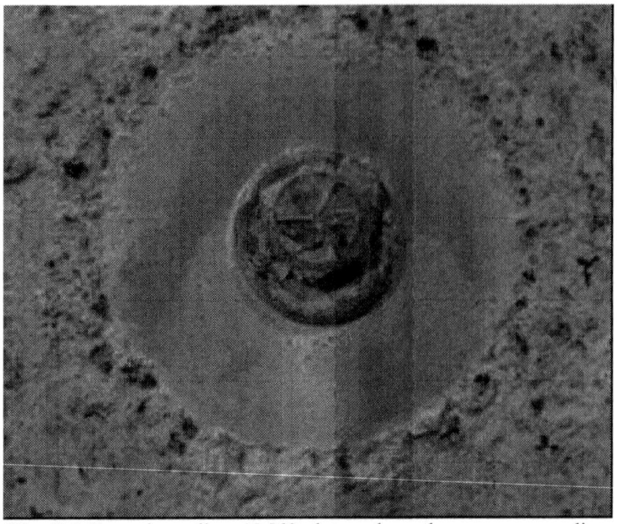

Fig. 7. Magnesium alloy AM60 plate, where the area surrounding the fastener was selectively cold sprayed with aluminum, after 1000 hours corrosion test, as per ASTM B117 [20].

While teams from all three countries will develop joint testing protocols and round-robin comparisons, there will also be opportunities for specializations inherent to the originating countries – e.g. U.S.- manufacturing processes, selection and system testing; Canada – environmentally-assisted corrosion and cold-spraying; China – novel surface treatment processes and testing.

Task 5. Low-Cost Extrusions and Forming Techniques

While high-pressure die casting is the dominant process for current magnesium automotive applications, wrought magnesium alloys and their manufacturing processes are receiving increasing attention from academia and industry. As magnesium is expanding into critical automotive body applications, there is a great need for developing magnesium extrusions and sheet products and manufacturing processes to provide improved mechanical and physical properties, crash performance and corrosion resistance. The primary reason that magnesium extrusions are not used in automotive applications is the low extrusion speeds. For example, the maximum extrusion speed of the workhorse commercial magnesium wrought alloy AZ31 is only about a half of that of aluminum extrusion alloy 6063, which makes magnesium extrusions much more expensive largely due to the higher material and processing costs [21]. Mg-Al-Mn based AM alloys are excellent die cast alloys which offer higher ductility compared to Mg-Al-Zn based AZ alloys. Recently, AM30 (Mg-3%Al-0.4%Mn) alloy was found to have about 20% faster extrusion speed than AZ31 (Mg-3%Al-1%Zn), with similar strength and slightly improved ductility at room temperature and elevated temperatures up to 200°C [22].

The key in this extrusion task is to develop/select magnesium extrusion alloys (AM30, AZ31, AZ61, etc.) with high extrusion speeds and improved mechanical properties. The extrusion process needs to be optimized for high speed (low cost) and better microstructure. New extrusion alloys and grain-refining techniques [23] will also be exploited for improved mechanical properties and extrudability. Extruded magnesium tubes will need to be bent to form structural parts in the front end. Rotary draw bending [24] and stretch bending processes will be investigated for magnesium tube bending. Moderate temperature bending might be needed. Gas forming [25] of magnesium tubes will be used to evaluate the tube formability. Fig. 8 illustrates the examples of magnesium tube bending and warm gas forming. Efforts are also directed to develop the simulation capability for magnesium extrusion and process-structure-property relationships in support of the integrated computational materials engineering (ICME) task.

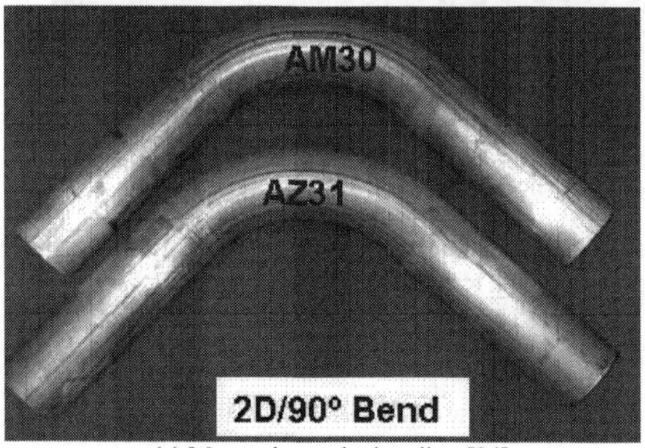

(a) Magnesium tube bending [24]

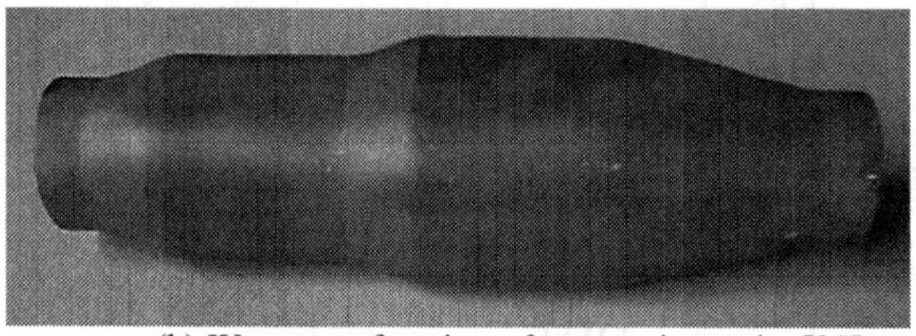

(b) Warm gas forming of magnesium tube [25]

Fig. 8. Magnesium tube bending and warm gas forming.

Task 6. Low-Cost Sheet and Forming Processes

Sheet is usually the last product form that comes to mind when considering magnesium automotive components. This is because of the high-cost and low formability of current magnesium sheet. Recent developments in the twin-roll strip casting of magnesium sheet are promising in terms of potential for low-cost, as supported by a recent USDOE sponsored cost modeling study [26]. In addition, magnesium ductility/formability has been shown to increase rapidly above about 230°C, Fig. 9 [27]. It has been demonstrated that magnesium sheet can be formed by warm stamping or superplastic forming [28, 29].

A major goal of the sheet and forming task is to procure, select and develop low-cost magnesium sheet based on the twin-roll continuous casting (CC) process, with formability and mechanical properties comparable or better than those of the current, high-cost sheet produced by the direct-chill (DC) ingot process. Sheet materials will be procured from several DC and CC sources, as well as developed by our collaborators in China and Canada. A few new sheet alloys, other than the current staple AZ31, will also be explored. The materials will be evaluated for microstructure, texture, warm formability and post-form mechanical properties. Research is also carried out to establish the process-structure-property relationships of magnesium sheet in support of the ICME efforts.

Task 7. High-Integrity Body Casting Development

Magnesium automotive components are mainly produced by high pressure die casting (HPDC). HPDC is a near-net shape manufacturing process in which molten metal is injected into a metal mold at high speed and allowed to solidify. This process, however, creates inherent defects, typically gas porosity in the castings due to the entrapment of air in the molten metal as a consequence of the high speed injection of the molten metal into the die cavity. The presence of gas porosity in castings is detrimental to their mechanical properties and pressure tightness. In addition, the formed pores, particularly those located adjacent to the casting surface, tend to expand during heat treatment, leading to the formation of blisters on the casting surface. The full potential of magnesium alloys and castings is, therefore, not realized.

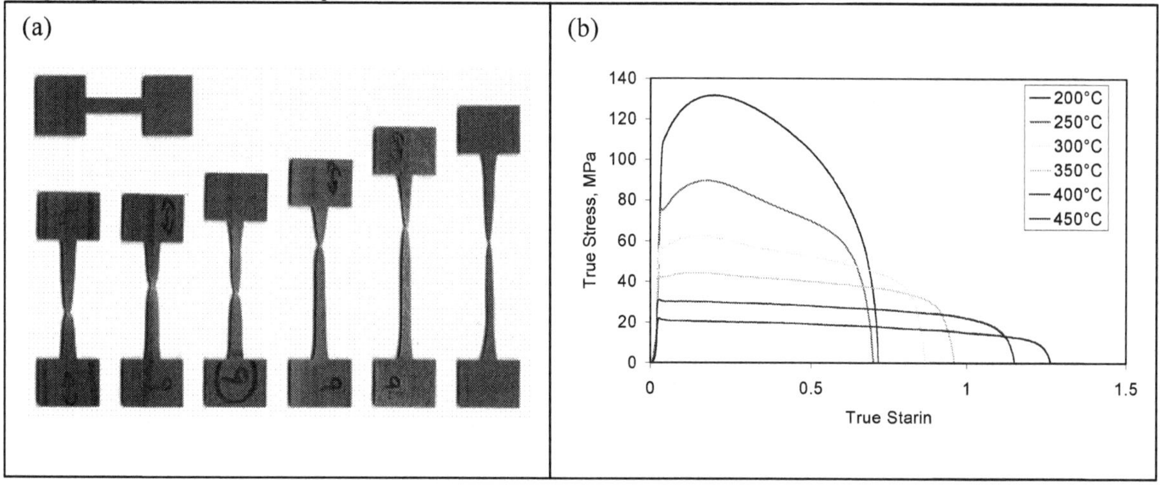

Fig. 9. Tensile test results of AZ31-H24 sheet material (2 mm gage) under constant speed pulling with initial strain rate of $1x10^{-2}$ s^{-1} (a) failed tensile bars, (b) true stress-strain curves at temperatures from 200 – 450°C [27].

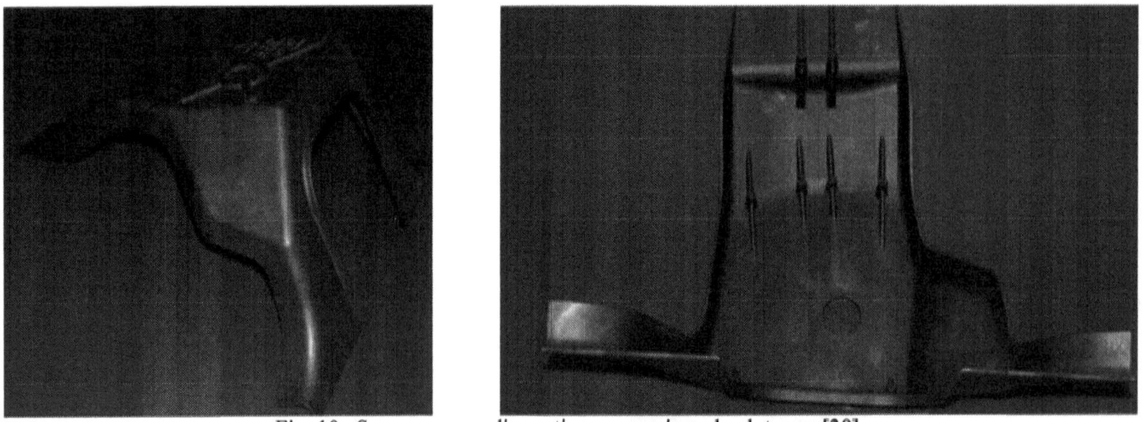

Fig. 10. Super-vacuum die casting magnesium shock tower [30].

Current aluminum body castings (about 3-5 mm wall) use special alloys and licensed vacuum die casting processes. These castings are very expensive due to the dedicated alloys and processes. The focus of this task is to develop high-integrity vacuum die casting processes for magnesium alloys to produce cost-effective magnesium castings that are heat-treatable and weldable. Porosity reduction and ductility improvement and thin-wall capability (about 2-2.5 mm) are some of the technical challenges.

Four different casting processes, vacuum die casting, squeeze casting, low pressure die casting and thixomolding, are being evaluated for their ability to produce high integrity magnesium castings. Fig. 10 shows a magnesium shock tower successfully cast at Contech using a super-vacuum process developed by the US casting team [30]. In addition, the heat treatment process will be studied to improve the mechanical properties further. As part of the task, the process-structure-property relationships of magnesium castings are being studied to achieve the ICME goals.

Task 8. Welding and Joining

Welding and joining is the very important final step in assembling the front end module from the cast and wrought magnesium components and to join the module to the steel body-in-white. There is a wide range of joining techniques that can be potentially used in for Mg-to-Mg and Mg-to-Al joining [31-34]. Fig. 11 shows a friction stir weld of AZ31B-H24 sheet with limited hardness drop and grain size increase in the weld and the heat-affected zones [34]. However, there have been no welded magnesium structures in automotive applications. The Welding and Joining team has structured an international arrangement to address multiple joining issues and technologies for Mg-to-Mg, Mg-to-Al, and Mg-to-Steel joining. Tasks have been divided between countries as follows:

- Canada: Laser welding, resistance spot welding (RSW), friction stir welding (FSW), and RSW weld bonding.
- China: Arc-welding (TIG, MIG, Plasma arc, A-TIG), Laser/TIG hybrid, and laser weld bonding.
- US: Surface pretreatment, adhesive bonding, self-piercing rivets (SPR) (including riv-bonding), Mechanical fasteners (threaded fasteners, blind rivets), and FSW.

All countries are also planning to include identification and/or development of nondestructive evaluation (NDE) methods suitable for determining the integrity of the joints produced by the respective joining technologies.

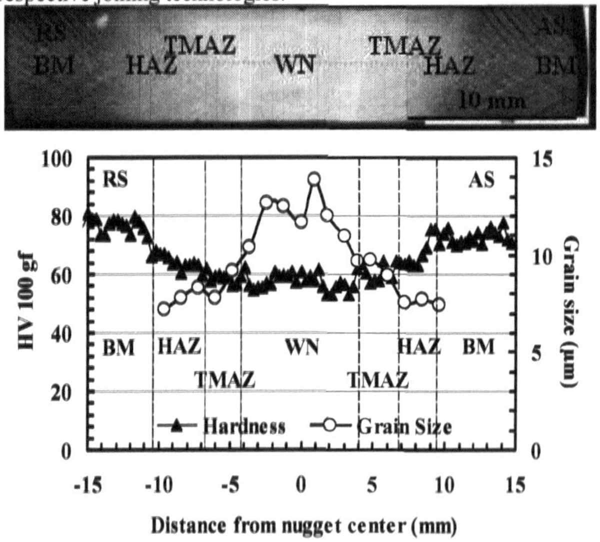

Fig. 11. Friction stir welding of AZ31B-H24 sheet showing limited hardness drop and grain size increase in the weld and the heat-affected zones [34].

Task 9. Integrated Computational Materials Engineering

Integrated Computational Materials Engineering (ICME), a new paradigm within the materials profession, offers a means to unify analysis of manufacturing, design and materials into a holistic system. A central component of ICME is the development and utilization of advanced material models which capture our quantitative knowledge of processing-structure-property relationships in a form which can be used by the broader engineering community. The goal of this task is to develop this key enabling technology for advanced lightweight magnesium processes and alloys for body structure and other body applications. This task is carried out in an accompanying Canada-China-USA collaborative project described in detail in another publication [35].

Summary

The first-of-its-kind Canada-China-USA collaborative project "Magnesium Front End Research and Development" has been formalized and launched in early 2007. The project seeks to explore emerging technologies and conduct fundamental research in the materials, manufacturing and performance of structures. This project brings together a unique team of international scope, from the United States, China and Canada to explore certain enabling technologies which could, if successful, permit the integration of magnesium-based subsystems into optimized assemblies for vehicle body structure applications using the front end structures as a test bed.

Acknowledgments

The successful launch of this project is due to the dedicated efforts of a large number of team members, especially the USAMP core team and all task leaders in Canada, China and the United States. We would like to acknowledge the sponsorship of the Natural Resources Canada, the Chinese Ministry of Science and Technology, the United States Department of Energy and the United States Automotive Materials Partnership (Chrysler, Ford and GM). We would also like to acknowledge the support and guidance provided by the project steering committee members, Dr. Joe Carpenter, Dr. Jenny Jackman, Dr. Jason Lo, Dr. Phil Sklad, Dr. Qi'an Wang and Dr. Xinmin Zhang; and their technical advisors, Dr. John Allison, Prof. Peter Frise, Mr. Swamy Kotagiri, Prof. Zheng Liu and Mr. Steve Logan. The continuing support of our respective organizations is also gratefully acknowledged. Alan Luo would like to acknowledge the dedicated effort of Dr. Robert McCune, the USAMP project technical administrator.

This project is supported in part by the US Department of Energy National Energy Technology Laboratory under Award Number DE-FC26-02OR22910. The views and opinions of authors expressed herein do not necessarily state or reflect those of the United States Government or any agency thereof.

References

1. J.A. Carpenter, J. Jackman, N. Li, R.J. Osborne, B.R. Powell and P. Sklad, "Automotive Magnesium Research and Development in North America", Materials Science Forum, 2007, vols. 546-549, pp. 11-24.
2. W. Shi, X. Zhang and Q. Wang, "Magnesium Research and Development in China", presentation at Magnesium Technology Symposium, TMS Annual Meeting, Orlando, FL, USA, February 25 – March 1, 2007.
3. K.U. Kainer, "Magnesium – Alloys and Technologies", Wiley-VCH, Weinheim, Germany, 2003.
4. A.A. Luo, "Magnesium: Current and Potential Automotive Applications", JOM Journal of Metals, 2002, 54, (2), pp. 42-48.
5. G.S. Cole, "Issues that Influence Magnesium's Use in the Automotive Industry", Magnesium Alloys 2003, Materials Science Forum, Trans Tech, Switzerland, 2003, vols. 419-422, pp. 43-50.
6. S. Schumann and H. Friedrich, "Current and Future Use of Magnesium in the Automobile Industry", ibid, pp. 51-56.
7. A.A. Luo, "Recent Magnesium Alloy Development for Automotive Powertrain Applications", ibid, pp. 57-66.
8. T. Kaneko and M. Suzuki, "Automotive Applications of Magnesium Alloys", ibid, pp. 67-72.
9. J.S. Balzer, P.K. Dellock, M.H. Maj, G.S. Cole, D. Reed, T. Davis, T. Lawson and G. Simonds: "Structural Magnesium Front End Support", SAE Technical Paper 2003-01-0185 (SAE International, Warrendale, PA, 2003).
10. M. Hoeschl, W. Wagener and J. Wolf, "BMW's Magnesium-Aluminium Composite Crankcase, State-of-the-Art Light Metal Casting and Manufacturing", SAE Technical Paper 2006-01-0069, SAE International, Warrendale, PA, 2006.
11. C.J. Duke and S. Logan, "Lightweight Magnesium Spare Tire Carrier", Proceedings of the 64th Annual World Magnesium Conference, International Magnesium Association, Wauconda, IL 60084, USA, 75-80.
12. BMW Group, "Aluminum Use in the New BMW 5&6 Series: Strategy and Experiences", presentation at the Aluminum Association Automotive Seminar, Livonia, MI, USA, Nov. 16, 2004.
13. A.A. Luo and R.C. McCune, "Magnesium Front End - AMD 603 and AMD 604", Automotive Lightweighting Materials, The United States Department of Energy FY 2006 Progress Report, January 2007.
14. M.H. Maj and T. N. Meyer, "Ultra-Large Castings of Aluminum and Magnesium", Automotive Lightweighting Materials, The United States Department of Energy FY 2005 Progress Report, August 2006.
15. Z. Liu, presentation at the "Magnesium Front End Research & Development" crashworthiness team web-meeting, July 19, 2007.
16. N. Atalla, "Numerical Modeling of Acoustic Performance of Magnesium Alloy flat Panels in Automobiles", Progress report 1, PWGCS, 2007.
17. S. Logan, A. Kizyma, C. Patterson and S. Rama, "Lightweight Magnesium Intensive Body Structure", SAE Technical Paper 2006-01-0523, SAE International, Warrendale, PA, 2006.
18. S. Begum, D.L. Chen S. Xu and A.A. Luo, "Low Cycle Fatigue Properties of An Extruded AZ31 Magnesium Alloy", paper to be submitted for publication, Oct. 2007.
19. B. Yuan, S. Yuan, B.-L. Jiang and F. Yuan, "Corrosion Resistance of Magnesium Alloy with Complex Surface-Modifying Micro-Arc Oxidation Coating and Organic Coating Studied by Neutral Salt Spraying", Materials Protection, 2006, 9.
20. J. Villafuerte and W. Zheng, "Corrosion Protection of Magnesium Alloys by Cold Spray" Advanced Materials and Processes, September 2007, 53-54.
21. A.A. Luo and A.K. Sachdev, "Wrought Magnesium Research for Automotive Applications", Magnesium Technology 2006, eds. A.A. Luo, N.R. Neelameggham, and R.S. Beals, TMS, Warrendale, PA, 2006, 333-339.
22. A.A. Luo and A.K. Sachdev: "Development of a New Wrought Magnesium-Aluminum-Manganese Alloy AM30", Metallurgical and Materials Transactions A, 2007, 38A, 1184-1192.
23. M. Yang, F. Pan, R. Cheng and A. Tang, "Effects of Solutionized Al–10Sr Master Alloys on Grain Refinement of AZ31 Magnesium Alloy", Journal of Alloys and Compounds, 2007, in press.

24. A.A. Luo and A.K. Sachdev, "Development of a Moderate Temperature Bending Process for Magnesium Alloy Extrusions", Materials Science Forum, 2005, 488-489, 447-482.

25. P. Martin, D. Baragar, K.P. Boyle, A.A. Luo, J.J. Jonas, S. Godet, K.W. Neale, "Elevated Temperature Property Measurements for Warm Forming Aluminum Alloy Tubes", in Proceedings of the 2nd International Light Metals Technology Conference 2005, Ed. Helmut Kaufmann, June 8-10, 2005, St. Wolfgang, Austria. 51-56.

26. W.H. Hunt Jr. and D.R. Herling, "Cost Assessment of Emerging Magnesium Sheet Production Methods", United States Department of Energy Report, PNNL-15368, September 2005.

27. R. Verma, unpublished work, GM R&D, June 2007.

28. P.E. Krajewski, "Elevated Temperature Forming of Sheet Magnesium Alloys", in Magnesium Technology 2002, ed. H.I. Kaplan, TMS, Warrendale, PA, 2002, pp. 175-179.

29. R. Verma and J.T. Carter, "Quick Plastic Forming of a Decklid Inner Panel with Commercial AZ31 Magnesium Sheet", SAE Technical Paper 2006-01-0525, SAE International, Warrendale, PA, 2006.

30. USAMP AMD604 Casting Task Team, "Magnesium Front End Research & Development" project casting trial, Contech, Dowagiac, MI, October 23, 2007.

31. A.A. Luo, "Wrought Magnesium Alloys and Manufacturing Processes for Automotive Applications", SAE Technical Paper 2005-01-0734, SAE International, Warrendale, PA, 2005.

32. G. Song, L. Liu and P.C. Wang, "Overlap Welding of Magnesium AZ31B Sheets Using Laser-Arc Hybrid Process", Materials Science and Engineering A, 2006, 429 (1-2), 312-319.

33. L. Liu, Z. Zhang and G. Song, "Mechanism and Microstructure of Oxide Fluxes for Gas Tungsten Arc Welding of Magnesium Alloy", Metallurgical and Materials Transactions A. 2007, 38 (3), 649-658.

34. X. Cao and M. Jahazi, "Use of Laser and Friction Stir Welding for Aerospace Magnesium Alloys", Proceedings of the 3rd International Conference on Light Metals Technology 2007, September 24-26, 2007, Saint-Sauveur, Quebec, Canada.

35. J. Allison, B. Liu, K. Boyle, Randy Beals and L. Hector, Integrated Computational Materials Engineering (ICME) for Magnesium: An International Pilot Project", presentation in Magnesium Technology 2008, TMS Annual Meeting, March 9-13, 2008, New Orleans, Louisiana, USA, 2008.

From *Magnesium Technology 2010*, S.R. Agnew, N.R. Neelameggham, E.A. Nyberg, and W.H. Sillekens, Editors

Integrated Computational Materials Engineering for Magnesium in Automotive Body Applications

John E. Allison[1], Baicheng Liu[2], Kevin P. Boyle[3], Lou Hector, Jr[4], Robert McCune[5]
[1]Ford Motor Company; Dearborn MI, USA
[2] Tsinghua University; Beijing, China
[3]CANMET Materials Technology Laboratory; Ottawa, ON, Canada
4 General Motors Corp; Warren, MI, USA
[5] Robert C. McCune and Associates; W Bloomfield, MI, USA

Keywords: Integrated Computational Materials Engineering, ICME, Magnesium Alloys, Computational Models, Automotive Applications

Abstract

This paper provides an overview and progress report for an international collaborative project which aims to develop an ICME infrastructure for magnesium for use in automotive body applications. Quantitative processing-microstructure-property relationships are being developed for extruded Mg alloys, sheet-formed Mg alloys and high pressure die cast Mg alloys. These relationships are captured in computational models which are then linked with manufacturing process simulation and used to provide constitutive models for component performance analysis. The long term goal is to capture this information in efficient computational models and in a web-centered knowledge base. The work is being conducted at leading universities, national labs and industrial research facilities in the US, China and Canada. This project is sponsored by the U.S. Department of Energy, the U.S. Automotive Materials Partnership (USAMP), Chinese Ministry of Science and Technology (MOST) and Natural Resources Canada (NRCan).

Introduction

A new field of study is evolving within the global materials profession called Integrated Computational Materials Engineering (ICME) (1,2). ICME has been defined as "the integration of materials information, represented in computational tools, with engineering product performance analysis and manufacturing-process simulation"(2). It promises to revolutionize the way the engineering community deals with materials and the way the materials community provides input to the engineering and scientific communities. It involves development of unified materials models which integrate information across length and time scales and across knowledge domains and enables integration of analysis of manufacturing, design and materials into a holistic system. Integrated Computational Materials Engineering (ICME) involves the development of materials models which integrate quantitative knowledge of processing-structure-property relationships for use by the engineering community. The development of processing-structure-property relationships has been a unifying paradigm in materials science and engineering since its inception. However this important paradigm has not resulted in the establishment of an effective computational materials engineering toolset. The focus of ICME is on developing such tools. In addition to increasing the efficiency of the engineering process, ICME has significant potential for accelerating development of new materials.

Significant opportunity exists for use of Mg in lightweight automotive body architectures. However to realize its full potential, development of new Mg alloys is required. In addition, improved understanding of the influence of manufacturing on microstructure and properties is required to more efficiently develop and optimize Mg components. Currently commercial use of Mg in high volume automotive applications are limited to two primary alloy families (AM and AZ) and manufacturing with the high pressure die casting process. Compared to most engineering metals, research on alloying and quantitative processing-structure property relationships in Mg alloys is in the early stages. ICME offers an efficient means to develop and deliver to the engineering community these quantitative processing-structure-property relationships and new alloys.

The ICME collaborative research effort described herein is a five year project which was initiated in 2007 as part of a larger program to develop enabling technologies for Mg in body applications. This program is sponsored by the U.S. Department of Energy, the U.S. Automotive Materials Partnership (USAMP), Chinese Ministry of Science and Technology (MOST) and Natural Resources Canada (NRCan). The goal of the ICME project is to establish an international ICME infrastructure (including contributor network, knowledge-base, and cyberinfrastucture) for Mg alloys in automotive structural body applications. This paper provides an overview of work being done at over fifteen institutions in China, Canada and the USA. It provides only highpoints in the specific areas of the project, with the responsible institutions listed in the headings for specific task areas. Details of this progress can be found in companion papers in this volume.

Calculated Phase Diagrams (CALPHAD), Diffusion Infrastructures and Database:

Northwestern University (USA); Northeastern University (China); Materials Informatics, Inc (USA); ThermoCalc (USA); University of Central Florida (USA); Oak Ridge National Laboratory

An essential aspect of ICME is an accurate means of calculating phase diagrams and diffusion of alloying species. Databases for calculating phase diagrams are often proprietary, thereby limiting the speed with which new information can be incorporated and utilized. In this program a novel approach to developing an open-source CALPHAD database and to optimize phase diagrams is proposed. This approach is called, the Extensible, Self-optimizing Phase Equilibrium Infrastructure for Mg Alloys (ESPEI). Input data for this open source database will originate from both first principles computations and experimentally-determined thermophysical properties.

Data infrastructures for storing input data used for thermodynamic modeling and output data have been established and tested, and the feasibility of this data structure to be implemented by the SQL (Structured Query Language) database software has been demonstrated. Fig 1 shows the overall structure of software/database designed for this project. First the available measured and predicted crystallographic, thermochemical, and phase-equilibrium data is collected and stored. Commonly used thermodynamic models are also stored in the database. These data and models are basic for thermodynamic modeling. Secondly, rules for optimization quality control, e.g. the weight of each thermodynamic datum and the optimization order during thermodynamic modeling, are established. These rules are heuristic and are treated as part of the database. Thirdly, the key data created in the optimization processes is saved in the database, in order to redo or continue a selected thermodynamic modeling. Finally, the modeled results, i.e. the self-consistent thermodynamic functions (e.g. the TDB files in ThermoCalc software) are stored. The GUI (graphical user interface) provides options for connecting the entire database and changing partial data through the database management. The GUI allows the user to model an alloy system (e.g. Al-Mg, Al-Mg-Mn, etc.) based on available thermodynamic and crystallographic data.

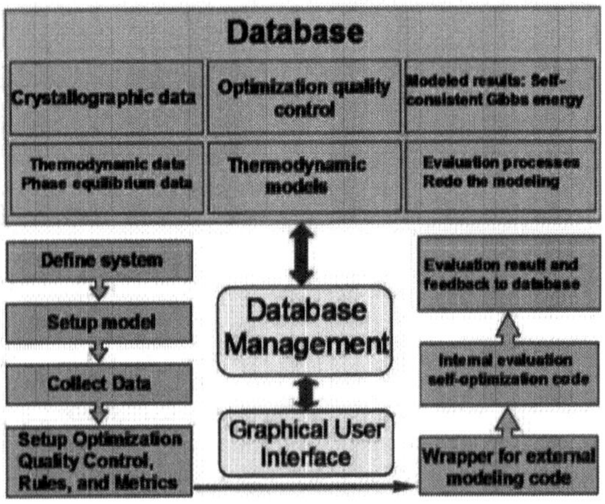

Figure 1. The ESPEI software/database structure.

The initial first-principles portion of this project focused on phases relevant to precipitation in the AZ91 (Mg-Al-Zn) system. Work has concentrated on phase stability, defects, and site substitution in the $Mg_{17}Al_{12}$ (β) phase, which is significant for this particular alloy. Efforts have been focused on calculating the effect of atomic vibrations (i.e. temperature) on the formation of native point defects in β-$Mg_{17}Al_{12}$ from the quasi-harmonic approximation. Defect formation energies at 0K were previously calculated and show that the lowest defect formation energies are those of anti-sites, $\Delta Hf(Al_{Mg})$ and $\Delta Hf(Mg_{Al})$, in β-$Mg_{17}Al_{12}$, and they are found to be very symmetric, with the Mg and Al anti-sites possessing similar defect energies. This treatment was extended by investigating the vibrational entropy and enthalpy of anti-site defects in β-$Mg_{17}Al_{12}$, and it was determined that the formation of the Al_{Mg} anti-site is not strongly stabilized by the vibrational entropy. Thus, the observed asymmetry in the off-stoichiometry of this phase in the Mg-Al phase diagram is not explained by simple native point defect thermodynamics, but must involve some more exotic effect, such as multi-defect complexes.

The calculation of defect formation free energies (e.g. 'anti-site') for a phase such as the β-$Mg_{17}Al_{12}$ involves both calculation of the enthalpy of formation (typically at 0 K) and also a formation entropy which will have a temperature dependence determined from vibrational modes. The formation entropies of Mg_{Al} and Al_{Mg} (anti-sites) at 300K are shown in Fig 2 along with the formation entropies of the intermetallic compounds, β-$Mg_{17}Al_{12}$, γ-$Mg_{23}Al_{30}$, and δ-$Mg_{89}Al_{140}$. In Fig. 2, the reference states are hcp Mg and fcc Al, respectively.

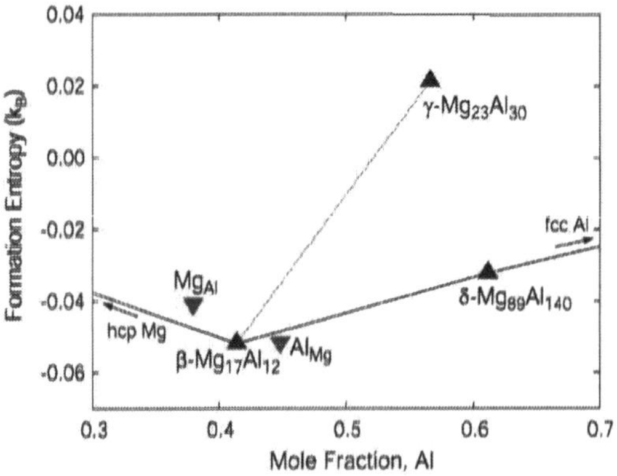

Fig 2. First-principles calculated formation entropies for the β-$Mg_{17}Al_{12}$ phase, along with the formation entropies of anti-site defects Al_{Mg} and Mg_{Al} at 300K.

In parallel with the development of ESPEI and a Mg-Al-Zn (Mn) first-principles database, efforts are also ongoing to develop experimentally determined phase diagrams for the Mg-Al-Zn and Mg-Al-Mn systems for assessing the calculated diagrams. In addition, efforts have been initiated to develop a diffusion database with efforts focused on diffusion of Al, Zn and Mn in Mg. The impurity diffusion coefficient of Al in pure Mg was measured at 300° C.

ICME for Mg Extrusions:

Mississippi State University (USA); Lehigh University (USA)

The extrusion effort is focused on development of quantitative processing-structure-property relationships in extruded shapes of Mg alloys AM30 and AZ61. A central focus is on understanding and modeling texture development in complex extrusions. This is being accomplished using experimental measurements, internal state variable constitutive models, and crystal plasticity.

A new, nested internal state variable (ISV) theory has been implemented that couples deformation and internal defects. A methodology related to a higher order tensorial rank of damage for void nucleation and void growth has been developed. This theory has been implemented in a user subroutine called DGM1.0. At lower scales, kinematics, kinetics, and hardening rules for twinning deformation have been implemented in a crystal plasticity code.

The extrusion strain and temperature history is being simulated using the commercial codes HyperXtrude®, Abaqus® and DEFORM 3-D®. To validate these simulation tools, a lab-scale extrusion experimental facility has been developed and experiments conducted on AM30 and AZ61 (Fig 3). Efforts in

DEFORM-3D have focused on the simulation of the process parameters such as stress, strain, strain rate and temperature using a modified Barnett's flow stress model and the geometry of the industrial-scale extrusion process. The numerical predictions are verified indirectly by material characterization performed on the billet material and extrudate material. This approach with two independent simulation projects allows for direct comparison between the DEFORM® and HyperXtrude® results.

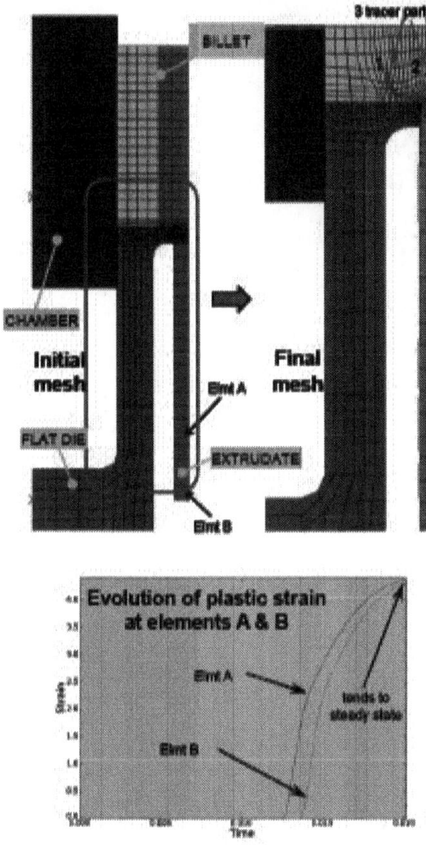

Figure 3: Finite-element model of preliminary lab-scale indirect extrusion experiments using ABAQUS.

Comprehensive experimental characterization of a variety of extruded components are being developed to extend and validate these models. Fig 4 presents results of microstructural studies (texture and grain morphology) performed on an extruded component of AM30. Besides being used for modeling purposes, such information helps to gain insights into how the process parameters may relate to the extrudate mechanical properties (e.g., anisotropy), an aspect strongly linked to product performance.

ICME for Mg Sheet Forming:

University of Virginia(USA); Mississippi State University (USA); CANMET MTL (Canada); General Motors Corp (USA); Central South University (China)

The sheet forming task is focused on Mg alloy AZ31 for developing processing-structure-property relationships for warm stamping (200–350°C) simulations of Mg sheet. Plasticity models that account for the effect of twinning and dynamic recrystallization (DRX) are being developed within phenomenological, internal state variable (ISV) and micromechanical crystal plasticity frameworks. The models are calibrated and validated with data from various concurrent mechanical and warm forming tests.

The phenomenological and ISV models are being implemented in the forming simulations codes PAM-STAMP2G® and Abaqus/Explicit®. An example of warm stamped AZ31 is shown in Fig 5. The result shows excessive strain localization in the pan corners (blue thickness contours) which is consistent with experiments. Various approaches which can be readily implemented within a finite element context are being considered including the Sellars-Tegart relation that describes steady-state behavior and the Bammann-Chiesa-Johnson (BCJ) ISV model (3). The BCJ model is incorporated into the DGM 1.0 subroutine and accounts for strain hardening behavior and damage evolution.

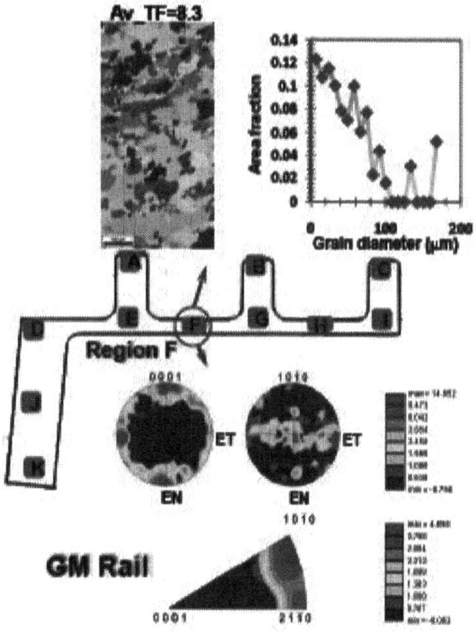

Figure 4: Microstructural characteristics at specific locations in a complex extrusion. The material is Mg alloy AM30.

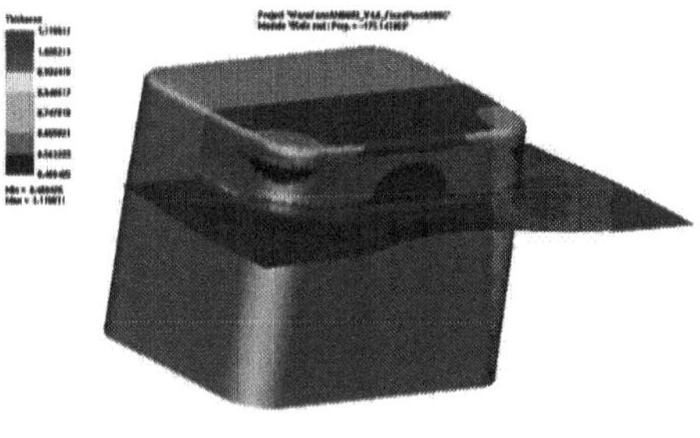

Figure 5. PAM-STAMP 2G simulation of warm stamping of AZ31. Contours depict sheet thickness variations.

A survey of available empirical and semi-empirical models, which take account of strain softening due to DRX has revealed a number of approaches in which the volume fraction recrystallized, X_D, is described by analogy with the Avrami relationship used to describe the nucleation and growth kinetics of phase transformations and static recrystallization.

$$X_D = 1 - \exp\left[-k \left(\frac{\varepsilon - \varepsilon_c}{\varepsilon_p} \right)^{n_d} \right],$$

where ε_c is the critical strain at which DRX begins, often assumed to be 80% of the peak strain ε_p (strain at which the flow stress reaches a peak) and the values k and n_d are material parameters which describe the kinetics. Other relationships, such as the power law dependence of the dynamically recrystallized grain size on the Zener-Holloman parameter seem to have already been well established for alloy AZ31 in the literature (see Fig. 6.)

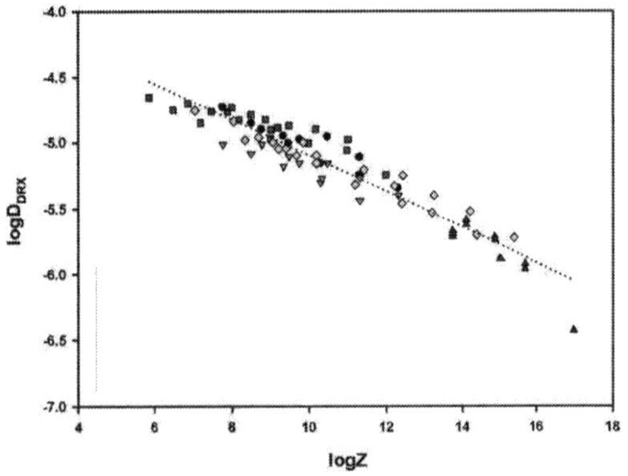

Fig 6. Relationship between the dynamically recystallized grain size, DRX, and the Zener-Hollomon parameter, Z, for AZ31 compared with literature data. A power law fit to the data points is depicted by the dotted line.

ICME for Super-Vacuum, High-Pressure Die Cast Mg:
Tsinghua University (China); The University of Michigan (USA); Ford Motor Co (USA); MAGMA Foundry Technologies, Inc. (USA), EKK, Inc,. (USA).

Super-Vacuum, High-Pressure Die Casting (SVDC) of Mg is a novel process being developed in the companion USAMP Mg Front-End Research and Development (MFERD) program. It offers significant potential for improved properties and for enabling the use of heat treatable Mg alloys in high-pressure, die cast geometries. This opens up a new vista of alloy development for cast Mg alloys. Alloy AZ91 was selected as a model alloy for this task due to its ease of casting as well as its potential for property improvement via heat treatment. Material was provided in the form of SVDC shock towers shown in Fig 7. Simulations of the SVDC process were performed using commercial software packages provided by MAGMA Foundry Technologies and EKK. To ensure accurate microstructural predictions, a comprehensive set of universal interfacial heat-transfer coefficients have been developed for the SVDC process.

Research efforts have focused on characterization of tensile properties and total strain-controlled, low-cycle fatigue behavior as a function of microstructure at the locations noted in Fig 7. The goal is to understand the influence of microstructure, porosity, and other casting-related features on fatigue life. Quantification of the role of heat treatment on monotonic stress strain behavior was also completed. The response to heat treatment on yield strength is shown in Fig 8 at an aging temperature of 168C. The aging response results from precipitation of beta phase platelets and the volume fraction of this phase was characterized by x-ray diffraction (Fig 9) and parallels the aging response shown in Fig 8. A yield strength model has been developed to describe this response.

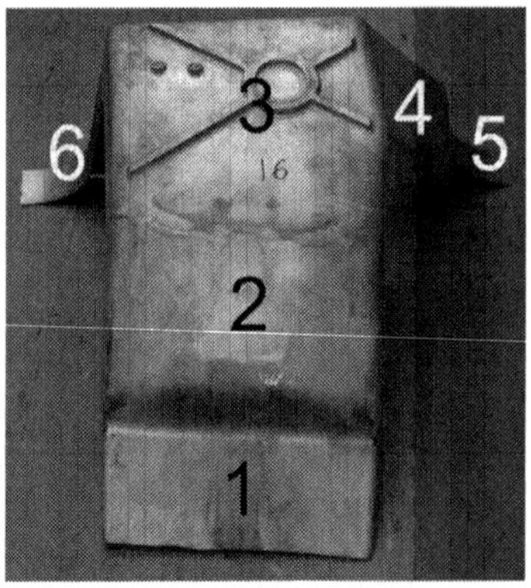

Figure 7. Super Vacuum High Pressure Die Cast Mg (AZ91) Shock Tower.

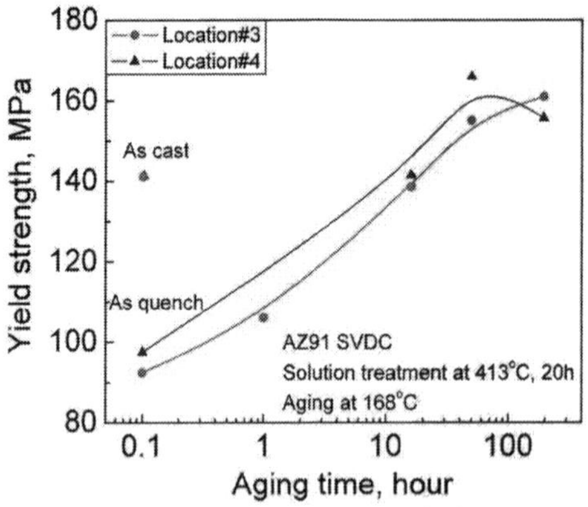

Fig 8. Yield strength response to solution treatment and aging of SVDC AZ91.

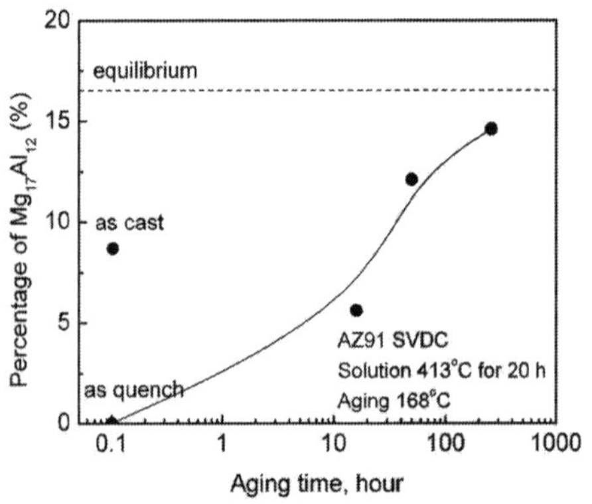

Figure 9. The influence of aging time on evolution of beta phase as determined by x-ray diffraction.

A modified cellular automaton (MCA) model has been developed for simulating the microstructure evolution and dendrite morphology of cast Mg alloys (4). The growth kinetics of dendrite tips was determined by the difference between local equilibrium composition and local actual composition obtained by solving the solute transport equation. Two sets of meshes, a hexagonal mesh and an orthogonal mesh, were used in the model to perform the simulation. The hexagonal mesh was used to perform CA calculation reflecting the texture of Mg alloy dendrites, while the orthogonal mesh was used to solve the solute transport equations. The model was applied to simulate single dendrite evolution and columnar dendrites growth of AZ91D Mg alloy, as well as multi-grain growth of Mg-10Gd-2Y-0.5Zr (wt%) Mg alloy. Permanent mold step-shaped castings of the two Mg alloys were poured and metallographic examinations were carried out for validating the present model. For example, the simulation of single equiaxed dendrite growth of AZ91D was studied under cooling rate of 10 K/s. At the beginning of simulation one cell was allowed to nucleate at the center of the calculation domain. It was assumed that AZ91D could be simplified to Mg-Al binary system.

Figure 10(a) shows the morphology evolutions and solute field of single AZ91D dendrite. To validate the simulation results, permanent mold step-shaped sample castings of AZ91D were poured and specimens for metallographic examination were obtained from each step. One of the metallographic results obtained from the step with the thickness of 35 mm, is shown in Figure 10(b). It can be seen that the simulation result and metallographic result are in good agreement.

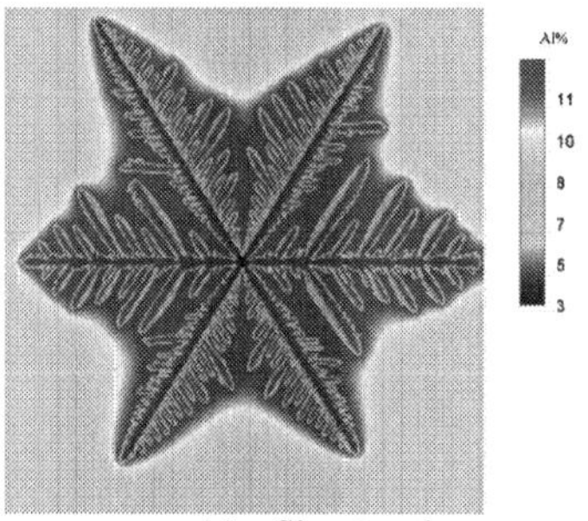

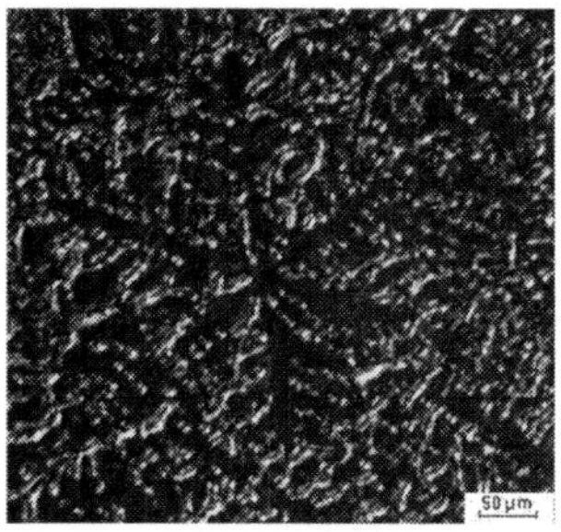

(a) Simulated (b) experimental

Fig. 10. Simulated result of single equiaxed AZ91D dendrite and metallographic result

Mg ICME Cyberinfrastructure:

Mississippi State University (USA); Pacific Northwest Labs, TMS

The ICME for Mg team participating in this program represents a geographically-distributed, collaborative community. An important enabler for such geographically-distributed ICME development teams is a web-centered knowledge-base referred to as "ICME Cyberinfrastructure" (2). ICME Cyberinfrastructure has been defined as "the Internet-based, collaborative materials science and engineering R&D environment that supports advanced data acquisition, data and model storage, data and model management, data and model mining, data and model visualization, and other computing and information processing services required to develop an ICME capability"(2).

The development of a Mg ICME Cyberinfrastructure is an integral part of this program. Development of specific services supporting an online repository of material properties, design optimizations, and multiscale simulations is ongoing. A repository of material properties has been developed and made available to test users. The repository comprises a database of experimental data (stress-strain, strain-life, and microstructure images), a database of material constants (for damage and multistage fatigue models), and is seamlessly integrated with online tools for fitting the internal state variable model, DGM 1.0, as well as for fitting a multi-stage fatigue model and microstructure analyzer. These tools are designed to extract key material constants from experimental data. Seamless integration means full support for searching experimental data directly from the tools interfaces, transferring analysis results from one tool to another (e.g., microstructural statistics to damage or fatigue model), and storing the analysis results in the material constants database so that they can be retrieved at a latter time, perhaps by a different user, and formatted according to specification of a specific finite element analysis (FEA) application. After system development and testing is completed and experimental data is curated, the intent is to make this Mg ICME Cyberinfrastructure available to the technical community under the auspices of TMS.

Summary

An international collaborative program developing an ICME infrastructure for Mg for use in automotive body applications was initiated in 2007. This paper provides an overview of this program. Quantitative processing-microstructure-property relationships are being developed for extruded Mg alloys, sheet-formed Mg alloys and high pressure die cast Mg alloys. These quantitative relationships are being incorporated into models that can be linked with manufacturing simulation to predict relevant properties and component response. The necessary infrastructure for phase diagram calculation, diffusion, and an ICME Cyberinfrastructure are integral components of this program.

Acknowledgments

The success of this project is due to the dedicated efforts of team members at Ford Motor Company, General Motors Corporation, Tsinghua University, Natural Resources Canada, The University of Michigan, The University of Virginia, Northwestern University, Mississippi State University, Lehigh University, Northeastern University (China), Materials Informatics, Inc., Oak Ridge National Labs, Magma Foundry Technologies, Inc., EKK, ThermoCalc Software, Inc., Pacific Northwest Labs and The Minerals, Metals and Materials Society (TMS).

The material is based upon work supported in-part by the Department of Energy National Energy Technology Laboratory under Award Number DE-FC26-02OR22910. This report was prepared as an account of work sponsored by an agency of the United States Government. Neither the United States Government nor any agency thereof, nor any of their employees, makes any warranty, express or implied, or assumes any legal liability or responsibility for the accuracy, completeness, or usefulness of any information, apparatus, product, or process disclosed, or represents that its use would not infringe privately owned rights. Reference herein to any specific commercial product, process, or service by trade name, trademark, manufacturer, or otherwise does not necessarily constitute or imply its endorsement, recommendation, or favoring by the United States Government or any agency thereof. The views and opinions of authors expressed herein do not necessarily state or reflect those of the United States Government or any agency thereof. Such support does not constitute an endorsement by the Department of Energy of the work or the views expressed herein.

References

1. John Allison, D. Backman and L. Christodoulou, "Integrated Computational Materials Engineering: A New Paradigm for the Global Materials Profession", JOM, The Mineral, Metals and Materials Society, Warrendale, PA, November 2006, pp. 25-27.

2. Tresa M. Pollock, John E. Allison, Daniel G. Backman, Mary C. Boyce, Mark Gersh, Elizabeth A. Holm, Richard LeSar, Mike Long, Adam C. Powell, IV, John J. Schirra, Deborah Demania Whitis, Christoper Woodward, "Integrated Computational Materials Engineering: A Transformational Discipline for Improved Competitiveness and National Security", The National Academies Press, 500 Fifth Street, N.W. Washington, DC 20001, 2008.

3. D. J. Bammann, M. L. Chiesa, M. F. Horstemeyer and L. I.,Weingarten, "Failure in Ductile Materials Using Finite Element Methods," Structural Crashworthiness and Failure, eds. T. Wierzbicki and N. Jones, Elsevier Applied Science, The Universities Press (Belfast) Ltd, 1993.

4. Liang Huo, Zhiqiang Han and Baicheng Liu, "Modeling and Simulation of Microstructure Evolution of Cast Mg Alloy", International Conference on Processing & Manufacturing of Advanced materials, August 25-29, 2009, Berlin, Germany

From *Magnesium Technology 2007*, R.S. Beals, A.A. Luo, N.R. Neelameggham, and M.O. Pekguleryuz, Editors

A Lightweight Automobile Body Concept Featuring Ultra-Large, Thin-Wall Structural Magnesium Castings

Stephen Logan[1]
[1]DaimlerChrysler Corporation, 800 Chrysler Dr. Auburn Hills, MI 48326, U.S.A.

Abstract

This paper describes a lightweight automobile body concept utilizing a hybrid materials approach with emphasis on a magnesium and aluminum structure to support a high fuel-efficiency vehicle project. This approach resulted in a more than 40% weight reduction over a conventional steel body-in-white while achieving significantly improved structural performance as evaluated through CAE simulations. A business case analysis was conducted and showed promising results. One concept vehicle was built for the purpose of demonstrating concept feasibility.

Introduction

Producing profitable, high fuel-efficiency vehicles that a large number of people will want to buy has long been a goal of the automobile industry. With worldwide energy demands increasing dramatically over the past few years and the trend indicating more of the same for the future, the priority on this goal is likely to increase. Unfortunately, most high fuel-efficiency offerings have tended to be small and underpowered. Additionally, it has been difficult to make a successful business case with such vehicles.

Seven years ago, the Liberty and Technical Affairs Group of DaimlerChrysler conducted a program to develop an engineering proof-of-concept vehicle to demonstrate how high fuel efficiency could be accomplished in a marketable product based on an existing vehicle package and design. The vehicle chosen as the baseline for this study was the then new Chrysler Sebring sedan. This was the type of high volume mid-size family sedan that could significantly impact CAFÉ performance. The V-6 option was chosen for all performance targets since it was the higher selling and most popular version of the Sebring.

In order to maintain a viable business case for the program, variable cost and investment costs had to be kept low enough so that a business case based on a five year plan for 250,000 vehicles per year would meet corporate requirements for an acceptable return.

To realize the aggressive fuel efficiency and acceleration objectives for the program, several performance simulations were conducted with the proposed powertrain system. As a result, it was determined that the total weight of the vehicle would have to be reduced by 20% and the aerodynamic drag of the vehicle would have to be reduced by 25%.

In order to achieve the required 20% total vehicle weight reduction, a goal of 40% weight reduction was established for the body. This goal was established through an iterative process to identify where the largest opportunities existed and where the required material and design changes could be made while fitting into the overall vehicle program requirements.

Lightweight Body Development

Design Description

The body structure consisted of extremely large thin-wall magnesium die-castings to improve integration and reduce plant complexity, combined with aluminum extrusions in key energy management areas, and aluminum closeout panels in the floor where no significant stamping challenges would be required and where the rectangular shape of the panels would reduce offal and cost of the more expensive sheet aluminum material. The minimum thickness for the large magnesium components was specified as 2 mm based on the literature [15] and recommendations from die-casting suppliers.

The magnesium alloy chosen was AM60B as this alloy was determined to have the best combination of strength, elongation, castability, cost, and corrosion resistance for body components.

Since the reduced aerodynamic drag requirements necessitated a smooth underbody and because the preliminary acoustic analysis showed that a very light weight underbody, regardless of material, would transmit significantly more road and engine noise into the occupant compartment than a heavier underbody unless the acoustic frequencies could be broken up and damped, the underbody was designed as a two layer construction with an expandable low density (0.52 g/cm^3) urethane stiffening/NVH foam between the two layers of metal.

By using this two layer underbody concept, the aerodynamic and NVH goals were met with no additional aerodynamic panels or acoustic damping materials required. The acoustic transmission loss was verified through laboratory testing of 1 m^2 samples of floor sections and Statistical Energy Analysis (SEA) using data from the test samples to predict overall vehicle acoustic performance. Additionally, the continuous integrated box sections created a very stiff underbody that could be optimized for improved performance in critical load transfer paths without adding extra reinforcements.

This underbody part break up, which results in significantly fewer components than a conventional steel underbody, is shown in Figure 1. The assembled underbody is shown in Figure 2, while Figure 3 shows a cutaway of the underbody, Figure 4 shows a section through the cutaway, and Figure 5 shows the section with the addition of the structural foam and the aluminum closeout panels.

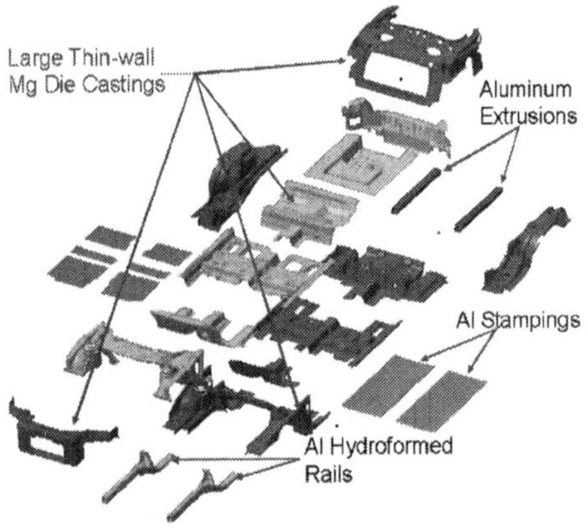

Figure 1 Underbody Component Breakup

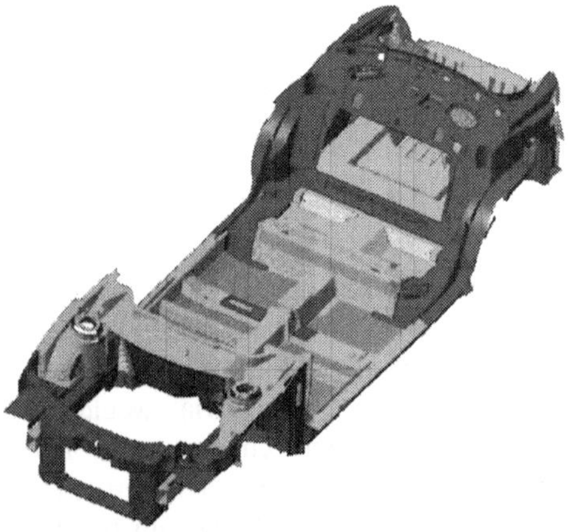

Figure 2 Assembled Underbody

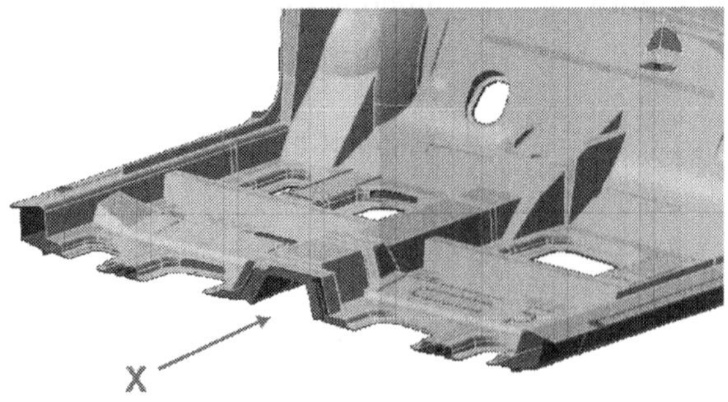

Figure 3 Cutaway View of Underbody without Aluminum Panels

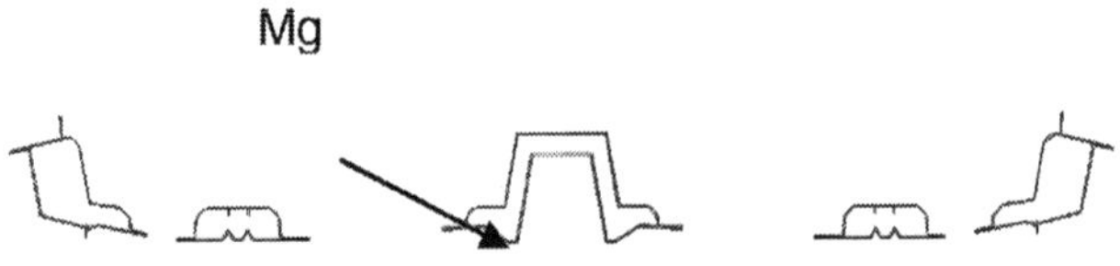

Figure 4 Cross Section View Along X Showing Box Sections

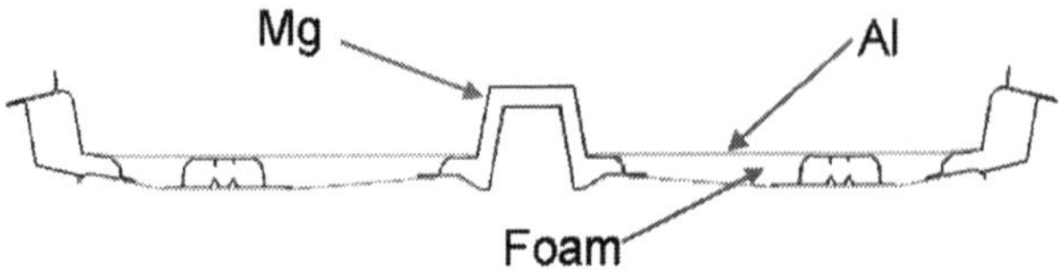

Figure 5 Cross Section View Along X Showing Box Sections, Aluminum Closeout Panels, and Structural Foam

For the body side assemblies, as shown in Figure 6, with the exception of the B-pillar reinforcement, which was cast from AM60B magnesium because of its strength and ability to integrate components, the body side apertures and reinforcements were stamped from 5000 series aluminum because of its high ductility for meeting side impact requirements and because of its formability compared to 6000 series aluminum. The body side assemblies are shown being installed to the underbody in Figure 7, and the stamped aluminum roof inner panel is shown assembled to the body sides in Figure 8.

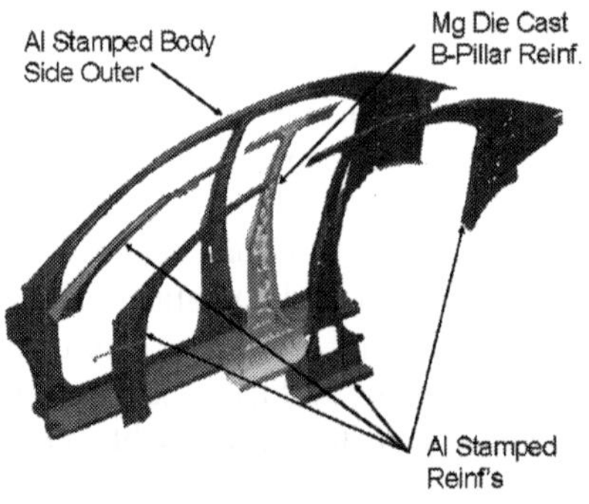

Figure 6 Body Side Assembly Component Breakup

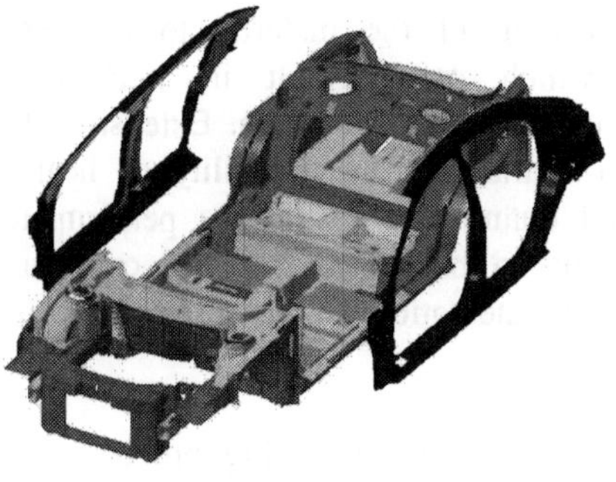

Figure 8 Completely Assembled Body Structure before Composite Panels

Class A exterior panels were chosen to be polymer composite materials with an extruded color film in order to eliminate the need for a paint shop.

Since the study was still conceptual in nature, and it was expected that no die-caster in the world would have the facilities and equipment in place to support such an extensive program at that point in time, a decision was made that casting size would be dictated by technical process feasibility, rather than the availability of existing machinery and equipment. Consequently, some of the castings used in this design would not be castable at any existing die-casting facility at this time. For a true production intent automobile body structure, with the current level of die casting infrastructure, the part breakup would have to be revised to break some of the parts into two.

Manufacturing

Production Intent

From the start of the lightweight body program, the intent was to identify and implement production feasible manufacturing and assembly processes that would closely resemble the normal sequence of processes in current automotive body shops and assembly plants wherever feasible. This would allow a business case to be developed that could be compared to current manufacturing processes. Thus, specific component manufacturing processes, joining processes and assembly processes had to be defined based on the knowledge and experience of using standard locating schemes and maintaining the same sequence of current assembly operations.

For manufacturing of the large structural components, the high-pressure die casting process was chosen. This is the most widely used process for casting magnesium and is the only process currently available with an infrastructure capable of supplying large thin-wall die castings in high volumes.

Since die-cast magnesium body parts are not conducive to spot welding, self-piercing riveting (SPR) coupled with adhesive bonding was chosen as the method of assembling the large castings and stampings to each other. This process not only closely resembles spot welding, but was also believed to be feasible for joining cast magnesium parts, due to studies at TWI [10,11], although background literature and experience were not as extensive as desired.

Magnesium sheet stamping was also considered but due to the high cost of magnesium sheet material, the business case did not support production intent feasibility. Instead, aluminum stampings were used in those applications where sheet metal stampings were considered to be favorable to thin-wall castings.

One of the decisions made at the beginning of the program was to reduce capital cost of a proposed greenfield assembly plant for the lightweight body concept by eliminating the paint shop. This led to the need for pre-coated metal components and the identification of joining processes compatible with the pre-coated components. All class A surfaces were made of polymer composite panels with extruded color film, which were attached to the metal structural components. Therefore, there was no need for any die cast body structural components to be of class A surface quality.

In addition to the design for production intent study, a detailed plant floor layout for component manufacturing as well as for the body shop was initiated. After the first phase of the study, which provided a conceptual layout and material flow for the entire body shop, including a magnesium casting and powder coating complex that feeds into the body assembly area, the study was curtailed due to budgetary constraints. Nevertheless, a manufacturing simulation software tool was used to virtually simulate step-by-step tasks at two body assembly stations: 1. Assembly of the front hydroformed rails to the cast magnesium structure and 2. Assembly of the body side aperture. This provided a level of confidence that volume production of the magnesium intensive body structure would be feasible.

<u>Casting</u>

For manufacturing of large castings, the high-pressure die casting process was chosen. This process was the only process suited to high volume production of large thin-wall magnesium components at the start of the program. The short cycle times help achieve a successful business case for high volume production, while the high pressure process helps to achieve the rapid part filling required to completely fill large thin-wall components before the material solidifies.

While there have been some concerns with the ability of the high-pressure die casting process to achieve consistent mechanical properties, the process has nevertheless been used in the past for producing numerous critical components in different applications and the quality can be improved with addition of vacuum assist if necessary. Also, recent work on heating or insulating the shot sleeve promises to improve the integrity of die-casting for large components as well [17].

One of the main advantages of casting the parts was to integrate several components together and reduce joining and assembly of the body components. With this intent, all body and structural components were designed so that they could be cast in a high-pressure die casting machine. The intent was to be able to manufacture most of the body and structural components by using 4000-ton die casting machines which were already available in the industry. Most, but not all of the components were designed to be die cast on available 4000-ton machines. Since the program was still conceptual in nature, some of the largest parts were allowed to exceed this limitation for the purposes of the study.

Joining

The selected joining process was a combination of adhesive joining and self-piercing riveting (SPR).

An example of a self-piercing rivet and its application are shown in Figure 11. The type of rivet used for the magnesium applications was a steel rivet with a zinc, tin, and aluminum alloy coating for galvanic corrosion protection. While aluminum fasteners were considered as a means of minimizing the risk of galvanic corrosion, preliminary evaluation showed that the aluminum rivets were too soft to form properly in the magnesium applications.

Figure 11 Self Piercing Rivets Before and After Application through Two Layers of Material

Additional research in collaboration with TWI (The Welding Institute) showed that optimized self-piercing riveting of the cast magnesium would benefit from localized heating of the material to prevent micro-cracking, which could result in weakened joints, especially in durability evaluation. Extensive effort was devoted to evaluating the feasibility of heat assisted process and quantifying its joining performance [11]. However, due to budget and timing constraints, the hardware implementation was not used in the prototype build

For the production intent joining process, a two-part epoxy adhesive was considered early in the program. Two-part adhesives have lower setting times and this can be an asset in high volume production as fewer fixtures may be required to maintain part geometry during handling. Additionally, since the program plan called for pre-coated composite outer panels to eliminate the need for a paint shop in the plant, the ability of a two-part epoxy to cure to its full strength without requiring the heat of a paint oven was considered to be beneficial.

However, during the program it became apparent that two-part epoxy adhesives did not provide the level of toughness (required for energy management applications) available in some of the one-part adhesives. Consequently, the decision was made to use a one-part epoxy adhesive instead of a two-part epoxy for most of the body structure. However, the use of a one-part epoxy adhesive would require a high temperature cure of at least 170°C for a minimum of 30 minutes to cross-link and obtain full strength, and this would require an oven to be added to the body shop.

Because of the desire to eliminate the paint shop from the assembly plant, the adhesive would have to be applied on pre-coated magnesium castings. Studies were conducted to determine the performance of the joint with and without coating on magnesium. This study confirmed that applying adhesive over coated magnesium castings did not significantly affect the joint performance. With proper control of pre-treatment and powder coating, the adhesive bonded joints using pre-coated magnesium castings were as good as or better than applying the chosen epoxy adhesive to bare magnesium components.

Corrosion

As mentioned earlier, several corrosion protection coatings were considered and evaluated. Alodine 5200 pre-treatment coupled with an epoxy powder coating was chosen as the corrosion protection for all body components.

Since a large die cast magnesium front of dash structure was just being launched on the Dodge Viper, and a die cast magnesium door inner panel was just going into production on the Mercedes SL Coupe', some of the corrosion protection knowledge from those programs was also applied to the lightweight body design.

As the initial plan was to bring in pre-coated magnesium components, which would later be joined through adhesive bonding and self-piercing riveting, the effect of this joining process on the corrosion protection coating was also studied. In this study, the process capability of the joining process was investigated through a Design of Experiments to understand the process capabilities and also the major factors that affect the coating on the pre-coated magnesium components.

From these studies, it was confirmed that the corrosion protection coating would be compromised at the joints by the self-piercing rivets. This observation drove design changes to protect these compromised areas by covering them with other body parts (stamped aluminum sheets) which also formed the structure of the underbody. Most exposed joints in the underbody were originally designed so that the flared end was in the ductile aluminum sheet rather than in the less ductile magnesium casting. After confirming that the self-piercing riveted joints were more water-tight at the head than at the flared tips, designs were reviewed for accessibility of the robot arm and the SPR C-clamp so as to drive the rivets with the heads exposed to the environment. This would minimize exposure of the corrosion protection compromised areas of the joints to the environment.

Prototype Build

In addition to a thorough production intent manufacturing study, several of the assembly methodologies were also verified during the prototype build. This was accomplished by using fixtures to establish principal locating points and verifying the feasibility of production intent body assembly. Several sizes of C-Clamps were staged to closely represent and perform the self-piercing riveting when the body parts were assembled in the correct sequential manner on the fixture. As most prototype parts were sand cast, the exercise proved to be challenging as geometric tolerancing was not feasible when handling thin-wall sand cast magnesium components that also had gone through some weld repairs, heat treatment, and corrosion protection coating prior to assembly. Fixtures were built to bring the prototype build as close to the production build intent as possible.

<u>Casting</u>

All prototype magnesium parts were sand-cast in the AZ91E alloy and heat-treated to a T-4 condition to obtain more ductile properties that would be close to die cast AM60B. Even after heat treatment, strength properties were only about 65% of die cast AM60B. To minimize geometric distortions during the heat treatment, all parts were placed on geometric fixtures during the heat treatment process (including the cool down period).

Since such large thin wall sand casting posed some challenges, several special techniques were required to produce these large sand castings. Molten metal was introduced into the mold simultaneously through several sprues and gates. Filters were used to minimize inclusions and slag in order to achieve good properties in the final prototype parts, which were eventually assembled to make the prototype body. While cold shuts and voids were minimized, some weld repairs were still required on several of the largest parts.

All sand-cast magnesium prototype parts were X-rayed in critical areas for defect identification before heat treatment and powder-coating. Although weld repairs were limited, they did nevertheless affect the geometric tolerance of the final parts. This in turn affected the assembly of the prototype vehicle. On a few occasions, the protective coating had to be removed to grind non-conforming parts to enable assembly. In some of these cases, corrosion protection was later applied by hand prior to assembly. All sand-cast parts were sand blasted for better finish and to enhance coatability.

Figure 12 shows the largest sand-cast part that was used on the magnesium intensive body structure prototype. This part combined the functions of 13 stamped steel parts from the steel production baseline vehicle. Due to difficulties in achieving consistent 2 mm wall thickness with sand casting of such large parts, the weight of the actual prototype part was 9.55 kg compared to the production intent weight of 7.95 kg. Similar weight discrepancies were observed on several of the prototype castings.

Figure 12 Largest Magnesium Prototype Sand Casting - Upper Trunk Housing

Joining

Although a two-part epoxy adhesive was briefly considered for high volume production, a one-part structural epoxy adhesive was always the intent for the prototype build. The reason for the decision to use a one-part adhesive during the prototype build was to prevent the adhesive from curing until the entire body was assembled. Since a prototype build process usually identifies a number of dimensional discrepancies from the production intent design and these part discrepancies require some time to modify or rework, then the extra curing time was necessary to provide the flexibility to move parts around, to modify them, even remove them and replace them later if required. The specified one-part epoxy adhesive would not cure until it was exposed to a temperature of at least 170°C for a minimum of 30 minutes.

Once the parts were properly positioned, the self-piercing rivets were applied as shown in Figure 13 to keep them properly positioned together until the entire body structure was assembled and ready for the adhesive to be cured in an oven. No special prototype tools were required for self-piercing riveting since this was the production manufacturing intent. During the build, several joints were successfully realized where the rivet had to penetrate through three and even four layers.

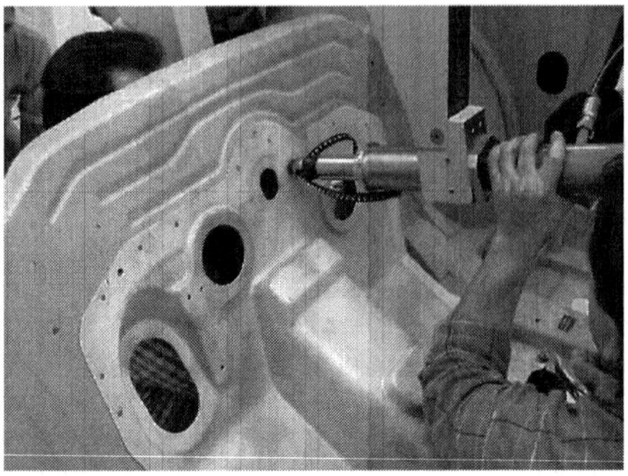

Figure 13 Prototype SPR Assembly Demonstration on Bare Magnesium Components

One unexpected finding was that the AZ91E sand cast magnesium was ductile enough so that most of the riveted joints did not result in any visual cracking at the flared end as can be seen on the test coupon shown in Figure 14.

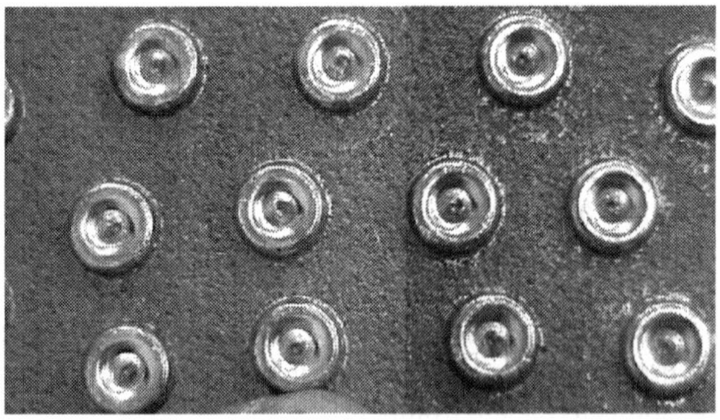

Figure 14 Back Side of Self-Piercing-Riveted Joints in Sand Cast Magnesium AZ91E with T-4 Heat Treat

The prototype body assembly fixture was designed to allow access to all joints from each side of the body. Figures 15-18 show some of the stages of assembly during the prototype build. Note in Figures 17 and 18 that the entire body-in-white was built on a fixture.

Since the sand casting process is not as dimensionally accurate as high-pressure die casting processes and since some areas did not fill completely and had to be welded, several of the prototype parts were out of dimensional tolerance. This created interference fit conditions at several joints and large gaps at others. The latter condition required bringing joint flanges together through a distance of over 20 mm. Nevertheless, these joints were performed with no failures, demonstrating that the sand cast body components were more ductile and workable than expected.

Figure 15 Front Firewall/Floor Assembly – Self Piercing Rivets are Visible - Larger Holes Are for Structural Foam Application

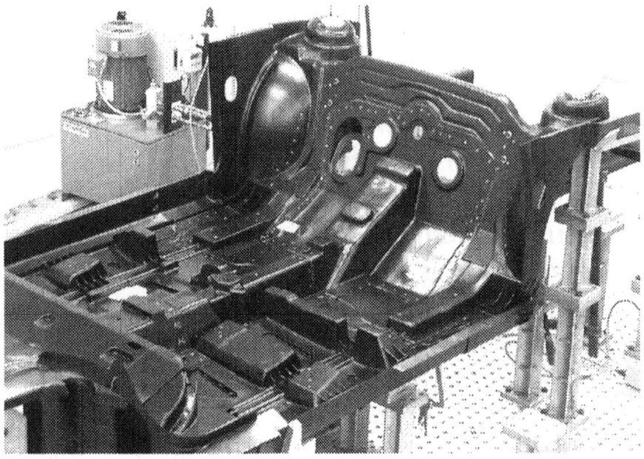

Figure 16 Firewall Assembly and Lower Center Floor – Note Ribs for Additional Stiffness and Strength in Key Areas

Figure 17 Entire Underbody Assembly – Only Thirteen Components

Figure 18 Body Structure Prior to Heat Curing of Adhesive

To cure the single part adhesive, the entire body structure and assembly fixture were staged in the curing oven and baked for over 6 hours. This was considerably longer than the specified 30 minutes, but the decision was made to be conservative since there would be considerable time required for the oven to achieve temperature and then for the body to achieve temperature as well in this one-off prototype process.

The body was left on the fixture to cool down to room temperature. As the fixture was made of aluminum with a coefficient of expansion very close to that of magnesium, any distortion due to differences in thermal expansion were expected to be negligible. This proved to be the case when the fully cured, adhesively bonded body was easily removed from the fixture for further processing as shown in Figure 19.

Figure 19 Complete Body Structure after Oven Curing

Corrosion

The conversion pre-treatment and epoxy powder coating were applied to all cast magnesium parts. Due to the large size of some of the magnesium parts, the powder coat was sprayed on by hand. All parts were baked after powder coating, and the part distortion during this baking process was kept to a minimum. Because the powder coating was sprayed by hand, the coating thickness was not always consistent. In some areas, the coating was observed as thick as 2.0 mm. This inconsistency interfered with smooth assembly of some of the prototype components, which required grinding to remove some of the thick layer of coating in order to fit together properly. This grinding posed no performance problems for the one-off prototype. However, if the vehicle had been exposed to a corrosive environment, the removal of powder coating would not have been acceptable.

Other Assembly Operations:

Once the fully cured body structure was taken off the fixture, the outer composite body panels were mounted and other assembly operations including wiring installation, powertrain installation, chassis installation, and interior installation were completed as with any prototype vehicle. Some rework was required for certain body components in order to achieve the showroom fit and finish that were expected of the final prototype vehicle. Once all body panels were mounted, finessing and rework completed, etc., the vehicle was painted to achieve the appearance shown in Figure 20.

Figure 20 Completed Vehicle after Paint Application

Conclusion

After investigation of several alternative materials and body constructions including high strength steels in conventional unitized body construction, aluminum unitized body construction, aluminum tubular space frame construction, and a composite intensive construction, a hybrid design with a primarily die-cast magnesium structure utilizing sheet aluminum floor pans, aluminum crush rails, and composite exterior Class A panels was derived.

The following comparisons are made between the lightweight body concept and the conventional steel body-in-white.

- Reduced part count of nearly 80% - 44 vs. 204.
- Total body in white weight (including composite exterior panels) reduced by 40% - 270 kg vs. 445 kg.
- Body structure weight reduced by 43% - 181 kg vs. 318 kg.
- Bending frequency remained constant at 36 Hz.
- Torsional frequency improved by more than 20% to 45 Hz.
- Reduced investment cost of 28%.
- Increased variable cost of less than 2%.
- Met or exceeded all NVH and energy management goals.

While only one concept vehicle was built using sand castings in place of die-castings to keep the project cost to a manageable level, and this vehicle was not subjected to any destructive tests such as frontal impact, roof crush, etc., the vehicle was thoroughly simulated with CAE analysis to predict performance in all categories and exceeded requirements in all of them. Additionally, although accurate NVH comparison was impossible due to the prototype build quality, one-off powertrain, suspension, tires, etc., a preliminary Statistical Energy Analysis (SEA) showed that the two layer underbody construction should result in a vehicle as quiet as the much heavier steel vehicle it would replace, and the on-road subjective evaluations of the vehicle revealed what seemed to be a very tight and solid body construction with no discernable BSR issues attributable to the body construction.

The prototype body weighed more than predicted by ~35 kg. Detailed analysis of the prototype build process revealed that the sand castings exceeded their design weights by a total of 10 kg (~10%) due to inability to cast all of the parts as thin as the production intent die-castings. Additionally, approximately 25 kg of body filler, primer, paint, etc. was required to give the composite exterior panels a show car quality appearance. This is not unusual for hand built prototype vehicles.

While additional research and development work would be required to fully validate the concept for high volume production, the work conducted on this project indicates that a lightweight body structure with extensive use of large thin-wall magnesium die castings can meet all current structural and energy management requirements at a competitive cost to conventional steel body structures while achieving significant weight savings.

Figure 20 Completed Lightweight Vehicle

Acknowledgments

The authors gratefully acknowledge the contributions of the entire JSX team of the former Liberty & Technical Affairs group of DaimlerChrysler Corp. for their work in developing this concept and studying its feasibility for real world applications as well as the DaimlerChrysler Cast Metals Engineering group for generously providing their technical expertise in designing for magnesium die casting and for resolving processing issues with the thin-wall sand castings in order to ensure that good quality components were delivered for the prototype build.

Additionally, the authors express gratitude to Tom Moore, former Vice President of the Liberty & Technical Affairs group for giving the approval to go forward with this program and to Tom Edson, Director for AVE-AMMT and Mark Chernoby, Vice President for supporting the publication of this paper.

References

1. Ni, C.M., "Crush Strength Analysis of Lightweight Vehicle Frame Components", SAE Technical Paper Series, 810232
2. Aune, T.K., et al, "Mechanical Properties of Energy Absorbing Magnesium Alloys", SAE Technical Paper Series 930418
3. Caldwell, T.W., Ward, H.M., "Automotive Applications for Magnesium Die Castings", SAE Technical Paper Series, 800415
4. Aune, T.K., et al, "Behavior of Die Cast Magnesium Alloys Subject to Rapid Deformation", SAE Technical Paper Series, 2000-01-1116
5. Simonds, G., Wang, J., "Structural Magnesium Die Castings. Developing a Robust Solution", 59th Annual World Magnesium Conference Proceedings
6. Grebetz, J.C., "A Comparison of the Impact Characteristics of Several Magnesium Die Casting Alloys", SAE Technical Paper Series, 930417
7. Dierks, M.W., et al, "Enhanced Mechanical Properties of Die Cast AM Series Magnesium; Through Part Design, Die Design and Process Control", SAE Technical Paper Series, 940410
8. Smith, L.S., "Joining of Magnesium Alloys to Dissimilar Materials", TWI Report 5700/3A/2000, (for a group of sponsors), March 2000
9. Smith, L.S., "Joining Magnesium Alloy Die Castings to Themselves and to Aluminum Sheet", TWI Report 13865/1/03, (single client project for DaimlerChrysler Corporation), June 2003
10. Hawke, D.L., et al, "Field Corrosion Performance of Magnesium Powertrain Components in Light Truck", SAE Technical Paper Series, 890206
11. Hawke, D.L., Olsen, A., "Corrosion Properties of New Magnesium Alloys", SAE Technical Paper Series 930751
12. Schreckenberger, H, et al. "The Magnesium Hatchback of the 3-Liter Car: Processing and Corrosion Protection" SAE Technical Paper Series, 2000-01-1123
13. Porro, R., "The Innovative Use of Magnesium in Car Design and an Analysis of Cost vs. Weight Savings", SAE Technical Paper Series, 980084
14. NHTSA FMVSS 214 test results database [http://www-nrd.nhtsa.dot.gov/]
15. Gjestland, H., et al, "Optimizing the Die Casting Process to Achieve Reliability in Automotive Applications" SAE Technical Paper Series 2005-01-0333
16. Juchmann, P.,Wolff, S., "Magnesium Sheet Components for Ultralight Construction", International Magnesium Association 2002 59th Annual World Magnesium Conference
17. Balzer, J.B., et al, "Structural Magnesium Front End Support Assembly", SAE Technical Paper Series 2003-01-0186
18. Blanchard, P. J., et al, "The Application of Magnesium Die Casting to Vehicle Closures", SAE Technical Paper Series 2005-01-0338

MAGNESIUM CASTINGS IN AERONAUTICS APPLICATIONS – SPECIAL REQUIREMENTS

Achim Wendt[1], Konrad Weiss[1], Arye Ben-Dov[2], Menachem Bamberger[3] , Boris Bronfin[4]
[1]RWP GmbH, Am Muensterwald 11; 52159 Roetgen; Germany
[2]Israel Aircraft Industries Ltd. ; 70100 Ben Gurion International Airport; Israel
[3]Technion- Israel Institute of Technology, 32000 Haifa, Israel
[4]Magnesium Research Institute; Kroitzer 1; 84111 Beer Sheva, Israel

Keywords: Magnesium; casting; aircraft; alloy development

Abstract

Although weight reduction in aircrafts is a fundamental matter, use of magnesium alloys is not widespread in airplanes due to "traditional" problems with corrosion resistance. Though there is creation of new magnesium alloys with improved corrosion resistance, these alloys are mostly designed for automotive industry. Alloys for aerospace industry must combine high performance regarding mechanical properties and corrosion resistance. A review is given here on the use of magnesium in aircrafts in the past. The paper then presents requirements of modern aircraft industry from magnesium castings regarding mechanical properties as well as corrosion resistance. Today the drawback is a lack of sufficient number of available alloys fulfilling the needs of modern aerospace industry. The methodology used in an international research project (IDEA) to develop new magnesium alloys for special use in aircrafts is described here, as well as the experimental and virtual methods utilised in this development.

Introduction

There is a need nowadays to lower the environmental load in traffic and industrial application and development in all industrial applications. It also widely affects the goals of European research policy. For example, the Work programme of the thematic priority "Aeronautics and Space" of the 6th EU Framework Programme (FP) states that "World aeronautics is entering a new age of aviation – the age of sustainable growth– characterized by the need of more affordable, cleaner, quieter, safer and more secure air travel. European aeronautics is committed to play a prime role in shaping aviation for this new age. Research and technology development will be essential in responding to this challenge"[1]
and "Consequently, the aeronautics research work programme is set against the two top-level objectives identified in the Strategic Research Agenda[2] and the Vision 2020 Report[3]:
To meet society's needs for a more efficient, safer and environmentally friendly air transport.
To win global leadership for European aeronautics, with a competitive supply chain, including small and medium size enterprises."

More concretely, this means that the EC supports research and development for improving the environmental impact with regard to emissions and noise, responding to the challenge of meeting the society's demand for sustainable transport as well as efforts for strengthening the competitiveness of the European manufacturing industry in the global market, responding to the challenge of delivering more economical, performing and better quality products and services. Thus, the "Aeronautics and Space" programme of the EC offers an opportunity to support aerospace industry and supplying industries like foundries as well.

Knowing these objectives of the 6th FP, it is quite obvious that magnesium could play an important role in lowering aircraft weight and thus also fuel consumption in air traffic if it were possible to massively introduce Magnesium castings to aeronautics industry. With this background a consortium was formed which intends to demonstrate the potential of Magnesium castings for essential weight reduction in aircrafts. This consortium, consisting of 14 partners from Finland, France, Germany, Hungary, Israel, Slovenia, Spain and the UK, worked out a corresponding research programme called IDEA. The acronym stands for "Integrated Design and Product Development for the Eco-efficient Production of Low-weight Aeroplane Equipment".

Use of Magnesium in Aircrafts in the Past

The challenge of weight reduction has always been one of the main factors affecting aeronautic design. Accordingly, Magnesium has been widely used in the history of aircraft manufacturing

Already in 1943 Northrop manufactured its XP56, *Figure 1*, which was an all-magnesium aircraft, similar to the F-80C built by Lockheed in 1950, *Figure 2*. Several reasons prohibited the widespread use of magnesium in aircrafts. Apart from the flammability risk, high corrosion tendency and lack of availability were also the reasons. Since the nineties of the 20th century the demand for reduced fuel consumption in street traffic boosted the development of magnesium alloys and magnesium castings for automotive applications.

Figure 1. Northrop XP56, Magnesium structure, 1949. Fuselage made from magnesium.

Figure 2. Lockheed F-80C, 1955, magnesium.

In aeronautics industry, however, components from magnesium are only rarely used. Boeing and Airbus, for example, do not accept structural Magnesium parts at all in their airplanes. The share of Magnesium in modern aircrafts is negligible, the share of castings as well. In the Boeing 757, e.g., aluminium makes 78% of the material, steel 12%, titanium 6%, composites 3% and others 1%. The share of castings is about 1%, which is negligible compared to a share of extrusions (28%), sheets (20%), forgings (18%), plates (16%) and bars (13%). In an Airbus 320 we have 65.5% aluminium, 9.3 % steel, 7.2% titanium, 12.5% composites and 5.5% others. Castings again make up a share of 1%. Accordingly the task to introduce Magnesium castings into aeronautics industry is a twofold challenge:

- The acceptance of casting as a manufacturing method must be increased.
- The acceptance of magnesium as a convenient material must be established.

Background

Aircraft costs are often regarded in terms of price per unit weight. Thus, there is a straight relationship between weight saving and cost. Data presented at the Mg 2000 conference [4] revealed that in aeronautics industry the value of a pound in weight saved ranges between 300 and 3000 US Dollars depending on the aircraft. For a business jet like I.A.I.'s – G-200 (Gulfstream) the value is 700 $ per lb.This is 100 to 1000 times more than in automotive industry. This causes an obvious demand for weight saving in aeroplane design and manufacture. However, weight saving has become quite a complicated task since a number of constraints and boundary conditions have to be considered. This might include material properties like corrosion resistance or flammability and properties of the component like stiffness. Production and material costs must also be considered.

Although the weight saving potential of magnesium is well known, and although we have seen that Magnesium has already been used in aircrafts, it does not play an important role in current aeronautics industry. The main arguments one can hear against magnesium are

- Magnesium has a high flammability
- Magnesium has poor corrosion properties
- Magnesium has poor mechanical properties
- The availability of Magnesium is too low for mass production

However, the flammability risk of up-to-date Magnesium alloys is so low that plastics used in aircrafts ignite at lower temperatures than Magnesium parts. The availability of Magnesium is not a problem either. The Dead Sea alone, for example, contains Magnesium enough to supply the world demand at the current rate of approximately 360,000 tons per year for 18,000 years. Corrosion and mechanical properties, however, should still be improved as the comparison of mechanical properties in *Table 1* shows. In general, Magnesium alloys show lower tensile and yield strength than Aluminium alloys.

Table 1. Mechanical Properties of aluminium and magnesium alloys.

	Alloy	TYS [Mpa]	UTS [Mpa]	E [%]
Mg, die cast	AZ91D	160	240	5
	AM60B	130	235	13
	AM50A	125	230	15
Al, die cast	A357	215	255	3
Requirement	Separate specimen	220	290	3

Obviously there are not enough magnesium alloys available which fulfil the demands of the aeronautics industry regarding, e.g., fatigue, strength, and vibration properties. In fact, currently approximately 20 Mg alloys are available, and more than 97% of the castings are made of AZ91, AM50, AM60, WE54 and AS21 alloys. Until now it is believed that the potential increase of Mg alloys' application is in the automotive industry and to be more specific, merely in components subjected to elevated temperatures like gear box, oil panel and engine. Thus, there is a need to increase the number of Mg-alloys available for aerospace applications with specific requirements to strength, damping properties, corrosion resistance etc. Since there is an obvious lack of knowledge about the characteristics and advantages of the magnesium and its alloys, and a lack of approved standards for magnesium components, the IDEA project aims to

- Develop further Mg-alloys especially for aeronautics applications.
- Optimise the relevant casting methods.
- Inform aviation designers broadly on the usability of Mg-alloys and Mg-castings.
- Contribute to standardisation.

By this it shall finally be demonstrated that the substitution of many aircraft components by Magnesium castings is a realistic measure to reduce the weight of aircrafts.

In order to reach these goals, in a first step convenient aircraft components for these demonstrations are selected and the special requirements of aeronautics industry are defined.

Requirements of Aerospace Industry from Mg castings

As we have seen, castings in general, and magnesium castings especially, are seldom used in aircrafts. Boeing and Airbus do not allow structural parts of magnesium at all. Accordingly, the first choice for demonstration objects is non-structural parts like electronic packaging or seat components. The components considered in the IDEA project are parts of the Gulfstream G200, a business jet, built by Israel Aircraft Industries Ltd (IAI), *Figure 3*.

The G200 contains several possible demonstrator components like an electronic rack and a housing, which is even a secondary structural part. These two parts are similar to the ones shown in

Figure 4. Elements of seat frames and other parts are also under consideration.

Figure 3. Business jet Gulfstream G200.

The first approach of aeronautic designers and stress engineers to formulate requirements of magnesium parts can be simply defined as: replace Aluminium with Magnesium without any change in mechanical properties and gain all the benefit of the weight reduction.

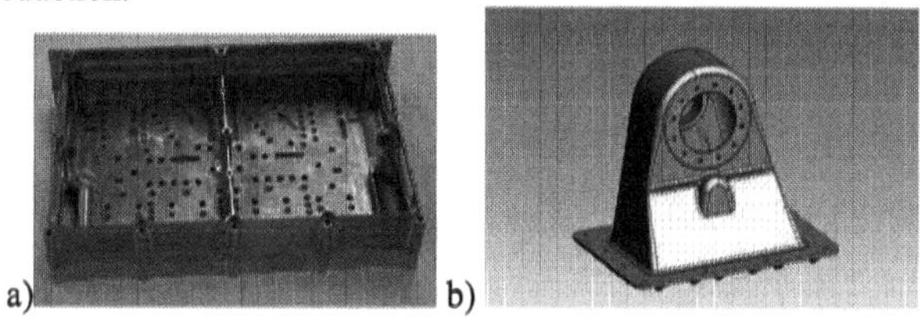

Figure 4. Typical parts that can be substituted by magnesium castings:(a) an electronic rack and (b) a secondary structural part.

Of course Mg is not and will never be Al simply because it is a different material. Therefore it is necessary to define the desired properties for developing Mg alloys for aeronautics applications. In order to be more accurate and realistic, we defined a range of properties that are a compromise between the aeronautical design requirements and the state of the art magnesium alloys' development. In the IDEA consortium and especially at IAI we see two main applications for Mg castings: electronic packages and secondary structural parts.

The results of our considerations can be summarised as follows. For electronic packaging made from magnesium we define limiting values for mechanical properties and physical properties as follows

Mechanical properties: UTS: 29 kg/mm^2, yield stress: 22 kg/mm^2, elongation= 3%

Physical properties: Thermal conductivity= 100W/m*K,

Electrical resistivity: 65 nΩ*m

Short circuit test: 250 Ampere without any severe damage.

For secondary structural parts the same mechanical properties apply. In addition, fatigue requirements are defined as in *Table 2.*

For both electronic packaging and secondary structural parts galvanic corrosion tests are to be defined. This will be done by characterisation of certain common cases, which are typical for aeronautic applications and aeronautic design. Several fasteners types and several protection systems will be tested in various environments and compared to each other as well as to a common design practice case.

Table 2. Axial fatigue for smooth test specimens.

Cycles	10^4	10^5	10^6
Percent of UTS R=0.2	88	66	50
Percent of UTS R=-1	61	44	28

Methodolgy

Apart from the alloy properties, the finally produced castings must also fulfil certain application dependent requirements regarding stiffness and deformation. Thus, the project follows two lines in parallel.

The first path aims to improve existing alloys by modification of the composition so that the required properties are met. This part consists of theoretical and practical alloy development. The effect of modification to the alloy properties will be modelled and experimentally investigated. Special effort is put to improve the corrosion properties of the alloys and of the final components by further developing coating technology for magnesium.

In parallel, the relevant magnesium casting technologies will be optimised to achieve high quality prototypes. The prototypes will undergo all the regular tests in aeronautics.

Since one of the objectives of the IDEA consortium is to boost the use of magnesium castings in aircrafts significantly the consortium will not only document the project results but also publish a design manual for magnesium castings in aircrafts. The manual is addressed especially to aircraft design engineers and shall give concrete advice where and how to use Magnesium castings.

Alloy development & surface treatment

The objectives related to alloy development and surface treatment are to develop a number of new Mg-alloys, which meet the end-users requirements regarding corrosion resistance, mechanical properties, creep and coatability. These alloys will be used for developing prototypes and demonstrators. The properties of the new alloys will be quantitatively described and compared to those of currently available commercial alloys. Based on this report, for each casting process (HPDC, sand casting, investment and Kern precision casting), the alloy which yields optimum casting properties will be selected and still further optimised. Alloy development is performed utilising computational thermodynamic and statistical methods.

In order to improve corrosion resistance several methods will be investigated. Apart from alloy modifications, formation of passive surface films, protection by a conversion layer, protection by anodisation processes and by paint layers will also be evaluated. In addition a metallic electro-deposition on conductive polymers is planned.

Mechanical, physical and corrosion properties are being determined for the commercial alloys and the new alloys as well.

Casting Technology and Simulation

The combination of state of the art alloys and processes will enable the layout design and production of high quantities with superior performance Mg components with predictable properties according to specified standards.

The IDEA project aims at improving and optimising casting technologies for the production of structural interior components like seats and the production of thin-wall castings (electronics housing and casings). The casting processes that have been selected are sand casting, investment casting, gravity die-casting, Kern Precision casting and high-pressure die-casting.

The castings will first be performed with commercial alloys like AM50, AM60 and AZ91. Optimum casting conditions will be determined and defined so that reference castings with the best possible properties are produced. The same castings will then be produced with the new alloys and optimum casting conditions. Properties of these castings, e.g. stiffness, will be compared with those of the previously cast components in order to document the progress achieved with the new alloys. When alloys and the casting processes are optimised, prototypes will be produced for demonstrating the capabilities of magnesium castings.

Simulation of the casting process and finite element analysis (FEA) play a prominent role in the casting development. Apart from traditional casting simulation, the local distribution of mechanical properties will also be modelled, as well as standard test procedures. Before simulating and producing prototypes, all investigations will be performed utilising a test casting, a test plate with varying wall thickness, which has been developed by Volkswagen AG, *Figure 5*. For each of the new alloys, solidification, microstructure, defects, stress and deformation will be simulated and verified. *Figure 6* illustrates a selection of possible results. Utilising the results of these tests and further measurements, material laws will be deduced for implementation into simulation software.

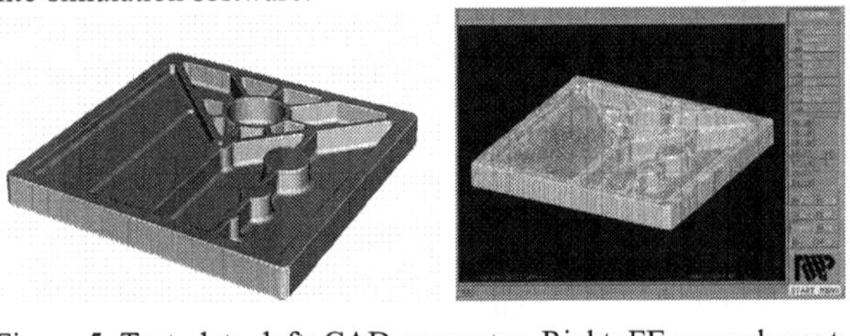

Figure 5. Test plate, left: CAD-geometry. Right: FE-enmeshment.

A further major result of this project will be the establishment of a database of comprehensive process parameters for conventional casting alloys. This database will be the infrastructure for the development of process and casting simulation capabilities. It will, e.g., enable the determination of local mechanical properties. Thus, the work packages related to casting technology and simulation provide optimised casting methods and a set of virtual design and test tools for magnesium castings for aeronautic applications.

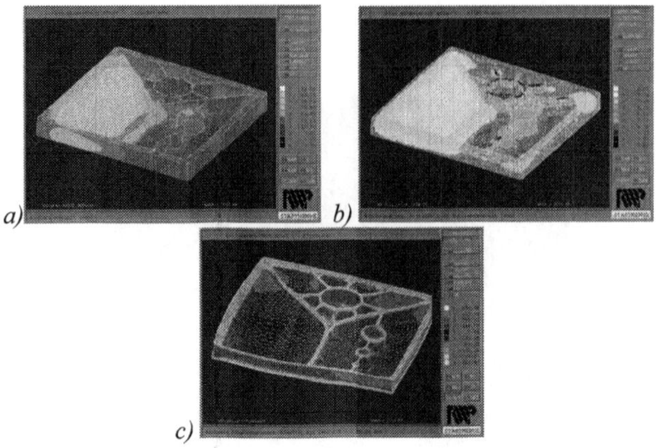

Figure 6. Results that will be used for evaluation of new developed Magnesium alloys: a) solidification behaviour, b) defect formation (porosity), c) stress and deformation.

Design Manual

One important objective of the IDEA project is to influence the way civil aircraft designs perceive the use of magnesium alloys. At the moment, though no standards that prohibit use of magnesium alloys for aerospace application exist (except for some very specific uses), many aircraft designers are reluctant to use this material. Some designers even claim that they think that the use of magnesium contradicts some standards though they could not specify them. It is evident that the ambiguity concerning the use of magnesium alloys in the aerospace industry must be removed.

The Design Manual provided by the project will encourage aviation designers to use Mg-components for aircrafts and guide them safely when choosing Mg-components. The manual will contain comprehensive information on the raw material, phase selection through appropriate production technologies, processes definition and up to a finished product and the complete assembly including recommendations for product marking, long range corrosion prevention and dissimilar metals. The design manual will present the selection of the different alloys, classified by their potential uses and the manufacturing techniques.

The manual will also contain a section on the alloys currently on the market, under what standards they are produced and what are their characteristics, and how the new alloys differ from the commercial ones.

The design manual for cast magnesium components will assure that the findings of this project, the design and construction rules, will regularly be applied to components produced after the end of this project and thus be a contribution to sustainable improvement of aircraft components with respect to product performance, environmental load and economic manufacture.

Conclusions and outlook

As has been shown here, the outcome of the IDEA project will be a demonstration of the technical and economic advantages of using magnesium castings in aircrafts. However, developing new alloys and demonstrating the convenience and advantages of magnesium castings alone is not sufficient. Aircraft manufacturing is widely standardised, and convincing especially large aircraft manufacturers to use magnesium castings can only be successful if standards are developed to deal with magnesium castings in aircrafts.

One important objective of the IDEA project is to influence the way civil aircraft designs perceive the use of magnesium alloys. All the guidelines and recommendations compiled in the Design Manual shall be coordinated, examined and approved by the CAA (Civilian Aviation Authorities) and will direct the aviation designer to perform the design as per the standard. Furthermore, the fact that the project will end up with demonstrations will cause designers to realise that the use of Magnesium in the aerospace industry is possible and is in line with current standards.

Standards in aviation industry are usually set by ICAO, the International Civil Aviation Organisation. One of ICAO's main activities is standardisation, establishment of International Standards, Recommended Practices and Procedures covering the technical fields of aviation, licensing of personnel etc. After a standard is adopted it is put into effect by each ICAO Contracting State in its own territories. As aviation technology continues to develop rapidly, the standards are kept under constant review and amended as necessary. Nevertheless, including new products in the standardisation may last up to ten years or more. Thus, while staying in touch with CAA and ICAO, the IDEA consortium takes measures to include magnesium castings into future standards.

References

1. Work Programme Thematic Priority 3 „Aeronautics and Space" of the 6th EU Framework Programme, Brussels, 2003.

2.ACARE. Strategic Research Agenda. Executive Summary, Volume 1, Volume 2. October 2002. See www.acare4europe.org

3. European Aeronautics – a Vision for 2020. Report of the Group of Personalities established by Commissioner Busquin. Brussels. January 2001. ISBN 92-894-0559-7

4. How to increase the use of Mg in aerospace applications, F.H. Froes, D. Eliezer and E. Aghion, Magnesium 2000 Proceedings of the Second Israeli International Conference on Mg Science and Technology, Edited by E. Aghion and D. Eliezer, 22-24 February 2000, Dead Sea, Israel, pp.43-49

MAGNESIUM ALLOYS IN U.S. MILITARY APPLICATIONS: PAST, CURRENT AND FUTURE SOLUTIONS

Suveen N. Mathaudhu[1] Eric A. Nyberg[2]
[1] U.S. Army Research Laboratory, Weapons and Materials Research Directorate, Building 4600, Attn: RDRL-WMM-B, Aberdeen Proving Ground, MD, 21005, USA
[2] Pacific Northwest National Laboratory, Energy Materials and Manufacturing, MSIN K3-80, P.O. Box 999, Richland, WA, 99352, USA

Keywords: Magnesium, Military, Army, Aircraft, Vehicle

Abstract

Since the 1940's Mg-alloys have been used for military applications, from aircraft components to ground vehicles. The drive for usage was primarily availability and lightweighting of military systems. But the promise of widespread military usage was not met largely based on corrosion and flammability concerns, poor mechanical behavior and inferior ballistic response. This review paper will cover historical, current and potential future applications with a focus on scientific, engineering and social barriers relevant to integration of Mg-alloy. It will also present mechanical and physical property improvements solutions which are currently being developed to address these issues.

Introduction

"When victory is ours that extraordinary weight-saving metal - hundreds of millions of pounds annually - will be available for innumerable industrial and domestic purposes. Magnesium will lighten the tasks of man in countless ways as yet undreamed of, except in the minds of far-seeing engineers and [...] who are already planning the future." - 1942 Dow Chemical Ad

This advertisement was motivated by the frequent usage of Mg-alloys for aircraft for WWII efforts. This was at the peak in Mg-alloy usage, when thousands of tons of Mg-alloy sheet (for brevity, unless a specific designation is given, Mg-alloys will be referred to as "magnesium"), extrusions and castings were being produced. Now, nearly 70 years later, the usage of magnesium has not yet met its potential for industrial, civilian or military applications. The auto industry has quadrupled its use of magnesium in the last decade, but concurrent development and military applications are few. In this paper, we will outline some historical applications and research, address historical barriers to magnesium usage by the military, highlight current applications and developments, and point to future military applications and opportunities.

Historical Applications and Research

World War II Era Applications

Owing to its low density and availability, magnesium was heavily exploited during World War I (1914 – 1918) and World War II (1939 – 1945). World War I applications consisted largely of tracer rounds and incendiary flares and munitions (applications which continue to the present time). During World War II, it became critical to develop the speed, range and capacity to deliver troops and materiel to distant theater locations. Given the limited power of early reciprocating and jet engines, it was necessary to reduce weight by usage of magnesium to accommodate increased payload and range. In fact, from 1941 to 1944, 13 government-owned plants were established to supply military demand (see Figure 1). Interesting applications, such as a 8.5 feet by 25 feet aerial delivery platforms composed of AZ31 and ZK60 support beams and AZ31 sheet top decks were developed. [1] The platform was capable of handling loads up to 17,000 lbs, with little or no damage during drops in a variety of terrains. By far, the largest use were sheet, plate, forgings, extrusions, tubular products and castings for military aircraft fuselage and wing skins, structural frame members, body panels, decking/flooring, brackets, engine compressor casings, gear box housings and landing gear wheels. [2, 3]

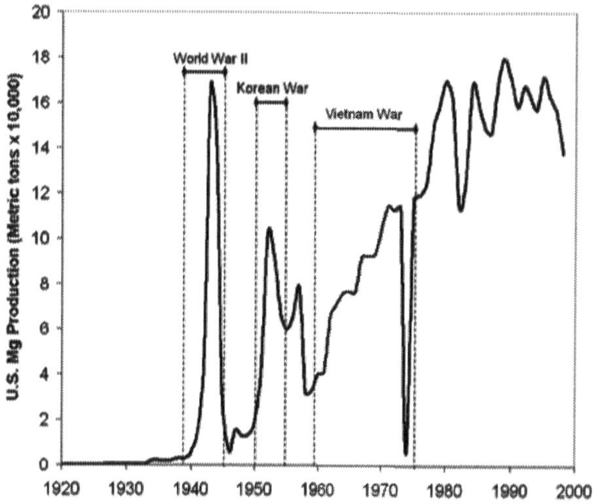

Figure 1. U.S. magnesium metal production by year (1920-1998) showing production spikes during wartime. (data from the U.S. Geological Survey, Historical Statistics for Mineral and Material Commodities in the United States, Data Series 140, 2007).

There were many aircraft at the time which extensively used magnesium components, but a key example is the design of the B-36 bomber, known as the "Magnesium Wonder of the World". [1] (see Figure 2) The B-36 incorporated 5,555 kg (12,200 lbs) of magnesium sheet, which covered 25% of the exterior, 700 kg (1,500 lbs) of magnesium forgings and 300 kg (660 lbs) of magnesium castings. Subsequent to the B-36 came the B-47 which used 5,500 kg (12,000 lbs) of magnesium components.

Korean and Vietnam Era Applications

Following World War II, the demand for magnesium drastically diminished. US production declined from 184,000 tons in 1943 to 5,300 tons in 1946. Better aircraft design and more powerful engines lessened the need for magnesium components, and completely eliminated the need in some cases. But the Korean War (1950 – 1953) renewed interest, and the government-owned magnesium plants were brought online by the end of 1951 (see Figure 1). The range of applications varied from the deep-drawn AZ31 MB-2 survival kit (15,000 manufactured between 1947-1960)[1] to the Sikorsky H-19 *Chickasaw* (see Figure 3), which was a large rotary wing aircraft designed for transporting and locating injured troops and landing in small areas. The H-19 had the highest percentage (by weight) of magnesium castings and sheet of any aircraft then in service (17 percent). [4]

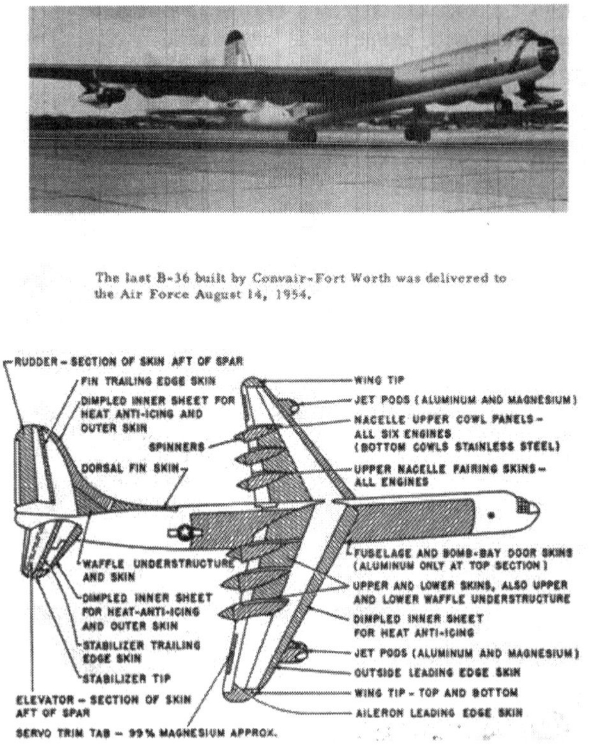

Figure 2. Schematic of some magnesium components used in the B-36 bomber (image courtesy of Robert E. Brown, Magnesium Assistance Group, Inc.).

Figure 3. Sikorsky UH-19B Chickasaw at the National Museum of the United States Air Force. (Photograph courtesy of the National Museum of the United States Air Force).

The 1950s also saw the use of magnesium for missile and ordnance components. A 105mm armor piercing round which used a magnesium sabot was developed, and is still used today. The Falcon GAR-1, one of the first air-to-air missiles, was first fielded in 1956, and had a 90% magnesium structure. The stabilizer frames were AZ91B die-castings, the rudders ZK60A-T5 forgings, the body was 0.040 in thick AZ31B-H24 sheet and ZK60A-T5 tubing. [2]

Land vehicles were not to be left out either. The mid-1950s and 1960s saw the development of the M-274 "mule" light cargo personnel and weapons carrier (see Figure 4) for use during the era of the Vietnam War (1959-1975). The M-274 weighed only 870 lbs, and could transport up to 1000 lbs for 90-150 miles. Its lightweight was due in part to cast magnesium axle housings and wheels and an extruded magnesium platform. Variants of the M-274 were manufactured from 1953 to 1970, and of the 11,240 made approximately 5,000 were left in Vietnam or destroyed. The Army continued use of the M-274 until the 1980s with the introduction of the HMMWV line of transports.

Figure 4. 1969 Baifield Industries M274 A5 USMC Reg# 324615. (Photograph courtesy of Charles Witt, webmaster, http://www.m274armymules.com).

Another Vietnam-era land (and sea) vehicle which made use of magnesium components was the M116 "Husky" amphibious personnel carrier (1957-1973). The 17-blade radiator cooling fan and air exhaust louver, along with floor and roof bows were made of magnesium. The 60lb floor covered 24.8 square feet, and was composed of 11.1 inch wide by one inch thick AZ31B extrusions. The M113 Bradley (see Figure 5) armored personnel carrier (1960-1984) had similar flooring, and was even prototyped with a magnesium-lithium alloy (14% Li and 2% Al) hull. [5,6] Lastly, the cargo version of the M113, the XM474E2 missile transporter (see Figure 6), had two 11 foot long by 10 inch wide extruded magnesium channels, each weighing only 85 lbs, and supporting a combined weight of 16,000 lbs of cargo.

Early Ballistic Research and Development

At each stage of magnesium usage, there were parallel efforts in armor research and development. The Watertown Arsenal Laboratory (an ancestor organization of the U.S. Army Research Laboratory) conducted extensive ballistic characterization of

From *Magnesium Technology 2010*, S.R. Agnew, N.R. Neelameggham, E.A. Nyberg, and W.H. Sillekens, Editors

"Dowmetal" as an aircraft armor material during the early 1940s in support of war efforts. Development of Dowmetal (refined magnesium with aluminum and manganese, now known as the "AM" series) was begun in 1919, shortly after World War I.

Figure 5. M113 Bradley armored personnel carrier at the U.S. Army Ordnance Museum. (Photograph courtesy of Edward Heasley, Curator, U.S. Army Ordnance Museum).

Figure 6. XM474E2 missile transporters with Pershing missile components and equipment. (Images courtesy of U.S. Army Artillery and Missile School Library).

These alloys were to offer better yield strength, hardness, corrosion resistance than prior alloys. Yet, in testing, the perforation resistance was found to be lower than face hardened steel alloys or "Duralumin" (age-hardenable aluminum alloys). [7-9] At this stage, the magnesium did not have suitable properties for armor. Later developments, such as the Mg-Li alloy used for the prototype M113 (developed from 1960-1963) [5] improved mechanical properties, corrosion resistance and ballistic response, but were never able to take a foothold in application due to a combination of perceptions (flammability, corrosion and poor strength), rapid development of aluminum alloys, and continued power capacity in transportation systems. Further improvements, such as novel ZK60A alloys investigated in the 1970s were met with similar response [10].

Historical Barriers to Magnesium Usage

Poor mechanical and ballistic properties alone were not enough to sink the usage of magnesium. The negative response was likely due to the real and perceived corrosion and flammability problems which revealed themselves in the 1950s and 1960s. The flammability issue was exacerbated by widespread knowledge of magnesium use in incendiary weapons, but mostly by the common high school magnesium-strip lighting experiments. The corrosion issue could not be attributed solely to perception. It is well known that magnesium has poor general and galvanic corrosion resistance, which does not favor high maintenance intervals and long product lifetime. Even with the development of high purity casting by protective melt cover gases, and barrier coatings and treatments (such as Dow 17® anodizing), normal usage resulted in wear, abrasion and mechanical damage which initiated corrosion. Poor engineering design involving joining of dissimilar metals and crevices for moisture collection and misapplication exacerbated the problem. A well known example is the pitting corrosion of magnesium wheels. This forced the Air Force to replace all magnesium aircraft wheels with aluminum counterparts.

In 1987, the U.S. Army Materials Technology Laboratory Center for Excellence for Corrosion and Prevention and Control (a more recent antecedent of the U.S. Army Research Laboratory) sponsored a workshop on the viability of magnesium in military hardware. [11] Corrosion issues dominated the workshop. A number of issues concerning the prospects for magnesium hardware were addressed. Engine parts were identified as to where magnesium would be suitable material for military hardware. Difficulty of maintaining corrosion free parts in the filed was seen to be the biggest roadblock for usage. Given that field maintenance was perceived to be the primary hindrance, the importance of proper process control, a good product, protected parts, and strict field maintenance were key factors in overcoming such objections.

Current Applications and Research

Knowledge of the Past Leads to New Developments

The experience of the five decades brought scientist and engineers to a point where the major limitations of magnesium were clear: perceived reactivity, low corrosion resistance, and low tensile strength. Poor low temperature formability, low fracture toughness, poor elevated temperature creep response and cost issues were also known to be minor limitations. The flammability myth has largely been dispelled. For magnesium to melt, it must be either exposed to high temperatures (greater than 600°C / 1112°F) [12] for long periods of time, or be in a particulate form with a high surface area to volume ratio, such as powders, shavings and ribbons. It is extremely difficult to ignite a bulk piece of magnesium unless exposed to a fire sustained to a point where it causes melting.

Corrosion will always be a problem, but designers are now well aware of the galvanic and general corrosion issues. High purity alloys which limit use of Fe, Co, Ni and Cu have drastically improved corrosion resistance. Modern solutions which limit magnesium contact with corrosive environments include electrochemical plating, conversion coatings, anodizing, gas-phase deposition processes, laser surface alloying/cladding and organic coatings. [13] Coating solutions such as Tagnite® anodizing, plasma gel anodizing, Gardobond® X4729 composite coating have provided superior corrosion, abrasion and wear resistance. [14] Since 1996, the Tagnite coating® has protected magnesium components from on the Sikorsky CH53 Naval helicopter, the Pratt and Whitney F-22, the Marine Expeditionary Force Vehicle (EFV), and Boeing Apache helicopter. Novel reclaiming techniques have also been recently developed to address corrosion in fielded vehicles. The U.S. Army Research Laboratory has developed cold-spray technology to be a cost effective and technically feasible solution for field repair and surface protection of Mg components. [15] In addition to alloy design and coatings, engineers now mitigate corrosion by proper joint design, insulation of dissimilar metals, manufacturing process control and suitable maintenance schedules.

The strength issue has largely been addressed with the arrival of rare-earth-containing alloys (AE, AX, AJ, WE and Elektron® series). Development of WE alloys (Mg-Y-rare earth) have produced alloys with comparatively high yield and tensile strength, excellent corrosion resistance and largely retained strength properties at elevated temperatures (up to 250°C). Recently developed alloys, such as Elektron 21 and Elektron 675 have extremely high strengths on par with many structural aluminum alloys. When combined with increased inherent corrosion resistance, and the novel solutions previously discussed, we can see how current applications have been realized. Table 1 gives the tensile properties of some of high-strength, corrosion resistant alloys, and compares them with those used in prior military vehicles and aircraft (AZ31B and ZK60A). Strength issues have also addressed through innovative processing techniques in both the cast and wrought realms. Thixomolding [16], twin-roll casting [17], rapid solidification [18] and equal channel angular extrusion [19,20] have been extensively researched to refine grain size and thus strengthen magnesium alloys.

Table 1. Tensile properties of various Mg-alloys

Alloy Designation	Tensile Strength (MPa)	0.2% Proof Stress (MPa)	Elong. to Failure (%)	Magnesium Elektron Datasheet*
AZ31B-H24	235	125	7	482
ZK60A-T5	290	180	6	486
AZ91E-T6	270	170	4.5	456
Electron 21	280	170	5	455
WE54-T5	300	200	10	480
WE43-T5	280	195	10	478
Elektron 675	410	310	9	102

* http://www.magnesium-elektron.com/.

Modern Era Application

A novel application of magnesium, illustrating the improved strength is a two-piece run-flat insert which was used in the original military Humvees (see Figure 6). These inserts were designed to keep a vehicle moving at 30 mph for 30 miles on one flat tire. They were later replaced by a rubber run-flat insert due to magnesium metal flanges cutting into the tires and shortening their lifespan. [21]

The most prominent current application is that of magnesium transmission and gear housings in over 3000 Army and Navy aircraft such as the Sikorsky Black Hawk. High strength alloys and corrosion resistance has made their usage commonplace. Similar alloys have been selected for use in the General Dynamics Marine Corps amphibious Expeditionary Fighting Vehicle (EFV). The EFV transmission housing is comprised of over 15 different magnesium parts made from Elektron 21, ZE41 and AZ91E. [12]

What about Ballistic Armor Properties?

Another lesson that scientists and engineers learned was that high strength properties and good corrosion resistance don't necessarily indicate adequate ballistic response and perforation resistance. The renewed interest has led to recent ballistic evaluations of both old (AZ31B-H24) and new (Elektron 675, Electron 21, etc.) wrought alloys to ascertain perforation resistance. [22-24] General purpose AZ31B-H24, has moderate strength and good corrosion resistance, and is in no way process optimized for ballistic performance, yet it outperformed 5083-H131 against armor piercing projectiles on a per-weight basis [23]. These promising results lead to the recent introduction of the first magnesium armor materials specification guide (MIL-DTL-32333 (MR)) to serve as a baseline for further advances in magnesium armor developments in materials such as Elektron 675 and WE 54. [25]

Figure 6. Humvee wheels with two-piece magnesium run-flat inserts (photograph courtesy of Don Weston, Mid-Kansas Military Vehicle Preservation Association).

Future Solutions

Renewed Military Interest and Applications

The increased performance and momentum have generated renewed interest in magnesium applications from the Department of Defense. A workshop was held in 2007 as part of the Army Research Laboratory Materials Center of Excellence program (International Magnesium Workshop). [26] The objective was to assess the global state-of-the-art in magnesium research and to exchange information on recent developments in processing, microstructure and corresponding properties. The advent of technologies such as nanostructured and ultrafine-grained materials, and integrated computational materials engineering

(ICME) and design of alloys [27, 28] and processing have opened the door to completely new design frameworks, and give indicators of the potential for extensive magnesium applications, from personnel protection to vehicle structures.

Many efforts are now ongoing to develop magnesium technology for such applications. The U.S. Army Research Laboratory has entered into a cooperative agreement with Magnesium Elektron North America, Inc. (MENA) to develop industrial manufacturing capabilities for production of magnesium alloys for armored ground vehicle applications. [29] High-strength WE43 and Elektron 675 alloys are being investigated under the agreement. Various alloys are also being investigated for composite helmet shells, where the light weight, specific stiffness and high damping, combined with suitable ballistic resistance may offer marked improvements in performance. [30] The formability of sheet and plate WE43 and Elektron 675 alloys is being studied, and generic helmet forming demonstrations are promising (see Figure 5).

Research like this on the front end is being matched by the U.S. Army Tank and Automotive Research, Development and Engineering Center (TARDEC) demonstration applications at the back end. The Fuel Efficient ground vehicle Demonstrator (FED) Program seeks to reduce fuel consumption in theater by design of a tactical vehicle with significantly greater fuel economy than a M114 HMMWV. In their consideration of high risk / high payoff technologies, they have sought to incorporate magnesium engine block and housing components.

Figure 5. Advanced concept helmet fabricated from WE43. (photograph courtesy of Shawn Walsh, U.S. Army Research Laboratory).

Successful Department of Energy Magnesium Programs

The U.S. Department of Energy has invested in similar projects over the last decade in support of the automobile industry, and a number of these projects have been coordinated via Pacific Northwest National Laboratory (PNNL). Today the Corvette uses a magnesium engine cradle developed under the "Structural Cast Magnesium Development" (SCMD) project. [31] The Mg Powertrain Cast Components (MPCC) program demonstrated the feasibility and benefits of using Mg in high-pressure die-cast powertrain components. [32] The Ultra-Large Casting for Lightweight Vehicle Structures project explored semi-solid molding and/or multi-port hot runner technologies for very-large structures to overcome the drawbacks of high-pressure die casting. [33] This was an outgrowth of an earlier Department of Energy project with Alcoa and a 2003 American Foundry Society benchmarking study. Finally, the Magnesium Front End Research and Development program is a multi-country effort to develop the enabling technologies and knowledge base that will lead to a vehicle front end body structure that is 50-60 percent lighter, equally affordable, more recyclable and of equal or better quality when compared to today's vehicles. [34] These and other success stories have demonstrated successful applications of magnesium alloys, and should inspire future military application as well.

Conclusion

With recent increased fuel costs and logistical burdens has come increased recognition of the need to develop and implement lightweighting strategies for the full range of military platforms (land, sea, and air) without a loss in platform functionality. It is hoped that the potential for significant improvements in the corrosion, mechanical and physical properties of magnesium will continue to drive design of ultra-lightweight metal structures and components, thus reducing large logistical burdens, minimizing operational constraints and liabilities, and reducing vulnerabilities. But for this to happen, the military must begin critically asking itself the same questions posed by the U.S. Army Materials Technology Laboratory workshop more than 20 years ago [11]:

1. Where is magnesium the suitable material for military hardware?
2. What is holding back the use of magnesium?
3. What can be done to overcome objections to the use of magnesium?
4. What are the prospects for success in such activities?
5. How much time will be required to establish the basis for greater use of magnesium?

In answering these questions, the designers would do good to keep the following in mind:

"The market for magnesium may be expected to develop along lines similar to those along which aluminum developed, but whether to anything like the same extent is entirely problematic. The market is somewhat unacquainted with many of the special qualities of the metal, and increased sales are largely a matter of education and research whereby a demand will be created and developed hand in hand with production. [...] The importance of a metal with properties similar to that of aluminum, but nearly half as light in these days of enormous automobile and aircraft expansion hardly needs much argument." - 1919 United States Bureau of Mines Bulletin [35]

Acknowledgements

Acknowledgments and special thanks are given to Jonathan Montgomery, James McCauley, Robert Dowding, Kyu Cho, Shawn Walsh, Tyrone Jones and Ernest S.C. Chin of the U.S. Army Research Laboratory, Robert E. Brown of Magnesium Assistance Group, Inc., Scott Shook of Applied Magnesium - International Group, Bruce Davis and Rick DeLorme of Magnesium Elektron, N.A. and Neale Nealameggam of U.S. Magnesium for discussions on military applications. The findings in this report are not to be construed as an official Department of the Defense, U.S. Army or Department of Energy position unless so designated by other authorized documents. Citation of manufacturers or trade names does not constitute an official endorsement or approval of the use thereof. The views and conclusions contained in this document are those of the authors and should not be interpreted as representing the official policies, either expressed or implied of ARL, PNNL or the U.S. Government.

References

1. B.B. Clow, "Magnesium: The Lightest One," *Metallic Materials for Lightweight Applications: Proceedings of the 40th Sagamore Army Materials Research Conference,* eds. Martin G.H. Wells, Eric B. Kula and John H. Beatty (Washington, D.C., Government Printing Office, 1993), 235-239.
2. H.E. Friedrich and B.L. Mordike, eds., *Magnesium Technology: Metallurgy, Design Data, Applications* (New York, NY: Springer, 2006), 610-620.
3. I.J. Polmear, "Magnesium and Magnesium Alloys" *ASM Specialty Handbook*, eds. M.M. Avedesian and H. Baker (Materials Park, OH: ASM International, 1999), 5.
4. L.S. Tagg, "On the Front Line of R&D: Wright Patterson Air Force Base in the Korean War, 1950-1953" Aeronautical Systems Center History Office. 2001. Wright Patterson Air Force Base. 9 September 2009, http://www.ascho.wpafb.af.mil/korea/contents.htm
5. J.F. Pashak, "Production and Fabrication of LA142 Pellets" (Report Department of the Army/AD409845, Dow Metal Products Company, 1963).
6. T.G. Byrer, E.L. White and P.D. Frost, "The Development of Magnesium-Lithum Alloys for Structural Applications" (Report NASA/NAS8-5049, Battelle Memorial Institute, 1964).
7. J.F. Sullivan and M.A. Brough, "Resistance of a Magnesium Alloy, Dowmetal (Type J-1), to Perforation by Fragment-Simulating Projectiles" (Memorandum Report/AD-A954953, Watertown Arsenal Laboratory, 1945).
8. J.F. Sullivan, "Aircraft Armor - Perforation Resistance Characteristics of Magnesium Alloys, (Dowmetal Type FS, J-1)" (Experimental Report/AD-A954259, Watertown Arsenal Laboratory, 1944).
9. J.F. Sullivan, "Aircraft Armor - Ballistic Characteristics of a Magnesium Alloy (Type FS)" (Memorandum Report/AD-A954458, Watertown Arsenal Laboratory, 1943).
10.
11. S. Isserow and R.S. Busk, "Workshop on the Viability of Magnesium in Military Hardware" (Special Publication/MTL-SP-87-7, U.S. Army Materials Technology Laboratory, 1987).
12. N. Jeal, "High-Performance Magnesium," *Advanced Materials & Processes*, 9 (2005), 65-67.
13. I. Ostrovsky and Y. Henn, "Present State and Future of Magnesium Application in Aerospace Industry" (Paper presented at ASTEC'07 International Conference– New Challenges in Aeronautics, Moscow, Russia, 19-22 August 2007), 1-5.
14. J.E. Gray and B. Luan, "Protective Coatings on Magnesium and its Alloys – a Critical Review," *Journal of Alloys and Compounds*, 336 (2002), 88-113.
15. V.K. Champagne, "The Repair of Magnesium Rotorcraft Components by Cold Spray," *Journal of Failure Analysis and Prevention*, 8 (2008), 164-175.
16. J. Mihelich and R.F. Decker, US Patent No 5.711.366, 1998 and US Patent No 5.819.839, 1998.
17. H. Watari, T. Haga, N. Koga, and K. Davey, "Feasibility Study of Twin Roll Casting Process for Magnesium Alloys," *Journal of Materials Processing Technology*, 192-193 (2007), 300-305.
18. C.A. Lavender, M.T. Smith and M.R. Garnich, "Development of Rapidly Solidfied Magnesium Alloys" (Contractor Report / BRL-CR-682, Ballistic Research Laboratory, 1992).
19. S.R. Agnew, P. Mehrotra, T.M. Lillo, G.M. Stoica, and P.K. Liaw, "Texture Evolution of Five Wrought Magnesium Alloys during Route A Equal Channel Angular Extrusion: Experiments and Simulations," *Acta Materialia*, 53 (2005), 3136-3146.
20. S.X. Ding, W.T. Lee, C.P. Chang, L.W. Chang, P.W. Kao, "Improvement of Strength of Magnesium Alloy Processed by Equal Channel Angular Extrusion", *Scripta Materialia*, 59 (2008) 1006-1009.
21. M. Green and G. Steward, *Humvees at War* (St. Paul, MN: Zenith Press, 2005), 39.
22. T. Jones and R. D. DeLorme, "Development of a Ballistic Specification for Magnesium Alloy AZ31B" (Technical Report / ARL-TR-4664, Army Research Laboratory, 2008).
23. T.L. Jones, R.D. DeLorme, M. S. Burkins and W.A. Gooch, "Ballistic Evaluation of Magnesium AZ31B" (Technical Report / ARL-TR-4077, Army Research Laboratory, 2007).
24. T. Jones and K. Kondoh, "Initial Evaluation of Advanced Powder Metallurgy Magnesium Alloys for Armor Development" (Technical Report / ARL-TR-4828, Army Research Laboratory, 2009).
25. F.T.M. van Wegen and E.P. Carton, "New Lightweight Metals for Armors" *Proceedings of the 24th International Symposium on Ballistics,* ed. S. Bless and J. Walker, CD Proceedings, 2008, 1-9.
26. E. Ma, K.T. Ramesh, R. Dowding and J.W. McCauley, "International Magnesium Workshop" (Scientific Report / ARL-SR-162, Army Research Laboratory, 2008).
27. H. Zhang, Y. Wang, S. Shang, L.Q. Chen and Z.K. Liu, "Thermodynamic Modeling of Mg-Ca-Ce System by Combining First-Principles and CALPHAD Method," *Journal of Alloys and Compounds*, 263 (2008) 294-301.
28. T. Uesugi and K. Higashi, "Materials Design for High-Strength Mg-based Alloys by Understanding from Ab-Initio Calculation," *Materials Science Forum*, 488-489 (2005), 131-134.
29. K. Cho, T. Sano, K. Doherty, C. Yen, G. Gazonas, J. Montgomery, P. Moy, B. Davis and R. DeLorme, "Magnesium Technology and Manufacturing for Ultra Lightweight Armored Ground Vehicles" (Reprint from 2008 Proceedings of the 2008 Army Science Conference / ARL-RP-236, Army Research Laboratory, 2009).
30. S.M. Walsh, B.R. Scott, T.L. Jones, K. Cho and J. Wolbert, " A Materials Approach in the Development of Multi-Threat Warfighter Head Protection," Proceedings of the 2008 Army Science Conference, Orlando, FL, December 2008, 8.
31. R.J. Osborne and D.E. Penrod, "Structural Cast Magnesium Development" (Annual Report / Contract No FC26-02OR22910, Department of Energy, 2005).
32. B.R. Powell, "Magnesium Powertrain Cast Components" (Annual Report / Contract No. FC26-02OR22910, Department of Energy, 2006).
33. M. Maj and E. McCarty, "Ultra-Large Casting For Lightweight Vehicle Structures" (Annual Report / Contract No. FC26-02OR22910, Department of Energy, 2008.)
34. E. Nyberg, A. Luo, K. Sadayappan and W. Shi, "Magnesium for Future Autos," *Advanced Materials & Processes*, (2008), 35-37.
35. "Make Our Own Magnesium", *the New York Times*, 4 July 1920.

HIGH-CAPACITY HYDROGEN-BASED GREEN-ENERGY STORAGE SOLUTIONS FOR THE GRID BALANCING

F.D'Errico[1], A.Screnci[2]
[1]Politecnico di Milano, Department of Mechanical Engineering, Via La Masa 34, 20156 Milan, Italy
[2]Mc Phy Energy , Zone d'activité 26190 La Motte-Fanjas (France)

Keywords: Magnesium, nanostructure, hydrogen storage, sustainability, renewable sources

Abstract

One of the current main challenges in green-power storage and smart grids is the lack of effective solutions for accommodating the unbalance between renewable energy sources, that offer intermittent electricity supply, and a variable electricity demand. Energy management systems have to be foreseen for the near future, while they still represent a major challenge. Integrating intermittent renewable energy sources, by safe and cost-effective energy storage systems based on solid state hydrogen is today achievable thanks to recently some technology breakthroughs. Optimized solid storage method made of magnesium-based hydrides guarantees a very rapid absorption and desorption kinetics. Coupled with electrolyzer technology, high-capacity storage of green-hydrogen is therefore practicable. Besides these aspects, magnesium has been emerging as environmentally friend energy storage method to sustain integration, monitoring and control of large quantity of GWh from high capacity renewable generation in the EU.

Introduction

Renewable Energy Sources (RES), such as Wind and Solar Energy, biomass, and hydropower have been relatively highly costly, but advances over the past few decades have made some renewable resources much more price-competitive. While the Organisation for Economic Co-operation and Development (OECD) countries' global primary energy demand is expected to show only a slightly grow with respect to other countries worldwide (with an overall 36% increase by 2035), their demand for RES will show a relatively large increase, thus calling for appropriate solutions with a systematic approach.

The problems of climate change, air pollution and energy insecurity requires a large scale conversion to clean, perpetual, and reliable energy at low cost together with an increase in energy efficiency. That is why, for the next future, RES will enter more and more the current mainstream, also consistently with EU strategic policies and initiatives in the energy segment (such as the European Electricity Grid Initiative (EEGI), the European Strategic Energy Technology (SET) Plan, etc.).

To this purpose the European Strategic Energy Technology (SET) Plan has identified fuel cells and hydrogen among the technologies needed for Europe to achieve the targets for 2020 as well as to achieve the long-term vision for 2050 towards decarbonisation.

A possible scheme for the future evolution of the energy distribution usually refers to the so-called SMART GRID that consists widely in the intelligent (i.e. dual side communication) integration of systems for distributed generation, centralized renewable power generation, non-renewable but cleaner centralized power generation for RES peaks stabilizing and information from the users' side to efficiently manage - in real-time - electricity peripheral flows in order to avoid mismatching between demand side (users) and supply side (power production).

All these features seek to build IT framework to overcome constraints due to "passive" architecture of current grid. Peaks loads efficient management is crucial for give chance to further increase for share of power generation via-renewable sources.

One other important issue the International Energy Agency (IEA) identified as key is energy storage.

Energy storage is considered one of the promising means both to feed the grid by RES and also to balance the grid energy flow according to the demand needs. It is also clear that advanced storage solutions, of course coupled with other infrastructures and technology initiatives (such as the ones above mentioned on Smart Grids) can power off-grid utilities, such as a green mobility system with electric vehicles, that are powered by stored electricity once the grid can feed exclusively the requested demand.

Energy storage is therefore one of the most promising solution to support the development and usage of RES, as well as introduce other benefits as it enables:

- to capture surplus renewable energy, that cannot be utilized by the grid due to low load demand or transmission constraints;
- to provide supplementary or backup power in correspondence of RES reduction or absence;
- to effectively alleviate power quality issues (e.g. interferences arising from the connection of wind turbines to the grid), thus improving the electrical quality of the RES power output.

Existing technologies for energy storage

Electricity storage systems can be categorized as mechanical, electromagnetic and electrochemical storage devices. Various technologies are at varying stages of technological maturity and they usually are exploited by their storage capacity (stored energy per mass or volume), or energy density, power (energy output per time), storage period (how long the energy should be stored) and size. Many of these key parameters actually can vary over a huge scale: from latent heat storage to prevent laptops from getting too hot (stored energy in the range of a few Wh) to the heat and cold thermal underground storage system underneath the German Reichstag in Berlin (stored energy in the range of some GWh).

Generally, energy density, power and energy ratings are essential criteria for any surveying process and synthetic selection maps have been developed to support selection. In Figure 1, currently electricity storage technologies are mapped over power and energy dimensioning basic features.

Together with above energy storage solutions today available, the "hydrogen battery" method is based on hydrogen energy storage (HES) route that consists in storing electricity into form of hydrogen (a high density energy carrier) produced by no-emission water electrolysis process powered by excess of electricity from RES. In Table I, some basic features of storage technologies are reported, as alternative solutions to HES systems.

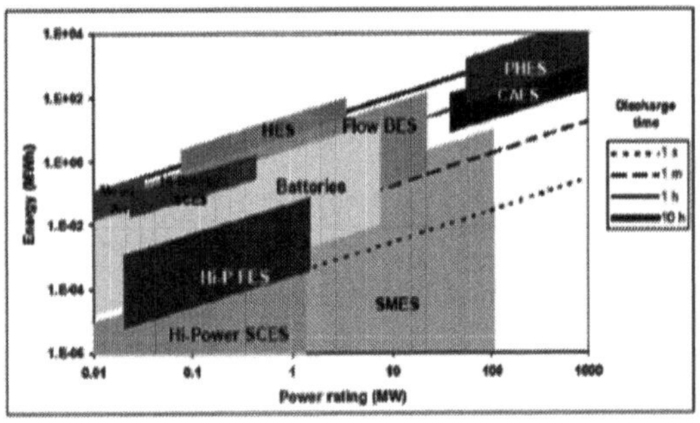

Figure 1 - Map of current available electricity storage technologies on energy vs. power comparison diagram (Source: Gonzales, 2004).

Table I – Main features, principles and drawbacks of the storage methods mapped in Figure 1.9, with the exception of HES.

Technologies	Main features, advantages and drawbacks
Compressed air energy storage (CAES)	Utilizes off-peak electricity to compress air that is then stored in an airtight reservoir (typically an underground geological formation such as a salt or limestone cavern) thus releasing it from the reservoir to run it through a gas-fired combustion turbine. Main advantage of this method is a high efficiency around 70% (Robyns, 2005), technology maturity and high power and energy capacity[1], but low energy density, in the order of 12 kWh/m^3 (Multon, 2003).
Pumped hydro energy storage (PHES)	Utilizes off-peak electricity to pump water from a lower reservoir into a higher reservoir then producing electricity by releasing the potential energy stored in elevated water through turbines (in the same manner as conventional hydro facilities). As CAES, PHES have a high conversion efficiency of about 65–80% (Multon, 2003), meanwhile, as for the large energy storage capacity of CAES, the main shortcoming of this technology is the need for a site with different water elevations.
Flywheel energy storage (FES)	Store kinetic energy in a rotating mass with minimized friction losses to improve efficiency. They have a great cycling capacity (a few 10,000 to a few 100,000 cycles) . While the instantaneous efficiency is around of 85%, the main drawback is an overall efficiency that would drop to 78% by progressively release of kinetic energy to after 5 h, and 45% after one day. For a long-term storage, this type of apparatus is therefore not foreseeable.
Batteries	Conventional chemical storage achieved through accumulators (lead–acid, nickel–cadmium, etc.) are high energy densities (up to 150 and 2000Wh/kg for lithium), while they have relatively low durability for large-amplitude cycling (a few 100 to a few 1000 cycles).
Flow batteries (Flow BES)	A two-electrolyte system in which the chemical compounds used for energy storage are in liquid state, in solution with the electrolyte. They have a relative high efficiency about 75% (Couffin, 2004–2005) and want to overcome the limitations of standard battery system in which the electrochemical reactions create solid compounds that are stored directly on the electrodes on which they form. They are at development stage; in 2003 Regenesys Technologies (UK) developed a system with a storage capacity of 15MW /120MWh.
Superconducting magnetic energy storage (SMES)	Stores energy in the magnetic field created by the flow of direct current in a coil of cryogenically cooled, superconducting material. Great instantaneous efficiency, near 95% (Anzano 1989) and a fast response time (under 100 ms). Major shortcomings represented by the refrigeration system and necessity for massive storage (5000–10,000MWh) of very large coils (several 100m in diameter) to generate enormous electromagnetic forces. Underground installation is foresee to limit infrastructure costs.
Super-Capacitors Energy Storage (SCES)	Store energy in the form of an electric field between two electrodes. Generally very durable (8–10 years), with very high efficiency around 95%, drawbacks are represented by a very high self-discharge around 5% per day and the low energy density 3–5 W·h/kg, against 30-40 W·h/kg for a lead acid battery.

[1] The first storage station using an underground compressed air reservoir has been in operation since November 1978 in Huntorf, near Bremen, Germany, while a larger installation has been set set in 1991 in MacIntosh, Alabama, (http://www.caes.net/mcintosh.html) began to deliver 100MW of power for 226 h, compressing ambient air and storing it at a pressure between 40 and 70 bars in a 2,555,000-m3 cavern, 700m deep in the ground.

Water electrolysis provide water decomposition when a direct current is passed between two electrodes immersed in water separated by a non-electrical conducting aqueous or solid electrolyte to transport ions and completing the circuit. Ideally, 39 kWh of electricity and 8.9 liters of water are required to produce 1 kg of hydrogen at 25 °C and 1 atmosphere pressure.

The energy efficiency of commercial water electrolyzer widely ranges from 50 to 70%, and it strongly depends also on how much system is efficient as external current density that feeds cell stack varies due to the variability and intermittency of RES power production.

As efficiency of the hydrogen generation system is one major drawback if compared with conventional battery, a very important key-point is that water electrolysis systems are cleaner in energy generation, modular as they can be easily grouped to produce hydrogen over wide scales ranging from few kilowatts to hundreds of megawatts and furthermore today commercial electrolyzers are capable of quick response to fluctuant electricity that is typically produced by intermittent and fluctuant wind turbines.

Regarding with modularity capabilities, power scale can be thought as unlimited, as a capillary distribution of these systems connected to the grid can build an infinite power capacity (Figure 2).

Storage System	Typical Output Rating (Megawatt)	Typical Storage Rating (Megawatt hours)	Physical Linkage Reqd.
Pumped Hydro	100	500- 10000	Yes
CAES	25	200	No
Battery	10	100	Yes
Hydrogen	10	Unlimited	No

Figure 2 – Summary of power and energy storage rating of HES compared with great capacity alternative energy storage technology (Price, 2000).

Behavior of water electrolyzer based HES in presence of high intermittent power source

Experiences have been performed by 2004 to provide information about potential operation of alkaline commercial water electrolyzer HySTAT® by Hydrogenics under emulated intermittent conditions generated from actual wind speed data [1]. In this test research a period of 14 h of a typical day with good wind conditions in the Sierra del Perdon's wind farm (Navarra region, Spain) was selected.

The evolution with time of the wind speed during this period and the electric power output expected the specific wind turbine used according to its power curve are reported in the Figure 2.

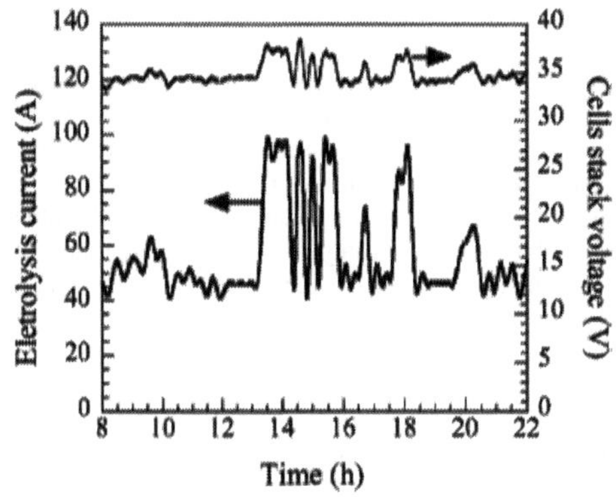

Figure 2- Evolution with time of the electrolysis current in a Hydrogenic HySTAT: total voltage established in the electrolyzer cell stack as a result of the passage of the electrolysis current (source: [1]).

The fast response of the stack voltage shown in Figure 2 demonstrated that electrolysis cell stack voltage readily matches a highly varying electric current-time profile, suggesting an essentially ohmic behavior governed by the internal resistances associated to the electrolyte and cells components. During test campaign electrolyzer worked reliably and at high efficiency, as it ranged 74-83% basing on the stack voltage. The hydrogen produced was at high purity as oxygen contents was always below 0.2 vol %.

The hydrogen produced by electrolyzer can be therefore stored to be released and used "on-demand" by various (also combined) user systems:

- an advanced modified turbine for power generation, advanced hydrogen fuelled turbines for power generation;
- a fuel cell stack to produce directly green-electricity, that functions in a "reverse mode" of the water electrolyzer, thus recombining hydrogen with oxygen to produce electricity and waste heat and pure water;
- as a green-fuel for either Fuel Cell vehicles or modified internal combustion engine vehicles;
- introducing this green-hydrogen by hydrogen merchants, thus providing an alternative to conventional hydrogen, normally produced by Steam Methane Reforming, SMR, at high carbon footprint.

Electricity storage technologies for the RES integrating purposes

Conventional network operation of distribution network, at least in some regions of Europe, is actually based on a "fit and forget" approach. This means that verifications are performed during the connection phase and the connection schemas adopted ensures any reasonable behavior of network users is allowed.

From a technical point of view, however, it would be possible to increase the amount of distributed generation points connected to the grid where RES plants could contribute to voltage regulation and to provide ancillary services.

An appropriately sized storage device would reduce uncertainties in the capability of the system of self-regulating, thus allowing a further increase of RES units connectable to the grid. As stated, the various storage technologies have a number of features, such as efficiency, size, dynamic behavior, etc. which are all relevant in the selection of the most suitable system for a particular application or set of applications.

Hydrogen-based energy storage systems deploy the highest energy density; thanks to high modularity, they are of unlimited storage capacity, but, conversely, when the storage pathway is completed by a fuel cell system for the back conversion of the electricity the round-trip efficiency[2] of the system is low (around 30%).

On the other hand, thanking to varying deploying architectures today possible to develop, these energy storage systems could be really promising in near-term.

Economic viability of large storage system has to be evaluated taking into account these major requirements:

- working in combination with RES, especially fluctuant RES
- serving for secure balancing of the grid
- reducing curtailing and allowing progressive and potentially unlimited installation of RES.

Accordingly, the optimal choice should be driven mainly by these following important driving key-factors:

- The charge and discharge response to fluctuant electricity;
- The capital cost of the facility, including those ones of the RES plant (charging equipment, storage equipment, and discharge equipment);
- The O&M costs for the facility, including those ones of the RES plant;
- The transmission power costs;
- The power and energy rate;
- The power quality;
- Safety and security for the grid;
- The round-trip system efficiency;
- Reversibility of charge/discharge cycle;
- Modularity and easiness of upscale;
- The dispatchability;
- The flexibility and adaptability to future trends in the electric sector;
- Environmental sustainability and safe operation
- Possible synergies

At present time, very few storage systems can solve power peaks curtailment over a longer term (10-12 hours) of functioning.

Magnesium as key-solution for safe and viable hydrogen solid storage for saving power curtailing

Both commercial water electrolyzers and fuel cell stacks have the major advantages of an instantaneous response: a) any electricity variability that flows into the water electrolyzer is used for producing hydrogen; b) any request of electricity by downstream users (i.e. demand) is matched by flowing variable hydrogen flux into the fuel cell stack.

On the other hand the major drawbacks of the supply-chain water electrolyzer→hydrogen storage→fuel cell stack is represented by:

[2] Round-trip efficiency is calculated throughout the first electricity conversion in hydrogen and the downstream back-conversion into electricity of hydrogen by use of fuel cell system.

- lost energy used during the storage phase, as hydrogen can be usually stored either as pressurized gas or liquefied cryogenically cooled to achieve adequate energy density;
- the low-efficiency of the fuel cell, that obviously affects the overall round-trip efficiency of the system: 70% for electrolyzer efficiency and not more than 50% for the fuel cell result generally in low round-trip efficiency of 35%.

On the other hand, one major advantage of using hydrogen is its high energy content, namely 39.4 kWh per kg of hydrogen[3]. But the actual energy density strongly depends on the storage methods used. Viability of the hydrogen storage systems are strictly related to the real capability of deploying a not expensive long-term safe storage of hydrogen.
The hydrogen produced can be stored in various forms: liquid (by cryogenic process, liquid hydrogen can be stored in tanks at 21.2 K at ambient pressure), compressed gas (high pressure gas cylinders usually operate at a maximum pressure of 20 MPa) or via solid–state methods. In any case the major objective of industrial hydrogen storage methods is reaching the highest volumetric density with full reversibility of hydrogen uptake and release. At ambient temperature and atmospheric pressure, 1 kg (equivalent to nominal 39 kWh) of the gas has a volume of 11 m^3. To increase hydrogen density in conventional storage methods it is necessary to spend energy either for compressing gas or lowering temperature for liquefying. With an un-conventional solid-state method, energy must be used to contrast repulsion deriving from the interaction of hydrogen with another material.
Generally speaking a solid-state hydrogen storage (HSS) method is intrinsically safe (no pressurized reactive hydrogen gas is present) and a long-term storage, since no boil-off problems can occur.
HSS into metals physically consists of diffusion of hydrogen into metal (absorption phase) that is finally bonded with host material forming stable metal hydrides. Therefore, hydrogen storage capability of metals is strongly influenced by the way this two-phase process develops. Hydrogen atoms[4] penetrate and diffuse into metal by preferential routes: they flow throughout grain boundaries, i.e. preferential pathways due to higher concentration of vacations. The host metal initially dissolves some hydrogen as a solid phase. As the hydrogen pressure together with the concentration of hydrogen in the metal is increased, interactions between hydrogen atoms become locally important, and nucleation and growth of the hydride phase can start. Therefore the "storing" phase develops by the formation of intermetallic compounds (interstitial hydrides, or complex metal hydrides) at any sites (interstitial sites present in crystal lattice host) available for hydrogen locating and bonding to metal element.
Further constraints that work against absorption phase are: a) the presence of oxide layers that spontaneously formed in air on metals and, b) such a "barrier effect" that impedes hydrogen atoms to more deeply penetrate inside bulk because top layers saturate. These represents actually physical (oxide layers) and chemical (saturated layers) barriers that can prevent a fully hydrogen absorption inside bulk material; these barriers are generally responsible of lower sorption kinetics and practical absorption rate (weight percentage of hydrogen absorbed into host metal) far from theoretical values.

Among metals that are more attractive for hydrogen storage, magnesium is considered a key material since that it is the eighth most frequent element on the earth, it has a good reversible storage capacity, achieving theoretically the 7.6 wt.%, it is relatively inexpensive and it has very low density. On the other hand, Mg major drawback is its high desorption temperature of over 300 °C; the high stability of magnesium-hydrogen bond is the reason of such a high enthalpy of hydride formation that is responsible of high desorption temperatures and efficient heat management systems. During the absorption phase magnesium hydrides form with heat release, while the desorption phase host material needs to be heated up for the hydrogen realease. An efficient heat management systems is therefore required for industrial systems.
One possible route to produce Mg hydride reactions with lowered reaction enthalpy (i.e. low temperatures for uptake and release phases) is to add elements which exhibit lower enthalpies of formation. Mg_2Ni is an example of this class of materials, as the enthalpies of formation for the ternary hydride Mg_2NiH_4 and the Mg_2Ni is of the order of -67 $kJmol^{-1}$ per H_2, therefore lower than the -78 $kJmol^{-1}$ per H_2 that is enthalpy for the MgH_2 formation. Unfortunately this route reduces drastically the storage capacity and increases the density of the modified Mg-based material.
In last two decades further breakthroughs in hydrogen sorption kinetics have been achieved by pre-treating magnesium by high-energy ball milling operations (Schulz, 1995; Zaluska 1999). Ball milling applies severe plastic deformation to material for a high refining microstructure (i.e. dramatic increase of grain boundaries) that finally becomes more reactive to hydrogen diffusion.
During the last decade advancements in kinetics sorption of hydrogen in magnesium-based host materials have been done by selection of additives and nano-structuring techniques for increasing grain boundary density [2-10].
Exploiting economically viable solid-state hydrogen by magnesium-based storing solutions is possible if hydrogen kinetics sorption is rapid, the storage capacity can reach the theoretical value of 7.6 wt.% of hydrogen (corresponding at around 400 kWh/m^3) and heavy and expensive additives are limited.

A major breakthrough in magnesium storage is represented by proprietary patented manufacturing process of nanostructured magnesium-based material for hydrogen storage created by Mc Phy Energy (Mc Phy), a French company based in Grenoble. Cheaper magnesium-based nanostructured materials produced by Mc Phy are capable of hydrogen kinetics absorption that are industrially exploitable. The modularity concept exploited by Mc Phy is schematically represented in the Figure 4.
Basing on solid-state technology, the Mc Phy solution does not require hydrogen from the water electrolyzer unit to be compressed, as the pressure for starting hydrogen absorption is of the same level of the hydrogen pressure at the electrolyzer outlet. And pressure of hydrogen desorbed by the tank is similar to the level (around 2 bars) required for fuel cell stack, a costly and energy dissipative compressor is therefore eliminated. These solid storage system are highly efficient, since about 3-5% are the losses in absorption-desorption cycles due to that part of energy used to heat-up tank material for achieving desorption. Storage modules require less energy than liquid or compressed hydrogen

[3] Energy content of 1 kg hydrogen (equivalent to 141.9 MJ) at Higher Heating Value (HHV).

[4] Molecular hydrogen is dissociated at the surface before absorption; two H atoms recombine to H_2 in the desorption process.

technologies, which generally use up to 33% of the hydrogen energy content for storage phase.

These storing systems are produced in modular sub-units that can be installed and grouped in parallel to increase system capacity. As hydrogen is stored in solid state, storage modules are furthermore safe to be transported as any non-hazardous goods. Raw materials used in manufacturing do not generate any hazardous waste at the end of their life cycle, and they are fully recyclable.

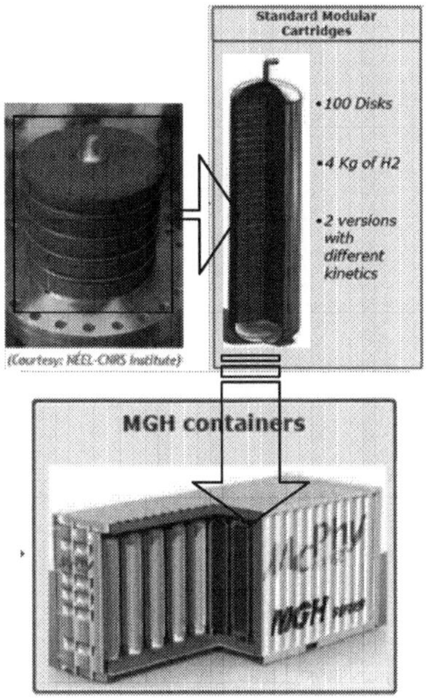

Figure 4 – Modularity concept realized by Mc Phy technology (courtesy of: Mc Phy Energy SA).

Dynamic behavior of the WE coupled with Hydrogen Solid Storage system

HSS needs to be optimally coupled with the WE, the upstream unit to assure high dynamic storage capability in transient and dynamic phase. This fact implies that HSS has to be capable to accept (absorb) in real time hydrogen at any rate that is produced by water electrolyzer that can function at varying conditions depending on external fluctuant electricity produced by RES.
For that, dynamic behavior of the HSS module has been extensively tested coupled with WE module that had to respond to an intermittent power source.
At CEA (Commissariat à l'Energie Atomique et aux énergies alternatives) facilities in Grenoble, a complete test campaign has been performed on a fully instrumented system constituted by a Mc Phy solid-state hydrogen storage tank, upstream coupled with a water electrolyzer and downstream connected to a fuel cell (figure 5). Extensive test campaign have done in order to analyze and optimize response of the system when connected with varying and fluctuant renewable source. Specifically the WE was directly linked to solar panel source emulator. Storage capability (i.e. sizing capacity for absorbing variable quantity of hydrogen) has been pointed out (figure 6).

Figure 5- Test bench facility set at CEA premises.

A load power profile had been modeled in order to feed electrolyzer simulating electricity production of a common solar farm over 8 hours of functioning, sunny in the morning and cloudy in the afternoon. A total 200 grams of hydrogen have been produced by the electrolyzer and constantly absorbed by McPhy tank in 8 hours.

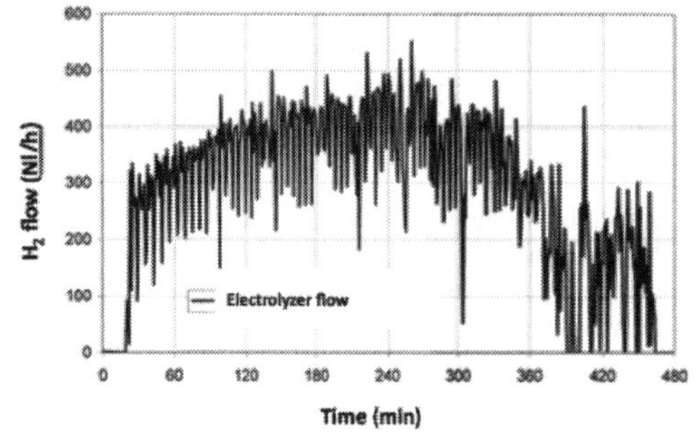

(a)

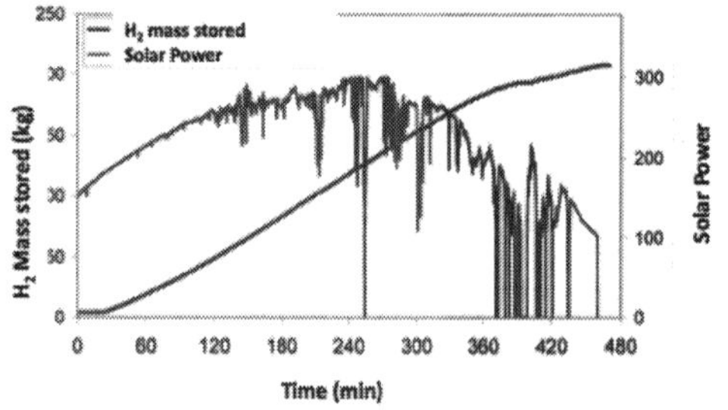

(b)

Figure 6 – Dynamic behavior diagrams showing: a) hydrogen flow produced by water electrolyzer fed by electricity produced by solar panels; b) hydrogen produced and stored in HSS module .

test results, water electrolyzer shown fully capability of accepting power variations from RES. Most important conclusions from test campaign were: a) electricity variations from power source feeding electrolyzer had no effects on hydrogen production; b) HSS module was well dimensioned to absorb all the hydrogen produced by water electrolyzer and it serves water electrolyzer unit in a fully stable mode.

Conclusions

Hydrogen storage via solid state (hydrides) is becoming a new challenge for magnesium based materials. Modular products commercially available are capable of sorption kinetics suitable to be coupled with up to date water electrolyzer technology.

Advancements have been done in the water electrolyzer technology specifically related to efficiency of such systems when they work at varying operating conditions, as high variable current density that is typical of intermittent and fluctuant wind sources. Behind this improvements, energy storage can be realized by new generation of hydrogen storage systems. New advancements in magnesium-based materials for hydrogen storage is leading to rapid improvements for high capacity "hydrogen batteries". Novel hydrogen solid storage modular units are suitable to be coupled with rapid and efficient electrolyzers to serve highly intermittent RES.

References

1. L.M. Gandía, R. Oroz, A. Ursúa, P. Sanchis and P.M. Diéguez, "Renewable hydrogen production: performance of an alkaline water electrolyzer working under emulated wind conditions", Energy Fuels, 21 (2007), 1699–1706.

2. G. Liang, J. Huot, S. Boily, A. van Neste and R. Schulz, Catalytic effect of transition metals on hydrogen sorption in nanocrystalline ball milled MgH2–Tm(Tm = Ti, V,Mn,Fe and Ni) systems. J Alloys Compd, 292 (1999), p. 247-252

3. R. Shulz, J. Huot, G. Liang, S. Boily, G. Lalande and M.C. Denis, et al. "Recent developments in the applications of nanocrystalline materials to hydrogen technologies. Mater Sci Eng, A267 (1999), 240-245

1. A. Zaluska, L. Zaluski and J.O. Stroem-Olsen, "Structure, catalysis and atomic reactions on the nanoscale: a systematic approach to metal hydrides for hydrogen storage", Appl Phys, A72 (2001), 157-165

2. R.C. Bowman and B. Fultz, Metallic hydrides: hydrogen storage and other gas-phase applications. MRS Bull, 27 (2002), 688-693

3. J. Huot, M. L. Tremblay and R. Schulz, "Synthesis of nanocrystalline hydrogen storage materials", J Alloys Compd, 356–357 (2003), 603-607

PART 2: ELECTROLYTIC AND THERMAL PRIMARY PRODUCTION

Neale R. Neelameggham

Magnesium is the only structural metal that can be obtained economically from both the lithosphere and hydrosphere. Magnesium [Mg] can be produced from oxidic resources of lithosphere or from magnesium ion raw materials originating from saline water. This has resulted in the development of two mainstream production technologies, unlike many other metals that use minor variations of a single process technology. This two-way production dates back to 1808, when Davy initially produced magnesium at a mercury cathode by electrolysis through moistened magnesium sulphate. He also demonstrated metallothermal conversion of red-hot magnesium oxide with potassium vapours collecting the magnesium into mercury.

The selected articles in this section can show only a glimpse of pertinent production techniques of the two routes for making magnesium metal that have been used since the 1990s:

1. The electrolytic method – magnesium chloride is the feed material to this process. The magnesium chloride is derived from magnesium oxide sources (by chlorination of magnesia or magnesite) or from chloride sources (by dehydration of brines or hydrous carnallite).
2. Metallothermic reduction – the dolomite feed material is calcined and then reduced, either by silicon or one of its alloys, or by other metals such as aluminum. One of the most important processes of this type, which uses ferrosilicon as the reductant, is known as the Pidgeon process.

This section starts with Byron Clow's "The History of Magnesium since World War II" (2002). Only a cursory mention of production of magnesium in China is noted in this article. Bob Brown, a former silicothermal producer from the 1960s, presents information on the Pidgeon Process in his "Lloyd M. Pidgeon – Magnesium Pioneer". The paper on magnesium production by the world's largest current electrolytic technology is presented by Wallevik and colleagues.

Little did the late Byron Clow expect that the thriving magnesium production industries noted in his article, such as Norsk Hydro magnesium, and many of the Western world magnesium production would become history a few years later. This happened when the dominance of Pidgeon process magnesium from China overtook and relegated the electrolytic technique to lower tonnage. Chinese Pidgeon process magnesium capacity is over 750,000 tons per year. The happenings in the present day befit the 1935 note by John Gann of Dow Chemical that stated "…our light metals occur only in the form of compounds so stable that their discovery, isolation, commercial production, and use were forced to await some of the modern advances in chemistry and engineering. Under such conditions, the evolution of a new industry is often a romance* in which scientific and industrial difficulties and near failures adds to the thrill of success."

---

* It should be noted that there is still a lot of room for innovation in magnesium metal production technology, and romance with it. Some of the areas are to reduce the solid–liquid or gaseous waste, reduce energy consumption, and make the production a sustainable process. Finding techniques to utilize hitherto unused magnesium bearing raw materials is another avenue wide open. These techniques will form the emerging readings in the future.

Yet electrolytic magnesium is thriving in the United States, Israel, Ukraine, and Russia, producing over 150,000 tons per year magnesium. Norsk Hydro's technology discussed in the paper by Wallevik *et al.* is in the process of relocating to China, which wants to utilize its hydrospheric magnesium from its saline lakes. The electrolytic production technology in operation at Dead Sea Magnesium closely resembles the technology discussed by Shekhovtsov and colleagues. The work by Lu and colleagues on a 5 kA pilot cell of using lanthanum chloride melt to electrolyse magnesium oxide at 750°C is a novel approach that is usable where the rare earth chlorides are available at a reasonable cost. The fundamental study by McLean and colleagues on cathode wetting confirms Norsk Hydro magnesium patent on improving magnesium cell performance by coating steel cathodes.

Readers are referred to the *JOM* article by Thayer and Neelameggham on the continuing electrolytic production of magnesium in the United States, titled "Improving the Electrolytic Process for Magnesium Production" (Vol. 53, 2001). Other laboratory level [less than 100 gm/day magnesium production as of 2012] descriptions of the solid oxide membrane electrolysis process can be found both in aforementioned *JOM* article, as well as in some of the *Magnesium Technology* proceedings.

Zang and Ding give a good comparison of parameters of production of magnesium by the Pidgeon process in China, and Mei and colleagues describe the vertical retort developed to increase the retort throughput being used by some of the magnesium producers in China.

Hu and colleagues describe a more economically technique of vacuum aluminothermic reduction of a mixture of calcined dolomite and calcined magnesite. Here the retort residue calcium aluminate is reacted with sodium carbonate to provide an aluminum cell feed in a cyclic fashion. This process is still in the pilot stages. Readers are referred to the article by Wadsley titled "Magnesium Metal by the Heggie-Iolaire Process" (*Magnesium Technology 2000*) for a history and background of aluminothermic magnesium. It is envisioned by the writer that aluminothermic method may become a sustainable technology approach for magnesium.

Hansgirg's carbothermal reduction of magnesia was in operation during 1940–1955 by Kaiser Industries. This technology had failed due to poor yields, and the need for costly inert gas recycle as a quenching medium. This process is being revived on a bench scale level in Australia, as described in the paper by Prentice and colleagues, and now in many other places.

History of Primary Magnesium Since World War II

Byron B. Clow
Executive Vice President (Retired)

International Magnesium Association
6731 Whittier Ave
Suite C-100
McLean, Virginia 22101

Abstract

Primary magnesium production just prior to WWII was around 32,000 tonnes with most of it coming from Germany and England with smaller amounts from France and the U.S.A. Peak production during the war was 232,000 tonnes with most of it coming from the 15 plants that had been built in the United States. By 1946 worldwide demand had fallen back to pre-war levels until 1980. The paper will identify the various plants built round the world since WWII and the impact of exports from Russia and China beginning in the early 90's. "The presentation will be by country in chronological order. We will not discuss the 15 or 20 potential projects going on around the world since this was well covered by Bob Brown at the Nashville Conference in the year 2000.

History By Country

Brazil: The Brazilian Air Force evaluated building an electrolytic plant during the 1970's but it did not materialize. In 1984 a ferro-silicon producer built a 10,000 tonne silico- thermic plant which is still in operation.

Canada: The original Pidgeon Process Plant built in Haley, Ontario in 1943 is still in operation today with a capacity of 7,000 tonnes.Alcan built a small electrolytic plant in Arvida Quebec in 1945 and closed it down in 1959. In 1989 Magnesium Company of Canada constructed a 12,500 tonne facility at High River Alberta, Canada. This was phase I with planned eventual capacity of 62,500 tonnes. It was based on a patented process for making anhydrous magnesium chloride cell feed. The raw material was magnesite from a nearby mine owned by others. It achieved production but was shut down in April 1991 for financial reasons & technical problems. Hydro Magnesium dedicated their 43,000 tonne electrolytic plant in Becancour in May of 1990 which is still operational today. The technology was developed at their plant in Norway. Noranda Magnesium started its electrolytic Plant in Danville, Quebec in mid 2000. The 60,000 tonne plant uses asbestos tailing for raw material which is a first.

Commonwealth of Independent States: The former Soviet Union had two operating magnesium plants in operation at the close of WWII and subsequently built three others. All are electrolytic using anhydrous Carnallite as cell feed. Carnallite is a double salt of potassium and magnesium chloride.

The technology was jointly developed by Solikamsk in Russia and the All-Union Aluminum Magnesium Institute. Raw material for four of the five plants came from a carnallite deposit near Solikamsk in the Ural mountains. Solikamsk started production in 1936 and is still operating. It is the oldest continuously operated magnesium plant in the world with current capacity of 20,000 tonnes. Avisma Titanium and Magnesium Works is located a few kilometers down the road from Solikamsk. It was started in 1945 and is still operational today. It's nominal capacity is 20,000 tonnes but this is dependent on the volume of titanium production. Magnesium is used to reduce titanium tetrachloride to titanium sponge leaving anhydrous magnesium chloride which is recycled to magnesium and chlorine. If titanium production is down you must reduce your magnesium production unless you have a separate outlet for the chlorine. The Zaporozia titanium magnesium complex in the Ukraine was opened in 1956 with 3,500 tonnes of magnesium available for sale. The facility was closed in 1993

The titanium magnesium complex in Ust Kamenogorsk in Kazakstan was brought on stream in 1965 with a magnesium availability of 12,000 tonnes over and above the requirements for titanium. They consumed around 6,000 tonnes internally to make magnesium powder. They closed the magnesium cells in 1993 and purchased magnesium from China for their titanium requirements. Recently, small amounts of magnesium from Kazakastan have been showing up on import statistics from the U.S. and Canada.

The plant at Kalush in the Ukraine started producing magnesium in 1986. It is basically a chemical plant with a 25 per cent magnesium chloride waste stream from a fertilizer operation. Carnallite is made from this waste stream for the electrolytic cell feed. Initial capacity was 18,000 tonnes which decreased over the years due to environmental and financial problems. It finally closed in 2000. A final source of primary magnesium was government stockpiles estimated to be 70,000 tonnes and released between 1992 and 1996.

France: A small electrolytic plant remained in operation for a few years following WWII but was forced to close due to the high cost of electric power. In 1963 three companies combined to form Magnetherm for the purpose of developing a patented new process for the production silico-thermic magnesium. They took over a plant at Marignac in the south of France which had electric furnaces for ferro-silicon production. The process was successfully developed and licensed to others around the world. The plant capacity at Marignac is 18,000 tonnes.

Israel: Construction started on the Dead Sea Magnesium plant in late 1993. The facility uses Russian technology using carnallite cell feed obtained from their nearby potash operations. The original plan was to build the plant in two phases of 25,000 tonnes. The first phase started up in late 1996 and has subsequently been increased to 33,000 tonnes. There is substantial magnesium research being conducted in Israel as well as Germany.

Italy: A silico-thermic plant was built in Bolzano in 1938 which operated continuously until its closure for economic reasons in 1993. It's capacity was 12,000 tonnes. An electrolytic pilot plant was built in Sardinia in 1957 but never became commercial. Italy has no primary magnesium production facilities today but they do have recycle capability.

Japan: All primary magnesium plants were closed following WWII. In 1954 Furukawa Magnesium built a silico-thermic plant in Oyama which had a capacity of 6,000 tonnes at the time of its closure in 1989 for economic reasons.. Ube Industries built a silico-thermic plant in Ube city during the 1960's using sea water as their source of magnesium instead of the customary dolomite. Their capacity was 9,000 tonnes at the time of their closure in 1994 for economic reasons. Japan Metal & Chemical built a 5,000 tonne plant in 1988 using the Magnetherm process under license from Pechiney. They closed in 1993 for economic reasons.

Norway: The Germans built an electrolytic plant in Porsgrunn in 1940 but very little production was achieved due to constant Allied bombing. Norsk Hydro rebuilt the plant in 1952 and it is still operating today with capacity of 45,000 tonnes. The new technology used in their Canadian plant was developed in Porsgrunn.

United Kingdom: Commercial production of magnesium in the U.K. dates back to 1864 when small amounts were produced for photographic purposes . The first commercial plant designed for industrial uses was built in 1938 by Magnesium Elektron, Ltd., at Clifton junction, Manchester, based on technology licensed from I.G. Farbenindustrie of Germany. The electrolytic plant was expanded to 9,000 tonnes and operated until 1947 when it closed because of the high cost of power. The government reactivated several electrolytic plants that had been placed on stand by following WWII for the Korean War, 1951-53. They were all dismanteled later. Magnesium Elektron built a Pidgeon process silico-thermic plant at Hopton in Derbyshire in the early 1960's to get around domestic high energy costs by importing energy as ferro-silicon. The project failed.

United States: Prior to WWII, Dow was the only producer of primary magnesium in the United States. During WWII there were 16 operating plants with a capacity of 232,000 tonnes although maximum production was 184,000 tonnes in 1943. Following the close of WWII all the plants were dismantled or closed except for Dows plants in Texas with combinded capacity of 45,000 tonnes. Several of the government owned plants that had been placed on stand by following WWII were reactivated during the Korean conflict but all were dismantled or sold following that emergency. Dow's seawater plant in Freeport, Texas continued to increase it's capacity to a peak of 120,000 tonnes in the 1970's. Due to aging facilites they cut back to 90,000 tonnes in 1981 and again to 60,000 tonnes in 1986. The remaining 60,000 tonnes were permanently closed in November 1998 following severe storm damage. The Alabama Metallurgical Corporation was formed in 1959 to build a 6,000 tonne silico-thermic Pidgeon process plant in Selma, Alabama. They had many difficulties but managed to survive until 1970. American Magnesium Company, headquarted in Tulsa, Oklahoma, constructed a plant in Snyder, Texas in 1968 to electrolytically produce magnesium from west Texas subterannean brines containing about 11 per cent magnesium chloride. The objective was a 30,000 tonne plant with Phase I being 12,500 tonnes. The plant had repeated difficulites with the electrolytic cells following which they signed a licensing agreement for Russian cell technology.

This was working well but a freak deluge of rain destroyed their evaporation ponds in the fall of 1980 from which they could not recover. In 1970 National Lead Industries constructed a 45,000 tonne electrolytic plant on the west shore of the Great Salt Lake in Utah. It's raw material was waters of the Great Salt Lake which contain about 0.4% - 0.5 magnesium. It is concentrated by solar evaporation, spray drying and chlorination. NL Industries sold the operation to Amax who subsequently sold it to Renco Metals who operate it today as Magnesium Corporation od America. The wholly owned subsidiary of the Aluminum Company of America, Northwest Alloys, dedicated their Magnetherm silico-thermic plant in Addy, Washington on May 13, 1976. The nine furnaces of Phase I had a rated capacity of 20,000 tonnes. Phase II was to add an additional 20,000 tonnes when power became available. Engineering improvements enabled them to raise the capacity to 45,000 tonnes with less than the original nine furnaces. Nevertheless, the plant was scheduled to be closed on October 1, 2001 for economic reasons due to imports.

Yugoslavia: a 5,000 tonne Magnetherm facility was constructed at the Bela Stena dolomite mine in the early 1980's and is still in operation on a limited basis.

Summary

Three major events have shaped the magnesium industry since WWII and are continuing to do so today. The first was the story of Dow Magnesium. They were the world's oldest and most dominant producer for 80 years until they folded their tent and left in 1998 for cost reasons. Over the years, Dow created vast amounts of technology which has been shared with the world and is the bedrock of today's advancing magnesium technologies. The second is the construction of four world class facilities of 40,000 tonnes or more in North America. NL Industries built a plant on the western shore od the Great Salt Lake in the early 1970's The basic electrolytic technology was that of I.G. Farbenindustrie in Germany. The current owners, Magnesium Corporation of America, are in a cell rebuild program to improve efficiency and expand capacity. On May 13, 1976, Northwest Alloys, a wholly owned subsidiary of the Aluminum Company of America, dedicated their Magnetherm licensed silico-thermic plant in addy, Washington which is about a one hour drive North of Spokane. In 1990 Hydro Magnesium dedicated their 43,000 tonne electrolytic facility in Becancour, Quebec, which is still operational and in mid 2000 Noranda Magnesium started their 60,000 tonne facility in Danville, Quebec to recover magnesium from asbestos tailings.

The third major event was the emergence of Russia and China as exporters of magnesium. When the former Soviet Union collapsed ten years ago it's effect on the magnesium world was that of a 40,000 tonne new plant being created overnight. Magnesium that had formerly gone into military markets in the U.S.S.R. was now being exported. China was a different case. They built numerous small silico-thermic Pidgeon process plants in response to market demand which was heavy in the mid 1990's Once installed it is still there even though market demand has softened. They have been exporting well over 100,000 tonnes per year. Exports from the C.I.S. and PRC now account for about 45 per cent of world market demand and will probably continue to do so for the forseeable future.

Extraction of magnesium from the world's unlimited resources for the metal is difficult because of its' reactivity with all of the other elements. It's value as an ultra light weight engineering solution will continue to drive venture capital in to the construction of new facilities.

LLOYD M. PIDGEON – MAGNESIUM PIONEER

December 2, 1903 – December 9, 1999

Robert E. Brown

Magnesium Assistance Group, Inc
226 Deer Trace
Prattville, Alabama 36067-3806

Abstract

Lloyd Montgomery Pidgeon was an unusual man in an unusual time. His contributions to the development of the magnesium industry have never been appreciated (or even known) by many of today's magnesium followers. Dr. Pidgeon, working with one technical graduate, achieved commercial development of a process to produce magnesium by reducing calcined dolomite with ferrosilicon, i.e. the silicothermic process. He also received patents for electrolytic magnesium processes. He worked with engineers to design and build six magnesium production plants in a very short period of time. The original plant at Haley, Ontario is still operating. Dr. Pidgeon received many technical honors, but was always quick-witted, with a humorous approach to life.

Background

Much of the content of this keynote address is derived from personal interviews with Dr. Pidgeon and from valuable information supplied by Ruth Pidgeon Bryson, his daughter. I personally had a chance to work briefly with Dr. Pidgeon in 1960 when he was retained as a consultant to a new silicothermic magnesium plant in Selma, Alabama. As plant metallurgist, I was part of a team that was struggling to learn how to operate a Pidgeon Process plant.

In 1994, I had the opportunity to personally spend a full day with Dr. Pidgeon and his wife Frankie, who had accompanied him to Selma in 1960. Neil and Ruth Bryson were hosts and guides during this visit which resulted in a biographical sketch printed in Light Metal Age magazine in April 1995.(1) Much of this paper is directly derived from that work.

Biography

Lloyd Montgomery Pidgeon was born in Markham, Ontario in 1903. In discussing his birth, Dr. Pidgeon said he was reminded of the old joke where the Englishman said he was born in Singapore because his parents happened to be there and he wanted to be close to them. Pidgeon's father was a Presbyterian clergyman. Dr. Pidgeon said that he does not remember Markham, but does recall living in St. Thomas, which was located near London, Ontario, on the Michigan Central Railroad. The family moved to Vancouver prior to World War I. His father, Dr. E. L. Pidgeon, was somewhat of a freethinker and was not happy in conservative Vancouver. He moved the family to Winnipeg, Manitoba, where he became a leader in the movement for Protestant church union in Canada. Dr. Lloyd Pidgeon received his B.A. at the University of Manitoba in 1925.

The family moved to Montreal in 1925. Dr. Pidgeon's father moved to a church near the McGill University campus. Lloyd Pidgeon then continued his advanced degree work at McGill University. His doctoral work was on cellulose. He was interested in the paper industry and a Cellulose Institute was planned for McGill. Receiving his Doctorate under the direction of Dr. Otto Maas in 1929, Dr. Pidgeon found that there were very few jobs available.

He sought and won a fellowship grant: The Sir William Ramsay chemical scholarship to Oxford. (Dr. Pidgeon pointed out that Ramsay was the discoverer of the rare gases, helium, xenon, neon, and krypton). At Oxford, Dr. Pidgeon actually worked on and received a B.Sc. Degree under the direction of Sir Alfred Egerton. His work was in the field of anti-knock compounds; investigations were conducted to actually determine just exactly what the "anti knock" mechanism did. This was his first exposure to working on high temperature processes. He joked that he only "professed metallurgy" since his Ph.D. was in chemistry. He left Canada in 1929 just as the Great Depression was beginning and returned in 1931.

Dr. Pidgeon was married to Frances Rundle in Winnipeg in 1928. The Rundle family was one of the pioneer settlers of Winnipeg and Western Canada. The couple lived together for 66 years. (Mrs. Pidgeon passed away in Dec 1994). After completing his studies at Oxford, Dr. Pidgeon returned to Canada and got a job with the National Research Council. His work was centered around experimentation with carbon black and on the utilization of Canadian natural gas (then becoming available in Alberta) to make carbon black for the reinforcement of rubber.

General A.G.L. McNaughton, an ex-military man was the head of the NRC. As war in Europe seemed near, the General felt that magnesium would be needed. He ordered the head of the chemistry department to put someone on magnesium research. Dr. Pidgeon was handed the undefined assignment.

Dr. Pidgeon began to work on the chlorination of brucite and subsequent electrolysis. He went on, "We had an associate committee on metallic magnesium which was set up with representatives from the Research Council, namely me, and a couple of fellows from the Mines Branch (normally responsible for metals) and a couple from the Armed Forces."

Based on literature research, two basic processes were chosen: the electrolysis of a fused salt and a distillation process that depended on the relative volatility of magnesium. Pidgeon advertised for a high-temperature electrochemist and hired Norman Phillips who immediately started to work on the electrolysis process based on the chlorination of brucite.

Dr. Pidgeon is given full credit in the published literature and technical reference books for developing and commercializing the magnesium production process that used silicon (in the form of ferrosilicon) to reduce calcined dolomite. [Note: Pidgeon reiterated over the years that he did not invent the process or put his name on it. Someone else did that.] It sounds very simple today in the brief encyclopedic descriptions. However, picturing an organic chemist with one assistant struggling with the total literature picture of magnesium production in 1938 is mind boggling. General McNaughton finally said that they were to develop a magnesium process that was energy efficient and would have to use readily available, native Canadian materials. The work with electrolysis was dropped and work was concentrated on the thermal reduction process.

The literature contained some references and patents of the work that was being done in Germany on thermal reduction. However, the process would depend on the ability to produce magnesium vapors and condense them. They would have to be produced by the high temperature reduction of MgO by a suitable reducing agent and the condensation of the evolved magnesium vapor into a dense, solid form.

Only two practical reducing agents were considered: carbon (Hansgirg Process) and silicon. Carbon, by far the cheapest, was ruled out because of its volatile oxide (carbon monoxide). Silicon was readily available as ferrosilicon. The reaction of ferrosilicon and calcined dolomite will produce magnesium vapor, the pressure of which will depend on the temperature of the system. The published vapor pressure over the charge was reported by Doerner of the U.S. Bureau of Mines to be 1.5 to 2.0 mm at 1100°C. This made it unlikely that there could ever be a commercial process developed in a temperature range that would permit alloy steel materials to be used for construction.

At this point, Dr. Pidgeon of the National Research Council in Canada and Glenn Bagley of Union Carbide in the U.S. conducted experiments which were designed to verify the vapor pressure of magnesium over the dolomite charge at 1100°C. They worked separately and independently with no knowledge of the other's existence. They both came up with the determination that the pressure over the charge was actually more in the order of 30 mm. This would make it possible to design steel reactors to produce magnesium. Pidgeon went with the horizontal tubular retort and Bagley went with very large vertical retorts (2).

Laboratory experimentation was conducted on a simple retort process by the NRC. Experiments concluded that it was possible to use silicon to reduce calcined dolomite and produce commercial magnesium at temperatures where an alloyed steel retort would have a reasonable life. After the pilot plant was built with 2-4" retorts, the first indications that alkali metals could become a problem were seen. These metals, mainly sodium and potassium, would not condense easily and would sometimes run out of the retorts. Special design modifications had to be made as the process developed to handle the sodium and potassium contents.

One of the big secrets of the Pidgeon Process success was the fact that the process was designed to be operated continuously at the reduction temperature. This prevented the large grain growth in the retorts that was so evident in the other processes such as the Bagley process that involved heating and cooling. Dr. Pidgeon also proved that it was also true that the condensing had to be done at a high temperature to get good crystalline deposits with no powder.

At this point, because the NRC did not go beyond the laboratory experimental stage, outside help was sought to continue process development. Two Canadian gold miners, Walter Segsworth (Mich. Tech '06) and Robert Jowsey, were looking for ways to help the Canadian war effort. They searched for a suitable source of dolomite and eventually found a very pure deposit in the Ottawa valley near Haley, Ontario. The promoters also included Thayer Lindsley (Falconbridge Nickel). They built a 5,000-tpy silicothermic plant at Haley, Ontario, adjacent to a very pure dolomite deposit. The company was called Dominion Magnesium and continues to operate today as Timminco. Dr. Pidgeon was appointed to be Director of Research at Dominion Magnesium in 1941.

Major C.J.P. Ball, the British magnesium pioneer said, "This successful translation of the Pidgeon Process from pilot plant to commercial production in such a relatively short period of time was a remarkable achievement, of great value to Canada and the USA, where five similar plants were built."(3)

During this time, Dr. Pidgeon was immersed in the rapid development of magnesium that was taking place throughout North America. He was contacted by many industrialists, including Henry Kaiser, for advice and consulting for silicothermic magnesium plant design. He worked with a large engineering firm, Singmaster and Breyer, to design and construct several of the plants built by the US Defense Plant Corporation. One of these plants was located at Luckey, Ohio and was run by National Lead. This plant was the most efficient of all the Pidgeon Process plants. (The plant manager was Edward Rowley, later Chairman of NL Industries and a leading proponent of the Great Salt Lake Magnesium project. The Luckey plant engineer was Robert Couch, later President of Amax Specialty Metals, who purchased the GSL Magnesium plant from NL Industries.)

Dr. Pidgeon was appointed Professor and head of the Department of Metallurgical Engineering at the University of Toronto in 1943. He was very successful in building a strong graduate school in Metallurgy and his department was one of the best in Canada. Many of his graduate students went on to leading positions in industry and academia. His appreciation of the physics of metals led to the growth of physical metallurgy. The department also expanded to include Material Science in 1965. He retired and was made Professor Emeritus in 1969.

Dr. Pidgeon was awarded major honors including the MBE (Member of the British Empire, awarded by King George VI), Officer of the Order of Canada, INCO Medal for contributions to Extractive Metallurgy, The Monel Medal of Columbia University for distinguished achievements in Mineral Technology, the Alcan Medal for contribution to the field of Metallurgy, the Falconbridge Innovation Award (with Timminco Metals).

He also received honors from many of the major metals organizations including The International Magnesium Association, The Canadian Institute of Mining and Metallurgy, The American Institute of Mining and Metallurgical Engineers. He is listed in Who's Who in Science, Who's Who in Canadian Science, Canadian Men of Science, and several other publications.

Silicothermic Production

The total amount of magnesium produced by the Pidgeon (horizontal retort) Process from the first plant at Haley, Ontario, Canada, through to about 1980 was an estimated 600,000 long tons of magnesium (4). This does not count production from the Italian process, the Brazilian process, the Magnetherm process, nor the Bagley vertical retorts in Spokane. These are technically silicothermic processes, but they use other furnace arrangements.

It appears from the production numbers available, that China has equaled this total magnesium production by the Pidgeon Process. Expansion in the number and capacity of the Pidgeon process plants in China is continuing.

Acknowledgments

The author is deeply indebted to Light Metal Age magazine for permission to use much of this material which was printed there first. Deepest thanks to Neil and Ruth Bryson for their warm hospitality and help in developing the original profile of Dr. Pidgeon. Special thanks to Ruth for her co-authorship of the original biographical sketch of Dr. Pidgeon. (1)

References

1. R.E. Brown and Ruth Bryson, "Magnesium Industry Legend", Light Metal Age, Volume 53, No.3 & 4, April 1995, pp. 44-45

2. L.M. Pidgeon and W.A. Alexander, "Thermal Production of Magnesium", Transactions of AIMME, Volume 159, 1944 pp. 315-352

3. C. J.P. Ball, "The History of Magnesium", Presidential Address to British Institute of Metals, 1956

4. Klagsbrunn, H.A., "Wartime Aluminum and Magnesium Production", Industrial and Engineering Chemistry, Vol. 37, No. 7, July 1945

Addendum

List of Canadian Patents by Dr. L.M. Pidgeon

1.	463416	Production of Calcium
2.	424665	Magnesium Producing Apparatus
3.	424664	Volatilizable Metal Recovery Apparatus
4.	424663	Volatilizable Metal Recovery Apparatus
5.	424662	Magnesium Producing Apparatus
6.	420245	Magnesium Production Apparatus
7.	420244	Magnesium Producing Apparatus
8.	420243	Magnesium Producing Apparatus
9.	420242	Ductile Magnesium Production
10.	420421	Volatilizable Metal Recovery Apparatus
11.	420240	Thermal Magnesium Production
12.	420239	Thermal Magnesium Production
13.	415765	Magnesium Producing Apparatus
14.	415764	Volatilizable Metal Recovering Apparatus
15.	412169	Anhydrous Magnesium Chloride
16.	361606	Aromatic Liquid and Carbon Black Production

Author Note

Dr. Pidgeon also has many U.S. Patents that followed the same line as the Canadian patents.

MAGNESIUM ELECTROLYSIS - A MONOPOLAR VIEWPOINT

Oddmund Wallevik[1], Ketil Amundsen[2], André Faucher[3], Thorvald Mellerud[4]

[1]Norsk Hydro ASA, Research Centre, N-3901 Porsgrunn, Norway
[2]Hydro Magnesium, N-0246 Oslo, Norway
[3]Norsk Hydro Canada Inc., 7000 Raoul-Duchesne, Bécancour (Québec), Canada G0X 1B0
[4]Hydro Magnesium, Ave. Marcel Thiry 83, B-1200 Brussels, Belgium

Abstract

Norsk Hydro has been continuously engaged in development of magnesium electrolysis all since the start of production in Porsgrunn, Norway, in 1951. The first technology was inherited from IG Farben. Later, Norsk Hydro has developed its own diaphragmless electrolyser, now being used for a number of years in the Norsk Hydro's plants in Porsgrunn as well as in Bécancour (Québec), Canada.

A presentation is made of the Norsk Hydro high-amperage monopolar electrolysis cell. Its performance is described, as basis for the conclusion that this type of cell presently is very competitive compared to other cell technologies, although it has a higher electrical energy consumption than bipolar cells.

Publications of performance data from magnesium plants in operation are scarce. Norsk Hydro hopes by presenting this paper to invite other producers to release comparable data.

History

Norsk Hydro has produced magnesium in Porsgrunn, Norway since 1951. The technology including the electrolysis was inherited from the German IG Farbenindustrie. Norsk Hydro has since then continuously developed the electrolysis, first by improving the IG technology, and then by developing its own diaphragmless electrolyser (DLE).

The first IG cells were operated at a current of 32 kA. In a new cell room built in the early 1960ies the cells were made for 62 kA, and an improved version were brought up to 80 kA. The last IG cell in Porsgrunn was stopped in 1989.

The Norsk Hydro development of a DLE cell was started in the late 1960ies, and two high-amperage prototypes were installed in Porsgrunn in 1974/75. The first high-amperage line was started in 1978 at a current of 260 kA, and the first full cell room brought on stream in Porsgrunn in 1983. Further developments have resulted in a current load well above 300 kA. A still higher amperage stage of more than 400 kA has been realised in the Norsk Hydro Canada plant in Bécancour, started up in 1989. Figure 1 shows a photo from the Bécancour cell room.

Presently, the annual production in Norsk Hydro high-amperage DLE cells amounts to some 85,000 tonnes of magnesium per year. About 50 % hereof is produced in Porsgrunn, based on liquid (molten) anhydrous magnesium chloride produced from dolomite and seawater, and the other 50 % in Bécancour based on solid (cold) anhydrous magnesium chloride produced from magnesite by the dehydration process developed by Norsk Hydro.

Figure 1. The Bécancour cell room

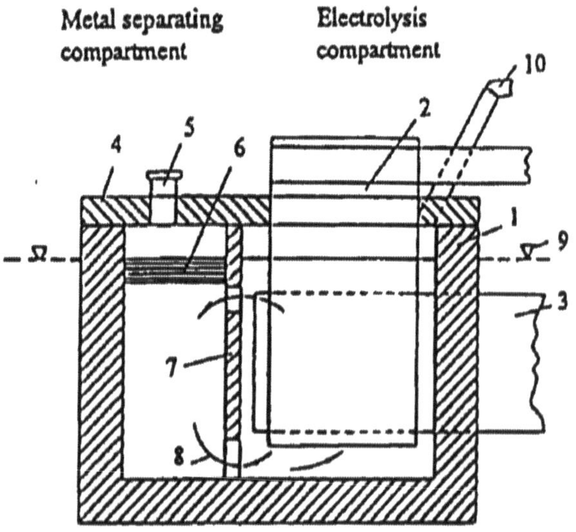

Figure 2. The Norsk Hydro diaphragmless electrolyser. 1) Refractory material. 2) Graphite anodes. 3) Steel cathodes. 4) Refractory cover. 5) Metal outlet. 6) Metal. 7) Partition wall. 8) Electrolyte flow. 9) Electrolyte level. 10) Chlorine outlet.

The Norsk Hydro monopolar cell

The main features of the Norsk Hydro cell are known from the description in the patent, ref. (1). Figure 2 shows the cell schematically.

The cell has a steel casing, lined with refractory materials. It is divided into two compartments, an electrolysis compartment and a metal separating compartment, separated by a partition wall. In the electrolysis compartment, densely packed graphite anode plates are installed from the top and double acting steel plate cathodes through the back wall.

The circulation of the electrolyte is parallel to the electrodes, bringing the metal to the separating compartment, from where it is extracted by vacuum operated vehicles and transported to the foundry. Chlorine gas is collected from one central pipe connection on top of the electrolysis compartment.

The anode tops are water cooled for longer life and higher cell productivity. The cell is equipped with special electrodes for adding AC electrical energy for temperature adjustment, or to keep the cell warm when the DC supply is off.

When liquid magnesium chloride feed is used, it is added to the metal separating compartment. When solid feed is used, it is fed into the electrolysis compartment countercurrent to the chlorine gas flowing out of the cell, to reduce the amount of air being brought into the cell with the feed.

Performance features

To assess the performance of alternative electrolysis cell technologies the following criteria are proposed:

- Amperage, current efficiency and cell capacity
- Specific energy consumption
- Cell life and rebuilding costs
- Chlorine quality
- Working environment and emissions
- Sludge formation
- Manpower productivity

An evaluation of each of these criteria is given below for the Norsk Hydro high-amperage DLE monopolar cell technology.

Amperage, current efficiency and cell capacity

The Norsk Hydro DLE cell is characterized by its high amperage load, exceeding 400 kA in the Bécancour plant. This is more than any aluminum smelter presently can apply, and represents an important feature for the economy of the cell room (for a monopolar type of plant).

The current efficiency of the electrolysis obviously depends critically on the purity of the feed material, in this case the quality of the anhydrous magnesium choride feed. In the Bécancour operation, the electrolysis is fed by solid, ambient temperature feed, produced by the Norsk Hydro dehydration process. This gives consistently a current efficiency in the range 89 - 91%.

The DLE cells are further characterized by long life (see below) and high on-line availability. Total availability over a 4 - 5 year life cycle of a cell, including cell rebuilding time, is typically 97%.

This gives a daily production in excess of 4 tonnes of Mg, and an average annual production close to 1,500 tonnes of Mg per cell.

Specific energy consumption

An important objective for Norsk Hydro's cell development has been to reduce its energy consumption. Measured on the main busbars, the cells are typically operating with a cell voltage of 5.3 volts, giving a DC energy consumption (with solid feed) of 13.0 kWh/kg Mg. Hereof 1.2 kWh/kg Mg is consumed in heating and melting of the feed. This compares favourably with modern aluminium electrolysis cells.

Figure 3 shows the development of energy consumption and daily production for Norsk Hydro cell types.

The net energy consumption of the plant will be further reduced by a system under development to recover heat from the cells. This will give a significant contribution to cover the thermal energy needs of the up-stream magnesium chloride dehydration process.

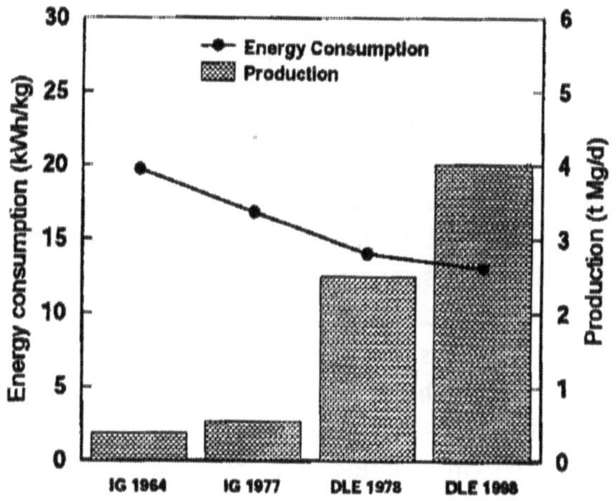

Figure 3. Energy consumption and daily production for various cell types.

Cell life and rebuilding costs

Another important performance parameter for the economy of the electrolysis is the average cell life (time between rebuilds). Norsk Hydro's experience with the DLE cells is today an average cell life of 4 - 5 years. Development work is on-going aiming to increase this to at least 6 years.

To achieve a high average cell life and low rebuilding costs it has been important to develop the life of anodes and cathodes to match the life of the cell itself. By development of the graphite quality as well as the tightness of the cells, anode life has been extended to match cell life. The graphite consumption is of the order of 2 - 3 kg per tonne Mg. Cathodes are today replaced only every third cell rebuild, and development work is on-going aiming to increase this to 4 cell lives.

As a result of these developments we today find that typical cell rebuilding costs for the Norsk Hydro DLE cells are in the range of 2 - 2.5 USc/lb Mg, and we aim by our development work to achieve rebuilding costs well below 2 USc/lb Mg.

Chlorine quality

The cells are kept tight to eliminate air ingress into the anode gas. This is important to reduce anode wear as well as for the downstream chlorine processing plant.

During normal operation, a concentration of 96 - 98 vol% Cl_2 is achieved in the anode gas.

Working environment and emissions

Substantial efforts have been put into improving the working environment in the electrolysis, focusing on reducing diffuse emissions of chlorine to the working atmosphere. This has been achieved by ensuring air tightness of the cells and vacuum operation for the anode gas system. The Norsk Hydro DLE cells can be operated with a satisfactory working atmosphere, the ambient air/working atmosphere for the operators containing Cl2 in the range 0 - 0.2 ppm . Typically, total air emissions from a cell room are of the order of 0.2 kg Cl2 per tonne Mg produced.

Sludge formation

It is of importance for the technical and economic as well as the environmental performance of the cells to reduce sludge formation to an absolute minimum. In the Norsk Hydro DLE cells we find that as long as the magnesium chloride feed is within specification and the cells are kept air tight, the sludge formation in the cells is negligible, and there is no need for sludging during a cell life.

Manpower productivity

The manning of the electrolysis plant is in generalized terms 1 operator per 1,000 tonnes Mg annual production

Conclusion

For any production of primary magnesium based on electrolysis, the efficiency of the electrolysis technology itself is critical for the total performance. Norsk Hydro has through many years of focused efforts developed its DLE cell technology, which now is considered key to Norsk Hydro's plants both in Porsgrunn and Bécancour.

It is Norsk Hydro's experience that for a large scale electrolysis plant the high-amperage monopolar cell, although having a higher electrical energy consumption than bipolar cells, is very competitive, all aspects considered.

It is also Norsk Hydro's experience that the performance of the electrolysis technology can not be seen isolated from the performance of the front end technology and the quality of the feed to the electrolysis. The good performance of the Norsk Hydro DLE cell must therefore be seen in context with the development of Norsk Hydro's up-stream magnesium chloride dehydration process.

Reference

1. K.A. Andreassen, Ø. Bøyum, H.K. Johnsen, L.B. Ognedal and P.R. Solheim: "Method and electrolyzer for production of magnesium", US Patent 4,308,116.

MAGNESIUM ELECTROLYTIC PRODUCTION PROCESS

G.Shekhovtsov, V.Shchegolev, V.Devyatkin,
A.Tatakin, I.Zabelin
State Research and Design Titanium Institute,
180 prospect Lenina, Zaporozhye, Ukraine
J.S.C. Russian National Aluminium and Magnesium Institute,
Saint-Petersburg

Abstract

Titanium Institute/VAMI propose two variants of magnesium electrolytic production process.
The first variant is based on two-stage preparation of magnesium raw material – carnallite ($KCl \cdot MgCl_2 \cdot 6H_2O$) for electrolysis. At the first stage, carnallite is dehydrated in fluidized bed dryers with on output of 400 t/day. Operation and control of the drying process has a high level of automation.
The second stage of carnallite dehydration is carried out in electric chlorinators with an output of 150-200 t/day. In chlorinators, carnallite is melted and chlorinets treated. Molten carnallite flows to the electrolysis cells. Electrolysis cells are connected in a flow line which operates as one highly productive electrochemical unit. By electrolyte flow, magnesium moves through electrolysis cells and accumulates in a separator cell from where it is extracted and transferred to the casting house for casting magnesium and magnesium alloys into ingots.
Electrolysis cells current intensity is ~ 200-300 kA.
Magnesium and magnesium alloys are refined in continuous refining furnaces with the capacity up to 100 t/day.
The process is adopted at magnesium and titanium-magnesium plants of Russia, Kazakhstan, Ukraine. The best modern projects are realised at DSM Magnesium Plant (Israel).
The second variant of magnesium electrolysis production process is based on high dehydration of carnallite in fluidized bed dryers by HCl injection into chambers together with combustion gases. HCl is gained from fuel burning in chlorine gas magnesium electrolysis cells. Solid highly dehydrated carnallite is charged into cells connected into flow lines having centralized magnesium collection.
The process has passed pilot-commercial tests and is ready for industrial realization.
Above mentioned carnallite processing variants can be used for different types of raw material: magnesite, magnesium chloride solutions, sea water, dolomite and carnallite.

1.Preparation of raw material for electrolysis

Double salt $MgCl_2{\cdot}KCl{\cdot}6H_2O$ is used as raw material for the production of magnesium by electrolysis. This salt can be of natural origin (Verkhne-Kamsk deposit in the Urals, salt deposits from evaporating basins of Dead Sea), or it can be artificial, when carnallite is obtained by synthesis from $MgCl_2$ and KCl solutions ("Oriana" Concern, Ukraine).
The raw material is prepared for electrolysis in two stages: the first stage is carried out in fluidized bed dryers with the feed of hot flue gases diluted by air. The dehydration is performed at temperatures of 300°-350°C. As the result, a free-flowing product – dehydrated carnallite – containing 5-7% H_2O and 1.2-1.9%MgO is produced.
The output of modern fluidized bed dryers is as high as 400 t/day of dehydrated carnallite with the consumption of fuel (fuel oil) from 90 to 105 kg per tonne of dehydrated carnallite.
The second stage of carnallite dehydration can be effected in two ways.
Variant 1 is used in CIS plants and includes the processing of dehydrated carnallite in chlorinators where it melts at a temperature of 530-540 °C, and then the products of magnesium chloride hydrolysis are chlorinated in the presence of carbon reductant (petroleum coke) at 770-800°C:

$$2MgOHCl + 2Cl_2 + C \Leftrightarrow 2MgCl_2 + 2HCl\uparrow + CO_2 \quad (1)$$
$$2MgO + 2Cl_2 + C \Leftrightarrow 2MgCl_2 + CO_2\uparrow \quad (2)$$

Variant 2 has been developed and tested in CIS on a pilot scale. This process involves the production of crystalline highly dehydrated carnallite (~0.3 wt.%MgO and~0.3 wt.%H_2O) in fluidized bed dryers. This concentration of the products of hydrolysis is attained due to a supply of hydrogen chloride to the fluidized bed. The method of raw material preparation is to be selected taking into consideration the chemical analysis of the raw material and potential chlorine utilization.

2.Magnesium production by electrolysis

Electrowinning of magnesium from carnallite and anhydrous magnesium chloride (electrolysis of anhydrous magnesium chloride in CIS is carried out using magnesium chloride from titanium production) is carried out in diaphragmless electrolysis cells .There are several variants of arranging the cells to ensure the minimum expenses when erected in cell rooms under reconstruction.

During the refinement of geometry and parameters of the electrolysis cell operation, the influence of current density, electrode height and length, interelectrode space, electrolyte temperature and raw material composition on the current efficiency for magnesium were investigated. In all cases, the electrolyte consisted of potassium, sodium and magnesium chlorides. Labour conditions in operating the cells were studied with the invitation of physicians. The cells were optimized with the use of computers.

In the experimental cell, with a current density of 0.5÷0.16 A/cm^2, the minimum consumption of the electric power across the cell shunt is 11÷10.5 kW·hr per kilogram of magnesium within a range of current densities from 0.16 to 0.25 A/cm^2, current efficiency for magnesium being equal to 79÷85%.

Based on the results of testing of pilot plants, pilot cells were developed, and, when tried out, they were widely introduced in cell rooms of operating commercial magnesium plants in CIS countries. In commercial operation, the cells were further improved during the batch process and in the continuous flow line.

2.1.Electrolysis cell performance during the bath process.

Table shows the performance of commercial electrolysis cells with a bottom entry of anodes at Joint Stock Company "AVISMA" (Berezniki, Russia) and with a top entry of anodes at UKTMP (Ust-Kamenogorsk, Kazakhstan). These are the best figures, attained in electrolysis cells operating with carnallite and magnesium chloride.

Table shows the advantages when carnallite is used as a feedstock of the bottom anode entry cells as to consumption of electric power. These electrolysis cells require less graphite consumption (18 kg/t as compared to 30 kg/t for the top anode entry cells). The bottom anode entry cells are more expensive (higher by ~ 30%). In the regions with higher electric power cost the bottom anode entry cell is preferable.

2.2.Electrolysis in a continuous flow line.

Time metering of the cells operation has shown that, during the batch process, about 70% of time is spent for such operations as movement of bridge cranes and electric wagons towards the cell for technical servicing (removal of sludge, magnesium and spent electrolyte). Besides, in the course of the cell operation, the level of the electrolyte varies over a range from 200 to 250 mm which is not optimal for the process. Other principal approaches which relieve the process for electrolysis of these drawbacks are attained by the use of continuous flow lines.

The flow line is schematically shown in Fig.1. It consists of a head cell for suppling molten carnallite and recycled electrolyte pumped by a pneumatic pump from a separator cell (PM) where magnesium produced in the electrolysis cell is collected.

Downstream, after the head cell, refining cells with the top anode entry are installed, followed by bottom anode entry flow cells. All the cells are interconnected by transfer channels. For magnesium chloride depletion to 6%(wt.), end cells with the bottom anode entry are set up after the separator cell

The process in the flow line enables a centralized feed of raw material to the head cell. The extraction of magnesium from the separator cell is centralized, too. Spent electrolyte can be pumped out of a single unit (end cell). Magnesium and raw material are transferred along the flow line by a stream of circulating electrolyte. Which is created by means of pneumatic pumps. During the cell shutdown for overhaul, by-pass channels which shunt the electrolysis cell under repair are used.

Fig.2 and 3 show the process parameters of the flow line operation.

When supplying solid highly dehydrated carnallite or magnesium chloride to the flow line, all the electrolysis cells have anodes introduced from the top.

Production of magnesium in the flow lines, as compared to the batch process, enables the increased labour productivity, increased amperage (by 15%), increased magnesium output (by 1.5÷2%), reduced power consumption (by 500-1000 kW·hr/t), extended service life of anodes and cells from 28 to 40-50 months.

TABLE. Performance of commercial electrolysis cells fed with carnallite and magnesium chloride

Ser No	Cell type	Raw material	Company	I, kA	η, %	U_{shunt}, V	W, kW·hr/kg Mg	Q_{Mg}, kg/day	Chlorine yield, kg/t
1	Bottom anode entry cell	carnallite	"AVISMA" (Russia)	90	77.2	4.71	13.5	728.5	2.72
2	Top anode entry cell	magnesium chloride	UKTMP (Kazakhstan)	200	80.5	4.83	13.2	1761	2.85
3	Top anode entry cell	carnallite	UKTMP (Kazakhstan)	166	75	5.04	14.6	1349	2.85

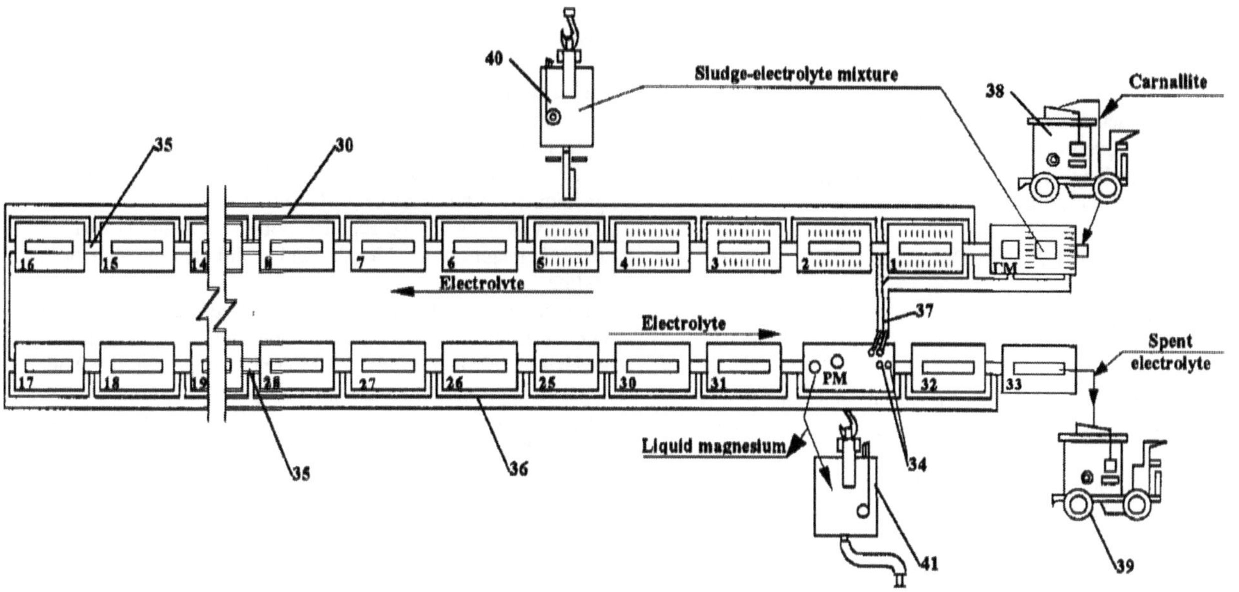

Figure.1: 1...5 – top anode entry (refining) cells; 6...31 – bottom anode entry (flow) cells; 32, 33 – end bottom anode entry cells; 34 – pumps for recycled electrolyte; 35,36 – transfer and bypass channels; 37 – channel for recycled electrolyte; 38 – carnallite ladle; 39 – spent electrolyte ladle; 40 – SEM ladle; 41 – magnesium ladle; ГМ – head cell; PM – separator cell.

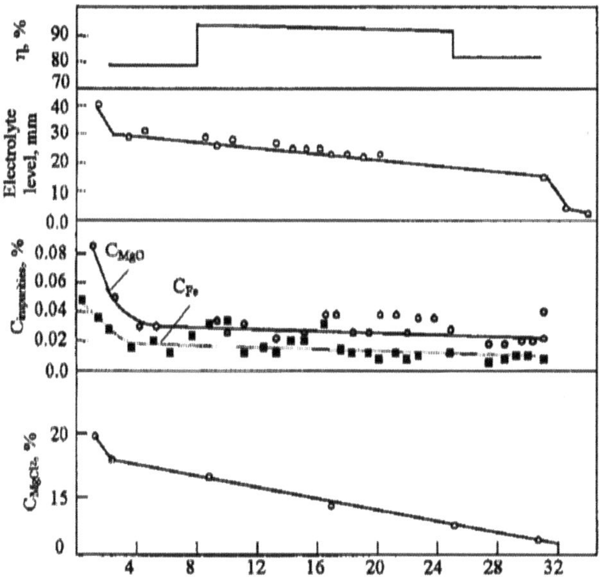

Figure.2: Change in concentration of magnesium chloride, impurities and electrolyte level in flow line electrolysis cells. Average current efficiency for magnesium in the groups of electrolysis cells.

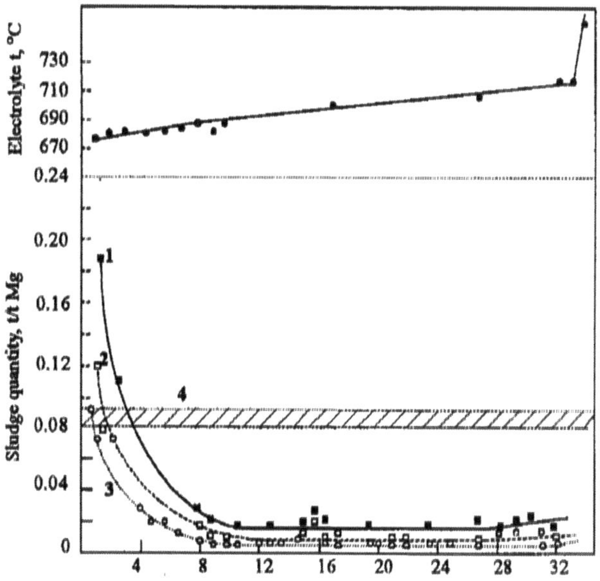

Figure.3: Quantity of sludge, removed from flow line electrolysis cells. Change of electrolyte temperature in flow line electrolysis cells.

1–summary quantity of removed sludge; 2–sludge pumped from sludge and electrolyte mixture; 3–sludge removed manually; 4–sludge quantity during batch process.

Magnesium extracted from the separator cell has a quality as per 99.8 ASTM. However, for convenience of magnesium collection and casting, the casting house is equipped with a continuous refining furnace. The furnace output can range from 50 to 100 t/day with the power consumption of 80 to 100 kW·hr/t Mg.

2.3.Raw material refining assured by the continuous flow line for electrolysis.

Electrochemically active impurities entering the flow line with the feedstock are removed in the head part of the flow line, particularly in the head cell and in the refining cells. These processes occur by the formulas:

$$MgOHCl + e \rightarrow MgO + 1/2\ H_2\uparrow + Cl^- \quad (3)$$

$$Ni^+ + e \rightarrow Ni(Mg) \quad (4)$$

$$Br^- - e \rightarrow 1/2\ Br_2 \quad (5)$$

$$Fe^{+2} + 2e \rightarrow Fe \quad (6)$$

In the flow line contaminated head cell magnesium can be separated from cleaner flow cell magnesium collected in the separator cell.

In a similar way, anode chlorine contaminated with bromine can be concentrated in the head part of the flow line and separated from the main portion of cleaner chlorine formed in the flow cells. Both grades of chlorine can be used for various purposes.

Above – cited process is used at the plants of CIS and the most up-to-date decisions are used at DSM plant (Israel).

3.Conclusions:

1.The advanced process for magnesium production by electrolysis from carnallite involves the application of highly efficient equipment such as fluidized bed dryers (400 t/day) chlorinators (150-170 t/day), electrolysis cells (180-200 kA), continuous magnesium refining furnaces (up to 100 t/day).

2.Electrolysis cells can be supplied both by anhydrous carnallite melt and magnesium chloride and solid highly dehydrated carnallite.

3.The electrolysis cells are interconnected into flow lines which assure a centralized collection of magnesium, an optimal level of electrolyte and, when supplied with carnallite melt, a distribution and purification of the feedstock. The power requirements with the molten magnesium chloride or carnallite feedstock is 12400-13200 kW·hr/t Mg.

4.Consumption of utilities to produce one tonne of marketable magnesium from carnallite;

4.1.When operating with carnallite melt:

electric power - 19000÷20000 kW·hr;

gaseous or liquid fuel – 900÷1150 kg (fuel oil);

4.2.When operating with solid highly dehydrated carnallite:

electric power - 17000÷18000 kW·hr;

gaseous or liquid fuel –1150 13150 kg (fuel oil);

cell output – by 15÷20% higher than that with carnallite melt.

PILOT EXPERIMENTS OF MAGNESIA DIRECT ELECTRYOLYSIS IN A 5KA MAGNESIUM REDUCTION CELL

Huimin Lu[1] , Wellton Jia[2], Chunfa Liao[1], Ruixin Ma[1] and Wenhui Yuan[1]
[1]University of Science and Technology, Metallurgl. Engrg. Sch.; No. 30 Xueyuan Rd.; Beijing 100083, China
[2]Bona Metals Research Institute; 21660 East Copley Drive, Suite390, Diamond Bar; CA 91765, USA

Keywords: **magnesia; $LaCl_3$-$MgCl_2$ system; molten salt electrolysis; pilot experiment**

Abstract

A new technique of magnesia direct electrolysis to produce magnesium with the $LaCl_3$-$MgCl_2$ system as support electrolyte published in Magnesium Technology 2004 has low energy consumption, high current efficiency and less pollution for the environment. In this paper, pilot experiments of the new technique of magnesia direct electrolysis in a 5kA magnesium electrolysis cell are conducted. The electrolyte is also the $LaCl_3$-$MgCl_2$ system. The experiment results indicate that the $LaCl_3$-$MgCl_2$ system is promising. These pilot experiments lay a good foundation for industrial comprehensive utilization of bischofite from Qinghai Lakes in China. In the meantime, the physico-chemical properties of $LaCl_3$-$MgCl_2$ bath system, electrolyte ionic species and electrode reactions of the magnesia direct electrolysis in the $LaCl_3$-$MgCl_2$ system are studied. The operational parameters of the new process are also compared with those of the common aluminum production process and $MgCl_2$ electrolysis process.

Introduction

In the paper "A New Technique of Magnesium Electrolysis with Bischofite from Qinghai Salt lakes in China" published in Magnesium Technology 2004, the technique on electrolysis of magnesia from bischofite of Qinghai salt lakes was investigated. Magnesia with 99.5% purity was prepared by ammonia process from bischofite. Magnesium was successfully produced in 1kA laboratory-scale cell with the $LaCl_3$-$MgCl_2$ system as support electrolyte system at 700°C, the current efficiency reached 90%, and the specific energy consumption was 11.5kWh/kg Mg. The new technique has the advantages of economy on energy and reduction of consumption, high efficiency and environmental amity. The experiment results have shown that the new magnesium electrolysis process with magnesia by ammonia process from bischofite of Qinghai salt lakes is practical and promising.

The background of the new technique is that the government of China has laid bischofite of Qinghai salt lakes as raw materials of magnesium electrolysis on a key development project of the Tenth Five-Year Plan of China [1]. Chaerhan Salt Lake of Qinghai province, located at the south central of Chaidamu basin, the exclusive production basis for potassium fertilizer in China, has extensive reserves of about 16.2 hundred million tonne magnesium chloride. In the production process of potassium fertilizer, 10 tonne magnesium chloride were secondarily produced accompanied by per 1 tonne potassium chloride produced. Chaerhan Salt Lake has actual production of about 70 ten thousand tonne potassium fertilizer per year; therefore about 700 ten thousand tonne magnesium chloride may be drawn from the mother liquor after potassium extraction. Bischofite ($MgCl_2 \cdot 6H_2O$) of 98% purity may be obtained from the mother liquor after carnallite extraction by natural and solar evaporation of Chaerhan salt lake bittern. This naturally crystal bischofite is in the nature of capitally crystallographic form, uniformly particle size (particle diameter 1.3mm on an average), little harmful impurities (major impurities being a small quantity of sodium, potassium and calcium). At present, the drained district of the mother liquor after potassium extraction has already large numbers of magnesium chloride deposits, the excavating cost is fairly low. Therefore, the bargain bischofite selected as the raw materials of production of metal magnesium by improvement of production process is an ultimate outlet road for low cost magnesium and the bischofite is also an inexhaustible ideal material for magnesium electrolysis [2].

For the magnesium electrolysis technique with bischofite as raw materials, production of anhydrous magnesium chloride from bischofite by a reasonable method to eliminate 6 pieces of crystal water is a key technology. However, production of anhydrous magnesium chloride from bischofite by warming is difficult because magnesium chloride will dissolve in the water per self and hydrolysis occurs concurrently [3], but production of magnesia from bischofite by warming or ammonia process is easy and has lower cost. Because the $LaCl_3$-$MgCl_2$ system were found that could dissolve sufficient magnesia and allow this dissolved magnesia to decompose electrolytically into magnesium and oxygen, magnesia in place of magnesium chloride could be used as a feed material which was similar to the Hall-Héroult process of aluminum production.

In this paper, pilot experiments of the new technique of magnesia direct electrolysis in a 5kA magnesium electrolysis cell are conducted on the basis of the previous studies. The electrolyte is also the $LaCl_3$-$MgCl_2$ system. The experiment results indicate that the $LaCl_3$-$MgCl_2$ system is promising. These pilot experiments lay a good foundation for industrial comprehensive utilization of bischofite from Qinghai Lakes in China. In the meantime, the physico-chemical properties of $LaCl_3$-$MgCl_2$ bath system, electrolyte ionic species and electrode reactions of the magnesia direct electrolysis in the $LaCl_3$-$MgCl_2$ system are studied. The operational parameters of the new process are also compared with those of the common aluminum production process and $MgCl_2$ electrolysis process.

Experimental raw materials

In the experiments, bischofite analyzed by XRD as raw material was taken from Qinghai Potassium Fertilizer Plant in China, its purity 98%, and major impurities being a small quantity of sodium, potassium and calcium.

So far, MgO has been produced from bischofite by these methods such as pyrolysis method, sodium salt method, lime base process, ammonia process and ammonium bicarbonate method [2]. By comparison, the pyrolysis method has the advantages as simple process and lower cost, but the disadvantages as bum product and unmanageable exhaust gas. High-grade product can be produced by both sodium salt method and ammonium bicarbonate method but high cost is obvious. In addition, lime base process has the disadvantage of bum product. However, by the ammonia process not only high-grade product but also lower cost can be obtained and the byproduct ammonium chloride is an important chemically raw material and commercial fertilizer.

In the experiments for producing MgO, the ammonia process was used. The project first prepared magnesium hydroxide by precipitating magnesium ion with ammonia, then by calcining magnesium hydroxide magnesia could be obtained. The molar ratio of magnesium ions and ammonia was 1:1.2, the magnesium ion content was 45g/L, the optimal reacting time was 20min and the reacting temperature was 55°C. The purity of magnesia produced was 99.5% and the recovery of magnesium from bischofite was 80%. The characters of this process are the high purity magnesia, the high magnesium recovery and ammonia recycling.

In fact, the amount of bischofite of Qinghai salt lakes used as the magnesium electrolytic material is little. Therefore, the great deal of the high-grade magnesia produced by ammonia process may be used as the material for super refractory production.

The electrolyte was compounded with dehydrated chemical grade $LaCl_3$ and $MgCl_2$ pro rata. In addition, dehydrated Analar grade chemicals NaCl and KCl (all>99.9mass %) were used in some experiments.

Physico-chemical properties of $LaCl_3$-$MgCl_2$ bath system

Liquidus temperature of electrolytes

Gradual cooling curve method was performed under an argon atmosphere to determine the liquidus temperature (T) of the $LaCl_3$-$MgCl_2$ system. The crucible and the lid used were made of platinum, a calibrated Pt/Pt10Rh thermocouple was employed as the temperature sensor, and super cooling of the melt was prevented by using a slow cooling rate chosen to be 0.7°C/min. The experimental relative error of the melting point measurement using NaCl and KCl as standard materials did not surpass ±0.8%. The corresponding fusibility for the $LaCl_3$-$MgCl_2$ system was listed in table 1 from the experimental readings of melting points of various salt mixtures of known composition. It was found that in the $LaCl_3$-$MgCl_2$ system there are electrolytes with wide range of low melting point. Some of them are even lower than 650°C melting point. The fusibility of the present $LaCl_3$-$MgCl_2$ system apparently has a minimum (at $\chi_{LaCl3} \approx 0.4$). It is apparent that across the $LaCl_3$ composition range of 0.2 to 0.6 mole fraction the melting point of the electrolyte is close to or below the melting point of the magnesium metal. The cell must operate above the melting point of the magnesium to ensure that the magnesium is produced as liquid. Table 1 suggests that over a wide range of electrolyte compositions the operating temperature does not need to be significantly higher than magnesium melting temperature. The fusibility of the $LaCl_3$-$MgCl_2$ system in table 1 suggests the optimum bath composition as bath electrolyte in the $LaCl_3$-$MgCl_2$ system may be around the 40:60 mole percent mixture of $LaCl_3$ and $MgCl_2$.

Table 1 Fusibility (T) of the $LaCl_3$-$MgCl_2$ system

χ_{LaCl3}	0	0.2	0.4	0.6	0.8	1.0
T in °C	714	648	630	656	810	850

Electrical conductivity

Wheatstone bridge method was adopted to measure the electrical conductivity (κ) of the molten $LaCl_3$-$MgCl_2$ electrolyte. The conductivity cells used here were quartz capillaries containing platinum wires of 1mm diameter. The cell constant was 102.4cm^{-1} determined using the molten NaCl at 831°C and the cell was calibrated using molten NaCl at 821°C. The result showed the cell constant varied little with temperature. The experimental relative error of the conductivity measurement did not surpass ±0.7%. Based on the experimental electrical conductivity values at different temperature corresponding to the various mixtures containing different proportions of $LaCl_3$ and $MgCl_2$, the measured conductivity was found to increase linearly with increasing temperature. Accordingly, the conductivity of the molten $LaCl_3$-$MgCl_2$ system was described by the linear expression $\kappa = A + BT$, where T represents the temperature in K and ranges from 873K to 1073K, A and B are constants. The coefficients A and B were determined by the least-squares in Table 2, which also includes the corresponding standard deviations (S_D). The calculated electrical conductivity values of the $LaCl_3$ composition range of 0.2 to 0.6 mole fraction in the molten $LaCl_3$-$MgCl_2$ electrolyte based on table 2 in 700°C are all higher than $1.4\Omega^{-1}\cdot cm^{-1}$. This is also one of the advantages of the molten $LaCl_3$-$MgCl_2$ electrolyte.

Table 2 The constants (A and B) of the temperature equation ($\kappa = A + BT$) of the electrical conductivity and the corresponding standard deviations (S_D) of the $LaCl_3$-$MgCl_2$ system

χ_{LaCl3}	A ($\Omega^{-1}\cdot cm^{-1}$)	10^4B ($\Omega^{-1}\cdot cm^{-1}\cdot K^{-1}$)	10^2S_D ($\Omega^{-1}\cdot cm^{-1}$)	κ700°C ($\Omega^{-1}\cdot cm^{-1}$)
0	0.5887	11	0.5436	1.659
0.2	0.8938	7.8	1.1231	1.6527
0.4	1.7956	-1.7	1.4571	1.5906
0.5	0.7786	8.6	2.5683	1.6154
0.6	-0.1298	16.05	2.7890	1.4319
0.8	0.6845	3.25	3.0241	1.0007

Bath density

The density of the molten $LaCl_3$-$MgCl_2$ electrolyte was determined by employing the Archimedean technique. A platinum sphere with a diameter of 16mm, suspended by a thin platinum wire on an automatic analytical balance, was used as the measuring body. The dependence of the sphere volume on the temperature was determined by calibration using molten NaCl and KCl. The accuracy of the density measurement was ±0.3%. The temperature dependence of the density was described by the linear equation $\rho = A + BT$, were T represents the temperature in K and ranges from 873K to 1073K, A and B are constants. The coefficients A and B were determined by the least-squares method.

These data are presented in table 3, which also includes the corresponding standard deviations (S_D).

Table 3 The constants (A and B) of the temperature equation (ρ = A + BT) of the density and the corresponding standard deviations (S_D) of the $LaCl_3$-$MgCl_2$ system

χ_{LaCl3}	A ($g \cdot cm^{-3}$)	10^4B ($g \cdot cm^{-3} \cdot K^{-1}$)	10^2S_D ($g \cdot cm^{-3}$)	ρ700°C ($g \cdot cm^{-3}$)
0	2.0276	-3.5	1.1244	1.7171
0.2	3.8123	-21	1.3456	1.7690
0.4	24.8064	-235.5	1.2456	1.8923
0.5	7.442	-53.125	1.4327	2.2729
0.6	3.6722	-12.2625	2.0214	2.4790
0.8	12.854	-105.2	1.2356	2.6180

Metal solubility

The gravimetric method was used to measure the solubility of magnesium in the $LaCl_3$-$MgCl_2$ melts. In the experiments, graphite crucibles made from a dense high-purity graphite electrode were used. The dimensions of the graphite crucible were diameter 50.30mm × height 70mm. The graphite crucible was placed in a closed steel crucible under an argon atmosphere to avoid air oxidation, and 60g of the $LaCl_3$-$MgCl_2$ mixture were melted in it at a predetermined temperature. After 20minute for melting of the mixture, 10±0.1g of pure magnesium in the form of small sticks (ϕ10mm) were added to the melt. The top of the crucible was closed off tightly with a steel cover. Magnesium and bath were kept in the crucible for 4h to reach the equilibrium state of metal dissolution when the metal loss became unchanged, and then the crucible was removed from the furnace. After cooling, the crucible was extracted and weighed. The metal loss was determined by weighing the metal before and after the experiment. All experiments were done in duplicate. The temperature was automatically controlled to a predetermined value. The solubility in the molten salts was determined from table 4 expressing the relationship between weight loss and time. In the experiments, time duration of 30, 60, 90, 120, 180 and 240minutes was allowed. The results are listed in table 4.

From table 4 it is seen that the metal solubility increases with an increases of $LaCl_3$ content in the molten electrolyte. It is also clear that an increase in the temperature will increase the metal solubility at a rate of 0.015mass%/10°C in the 650 ~ 750°C range when χ_{LaCl3} is 0.5 and the metal solubility at 700°C is 0.078mass%. The experiments show that the metal solubility in $LaCl_3$-$MgCl_2$ system is low. Thus, it will contribute to high current efficiency.

Table 4 measurement of metal solubility from the metal-loss in mass% /time in min

Time	30	60	90	120	180	240
χ_{LaCl3}0.4 (700°C)	0.022	0.043	0.051	0.063	0.064	0.066
χ_{LaCl3}0.6 (700°C)	0.042	0.052	0.084	0.092	0.094	0.095
χ_{LaCl3}0.8 (700°C)	0.102	0.122	0.133	0.143	0.145	0.144
χ_{LaCl3}1 (700°C)	0.123	0.132	0.143	0.160	0.164	0.162
χ_{LaCl3}0.5 (650°C)	0.029	0.040	0.055	0.069	0.070	0.068
χ_{LaCl3}0.5 (700°C)	0.037	0.048	0.062	0.077	0.079	0.076
χ_{LaCl3}0.5 (750°C)	0.045	0.056	0.070	0.085	0.086	0.088

Magnesia solubility

In the process of magnesium electrolysis when magnesia is used as the material, its solubility and the rate of dissolution are important. The size of magnesia used was in the range of 40~50μm. Table 5 lists the results of the solubility experiments. From the table it is apparent that the magnesia solubility in the $LaCl_3$-$MgCl_2$ system increase s with a increase of $LaCl_3$ content. It was found that the melts were saturated by magnesia in duration of 10min. The magnesia solubility in the $LaCl_3$-$MgCl_2$ system is about 10mass% when χ_{LaCl3} is 0.5 and the temperature at 700°C. Therefore, the $LaCl_3$-$MgCl_2$ system is competent as support electrolyte when magnesia is used as the material.

Table 5 Magnesia solubility and dissolution rate in molten $LaCl_3$-$MgCl_2$ mixture at different compositions

χ_{LaCl3}	T(°C)	5(min)	10(min)	60(min)
0.2	700	2.0	2.5	2.5
0.4	700	7.4	10.2	10.3
0.5	700	8.3	10.7	10.7
0.5	650	8.5	10.5	10.5
0.5	750	8.8	10.7	10.7
0.6	700	10.2	12.3	12.4
0.8	700	11.2	13.5	13.5
1.0	700	12.4	15.1	15.1

Electrolysis experiments in 5kA capacity cell

Experimental method

Electrolysis was carried out in a 5kA electrolysis cell, which is shown in Fig. 1. A graphite crucible was used, which served both as the container and as the anode, which density was 1.7g/cm^3. A steel bar was used as the cathode. The anode-cathode distance was 40mm. A corundum tube was placed around the cathode in order to protect magnesium liquid from being oxidized by anodic gas. The duration of electrolysis was 72h. The current efficiency (CE) was determined from the amount of magnesium metal produced by weighing after the experiments.

Fig.1 5kA magnesia electrolysis cell

Electrolysis experiment results from 5kA capacity cell

Based on the exploratory condition experiments in 1kA lab electrolysis cell and the physico-chemical properties of $LaCl_3$-$MgCl_2$ bath system, the technological conditions of magnesia electrolysis in 5kA electrolysis cell were as follows: the electrolysis temperature 973K, the anode-cathode distance 4cm, cathode working height 50cm, cathode diameter 51cm, cathodic density $0.81A\cdot cm^{-2}$, anodic density $0.539A\cdot cm^{-2}$, cell voltage 4.6v, duration of electrolysis 72h. In this pilot experiment, about 232kg of magnesia used and 138kg of magnesium was produced, the purity of magnesium produced was 99%, current efficiency (CE) was about 85% and the specific D.C. energy consumption was 11.87kWh/kg Mg.

In comparison with the existing magnesium chloride process, the magnesia electrolysis process has fewer steps, lower production cost, higher current efficiency, lower D.C. energy consumption, and produces CO_2 at the anodes. If the anode is made of inert materials, then O_2 will be produced at this anode when magnesia is used as electrolytic raw material. The technique belongs indeed to clean production process and is also the target that magnesium industry has sought for donkey's years. Comparison of electrolytic processes for aluminum and magnesium production is listed table 7.

Table 7 Comparison of electrolytic processes for aluminum and magnesium production

	Hall-Heroult Process	Dow Process	The New Process
Metal	Aluminum	Magnesium	Magnesium
Raw Material	Al_2O_3	$MgCl_2$	MgO
Electrolyte	Cryolite Fluorspar Al_2O_3	$MgCl_2$ NaCl $CaCl_2$ caF_2	$LaCl_3$ $MgCl_2$ MgO
Operating Temp. (°C)	950	700	700
Cathode	Carbon	Steel	Steel
Anode	Carbon	Carbon	Carbon
By-product	CO	Cl_2	CO
Current Efficiency	94	80	85
D.C. energy consumption (kWh/kg)	13.5	18	12

Electrolytic ionic species and electrode reactions

In the molten salt $LaCl_3$-$MgCl_2$ system, there may be some reactions when carbon anode takes part in [4~6]:

$MgO + LaCl_3 = MgCl_2 + LaOCl$ (1)
$MgCl_2 = Mg + Cl_2$ (2)
$LaCl_3 = La + 3/2Cl_2$ (3)
$MgO + C = Mg + CO$ (4)
$MgCl_2 + C + LaOCl = LaCl_3 + CO + Mg$ (5)

Therefore, theoretical decomposition voltages of MgO, $MgCl_2$, and $LaCl_3$ in 700°C calculated is 1.66v, 2.61v and 3.17v respectively. Because the solubility of MgO in $LaCl_3$-$MgCl_2$ system is 10mass%, MgO will decompose into Mg^{2+} and O^{2-} and be electrolyzed most easily.

In the $LaCl_3$-$MgCl_2$ system, MgO, $MgCl_2$, $LaCl_3$ and LaOCl are dissociated as follows:

$MgO = Mg^{2+} + O^{2-}$ (6)
$MgCl_2 = Mg^{2+} + 2Cl^-$ (7)
$LaCl_3 = La^{3+} + 3Cl^-$ (8)
$LaOCl + Cl^- = LaOCl^-_2$ (9)

When electrolysis starts, at the cathode, magnesium is produced by the reaction:

$Mg^{2+} + 2e = Mg$ (10)

At the graphite anode, the following phenomena occur:

$LaOCl^-_2\,(bath) = LaOCl^-_2\,(electrode)$ (11)

Dissociation of $LaOCl^-_2$ complexes, adsorption and discharge of O^{2-} ions, and formation of adsorbed CO_{ads}:

$LaOCl^-_2\,(electrode) + C = La^{3+} + 2Cl^- + CO_{ads} + 2e$ (12)
$O^{2-} + C = CO_{ads} + 2e$ (13)
$LaOCl^-_2\,(electrode) + CO_{ads} = La^{3+} + 2Cl^- + CO_2\uparrow + 2e$ (14)

Under the condition of carbon anode, the MgO electrolysis equation is as follows:

$2MgO + C = 2Mg + CO_2\uparrow$ (15)

Magnesium oxide is practically insoluble in any melt containing of chlorides and fluorides of alkali and alkali-earth metals at practical, cell-operating temperatures. However, it may be possible to dissolve magnesia in these melts if the oxide can be converted into a complex. The complex should not be so tightly bonded as to preclude the reduction of oxide-containing ionic species. This approach is similar to the dissolution of alumina in cryolite used in the production of aluminum in the Hall- Heroult process; alumina dissolves in these melts through complex formation [7]. It is also essential to dissolve magnesia in a melt in which magnesia can be electrolytically decomposed without decomposing melt components. In this research, the used electrolyte $LaCl_3$ was not consumed as cryolite in aluminum electrolysis, but magnesia solubility increased with an increase of $LaCl_3$ in the studied electrolyte, and reached 10% at 700°C. At the present time, we are actively improving the new process that is, preparing magnesia by ammonia process from bischofite of Qinghai Salt Lakes and magnesia electrolysis, at last establish the new technological system with bischofite as magnesium production material.

Conclusion

The physico-chemical properties of $LaCl_3$-$MgCl_2$ bath system, electrolyte ionic species and electrode reactions of the magnesia direct electrolysis in the $LaCl_3$-$MgCl_2$ system have been studied to find out the suitability of the system as a candidate electrolyte for the new technique on electrolysis of magnesia from bischofite of Qinghai salt lakes. It was found that there is broad low temperature range in the $LaCl_3$-$MgCl_2$ system, where the $LaCl_3$ mole fraction is 0.4~0.6, fusibility is lower than 650°C, electrical conductivity is greater than $1.4\Omega^{-1}\cdot cm^{-1}$, bath density is in $1.89\sim2.48g\cdot cm^{-3}$ range, magnesium solubility is lower than 0.20mass%, and magnesia solubility is greater than 10mass%. Magnesium was successfully produced in 5kA pilot electrolysis cell with the $LaCl_3$-$MgCl_2$ system as support electrolyte system at 700°C, the anode-cathode distance 4cm, cathodic density $0.81A\cdot cm^{-2}$, anodic density $0.539A\cdot cm^{-2}$, cell voltage 4.6v, the purity of magnesium produced was 99%, current efficiency (CE) was about 85% and the specific D.C. energy consumption was 11.87kWh/kg Mg. The new technique has the advantages of economy on energy and reduction of consumption, high efficiency and environmental amity.

Acknowledgement

The authors wish to thank the National Natural Science Foundation of China for the financial support to the project 20166003.

References

1 Fengwen Han. Light Metals(in Chinese), 249(7) (1999), 40-41.
2 Wankun Chen, Junwu Shi. Nonferrous Metals Progression – Light Metals (in Chinese). Central South University of Technology Press, Changsha, 1995, 165-232.
3 Riyao Xu. Magnesium Metallurgy (in Chinese), Second edition. Metallurgical Industry Press, Beijing, 1993, 10-22.
4 Ram A. Sharma. "A New Electrolytic Magnesium Production Process", JOM, 6 (1996), 39-43.
5 H. Mediaas, O. Tkatcheva, V. Dracopoulos, G. N. Papatheodorou, G J.Kipouros, and T. Østvold. Metallurgical and Materials Transactions B. 31B (2000), 631- 639.
6 K. J. Cathro, R. L. Deutscher, R. A. Sharma. Journal of Applied Electrochemistry. 27(1997), 404-413.
7 Zhuxian Qiu. Theory and Application of Aluminum Electrolysis (in Chinese). Chinese University of Mining Technology Press, Beijing, 1998, 192-225.

From Magnesium Technology 2009, E.A. Nyberg, S.R. Agnew, N.R. Neelameggham, and M.O. Pekguleryuz, Editors

CATHODE WETTING STUDIES IN MAGNESIUM ELECTROLYSIS

Kevin McLean[1] James Pettingill[2], and Boyd Davis[2,3]
[1]Dept. of Chemical Engineering, Queen's University, Kingston, ON K7L 3N6 CANADA
[2]Dept. of Mining Engineering, Queen's University, Kingston, ON K7L 3N6 CANADA
[3]Kingston Process Metallurgy Inc., Kingston, ON K7L 3N6 CANADA

Keywords: Magnesium, electrolysis, cathode, wetting

Abstract

The effects of cathode materials and electrolyte additives on magnesium wetting were studied with the goal of improving current efficiency in a magnesium electrolysis cell. The study consisted of static wetting and electrolysis tests, both conducted in a visual cell with a molten salt electrolyte of $MgCl_2$-$CaCl_2$-NaCl-KCl-CaF_2. The wetting conditions were tested using high resolution photography and contact angle software. The electrolysis tests were completed to qualitatively assess the effect of additives to the melt and were recorded with a digital video camcorder. Results from the static wetting tests showed a significant variation in wetting depending on the material used for the cathode. Mo and a Mo-W alloy, with contact angles of 60° and 52° respectively, demonstrated excellent wetting. The contact angle for steel was 132° and it ranged from 142°-154° for graphite depending on the type. Improvements to the cathode wetting were observed with tungsten and molybdenum oxide additives.

Introduction

Magnesium production has seen a drastic change in the past 20 years. Prior to the 1990s, electrolytic production in the western world, which boasted low energy and operating costs, was the main source of magnesium. Silicothermic magnesium, which has high labour and low capital costs, was only a small part of global magnesium production. However, labour intensive processes are not a barrier to countries such as China that have low labour costs, and therefore silicothermic magnesium has displaced electrolytic magnesium almost entirely over the past two decades. Chinese production of magnesium has increased significantly from 120700 tonnes in 1999 to 534000 tonnes in 2006 whereas electrolytic production in North America has almost dropped to zero with only one plant remaining in Utah (Canadian Minerals Yearbook, 2005; USGS Minerals Yearbook, 2006). This has caused an imbalance in the global magnesium market which has, in part, lead to the difficulty for magnesium to penetrate automotive markets. Electrolytic magnesium continues to be produced in many countries in a closed loop for titanium production using the Kroll process.

A resurgence of electrolytic magnesium can only be helped by improved efficiencies. The two main energy consuming areas that should be targeted are anhydrous magnesium chloride production and current efficiency. It is the latter issue that this work has investigated. Typical current efficiencies in magnesium electrolysis cells are about 85% with the bulk of the loss thought to be due to the recombination of magnesium and chlorine in the anode-cathode space (Wypartowicz, Østvold, & Øye, 1980). The relatively high recombination of metal and gas is a result of metal migration from the cathode into the electrolyte as a result of poor metal wetting on the surface of the cathode.

Improvements to the wetting of the magnesium on the cathode could have a substantial effect on the current efficiency of the process in three ways. First, wetted magnesium would be limited in its ability to enter into the bulk electrolyte and back react with chlorine gas. Second, a continuous metal pad of magnesium on the cathode would reduce the cathodic overpotential, which is estimated to be about 2.4 – 3.0 V (Kipouros & Sadoway, 1987). Third, the wetted magnesium will have a reduced likelihood of reacting with oxide and falling to the cell floor to form part of the sludge. For these reasons, this work was planned to investigate the impact of a number of cathode materials and electrolyte additives on magnesium wetting in molten salts.

Contact angles measure the wetting of a liquid on a solid surface. As seen below in Figure 1, low contact angles represent wetting, or spreading, with 0° being the perfect wetting scenario where the formation of a liquid film occurs on the solid surface. In contrast, high contact angles represent non-wetting, with 180° being the perfect non-wetting scenario, where the formation of a sphere forms on the solid surface.

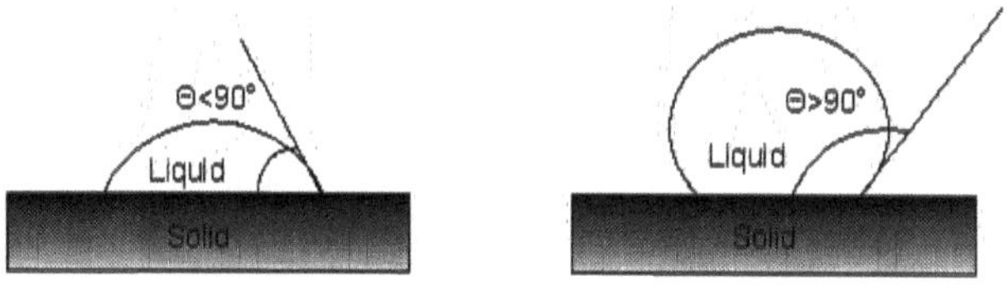

Figure 1: Contact angles representing wetting and non-wetting conditions

One method of measuring contact angles is the sessile drop technique (Gu, 2002). In this method, the contact angle is measured directly from the angle between the solid surface and the tangent to the drop. As determination of the tangent is subjective, high accuracy is difficult to achieve. Equipment including a digital camera, a lighting source, and contact angle software can be utilized to measure contact angles with this technique.

Previous reports and patents have examined various cathodic materials (steel, graphite, tungsten and molybdenum), additives to the electrolyte, and cathodes with coatings. A summary of the previous studies in Table 1 shows no wetting with graphite, poor wetting with steel, and good wetting with molybdenum and tungsten.

Studies by Anundskas et al. (1977) and Komura et al. (1973) concluded that magnesium does not wet graphite. The contact angles obtained in melts of $MgCl_2$ and $MgCl_2$-NaCl-$CaCl_2$-KCl were 133° and 146° respectively. It was proposed that the reason for the low wettability of graphite was the low surface energy of the soft graphite solid.

Anundskas et. al (1977) also measured the contact angle of magnesium on steel St 37 (max 0.17 wt.% C). With pure $MgCl_2$ and $MgCl_2$-NaCl (x_{MgCl2}>0.15), the contact angles were found to be 0°, representing perfect wetting. However, with a system of $MgCl_2$-NaCl-$CaCl_2$-KCl, which is close to commercial electrolytes, the contact angle was 144°C, indicating that the magnesium did not wet the steel. The exact composition of this melt is not known. The effect of the electrolyte composition clearly has a major effect on wetting.

Table 1: Summary of previous work and wetting conditions on various materials

	Previous work	Wetting conditions/Results
Graphite	Anundskas et. al (1977) Komura et. al (1973)	No wetting observed in any case Contact angle of 133° in $MgCl_2$ Contact angle of 146° in $MgCl_2$-$NaCl$-$CaCl_2$-KCl
Steel	Shi et. al (2001) Wallevik et al (1988) Anundskas et. al (1977)	Overall poor wetting, but better than graphite Contact angle of 0° in $MgCl_2$ and $MgCl_2$-$NaCl$ (x_{MgCl2}>0.15) Contact angle of 144° in $MgCl_2$-$NaCl$-$CaCl_2$-KCl
Mo/W	Wallevik et. al (1988) Beale (1978)	Wetting is improved over steel (qualitative) Films of magnesium were obtained on cathode surfaces

Similarly, Shi et. al (2001) examined the contact angles of magnesium on iron and stainless steel surfaces under an atmosphere of argon. With iron, the magnesium was not found to wet the surface, with an initial contact angle between 105-115° for temperatures of 700-725°C. The study also examined two types of stainless steels: SUS430 and SUS316. The initial contact angle of magnesium on the SUS430 was found to be 115° at 700°C. With SUS316, the magnesium was found to spread on the surface, with an initial contact angle of 70°.

Beale (1978) examined the effect of additions to the electrolyte on the cell efficiency. Molybdenum and tungsten compounds were found to improve the current efficiency; for example, MoO_3 additions to the melt resulted in increases of 5 and 8 % over the control tests.

Wallevik et. al (1988) examined the wetting of various cathodes following electrolysis. For pure molybdenum and tungsten cathodes as well as the steel and graphite cathodes that were coated with molybdenum, films of magnesium were obtained on the cathode surfaces, indicating excellent wetting. Figure 2 shows wetting with the molybdenum coated cathode (film formed on the surface) compared to poor wetting with the steel cathode (beads 3-6 mm in diameter were formed on the surface).

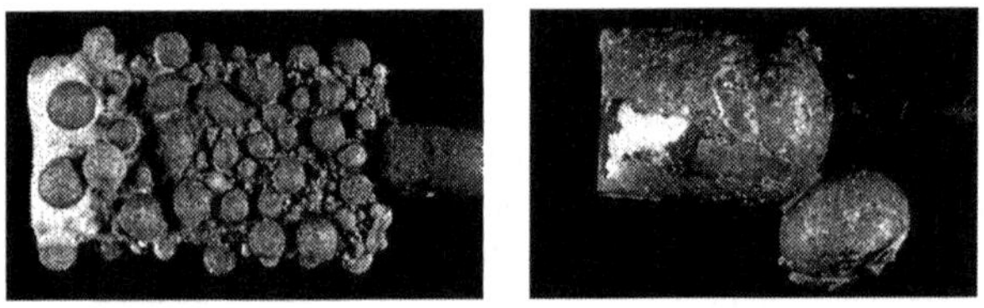

a) b)

Figure 2: Wetting Conditions for a) Steel Cathode b) coating of 0.35 mm molybdenum on steel (Wallevik et. al, 1988)

While studies and patents exist that have researched the effect of cathodic materials and additives to the electrolytic production of magnesium, there does not exist a comprehensive, systematic study. Previous studies have reported cell efficiencies or wetting conditions of cathodes following electrolysis. A comprehensive study that uses an electrolyte simulating a commercial cell, while involving wetting analyses and cathode/metal interactions inside the molten salt is lacking. Only recent advances with digital equipment and software could make such a study possible. Thus, the objective of the present study was: to analyze and compare the wetting of different cathode materials in-situ via static wetting tests; and perform electrolysis tests to view the production of magnesium and qualitatively assess various electrolyte additives.

Oxides and chlorides of W and Mo were chosen for use as additives to the electrolyte as it is clear that these compounds will preferentially electroplate. The decomposition voltages of WCl_6 and $MoCl_6$ in the molten salt are 0.404 and 0.317 V respectively – significantly lower than the 2.53 V for $MgCl_2$. This means that even small concentrations of W or Mo will electroplate on the cathode and produce a thin metal coating. In a review article by Predel (1997a), it was reported that Mg does not react with Mo and the phase diagram at the temperature of 675°C indicates the existence of liquid magnesium and solid molybdenum in separate phases. Predel (1997b) reviewed the Mg-W system and stated that the phase diagram for Mg and W is unknown, however, Mg does not react with W.

Experimental

A custom-made, 2.5 kW electric furnace was used for both the static wetting and the electrolysis tests. A Eurotherm 2416 temperature controller was used to set the furnace temperature. Two 3" diameter holes were present in the front and back of the furnace. The front hole was used as a viewing port and the frosted quartz window in the back allowed light to pass and provide a luminescent background. Two quartz windows in the viewing port prevented heat loss. A flat bottom, fire polished clear fused quartz crucible with an 85 mm inner diameter, 90 mm outer diameter and 355 mm in length was placed inside the furnace to hold the electrolyte mixture. The aluminum lid for the crucible, consisting of a top and bottom clamped together and sealed with an o-ring, had ports for gas in, gas out, thermowell, dispensing tube (static wetting tests) and cathode and anode leads (electrolysis tests). Water was supplied to the lid during experiments for cooling. Fused quartz tubes (4 mm inner diameter, 6.35 mm outer diameter) were used as thermowells and gas lances. A fume collection tube leading to a caustic scrubber was used to remove the off-gasses.

The electrolyte mixture consisted of the following salts: NaCl (ACS grade, Fisher Scientific), $MgCl_2$ (anhydrous, 99%, Alfa Aesar), $CaCl_2$ (anhydrous, 4-20 mesh, Fisher Scientific), KCl (ACS grade, 99-100.5 %, Alfa Aesar) and CaF_2 (99.7% metals basis, Alfa Aesar). The electrolyte composition, shown below in Table 2, was selected to replicate typical industrial electrolysis settings. For the static wetting tests, a charge of 687.5 g was used and for electrolysis tests a charge of 1250 g was added to the crucible.

Table 2: Electrolyte Composition

Salt	Wt. %	Weight (g)	Weight (g)
$MgCl_2$	20.0	125	250
$CaCl_2$	28.5	178	356
NaCl	48.0	300	600
KCl	2.0	12.5	25
CaF_2	1.5	9.4	19
Total	100	625	1250

For the static wetting tests, the following test plates were used to compare wetting: alumina, 1020 HR steel, 99.95 % pure molybdenum, a 70% molybdenum/30% tungsten alloy, an M2 steel alloy and four types of graphite. The typical composition of the M2 steel alloy is shown below in Table 3 and a summary of the types and properties of the graphite tested is shown in Table 4. The test plates were 2.5" in diameter and ranged from 0.1" to 0.25" in thickness. A 9/16" hole was made for the dispensing tube to pass through and 5/16" holes were made at the edges of the discs so that it could rest on a crib. The crib, made from bisque alumina, consisted of a 3" diameter base and 3 1/2" diameter pegs. A diagram of the crib and test plate is shown below in Figure 3. The alumina dispensing tube had a 0.5" outer diameter, a 0.375" inner diameter and was 18" long. Magnesium pellets (0.25" by 0.25", 4N purity, ESPI) were used to test the wetting conditions on the various test plates. A Nikon D70 digital camera was used to take pictures during the static wetting tests and the software Fta32 Video 2.0 by First Ten Angstroms was used to analyze the pictures and calculate contact angles.

Table 3: Typical Composition of M2 Steel Alloy
(as provided by Cincinnati Tool Steel Co.)

Carbon	0.830
Molybdenum	5.000
Tungsten	6.350
Chromium	4.150
Vanadium	1.900

Table 4: Properties of Graphite Tested
(as provided by SGL Carbon Inc)

	Benchmark Graphite		R6510	R6810	Grade HLM extruded	
	WG	AG			WG	AG
Ash value (ppm)	290		200	200		
Average grain size (μm)	220		10	20	760 (max grain size)	
Bulk density (g/cm^3)	1.69		1.83	1.82	1.73	
Compressive strength (MPa)	NA		125	100	40.7	38.6
Flexural strength (MPa)	28	14.5	60	45	19.3	12.4
Medium pore size (μm)	NA		1.5	2.5	NA	
Open porosity (Vol. %)	25		10	11	23	
Rockwell hard. $HR_{10/100}$	24		95	95	78	
Spec. electrical resist.(μΩm)	6	9	13	10.5	6.6	10.2
Thermal cond. ($Wm^{-1}K^{-1}$)	NA		100	130	195	128
Thermal expan. (20-200°C) (10^{-6}°C^{-1})	2.3		4.0	4.1	NA	
Young's modulus (GPa)	11.1		11.5	10	NA	

(WG = with grain, AG = against grain)

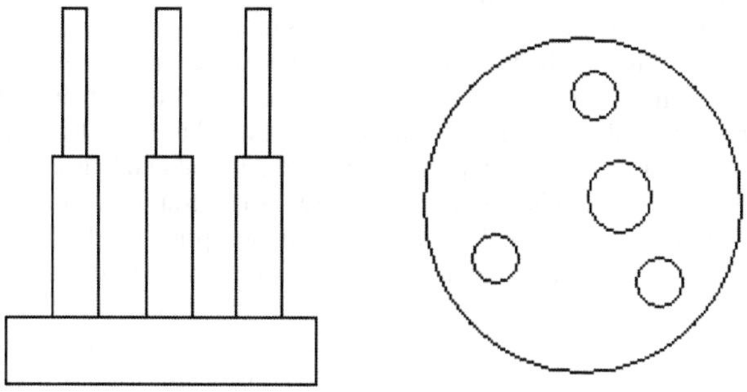

Figure 3: Bisque alumina crib (side-view)
and test plate (top view)

For the electrolysis tests, 1" by 1" and 3/8" thick low carbon steel (LCS) cathodes and graphite anodes were used. A threaded steel rod was used as the cathode lead and a 1/4" graphite rod was used as the lead to the anode. The additives to the melt consisted of MoO_3 (ACS grade, 99.5% min, Alfa Aesar) and WO_3 (milled, 99.8% metals basis, Alfa Aesar). A Panasonic digital video camcorder, model PV-GS15, was used to record the electrolysis tests.

Static Wetting Tests

The bisque alumina crib was constructed using a drill press to make holes in the base and a lathe was used to make the three pegs. Holes were made in the test plate with the drill press. For each experiment, the salts were weighed out, mixed and added to the crucible with the crib on the bottom. The crucible was then placed in the furnace and the temperature was set to heat up to 675°C. Once the salt mixture was completely molten, the melt was chlorinated with HCl gas flowing at 500 mL/min to remove the oxides. With this chlorination, oxide levels were maintained at approximately 200 ppm during the experimental trials.

During the chlorination process, the crucible lid was assembled with ports for the thermowell, gas in, gas out and dispensing tube. After approximately 90 min of chlorination, the test plate was lowered onto the crib with tongs and the crucible lid was placed on top of the crucible. The cooling water was connected to the lid and the thermowell and gas lance were placed in the appropriate ports. The dispensing tube was lowered to rest just below the test plate. Each magnesium pellet was cleaned using HCl and dropped into the melt via the dispensing tube. For each pellet that remained in view of the furnace's window, pictures were taken. These pictures were later analyzed using the Fta32 Video 2.0 software. Once the contact angle remained constant, the magnesium pellet was removed from the test plate by tapping the dispensing tube and the next magnesium pellet was dropped into the melt. This process usually continued until the melt became too cloudy and the pellets were no longer visible.

Electrolysis Tests

The preparation for the electrolysis tests involved making the anode, the graphite lead, and rounding the edges of the cathode. The back edges of a graphite block were sanded and a hole was drilled in the top and threaded to make the anode. The graphite lead was sanded and threaded. The back edges of the LCS cathode were sanded using a disc sander.

The methods described for the static wetting tests of weighing out the salts, heating the furnace and chlorinating the melt were the same for the electrolysis tests. For these trials, the lid was assembled with ports for the cathode and anode instead of the dispensing tube. The power source was set-up to the cathode and anode and a small voltage was applied prior to placing the cathode in the melt so that the LCS cathode would not corrode. Once the electrodes were inserted into the melt, a current of 5.4 A was applied and the electrolytic production of magnesium began to occur. The electrolysis tests were recorded with a digital camcorder. Current and voltage readings were taken at 5 min intervals using a multimeter.

Results and Discussion

Static Wetting Tests

Different types of materials were tested for the static wetting phase of the project. Alumina was used as a baseline for poor wetting. Pictures were taken showing the magnesium drops in contact with the various test plates. The contact angles in these pictures were then analyzed using the Fta32 Video 2.0 software. It was observed that the different materials showed different wetting conditions. As seen below in Table 5 and in the pictures shown in Figure 5, graphite, alumina and steel demonstrated poor wetting whereas the molybdenum and Mo/W alloy demonstrated good wetting, which is consistent with the previous studies discussed above.

Table 5: Static Wetting Test Materials and Results

Test Material	Results
Benchmark Graphite	Limited Wetting Contact angle approximately 152°
HLM Graphite	Limited Wetting Contact angle approximately 149°
R6510 Graphite	Limited Wetting, but slight improvement over other graphites Contact angle approximately 142°
R6810 Graphite	Limited Wetting Contact angle approximately 153°
Alumina	Limited wetting Contact angles ranged from 135-145°
1020 Steel	Poor wetting Contact angle approximately 133°
M2 Steel Alloy	Poor wetting – no improve. over steel Contact angle approximately 137°
Molybdenum	Excellent wetting Contact angle decreased below 60° 60 seconds after the drop made contact with test plate
Molybdenum/tungsten alloy	Excellent wetting Contact angle approximately 52°

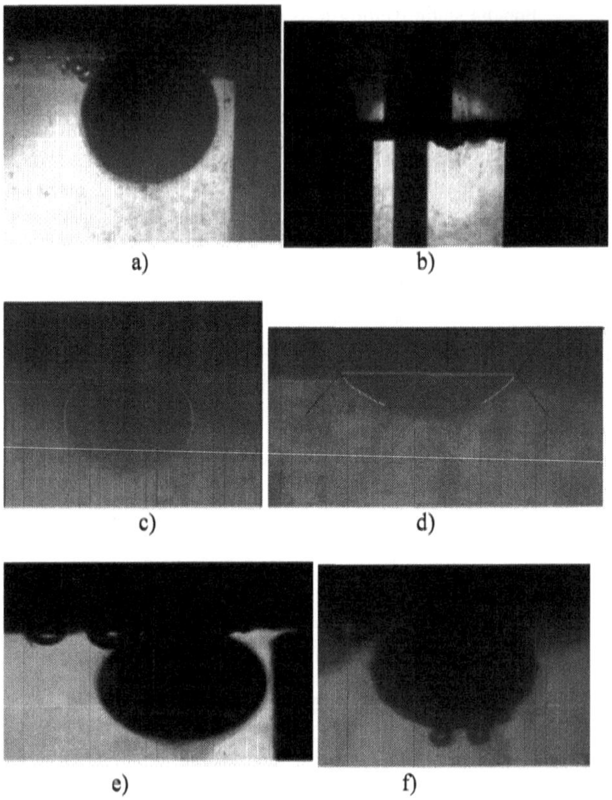

Figure 4: Static wetting tests with a) alumina, b) molybdenum, c) 1020 steel, d) Mo/W alloy, e) Graphite and f) M2 Steel Alloy

Four types of graphite were tested to determine the effect of pitch impregnation. Except for the benchmark graphite, the other types were pitch impregnated, which improves the strength and chemical resistivity, increases the density and decreases the porosity of the graphite. The range of contact angles obtained for graphite demonstrated limited wetting although the R6510 grade showed slight improvement over the other three types. This improvement may be a result of the pitch impregnation and the finer grain size of the type. With alumina, the contact angles obtained for several drops ranged from 135-145°. This range represents good reproducibility as the various magnesium drops may have varied slightly in shape, the drops may have landed on different locations on the test plate, and the oxide level in the melt may have increased slightly over time. With the 1020 steel, corrosion product in the electrolyte made the melt cloudy and as a result it was sometimes difficult to obtain accurate contact angles. The M2 steel alloy did not show improvement over the 1020 steel. This may be explained by the higher carbon content in the alloy or perhaps the molybdenum/tungsten content was not significant enough to positively affect the wetting. The pictures shown in Figure 5 demonstrate that the magnesium drops remain mostly spherical when in contact with graphite, alumina and steel.

With molybdenum and the Mo/W alloy, the wetting was significantly improved. As shown in Figure 5, it was observed that the magnesium drops did not remain spherical, but rather flattened in shape when in contact with these two materials. Upon contact with the molybdenum plate, the magnesium drop flattened and a violet haze was observed streaming from the drop. This

From *Magnesium Technology 2009*, E.A. Nyberg, S.R. Agnew, N.R. Neelameggham, and M.O. Pekguleryuz, Editors

indicates the possibility of a chemical reaction resulting from the contamination by oxide films, such as:

$$Mo + 2\,MgO \rightarrow MoO_2 + 2\,Mg \quad (1)$$

$$2\,Mo + 5\,MgO \rightarrow Mo_2O_5 + 5\,Mg \quad (2)$$

The contact angle of the drop decreased to below 60° approximately 60 seconds after it made contact with the molybdenum test plate. With the Mo/W alloy, the wetting was again significantly improved over the alumina and steel. The contact angle of the drop seen in Figure 4 was approximately 52°.

Electrolysis Tests

For the electrolysis tests, LCS cathodes were used and the effects of additives to the melt were examined. These additives included tungsten and molybdenum oxide. One experiment was also conducted using a rotatable anode to allow for the interaction of Cl_2 bubbles with Mg to be observed.

These electrolysis tests demonstrated the fine nature of the Mg particles produced and their transport into the electrolyte. As summarized in Figure 4, a fine mist of Mg from the cathode was present during electrolysis. The mist was surprising given that the generally perceived mechanism for magnesium production on the cathode was growth of small droplets, followed by some coalescence and detachment. The mist is visible at the top of the cathodes in pictures a) and b) of Figure 5 and reinforces the likelihood of significant back reaction of Mg and Cl_2 in the relatively narrow anode-cathode distance.

The rotatable anode in part c) of Figure 5 was tested to determine if the chlorine bubbles rising off the anode face would clean the cathode face if they were in close proximity to the cathode. This was indeed found to be the case, even with the relatively slow gas velocity. Once the anode was rotated close to the cathode, the chlorine bubbles came in contact with the cathode face and no accumulation of Mg on the face was observed - although a film may have continued to exist. This finding could have an impact on magnesium growth on wetted cathodes. This method also allowed the surface of the cathode to be refreshed during a run without physical intervention.

With the MoO_3 addition to the melt, the melt turned a dark purple and the electrolysis was difficult to view. The mist appeared only at the start of the electrolysis and not to the same extent as the LCS cathode without additives. This indicates that the magnesium was wetting the cathode.

As seen in the pictures of the cathodes following the electrolysis tests in Figure 6, the agglomeration of Mg on the cathode face was significantly improved with the WO_3 addition to the melt. This suggests that the W will electroplate onto the cathode, creating a thin metal coating that improves the wetting of magnesium.

Table 6: Electrolysis Test Conditions and Results

Cathode	Additive to melt	Results
LCS cathode		Fine dissemination of Mg from cathode ("mist")
LCS cathode	WO_3	Dendrite formation of Mg on cathode Mist like dissemination of Mg
LCS cathode	MoO_3	Some mist observed, but not to same extent as LCS cathode with no additives

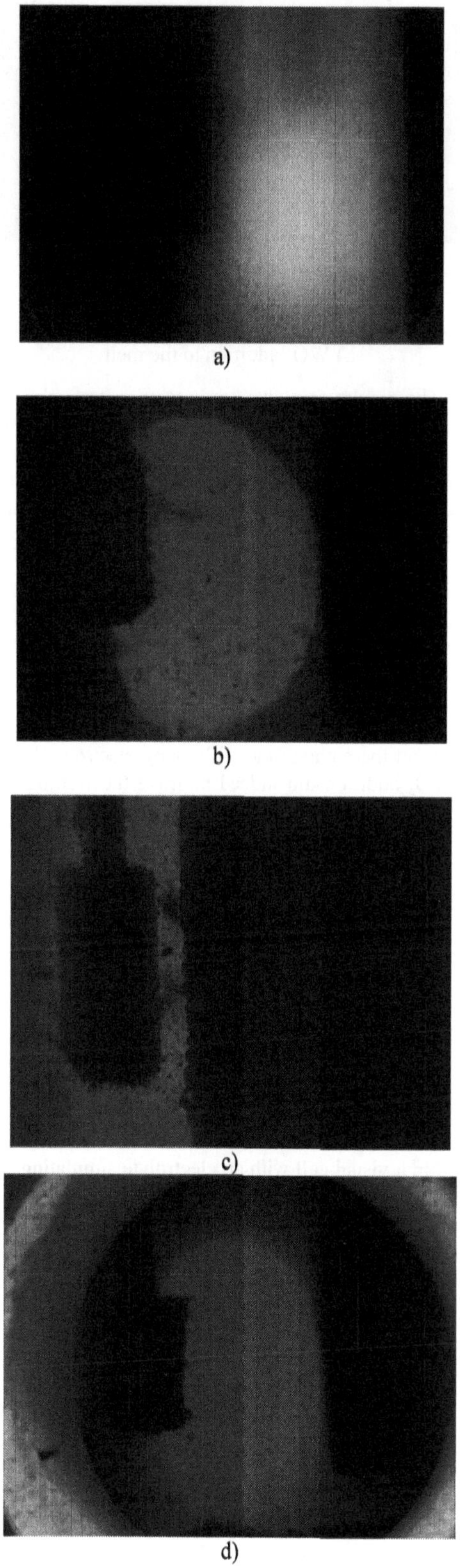

Figure 5: Electrolysis tests with a) LCS cathode, b) WO_3 addition to the melt, c) rotatable anode and d) MoO_3 addition to the melt

a) b)

Figure 6: Cathodes following electrolysis a) LCS cathode, b) WO_3 addition to the melt

The results from the static wetting tests show that the wetting of magnesium on the cathode surface can be significantly improved by using molybdenum and tungsten. As discussed above, improving the wetting would significantly increase the current efficiency in industrial cells and would contribute to stimulating resurgence in electrolytic magnesium production. However, due to excessive costs, switching from steel to molybdenum or tungsten cathodes in industrial cells would not be practical. Thus, tungsten and molybdenum oxide additions were made during electrolysis tests to determine if these metals would electroplate onto the cathode and form coatings that would equally improve the wetting. A steel cathode with a tungsten coating showed significant improvement in the agglomeration of magnesium on the cathode face following electrolysis, indicating that the WO_3 addition did indeed improve the wetting. The Mg mist from the cathode, an indicator of significant back reaction of Mg and Cl_2, was virtually eliminated with the MoO_3 addition. Therefore, the results from this study suggest that chemical additions to the melt (or metal coatings) would successfully improve the wetting on the cathode surface, thus reducing the extent of the back reaction and improving the current efficiency of electrolytic cells.

Conclusions

Digital equipment, including a high resolution camera and a digital video camcorder, and contact angle software were used in a comprehensive study to analyze cathode wetting conditions in magnesium electrolysis. The experiments were conducted in a visual cell with an electrolyte simulating industrial conditions. The wettability was analyzed in-situ allowing for the consideration of the cathode/metal interactions inside the molten salt. The static wetting tests performed in the study provided a semi-quantitative assessment for the value of cathode material modification. The wetting of magnesium on graphite was limited, with a range of 142-154° obtained for the contact angles depending on the type of graphite. With steel, poor wetting was observed with a contact angle of approximately 132°. The wetting on steel was not improved with the M2 steel alloy containing small percentages of molybdenum and tungsten. The wetting of magnesium was clearly improved with molybdenum and a Mo/W alloy, which produced contact angles of 60° and 52° respectively. The visual electrolysis tests have shown, for the first time, the fine nature of Mg produced at the cathode and its transport into the electrolyte. The mist present during electrolysis reinforces the likelihood of significant back reaction of Mg and Cl_2. Improvement to the cathode wetting was also observed with the tungsten and molybdenum oxide additives.

Future directions in this field should include wetting studies with steel alloys containing higher percentages of molybdenum and tungsten and steel cathodes coated with these elements. Other electrolyte additives could also be studied.

Acknowledgements

The authors thank the Natural Science and Engineering Research Council of Canada (NSERC) for financial assistance with this project, Dr. David Creber (RioTintoAlcan) for his financial and technical support, Dr. Neil Johnson from SGL Carbon for anode and cathode samples and technical support on graphite issues, and Dr. Georges Houlachi (Canadian Council on Electrochemical Technologies) for stimulating the project and encouraging industry-academic interactions.

References

A. Anundskas, K. Grjotheim, A.K. Schultz, H. Svendsen, & H.A. Oye, "Surface Properties Related to Magnesium Electrolysis" Proceedings of the AIME, TMS Light Metals Committee, 1977, pp. 513-25.

A.F. Beale, "Method for Magnesium Production Using Tungsten or Molybdenum." 1978, U.S. Patent 4,073,704.

Canadian Minerals Yearbook - Mineral and Metal Commodity Reviews, 2005. Retrieved September 24, 2007 from http://www.nrcan.gc.ca/mms/cmy/com_e.html

Fta32 Video 2.0, 2004. [Computer Software]. Portsmouth, Virginia: First Ten Angstroms.

Y. Gu, "Contact Angle Measurement Techniques for Determination of Wettability." *Encyclopedia of Surface and Colloid Science*. 2002.

G. J. Kipouros and D. R. Sadoway, "The Chemistry and Electrochemistry of Magnesium Production." *Advances in Molten Salt Chemistry 6*, 1987, pp. 187-209.

A. Komura, H. Imanaga, N. Watannabe, "Preparation of Magnesium by Molten Salt Electrolysis IX Dispersion of Magnesium into (Magnesium-Potassium) Chloride Melts." Vol. 41, No. 4, 1973, pp. 281-6.

B. Predel, "Mg-Mo (Magnesium-Molybdenum)." *Landolt-Börnstein - Group IV Physical Chemistry Vol. 5H*, 1997a.

B. Predel, "Mg-W (Magnesium-Tungsten)." *Landolt-Börnstein - Group IV Physical Chemistry Vol. 5H*, 1997b.

W. Shi, M. Kobashi, T. Choh, "Wetting of Alumina, Iron and Stainless Steel Substrate by Molten Magnesium." Z. Metallkd., V. 92, No. 4, 2001, pp. 382-85.

USGS Minerals Yearbook – Metals and Minerals, 2006. Retrieved January 10, 2008 from http://minerals.usgs.gov/minerals/pubs/commodity/myb/#M

O. Wallevik, O.J. Kaasa, G. Endal, "Cathode for Magnesium Production." 1988, U.S. Patent 4,784,742.

J. Wypartowicz, T. Østvold, H.A. Øye, "The solubility of Magnesium Metal and the Recombination Reaction in the Industrial Magnesium Electrolysis." *Electrochimica Acta, Vol. 25*, 1980, pp. 151-156.

THE PIDGEON PROCESS IN CHINA AND ITS FUTURE

Jing Chun Zang
Gold River Magnesium Plant
Ningxia Huayuan Magnesium Group
No. 50 Wenhuadong Street, Yinchuan, China 750004

Weinan Ding
Sinomag
1204 Floor 1 Landmark Tower, 8 Dongsanhuan Bei Lu,Beijing, China 100004

Abstract

Magnesium production in China has been growing steadily over the past 10 years. Most of the metal has been produced by the Pidgeon process. This process uses horizontal steel tubes called retorts, in furnaces and under vacuum. In the retorts mixtures of finely ground calcined dolomite and ferrosilicon formed into briquettes react to form magnesium vapors which are condensed and later remelted into ingots. The Pidgeon process was long thought to be uneconomic and obsolete. The Chinese have used the advantages of excellent raw material, location, large skilled labor supply, and low capital costs to produce magnesium by this process. The Chinese magnesium is being sold at the lowest prices in the world and lower than aluminum on a pound for pound basis.

Introduction

The first magnesium meal was produced by Fushun Aluminium Plant by the electrolytic process in 1958. The first metal produced by the Pidgeon process was in Nanjing in 1978, but the costs of production were much higher than the electrolytic process. In 1988, Shenyang Al and Mg Engineering & Research Institute completed the design and installation of the first Magnesium plant using the Pidgeon Process. It was located in Tongshan, Hubei Province with a design capacity of 500 metric tons per year.

After this plant was started, two Pidgeon plants were set up in Ningxia with a total capacity of 2000 metric tons per year. Additional Pidgeon plants were built in Shanxi and Henan provinces. The number of magnesium production plants using the Pidgeon Process was over 200 by 1997. The plant sizes varied from 100 metric tons per year to 3000 metric tons per year. The total Chinese capacity was more than 200,000 metric tons per year in 1999, while the actual production reached 160,000 metric tons. There were 3 special features that assisted in the development of the magnesium industry,: 1) Good raw material and large labor supply (no high technology, small investment); 2) Private and family businesses (not the normal state-owned plant); 3) World market oriented.

Magnesium produced in China has had a big impact on the world market. It is available in large quantities for an extremely low price compared with other magnesium. The quantity and price of Chinese magnesium has helped to increase the interest of many industries, particularly automotive. Rapid reduction in selling prices also caused many of the smaller Chinese magnesium plants to close. These closures were also due to the anti-dumping duties and the decrease in the prices in Europe and North America.

Chinese producers are consolidating their process knowledge and working with research and engineering institutions to further improve the process, improve productivity and quality with better service, to reduce costs, and to increase the production capacity.

The development of the Chinese magnesium industry has changed the structure of world supply and demand, facilitated and encouraged the technical development and new applications in many new fabrication and industry areas.

Present Situation

The Pidgeon Process is used in China to produce over 95% of all primary magnesium. About 40% of the world production is by the Pidgeon Process. Other countries that use the Pidgeon process are Canada, India, and North Korea. Today, there are about 130 plants running with Pidgeon process in China. Average capacity is over 1000 metric tons per year. There are 10 plants with over 4000 metric tons per year capacity and four plants with a capacity of over 10,000 metric tons per year. In the past several years, many of the small magnesium producers have merged together to become larger enterprises.

There are about 10 joint ventures for producing magnesium. The largest foreign investment in these ventures is from Japan and then North America. Participation from Europe has been small. Norsk Hydro is talking about alloy production in Xi'an, Ningxia with a capacity of 5000 mt per year.

In 2000, owing to the weak magnesium market price in the world and new anti-dumping charges in Europe, many small magnesium production in China are expected to close. However, Ningxia Huayuan Magnesium Group plans to increase its magnesium output capacity by 12,000 mt in 2001. Wenxi Yinguang will also add 10,000 mt within 2001, Taiyuan Tongxiang Magnesium will add 7,000 mt at end of 2000. The total magnesium production capacity is expected to be about 260,000 mt in 2001 and 300,000 mt in 2002, including alloys, anodes and powder/granules.

The Pidgeon Process has developed rapidly in China for several reasons:

1. There is a rapidly increasing demand for magnesium on the world market
2. There are convenient circumstances in China, flexible structure, global oriented production and marketing.
3. Excellent raw material. For example, Ningxia has 5 to 100 million tons of dolomite resources; 200,000 tons of locally produced ferrosilicon; 30 to 200 billion tons of high quality Anthracite coal; 2.2 million KW at US$0.03/Kwh; suitable dry climate, annual precipitation of 200-700 mm.
4. Qualified and experienced operating workers; reasonable labor costs; more mechanical/automatic equipment.
5. Plants require small initial investment and have a very short payback period.

Process Improvement History

In the past ten years of operating the Pidgeon process in China, the theory remains the same. However, some technical advantages have developed that have improved quality and reduced the cost of operation.

Technical Improvements in Pidgeon Process

A summary of the improvements that have been made and are being made include the following:

1. Heating of reduction furnace is improved with anthracite or coal powder instead of coal gas. Cost reduction of more than 10 million RMB (about $1.2 million) for one plant with capacity of about 6000 mt per year. It means that the cost of one metric ton of pure Mg metal shall be lowered by $450.
2. Vacuum for the process to be supplied by steam jet ejectors instead of mechanical pumps. The new process is more effective and faster (30 min instead of 2 hours to lower pressure to 10 Pa or under). It requires lower maintenance and no vacuum oil. Steam comes from surplus energy of reduction furnaces in a waste heat recovery boiler. More than US$150,000 shall be reduced in the installation of a 6,000 mtpy plant and 932 KW of electric capacity eliminated. The production costs per ton of pure magnesium shall be reduced by US$80 (for electricity and vacuum oil).
3. Coating the surface of the reduction tubes with special paint would extend the lifetime of the tube from 4 months to 8 months. That will cut costs by US$85-150 per ton of pure magnesium.
4. Weighing and making up the charges will be computer rather than manual, which will increase precision and uniformity of the charges, ending waste, especially of FeSi.
5. Casting ingots in a continuous conveyor by pumping reduces oxidation and burning loss, improved internal cleanliness and surface appearance.
6. Double quality inspections: Chemical and Spectrometer, raw materials & semi / end products.

Prospect of Pidgeon Process In China

Overview

As can be seen from the prior discussions, the development of the Chinese magnesium industry is very young. The production of magnesium metal by the Pidgeon Process is expected to have some impact on the world market.

The total Chinese magnesium production capacity will increase to 300,000 metric tons per year in the next 2-3 years. China's magnesium producers need to improve quality control and service. They need to work to eliminate any anti-dumping charges from other countries, and to help develop the domestic automobile market. Also, if the demand for magnesium in the European and North American markets increases, it will provide a bright opportunity for the Chinese magnesium sector.

Investors from Western Europe, North America, Japan and Taiwan are welcome to work together to develop the Mg sector using the Pidgeon process. The development would be under the fair participation and technology exchange. With the abundance of raw materials in China, the foundation for the magnesium industry is very solid.

Summary

In the last part of 2000, the magnesium price for Chinese magnesium dropped to US$1400 / mt, FOB Chinese port, because of the weakness of the Euro and purchasing panic on the announcement of increased anti-dumping duties in EC. Because there are a large number of small Chinese magnesium producers, independent and speculative traders, a defense against the challenger is very difficult. It can be seen that Chinese magnesium is too dependent on the international free market.

Fig. 1 List of Pidgeon Process Plants with Capacities of over 3000 metric tons per year.

Province (Total plants)	Plant	Capacity (metric tons/year)	Expansion projects for 2000-2001
Henan (45 plants)	Huaqi Mg Plant	8,000	2,000 mt
	Hebi Longxiang Mg Plnt	3,000	Alloy production
	Henan Wuhua Mg Co	3,000	No
	Fuda Mg Plant	3,000	No
	Yufeng Mg Plant	3,000	No
Shanxi (72 plants)	Taiyuan Tongxiang Mg	12,000	7,000 mt
	Wenxi Yinguang Mg	11,000	5,000 mt/2000 5,000 mt/2001
	Taiyuan Tongxiang Mg	15,000	No
	Guangling Jinghua Mg	8,000	No
	Tongbao	6,000	2,000 mt/ 2000
	Wenxi Hongfu Mg	3,000	2,000 mt/ 2001
	Pingding Maikelin Mg	3,500	No
	Taiyuan Longhe Mg	3,000	No
Ningxia (13 plants)	Gold River Mg Plant	9,000	12,000 mt/2001
	Huayuan Zhongwei Changle Mg	3,000	3,000 mt/ 2000
	Huinong Mg Plant	4,500	No
	Huayuan Shizuizhan Mg	8,000	No

Fig. 2 Technical and Economic Comparison between Chinese and Japanese Pidgeon Process.

Item	Unit	Chinese Process	Chinese Process	Japanese Process	Japanese Process
		1995	2000	Koga	Ube
Dolomite	Mt per Mt of Mg	14.0	11.0	12.56	11.52
Ferrosilicon (75)	Mt per Mt of Mg	1.35	1.10	1.20	1.16
Electricity	Kwh/ Mt of Mg	2000	888	1440	2560
Coal	Mt		10.5		
Heavy Oil	Mt				4.31
Coal Gas	Cu M/Mt of Mg	60,000			
Gas	Cu M/Mt of Mg			5090	
Burden/Mg ratio		7-8	6.20	6.25	6.30
Crown Recovery	Per Cent	75	85.5	81.3	83.5
Refinery Recov.	Per Cent	93	96	94.5	93

Table 1. Magnesium Process Comparisons
World Electrolytic vs. Pidgeon Process in Ningxia

Items	Electrolytic	Pidgeon in Ningxia
Investment per Metric Ton of Installed Capacity US$/Mt	Dead Sea Magnesium $15,300/mt	Huayuan Shizuishan $3,370/mt
	Noranda Magnesium $ 8,200/mt	Gold River Mag $1.020/mt
	Australian Average $ 6,100/mt	
Production Costs US$/metric ton	Australian Mag Corp $2,070 /mt	Huayuan Shizuishan $1,446/mt
	Dead Sea Magnesium $1,784	Gold River Mag $1,265/mt
	Norsk Hydro Norway $1,872/mt	
	Norsk Hydro Canada $1,630/mt	
	Russia $1,388/mt	
	Australian Average $1,344/mt	
Quality	CHN GB II-III Mg 99.8-99.90%	CHN GB I-II Mg 99.90-99.95%
Environmental Protection	Chlorine, closed workshop	Clean production, few dust and SO_2

Comments and Opinions

1. The capacity of Chinese magnesium industry shall be increased to meet the world demand. The price will be going down further to where the supply and demand of domestic supply shall be changed slightly. Under this assumption , the Mg supply and demand gap will be widened in the world market in 2000-2001. However, after this period, the outlet for this demand will increase.

2. The low price and large capacity will make it difficult for new projects already running or just being planned. But it will encourage new applications in different areas. If China can enter the WTO smoothly in 2001, we believe that western enterprises will participate in the magnesium industry in China.

3. Because of unreasonable anti-dumping duties, the export of Chinese magnesium shall be prevented from reaching many markets.

4. Use of magnesium in the domestic market by automotive, electronic, and metallurgical industries is an important factor. But it is presently a secondary role as they need qualified manpower, machinery , money and markets.

5. The price of alloys is directly affected by the price of pure magnesium. The price will continue to stay low for the next few months. Further effort to improve the quality and service is needed. Aluminum and magnesium prices are related. Magnesium is used in many aluminum alloys. A magnesium price of US$1,900 to 2,100/mt would be very good both for the users and the producers.

VERTICAL LARGER-DIAMETER VACUUM RETORT MAGNESIUM REDUCTION FURNACE

Xiaoming Mei, Alfred Yu, Shixian Shang and Tianbai Zhu

Nanjing Welbow Metals Co., Ltd.
1 Yunhai Road, Jingqiao
Lishui, Nanjing, Jiangsu 211224, China

Abstract

A new magnesium reduction technique has been developed to improve the Pidgeon reduction process. A demo-plant of 1000t magnesium per year succeeds in applying this new technique. Firstly, a new furnace is developed and a larger-diameter vertical settled vacuum retort is used instead of traditional horizontal retort. So the furnace can be designed with more compact structure to raise the magnesium output per furnace volume. Secondly, calcined dolomites and ferrosilicon is compressed into given unitary shape for enhancing heat and mass transfer during the reduction and shorten remarkably the reduction time. The shape is designed with reference to the numerical simulation result. Demo operation shows that, with application of the technology, significantly production capacity increases in the same furnace, reduction period decreases (only two thirds of the traditional reduction period), energy consumption decreases too, retort's life extends, operation becomes easy and the total production cost reduces.

Introduction

Among the techniques of magnesium production, Pidgeon reduction process is one of those that have been widely applied. The advantage of Pidgeon process is that it produces high purity magnesium that especially meets the requirement of automotive industry. Its shortage is that in the commonly used horizontal reduction furnace (Figure 1), heat and mass transfer conditions in the retort is very poor, so that its energy consumption is high, productivity is low, operating condition is intensive, unit investment is high, facility needs more area and it is not suited for large scale production and management.

Based on the situation of China resources, Pidgeon process will still be the main technique for producing magnesium in future. This paper will introduce a Vertical Larger-diameter Vacuum Retort Magnesium Reduction Furnace and its demo-plant of 1000t that has overcome the shortage of horizontal magnesium reduction furnace, and that will make larger-scale production by using Pidgeon process possible.

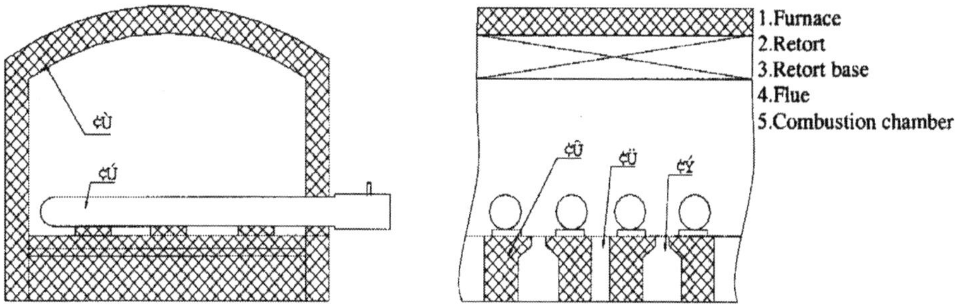

Figure 1: Typical horizontal retort reduction furnace

Enhancing Reduction Process

Horizontal retort

In the horizontal reduction furnace, calcined dolomite and ferrosilicon (>75%Si) are finely ground, mixed, and compressed into briquettes, then packed in paper bags and thrown into the retort. At a temperature of 1150℃~1200℃ and 5-20 Pa vacuum, after 12-14 hrs chemical reaction, magnesium is reduced. Figure 2 shows a section draft of a typical horizontal retort.

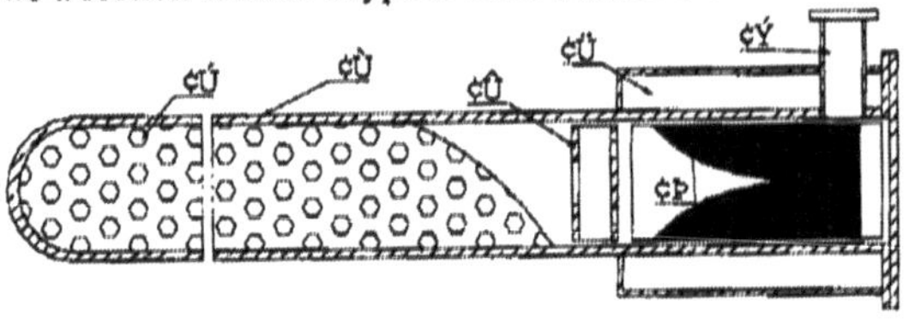

Figure 2: Typical horizontal retort

1—Retort, 2—briquettes, 3—Baffle, 4-- Cooling jacket, 5-- Vaccum pipe, 6-- Crystal magnesium

In the process, due to its structure, heat can only transfer in one direction, meanwhile the heat conductivity of retort contents is very low, so it warms up very slowly, thus cause the long reduction time cycle and vast energy consumption. Since the retort diameter is short, the material loading is little, so the productivity is not high.

Vertical retort

To overcome the shortages of horizontal magnesium reduction furnace, and make large-scale production by using Pidgeon process possible, heat and mass transfer of material in the retort must be enhanced. In view of the shortages of horizontal retort, vertical retort is introduced to shorten reduction time, reduce energy consumption and increase loading of unit retort through enhancing reaction process.

1—Retort,
2—Circularity Reduction stuff
3—Suspender & heat reflector

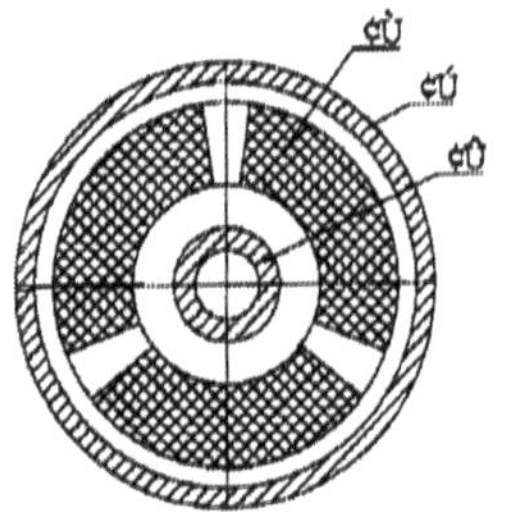

Figure 3: Section draft of vertical retort

It is shown in Figure 3 that calcined dolomite and ferrosilicon are mixed, crushed and compressed into circularity with slots. All circular stuff are overlapped on a suspender, and then put into furnace by using a lift. Due to slots in the circular stuff, the retort can heat the metal suspender through radiation quickly, and make the suspender become another heat source, thus change the heat transfer from one way to two directions, heat transfer in the stuff is thus enhanced. Meanwhile, slots in the stuff, spaces between suspender, stuff and retort also improve the mass transfer condition. So the stuff can be heated up quickly, and the reaction speed is faster.

Numerical optimization

A numerical model of heat and mass transfer in the vertical retort was set up. Figure 4 shows the two-dimensional model structure. There is radiation heat exchange between retort and stuff outer surface, between retort and suspender, between suspender and stuff inner surface, and between retort, suspender and slot sections. In the stuff, there is conduction heat transfer from surface to center, and also convection heat transfer and mass transfer while magnesium is reduced and magnesium vapor moves out. The model is to optimize the process, the dimensions of retort and circular stuff, and reaction cycle. Calculation was validated by experimental and operation data.

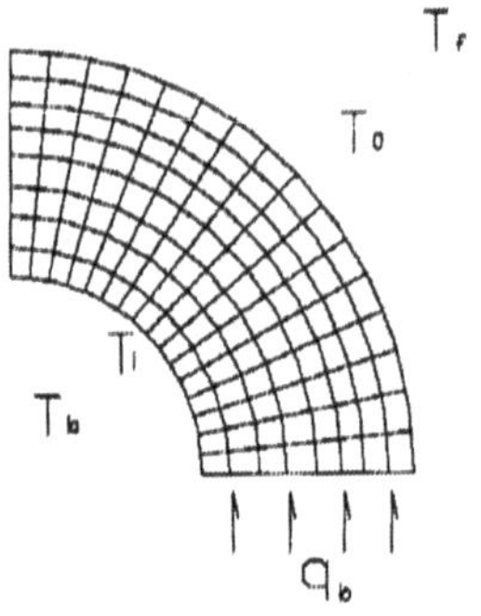

T_r—Retort temperature
T_o—Stuff outer temperature
T_i—Stuff inner temperature
T_b—Suspender temperature
Q_b—Heat flux

Figure 4: Model structure

Example of calculation results

Under the conditions of same/different thickness, loading and retort temperature, reaction processes of circular stuff with and without slots were all simulated. Table 1 is one of the results.

Table I Comparison for different shapes with same loading

	Horizontal retort	Vertical retort	
		Without slots	With slots
Retort dimension (mm)	D300x2700	D300x1800	D300x1800
Circularity thickness (mm)	/	60	60
Stuff Loaded (kg)	150	150	150
Reaction time (hrs)	12	12	7
Stuff magnesium ratio	7: 1	6: 1	5.5: 1

Vertical Retort Reduction Furnace

Based on the above analysis and study, a new reduction furnace was designed. The retort is set vertically in the new furnace, instead of horizontally, thus the structure of the furnace is entirely different from the old one. Figure 3 is a section draft of the new vertical retort reduction furnace. It is composed of retort (vertical), furnace and combustion chamber. It can burn gas, diesel, heavy oil or coal. Reduction stuff is first pressed into pellets, and then compressed into circularity with slots as shown in figure 2. Stuff is set on the suspender and then put into furnace for reduction. In designing the new furnace, numerical optimization was used for making the furnace flow field and temperature field even and reasonable, the furnace itself is more compact, thus compared with horizontal retort furnace, the new furnace has the following advantages: low Investment for same capacity, high productivity, low energy consumption, long retort's life-span is extended. Figure 5 is Section draft of vertical retort furnace.

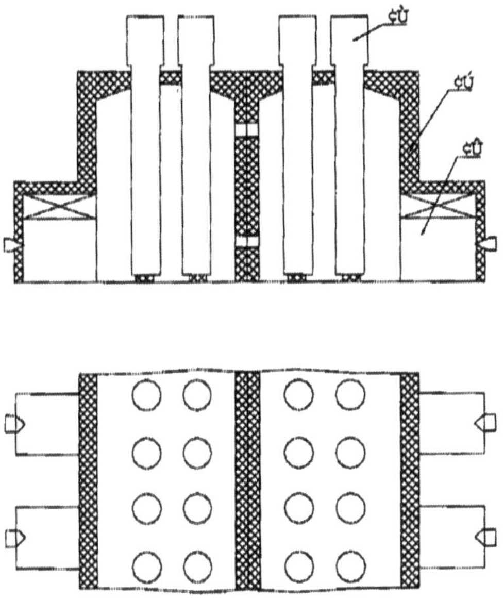

1—Vertical retort, 2—Reduction furnace, 3—Combustion Chamber

Figure 5: Section draft of vertical retort furnace

1.200MT/a Demo Plant

After careful study, a 1200mt/a demo magnesium plant was built in Nanjing Welbow Metals Co., Ltd. to validate its practicability of this new technique. Table 2 shows the investment, capacity comparisons between vertical retort furnace and horizontal retort furnace at same area location.

It is obviously that, for same capacity, investment of vertical reduction furnace is only 1/3.3 of horizontal furnace.

Table II Investment, capacity comparisons

Items	Vertical furnace	Horizontal furnace
Number of retorts	36	9
Area located (m*m)	8*6m	8*4m
Diameter of retort	400mm	300mm
Length of retort	1800mm	2700mm
Total investment (US$)	14,457	9,638
Construction period	same	same
Capacity (t/d)	2.16	0.43
Investment/capacity (US$/t)	22.31	74.71

Table 3 shows the operation economy comparison between this demo plant and a same capacity plant using horizontal retort furnace.

Table III Economy for producing 1t magnesium

No.	Items	Unit	Vertical furnace	Horizontal furnace
1	Dolomite	T	11	13
2	Ferrosilicon	T	1.15	1.35
3	Fluorite	T	0.17	0.19
4	Retort	Piece	0.15	0.2
5	Electricity	Kwh	2300	3000
6	Coal	T	9	12
7	Vacuum oil	Kg	10	10
8	Maintenance	US$	48	60
9	Labour	US$	100	145

Conclusions

1. By using vertical large diameter retort furnace in Pidgeon magnesium reduction process, compared with horizontal retort furnace, it makes the furnace more compact, the output of same volume furnace increases.

2. Enhancement of heat and mass transfer makes the retort contents warm up quickly and shorten the reduction period.

3. Productivity increases remarkably, meanwhile, energy consumption and production cost decrease.

4. Operation becomes easy, it is fit for large scale production.

References

1. Emley, E.F. Principles of magnesium technology. Pergamon Press. 1966
2. G.V. Raynor, The physical metallurgy of magnesium and its alloys. Pergamon press, P13-14. 1959.
3. Riyao Xu, Magnesium, China Metallurgy Press, 1986.

MAGNESIUM PRODUCTION BY VACUUM ALUMINOTHERMIC REDUCTION OF A MIXTURE OF CALCINED DOLOMITE AND CALCINED MAGNESITE

Wen-xin Hu, Nai-xiang Feng, Yao-wu Wang, Zhi-hui Wang
School of Materials and Metallurgy, Northeastern University, Shenyang, Liaoning 110004, China

Keywords: vacuum aluminothermic reduction; magnesium; $CaO{\cdot}2Al_2O_3$; alumina leaching

Abstract

A new method of magnesium production was proposed that using a mixture of calcined dolomite and calcined magnesite as raw materials with the molar ratio of MgO to CaO was 6:1 by vacuum aluminothermic reduction. The reduction process was studied by thermodynamic analysis and X-ray diffraction analysis of reduction residue. The reaction of reduction process was $CaO+6MgO+4Al=CaO{\cdot}2Al_2O_3+6Mg$. The effect of briquetting pressure, reduction temperature, time and CaF_2 (MgF_2) on reduction ratio of MgO was investigated. And the reduction residue that the main phase of $CaO{\cdot}2Al_2O_3$ was leached in alkaline solution for producing sodium aluminate-the raw material for special alumina. The results show that the reduction ratio is increased with increasing of the temperature, time, briquetting pressure in range from 40 to 100 MPa and addition of CaF_2 or MgF_2 in range from 0 to 3%. The alumina leaching ratio of reduction residue reached 88% at the conditions of leaching temperature 95 ℃ and the concentration ratio of Na_2CO_3 to NaOH was 100:75 in leaching solution.

Introduction

Magnesium is the lightest metal of the commonly used metals and can be applied widely. For example, there are metallurgical applications, chemical applications, structural applications, and automobile parts production [1,2]. It means huge demands for magnesium in the world. The raw materials of magnesium production such as magnesite and dolomite are widespread and high purity magnesium can be produced under low cost, simple and flexible conditions of vacuum thermal reduction process versus electrolytic production.

The Pidgeon process is a widely used process in vacuum thermal reduction process, and this technology has been in stage of maturity [3-5]. But because of the low reactivity of ferrosilicon alloy, higher reduction temperature (>1170 ℃) and longer reduction time (>7 hours) are required, which in order to reach high reduction ratio of MgO. The reduction ratio of MgO was only 73.9% at the conditions of temperature 1443 K, time 120 min [6]. More energy is needed for heating, the service life of Ni-Cr alloy reactors decrease greatly under vacuum and the conditions mentioned above, make higher cost of the Pidgeon process eventually; the residue after reduction (main phase is γ-$2CaO{\cdot}SiO_2$) is useless to other fields, such as cement industry with a reason that unreacted MgO has negative influence on the performance of cement [7,8]. According to the above information, there are two problems should be solved, firstly, the service life of reactors and the high cost of process; secondly, the emission of CO_2 and the waste of reduction residue.

A new vacuum thermal reduction method is proposed for solving the two problems about the Pidgeon process in the present paper. The new reduction process is aluminothermic reduction of a mixture of calcined dolomite and calcined magnesite with the molar ratio of MgO to CaO was 6:1. It is studied by thermodynamic analysis and the effects of reduction temperature, reduction time, vacuum of reduction system, briquetting pressure and addition of CaF_2 or MgF_2 on the reduction ratio of MgO, which for ensuring the optimal conditions of reduction process. Using alkaline solution to leach alumina from $CaO{\cdot}2Al_2O_3$, the main phase of reduction residue. The new magnesium production method in the present paper is in order to create a green, lower cost but higher economic benefit process.

Thermodynamic Analysis

Figure 1 is the X-ray diffraction analysis of reduction residue, the reduction condition of experiment is reduction temperature 1413 K, reduction time 120 min, pellets forming pressure of 50 MPa and vacuum of 4 Pa. It is noticed that $CaO{\cdot}2Al_2O_3$ phase with higher intensity; the other phases are from raw materials. Thus, the main reaction of reduction process occurred in reduction is

$$CaO\,(s) + 6MgO\,(s) + 4Al\,(l) = CaO{\cdot}2Al_2O_3\,(s) + 6Mg\,(g) \quad (1)$$

$$\Delta G^{o} = 1013.7 - 0.615T \quad (\text{kJ/mol}) \quad (2)$$

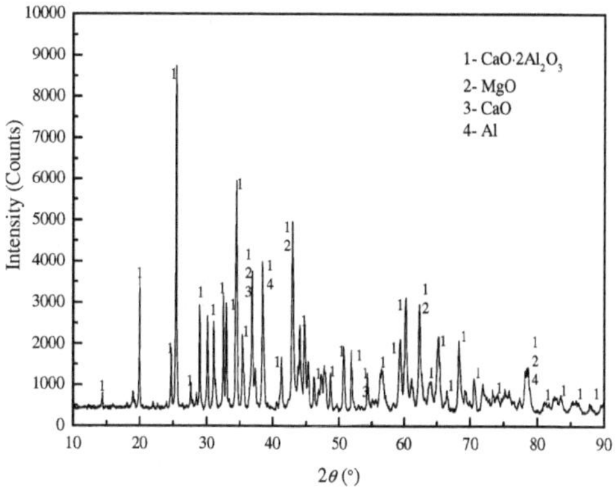

Figure 1. The XRD pattern of reduction residue.

The relationship between the equilibrium magnesium partial pressure (P_{Mg}) and reduction temperature is given by Equation 3.

$$\Delta G = 1013.7 - 0.615T + RT\ln\left[\frac{(\frac{P_{Mg}}{P^{o}})^{6}}{(\frac{\alpha_{Al}}{\alpha^{o}})^{4}}\right] \quad (\text{kJ/mol}) \quad (3)$$

Where α_{Al} is the activity of aluminum; α^{o} and P^{o} are values of standard state; R is the gas constant.

The equilibrium temperature of Equation 1 is 1648.3 K at reduction system pressure of 1 atm. But the reaction temperature decreases to 904.4 K at the experimental vacuum of 4 Pa. Therefore, the reaction Equation 1 could occur at temperatures of 1223, 1273, 1323, 1373, 1413, 1443 and 1473 K at vacuum of 4 Pa (during the reduction process, P_{Mg} is approximately equal to the pressure of vacuum reduction system [9], which displays on vacuometer).

Experimental

Materials

Dolomite and magnesite were obtained from Liaoning, China; aluminum powder (purity 99%), CaF_2, MgF_2, NaOH and Na_2CO_3 were supplied by Tian Jin Kermal Chemical Reagent CO., LTD. Table I and II show the chemical compositions of dolomite and magnesite using chemical analysis.

Table I. Chemical composition of dolomite.

Component	MgO	CaO	SiO_2	Al_2O_3
Content (%, ω)	23.62	29.86	0.22	0.10

Table II. Chemical composition of magnesite.

Component	MgO	CaO	SiO_2	Al_2O_3
Content (%, ω)	47.34	0.67	0.13	0.08

Apparatus

Figure 2 schematically shows the experimental apparatus for reduction. A furnace using silicon carbide rods was used to heat the reactor which was made of high-temperature alloyed steel; temperature control system was used to control the heating ratio of reduction system and shown the temperature which was measured by NiCr-NiSi thermocouple; vacuum system was made up of ZJP-70 roots vacuum pump and ZX-15 rotary-vane vacuum pump; condenser, the upper part of reactor was used to condense magnesium vapor and collected magnesium.

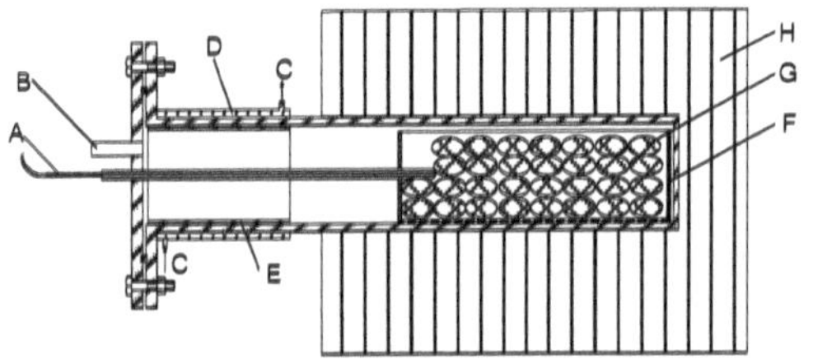

A- thermocouple B- vacuum tube
C- circulating water D- water-cooled jacket
E- condenser F- briquettes bucket
G- briquettes H- furnace

Figure 2. Schematic of experimental apparatus.

Procedures

The vacuum thermal reduction experiments were carried out by using a mixture of calcined dolomite and calcined magnesite. According to the experimental data of calcinations, dolomite and magnesite should be calcined at 1060 °C and 830 °C for 90 min respectively for ensuring the reactivity of MgO·CaO and MgO. The calcined dolomite and calcined magnesite were pulverized and sieved 150~180 μm out, mixed with the molar ratio of MgO, CaO to Al 6:1:4. The mixed charges were formed into pellets under different pressure, having a diameter of 25 mm and a thickness of about 10 mm. The pellets were charged into the reactor for reduction experiments. The magnesium was reduced by aluminum in the conditions of vacuum and high temperature. The magnesium was collected on the condenser. The reduction residue was leached under the conditions of leaching temperature 95 °C and the concentration ratio of Na_2CO_3 to NaOH was 100:75 by calculation of Equation 6. The flow sheet of process show is shown in Figure 3. After reduction and reactor cooling, the condensed magnesium product and the residue obtained were weighed and analyzed.

The reduction ratio of MgO (R) is calculated by the following formula:

$$R = \frac{m_r}{m_0} \times 100\% \tag{4}$$

Where m_0 is the initial quantity of magnesium in the raw materials and m_r is the quantity of magnesium on condenser.

The leaching ratio of Al_2O_3 according to the following formula:

$$\eta_{Al_2O_3} = \frac{\omega_l}{\omega_0} \times 100\% \tag{5}$$

Where ω_0 is the quantity of Al_2O_3 (elemental aluminum in reduction residue is convert to Al_2O_3), ω_l is the quantity of Al_2O_3 in leaching solution.

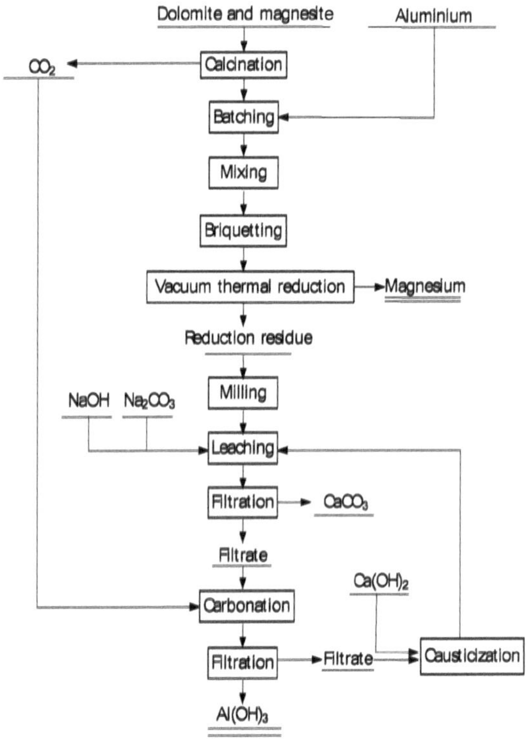

Figure 3. The flow sheet of reduction and leaching process.

Analysis of Experimental Results

Effect of Reduction Temperature on Reduction Ratio of MgO

Figure 4 shows the changes in reduction ratio of MgO at different temperatures. At the conditions of briquetting pressure of 50 MPa, reduction time 120 min and vacuum 4 Pa. With increasing temperature from 1223 K to 1413 K, the reduction ratio of MgO increased remarkably, from 44.47% to 82.06%. The reduction ratio of MgO increased slowly at temperature range from 1413 K to 1443 K. The final reduction ratio was 91.35% at 1473 K.

Reduction temperature is an important factor to the reaction rate; the reaction ratio was accelerated with temperature increasing. But at the stage of 1443 K, the reduction is close to the end which the reason of reduction ratio increased slowly. The high reaction rate and good results of reduction ratio of MgO needed high temperature. But the reasonable reduction temperature is chosen 1413 K by experimental data. The reason that decreasing the reduction temperature for saving cost of high-temperature alloy is one of the principal objectives of the present reduction process. At 1413K, the reduction ratio of MgO is higher than Pidgeon process at 1473 K [6]. The reduction ratio of MgO should increase by change other factors, such as reduction time.

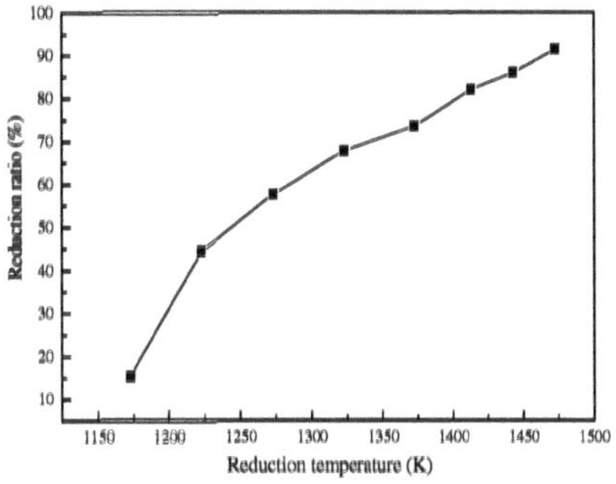

Figure 4. Effect of reduction temperature on reduction ratio of MgO.

Effect of Reduction Time on Reduction Ratio of MgO

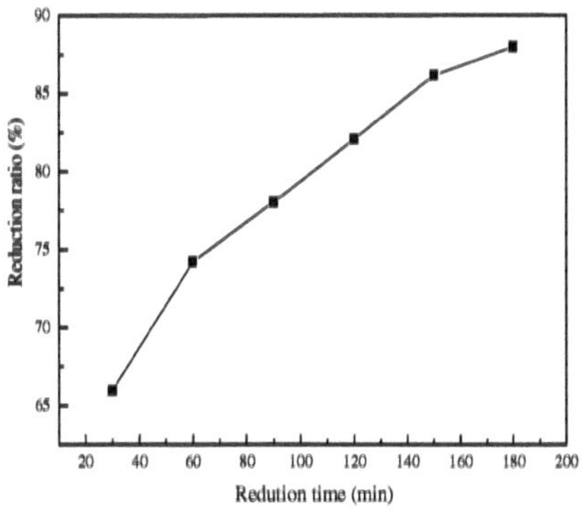

Figure 5. Effect of reduction time on reduction ratio of MgO (the initial reduction ratios according to experimental data are 40.9% in temperature-rise period from room-temperature to 1413 K).

The effect of reduction time on reduction ratio of MgO is presented in Figure 5. The reduction temperature was 1413 K, briquetting pressure 50 MPa and vacuum 4 Pa. The reduction ratio increased obviously at the initial reaction stage from 30 min to 150 min, the reduction ratio reached 86.18% at 150 min. After reduction time of 150 min, the reduction ratio of MgO increased slowly that only increasing 1.81% at 180 min. The reduction ratio increased almost no more. It should be the meaning of the end of reduction.

The more energy will lead to high reaction rate of reduction process. The energy which was supported to pellets by heat transfer from furnace to pellets by thermal radiation almost, and then from surface into the core of pellets with reduction time extended. In present paper, time of 150 min was not the best reduction time, but the time which the corresponding reduction ratio of MgO was closed to the maximum. It is one of key links for saving the cost of energy, and enhanced the effects of briquetting pressure and addition of CaF_2 or MgF_2 on reduction ratio of MgO.

Effect of Briquetting Pressure on Reduction Ratio of MgO

Figure 6 shows the effect of briquetting pressure on reduction ratio of MgO. The reduction temperature was 1413 K, reduction time 120 min and vacuum 4 Pa. For the briquetting pressure of 100 MPa, the reduction ratio proceeded the fastest and the reduction ratio was 88.96%. For the briquetting pressure of 50 MPa or 200 MPa, the reduction ratio was decreased remarkably. The reason of this phenomenon is from 50 MPa to 100 MPa, increasing briquetting pressure can increase the contact area between MgO-CaO and reduction agent; from 100 MPa to 200 MPa, the briquetting pressure make the gap is too small between MgO-CaO and reduction agent that magnesium vapor is difficult to escape from the pellets. Thus the appropriate briquetting pressure could balance progresses of contact between MgO-CaO and reduction agent, and the escape of magnesium vapor.

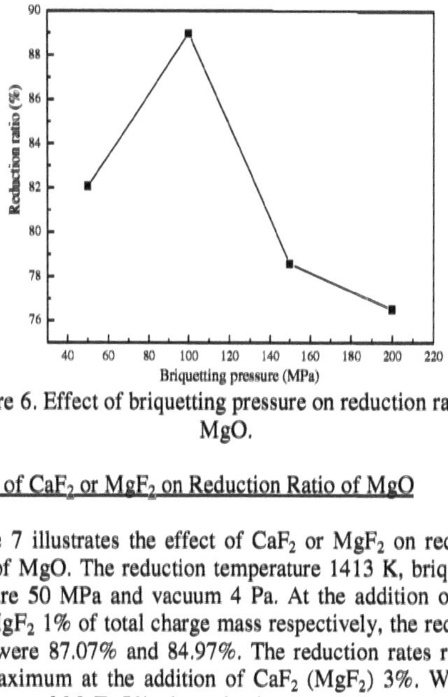

Figure 6. Effect of briquetting pressure on reduction ratio of MgO.

Effect of CaF_2 or MgF_2 on Reduction Ratio of MgO

Figure 7 illustrates the effect of CaF_2 or MgF_2 on reduction ratio of MgO. The reduction temperature 1413 K, briquetting pressure 50 MPa and vacuum 4 Pa. At the addition of CaF_2 and MgF_2 1% of total charge mass respectively, the reduction rates were 87.07% and 84.97%. The reduction rates reached the maximum at the addition of CaF_2 (MgF_2) 3%. With the additions of MgF_2 5%, the reduction rates slightly decreased that 95.59% and 94.95%. The addition of CaF_2 (MgF_2) could activate the surface of MgO-CaO and the reactivity of oxygen ion was increased which enhanced the reaction between MgO-CaO and aluminum.

The addition of CaF_2 (MgF_2) was an effective way to accelerate reduction progress, but the mass of addition should be appropriate as 3% of total charge mass according to the considerate of cost and the high reduction ratio of magnesium. The accelerating effect of MgF_2 to reduction was better than CaF_2 of addition of 3% or above.

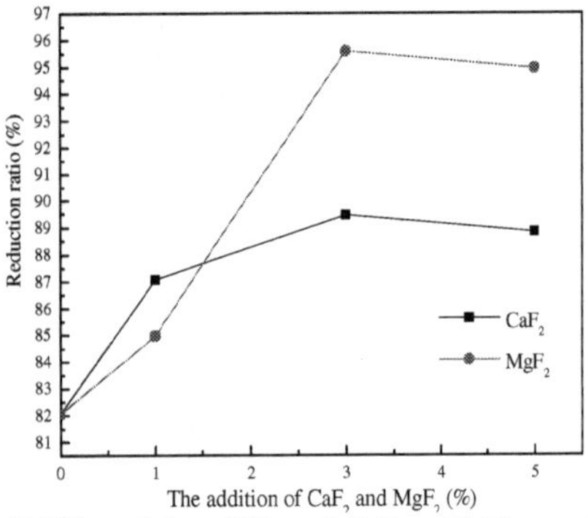

Figure 7. Effect of the addition of CaF_2 or MgF_2 on reduction ratio of MgO.

The reduction ratio of MgO was 97.01% at temperature 1413 K, time 150 min, briquetting pressure 100 MPa and 3% MgF_2.The magnesium produced in reduction as shown in Figure 8, the ratio of purity beyond 99.17% by chemical analysis.

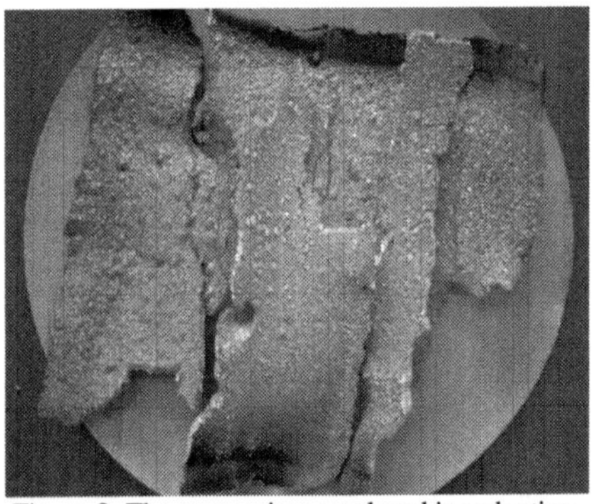
Figure 8. The magnesium produced in reduction.

Alumina Leaching from Reduction Residue

According to the XRD pattern shown in Figure 1, $CaO{\cdot}2Al_2O_3$ was the major phase in reduction residue. Using alkaline solution of a mixture of Na_2CO_3 and NaOH with the concentration ratio of 100:75 to leach $NaAl(OH)_4$ from $CaO{\cdot}2Al_2O_3$ at temperature 95 °C and time 120 min. The leaching ratio of Al_2O_3 reached 88%. $NaAl(OH)_4$ was converted to $Al(OH)_3$ by carbonation. CO_2 was collected at the stage of dolomite and magnesite calcinations. This step is contributed to reduce cost and reduced the emission of CO_2. The product of $Al(OH)_3$ is shown in Figure 9.

The phase of residue after leaching was shown in Figure 10. $CaCO_3$ could be isolated firstly and then be the raw material of CaO in cement industry. The leaching process is the following

$$CaO{\cdot}2Al_2O_3(s) + 2NaOH(l) + Na_2CO_3(l) + 7H_2O = CaCO_3(s) + 4NaAl(OH)_4(l) \quad (6)$$

and the reaction during the carbonation of $NaAl(OH)_4$ is

$$2NaAl(OH)_4(l) + CO_2(g) = 2Al(OH)_3(s) + Na_2CO_3(l) + H_2O \quad (7)$$

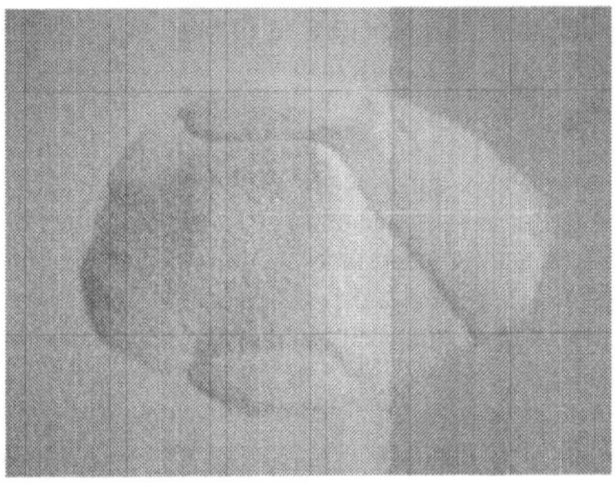
Figure 9. The product of $Al(OH)_3$.

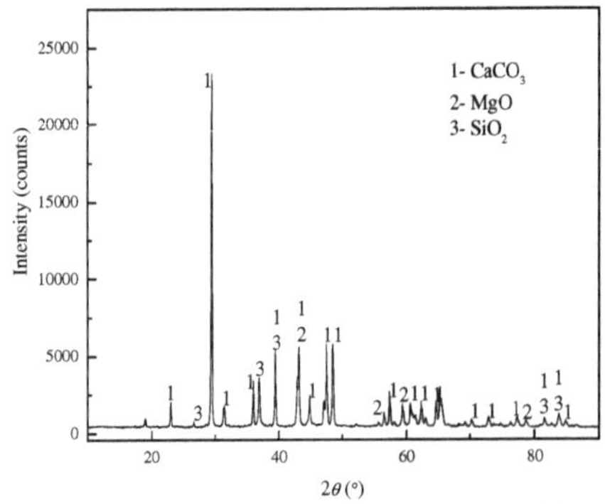

Figure 10. The XRD pattern of leaching residue.

Conclusions

(1) The reduction temperature, reduction time, briquetting pressure and the addition of CaF_2 (MgF_2) are important factors to reduction process. With increasing the above factors increased the reduction ratio of MgO. But briquetting pressure and the addition of CaF_2 (MgF_2) should be in the optimum ranges.

(2) According to the analysis of experimental results, the possible industrial operating conditions of reduction process are reduction temperature 1413 K, reduction time 150 min, vacuum 4 Pa, pressure for briquetting of raw materials 100 MPa and the addition of 3% MgF_2.

(3) The main phase of reduction residue is $CaO{\cdot}2Al_2O_3$ that can extract sodium aluminate and produce $Al(OH)_3$ by carbonation. The leaching ratio reached 88% under leaching temperature 95 °C of 120 min and the concentration ratio of Na_2CO_3 to NaOH was 100:75.

(4) Compare to the Pidgeon process, the new reduction technology presents in this paper can accelerate reduction process, $Al(OH)_3$ can be produced from reduction residue, CO_2 is collected at the stage of dolomite and magnesite calcinations and $CaCO_3$ in the leaching residue also can be used, which mean the new reduction technology reduce production cost effectively and almost none waste be discharged into environment.

References

1. D. Eliezer, E. Aghion, F.H. Froes, "Magnesium Science, Technology and Applications", *Advanced Performance Materials*, 5 (1998), 201-212.

2. G. Hanko, H. Antrekowitsch, P. Ebner, "Recycling Automotive Magnesium Scrap", *JOM*, 2 (2002), 51-54.

3. I.M. Morsi, K.A. Elbarawy, M.B. Morsi et al., "Silicothermic Reduction of Dolomite Ore Under Inert Atmosphere", *Canadian Metallurgical Quarterly*, 1 (2002), 15-28.

4. D. Minic, D. Manasijevic, J. Dokic et al., "Silicothermic reduction process in magnesium production", *Journal of Thermal Analysis and Calorimetry*, 2 (2008), 411-415.

5. G. C. Holywell, "Magnesium: The First Quarter Millennium", *JOM*, 7 (2005), 26-33.

6. W.-X. Hu, J. Liu, N.-X. Feng, "Analysis of New Vacuum Reduction Process in Magnesium Production", *Non-ferrous Mining and Metallurgy*, 4 (2010), 40-42.

7. F. Gao, Z.-R. Nie, Z.-H. Wang et al., "Environmental Assessment of Energy Usage Strategies for Magnesium Production Using the Pidgeon Process", *Journal of Beijing University of Technology*, 6 (2008), 646-650.

8. S.-Y. Yan, "The current status and analysis of making magnesium by means of Pidgeon Process in China", *Light Metals*, 6 (2005), 37-40.

9. J.R Wynnyckyj, L. M. Pidgeon, "Equilibria in the Silicothermic Reduction of Calcined Dolomite", *Metallurgical transactions B*, 2 (1971), 979-986.

CARBOTHERMAL PRODUCTION OF MAGNESIUM: CSIRO'S MAGSONIC™ PROCESS

Leon H. Prentice[1]; Michael W. Nagle[1]; Timothy R. D. Barton[1]; Steven Tassios[1]; Benny T. Kuan[2]; Peter J. Witt[2]; Keri K. Constanti-Carey[1]
[1]CSIRO Process Science and Engineering; Box 312; Clayton South, VIC, 3169, Australia
[2]CSIRO Mathematics Informatics and Statistics; Box 312; Clayton South, VIC, 3169, Australia

Keywords: Carbothermal Magnesium, Carbothermic Reduction, Shock Quenching, Reversion, Laval Nozzle, Supersonic Flow, Pyrophoric Powder, Computational Fluid Dynamics, Scale Up, Pilot Plant, Process Safety

Abstract

Carbothermal production has been recognized as conceptually the simplest and cleanest route to magnesium metal, but has suffered from technical challenges of development and scale-up. Work by CSIRO has now successfully demonstrated the technology using supersonic quenching of magnesium vapor (the MagSonic™ Process). Key barriers to process development have been overcome: the experimental program has achieved sustained operation, no nozzle blockage, minimal reversion, and safe handling of pyrophoric powders. The laboratory equipment has been operated at industrially relevant magnesium vapor concentrations (>25% Mg) for multiple runs with no blockage. Novel computational fluid dynamics (CFD) modeling of the shock quenching and metal vapor condensation has informed nozzle design and is supported by experimental data. Reversion below 10% has been demonstrated, and magnesium successfully purified (>99.9%) from the collected powder. Safe operating procedures have been developed and demonstrated, minimizing the risk of powder explosion. The MagSonic™ Process is now ready to progress to significantly larger scale and continuous operation.

Introduction

Magnesium is the lightest of the structural metals and is useful for a wide range of industrial applications [1]. The production of magnesium via carbothermal reduction is not a new concept; indeed, a process using this chemistry was patented as far back as 1932 [2]. The basic chemistry depends on the use of carbon as the reductant to produce magnesium and carbon monoxide from magnesia and carbon:

$$MgO_{(s)} + C_{(s)} \leftrightarrow Mg_{(v)} + CO_{(g)} \quad (1)$$

The well-known primary issue with the carbothermal process is reversion – the reaction above proceeds in the reverse direction very easily; this can be minimized by cooling the gaseous reaction products very rapidly. The largest industrial-scale operation of carbothermal technology was at Permanente, CA [3, 4]. Cooling of the gases was achieved either by heat transfer with surfaces [3] or shock-chilling by dilution with, for example, natural gas [4], nitrogen [5], molten magnesium chloride [6], or even molten magnesium itself [7]. Other versions of the chemistry focused on removing magnesium from the vapor phase by reaction in a solution [8, 9].

The idea of supersonic quenching is also not new. The reaction products (gases) are accelerated to several times the speed of sound (typically $Ma \approx 4$-5), achieving very rapid gas cooling. Useful reviews of the literature have been presented by the present authors [9-11]. Previous attempts appear to have failed primarily because of blockage of the supersonic Laval nozzle [12] under sustained operation. This paper describes one form of the technology, called MagSonic™, developed by CSIRO (Commonwealth Scientific and Industrial Research Organisation) in Australia.

Technology Description

Feed Preparation

Magnesite may be commercially mined at relatively high purity and calcined to produce magnesia (MgO) suitable for reaction. Options exist for additional purification of the magnesite or magnesia, which depend on purity and morphology requirements and may include acid leaching, flotation, and other processing. The MagSonic™ process has demonstrated itself remarkably robust, processing magnesia that has been only calcined (not purposely purified). The behavior of impurities in the carbothermal reduction process has been addressed previously [13], reporting that essentially all the impurity species are reduced before the carbothermal magnesium reaction begins. The species may be detected in the cyclone (used to separate the magnesium powder from the gas) by Laser-Induced Breakdown Spectroscopy (LIBS).

Feed preparation is completed by briquetting calcined magnesia with a carbon source. For much of the experimental program, powdered graphite has been used ($d_{50} \approx 12$ µm, >99.5% C, Merck Pty Ltd, Kilsyth, VIC 3137, Australia). Although briquetting graphite without a binder is difficult, briquettes of the size and shape suitable for piloting have been produced without a binder (Figure 1), and have been tested in the laboratory rig.

Figure 1: Magnesia-carbon briquette

Reaction

The reaction proceeds quickly above 1700 °C, and there is evidence of some reduction/vaporization as low as 1250 °C [13]. In experimental work to date, heating has been achieved by the use of an induction furnace (up to 50 kW), with sustained reaction at temperatures typically around 1800 – 1900 °C [14]. The reaction kinetics have been described by the present authors [14, 15], including achievement of sustained (60+ minutes) operation and confirmation of kinetic modeling for pilot plant and full-scale design.

Supersonic Quenching

The aerodynamics of supersonic Laval nozzle operation are well known, but the present operation presents a number of novel aspects. These include high temperatures, metal vapor condensation and solidification, and relevant boundary layer interactions. CSIRO has developed a number of novel computational fluid dynamics (CFD) tools that have been implemented and validated, providing significant information for scale-up and further development.

The effect of metal vapor condensation may be illustrated by Figure 2. Condensation adds significant heat to the stream, affecting temperature distributions, expansion shocks, and other aspects of the flow. Early CFD work identifying the condensation effect shown in Figure 2 was reported in [9] with further publication on these topics pending.

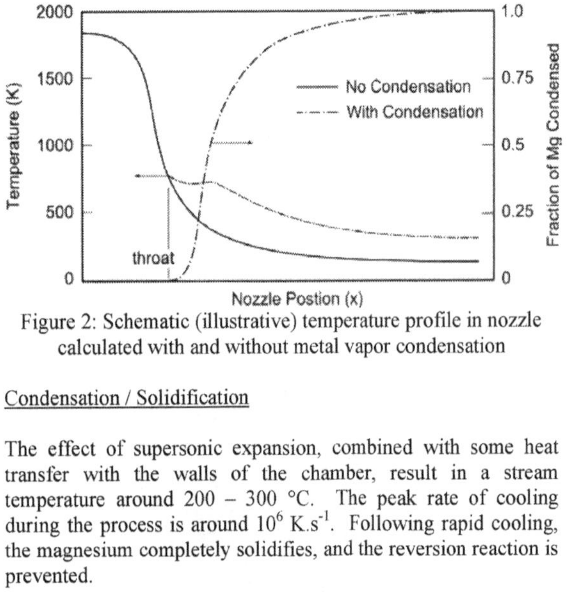

Figure 2: Schematic (illustrative) temperature profile in nozzle calculated with and without metal vapor condensation

Condensation / Solidification

The effect of supersonic expansion, combined with some heat transfer with the walls of the chamber, result in a stream temperature around 200 – 300 °C. The peak rate of cooling during the process is around 10^6 $K.s^{-1}$. Following rapid cooling, the magnesium completely solidifies, and the reversion reaction is prevented.

Separation

Separation may be achieved by a number of methods, the simplest of which is a cyclone. While some reduced efficiency at moderate temperatures may be experienced [16, 17], separation efficiencies above 90 %, often higher than 95 %, have been consistently achieved. The powder may be conveniently recovered, despite its very low bulk density (typically < 0.3 $g.ml^{-1}$), and flows easily.

Gas Handling

Vacuum systems are typically designed to handle one or two of the relevant process conditions – low pressure, high mass flow, high temperature – but rarely all three at once. As such, the vacuum system is a major user of energy, estimated as 0.46 MWh per tonne Mg at full scale [18]. The vacuum system must maintain the absolute pressure downstream of the nozzle below 50 % (but preferably below 10%) of its upstream value to maintain sonic flow. The lower limit of vacuum is determined primarily by the economics of operation; achieving a pressure below 0.5 kPa is likely to be prohibitively expensive, and unnecessary for optimum operation.

Purification

The experimental product powder is rendered safe either by controlled surface oxidation ('passivation') or by covering it with oil to prevent contact with air. The product has been purified from both media, either by direct sublimation from the passivated powder, or firstly by distilling off the oil at 400-500 °C at atmospheric pressure and subsequently distilling the magnesium metal at about 980 °C and 0.8 kPa absolute pressure in a graphite-lined steel retort. The resulting magnesium metal is distilled again and yields metal containing low levels of metallic impurities (200 ppm Al, 90 ppm Ca, 80 ppm Fe, 10 ppm Mn and 60 ppm Si).

Major Technical Challenges

High Temperature Operation and Control

The control of a kinetically-constrained gas-solid endothermic reaction at very high temperatures is non-trivial, though the total system pressure can be relatively easily maintained by the flow rate of an inert cover gas. For maximum efficiency the magnesium concentration in the vapor should be as high as possible, but any higher than a partial pressure of 50 kPa risks over-pressurizing the reactor. The equilibrium partial pressure of 50 kPa is reached at 1764 °C, but the reactor is likely to operate at 1900 – 2000 °C.

In terms of process control, reactor temperature and cover gas flow rate are controlled in the laboratory batch system. Continuous operation will likely operate at constant temperature, with the reaction rate controlled by the amount of reacting bed, and to a lesser extent by the inert gas flow rate. The authors have presented data on the modeling of the reaction kinetics [14, 15, 19], which are useful for sizing and control.

Preventing Nozzle Blockage

Blockage of the gas exit/quenching nozzle has been a significant problem for the carbothermal process [12]. Typically, magnesium and/or impurity metals condense on the surface and revert to oxides with high melting points, thereby accreting and blocking the nozzle. The CSIRO research team has overcome these issues and provisionally patented aspects of the method [11]. Despite demonstrated operation without blockage at laboratory scale, safe design includes the possibility of recurrence and must be included in risk assessments.

Pyrophoric Powder and Process Safety

Issues with pyrophoric magnesium powders are well known, and previous industrial operation has resulted in fatalities [20]. Studies have quantified the conditions for spontaneous ignition [21]. National Fire Protection Association standards 69 and 484 [22, 23] are also useful for magnesium powder explosion prevention and process safety. Given that magnesium powder produced by the MagSonic™ process is relatively fine and is certainly spontaneously combustible, and that operation is under vacuum, conventional measures such as deflagration venting, explosion suppression, or dust concentration control are not likely to be practically effective in ensuring safe operation.

Design for scale-up presently focuses on explosion containment, whereby equipment is designed to withstand a magnesium powder explosion without rupturing. Safety methods under development are relatively novel (and are the subject of future publications), but will ensure exceedingly high safety integrity levels are met for practical full-scale operation.

At laboratory scale, residual internally-adhered powder is oil-sprayed for safety. At larger scales, oil spraying is impractical and undesirable. Therefore, the metal powder must be rendered safe from a "dry" state. Options include a) batch passivation by controlled oxidation; b) melting of powder directly, or after briquetting, or by stirring into a liquid pool of magnesium; and c) distillation and condensation. The latter two options lead directly to the purification stage of the process.

Purification

The temperatures required to achieve reaction also means that the other metals in the feed are also reduced and contaminate the product to various extents. The product metal that is collected will require subsequent purification to achieve commercial magnesium metal purity.

Purification of magnesium by distillation is known technology [24, 25], but difficulties with fine powders are also known [3]. At Permanente the fine powder was briquetted with tar prior to vacuum distillation, but it is unlikely that a similar process will be used for MagSonic™ at full scale. The CSIRO team has overcome the difficulty of fine powder purification by dissolution into a molten metal, from which magnesium may be vacuum distilled. The molten metal may a metal with low vapor pressure (such as Sn, Pb, Cu), or indeed magnesium itself. Because of its lower oxygen content than Permanente's powder, incorporation into a melt, and subsequent distillation, are easier to achieve. The CSIRO project team has successfully purified the magnesium produced from the MagSonic™ rig and has data to size pilot plant equipment to purify magnesium.

Experimental Progress

Progress in the experimental program has been rapid following the initial stage (proof of concept) from 2004 - 2007. The charts below illustrate experimental progress.

The concentration of magnesium in the nozzle is an important factor for efficient economic operation. Dungan [3] reported that the (subsonic) "Permanente magnesium plant circulates to each reduction unit from a common system approximately 13,000 $ft^3.min^{-1}$ of natural gas containing up to 6 per cent carbon monoxide." This implies that the magnesium concentration in the quench section was relatively low. For economic operation of the MagSonic™ process, the concentration target has been 25 % Mg in the nozzle. This corresponds to less than 50 % inert gas, though it may also be recycled as in the Permanente plant.

Figure 3 shows that the 25 % target was consistently met towards the end of this experimental program. Not all experiments aimed for very high concentrations, as other goals of the program included kinetics and control operations and the achievement of 'pseudo-steady-state' control [15].

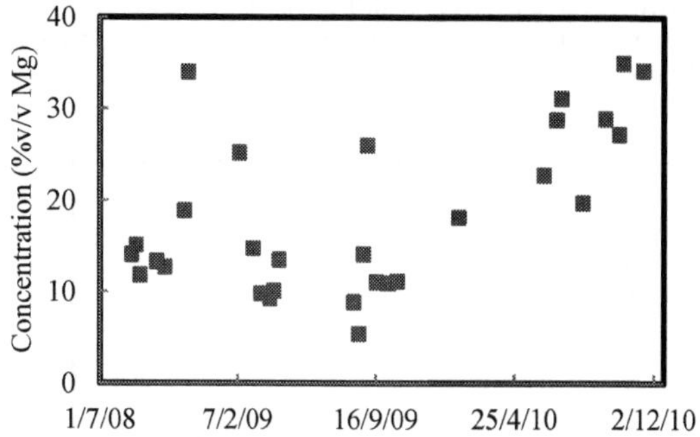

Figure 3: Peak magnesium concentration in nozzle entry gases over main experimental program

The data presented in Figure 3 are higher than reported for similar systems. Odle and McClaine [26] cite experimental work by Mineral Development International (MDI) in Birkerod, Denmark, for which the maximum Mg concentration appears to be around 0.25 %. Hori *et al* [27, 28] worked at reduced reactor pressure, which would also lead to a lower magnesium concentration (per unit volume) than that produced by the MagSonic™ process.

Continuous operation is also imperative. In the laboratory setting the process has been a batch operation, consuming MgO-C pellets. Operation of the reactor, nozzle, vacuum system, and powder recovery process were tested for durations up to around 2.5 hours of operation above 1700 °C (Figure 4). Towards the end of the experimental program, two hours of consistent operation could be reliably achieved, part of which may be considered pseudo-steady-state [15].

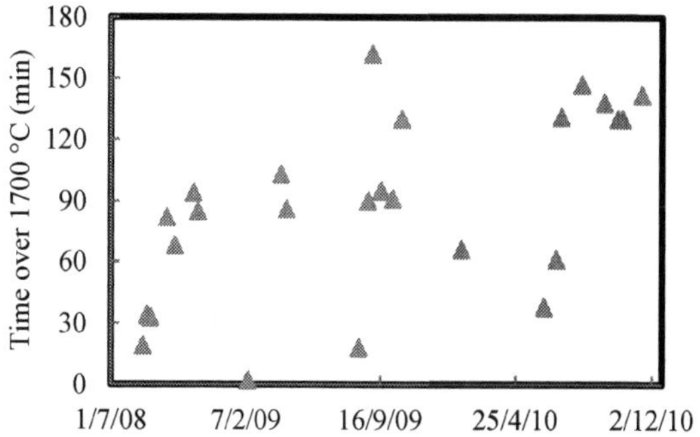

Figure 4: Experiment duration (minutes over 1700 °C) over main experimental program

Finally, another practical measure of continuous operation is the mass of reactants consumed. The laboratory equipment was upgraded in late 2008 to allow up to 2500 g of reactants to be loaded into the reactor. The charge consumption in each experiment is shown in Figure 5. Some failures are included (from, for example, nozzle blockage or operational issues), but since late 2009 consistently good and reliable experiments have been conducted. A 'routine' mass of 2000 g has been used for many of the latter experiments, all of which have gone to effective completion (>99.9 % MgO reacted).

The focus of each experiment was not necessarily maximum charge consumption. As mentioned above, occasionally the purpose is pseudo-steady-state control, or the investigation of alternative cover gases, or modifications to process conditions (e.g. downstream pressure), or simply collection of a batch of powder.

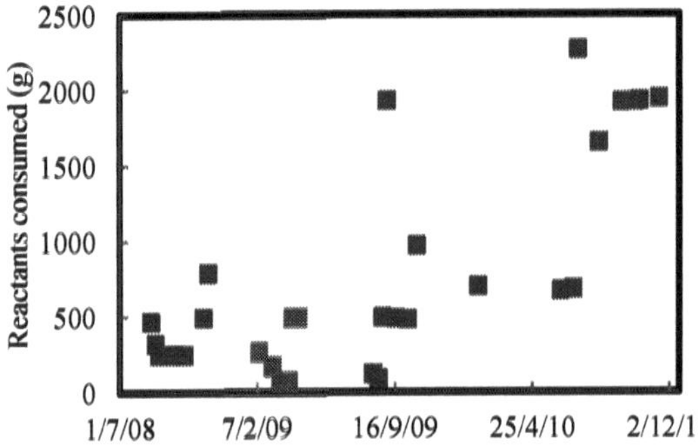

Figure 5: Consumption of reactants in each experiment (g) over main experimental program

The mass of magnesium produced in a batch may be compared with some data available in the literature. Hori *et al* [27, 28] report the consumption of around 1000 g of MgO and carbon, for two experiments listed. Odle and McClaine [26] report that 11 experiments 'processed about 90 to 285 grams of feed'. Engell *et al* [29], whose work may underpin Odle and McClaine's, give examples of consumption of 10-15 g reactants. Larger-scale work was undertaken by Cameron *et al* [30, 31], who reported feeding tens of kg per hour. By contrast, kinetic work by Sano's group [32, 33] operated at the mg scale. Nusheh *et al*'s recent work [34] was also at mg to µg scale. Donaldson and Cordes [12] did not provide data for magnesium experiments.

Future Development

The development of a novel process for the production of a mass-produced material necessarily involves significant scale-up factors. The MagSonic™ process has proceeded from the gram scale to the kilogram (and from Mg concentrations below one percent to upwards of 25 %), yet significant scale-up factors remain. For example, from Figure 4 and Figure 5 one may deduce a laboratory productivity around 500 g Mg per hour. Full scale may be around 30,000 tonnes Mg per year, which at 80% utilization is around 4.3 tons per hour, or nearly 10,000 times laboratory scale.

The next step in the development of MagSonic™ technology is a pilot plant, for which CSIRO has developed a relatively detailed preliminary design and costing. Such a plant would aim to operate continuously and produce both purified magnesium for industry evaluation and good data for design and operation control. Sustained continuous operation would also validate the environmental claims of MagSonic™ [18], provide data on process robustness, and lead to further scale up and process development.

Conclusion

CSIRO's MagSonic™ carbothermal magnesium technology has been under development in Australia since mid-2003. Recent advances have demonstrated that sustained operation can be achieved, and the laboratory equipment and experimental program have achieved concentration, reaction time, consumption, recovery, and purification targets. The technology is now ready to progress to the next stage, a continuous Pilot Plant.

References

1. Abbott, T., *Why Choose Magnesium?* in *4th International Light Metals Technology Conference (LMT 2009)*. 2009. Gold Coast, AUSTRALIA: Trans Tech Publications Ltd.

2. Hansgirg, F., *Production of Metallic Magnesium*. 1932, US1884993.

3. Dungan, T. A., *Production of Magnesium by the Carbothermic Process at Permanente*. in *Transactions of the American Institute of Mining and Metallurgical Engineers*. 1944:(308-314).

4. Byrns, A. C., *Carbothermic Process for Magnesium at Permanente*. Chemical Engineering Progress, 1947. **43**(4): p. 172-173.

5. Dean, K. C., Edlund, V. E., and Lawrence, A. G., *Quenching Carbothermic Magnesium with Nitrogen*. Light Metal Age, 1972. **30**(5-6): p. 21-22.

6. Avery, J. M., *Process for Recovering Magnesium*. 1983, EP75836-A.

7. Avery, J. M., *Method for Producing Magnesium*. 1981, US4290804.

8. Eckert, C. A., Irwin, R. B., and Graves, C. W., *Liquid Metal Solvent Selection: The MgO Reduction Reaction*. Industrial & Engineering Chemistry, Process Design and Development, 1984. **23**(2): p. 210-217.

9. Brooks, G., Trang, S., Witt, P., Khan, M. N. H., and Nagle, M., *The Carbothermic Route to Magnesium*. JOM, 2006. **58**(5): p. 51-55.

10. Brooks, G., Nagle, M., Tassios, S., and Trang, S., *The Physical Chemistry of the Carbothermic Route to Magnesium*. in *Magnesium Technology 2006*. 2006. San Antonio, TX, United States: Minerals, Metals and Materials Society, Warrendale, PA 15086, United States (25-31).

11. Tassios, S., Barton, T. R. D., Constanti-Carey, K. K., Nagle, M. W., and Prentice, L. H., *Manufacture of Metal E.G. Magnesium, Involves Performing Carbothermal Reduction of Metal Oxide, Preventing Reformation of Metal Oxide, and Cooling Stream Using Nozzle Heated with Unit Other Than Gas under Specific Condition.* 2010, WO2010012042-A1.

12. Donaldson, A. and Cordes, R. A., *Rapid Plasma Quenching for the Production of Ultrafine Metal and Ceramic Powders.* JOM, 2005. **57**(4): p. 58-63.

13. Prentice, L., Nagle, M., and Constanti-Carey, K., *Impurities in the Carbothermal Production of Magnesium: To 1500 °C.* in *High Temperature Processing Symposium.* 2009. Swinburne University, Hawthorn, Australia.

14. Prentice, L. and Nagle, M., *Mechanism and Kinetics of Reduction of Magnesium Oxide with Carbon.* in *Magnesium Technology 2009.* 2009. San Francisco, CA: The Minerals, Metals, and Materials Society (35-39).

15. Prentice, L. H., *Psuedo-Steady-State Control of High Temperature Gas-Solid Reaction.* in *Chemeca2011.* 2011. Sydney, Australia: Engineers Australia.

16. Bohnet, M. and Lorenz, T., *Separation Efficiency and Pressure Drop of Cyclones at High Temperatures,* in *Gas Cleaning at High Temperatures,* R. Clift and J.P.K. Seville, Editors. 1993, Blackie Academic and Professional: London, UK. p. 17-31.

17. Bohnet, M., *Influence of the Gas Temperature on the Separation Efficiency of Aerocyclones.* Chemical Engineering and Processing, 1995. **34**(3): p. 151-156.

18. Prentice, L., Wai Poi, N., and Haque, N., *Life Cycle Assessment of Carbothermal Production of Magnesium in Australia,* in *IMA 67th Annual World Magnesium Conference.* 2010, International Magnesium Association: Hong Kong, PRC. p. 77-82.

19. Prentice, L. H., *A Kinetic Model for the Carbothermal Production of Magnesium.* 2009, CSIRO Process Science and Engineering: Clayton, VIC (Internal Report).

20. *Permanente Squeaks Through,* in *Time.* February 8, 1943.

21. Yuasa, S., Kawashima, M., and Sakurai, T., *Spontaneous Ignition of Ultra-Fine Magnesium Powder without an Original Oxide Coat at Room Temperature in O_2/N_2 Mixture Streams.* Proceedings of the Combustion Institute, 2009. **32**(2): p. 1929-1936.

22. National Fire Protection Authority, *NFPA 69 - Standard on Explosion Prevention Systems, 2008 Edition.* 2007, NFPA: Quincy, MA.

23. National Fire Protection Authority, *NFPA 484 - Standard for Combustible Metals, 2009 Edition.* 2008, NFPA.

24. Revel, G., Pastol, J.-L., Rouchard, J.-C., and Fromageau, R., *Purification of Magnesium by Vacuum Distillation.* Metallurgical Transactions B, 1978. **9B**(December): p. 665-672.

25. Hideo, T., Noboru, K., Gotou, T., Akiyoshi, K., and Yuuji, K., *Manufacture of Metallic Magnesium.* 1979, JP54130413.

26. Odle, R. R. and McClaine, A. W., *Economic Evaluation of a Nozzle-Based Carbothermal Magnesium Process.* 2007, Metallurgical Viability Inc: Elkton, MD.

27. Hori, F., *Apparatus for Obtaining Mg and Ca through Carbon Reduction.* 1980, US4200264.

28. Hori, F., *Method for Obtaining Mg and Ca through Carbon Reduction.* 1979, US4147534.

29. Engell, J., Frederiksen, J., and Nielsen, K. A., *Method of Producing Metallic Magnesium, Magnesium Oxide, or a Refractory Material.* 1998, US5803947.

30. Warren, G. F. and Cameron, A. M., *Process for Producing Magnesium.* 1985, EP0146986.

31. Cameron, A. M., Lotens, J. P., Ouwehand, C., and Aurich, V. G., *Carbothermic Production of Magnesium,* in *Pyrometallurgy '87.* 1987, The Institution of Mining and Metallurgy: London, UK. p. 195-222.

32. Hong, L., Sohn, H. Y., and Sano, M., *Kinetics of Carbothermic Reduction of Magnesia and Zinc Oxide by Thermogravimetric Analysis Technique.* Scandinavian Journal of Metallurgy, 2003. **32**(3): p. 171-176.

33. Li, R., Wei, P., and Sano, M., *Kinetics and Mechanism of Carbothermic Reduction of Magnesia.* Metallurgical and Materials Transactions B: Process Metallurgy and Materials Processing Science, 2003. **34**(4): p. 433-437.

34. Nusheh, M., Yoozbashizadeh, H., Askari, M., Kuwata, N., Kawamura, J., Kano, J., Saito, F., Kobatake, H., and Fukuyama, H., *Effect of Mechanical Milling on Carbothermic Reduction of Magnesia.* ISIJ International, 2010. **50**(5): p. 668-672.

PART 3: MELTING, REFINING, RECYCLING, AND LIFE-CYCLE ANALYSIS

Neale R. Neelameggham

Magnesium is a structural commodity metal, yet it requires a tender touch in melting and casting due to its ignition point of just above 500°C, and molten metal not forming an oxidation retarding film without the help of cover fluxes or inhibitive gases. The articles in this section present basics of how to melt refine and recycle the metal and its alloys.

Cashion and Ricketts fundamental study on "The Use of SO_2 as a Cover Gas for Molten Magnesium" gives insight to the technology to follow. Sulphur and sulfur dioxide has been used for over a century as an important inhibitive cover gas for magnesium, and had lost its place to Sulfur hexafluoride (SF_6), a less noxious gas, during 1975–2005. The same low noxiousness and persistence of sulfur hexafluoride made it one of the worst greenhouse gases, with a global warming potential (GWP) of 23,700 tons of CO_2 equivalent from its seven atom molecule being the highest emissivity among known gases. Worldwide environmental control and carbon taxation forced the industry to seek other methods, including going back to known techniques used in an environmentally safe manner.

In their Canadian plant, Norsk Hydro's magnesium process flow sheet demonstrated lower dioxin levels than their pre-1980s Porsgrunn plant (Eklund *et al.*). As noted earlier, this process is no longer in service, but the fundamentals of this technology are still sound, but are not sustainable as they depend on high-cost petroleum-derived hydrogen gas to handle chlorine in the anode gas. This process was purported to be of lower costs in 2002, but had to close in 2006 due to higher raw material magnesite costs. This process depended on others to supply basic raw material, and the closure was similar to their Norwegian plant in 1991 (noted in the article).

The use of calcium as an ignition retardant in molten magnesium had been known for a while, but was not perfected until recently as noted in the article by Lee and Kim. The article shows calcium made *in-situ* by adding calcium oxide or calcium hydroxide that can effectively suppress ignition of magnesium alloy melt at the normal casting temperatures of 650 to 700°C. Their technique is known as the Eco-Mag technique.

Galovsky and Kuehlein discuss the fluxless recycling of high purity magnesium. Here the authors note that a cover gas of argon + 1%SO_2 or N_2+1%SO_2 has a more significant effect than whether the molten material is treated with bubbling the cover gas or using flux refining with proper settling.

Zhu and colleagues show that vacuum distillation is a feasible approach to recycling the metal from dirty, oily metal, sludge, etc. and Hu and colleagues present fundamentals on improving the heat transfer in the vertical retort process in a study done at Windsor, Ontario for application in Chinese magnesium production facilities.

Ramakrishnan and Koltun's 2004 study and the 2005 follow-up article by Koltun and colleagues has been cited as one of the earliest studies on the life cycle analysis (LCA) of magnesium production by many of the LCA analysts discussing magnesium's usefulness in the abatement of GHG by light weight automobiles. The authors compare the Pidgeon silicothermic process then in use in China and the hypothetical AM Process, of an electrolytic magnesium plant which was never built and operated. The next article describes the LCA of magnesium components used in automobiles. The Chinese producers have since been reducing energy consumption of the process by utilizing the hot retort furnace exhaust gases, as noted in the articles published during the 2009–2012 period, bringing their CO_2 emissions down considerably.

Aarstad and colleagues of the Norwegian University of Science and Technology measured the solubility of Fluorine in Molten Magnesium to be at the 10 to 100 ppm level and suggested using this principle in providing an environmentally effective cover in a controlled fashion. This fundamental study and its implications have not yet become commercialized or compared to the age old technique of using sulfur dioxide as the cover in a controlled fashion.

THE USE OF SO_2 AS A COVER GAS FOR MOLTEN MAGNESIUM

S. Cashion[†] and N. Ricketts[‡]

CRC for Alloy and Solidification Technology (CAST)

[†] Department of Mining, Minerals & Materials Engineering, The University of Queensland
[‡] CSIRO Manufacturing Science and Technology

Abstract

The two major problems with using sulphur hexafluoride (SF_6) as a cover gas component for protecting molten magnesium are its high cost and extremely high Global Warming Potential of approximately 24,000. Sulphur dioxide (SO_2) is a viable alternative to SF_6. In comparison, SO_2 is relatively inexpensive and has a Global Warming Potential of zero.

However, there has been some concern regarding the safety of SO_2 cover gases, particularly when used in die casting melting furnaces. During the 1940s and 1950s, numerous incidents were reported in magnesium die casting operations, where sulphur domes were used to protect the magnesium die casting alloys. This "Sulphur Dome Effect" was attributed to the presence of SO_2 above the melt, with the suggestion that the incidents were associated with formation of magnesium sulphate.

To optimise the conditions for use of SO_2, experiments were conducted to assess the performance of various SO_2-based cover gas mixtures for ingot casting. Ingots of pure and alloyed magnesium were cast under cover gases of SO_2 diluted with air, carbon dioxide, argon and nitrogen. A wide range of SO_2 concentrations was examined, and the ingots were ranked on the basis of visual appearance, dross formation and mould-coat interaction.

To determine the significance of the Sulphur Dome Effect, melts of magnesium alloys were exposed to cover gas mixtures of 0.5% SO_2 + dry air, CO_2 and N_2 at 660, 680 and 700°C. The scale that deposited on the crucible wall was then scraped with a mild steel spatula, simulating the de-sludging procedure used for crucible cleaning. Samples of the melt surface film and the crucible scale were collected for chemical analyses by X-ray Photoelectron Spectroscopy (XPS) and X-ray Diffraction (XRD).

For both pure and alloyed magnesium, scraping of the crucible scale produced significant sparking and burning, with little difference noted when changing the diluent gas or melt temperature. Chemical analyses revealed the presence of MgO and $MgSO_4$ in the surface film and the crucible scale. The highly unstable compound $Al_2(SO_4)_3$ was also detected in the crucible scale. These findings support the currently proposed mechanism for the "Sulphur Dome Effect", of a highly exothermic reaction between molten magnesium and magnesium sulphate.

Introduction

Following the success of elemental sulphur as an effective agent for inhibiting the burning of molten magnesium, sulphur dioxide (SO_2) was one of the first gases trialed in the move to protecting molten magnesium without the use of solid fluxes. The use of SO_2 was widely accepted until the work of Fruehling (1) showed that sulphur hexafluoride (SF_6) gas offered superior protection at lower concentrations. As SF_6 was non-toxic and non-corrosive, it was soon adopted as the industry standard. The magnesium industry is now facing increased pressure to eliminate the use of SF_6 for two main reasons:

a) SF_6 is becoming increasingly expensive.
b) SF_6 has been identified as an extremely potent greenhouse gas, with a 100 year Global Warming Potential (GWP) of approximately 24,000 relative to CO_2.

In the absence of a totally new protection technique, SO_2 has once again become the focus of attention as an alternative to SF_6. The advantages of replacing SF_6 with SO_2 are:

a) SO_2 is readily available and cheaper than SF_6.
b) SO_2 has no Global Warming Potential.
c) The success of SO_2 has already been demonstrated.

A potential concern with the use of SO_2 as a cover gas component was the discovery of a phenomenon known as the "Sulphur Dome Effect". This term has been used to describe incidents observed during die casting operations, where the magnesium was held in a melting furnace equipped with a sulphur dome. Figure 1 shows a schematic diagram of a furnace with a sulphur dome.

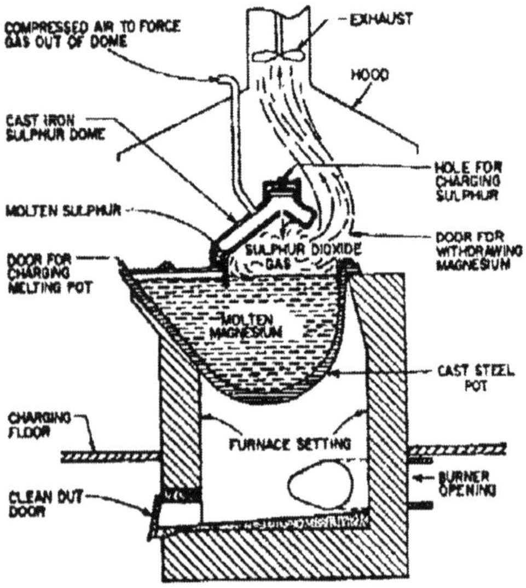

Figure 1: Schematic diagram of a furnace with a sulphur dome (reproduced from Doehler (2)).

To protect the crucible of molten magnesium, SO_2 gas was generated by charging sulphur into the sulphur dome. The heat from the crucible was sufficient to melt the sulphur, and the resulting SO_2 was forced out of the sulphur dome with compressed air. It was reported by Doehler (2) that sparking and explosions were sometimes observed when the molten metal contacted the scale on the crucible wall. These sulphur dome reactions were particularly prevalent during routine maintenance, when the crucible was de-sludged. The cause and mechanism of this effect is largely speculative, and much of the anecdotal evidence originates from the 1940s and 1950s.

The present work was initiated in order to gain a greater understanding of how SO_2 behaves in the presence of molten magnesium. It describes the results of ingot casting trials using SO_2 cover gases, and details a more up-to-date study of the "Sulphur Dome Effect".

Ingot Casting

Experimental

For production of the ingots, a vacuum ingot casting apparatus was used (Figure 2). A 7 kg cast iron mould was used for all trials. The mould was placed in a drawer that slides under a water cooled lid. The top of the drawer is sealed against the lid with an O-ring seal. For each experiment, the mould chamber is purged with the cover gas for approximately 20 minutes, and then evacuated to 0.75 bar. The molten magnesium is drawn, by the vacuum, from the crucible into the mould through a heated steel transfer tube. After evacuation and casting, the cover gas is re-introduced during solidification and cooling.

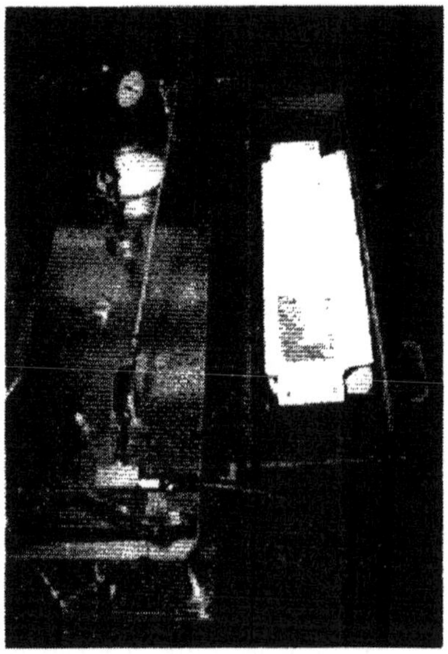

Figure 2: The vacuum ingot casting apparatus.

Results

Table I summarises the experimental conditions for each of the ingots, and Figure 3 shows examples of the ingots cast.

Table I: Experimental conditions for the ingot production.

		Cover gas	
Ingot	Alloy	SO_2 (l/min.)	Air (l/min.)
(a)	Pure	0.0125	10
(b)	AZ91	0.0125	10
(c)	AM60	0.0125	7.8

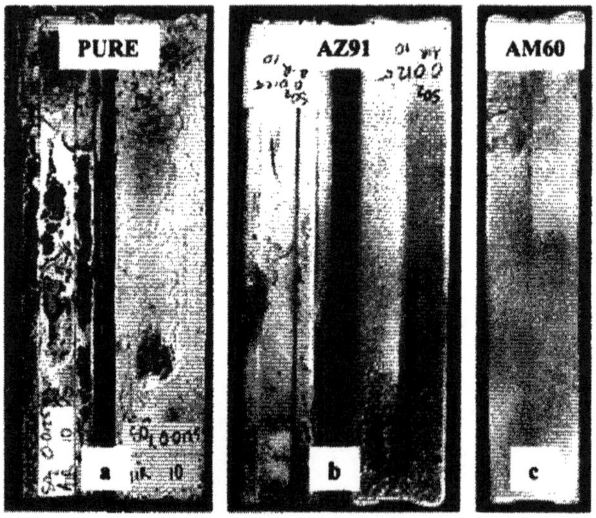

Figure 3: Ingots produced under SO_2 cover gas mixtures.

The pure magnesium ingot (Figure 3a, left) clearly shows a severe reaction between the base of the ingot and the mould coating. The top surface of the ingot (Figure 3a, right) has a high level of dross and shows evidence of burning around the filling point. The ingots of alloyed magnesium have very low levels of dross, minimal mould-coat interaction and a shiny appearance.

Changing the diluent gas from air to CO_2, argon or nitrogen made no universal improvement. The use of Ar/SO_2 and N_2/SO_2 cover gases produced ingots with very low dross levels, but a dull surface finish. Conversely, a CO_2/SO_2 cover gas imparts a shiny, metallic lustre to the ingot, but creates a high level of dross.

The Sulphur Dome effect

Experimental

Melts of magnesium alloy AZ91 and AM60 (approximately 25kg mass) were exposed to three cover gas mixtures of 0.5% SO_2 mixed with dry air, nitrogen and carbon dioxide respectively. All melts were contained in a mild steel crucible (diameter 245mm) with a loose-fitting mild steel lid. A cover gas flow rate of 1 litre per minute was used for all trials, with melt temperatures of 660°C, 680°C and 700°C. Samples of the air/SO_2 protective film formed on the melt surface were periodically removed, and analysed using XPS and XRD. As mentioned in the introduction, most of the sulphur dome reactions occurred during the cleaning and de-sludging of the crucible, whereby the scale that deposited on the crucible wall was disturbed. Therefore, the most important information to be obtained from these experiments was the modelling of the cleaning process and the observation of any interaction between the molten magnesium and the scale.

To recreate this scale, a portion of the melt was removed with a ladle, thus lowering the metal level in the crucible. As the level of the magnesium fell, a band of molten magnesium remained bonded to the crucible wall (shown schematically in Figure 4) and was therefore exposed to the cover gas mixture. Ladling of the melt continued until the melt level was just below a thermocouple attached to the crucible wall.

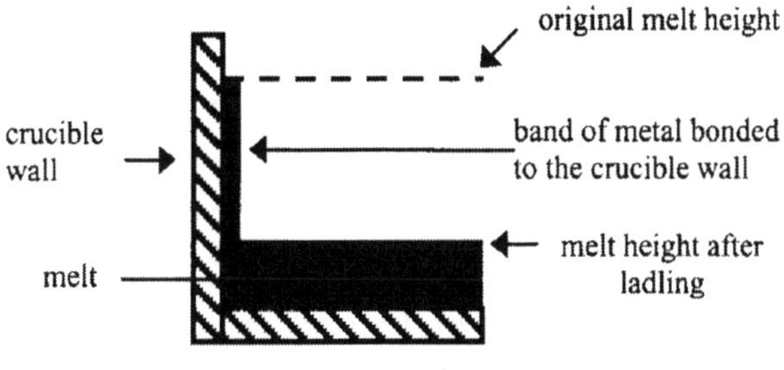

Figure 4: Re-creating scale formation on the crucible wall.

The band of metal was then exposed to the cover gas mixture for up to 5 hours. During this time, a layer of scale formed on the metal band, as shown in Figure 5.

To model the cleaning process, a mild steel spatula was then used to scrape the oxide layer away from the crucible wall. The resulting reaction was monitored with a video camera. Two thermocouples attached to the crucible wall recorded the change in temperature as the reaction proceeded.

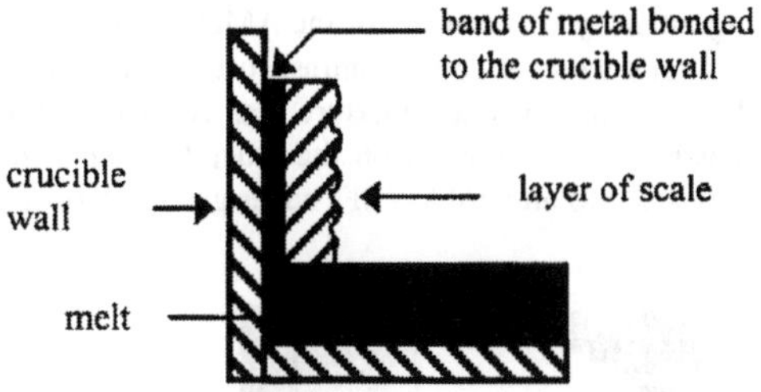

Figure 5: The layer of scale formed on the band of metal, after extended cover gas exposure.

Visual observations of the Sulphur Dome effect

Figure 6 shows the crucible after a portion of AZ91 alloy had been ladled out. The band of metal adhering to the crucible wall can clearly be seen in Figure 6a, while Figure 6b shows the oxidised metal band after exposure to the air/SO_2 cover gas.

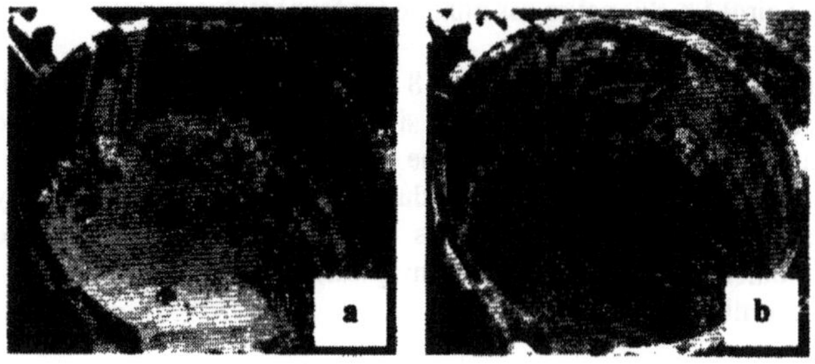

Figure 6: Metal band adhering to the crucible wall; a) – immediately after ladling, and b) – after 4 hours exposure to the cover gas.

Scraping of the scale deposit on the crucible wall produced a violent and sustained reaction (see sequence in Figure 7). Severe sparking was observed immediately upon scraping, followed by localised burning at the point of scraping and on the melt surface where burning deposits fell back into the crucible. In some instances, scraping at one location initiated a chain-reaction, with the entire ring of scale burning around the crucible.

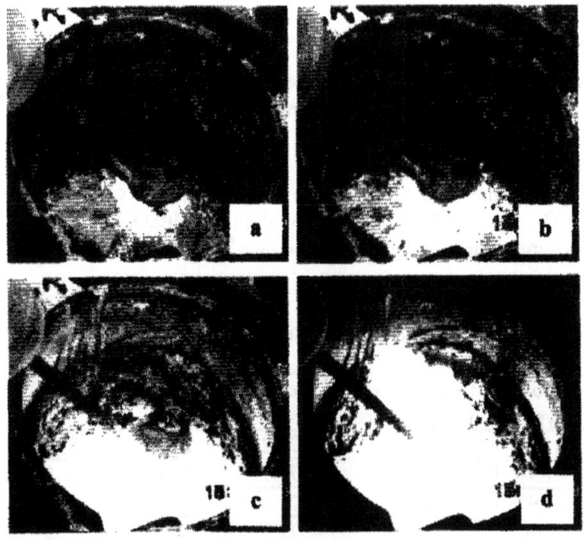

Figure 7: Reaction sequence observed upon scraping the crucible scale, taken from video footage recorded during the trial.

The sparking intensity was similar for the AM60 alloy. However, these observations were in direct contrast to the results obtained when the SO_2 in the cover gas mixture was replaced with SF_6. Figure 8 shows the scale formed on the crucible after a melt of AM60 was exposed to an air/0.5% SF_6 cover gas mixture for four hours.

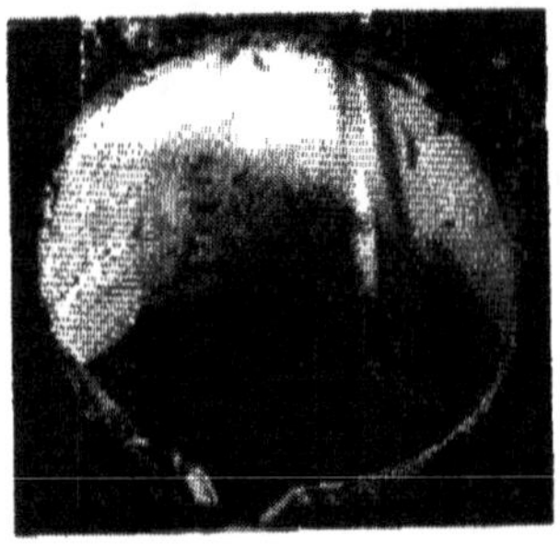

Figure 8: The scale formed after a melt of AM60 was exposed to an air/0.5% SF_6 cover gas mixture for four hours.

Comparing the scale in Figure 8 with that shown in Figure 6b, it can clearly be seen that there is significantly less build-up of scale after prolonged exposure to the air/SF_6 cover gas. Subsequent scraping of the SF_6 scale produced no burning, and only very minor thermite sparking was observed, resulting from the exothermic reaction between magnesium and iron oxide (rust) according to the equation:

$$Mg\,(l) + FeO\,(s) \longrightarrow Fe\,(l) + MgO\,(s) \qquad (1)$$

In another experiment, the crucible was emptied and thoroughly cleaned prior to the trial. The crucible wall was then spray-coated with an aqueous suspension of boron nitride, which is often used as a non-wetting agent for magnesium, and thoroughly dried. In this experiment, when some of the molten magnesium was ladled out, the boron nitride prevented any metal from adhering to the crucible wall. Thus, after exposure to the cover gas, there was no oxide build-up on the wall. Scraping of the wall resulted in only minor thermite sparking, and no violent sulphur dome reaction was observed.

Chemical analyses

From the XPS analyses of samples from the surface film formed on the melt, the main chemical species detected were Mg, O and S. The alloying elements Al, Mn and Zn were not detected in the film. The positions of the Mg, O and S peaks in the XPS spectra suggested that the surface film was composed of MgO and $MgSO_4$. For AZ91, the concentration of $MgSO_4$ increased as exposure to the cover gas increased (see Table II).

Table II: The composition of surface films exposed to an air/SO_2 cover gas mixture at 700°C.

alloy type	time of cover gas exposure	element concentration (atom%) O	Mg	S
AZ91	1 hour	70.2	22.1	7.7
AZ91	2 hours	70.7	17.8	11.5
AZ91	5 hours	71.6	14.6	13.8
AM60	1 hour	71.5	15.3	13.2
AM60	2 hours	73.5	14.1	12.4
AM60	5 hours	69.1	26.0	4.9

The scale from the crucible wall was collected and examine using XRD. As was the case for the surface film, both magnesiur oxide and magnesium sulphate were found to be presen. However, unlike the surface film analyses, aluminium sulphat was detected in the scale.

Temperature profiles

Figures 9 displays an example of the temperature profile at th crucible wall, obtained during the scraping of the crucible wa scale, after the scale had been exposed to the cover gas mixture fc 4 hours.

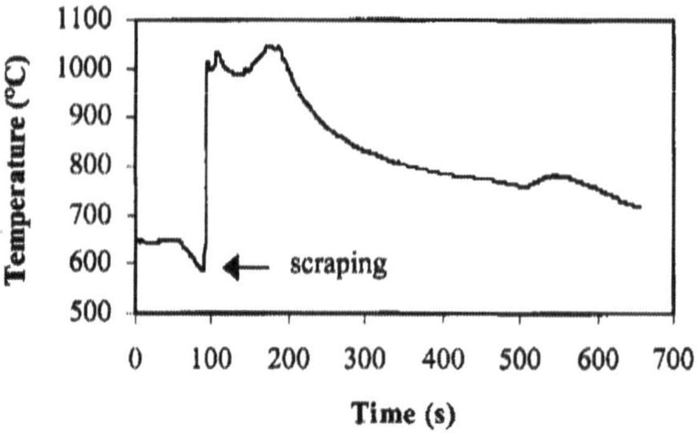

Figure 9: The temperature profile obtained after scraping th crucible wall scale exposed to a cover gas of air + 0.5% SO_2 at melt temperature of 700°C.

The plot is characterised by a sharp increase in temperatur corresponding to the onset of sparking and burning. Th temperature then gradually decreases as the sulphur dome reactic subsides. From a start temperature of 645°C, the plot reaches maximum temperature of 1045°C, giving a difference of 400° After 680 seconds, the temperature of the system is still above th start temperature.

Discussion

Ingot Casting

The ingot photographs show that an air/SO_2 cover gas provid the best results for the alloyed magnesium. No discernib difference was noted between the AZ91 and AM60 ingo Although the cover gas protected the pure magnesium ing effectively, the visual appearance of the ingot is probab unacceptable for commercial production. Importantly, good all ingots were obtained using a very low concentration of SO_2, whi helps negate the deleterious effects of corrosion and toxicity th come with SO_2 use. All of the ingots exhibit significant dro around the filling point, however this is mainly due to the casca nature of the mould filling, which creates turbulence within t molten metal. The occurrence of dross at this location can eliminated with the use of pouring wheels or bottom-filli techniques.

The Sulphur Dome effect

It has been proposed by Russell (3) that the "Sulphur Dor Effect" is a result of the reaction:

$$4Mg\,(l) + MgSO_4\,(s) \longrightarrow MgS\,(s) + 4MgO\,(s) \qquad (2)$$
$$\Delta H^\circ = -352.86 \text{ kcal/mol}$$

The presence of $MgSO_4$ and MgO in the surface film and crucible scale supports this theory. The visual observations, combined with detection of $Al_2(SO_4)_3$ in the crucible scale also supports the measurement by Russell (3) of a highly energetic reaction between $Al_2(SO_4)_3$ and magnesium:

$$12Mg\,(l) + Al_2(SO_4)_3\,(s) \longrightarrow Al_2S_3\,(s) + 12MgO\,(s) \qquad (3)$$
$$\Delta H^\circ = -1026.7 \text{ kcal/mol}$$

No MgS was detected in surface films or the crucible scale. However, it is possible that any MgS present may have oxidised to $MgSO_4$ in the time between sampling and analysing the film.

The temperature profiles did not provide consistent results for each cover gas composition, and no discernible trends were noticed. The examples provided clearly indicate the occurrence of a highly exothermic reaction, resulting from the disturbance of the crucible scale.

The most surprising result was the absence of any reaction when molten metal was splashed against the scale or during charging of the melt with fresh magnesium. This can be explained in terms of the phase structure of the scale. Cashion (4) showed that, when magnesium is protected by an air/SO_2 cover gas mixture, the surface film formed has a distinctly layered structure. This is shown schematically in Figure 10 (a). If this layered growth mechanism is assumed for the wall scale, it can be seen that the reactive $MgSO_4$ is "shielded" by the outer layer of MgO, as displayed in Figure 10 (b). Therefore, the sulphur dome reaction will not occur until the MgO layer is broken (i.e by scraping the wall scale), allowing the molten magnesium to contact the $MgSO_4$.

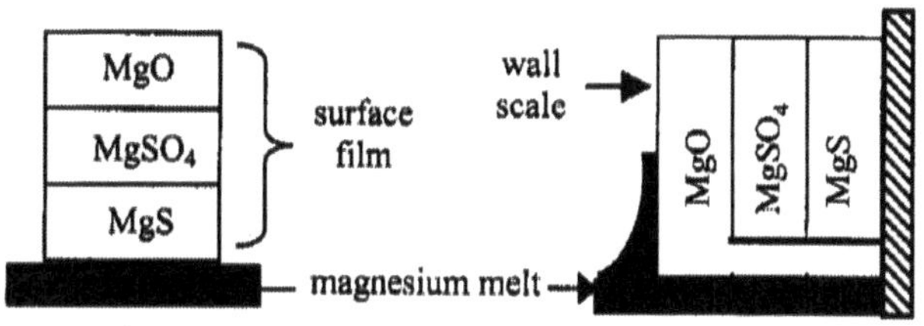

Figure 10: Layered structure of; (a) - surface film, and (b) - wall scale, formed under an air/SO_2 cover gas.

Conclusions

This work has shown that a low concentration of SO_2 in air provides an effective cover gas for the casting of ingots of AZ91 and AM60 magnesium alloy. For the apparatus used in this work, the use of air/SO_2 cover gas mixtures for pure magnesium did not produce ingots that were acceptable in terms of visual appearance and dross level. In light of the advantages of SO_2 detailed in the introduction, the use of air/SO_2 cover gases would be a suitable replacement for air/SF_6 mixtures.

However, under certain operating conditions, the scraping of a crucible scale that had been exposed to a cover gas containing SO_2 caused a highly energetic reaction. Sustained sparking was observed, with subsequent burning of the scale. Sections of the burning scale detached from the crucible wall and fell into the magnesium melt. As the experiments were performed on a laboratory scale, the severity of the reaction could be kept to a safe level. However, for a commercial-sized operation, such a reaction could have serious safety implications.

It would appear that the occurrence of this reaction depends on the formation and subsequent disturbance of an oxidised scale. If techniques can be employed to minimise the growth of such a scale, for example sprayed crucible coatings, it should be possible to safely use cover gases containing SO_2 for protection of magnesium in a melting furnace environment. An added benefit of using such sprayed coatings may be increased crucible life.

Acknowledgments

The authors would like to thank Mr. Craig Korn (CSIRO) for conducting the scraping trials for all experiments, Dr. Barry Wood (UQ) for the XPS analyses and Mr. Lambert Bekessy (UQ) for the XRD analyses. The authors wish to acknowledge the financial assistance provided by the CRC for Alloy and Solidification Technology (CAST) and the Australian Magnesium Corporation (AMC).

References

1. Fruehling, J.W., "Protective Atmospheres for Molten Magnesium", Ph.D. Thesis, University of Michigan, 1970.

2. Doehler, H.H, "Die Casting", International Edition, McGraw Hill, New York, January 1951.

3. Russell, A.S., "Sparking Magnesium", Metal Progress, vol. 59, pp 228-231, 1951.

4. Cashion, S.P., "The Use of SF_6 for Protecting Molten Magnesium", Ph.D. Thesis, The University of Queensland, 1998.

AN IMPROVED PROCESS FOR THE PRODUCTION OF MAGNESIUM

Hans Eklund, Hydro Magnesium, N-0246 Oslo, Norway
Per Bjørn Engseth, Hydro Magnesium, N-3907 Porsgrunn, Norway
Birger Langseth, Norsk Hydro ASA, Research Centre, N-3907 Porsgrunn, Norway
Thorvald Mellerud, Hydro Magnesium, Av. Marcel Thiry 83, B-1200 Brussels, Belgium
Oddmund Wallevik, Norsk Hydro ASA, Research Centre, N-3907 Porsgrunn, Norway

Abstract

In the late 60's, Norsk Hydro started to develop a process for production of anhydrous $MgCl_2$ based on dehydration of $MgCl_2$ brines. The development of the dehydration process has continued, and today an improved third generation process is ready for implementation in brown-field expansions or green-field projects. Several elements of the improved process have already been installed in the Norsk Hydro Bécancour plant. Compared to the previous generation, this new technology offers a minimized energy consumption of 6 kWh[1]/kg Mg in a large scale mass and energy integrated plant, fully automated and continuous operation with a typical productivity of 1 t Mg/man hour. The Norsk Hydro dehydration technology (considered BAT in EU) offers an environmentally sound solution with minimal emissions of chlorine and chlorinated hydrocarbons, high productivity and improved quality of the anhydrous $MgCl_2$ feed, ensuring a significantly improved performance of the electrolysis.

History

Norsk Hydro has produced primary Magnesium in Porsgrunn, Norway since 1951. This production is based on the IG Farben carbo-chlorination process for production of anhydrous $MgCl_2$ as feed- stock to the electrolysis. An improved version of the process is in operation at the Porsgrunn plant.

The availability of vast amounts of $MgCl_2$-brines, generated as a by-product from the Potassium industry in the two former German states initiated a research and development project at Norsk Hydro in the middle of the 60's. The raw brine imported from Germany contained trace amounts of constituents detrimental to the performance of the Norsk Hydro electrolysis cell technology /1/.

After an initial screening of possible ways to produce anhydrous $MgCl_2$ from the purified brines, the main route with sequential dewatering of the brine with the final step of water removal to be carried out in HCl gas atmosphere was endeavored upon /2/. The process steps involved were evaporation, particle generation by prilling, air drying and finally drying in hot HCl gas atmosphere to generate water-free $MgCl_2$-prills.

The basic elements of the evaporation and prilling steps were based on Norsk Hydro fertilizer experience. The drying steps in air and HCl were operated in pilot scale, which yielded important information about drying chemistry, kinetics and not the least information about materials technology in the demanding process environment of hot HCl gas, water and acidic $MgCl_2$ solutions. Many difficulties were encountered and were overcome in a several yearlong pilot plant testing program. Finally, an industrial process concept was demonstrated and a decision was made in the early 70's to build a full scale demonstration plant with an annual design capacity of 15 000 t Mg. The by-product chlorine gas was to be purified and liquefied before being exported to a nearby Norsk Hydro Vinyl Chloride complex.

The plant was commissioned in 1978. The first years of operation were plagued with numerous problems related to process chemistry and equipment failure/corrosion from the harsh process conditions. A fire in the plant's thermal oil system caused severe damage in 1981.

When the plant was shutdown in 1991, due to high raw material costs, following reunification of Germany and unfavorable Magnesium market conditions at the time, the plant availability was above 92 % (basis: 8760 hrs), and the production some 30 % above initial design. Continuous optimization efforts had reduced the energy consumption from an initial figure of 25 kWh[ii]/kg Mg to 19 kWh/kg Mg.

The Norsk Hydro Bécancour plant with a nameplate capacity of 40 000 t Mg/yr, split on two dehydration trains was started-up in 1989. The process concept includes the same process unit operations as the Porsgrunn plant. Also the start-up of this plant was not without problems, and several process modifications were required before the targeted plant performance was established.

At a production level of 70 kt, the Bécancour dehydration plant will have an estimated energy consumption of about 11 kWh/kg Mg.

With the experience from start-up of the Bécancour plant on hand; Hydro Magnesium initiated a basic process research program in 1993 to further develop the Norsk Hydro dehydration technology. The main objectives with the project, designated Mg-2000, were continued reduced energy consumption and reduced investment and maintenance costs through process simplifications. Compared to the previous technology, the Mg-2000 concept has a simplified evaporation step, the prilling and air drying steps are replaced by a new fluidized bed granulation system /3/, a larger unit size HCl drying bed is applied, and energy integration with other plant systems is explored to a maximum extent.

The granulation concept has been demonstrated in a pilot plant installation. The development program for the Mg-2000 project is now completed and ready to be included into the design basis of a major brown- or green-field expansion project. The energy consumption of the Mg-2000 dehydration process for a 90 kt Mg/yr single train plant is estimated at 6 kWh/kg Mg.

Process description

Introduction

Raw materials for electrolytic Magnesium plants can be either magnesite, dolomite/sea water, brucite or $MgCl_2$ brine. The latter is only feasible if chlorine or HCl acid can be used for other purposes. Some electrolyte must be bled out due to accumulation of salts (1 – 4 %). Make up for these losses can be externally purchased HCl acid or $MgCl_2$ brine.

The capacity of such a plant can vary, one train Dissolving/Dehydration plant can have capacities between 15 to 90 kt Mg per year. A block diagram for the new Norsk Hydro Magnesium process is shown in Figure 1. A more detailed block diagram of the Dehydration process is presented in Figure 2.

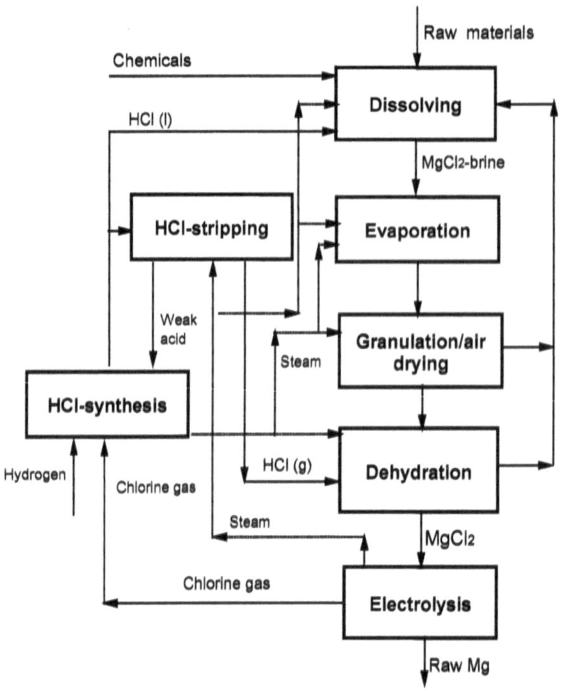

Figure 1: Overall Block Diagram, Norsk Hydro Magnesium Plant

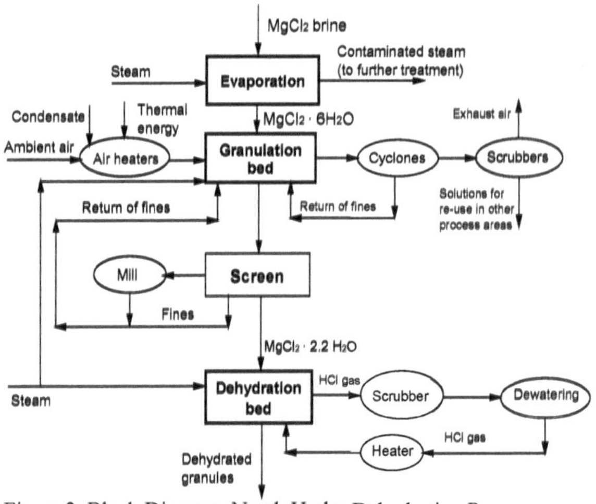

Figure 2: Block Diagram, Norsk Hydro Dehydration Process

Description of the various process steps

In order to have a suitable feed for the dehydration plant, a dissolving and brine purification plant is required for processing the various raw materials. Several HCl and $MgCl_2$ streams from other plant areas are returned to the dissolving plant. The purified brine is either sent to storage or routed to the evaporation step.

Brine evaporation can be done in a system consisting of a three-stage plate evaporator system. Steam produced in other areas of the Mg process is used as energy carrier. The brine entering and leaving the evaporation has a $MgCl_2$ concentration of approx. 34 % (wt) and 47 % (wt) respectively. A concentration of 47 % by weight corresponds to 6 moles H_2O per mole $MgCl_2$.

In the granulation step, particles with a water content of 2.2 moles H_2O per mole $MgCl_2$ are formed in a fluidized granulation bed /3/. The brine is sprayed into the fluidized bed by the use of nozzles submerged in the fluidized bed. The seed particles (return from mill, screen and cyclone) are fed into the system below the nozzles. Natural gas is used to indirectly heat the fluidizing air, and steam recovered in HCl synthesis is used to supply heat to panels submerged in the fluidized bed. The granulation step also includes an air drying step. In total, 3.8 moles H_2O are removed per mole $MgCl_2$. Most of the dust in the exhaust air from the granulation bed is removed by cyclones and returned to the granulation bed. The remaining dust and HCl produced during the drying process are removed in scrubbers, and the acidic $MgCl_2$ solutions from the scrubbers are returned to other process areas. The granule flow from the granulation bed is screened. The oversized particles are crushed and along with the fine fraction from the screen, returned as seed to the granulation bed. The product fraction (0.7 - 2.5 mm particles) enters the dehydration step.

The particles from the granulation bed are dried to anhydrous granules in the dehydration step. The dehydration step consists of a multistage fluidized bed, using HCl gas as fluidizing medium. The HCl gas leaving the multistage fluidized bed is purified, water is removed and the gas is heated in an electrical calorifier before being recirculated back to the fluidized bed. Energy is also supplied to the system by the use of panels and coils in the fluidized beds. Steam is used for energy supply to panels on the upper trays of the bed. High temperature coils, such as thermal oil, molten salt or electrical coils, are used for the lower trays. A bleed stream from the HCl circuit is utilised to increase the brine concentration in another process step. The anhydrous granules from the HCl dryers are vented and then cooled in a fluidized bed. Cooling coils are used in the cooling bed. The cooled granules are pneumatically transported to storage bins.

Pneumatic conveyors transport the granules from the storage bins to electrolysis where $MgCl_2$ is split into raw magnesium and chlorine gas. A description of the Norsk Hydro electrolysis technology is presented in /1/. The chlorine gas is purified and then routed to HCl synthesis. Heat recovered from the electrolytic cells is utilised for steam production.

In the HCl synthesis plant, HCl gas is absorbed in water and weak acid streams from other part of the plant, and highly concentrated HCl acid is produced. The acid is partly used in the HCl gas production unit where HCl gas is produced as make-up for the Dehydration plant. All the steam used in panels in the granulation bed and in the HCl dryer is produced in the HCl synthesis plant /4/.

Performance data.

The following criteria are proposed to describe the performance of alternatives for production of dry Magnesium chloride suited for the Norsk Hydro high-amperage diaphragm-less electrolyser:

- Specific energy consumption
- Manpower productivity
- Product quality
- Availability and maintenance cost
- Emissions

An evaluation of these criteria is given below for the Norsk Hydro Mg-2000 technology.

Specific energy consumption

Specific energy consumption is an important parameter both in terms of process efficiency and production cost. Reduced energy consumption has therefore been an important issue during the development of the new technology.

Figure 3 shows the development of energy consumption and production capacity for the Norsk Hydro Dehydration process (Gen stands for generation). Gen 2 (Future) shows estimated data for an expanded Bécancour plant with 72 000 t Mg/yr production, using the present technology, while Gen 3 is a feasibility study estimate for the latest technology (Mg-2000), applied to a 90 000 t Mg/yr plant located in Porsgrunn, Norway.

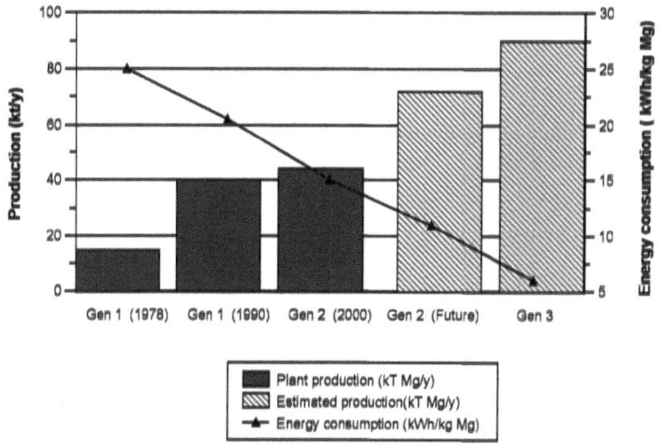

Figure 3: Energy consumption and production for Norsk Hydro dehydration technology

Regarding the energy consumption for the Mg-2000 concept, the reduction is mainly due to close energy integration with HCl-synthesis and electrolysis, and the introduction of the granulation technology.

Manpower productivity.

The new Dehydration process is a highly automated plant running a production of 90 000 t Mg/yr through one train. The operation is continuous, with a production manning of five operators per shift. The output of the dehydration plant is in the range of 1 ton Mg per working hour.

Product quality.

The dry magnesium chloride produced in the Dehydration plant must fulfil the requirements for the Electrolysis described in /1/. This demands that the cell feed produced in the Dehydration plant must meet the requirements needed to achieve current efficiency above 90 %, cell life above 6 years and very low sludge formation. Consequently, the content of water, other oxygen containing species, and other impurities must be very low. The production in Bécancour has proven that these goals are consistently met in commercial operation.

Availability and maintenance cost.

Based on experience from the first Dehydration plant in Porsgrunn, and the Bécancour plant, the plant availability is expected to be above 8000 hrs per year, or 91 %. To achieve this, material selection and in-depth process understanding are crucial elements. The main stoppage time is due to two annual shutdowns for maintenance work.

Including both the maintenance personnel mentioned above and maintenance materials, the total maintenance cost for a 90 kt Mg/yr dehydration plant will be below 1.5 US c/lb Mg.

Emissions

Emissions from the new Dehydration process to air and water will be in the same range or lower than what is achieved in the Bécancour plant, and which is regarded as BAT today. A comparison of emission rates between the existing Chlorination step, which is used in Norway today, and the new Mg-2000 dehydration process is provided in the following table:

Relative emissions	**Units**	**Chlorination**	**Mg-2000 dehydration**
Dioxins to water	µg /t Mg	22	< 0.1
Dioxins to air	µg/t Mg	44	< 10
CHC * to water	g/t Mg	0.03	< 0.005
Cl_2/HCl to air	kg/t Mg	3.2	< 0.5
SO_2 to air	kg/t Mg	5.0	< 0.2

* Sum Hexachlorobenzene (HCB), Pentachlorobenzene (PCB) and Octachlorostyrene (OCS).

References

1. O. Wallevik, K. Amundsen, A. Faucher, Th. Mellerud. "Magnesium Electrolysis - A Monopolar Viewpoint". Paper presented at the TMS 2000, Nashville, Tennessee, March 12-16, 2000.

2. Ø. Bøyum, K. E. Eriksen, P. Solberg, K. W. Tveten. "Process for the preparation of anhydrous $MgCl_2$ prills". US patent no.: 3,742,100. October 27, 1970.

3. B. Langseth, A.B. Frigstad, L. K. Grønstad. "Method of producing magnesium chloride granules". Norwegian patent no.: 309260, October 11, 1996. International application no.: PCT/NO97/00275. Publication no.: WO98/16306, October 10, 1997.

4. H. R. Eklund, P. B. Engseth, B. Langseth, T. Pettersen, B. Rønnevig, J. Sandviknes. "Method and apparatus for synthesis of HCl". Norwegian application no.: 19994856, October 6, 1999. International application no.: PCT/NO00/00313. Publication no.: WO01/25143. Sept. 22, 2000.

[i] kWh refers to both kWht (thermal) and kWhe (electrical)
[ii] kWh/t Mg refers only to the dehydration process

EFFECT OF $Ca(OH)_2$ ON OXIDATION AND IGNITION RESISTANCES OF PURE MG

Dong-In Jang and Shae K. Kim
Eco Materials & Processing Department, Korea Institute of Industrial Technology, 7-47 Songdo-dong, Yeonsu-gu, Incheon 406-840, Korea

Keywords: Mg alloy, CaO, TGA, AES, DTA

Abstract

CaO added Eco-Mg alloy has the potential to maximize the environmental benefits provided by lightweight, unlimited, and recyclable Mg alloy by eliminating global warming SF_6 or other protective gases as well as Be addition. It is possible to ensure the safety during manufacturing and application, especially without sacrificing process abilities and mechanical properties and increasing the cost of Mg alloy. However, the one limitation of CaO is prone to moisture absorption during storage. Instead of CaO, it is attempted to use $Ca(OH)_2$, which does not absorb moisture during storage, for Eco-Mg alloy. This paper discusses the effect of $Ca(OH)_2$ on oxidation and ignition resistances of pure Mg and to compare the results with them of CaO addition. The purpose of this study is to investigate effects of CaO and $Ca(OH)_2$ on pure Mg through microstructure observation, ignition test and phase analysis. With increasing $Ca(OH)_2$ content, the hardness of $Ca(OH)_2$ added Mg alloy increased by grain refinement. From oxidation test by TGA, the oxidation behavior of $Ca(OH)_2$ added Mg was comparable to that of CaO added Mg alloy for the previous study. Consequently, it seems that reduction of fluidity and mold adhesion could be minimized by adding small amount of $Ca(OH)_2$ which is cheap and easy to be handled due to its stability in application for Eco-Mg alloy.

Introduction

Mg alloys are widely used as structural materials for transportation industry due to their properties such as low density, good castability and high specific strength [1-2]. However, molten Mg and Mg alloys are ignited without the melt protective gases during melting and casting process due to their high reactivity [3]. Melt protective gases such as SF_6 and HFC-134a have been used in order to prevent the ignition of molten Mg during melting and casting. However, the use of these gases is limited because of theirs significant impact on global warming potential. If gas concentration increases during melting, it has also bad influence on corrosion resistance of steel based equipments. In order to reduce the use of melt protective gases, many researchers have performed studies to improve the ignition resistance of Mg through alloying to Mg alloys. It is well known that Ca is used to improve the ignition and oxidation resistance of Mg. However, Ca is difficult to be handled due to their high reactivity under an ambient atmosphere [4-8].

CaO added Eco-Mg alloy has the potential to maximize the environmental benefits provided by lightweight, unlimited, and recyclable Mg alloy by eliminating global warming SF_6 or other protective gases as well as Be addition [9-15]. It is possible to ensure the safety during manufacturing and application, especially without sacrificing process abilities and mechanical properties and increasing the cost of Mg alloy. However, the one limitation of CaO is prone to moisture absorption during storage. Instead of CaO, it is attempted to use $Ca(OH)_2$, which does not absorb moisture during storage, for Eco-Mg alloy.

This paper discusses the effect of $Ca(OH)_2$ on oxidation and ignition resistances of pure Mg and to compare the results with them of CaO addition. Pure Mg was used instead of Mg alloys to minimize the effects of other elements. The oxidation resistance was investigated through TGA (Thermogravimetric Analysis). The AES (Auger electron spectrometry) depth profile was performed to evaluate the oxidation and ignition resistances in terms of surface oxide stability. The ignition resistance was examined by DTA (Deferential Thermal Analysis) under dry air atmospheres.

Experimental Procedures

Pure Mg was melted in a steel crucible in an electric resistance furnace under SF_6 and CO_2 gases. The desired fraction of CaO or $Ca(OH)_2$ particles was added into molten pure Mg. The CaO and $Ca(OH)_2$ added melts were cast into a steel mold preheated at 200°C. Chemical compositions were determined by ICP-mass spectrometer. The chemical compositions are given in Table 1. In order to observe the change of the microstructures by CaO and $Ca(OH)_2$ addition, the cast billets were polished and observed using an Olympus PME3 microscopy. The variation of hardness was investigated by Rockwell hardness tester (15T scale). The obtained values of hardness were based on the average of 7 tests.

Table 1. Chemical Composition of CaO and $Ca(OH)_2$ added Mg.

Added (wt.%)	Measured (wt.%)	Added (wt.%)	Measured (wt.%)
0.1CaO	0.10	0.1$Ca(OH)_2$	0.20
0.5CaO	0.35	0.5$Ca(OH)_2$	0.37
1.0CaO	0.82	1.0$Ca(OH)_2$	0.70
2.0CaO	1.67	2.0$Ca(OH)_2$	1.70

Oxidation test was performed with TGA (Thermogravimetric Analysis) to investigate the effect of $Ca(OH)_2$ on oxidation resistance of pure Mg. TGA was performed at 500°C for 5 hrs. After TGA, the surface was analyzed by AES (Auger Electron Spectroscope). Mapping Mg, Ca and O by EDS (Energy Dispersive X-ray Spectroscopy) was performed to investigate the existence of $Ca(OH)_2$ in pure Mg. Ignition test was performed to investigate the effect of $Ca(OH)_2$ on the ignition resistance of pure Mg. DTA (Deferential Thermal Analysis) was performed under ambient atmosphere. Although there were the previous results, all the tests were also done to compare the effect of CaO with $Ca(OH)_2$.

Results and Discussions

Fig. 1 shows the changes of surface appearance for pure Mg, 0.35wt%.CaO added Mg alloy and 0.3wt%.$Ca(OH)_2$ added Mg alloy during casting without a protective gas under ambient atmosphere. In the case of pure Mg as shown in Fig. 1(a), the tarnished surface appeared just after being cast. In 30 seconds as shown in Fig. 1 (b), burning oxides covered all the surface of pure Mg melt. As shown in Fig. 1 (c) and Fig. 1 (d), burning continually proceeded and white flame occurred with lots of heat generation on the surface of pure Mg. The surface was totally covered with burning and ignition oxides. On the contrary, there was no burning oxides and shiny surface appeared in 0.35wt.%CaO added Mg alloy, although it was cast without protective gas, as shown in Fig. 1 (e) through Fig. 1 (h). Although not so severe, some of burning oxides appeared in 0.3wt.%$Ca(OH)_2$, as shown in Fig. 1 (j) through Fig. 1 (l). 0.3wt.%$Ca(OH)_2$ added Mg alloy was analogous to 0.35wt%.CaO added Mg alloy in terms of Ca weight percentage. The continuous oxidation and ignition of pure Mg melts was occurred because of the porous surface MgO oxide films which could not act as a protective layer to prevent further oxidation and ignition.

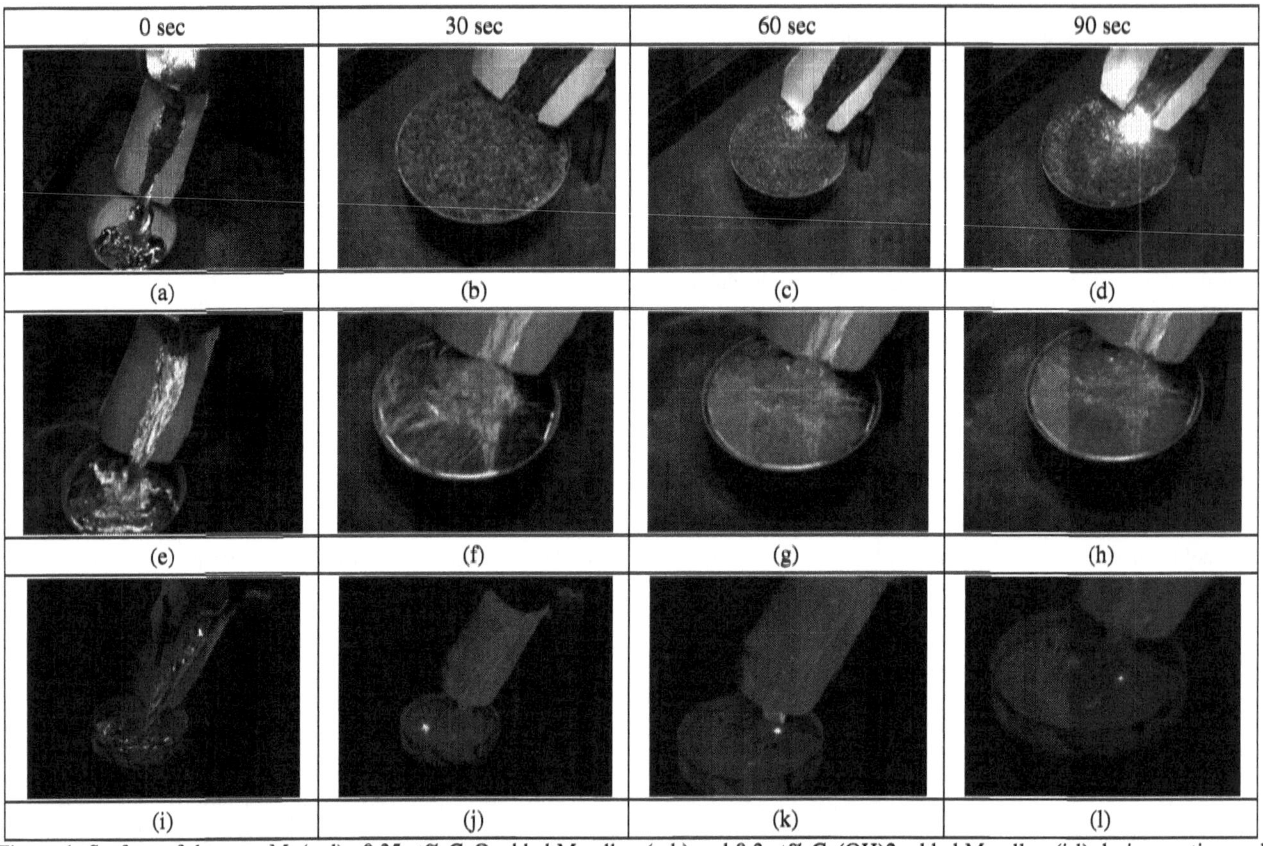

Figure 1. Surface of the pure Mg(a-d), 0.35wt%.CaO added Mg alloy (e-h) and 0.3wt%.Ca(OH)2 added Mg alloy (i-l) during casting and solidification; (a), (e) and (i): 0sec., (b), (f) and (j): 30sec., (c), (g) and (k): 60sec. and (d), (h) and (l): 90sec..

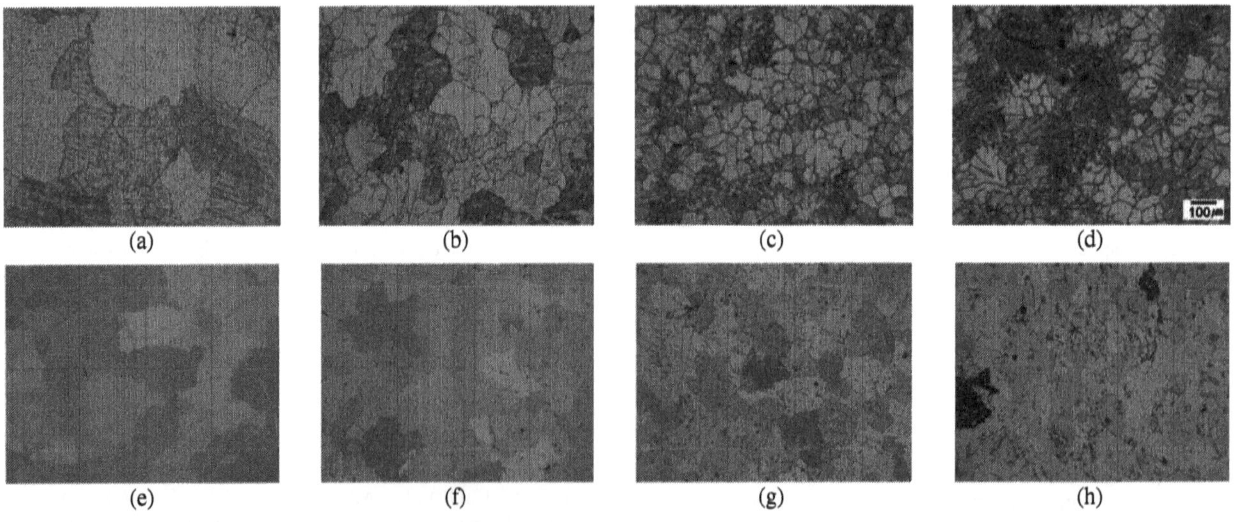

Figure 2. As-cast microstructures of (a) Mg-0.10wt.%CaO, (b) Mg-0.35wt.%CaO, (c) Mg-0.82wt.%CaO, (d) Mg-1.67wt.%CaO, (e) Mg-0.20wt.%$Ca(OH)_2$, (f) Mg-0.37wt.%$Ca(OH)_2$, (g) Mg-0.70wt.%$Ca(OH)_2$, and (f) Mg-1.70wt.%$Ca(OH)_2$ alloys.

Fig. 2 shows the microstructures of CaO and $Ca(OH)_2$ added Mg alloys. With increasing CaO content, there was a visible variation on the microstructures of CaO added Mg alloys and the grains were refined even in the case of 0.35wt.%Ca added Mg alloy. It could be explained that the grains were refined because Mg_2Ca phase on the grain boundary was formed even below Ca solubility (1.35wt.%) [16]. The microstructures of $Ca(OH)_2$ added Mg alloy were different from those of CaO added Mg alloy. The grains in $Ca(OH)_2$ added Mg alloys were refined by Mg_2Ca with 0.70wt.% $Ca(OH)_2$ addition.

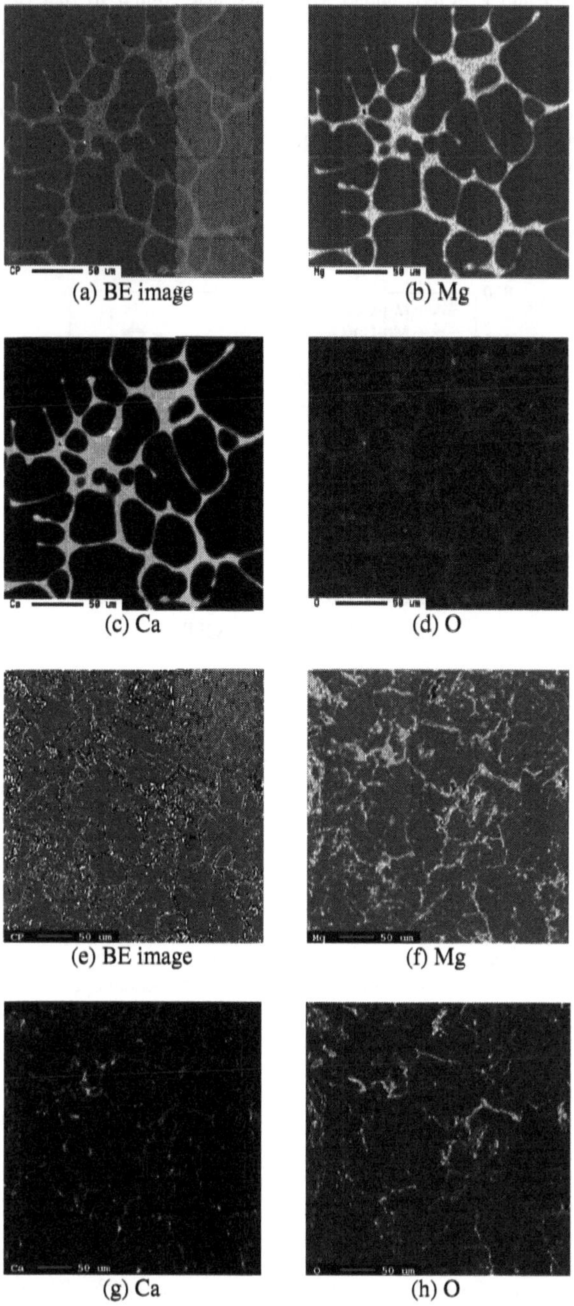

Figure 3. EPMA mapping results of Mg, Ca and O for Mg-1.67wt.%CaO (a-d) and Mg-1.70wt.%$Ca(OH)_2$ (e-h) alloys.

Fig. 3(a)-3(d) show EPMA mapping result of Mg, Ca and O for 1.67wt.%CaO added Mg alloy. Ca was observed only on the grain boundary and existed as Mg_2Ca. Fig. 3(e)-3(h) show EPMA mapping result of Mg, Ca and O mapping for 1.70wt.%$Ca(OH)_2$ added Mg alloy. The concentration of Ca was the same as 1.67wt%CaO added pure Mg and O was not detected.

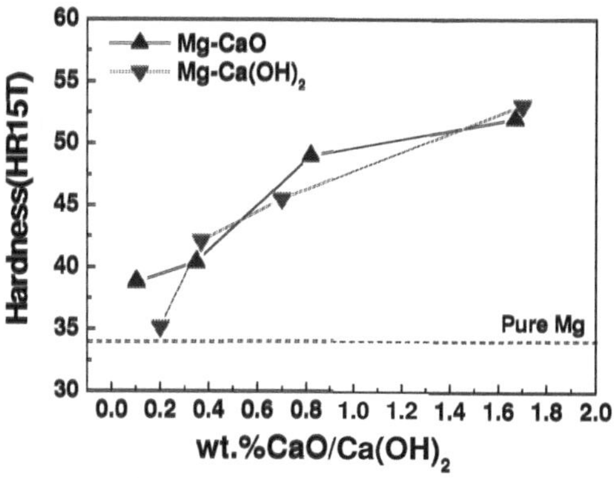

Figure 4. Results of Hardness (HR15T) in CaO and $Ca(OH)_2$ added Mg alloys.

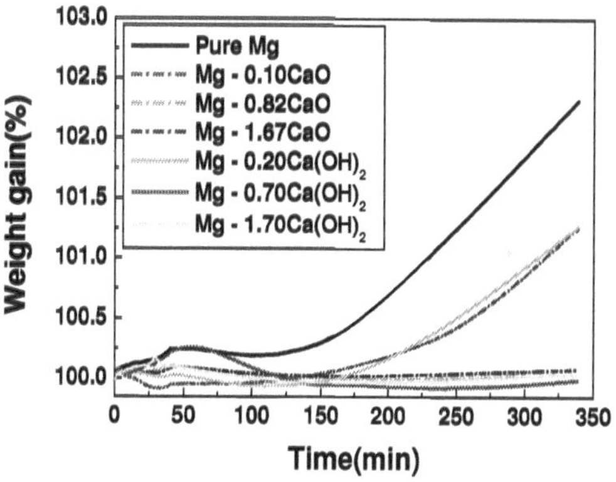

Figure 5. Weight gain curves of pure Mg, CaO added Mg alloy and $Ca(OH)_2$ added Mg alloy oxidized during TGA at 500°C for 5 hrs.

Fig. 4 shows the dependence of hardness on the mass fraction of CaO and $Ca(OH)_2$ addition. The hardness of CaO and $Ca(OH)_2$ added Mg alloys was increased by grain refining.

Fig. 5 shows the weight gain of pure Mg, CaO added Mg alloys and $Ca(OH)_2$ added Mg alloys as a function of oxidation time at 500°C. The results indicate that pure Mg showed a parabolic law due to the loose and porous structure of MgO. On the contrary, the oxidation rate was reduced with respect to CaO or $Ca(OH)_2$ content for CaO added Mg alloys and $Ca(OH)_2$ added Mg alloys at the same temperature.

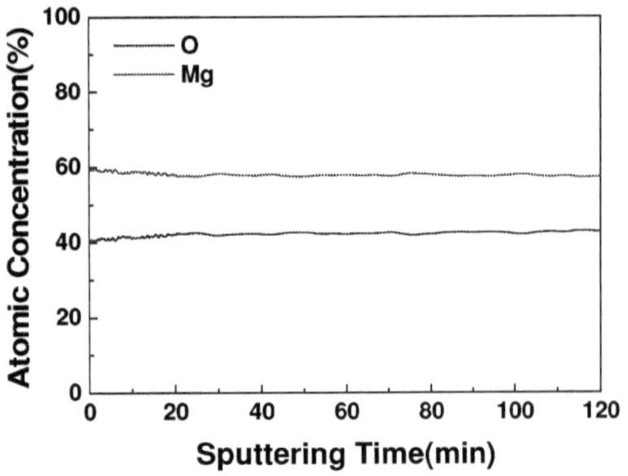

(a)

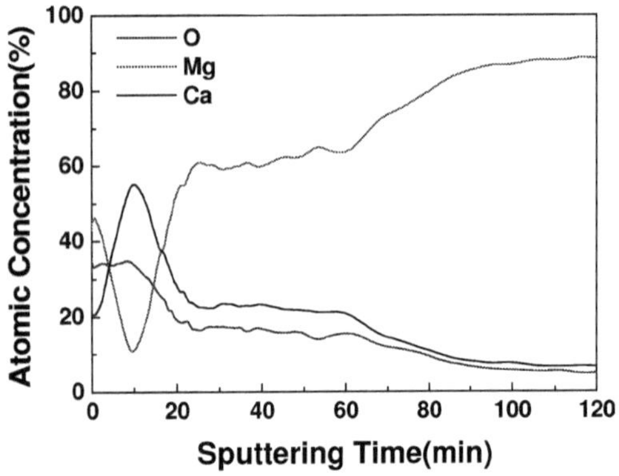

(b)

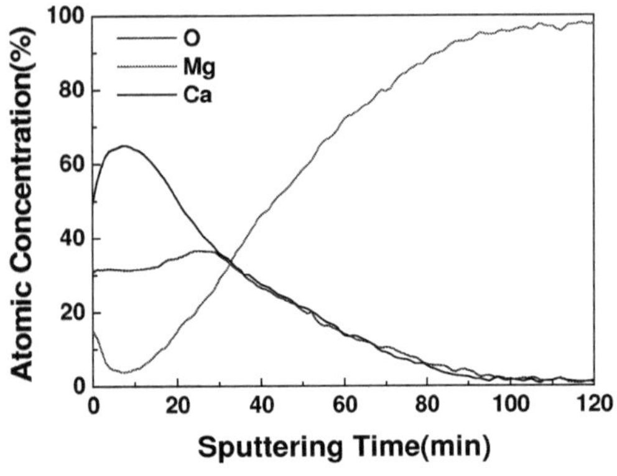

(C)

Figure 6. AES depth profiles sputtered from the surface of (a) pure Mg, (b) CaO added Mg alloy and (c) $Ca(OH)_2$ added Mg alloy oxidized during TGA at 500°C for 5 hrs.

Fig. 6 shows the AES depth profile sputtered form the surface. In pure Mg, it was confirmed that there was a thick MgO oxide layer on the surface because the concentration of Mg and O were constant for sputtering time. It could be explained though Pilling-Bedworth ratio, R [17]. Pilling-Bedworth ratio means the ratio of oxide volume to metal volume. Al is greater than 1, which explains the protective effect of the Al oxide. Because the Pilling-Bedworth ratio for Mg is 0.81, MgO oxide film on surface of molten Mg could not act as a protective layer to prevent ignition and oxidation of molten Mg. In CaO and $Ca(OH)_2$ added Mg alloys, there were the thin oxide layers mixed with MgO and CaO. The mixed MgO and CaO oxide layer, generated by CaO or $Ca(OH)_2$ addition, exhibits protective behavior below about, while weight gain by oxidation is accelerated with respect to temperature above 500°C because of porous MgO surface oxide film at high temperature. The behavior of oxidation over 500°C will be discussed in terms of ignition. The oxidation behavior of CaO and $Ca(OH)_2$ added Mg alloys was similar to that of Ca added Mg studied by B. S. You et al [4, 5]. Although it could not be explained by Pilling-Bedworth theory, it has been reported that the oxidation resistance of Ca added Mg alloys was suppressed due to the dense film which consisted of porous CaO (0.78) and MgO (0.83) [12-13]. Further study should be performed to examine the ignition suppression mechanism by Ca, CaO, and $Ca(OH)_2$ addition.

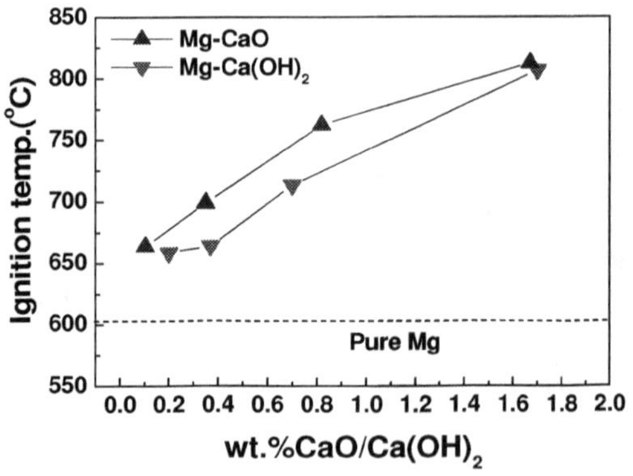

Figure 7. Results of ignition temperature under an ambient atmosphere in Cao and CaO added Mg alloys.

Fig. 7 shows the result of ignition temperature under an ambient atmosphere by DTA. With increasing CaO and $Ca(OH)_2$ contents, the ignition temperatures of CaO and $Ca(OH)_2$ added Mg alloys increased. The ignition temperature of 0.10wt.%CaO added Mg alloy was much higher than that of pure Mg. With increasing CaO content, the ignition temperature of CaO added Mg alloys increased linearly. The ignition temperature results of $Ca(OH)_2$ added Mg alloys exhibited similar tendency, compared with those of CaO added Mg alloys. The oxidation and ignition of pure Mg were occurred because of porous MgO surface oxide film at high temperature. MgO oxide film on surface of pure Mg could not act as a protective layer to prevent the oxidation at 500°C and ignition at higher temperatures over 500°C. The oxidation of CaO and $Ca(OH)_2$ added Mg alloys could be suppressed by the thin and dense oxide layer of the mixed MgO and CaO at 500°C. However, the ignition of CaO and $Ca(OH)_2$ added Mg alloys was occurred because the thin and dense oxide layer of the mixed MgO and CaO could no longer act as the protective layer at at higher temperatures over 650°C. The result might indicate that small CaO or $Ca(OH)_2$ addition was of interest for non-SF_6 processing and fire-retardation of process burrs and machined chips of cast and wrought Mg alloys.

Conclusions

The study to investigate the effect of $Ca(OH)_2$ on the oxidation and ignition behaviors of pure Mg was carried out. With respect to $Ca(OH)_2$ content, the hardness of $Ca(OH)_2$ added Mg alloys increased by grain refinement. From the oxidation result by TGA, the oxidation behavior of $Ca(OH)_2$ added Mg alloys was superior to that of pure Mg and comparable to that of CaO added Mg alloys. Consequently, it seems that reduction of fluidity and mold adhesion could be minimized by adding small amount of $Ca(OH)_2$ which is cheap and easy to be handled due to its stability in application for Eco-Mg alloy.

Reference

1. I. J. Polmear, "Magnesium alloys and applications," *Materials Science and Technology*, 10 (1994), 1-16.
2. I. J. Polmear, "Recent Developments in Light Alloys," *Materials Transactions, JIM*, 37 (1996), 12-31.
3. R. Schmid-Fetzer et al., "Focused Development of Magnesium Alloys Using the Calphad Approach," *Advanced Engineering Materials*, 3 (2001), 947-961.
4. B. S. You et al., "The effect of calcium addtions on the oxidation behavior in magnesium alloys," *Scripta materials*, 42 (2000) 1089-1094.
5. B. S. You et al., "Oxidation Behavior of Molten Magnesium Containing Calcium," J. Kor. Inst. Met. & Mater., 39 (2001), 446-450.
6. B. H. Choi et al., "Characterization of Protective Oxide Layers Formed on Molten AZ91D Alloy Containing Ca and Be," *METAL AND MATERIALS International*, 12, (2006), 63-67.
7. X. Q. Zeng et al., "Study on ignition proof magnesium alloy with beryllium and rare earth additions," *Scripta Materialia*, 43 (2000) 403-409.
8. M. Sakamoto and S. Akiyama, "Suppression of ignition and burning of molten Mg alloys by Ca bearing stable oxide film," *Journal of Material Science Letter*, 16 (1997) 1048-1050.
9. S. K. Kim et al., "The behavior of CaO in Mg alloys," *Proc. Magnesium Technology* 2005, (2005), 285-289.
10. J. K. Lee et al., "Development of CaO added wrought Mg alloy for cleaner production," *Proc. Magnesium Technology* 2006, (2006), 185-189.
11. J. K. Lee et al., "Effect of CaO addition on ignition behavior in molten AZ31 and AZ91D Magnesium alloys." *Rare Matals*, 25 (2006) 155-159.
12. S. H. Ha et al., "The behavior of CaO and Ca in pure Mg", *Rare Matals*, 25 (2006) 150-154.
13. S. H. Ha et al., "Effect of CaO on oxidation resistance and microstructure of pure Mg", *Materials Transactions, JIM*, 49 (2008), 1081-1083.
14. D. I. Jang et al., "Effect of CaO on AZ31 Mg Strip Castings", *Materials Transactions, JIM*, 49 (2008), 976-979.
15. J. K. Lee et al., "Development of Novel Environment-friendly magnesium alloys", *Advanced Materials Research*, 47-50 (2008) 940-943.
16. T. B. Massalski et al., "Binary Alloy Phase Diagram," American Society for Metals, (1986), 618-621.
17. N. B. Pilling , "The oxidation of metal in high temperature," *Journal International Metals*, 29 (1923) 529-582.

A NEW CONTI-PROCESS FOR THE FLUXLESS RECYCLING OF HIGH PURITY MAGNESIUM

U. Galovsky, M. Kühlein

ARC Leichtmetall Kompetenzzentrum Ranshofen GmbH, LKR, Austria

Abstract

As the production of magnesium die-castings for automotive applications increases, the recycling of inhouse scrap and post consumer scrap plays a more and more important role in the supply of magnesium in the long term.

An innovative recycling concept for low-cost recycling of magnesium scrap is presented. The operation of this new fluxless conti-process for the recycling of return material class 1 is introduced. The components of the remelting unit are explained.

Recycled material from flux-based processes is compared with fluxless recycled material in terms of mechanical properties.

It is shown that the used gas atmosphere in the furnace during melting process has an influence on the inclusion content. The mechanical properties differ markedly depending on the atmosphere and the ratio of surface to volume of the used scrap.

Introduction

Die casting produces considerable quantities of scrap such as runners, overflows, flash, rejects etc. This leads to increasing volumes of magnesium alloy scrap, which must be recycled for a cost-competitive, sustainable die casting process. Remelting of class I scrap in the die casting shop – the so-called in-house remelting process – reduces the cost noticeably. By sparging fine bubbles of an inert gas through the melt, it is possible to float most of the inclusions to the top and remove them by skimming [1,2]. Contrary to other die casting metals like aluminium and zinc remelting of magnesium scrap requires careful handling. Magnesium and magnesium alloys tend to oxidize rapidly and to burn in air. Therefore, the melts need protection from oxidation. Until recently a flux of mixtures of salts such as $MgCl_2$, KCl, $BaCl_2$, CaF_2 and MgO was used to cover the surface of the bath without any hesitation [3]. But the introduction of high purity die casting magnesium alloys has made the production of high quality metal from scrap material to a critical issue, and the use of salt became a quality problem.
For many years the melt surfaces have been protected with rarefied mixtures of SF_6 in dry air/CO_2 [4,5]. But studies [6,7] showed that SF_6 has a high global warming potential (GWP) relative to CO_2. Therefore the consumption of SF_6 should be reduced. It is a challenge for the industry to look for alternative protection methods. It was shown that concentration levels of SO_2 as low as 0,5 % in air are an acceptable alternative to SF_6 [8]. Opposite to SF_6 the GWP of SO_2 is low, but it is toxic and demands safety devices in the working environment [9,10].

Economical aspects

The magnesium production of 1998 was in the range of 400,000 tonnes. About half of this was used in the aluminium industry, a quarter of for magnesium casting and the rest for steel production and others. Considerable increase in the magnesium industry is expected. In 2005 about 500,000 tonnes of magnesium use is predicted, mainly due to a significant increase of magnesium die casting. It is estimated that the share of die casting will cover as many as 40 percent of the annual production in 2005. Therefore in five years the magnesium volume used in the die casting industry will be more than twice of the processed magnesium in 1998 [11].

Mg used in	1998	2005
Aluminium alloys	195,000 tonnes	225,000 tonnes
Die casting	94,000 tonnes	214,000 tonnes
Steel industry	74,000 tonnes	4,000 tonnes
Others	27,000 tonnes	30,000 tonnes

Table 1: Estimated magnesium consumption for 2005

In a die casting foundry clean process scrap consisting of biscuits, runners, overflows and warm-up shots is produced. Depending on the part design and process used, this can amount to 25 – 50 percent of the shot weight. The dramatic growth of the magnesium die casting production volume leads to an increasing amount of magnesium scrap.
So far the major part of recycling work was done in the primary magnesium industry. However, long and expensive transportation and storages are necessary. Recycling of this scrap in the die casting company could be much more economical. The estimated cost for external recycling is about US$1,000/tonne depending on the transportation distance. On the other hand the cost of inhouse recycling is in the range of US$415/tonne. The total saving potential for the magnesium die casting industry in 2005 is in the order of US$62 million.
An average medium sized die cast plant has to cope with approximately 1,500 tonnes of inhouse scrap per year. Calculated with the saving potential above an annual saving of more than US$850,000 results. This means that investment cost for a modern inhouse recycling unit can be earned within one year. All economical and environmental aspects (think about transportation by trucks) are in favor of inhouse recycling versus recycling at the recycling plant.
It is now necessary to take a look at the recycling process and judge the quality of recycled materials.

Recycling Processes

Basically there are two different types of remelting techniques for magnesium scrap:

- With flux in a batch wise process
- Fluxless in a continuous process

In the batch process it is possible to interrupt after each batch, for instance for changing the alloy. Nearly all types of scrap from the diecasting shop can be recycled, if the material is dry and free of oil and lubricants. Such plants consist of a conveying belt, one, two or more crucible furnaces, one or more holding furnaces and an ingot casting conveyor.
Most of the inhouse scrap of the die casting foundry consists of runners, rejects and overflows. If the foundry uses various alloys such as AZ91 and AM60, these can be easily separated. This class I scrap can be remelted in a continuous fluxless process.
The fluxless conti-process is more sensitive and requires attention and care during processing. Especially the melt-analysis and, as far as necessary the addition of alloying elements, should be done carefully. Certainly no contamination of the remelted magnesium alloy with flux inclusions will appear and, therefore, corrosion problems will be reduced.
In the following the main components of the new conti-unit for the fluxless recycling of return material and inhouse scrap class I are described. The recycling unit of company Rauch Fertigungstechnik Ltd., Austria (figure 1) installed in LKR's laboratory in Ranshofen is used for remelting magnesium scrap and casting it into ingot or billet shape. This fluxless continuous process uses protective gas based on the addition of SO_2 and allows to remelt 500 kg of class I scrap per hour. It can be used in foundries as an in-house recycling unit.
In this pilot plant low amounts of dross and sludge were produced, in fact about 4 % during fluxless recycling process.

Figure 1: Recycling pilot plant of company Rauch Fertigungstechnik Ltd., Austria, installed at the Lightmetal Competence Center (LKR) in Ranshofen, Austria

The continuous feeding of the magnesium melting furnace is done with an electric conveying belt. The fall shaft of the magnesium melting furnace is equipped with a safety flap. Before transfer of the melt from the melting to the holding furnace in an electrically heated transfer pipe the alloy composition can be adjusted. In the electrically heated holding furnace the melt is also protected against oxidation with a cover gas. If necessary, refining could take place in this furnace flushing with inert gases.
The holding furnace is equipped with two electrically heated transfer pipes and can be positioned at the ingot casting conveyor and a vertical direct chill casting unit for billet casting.

Experimental procedure

Laboratory tests under different atmospheres were done for determination of mechanical properties. For this investigations class I scrap of the AZ- and AM-alloy families was remelted in an electrical resistance furnace equipped with a cylindrical steel crucible containing 30 kg of scrap. There were two types of scrap:

- Coarse scrap: runners, rejects, AZ91
- Fine scrap: overflows, flashes, AM60

The scrap was supplied by different HPDC foundries and the alloy composition of coarse and fine scrap varied widely. In addition to that both scrap types differ markedly in the ratio of surface (S) to volume (V):

$$S/V \text{ (coarse scrap)} << S/V \text{ (fine scrap)}.$$

Flushing gas tests were performed with a pulsating device (volume flow 5 l/min.), distributing small bubbles of protective gas mixtures in the melt, see table 2.

Test portions of the cleaned melt were extracted from the crucible by use of a cylindrical lifting device.
Table 2 gives an overview of the combination of protecting gases and flushing gases applied to melts from coarse and fine magnesium scrap.

Trial	Melt protection	Flushing gas	Scrap type
A1	N_2 + 1% SO_2	N_2 + 1% SO_2	AZ91 - coarse
A8	N_2 + 1% SO_2	Ar + 1% SO_2	AZ91 - coarse
A3	Ar + 1% SO_2	Ar + 1% SO_2	AZ91 - coarse
A5	Ar + 1% SO_2	Ar + 1% SO_2	AZ91 - coarse
B4	Flux Emgesal 5	N_2	AZ91 - coarse
B7	Flux Emgesal 5	N_2	AZ91 - coarse
B9	Flux Emgesal 5		AZ91 - coarse
B12	Flux Emgesal 5		AZ91 - coarse
A2	N_2 + 1% SO_2	N_2 + 1% SO_2	AM60 - fine
A9	N_2 + 1% SO_2	Ar + 1% SO_2	AM60 - fine
A4	Ar + 1% SO_2	Ar + 1% SO_2	AM60 - fine
A6	Ar + 1% SO_2	Ar + 1% SO_2	AM60 - fine

Table 2: Parameter variations for fluxfree remelting laboratory tests

Table 3 shows the alloy composition for the alloys used in this study and provides the range of alloying elements given by ASTM B93 standards.

During a typical industrial procedure samples were taken to investigate the influence of flux and flushing gas on the mechanical properties.

<u>Results</u>

Chemical analyses show that coarse and fine scrap originate from different batches. The chemical composition (see table 3) of the coarse scrap is within the compositional limits of AZ91D, except for the Ni-content. The chemical composition of the fine scrap is nearly within AM60 specification.

Figure 2 illustrates the influence of applied gas atmosphere on the ultimate tensile strength R_m and elongation to fracture A5 for coarse AZ91 and fine AM60 scrap in unflushed condition. A significant difference in R_m and A5 could be seen for the fine AM 60 scrap, when either Ar + 1% SO_2 or N_2 + 1% SO_2 were used, with Ar + 1% SO_2 yielding the significantly better results. For the coarse AZ91 scrap no clear effect was observed. There was no difference in yield strength $R_{p0,2}$ utilising either Ar + 1% SO_2 or N_2 + 1% SO_2.
The average yield strength $R_{p0,2}$ of the specimens remelted from coarse AZ91 scrap is in the range of 100 MPa. The specimens taken from the remelted fine AM60 scrap present a lower yield strength $R_{p0,2}$ of about 70 MPa.

Trial	Al [wt.%]	Zn [wt.%]	Mn [wt.%]	Si [wt.%]	Cu [wt.%]	Ni [wt.%]	Fe [wt.%]
A1	8,61	0,633	0,142	0,0308	0,0078	0,0021	0,0011
A3	8,40	0,614	0,133	0,0381	0,0095	0,0011	0,0009
A5	8,40	0,594	0,196	0,0585	0,0130	0,0013	0,0058
A8	8,41	0,622	0,128	0,0552	0,0127	0,0016	0,0019
B7	8,72	0,751	0,180	0,0286	0,0059	0,0008	0,0035
B9	9,02	0,780	0,168	0,0216	0,0040	0,0007	0,0042
B12	8,87	0,761	0,186	0,0208	0,0039	0,0007	0,0031
AZ91D ASTM B93 [12]	8,3 – 9,7	0,35 – 0,9	0,15 – 0,5	max. 0,1	max. 0,03	max. 0,002	max. 0,005
A2	6,24	0,292	0,249	0,0180	0,0036	0,0007	0,0116
A4	5,83	0,203	0,185	0,0148	0,0029	0,0006	0,0008
A6	4,69	0,074	0,460	0,0202	0,0008	0,0007	0,0069
A9	5,75	0,129	0,275	0,0240	0,0025	0,0009	0,0035
AM60B ASTM B93 [12]	5,5 – 6,5	max. 0,22	0,25 – 0,6	max. 0,1	max. 0,01	max. 0,002	max. 0,005

Table 3: Chemical composition of specimens from remelted scrap

Prior to taking these samples sludge was removed from the melt surface. The samples with a diameter of 30 mm and length of 150 mm were rapidly cooled in water.

The chemical composition of each sample was chemically analysed by using an atomic emission spectrometer SPECTROLAB. Tensile tests were performed with round specimens of 6 mm diameter and a measure length (l_0) of 30 mm on a Zwick Universalprüfmaschine Z100 calibrated to EN 10002 standard.
Recycling was also carried out by a recycling unit in industrial scale using flux for protection. Class I scrap of AZ91D melting was done in a 720 kg capacity stainless steel crucible enclosed in a gas fired furnace. The addition of flux EMGESAL 5 protected the melt against oxidation.

It is important to note that the data shown in figure 2 do not change markedly when flushing is applied. Thus the authors assume that the gas distributor has to be modified, which is part of an ongoing study.

The influence of recycling procedure on mechanical properties is shown in figure 3. Trial A1, A8, A3 and A5 are remelted in the fluxless process using an Ar + 1% SO_2 or N_2 + 1% SO_2 atmosphere. Batches B4, B7, B9 and B12 were remelted with flux Emgesal 5. Batches B4 and B7 were additionally flushed with nitrogen. Using flux results in slightly lower tensile strength R_m, but clearly lower elongation to fracture A5. Flushing with nitrogen improves apparently neither tensile strength nor elongation. The yield strength $Rp_{0,2}$ is about 100 MPa.

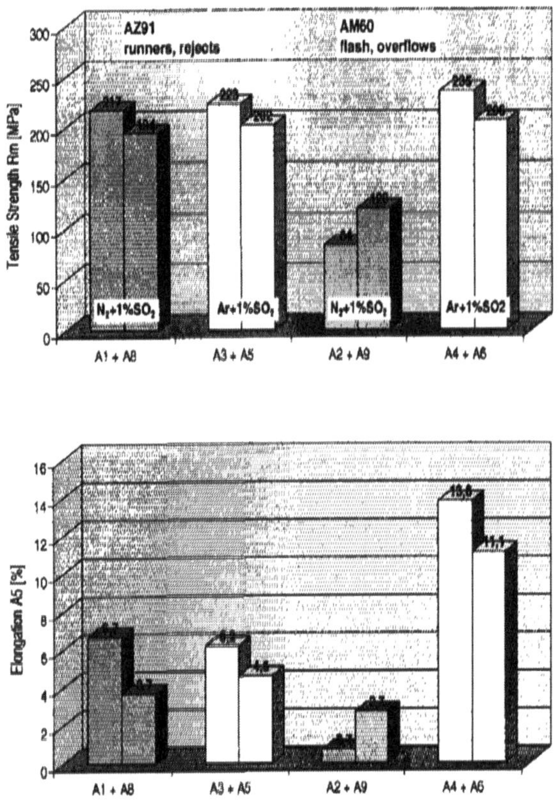

Figure 2: Tensile strength R_m and elongation of fracture A5 of AZ91 and AM60 scrap remelted under different cover gas mixtures. Two different batches.

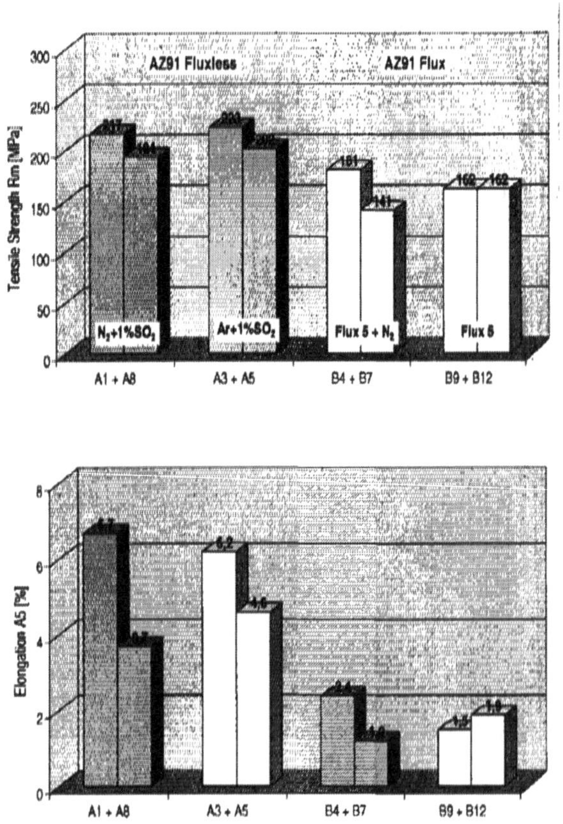

Figure 3: Tensile strength R_m and elongation of fracture A5 of AZ91 scrap remelted fluxless and with flux Emgesal 5 and flushing with nitrogen.

The fracture surfaces of the tensile specimens were examined in a scanning electron microscope. Specimens of fine AM60 scrap remelted in the fluxless process under Ar + 1% SO_2 (A4 and A6) present quite smooth, light coloured and glistening fracture surfaces. No inclusions are visible, see figure 4.

But the fracture surfaces of the specimens remelted in the fluxless process under N_2 + 1% SO_2 atmosphere (A2 and A9) are fissured and there is a great number of grey or yellowish inclusions. Figure 5 shows the surface of a yellowish inclusion.

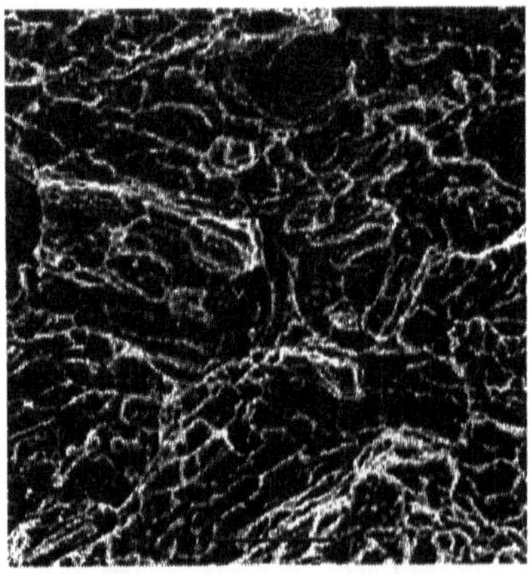

Figure 4: SEM of A4 matrix – remelted in the fluxless process under Ar + 1% SO_2 atmosphere

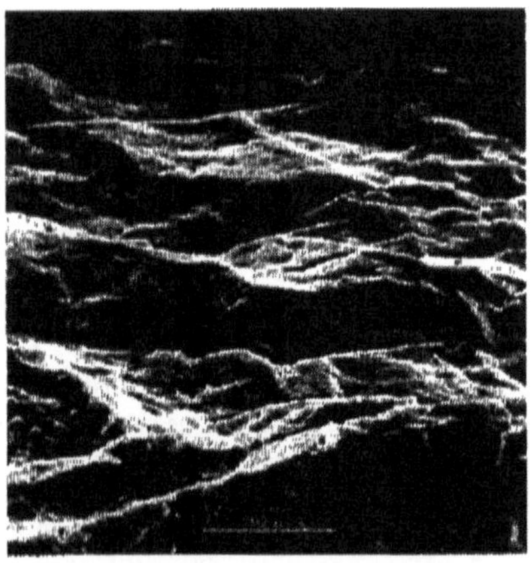

Figure 5: SEM of A2 yellowish inclusion– remelted in the fluxless process under N_2 + 1% SO_2 atmosphere

According to qualitative EDX analyses this inclusion has evidently higher concentrations of oxygen, calcium and sulphur than the surrounding matrix. The surface topography differs markedly from the surrounding matrix, as well.
The surface of a grey inclusion is shown in figure 6. EDX analysis of this inclusion shows a high concentration of oxygen.

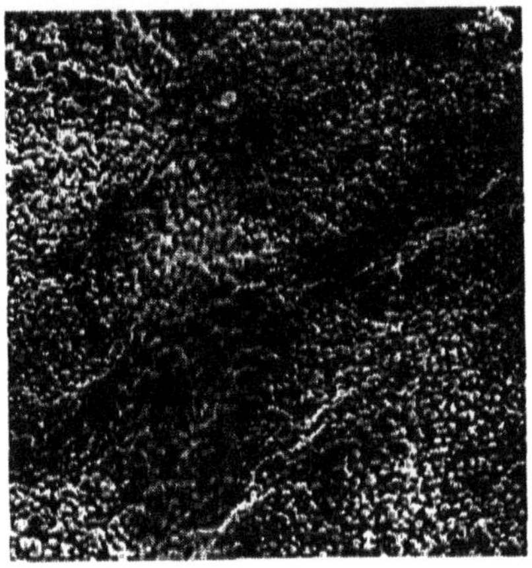

Figure 6: SEM of A2 grey inclusion– remelted in the fluxless process under N_2 + 1% SO_2 atmosphere

Summary

Two different types of scrap (fine AM60 and coarse AZ91) were remelted in the fluxless process under different gas atmospheres. The application of an Ar + 1% SO_2 atmosphere during remelting of fine scrap affects the quality of the melt positively. Significantly less inclusions could be observed in the resulting samples. No such effect could be seen for the coarse scrap, where Ar + 1% SO_2 yielded similar results as N_2 + 1% SO_2. The yield strength of both types of scrap remained independent of the protective gas composition and agreed well with typical data presented in literature for these alloys [12].
The special observation of this investigation was the significant effect of gas composition on the fracture elongation A5 and ultimate tensile strength Rm, respectively, for fine scrap of AM60.
Samples tested after recycling in the batch process with flux yielded significantly lower values of tensile elongation.

Fluxless remelted ingots show no significant difference regarding surface quality and fracture topography to ingots recycled in salt based process. Micrographs of the ingots reveal similar porosity and oxide content. The fluxless recycling process produces only small quantities of salt-free sludge and dross.

Acknowledgements

The authors would like to thank the Technologie Impulse GmbH (TIG) and the State of Upper Austria for supporting this research work within the frame of the Austrian Kplus – Programme. Special Thanks to Gernot Schmidt of Linde Germany and Alfred Sigmund of Rauch, Gmunden Austria supporting this research project actively.

References

1. S. E. Housh, V. Petrovich, Magnesium refining: a fluxless alternative, International Congress & Exposition Detroit, Michigan Feb. 24-28, 1992 No. 920071

2. V. W. Petrovich; J. S. Waltrip, Fluxless refining of magnesium scrap, Proceedings from the 45th Annual World Magnesium Conference pp 11-18, International Magnesium Association, May 1988

3. J.D. Hanawalt, SF_6 – Protective atmosphere for molten magnesium, 8th SDCE Int. Die Casting Expo. & Congr., Paper No. G-T75-111, Detroit 1975

4. S. L. Couling, F. C. Bennet, T. E. Leontis, Fluxless melting of magnesium, AIME Light Metals, 1977, Vol. 1 545 - 560

5. S. L. Couling, T. E. Leontis, Improved protection of molten magnesium with air/CO_2/SF_6 gas mixtures, Batelle Columbus Labs, Columbus, Ohio, Light Metals 1980, 997 – 1009

6. F. Stordal, B. Innset, A. S. Grossmann, SF_6 as a green house gas, Norwegian Institute of Air Research, 1993.

7. M.K.W. Ko, N.D. Sze, W.-C. Wang, G. Shia, A. Goldman, F.J. Murcray, D.G. Murcray, C.P. Rinsland, Atmospheric sulfur hexafluoride: sources, sinks and Greenhouse Warming, Journal of Geophysical Research, Vol. 98, No. D6, 1993, 10499-10507

8. H.Gjestland, H. Westengen, S. Plahte, Use of SF6 in the magnesium industry: an environmental challenge, Proceeding of the Third International Magnesium Conference, Manchester, England 10.-12. April 1996, 33-41

9. H. Gjestland, D. Magers, Progress to eliminate SF6 as a protective gas in the magnesium diecasting, 6. Magnesiumguss Abnehmerseminar und Automotive Seminar, 30.9.-1.10.98

10. W. Schubert, H. Gjestland, Use of SO_2 as protection gas in magnesium diecasting, Magnesium Alloys and their Applications, ed. by K.U. Kainer, Wiley-VCH Verlag GmbH, Weinheim 2000, ISBN 3-527-30282-4, pp.760-766, Magnesium 2000 27.-28.9.2000 München

11. E. Rauch, U. Galovsky, A. Sigmund, Technische und betriebswirtschaftliche Aspekte des Inhouse-Recycling von Mg-Legierungen, Tagungsband Ranshofner Leichtmetalltage 2000, ISBN 3-902092-00-9, pp.147-154

12. ASM Speciality Handbook, Magnesium and Magnesium Alloys, edited by Michael M. Avedesian and Hugh Baker, 1999

INNOVATIVE VACUUM DISTILLATION FOR MAGNESIUM RECYCLING

Tianbai Zhu[1], Naiyi Li[2], Xiaoming Mei[1], Alfred Yu[3], and Shixiang Shang[1]

[1] Nanjing Welbow Metals Company, Nanjing, Jiangsu 211224, China
[2] Ford Motor Company, MD 3135, Dearborn, MI 48121, U.S.A.
[3] Nanjing Welbow North America Office, Windsor, Ont. N9C 4C7, Canada

Abstract

Magnesium recycling now becomes a very important subject as magnesium consumption increases fast around the world. All commonly used magnesium die-casting alloys can be recycled and recovered to the primary metal quality. The recycled materials may be comprised of biscuits, sprues, runners, flash, overflows, dross, sludge, scrap parts, and old parts that are returned from service. An innovative magnesium recycle method, vacuum distillation, is developed and proved out to be able to recycle magnesium scraps, especially machining chips, oily magnesium, smelting sludge, dross or the mixture. With this process at a specific temperature and environment condition, magnesium in scraps can be gasified and then solidified to become crystal magnesium crown. This 'recycled' magnesium crown is collected and used as the raw material of magnesium alloys. The experimental results show the vacuum distillation is a feasible and plausible method to recycle magnesium. Further, the cost analysis will be addressed in this paper.

Introduction

All commonly used magnesium die-casting alloys can be recycled. The different types of metal returns may be restored to primary metal quality by selecting the correct recycling processes. These return materials may be comprised of biscuits, sprues, runners, flash, overflows, dross, sludge, scrap parts, and old parts that are returned from service. Of these categories, the scrap parts, biscuits, sprues, and runners are the easiest to recycle since they are relatively clean and free of oxides. Machining chips, flash, overflows, dross, and sludge will contain higher quantities of oxides and require special precautions during recycling. Old parts returned from service will in many cases be covered with oil or paint that must be removed before melting. In addition, care must be taken to avoid contamination of the metal from any bushings etc. in the castings.

A variety of recycling processes are in current use, based on the use of refining fluxes as well as fluxfree methods. Hydro Magnesium [1,8] operates two recycling units located at the primary production sites in Porsgrunn, Norway and Bécancour, Canada. The return material is first melted under flux protection and later refined in the same type of refining and casting furnace as is used for primary metal. Hydro Magnesium claims the resulting product meets all the specifications for primary metal and the properties of the resulting castings are indistinguishable from those derived from primary metal sources. Chrysler Corporation did qualify the utilization of the refined recycled process for their automotive component applications as the results of chemical analysis, optical and electron metallography, image analysis and corrosion testing of the ingot materials showed the electrolytic and refined recycled forms of the alloys to be indistinguishable. [2]

Today, a few die-casting shops do their own recycling. Generally, the capital cost of such an operation, including environmental control, is considered quite high and might be an unattractive investment. Special equipment is also required to insure the quality of metal coming from an in-house recycling system. For developing applicable regulations or permit requirements in North America, there are at least four separate but related regulatory agencies of legislation that impact residual magnesium handling. [3,7]

In this paper, we will introduce an innovative magnesium recycle method, vacuum distillation, is developed and proved out to recycle magnesium scraps, especially for machining chips, oily magnesium, smelting sludge, dross or the mixture. With a specific temperature and environment condition, magnesium in scraps can be gasified and then solidified to become crystal magnesium crown. This 'recycled' magnesium crown collected is therefore to be used as alloys' raw material. The experimental results with various magnesium alloys show that recycling magnesium via vacuum distillation is a feasible and plausible method. Further, the cost of this method is reasonable, and the composition of scraps recycled has a direct relationship with its recovery ratio.

Materials

Before we recycle magnesium alloys, we need to review some basic information of magnesium alloys. In the ASTM standard magnesium alloys are designated by an alphanumerical code. Letters denote the various main alloying elements. Following the letters, numbers indicate the percentage of each element. A final letter may designate a special version of the alloy. For example, alloy AZ91D contains 9 wt% aluminum and 1 wt% zinc. The letter D indicates a special high purity version.

In the European Standard magnesium alloys are designated as illustrated in the Figure 1.

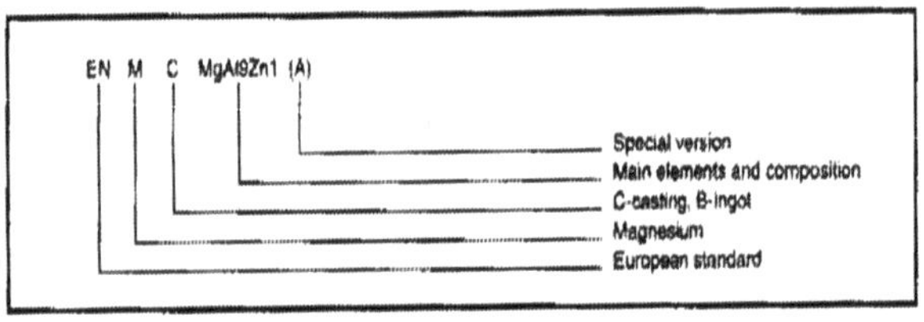

Fig. 1 MgAl9Zn1 in Europe as AZ91D in ASTM standard

The various alloying elements in die casting alloys are designated by letter codes as follows:

Aluminum	A
Zinc	Z
Manganese	M
Rare Earth Elements	E
Silicon	S

The current commercial magnesium die casting alloys all contain aluminum as the main alloying element. Aluminum improves the mechanical strength, corrosion properties and castability. Ductility and fracture toughness are gradually reduced with increasing aluminum content. This has led to the introduction of a series of alloys with reduced aluminum contents, the AM-series, which is used extensively for automotive safety related components.

Zinc is added to the most common die casting alloy AZ91 at a level around 0.7 wt%, resulting in a minor improvement of strength and corrosion properties.

Manganese is added to control the iron content of the alloys. The level of manganese additions varies from one alloy to the other, depending upon the mutual solubilities of iron and manganese in the presence of other alloying elements. A basic requirement for the high purity alloys is that the iron content of die cast parts is limited to a maximum of .005 wt%. Other impurities like nickel and copper must also be strictly controlled.

Efforts to improve the elevated temperature creep properties have resulted in the introduction of alloys containing silicon or rare earth elements. These alloying elements form intermetallic constituents which stabilize the grain boundaries. In these alloys aluminum has to be kept at relatively low levels.

Beryllium is added to the die casting alloys to levels in the range 5~15 ppm to reduce the oxidation rate in the molten condition. The detailed chemical compositions of the current magnesium die casting alloys can be seen in the literature. [4]

The Principle of Vacuum Distillation

Pidgeon reduction process [5] has been widely and successfully applied in the magnesium production. In this section, we will cover the principle of vacuum distillation that magnesium can be distilled from raw stuff under a specific vacuum and temperature. This principle is based on a) magnesium has a vapor pressure under a condition of specific vacuum and temperature, and b) impurities in magnesium have higher boiling point. Therefore, it is doable to use vacuum distillation to recycle magnesium.

The relationship of the magnesium vapor pressure and temperature can be shown as follows [6].

$$\log P = (3.27 - 2.95*10^{-3}T - 7.64*10^{-3}T + 8.74*10^{-7}T^2 + 2.5\log T)*133.3 \qquad 570<T<924K$$

$$\log P = (11.61 - 7.61*10^{3}T^{-1} - 1.02\log T)*1.33 \qquad 924<T<1380K$$

where P is the magnesium vapor pressure in *Pa* and T is the surrounding temperature in *K*. The graphical relation between P and T is shown in Figure 2 with data given in Table I.

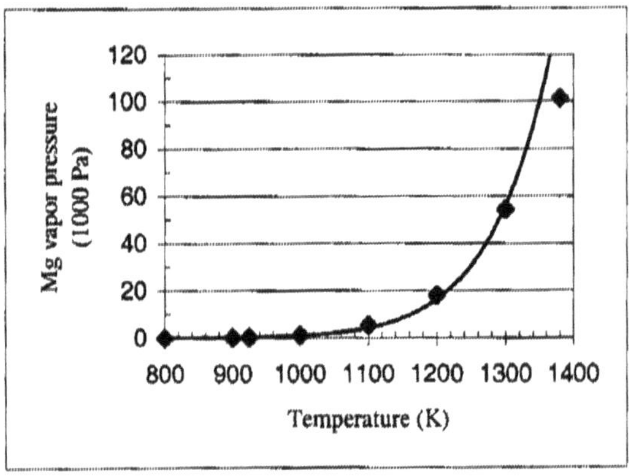

Fig 2. Magnesium vapor pressure chart

Table I Magnesium vapor pressure and temperature

Data No	Temperature, K	P_{Mg}, Pa
1	600	/
2	700	/
3	800	19.3
4	900	221
5	924	351
6	1000	1161
7	1100	5305
8	1200	17892
9	1300	54306
10	1380	101308

Magnesium collecting method used in vacuum distillation is the same as in Pidgeon process. A cooling zone is configured in the vacuum distilling facility, where magnesium vapor of scraps will condense when temperature is 800K or lower. Magnesium is then collected in the crystal "crowned" status.

Impurity in magnesium scraps includes Al, Zn, Mn, Fe, Si, Cu, Ni and oxides. Among those elements, except Zn, they all have high boiling point that means these elements can't be distilled under the temperature when magnesium is gasified. For recycling magnesium, this is the same principle as Pidgeon process uses.

Vacuum Distillation Experiment

The experimental facilities include retort, vacuum pump and furnace as shown in Figure 3 and Figure 4.

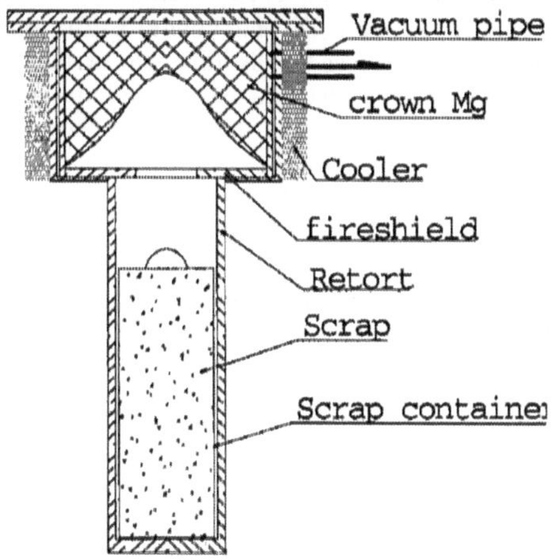

Fig 3 Structure of retort

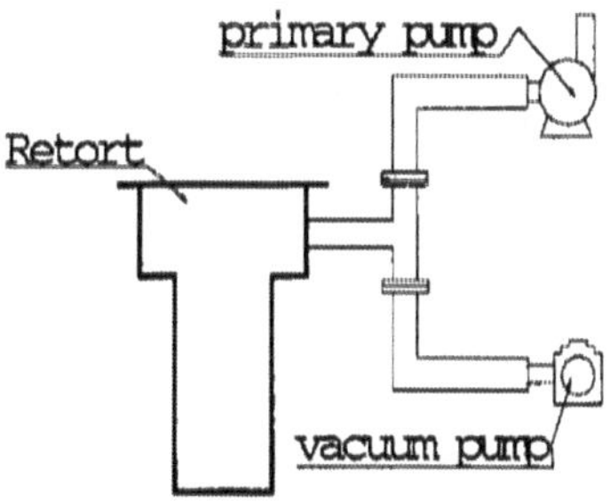

Fig 4 Retort and vacuum pump

Both magnesium scraps and virgin magnesium alloys were used as the experimental materials [9].

Scraps:

Type 1A: Magnesium chips and pieces with a little bit oil and water. Magnesium scarp 500kg was sampled and melt in a closed pot, then cast into ingots. The recovery ratio is 86.4%. Its chemical composition is as shown in Table II.

Table II Chemical composition of magnesium scraps Type 1A

Code	Al	Zn	Mn	Fe	Si	Cu	Ni
1A	8.43	0.57	0.31	0.007	0.12	0.04	0.0062

Type 1B: Crushed pieces of magnesium parts with painting, dust and water. Magnesium scarp 500kg was sampled and melt in a closed pot, then cast into ingots. The recovery ratio is 89.2%. Its chemical composition is as shown in Table III.

Table III Chemical composition of magnesium scraps Type 1B

Code	Al	Zn	Mn	Fe	Si	Cu	Ni
1B	9.78	0.62	0.19	0.016	0.12	0.17	0.0041

Virgin alloy ingots:

Type 2A: AZ91D Ingots (meet ASTM B93 Designation)
Type 2B: AM60B Ingots (meet ASTM B93 Designation)
Type 2C: AZ31 Ingots (meet ASTM B93 Designation)
Their chemical compositions are shown in Table IV

Table IV Chemical compositions of Virgin alloy ingots

Code	Al	Zn	Mn	Fe	Si	Cu	Ni
2A	8.91	0.61	0.27	0.002	0.024	0.001	0.0003
2B	6.02	0.01	0.31	0.002	0.021	0.001	0.0005
2C	2.89	0.75	0.33	0.001	0.030	0.001	0.0006

The experimental process steps are as follows,

a. Control furnace temperature within 850±10•.
b. Add 60kg experimental material into a retort per charge.
c. Exhaust 10 minutes
d. Vacuum and hold furnace temperature
e. Keep stable process for 6 hours
f. Take out crystal magnesium and residue, weigh them after cool and record results
g. Repeat a~f 5 times for the same material, respectively

The recovery results of all the experimental materials under same conditions are shown in Table V.

Table V Test results of magnesium shown in Tables II~IV

Material type	1A	1B	2A	2B	2C
Qty added (kg)	300.1	302.1	297.5	299.4	298.7
Crystal Mg (kg)	219.8	217	221	247.1	272.1
Residues (kg)	78.2	81.2	74.2	51.7	27.8
Mg recovery (%)	73.24	71.83	74.29	82.53	91.09
Total recovery (kg)	298	298.2	295.2	298.8	299.9
Total recovery (%)	99.30	98.71	99.23	99.80	100.4
Note		1.4kg dust			

Note that the recovery ratio over 100% for 2C may be caused by the weight increase of MgO in the magnesium crystal. Further, the magnesium crystal and the residues of all the distilled experimental materials were sampled and analyzed. The chemical compositions are shown in Table VI and Table VII.

Table VI Chemical composition of all the crystal magnesium

Material distilled	Al	Zn	Mn	Fe	Si	Cu	Ni
1A	0.03	0.73	0.003	0.001	0.004	0.001	0.0003
1B	0.03	0.82	0.002	0.001	0.003	0.001	0.0004
2A	0.01	0.81	0.005	0.001	0.003	0.001	0.0004
2B	0.03	0.01	0.003	0.002	0.004	0.001	0.0004
2C	0.03	0.79	0.002	0.001	0.003	0.001	0.0002

Table VII Chemical compositions of all the residues

Material distilled	Al	Zn	Mn	Fe	Si	Cu	Ni
1A	31.6	0.14	1.12	0.031	0.410	0.073	0.0140
1B	34.7	0.07	0.66	0.071	0.340	0.590	0.0120
2A	33.2	0.09	0.92	0.009	0.061	0.002	0.0010
2B	33.9	0.01	1.71	0.010	0.052	0.001	0.0012
2C	32.5	0.12	1.92	0.007	0.030	0.002	0.0010

The results shown in Tables VI and VII indicate that impurity is very low in all crystal magnesium except Zn in 1A, 1B, 2A and 2C, and impurity in the raw material all gather in the residues. It clearly shows that this distillation process can purify magnesium scraps and help magnesium alloy production.

Although the experiment results show that the overall recovery of every scrap is almost 100%, the recovery of magnesium crystal varies greatly. The main chemical compositions of all the experimental materials and their recovery are listed in Table VIII.

Table VIII Chemical composition and magnesium recovery

Matl distilled	Main composition (%)					Mg Added (kg)	Mg Recovered (kg)	Mg recovery (%)
	Al	Zn	Mn	Si	Mg			
1A	8.43	0.57	0.31	0.12	90.6	271.8	218.3	80.3
1B	9.78	0.62	0.19	0.12	89.3	269.7	215.3	79.8
2A	8.91	0.61	0.27	0.024	90.2	268.3	219.2	81.7
2B	6.02	0.01	0.31	0.021	93.6	280.4	246.9	88.1
2C	2.89	0.75	0.33	0.03	96.0	286.8	271.6	94.7

The crystal magnesium recovery may depend on both Al and Mg content in the experimental materials. The correlation based on the Table VIII is shown in Figure 4.

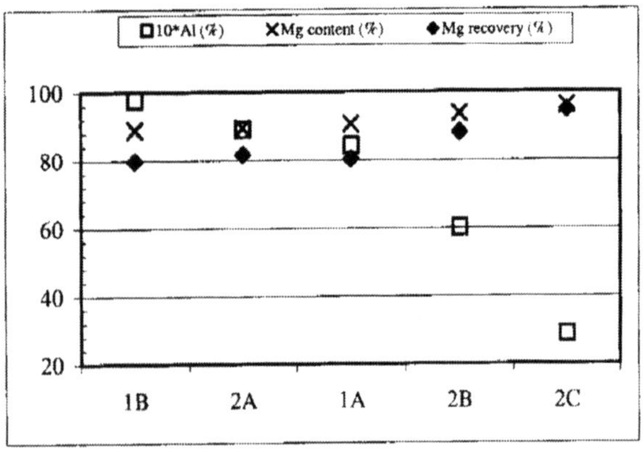

Fig 4 Recovery of magnesium crystal from various Mg alloys

From Figure 4, it is clear that the crystal magnesium recovery increases as the magnesium content in the experimental materials increases. Yet, the relationship between the recovery and the aluminum content in the experimental materials is the reverse. The possible explanation is that more metallic compounds are produced as the aluminum content increases in the experimental materials, so it requires more energy and time to break such compounds and gasify magnesium. To confirm this point and to explore the peak-point aluminum in the residues, we have the following tests to identify the maximum magnesium recovery with the high aluminum content alloy.

Maximum Recovery

Generally, the recovery depends on two factors, the distilling time period and the distilling temperature. However, does the maximum recovery exist? If does, how to reach it? By using the design of experiment method, we vary the distilling temperature to be at 900±10•, 950±10•, 1000±10••1100±10• and 1200±10•, respectively for a 6-hour distilling time period. In addition, for the distilling temperature at 850±10•, we vary the distilling time period to be at 8, 10, 12, 15, 18, 22, 26, and 30 hours, respectively. To assure the consistency, we use virgin AZ91D for all the tests. Its chemical compositions are the same as the material 2A in Table IV. The testing process is the same as those described in the prior section. During the distilling process, the increasing magnesium pressure may affect the vacuum status in the test retort. To maintain its vacuum status, we only added scraps of 40kg in the retort per charge. For the repeatability, we had conducted two tests in parallel under the identical conditions.

The distillation results for various temperatures and fixed 6-hour period are shown in Table IX and Figure 5. The magnesium recovery increases slightly as the distilling temperature increases, but there is almost no change as the temperature is 950• and higher. The aluminum content in the residues shows the same trend. It shows that under these industrial temperatures, the recovery of magnesium could reach the peak-point, the maximum.

Table IX Distilled results at various temperatures

Distilling temp (•)	850	900	950	1000	1100	1200
Qty. of material (kg)	81.2	80.4	80.3	79.7	81	80.3
Crystal Mg recovered (kg)	58.8	58.7	61.2	61.1	62.5	61.9
Crystal Mg recovery (%)	72.4	73.0	76.2	76.7	77.2	77.1
Al content in residues (%)	32.6	33.1	37.3	38.3	38.5	38.7

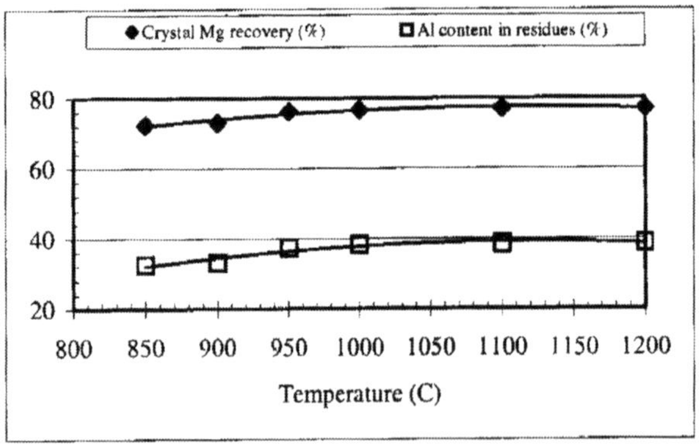

Fig 5 Mg recovery and Al content in the residues at various temperatures for distilling 6 hours.

The distillation results for various periods at the fixed temperature, 850±10•, are shown in Table X and Figure 6.

Table X Distilled results for various periods

Distilling time (hr)	8	10	12	15	18	22	26	30
Quantity of material (kg)	79.7	80	80.5	80.2	81	80.2	79.9	80.9
Crystal Mg recovered (kg)	57.9	58.1	61.4	61.2	62.7	62.5	62.8	63.1
Crystal Mg recovery (%)	72.7	72.6	76.3	76.3	77.4	77.9	78.6	78.0
Al content in residues (%)	33.0	33.4	36.5	37.9	39.1	38.6	39	40.1

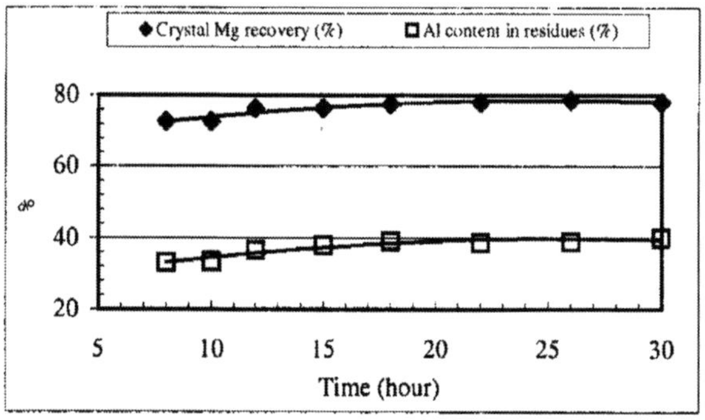

Fig 6 Mg recovery and Al content in the residues under different distilling times at 850•

It is shown in Figures 5 and 6 that both the magnesium recovery and the aluminum content in the residues have its peak-point or the maximum, respectively. Therefore, we can optimize the recycling process to get the maximal recovery which will be addressed somewhere else.

Industrial Demonstration

The experiments reported here were conducted at the Nanjing Welbow Metals Co., Ltd. It has an annual capacity of 600MT the magnesium vacuum distillation recycling line, which was rebuilt from a part of its Pidgeon reduction furnaces. The magnesium and the residues produced from the vacuum distillation process are directly used in magnesium alloy and aluminum alloy production. The impact is significant, not only technically also financially. For example, in a 2000MT/year vacuum distillation facility, the normalized production cost of crystal magnesium from recycling scraps of AZ91D is shown in Table XI.

Table XI Production cost per MT magnesium, US$

Item	Unit Price, $	Multiplier	Cost, $	Cost, %
Scraps AZ91D	0.84	1370 kg	1155	83.2
Oil	0.193	800 kg	154	11.1
Electricity	0.06	300 kw-h	18	1.3
Retort (stainless steel)	4.33	6 kg	26	1.9
Labor	24	1	24	1.7
Others	12	1	12	0.9
			1389 Total	100

In general, the vacuum distillation will leave 20~25% of scraps to remain as the residues of which the main components are 30~40% Al, a small quantity of Mn and Si, etc. Reasonable usage of the residues will enhance the effect of the vacuum distillation.

At the present, the residues are used as additive in aluminum alloys production or are used to produce AM50 alloy directly. Its applications are still relatively new to us. Future research tasks should include maximizing the utilization of the residues. With the total recovery of the vacuum distillation at 95% or higher, it will become a low-cost recycling process. The cost of recycled product will then mainly depends on the prices and quantities of scraps.

Further Tasks

The present vacuum distillation process for magnesium recycling, similar to that of Pidgeon magnesium production, is not a continuous process and its productivity is not high. Our immediately goal is to develop a low-cost recycling process, i.e., a new and continuous vacuum distillation method for recycling magnesium scraps.

Further, when recycling the oily, painted and wet scraps, the gas mixture vented in the preliminary exhaust must be filtered and treated to meet the industrial regulations before emission. However, it is unnecessary to treat the exhaust gas once the vacuum is reached.

Conclusions

1. The vacuum distillation process is developed and can be used to recycle all kinds of magnesium scraps. It is appropriate for recycling magnesium with high impurities, machining magnesium chips, oily magnesium, smelting sludge, dross or the mixture that would not be treated by the remelting method as its recovery is very low.

2. The aluminum content in the scraps may affect the crystal magnesium recovery in the vacuum distillation process. When the aluminum content in the residues is higher than 40%, magnesium could not be distilled under the practical industrial temperature.

3. The overall magnesium recovery in the vacuum distillation is higher than that in the melting recycling. Its economical impact will be significant with the maximized use of the crystal magnesium and the residues.

4. Compared with the direct remelting, the advantage of the vacuum distillation for recycling magnesium is that it separates Mg from the impurities and the rest of components such Si, Cu and Ni in the scraps. In addition, this process doesn't use any protection gas, e.g. no SF_6 is involved.

Acknowledgement

The work presented is a part of the "Magnesium Manufacturing & Product Development group" projects in Manufacturing Systems Department, Ford Research Laboratory, Ford Motor Company. The experimental works were conducted at the Nanjing Welbow Metals Company.

<u>References</u>

1. Hydro Magnesium, "Melting and Handling Magnesium for Die Casting," 1996

2. J. G. Grebetz and D. L. Albright, "Recycled Magnesium Alloys for High Ductility Automotive Applications," SAE 960413, 1996

3. J. S. Waltrip, "Magnesium Scrap and Residual Magnesium: Trash or Treasure?" 47th Annual World Magnesium Conference Proceedings, International Magnesium Association, pp.29-33, 1990.

4. M. Avedesian and H. Baker ed. <u>Magnesium Handbook</u>, ASM, 1999.

5. E. F Emley, <u>Principles of Magnesium Technology</u>, Pergamon Press, 1966.

6. G. V. Raynor, The Physical Metallurgy of Magnesium and its Alloys, Pergamon Press, pp13-14, 1959.

7. L. Riopelle, "The Recycling of Magnesium Makes Cents," JOM, vol. 48, no. 10, pp. 44-46, Oct. 1996.

8. D. L. Albright, "Technology Paths to Magnesium Alloy Recycling," Die Casting Management, vol. 11, no. 6, pp. 24-28, Dec. 1993.

9. J. C. Grebetz and A. G. Haerle, "Qualification of Recycled Magnesium Type 1 Die Casting," A Global Vision for Magnesium, Coronado, CA, pp. 17-19, May 1998.

A Numerical Understanding of Vertical Magnesium Thermal Reduction Processes

Henry Hu[1], Alfred Yu [1]and Naiyi Li[2]

[1]Mechanical, Materials and Automotive Department
University of Windsor
401 Sunset Ave.
Windsor, Ontario
Canada N9B 3P4
Email: huh@uwindsor.ca & yu11@uwindsor.ca

[2] Manufacturing System Department
Ford Research Laboratory
Ford Motor Company
2101 Village Road, P.O. Box 2053
Dearborn, MI48121-2053 USA
Email: nli1@ford.com

Abstract

A mathematical model has been developed to simulate the occurrence of heat transfer in vertical retorts. The simulations were run to determine the effect of varying parameters, such as the diameter and thickness of the compound, and slot angle, on the magnesium reduction time. The model predicted the temperature distributions, the heating curves, the recovery ratio of magnesium, and the total process time. The predictions were used to optimize the magnesium reduction process parameters, which consist of dimensions of the retort, shapes of charged materials, and reduction cycle time. The computed results show that the utilization of the optimized process parameters leads to a decrease in reduction time and energy consumption, and an increase in production capacities and recovery rates. Consequently, the magnesium thermal reduction process is significantly improved in the vertical retort.

Introduction

Magnesium consumption has been growing considerably in the past decade, [1]. Of the total magnesium production, 30% is produced by the thermal reduction or so-called Pidgeon process. The Pidgeon process is still anticipated to be the main technology for producing magnesium in future [2]. One of the advantages of the Pidgeon process is that it is capable of producing high purity magnesium demanded by both the automotive and aerospace industries. However, the horizontal Pidgeon reduction process currently deployed in China possesses a characteristic, i.e., poor inefficiency of heat and mass transfer occurred in the reduction process. This inefficiency always leads to low productivity, high energy consumption, labor-intensive operation, large floor space and high capital investment. Hence, this horizontal process confines its production scale [3].

To overcome the shortcomings of the horizontal magnesium reduction process, a novel vertical magnesium reduction process for large-scale production has been developed [3], in which the heat and mass transfer in the retort is enhanced. The newly developed process shortens reduction time, reduces energy consumption and increases productivity.

In order to understand the heat transfer mechanism in the vertical production process, a numerical model of heat transfer was formulated based on the control-volume finite difference approach to simulate and optimize the process [4]. The computed results show that for a certain amount of reactant materials, the design of the open compounds reduces considerably the reduction time to a level less than half of that required by the closed compounds (Figures 1 and 2). An increase in the compound thickness results in extended reaction times for both designs of the closed and open compounds. However, the effect of the thickness on the reaction time for the open compounds is not as significant as that for the closed compounds.

In order to further develop the vertical retort technology with increased productivities, the influence of slot angles and compound diameters on the process needs to be investigated. The present study was focused on the simulation of the vertical retort with various design of slot angles and different sizes of compound diameters by using the developed mathematical model[4]. The paper discusses the effect of varying slot angles and compound diameters on the process.

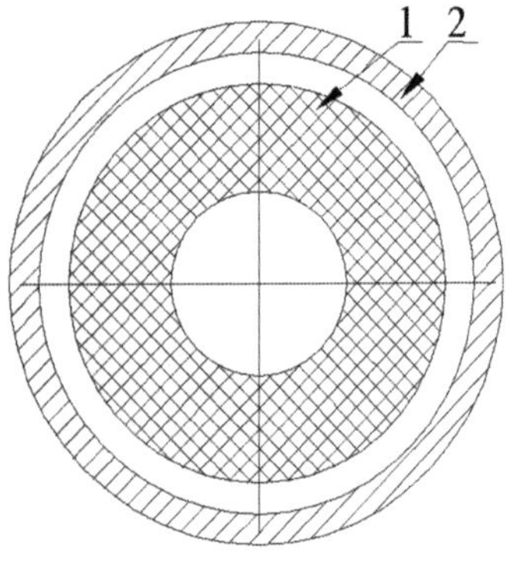

(a)

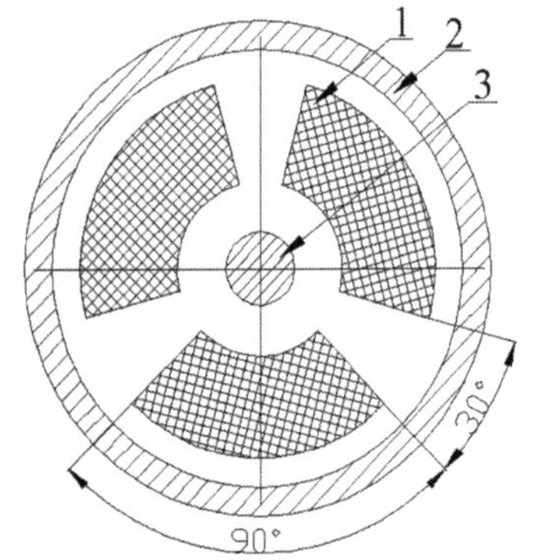

(b)

1—Reactant compound, 2—Retort, 3—Suspender (reflection shell)

Figure 1: A cross section of vertical retort containing (a) closed, and (b)open cylindrical reactant compounds.

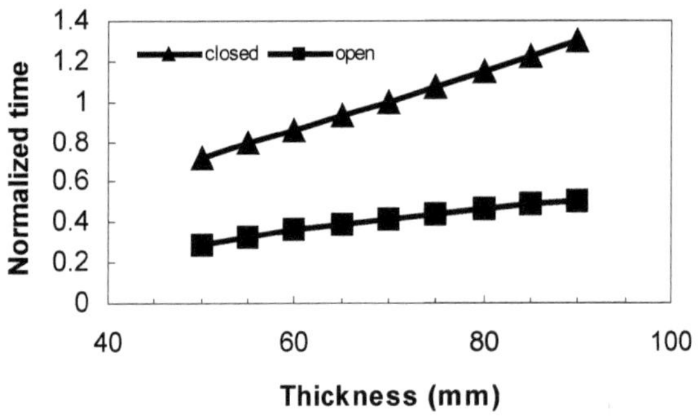

Figure 2: Variation of normalized reaction times for both closed and open compounds with different thickness.

Numerical model

As it has been described in reference 4, the numerical model of heat transfer was formulated based on the control-volume finite difference approach to simulate and optimize the process. The governing equation for this 2-D model in a polar coordinate system can be represented as [5]

$$\rho c\frac{\partial T}{\partial t} = k\left[\frac{1}{r}\frac{\partial}{\partial r}(r\frac{\partial T}{\partial r}) + \frac{1}{r^2}\frac{\partial^2 T}{\partial \theta^2}\right] + S$$

(1)

where the source term (S) is the reaction heat calculated from the following [6]:

•H=115,600-5.1T (2)

And the overall chemical reaction during the reduction of magnesium is described by [7]

$2MgO.2CaO+Si\ (Fe) = Ca_2SiO_4+2Mg+(Fe)$ (3)

Heat transfer coefficients applied on boundaries are computed as follows [8, 9]:

$$q_{12} = \frac{\sigma(T_1^4 - T_2^4)}{(1-\varepsilon_{12})/\varepsilon_{21}A_1 + 1/A_1F_{12} + (1-\varepsilon_{21})/\varepsilon_{12}A_2}$$

(4)

and,

$$h_1 = \frac{q_{12}}{A_1(T_1 - T_2)}$$

(5)

The radiation shape factors are calculated by the first summation principle [10]:

$$\sum_{j=1}^{N} F_{s-j} = 1$$

(6)

and the second summation principle

$$F_{1-(2,3)} = F_{1-2} + F_{1-3}$$

(7)

Due to the homogeneity of the mixed reactant materials in the compounds, thermal physical properties used in the computation are the stoichiometrical property values of calcined dolomite and ferrosilicon until the reaction is completed (Table I). After the reaction, the stoichiometrical

From *Magnesium Technology 2002*, H.I. Kaplan, Editor

thermal physical properties of calcium orthosilicate and iron are used for the computation as listed in Table I.

Table I. Thermophysical properties

Properties	Reactant (before reaction)	Products	Retor t	Suspender (reflection shell)
Thermal Conductivity k (W/m ^{0}C)	0.16	0.13	/	/
Specific Heat C_p (J/kg ^{0}C)	1200	1000	/	/
Density ρ (kg/m^3)	1800	1500	/	/
Emissivity•	0.8	0.8	0.8	0.8

In the process, the governing equations are discretized with the control volume finite difference procedure[10]. In this procedure, the partial differential equation is integrated over the control volume to obtain the discretization equation. The power law scheme is used for formulating the discretization equation from the governing equation in order to acquire a set of algebraic equations. Owing to the non-linear nature of the problem, the iterative methods are needed to solve the algebraic equations. A method used in this procedure is known as the line-by-line method which is a combination of the tridiagonal matrix algorithm (TDMA) and the Gauss-Seidel method. The convergence for each iteration is assured with the time step according to the stability criterion $t<\bullet r^2/2\bullet$.

Result and Discussion

1. Effect of slot design

The geometric dimensions of open compounds for computing the effect of slot angle are listed in Table II with reference to Figure 3.

Table II: Geometric dimensions of the closed and open compounds

Item	Standard compound	Compounds with varied slot angle	Compounds with varied diameters	5-slot compound
Retort radius (inner) R_r /mm	150	150	150-300	300
Compound radius(outer) R_d /mm	130	130	130-260	260
Compound radius(inner) R_i/mm	60	60	60-190	190
Suspender (reflective shell) radius R_s /mm	25	25	25-155	155
Slot angle α /^{0}	30	15-45	30	20
Slots number N	3	3	3	5

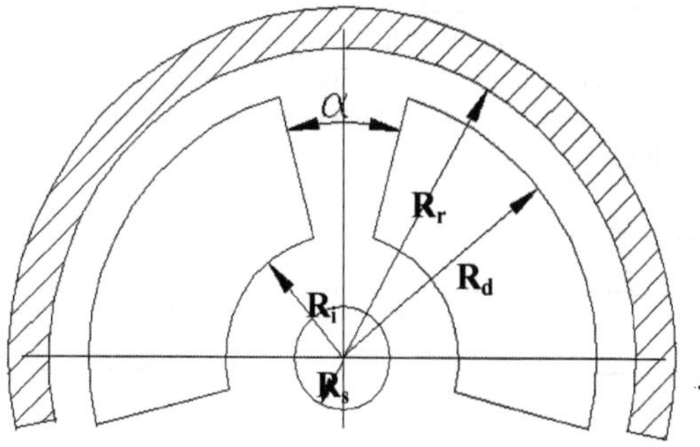

Figure 4 shows the variation of reaction time of circularity with different slot angles.

The predictions indicate that the absolute reaction time decreases slightly as the slot angle increases due to the enhanced radiant heat transfer by relatively large side surface area. By using the normalized reaction time for the purpose of comparison, however, it is evident that the increase in slot angles extends reaction times.

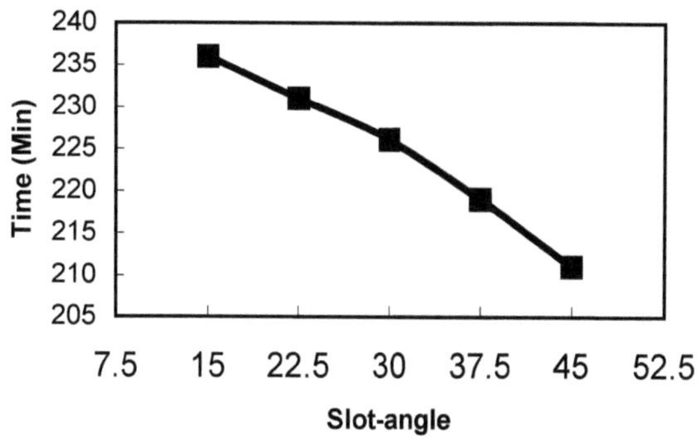

Figure 4: Variation of reaction time for compounds with different slot angles

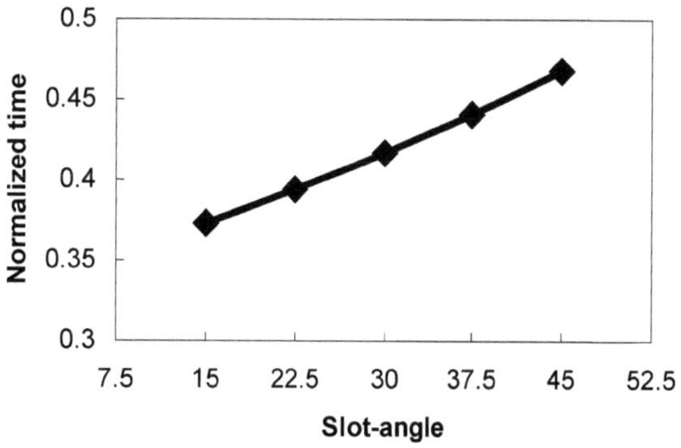

Figure 5: Variation of normalized reaction time for compounds with different slot angles

This is mainly because the enhancement of heat transfer resulting from the increase in slot angles is offset by the decrease in compound mass of each load.

The simulation was also carried out on a 520mm-diameter compound with different numbers of slots as shown in Figure 8. The results listed in Table III indicate that there is no large difference in reaction times between two designs of the compounds. This alternatively proves that the variation of slot angles has limited influence on the total reaction time as long as the number of slots does not change considerably.

Table III: Reaction time for 3-slot and 5-slot compounds with a diameter of 520mm

	Reactant load*	Reaction time (min)
3-slot compound	2.36	222
5-slot compound	2.28	210
Standard	1	226

*note: load is normalized with standard compound

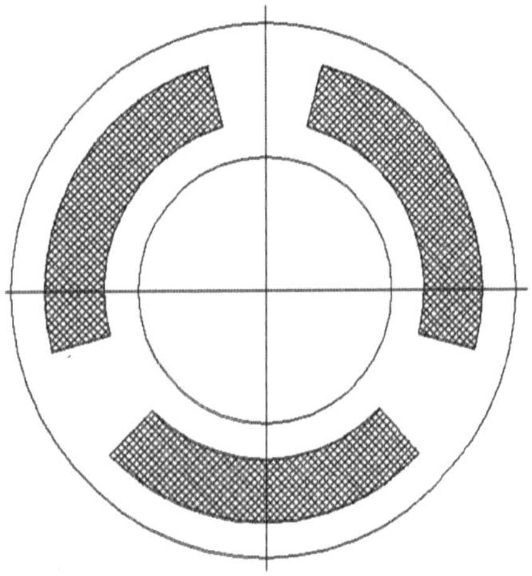

(a)

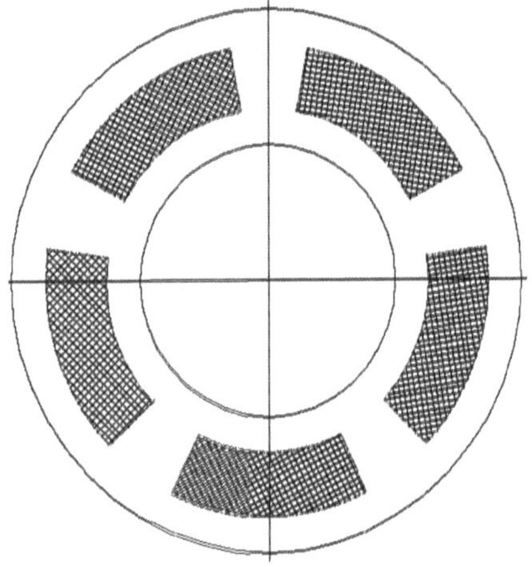

(b)

Figure 6: 520mm diameter reactant compounds with (a) 3 slots, and (b) 5 slots

2. Effect of compound diameters

The geometric dimensions of compounds for computing the effect of diameter are also listed in Table II with reference to Figure 3. During the numerical study, the compound diameters vary from 260mm (standard) to 520mm with the same thickness of 70 mm while the gap distances between the inner surface of the retort and the outside surface of compound, and between the inner surface of compound and the outside surface of reflective shell (suspender) are kept constant.

Figure 7 shows that reactant loads increase linearly with compound diameters. However, the increase in the compound diameter has little influence on the reaction time as depicted in Figure 8. This observation indicates that the reaction time does not alter significantly with the variation of compound diameters when the shape of reactant compounds remains unchanged. This is mainly because that, as the compound diameter increases, the heat transfer to the compound from both the inside surface of the retort and the suspender are enhanced by relatively high shape factors due to the enlarged surface area.

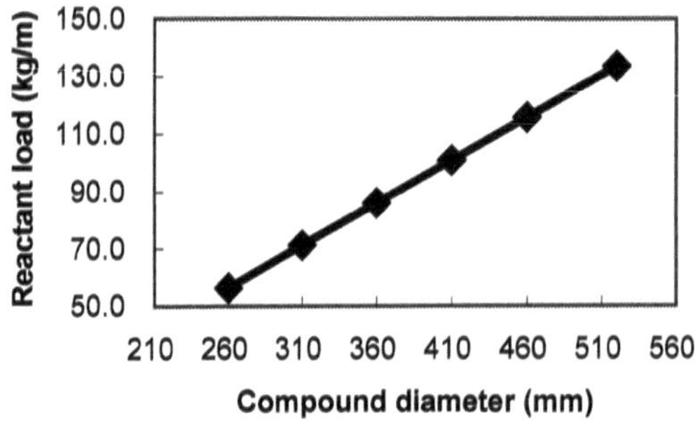

Figure 7: Variation of load for compounds with different diameters

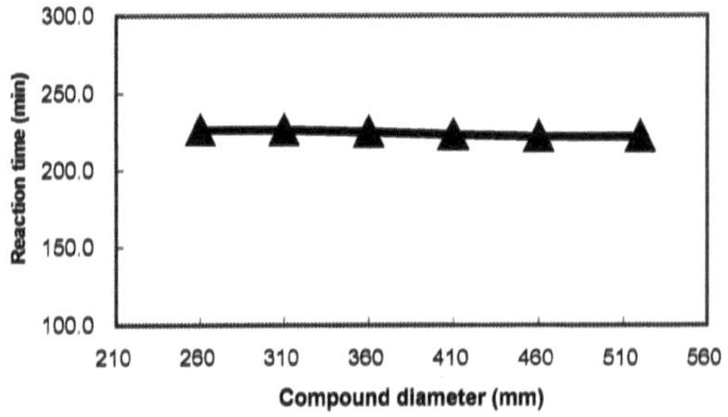

Figure 8: Variation of reaction time for compounds with different diameters

Conclusions

A developed model has been used to understand the effect of slot designs and compound diameters on reaction times in vertical magnesium thermal reduction processes. The predicted results indicate that, for a certain thickness of compounds, the slot design including slot angles and number, and the compound diameter does not affect the reaction time significantly.

Nomenclature

A—Area (mm^2)

C_p—Specific Heat(J/kg K)
F—Radiation shape factor
ΔG—Free energy of reaction (kJ)
ΔH— Heat of reaction (kJ)
h—Heat transfer coefficient (W/m^2 ^{0}C)
k—Thermal conductivity (W/m ^{0}C)
N—Slots number
q —Heat flux (kJ)
R_r—Retort inner radius (mm)
R_o, R_i—Compound outer and inner radius (mm)
R_s—Suspender radius (mm)
T—Temperature (^{0}C)

Greek Symbol

α—Slot angle (^{0})
ρ—Density (kg/m^3)
□—Emissivity
σ—Stefan-Boltzmann constant (W/m^{2}K^4)

Acknowledgement

The authors would like to take this opportunity to thank the Natural Sciences and Engineering Research Council of Canada and the University of Windsor for supporting this work.
The work presented is also a part of the "Magnesium Manufacturing & Product Development group" projects in Manufacturing Systems Department, Ford Research Laboratory, Ford Motor Company.

References

1. R. Osborne, G. Cole, Bruce and D. Penrod, "USCAR Project on Magnesium Structural Castings," Proc. 57th Annual World Magnesium Conference, Vancouver, Canada,2000.
2. JingChun Zang and Weinan Ding, "The Pidgeon Process in China and Its Future", Magnesium Technology 2001, TMS, 2001.
3. Xiaoming Mei, Alfred Yu, Shixian Shang and Tianbai Zhu, "Vertical Large-Diameter Vacuum Retort Magnesium Reduction Furnace", Magnesium Technology 2001, TMS, 2001.
4. Alfred Yu and Henry Hu, "Numerical optimization of magnesium reduction in a modified Pidegon process ", Symposium of Computational Modeling of Materials, Minerals and Metals Processing, TMS, 2001.
5. D.R.Poirier and G.H.Geiger, Transport Phenomena in Materials Processing, TMS, 1994.
6. O.Kubaschewski, Metallurgical Thermochemistry. Pergamon Press, 1967.
7. Emley, E.F. Principles of magnesium technology, Pergamon Press. 1966.
8. E.R.G. Eckert, Heat and Mass Transfer, McGraw-Hill, New York, 1959.
9. Lindon C. Thomas, Heat Transfer—Professional Version, Capstone Publishing Corporation, 1999.
10. S.V. Patankar, Numerical Heat Transfer and Fluid Flow, McGraw-Hill, New York, 1980

A COMPARISON OF THE GREENHOUSE IMPACTS OF MAGNESIUM PRODUCED BY ELECTROLYTIC AND PIDGEON PROCESSES

Subramania Ramakrishnan, Paul Koltun
CSIRO Manufacturing & Infrastructure Technology
Locked Bag No. 9, Preston, Vic. 3072, Australia

Keywords: Magnesium production, Electrolytic magnesium, Pidgeon process, Lifecycle analysis, Greenhouse impact

Abstract

With a focus on the global warming impact, this paper deals with the cradle-to-gate life cycle study of the following two practical production systems for producing magnesium ingots: (i) Magnesite ore is processed using the Australian Magnesium process to produce anhydrous magnesium chloride, which is then electrolysed to produce magnesium; and (ii) Dolomite ore is calcined to produce magnesium oxide, which is then thermally reduced with ferrosilicon using the Pidgeon process, based on the current practice used in China for magnesium production

.

The global warming impact of the ingots produced by the Australian Magnesium process is ~ 24.3 kg CO2-eq/kg of Mg ingot, assuming to HFC-134a is used as the blanketing gas, and electricity and steam are co-generated using Australian black coal. The impact of the magnesium ingots produced in China using the Pidgeon process is ~ 42 kg CO2-eq/kg Mg ingot. This value is based on the current usage of sulphur powder for blanketing molten magnesium while casting. The values of GHG impact for the two systems are discussed in light of energy consumption and process contributions.

Introduction

A major agenda behind the growing use of magnesium for light weighting manufactured products, particularly automobiles [1, 2], is the reduction of environmental impacts of products over their material life cycle. Application of the emerging Life Cycle Assessment (LCA) methodology to the magnesium product value chain will assist in designing, developing and manufacturing magnesium-based products that have lower environmental impacts over the entire product life cycle, which extends from material production, through product manufacture and use, to final disposal and possible recycling. Life cycle environmental data on magnesium production is needed to conduct any LCA study of magnesium products. Currently, only very limited LCA data is available in the public domain [3].

Magnesium is produced commercially either by electrolysing the magnesium chloride derived from raw materials [4], or by using a thermal process, known as the Pidgeon process [5, 6], using dolomite as the raw material. The Pidgeon process is used mainly in China [7], which now supplies ~ 40% of the world demand for magnesium. The remaining is produced mostly via the electrolytic route. However, there are many variations in the way the electrolytic route is applied depending upon the raw material used, preprocessing of the raw material to obtain the feedstock for the electrolytic cell, and design of the cell. In this paper, we have chosen for the cradle-to-gate life cycle study a specific electrolytic route that uses the Australian Magnesium (AM) process [8], and the Pidgeon process as it is used in China.

.

The cradle-to-gate life cycle study of the two systems extends from raw material extraction at the cradle end to the production of magnesium ingots at the gate end of the systems.

The goal of the life cycle study is to compare the global warming (or greenhouse gas emission, GHG) impacts of the magnesium ingots produced by the two processing routes in terms of kg CO_2 –eq for a functional unit of 1kg of magnesium ingots.

Electrolytic Magnesium Production using the AM Process

Australian Magnesium Corporation Limited intends to produce magnesium and magnesium alloy ingots in Stanwell, Queensland, Australia using magnesite as the principal raw material. The overall AM process is shown in Figure 1. Magnesite ore is leached with hydrochloric acid to produce magnesium chloride brine. The brine is treated to precipitate unwanted impurities, filtered to remove solid waste residues, and then dehydrated using proprietary technologies to produce clean and dry magnesium chloride powder that is suitable for electrolysis. The magnesium chloride powder is electrolysed in an Alcan electrolytic cell to produce molten magnesium metal. The molten magnesium is alloyed with elements, such as aluminium and zinc, and then cast into ingots. The chlorine evolved at the electrolytic cell is recycled to produce hydrochloric acid, which is used for leaching the ore. The nominal system considered for the study includes the following design features:

- Production rate of 90,000 metric tons per year to produce ~97,000 tons per annum of ingots of pure magnesium and several magnesium alloys
- Co-generation of steam and electricity using Australian black coal;
- Use of a proprietary gas mixture, containing HFC134a and a diluting gas [10], as the blanketing gas for protecting molten magnesium from oxidation.

We have considered nine unit process stages (including the seven shown in Figure 1) in the product system considered for the life cycle study. The life cycle data required for the study were derived from the process information and data obtained from Australian Magnesium Corporation Limited. Thermodynamic calculations of mass and energy balance for all process stages were made to verify the data supplied.

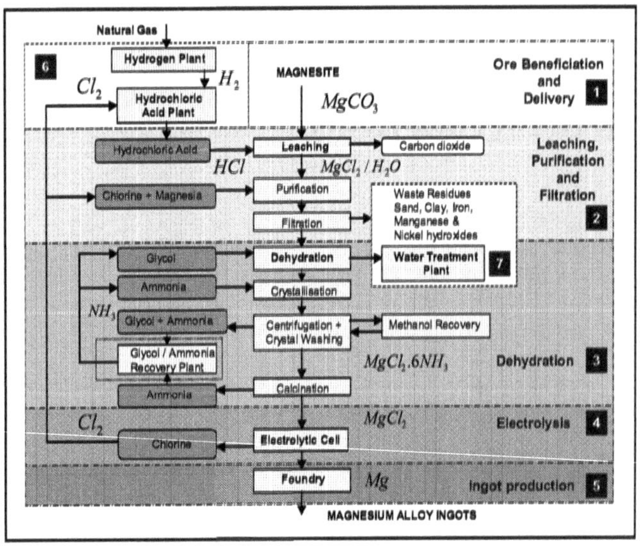

Figure 1. Electrolytic magnesium production using AM process (Numbers in filled squares show unit process stages)

Process stage AM1: Mining ore, beneficiation and delivery

Magnesite ore is mined and beneficiated at the Kunwarara mine site to produce a clean ore feed stock. The beneficiated ore is then transported by heavy 40-ton diesel trucks over ~ 100km to the magnesium plant at Stanwell. Typically, an amount of 365.7×10^{6} kg per annum of ore is required to produce 90×10^{6} kg per annum of magnesium. The ore contains $MgCO_3$ (~90%), $CaCO_3$ (~3%), SiO_2 (~6.9%) and other minor impurities. Electrical energy for mining equipment, and energy from diesel for transportation are used.

Process stage AM2: Leaching, purification and filtration

The ore received from the mine is leached by mixing it with hydrochloric acid in agitated tanks at the magnesium plant to produce predominantly magnesium chloride solution, as given by the chemical equation 1. Depending upon the amount of calcium carbonate present in the ore, some calcium chloride is also produced.

$$MgCO_3 + 2HCl \rightarrow MgCl_2 + CO_2\uparrow + H_2O \qquad (1)$$

The brine obtained from leaching is treated with magnesia and chlorine to remove impurities, such as iron, nickel and manganese. The purified brine is filtered to remove unreacted solids and precipitated impurities to obtain clear, clean magnesium chloride brine that also contains some calcium chloride. The filter residue of clay and sand is disposed of at the mine site. A small amount of electrical energy is used in the process.

Process stage AM3: Dehydration

This is a crucial process stage that is based on proprietary technologies in which the magnesium chloride obtained from the leaching process stage is converted into pure anhydrous magnesium chloride crystals for feeding into electrolytic cells.

The brine from the leaching circuit is mixed with ethylene glycol; the water is boiled off in distillation columns to give magnesium chloride in glycol. The magnesium chloride in the glycol mix is reacted with ammonia to grow fine white crystals of magnesium chloride hexammoniate. Magnesium chloride hexammoniate is then separated from ethylene glycol using centrifuges. The crystals are washed with methanol to drive away any remaining glycol. The remaining calcium chloride in the brine is removed by treating with magnesia and carbon dioxide, and also by steam stripping. Crystals of magnesium chloride hexammoniate are heated in fluidised beds to dry the crystals and drive off ammonia. This calcination process leaves a dry white anhydrous powder of magnesium chloride for feeding into the electrolytic cell.

Ammonia is separated from glycol by cooling the mixture to the ambient temperature. The methanol from centrifugation is separated from water by heating up the mixture, and then condensing the vaporised methanol.

Thermal energies from steam and natural gas, and some electricity are used by this process stage.

Process stage AM4: Electrolysis

Anhydrous magnesium chloride is fed into the Alcan electrolytic cell. The magnesium chloride dissolves into the electrolyte, which is maintained at 660 °C. The passage of electric current splits the magnesium chloride into pure magnesium, and chlorine gas which is liberated at the anode. The molten magnesium is tapped from the cell in batches. The chlorine gas liberated in the process is returned for hydrochloric acid production (process stage AM6).

Electrical energy is consumed, both for maintaining the temperature of the electrolyte, as well as for electrolysis.

Process stage AM5: Ingot production

The molten metal tapped from the electrolytic cell is mixed with other alloying metals to form the required alloy. The molten alloy is then cast as ingots. Natural gas and electricity are used.

A cover gas is used to prevent the molten metal from oxidising. In the nominal product system, a mixture of HFC-134a at 1.1g diluted by air at 0.1kg is used per 1kg of ingot produced.

Process stage AM6: Hydrochloric acid production

Sodium chloride, supplied as anhydrous crystals, is dissolved in water; the solution then undergoes electrolysis to produce chlorine. Chlorine, obtained at the anode of the electrolytic cell, is also used to make up for the lost chlorine. Approximately 50% of the sodium hydroxide obtained from electrolysing sodium chloride is used as a by-product.

Some of the hydrogen librated during the electrolysis of sodium chloride is used for acid production. However, ~ 97% of the hydrogen required for making hydrochloric acid is produced from natural gas, which is mixed with steam in the presence of a catalyst at high temperature. This process produces hydrogen and carbon monoxide. Then, steam is added to the mixture of gases in the presence of a catalyst at a temperature above 350 ° C to extract further hydrogen and to oxidise carbon monoxide to carbon dioxide. Carbon in the natural gas used in the process is liberated as carbon dioxide.

Hydrogen is fumed in chlorine to form hydrogen chloride, which is absorbed by the water inside absorption columns. Chlorine, obtained at the anode of the electrolytic cell, is used to close the chlorine loop.

Process stage AM7: Water and waste treatment

Water discharged from different processes is treated in a special biological treatment plant and disposed of to the environment. Solid wastes generated by the plant, including spent cell anodes and refractories, and dross and sludge are treated on site for disposal off site.

Process stage AM8: Auxiliary services

The magnesium plant uses many auxiliary services. These include: water reticulation; cooling water systems; steam and condensate; supply of compressed air, inert gas and fuel; electricity distribution; infrastructure etc. Electrical energy is used for providing auxiliary services.

Process stage AM9: Co-generation of steam and electricity

Both steam and electricity are obtained for magnesium production from the Stanwell power plant. Coal is used as the fuel to generate energy at the Stanwell power plant.

In order to take into account cogeneration, we have included co-generation as a process stage within the product system [9] by extending the boundary of the product system to include the co-generation cycle of the Stanwell power plant. When co-generation cycle is used in a thermal power station, an increased thermal efficiency in producing electricity may be attained.

The Pidgeon Process in China

The Pidgeon process is based on reducing the oxide of magnesium using silicon by supplying the thermal energy required for the reaction [5]. The practical application of this principle requires the reduction of calcined dolomite ore with ferrosilicon at high temperatures. The core chemical reaction in the process is as follows:

$$2\,(MgO,CaO)_{(s)} + (Fe)Si_{(s)} = 2Mg_{(g)} + 2CaO, SiO_{2(s)} + (Fe)_{(s)} \tag{2}$$

The reaction is conducted in retorts at temperatures in excess of ~1100°C, and magnesium is released as a vapour, which may be condensed to obtain pure magnesium.

The Pidgeon process, as it is used in China today [7], bears many similarities to the original plant designs, proposed in the 1940s by their developers in Canada [6]. However, there are variations in the material and energy intensities across the magnesium industry in China depending upon the size of plants, their location, and the adoption of new technologies.

We have included six process stages by considering a representative product system for magnesium production in China, as shown in Figure 2. We have obtained some data from a consulting company, Clark & Marron Pty. Ltd, that has a first hand experience on the magnesium industry in China. Much of the information and data obtained compare favourably with the data provided in the publication of Zang [7] on the Pidgeon process in China

Process stage CM1: Mining and transportation of dolomite and coal

The principal raw material is dolomite ore, which is mined and transported to the magnesium plant. An amount of 11.6kg of ore is used to produce 1kg of magnesium. The ore is predominantly dolomite ($CaCO_3 . \bullet MgCO_3$) with the proportion of MgO in the ore being ~20%. Hence, the theoretical mass proportion of impurities on the basis of molecular weights of the minerals is 8.6%. The efficiency of recovery of magnesium from the available magnesium in the ore is 71.5%.

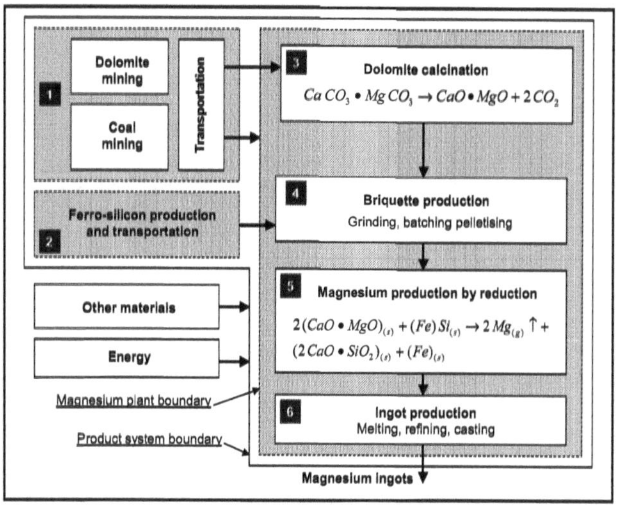

Figure 2. Magnesium production using the Pidgeon process in China (Numbers in filled squares show unit process stages)

The consumption of coal for producing 1kg of pure magnesium is 10.5kg. The coal used has a calorific value of 25MJ/kg coal. An amount of 0.11 MJ of electrical energy is used for delivering coal on the ground, and for using other auxiliaries, such as air fans etc. We have assumed that both coal and the dolomite ore are transported to the magnesium plant by 40-ton diesel trucks.

Process stage CM2: Ferrosilicon production and transportation

Ferrosilicon is used at the rate of 1.19kg/kg of magnesium. The ferrosilicon contains 75% silicon and 25% iron. Ferrosilicon is produced in a three-phase submerged arc furnace by the carbothermic reduction of quartz in the presence of high quality scrap steel. The electrical energy consumed to supply 1.19 kg of ferrosilicon to produce 1kg of magnesium is 44.98MJ. The transportation of ferrosilicon to the magnesium plant takes place over a long distance of ~ 700km.

Process stage CM3: Dolomite calcination

The dolomite ore that is transported to the magnesium plant is calcined by the addition of heat in furnaces at ~1300°C (equation 3).

$$(CaCO_3 \bullet MgCO_3)_{(s)} + heat \rightarrow (CaO \bullet MgO)_{(s)} + 2CO_{2(g)} \uparrow \quad (3)$$

Energy for calcining is derived from the combustion of coal. The calcining process yields 'dolime'.

From the 11.6kg of ore with impurities supplied to the process, the theoretical yield of dolime from the process is approximately 5.36kg, assuming 5% loss of dolime. The amount of coal supplied for the process is 2kg for 1kg of magnesium produced.

Process stage CM4: Briquette production

The calcined dolomite is then finely ground and mixed in a specific ratio with finely ground ferrosilicon containing 75% of silicon. Calcium fluoride at a nominal rate of 0.2kg per 1kg Mg may be added to the mixture as a catalyst. The mixture of calcined dolomite, ferrosilicon and calcium fluoride is formed into briquettes, which are typically 12mm in diameter and 50mm in length, and have a mass density of 2300kg/m^3.

Process stage CM5: Magnesium production by reduction

The reduction process is conducted in refractory brick furnaces, which are fired using coal. The capacity of a furnace may be in the range of 10 - 30 retorts. The retorts are charged with briquettes from the previous stage, the approximate charge per retort being 128 kg. The temperature inside the retort is held at 1160-1165°C, and a vacuum pressure of 1.5 - 2.0 Pa is maintained for condensing magnesium vapour at the exit end of the retort.

As the reduction reaction proceeds, magnesium vapour condenses in crystalline form on a recyclable, water-cooled steel. The condensed magnesium, known as the "crown" magnesium, produced in each retort at the completion of the reaction weighs ~ 20 kg. The sleeve, together with the condensed magnesium, is removed from the retort while the metal is still hot. The metal crown is subsequently removed from the steel sleeve.

The estimated magnesium recovery from the reduction process is 4.34/5.09 = 0.852, a value that agrees with the published value of Zang [7]. The efficiency of usage of ferrosilicon is 0.812/0.95 = 0.854.

The amount of coal used for the reduction process is approximately 6.8kg per 1kg of pure Mg. A small amount of electrical energy of 0.55MJ is also used by the process stage to assist water-cooling and other auxiliary processes.

Process stage CM6: Ingot production

The magnesium crowns extracted from the retorts are melted in melting furnaces, and refined by heating the melt above the melting point of 650 °C. Alloying elements may be added when magnesium alloys are produced. The molten metal is then transferred from the melting furnace and poured into ingot moulds to produce magnesium ingots.

Coal at 0.3kg/kg Mg is used for melting and refining magnesium crowns.

The current practice in China for protecting molten metal from oxidation is to disperse manually sulphur powder over exposed molten metal surfaces in moulds. The amount of sulphur used for melt protection is 0.05kg/kg Mg.

Calculated Impacts

Life cycle inventory and impact analyses of the two systems were conducted by using SimaPro, well known software for conducting LCA studies based on international standards.

As the focus of this study is the global warming impact category, all the greenhouse gases emitted by the product system are quantified. The mass of each greenhouse gas emitted is multiplied by its Global Warming Potential (GWP) to determine the contribution of the gas to the global warming or greenhouse impact. The global warming impact of the product system is then obtained by totaling the contributions of all the greenhouse gases in the calculated inventory. The estimated global warming impact from the product system on a functional unit basis is expressed as kg CO_2-eq/kg Mg ingot.

Uncertainty analyses were conducted using the Monte Carlo technique to estimate the uncertainty in the estimated global warming impacts by assuming values for the lower and upper bounds for the uncertainties in the various process parameters of the two systems.

In Table I, we have shown the calculated results of global warming impact, cradle-to-gate life cycle (or primary energy) and process energy for the two systems. For estimating the primary energy, we have used values appropriate to the region for conversion efficiencies and calorific values of the fossil fuels used [9,11,12]. Figures 3(a) and (b) show the distribution across process stages of the cradle-to-gate global warming impact and energy of the two systems.

Discussion

The difference between the global warming impacts of the two systems (Table I) may be discussed in the light of the differences between the core processing stages that produce magnesium and the other process stages that support the core processing stage in each system (Figure 3(a) and (b)).

Core processing stages

In the electrolytic route using the AM process, the core processing stage is the electrolysis (AM4) that extracts magnesium from magnesium chloride. This process consumes ~ 46.8 MJ/kg Mg ingot of energy, mostly in the form of electricity, which is produced in Australia using black coal. The primary energy supplied to the process stage is 121 MJ/kg Mg ingot. Hence the contribution of this process stage to the total greenhouse impact of the system is 12.1 kg CO_2 -eq/kg Mg ingot (~50%).

In the case of the Pidgeon processing route followed in China, the core processing stage is the thermal reduction (CM5) of magnesium oxide to magnesium using thermal energy and ferrosilicon. An amount of 181 MJ/kg Mg ingot of thermal energy derived from coal is supplied to this process. The GHG impact of the reduction process stage is 15.9 kg CO_2 -eq/kg Mg ingot (~38%).

We may observe that the contribution of the core processing stage to the total impact is significant in both systems. We may also observe that the major difference between the impacts of the two systems is not mainly due to the differences in the impacts of the two core processing stages.

Table I. Calculated results of impacts for the two systems

	Electrolytic route using AM process	Pidgeon process in China
Global warming impact (kg CO_2 -eq/kg Mg ingot)	24.3 within uncertain range: 20.4 – 26.4	42 within uncertain range: 37 - 47
Process energy (MJ/kg Mg ingot)	174 Steam: 42.3%; Electricity: 34.1%; Natural gas: 23.1%; Diesel: 0.5%	284 Coal: 83%; Electricity: 16%; Diesel: 1%
Cradle-to-gate life cycle energy (MJ/kg Mg ingot)	275	355
Melt protection	Cover gas: HFC-134a (GWP: 1600)	Sulphur powder Non-GHG

Supporting processing stages

The dehydration process stage (AM3) that produces the anhydrous magnesium chloride contributes 9.2 kg CO_2 -eq/kg Mg ingot (~37%) to the global warming impact of the electrolytic route using the AM process. The process consumes 102.4 MJ/kg Mg ingot of energy at the process level of which ~ 72MJ/kg Mg ingot is derived from steam, which is co-generated with electricity.

The dehydration (AM3) and leaching (AM2) stages together serve the purpose of producing the feedstock required for electrolysis. The combined impact of the two stages is 11.1 kg CO_2 -eq/kg Mg ingot (46%), of which 3.2 kg CO_2 -eq/kg Mg ingot is recovered using cogeneration (AM9 in Figure 3a). Thus, the combined contribution of the process stages that contribute towards feed preparation is 7.9 kg CO_2 -eq/kg Mg ingot.

In the case of the Pidgeon process in China, the calcination process (CM3: 10.1 kg CO_2 -eq/kg Mg ingot; 24%), together with briquette production (CM4: 0.2 kg CO_2 -eq/kg Mg ingot), serves a similar role to that of the dehydration and leaching stages in the electrolytic system in preparing the feedstock for magnesium production.

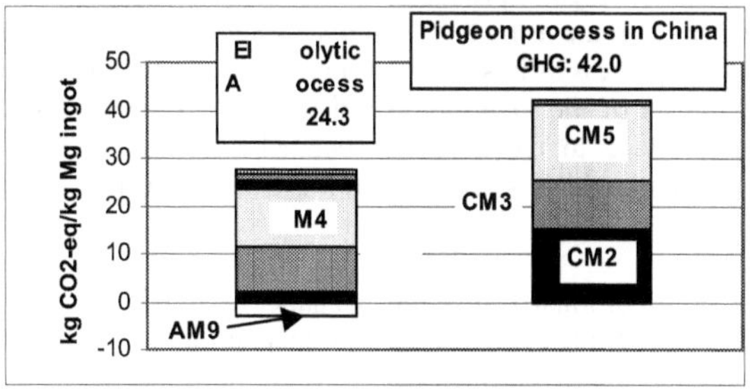

(a)

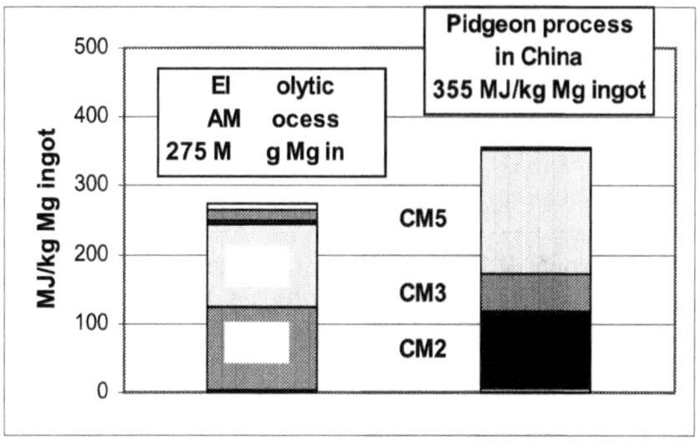

(b)

Legend

AM3 – Dehydration; AM4 – Electrolysis; AM9 – Cogeneration.

CM2 – Ferro-silicon Production; CM3 – Dolomite Calcination; CM5 – Mg Production by Reduction.

Figure 3(a). Calculated global warming impacts of magnesium ingots produced by the two systems

Figure 3(b). Cradle-to-gate life cycle energy used by the two systems

An essential ingredient for reducing magnesium oxide using the Pidgeon process is ferrosilicon. According to our calculations, the ferrosilicon production stage (CM2) produces a global warming impact of 14.7 kg CO_2 -eq/kg Mg ingot as it consumes an amount of 45MJ/kg Mg ingot of electricity, which is produced predominantly using coal.

It may be seen from Table I that the difference in the global warming impacts of the two systems considered in this study is ~17.7 kg CO_2-eq/kg Mg ingot. Our study reveals that a significant proportion of this increased GHG emission from the Pidgeon process, as it is used currently in China, is due to the impact associated with the production of ferrosilicon, which is required for the reduction of magnesium oxide.

Conclusion

We have estimated the life cycle global warming impacts of magnesium ingots produced via two processing routes: an electrolytic route using the Australian Magnesium process and the Pidgeon process as it is used in China. The calculations show that the impact of the ingots produced using the Pidgeon process is higher owing to the impact associated with the production of ferrosilicon, which is an essential ingredient used in the Pidgeon process.

The values of impacts given in this paper need to be considered in the context of the entire life cycle of magnesium products to foster continuous improvements over the whole value chain.

Acknowledgements

We wish to thank Cooperative Research Centre for CAST for supporting this work. We also thank Jim Marron of Clark & Marron Pty Ltd. for providing information and data on magnesium production in China

References

1 Brown R.E., "Magnesium industry growth in the 1990 period," *Magnesium Technology 2000*, ed. H.I. Kaplan, J.N. Hryn and B.B. Clow (Warrendale, PA: The Minerals, Metals & Materials Society, 2000), 3-12.

2. B.L.Mordike and T. Ebert, "Magnesium: properties - applications – potential," *Materials Science and Engineering*, A302 (2001), 37-45.

3. L.K. Sivertsen, O. Haagnesen and D. Albright, A review of the life cycle environmental performance of automotive magnesium, 2003, *SAE Technical Paper Series*, SAE International, 2003-01-0641.

4. R. W. Stanley, M. Berube, C. Celik, Y. Oosaka, J. Pearcy and M. Avedesian, "The Magnola process for magnesium production," *Proceedings of the International Magnesium Association 54: Magnesium Trends*, (1997), 58-65.

5. L.M. Pidgeon and W.A. Alexander, 1944. "Thermal production of magnesium – pilot plant studies on the retort ferrosilicon process," *Transactions American Institute of Mining and Materials Engineers*, 159 (1944), 315-352.

6. A. Mayer, 1944. "Plant for production of magnesium by ferrosilicon process," *Transactions American Institute of Mining and Materials Engineers*, 159 (1944) 363-376.

7. J.C. Zang, 2001. " The Pidgeon process in China and its future," Magnesium Technology 2001, ed. J. Hryn , Magnesium Technology 2001, (Warrendale, PA: The Minerals, Metals & Materials Society, 2001), 7-10.

8. G.L. Baker, "The Stanwell magnesium project," *Magnesium Industry*, 5 (2001), 18-24.

9. P. Koltun , S. Ramakrishnan and A. Tharumarajah, "A methodology for the inclusion of cogeneration in product life cycle assessment," *The Third Australian Conference on Life Cycle Assessment: Life Cycle Decision Making for Sustainability*, Gold Cost, Queensland, Australia, July 2002.

10. N. Ricketts, S. Cashion and R. Bailey, 2003. "Industrial trials with the AM-cover gas system for magnesium melt protection," *Proc. 1st Int. Light Metals Technology Conf.*, Brisbane, Australia, ed. A. Dahle (2003) 275-280.

11. C. Kroeze, J. Vlasblom, J. Gupta, C. Boudri and K. Blok, 2004. "The power sector in China and India: greenhouse gas emissions reduction potential and scenarios for 1990-2020," *Energy Policy*, 32 (2004), 55-76.

12. B. He and C. Chen, 2002. "Energy ecological efficiency of coal fired plant in China," *Energy Conversion and Management*, 43 (2002), 2553-2567.

LIFE CYCLE ENVIRONMENTAL IMPACT OF MAGNESIUM AUTOMOTIVE COMPONENTS

P. Koltun[1], A. Tharumarajah and S. Ramakrishnan
CSIRO Manufacturing & Infrastructure Technology
PO Box 56. 9, Highett, Vic. 3190, Australia

Keywords: Magnesium production, Electrolytic magnesium, Lifecycle analysis, Greenhouse impact, Converter housing

Abstract

The development of magnesium applications for automotive industries is receiving significant attention. One aspect of this attention is the assessment and reduction of the cradle-to-grave environmental impact of magnesium components.

The study investigates detailed impact of magnesium converter housing for automotives starting from the production of magnesium ingots to the manufacture and assembly, use and recycling.

Extensive sensitivity analysis is conducted to examine the impact of key process parameters that can improve the environmental performance of the component. The parameters include: cover gases other than SF_6, improvements to product yield and use of secondary magnesium. From this analysis, a number of environmental performance scenarios are proposed and used to compare the impact of similar functional components made using of magnesium produced in China, aluminium and iron.

The investigations clearly show a significant reduction in the greenhouse gas impact may be achieved from the lighter weight of the magnesium components. Also, process improvements to reduce the impact improve the break-even distances in the use of automobiles at which magnesium becomes comparable with other competing metals.

Introduction

The environmental drive towards reducing emissions from automobiles has resulted in the growing use of lightweight magnesium components to achieve weight reduction. This drive calls for a clear understanding of the environmental impacts associated with the light-metal value chain. With a focus on this important domain, this paper addresses the greenhouse impact associated with manufacturing, using and recycling magnesium automotive components, and discusses some issues for improving the environmental performance of the value chain.

The life cycle stages of magnesium components broadly includes production of magnesium alloy from ore, manufacture of a representative magnesium component, assembly, use of automobiles, and recycling of the end of life vehicles (ELV) that are discarded after use. Many researchers have paid attention [1,2] to quantify the greenhouse impact of producing magnesium metal by electrolytic means. However, environmental issues related to manufacturing of components, and the advantages to be gained from using manufactured magnesium components in automobiles, have received only limited attention. In this paper, we have made an attempt to improve our understanding of the cradle to grave environmental impact by adding the manufacturing, use and recycling stages so as to extend the life cycle boundary from cradle-to-grave.

We use life cycle assessment (LCA) that is currently being recommended by international standards [3] to quantify the environmental impacts associated with a product, namely converter housing (CH). Using the results of the "cradle-to-grave" life cycle study, we have conducted sensitivity analysis on selected parameters that represent options for reducing GHG emissions. The parameters considered include reduction in energy use, improvements to product yield, increasing the use of secondary magnesium and light-weighting. The analysis is extended to compare the GHG emissions of using magnesium with other competing metals, such as iron and aluminium.

Goal and Scope of the Life Cycle Study

The principal goal of the current LCA study is assessment of different processes, materials, and systems that are major contributors to environmental impact through whole life cycle of magnesium automotive components and identification of possible improvements with the aim to reduce such impacts. The impact category that has been selected to address the above goal is the global warming impact that deals with the emission of greenhouse gases (GHG) from the product system.

In order to understand the impact of magnesium components that could be used in automobiles we have chosen for the LCA study a converter housing made from magnesium alloy AZ91, which houses the torque converter used in cars with automatic transmission (see Figure 1, also, the mass of the magnesium converter housing of 3.1kg is obtained from [4]).

The functional unit of this product is chosen to be the product itself, i.e. CH. The life cycle impact of a chosen magnesium component is therefore expressed in terms of impact units per component, and any comparison of the impact for magnesium with the impact of another material is made on the basis of the impact of the functional component (e.g. kg CO-eq /Mg CH vs. kg CO-eq/Al CH).

The life cycle product system for magnesium relevant to automotive applications may be viewed as a generic product

[1] Corresponding author: Paul Koltun, CSIRO Manufacturing & Infrastructure Technology, PO Box 56, Highett, Vic 3190, Australia; Ph: 61 3 9252 6599; Fax: 61 3 9252 6248;
E-mail: paul.koltun@csiro.au

system that is a collection of the four life cycle stages shown in Figure 2: Primary magnesium production, Manufacturing and assembly of components into vehicles, Use of vehicles, and finally, Secondary magnesium production by open loop recycling.

Each life cycle stage in the generic product system consists of several unit processes. The details of the processes that are used for making the component are described below.

Figure 1. Converter housing (CH)

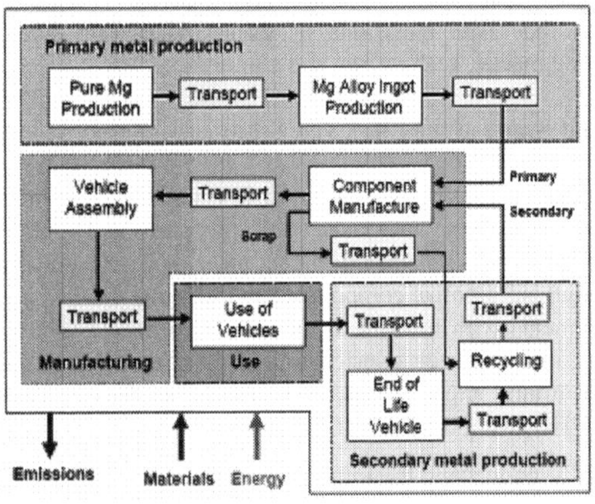

Figure 2. Generic product system considered for LCA study

Description of Processes in Product Systems

Primary Magnesium Ingot Production

This stage deals with the production of ingots of magnesium alloys from raw materials. We have considered production of magnesium alloy ingots in Australia by an electrolytic process [5] by using AMC proprietary technologies.

The process stages related to the production of magnesium alloy ingots in the generic product system (Figure 2) includes mining of ore, all the processes that are used for making pure magnesium, and the processes for making ingots of magnesium alloys from pure magnesium. The details of the various processes associated with the production of magnesium alloy ingots made by two different production routes have been presented in [6].

Manufacturing and Assembly of Converter Housing

Magnesium alloy ingots (both primary and secondary) are transported to component manufacturing, where alloy ingots are transformed into components for use in cars. The components are transported to vehicle assembly plants, where they may be further processed and assembled into cars. It is assumed that a cover gas that contains sulphur hexafluoride (SF_6) is used in the manufacturing plant for protecting molten magnesium form oxidation. All manufacturing scrap is recycled externally, i.e. magnesium scrap from manufacturing plants is transported to recycling centre, where secondary magnesium is produced.

The core manufacturing process that is used for manufacturing the magnesium converter housing is the HPDC process. (Figure 3 shows the product system associated with the manufacturing of magnesium CH). From the mass flow of magnesium shown in the figure, it may be seen that an amount of 5.709kg of magnesium alloy ingots is required to produce a converter housing of mass 3.1kg. It is assumed that the metal received by the manufacturing plant is made up of 70% of primary metal and 30% of secondary metal [7]. Further, the processes and electrical energy source are assigned to the currently adopted HPDC technology in USA for similar magnesium products and average electrical energy source in USA, respectively.

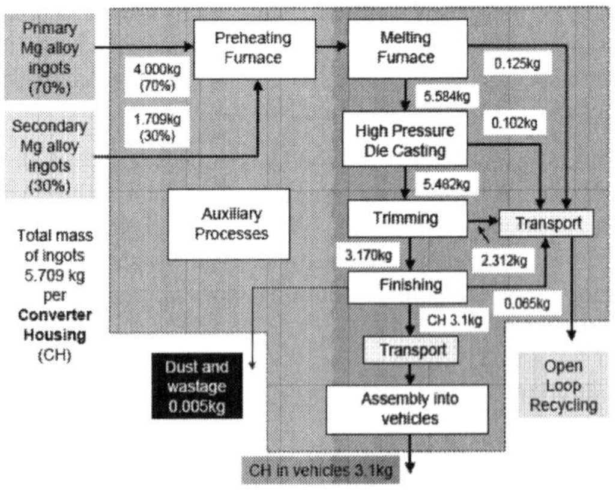

Figure 3. Manufacturing and assembly of CH

The various processes used in manufacturing magnesium converter housings and assembling them into vehicles (Figure 3) are described below.

Preheating and melting magnesium ingots. Ingots of magnesium alloy received by the die-casting plant are initially preheated to a temperature of 180°C using an electrical furnace, similar to patented "Ferrywheel" furnace [8], having a thermal efficiency of 50%.

The preheated magnesium alloy ingots are charged into melting furnace, which is located at the die-casting machine while casting magnesium. We have assumed that the melting furnace used in manufacturing is similar to the Rausch MM550 that consumes 1.3MJ of electrical energy per 1kg of magnesium melt [8] with a thermal efficiency of 62%. For calculating the GHG impact of consuming electrical energy in the die-casting operation, we have assumed an emission factor of 0.207 kg CO-eq/MJ of electrical energy on the basis of the average values including grid loss for producing electrical energy in the USA [9].

SF_6 cover gas is used while handling molten magnesium to prevent the oxidation of the metal. An average consumption of 0.96g of SF_6 per 1kg of molten metal is assumed. Approximately 10% of the SF_6 supplied is absorbed by molten magnesium.

Die-casting, trimming and finishing processes. Molten metal from the melting and holding furnace is transferred to the shot sleeve of the HPDC machine to produce the raw casting. We have used process data that are typical to the manufacturing of die-cast automotive components in the USA in estimating the materials and energy input for making the converter housing.

The output of die-casting machine contains the final desired product (3.1kg of CH) and additional solidified metal (2.482kg) in the runners and gates. The excess material is trimmed using a trimming machine, and sent to recycling centres for secondary metal production.

Auxiliary processes. There are many auxiliary and subsidiary processes that support the core processes in a die-casting plant. These include processes, such as systems that provide compressed air, plant for treating water, and others that form the essential infrastructure. These processes consume energy and use many materials and physical products, such as process consumables, clothing etc.

Assembly of converter housings into vehicles. The converter housings manufactured at the component manufacturing plant are transported to vehicle assembly plants, where they are assembled into vehicles. It is assumed that the transportation of converter housings is carried out using 40-ton diesel trucks over an average distance of 300kms.

Table 1 gives the process data used for the LCA study.

Secondary metal production

In this study, secondary magnesium production is treated as open-loop recycling, as shown in the generic product system (Figure 4). The open loop recycling stage involves transporting, shredding and melting of scrap; cleaning and alloying of the melt material; and casting molten metal as ingots.

The production of secondary magnesium by recycling assumes the current practice of magnesium recycling industry in the USA. New scrap sent from component manufacturing plants (about 60% of all scrap), and old scrap received from recycled cans (mostly) and other magnesium containing products (such as end of life of vehicles), are used by recyclers to produce secondary magnesium alloy ingots that meet the quality standards of manufacturers [7].

The operations associated with secondary magnesium production include (shown in Figure 4): recovery, preparation and handling of process materials; shredding and de-coating activities. The shredded scrap is then melted in a furnace. After melting, the melt is cleaned by adding flux. Then it is transferred to a holding furnace over a launder. Alloying takes place in the holding furnace.

On the basis of the assumption that the ratio of primary to secondary magnesium ingots received is 70:30%, the mass of secondary magnesium ingots delivered to the manufacturing plant is 1.709kg for making a converter housing of mass 3.1kg. Details of energy and material consumption in secondary magnesium production are given in Table 2.

Converter Housing: Manufacturing & Assembly			
Process	Main materials (kg/CH)	Principal products (kg/CH)	Energy input (MJ/CH)
Preheating and melting magnesium ingots	Mg ingots 5.709 (Primary: 4.000; Second.: 1.709)	Molten Mg alloy: 5.584	El. energy: 1.92 + 7.42 (melting & holding)
	Cover gas SF6: 0.00549 dry air: 0.0608		
Casting, trimming and finishing	Molten Mg alloy: 5.584	Converter housing: 3.1	El. energy: 8.52
	Water: 7.5		
	Aluminium shot: 0.0002		
	Lubricant: 0.013		
Auxiliary processes	Cotton, rubber, Fellpro, etc.		El. energy: 3.2 + Thermal energy: 17.73 (gas and fuel)
	Water from all processes: 6.9		
Transition to assembly plant	Diesel: 0.054		Energy from diesel fuel: 2.28
Assembly into car			El. energy: 1.51

Table 1. Details of processes for manufacturing CH

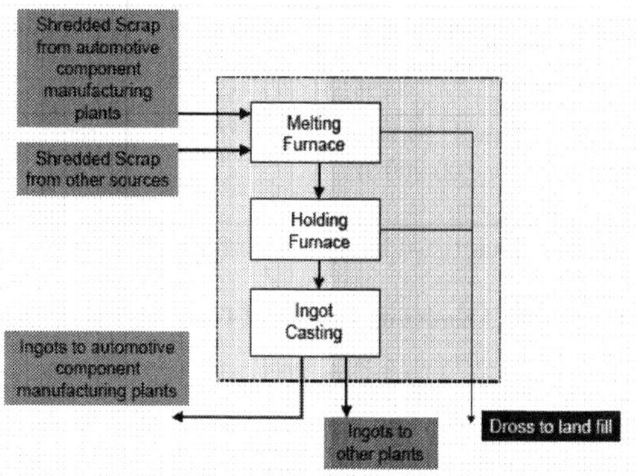

Figure 4 Open loop recycle of magnesium scrap for secondary metal production

Use in cars.

During use phase, the fuel consumption (and hence the emission) is assigned to the components on basis of the mass contribution to the whole mass of the car. A mid-size car (with a mass $\mathbf{M_V}$ = 1400kg and fuel consumption of $\mathbf{K}$ = 8.5 litres per 100km of driving [10]) is considered for the study, and it is assumed that the car reaches the end of its life after a driving distance of 200,000kms.

Converter housing: open loop recycling				
Process	**Materials input (kg/CH)**	**Principal products (kg/CH)**	**Energy input (MJ/CH)**	**Comments**
Recycling Mg (open-loop)	Mg scrap: (2.312+1.16) x 0.55 Fluxes: 0.515 SF6:	Mg alloy ingots: 1.709 (for CH) 1.36 (for other products)	El. Energy: 6.48 Nat. gas: 4.16	Energy and materials allocated for CH. Dross and sludge to landfill:0.65

Table 2. Details of open-loop recycling for CH

Assuming that the average fuel consumption to be linearly dependent on the mass of the car, the amount of fuel consumed over the life of the car in moving 1kg of the magnesium component used is given by,

$$F = K \times L / M_V = (8.5/100) \times 200000 / 1400 =$$
$$= 12.14 \text{ litres of fuel/kg of component.}$$

Using the data for GHG emission from cars of 2.85 kg CO2 eq/litre [10], we get the life cycle GHG emission from using a component be 34.60 kg CO-eq/kg of component.

In calculating the GHG emission resulting from the use of a component in a car, we have accounted for the mass of the support structure that carries the component in the car. According to [11, 12] this secondary weight can save up to 50% of the primary weight save, though we have made a conservative assumption of 25%. Table 3 gives details of using the converter housing in a car.

Converter Housing: use in cars			
Process	Materials input (kg/kg of product)	Energy input (MJ/kg of product)	Comments
Use CH in car	Petrol: 34.14kg for 200,000km	Energy from petrol: 1623.2MJ	Mass of replacement Mg: 3.1 X 1.25 kg

Table 3. Details of using converter CH in car.

Life Cycle GHG Emission Impact of Converter Housing

A life cycle model was created for the product system shown in Figure 2 using SimaPro LCA software [11]. The model was analysed to estimate life cycle GHG emission inventories and global warming impacts for different scenarios for the various processes that make up the product system. In estimating the global warming impact, the global warming potentials (GWP) of the various greenhouse gases emitted were used to quantify the impact in terms of kg CO-eq /kg of the functional unit under consideration. Aggregated global warming impacts for the generic processing stages were obtained as a result of the study.

The basic model has been considered to quantify the life cycle impact under nominal conditions. The basic system is characterised by the following: Primary magnesium alloy ingots are produced in Australia using the AM process, and transported to the USA. (AM-cover is used as a cover gas while casting ingots). Secondary magnesium alloy ingots are produced in the USA using the scrap produced in the USA. (SF_6 is used as a cover gas while casting secondary ingots). Manufacturing of converter housings is done in the USA using a mix of 70% primary magnesium (from Australia) and 30% secondary magnesium (from the USA). The converter housings are assembled into vehicles in the USA. The car with the manufactured converter housing is used in the USA over a total driving distance of 200,000kms Impacts under conditions that deviate from the nominal condition are assessed in sensitivity analyses.

Figures 5 and 6 present the GHG impact and the primary energy consumption of the converter housing under nominal system conditions, respectively. The overall "cradle-to-grave" life cycle GHG impact for the CH is 349.8 kg CO2-eq/CH. It can be seen (Figure 5) that the primary production, manufacturing and use stages of the life cycle contribute significantly towards the GHG impact.

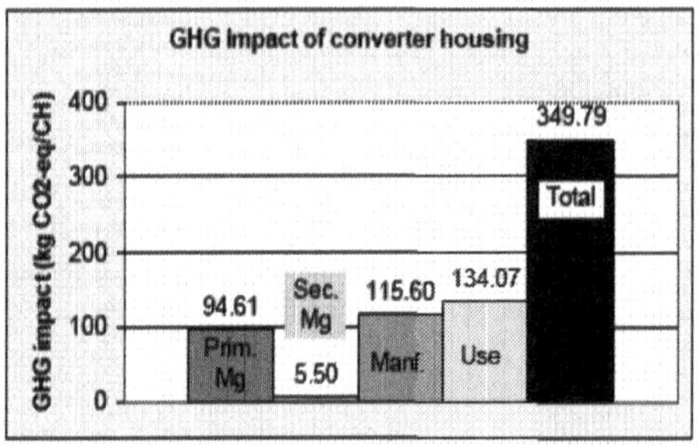

Figure 5. GHG impact of CH over the life cycle.

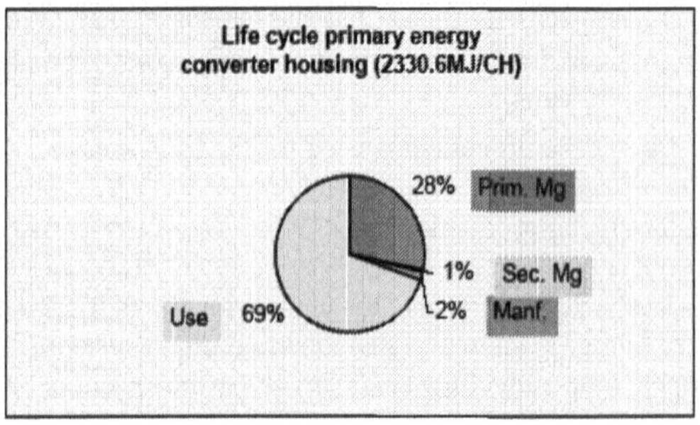

Figure 6. Primary energy consumption of CH over the life cycle

The primary energy over the life cycle of the CH is 2330.6MJ/CH, of which most of the contribution is from the primary magnesium production and use stages. The manufacturing stage uses little primary energy. The enhanced GHG impact of manufacturing is primarily due to the use of SF_6, which has a very high GWP of 22,200.

Sensitivity analysis of the impact of converter housing

The sensitivity analysis of the life cycle environmental impact of the CH has been conducted for the following important factors: usage of cover gas in manufacturing and recycling stages, product yield in manufacturing stage, source of primary magnesium alloy ingots and the proportion of secondary metal used by the product system. The calculated results are shown in Figures 7 and 8. It can be seen that the GHG impact is significantly reduced by the use of AM-Cover gas for manufacturing and recycling as the gas contains HFC-134a having a GWP of 1,600. Significant reductions in the GHG impact can also be achieved by increasing the product yield in manufacturing from the nominal value of 54%, and also by increasing the proportion of secondary magnesium used for manufacturing from the nominal value of 30%.

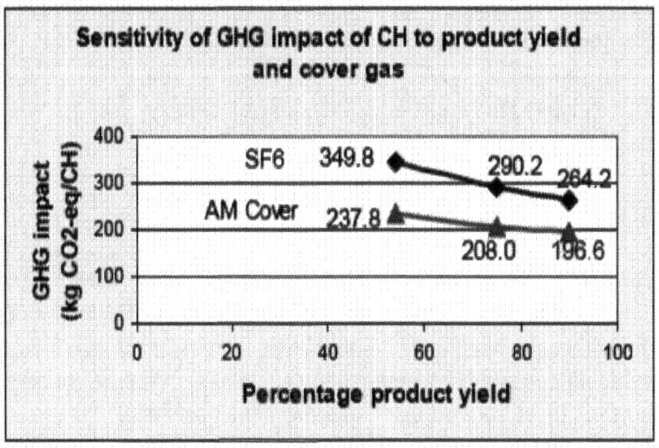

Figure 7. Sensitivity of GHG impact of CH to product yield and the cover gas used in manufacturing and recycling

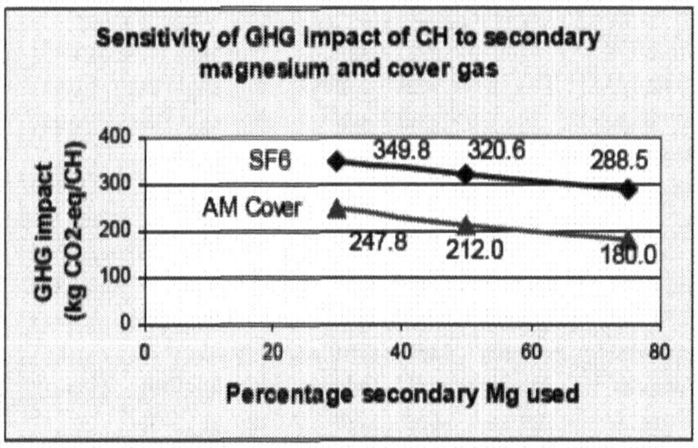

Figure 8. Sensitivity of GHG impact of CH to product percentage of secondary magnesium and the cover gas used in manufacturing and recycling

Discussion of Results

Cover Gas: The results of the study clearly show that the use of a cover gas that contains SF_6 (GWP: 22,200) to protect molten magnesium in the manufacturing and recycling stages of the magnesium life cycle produces a high GHG impact. The AM-Cover gas that contains HFC-134a (GWP: 1,600) has a significant influence in reducing the GHG impact of magnesium products.

Product yield in manufacturing: The product yield obtained in the high-pressure die-casting operation used for manufacturing products such as the CH is as low as 55%. This results in not only a higher GHG impact, but also a large amount of scrap that needs to be sent out the manufacturing plant for recycling. It is essential that the product yield from die-casting operations is increased to reduce the GHG impact of magnesium.

Secondary metal content: We have assumed in this study that the secondary magnesium ingots obtained from recycling constitute 30% of the total magnesium ingots used for manufacturing magnesium components. If this is increased to 75%, the GHG impact over the life cycle of magnesium could be reduced by as much as 25% from the nominal impact calculated in the study. One must note that in-house recycling of scarp can further reduce the GHG impact of secondary magnesium, such as through savings in transport and sorting.

Light weighting cars: The study also highlights the emission reduction in the use stage of a light-weighted passenger car that is built with magnesium components. To illustrate the comparative performance of magnesium, we have compared the impact of nominal product system with the impacts of five other competitive materials product systems, as shown in Table 4.

#	Product System	Primary	Second	Manuf.	Cover Gas	Ref..
1	Nominal	AU. Mg 70%	Open-loop US 30%	US	SF6	This study
2	China Mg	China Mg 70%	Open-loop US 30%	US	SF6	This study, [14]
3	Iron	AU 50%	AU. 50%	AU.	N/A	[9]
4	AU. Al	AU. Al 70%	AU. AL 30%	AU.	N/A	[9,16]
5	US Al	US Al.40%	USA Al.60%	US	N/A	[9,13]
6	AU. Mg & AM-Cover	AU. Mg 70%	Open-loop US 30%	US	AM-Cover	This study

Table 4. Product systems considered for comparison impacts

The GHG impacts of the above six product systems are compared in Figure 9. The impact values shown in this figure assume a life cycle total driving a distance of 200,000kms. It can be seen from this figure that the largest GHG impact of the CH results from the Chinese magnesium product system (product system 2) that uses SF_6 as the cover gas for both manufacturing and recycling stages. The nominal system (product system 1) that is based on Australian magnesium and uses SF_6 gives the second highest GHG impact. A much lower impact results (product system 6) if AM-Cover gas is used instead of SF_6 with Australian magnesium. The lowest impact is produced by CH made from US aluminium due to high recycling ratio (60% of secondary aluminium).

Figure 10 gives the GHG impact of the six product systems as a function of the driving distance. This figure provides information on the break-even driving distance. In the case of the product system 6 that uses Australian magnesium and AM-Cover breaks even with the CH made from iron (product system 3) at a driving distance of ~ 100,000kms. Also, it may be seen from Figure 10 that such a converter housing will be only as good as converter housing made from Australian aluminium. However, if a component with a higher mass is substituted, then the break-even distance will be much smaller. For example, a magnesium component of 10 kg mass under the same condition will have break-even distance with iron of ~20,000 km.

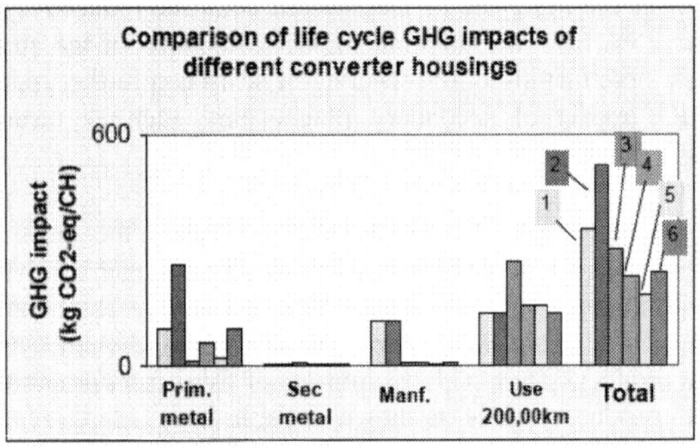

Figure 9. Comparison of GHG impact of different materials product systems
1 – Nominal; 2 – Chinese Mg; 3 – Iron; 4 – AU Al; 5 – US Al; 6 – AU Mg and AM-cover

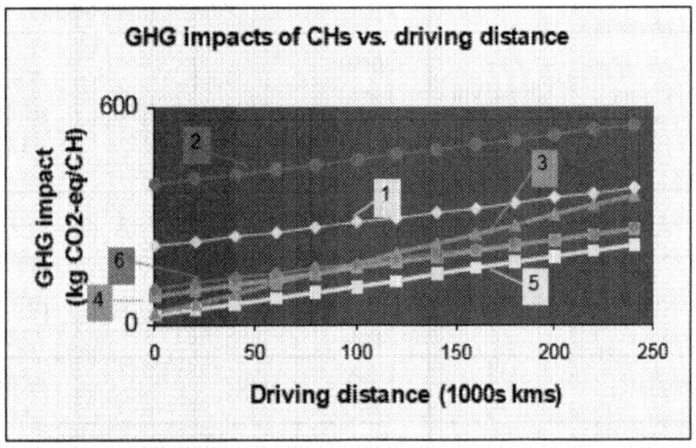

Figure 10. GHG impact of different product system for CH vs. driving distance
1 – Nominal; 2 – Chinese Mg; 3 – Iron; 4 – AU Al; 5 – US Al; 6 – AU Mg and AM-cover

Conclusions

In the conducted study we have estimated the GHG impacts over the entire life cycle of magnesium CH made from Australian magnesium.

The study clearly shows that the use stage of a passenger car contributes significantly to the total GHG impact of magnesium components over their entire life cycle. A significant reduction in the GHG impact during the use stage, and hence over the entire life cycle, may be achieved from a significant mass reduction of a car by using magnesium components. The magnesium CH can only provide improvement if SF_6 will be substituted by other cover gases with significantly lower GWP. Further, a simple substitution of magnesium without significant reduction in GHG is not competitive environmentally with other metals as demonstrated in the case of Chinese magnesium.

Sensitivity analyses conducted clearly indicate the reduction in GHG impact when AM-Cover gas, instead of SF_6, for protecting molten magnesium from oxidation during the manufacturing and secondary metal production stages is used. Higher percentage use of secondary magnesium and increase of yield in die-casting have substantial reductions as well.

This study has dealt only with the GHG impact of magnesium components, and comparisons with other metals are based on the GHG impact too. However, other impact categories need to be considered if magnesium components are to be compared with components made from non-metallic materials.

Acknowledgements

We wish to thank the Cooperative Research Centre for Cast Metals Manufacturing (CAST) for supporting this work. We also thank Dr. Nigel Ricketts and Mr. Rob Bailey for providing information and data for this study.

References

1. Brown R.E., "Magnesium industry growth in the 1990 period" *Magnesium Technology 2000*, ed. H.I. Kaplan, J.N. Hryn and B.B. Clow, The Minerals, Metals & Materials Society, (2000), pp. 3-12.
2. Albright, D.L. and Haagensen, J.O., "Life cycle inventory of magnesium", *in Proceedings of the International Magnesium Association*, Toronto, Canada, (1997) pp. 32-37.
3. Standards Australia, AS/NZS ISO 14000: Australia/New Zealand Standard "Environmental management – life cycle assessment – principles and framework", Australia, (1998) p. 12.
4. Beggs P., Brynes D., Song G., Qian M., Song W., Christodoulou P., Brandt M., "Design and Development of Magnesium Alloys", *BTRA/CAST Joint Project Phase 1 – Project Review Report*, CAST Report Number: CASTREP/2000135, (2000)
5. Stanley R. W., Berube M., Celik C., Oosaka Y., Pearcy J., and Avedesian M. "The Magnola process for magnesium production" *Proceedings of the International Magnesium Association 54: Magnesium Trends*, (1997), 58-65.
6. .Ramakrishnan S, Koltun P. "Comparison of the Greenhouse Impacts of Magnesium Produced by Electrolytic and Pidgeon Processes". *TMS (The Minerals, Metals & Materials Society)*, (2004).
7. Kramer D. A. "Magnesium Recycling in the United States in 1998" *U.S. Geological Survey, Circular 1196-E*, http://pubs.usgs.gov/circ/c1196e (Online Only).
8. Dorsam H., Westofen S. "Magnesium Melting, Casting and Remelting in Foundries", *Magnesium Technology*, (2002) p. 725-737
9. Amersfoort B.V., *SimaPro Database*, Pre Consultants, The Netherlands, (2003).
10. Moore D.M. "Environment and Life Cycle Benefit of Automotive Aluminium" *Canadian Technology Development*, http://www.alcan.com, Toronto, Canada, Oct., (2001).
11. Antrekowitsch H., Hanko G. "Recycling of Different Type of Magnesium Scrap", *Magnesium Technology, TMS, The Minerals, Metals & Materials Society*, (2002)
12. http://www.rauch-ft.com/e001home.htm: *Magnesium machine furnace MMO and MMOSL*, (2003).
13 International Primary Aluminium Institute, "Aluminium Applications and Society Life Cycle Inventory of the World Wide Aluminium Industry with Regard to Energy Consumption and Emissions of Greenhouse Gases", *Report 1 – Automotive*, May (2000).
14.Gjestland H., Magers D. "Practical Usage of Sulphur Hexafluoride for Melt Protection in the Magnesium Die-Casting Industry", *Norsk Hydro, Magnesium Technology* (2001).

SOLUBILITY OF FLUORINE IN MOLTEN MAGNESIUM

Kari Aarstad[1], Martin Syvertsen[2], and Thorvald Abel Engh[1]

[1]Norwegian University of Science and Technology,
Dept. of Materials Technology and Electrochemistry, N-7491 Trondheim, Norway
[2]SINTEF Materials Technology, N-7465 Trondheim, Norway

Abstract

The solubility of fluorine in molten magnesium was measured at temperatures between 700°C and 950°C. The experimental approach involved equilibrating molten magnesium in a MgF_2 crucible under argon. Samples of the equilibrated melt were taken using alumina tubes and a syringe. The fluorine solubility was determined employing the "Sintalyzer method", an electrochemical method developed at SINTEF, Norway. The fluorine content in liquid magnesium was described by an exponential function of inverse temperature and varied between 10 and 100 ppm by weight in the measured temperature interval. The standard Gibbs free energy of the fluorine dissolution reaction: ½ F_2 (gas) = $\underline{F}$ (in mass %), in the temperature range 700°C to 950°C, was determined to:

$$\Delta G^\circ_3/2 = (-\,473\,000 \pm 3250) + (53 \pm 3)T$$

Introduction

During experiments with molten magnesium protected with fluorine containing gases, a residual protection effect when the supply of gas is cut off, has been observed. That is, the molten metal does not oxidise the moment the gas flow is cut off. Previous work has shown that fluorine is an active element in the protection of the melt, by helping to build a stable surface film (1,2). It is hence possible that the observed residual effect can be attributed to a limited solubility of fluorine in the molten magnesium, which may supply the surface with necessary fluorine to remain protective for some time.

An extensive literature search carried out by Nayeb-Hashemi and Clark (3) for the publication of their book on phase diagrams, revealed no publications on the solubility of fluorine in magnesium. Since then (1988), no new publications on the topic have been found.

SF_6 is today the most commonly used protective gas for molten magnesium to prevent uncontrolled oxidation of the melt. Due to its extremely high Global Warming Potential (the GWP of SF_6 is 23 900 relative to CO_2 on a 100 years time horizon (4)), SF_6 will however be phased out from the market in the foreseeable future. Hence, if for the purpose of magnesium melt protection, it is possible to introduce fluorine into the melt in a different way than using a protective gas, a number of environmental problems related to gases and their dissociation/reaction by-products, would be eliminated.

With this background, the work on measuring the solubility of fluorine in molten magnesium was started.

The following equilibrium reactions for the solution of F in Mg, with their associated standard Gibbs free energy values (5,6), should be considered:

$$MgF_2 = Mg + F_2\,(g) \qquad \Delta G^\circ_1 = 1\,120\,000 - 171.2T \pm 2000$$
$$Mg = Mg\,(l) \qquad \Delta G^\circ_2 = 8\,960 - 9.71T \pm 400$$
$$F_2\,(g) = 2\underline{F} \qquad \Delta G^\circ_3 \qquad (1)$$
$$MgF_2 = Mg\,(l) + 2\underline{F} \qquad \Delta G^\circ$$

The experiments should give ΔG° and from there it is possible to find ΔG°_3.

The equilibrium can be written as

$$\frac{a_F^2 \cdot a_{Mg(l)}}{a_{MgF_2}} = e^{-\Delta G^\circ / RT} \qquad (2)$$

with

$$a_{Mg(l)} = 1 \qquad (3)$$

$$a_{MgF_2} = 1 \qquad (4)$$

and setting the activity coefficient f_F equal to 1:

$$a_F = [\%F] \qquad (5)$$

Equation (2) then reduces to:

$$[\%F] = e^{-\Delta G^\circ / 2RT} \text{ (by mass)} \qquad (6)$$

Experimental

To study the solubility of F in Mg, the chosen experimental approach was to equilibrate pure liquid magnesium in a solid magnesium fluoride crucible. Fluorine in magnesium was measured at temperatures from 700°C to 950°C. The experiments were performed in an argon atmosphere.

The crucible

MgF_2 from Merck, quality 5836, was mixed with polyvinyl alcohol. The polyvinyl alcohol acts as an adhesive, and 2% by weight was dissolved in 250 ml water and heated to 80°C. The solution was poured into a bottle together with 250g magnesium fluoride. The bottle was placed on a ball grinder for two hours to make sure that the solution was mixed completely. The solution was then left in a vacuum furnace at 80°C over night to evaporate the water. What remained was crushed and pressed in a rubber mould at 2000 bar.

It was necessary to sinter the crucible to give strength. Therefore, the crucible was held at 1050-1100°C for about an hour. The size of the crucible after sintering was about 50 mm in outer diameter and a height of 80 mm.

The furnace

Figure 1 shows a sketch of the furnace set-up. The furnace is a resistance element tube furnace. The crucible was charged with 60-80 g 99.99% pure magnesium and placed in the middle of the furnace. This gave a melt depth of approximately 50 mm. There were radiation shields both above and below the crucible to obtain a uniform temperature in the melt.

Before the experiments were started, the furnace was evacuated to 1 mbar. The chamber was filled with argon (purity 99.99 %), evacuated once more, filled with argon again so that the pressure was a little higher than atmospheric pressure. The off-gas valve was subsequently opened and argon allowed to flow through the furnace. The gas flow was controlled by a Bronkhorst flowmeter and set to approximately 70 ml/min.

The temperature in the melt was measured with a thermocouple placed inside an alumina tube lowered into the melt.

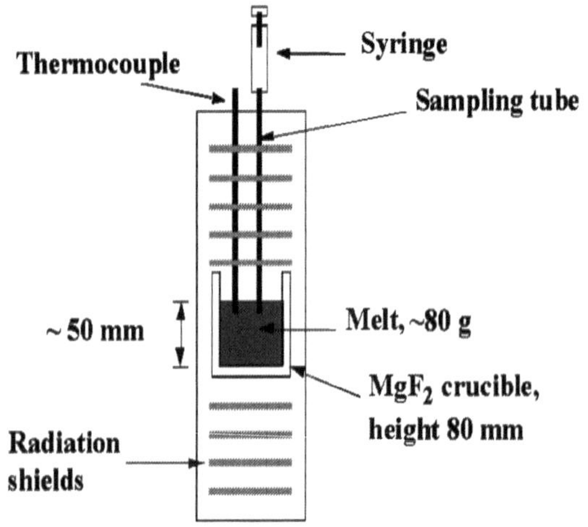

Figure 1: The inside of the furnace used.

Sampling

The melt was sampled using an alumina tube with inner and outer diameters 3 and 5 mm, respectively.

The flow of argon was increased before the lid of the sampling hole was removed and the sampling tube was lowered into the furnace. This was done to create an excess pressure inside the furnace so that air is not drawn into the furnace during sampling.

Samples were taken by placing a syringe at the end of an alumina tube. The tube was carefully lowered into the melt, and liquid metal drawn into the tube. Samples were not taken close to the bottom of the crucible to avoid sampling potential fluoride particles. Therefore, samples were taken at least 2 cm above the bottom of the crucible, except in special cases with high temperature where the evaporation of magnesium was so high that there was not much metal left. The tube was then quickly taken out of the melt and cooled with helium in a cooling zone at the top of the furnace. The sample was allowed to cool for a few minutes before the alumina around the sample was removed with a hammer.

For each temperature, samples were taken after both short (4-6h) and long (≈24h) holding times to assure that the melt had attained equilibrium. Samples were taken at the particular temperatures both on the way up in temperature and on the way down.

Analysis

The samples were analysed with the "Sintalyzer". This method is an electrochemical method developed at SINTEF originally for the aluminium industry (7). In this analysis method, the samples are dissolved in a buffer and ion selective electrodes used to measure potentials. Controlled amounts of standard fluorine are added and new potentials are registered. Plotting these results gives a regression line and the original fluorine concentration can be determined.

The uncertainty in these measurements is 10% for values higher than 5 mg F/kg Mg (ppm) and 20% for values less than 5 ppm.

It should be noted that all the fluorine in the samples will contribute in the analysis, whether it is in solid solution or in a chemical compound e.g. magnesium fluoride.

Results

The content of fluorine in magnesium, in equilibrium with MgF_2, is plotted as a function of inverse temperature in Figure 2. The uncertainty in the analysis of the measurements is plotted as error bars in the graph. The results from both the short (4-6 h) and the long time (17-26 h) exposures are given in this graph.

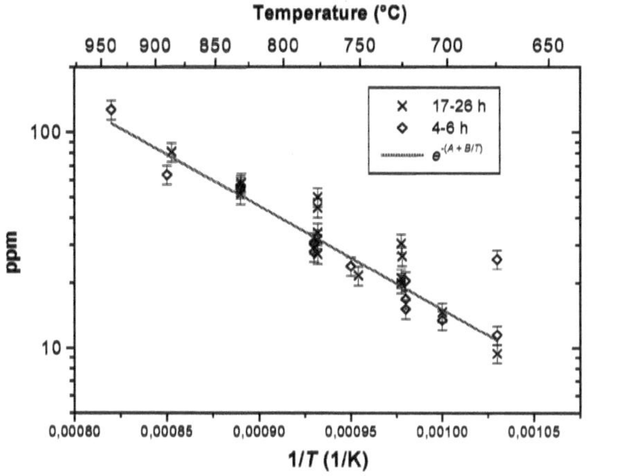

Figure 2: Fluorine in molten magnesium in equilibrium with MgF_2 as a function of temperature.

It can be seen from Equation (6) that the fluorine concentration in parts per million can be described by the exponential function:

$$[\text{ppm F}] = e^{-(A+B/T)} \tag{7}$$

Equation (7) is fitted to all the measurements by a non-linear curve fitting method (8), with the uncertainty of the measurements as weight. This fitting gives

$A = -13.74 \pm 0.34$
$B = 11\,030 \pm 358$

Combining Equations (6) and (7) gives

$$A + B/T - \ln 10^{-4} = \Delta G°/2RT$$

or

$$\Delta G° = 2RT(A + B/T - \ln 10^{-4}) \tag{8}$$

Then

$$\Delta G° = (183\,400 \pm 6000) - (75 \pm 6)T$$

Taking the mean squared uncertainties into account:

$$\Delta G°_3 = (-945\,560 \pm 6500) + (106 \pm 6)T$$

Discussion

Some points have to be considered when critically evaluating the results.

Firstly, it is possible that magnesium fluoride particles from the crucible could have been drawn into the alumina tube during sampling and hence affected the amount of fluorine detected in the analysis. To investigate this, the metal that was left in the crucible after one of the experiments was finished, was sliced vertically along the centre axis and polished and prepared for electron micro-probe analysis. Figure 3 shows a micrograph taken from the area where a few MgF_2-particles were found. The fluoride particles appear darker than the matrix. As one can see in the lower left corner, some magnesium fluoride is torn off from the crucible when the metal is taken out. Not many particles were found, and the few that were found were close to the bottom of the melt or the crucible wall. The uppermost particle was found at a height approximately ½ mm from the bottom. Above this height, no fluoride particles were registered. So, unless there is a strong convection in the crucible, it is not likely that magnesium fluoride particles were drawn into the test tube, keeping in mind that the samples were taken at least 2 cm above the bottom.

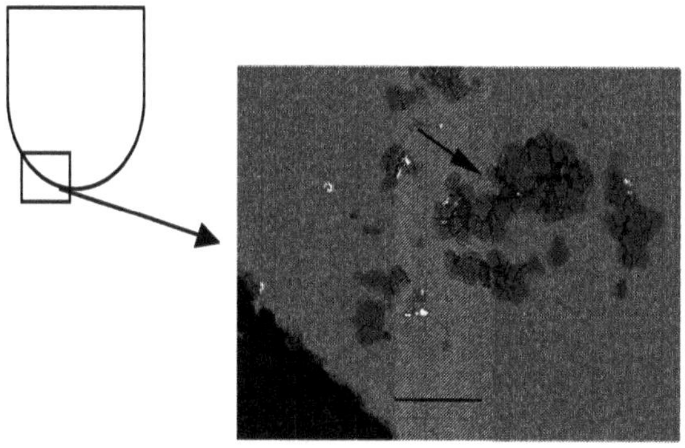

Figure 3: Magnesium fluoride particles in the remaining metal in the crucible.

At high temperatures the evaporation of magnesium becomes significant. This may give a higher content of fluorine than the equilibrium concentration since magnesium may evaporate at a relatively higher rate than fluorine. However, this is probably not a significant effect since no discrepancy between measurements at the same temperature when the temperature was rising or decreasing, could be detected.

In future experiments, the effect of alloy addition such as Al and Zn, and the presence of iron on the fluorine content will be studied.

Conclusion

The solubility of fluorine in molten magnesium was measured at temperatures between 700°C and 950°C. The experimental approach involved equilibrating molten magnesium in a MgF_2 crucible under argon. The fluorine content in liquid magnesium was described by an exponential function of inverse temperature and varied between 10 and 100 ppm by weight in the measured temperature interval. The standard Gibbs free energy of the fluorine dissolution reaction: ½ F_2 (gas) = $\underline{F}$ (in mass %), in the temperature range 700°C to 950°C, was determined to:

$$\Delta G°_3/2 = (-473\ 000 \pm 3250) + (53 \pm 3)T$$

The standard Gibbs free energy of the reaction between dissolved fluorine in molten magnesium in equilibrium with MgF_2: $MgF_2 = Mg\,(l) + 2\underline{F}$, is given by

$$\Delta G° = (183\ 400 \pm 6000) - (75 \pm 6)T$$

Given the relatively high solubility of fluorine in liquid magnesium, the potential of protecting molten magnesium alloys by adding fluorine directly to the melt in an effective manner in combination with the use of an environmentally friendly cover gas, is presently investigated

Acknowledgements

The authors gratefully acknowledge the financial support provided for this work by the Norwegian Research Council and the International Magnesium Association.

References

1.G. Pettersen et al., "Characterisation of the Surface Films Formed on Molten Magnesium in Different Protective Atmospheres", (2000) To be published

2. S.P. Cashion, "The Use of Sulphur Hexafluoride (SF_6) for Protecting Molten Magnesium" (Ph.D. thesis, The University of Queensland, 1998), 80.

3. A.A. Nayeb-Hashemi and J.B. Clark, ed., Phase Diagrams of Binary Magnesium Alloys (Ohio: ASM International, 1988), 117.

4. M. Maiss and C.A.M.Brenninkmeijer, "Atmospheric SF_6: Trends, Sources, and Prospects", Environmental Science & Technology ,32 (20) (1998), 3077-3086.

5. Thorvald Abel Engh, Principles of Metal Refining (New York: Oxford University Press, 1992), 407-425.

6. E.T. Turkdogan, Physical Chemistry of High Temperature Technology (London: Academic Press, 1980), 14.

7. "SINTALYZER for Fluorine Analyses", brochure from SINTEF Inorganic Process Chemistry and Analysis, may 1999.

8. W.H. Press et al. Numerical Recipes in C: The Art of Scientific Computing (2nd ed.) (Cambridge: Cambridge University Press, 1992), chapter 15.

PART 4: CASTING AND SOLIDIFICATION

Wim H. Sillekens

Developments in the field...[1]

Melt-metallurgical processing is the main route for manufacturing magnesium (semi)-finished products. This applies to shape castings that account for the vast majority of all magnesium structural parts, but also to the initial step in wrought products' manufacturing. Thus magnesium technology has long since been focusing on casting, with leitmotivs being the pursue of alloys with better workability and in-service properties as well as of processes yielding enhanced productivity and product quality. Meanwhile a mature industrial supply chain for magnesium castings has been established.

One general aspect to be dealt with in the processing of a magnesium melt is in its high affinity to oxygen, which implies both a safety hazard and the likelihood of excessive oxide formation and associated risk of melt contamination. While early solutions consisted of molten salts covering the melt, the use of sulphur hexafluoride gas (or more precisely, of mixtures of inert carrier gases like nitrogen or argon with typically 0.2–0.5% SF_6) in closed furnaces has been the prevailing means of melt protection over the last decades. With SF_6 being an extremely potent greenhouse gas, however, its industrial use is presently being phased out under (European) legislation, which has triggered developments to come with more environmentally friendly alternatives. Mixtures in which SF_6 is substituted by SO_2 are now recommended practice although this is regarded a temporary solution owing to their toxicity and corrosiveness. Further developments are thus due with the overall trend being toward abandoning fluorinated gases at all.

A further aspect associated with casting and solidification is the use of appropriate alloy compositions. Alloys of the AZ and AM series (with typical representatives being AZ91 for the former, and AM50 and AM60 for the latter) can be regarded as the workhorses of the casting materials. Both alloy systems have been developed for die casting and give good mechanical properties at room temperature. With a major application area for castings being in the drive-train of vehicles with associated temperatures of typically 120°C for oil-containing casings and up to 150°C for actual engine components, creep-resistant magnesium alloys are another important class. Early alloys from the AS series (especially AS21 and AS41) were used extensively for die casting of VW Beetle components. A successor is the AE series, where AE42 has long been the reference yet without extensive application. For gravity castings, the WE series is well established, with WE43 as its main representative being used in such niche products as helicopter transmission casings. Alloy development for shape castings has been and is still strongly directed toward enhancing high-temperature properties. Improvements are being sought by modifying the AZ and AM series with a diversity of elements including calcium, rare earths, silicon, strontium, and tin. These additions target the introduction of stable precipitates (intermetallic compounds) within the grains and at the grain boundaries so as to hamper grain-boundary sliding at elevated temperatures and thus to limit creep of the thermally and mechanically loaded components. Especially rare earths have been demonstrated to be effective in this respect. As an outcome of these developments, a number of novel (proprietary) alloys have been introduced to the market

---

[1] Based on: Sillekens W.H., Hort N.; "Magnesium and magnesium alloys" – chapter 3 in *Structural Materials and Processes in Transportation* (eds. Lehmhus D., Busse M., Herrmann A.S., Kayvantash K.), Wiley-VCH, Weinheim, Germany (2013).

during the last decades, among which AS31, AJ62, MRI-153M, MRI-230D, and AE44 for die casting, and MRI-201S, Elektron 21, and AM-SCI for gravity casting.

As for actual casting operations, a distinction can be made between die casting, gravity casting, and other processes (mainly for wrought alloys). The state of affairs in these fields can be outlined as follows.

- **Die casting.** High-pressure die-casting equipment for magnesium is essentially the same as for aluminum and in that sense to be characterized as mature. Revolutionary developments are not anticipated anymore, yet there is still a trend toward higher performance (e.g., clamping forces) which expands production capabilities in terms of product size and quality as well as in the alloys that can be processed. Meanwhile the industrial knowledge base for high-pressure die casting in view of such aspects as mold design has been and is still progressing, among others supported by process-simulation tools. One particular development relates to vacuum-assisted casting that was adopted from aluminum high-pressure die casting. This is an interesting option for safety-relevant components with high mechanical demands; however, the industrial application appears to be limited as yet in view of the additional costs. As a further variant, squeeze casting of magnesium alloys has been subject of study at several research organizations, although this seems not to have resulted in commercial applications as yet. Semi-solid casting covers a class of processes that revolve around the use of partially molten or partially solidified material as the feed for casting; in their actual implementation, however, these processes—including thixocasting, New Rheo-Casting, and injection molding (or thixomolding)—are distinctly different. Both thixocasting and New Rheo-Casting have as yet not seen sustained industrial application, but injection molding has established itself over the past few decades as an industrial manufacturing technique for mainly the consumer electronics market (casings).
- **Gravity casting.** Sand casting and permanent-mold casting are both traditional means of casting that are apt mainly for low-to-medium volume production. Used often for niche applications, developments are correspondingly (company)-specific.
- **Other processes.** Casting is also the first step in wrought products' manufacturing, where so-called (semi-continuous) direct-chill casting is the prevailing method for series production of strands and slabs, which are subsequently processed into extrusions, forgings, and sheets. Here, process engineering is concerned with such issues as reaching high productivity while simultaneously avoiding hot cracking and realizing a fine and homogeneous microstructure. The trend to integrate the downstream processes in the so-called twin-roll casting process in order to enhance cost-effectiveness is now attracting ample attention for magnesium as well. This process is meanwhile under development at several institutes and companies worldwide, with industrial production being in the start-up phase and twin-roll cast sheet becoming commercially available. One particular challenge is the manufacturing of wide strip—with widths of up to 2,000 mm suitable for automotive sheet—in which tip design for a balanced flow of the melt between the rolls is a critical aspect.

...and their relationship with this book

The papers in this part represent a fair, yet exemplary selection of the research topics on the subject as outlined above. Being more or less thematically grouped, they include recent and less recent reports on generic issues like castability and solidification (by Bowles *et al.*, StJohn *et al.*, and Saddock *et al.*), possibly incurred defects such as hot cracking and porosity (by Cao *et al.*, and Lee and Gokhale), mechanical performance of die castings (by Chadha *et al.*), semi-solid processing (by Czerwinski, and Moore *et al.*), and twin-roll casting (by Park *et al.*, and Duygulu *et al.*).

Two major aspects associated with industrial casting operations are treated elsewhere in this volume in a more generic setting: melt protection (in Part 3) and alloy development (in Part 5).

CASTABILITY OF MAGNESIUM ALLOYS

A.L. Bowles[1], Q. Han[1], J.A. Horton[1]
[1]Oak Ridge National Laboratory, Metals and Ceramics Division. Oak Ridge, TN, 37831-6115, USA

Keywords: Magnesium, Castability

Abstract

There is intense research effort into the development of high pressure die cast-able creep resistant magnesium alloys. One of the difficulties encountered in magnesium alloy development for creep resistance is that many additions made to improve the creep properties have reportedly resulted in alloys that are difficult to cast. It is therefore important to have an understanding of the effect of alloying elements on the castability. This paper gives a review of the state of the knowledge of the castability of magnesium alloys.

Introduction

The issue of castability is important in an industrial situation; castability is an ill-defined term, but in general it gives the ability to determine if an alloy is suitable to cast into a certain end-product with satisfactory quality. The ease of casting an alloy (the *castability*) will depend significantly on the die design, pouring system (shot velocity or pouring velocity, metallostatic head and gating system) and the mould/die materials as well as the metallurgical aspects, and this should be borne in mind when devising experiments and interpreting data.

Approximately 90% of magnesium alloys are based on the Mg-Al system; these alloys have good room temperature mechanical properties, corrosion resistance (in high purity forms) and excellent castability in a wide range of processes. The major disadvantage of this alloy system is the poor high temperature strength and creep properties. It has been recognised that the presence of aluminium in these alloys plays a major role in the poor creep behaviour as a result of the dynamic precipitation of grain boundary discontinuous β phase, $Mg_{17}Al_{12}$ [1, 2].

The use of magnesium alloys in high temperature applications (up to ~175°C) is envisaged for the very near-future, and as a result, much effort has been placed on the development of creep resistant magnesium alloys. Indeed, a number of commercial and experimental alloys have been developed. The development of creep resistant magnesium alloys has essentially followed two paths: either the elimination of aluminium or the addition of elements to Mg-Al alloys to preferentially form Al-containing phases, thereby reducing the level of aluminium in solution. One of the difficulties encountered in alloy development is that many additions made to improve the creep properties have reportedly resulted in alloys that are difficult to cast [3-5]; especially when this is combined with the reduction or removal of aluminium which is known to impart excellent castability.

It is recognised that the issue of castability must be addressed for successful alloy development. There is evidence that suggests aluminium, Zinc, silicon and rare earths (RE) increase the castability[1] [6-14], while calcium decreases it (incidentally it is also reported that RE additions decrease the castability [11, 15]). It is therefore important to have some understanding of the effect of alloying additions on the castability of magnesium alloys.

In contrast to magnesium alloys, much information exists on the castability of aluminium alloys, and a number of excellent review papers have been published (see for example refs. [16, 17]). With this paper we attempt to highlight the state of knowledge on the castability of magnesium alloys; specifically the metallurgical aspects, and direct attention towards areas that require further investigation.

Castability

When casting an alloy, the metal charge is first raised to the casting temperature (some temperature above the liquidus temperature - superheated). During the melting and casting operation the molten alloy is contained in some material (crucible, furnace, mould or die); exposed to the atmosphere, in contact with a flux or cover gas and racked or stirred, poured or siphoned into a die or mould with a coated tool – all these interactions with other materials (atmospheres) may cause reactions resulting in the formation of unwanted precipitates, inclusions and oxides, and the pick up of certain elements. During the actual casting operation, the metal feeds into the mould or die cavity and immediately begins to solidify. In the first stage of filling there is the bulk flow of liquid; the solid fraction is low, and any particles are essentially free to move about in a liquid of low viscosity. The filling of thin sections will be controlled by the surface tension and large particles (inclusions/oxides) may block the flow. As solidification progresses, the viscosity will increase (due to the temperature drop, increase in solid fraction and change in the liquid composition) and the morphology and volume fraction of any phases, as well as the latent heat evolved during phase transformations will affect the flow of the alloy. While the temperature continually decreases, the solid fraction increases and bulk flow becomes restricted, during this stage of filling the primary dendrites come in contact and the dendrite coherency point is reached. As solidification continues, interdendritic feeding begins to take place and the dendrites are impinging on one an other and are no longer free to move. At this stage (when flow occurs primarily through the interdendritic channels) the size and shape of the primary dendrites affects the permeability of the solid network, intermetallic phases may block the flow between dendrites as may inclusions or oxides, and porosity formation occurs.

High strains may develop during solidification due to contraction and thermal gradients [18], the stress on the dendrite network in the mushy zone may exceed its strength resulting in hot tearing, and this is especially significant in regions isolated from liquid flow (dramatic changes in section thickness).

Factors Effecting Castability

Considering only the alloy properties, the factors that affect the castability are: porosity formation (gas and shrinkage), hot tearing susceptibility, mould filling ability and die soldering; these characteristics can be related to (but not limited to): phase transformations (composition, morphology and volume fraction), morphology of the primary phase (grain size and volume fraction), viscosity (as a function of temperature, shear rate and composition of the liquid during solidification), solidification mechanism (strength of the mushy zone, permeability, eutectic solidification), freezing range, latent heat, surface tension, gas solubility gap (difference in gas solubility between liquid and solid), propensity to form inclusions and reactions with the melt (for instance die soldering). In the following sections a brief review of these factors is given with reference to the behaviour of magnesium alloys.

Porosity

Porosity is a combination of gas and shrinkage porosity. Both can be affected by alloying elements. Alloying additions may change the solubility of hydrogen, volumetric shrinkage (composition and volume fraction of phases) and affect the solidification mechanism (by altering the solid fraction, formation of precipitates (V_f, morphology and composition), leading to changes in the interdendritic feeding during solidification). Other secondary effects may include changes in the grain size and the formation of pore nuclei such as oxides.

Magnesium alloys do not suffer significantly from hydrogen porosity as the difference between the solid and liquid solubility of hydrogen in magnesium alloys is not as large as in aluminium alloys. The hydrogen solubility in magnesium has been related to the Al content in Mg-Al alloys, with the solubility decreasing with increasing aluminium content [19].

The total amount of porosity is also reported to be related to the aluminium content in binary Mg-Al alloys [20, 21]; the maximum in porosity occurs at around 9wt% Al (12wt% for ref. [21]) which is a composition slightly below the maximum freezing range, corresponding to the first appearance of the eutectic (according to the equilibrium phase diagram). Schaffer *et al.* [20] suggest that the maximum in porosity occurs at this composition as it is the "worst combination of mushy zone size, interdendritic feeding permeability and eutectic volume fraction". Mg-Zn and Mg-Sn alloys have also been found to show a maximum in porosity at the composition of the maximum freezing range [21].

[1] Tin is also reported to increase the castability [6-8]; tin is currently not a common alloy addition.

Hot Tearing/Cracking

Hot tears (sometimes referred to as hot cracks) are casting defects that can affect the pressure tightness, mechanical properties, and the appearance of castings. Hot tearing occurs in areas of low strength and high strain. This defect commonly occurs in areas of dramatic changes in section thickness, where regions are not fully solidified (low strength) and that are also isolated from liquid feeding such that the local stresses developed exceed the strength of the partially solidified metal.

Dodd *et al.* [22] studied the hot tearing susceptibility of a number of magnesium alloys (Mg-Al, -Zn, -Al-Zn, -Zn-Zr, -RE-Zr, -RE-Zn-Zr, -Th-Zr, -Th-Zn-Zr, -Mn and RE-Mn). Binary Mg-Al were found to be the most hot tear resistant as were Mg-Al-Zn alloys (although the addition of Zn impairs the hot cracking resistance); RE containing alloys were slightly worse and the worst alloys were Mg-Zn-Zr and Mg-Zn-Th-Zr. Hot tearing susceptibility is related to several parameters. Consider the grain size - the larger the grain size the more susceptible alloys tend to be (although note that the alloys reported to be the most susceptible to hot tearing by Dodd *et al.* were those grain refined by the addition of Zr, indicating some other factor is more important in that case). The freezing range of an alloy is also plays a role; larger freezing range alloys exhibit a greater tendency for hot cracking (consider Mg-Al to Mg-Zn [23]) and the amount of eutectic present (amount of liquid available near the end of solidification) should also be considered [22].

In Mg-Al alloys the maximum in hot tearing was reported to occur around 1%wt Al [22], which is close to the peak in the non-equilibrium freezing range (according to the Scheil equation) which is commonly associated with the maximum in hot tearing susceptibility [24]. Most common Mg-Al containing alloys have more than 2wt% Al and normally contain around 4-6wt% which provides the best combination of strength and castability. Hence commercial Mg-Al alloys are not considered prone to hot cracking.

In more recent work, Pekguleryuz *et al.* [5] compared the hot tearing susceptibility of a range of commercial and experimental alloys in a constrained rod test. Rating from best to worst they found AJ62x (6Al, 2Sr), aluminium alloy A380, AZ91 (E then D), AS21x (+RE), AM60, AJ50x (0.2Ca, 0.5Sr), AS21, AS41, with AE42 the worst (which is similar to the results of Dodd *et al.* [22]). From their results, in general, the hot cracking resistance improves with increasing aluminium content, with the alloys containing silicon and rare earths faring worse.

Calcium, a recently favored addition to magnesium alloys for creep resistance, is known to increase the level of hot cracking [3, 25, 26], although the effect appears to be composition dependent. At low Ca levels (0.9wt% Ca), Mg-Al-Ca-Sr alloys suffer significantly from hot tearing, die sticking, cold shuts, staining and soldering, while at higher calcium levels hot cracking and die sticking problems improve significantly [3, 25].

The hot tear behaviour of Mg alloys generally conforms to accepted theories of solidification cracking, but alloying effects are poorly understood. Many apparently contradictory observations exist regarding alloying elements. Some of these may be the result of interaction effects. For instance the behaviour of a major alloying element in a binary system may be quite different from its behaviour in a ternary or higher order system. This might be due to changes in the thermodynamics of the system, liquid viscosity, solid phase nucleation details, etc. Clearly, much more work is needed to thoroughly understand solidification cracking behaviour of Mg alloys.

Melt reactions

The castability is also affected by reactions with the molten metal, for example die soldering, oxide formation and reactions with containment materials, lubricants and mould coatings.

For high pressure die casting, magnesium alloys are not as susceptible to die soldering as aluminium alloys [27]. Die soldering or sticking of magnesium alloys appears to be related to the Al and Mn content of the melt and the formation of Al-Fe-Mn intermetallics [27]. For compositions above ~7wt%Al soldering problems increase, as the solubility of iron in magnesium decreases at higher Al contents; Mn also reduces the solubility of iron in magnesium. Some magnesium alloys are reported to be more susceptible to soldering than others. For instance, some problems have been reported with die soldering with the alloy composition AJ52x (5Al-2Sr), which can be alleviated by maintaining a high die temperature (~300°C) [5] although the reason for this is not clear. Calcium containing alloys are also known to have some problems with die soldering [6, 25, 26, 28, 29]. Powell *et al.* [25] note that die soldering occurred only in AXJ alloys (Al-Ca-Sr) with Ca levels below 2%, although Pekguleryuz and Renaud [30] note that in Mg-6Al-XCa (0.6,0.8 and 1 wt%) die soldering becomes a problem when the composition reaches 1wt% Ca. Similarly, in Mg-Al-Ca-Sr alloys [3], die soldering was reported to be more pronounced at 1wt% Ca, and reduced when the Ca content was above 2 wt%. It should also be noted that die soldering in Ca containing alloys is reported to be reduced with increased Al content [6]. Bakke *et al.* [6] report that Sr has little effect on die soldering in Mg-Al-Ca-Sr alloys; this appears in contradiction to the results for AJ type alloys (Mg-Al-Sr) which do suffer from die sticking. RE containing alloys, like AE42 have also been reported to suffer from die soldering [31]. The role of alloy composition apart from aluminium and manganese in die soldering has not been fully explained.

The cleanliness of melt is also a factor when considering the castability of an alloy and in this sense cleanliness refers to the number of contaminates in the melt, including oxides and inclusions (particles that result from melt reactions with containment materials, mould coatings etc). A "dirty" melt can result in casting difficulties as the viscosity is effectively increased by an increase in the solid fraction. Additionally if oxides are present they have a tendency to form as films and these can block filling channels and result in incomplete filling [32].

Magnesium oxidizes easily; magnesium melts are generally protected by a cover gas to limit heavy oxidation. Magnesium alloys have been reported to have higher oxide inclusion contents compared to aluminium alloys [33]. Hu *et al.* [33] report that as the weight fraction of MgO increases the fluidity length decreases in AZ91D, likely to be a result of an increase in the effective viscosity and the blocking of flow due to large solid particles. The presence of alloying elements can also affect the formation of oxides, for instance Ca and RE's have a high affinity for oxygen. Mg-Ca melts have been reported to form a heavy oxide/dross layer on the surface [34] during melting and casting, which retards the oxidation of the melt. The formation of other oxides apart from MgO gives the potential for the entrainment of a larger number of inclusions for alloys containing elements with a high affinity for oxygen [33], however the effect of reactive elements on the amount of oxide inclusions has not been studied in magnesium alloys.

Little information exists in the literature about reactions between magnesium alloys and common containment materials (crucibles and furnace lining), however Cao et *al.* [35] note the uptake of iron into magnesium melts in mild steel crucibles, and Emley [36] also reports iron pick up in "conventional steel" crucibles. This indicates that some steel crucible materials should be avoided to reduce the pick up of iron and the potential re-precipitation of Fe-containing intermetallics in the melt.

<u>Fluid flow in the die/mould</u>

<u>Viscosity</u>

Viscosity is a fundamental physical parameter of liquids and is a function of temperature and can also be a function of shear rate (for example thixotropic materials). The viscosity (and the effect of alloying elements) is likely to be important in solidification feeding [37], where fluids of low viscosity flow more easily; this is especially important to feeding in the mushy zone, note also that as solidification progresses the remaining liquid becomes enriched in solute so that the composition, and hence viscosity is continually changing. Aluminium is reported to decrease the viscosity of magnesium alloys [38]; there is no other information in the literature on the effect of other common alloying additions on the viscosity.

<u>Surface tension</u>

When filling thin sections the heat transfer between the mould and metal, and the surface tension act to halt filling (if the melt wets the substrate, surface tension will actually aid filling). As we are considering only the metallurgical character of an alloy, the surface tension is an important consideration that can be affected by alloying. For the mould to fill the external pressure acting on the metal must be greater than the effective pressure due to surface tension [39]. A high surface tension inhibits filling and therefore reduces the castability of an alloy in thin sections. For aluminium alloys a number of alloying elements are known to reduce the surface tensions (Li, Bi, Pb, Mg, Sb, Ca and Sn [37]) while others increase the surface tension (Ge, Zn, Ag, Fe, Mn and Cu [37]). For magnesium alloys, Al has been reported to decrease the surface tension [38]; the effect of other alloying elements has not been studied.

Fluidity

When the term fluidity is used in reference to castability it describes the measure of how far a molten alloy will flow in a specific mould; normally using a spiral mould or vacuum suction technique. The fluidity is then expressed as the flow length.

Fluidity spirals are commonly permanent mould or sand castings, but may also be high pressure die castings. When interpreting "fluidity lengths", in general only castings made under the exact same conditions can be compared because the mould material (and any coatings), die design, metal temperature (constant temperature or constant superheat when comparing alloys), pouring speed and cleanliness of the metal will all affect the results. In general terms, short freezing range alloys have higher fluidity than long freezing range alloys. The fluidity length of an alloy will depend on the viscosity (liquid composition, solid fraction and morphology of the solid phases), freezing range and solidification mode, oxides or inclusions that block flow, and surface tension (especially when filling thin sections).

Aluminium is known to impart excellent fluidity to magnesium alloys [14, 40]; the effect of other alloying elements is less well-known. Niesse *et al.* [40] used a Ragone test (vacuum suction) to examine the fluidity of Mg-Al, -Zn, -Al-Zn, -Zn-Zr, -Th-ZR, -RE, -RE-Zr alloys. The fluidity decreases from that of the pure metal as solute is added, reaching a minimum at the non-equilibrium freezing range and then increasing again towards the eutectic composition (a second maximum does not always occur at the eutectic composition, for example in Al-Si alloys, the maximum is at a slightly hypoeutectic composition; this is normally attributed to the high latent heat of formation of Si).

The addition of RE's to Mg-Al alloys has been reported to increase the fluidity length [8, 9, 14]; with the effect more pronounced at lower Al contents than at higher levels; it should be noted that the effect of the addition of Al has a much larger effect on fluidity than the addition of RE's [14]. The increase in the fluidity length with increasing RE content has been attributed to the reduction in the freezing range of the alloys and the high latent heat[2] of RE containing precipitates ($Al_{11}RE_3$ ~41 kJ/mol [14]) which may act to slow the solidification. RE additions also have a grain refining effect and affect the volume fraction and morphology of the eutectic β ($Mg_{17}Al_{12}$) phase [9], which can alter the interdendritic feeding paths resulting in an increase in the fluidity. Lü *et al.* [9] also note that RE's have a high affinity for oxygen, this may result in the entrainment of more oxides resulting in a increase in the apparent viscosity, and decrease the fluidity of an alloys [33].

Silicon is also reported to increase the fluidity length of Mg-6Al-XSi alloys [14] with an effect larger than the same addition of RE's (in weight %). Lü *et al.* [14] discuss the microstructure and solidification mode in Mg-Al-RE and Mg-Al-Si alloys and note that magnesium alloys exhibit "mushy zone solidification" and point to the importance of solidification in the mushy zone, where interdendritic feeding occurs as an important measure of an alloys fluidity. During the initial stages of feeding the fluidity is controlled by the viscosity of the alloy, as solidification progresses the viscosity will increase (due to the temperature drop, the increase in solid fraction and composition change of the liquid) and the morphology and volume fraction of any phases, as well as the latent heat will affect the fluidity. As solidification continues, interdendritic feeding begins to take place and the dendrites are impinging on each other and are no longer free to move; the maximum fluidity length will be reached in this stage of feeding (prior to complete solidification), as the flow of liquid is significantly restricted.

Summary

Castability is not a single property of an alloy; many factors play a role. The castability issues that occur most frequently for magnesium alloys are the amount of porosity, die soldering, hot cracking and ability to cast very thin wall sections. These factors are affected by alloying additions. An understanding of the effects of common alloying elements for magnesium is lacking except perhaps in the case of aluminium, which is known to produce castable alloys. Aluminium additions above 1wt% result in alloys which are not prone to hot cracking or die soldering and appear not to suffer from excessive porosity or oxide formation;

[2] The latent heat of formation for $Mg_{17}Al_{12}$ is 10 $kJmol^{-1}$, Mg_2Si 80 $kJmol^{-1}$ [14], Mg_2Sn 81 $kJmol^{-1}$ and Mg_2 n 14 $kJmol^{-1}$ [41].

interestingly it seems that the addition of aluminium to alloys with unfavorable castability can result in an improvement (presumably with some negative effect on the creep behaviour). Calcium, Zinc and rare earths tend to promote hot cracking and die soldering and may adversely affect the oxide content. While much anecdotal evidence for the affect of alloying additions in magnesium alloys exists, there are few systematic, comprehensive studies of the castability of magnesium based alloys.

Acknowledgments

Research sponsored by the U.S. Dept. of Energy, Assistant Secretary for Energy Efficiency and Renewable Energy, Office of FreedomCAR and Vehicle Technologies, as part of the High Strength Weight Reduction Materials Program, and by the Division of Materials Sciences and Engineering at the Oak Ridge National Laboratory, operated by UT-Battelle, LLC, under contract DE-AC04-00OR22725.

References

1. M.S. Dargusch, G.L. Dunlop and K. Pettersen. in *Magnesium Alloys and Their Applications*. 1998. Wolfsburg, Germany: Werkstoff-Informations GmbH. p. 277-282.
2. M. Dargusch, *The Role of Microstructure in the Creep of Die Cast Magnesium Alloys*, PhD Thesis. 1998, Brisbane: The University of Queensland, Australia
3. B.R. Powell, V. Re hets, A. Luo, A., *et al.*, General Motors Corporation, USA Patent US 6264763, 2001.
4. A.P. Druschitz, E.R. Showalter, J.B. McNeill, *et al.* in *Magnesium Technology 2002, Feb 17-21 2002*. 2002. Seattle, WA, United States: Minerals, Metals and Materials Society. p. 117-122.
5. M.O. Pekguleryuz, P. Labelle, D. Argo, *et al.* in *Magnesium Technology 2003*. 2003. San Diego, USA: TMS. p. 201-206.
6. P. Bakke, K. Pettersen and H. Westengen. in *2003 Magnesium Technology, Mar 2-6 2003*. 2003. San Diego, CA, United States: Minerals, Metals and Materials Society. p. 171-176.
7. K. Pettersen, P. Bakke and D. Albright. in *Magnesium Technology 2002, Feb 17-21 2002*. 2002. Seattle, WA, United States: Minerals, Metals and Materials Society. p. 241-246.
8. W. Qudong, L. Yi hen, Z. Xiaoqin, *et al.*: Mater. Sci. Eng A Vol. 271 (1999): p. 109-115.
9. Y. Lü, Q. Wang, X. Zeng, *et al.*: Materials Science and Engineering A Vol. 278 (2000): p. 66-76.
10. G.S. Foerster: Light metals Age Vol. 30 (1972): p. 11-13.
11. R. Beals, L. Kopka, J. Allison, *et al.* in *Magnesium Technology 2004, Mar 14-18 2004*. 2004. Charlotte, NC., United States: Minerals, Metals and Materials Society. p. 11-17.
12. B. Bronfin, M. Katsir and E. Aghion: Materials Science and Engineering A Vol. 302 (2001): p. 46-50.
13. A. Katsman, S. Cohen, G. Goren-Muginstein, *et al.* in *Magnesium Technology 2004, Mar 14-18 2004*. 2004. Charlotte, NC., United States: Minerals, Metals and Materials Society. p. 297-300.
14. Y. Lü, Q. Dong, W. Ding, *et al.*: Z. Metallk. Vol. 91 (2000): p. 477-482.
15. B.L. Mordike and T. Ebert: Mater. Sci. Eng A Vol. 302 (2001): p. 37-45.
16. J.A. Taylor: Cast Metals Vol. 8: p. 225-252.
17. E.N. Pan and J.F. Hu: AFS Trans. Vol. 48 (1997): p. 413-418.
18. D.G. Eskin, Suyitno and L. Katgerman: Progress in Materials Science Vol. 49 (2004): p. 629-711.
19. E. Øverelid, P. Bakke and T.A. Engh: Light Metals Vol. (1997): p. 141-154.
20. P.L. Schaffer, Y.C. Lee and A.K. Dahle. in *Magnesium Technology 2001*. 2001. New Orleans, USA: TMS. p. 87-94.
21. E.J. Whittenberger and F.N. Rhines: Journal of Metals Vol. (1952): p. 409-420.
22. R.A. Dodd, W.A. Pollard and J.W. Meier: Trans AFS Vol. 65 (1957): p. 100-117.
23. T.W. Clyne and G.J. Davies: Brit Foundry Vol. 74 (1981): p. 65-73.
24. M. Rappa , J.-M. Drezet and M. Gremaud: Metall. Mater. Trans. A Vol. 30 (1999): p. 449-455.
25. B.R. Powell, A. Luo, A., L. Basant, *et al.* in *Magnesium Technology 2002*. 2002. Seattle, USA: TMS. p.
26. M.O. Pekguleryuz and A. Luo, A., Inst la Technologie Du Magnesium, World Patent WO9625529, 1996.
27. C. Tang, M. Jahedi and M. Brandt: Advanced engineering materials Vol. 5 (2003): p. 903-906.
28. W. Schmidt: Giesserei Vol. 57 (1970): p. 535-538.
29. J.J. Berkmortel, H. Hu, J.K. Kearns, *et al.*: SAE Technical Paper - 2000-01-0111 Vol. (2000).
30. M.O. Pekguleryuz and J. Renaud. in *Magnesium Technology 2000*. 2000. Nashville, USA: The Minerals, Metals and Materials Society. p. 279-284.
31. A. Luo, A. and M.O. Pekguleryuz: J. Mater. Sci. Vol. 29 (1994): p. 5259-5271.
32. J. Campbell: *Castings*.(Butterworth-Heinemann 1998).
33. H. Hu and A. Luo, A.: JOM Vol. 10 (1996): p. 47-51.
34. B.-S. You, W.-W. Park and I.-S. Chung: Scr. Mater Vol. 42 (2000): p. 1089-1094.
35. P. Cao, M. Qian, D. St. John, *et al.*: Mater. Sci. and Tech. Vol. 20 (2004): p. 585-592.
36. E.F. Emley: *Principles of Magnesium Technology*. (Pergamon Press 1966).
37. C.R.J. Loper: AFS Trans. Vol. 147 (1992): p. 533-538.
38. A.M. Kovol'kov: *Casting properties of metals and alloys*.(Consultant Bureau Enterprises 1963).
39. D.M. Stefanescu: *Science and engineering of casting solidification*.(Kluwer Academic/Plenum Publishers 2002).
40. J.E. Niesse, M.C. Flemings and H.F. Taylor: AFS Trans. Vol. 65 (1957): p. 207-211.
41. C.J. Smithells and E.A. Brandes: *Smithells metals reference book*. 6th ed.(Butterworths 1983).

SOLIDIFICATION OF CAST MAGNESIUM ALLOYS

D.H. StJohn[1], A.K. Dahle[1], T. Abbott[2], M.D. Nave[3], and Ma Qian[1]
CRC for Cast Metals Manufacturing (CAST), UDP No. 55, University of Queensland, 4072, Australia
[1]University of Queensland, Division of Materials Engineering, Brisbane Qld 4072, Australia
[2]Monash University, School of Physics & Materials Engineering, Melbourne Vic 3168, Australia
[3]Deakin University, School of Engineering & Technology, Geelong Vic 3217, Australia

Abstract

A description of the key solidification steps in the formation of the as-cast microstructure of magnesium alloys is presented. The focus is on the two common magnesium alloy groups: Mg-Al alloys and Mg-Zn-rare earth alloys. The key elements described are: nucleation (including grain refinement), growth of the primary phase and the formation of the eutectic phases. In addition the effect of casting process (e.g. high-pressure die-casting and sand casting) on the outcomes from solidification are discussed. This includes consideration of the formation of banded defects during solidification in the dynamic environment of high-pressure die casting.

Introduction

Each year magnesium alloys are finding new applications in the automotive industry despite the fact that there is continual pressure to reduce the cost of producing automotive components [1]. Magnesium has a number of advantages particularly in allowing parts consolidation which results in a lower overall cost. However, the need to achieve consistent properties is becoming paramount as applications move to components requiring a higher level of structural integrity. Therefore, a better understanding of the solidification process during the casting of components is required to ensure high quality and low cost [2]. Casting is still the dominant process for producing magnesium components. By far the most popular process is high pressure die casting (HPDC) which is used for large thin section castings such as instrument panels and seat frames. Thixoforming is used for small components such as mobile phone and computer cases and development work is underway to extend this process to larger automotive components. Sand casting has been the main casting method for large aerospace castings such as helicopter gear box housings.

The properties achieved depend to a large extent on the alloy used and the resulting microstructure achieved. The scale of the microstructure can vary by a large magnitude due to this very wide range of casting methods. There are two main types of magnesium alloys. Magnesium – aluminium alloys (e.g. AZ91, AM60 and AM50) are predominantly used for HPDC and thixoforming of thin section castings of electronic components and structural automotive castings such as instrument panels and steering wheels. The other group of alloys is magnesium - zinc alloys (e.g. ZE41, ZK60) that can be alloyed with zirconium to provide a fine grain size and rare earth elements to provide creep resistance. Applications include engine components and extrusion billet.

The key solidification processes that control the final microstructure are the initial nucleation events, the growth of these nuclei into primary dendrites and finally eutectic solidification. The slow cooling rates encountered in sand casting tend to produce coarse microstructures that have low ductility. To improve ductility the grain size needs to be decreased by the addition of grain refiner. Alternatively, the component can be heat treated which increases ductility but also lowers the yield strength. In contrast, HPDC creates a very dynamic environment for solidification and solidification begins in the shot sleeve even before the melt enters the cavity. The mush enters the die cavity at high speed and solidification progresses to high solid fraction while the mush continues to fill the cavity. This can result in banded defects that may be detrimental to mechanical properties. Thus solidification processes need to be considered in the light of the casting process being used. This paper briefly describes the response of magnesium alloys to the key elements of the solidification process and also considers solidification in the dynamic environment encountered in HPDC and to some extent in Thixoforming and squeeze casting.

Nucleation and Grain Refinement

In HPDC nucleation is not usually an issue as the high chill of the mould materials ensures many small grains of primary magnesium phase are formed and these are carried into the melt to generate a fine microstructure. In sand casting, on the other hand, grain refinement is required to provide uniform properties and a high yield strength. Mg-Al alloys are difficult to adequately grain refine as there is not a simple and reliable grain refining method. Attempts to find a suitable refining technology are the subject of ongoing research and a review of these methods was recently published by Lee et al. [3]. This paper will focus on the non-aluminium containing alloys where zirconium is a very effective grain refiner.

Zirconium (Zr) is a potent grain refiner for magnesium alloys that contain little (impurity level) or no Al, Mn, Si, Fe, Ni, Co, Sn and Sb (zirconium forms stable compounds with these elements) [4]. When added to these magnesium alloys, where the maximum solubility of Zr in molten magnesium is approximately 0.6%, the use of Zr can readily cause the average grain size to decrease by 80% or more for normal cooling rates. This exceptional grain-refining ability makes Zr an important alloying element for magnesium alloys containing zinc, rare earths, thorium, calcium, or a combination of these elements.

The most characteristic feature of the microstructure of a magnesium alloy containing more than a few tenths per cent soluble zirconium is the Zr-rich cores that exist in most magnesium grains [4], particularly when observed using a scanning electron microscope (SEM) in the backscattered (BES) mode (See Fig. 1a). It is generally accepted that solidification of Mg-Zr alloys containing Zr close to 0.6% occurs by a peritectic mechanism [4,5] and the Zr-rich cores are products of this peritectic reaction. Observations of the Zr-rich cores on polished sections cut at different angles from a Mg-0.56Zr alloy cone sample (including transverse, longitudinal and a 45° diagonal) demonstrated that these Zr-rich cores generally appear with a nearly spherical or elliptical form and many contain a tiny particle in their central regions. Electron microprobe analyses of a total of 50 such tiny particles observed in the Mg-0.56Zr alloy showed that the majority were almost pure Zr particles. On the other hand, examination of the distribution of Zr in areas that surround the Zr-rich cores showed little presence of Zr in most traversed areas, indicating that almost all of the Zr in solid solution is concentrated in the Zr-rich cores. The distribution of Zr in these cores was found to be inhomogeneous and vary in a wide range, e.g., from 0.5% to 3% (the maximum solubility of Zr in solid magnesium is ~ 3.8%). The Zr-rich cores observed in different Mg-Zr alloys are usually less than 20 μm in diameter for normal cooling rates and the tiny particles inside are usually less than 2 μm.

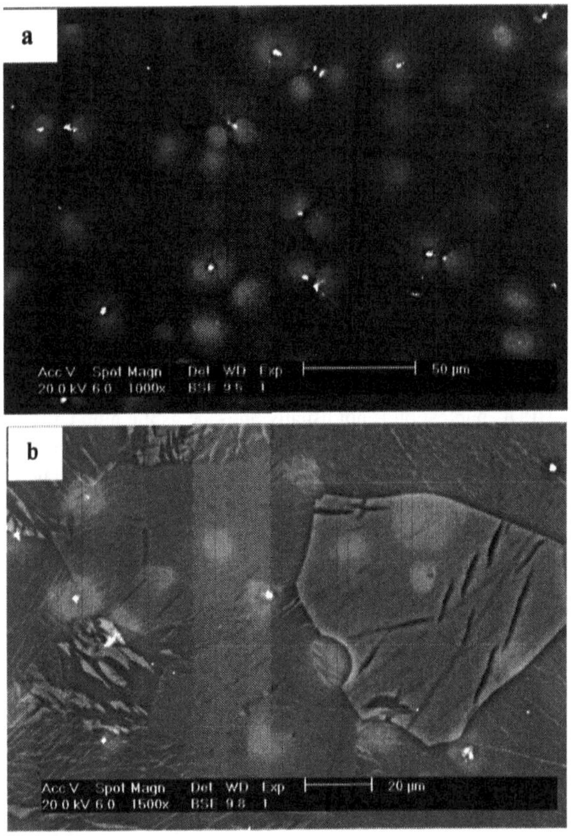

Figure 1: a) A typical view of the zirconium-rich cores in a Mg-0.56Zr alloy (backscatter SEM), and b) SEM backscatter image showing that each magnesium grain may contain a number of Zr-rich cores (Mg-0.56Zr alloy).

Despite the general agreement on a peritectic mechanism for the formation of the characteristic Zr-rich cores, it is still far from clear exactly how the subsequent grain growth occurs once a Zr-rich core has formed. It has been found that many magnesium grains in a Zr-containing Mg alloy contain a number of Zr-rich cores rather than a single core, as shown in Fig. 1b, and these Zr-rich cores can exist anywhere inside a grain rather than just in the centre of the grain. These are an important microstructural feature that must be understood in order to fully understand the solidification of Mg-Zr alloys.

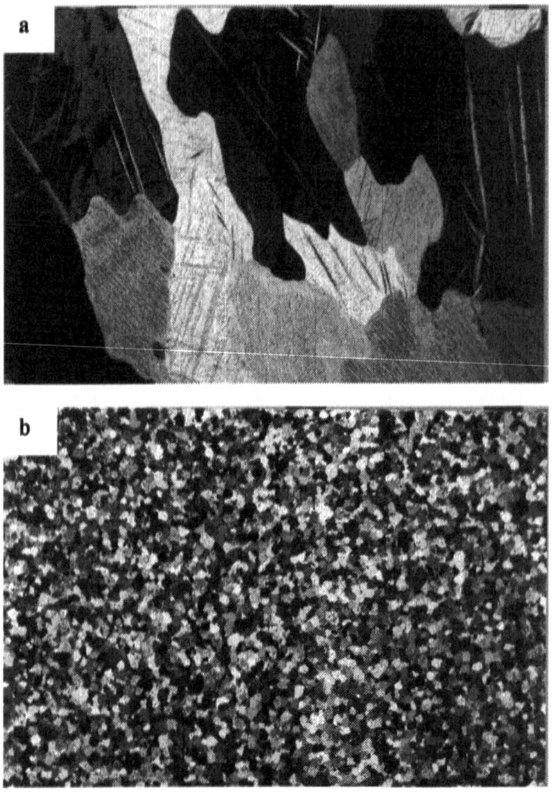

Figure 2: Grain refinement with zirconium: (a) pure magnesium and (b) after the addition of 1% Zr at 730°C. Micrographs are of the same magnification.

Grain refinement of magnesium alloys by Zr is noticeable at very low levels of soluble Zr (~0.04%) [3] and, in general, improves with increasing soluble zirconium towards the maximum soluble Zr content (~0.6%). According to the early work by Sauerwald [4], only Zr that is dissolved in the liquid magnesium at the time of pouring is effective in grain refinement. Insoluble Zr such as undissolved Zr particles was believed to be irrelevant to grain refinement, despite both α-zirconium and magnesium having the same type of crystal structure and nearly identical lattice parameters. This view appeared to be widely accepted until it was recently shown by Tamura et al. [6] that undissolved Zr particles actually play an important role in the grain refinement of magnesium alloys. More recent work [7] has confirmed the beneficial role of undissolved Zr particles in grain refinement and showed that grain refinement of magnesium alloys by Zr is dictated by both soluble and insoluble zirconium. As such, an ideal Zr alloying process should end up with both high soluble and high total Zr at the time of pouring in order to achieve the best grain refinement. Figure 2 shows an example of grain refinement of pure magnesium with 1% Zr addition

The above findings confirm that the mechanism of grain refinement in magnesium alloys is identical to that in aluminium alloys where both potent nucleant particles and a strongly segregating solute element must be present [8]. Considering the work on grain refinement of aluminium alloys it is clear that the nucleant particles present in Mg-Al alloys are being poisoned and thus it is difficult to obtain sufficient refinement unless this poisoning effect is overcome. Understanding the poisoning mechanism is a key to developing a suitable grain refiner for Mg-Al alloys.

Dendritic Growth

Figure 3 shows a typical microstructure of Mg-Al alloys. Because magnesium has a hexagonal close-packed structure, the dendrite arms branch with a six-fold symmetry. Well-developed primary α-Mg dendrites with secondary arms (A) showing sixfold symmetry are clearly visible in Figure 3. The eutectic of $Mg_{17}Al_{12}$ (B) and α-Mg solid solution (C) is located in the interdendritic regions. According to the Mg-Al equilibrium phase diagram, the eutectic phase ($Mg_{17}Al_{12}$) is expected to appear when the aluminium content reaches ~13 wt%. However, the eutectic phase appears in alloys containing as little as 2 wt% Al for non-equilibrium cooling conditions normally encountered in castings.

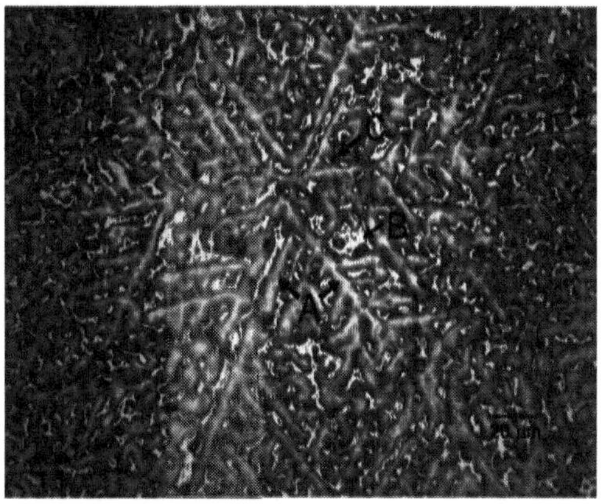

Figure 3: Micrograph of fully developed dendrites in a Mg-15 wt% Al alloy permanent mould casting. The magnesium dendrites have a characteristic sixfold symmetric shape (A). The white phase between the dendrites is secondary eutectic phase $Mg_{17}Al_{12}$ (B) and the dark regions between the dendrites are Al-rich Mg solid solution (C).

The growth of magnesium dendrites occurs according to the usual principles of primary phase solidification. Solute elements with a distribution coefficient less than unity are rejected ahead of the dendrites and this can cause changes in the microstructure. For example, a small addition of aluminium to pure magnesium leads to a morphological change of the primary phase from a cellular to a dendritic structure. Rosette-like globular equiaxed grains form with Al-rich solid solution between the dendrite arms. As the aluminium content is increased further to 5 wt%, dendrites with pools of eutectic phase between the dendrite arms start to develop and, when the aluminium content is further increased, a fully developed dendritic structure with sharp tips is observed. The concept of growth restriction factor (GRF) seems to be applicable for magnesium alloys and the elements with the largest segregation potential, in decreasing order, are Zr, Ca and Si.

Eutectic Growth

Magnesium forms eutectic systems with a wide range of alloying elements [9] and the microstructures of the common casting alloys, AZ91, AM50/60, AS21/41, AE42, as well as recently developed alloys such as ZA85, contain small amounts of eutectic [10, 11]. While alloys containing any proportion of eutectic may be produced [12], commercial casting alloys contain very low volume fractions of eutectic compared with cast irons and aluminium-silicon alloys. Pure magnesium is very weak and low to moderate levels of alloying provide solid-solution strengthening and strengthening from the presence of massive second-phase particles formed during eutectic solidification. Addition of some elements also allows precipitation hardening. While higher levels of alloying addition should improve castability by increasing the volume fraction of eutectic, the dendrite-eutectic interface provides an increasingly convenient crack path around the more ductile magnesium dendrites [13] resulting in a sharp decrease in the ductility of the alloys [9]. Large additions of heavy and/or expensive elements also detract from the economic viability of the alloys by significantly increasing their density and cost. For these reasons, addition levels are generally below equilibrium maximum solid-solubilities and it is only the non-equilibrium cooling in practical casting processes that causes eutectic formation in commercial alloys.

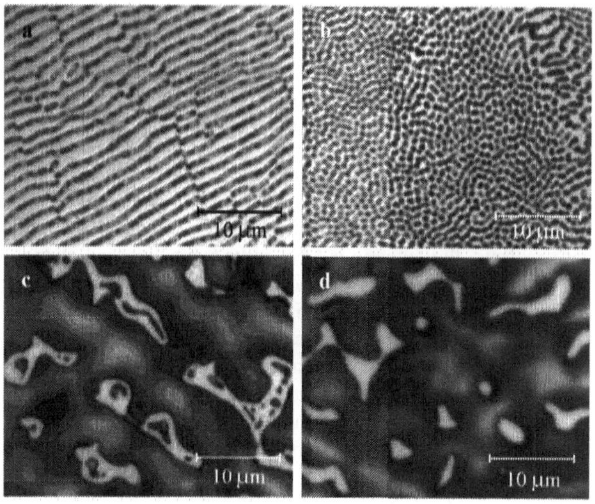

Figure 4: (a) Lamellar, (b) fibrous, (c) partially divorced and (d) fully divorced morphologies in Mg-Al alloys of various compositions [12,14].

Several eutectic morphologies that may be produced in magnesium alloys are shown in Fig. 4. For alloy compositions and casting conditions used commercially, the eutectic morphology is generally fully or partially divorced. In the fully divorced morphology (Figure 4d), the "eutectic" Mg coats the primary dendrites and the intermetallic phase exists as massive particles or a thin layer between the coated dendrites. The partially divorced eutectic (Figure 4c) is similar, however not all the "eutectic" Mg coats the primary dendrites – a small amount solidifies as islands inside the intermetallic phase. Fully divorced morphologies are usually formed in die-casting, while partially divorced morphologies are more likely at lower cooling rates. In permanent mould cast Mg-Al [12] and Mg-Al-Zn [14] alloys, a reduction in aluminium content from 15 wt% Al to 9 wt% Al also favours a more divorced morphology. This trend is expected to continue to lower aluminium contents, but is not always evident among AZ, AM and AS alloys due to the different ternary additions. It is worth noting that as the morphology of the eutectic is quite sensitive to small changes in alloy content and cooling conditions, different morphologies are sometimes observed in the same alloy cast under similar conditions and even in nearby regions of the same casting.

The general similarities in the eutectic morphologies of magnesium casting alloys and the variation of these morphologies with cooling rate and alloy content can be explained in the following manner [12,14]: The predominance of fully and partially divorced morphologies is primarily a consequence of the high volume fraction of primary dendrites, which restricts eutectic solidification to small regions between dendrite arms and adjacent dendrites. When the interdendritic regions are small, a relatively small undercooling for nucleation and growth of the intermetallic phase is sufficient to cause solidification of practically all of the interdendritic liquid to occur outside the coupled zone, producing fully or partially divorced morphologies. Furthermore, if a large number of regions of interdendritic liquid become isolated from one another, nucleation of the intermetallic phase is required in each of these regions and these regions are more likely to solidify with a fully divorced morphology, since the undercooling required for nucleation is larger than the undercooling required for growth. Consequently, fully divorced morphologies are more likely in alloys with low levels of alloying additions, since the interdendritic regions tend to be smaller and more isolated when "eutectic" solidification commences.

Generally, high cooling rates produce more highly branched dendrites and trap the eutectic liquid into smaller, more isolated spaces. High cooling rates also increase the average undercooling required for nucleation of the intermetallic phase and for both of these reasons, the eutectic is frequently more divorced at higher cooling rates [10,12,14]. Although higher cooling rates are expected to produce more coring during dendrite growth and therefore a larger volume fraction of eutectic, which should produce less divorced eutectic morphologies, in most cases this effect appears to be counteracted by the mechanism just described. Since eutectic morphology is highly sensitive to the morphology of the primary dendrites [12,14], grain refinement levels and columnar-to-equiaxed transitions in the primary phase may also affect the morphology of the eutectic.

The effect of ternary elements on eutectic morphology is quite complex. Different elements may affect eutectic growth in very different ways. It has been established that the addition of small amounts of zinc to Mg-Al alloys increases the probability of forming a fully divorced eutectic [14]. There are several possible reasons for this effect, but the most likely reasons are the promotion of a more highly-branched dendrite structure (primarily due to the low partition coefficient of Zn in Mg) and changes in the equilibrium and non-equilibrium phase boundaries [14].

Besides the detrimental effect of the divorced Mg-$Mg_{17}Al_{12}$ eutectic on the ductility of Mg-Al alloys, the presence of $Mg_{17}Al_{12}$ particles along the grain boundaries and discontinuous precipitation of $Mg_{17}Al_{12}$ during high temperature service may contribute to their low creep strength by assisting intergranular cavitation and grain boundary sliding [15,16]. Compared with AZ and AM alloys, creep strength is higher in AS alloys, which have only small amounts of $Mg_{17}Al_{12}$ and larger amounts of the more stable Mg_2Si in the grain boundary regions. Creep strength is further improved in AE42 [17], where the Mg-$Mg_{17}Al_{12}$ eutectic reaction is avoided completely and the $Mg_{17}Al_{12}$ is replaced with Al_4RE and $Mg_{12}RE$ [18], respectively. Minor additions of Bi and Sb to AZ91 provide remarkable improvements in creep strength, which are partly due to the suppression of discontinuous precipitation and the presence of rod-shaped Mg_3Bi_2 and Mg_3Sb_2 particles that are stable at high temperatures and often straddle grain boundaries [16]. It should be noted, however, that grain boundary sliding is only one of a number of creep mechanisms active in magnesium alloys and the dominant mechanism varies with alloy composition and service conditions (e.g. Table 3 in [18]).

Dynamic Solidification and Defect Formation

High-pressure die cast magnesium alloys display several characteristic macro- and microstructural features including:

- Distinctive bi-modal grain size distribution
- Skin
- Defect bands

In cold chamber machines it is common for some solidification to occur in the shot sleeve prior to the commencement of the shot. The solid phase formed in this way is able to grow considerably larger than the material solidified during die filling, giving rise to a bimodal distribution of primary magnesium grains. The grains formed in the shot sleeve are referred to as externally solidified grains. The amount of these grains varies depending upon the metal temperature and other factors, and can range from zero up to about 20% [19].

Another common feature of pressure die cast magnesium is the presence of a surface layer or skin. The skin is characterised by a higher volume fraction of second phases, $Mg_{17}Al_{12}$ in the case of Mg-Al alloys and, often, a finer microstructure [20]. The formation of the skin suggests the segregation of solute elements to the surface that may occur due to the semi-solid state of the metal as it flows in the cavity. The depletion of solid phase adjacent to surfaces is a common phenomenon for flowing suspensions [21]. Alternatively the higher $Mg_{17}Al_{12}$ content may be a consequence of a faster cooling rate near the surface without enrichment in solute.

Possibly the most visually apparent microstructural feature of magnesium pressure die castings are bands of defects which, in most cases, form parallel to the casting surface. The make up of these bands seems to vary considerably and may consist of cracks, porosity (shrinkage and/or gas porosity) and/or segregation. Examples of defect bands are shown in Fig. 5. The bands may be single or multiple and can sometimes intersect with the surface. According to a mechanism proposed by Dahle and StJohn [22] the solid phase, within the semi-solid mixture flowing in the die cavity, migrates away from the walls and towards the centre. While this effect raises the solid fraction towards the centre, heat extraction through the die results in increased solid fraction near the surface. As a consequence there is an intermediate zone with relatively high liquid fraction. As solidification and filling proceeds, the centre and surface regions begin to develop some mechanical strength resulting in a localisation of shearing in the high liquid fraction region. This mechanism should result in the central region having a lower solute content than the outer region, which has been reported to be the case in at least one instance [19].

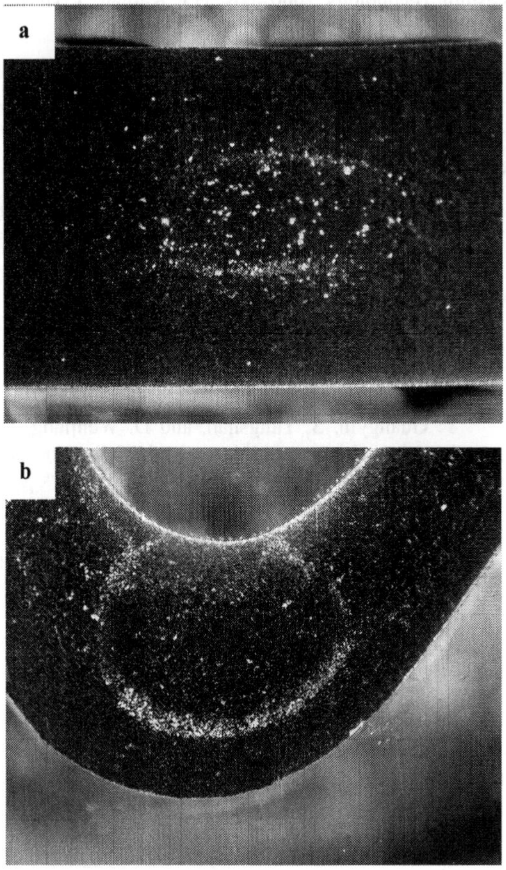

Figure 5: Example of defect bands (a) in a 2mm thick cast tensile bar, (b) in a commercial casting (3mm thick).

Mao [23] observed bands of pores that appeared spherical suggesting they were due to gas porosity. The porosity was thought to have formed due to a sudden drop in pressure, possibly as a result of a small back movement of the plunger tip during intensification. This mechanism would appear only to explain instances of gas porosity and not shrinkage porosity or segregation.

Rodrigo and Ahuja [24] investigated the effects of casting parameters on the formation of defect bands. Although they did not propose a formation mechanism they did assess the validity of the above mechanisms in light of their experimental findings. The model proposed by Dahle and StJohn [22] was disputed by Rodrigo and Ahuja [24] on the basis that the presence or absence of externally solidified grains did not influence the appearance of the pore bands and that when externally solidified grains were present they often appeared on either side of the defect bands. However, Dahle and StJohn's theory [19,22,25] also applies to the situation where externally solidified grains are not present so it is not clear whether this is in fact contradictory evidence. The theory proposed by Mao [20] was also disputed by Rodrigo and Ahuja [24] as the observed porosity was due to shrinkage rather than gas.

A patent application by Murray and Cope [26] suggests an alternative mechanism for the formation of these bands. Although they do not explicitly refer to bands, they claim, from microstructural examination, that in the runner and in sections of the casting where flow is uni-directional, flow occurs through a cylindrical section much smaller than the physical cross-section. The size of the cylindrical section appeared to be determined by the flow velocity, which typically fell in the range of 140 - 165m/s. It was claimed that solidification initially occurred in the runner, reducing the effective cross section and increasing the velocity. Once a velocity of about 150m/s was attained, viscous heating prevented further solidification. It was claimed that because viscosity rises rapidly for solid percentages above 50%, the viscous heating process would stabilise the solid fraction of material entering the cavity. As this phenomenon was also observed in sections of the cavity where flow was uni-directional, this suggests that the bands of defects occur where solidification was temporarily arrested by viscous heating. Rodrigo and Ahuja [24] also claim that defect bands are more defined in regions where flow is largely unidirectional.

The observation of banded defects is far more common in magnesium alloys than in aluminium alloys. Also, HPDC of magnesium alloys is able to fill long thin sections while aluminium is not. The reason for this is not clear. However, the two behaviours may be linked. A possible explanation can be proposed by considering the differences in solidification behaviour between aluminium and magnesium alloys. AZ91 has a freezing range of 160°C which is five times larger than Al-7wt%Si (e.g. A356) at about 30°C. Also, the volume fraction of eutectic in AZ91 is about 25% compared with about 50% for A356. By considering this and assuming thermal properties are similar, it can be envisaged that AZ91 will flow significantly longer distances before the semisolid microstructure locks up with eutectic solidification. This extra time allows for rearrangement of the primary magnesium grains and the formation of a liquid band to lubricate the flow of the semisolid mush into the die cavity. This segregated solute enriched band allows flow to continue until the temperature drops to near the eutectic temperature. Hence the bands are clearly visible in the microstructure.

Defect bands occur in a range of forms, such as cracks, bands of porosity (gas and/or shrinkage) and segregation. These all have superficial similarity, however it is not clear whether they form by the same process or whether multiple mechanisms are involved. Further investigations are necessary before their formation mechanism is fully understood.

Summary

The solidification of magnesium alloys differs in a number of essential ways from the solidification of aluminium foundry alloys. Magnesium alloys have a much lower volume fraction of eutectic and the composition of most alloys is within the range of the maximum solid solubility of the major alloying elements such as aluminium and zinc. This means that commercial alloy microstructures usually have divorced eutectics that form due to non-equilibrium solidification. This can be a problem for structural alloys as it decreases ductility, but the high solidification rates in HPDC or heat treatment result in improved ductility. On the other hand, divorced eutectic formation is likely to be an advantage for creep resistant alloys as the stable intermetallic phase present in the grain boundaries increases the microstructure's resistance to deformation at elevated temperatures. Grain refinement of sand cast alloys also assists in improving the distribution of the intermetallic phase and this is used to good effect in non-aluminium containing magnesium alloys by adding zirconium. Zirconium provides very potent nucleant particles and high segregating solute, both of which ensure very good grain refinement. Unfortunately a similarly effective grain refiner for aluminium containing alloys has not been developed and is the subject of ongoing research.

Another difference compared to aluminium alloys is the large freezing range. This may enhance the formation of banded defects in HPDC and thixocast alloys. These defects may reduce the ductility of the cast components but may also be responsible for the ability of magnesium alloys to fill long thin sections. A better understanding of the rheological behaviour of partially solidified alloys is required in order to reduce the negative effects of banded defects on mechanical performance while optimising the filling behaviour of magnesium alloys during HPDC.

Acknowledgements

CAST was established under the Australian Government's Cooperative Research Centres Scheme.

References

1. A.K. Dahle, D.H. StJohn and G. Dunlop, "Developments and Challenges in the Utilisation of Magnesium Alloys", Materials Forum, 24 (2000), 167-182.

2. A.K. Dahle et al., "Development of the As-Cast Microstructure in Magnesium-Aluminium Alloys", J. Light Met., 1 (2001), 61-72.

3. Y.C. Lee, A.K. Dahle and D.H. StJohn, "The Role of Solute in Grain Refinement of Magnesium", Met. Mat. Trans., 31A (2000), 2895-2906.

4. E.F. Emley, Principles of Magnesium Technology, (Oxford: Pergamon Press, 1966), 126-156; 254-254.

5. A.A. Nayeb-Hashemi and J.B. Clark, Phase Diagrams of Binary Magnesium Alloys, (ASM, Metals Park, OH, USA, 1988) 365-369.

6. Y. Tamura et al., "Grain Refining Mechanism and Casting Structure of Mg-Zr Alloy," J. Jap. Inst. Light Met., 48 (1998), 185-189.

7. M. Qian et al., "Settling of Undissolved Zirconium Particles in Pure Magnesium Melts," J. Light Met., 1 (2001), 157.

8. M. Easton and D.StJohn, "Model of Grain Refinement Incorporating Alloy Constitution and Potency of Heterogeneous Nucleant Particles", Acta Mat., 49 (2001) 1867-1878.

9. I.J. Polmear, Light Alloys: Metallurgy of the Light Metals, 3rd ed. (London: Arnold, 1995).

10. A. Luo, "Understanding the Solidification of Magnesium Alloys," 3rd Int. Magnesium Conference, ed. G.W. Lorimer, (The Inst. of Mat., Manchester, UK, 1996), 449-464.

11. M. Vogel et al., "Quasi-Crystalline Grain-Boundary Phase in the Magnesium Die-Cast Alloy ZA85," Scripta Materialia. 45 (2001). 517-524.

12. M.D. Nave, A.K. Dahle, and D.H. StJohn, "Eutectic Growth Morphologies in Magnesium-Aluminium Alloys," Magnesium Technology 2000, eds. H.I. Kaplan, J. Hryn and B. Clow, (TMS, 2000), 233-242.

13. S. Lee, S.H. Lee, and D.H. Kim, "Effect of Y, Sr and Nd Additions on the Microstructure and Microfracture Mechanism of Squeeze-Cast AZ91-X Magnesium Alloys," Met. Mat. Trans., 29A (1998), 1221-1235.

14. M.D. Nave, A.K. Dahle, and D.H. StJohn, "The Role of Zinc in the Eutectic Solidification of Magnesium-Aluminium-Zinc Alloys," Magnesium Technology 2000, eds. H.I. Kaplan, J. Hryn and B. Clow, (TMS, 2000), 243-250.

15. M. Regev et al., "Creep Studies of Coarse-grained AZ91D Magnesium Castings," Mat. Sci. Eng., A252 (1998), 6-16.

16. Y. Guangyin, S. Yangshan, and D. Wenjiang, "Effects of Bismuth and Antimony Additions on the Microstructure and Mechanical Properties of AZ91 Magnesium Alloy," Mat. Sci. Eng., A308 (2001), 38-44.

17. M.S. Dargusch, G.L. Dunlop, and K. Pettersen, "Elevated Temperature Creep and Microstructure of Die Cast Mg-Al Alloys," Magnesium Alloys and their Applications, eds. B.L. Mordike and K.U. Kainer, (Wolfsburg, Germany: Werkstoff-Informationsgesellschaft mbH, 1998), 277-282.

18. I.P. Moreno et al., "Microstructure and Creep Behaviour of a Die Cast Magnesium-Rare Earth Alloy," Magnesium Technology 2002, ed. H.I. Kaplan, (TMS, 2002), 111-116.

19. A.K. Dahle et al., "Formation of Defect Bands in High Pressure Die Cast Magnesium Alloys", J. Light Met., 1 (2001), 99-103.

20. H. Mao et al., "Microstructural Characteristics of Die Cast AZ91D and AM60 Magnesium Alloys" SAE paper, (1999) No. 999-01-0928

21. P.D. McCormack and L. Crane, Physical Fluid Dynamics (Academic Press, New York, 1973), 407.

22. A.K. Dahle and D.H. StJohn, "Rheological Behaviour of the Mushy Zone and Its Effect on the Formation of Casting Defects during Solidification" Acta Mat., 47 (1999) 31-41.

23. H. Mao, "Microstructural Features of Cold Chamber Die Cast AM60B" (MSc thesis, Ohio State University, 1999).

24. P.D.D. Rodrigo and V. Ahuja, "Effect of Casting Parameters on the Formation of 'Pore/Segregation Bands' in Magnesium Die Castings", Magnesium 2000 – 2nd Israeli Conf. on Magnesium Science and Technology (2000) 97-104.

25. A.K. Dahle and D.H. StJohn, "The Origin of Banded Defects in High Pressure Die Cast Magnesium Alloys", NADCA Transactions, 1999, 203-210.

26. M.T. Murray and M.A. Cope, Magnesium Pressure Casting. 1999: International Patent WO 99/28065.

Solidification and Microstructure of Mg - Al - (Ca, Sr, Ce, La) Ternary Alloys

N.D. Saddock[1], A. Suzuki[1], K. Wu[2], S.C. Wildy[3], Y.A. Chang[2], T.M. Pollock[1], J. W. Jones[1]

[1]Department of Materials Science and Engineering, University of Michigan, 2300 Hayward St., Ann Arbor, MI 48109, USA
[2]Department of Materials Science and Engineering, University of Wisconsin, 215 Materials Science and Engineering Building, 1509 University Avenue, Madison, WI 53706, USA
[3]Department of Materials, University of Oxford, Parks Road, Oxford, OX1 3PH, UK

Keywords: Thermal Analysis, Solidification, Scheil Model

Abstract

Solidification behavior and as-cast microstructure have been analyzed in a series of ternary Mg alloys consisting of Mg – Al – X: (Ca, Sr, Ce, La). Thermodynamic calculations of the Mg - Al - Ca , Mg – Al –Sr, and Mg – Al – Ce ternaries were conducted to predict the solidification path of the Mg – 4 Al – 4 X and Mg – 4 Al – 2 X systems. Thermal analyses of the solidification process and accompanying electron microscopy studies of as-cast microstructure were conducted on the same alloys. Scheil calculations and the experimental measurements of liquidus and solidus temperatures are compared. All microstructures consisted of primary α-Mg grains with intermetallic phases along the grain boundaries. Additions of Ca and Sr widen the solidus to liquidus gap to a much greater degree than equivalent levels of Ce and La.

Introduction

Weight reduction of automobiles for increased fuel efficiency and performance has stimulated interest in magnesium alloys for multiple automotive applications. Magnesium alloy systems capable of replacing existing aluminum and steel powertrain components are currently being developed by industry [1-6]. These systems include Mg – Al based alloys with Ca and/or Sr additions. There is also interest in the influence of rare earth (Ce, La) additions on microstructure and mechanical properties [7-9]. Information on the effects of individual alloying additions on solidification and microstructure is limited with only a few fundamental studies [10, 16-20] on solidification paths, diffusion, phase stability, and elevated temperature mechanical deformation.

Since high temperature die-cast Mg alloys are likely to be used in their as-cast state, understanding the solidification behavior of these alloys is particularly important. Furthermore, better predictive capabilities of the influence of individual alloying elements on liquidus and solidus temperatures, phases formed and castability have the potential to provide an important tool for future alloy design and development. In this paper we describe the initial results of a study to examine, experimentally and through thermodynamic modeling, the role ternary alloy additions on the development of microstructure in magnesium alloys for elevated temperature automotive applications.

Experimental

Master alloy ingots were cast by Norsk Hydro with nominal compositions of Mg – 4 Al – 4 X, where X represents Ca, Sr, Ce, or La. Results of chemical analyses (mass spectroscopy, flame atomic absorption spectrometry, and ICP-AES) of the as received ingots are given in Table 1. Small samples (20~25 g) from these ingots were remelted under controlled conditions to study solidification behavior and microstructure development.

The apparatus constructed for remelting samples for solidification studies was similar to the unit described by Riddle [19]. A schematic view of the apparatus is illustrated in Figure 1. The system consisted of an induction coil placed around an environmental chamber that was backfilled with a cover gas of 1% sulfur hexafluoride with a balance Argon. A single type K thermocouple, 0.040" in diameter is inserted in the center of the melt for detecting transformations. A mild steel crucible with no refractory coating is used for the melt cup. Temperature measurement is calibrated using distilled H_2O as an ice / boiling water bath, and the melting temperatures of pure tin, lead, magnesium, and aluminum as references. During solidification, the temperature of the melt is recorded every 500 ms using a Dataq DI – 1000TC device connected to a PC with WinDaq data acquisition software. Liquidus and solidus temperatures are then estimated from the plots of DT/dt vs. time, and are based on a measurable deviation from linearity, when extrapolating from the all-liquid to all-solid ranges. An example of a typical experimental cooling curve and the derivative is shown in Figure 2. It should be noted that no significant undercooling was observed for any of the alloys examined.

The resulting microstructures were characterized using a Hitachi S-3200N SEM equipped with a Kevex EDS System. Sample preparation was completed as outlined in [10] except that the samples were mounted in a cold setting epoxy resin.

Volume fractions of eutectic were determined using the standard point count method, with a fine grid containing 552 markers. Grain size measurements were completed by a line intercept method using ten randomly placed lines per micrograph and calculating the average grain size.

Thermodynamic Modeling

Thermodynamic calculations were carried out by means of the CALPHAD method, using a thermodynamic database modified from that developed by R. Schmid-Fetzer's group. Their assessments for Mg-Al-Ca [11] and Mg-Al-Ce [12] have been adopted. The liquid phase and terminal solid solutions were described by the substitutional solution model. Several intermediate phases show appreciable ternary solubilities and were thus treated as line-compounds. They are $CaMg_2$, Al_2Ca and Al_3Ca_8 in Mg-Al-Ca system, which were described by $Ca(Al,Mg)_2$ and $Al_3(Ca,Mg)_8$ [11]. Meanwhile, the isostructural

Table 1: Chemical composition of as received ingots. Values in weight percent.

Alloy	Al	X	Zn	Mn	Si	Fe	Cu	Ni	Mg
Mg - 4Al	4.16		0.001	0.016	0.0033	0.0215	0.0007	0.0003	Bal.
Mg - 4Al - 4 Ca	3.62	3.71	0.001	0.015	0.003	0.0232	0.0014	0.0003	Bal.
Mg - 4Al - 4 Sr	3.94	4.27	0.001	0.012	0.0029	0.0187	0.0006	0.0001	Bal.
Mg - 4Al - 4 Ce	3.99	3.96	0.001	0.014	0.0032	0.0200	0.0005	0.0001	Bal.
Mg - 4Al - 4 La	3.95	4.04	0.001	0.013	0.0029	0.0236	0.0006	0.0001	Bal.

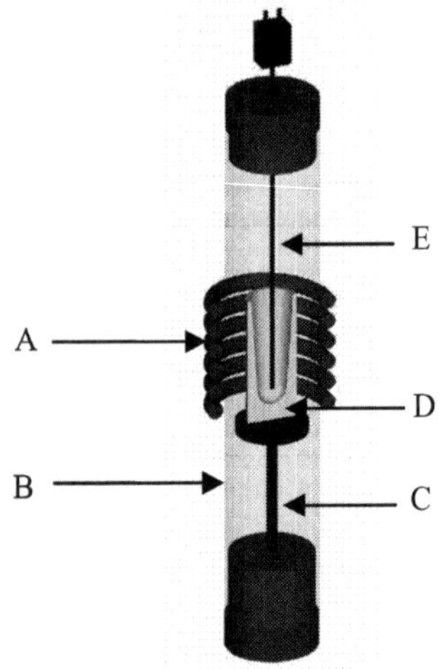

Figure 1: Schematic of melting apparatus: A) Cross Section of Induction Coil, B) Quartz Environmental Chamber, C) Crucible Stand, D) Cross Section of Crucible, E) Thermocouple.

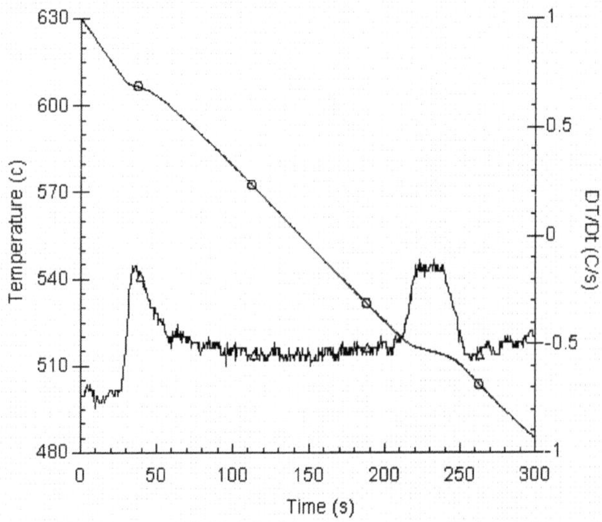

Figure 2: Thermal measurements during solidification for Mg – 4 Al – 4 Ca.

phases $CeAl_2$ and $CeMg_2$ were rationalized as a continuous solid solution $Ce(Al,Mg)_2$ with a low-temperature miscibility gap to match the solubility limit [12]. Other intermetallic compounds were treated as stoichiometric binary phases.

The Mg-Al-Sr system, which is yet to be assessed, was calculated preliminarily by the Muggianu extrapolation method based on Mg-Al, Al-Sr and Mg-Sr sub-binaries. As a first approximation, the ternary solubilities of binary phases were neglected.

The solidification simulations were conducted with the Scheil method [13], assuming no diffusion in the solid, uniform liquid composition, planar and locally equilibrated solid-liquid interfaces, and equally dense solid and liquid. The model is extended to multi-component and multiphase systems, and is solved numerically as the input parameters, e.g., the liquid and solid phase compositions as well as the actual partition coefficients, are provided stepwise by incorporating the aforementioned thermodynamic database, so that more accurate simulations are obtained.

The Pandat® thermodynamic software was used to perform all the calculations [14,15]. Calculated ternary phase diagrams for the Mg – Al – Ca, Mg – Al – Sr, and Mg – Al – Ce systems can be seen in Figure 3.

Results and Discussion

Effect of Ternary Additions on Liquidus and Solidus Temperatures

Eight ternary alloys were examined in the manner described above. The liquidus and solidus temperatures determined for each alloy are shown graphically in Figure 4, along with the corresponding predictions using the Scheil method for all but the La-containing alloys. The data are tabulated for reference in Table 2. The Scheil calculations for solidification for each alloy are shown in Figure 5.

In general the calculated liquidus temperatures were in good agreement with the measured values, whereas the calculated solidus temperatures were lower than those estimated from the cooling curves, particularly for the Ce containing alloys. This is likely due to the prediction that a small fraction of liquid in the Mg – Al – Ce system approaches a low melting point ternary eutectic in the Scheil solidification. The Ca and Sr containing ternary alloys had significantly wider solidification ranges (T_{liq}-T_{sol}) than the alloys containing Ce and La. As shown in Figure 4, the solidus to liquidus gaps for the 4% and 2% Ca alloy are 95 °C and 100 °C, respectively, and that for the 4% and 2% Sr alloy they are 65 °C and 92 °C, respectively. Alloys containing Ce and La have considerably smaller melting ranges than observed for the Ca and Sr alloys and essentially no dependence on composition. These ranges were 24 °C and 26 °C for the 4 wt. % Ce and 2 wt.

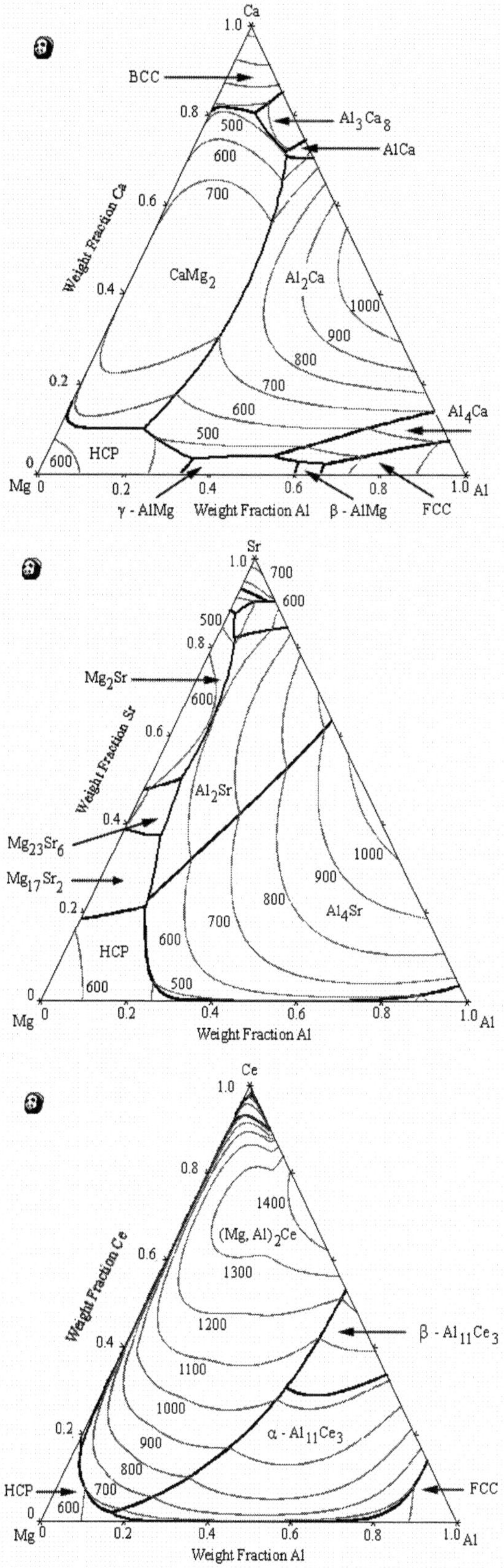

Figure 3: Liquidus projections for the Mg – Al – Ca, Sr, Ce systems.

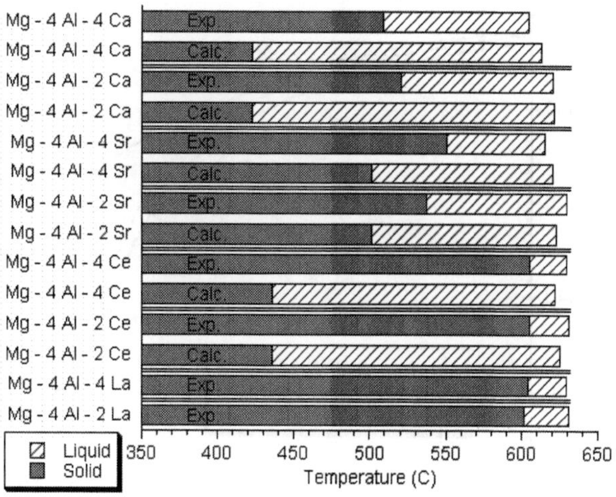

Figure 4: Liquidus and solidus temperatures for experimental (exp.) and Scheil calculations (calc.).

Table 2: Tabulated liquidus and solidus temperatures for experimental and Scheil calculations. (°C).

	Experimental		Calculated	
	Liquidus	Solidus	Liquidus	Solidus
Mg - 4 Al - 4 Ca	604	509	613	422
Mg - 4 Al - 4 Sr	615	550	620	501
Mg - 4 Al - 4 Ce	629	605	622	436
Mg - 4 Al - 4 La	629	604		
Mg - 4 Al - 2 Ca	620	520	621	422
Mg - 4 Al - 2 Sr	629	537	623	501
Mg - 4 Al - 2 Ce	631	605	625	436
Mg - 4 Al - 2 La	631	602		

% Ce, respectively. They were 25 °C and 29 °C, respectively, for the 4 wt. % La and 2 wt. % La, respectively. The largest deviation between measured and predicted solidification range occurs in the Ce-containing alloys. Finally, these differences in the influence of Ca, Sr, Ce, and La on freezing range suggest that castability could be influenced by controlling the balance of these elements to optimize fraction solid as a function of temperature.

Microstructural Analysis

As-cast microstructures for the Mg – 4 Al – 4 X (Ca, Sr, Ce, La) and Mg – 4 Al – 2 X (Ca, Sr, Ce, La) systems are shown Figure 5. The microstructures for all alloys are characteristic of a final eutectic reaction that results in Ca, Sr, Ce or La rich phases along the grain boundaries where the final liquid solidifies. Note that volume fraction of eutectic varies with composition and appears to be greatest for the Mg – 4Al – 4 La alloy. Other features of microstructure observed experimentally and predicted by thermodynamic calculations are discussed for each system individually.

Magnesium – Aluminum – Calcium

Alloys containing 2 wt% and 4 wt% have comparable microstructure with dendritic and globular dendritic primary α –

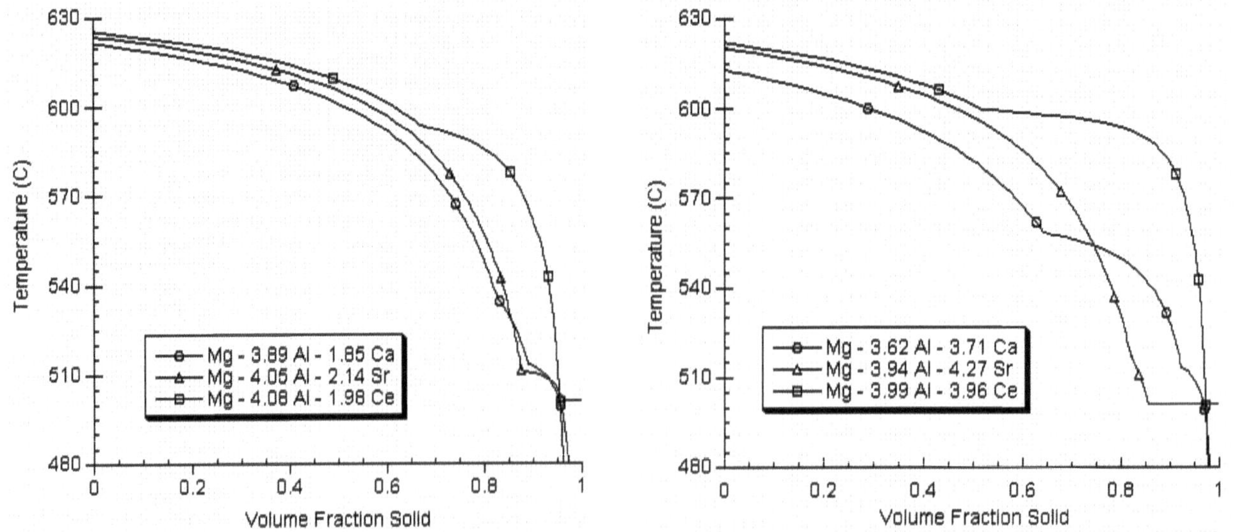

Figure 5: Calculated volume fraction solid for (a) Mg – 4 Al – 2 X: (Ca, Sr, Ce, La) system and (b) Mg – 4 Al – 4 X: (Ca, Sr, Ce, La).

Mg grains, between 20-45 μm, surrounded by a eutectic compound. The crystal structures of the eutectic intermetallic compounds have been identified and reported in detail elsewhere [20]. Scheil simulations predict formation of the C15 phase along grain boundaries, while experiments reveal the presence of a closely related C36 phase [20].

Magnesium – Aluminum – Strontium

Microstructural comparisons of the two alloys again reveal globular dendritic α - Mg grains. The morphology of the eutectic phase is similar to the eutectic phase of the calcium containing alloy, though different intermetallic compounds are formed. An increase in volume fraction of the eutectic is also observed with the 4 wt. % strontium alloy versus the 2 wt% strontium alloy. The eutectic compound has not been characterized by TEM but predictions from modeling suggest the Al_4Sr and $Mg_{17}Sr_2$ phases. Chartrand [21] predicts similar equilibrium phases. Pekguleryuz and Baril [22] have tentatively identified the $Mg_5Al_{17}Sr_3$, $Mg_{13}Al_{10}Sr$ and $Mg_{13}Al_3Sr$ phases in AJ52 and in AJ51, though could not conclusively confirm them since the crystal structure is not yet known for any of these compounds.

Magnesium – Aluminum – Cerium

The intermetallic phases in these ternary alloys are finer and less continuously distributed along the grain boundries. The grain boundary intermetallics are predicted by modeling to be $Al_{11}Ce_3$ and $(Mg,Al)_2Ce$. Grobner [12] also predicts similar phases in ternary Mg – Al – Ce alloys.

Magnesium – Aluminum – Lanthanum

Thermodynamic data was unavailable for the lanthanum binary phase diagrams so the Mg – Al –La system has not yet been modeled thermodynamically. This system is significant due to the large fraction of lanthanum in the mischmetals used for alloying rare earth containing creep resistant magnesium alloys. Additions of La had less effect on liquidus and solidus temperatures, compared to other ternary additions under investigation. Lanthanum had the largest effect on volume fraction eutectic with a significant increase in volume fraction from 2 wt. % to 4 wt. % of all the alloy systems studied. The microstructure of the alloy again consisted of α dendrites surrounded by eutectic with a lamellar plate shaped morphology. Electron microscopy studies of the phases formed in this alloy during solidification are in progress.

Conclusions

The microstructure and solidification behavior of the four Magnesium – Aluminum based ternary systems have been investigated. Solidification paths were modeled using Pandat® thermodynamic software. Analyses indicates that:

1. There is good agreement between thermodynamically predicted and experimentally measured liquidus temperatures in alloys containing 2-4 wt. % of Ca, Sr, Ce, or La.

2. For comparable levels of alloying additions, Ca and Sr containing alloys have larger freezing ranges than Ce and La containing systems.

3. The microstructure of all ternary alloys consisted of globular dendritic primary α grains with intermetallic compounds located along grain boundaries.

Acknowledgements

The authors acknowledge the support of NSF FRG Grant No. DMR – 0309468. The authors also would like to thank C.J. Torbet of the University of Michigan and Daryll Albright and Per Bakke of the Norsk Hydro Company for their assistance.

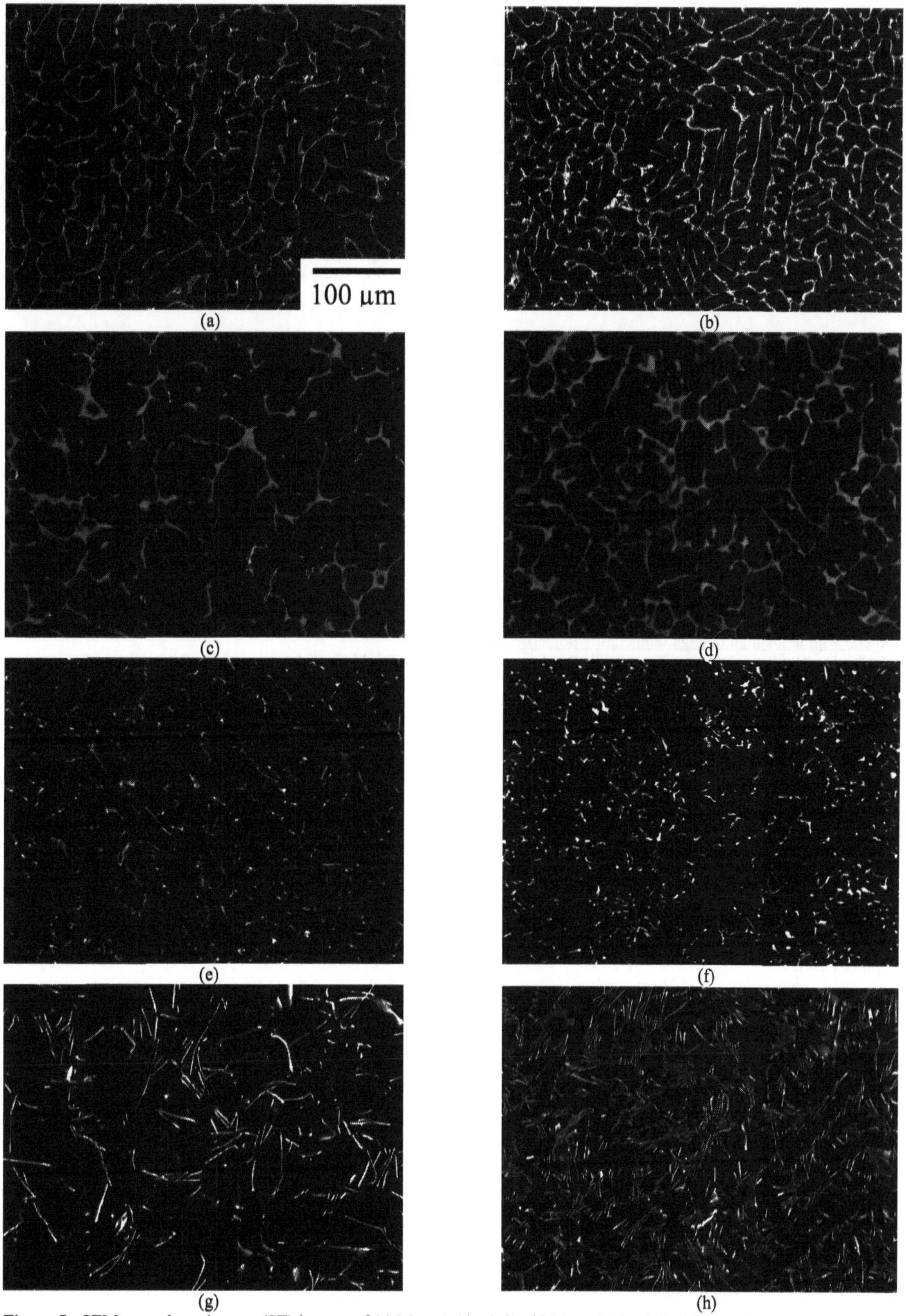

Figure 5: SEM secondary electron (SE) images of (a) Mg – 4 Al – 2 Ca (b) Mg – 4 Al – 4 Ca (c) Mg – 4 Al – 2 Sr (d) Mg – 4 Al – 4 Sr (e) Mg – 4 Al – 2 Ce (f) Mg – 4 Al – 4 Ce (g) Mg – 4 Al – 2 La (h) Mg – 4 Al – 4 La

References

[[1] B.R. Powell, V. Rezhets, A.A. Luo, B.L.Tiwari, Low-Cost, Creep-Resistant Magnesium Alloy for Powertrain Components, US Patent No. 6,264,764. July 24, 2001.

2] B.R. Powell, A.A. Luo, V. Rezhets, J.J. Bommarito, B.L. Tiwari, Development of Creep-Resistant Magnesium Alloys for Powertrain Applications: Part 1 of 2, SAE Tech. Paper 2001-01-0422, March 6, 2001.

[3] A.A. Luo, M.P.Balogh, B.R. Powell, Tensile Creep and Microstructure of Magnesium-Aluminum-Calcium-Based Alloys for Powertrain Applications: Part 2 of 2, SAE Tech. Paper, 2001-01-0423, March 6, 2001.

[4] A.A. Luo, M.P. Balogh, B.R. Powell, Creep and Microstructure of Magnesium-Aluminum-Calcium Based Alloys, Met. Mat. Trans. A., 2002, pp 567-574.

[5] E. Baril, P. Labelle, A. Fischersworring-Bunk, AJ(Mg-Al-Sr) Alloy System Used for New Engine Block, SAE Tech. Paper, 2004-01-0659, March 8, 2004.

[6] B. Brofin, E. Aghion, S. Schumann, P. Bohling, K. U. Kainer, Magnesium Alloy for High Temperature Applications, US Patent No. 6,139,651, October 31, 2000.

[7] I. A. Anyanwu, Y. Gokan, A. Suzuki, S. Kamado, Y. Kojima, S. Takeda, T. Ishida, Effect of substituting cerium-rich mischmetal with lanthanum on high temperature properties of die-cast Mg-Zn-Ca-RE alloys, Mat. Sci. Eng. A 380, 2004, pp 93-99.

[8] P. Bakke, H. Westengen, S. Sannes, D. Albright, Powertrain Components – Opportunities for the Die Cast AE Family of Alloys, SAE Tech. Paper, 2004-01-0655, March 8-11, 2004.

[9] J. Berkmortel, H. Hu, J. E. Kearns, J.E. Allison, Die Castability Assessment of Magnesium Alloys for High Temperature Applications: Part 1 of 2, SAE Tech. Paper, 2000-01-1119, March 6-9, 2000.

[10] Y. W. Riddle, L. P. Barber, M. M. Makhlouf, Characterization of Mg Solidification and As-Cast Microstructures, Mag. Tech. Conf. Proceedings, March, 14-18 2004, pp 203-208.

[11] J. Groebner, D. Kevorkov, I. Chumak, R. Schmid-Fetzer, Experimental investigation and thermodynamic calculation of ternary Al-Ca-Mg phase equilibria, Z. Metallk., 94, 2003, pp 976-982.

[12] J. Grobner, D. Kevorkov, R. Schmid-Fetzer, Thermodynamic modeling of Al-Ce-Mg phase equilibria coupled with key experiments, Intermetallics., 10. 2002. pp 415-422.

[13] E. Scheil, Z. Metallk., 34, 1941, pp.70-72.

[14] Pandat-phase diagram calculation engine for multicomponent systems. CompuTherm LLC, 437 S. Yellowstone Dr., Suite 217, Madison, Wisconsin, USA(2000).

[15] S-L. Chen, S. Daniels, F. Zhang, Y.A. Chang, W.A. Oates, R. Schmid-Fetzer, On the calculation of multicomponent stable phase diagrams, J. Phase Equilibria, 22.2001. pp. 373-378.

[16] S. R. Agnew, O. Duygulu, TEM Investigation of Dislocation Mechanisms in Mg Alloy AZ31B Sheet, Mag Tech. Conf. Proceedings, March 14-18 2004, pp 61 - 65

[17] G. Chadha, J.E. Allison, J. W. Jones, The Role of Microstructure and Porosity in Ductility of Die Cast AM50 and AM60 Magnesium Alloys, Mag. Tech. Conf. Proceedings, March 14-18, 2004, pp 181 – 186.

[18] C. Brubaker, Z.K. Liu, Diffusion Couple Study of the Mg-Al System, Mag. Tech. Conf. Proceedings, March 14-18, 2004, pp 229 – 234.

[19] Y. W. Riddle, M.M. Makhlouf, Characterizing Solidification by Non-Equilibrium Thermal Analysis, Mag Tech. Conf. Proceedings, March 2-6, 2003, pp 101 – 106.

[20] A. Suzuki, N.D. Saddock, J.W. Jones, T.M. Pollock, Structure and transition of eutectic $(Mg,Al)_2Ca$ Laves phase in a die-cast Mg-Al-Ca base alloy, *Scripta Mater.*, 51, 2004, in press pp 1005-1010.

[21] P. Chartrand, A.D. Pelton, Critical Evaluation and Optimization of the Thermodynamic Properties and Phase Diagrams of the Al-Mg, Al-Sr, Mg-Sr, and Al-Mg-Sr Systems, Journal of Phase Equilibria., 15. 1994. pp 591 – 605.

[22] M.O. Pekguleryuz, E. Baril, Creep Resistant Magnesium Diecasting Alloys Based on Alkaline Earth Elements, Mat. Trans., 42, 2001, pp 1258-1267.

HOT CRACKING SUSCEPTIBILITY OF BINARY Mg-Al ALLOYS

Guoping Cao, Sindo Kou and Y. Austin Chang
Department of Materials Science and Engineering
University of Wisconsin, Madison, WI 53706

Keywords: Mg-Al alloys, hot cracking, casting

Abstract

The susceptibility of binary Mg-Al alloys in permanent mold casting was tested, including Mg-0.25Al, Mg-0.6Al, Mg-1Al, Mg-2Al, Mg-4Al and Mg-8Al (all in wt %). A steel mold was used for constrained rod casting (CRC), in which rods of various lengths were cast with sudden enlargement at both ends of each rod to prevent it from free contraction during solidification and thus induce hot cracking. The cooling rate near the location of cracking was measured. The hot cracking susceptibility was evaluated based on the widths and locations of the cracks observed. The curve of crack susceptibility vs. Al content was determined and compared with previous curves based on ring casting in steel molds and constrained rod casting in sand molds. The Scheil model of solidification was used to calculate the curves of temperature vs. fraction solid (T-f_S) of the alloys. The crack susceptibility curve based on a simplified hot cracking criterion and the T-f_S and cooling curves was compared with that based on CRC.

Introduction

Hot cracking, also referred to as hot tearing or solidification cracking, occurs at high temperatures during solidification. Dodd et al. [1] studied the hot cracking susceptibility of binary Mg-Al alloys up to 8% Al (by weight) by ring casting in a steel mold and found the peak susceptibility at 0.75% Al. Rosenberg et al [2] studied hot cracking of binary Mg-Al alloys up to 17% Al by restrained rod casting in a sand mold, and observed the peak susceptibility around 2.5% Al. Argo et al. [3] and Pekguleryuz et al. [4] studied hot cracking of high temperature creep resistant Mg-Al-Sr alloys using constrained rod casting (CRC) in a steel mold. Wang et al. [5] detected the onset of hot cracking of the commercial magnesium alloy AZ91 in a sand mold by using a load cell and a thermocouple. Powell et al. [6] studied the hot cracking of Ca-containing Mg alloys in die casting.

Experimental Procedure

A mild steel crucible was used for melting. It had an opening at the bottom wall for discharging the liquid metal into the mold. The opening was closed with a stopper during melting. The inside of the crucible was coated with BN the first time it was used. About 600g of pure magnesium (99.8%) was melted in the crucible under a protective gas of CO_2+0.7%SF_6 at the flow rate of 1.6 L/min. Pure aluminum (99.99%) was added to the pure magnesium melt when the melt temperature reached 700°C. After about 20 minutes, the melt was stirred for 2 minutes manually. The melt was held at the pouring temperature for about 30 minutes before casting. The pouring temperature was set at 125°C above the liquidus temperature of the alloy. During pouring, a CO_2+2%SF_6 gas was directed at the flowing Mg melt to protect it from oxidation and possible combustion.

The steel mold for constrained rod casting is shown in Fig. 1. It allowed four rods of 9.5 mm diameter to be cast at the lengths of 51, 89, 127, and 165 mm, with a ball of 19 mm diameter at the end of each rod acting as an anchor to restrain the rod from free contraction during casting. The mold was coated with BN, about 40 μm thick. The mold preheating temperature, usually 150 to 400°C [1, 3], was 335°C and was achieved with a resistance heater. The mold was opened 8 minutes after pouring to remove the casting from the mold.

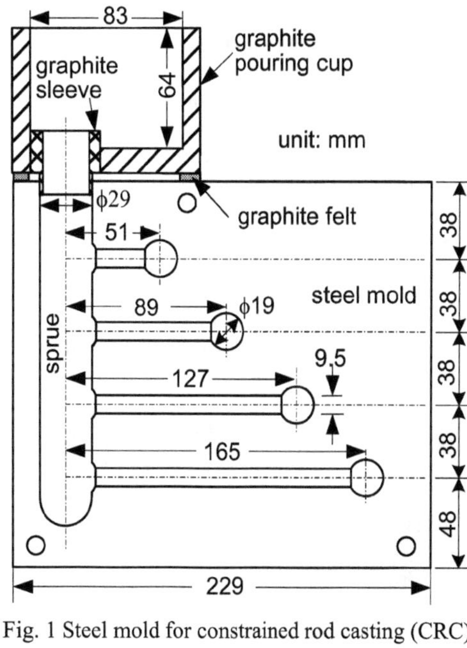

Fig. 1 Steel mold for constrained rod casting (CRC).

The CRC mold in the present study differed from that of Argo et al. [3] for testing the hot cracking susceptibility of Mg-Al-Sr alloys. The pouring cup of Argo et al. [3] was internal, that is, part of the steel mold. It was a funnel-shaped cup widening from the top of the sprue to the top of the mold to facilitate pouring. In the present study, however, an external graphite pouring cup was connected to the steel mold with a graphite sleeve. The graphite felt, about 4 mm thick, between the cup and the mold acted as a cushion and allowed the pouring cup to move freely downward toward the mold as the remaining melt in the cup solidified and contracted. This kept solidification in the cup from inducing severe shear strains at the sprue end of the rods.

The cooling rate during solidification was measured in a similar steel mold, where the tensile force developing during solidification in the bottom rod due to hindered contraction could be measured [7]. A K-type thermocouple with a closed stainless steel sheath of 0.5 mm (0.02 in) outer diameter was positioned at the sprue end of the bottom rod through a 1.4 mm-diameter hole in the left wall of the steel mold. The hole remained open throughout casting and thus allowed free movement of the thermocouple relative to the mold wall in order not to induce additional strains in the solidifying material. To keep the thermocouple from being moved by the liquid metal during pouring, it was covered with a stainless tube of 0.7 mm inner diameter and 1.1 mm outer diameter up to 5 mm away from the tip. After pouring the melt into the mold, the temperature and load were recorded with a computer data acquisition system, National Instruments SC-2345. The sampling rate was 20 per second.

A digital photo system was used to take pictures of the cracks in the resultant castings, consisting of a digital camera, close-up lens, and bellows. The digital images were used to measure the widths of the cracks with a computer using computer software. The cracks were enlarged on the computer monitor, and their widths measured with tools available in the commercial software Visio 2000 [8].

Results and Discussion

Figure 2 is a typical casting showing, from the top down, the material solidified in the graphite pouring cup, the graphite sleeve, the material solidified in the sprue, and the rods attached to it. The spikes connected to the balls came from the liquid metal that solidified in the vents (not shown in Fig. 1). Figure 3 shows examples of cracks at the sprue end of the rods. However, cracking at the ball-end or in the middle of the rods, which is more difficult, sometimes occurred. In any case, the maximum crack widths were determined with the help of computer software. As shown in Fig. 4, the hot cracking susceptibility (*HCS*) was defined as follows:

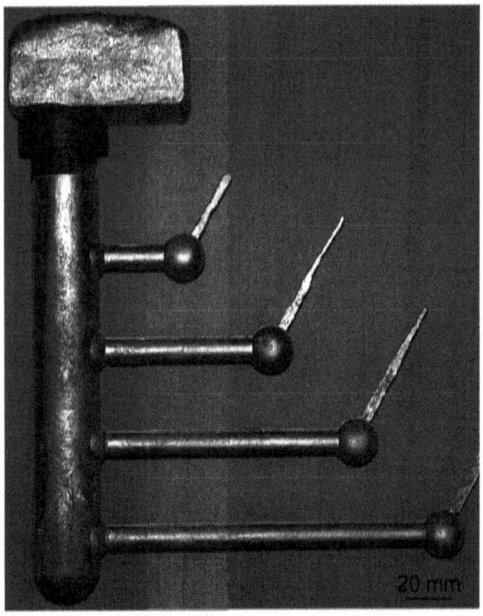

Fig. 2 Photograph of a casting made with the CRC mold.

Fig. 3 Close-up views of cracks in the bottom rods near the sprue: (a) Mg-2Al; (b) Mg-4Al; (c) Mg-8Al.

$$HCS = \sum (w_{crack} \times f_{length} \times f_{location}) \quad (1)$$

where, as in the study of Argo et al [3], f_{length} is the crack length factor (4 for the longest rod, 8 for the second longest rod, 16 for the third longest rod, and 32 for the shortest rod), and $f_{location}$ the crack location factor (1 for cracking at the sprue end, 2 at the ball end and 3 in the middle of the rod). However, the crack width factor f_{crack} (1 for short hair line, 2 for full hair line, 3 for crack, and 4 for half broken rod) of Argo et al. [3] was not used. Rather, the maximum crack width (measured in mm) w_{crack} was adopted.

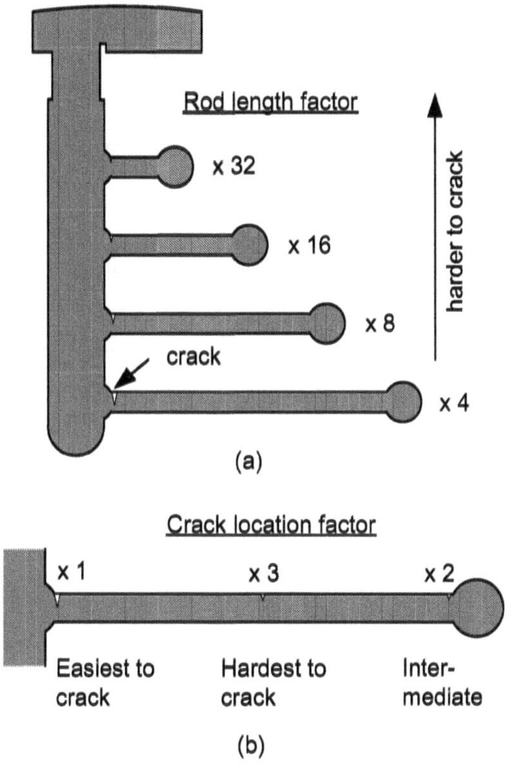

$$HCS = \sum (w_{crack} \times f_{length} \times f_{location})$$

(c)

Fig. 4 Determining crack sensitivity from widths of cracks in CRC: (a) rod length factor; (b) crack location factor; (c) hot cracking susceptibility.

The results of the crack susceptibility test are shown in Fig. 5. The susceptibility rose sharply from pure Mg, reached its maximum at Mg-1Al, decreased gradually with further increase in the Al content, and diminished beyond 8% Al. The wide susceptibility scatter at Mg-0.25Al is likely to be associated with the sharp rise before the maximum susceptibility. The crack susceptibility curve of Dodd et al. [1] from ring casting in a steel mold is shown in Fig. 6 for comparison. Based on the measured total crack length, the maximum susceptibility was found to be at about 0.75% Al, close to the maximum susceptibility at 1% Al shown in Fig. 5. However, the crack susceptibility curve appears much narrower than that in Fig. 6, that is, the crack susceptibility drops off with increasing aluminum content much more rapidly. Perhaps, hot cracking is more sensitive to composition variations in ring casting than in constrained rod casting. The crack susceptibility curve of Rosenberg et al. [2] from constrained rod casting in sand molds is also shown in Fig. 6. The maximum crack susceptibility appears around 2.5% Al (but can be anywhere between 1 and 5% Al considering the resolution of the distribution of the data points). The crack susceptibility curve appears much wider than that in the present study (Fig. 5) and that of Dodd et al. [1], perhaps because of the differences between hot cracking in metal and sand molds.

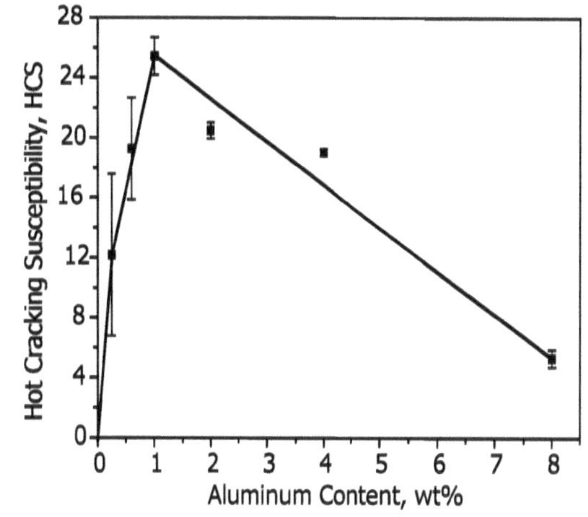

Fig. 5 Hot cracking susceptibility curves of binary Mg-Al alloys based on CRC testing.

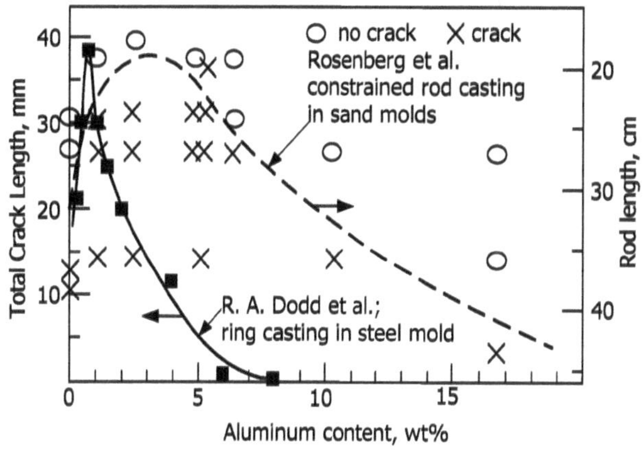

Fig. 6 Hot cracking susceptibility curves of binary Mg-Al alloys based on ring-casting and sand casting. From Dodd et al. [1] and Rosenberg et al. [2].

As shown in Fig. 5, the maximum crack sensitivity (at 1 wt% Al) occurs somewhere between pure magnesium and highly alloyed magnesium, say Mg-8%Al. Such a maximum can be explained qualitatively based on the principles of solidification. Pure magnesium is not susceptible to hot cracking because there is no grain-boundary liquid to cause hot cracking. In highly alloyed magnesium, on the other hand, there can be much grain-boundary liquid available to backfill and "heal" incipient cracks. Somewhere in between the two, however, the amount of grain-boundary liquid can be sufficient to form thin continuous intergranular films to separate grains but insufficient to heal cracks, thus resulting in a maximum susceptibility to hot cracking.

The crack susceptibility will be discussed in light of the cooling curve and the curve of temperature vs. fraction solid (T-f_S). Figure 7 shows the cooling curve (temperature vs. time) recorded during the casting of alloy Mg-4Al at the sprue end of the bottom rod. A rapid rise in temperature was recorded immediately after pouring and the temperature started falling in about 0.5 sec. Time zero corresponds to the moment the temperature at the thermocouple position started falling, that is, when cooling began.

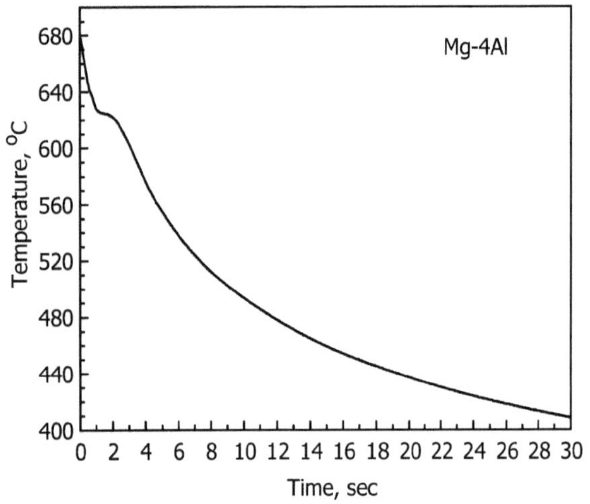

Fig. 7 Cooling curve recorded during casting of Mg-4Al.

For convenience of discussion, the Mg-rich side of the Mg-Al binary phase diagram [9] is shown in Fig. 8. As shown, the eutectic temperature is 437°C and the liquidus line is not a straight line. The curves of temperature (T) vs. fraction solid (f_S) for the binary alloys studied are shown in Fig. 9. These T-f_S curves, calculated using the computer code Pandat [10], which is a software package for calculating multicomponent phase diagrams, solidification paths and thermodynamic properties. In the computation, the compositions of the solid and liquid phases at each temperature are calculated. The widely used Scheil solidification model was used, that is, assuming equilibrium at the solid/liquid interface, no solid-sate diffusion, and uniform composition in the liquid. However, unlike a simple Scheil equation for binary alloys, the equilibrium partition coefficient (k) and liquidus slope (m_L) were both calculated at each temperature instead of being held constant. The effect of the cooling rate on the equilibrium at the solid/liquid interface is not expected to be significant because the measured cooling rate was only on the order of about 30°C/s at the liquidus temperature. For instance, from Fig. 7 for alloy Mg-4Al, the cooling rate was 32 °C/s at the liquidus temperature 630 °C. This relatively low level of cooling rate is not likely to cause undercooling. On the other hand, this cooling rate is not so slow as to allow much solid-state diffusion during solidification, either. Thus, the Scheil model is expected to be reasonable.

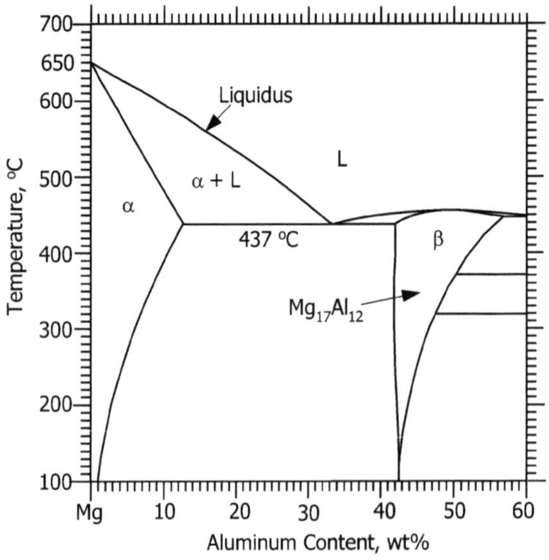

Fig. 8 Mg-rich side of binary Mg-Al phase diagram [9].

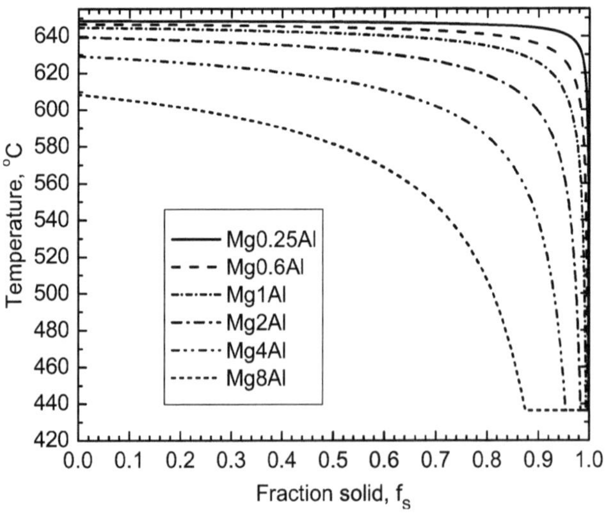

Fig. 9 Temperature vs. fraction solid for binary Mg-Al alloys based on Scheil model. Curves calculated using Pandat of CompuTherm LLC [10].

Suyitno et al. [11] recently reviewed several criteria for hot cracking, including those based on nonmechanical aspects, those based on mechanical aspects and those that combine these features. For simplicity, the criterion of Clyne et al. [12] will be considered here. They proposed to use the following cracking susceptibility coefficient (CSC') to evaluate the susceptibility of an alloy to hot cracking:

$$CSC' = \frac{t_V}{t_R} \qquad (2)$$

where t_V is the time period the solidifying alloy is vulnerable to hot cracking, and t_R the time available for stress relief processes.

The former, corresponding to the time period of interdendritic separation, was assumed to be the time period for the fraction solid to increase from 0.9 to 0.99, that is, $(t_{0.99} - t_{0.9})$. The latter, corresponding to the time period of mass and liquid feeding, was assumed to be the time period for the fraction solid to increase from 0.4 to 0.9, that is, $(t_{0.9} - t_{0.4})$. In other words,

$$CSC' = \frac{(t_{0.99} - t_{0.9})}{(t_{0.9} - t_{0.4})} \quad (3)$$

From the cooling curve and the T-f_S curve of each alloy, the values of $t_{0.4}$, $t_{0.9}$ and $t_{0.99}$ were determined, and the resultant *CSC'* vs. composition curve was plotted and shown in Fig. 10. The agreement between the curve and the CRC hot cracking susceptibility curve was reasonably good. However, the validity of assuming that hot cracking occurs between $t_{0.9}$ and $t_{0.99}$ needs to be checked, as discussed elsewhere [7].

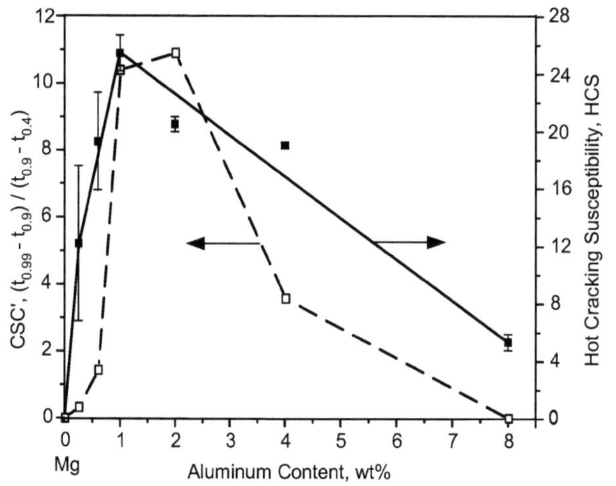

Fig. 10 Comparison between hot cracking susceptibility curves based on the cracking susceptibility coefficient (*CSC'*) and CRC testing.

Conclusions

The hot cracking susceptibility of binary Mg-Al alloys has been tested from 0.25 to 8.0 wt% Al by constrained rod casting (CRC) with a steel mold. The curve of the crack susceptibility vs. the Al content shows that alloys containing 1 to 2 wt% Al are most susceptible to hot cracking. This is close to the 0.75% Al of Dodd et al. [1] and 2.5% Al of Rosenberg et al [2]. However, the crack susceptibility curve is wider than that of the former and narrower than that of the latter, possibly due to the differences in the casting process and the mold. The crack susceptibility curve based on the hot cracking criterion of Clyne et al. [12] and the T-f_S and cooling curves compares reasonably well with that based on CRC.

Acknowledgements

This work was supported by NSF FRG Grant No. DMR-0309468.

References

[1] R. A. Dodd, W. A. Pollard, and J. W. Meier, Trans. AFS, 65 (1957) 100-116.
[2] R. A. Rosenberg, M. C. Flemings, and H. F. Taylor, Trans. AFS. 68 (1960) 518-528.
[3] D. Argo et al., Die Casting 2001, October 29 to November 1, 2001, Cincinnati, Ohio, pp. 181-186.
[4] M. Pekguleryuz et al., Magnesium diecasting alloy AJ62x with superior creep resistance, ductility and diecastability, ed. H. I. Kaplan, Magnesium Technology 2003, TMS, Warrendale, PA, 2003, 201-206
[5] Y. Wang et al., Materials Letters, 1-2 (2002) 35-39.
[6] B. R. Powell et al., The die castability of calcium-containing magnesium alloys: thin wall computer case, ed. H. I. Kaplan, *Magnesium Technology 2002,* TMS, 2002, 123-129.
[7] G. Cao and S. Kou, Materials Science and Engineering, in press.
[8] Visio 2000, Visio Corporation, Seattle, WA, 2000.
[9] American Society for Metals. 1986. *Binary Alloy Phase Diagrams vol. 1*: p. 130, Metals Park, O.H., American Society for Metals.
[10] Pandat – Phase Diagram Calculation software package for Multicomponent Systems, Computherm LLC, Madison, WI 53719, 2001.
[11] Suyitno, W. H. Kool, and L. Katgerman, Metallurgical and Materials Transactions A, 36A (2005), 1537-1546.
[12] T. W. Clyne and G. J. Davies, Br. Foundryman. 4 (1981), 65-73.

PHENOMENA OF FORMATION OF GAS INDUCED SHRINKAGE POROSITY IN PRESSURE DIE-CAST MG-ALLOYS

Soon Gi Lee[1] and Arun M. Gokhale[1]
[1]School of Materials Science and Engineering; Georgia Institute of Technology; 771 Ferst Drive NW; Atlanta, Georgia, 30332-0245, USA

Keywords: Mg-alloys, Pressure die-casting, Casting defects, Porosity

Abstract

It is usually believed that formation of gas and shrinkage porosity has different dependencies on the process conditions, and consequently, the gas and shrinkage pores are usually regarded as independent microstructural features. In this contribution, it is shown that in high-pressure die-castings, gas pores can lead to the formation of shrinkage porosity under specific conditions. This is because the air/gas in the gas pores is an efficient heat-insulating medium. Therefore, presence of gas porosity can retard heat transfer in the liquid melt as compared to similar regions without gas porosity. Consequently, the local solidification rate is lower in such regions, leading to shrinkage porosity formation. Two and three-dimensional (3D) microstructural observations using digital image analysis and numerical simulation for heat transfer support such a hypothesis. Finite element (FE)-based simulations have also been performed on the 3D microstructures to reveal the variations of local stresses caused by gas induced shrinkage porosity.

Introduction

During the recent years, applications of cast Magnesium alloys have considerably increased in the automotive as well as numerous other industries due to their attractive properties such as high strength to weight ratio [1,2]. Majority of Mg-alloy cast components are manufactured using high-pressure die-casting (HPDC) processes, which inevitably lead to the formation of gas (air) and shrinkage pores that deteriorate the fracture sensitive mechanical properties (ductility, fatigue life, fracture toughness, etc) of the castings [3-7]. Gas pores typically have equiaxed round shapes, whereas shrinkage pores have "crack-like" morphologies. It is usually *believed* that the mechanisms of formation of gas and shrinkage porosity are distinctly different and independent, and consequently, the gas and shrinkage pores can be regarded as independent microstructural features. This is the basic under pinning of the majority of approaches for improvement of fracture sensitive mechanical properties of the HPDC Mg-alloys via control of process parameters such as molten metal temperature, intensification pressure, gate velocity, and casting geometry that affect porosity [8-12]. Conventional shrinkage porosity can be present in the regions that have gas pores as well in the regions that do not contain any gas pores. Further, when conventional shrinkage porosity is present in the regions surrounding gas pores, it can be topologically connected to the corresponding gas pores. In the present work it is shown for the first time that, under specific conditions, in high-pressure die-castings, presence of gas pores can facilitate formation of shrinkage porosity, namely gas induced shrinkage porosity. The gas pores are required for formation of such gas induced shrinkage porosity. Therefore, *gas induced shrinkage porosity* (unlike conventional shrinkage porosity) is present only in the regions containing gas pores However, unlike conventional shrinkage porosity, the *gas induced shrinkage pores* are not topologically connected to the corresponding central gas pores around which they form. The gas pores facilitate formation of gas induced shrinkage porosity around them because the air in the gas pores is an efficient heat-insulating medium, and consequently, the presence of gas pores can decrease the heat transfer rate in the liquid melt as compared to similar regions without gas porosity, leading to lower the local solidification rate in regions containing gas pores, which induces formation of shrinkage pores around the gas pores. To present direct experimental metallographic evidence of the presence of gas induced shrinkage porosity, it is essential to show that such porosity (i) surrounds gas pores, and (ii) is not topologically connected in the three-dimensional microstructural space to the gas pore(s) that it surrounds. In this contribution, we present such experimental evidence via reconstruction of high-resolution three-dimensional microstructure of HPDC AM60 alloy. To further support the hypothesis of the formation of gas induced shrinkage porosity heat transfer simulations are performed to show that regions around gas pores have lower local solidification rate, which can induce formation of shrinkage pores around gas pores. Finally, finite element (FE) based computations are used to show that presence of gas induced shrinkage pores can substantially increase the local stress concentration, which can deleteriously effect the mechanical properties of the cast alloy.

Experiments and Simulations

Materials and Processes

The experiments were performed on AM60 Mg-alloy (nominal 6% Al, 0.25 to 0.5% Mn, 0.22 % (max.) Zn, Fe, Si, Cu, and Ni impurities). The rectangular plates of 200 mm × 100 mm × 4 mm dimensions were high-pressure die-cast in a cold chamber machine at Gibbs Die Castings Co., Evansville, IN. Four sets of plates were high-pressure die-cast using combinations of the process parameters given in Table I. The parameters such as vent area, casting geometry, gate dimensions, die temperature, etc were kept the same during high-pressure die-casting of all the plates. This design of experiments permits "deconvolution" of the individual effects of process parameters on the porosity attributes in the cast microstructures.

Table I. Process parameters for high-pressure die-castings.

Plate ID	Gate Velocity (m/s)	Intensification	Liquid Melt Temp. (°C)
I	5.2	On	705
G	2.7	Off	705
M	5.2	Off	682
W	5.2	Off	705

Metallography

The metallographic specimens were sectioned along the planes containing the thickness direction of the plates. These samples were mounted in standard metallographic mounts, and then they were ground using 320 grit to 1000 grit abrasive SiC papers. Immediately after the last grinding step, the specimens were washed with water, rinsed with methanol, and dried. The fine polishing was conducted using 6μm and 1μm oil based diamond compounds. The final steps in polishing involved the use of 0.3μm alumina and 0.05μm colloidal silica. Glycerol and oil base lubricants were used during these polishing steps to avoid the formation of a surface film. The specimens were immediately rinsed with methanol and dried. The specimens were examined in the as-polished condition (without etching).

Reconstruction and Visualization of 3D Microstructures

Microstructural observations carried out in the 2D metallographic planes through 3D microstructural domains of interest do not contain all the information concerning the topological of features in the 3D microstructure. In particular, 2D observations are not sufficient to obtain the information concerning complex topological connectivity of pores in HPDC Mg-alloys. In the present work, the 3D microstructure visualization has been carried out using 102 montage serial sections; each montage serial section containing 9 contiguous microstructural fields grabbed at 500× with a 2μm section depth. The 3D microstructure was reconstructed from these serial sections and visualized using image analysis software Voxblast 3.10. The details of serial sectioning technique and 3D microstructure reconstruction are given elsewhere [14-16]. The resolution of reconstructed 3D microstructural image is 2 μm in the Z-direction (which is the distance between serial sections), and 0.2 μm in the X- and Y-directions (i.e., in the planes of the serial sections).

3D Finite Element Based Analysis to Compute Local Stress Distributions

The local stress distributions around the gas pores in the cast microstructure that have gas induced shrinkage porosity around them may not necessarily be same as those around the gas pores in the microstructural regions that do not contain gas induced shrinkage porosity. To gauge the extent of these differences in the local stress distributions, the digital image of the 3D microstructure reconstructed from serial sections is implemented in a 3D finite elements (FE) computational framework. The reconstructed 3D microstructure in STL (stereo lithography) format is exported into Harpoon (CEI), commercial automatic meshing software to mesh the geometry with tetrahedral elements. Only the external volume of the STL file is meshed leaving porosity free of elements. The FE mesh is then exported to commercial finite element code ABAQUS to simulate the elastic mechanical response due to an applied uniaxial tensile load.

Numerical Simulation for Transient Heat Transfer

The heat transfer simulations were performed to study the effect of the presence of gas pore on the local heat transfer rate that subsequently determines the local solidification rate of the casting using the initial conditions and boundary conditions given in Table II. These numerical simulations were carried out using commercial software ANSYS 7.1. A gas pore of 100μm radius was simulated in the cast plate to investigate the effects of the presence of gas porosity on the temperature distribution surrounding it as compared to the region without gas porosity. The initial liquid melt temperature was computed at the time when the liquid melt completely filled the die cavity from die casting simulation using CFD modeling software, FLOW 3D 8.2. These simulations involved the following assumptions and simplifications.

- The simulations were performed on casting of pure Magnesium (rather than AM60 alloy) as the required auxiliary physical data are not available for commercial AM60 alloy
- A temperature of 703°C has been assumed at the interface between liquid melt and a gas pore.
- Thermal convection and radiation effects within the gas pore have been ignored.

The simulation begins at the instant where the liquid metal has completely filled the mold cavity, therefore, liquid metal flow is not incorporated in the simulations.

Table II. Modeled geometry and initial and boundary conditions.

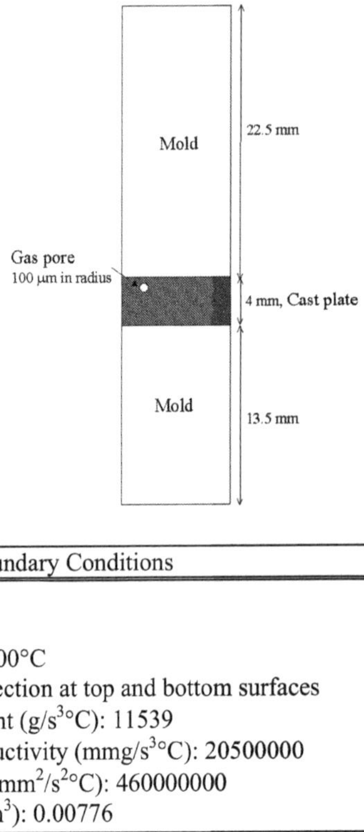

Initial and Boundary Conditions		
Mold		
H13 die steel		
Initial temp: 100°C		
Thermal convection at top and bottom surfaces		
Film coefficient ($g/s^3{}^\circ C$): 11539		
Thermal conductivity ($mmg/s^3{}^\circ C$): 20500000		
Specific heat ($mm^2/s^2{}^\circ C$): 460000000		
Density (g/mm^3): 0.00776		
Liquid Melt		
Pure Mg		
Initial temp: 703°C		
Temp (°C)	Thermal Conductivity	Enthalpy
127	1.48×10^8	1.07×10^{10}
300	1.45×10^8	3.20×10^{10}
527	1.43×10^8	5.80×10^{10}
703	1.43×10^8	12.02×10^{10}
Gas Pore (air)		
Temp: 703°C		
Heat transfer in liquid melt only by thermal conduction		

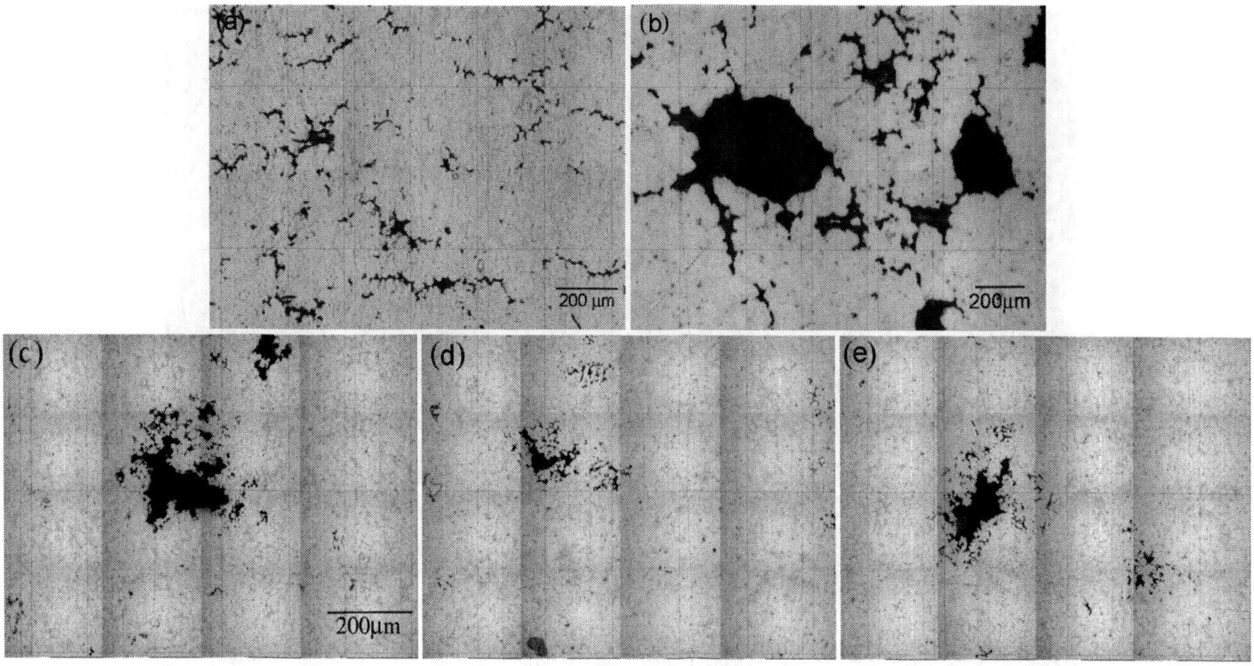

Figure 1. (a) Conventional shrinkage porosity in a region not containing gas pores and (b) conventional shrinkage porosity around gas pores, (c), (d), and (e) gas induced shrinkage porosity clustered around gas pores.

Results and Discussion

Two-Dimensional Metallographic evidence of gas induced shrinkage porosity

Figure 1c, d, and e depict several magnified images of large gas pores and shrinkage porosity clustered around these gas pores. However, the shrinkage pores in Figure 1c to 1e do not appear to be connected the corresponding central gas pores around which they are clustered. It will be shown in the next section that such shrinkage pores are not connected to corresponding gas pores in the third-dimension as well. Figure 1a illustrates conventional shrinkage porosity that has formed in the regions not containing gas pores. Figure 1b shows conventional shrinkage porosity around gas pores. However, as opposed to gas induced shrinkage porosity in Figures 1c to 1e, the conventional shrinkage porosity in Figure 1b is indeed connected to the corresponding gas pores.
Thus gas induced shrinkage porosity is clustered around gas pores whereas conventional shrinkage porosity is present in the regions that may not (Figure 1a) or may (Figure 1b) contain gas pores. It is interesting to note that the "sponge" type gas induced shrinkage porosity is not observed in the regions that do not contain gas pores. The air in the gas porosity is an efficient heat-insulating medium; therefore the presence of gas porosity can retard the heat transfer in the liquid melt as compared to similar regions without gas porosity. Consequently, the local solidification rate is lower in such region, leading to shrinkage porosity formation. In the following section, quantitative microstructural clustering data are presented to characterize the spatial clustering and correlations between the gas pores and the associated gas induced shrinkage porosity.

Reconstruction and Visualization of 3D Microstructure

It is very essential to reconstruct the 3D microstructure to understand the complex real 3D geometry and topological connectivity of porosity. In the present work, the 3D microstructure visualization has been done using 102 montage serial sections; each montage serial section containing 9 contiguous microstructural fields grabbed at 500× with a 2μm section depth. The resolution of reconstructed 3D microstructural image is 2 μm in the Z-direction (which is the distance between serial sections), and 0.2 μm in the X- and Y-directions (i.e., in the planes of the serial sections). Figure 2 shows a stack of aligned 20 montage serial sections of the cast microstructure. Figure 3a shows surface rendered 3D microstructure depicting only the pores. This 3D image contains the central large gas pore and the surrounding shrinkage porosity.

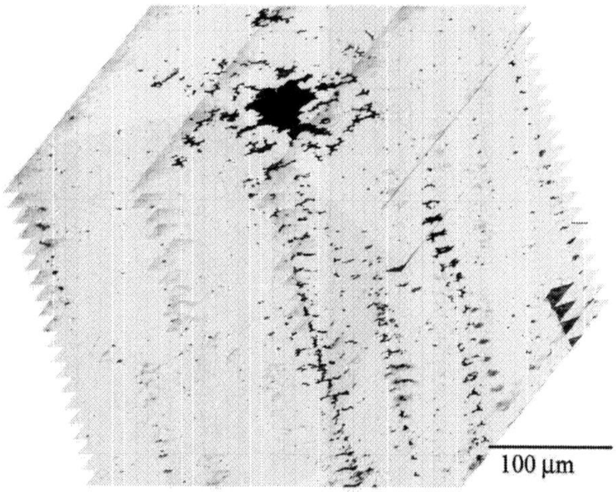

Figure 2. Stack of 20 montage serial sections for the specific region containing a large gas pore and surrounding gas induced shrinkage pores.

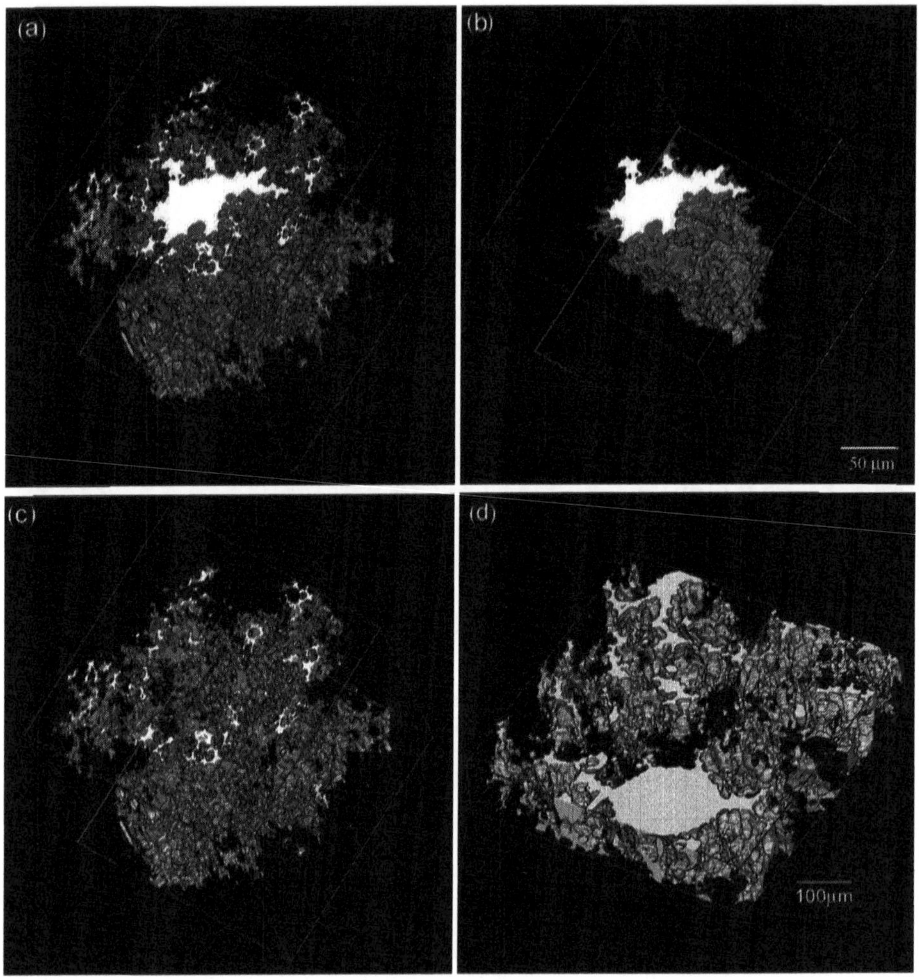

Figure 3. (a) Surface-rendered reconstructed 3D microstructure of gas induced shrinkage porosity clustered around central gas pore, (b) view of the center gas pore from the porosity shown in Figure 3a, (c) view of gas induced shrinkage porosity in Figure 3a; the Figure 3b and 3c can be combined to produce the image in Figure 3a, and (d) 3D microstructure showing conventional shrinkage porosity usually connected to gas porosity.

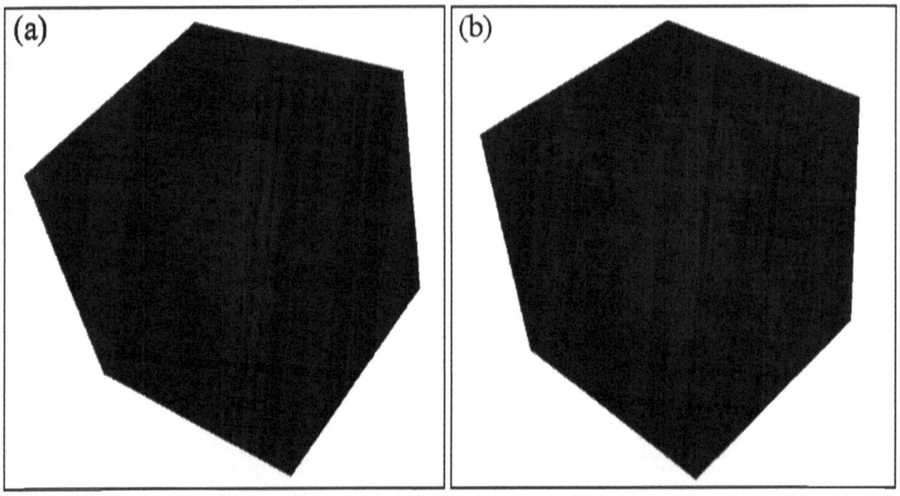

Figure 4. (a) Local stress distributions for only gas pore and (b) gas and shrinkage pores.

It needs to be shown that the shrinkage porosity in Figure 3a is not topologically connected to the central gas pore. To show this, the central gas pore is separately shown in Figure 3b. This image contains only those bright voxels (3D pixels) that are connected together, i.e., one can traverse from any one bright voxel in this image to any other bright voxel without ever entering any dark voxel. Note that this segmentation of the central gas pore is done *automatically* via image analysis. Such separation is possible *if and only if* corresponding shrinkage porosity is not connected to the gas pore. *The image analysis software does not permit such separation if the shrinkage porosity were connected to the central gas pore.* The image processing also permits separate visualization of the gas induced shrinkage porosity in the three-dimensions as shown in Figure 3c. Such clear separation of gas and shrinkage porosity is not possible if the shrinkage porosity is connected to the gas pore. Figure 3d shows 3D microstructure for conventional porosity image depicting the complexity of the porosity shape and random spatial arrangement of the gas (air) and shrinkage porosity [16]. In Figure 3d, most of gas (air) and shrinkage pores are connected, except a few that is in very small size ranges, which is very different from porosity distribution in Figure 3a. *These 3D microstructural observations clearly reveal the presence of gas induced shrinkage porosity, i.e., microstructure contains shrinkage pores that are clustered around gas pores but not topologically connected to the corresponding gas pore(s).* This type of gas induced shrinkage porosity is of concern not only because of an increase in total amount of porosity, but also because it leads to higher local stress concentrations around the gas pores as demonstrated in the next section.

Three-Dimensional Finite Element Based Analysis

The surface rendered 3D microstructure in STL (stereo lithography) format is exported into Harpoon (CEI), commercial automatic meshing software to mesh the geometry with tetrahedral elements. Gas and shrinkage porosity were combined to study the effects of gas induced shrinkage porosity on the local stress distribution. The FE meshes are then exported to ABAQUS, commercial finite element code and analyzed for the elastic mechanical response to a tensile load. Figure 4 shows the local von Mises stress distribution in the microstructural volume due to an applied uniaxial tensile load. These figures clearly demonstrate that the local stress distribution is significantly affected by the presence of gas induced shrinkage porosity: the local Von Mises stresses are higher in the regions around gas induced shrinkage porosity. This is the reason why the gas induced shrinkage porosity can adversely affect the fracture related mechanical properties of the HPDC Mg-alloy casting. In the next section, heat transfer simulations are presented to show the effect of gas pores on local heat transfer and solidification rate that is responsible for formation of gas induced shrinkage pores.

Numerical Simulation for Transient Heat Transfer

Figure 5 shows the temperature distribution in cast plate and depicts that the local temperature distribution is changed due to the presence of gas pore. Compared to the center region, the temperature near the large gas pore is much higher, and consequently that region is expected to solidify last leading to the formation of shrinkage porosity. This simulation is very simplified in that thermal radiation and convection within the gas pore are assumed to be negligible and other thermal effects of eutectic phases are also not considered. However, this result shows the feasible effect of gas pore in the liquid melt on the local heat transfer and solidification rate.

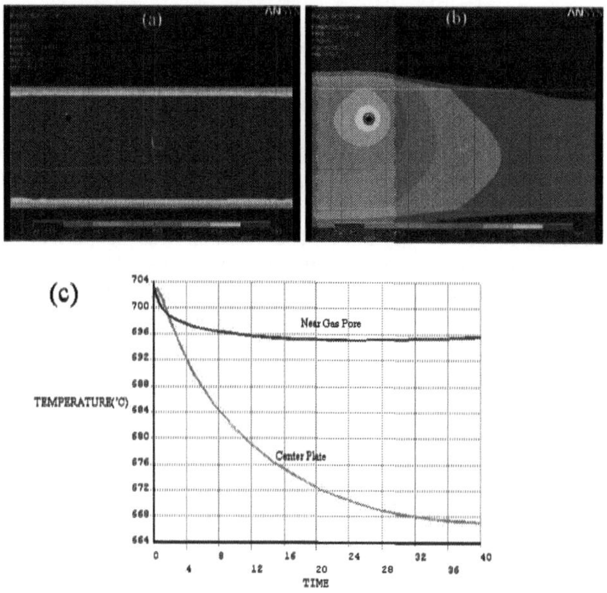

Figure 5. (a) Temperature distribution at time=0sec, (b) at time=22sec, and (c) Comparison of temperature profiles between near the gas pore region and center of the plate region.

Formation of gas induced shrinkage porosity has some distinctive dependencies on several factors essentially affecting the heat transfer and local solidification rate during the casting such as component thickness, alloy compositions, and process conditions. More detail study on the effects of these factors is currently under investigation.

Summary and conclusions

The metallographic evidence is presented for existence of gas induced shrinkage porosity using 2D and 3D microstructural observations. The metallographic evidence is further substantiated using heat transfer simulations. Further, the importance of gas induced shrinkage porosity on the local stress distribution is studied. These experimental data and numerical simulations lead to the following important observations and conclusions.

- The presence of gas porosity in the casting components can lead to the formation of a new type of shrinkage porosity, gas induced shrinkage porosity, under specific conditions. A numerical simulation for heat transfer during the casting process supports that the presence of gas porosity can retard the heat transfer and change the local solidification rate leading to the formation of gas induced shrinkage porosity.
- The visualization of the reconstructed 3D microstructures yields the real morphologies of the gas and shrinkage porosity, their connectivity, and spatial correlations, which further support conclusions drawn from the numerical analysis and 2D microstructural observations
- It is illustrated that the reconstructed 3D microstructures can serve as a representative volume element in the FE-based computational simulations of mechanical

response. The FE simulations show that presence of shrinkage porosity around gas pores can significantly increase the local stresses in such regions that can adversely affect the fracture sensitive mechanical properties of HPDC castings.

Acknowledgement

The authors thank Richard Osborne and Don Penrod for discussions, encouragement, and support, and Mike Evans of Gibbs Die Casting Company for providing the HPDC plate castings of AM60 Mg-alloy. The research was supported by research grants from Structural Cast Magnesium Development (SCMD) program of USCAR Project and the U.S. National Science Foundation (grant no. DMR-0404668), and American Foundry Society. The financial support is gratefully acknowledged. Any opinions, findings, and conclusions or recommendations expressed in this contribution are those of the authors and do not necessarily reflect the views of the funding agencies. The authors thank Mr. Kent Misegades, President, CEI for providing with an evaluation copy of Harpoon.

References

1. B. L. Mordike and T. Elbert: *Materials Science Engineering*, 2001, Vol. A302, p. 37-45.
2. B. R. Powell, L. J. Ouiment, J. E. Allison, J. A. Hynes, R. S. Beals, L. Kopka, and P. P. Reid: *Proceedings of Magnesium Technology* 2004, A. A. Luo, eds. TMS, Warrendale, PA, 2004, p. 3-10.
3. A. Balasundaram and A. M. Gokhale: *Materials Characterization*, 2001, Vol. 46, p. 419-426.
4. A. K. Dahl, S. Sannes, D. H. St. John, and H. Westengen: *Journal of Light Metals*, 2001, Vol. 1, p. 99-103.
5. B. J. Coultes, J. T. Wood, G. Wang, and R. Berkmortel: *Proceedings of Magnesium Technology 2003*, H. I. Kaplan, eds, TMS, Warrendale, 2003, p. 45-50.
6. A. M. Gokhale and G. R. Patel: *Proceedings of Magnesium Technology 2001*, J. Hryn, eds, TMS, Warrendale, PA, 2001, p. 195-199.
7. Y. J. Huang, B. H. Hu, I. Pinwill, W. Zhou, and D. M. R. Taplin: *Materials and Manufacturing Processes*, 2000, Vol. 15, p. 97-105.
8. W. P. Sequeria, M. T. Nurray, G. L. Dunlop, and D. H. StJohn: *Proceedings of Automotive Alloys 1997*, TMS, Warrendale, PA, 1997, p. 169-187.
9. D. F. Allsop and D. Kennedy: *Pressure Die Casting*, The technology of the Casting and the Die, Pergamon Press, Oxford, UK, 1983, p. 216-219.
10. A.M. Gokhale and G. R. Patel: Aluminum 2002: *Proceedings of TMS Symposium on Automotive Alloys 2002*, Michael Skillinberg and S. Das, eds, TMS, Warrendale, 2002, p. 65-73.
11. E. M. Gutman, Y. B. Unigovski, A. Eliezer, E. Abramov, and E. Aghion: *Materials Technology*, 2001, Vol. 16, p. 126-132.
12. H. T. Gjestland, S. Sannes, H. Westerngen, and D. Albright: *Die Casting Engineer*, July 2004, p. 56-65.
13. S. G. Lee, A. M. Gokhale, G. R. Patel, and M. Evans, submitted to *Metallurgical and Materials Transactions B.*
14. A. Tewari, A.M. Gokhale, and R. M. German: *Acta Materialia*, 1999, Vol 27, p. 3721.
15. A. M. Gokhale and A. Tewari: *Journal of Microscopy*, 2000, Vol 200, p. 277.
16. Soon Gi Lee, Arun M. Gokhale, and Arun Sreeranganathan, submitted to *Materials Science and Engineering A*.

THE ROLE OF MICROSTRUCTURE AND POROSITY IN DUCTILITY OF DIE CAST AM50 AND AM60 MAGNESIUM ALLOYS

Gurjeev Chadha[1], John E. Allison[2], J. Wayne Jones[1]
[1]Department of Materials Science and Engineering, University of Michigan, 2300 Hayward St., Ann Arbor, MI 48105, USA
[2]Materials Research Laboratory, Ford Motor Company, 20000 Rotunda Drive, Dearborn, MI 48124 USA

Keywords: die cast, magnesium alloys, fracture, ductility, porosity, voids, microstructure, failure, bending, archimedes.

Abstract

The deformation and fracture behavior of die-cast AM50 and AM60 magnesium alloys have been examined to determine the processes leading to fracture in bending and tension and to elucidate the influences of microstructure and porosity on ductility. Damage accumulation in terms of crack initiation, growth, linking and eventual failure was quantified as a function of strain for as-cast plates with section thicknesses of 2, 6 and 10 mm. Microstructure and porosity distribution are dependent on section thickness, with bands of high porosity resulting from the nature of the die filling process. In both tensile and three-point bend studies, damage accumulation and subsequent ductility at fracture are strongly influenced by the heterogeneous porosity distribution, with early crack formation occurring in regions of highest porosity. A model is presented that relates the observed ductility of these alloys with porosity distribution, crack size, crack tip radius, and elastic modulus.

Introduction

Magnesium alloys have been receiving considerable attention from the automobile industry because of their combination of good mechanical properties and low density make them important alloys for automobile weight reduction efforts [1-3]. For applications demanding good ductility the Mg-Al-Mn (AM series) alloys are used, whereas, for strength sensitive applications the AZ91D and other alloys are preferred [1,3,4].

The present study investigates two of the prominent high pressure die cast alloys, AM50 and AM60 that are used in applications requiring good ductility. The mechanical properties of these alloys are well characterized with much data correlating mechanical properties with bulk porosity reported [5-7]. However, an understanding of the effect of section thickness variation on the mechanical properties and the role of microstructure on ductility is incomplete [3,6,8-11]. The high pressure die casting process is characterized by rapid die filling, cooling and non-equilibrium solidification of the metal in the die. This leads to formation of bands where porosity is distributed heterogeneously with deleterious effects on strength and ductility [12-14]. This study examines the phenomenon of damage accumulation during deformation, with attention to the role of heterogeneous regions of porosity on the failure process. Section thicknesses of 2 mm, 6 mm and 10 mm for both AM50 and AM60, were examined. These thicknesses were selected because they are representative of the section thickness variation in die cast automotive components.

Material

Plates were die cast by Spartan Light Metals using a 700 ton cold chamber high pressure die casting unit. The melt temperature was 677 °C and the die temperature was 95 °C. All die cast plates were 126 mm long and 26 mm wide. Three thicknesses (2, 6, and 10 mm) were examined for both AM50 and AM60 (Figure 1). The composition of these alloy plates was determined using inductively coupled plasma technique and the results are shown in Table I.

Microstructure was investigated by optical and scanning electron microscopy on metallographically prepared samples. The final polishing step used a 0.05 μm alumina powder suspension in isocut fluid on a Chemomet® pad, followed by etching in acetic glycol. Bulk porosity measurements were performed on regions of interest for all the plates using Archimedes method. The liquid medium used in this study was chemical grade 2-propanol solution, under stable ambient temperature conditions.

Table I: Composition (in wt %) for all Three Die Cast Plates for AM50 and AM60

Elemental Comp. (in wt %)	AM50			AM60		
	2 mm	6 mm	10 mm	2 mm	6 mm	10 mm
Al	5.16	5.04	4.95	6.07	6.07	6.12
Mn	0.36	0.35	0.35	0.34	0.33	0.33

Mechanical Testing

Tensile tests and progressive strain tests were performed on specimens machined from rectangular plates to understand the phenomenology of damage evolution. Bending as a loading mode has not been extensively studied in these alloys [15]. Three-point bend tests were done to investigate and elucidate how microstructure controls ductility in these alloys under this important loading mode.

Tensile Testing

Specimens were machined from flat plates for all three section thicknesses of 2, 6 and 10 mm for both AM50 and AM60 (Figure 1). At least 8 tensile tests were conducted for each thickness at a strain rate of $5x10^{-4}$ s^{-1} on an Instron 5582 at room temperature. Strain levels were measured using automated data capture with a 25.4 mm gage length extensometer.

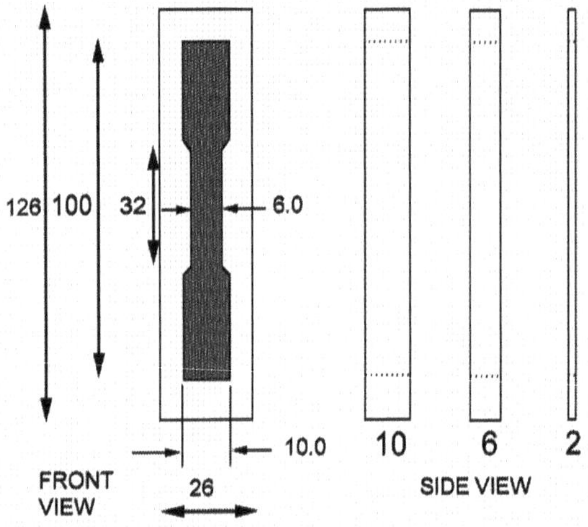

Figure 1: Tensile Test Specimens machined out from the plates. A common front view and side views for all three thicknesses are shown here (all dimensions in mm).

Progressive Straining

Surface observations of damage accumulation were made on surfaces perpendicular to the thickness direction at increasing strain levels. The 6 mm thick plates were selected for this study because they showed higher ductility and low scatter. It was assumed that the microstructure did not vary significantly from plate to plate in either alloy. Based on this assumption, different samples were strained to each of the strain levels and sectioned for examination.

The incremental strain levels selected for detailed observation were approximately 1%, 3% and 5%. The gauge sections were cut from the three tensile specimens and the side-surface perpendicular to the thickness direction was metallographically polished and observed in a Phillips XL30FEG SEM in the back scattered mode (Figure 2). This approach provided a more comprehensive phenomenological understanding of the damage accumulation steps and a better understanding of the behavior of the porosity segregation zone and skin region under loading.

Three Point Bend Studies

Three point bend tests were performed on the 6 mm and 10 mm thick plates in the as-cast condition. A simple bend test fixture, shown in Figure 3 was used in an INSTRON 5582 for this study. Failure in bend tests was defined as a 5% drop in the load levels, unlike for tensile tests where fracture terminated the tests. Tests for both 6 mm and 10 mm section thicknesses (both alloys) were stopped after the 5% load drop showed presence of surface cracks on the tensile side of the plate surface.

The 2 mm plate did not fail in this manner due to the limitations posed by the test fixture causing tests to be terminated well before failure.

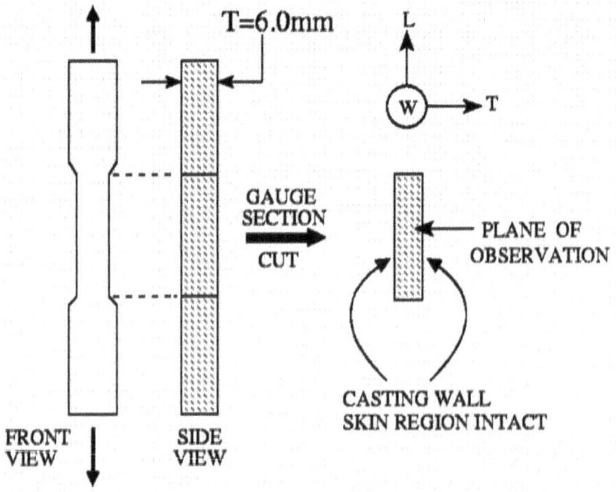

Figure 2: Schematic showing sample preparation for observing progressively strained samples at 1%, 3%, and 5%. Plane of observation denotes the face metallographically polished and examined.

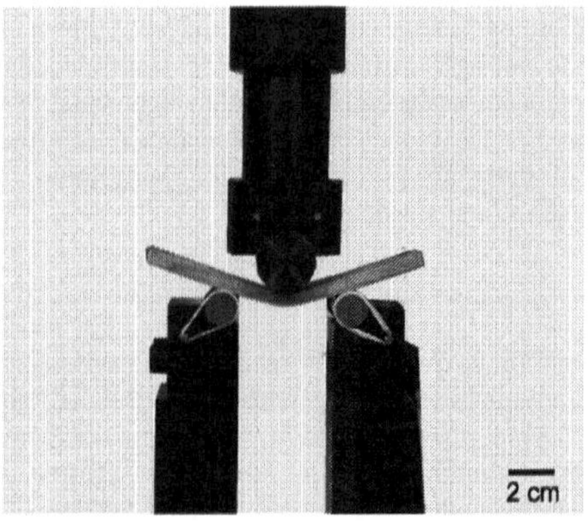

Figure 3: Three point bend setup

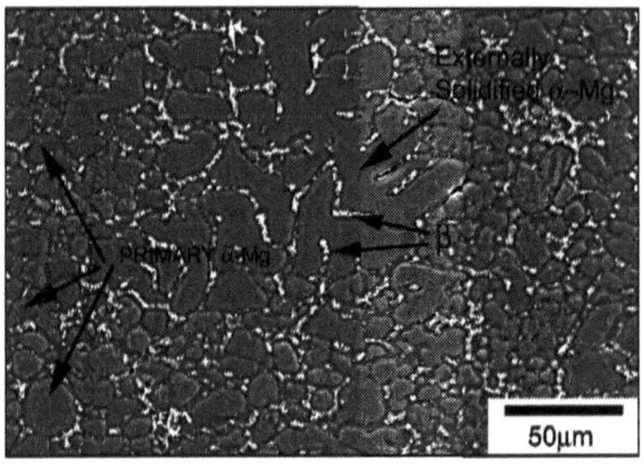

Figure 4: Photomicrograph showing the primary microstructural phases in these alloys

Results and Discussion

Microstructure

Microstructure of these alloys, shown in Figure 4 consists of primary α–Mg dendrite cells, externally solidified α-Mg dendrites, and a divorced-eutectic (α–Mg + β-$Mg_{17}Al_{12}$).

Heterogeneous bands of microporosity were observed in the plate interior and a low porosity, fine grained region, normally termed the skin was observed at all casting surfaces (Figure 5). The nature of pores observed in all castings was primarily shrinkage, either as large shrinkage pores or small interdendritic micropores, concentrated at the center for all the castings. Hence, the word "porosity" is used without differentiating between shrinkage, microshrinkage, and gas related porosity.

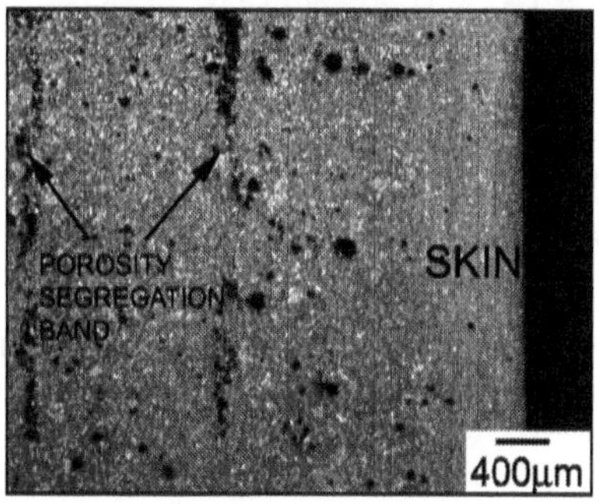

Figure 5: Porosity segregation band as seen AM50-6 mm plate, note the pore free skin region

The volume fractions of porosity measured on gauge sections using Archimedes Method are given in Table II. The values shown are averages of 3 sections cut from unstrained samples for each thickness. As expected, the volume fraction of porosity increased with increasing section thickness as longer solidification time for higher plate thicknesses results in inefficient inter-dendritic feeding of metal. However, as stated earlier, all the porosity was concentrated at the center region of the castings with little or no porosity in the skin region.

Table II: Average Bulk Porosity Levels in the Gauge Section for all Three Die Cast Plates for AM50 and AM60

Plate Thickness (mm)	Porosity (vol %)	
	AM50	AM60
2	0.9 ± 0.1	1.0 ± 0.2
6	1.3 ± 0.2	1.5 ± 0.3
10	3.3 ± 0.2	3.6 ± 0.2

The center regions of all three thicknesses for the AM50 plates are shown in Figure 6. Size of pores increases significantly as the section thickness increases which relates well with the increasing volume fraction of porosity. The 2 mm and 6 mm AM60 plates had no large pores as observed in AM50 plates which had a pore rich center region (compare Figure 6(b) and 7). This suggested that the porosity was of a more distributed nature in the AM60 2 mm and 6 mm plates. Porosity measurements were also made on the whole cross-section for the 6 mm and 10 mm plates to correlate failure strains in bending with the volume percentage of porosity. Sections were taken at half length of the plates, corresponding to the region below the center-pin in bend tests. Similar volume fractions of porosity were recorded for both AM50 and AM60, 1% for the 6 mm plate and 2.5% for the 10 mm plate.

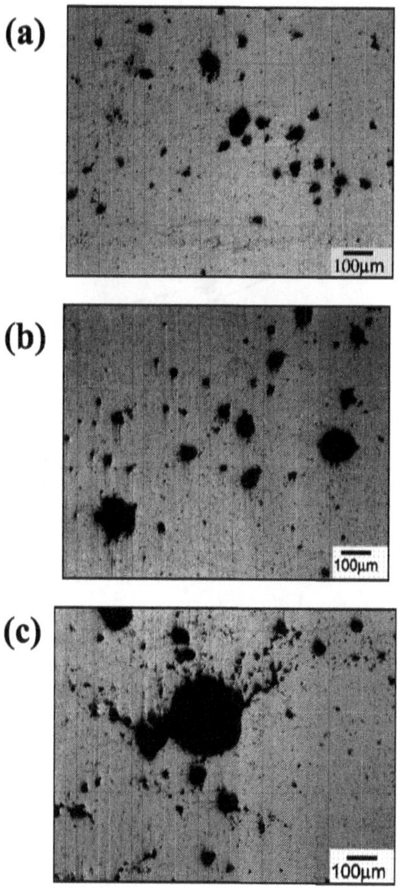

Figure 6: Center region of the (T) x (W) cross-section in AM50 alloys plates (a) 2 mm, (b) 6 mm, and (c) 10 mm

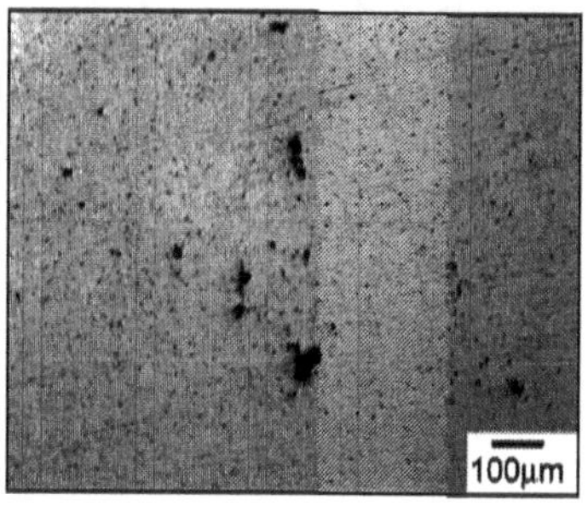

Figure 7: Small Pores seen in AM60-6 mm plates compared to AM50-6 mm, porosity is more distributed in AM60 plates.

Mechanical Properties

Both strength and ductility decrease linearly as the section thickness increases for both AM50 and AM60. Ductility variation with increasing plate thickness is shown in Figure 8. The relative decrease in mechanical properties normalized with respect to the 2mm thick plate properties is shown for both AM50 and AM60 in Figure 9. The mechanical properties in this study (for both AM50 and AM60) are slightly lower than that reported for these alloys in the literature and is likely a direct effect of the loss of the fine grain skin regions on each side of the tensile test specimens. Also, the ductility of the AM60 die cast plates is higher than that for the

AM50 plates for the 2mm and 6mm thicknesses, although AM50 would be expected to have a higher ductility [1,4]. The difference may be a direct consequence of the more highly localized regions of porosity in the AM50 castings (Figures 5 and 6)) compared to the more uniformly distributed porosity in the AM60 plates (Figure 7).

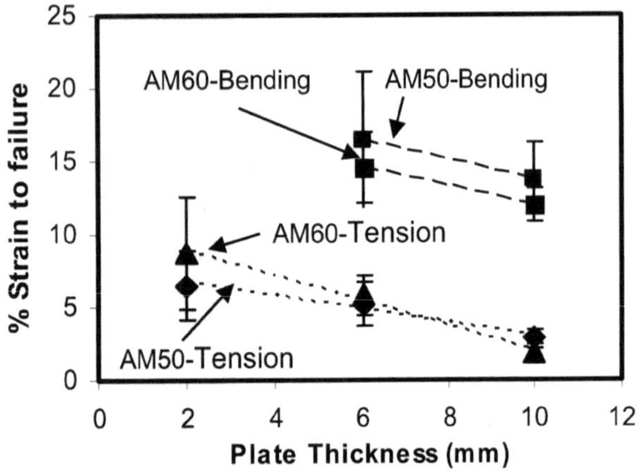

Figure 8: Variation of tensile ductility and bend failure strains with plate thickness for both AM50 and AM60

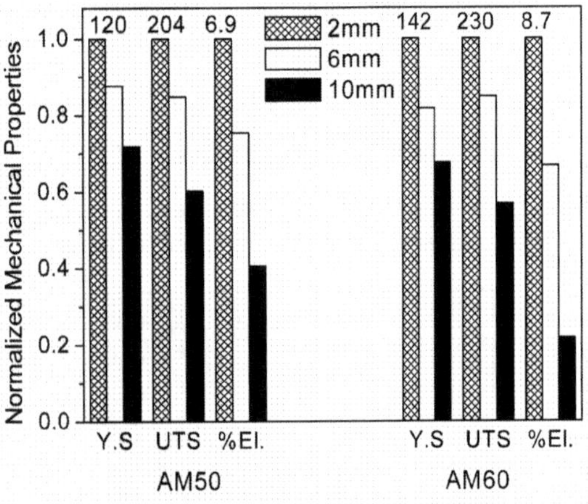

Figure 9: Relative Decrease in YS, UTS, and %Elongation with increase in section thickness normalized w.r.t. 2 mm thick plate properties (Numbers on top correspond to 2 mm plate specimens).

The variation of strain to fracture with bulk porosity parameter is shown in Figure 10. For both AM50 and AM60 die cast plates, strain to fracture decreases with increasing volume fraction of porosity. Again it should be noted that AM60 tensile specimens show higher ductility than AM50 for porosity fractions less than 2.0%, which corresponds to the 2 mm and 6 mm thick plates (Table II). This indicates that the role of other microstructural features (or porosity attributes) in limiting ductility in these alloys at low porosity may be important.

Higher ductility, measured at a 5% load drop, is observed for both AM50 and AM60 alloy plates under bending compared to tension for both 6mm and 10mm thick plates (Figure 8 and 10). This is not surprising because the region under highest strain and

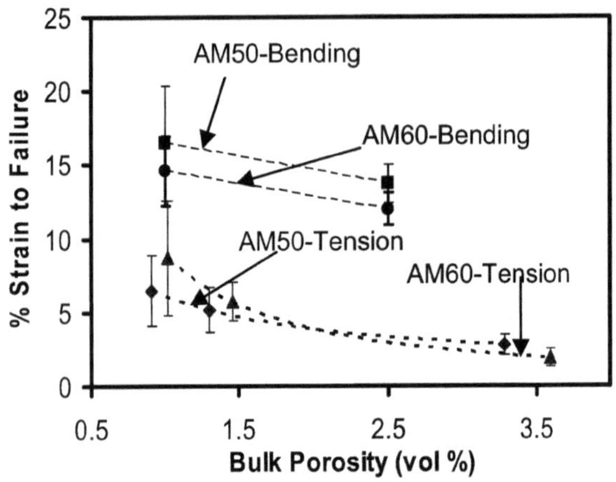

Figure 10: Variation of tensile ductility and bend tests strains to failure with bulk porosity for both AM50 and AM60

stress is a low defect region with almost neglible porosity. The strains to failure were calculated using the radius of curvature of bend tested plates. This gave an approximate lower bound of the actual plastic strains at failure for both 6mm and 10mm thick plates. However, even this strain value is almost 3 times larger than the tensile ductility observed for these alloys.

Phenomenology of Damage Evolution

The progressive straining tests clearly elucidated the response of macro-features such as the porosity segregation bands and the pore-free skin region to damage accumulation during straining. Damage accumulation, qualitatively defined by number and size of cracks observed were higher in AM50 than in AM60 at each strain level (Figure 11). This could directly explain the slightly lower ductility values observed for AM50 6 mm thick plates. The results presented here elucidate phenomenon occurring at a macro scale, a more detailed study of crack initiation and void linkage will be part of a separate communication.

It was observed that damage accumulation occurred throughout the gage length of the specimens at multiple locations and was localized primarily in the porosity segregation zone. Similar observations were also made for AM60 but the intensity of damage, again qualitatively defined in terms of the number and size of cracks seen on the surface after 1% and 3% straining was lower compared to AM50. For the 5% strained sample a similar level of damage was recorded and again localized in the porosity segregation bands that are a result of rheological characteristics of die filling process [13].

In order to understand the phenomenology of deformation during bend failures, sections were taken from failed plates to observe cracks and how they propagate through thickness for both 6 mm and 10 mm plates. For 6 mm thick plates, in both AM50 and AM60 failure took place at the outer skin region and propagated inwards. Multiple cracks were also formed in the low strength porosity band (Figure 12(a)) but they did not cause failure. For the 10 mm thick plates for both AM50 and AM60 sharp cracks from pores grew and joined to form a longer crack in the porosity rich interior which then propagated towards the outer skin causing failure instead of failure initiating at the skin. This is explained by the high concentration of pores in the interior of 10 mm thick plates compared to the 6 mm plates which made the interior crack earlier than the edge in this case (Figure 12 (b)).

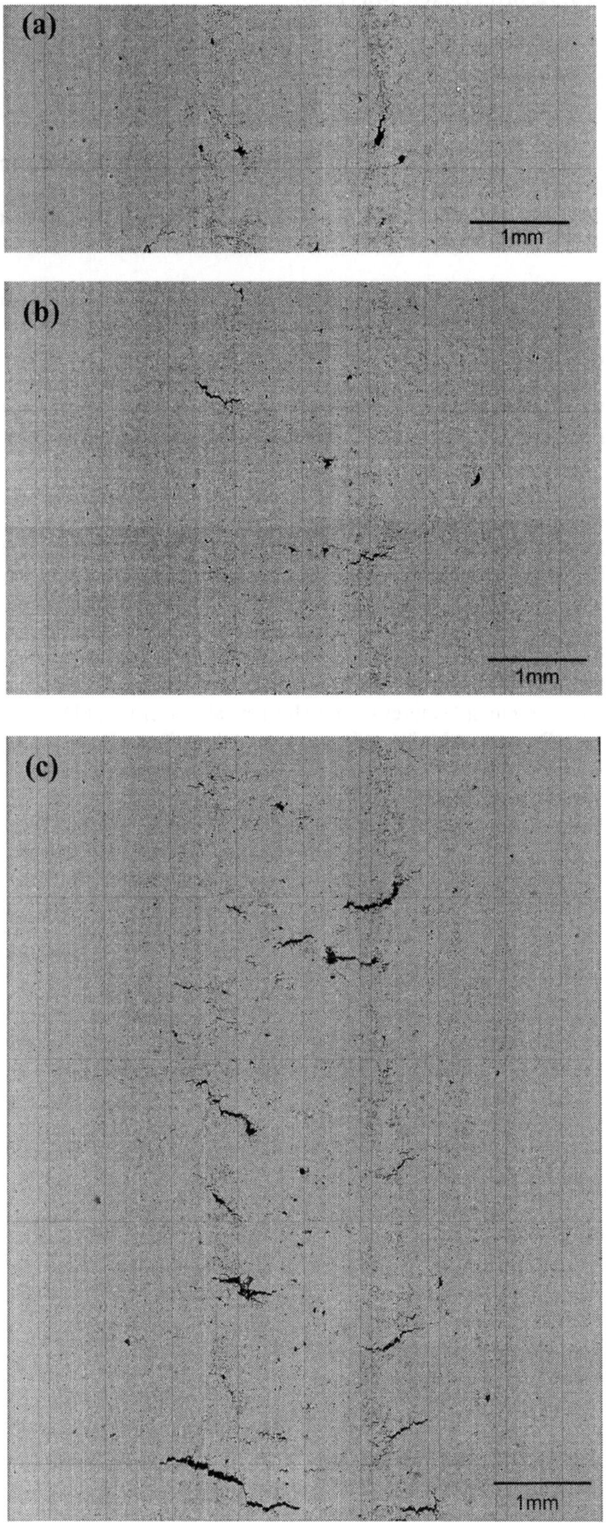

Figure 11: Photomicrographs showing damage accumulated in 3 different AM50 test specimens after different strain levels of (a) 1% (b) 3% and (c) 4.5%.

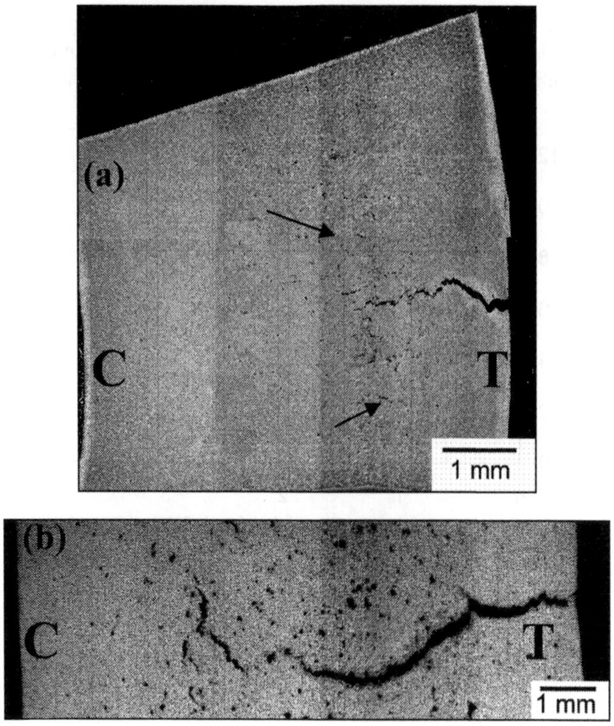

Figure 12: Representative photomicrographs showing different active zones of failure in (a) 6 mm AM50 plate with failure occurring at the outer fiber and a (b) 10 mm AM60 plate showing cracks propagating from the plate interior to outside fiber (Note: C denotes Compressive side and T: Tension side in bend specimen)

Model

Based on these phenomenological observations it is clear that cracks are initiated even at low strains due to micropore coalescence in the heterogeneous regions of porosity. Cracks are formed at multiple locations and eventually a few of these grew, coalesced and led to eventual failure. This understanding has been applied to a modified Brown-Embury model that analytically predicts fracture strain from information such as, crack size, crack-tip radius, volume fraction of porosity, and elastic modulus of the alloy [16]. The model assumes cracks will propagate until the distance between the two cracks is equal to the critical plastic zone size. Irwin's approach to elastic-plastic deformation is used to obtain this analytical expression for ductility.

$$\varepsilon_f^P = \frac{2\sigma_y^P r}{E_P \rho}\left[\left(\frac{2\pi}{3f_P}\right)^{1/2} - \frac{\pi}{2}\right] \quad (1)$$

ε^P_f: Fracture strains of porous alloy, σ^P_y: Yield stress of porous alloy, r: mean pore radius, ρ: crack tip radius, f_P: Volume fraction of pores, E_P: Elastic Modulus of porous alloy.

The analytically obtained values agree well with the tensile test data for both AM50 and AM60 (Figure 13). Importantly, the shape of trend predicted by the model (dotted line) follows closely the experimentally obtained non-linear correlation between ductility and the volume fraction of porosity. The values of modulus of elasticity and yield stress were obtained from tensile test. The value of r/ρ was taken as 100 for both AM50 and AM60 plates in these calculations. These values offer a good fit considering the fact that cracks sizes (r) of 300-400 μm and crack tip radius (ρ) of 3-4 μm are reasonable estimates for these alloys.

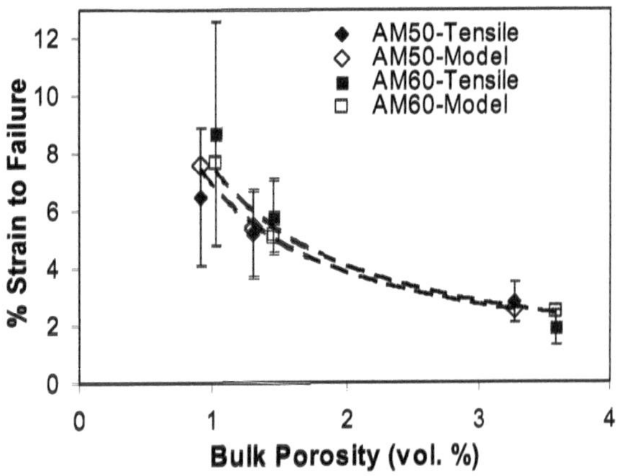

Figure 13: Tensile Ductility as a function of pore volume fraction. Compare values obtained from tensile testing and the analytical model (dotted line) for both AM50 and AM60

Conclusions

a) Increases in section thickness lead to a considerable decrease in ultimate tensile strength, yield strength and ductility in both tensile and bend loading conditions.
b) Heterogeneous distributions of porosity resulting in the formation of porosity segregation bands have the most deleterious effect on properties. They limit the ductility in these alloys by forming sharp cracks due to micro-porosity coalescence during early stages of straining which then grow in size to meet other cracks leading to eventual failure.
c) Significant improvement in strain to fracture occurs in bending and results from the lower strains in the high porosity regions, which are closer to the neutral axis.
d) There is a corresponding change in the damage accumulation sequence in bending between 6 mm and 10 mm thick plates which is a direct result of the size and distribution of the heterogeneous porosity. The 6mm plates failed at the outer skin region, unlike the 10 mm plates which failed in the pore rich interior.

Acknowledgement

Support for this research came from the Ford Motor Company and is gratefully acknowledged.

Reference

1. D. J. Sakkinen, "Physical Metallurgy of Magnesium Die Cast Alloys", *SAE Technical Paper*, 940779, SAE, Detroit, MI, (1994).

2. B.L. Mordike and T. Ebert, "Magnesium: Properties-applications-potential", *Materials Science and Engineering*, A302 (2001), 37-45.

3. H. Mao et.al., "Microstructural Characteristics of Die Cast AZ91D and AM60 Magnesium Alloys", *SAE Technical Paper*, 1999-01-0928, SAE, Detroit, MI, (1999).

4. T.Kr. Aune et.al., "Properties of Die Cast Magnesium Alloys of varying Aluminum Content", *SAE Technical Paper*, 900792, SAE, Detroit, MI, (1990).

5. C.D. Lee, "Effect of Microporosity on Tensile Properties of As-Cast AZ91D Magnesium Alloy", *Metals and Materials International*, 8(3) (2002), 283-288.

6. A.L. Bowles et.al., "Ductility and the Skin Effect in High Pressure Die Cast Mg-Al Alloy", *Magnesium Technology 2001*, ed. J. Hryn (New Orleans, LA: The Minerals, Metals and Materials Society TMS, 2001), 161-168.

7. Z. Liu et.al., "The Influence of Porosity of Magnesium Die Casting Alloys on its Mechanical Properties", *Metall*, (54) (2000), 122-125.

8. T. Abott, M. Easton, and W. Song, "Mechanical Behavior of Cast Magnesium Alloys", *Materials Science Forum*, 419-422 (2003), 141-146.

9. A. Stich, "Dimensions Strategy for High Stress Cast Magnesium Components", *Magnesium 2000*, ed. E. Aghion and D. Eliezer (Dead Sea, Israel: Proceedings of the Second Israeli International Conference on Magnesium Science and Technology, 2000), 27-34.

10. D. Rodrigo et.al., "Effects of Section Size and Microstructural Features on the Mechanical Properties of Die Cast AZ91D and AM60B Magnesium Alloy Test Bars", *SAE Technical Paper*, 1999-01-0927, SAE, Detroit, MI, (1999).

11. G. Schindelbacher and R. Rosch, "Mechanical Properties of Magnesium Die Casting Alloys at Elevated Temperatures and Microstructures in Dependence of Wall Thickness", *Werkstoff-Informationsgesellschaft*, ed. B.L. Mordike and K.U. Kainer (Wolfsburg, Germany: Magnesium Alloys and their Applications, 1998), 247-252.

12. S. Sannes et.al., "Die Casting of Magnesium Alloys-The Importance of Controlling Die Filling and Solidification", *SAE Technical Paper*, 2003-01-0183, SAE, Detroit, MI, 2003.

13. A.K. Dahle et.al., "Formation of Defect Bands in High Pressure Die Cast Magnesium Alloys", *Journal of Light Metals*, 1 (2001), 99-103.

14. P.D.D. Rodrigo and V. Ahuja, "Effect of Casting Parameters on the Formation of Pore/Segregation Bands in Magnesium Die Castings", *Magnesium 2000*, ed. E. Aghion and D. Eliezer (Dead Sea, Israel: Proceedings of the Second Israeli International Conference on Magnesium Science and Technology, 2000), 97-104.

15. B.J. Coultes et.al., "Mechanical Properties and Microstructure of Magnesium High Pressure Die Castings", *Magnesium Technology 2003*, ed. H.I. Kaplan (San Diego, CA: The Minerals, Metals and Materials Society TMS, 2003),45-50.

16. J.Y. Jung, "Prediction of tensile ductility in porous solids", *Philosophical Magazine A*, *82*(11) (2002), 2263-2268.

SEMISOLID PROCESSING AND ITS APPLICATION TO MAGNESIUM ALLOYS

Frank Czerwinski
Development Engineering, Husky Injection Molding Systems Ltd., Bolton, Ontario, L7E 5S5 Canada

Keywords: semisolid processing, magnesium alloys, injection molding, Thixomolding®

Abstract

Since its discovery over thirty years ago, semisolid processing is mainly applied to alloys with relatively low melting temperatures, particularly aluminum. Although historically an interest in magnesium reaches back to the early 1970's, compared with aluminum, investigations of semisolid magnesium alloys are still scarce. This report presents the origin of semisolid processing, rheological behaviour of semisolid slurries and the key technologies, available in today's industry, based on thixo- and rheo-routes. Particular attention is being paid to emerging techniques of injection molding. The major requirements imposed on potential alloys and typical components, manufactured commercially, are characterized.

Introduction

Magnesium alloys can be transformed into specific products while either in liquid or solid states. Since their hexagonal crystal structure has relatively poor formability at room temperature, preheating is usually required during solid state forming operations including forging, extrusion and rolling. Due to limited solid state workability, casting is often preferred where complex shape components can be produced by nearly all of the conventional gravity and pressure casting methods: sand, permanent and semi-permanent mold and shell, investment casting and die casting. The choice of a method depends upon factors such as the configuration of the design, the application, the properties required, the total number of parts and the castability of the alloy. In indirect squeeze casting, which shows some similarities to low pressure die casting, the gates are replaced with thick ones and rapid die filling is replaced by a much slower rate to avoid turbulence. For magnesium alloys, the turbulence free die filling is difficult to obtain due to quick solidification caused by low heat content.

In recent years, alternative techniques have been commercialized. The most distinct is semisolid processing where net-shape components are manufactured at temperatures between solidus and liquidus. So far, semisolid technologies are firmly established for aluminum alloys. Although some techniques are commercially applied to magnesium, a majority of them were tested only during laboratory trials. The fundamentals of semisolid processing and its application for magnesium alloys are presented in this paper.

Origin of semisolid metal processing

Semisolid metal processing (SSP), sometimes referred to as semisolid metallurgy (SSM), was initiated at The Massachusetts Institute of Technology in 1971 by Spencer, Flemings and co-workers [1]. During experiments with Sn-15Pb alloy they found that applying shear during solidification substantially reduced the stress measured. In fact, the stress at a given temperature below the liquidus was orders of magnitude less than when the alloy was cooled to that temperature without shear. Important for SSP, the phenomenon of thixotropy, however, was discovered about a half of century earlier in 1923 by Schalek and Szegvari [2] in non-metallic systems. The essence of their finding was the remarkable property of aqueous iron oxide gels which were becoming completely liquid through gentle shaking alone to such an extent that the liquefied gel was hardly distinguishable from the original sol. Previously, these kinds of physical changes had only been known to occur by modifying the temperature when gels would melt on heating and then re-solidify on cooling. The term thixotropy was introduced by Peterfi [3] in 1927 as a combination of two Greek words: *thixis*, meaning stirring or shaking and *trepo*, meaning turning or changing. During following years, various systems were studied, including clays, oil suspensions, creams, drilling mud, flour doughs, flour suspensions, fibre greases, jellies, paints and starch pastes. The progress made over the past decades was reviewed by Barnes [4].

Rheological behaviour of semisolid slurries

According to Bauer and Collins [5], the system is described as thixotropic when a reduction in magnitude of its rheological properties, such as elastic modulus, yield stress and viscosity occurs reversibly and isothermally with a distinct time dependence on application of shear strain. The most frequent structural changes that produce thixotropy are those where the structure breaks down under high shear rate but recovers under low shear rate or when at rest. Therefore, thixotropy arises from changes in flock structural arrangement due to forces acting between suspended particles and breakdown due to the shear rate. It was revealed in 1930's that thixotropy was more pronounced in systems containing non-spherical particles because rotation and movement allowed for their alignment and dis-alignment in a three-dimensional structure [6].

Semisolid metal processing represents one of the many engineering applications which utilize thixotropy. It should be pointed out that, although metallic alloys sheared below their melting point are described as thixotropic, in fact, they exhibit mainly shear-thinning behaviour. Shear thinning, also referred to as pseudoplasticity, describes a decrease of viscosity with an increase in shear rate. Both terms are frequently confused since in real systems it is difficult to separate the time dependent and time-independent behaviour of the fluid. An explanation of pseudoplastic behaviour is the formation of agglomerates/clusters of favorably oriented solid particles at low shear rates which

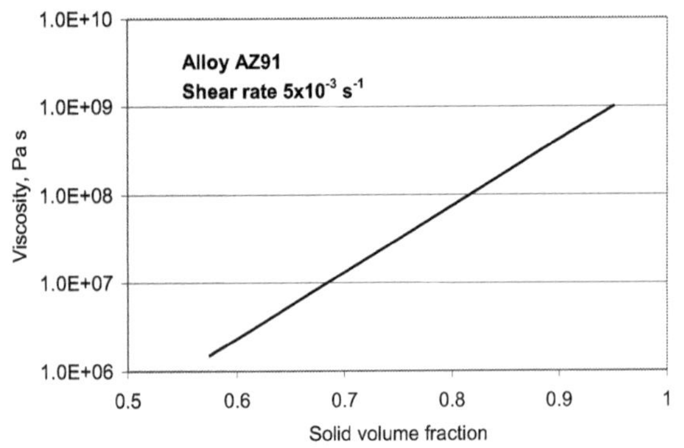

Fig. 1. The apparent viscosity as a function of the solid fraction for AZ91 magnesium alloy [7].

increase viscosity. It is assumed that changes are reversible and high shear rates break down agglomerates, reducing viscosity.

The rheological behaviour of AZ91 magnesium alloy were investigated by Gebelin et al. [7] and the apparent viscosity of the alloy over different shear rates were measured using compression and backward extrusion methods. It was found that from 10^{-3} to 10^{3} s^{-1} the material exhibited the shear thinning behaviour and the apparent viscosity (η) decreased with increasing shear rate (γ) according to a power law equation:

$$\eta = K\gamma^{n} \qquad (1)$$

K in equation (1) is the solid fraction dependant factor. The exponent n was found to equal –0.65 in the low shear range (compression) and to be close to –0.95 in the high shear rate for backward extrusion. When combining all data the average exponent of the shear ratio equaled –0.87, which is similar to values reported by Couette viscometer experiments (Fig. 1).

Benefits of semisolid technologies

The well known processing advantages related to reduced operating temperature are universal for all alloys used: lower energy consumption, no handling of liquid metal, longer die life, better yield from the raw material due to lower oxidation, evaporation and other losses related to melt overheating. The second part of the benefits arises from an improvement in the product's properties. The frequent assumption that the benefits of SSP arise exclusively from the flow behaviour of the partially solidified metal is, therefore, a simplification.

The influence of the semisolid slurry on the component's properties is complex and the major factors are listed in Fig. 2. In general, the slurry affects product integrity throughout the reduction of porosity. The turbulent flow of a liquid alloy into a mold can result in the entrapment of air and mold gases into the melt which in turn may translate to micro and macro porosity. Smooth flow of the semisolid slurry allows for minimizing these defects (Fig. 3a). In contrast, a reduced liquid content within the semisolid slurry reduces shrinkage porosity, which not only improves the material properties, but also allows achieving near-net shape capabilities (Fig. 3b). The high internal integrity makes

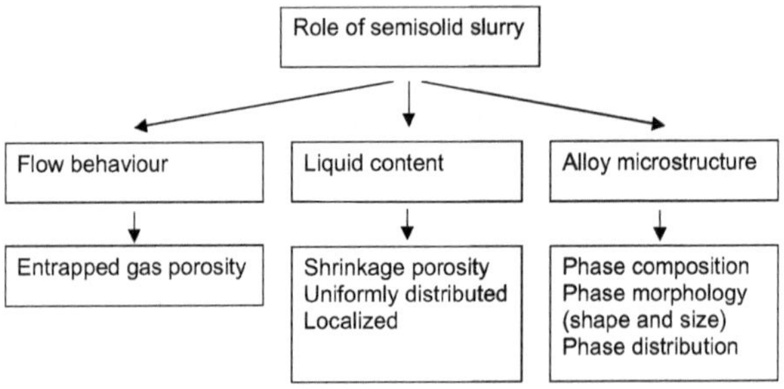

Fig. 2. The role of the semisolid slurry in controlling the integrity and microstructure of components, manufactured by SSP.

a component heat-treatable. As was the case for the general benefits mentioned above, the enhanced component's integrity has a universally positive effect for all alloys.

The presence of globular solid particles within the alloy slurry essentially changes an alloy's microstructure after solidification. The morphological changes are portrayed in Fig. 3c. The new features, typical for semisolid processing, are unmelted globules with a size less than 100 μm. The globules are surrounded by the solidification product of the liquid fraction. Their presence leads to a segregation of chemical elements which translates to a new distribution of phases and microstructural components. It is clear that the presence of an α-Mg rich unmelted solid will affect the composition of liquid and its tendency to the formation of the intermetallic phases, e.g. the $Mg_{17}Al_{12}$ compound in Mg-Al or Mg-Al-Zn alloys. This phase controls properties of magnesium alloys, including strength, ductility and corrosion resistance [8,9]. The influence of globular solids on the direction of changes in alloy properties is not obvious and should be evaluated for individual alloys. The present literature does not separate the structural factor and benefits of the component's integrity for the improvement in properties.

Technologies available commercially and their applicability to magnesium alloys

A number of processes have been developed and commercialized to take advantage of the behaviour of semisolid slurries and they can be divided into two fundamentally different basic routes: the rheo- and thixo-route (Fig. 4). During utilizing rheo-processing, the starting fully molten precursor is pre-solidified under controlled agitation conditions and then transferred into the mold. The thixo route involves two stages: (i) billet preparation, which consists, in fact of a portion of rheo-route; and (ii) billet's re-heating and component forming. As shown in Fig. 4, there are also possible hybrids, combining features of both the thixo and rheo route.

There are several commercial technologies based on the above routes. Thixoforming is a general term used for net shaping. If the process is performed in a closed die, it is called thixocasting. When a mold is open during this operation, the process is called thixoforging. Both technologies involve a step of preheating the previously prepared billet to the temperature of the semisolid range. Its purpose is to provide the material with the precisely controlled solid fraction of fine spherical particles uniformly

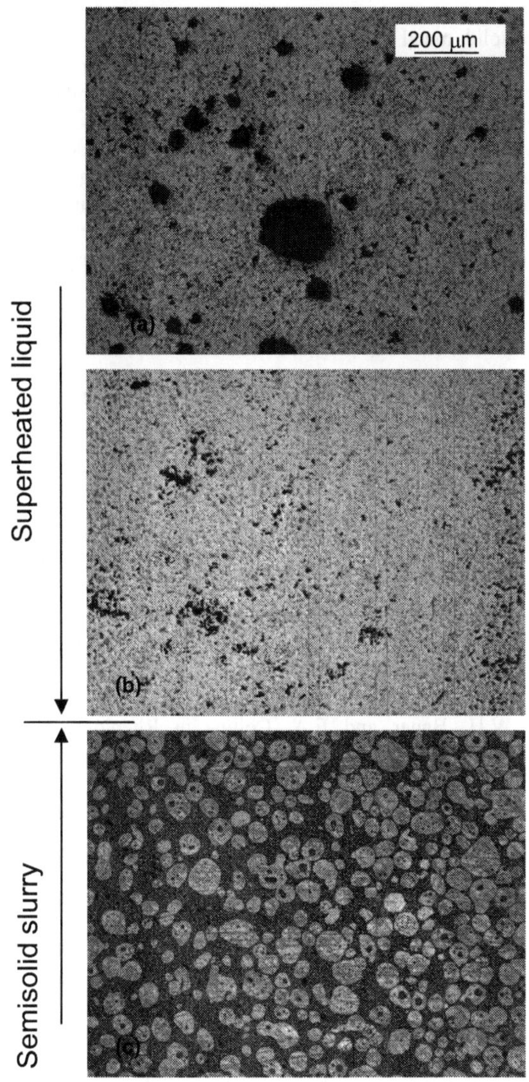

Fig. 3. The structural integrity of the superheated liquid cast and semisolid processed components: (a) AM60B, die casting, entrapped gas porosity; (b) AM60B, die casting, solidification shrinkage porosity; (c) AZ91D, Thixomolding, globular structure.

distributed in a liquid matrix, having a lower melting temperature. Both conventional and induction heating methods are used.

Historically, thixoforming methods were the first implemented commercially, but due to cost disadvantage, they are being recently dominated by rheocasting technologies of several modifications. The major difference among them is the way of melt agitation, applied to avoid the formation of dendritic forms of growth.

A distinct route of SSP is represented by injection molding. The one-step processing is possible due to the specific structure of coarse particulates, used as a feedstock, which transform to thixotropic slurries under the sole influence of heat [10,11]. In principle, all techniques are applicable to magnesium alloys. The

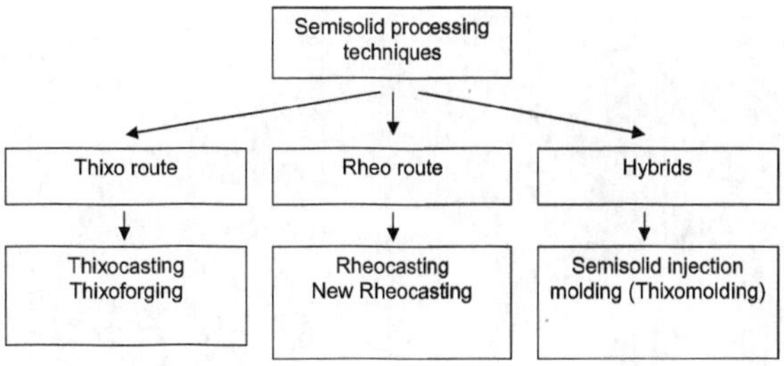

Fig. 4. The basic classification of semisolid processing techniques.

specific nature of magnesium and its extremely high affinity to oxygen, favours injection molding.

Semisolid injection molding

An injection molding is at present the leading semisolid technique applied commercially to magnesium alloys. To execute the processing, a special system was designed and built by Husky. The new generation unit, with a clamp force of 500 tons and based on Hylectric and Quadlock machines, is shown in Fig. 5. It is a part of the system which includes, additionally, a high-speed top-entry robot for part removal and die lubricant application, material loading equipment, a part cooling conveyor and mist filtration, as well as mold lubricant mixing and distribution.

The essential part of the unit is the barrel having a diameter of 70 mm, a length of approximately 2 m, and its temperature profile is maintained by coaxial style heating coils embedded into grooves, machined in the outer diameter of the barrel. The barrel heaters are grouped into a number of independently controlled zones. Molds are preheated typically to 200°C and the alloy is injected at a screw velocity up to 6 m/s. As described previously [12], the hopper is filled with cold feedstock which is heated and conveyed within the machine barrel while protected against oxidation by argon gas. Upon accumulating the required shot size in the front of the non-return valve, the screw is accelerated forward, injecting the alloy through the sprue and gates into the mold cavity. The slurry velocity during injection depends on the size of the mold's gating system.

Alloy Selection

The general rules of material selection from a process-ability point of view are the same as for other semisolid techniques. The knowledge of the liquid fraction versus temperature curve permits selection of optimal compositions of alloys for semisolid processing. The basic condition is a sufficiently wide separation between liquidus and solidus where changes in temperature allow controlling the solid/liquid ratio. The physicochemical selection criteria were developed for aluminum alloys based on thermodynamic modeling and scanning calorimetry experiments [13]. The fraction liquid-temperature curve can be generated by thermodynamic modeling, which reduces timing and cost. Based on that analysis it was found that eutectic Al-Si alloys are not practical for semisolid technology. The reason is that at 50% solid the slope is too steep and the width of the mushy zone is too narrow. The high value of $dF/dT(T1)$ causes complications during reheating since small temperature variations will cause considerable changes in the solid/ liquid ratio (F-fraction liquid,

Fig. 5. The general view of the Husky injection molding system with a clamp force of 500 tons.

T-temperature, T1-temperature at which the slurry contains 50% liquid.

Although the future family of thixomolded alloys will include magnesium based, aluminum based and zinc based grades, the present commercial interest is limited to the first group. Among those, alloys used in die-casting are explored most frequently. There are three major families of magnesium alloys with additions of aluminum, zinc, manganese and silicon in various configurations. The most common alloy from this group, AZ91D, used in many of structural and nonstructural components, exhibits good mechanical and physical properties in combination with excellent castability and saltwater corrosion resistance. For applications requiring greater ductility than available with AZ91D, the high purity die cast alloy AM60B is used, which in turn represents better elongation and toughness. For high temperature applications, where creep resistance is important, alloys with additions of rare earth elements such as AJ52, AJ62 with strontium, MEZ with cerium or Mg-Al alloys with calcium, are used.

Product Examples

The components manufactured exhibit net, or near net-shape, which eliminates or reduces the finishing requirements, as well as improving overall quality. When compared to thermoplastics, magnesium alloys offer better stiffness, heat dissipation and electromagnetic shielding. Other advantages include dimensional stability, low porosity and tighter part tolerances with reduced shrinkage, residual stress and component distortion. The process is excellent for parts requiring extremely thin walls of the order of 0.5 mm. The present world market is focused on consumer electronics, personal entertainment and home hardware. The most popular examples include laptop computer cases, keypads for cell phones and cases for digital cameras and camcorders. There is also interest expressed by the automotive industry in using this technique for manufacturing structural automotive components.

Concluding remarks

Semisolid metal processing is establishing itself as the viable technology for manufacturing of high integrity components. Among several semisolid technologies available today, injection molding represents the highly developed method with well advanced commercialization. Since the alloy processed is tightly closed inside the unit's barrel, injection molding is especially applicable for magnesium alloys, prone to oxidation and ignition. It is anticipated that processing capabilities of injection molding will expand to structural components with superiour properties for highly demanding markets, e.g. automotive.

References

1. D. B. Spencer, R. Mehrabian and M.C. Flemings, Metall. Trans, Vol. 3, 1972, 1925.
2. E. Schalek and A. Szegvari, Koll. Z., Vol. 32, 1923, 318, 326.
3. T. Peterfi, Arch. Entwicklungsmech. Organ., Vol. 112, 1927, 680.
4. H.A. Barnes, J. Non-Newtonian Fluid Mech., Vol. 70, 1997, 1.
5. W.H. Bauer and E.A. Collins in Rheology:Theory and Applications, Vol. 4, Academic Press, New York, 1967.
6. J. Pryce-Jones, JOCCA, Vol. 17, 1934, 305.
7. J.C. Gebelin. M. Suery and D. Favier, Mater. Sci. Eng. A, Vol. 272, 1999, 134.
8. F. Czerwinski, A. Zielinska-Lipiec, P.J. Pinet and J. Overbeeke, Correlating the Microstructure and Tensile Properties of a Thixomolded AZ91D Magnesium Alloys, Acta Materialia, Vol. 49, 2001, 1225-1235.
9. F. Czerwinski, The Oxidation Behaviour of AZ91D Magnesium Alloy at High Temperatures, Acta Materialia, Vol. 50, 2002, 2639-2654.
10. F. Czerwinski, On the Generation of Thixotropic Structures during Melting of Mg-9%Al-1%Zn Alloy, Acta Materialia, Vol. 50, 2002, 3265-3281.
11. F. Czerwinski, The Microstructural Development of Mg-9PctAl-1PctZn Alloy during Injection Molding, Metallurgical & Materials Transactions, Vol. 33A, 2002, 2963-2972.
12. F. Czerwinski, Assessing Capabilities of Thixomolding System in Semisolid Processing of Magnesium Alloys, International Journal of Cast Metals Research, Vol. 16, 2003, 389-396.
13. A.A. Kozakow, Alloy Compositions for Semisolid Forming, Advanced Materials and Processes, Vol. 157, no 3, 2000, 31-33.

FATIGUE BEHAVIOR OF THIXOMOLDED® MAGNESIUM AZ91D USING ULTRASONIC TECHNIQUES

A. R. Moore[1], C. J. Torbet[2], A. Shyam[2], J. W. Jones[2], D. M. Walukas[3], R. F. Decker[3]

[1]General Motors Research & Development Center, 30500 Mound Road, Warren, MI 48090-9055, USA
[2]The University of Michigan, Materials Science & Engineering, Ann Arbor, MI 48109, USA
[3]Thixomat, Inc., 620 Technology Drive, Ann Arbor, MI 48108, USA

Keywords: Thixomolding®, magnesium alloys, ultrasonic fatigue, probability of failure, cumulative distribution function

Abstract

The fatigue behavior of the Thixomolded® magnesium alloy AZ91D has been examined using ultrasonic fatigue testing techniques at frequencies of approximately 20 kHz and for lifetimes as long as 10^9 cycles. An apparent endurance limit of approximately 65-70 MPa is observed. Comparison with the fatigue behavior of AZ91 produced by conventional die casting indicates that the Thixomolded® material has an endurance limit substantially higher than that of die cast material. This is attributed primarily to reduced porosity associated with the thixotropic processing technique. Fractographic analyses indicate that fatigue cracks leading to failure initiate at internal porosity in approximately 75% of tests and at the as-molded surface in the remainder of tests. Fractographic studies indicate that, in general, longer fatigue lifetimes are associated with smaller fracture initiation sites. It was also observed that a bimodal distribution of fatigue lives occurred, especially at lower stresses. This was analyzed using cumulative distribution functions, which confirmed the existence of dual failure populations.

Introduction

Magnesium alloys have recently received much attention as possible structural materials in a variety of engineering applications. The automotive industry frequently uses magnesium for components that are replacing parts previously manufactured from steel and aluminum [1]. The major attraction of magnesium alloys is that they have a low density of approximately 1.8 g/cm^3, and a corresponding high strength-to-weight ratio. This translates to better fuel economy and cost savings in the long run. Typical automotive applications of magnesium alloys include camshaft covers, clutch and transmission housings, and intake manifolds [1]. These components are frequently subjected to cyclic stresses greater than 10^8 cycles during automobile operation, indicating that evaluation in the very high cycle regime is necessary in order to produce durable vehicle components. The primary goal of this study was to investigate the fatigue behavior of the magnesium alloy AZ91D at ultrasonic frequencies.

Typical closed loop servohydraulic fatigue test machines operate at a maximum cyclic frequency between 20 and 40 Hz. Recent development of the ultrasonic fatigue testing technique has dramatically increased the rate at which fatigue tests can be performed [2,3]. This method utilizes pulses at approximately 20 kHz to test samples in a variety of loading conditions. Ultrasonic testing is readily used to analyze behavior in the conventional fatigue testing regime below 10^7 cycles, but also affords testing capabilities in the very high cycle regime beyond 10^9 cycles where typical test times are one to two days, rather than several months required at conventional frequencies.

In ultrasonic fatigue testing, an electrical power supply generates a sinusoidal 20 kHz electrical signal which is transformed into mechanical vibration by means of a piezoelectric transducer. These vibrations pass through a mechanical amplifier, or ultrasonic horn, and into the test sample [4]. A condition of resonance is established within the specimen, and loading occurs as a result of the imposed mechanical vibrations. An in depth review of the use of ultrasonic techniques for fatigue studies can be found in Ref. [5].

The ultrasonic fatigue testing technique has been successfully employed for numerous material systems. Studies have shown that this approach can effectively replicate conventional servohydraulic fatigue test results for aluminum alloys [6-8]. These findings indicate a negligible effect of loading frequency on fatigue behavior at room temperature. Other ambient temperature studies have also shown a negligible frequency effect on fatigue behavior for ferrous alloys [9], for nickel-base superalloys [10], and for several titanium alloys [11,12]. Ultrasonic fatigue testing has also been implemented in the study of magnesium, where the capability of this technique to model conventional fatigue has been demonstrated for several alloys [13].

Several studies have been conducted in order to understand the fatigue behavior of conventionally die cast AZ91 alloy derivatives at conventional frequencies. Bag and Zhou [14] have studied AZ91D fatigue behavior at R=0.1, and have determined a 10^7 fatigue strength at a stress amplitude of approximately 25 MPa. Goodenberger and Stephens [15] have reported the 5×10^5 fatigue strength at a stress amplitude of approximately 135 MPa for AZ91E-T6 alloy at a load ratio of -1.

Mayer *et al.* have used ultrasonic techniques to study the fatigue behavior of AZ91 hp produced by die casting and by a gravity casting technique (*in-vacuo*) known as Vacural die casting [13]. Die cast AZ91 hp was determined to have a fatigue limit of 45 ± 10 MPa, while Vacural cast AZ91 hp exhibited a 10^9 fatigue strength of approximately 65 MPa, but showed no endurance limit up to 10^9 cycles. The variation in fatigue behavior is most likely the result of differences in microstructure that stem from deviations in processing techniques.

In recent years, thixotropic processing has been used to refine microstructures and significantly reduce porosity. One such technique is called Thixomolding® and resembles an injection molding process for metals. In the Thixomolding® process, magnesium alloys are heated to a semisolid state at temperatures between their liquidus and solidus. In this temperature regime, the alloy is molten with the exception of the larger un-melted α-magnesium particles which remain solid. These particles are the definitive feature of Thixomolded® microstructures, and are expected to impart enhanced mechanical properties compared to die cast components [16]. In addition, the Thixomolding® process affords lower levels of porosity than other casting techniques because of the planar flow of injected semisolid material. This also contributes to the enhancement of mechanical properties.

This work is an attempt to analyze the fatigue behavior of thixotropically processed magnesium alloy AZ91D, and to relate the fatigue properties to the constituents present within the microstructure.

Experimental Procedure

Material

The alloy AZ91D consists of approximately 9% Al and 1% Zn. The microstructure of Thixomolded® AZ91D obtained by scanning electron microscopy is shown in Fig. 1. This material consists of a fine α-magnesium grain structure, with large un-melted α-magnesium particles, and a divorced eutectic of α-magnesium and the intermetallic phase $Mg_{17}Al_{12}$ [16]. This image clearly shows that the large un-melted α-magnesium particles are for the most part equiaxed, non-dendritic structures, as reported in the literature [16,17]. Test specimens in this work consisted of approximately 5% un-melted α-magnesium particles. Porosity has been characterized using the montage of several optical micrographs displayed in Fig. 2, along with the digitally thresholded porosity distribution image of this montage. The thresholded image was obtained in a manner similar to that described in Ref. [18]. Approximately 1.75% porosity is measured, as quantified by computational analysis and confirmed by Archimedes density measurements. This level of porosity is somewhat high for Thixomolded® components, and is attributed to the comparatively large diameter of test samples when compared with the section thickness of typical Thixomolded® components.

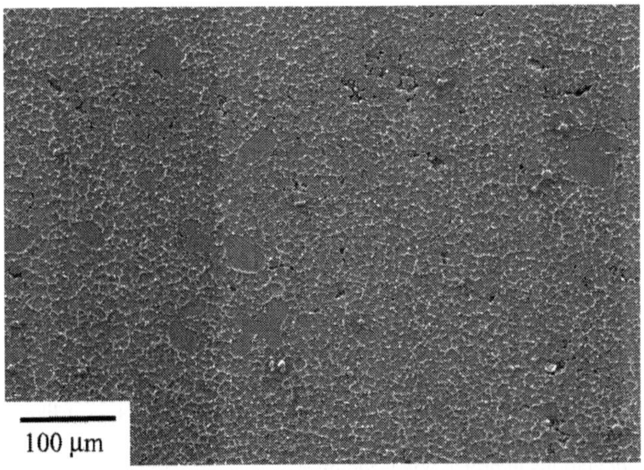

Fig. 1: SEM image of Thixomolded® AZ91D microstructure showing fine α-magnesium grain size (dark), large un-melted α-magnesium particles (dark), and the divorced eutectic matrix (bright)

Fig. 2: Montage of optical microscope images to view porosity (top) and corresponding digitally thresholded image used to quantify porosity (bottom)

Test samples of AZ91D with nominal composition were produced at Thixomat Inc., Ann Arbor, Michigan. A Japan Steel Works Thixomolding® machine model JLM220MG was used. AZ91D chips several millimeters in size were partially melted at a temperature of approximately 605 °C and injected at a velocity of 80 inches per second into a die preheated to approximately 180 °C. The die consisted of two chambers, from which parts were extracted and prepared for machining. The Thixomolded® blank dimensions are shown as the dashed lines in Fig. 3, along with the specific machining dimensions for ultrasonic fatigue test samples. The gauge section was not altered during the machining process, and hence as-molded surfaces were maintained.

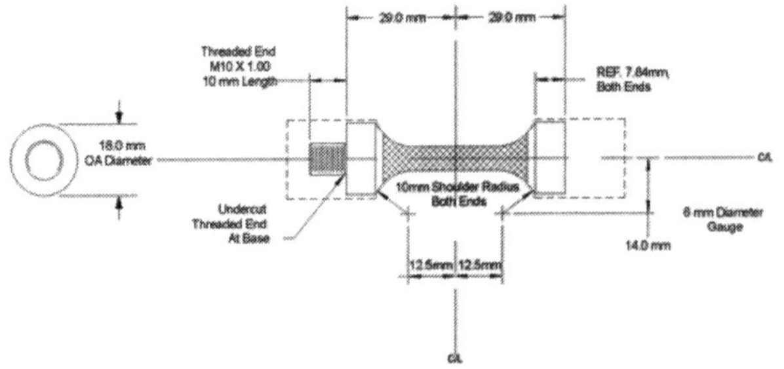

Fig. 3: Depiction of the specimen geometry used in ultrasonic fatigue tests, where the cross-hatched area indicates the as-molded surface

Ultrasonic Fatigue

Strain gauges were used on each test to determine the maximum strain amplitude, and corresponding stress amplitudes were calculated from the strains using an elastic modulus of 45 GPa. Fatigue studies were conducted for fully reversed loading (R= - 1) at stress amplitudes ranging from 70 to 95 MPa. At these stresses specimen heating would normally occur at ultrasonic frequencies. This was mitigated by cycling at 20 kHz in pulses, where the pulse/pause ratio was adjusted depending on the stress level. In the present study pulse/pause times ranged from 5 ms/150 ms to 40 ms/80 ms. Forced air cooling methods were also used to prevent heating. Sample temperature was maintained below 35 °C for tests performed at and below 85 MPa, and below 50 °C for tests performed at 90 and 95 MPa. These higher stresses resulted in increased temperatures as a result of some initial plastic deformation.

A confirmation check of resonance features was performed via computer modeling. A program was created in JAVA™ to model the strain and displacement resulting from the propagation of planar tension/compression waves. This model was plotted as a function of length from the center of the gauge section. The results of this analysis for AZ91D samples with the geometry described above are presented in Fig. 4. This plot indicated that the strain reached a maximum value at the center of the gauge section, and decreased with increasing distance from the center. This also shows that the strain in the gauge section is constant within about 3% error, which correlates with experimental results. On the contrary, there is a displacement node at the center of the specimen while its ends are anti-nodes. This plot represents one half of the test sample, and the other half would be a mirror image of these profiles.

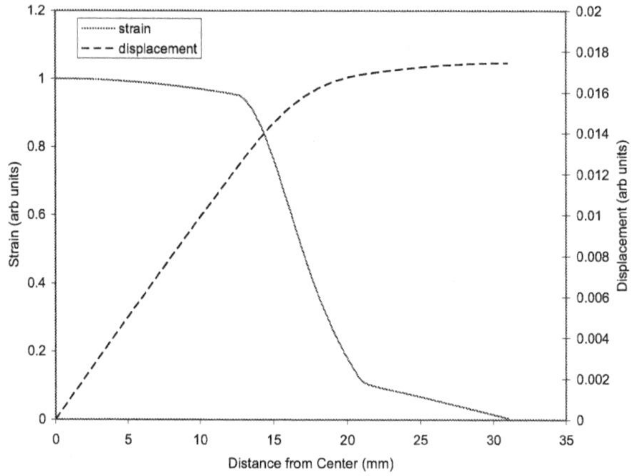

Fig. 4: Plot of calculated strain and displacement versus distance from center of gauge section, indicating that the strain is constant in the gauge section within 3% error

Tests were completed when a sample survived 10^9 cycles or when fatigue failure occurred. A sample which survived 10^9 cycles was declared a run out. Fracture surfaces were analyzed with a Phillips XL30 FEG Scanning Electron Microscope in order to study the details of the fracture origin. The location of crack initiation was determined for all samples that fractured by fatigue failure.

Results and Discussion

S-N Data

Test data has been collected for approximately fifty test samples in a fully reversed loading condition at varying stress levels. These results have been tabulated and plotted in the form of an S-N diagram. The S-N data for Thixomolded® AZ91D with approximately 5 % un-melted α-magnesium is shown in Fig. 5. This plot depicts stress amplitude versus the number of cycles until failure (or run out) on a semi-logarithmic plot. Run outs are located at approximately 10^9 cycles, and are indicated by arrows pointing diagonally upward. The results indicate that the material follows a steeply sloped curve for higher stress values, and then becomes more horizontal at a stress amplitude around 70 MPa. The data also indicates that there is a bimodal distribution of results, visible in the three orders of magnitude difference between data points at the lower stress levels.

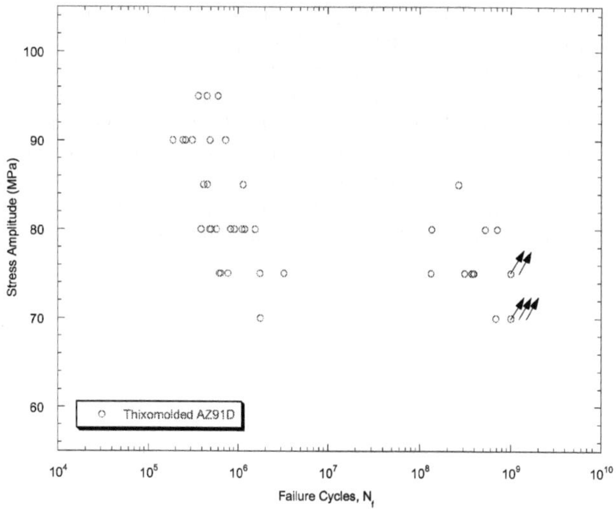

Fig. 5: S-N plot of Thixomolded® AZ91D with approx. 5 % un-melted α-magnesium

Ultrasonic fatigue test results exist for the AZ91 alloy produced by other manufacturing techniques. As previously noted, Mayer *et al.* [13] have studied the fatigue behavior of AZ91 produced by a conventional die casting process, and by the Vacural vacuum casting process. These tests were performed at ultrasonic frequencies, and at R= -1.

A comparison between S-N data of Thixomolding® and die casting production techniques is shown in Fig. 6. This plot illustrates the marked difference in fatigue behavior of the two processing techniques. The Thixomolded® samples afford a longer fatigue life at a given stress level than their die cast counterparts. Also, the 10^9 cycle fatigue strength is about 70 MPa in Thixomolded® samples versus 45 MPa in die cast samples.

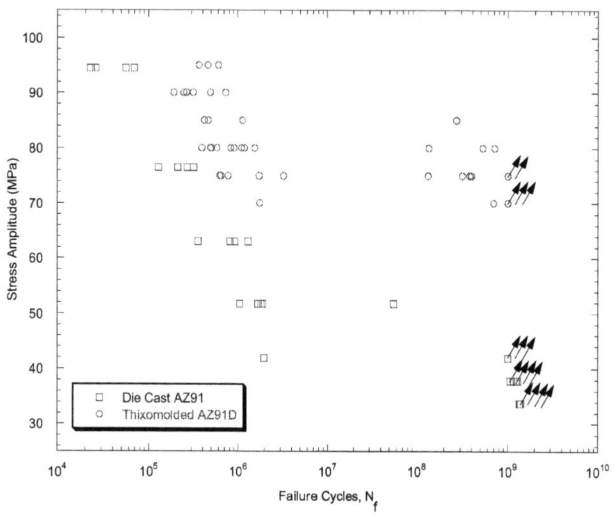

Fig. 6: S-N plot comparing Thixomolded® AZ91D to die cast AZ91

The Thixomolded® material was also compared against AZ91 produced by the Vacural technique. Results of this comparison are shown in Fig. 7. This plot indicates that the two processing methods produce materials with comparable fatigue properties. However, the 10^9 cycle fatigue strength of Thixomolded® samples appears to be slightly higher than that corresponding to Vacural cast samples. The Vacural die cast results do not show a bimodal fatigue failure regime observed in the present study. These results indicate that the fatigue behavior, as expected, is primarily a factor of the size and distribution of the porosity within the specimen, and less a factor of the microstructural constituents.

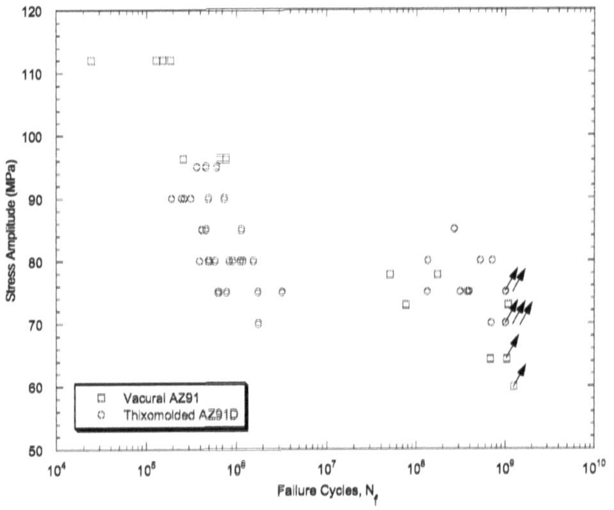

Fig. 7: S-N plot comparing Thixomolded® AZ91D to Vacural cast AZ91

Fracture Surface Investigation

In the majority of samples, internal porosity was determined to be the crack initiation site, as previously discussed. Gas and/or shrinkage pores acted as sites for crack initiation. Fractographic examination of samples that failed from surface locations revealed no noteworthy defect whose size could be quantified.

Sample orientation relative to the test machine was recorded to check for the presence of possible bending effects. This was done by visually locating the fracture origin on the failure cross-section and creating a radial type plot. Fig. 8 shows the relative failure origin location for all fatigue failures. When studied with respect to angular orientation, a fairly random distribution of failures across the fracture surfaces is observed. However, a definite trend in failure location exists with respect to radial position from the specimen center, as the majority of failures originated approximately 1 mm from the outer as-molded surface. This general location is indicated by a dashed line in Fig. 8 and displayed an increased amount of local porosity.

In addition, Fig. 8 shows that approximately 25% of samples failed at surface locations. Each of these surface failures occurred on the parting line (location where two halves of the die meet during casting) for that respective sample. No surface defects besides the parting line led to fatigue failure. The gauge section of all samples tested were in the as-cast condition, and hence light polishing of the gauge section surface may have further reduced the number of surface failures and extended lifetimes.

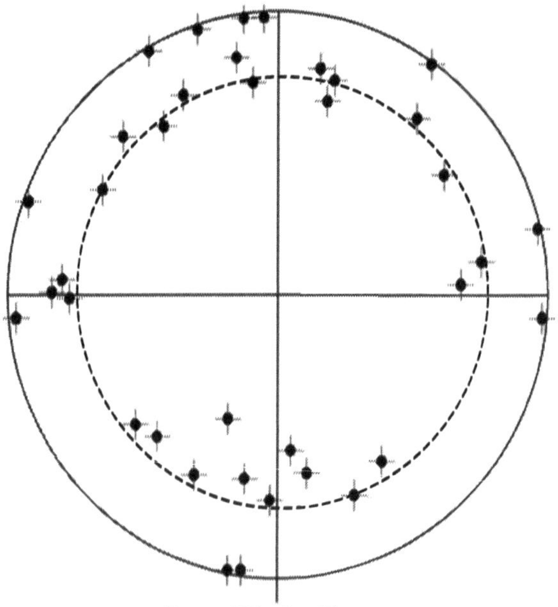

Fig. 8: Schematic depicting the relative locations of failure origins on the fracture surface, and approximate area where increased local porosity existed (dashed line)

Fatigue failures initiating at internal porosity were examined and the initiating pore feature was outlined using image analysis software. A total of 15 samples that failed at stress levels of 75 and 80 MPa were examined in this fashion. The area of the initiation sites was then determined. Fig. 9 is an image of a fracture initiation site. This image shows a region of shrinkage porosity, with an area of approximately 1.72 x 10^5 μm^2 (average diameter of 415 μm). Similar characterization was performed for the 14 other samples at these stress levels.

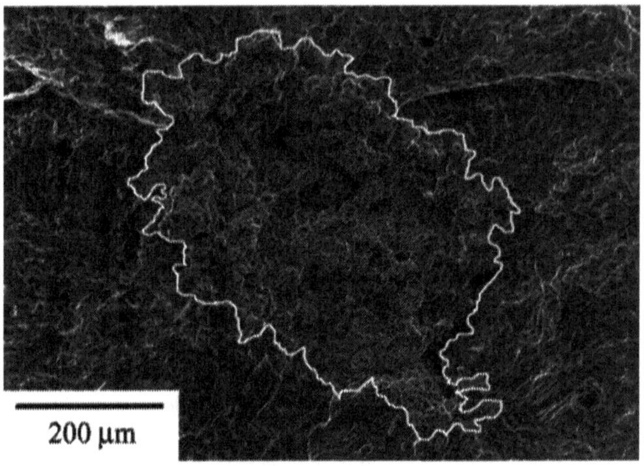

Fig. 9: SEM image of fatigue crack initiation site with defect area of 1.72 x 10^5 μm^2 and average equivalent diameter of 415 μm

Analysis was performed in accordance with methods defined by Murakami *et al.* [19], wherein the square root of the initiation feature area was taken as a measure of the origin size. The origin size was plotted against fatigue lifetime, and the results are presented in Fig. 10. This data indicates that there is a general increase in fatigue lifetime with a corresponding decrease in origin size. Some scatter exists within this data, which can possibly be attributed to other factors including the proximity of the failure initiation site to the surface of the sample.

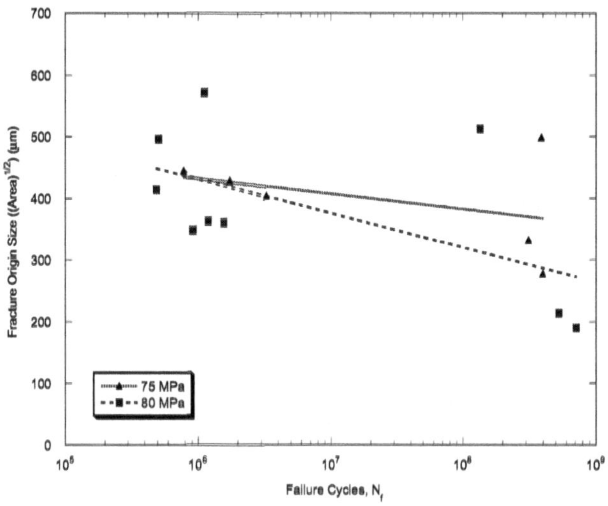

Fig. 10: Plot of fracture origin size versus number of cycles to failure

Analysis of Dual Failure Populations

As shown in Fig. 5, considerable variability in fatigue life at select stress levels was found to exist for the Thixomolded® material. Fatigue lifetimes at stress levels of 75 and 80 MPa varied by as much as three orders of magnitude. This behavior can be rationalized by recognizing the existence of two distinct failure populations.

A possible explanation for this bimodal distribution of fatigue lifetimes could stem from the difference between surface and interior failure initiation sites. At these stress levels, both internal and surface originating failures led to similar fatigue lifetimes in the lower-cycle regime. In addition, no surface originating failures survived into the regime above 10^8 cycles. However, if surface failures were removed from this analysis a bimodal distribution of lifetimes still exists, indicating that variability cannot be explained solely on the basis of surface versus internal crack nucleation.

In order to further analyze these two distinct failure populations, cumulative probability of failure (POF) plots were created at selected stress levels with respect to the fatigue lifetime. As in a recent study [20], the failure populations have been defined as Type I (lower-life regime) and Type II (higher-life regime) for further investigation. Type I and Type II failures were separated and plotted as individual distributions in lognormal probability space (Fig. 11). The linear behavior of both sets of data in Fig. 11 indicates that the lifetimes for both populations can be individually described by two distinct lognormal distributions. This provides the motivation for modeling this data by combining two cumulative distribution functions (CDFs).

Jha *et al.* [20] have shown the ability to create a step-like CDF by superimposing weighted probabilities in order to successfully model an entire population. A CDF that results from superimposing Type I and Type II failures was created using the equation:

$$CDF(N) = p_{TypeI}CDF_{TypeI}(N) + p_{TypeII}CDF_{TypeII}(N) \quad (1)$$

where N is the number of cycles and p_{TypeI} and p_{TypeII} are the probabilities of Type I and Type II failures, respectively. The probabilities were calculated so that the failures resulting from the two populations are independent and make up the entire population. Further details can be found in reference [20].

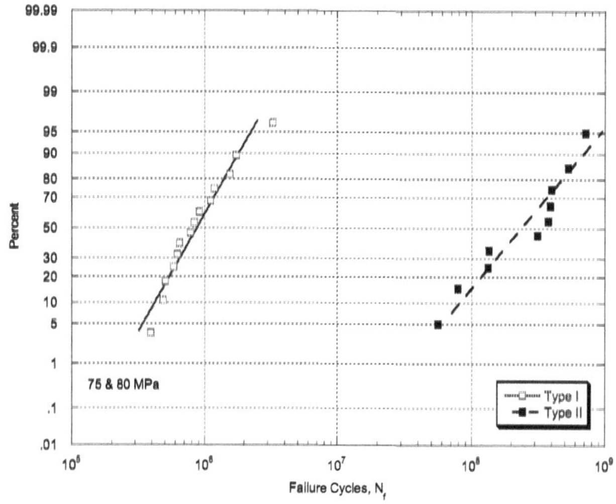

Fig. 11: POF plot showing Type I and Type II failures plotted as separate distributions for combined stresses of 75 and 80 MPa

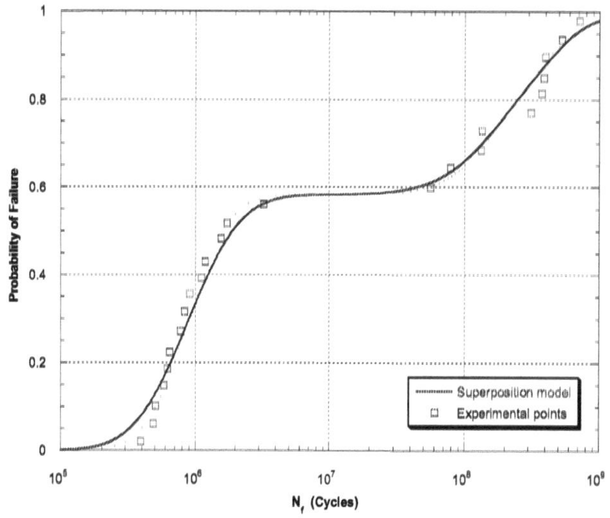

Fig. 12: POF plot representing the superposition model fit of experimental Type I and Type II failures at combined stresses of 75 and 80 MPa

This superposition technique of CDFs has been used to analyze experimental fatigue data, the results of which can be seen in Fig. 12 for combined stress levels of 75 and 80 MPa. Experimental points are plotted in lognormal probability space and are grouped into two distinct populations. In addition, the superposition model of CDFs is plotted, which represents the entire distribution of lifetimes very well. This indicates that two distinct failure populations corresponding to different lifetime regimes can be identified in this material.

Conclusions

An investigation of the fatigue properties of the Thixomolded® magnesium alloy AZ91D was performed using ultrasonic fatigue techniques. With this technology, fatigue behavior in the very high cycle regime was obtained. An endurance limit of approximately 70 MPa was determined for this alloy, which is significantly higher than that corresponding to AZ91 produced by conventional die casting methods. Comparison with AZ91 produced by a vacuum casting method revealed similar results, indicating that fatigue behavior is primarily dependent on the size and distribution of internal porosity.

Fractographic examination indicated that internal porosity was the primary feature leading to failure, and was present in the form of gas and shrinkage type pores. Fracture origin size was studied, and it was determined that in general fatigue lifetime decreases as fracture originating pore size increases.

POF plots revealed that failures could be separated into two distinct groups while maintaining linear behavior in lognormal probability space, indicating that experimental data could be described by two distinct distributions. A superposition of two CDFs modeled this data very well, and made evident the existence of dual failure populations in this material.

Acknowledgements

This work was supported, in part, by NSF Grant DMR 0211067. Test samples were supplied by Thixomat, Inc., Ann Arbor, MI. The authors acknowledge helpful discussions with Herwig Mayer, James M. Larsen and Sushant Jha.

References

1. A.A. Luo, "Materials Comparison and Potential Applications of Magnesium in Automobiles," *Magnesium Technology 2000*, ed. H.I. Kaplan et. al., (Warrendale PA: The Minerals, Metals and Materials Society, 2000), 89-97.

2. L.E. Willertz, "Ultrasonic Fatigue," *Int. Mat. Rev.*, 25 (2) (1980), 65-78.

3. ASM: *ASM Handbook*, Vol. 8 (ASM International, Materials Park, OH, 1985).

4. S.E. Stanzl-Tschegg, "Fracture Mechanisms and Fracture Mechanics at Ultrasonic Frequencies," *Fatigue Fract. Engng. Mater. Struct.*, 22 (1999), 567-579.

5. H. Mayer, "Fatigue Crack Growth and Threshold Measurements at Very High Frequencies," *Int. Mat. Rev.*, 44 (1) (1999), 1-34.

6. H. Mayer et. al., "Influence of Loading Frequency on the High Cycle Fatigue Properties of AlZnMgCu1.5 Aluminum Alloy," *Mater. Sci. Engng.*, A314 (2001), 48-54.

7. S.E. Stanzl-Tschegg and H. Mayer, "Fatigue and Fatigue Crack Growth of Aluminum Alloys at Very High Numbers of Cycles," *Int. J. Fatigue*, 23 (2001), S231-S237.

8. M.J. Caton et. al., "Demonstration of an Endurance Limit in Cast 319 Aluminum," *Metall. Mater. Trans. A*, 34A (2003), 33-41.

9. Q.Y. Wang et. al., "Gigacycle Fatigue of Ferrous Alloys," *Fatigue Fract. Engng. Mater. Struct.*, 22 (1999), 667-672.

10. S.A. Padula II et. al., "High Frequency Fatigue Crack Propagation Behavior of a Nickel-Base Turbine Disk Alloy," *Int. J. Fatigue*, 21 (1999), 725-731.

11. R.O. Ritchie et. al., "Thresholds for High-Cycle Fatigue in a Turbine Engine Ti-6Al-4V Alloy," *Int. J. Fatigue*, 21 (1999), 653-662.

12. M. Papakyriacou et. al., "Influence of Loading Frequency on High Cycle Fatigue Properties of B.C.C. and H.C.P. Metals," *Mater. Sci. Engng.*, A308 (2001), 143-152.

13. H. Mayer et. al, "Influence of Porosity on the Fatigue Limit of Die Cast Magnesium and Aluminum Alloys," *Int. J. Fatigue*, 25 (2003), 245-256.

14. A. Bag and W. Zhou, "Tensile and Fatigue Behavior of AZ91D Magnesium Alloy," *J. Mater. Sci. Lett.*, 20 (2001), 457-459.

15. D.L. Goodenberger and R.I. Stephens, "Fatigue of AZ91E-T6 Cast Magnesium Alloy," *J. Engng. Mater. Tech.*, 115 (1993), 391-397.

16. F. Czerwinski et. al., "Correlating the Microstructure and Tensile Properties of a Thixomolded AZ91D Magnesium Alloy," *Acta Mater.*, 49 (2001), 1225-1235.

17. F. Czerwinski, "Size Evolution of the Unmelted Phase During Injection Molding of Semisolid Magnesium Alloys," *Scripta Mater.*, 48 (2003), 327-331.

18. A. Balasundaram and A.M. Gokhale, "Quantitative Characterization of Spatial Arrangement of Shrinkage and Gas (air) Pores in Cast Magnesium Alloys," *Mater. Charact.*, 46 (2001), 419-426.

19. Y. Murakami and M. Endo, "The √Area Parameter Model for Small Defects and Nonmetallic Inclusions in Fatigue Strength: Experimental Evidences and Applications," *Theoretical Concepts and Numerical Analysis of Fatigue*, (Birmingham UK: 1992), 51-71.

20. S.K. Jha et. al., "Dual Fatigue Failure Modes in Ti-6Al-2Sn-4Zr-6Mo and Consequences on Probabilistic Life Prediction," *Scripta Mater.*, 48 (2003), 1637-1642.

DEVELOPMENT OF WROUGHT Mg ALLOYS VIA STRIP CASTING

Sung S. Park, Jung G. Lee, Hak C. Lee and Nack J. Kim

Center for Advanced Aerospace Materials
Pohang Univ. of Science & Technology, Pohang 790-784, Korea

Keywords: Strip casting, Wrought Mg alloy, AZ31 alloy

Abstract

Development of wrought Mg alloy, particularly in sheet form, are essential to support the growing interest for lightweight components in the automotive industry. However, development of Mg alloy sheets has been quite slow due to the complexity of sheet production originated from limited deformability of Mg. In this respect, strip casting, a one-step processing of flat rolled products, can be an alternative for the production of Mg alloy sheets. In this study, AZ31 and experimental alloys are strip cast into 2 mm thick strips. The microstructure of the as-cast AZ31 alloy strip consists of columnar zones near the roll side and equiaxed zones in the mid-thickness region. On the other hand, as-cast MX1 alloy strip shows equiaxed dendritic structure through the thickness of strip. The cooling rate estimated from the secondary dendrite arm spacing is around 10^2 K/s. These alloys were subjected to various thermomechanical treatments and their tensile properties were evaluated. Strip cast AZ31 alloy in H24 condition has equivalent yield and tensile strengths with similar ductility compared to commercial ingot cast AZ31-H24 alloy, indicating that strip casting is a viable process for the fabrication of Mg alloy strips. The experimental MX alloys have a large volume fraction of fine dispersoid particles in the microstructure, resulting from the beneficial effect of strip casting on microstructural refinement. It has been shown that the experimental MX0 alloy has superior tensile properties compared to commercial ingot cast AZ31-H24 alloy, suggesting the possibility of the development of new wrought Mg alloy sheets by strip casting.

Introduction

The main advantages of Mg are that it has very low density of 1.74 g/cm^2 which is 30% lighter than Al and it has nearly inexhaustible natural resources. Mg alloys normally show a very good castability and machinability as well as excellent specific strength and stiffness. Therefore near net shape die-casting technology has already been established for fabricating various components such as instrumental panel, gear box housing or seat frames [1-9]. However, there is a lack of competitive Mg wrought products, especially sheet materials, which are much needed for numerous weight-sensitive applications. According to the comparison of mass savings potential of various materials, component mass can significantly be reduced by using Mg as opposed to steel and Al alloys particularly where bending stiffness is the design limit. Hence the advantages of Mg alloy sheets can be utilized particularly in large thin-walled areas such as automotive body components [10, 11]. The substitution of conventional sheet materials by Mg alloy sheets would lead to essential weight savings [12]. However, application of Mg alloy sheets in transportation systems is rather limited since few Mg alloy sheet products fulfill the properties and cost requirements of automobile industry [10].

In this respect, it is believed that the strip casting process can be an alternative for the production of Mg alloys sheets. Strip casting process combines casting and hot rolling into a single step, having an advantage of one-step processing of flat rolled products. Besides being such a cost-effective process, strip casting also has beneficial effects on microstructure such as reducing segregation, improving inclusion size distribution and refining microstructural and textural homogeneity. The current interest in strip casting is on the production of steel products such as stainless steels [13-14] and Al alloys with limited alloy contents [15]. Recently, attempts have been made to fabricate the highly alloyed Al alloys by strip casting [16] and it shows that strip casting can be applied for the production of highly alloyed Al alloys.

The present research is aimed at studying the feasibility of strip casting for the production of Mg alloy sheets and the development of new wrought Mg alloys via strip casting. AZ31 and experimental alloys were strip cast and their microstructure and mechanical properties were discussed as compared to those of conventional alloys.

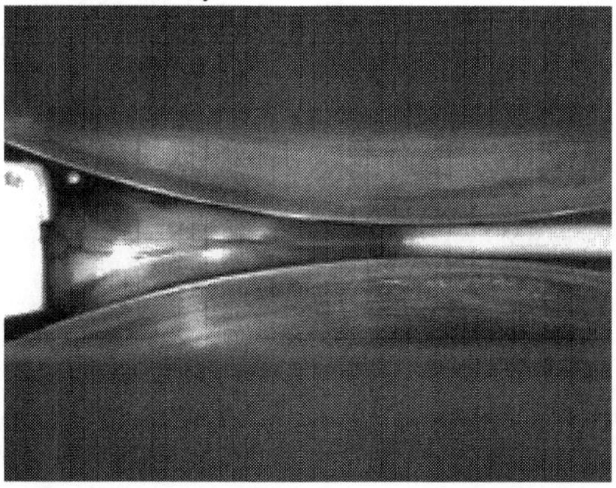

Figure 1. Photograph of the roll gap area showing the actual fabrication of strips.

Twin Roll Strip Casting

The horizontal twin roll strip caster used in the present study is basically same as the one used by the authors for the production of Al alloy strips [14]. The caster consists of induction melting furnace, tundish, nozzle, and rolls. The molten metal is transferred to a tundish where the temperature is stabilized and controlled before being shot through a nozzle into the rolls. Roll is made of Cu-0.7 wt.% Be and cooled by the circulating water through the holes inside the rolls. Since the water cooled Cu is used as a roll material, the present strip caster can give a faster solidification rate than the one made of steel. Previous studies on Al alloys using the same strip caster shows that a solidification rate of 10^2-10^3 K/s can be achieved by the present strip caster [16]. One of the possible drawbacks using the Cu as a roll material is the amount of separating force the rolls can withstand due to relatively low strength of Cu. After the strip exits the roll gap, the strip is usually air cooled.

Fig. 1 shows the roll gap area of the horizontal strip caster employed in producing Mg alloy strips. A slot type ceramic nozzle is used to achieve a uniform flow across the width. As the molten metal feeds through two rotating rolls, there is a formation of solidified shell near the contact region between melt and rolls. At the same time, its thickness is reduced by the rolling action of the rotating rolls. As can be seen from the difference in gray scale, the mid-thickness area is in semi-solid state before it reaches the roll nip. Usually the heat transfer coefficient reaches its highest value at roll nip due to the increased roll separating force. Solidification can end before or after strip passes the roll nip depending on several processing variables [17].

Experimental Procedure

AZ31 alloy (Mg-2.9wt.% Al-0.94wt.% Zn-0.26wt.% Mn) and experimental MX alloys (MX0 and MX1) were studied in the present study. The reason for choosing AZ31 alloy in the present study is that it is the most common wrought Mg alloy. The alloys were remelted under an inert atmosphere of CO_2 and SF_6 mixture. After melting, the molten alloy was transferred into preheated tundish whose temperature was controlled to select the superheating of melt. The degrees of superheating for AZ31 and MX alloys were 118 °C and 70 °C, respectively. Roll gap was set at 2 mm and roll speed was in the range of 4 ~ 4.5 m/min. The cast AZ31 and MX alloys were homogenized at 420 °C for 1 hr and 330 °C for 10 hr, respectively. The homogenized strips were subjected to thermomechanical treatment to compare with commercial ingot cast alloys. The AZ31 alloy strip was warm rolled at 200 °C with a total reduction of 50 %, followed by annealing at 150 °C for 60 min to give H24 condition (strain hardened and partially annealed). The MX alloys strips were hot rolled at 300 °C with a total reduction of 40 ~ 50 %, followed by annealing at 330 °C for 30 min. After annealing, MX alloys were artificially aged at 150 °C to give T6 condition. Specimen blanks for tensile specimens were cut from the midsection of strip parallel to the casting direction. Tensile properties were measured by using flat tensile specimens with 12.6 mm gage length, 1 mm gage thickness and 5 mm gage width. Strain rate used was 6.4×10^{-4} s^{-1}. Microstructural observation of the strip was conducted on the cross sections by optical microscopy (OM) and scanning electron microscopy (SEM).

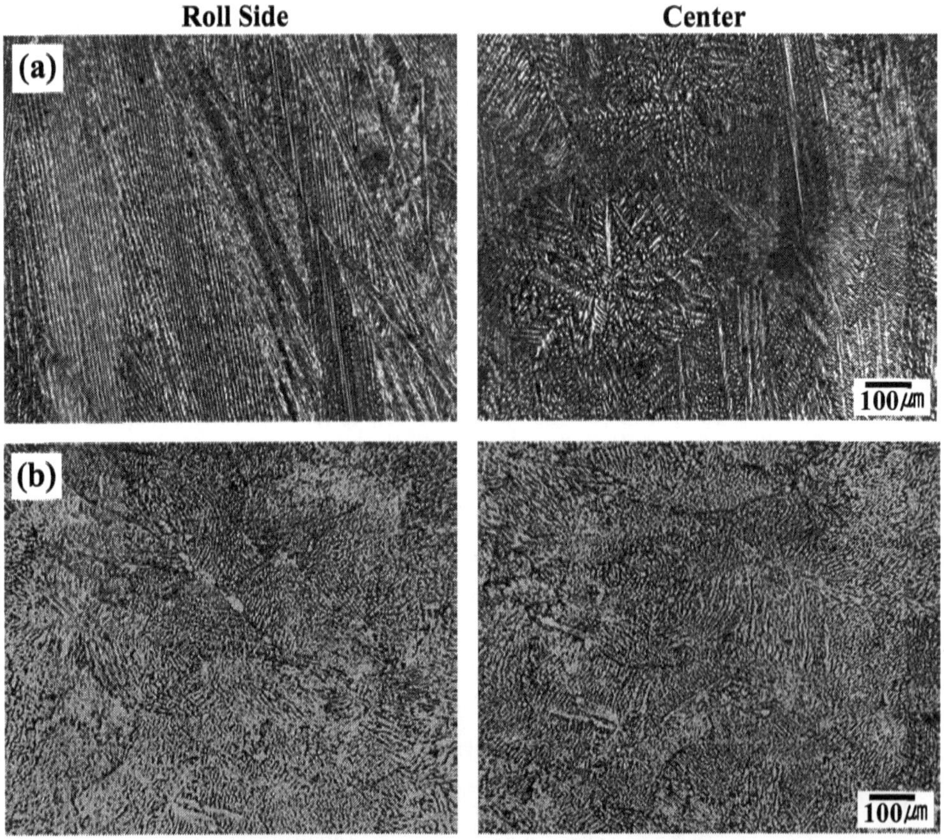

Figure 2. Cross-sectional optical micrographs of the strips: a) AZ31 alloy and b) MX1 alloy

Results

Microstructure

Fig. 2 (a) shows the cross sectional as-cast microstructures of AZ31 alloy. Columnar dendrites are observed in upper and lower surfaces and both columnar zones grow along the direction of heat transfer. The columnar to equiaxed transition occurs in mid-thickness region. Distinct macrosegregation through the thickness of strip was not detected. Secondary dendrite arm spacing (SDAS) was measured to estimate the cooling rate during strip casting. It shows that the SDAS of the strip varies from 2.6 to 3.9 μm for the AZ31 alloy. As expected, the center of the strip shows the largest SDAS, indicating that it experiences the slowest cooling rate. The cooling rate was estimated from the measured SDAS by using the following relationship [19];

$$SDAS = 5.3 \cdot t_f^{0.43}$$
$$R = \Delta T_f / t_f$$

where the t_f is the local solidification time (s), R is the cooling rate (K/s) and ΔT_f is the non-equilibrium temperature range of solidification, which is approximately 66 K for the AZ31 alloy. It shows that the cooling rate experienced in the AZ31 alloy strip ranges from 130 K/s to 350 K/s. These values of cooling rate are faster than those experienced during conventional ingot casting (~10 K/s), but slower than those obtained in our previous study of Al alloys [16].

As is the case of SDAS, grain size of AZ31 strip after homogenization also varies through the thickness of strip, ranging from 34 to 38 μm. Variation of grain size through the thickness after homogenization follows the same trend observed in the variation of SDAS. Nevertheless, it is worth mentioning that the grain size of the strip cast alloy is much finer than 250 μm which is the typical grain size of ingot cast alloy. Within Mg matrix, Al-Mn particles are formed sparsely. In contrary to AZ31 alloy, MX1 alloy strip shows equiaxed dendritic structure through the thickness of strip (Fig. 2 (b)). The microstructure is quite homogeneous in as-cast MX1 alloy strip. The microstructure of MX alloy is somewhat similar to that of the previously studied AZ91 alloy strip [18, 20].

Fig. 3 shows the optical micrographs of the homogenized and rolled AZ31 alloy strip. By warm rolling at 200 °C, Mg grain size is reduced by dynamic recrystallization since the recrystallization temperature of AZ31 alloy is about 205 °C [21]. After partial annealing at 150 °C, the elongated Mg grains of as-rolled alloy change to somewhat equiaxed shape and the grain size becomes 6 μm. The microstructure of strip cast alloy after thermomechanical treatment is similar to the commercial ingot cast AZ31-H24 alloy (Fig. 3 (d)). Fig. 4 shows the optical and SEM micrographs of MX1 alloy strip after solution treatment (T4). Grain size of MX alloys ranges from

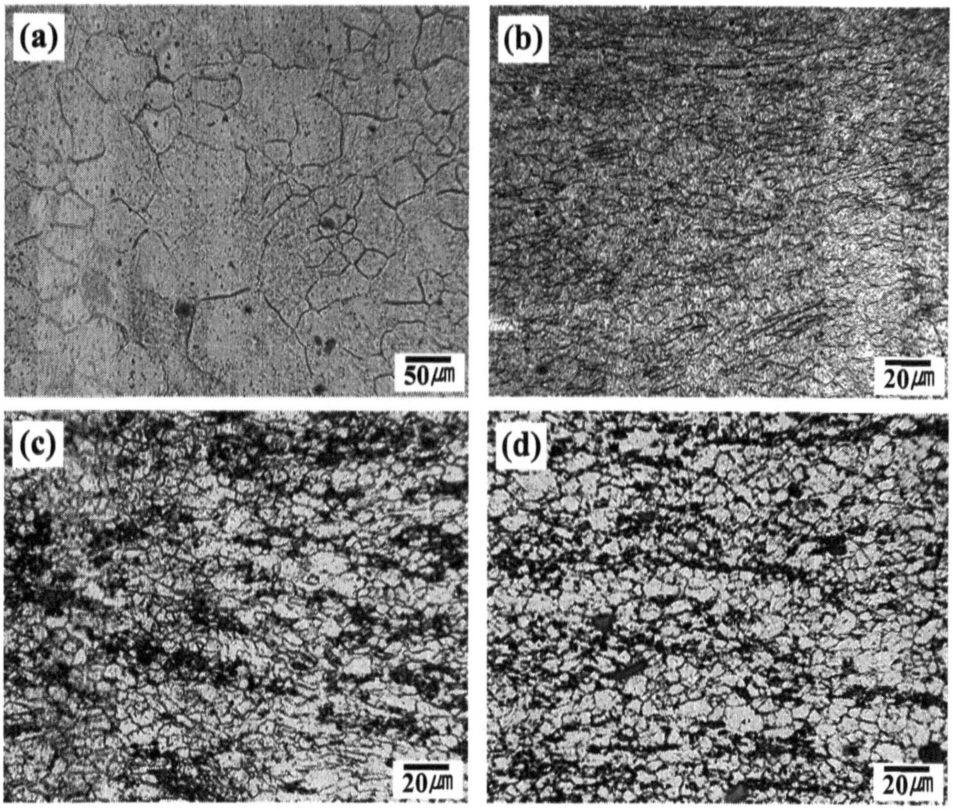

Figure 3. Optical micrographs of (a)~(c) strip cast and (d) ingot cast AZ31 alloy: a) homogenized b) as-rolled (c) annealed (d) commercial sheet.

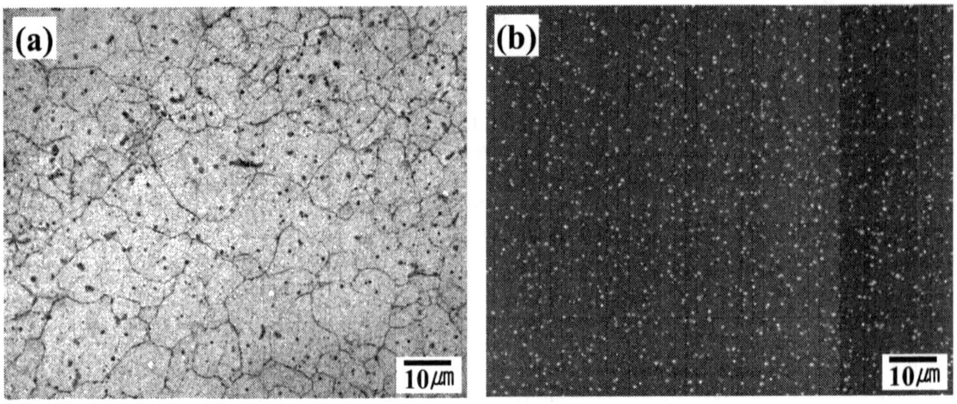

Figure 4. Optical (a) and SEM (b) micrographs of MX1 alloy after solution treatment.

12 μm to 15μm, which is finer than that of AZ31 alloy. It should be noted that formation of dispersoid particles in MX1 alloy is more evident than in AZ31 alloy. The size of dispersoid particles in MX1 alloy ranges from 10^2 to 10^3 nm with the volume fraction of 3.8 %. The MX0 alloy shows a smaller volume fraction (1.5%) of dispersoid particles than MX1 alloy. Fig. 5 shows the microstructure of annealed MX alloy strip after hot rolling. It shows that the grain size is reduced by recrystallization from 12 μm to 2.6 μm. As is the case of AZ31 alloy, the annealed MX alloys show more uniform microstructure than the solution treated ones.

Mechanical Properties

Tensile properties of the strip cast alloys are listed in Table 1 along with those of commercial AZ31 sheet in H24 condition for comparison purpose. It shows that the strip cast AZ31 alloy has equivalent strength and ductility to those of commercial ingot cast alloy in H24 condition. It is expected that the Al-Mn particles give little influence on tensile properties of AZ31 alloy since its volume fraction is too small. The production of Mg alloy sheet usually involves several hot rolling steps, which are high cost operations. In the case of commercial AZ31 alloy, high temperature rough rolling of about 95 % reduction is usually conducted prior to the thermomechanical treatment needed for H24 temper. The strip cast AZ31 alloy, however, can directly be subjected to H24 treatment, giving similar strength and ductility values to those of the commercial alloy.

Tensile properties of the MX alloys are also listed in Table 1. There are large increases in strengths of both MX0 and MX1 alloys by aging treatment. Despite the large increases in strengths of the MX alloys by aging treatment, there are only slight decreases in ductility values by aging treatment. Yield and tensile strengths of the MX alloys in T6 condition are higher than those of AZ31 alloy. In particular, the MX0 alloy in T6 condition has higher yield and tensile strengths with equivalent ductility as compared to those of commercial AZ31-H24 sheet. The MX1 alloy has much higher yield and tensile strengths but with much reduced ductility than the MX0 alloy in both T4 and T6 conditions. Higher strength of the MX1 alloy than that of MX0 alloy is mainly due to the larger volume fraction of dispersoid particles in the former than in the latter.

Discussion

There are several factors which control the solidification microstructure, e.g., freezing range and degree of superheating (temperature gradient). The freezing range of alloy and temperature gradient simultaneously change the length of dendrite [22]. For instance, a wide freezing range and low temperature gradient causes long and thin dendrites which are easily fragmented by liquid convection. In this

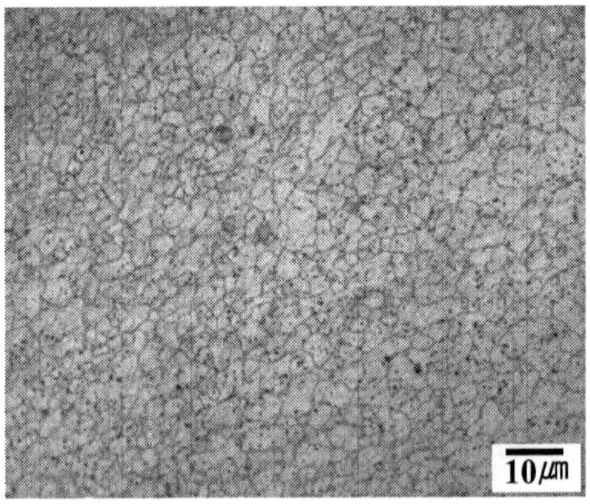

Figure 5. Optical micrograph of annealed MX alloy strip after hot rolling.

Table 1. Tensile properties of the strip cast alloys. Properties of ingot cast alloy are also shown for comparison purposes.

Alloy	Condition	YS [MPa]	UTS [MPa]	Elongation [%]
AZ31	Strip cast (H24)	236	284	15.3
	Ingot cast (H24)	236	283	16.8
MX0	Strip cast (T4)	134	271	18.4
	Strip cast (T6)	256	310	16.2
MX1	Strip cast (T4)	180	290	8.7
	Strip cast (T6)	319	360	6.3

situation, equiaxed dendritic structure occurs. In the previous study of AZ91 alloy, equiaxed dendritic structure was observed through the thickness of as-cast strip [20]. On the other hand, the columnar dendritic structure as shown in Fig. 1 is favored in the alloy having a narrow freezing range and by a high temperature gradient. The freezing range of AZ31 alloy is smaller than that of other Mg alloy such as AZ91 since AZ31 alloy has smaller alloying content than AZ91 alloy. The transition from columnar to equiaxed structure arises from the loss of melt superheat in the last solidified center region.

In addition to the solidification microstructure, macrosegregation was also affected by freezing range of alloys. In AZ31 alloy, a noticeable macrosegregation is not found through the thickness of the strip. In the previous study of AZ91 alloy, however, there is a considerable macrosegregation of Al deviated from the average concentration by 20 %. Generally the cause of macrosegregation in casting is understood to be the liquid movement which ascribes to thermal contraction, solidification shrinkage and density difference within the mushy (liquid-solid) zone [22]. However the liquid movement in strip casting is mainly due to roll separating force which squeezes the solute-rich liquid from the mid-thickness region to the outer surface. Therefore, so-called negative segregation occurs in strip casting, which is undoubtedly more evident in the alloys with wider freezing ranges. This negative segregation can be reduced by controlling the roll separating force which is a function of several processing variables such as melt feeding, roll speed or roll gap. By doing so, macrosegregation of MX alloys having similar freezing range to AZ91 alloy was cut down by half as compared to that of AZ91 alloy.

Usually, commercial AZ31 sheet is produced in H24 temper to obtain optimum combinations of strength and ductility, which are intermediate between those of the fully strain hardened and the fully annealed conditions. Strain hardening is the major source of strength for AZ31 alloys since it has a small amount of solute contents and the contribution of usual precipitation hardening is rather small. Since most of forming operations on wrought Mg alloy are done at elevated temperatures, the need for fully annealed condition is less than that for other metals. Thermomechanical treatment to give H temper for Mg alloys is usually conducted at elevated temperatures to promote non-basal deformation mode. The low cold workability of Mg alloys results from their hexagonal crystal structure having limited slip system at room temperature. It has been reported that the critical resolved shear stress (CRSS) for basal plane slip in Mg single crystal exhibits 100 times lower value than that for non-basal plane slip near room temperature. At higher temperatures, further deformation modes (pyramidal / prismatic slip) become operative. Partial annealing of fully strain hardened AZ31 alloy at 150 °C results in the changes in microstructure by substructural changes in dislocation density and rearrangement of dislocations into polygonization [23]. In the case of MX alloys, thermomechanical treatment to give H temper condition was not conducted since these alloys derive their strength mainly from dispersion and precipitation hardening. Thermomechanical treatment for MX alloys was set up to induce dynamic recrystallization, which results in refinement of grain size. Such refined grain sizes of MX alloys do not change by annealing treatment due to the presence of dispersoid particles which impede the grain boundary movement.

One of the main advantages of strip casting is the refinement of intermetallic compounds, which usually exist as harmful coarse particles in conventionally processed alloys. Presence of fine intermetallic particles in the microstructure is quite beneficial for mechanical properties. It has been shown that the refinement of Al-Mn particles in Al alloys, which are generally known as deleterious particles, results in improvement in both strength and ductility [24]. Al-Mn particles retards the motion of dislocations to increase strength and at the same time the blocked dislocations change the slip systems by cross-slip to maintain good ductility [24]. Previous study of strip cast Al alloys also showed the similar results [16]. As mentioned previously, fine dispersoid particles inhibit the grain growth during high temperature heat treatment. They can also form intense deformation zones, which act as sites for recrystallization during rolling and subsequent annealing. Based on these respects, the compositions of MX alloys were designed to form dispersoid particles during strip casting. The strength of the MX0 alloy is about 20 MPa higher than that of commercial AZ31 sheet with a similar ductility level. Moreover the strength of MX1 alloy is comparable to those of 7000 series Al alloys on specific strength basis. Nevertheless the origin of relatively poor ductility of the MX1 alloy is somewhat obscure and further study is needed to improve the microstructural homogeneity and ductility of strip cast alloys.

Summary

1. Roll side of the as-cast AZ31 alloy strip consists of columnar zones, which change to equiaxed zones in the mid-thickness region. On the other hand, MX1 alloy strip shows equiaxed dendrite structure through the thickness of strip.
2. Secondary dendrite arm spacing varies through the thickness of strips, having the smallest value at the roll side and the largest value at the center of strips. The cooling rate estimated from the secondary dendrite arm spacing is around 10^2 K/s.
3. Strip cast AZ31 alloy in H24 condition has equivalent yield and tensile strengths with similar ductility compared to commercial ingot cast AZ31-H24 alloy.

4. Strip cast experimental MX0 alloy in T6 condition, utilizing dispersion and precipitation hardening, has superior tensile properties compared to commercial ingot cast AZ31-H24 alloy.

References

[1] J. Enss et al., "Properties and Perspectives of Magnesium Rolled Products," *Magnesium Alloys and their Applications*, ed. K.U. Kainer (Weinheim, Germany, Wiley-VCH, 2000), 590-591.

[2] A.A. Luo et al., "Magnesium Castings for Automotive Applications," *JOM*, July (1995), 28-31.

[3] D.H. Kim, K.S. Shin, and N.J. Kim, "Recent Research and Development Activities of Mg Alloys at CAAM," *Advanced Performance Materials*, 5 (1998), 319.

[4] M.S. Yoo et al., "Tensile and Creep Properties of Squeeze Cast Mg Alloys with Various Second Phases," *Materials Science Forum*, 419-422 (2003), 419-424.

[5] D.H. Kim et al., "Precipitation Behavior of $Mg_{17}Al_{12}$ in Monolithic and Al_2O_3 short Fiber Reinforced Mg-Al-Zn Alloys," *Metals & Materials*, 3 (4) (1997), 211-215.

[6] S.H. Lee et al., "Effect of Applied Pressure on Mechanical Properties of Squeeze Cast Mg Matrix Composites," *Metals & Materials*, 1 (1) (1995), 37-46.

[7] S.W. Kim et al., "Invest Casting of AZ91HP Magnesium Alloy," *Metals & Materials*, 6 (3) (2000), 275-279.

[8] W.W. Park et al., "Age Hardening Phenomena of Rapidly Quenched Non-Combustible Mg-Al-Si-Ca and Mg-Zn-Ca Alloys," *Metals & Materials International*, 7 (1) (2001), 9-13.

[9] C.D. Lee, C.S. Kang, and K.S. Shin, "Nonstoichiometric Precipitation in As-Cast Mg-xAl-1Zn Alloy," *Metals & Materials*, 6 (6) (2000), 497-503.

[10] W. Sebastian, K. Droder, and S. Schumann, "Properties and Processing of Magnesium Wrought Products for Automotive Application," *Magnesium Alloys and their Applications*, ed. K.U. Kainer (Weinheim, Germany, Wiley-VCH, 2000), 602-603.

[11] A.A. Luo, "Magnesium: Current and Potential Automotive Applications," *JOM*, February (2002), 42-50.

[12] E. Doege and K. Droder, "Sheet Metal Forming of Magnesium Wrought Alloys-Formability and Process Technology," *Journal of Materials Processing Technology*, 115 (2001), 14.

[13] D.K. Kim et al., "Microstructural Evolution in a Strip-cast Ni-base Superalloy," *Metals & Materials*, 2 (4) (1996), 211-217.

[14] S.G. Pyo, J.J. Kim, and N.J. Kim, "Solidification Microstructure of Strip Cast 304 Stainless Steel," *J. Kor. Inst. Met. & Mater.*, 37 (10) (1999), 1204-1212.

[15] M. Yun, S. Lokyer, and J.D. Hunt, "Twin Roll Casting of Aluminum Alloys," *Materials Science and Engineering A*, 280 (2000), 116-123.

[16] Y.S. Park, S.B. Lee, and N.J. Kim, "Effect of Mn addition on Microstructure and Mechanical Properties of Strip Cast Al6061 Alloy," *Light Materials for Transportation Systems*, eds. N.J. Kim, C.S. Lee and D. Eylon (CAAM, POSTECH, 2001), 553.

[17] Y. Fujita et al., "Solidification and Roll-bonding of Shell in Twin-roll Casting Process," *ISIJ International*, 29 (1989), 495-502.

[18] S.S. Park, Y.S. Park, and N.J. Kim, "Microstructure and Properties of Strip Cast AZ91 Mg Alloy," *Metals and Materials International*, 8 (6) (2002), 551-554.

[19] S.Y. Eum, H.M. Park, and T.W. Nam, " The Effect of Misch Metal Addition on the Microstructure and Mechanical Properties of Rapidly Solidified AZ91 Alloy," *Journal of the Korean Foundrymen's Society*, 20 (1) (2000), 31.

[20] S.S. Park et al., "Fabrication of Mg Alloy Strips by Strip Casting," *Materials Science Forum*, 419-422 (2003), 599-604.

[21] H. Baker et al., eds., Metals Handbook, Vol. 2 (Metals Park, OH: American Society for Metals, 1979), 555.

[22] J.D. Verhoeven, *Fundamentals of Physical Metallurgy* (New York, NY: John Wiley & Suns, Inc., 1975), 283-294.

[23] V. Masseria et al., eds., Metals Handbook, Vol. 4 (Metals Park, OH: American Society for Metals, 1981), 708-709.

[24] S.W. Nam, "The effect of Mn on the Mechanical Behavior of Al Alloys," *The First KIM-JIM Joint Symposium*, ed. S.W. Nam (1999), 37-40.

From *Magnesium Technology 2009*, E.A. Nyberg, S.R. Agnew, N.R. Neelameggham, and M.O. Pekguleryuz, Editors

DEVELOPMENT OF 1500MM WIDE WROUGHT MAGNESIUM ALLOYS BY TWIN ROLL CASTING TECHNIQUE IN TURKEY

Ozgur Duygulu[1], Selda Ucuncuoglu[1], Gizem Oktay[1], Deniz Sultan Temur[1], Onuralp Yucel[2] and Ali Arslan Kaya[1,3]

[1]TUBITAK MRC, Materials Institute, P.O.B. 21, Gebze, Kocaeli 41470, TURKEY
[2]Istanbul Technical University, Department of Metallurgical and Materials Engineering, Istanbul 34469, TURKEY
[3](Currently) Mugla University, Engineering Faculty, Department of Metallurgy and Materials, Kotekli, Mugla 48000, TURKEY

Keywords: Magnesium alloy, Twin roll casting, AZ31, AZ61, AZ91, AM50, AM60.

Abstract

Magnesium alloy AZ31, AZ61, AZ91, AM50 and AM60 sheets were produced by twin roll casting first time in Turkey. Sheets of 4.5-6.5mm thick and 1500mm width were successfully achieved. Microstructure of the sheet was analyzed by optical microscope, scanning electron microscope (SEM) and transmission electron microscope (TEM). Semi-quantitative analyses were performed by SEM-EDS. In addition, X-ray studies were performed for both characterization and texture purposes. Mechanical properties were investigated by tensile tests and also hardness measurements. Homogenization and annealing heat treatments were performed on the produced sheets.

Introduction

There is increasing interest in using magnesium alloys for various applications especially for electronics and automotive industries [1-6]. Application of wrought magnesium alloys especially in the form of sheet is limited due to the price of conventional rolling product. However, the demand for decreasing the magnesium sheet prices is high and can be met through twin-roll casting [7].

With conventional magnesium sheet production, the starting material is a slab with a thickness of about 100-250mm. The magnesium alloy slab has to be rolled many times with small amounts of deformation passes at each step and intermediate annealing treatments to produce the sheet material [8]. This multi-step process is not cost-effective in terms of time and energy. Therefore, twin roll casting (TRC) is focused on due to its potential for relatively low-cost production.

Universities, companies and institutes have performed some laboratory as well as pilot scale industrial trials using TRC technology. Investigations on strip casting have been mainly in Australia [9,10], South Korea [11-15], Germany [16-20], Japan [21-24], China [25-27] and Canada [28]. In CSIRO (Australia), magnesium alloy strips (AZ31, AZ61, AM60, and AZ91 alloys), having a width of 100-600mm and thickness of 2.3-5mm have been produced [9-10]. POSCO Magnesium (South Korea) produces 530-600mm wide, 0.4-4.3mm thick AZ31 sheet or coils [11]. Thyssen Krupp MgF Magnesium Flachprodukte GmbH (Germany) is producing strips up to 700mm wide and 4.5-7mm thick [16-20]. In Japan, AZ31, AZ61, AZ91, AM50 and AM60 strips are produced with 4-5 mm thickness and up to 250mm width [21-28]. In China, 1.0-2.0mm thick, 150mm wide; and 0.5-8mm thick, 600mm wide AZ31 strips are produced [8, 29-31].

Recently, an industrial scale magnesium twin roll casting plant has been set up in Turkey. The system is incorporated with a gas-fired chamber furnace with a maximum melting capacity of approximately 3200Kg magnesium. Water-cooled steel twin-rolls of 1600mm width and 1125mm diameter with 15° tilt angle, a pinch roll unit, a cross-cut shear and a coiler are incorporated into the system.

1500mm wide magnesium alloy AZ31, AZ61, AZ91, AM50 and AM60 sheets of 4.5-6.5mm thickness have been successfully casted. To our knowledge, these are the widest magnesium alloy sheets that have been produced by this method in the world so far. The production has been continued for 10's of meters and the sheet was coiled or sheared to test the system in process. The surfaces of the sheets were of good quality. No cracks or voids were observed other than some minor problems along the edges.

Experimental Procedure

Commercially available ingots were first heated and charged into the melting furnace. N_2, SF_6 gases and Al-Be alloy was used to protect the molten alloy during melting. Nominal bath temperature was approximately 700°C in the melting furnace. 1500mm wide ceramic tip was used. The tip and the rolls were heated by gas flame before casting process. The roll speed was 1.5-3m/min. The temperature was controlled by a FLIR Systems ThermaCAM thermal camera and a thermocouple.

For chemical compositions of the produced magnesium alloy sheets, Spectrolab M8 optical emission spectrometer is used. The chemical compositions of magnesium alloy sheets are listed in Table 1.

Table 1. The chemical compositions of magnesium alloy sheets obtained by optical emission spectrometer (wt%).

Sample	Al	Zn	Mn	Mg
AZ31	2.7	1.03	0.31	bal.
AZ61	6.0	0.84	0.23	bal.
AZ91	8.3	0.74	0.18	bal.
AM50	5.4	0.08	0.29	bal.
AM60	6.2	0.08	0.27	bal.

Metallographic samples were prepared to observe the microstructure in the length, width and thickness directions. The samples were mounted and polished starting with SiC paper, followed by diamond slurry and colloidal silica. Etching was done using acetic picral acid solution (4.2g picric acid, 10ml acetic acid, 70ml ethanol, 10ml water) for AZ alloys, and acetic glycol solution (20 ml acetic acid, 1 ml HNO_3, 60 ml ethylene glycol and 20 ml water) for AM alloys face up for 5 seconds immediately followed by washing with ethanol. Heyn Lineal Intercept procedure (ASTM E112-96 standard) is applied for measuring the size of the grains.

In order to evaluate the deformation behavior of the sheets, tensile tests were preformed. Tensile samples were machined via EDM (Electric discharge machining) in accordance with ASTM standard B557M [33]. The samples were subsize specimens. A Zwick Z250 tensile testing machine at a strain rate of $1x10^{-3}$ s^{-1} was used for tensile testing.

Vickers microhardness measurements were performed with a 200g applied load using a Zwick ZHV10 device.

For X-ray diffraction (XRD) studies, specimens were polished and a Shimadzu XRD-6000 diffractometer producing Cu-Kα radiation at 40kV and 30mA was used.

Texture measurements were performed by PANalytical X'Pert Pro operated at 45kV and 40mA after specimens were polished to third quarters of sheet thickness.

Scanning electron microscope (SEM) observations were carried out using a JEOL-JSM-6335 FEG operated at 20 kV equipped with Oxford EDS system and Inca software.

For transmission electron microscopy (TEM) studies, specimens were cut; 3mm diameter discs were punched and ground to a thickness less than 80μm. Punched specimens were dimple ground with Gatan 656 Dimple Grinder. Specimens were ion polished at 3kV with Gatan 691 Precision Ion Polishing System (PIPS). The foils were examined with a JEOL 2100 TEM with a LaB_6 filament operated at 200kV and equipped with an Oxford EDS system. Images were taken by Gatan Model 694 Slow Scan CCD Camera with Gatan Digital Micrograph software.

Results

Photographs from the production of 1500mm wide magnesium alloy AZ31 sheet by twin roll strip system and coiling are shown in Figures 1,2 and 3. Figure 4 shows the thermal camera view of the sheet during twin-roll casting. As can be seen, the thermal temperature gradient is homogenous thorough width of the sheet.

Figure 1. Photograph of 1500mm wide magnesium AZ31 sheet during twin roll casting process.

Figure 2. Photograph of 1500mm wide magnesium AZ31 sheet during coiling process.

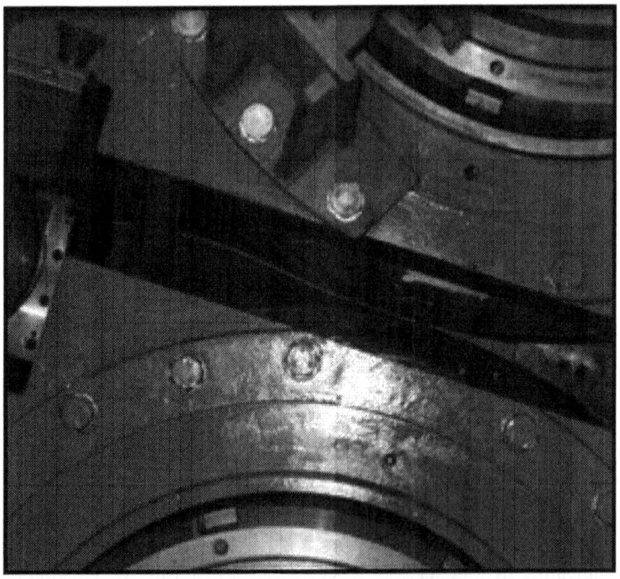

Figure 3. Photograph of magnesium AZ31 sheet during twin roll casting process.

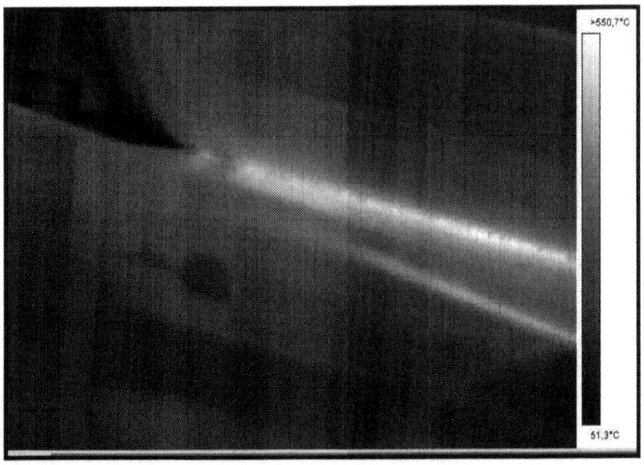

Figure 4. Thermal camera view of the sheet during twin-roll casting.

In Figure 5, the XRD spectrums of as twin-roll cast AZ31, AZ61, AZ91, AM50 and AM60 magnesium alloy sheets are given. It is observed that peak intensity levels are similar for all Mg alloys and second phase peaks are not significant other that $Mg_{17}Al_{12}$ phase for AZ91 alloy.

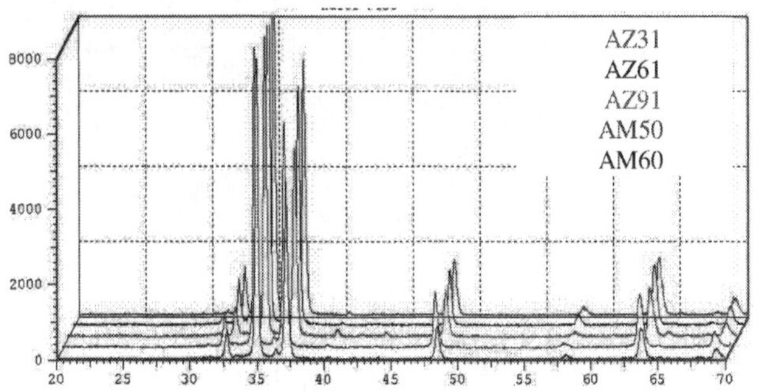

Figure 5. The XRD spectrums of as twin-roll cast AZ31, AZ61, AZ91, AM50 and AM60 magnesium alloy sheets.

Figure 6 shows plane view microstructures of TRC Mg alloys in the as-cast condition. Dendritic structure was observed and dendritic grain size was approximately 200-300μm. As seen from Figure 7, a gross macrosegregation was not detected for the TRC Mg alloys.

The as cast Mg alloy sheets were homogenized between 350-450°C for 1-2-4-6hr in laboratory furnace. In Figure 8, microstructures are given for all Mg TRC sheets after homogenization heat treatments at 400°C for 2hr. Grain sizes of less than 20μm are achieved. Homogenization processes are under investigation for longer times for industrial scale sheets.

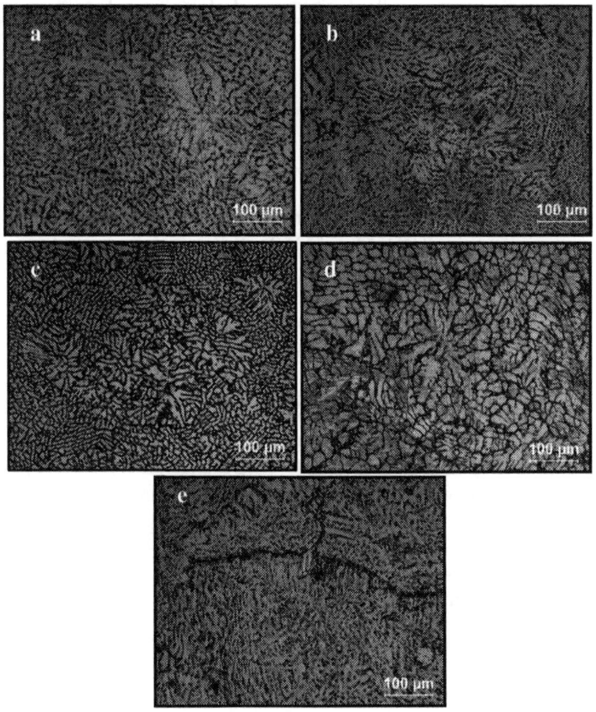

Figure 6. As-cast plane views of **a)** AZ31, **b)** AZ61, **c)** AZ91, **d)** AM50 and **e)** AM60 magnesium alloys.

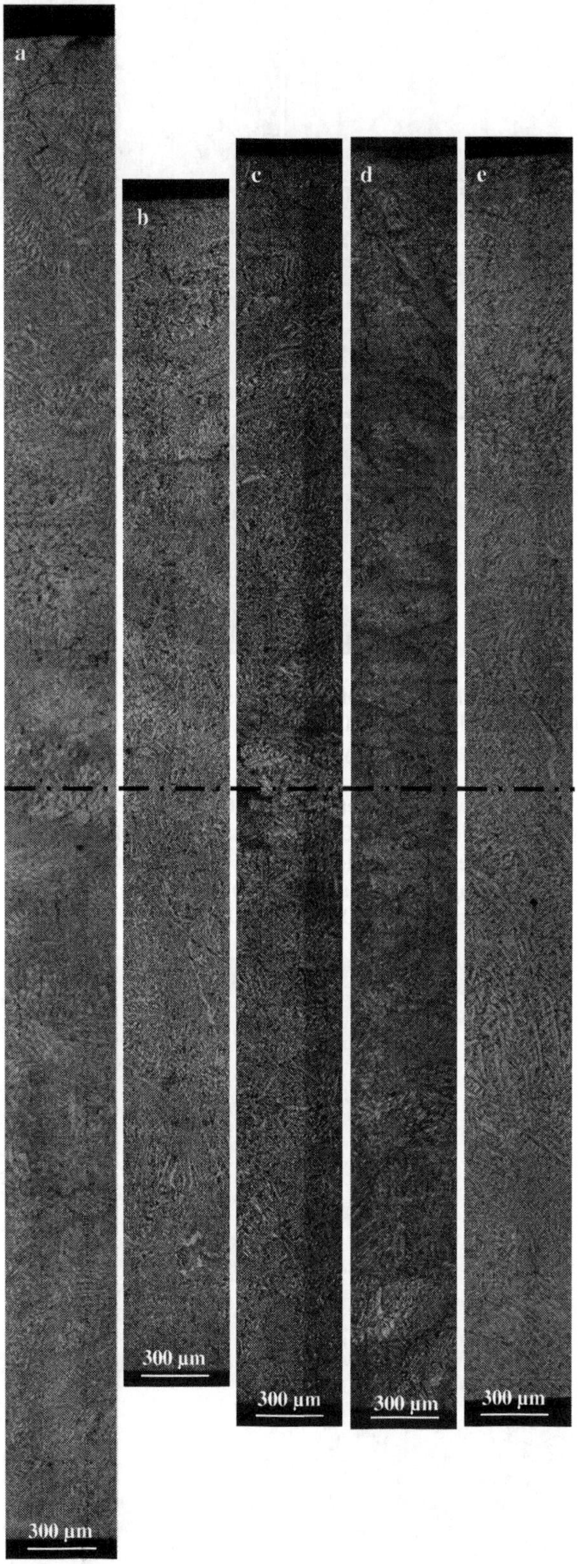

Figure 7. As-cast cross-section microstructure views of **a)** AZ31, **b)** AZ61, **c)** AZ91, **d)** AM50 and **e)** AM60 magnesium alloys (the dashed line shows the centerline).

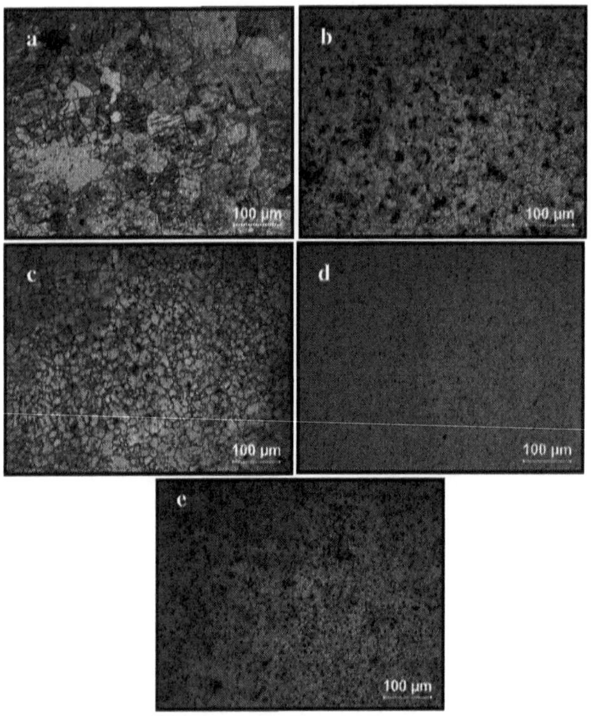

Figure 8. Plan view microstructures after homogenization at 400°C for 2hr: **a)** AZ31, **b)** AZ61, **c)** AZ91, **d)** AM50 and **e)** AM60 magnesium alloys.

General view TEM picture of TRC Mg AM50 alloy after homogenization at 400°C for 2hr is given in Figure 9 as a representative TEM microstructure. Al-Mn precipitates are observed in all TRC Mg sheets which is shown in more detail in HRTEM image (Figure 10) with its EDS analysis.

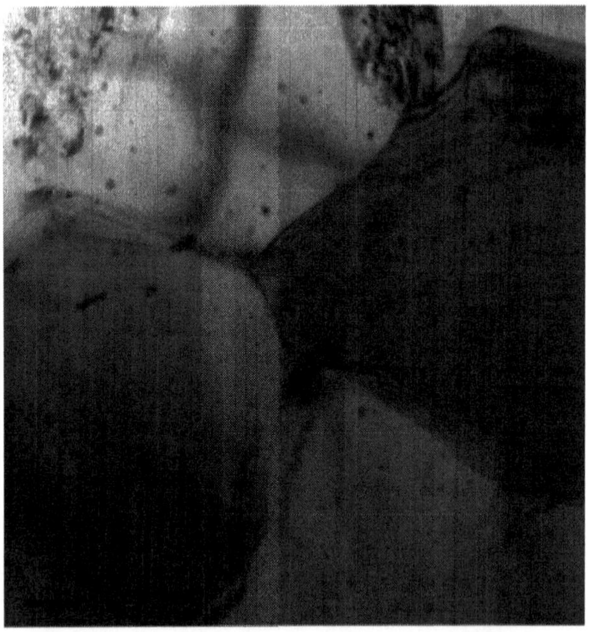

Figure 9. General view TEM picture of TRC Mg AM50 alloy after homogenization at 400°C for 2hr.

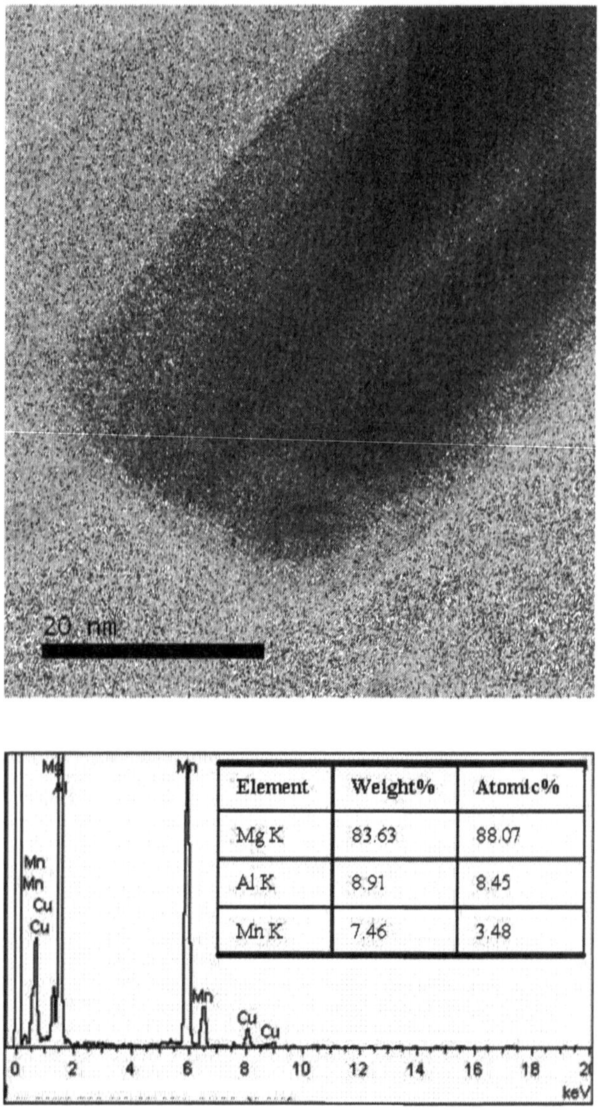

Figure 10. HRTEM picture and EDS analysis of Al-Mn precipitate in TRC Mg AM50 alloy after homogenization at 400°C for 2hr (Cu peaks are from TEM holder).

SEM studies were also performed on the samples for characterization of the microstructure and investigating the features such as segregation, inclusions, oxides, cracks, particles and precipitates by SEM-EDS. In Figure 10, one example SEM picture is shown from the centerline segregation zone in the TRC AZ31 alloy. By semi-quantitative EDS analysis it is observed that, the amount of Al, Zn and Mn elements are increased with some oxide in the segregation zone. However, significant amount of segregations or inclusions in big dimensions or with too much change in element content is not observed for none of the produced TRC Mg alloys.

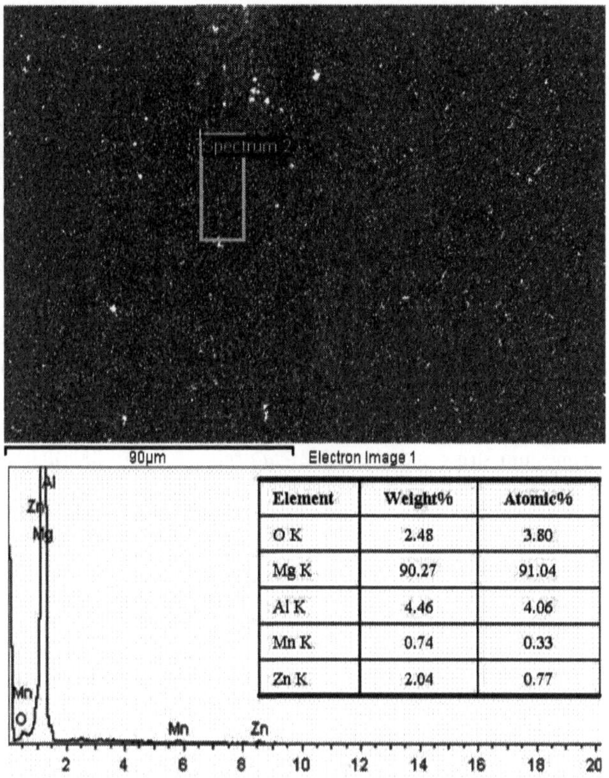

Element	Weight%	Atomic%
O K	2.48	3.80
Mg K	90.27	91.04
Al K	4.46	4.06
Mn K	0.74	0.33
Zn K	2.04	0.77

Figure 11. SEM picture and EDS analysis of segregation zone of TRC Mg AZ31 sheet.

Figure 12 and 13 shows flow curves of as TRC and homogenized TRC Mg alloys at 400°C for 2hr respectively. The low ductility observed for some as-cast Mg alloy sheets are thought to be related with their inhomogeneous microstructures, since ductility values reached more than 20% elongation after homogenization. The conditions employed for thermomechanical processing showed that the microstructure and mechanical properties of the produced sheets can be improved significantly. Moreover, SEM fractographs showed mostly ductile fracture morphology. Hardness measurements also gave quite reasonable hardness values compared to the literature [15].

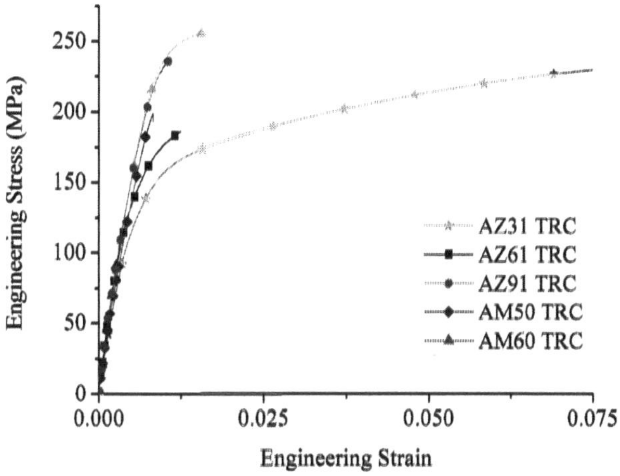

Figure 12. Tensile test results for the as TRC 1500mm wide magnesium alloy AZ31, AZ61, AZ91, AM50 and AM60 sheets.

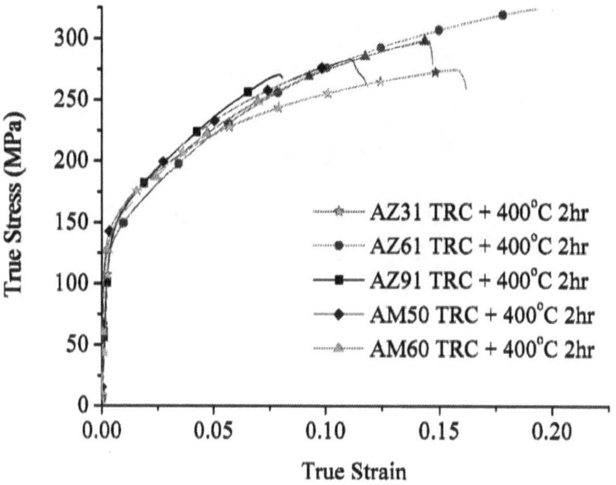

Figure 13. Tensile test results for the homogenized TRC Mg alloys at 400°C for 2hr.

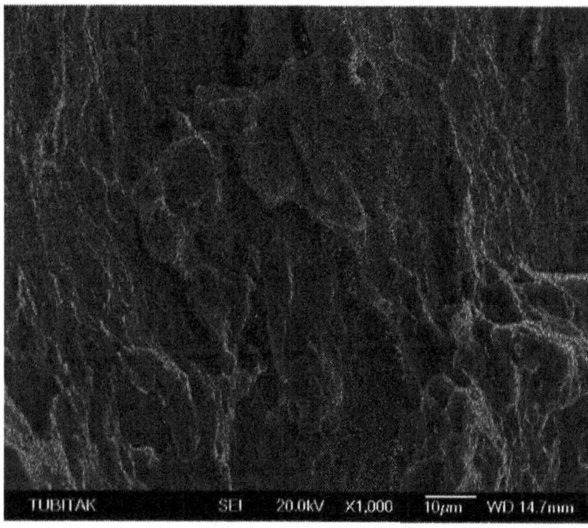

Figure 14. SEM fracture surface of tensile tested sample.

Table 2. Hardness measurements of magnesium alloys.

Sample	As Cast	After Homogenization
AZ31	72	56
AZ61	72	62
AZ91	83	73
AM50	70	54
AM60	76	59

(0002) pole figures (Figure 15) of the as TRC AZ31 and AZ61 alloy sheets show that texture is almost random. Since texture and anisotropy is critical for forming operations starting texture is important. Therefore, having a random texture will aid future rolling and forming processes.

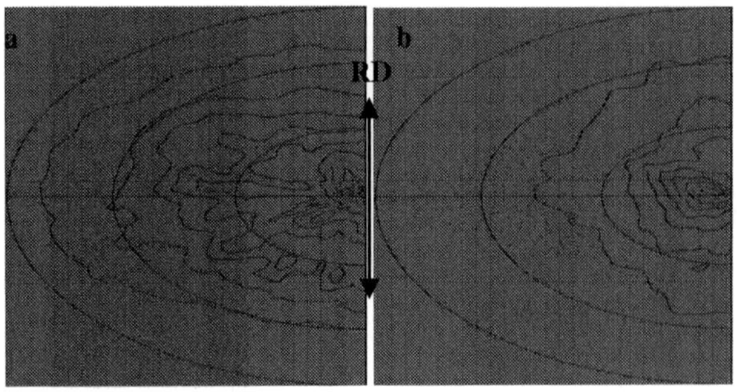

Figure15. (0002) pole figures of as TRC **a)** AZ31 and **b)** AZ61 alloy sheets.

Summary and Conclusions

Magnesium alloys AZ31, AZ61, AZ91, AM50 and AM60 sheet castings of 4.5-6.5mm thick and 1500mm width are easily achievable by TRC method. The conditions employed for thermomechanical processing showed that the microstructure and mechanical properties of the produced sheets can be improved significantly. Further processing for thinner sheets is attempted using the same TRC plant in Turkey [34].

References

[1] M.M. Avedesian and H. Baker, *Magnesium and Magnesium Alloys, ASM Specialty Handbook* (Ohio, ASM International, 1999).

[2] E.F. Emley, *Principles of Magnesium Technology* (Pergamon, Oxford, 1966).

[3] E. Beck, *The Technology of Magnesium and Its Alloys* (London, Hughes, F.A., and Co. Limited, 1943).

[4] E. Aghion and D. Eliezer, *Magnesium Alloys, Science, Technology and Applications* (Haifa, Israel, 2004).

[5] I. J. Polmear, *Light Alloys, Metallurgy of the Light Elements* (London, Arnold, 1995).

[6] K.U. Kainer, *Magnesium alloys and technologies* (Weinheim:Wiley-VCH, Verlag GmbH & Co., 2003).

[7] F. Moll et al., "Application of Mg sheets in car body structures" (The paper presented at the Magnesium, Proceedings of 6th International Conference Magnesium Alloys and their Applications, 2003), 936-942.

[8] P. Ding et al., "Status and development of magnesium alloy thin strip casting," *Materials Science Forum*, 546-549 (2007), 361-364.

[9] D. Liang and C.B. Cowley, "The twin roll strip casting of magnesium," *Journal of Materials Processing Technology*, 56(5) (2004), 26-28.

[10] D.H. StJohn, "Overview of current international magnesium research and recent CAST CRC developments," *Advanced Materials Research*, 29-30 (2007), 3-8.

[11] POSCO Co. Ltd web site: www.poscomagnesium.com.

[12] S.S. Park et al., "Development of wrought Mg alloys via strip Casting" (Paper presented at the Magnesium Technology 2004, TMS, Charlotte, North Carolina, U.S.A., March 2004) 14-182.

[13] S.S. Park et al., "Development of Mg alloy sheets via strip casting," *Materials Science Forum*, 475-479 (2005), 457-462.

[14] S.S. Park et al., "Microstructure and mechanical properties of twin-roll strip cast Mg alloys," *Materials Science Forum*, 539-543 (2007), 119-126.

[15] S.B. Kang et al., "Effect of reheating and warm rolling on microstructure and mechanical properties of twin roll strip cast Mg-4.5Al-1.0Zn-0.4Mn-0.3Ca alloy sheet" (The paper presented at the Magnesium Technology 2008, TMS, New Orleans, Louisiana, U.S.A., 2008) 147-152.

[16] ThyssenKrupp Magazine, 2004, web site: .thyssenkrupp.com

[17] R. Kawalla et al., "Development of a strip-rolling technology for Mg alloys based on the twin-roll-casting Process" (The paper presented at the Magnesium Technology 2008, TMS, New Orleans, Louisiana, U.S.A., 2008), 177-182.

[18] F.W. Bach, Rodman M., and Rossberg A., "High quality magnesium sheets for automotive applications," *Advanced Materials Research*, 6-8 (2005), 665-672.

[19] B. Engl, "Future aspects of magnesium sheet materials using a new production technology" (The paper presented at the 62nd Annual World Magnesium Conference, Berlin, Germany 22-24 May 2005), 27-34.

[20] R. Kawalla et al., "New technology for the production of magnesium strips and sheets," *Metalurgija,* 47 (3) (2008), 195-198.

[21] Y. Nakaura and K. Ohori, "Properties of AZ31 magnesium alloy sheet produced by twin roll casting," *Materials Science Forum*, 488-489 (2005), 419-426.

[22] H. Watari et al., "Semi-solid manufacturing process of magnesium alloys by twin-roll casting," *Journal of Materials Processing Technology*, 155-156 (2004), 1662-1667.

[23] T. Haga, H. Watari, and S. Kumai, "High speed twin roll casting of Mg alloy strip by a vertical twin roll caster," *Journal of Achievements and Manufacturing Engineering*, 15(1-2) (2006), 186-192.

[24] H. Watari et al., "Effects of production parameters on characteristics of magnesium alloy sheets manufactured by twin roll casting," *Journal of Achievements and Manufacturing Engineering*, 16(1-2) (2006), 171-175.

[25] H. Watari et al., "Twin roll casting of magnesium alloys with high aluminium contents," *Journal of Achievements and Manufacturing Engineering*, 18(1-2) (2006), 419-422.

[26] H. Watari et al., "Mechanical properties and metallurgical qualities of sheets manufactured by twin-roll casting," *Key Engineering Materials*, 345-346 (2007), 65-168.

[27] H. Watari et al., "Feasibility study of twin roll casting process for magnesium Alloys," *Journal of Materials Processing Technology*, 192-193 (2007), 300-305.

[28] H. Watari et al., "Development of manufacturing process of wrought magnesium alloy sheets by twin roll casting," *Journal of Achievements in Materials and Manufacturing Engineering*, 20 (1-2) (2007), 515-518.

[29] C. Yang et al., "The microstructure and processing in twin roll casting of magnesium alloy strip," *Materials Science Forum*, 488-489 (2005), 427-430.

[30] H. Di et al., "New processing technology of twin roll strip casting of AZ31B magnesium strip," *Materials Science Forum*, 488-489 (2005), 615-618.

[31] B. Jiang et al., "Rolling of AZ31 magnesium alloy thin strip," *Materials Science Forum*, 546-549 (2007), 365-368.

[32] J.A. Jackman et al., "Overview of key R&D activities for the development of high-performance magnesium materials in Canada," *Materials Science Forum*, 488-489 92005), 21-24.

[33] R.F. Allen et .al., eds., *Annual Book of ASTM Standards,* section 3, (Metals Test Methods and Analytical Procedures, ASTM, Philadelphia, PA, 2001).

[34] O. Duygulu et al., "Development of Rolling Technology for Twin Roll Casted 1500mm Wide Magnesium AZ31 Alloy," *Submitted to Magnesium Technology 2009 TMS, San Fransisco, U.S.A. 2009*

PART 5: ALLOY AND MICROSTRUCTURAL DESIGN

Suveen N. Mathaudhu

The alloy and microstructural design of Mg-alloys has been studied for the past century, but these aspects have historically occurred at a sluggish pace when compared to Fe-, Ni- and Al-based alloys over the same time period. For the most part, only a few major alloy families, such as Mg-Al-Zn (AZ) and Mg-Zn-Zr (ZK) alloys, have seen application and usage in casting and wrought components. However, the recent renaissance in Mg-alloy research has fueled the increased demand for lightweight transportation components and has renewed interest in microstructural modification and alloying for improved performance in the diverse and extreme environments that the parts would be exposed to. Many of these advances have been historically reported first in the *Magnesium Technology* proceedings, and here we present three key areas that have seen great progress: (1) grain refinement and solute additions for high strength, (2) alloying for creep resistance and high temperature applications, and perhaps most importantly, (3) property enhancement through the introduction of rare-earth elements.

The first key area of importance addresses the critical need to improve the strength for the rigorous environments seen in automobile and aerospace applications. The first paper in this section by Lee and colleagues provides an excellent review on the effects of solute and particle effects on the grain refinement, including the mechanisms of refinement. Key attention was paid to the addition of Al, Zr, Si, Ca, and Sr solutes, which have since turned into critical alloying components of many new high strength Mg-alloys. This paper, and the next one in the section (Klösch *et al.*) also address the effects of the addition of small particles, such as ZrB_2 on the grain refinement, with marked improvements in mechanical properties. The next two papers (Blake and Caceres, Sasaki *et al.*) represent early reports on the effect of specific solutes in improving the mechanical strength, via modification of the strain hardening rate, and enhanced age hardening, respectively. These four papers embody transformational approaches to increasing the strength beyond the conventional limits of traditional alloy systems, and in many ways forecast and led rapid global developments in grain refinement and alloy modification for enhance mechanical properties.

The second key area addresses the historically low temperature limitations of Mg-alloys, owing to their low melting temperatures (~650°C for pure Mg). Such limitations have prevented application of Mg alloys in powertrain applications. Powell and colleagues reported on the mechanism for loss in creep strength of Mg-Al-RE (AE42) alloys that were originally developed by Dow for engine and transmission blocks, but suffered rapid property deterioration above 150°C. Importantly, they disproved the prevalent theory which postulated rare earth solutes in Mg as the predominant mechanism of creep resistance. The shortcomings of the AE and AS (Mg-Al-Si) systems for creep resistance fostered the development of new families of alloys, such as the Mg-Al-Sr alloy reported by Pekguleryuz and Baril that demonstrated enhanced creep resistance, mechanical properties, and—surprisingly—corrosion resistance. Suzuki et al. subsequently reported similar alloy investigations and improvements for Mg-Al-Ca alloys, and an excellent overview on the breadth and applicability of creep resistance alloys is presented by Pekguleryuz and Kaya. On the far end of the temperature spectrum, Lee and Kim reported on the critical evaluation of "fire-proof" Mg-alloys through the addition of a small percentage of CaO into conventional Al-containing alloys. These alloys showed incredible resistance to ignition, forecasting implementation in many environments.

Perhaps the most significant enhancement in Mg-alloy development has been the introduction of rare-earth containing alloys that have significantly opened up the design space through improvements in strength, plasticity, high-temperature properties, and flame resistance. Bakke and Westengen provided an early report on the role of rare-earth elements, along with Sr, Ca, and Si (as seen in prior papers in this section) on the formation of creep resistant microstructures and secondary phase inclusions in AE alloys. Their results point to grain boundary stabilizing dispersoid phases composed of Al/RE precipitates. This ability to engineer and control precipitates was also demonstrated by Rosale and colleagues on a Mg-Zn-Y alloy where low-aspect ratio β'1 (Mg4Zn7) precipitates were shown to form on twin boundaries, and to be very different from the previous rod-like precipitates seen in this alloy. The impact and value of the ability to tailor and control microstructures and precipitates by rare earth additions, as demonstrated in all the papers selected for inclusion in this section, is brought to light in the last paper. Okamoto and colleagues reported on the possibility of using Mg-Zn-(Y,Gd) alloys for engine pistons; possibly one of the harshest environments seen in an engine. Castings of Mg-2Zn-11Y-5Gd-0.5Zr alloys were gravity cast, T6 heat-treated, and compared with aluminum alloy A336, which is a commonly used piston alloy. While the fatigue strength of the Mg-alloy was 27% lower than the A336 at room temperature, it proved to be 35% higher at the piston operating temperature of 300°C, forecasting the application of Mg-alloys in entirely new and extreme environments.

From *Magnesium Technology 2000*, H.I. Kaplan, J.N. Hryn, and B.B. Clow, Editors

GRAIN REFINEMENT OF MAGNESIUM

Y.C. Lee, A.K. Dahle and D.H. StJohn

CRC for Alloy and Solidification Technology (CAST)
Department of Mining, Mineral and Materials Engineering
The University of Queensland
Brisbane Qld 4072, Australia

Abstract

Grain formation during solidification of magnesium and Mg-Al alloys has been studied with a focus on grain refinement mechanisms, solute and particle effects. The variation in grain size with increased aluminium content in hypoeutectic Mg-Al alloys showed a continuous decrease in grain size up to 5 wt% Al, and a stabilisation at higher Al contents (above 5 wt%). Strontium additions to both low- and high-aluminium content magnesium alloys showed that Sr had a significant grain refining effect in low-aluminium containing alloys. However, strontium had a negligible effect on grain size in the Mg-9Al alloy. Additions of Zr, Si, or Ca to pure magnesium produced significant grain refinement, probably because these elements have high growth restriction effects during solidification. An attempt was made to identify the grain refinement effect of particles added directly to the melt that are considered to be powerful nucleants in Al based alloys (TiC) and in Mg based alloys (AlN, Al_4C_3). Most of these particles produced grain refinement, probably because of enhanced nucleation due to the small lattice disregistry between their crystal structures and that of magnesium. However, it is not clear whether the grain refining mechanism of the effective particles was catalysis of primary crystal nucleation or simply restriction of crystal growth during solidification.

1 Introduction

Unlike Al-based alloys, casting is the most commonly used processing method for Mg-based alloys (1). Most of the production is normally by high pressure die casting, but other casting techniques are also being used or developed for use. An important area of concern is the ability to control the grain size of the alloy in the final cast products, particularly in casting processes where the cooling rate is low.

The main advantage of fine grain size is enhanced mechanical properties such as tensile and yield strength. This relationship can be explained by the Hall-Petch equation ($\sigma_y = \sigma_0 + Kd^{-1/2}$ where σ_0 is yield stress of a single crystal, K is constant and d is grain size). The effect of grain size on mechanical properties is more significant in alloys which have a H.C.P structure, such as Mg and Ti, than in many other alloys with different crystalline structures (F.C.C or B.C.C), since there are fewer slip systems in the H.C.P structure (2). Other advantages of a fine grain size include a reduction in hot tearing tendency and in the size of defects such as porosity. Heat treatment is more effective in Mg-Al alloys with a finer grain size because of the smaller diffusion distance for the dissolution of $Mg_{17}Al_{12}$ phase which is present on the as-cast grain boundaries. In addition, grain refinement improves resistance to corrosion as well as machinability of cast products (2).

A universally reliable grain refiner system does not exist for the wide range of Mg-based alloys as it does for Al-based alloys, where well established, reliable grain refiners are commercially available. While there are several grain refinement methods available for Mg alloys, none is regarded as a reliable method for a broad range of Mg-based alloys. The method used depends on the alloy system, and even on the specific composition of the alloy. Zr additions are considered to be the most reliable and effective grain refiner for some magnesium alloys, but have limited applicability. The mechanisms underlying the various grain refinement methods for magnesium have not yet been established, and this lack of understanding creates confusion and leads the industry to rely on empirical approaches for grain refinement. Because of the significance of grain refinement in the development of improved performance magnesium alloys, the current work aims to improve understanding of the factors controlling grain refinement in Mg alloys.

Work on the mechanisms of grain refinement has been concentrated on the popular Al-Ti-B grain refiners for aluminium alloys and found that the presence of powerful nucleants in the melt is not the only requirement for the formation of fine equiaxed grains and subsequent grain refinement (3). A certain degree of driving force, which depends on the nucleating potency of the particles, is also required to activate those nucleants, and the driving force is usually provided by undercooling (thermal, constitutional and capillarity undercooling) during the solidification (4). As constitutional undercooling plays a more significant role in casting of alloys, the role of solute elements on grain formation and subsequent grain refinement in Al-based alloys has been investigated and well documented (5-7).

The current understanding incorporates the role of both the solute present in an alloy and the potency of nucleant particles. These two aspects are considered separately in this investigation.

The effect of solute additions on grain size can be understood in terms of the growth restriction factor (GRF). GRF equals $\sum_i m_i C_{0,i}(k_i - 1)$ where m_i is the slope of the liquidus line (assumed to be a straight line), k_i is the distribution coefficient and $C_{0,i}$ is the initial concentration for element i (8). It should be noted that this equation is only applicable for alloying elements present in dilute concentration levels (8). Addition of alloying elements or solute generates constitutional undercooling in a diffusion layer ahead of the advancing solid/liquid interface which restricts the growth of the crystals. The growth of crystals is largely controlled by diffusion of the rejected solute at the interface since it changes the equilibrium freezing temperature. Creation of the constitutionally undercooled region in front of the interface also increases the number of nucleation events because nucleants in the melt are

number of nucleation events because nucleants in the melt are more likely to survive and be activated in the constitutionally undercooled zone.

The GRF values for various alloying elements in magnesium alloys have been determined using binary phase diagrams and are presented in Table 1. The GRF values of Zr, Ca, and Si are relatively large compared to those of other elements such as Al, Y, Sn etc., indicating more powerful growth restriction, and hence stronger grain refining ability with small addition levels of these elements. Based on these calculations the experimental work focused on the three elements with the highest GRF and also Al because it is the most common alloying element with magnesium and Sr as it has been shown to produce refinement (1, 9).

Table 1 Slope of the liquidus line (m), equilibrium distribution coefficient (k) and growth restriction parameter m(k-1) for various alloying elements in magnesium (based on the assumption that the liquidus and solidus are straight lines).

Element	m	k	m(k-1)	System
Zr	6.90	6.55	38.29	peritectic
Ca	-12.67	0.06	11.94	eutectic
Si	-9.25	≈0.00	9.25	eutectic
Ni	-6.13	≈0.00	6.13	eutectic
Zn	-6.04	0.12	5.31	eutectic
Cu	-5.37	0.02	5.28	eutectic
Ge	-4.41	≈0.00	4.41	eutectic
Al	-6.87	0.37	4.32	eutectic
Sc	4.02	1.99	3.96	peritectic
Sr	-3.53	0.006	3.51	eutectic
Ce	-2.86	0.04	2.74	eutectic
Yb	-3.07	0.17	2.53	eutectic
Y	-3.40	0.50	1.70	eutectic
Sn	-2.41	0.39	1.47	eutectic
Pb	-2.75	0.62	1.03	eutectic

Investigations reported in the literature related to various methods of grain refinement have produced results that appear to contradict the above GRF-based hypothesis. For example, grain refinement achieved by the superheating method is suppressed by the presence of certain other alloying or trace elements such as Ti, Zr and Be (10-12). Alloying elements such as Zr and Be also interfere with the Elfinal process, causing grain coarsening by forming intermetallic compounds with iron (13). Carbon addition methods have become the major grain refinement technique for magnesium alloys containing Al because of the practical advantages derived from the low operating temperature, increased melt volumes, and grain refining power equivalent to the Superheating and Elfinal methods. The alloys must contain above 0.5 wt% aluminium for successful grain refinement, while Zn and Mn have negligible effects (14). Zr, Be and Ti cause grain coarsening as they do in the Superheating process, even when present in very dilute concentrations.

On the other hand, Zr can be used for grain refinement of pure Mg and Mg alloys containing Zn, rare earth elements, Ag or Th (15). This method of grain refinement is capable of producing extremely fine grains (significantly finer than those produced by other methods). The mechanism of Zr grain refinement in magnesium is not well understood, but the most commonly accepted hypothesis relates to lattice disregistry between Mg and Zr. Zr has the same crystal structure as Mg (H.C.P.), and the lattice parameters of Zr (a = 0.323 nm, c = 0.514 nm) are very close to those of Mg (a = 0.320 nm, c = 0.520), suggesting that Zr particles, elemental or intermetallic, could be potent nucleants in magnesium. However, Zr is soluble in magnesium to about 0.58 wt% Zr in the melt.

A major problem with this method is the limited choice of alloy systems to which Zr can be applied. For example, magnesium alloys that contain Al can not be grain refined by Zr additions (15). In spite of this problem, however, Zr addition is the most commonly used grain refinement method for alloys such as Mg-Zn or Mg-Zn-Y.

It has recently been reported that Sr is an effective grain refiner of Mg-Al alloys. The effect of Sr additions was investigated by Gruzleski and Aliravci (9) who reported that Sr produced significant grain refinement in AZ91 alloy (in this experiment the base material had been pre-grain refined by carbon addition). Similar results have been reported by Nussbaum et al. (16), even though their base material was not pre-grain refined. They also found that as the amount of Sr was increased, a moderate amount of Sr precipitates, such as Al_4Sr and Mg_2Sr, was formed, and although it has not yet been determined whether these precipitates are potent nucleation sites, the possibility of nucleation by these particles can not be excluded.

The role of particles in effecting refinement of magnesium alloys is also unclear as evidenced above in relation to Zr and Sr additions, and therefore requires further investigation. Emley (10) has suggested that grain refinement by the Elfinal method involves nucleation by Fe-Mn-Al compounds. With regard to carbon treatments, aluminium carbide (Al_4C_3) or a combination of Al_4C_3 and AlN particles are proposed to form during the carbon treatment and act as the major nucleating particles (10, 11). As a result, AlN and Al_4C_3 particles have been proposed to be excellent nucleants for Mg-based alloys (17, 18). Recently, grain refinement of AZ91 by addition of SiC particles has been investigated (19, 20). It was found that significant grain refinement results, possibly from enhanced heterogeneous nucleation that occurs because of a relatively small lattice disregistry between SiC and Mg. Luo (20) found that some of the SiC particles were pushed out to the grain boundaries and therefore proposed that the grain refinement effect of SiC may also be attributed to a reduction of the growth rate of the primary phase.

Based on the above, AlN, Al_4C_3 and SiC particles were selected to be deliberately added to the melt to determine their relative effect on grain size. In addition, the effect of TiC particles is also investigated as TiC is a relatively well-known grain refiner for various Al-based alloys (21, 22), however its effect on grain size of Mg-based alloys has not previously been investigated.

2 Experimental

Binary Mg-Al alloys were prepared in an electric resistance furnace using commercial purity Mg ingot (99.7%) and Al ingot (99.8%). Melting and casting were conducted under a protective atmosphere of 0.5% SF_6 in air using mild steel crucibles which were coated with a thin layer of boron nitride to prevent direct contact of the melt with the crucibles. The addition of grain refiner or master alloy was carried out at 700°C and the melt was held at that temperature for approximately 15 minutes. The melt was stirred for 30 seconds, 5 minutes prior to pouring and then cast into steel moulds which had been preheated to 200°C. The moulds consisted of cubic blocks with 100 mm sides, each with a cylindrical cavity of 25 mm radius and 70 mm depth in the centre.

The effect of additions of strontium, zirconium, silicon, calcium and aluminium on grain size was investigated. Two different types of Sr-Al master alloys were used: a 90 wt% Al-10 wt% Sr master alloy rod for low Sr additions (0.01 and 0.1 wt% Sr), and a 90 wt% Sr-10 wt% Al master alloy for higher Sr addition levels (>0.1 wt% Sr). A Mg-Zr master alloy containing 33 wt% Zr was used for Zr additions, while the Si and Ca additions were made using commercially available Si chips and granular Ca respectively. The effect of additions of TiC, AlN, Al_4C_3 and SiC particles on grain size in Mg-1Al was investigated by submerging pellets containing 1 vol% of the particles. The particles were mixed with pure magnesium chips in a tumble mixer for 20 minutes in order to achieve a homogenous mixture, and then compressed to form pellets. The pellets were submerged in the melt and then stirred by hand with a small stirrer for 30 seconds.

To reveal the grain boundaries, each sample was sectioned in half vertically or transversely and subjected to solution heat treatment in an air furnace. The samples were held in the furnace for 30 minutes at 413°C, then air-cooled. Alloys containing less than 5 wt% Al were etched with acetic-picric acid, and those containing more than 5 wt% Al were etched with citric acid. The etching was performed at room temperature for 2-3 seconds. Because of the different cooling rates between the centre and edge of the samples, 2 optical micrographs were taken for grain size measurement from each of three different sections (edge, half radius and centre) of each sample. Grain size measurements were performed using the lineal intercept method described in ASTM standard E 112-88, and more than 50 intercepts were counted for each grain size determination. Chemical composition of the samples was determined with the ICP-AES method.

3 Results and Discussion

3.1 Effect of aluminium on grain size

The addition of Al resulted in some grain refinement as shown in Figure 1. Significant grain size reduction was observed upon the initial additions of Al (1 and 3 wt%), and the reduction in grain size continued as the Al concentration was increased up to about 5 wt%. However, increasing the Al content beyond 5 wt% produced no further significant refinement of the grain size. As shown in Figure 1, the grain size of AZ91 was comparable to that of the binary Mg-9Al alloy. As the aluminium content increased, the variation in grain size in the different sections of the sample (edge, half radius and centre) was reduced, together with a suppression of the columnar structure at the walls. While pure magnesium exhibited long columnar grains, this was suppressed by about 75% and the columnar grains were also much narrower when 3 wt% Al was added. No columnar structure was detected at Al concentrations exceeding 5 wt%. A change in growth morphology, from cellular to dendritic, was also observed with increasing Al content. This morphological transition was apparent at aluminium contents between 5 and 7 wt%, and dendrites with sharp tips were fully developed at 9 wt% Al.

3.2 Effect of strontium on grain size

The effect of Sr additions (ranging from 0.01 wt% to 1 wt%) on grain size was investigated in both low (3wt%) and high (9wt%) Al content alloys. The results for the Mg-3Al alloy (Figure 2) showed a similar pattern to that observed with Al additions above (i.e. a large change in grain size with initial Sr additions and a small change upon further additions). The addition of 0.01wt% Sr had a dramatic grain refining effect. However, a further addition of Sr to 0.1 wt% caused a slight increase in grain size (or at least did not show any further significant grain refinement). A further addition of Sr to 0.3 wt% resulted in a decrease in grain size, and this trend continued with increasing strontium contents up to 1 wt%.

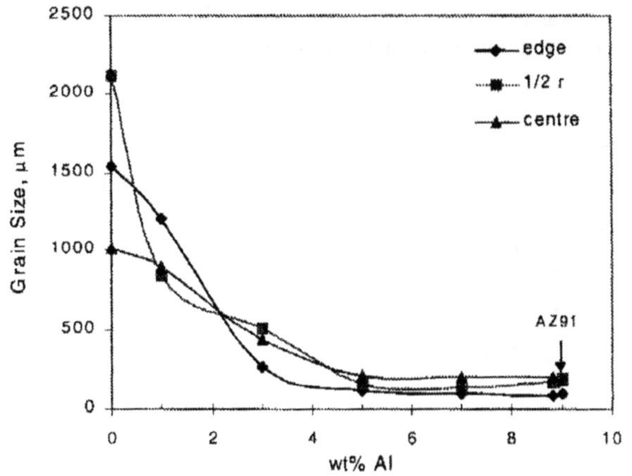

Figure 1 Variation in grain size of pure magnesium with increasing Al content.

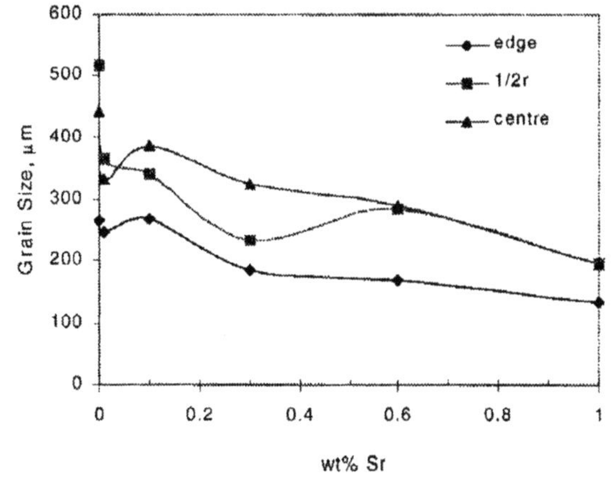

Figure 2 Variation in grain size of binary Mg-3 wt% Al alloy with increasing Sr content.

Sr additions showed a somewhat different effect on grain size of the binary Mg-9 wt% Al alloy compared to the Mg-3wt% Al alloy described above. Figure 3 shows that there were significant fluctuations in grain size with different levels of strontium additions. Additions of 0.01 wt% and 0.1 wt% Sr resulted in an increase in grain size, followed by slight reductions upon further strontium additions up to 1 wt% Sr. However, the extent of change in grain size over this range of Sr additions was within experimental error, and no significant difference was observed when the grain size of the alloy refined with 1 wt% Sr was compared with that of the unrefined binary Mg-9Al alloy.

3.3 Effect of silicon on grain size

From the grain size results shown in Figure 4, it can be observed that a small addition of Si produced a strong grain refining effect in pure magnesium. After a small addition of Si (0.07 wt%), long columnar grains were still present. However, when the Si level was increased to 0.15wt%, the columnar grains were suppressed and fine equiaxed grains were formed throughout the sample. No further significant grain size reductions were found for subsequent increases in Si level up to 0.5 wt%. Attempts to increase the silicon content beyond 0.5 wt% were problematic and although each melt was held at 700°C for two hours with constant stirring, complete dissolution of the Si chips was not achieved.

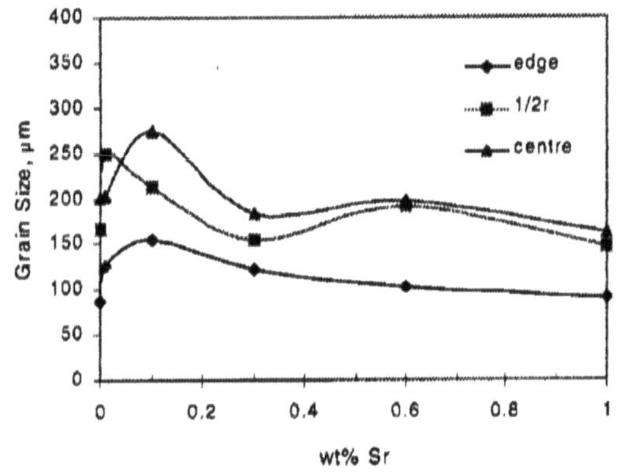

Figure 3 Variation in grain size of binary Mg-9 wt% Al alloy with increasing Sr content.

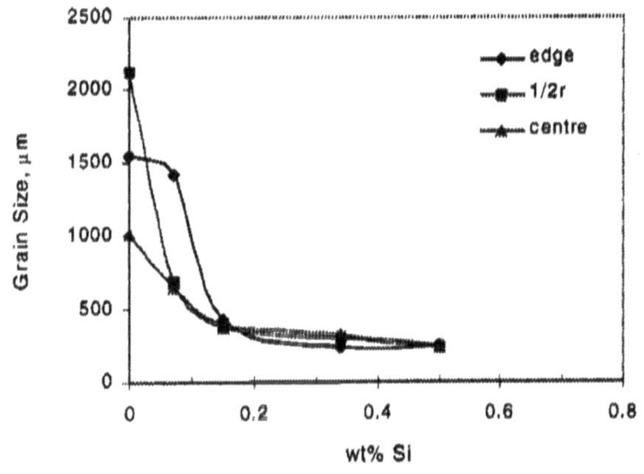

Figure 4 Variation in grain size of pure magnesium with increasing Si content.

3.4 Effect of zirconium on grain size

Grain size measurements in the Mg-Zr alloys were only performed at the centre of each sample because well developed equiaxed grains were distributed homogeneously throughout the samples, except for the low Zr content sample (0.04 wt% Zr) which exhibited pockets of small equiaxed grains surrounded by large columnar grains. The variations in grain size of pure magnesium with Zr additions are shown in Figure 5, and corresponding macrographs are presented in Figure 7. As can be observed in Figure 5, the trend in grain size with addition level is very similar to that observed with Al and Si additions. Initially, grain size decreased rapidly as the Zr content increased. Further additions of Zr (up to 0.15 wt% Zr) resulted in a more gradual reduction in grain size. Once the Zr level was increased to about 0.32 wt%, the magnitude of grain size reduction was minimal and the grain size appeared to be reaching the saturation level. It should be noted that fine equiaxed grains were only observed in the centre of the sample at 0.04 wt% Zr and the size of these fine grains are therefore not representative of the overall sample grain size. The Zr concentration was targeted to vary from 0.1 wt% to 1 wt%, however chemical analysis results showed poor recovery and consistently low Zr concentrations.

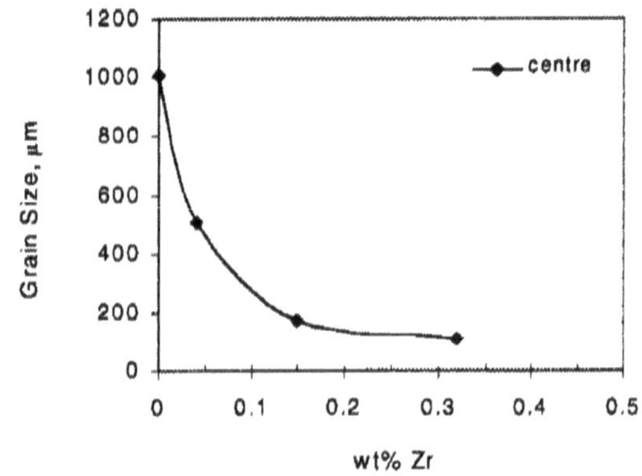

Figure 5 Variation in grain size of pure magnesium with increasing Zr content.

3.5 Effect of calcium on grain size

Significant reductions in grain size were observed with Ca additions to pure magnesium. As can be observed in Figure 6, grain size behaviour with increased Ca level displayed a similar trend to that found for the other elements studied (i.e. a large reduction in grain size with the initial additions and then a relatively minor effect on grain size with subsequent additions, followed by stabilisation of grain size). Grain size seems to have reached a saturation level after the addition of 0.4 wt% Ca, with further additions having an insignificant effect on grain size.

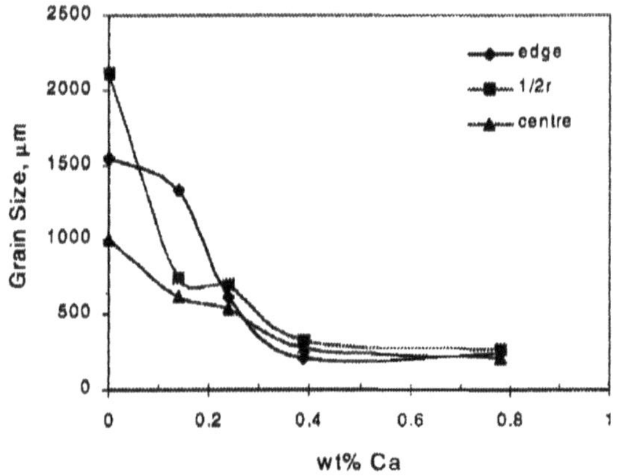

Figure 6. Variation in grain size of pure magnesium with increasing calcium content.

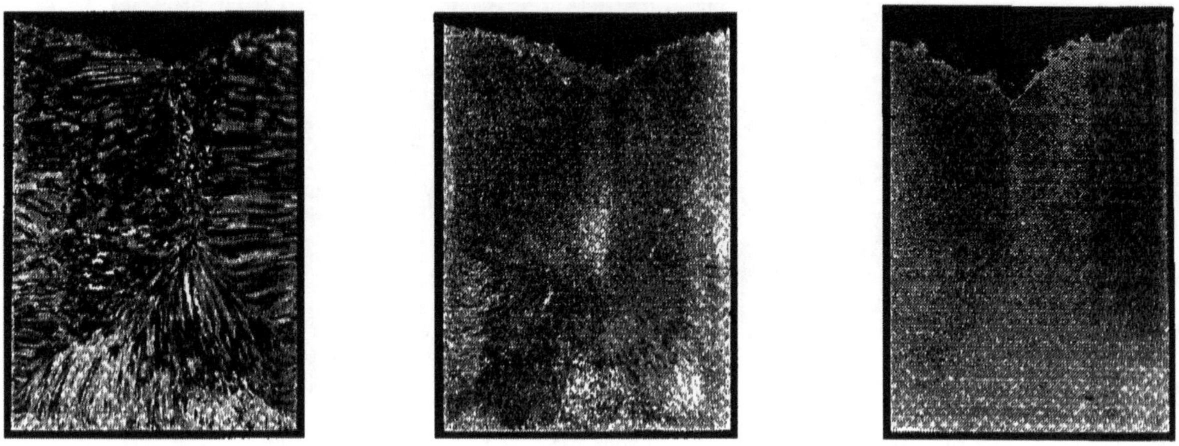

a) Mg-0.04 wt% Zr b) Mg-0.15 wt% Zr c) Mg-0.32 wt% Zr

Figure 7 Macrographs of pure magnesium with various Zr contents. In the Mg-0.04 wt% Zr alloy, a large pocket of equiaxed grains is surrounded by columnar grains. As the Zr content increases, fine equiaxed grains are formed throughout the samples.

3.6 Summary of Solute Effects on grain size

Based on GRF calculations, addition of Zr, Ca or Si should lead to a more significant reduction in grain size than a similar addition of Al. Addition of less than 1 wt% of these elements was indeed found to produce more effective grain refining than additions of relatively large amounts of Al (up to 9 wt%). Zr has the highest value of GRF among the common alloying elements in magnesium, indicating significant grain refinement potential with its strong segregating power. This is consistent with the experimental results reported above.

An interesting result was observed for Zr additions. The results of chemical analyses showed that the maximum Zr concentration achieved in the experimental samples was 0.32 wt%, which is far below the peritectic point (0.56 wt%). It therefore seems unlikely that Zr is present in the form of nucleant particles, and the grain refinement achieved by the addition of Zr may instead be caused by the high growth restriction of Zr as solute. Alternatively, there may be undissolved Zr particles present, in elemental form or as intermetallic compounds, which may act as nucleants, even below the peritectic point. This possibility is supported by a comparison of the minimum (saturation level) grain size of the samples. The experimental results show a minimal reduction in grain size with further additions of alloying elements after the saturation level is reached. Therefore it is reasonable to assume the minimum grain size achievable with addition of Zr to be about 100 μm (actual grain size with 0.32 wt% Zr), while it is about 200 μm for aluminium additions under the casting conditions used in the current work. Because grains are actually three-dimensional, there is a cube-root relationship between grain volume and grain size measurement. Because the saturation grain size achieved with zirconium addition is approximately half of that established for additions of aluminium, there must be about eight times as many nucleant particles activated in the Zr-refined sample compared with that in the Al sample. Since only a small proportion of nucleants present in a melt are activated, the difference in the actual number of nucleants present in the Zr and Al samples may be enormous.

3.7 Effect of particle additions on grain size of Mg-Al alloys

Various particles which are known to be good grain refiners for aluminium or magnesium alloys, such as TiC, SiC, AlN and Al_4C_3, were added to a Mg-Al melt in pellet form and the resultant grain sizes were compared. The resultant grain sizes achieved by the addition of these different particles to a Mg-1Al alloy are shown in Figure 8. Corresponding macrographs for the particle additions are shown in Figure 9. All particle additions produced some grain refinement, while SiC and Al_4C_3 particles performed slightly better than AlN and TiC particles

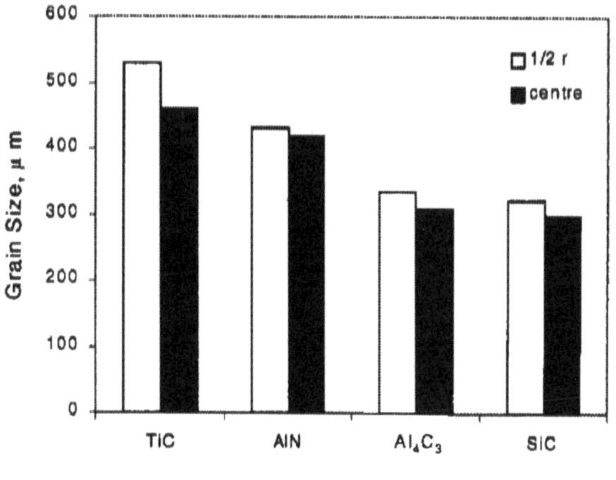

Figure 8 Grain size of binary Mg-1Al alloy after addition of 1 vol% of various particles. The grain size of the unrefined alloy was around 550 μm. These experiments were undertaken with pure magnesium obtained from a different source. The grain size obtained for Mg-1%Al was less than obtained with the other magnesium used in the solute experiments. Differences in impurity content between the two sources of magnesium may explain this difference in grain size.

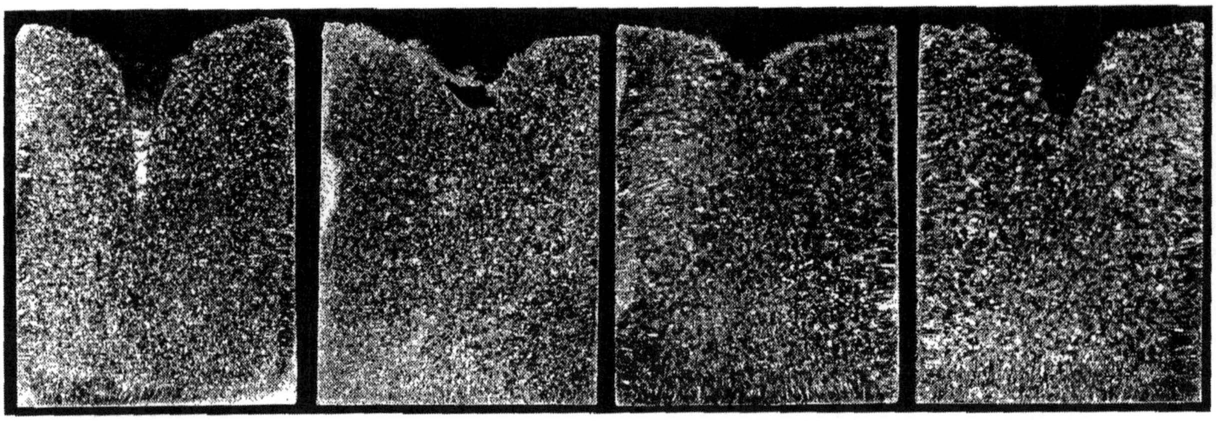

a) 1 vol% SiC b) 1 vol% Al_4C_3 c) 1 vol% AlN d) 1 vol% TiC

Figure 9 Macrographs of Mg-1 wt% Al binary alloys after the addition of various particles.

Previous studies of the behaviour of SiC particles during solidification of alloy AZ91 (19, 20, 23) have shown that the SiC particles were found both within the primary magnesium dendrites as well as in intergranular regions. Particles within the primary dendrites may be an indication of heterogeneous nucleation potential. To date there have been no reported experimental observations on whether the primary magnesium can nucleate directly on SiC particles. As pointed out by Luo (20), the possibility of nucleation on the SiC (β-SiC) particle may increase with good lattice matching (low disregistry) with Mg. In classical nucleation theory, nucleation can be greatly facilitated when the foreign particle is wet well by the liquid (i.e low wetting angle) and also displays a small lattice mismatch with the nucleating solid phase (4). The lattice mismatch between magnesium and the particles investigated in the present work has been calculated to provide some estimate of the nucleating potency of the particles. The results are presented in Table 2.

Table 2 Crystal structure of the particles and their lattice parameters (24), and the calculated lattice disregistry between the particles and magnesium.

Name	Structure	Lattice Parameter (nm)	Disregistry
AlN	Hexagonal	a = 0.31 c = 0.497	3.4 %
Al_4C_3	Hexagonal	a = 0.333 c = 0.249	3.8 %
SiC	Cubic (ZnS)	a = 0.435	4 %
TiC	Cubic (NaCl)	a = 0.303	4.8 %

All particles have a relatively small lattice disregistry with Mg suggesting that they may provide potential heterogeneous nucleation sites with good wetting. However, the grain size cannot be predicted by considering the lattice mismatch alone since the size and the distribution of the particles in the melt are also critical factors that must be considered. The size of the particles used in the present work was less than 325 mesh (45 μm), and the particle size distribution varies depending on the manufacturing method used to produce the particles.

It is difficult to achieve a homogenous distribution of particles throughout the melt and a large amount of agglomerated particles was found in all samples. In particular, agglomerates were found close to the bottom of the samples indicating significant settling of the particles. Thus the nucleating potency of the particles can not be directly evaluated and compared by particle additions to the melt, as differences in the stirring procedure and the particle settling and agglomeration characteristics could influence the results.

However, in addition to the direct nucleation of Mg on the particles, grain refinement can also be expected to be induced by particles found in the interdendritic regions. Pushing of the particles by the growing crystals occurs because of various factors such as growth rate, surface tension, heat transfer coefficients of particles and metal and fluid flow characteristics of the melt (25). Thus the particles may play a significant role in terms of grain refinement by restricting growth during solidification. Since an enrichment of particles was observed at the grain boundaries of all samples produced in the present work, the possible contribution of growth restriction by these particles to grain refinement should be considered.

The superior performance of SiC particles in terms of grain refinement could be further explained by the reaction of the particles with the matrix. The presence of Mg_2Si between the SiC particles and the matrix interface was observed, suggesting that the particles react with aluminium in the melt to produce Al_4C_3 and Si (19). Both Al_4C_3 and Si have a significant effect on grain size, as confirmed by the present work, and improved grain refinement could have resulted from the liberation of these reaction products in a chemical reaction. However, it is also possible that all liberated Si is tied up in Mg_2Si precipitates.

Conclusions

The addition of certain alloying elements (Al, Si, Ca, Zr) to pure magnesium results in a reduction in grain size. A significant reduction occurs with initial additions of these elements, and the degree of grain size reduction decreases with further additions.

- The addition of aluminium produces moderate grain refinement, mainly due to growth restriction.
- The addition of strontium produces significant grain refinement in low-Al content alloys (Mg-3 wt% Al). However, addition of Sr to a high Al content (Mg-9 wt% Al) binary alloy does not produce significant grain refinement.
- Small additions of Si, Ca, and Zr to pure magnesium produce significant grain refinement due to their strong segregating power. Zr has the highest value of the growth restriction factor among the elements investigated in this work and produces the finest grain size.
- The addition of Al_4C_3, SiC, AlN and TiC particles to a Mg-1Al alloy produced some grain refinement. All of these particles have a relatively small lattice disregistry with magnesium. The grain refinement obtained by the addition of AlN and TiC particles was not as strong as that obtained by Al_4C_3 and SiC additions.

Acknowledgment

The CRC for Alloy and Solidification Technology (CAST) is acknowledged for financially supporting the project. CAST was established under the Australian Federal Government's Co-operative Research Centres Scheme.

References

1. I.J. Polmear, Light Alloys: Metallurgy of the Light Metals, (London, UK: Arnold, 1995) 197-209.

2. J. Campbell, "Castings", (Oxford, UK: Butterworth-Heinemann, 1997) 261-262.

3. M.A. Easton, "Grain Refinement Mechanisms in Aluminium and its Alloys and the Effect of Grain Refinement on Castability," (Ph.D. thesis, The University of Queensland, 1999), 57-85.

4. W. Kurz, and D.L. Fisher, Fundamentals of solidification, (Aedermannsdorf, Switzerland: Trans Tech Publicatons, 1989) 21-42.

5. J.E.C. Hutt, D.H. StJohn, and A.K. Dahle, "Equiaxed Solidification of Al-Si Alloys," Materials Science and Technology, 15 (1999) 495-500.

6. Y.C. Lee, A.K. Dahle, and D.H. StJohn, "The Effect of Grain Refinement and Silicon Content on Grain Formation in Hypoeutectic Al-Si Alloys," Materials Science and Engineering A A 259 (1999), 43-52.

7. M.A. Easton, and D.H. StJohn, "Grain Refinement of Aluminium Alloys II. Conformation of, and a Mechanism for the Solute Paradism," Metallurgical and Materials Transactions A 30A, (1999), 1625-1633.

8. A.K. Dahle, and L. Arnberg, "On the Assumption of Additive Effect of Solute Elements in Dendrite Growth," Materials Science and Engineering A, A225, (1997) 38-46.

9. J.E. Gruzleski, and C.A. Aliravci, "Low porosity, Fine Grain Sized Strontium Treated Magnesium Alloy Castings," United States Patent No 5,143,564 (1992)

10. F. Emley, Principles of Magnesium Technology (Oxford, UK: Pergamon, 1966), 260.

11. L.A. Davis, and L.W. Eastwood, "Grain Refinement of Magnesium Casting Alloys," Trans AFS, 53 (1945) 352-362.

12. N. Tiner, "Superheating of Magnesium Alloys," Metals Technology, 12 (1) (1945) 1-19.

13. E.L. Glasson and E.F Emley, "Heterogeneous Nucleation in the Solidification of Aluminium and Magnesium Alloys," The Solidification of Metals, (London, UK: The Iron and Steel Institute, 1967) 1-9.

14. V.B. Kurfman,, "Nucleation Catalysis by Carbon Addition to Magnesium Alloys," Trans AIME, 221 (1961) 540-546.

15. D.O. Karlsen, D. Oymo, H. Westengen, P.M.D Pinfold, and S.I. Stromhaug, "Development of Grain Refiner for Aluminium Containing Magnesium Alloys," Light metals Processing and Applications, Canadian Inistitute of Mining, Metallurgy and Petrolium Eds, (Canadian Inistitute of Mining, Metallurgy and Petroleum, Quebec, Canada, 1993) 397-408.

16. G. Nussbaum, P. Bridot, T.J. Warner, J. Charbonnier, and G. Regazzoni, "New Mg-Al Based Alloys with Improved Casting and Corrosion Properties," Magnesium alloys and their applications, B.L. Mordike, and Hehmann Eds., (Oberursel, Germany: DGM informationsefesellshaft, 1992) 351-368.

17. C.E. Nelson, "Grain Size Behaviour in Magnesium Casting Alloys," Trans AFS, 56 (1948), 1-23.

18. R.T. Wood, "Grain Refinement of Magnesium Castings," The Foundryman, (1953) 98-99, 256-261.

19. A. Luo, and M.O. Pekguleryuz, "Solidification Processing and Microstructure of Cast Magnesium MMCs," Trans AFS, 102 (1994) 313-320.

20. A. Luo, "Heterogeneous Nucleation and Grain Refinement in Cast Mg(AZ91)/SiC Metal Matrix Composite," Canadian Metallurgical Quarterly, 35 (4) (1996) 375-383.

21. K. Sato and M.C. Flemings, "Grain Refinement of Al-4.5 Cu Alloy by Adding an Al-30TiC Master Alloy," Metallurgical and Materials Transactions A 29A (1998) 1707-1710.

22. D.G. McCartney,"Grain Refining of Aluminium and Its Alloys using Inoculants," International Materials Reviews 34 (1989) 247-260.

23. A. Luo, "Development of Matrix Grain Structure During the Solidification of a Mg(AZ91)/SiC Composite," Scripta Metallurgica et Materialia., 31 (1994) 1253-1258.

24. E.A. Brandes, ed., Smithells Metals Reference Book 6th ED (London, UK: Butterworth, 1983), 6-1~6-35.

25. P.K. Rohatgi, "Interfacial Phenomenon in Cast Metal-Ceramic Particle Composites," Interfaces in Metal-Matrix Composites (Warrendale, USA: The Metallurgical Society/AIME, (1986), 185-202.

PRELIMINARY INVESTIGATION ON THE GRAIN REFINEMENT BEHAVIOR OF ZrB_2 PARTICLES IN Mg-Al ALLOYS.

G. Klösch[1], B. J. McKay[2], P. Schumacher[1,2]
[1] Austrian Foundry Research Institute; Parkstrasse 21, Leoben, Styria, A-8700, Austria
[2] University of Leoben; Chair of Casting Research, Franz-Josef Strasse 18, Leoben, Styria, A-8700, Austria

Keywords: Grain Refinement, Magnesium Alloys, Nucleation

Abstract

This paper investigates the effect of ZrB_2 particles on the grain refinement of an Mg-Al alloy. A comparison, with respect to grain size, of the effect of ZrB_2 particles formed in situ and commercially obtained synthetic ZrB_2 particles, which are subsequently added to the melt, was made. In situ ZrB_2 particles were formed by reacting conventional Al-Ti-B and Al-Zr alloys. Samples were taken in accordance with the TP1 test procedure and the resulting grain size of the primary Mg measured using the intersect method. An SEM equipped with EDS was employed to elucidate the effect of the Zr. Results show that the ZrB_2 successfully grain refines the Mg-Al alloy resulting in ultimate grain sizes of 100 and 60 µm for the in situ and synthetic ZrB_2 particles respectively. Mg-Al alloys can be successfully grain refined using ZrB_2 heterogeneous particles and the resultant effect should be beneficial in improving the mechanical properties of the alloy.

Introduction

Mg alloys are used for a wide variety of applications. Mg castings have become more prevalent due to their high specific strength and low weight in transport industries especially automotive [1]. The high demand in the automotive industry for weight savings has resulted in successful applications for Mg products which include seat frames, inner door frames, instrument panels, steering wheels, cylinder head covers etc. The majority of the Mg alloys produced are used for die-casting but over the last few years the demand for Mg wrought products has increased [2].

One of the most effective approaches to improve the mechanical properties of Mg alloys is to reduce the grain size of the primary Mg. Due to the hexagonal close packed lattice structure of the Mg and the poor plastic deformation rates at room temperature a fine grain size improves the plastic deformation behavior. The grain size of castings can be influenced by modifying casting parameters or by adding alloying elements, which contain nucleants. For Al alloys there are reliable grain refiners commercially available such as TiB_2 and TiC based master alloys. Unfortunately this is not the case for all Mg alloys [3].

Nucleation

For non Al containing alloys peritectic Zr additions (Zr > 0.6 wt. %) have been found to be beneficial in decreasing the grain size effectively. Zr particles act as nucleation sites and free Zr can act as a growth restrictor. The lattice parameters of Zr are very close to those of Mg (a = 0.320 nm, c = 0.520 nm) and if undissolved Zr particles exist, they may readily act as nucleants. Another problem is the correct particle size because Zr particles settle relatively fast in molten Mg. The driving force is the difference between the densities of Zr and the molten Mg ($\delta_{Zr}/\delta_{Mg} \approx 4$). The optimum overall process particle size lies between 2 and 5 µm. For classical nucleation and typical values for the interfacial energy ($\gamma_{sl} = 0.558$ J/m^2), entropy of fusion ($\Delta S_f = 0.68506 \cdot 10^6$ J/m^{3}K^{-1}) and undercooling $\Delta T = 1°C$ a critical radius is on order of 1.3 µm. The particle substrate diameter should lie in the same range (Eqn 1) as twice the critical nucleus radius.

$$r_{crit} = -\frac{2 \cdot \gamma_{sl}}{\Delta S_f \cdot \Delta T_m} \qquad \text{(Eqn. 1)}$$

r_{crit}	critical nucleation radius
γ_{sl}	surface interfacial energy
ΔS_f	entropy of fusion per unit volume
T	temperature

A particle size of ~ 3 µm is therefore a good compromise between settling velocity and nucleant potency and efficiency [4].

For effective heterogeneous nucleation the nucleant, the lattice mismatch, and the chemical interaction, such as a peritectic reaction, all play an important role in the effect on the overall interfacial energy (γ_{sl}) [5]. Table 1 shows the lattice parameters for Zr, ZrB_2 and Mg and the calculated lattice mismatch. ZrB_2 is a promising candidate to act as an effective nucleant for Mg due to the small mismatch.

Table 1: ZrB_2 particles and the lattice match to Mg [6].

Particle	Melting point [°C]	Lattice Parameters [nm] a	Lattice Parameters [nm] c	Mismatch compared to Mg [%]
Zr	1855	0.323	0.514	< 1.00
ZrB_2	3038	0.31650	0.35200	1.33
Mg	650	0.32075	0.52076	-

Growth

The role of solute elements on grain size is as important as that of the nucleant. After initial nucleation the growth velocity determines the rate of heat released by recalescence. This decreases the activation undercooling for the available nucleating particles and reduces the number density of growth centers. The solute elements generate a constitutional undercooling in the diffusion layer ahead of the advancing solid/liquid interface, which restricts the grain growth. Therefore in front of the interface, nucleants are more likely to survive and be activated. The growth restriction factor (Q) for equiaxed growth can be estimated from binary phase diagrams:

$$Q = \sum_i m_i C_{0,i}(k_i - 1) \qquad \text{(Eqn. 2)}$$

Where m_i is the slope of the liquidus line, k_i the distribution coefficient and $C_{0,i}$ the initial concentration of element i [3, 7, 8]. From table 2 it can be seen that Zr and Fe are very good growth restrictors. Due to the corrosion behavior of Mg, the addition of Fe is not a suitable solution for alloys. However, only solute elements in the melt are freely available for growth restriction. Elements such as Al, Mn, Si, Fe, Ni, Sn and Sb are said to form stable compounds with the Zr [10]. In the case of Al containing alloys, Zr forms Al_3Zr and thereby generates a poisoning effect of the grain refinement mechanism [9].

Table 2: Growth Restriction Parameter (m(k-1)) for various alloying elements in Mg [10, 11].

Element	Q [m(k-1)]
Fe	52.56
Zr	38.29
Si	9.25
Zn	5.31
Cu	5.28
Al	4.32
Sn	1.47
Mn	0.15

Grain refiner addition

There are potentially two different routes to introduce grain refining particles into an Mg melt. Suitable particles may be mixed into the melt using conventional stirring. For successful nucleation these particles must be in direct contact with the melt and not covered by an oxide layer. However, the strong affinity of Mg to oxygen may reduce oxides formed prior to addition. The second route is to produce grain refining particles in situ i.e. within the melt, to act as nucleation sites. Particles, such as TiB_2 available from conventional grain refiner master alloys developed for Al alloys, can be transformed by Zr additions in Mg alloys to form ZrB_2. From the Gibbs energies of formation TiB_2 ($-\Delta G$ = 310.4 kJ at 727°C) should be transformed and ZrB_2 ($-\Delta G$ = 309.3 kJ at 727°C) should be formed because of the lower free Gibbs energy of formation for ZrB_2 [12].

The TP1 – test is commonly used in the Al industry to simulate the grain size of billets produced by DC-casting. This test can also be applied to Mg castings. For Mg e.g. AZ31, typical grain sizes are about 300 µm without using grain refiners. Of particular interest is a grain refining agent, which works with both Al and non Al containing alloys.

The aim of this work is to access the grain refiner efficiency of in situ formed ZrB_2 particles and added synthetic ZrB_2 particles in Mg melts containing Al.

Experimental Methods

In situ formation of ZrB_2

For the in situ formation of ZrB_2, two master alloys were used. The first was an Al-Ti-B master alloy. This was a typical master alloy used in refining Al. The second was an Al-Zr alloy, which contained 10 wt. % Zr. Table 3 shows the chemical composition of the Al-Zr master alloy. For the production of the new master alloy an induction furnace and a clay bonded crucible were used. 500 g of each alloy was added. The alloys were heated up to 800°C under an Ar atmosphere to protect the melt surface from excessive oxidation. After 30 minutes of continuous stirring the alloy was poured into a graphite coated steel ingot. After pouring, the alloy was examined metallographically, chemically by induced coupled plasma spectroscopy (ICP-AES method) and by a scanning electron microscope (SEM) equipped with EDS. Table 4 shows the chemical composition of the produced master alloy.

Table 3: Chemical composition of the Al-Zr master alloy (balance: Al).

Si [%]	Fe [%]	Cu [%]	Cr [%]	Sn [%]	Ti [%]	Ca [%]	Zr [%]
0.095	0.213	0.016	0.012	0.340	0.032	0.005	10.00

Table 4: Chemical composition of the produced grain refiner alloy.

	Zr [%]	Ti [%]	Al [%]
Chemical composition	5.1	1.3	balance

Grain refinement of Mg using the produced master alloy

Two tests with different Zr contents of 0.17 wt. % and 0.25 wt. % were performed. Pure Mg (table 5) was used and melted in an electrical resistance furnace using a graphite coated mild steel crucible. A schematic of the experimental setup is shown in figure 1. The overall Al content of the first melt was 2.7 wt. % (0.17 wt. % Zr) and for the second melt 3.7 wt. % (0.25 wt. % Zr) which resulted in Q-values of 18.17 and 25.55 respectively. The melt temperature was 730°C. The TP1 test apparatus (figure 2) was used for cooling the Mg samples. According to the TP1 test procedure described in the "Standard Test Procedure for Al Alloy Grain Refiners 1990, TP-1" [13] the samples are taken before and after adding the previously produced grain refiner additions. The graphite coated conical sampling ladle (fabricated of mild steel) was preheated to a temperature of 320°C (figure 3). To protect the melt surface form the excessive oxidation a gas mixture of Nitrogen and 1.25 vol. % SF_6 was used.

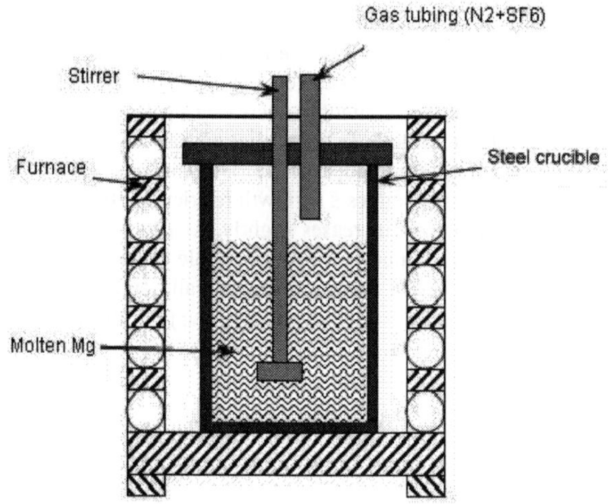

Figure 1: Schematic of experimental setup for melting.

Figure 2: TP1 test apparatus.

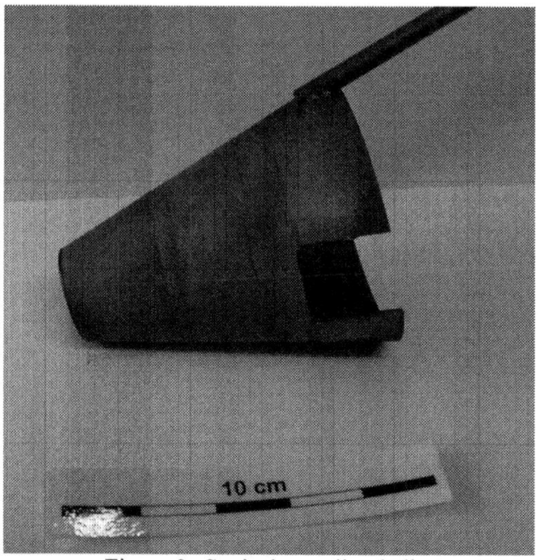

Figure 3: Conical sampling ladle.

Table 5: Chemical composition of the Mg alloy resulting in an overall Q-value of 0.50.

Al [%]	Si [%]	Mn [%]	Fe [%]	Zn [%]	Cu [%]	Ni [%]	Be [ppm]
<0.005	0.03	0.007	0.003	0.005	<0.002	<0.001	<1

One sample was taken before the addition of the master alloy to the melt and subsequent samples were taken at intervals of 1, 2, 5, 10, 20, 40, 60, 80, 100, 120, 180 and 185 minutes. After 120 minutes the continuous stirring was stopped. At 180 minutes a sample was taken and the stirrer restarted. After 5 minutes of stirring a sample was taken at 185 minutes to investigate if grain refiner performance is restored.

Grain refinement of Mg using synthetic ZrB_2 particles

The commercially produced synthetic particles were supplied by TREIBACHER Industrie AG with an average diameter of ~ 4 µm and a density of 6.1 g/cm^3.

Two series with different synthetic particle contents were cast. Approximately. 2 wt. % Al (99.9 %), which results in a Q factor of 9.12, was added to each Mg melt. The metal was melted in a resistive furnace using a graphite coated steel crucible. The melt temperature was constant at 730°C in all tests. A mixture of 1.25 vol. % SF_6 in N_2 was used for the melt protection.

Approximately 6 kg of Mg was used per test. 0.1 wt. % and 0.5 wt. % of the ZrB_2 powder was added in each test under continuous stirring of the melt. The stirrer was made of mild steel and constructed to distribute the particles throughout. Samples were taken at 1, 2, 5, 10, 20, 40, 60, 80, 100 and 120 minutes after the ZrB_2 addition to the melt, in accordance with the TP1 test procedure. All conical samples were sectioned at the same height of 38 mm and only the 12 x 12 mm mid section was used for metallographic examination (figure 4). After the preparation the samples were etched with picric acid and the grain size measured (horizontally and vertically) with the intersect method as described in the ASTM standard E 112-88, with more than 100 intercepts counted for each grain size determination.

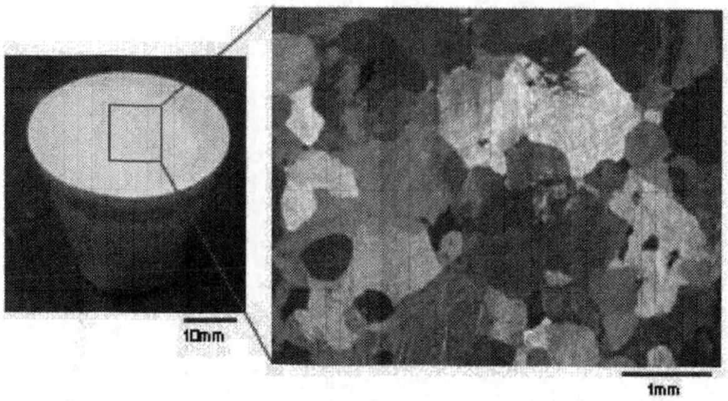

Figure 4: TP1 sample, the red box indicating the area of investigation.

Results and discussion

In situ formation of ZrB_2

The produced grain refiner alloy composed of Al-Zr and Al-Ti-B was examined with a scanning electron microscope, in order to locate ZrB_2 particles present in the samples. Because TiB_2 was not as stable at elevated temperatures than ZrB_2, TiB_2 should be transformed to ZrB_2 [14 - 16]. Figure 5 shows a scanning electron micrograph, in which two small particles are visible as bright rectangular particles in the vicinity of a larger grey $Al_3(ZrTi)$ phase (figure 5). EDS spectrums from these small bright particles revealed a large Zr peak and a smaller B peak (figure 6). This suggests that the particles are ZrB_2.Analysis from the surrounding matrix indicates the presence of Al and little Ti. Because of the excellent crystallographic match with the Mg these in situ formed ZrB_2 particles should readily act as heterogeneous nucleants. Furthermore, the in situ formation of the particles has the advantage, that there is no oxide on the surface of the particles.

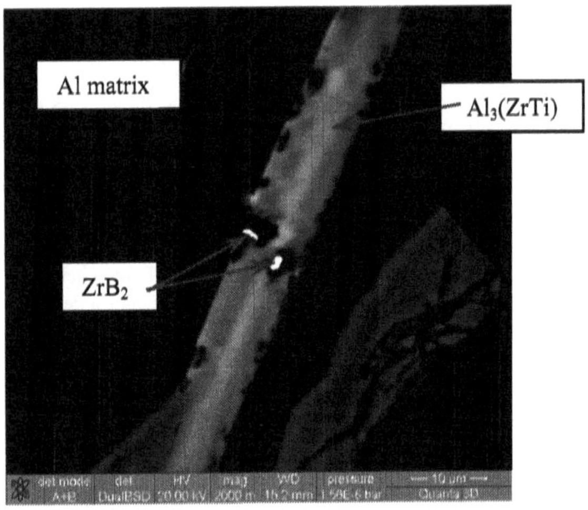

Figure 5: SEM BSE image, using Z contrast, of ZrB_2 particle (bright) and AlZrTi phases (grey).

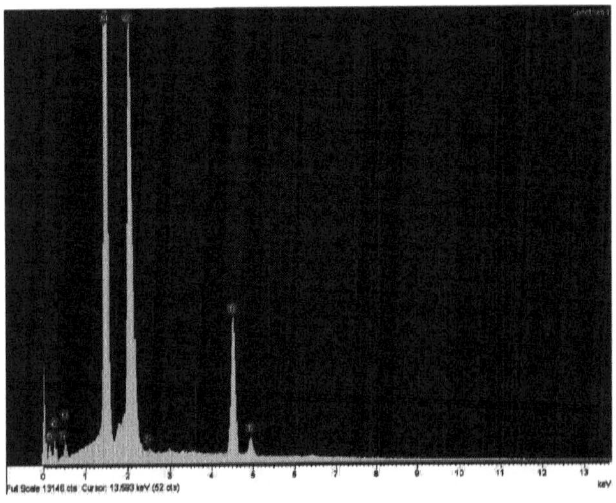

Figure 6: Corresponding EDS spectrum from ZrB_2 particle (bright) in figure 5.

Thereby the crystallographic match of the particle with Mg will not be influenced by oxides on the surface of the nucleant. As it can be seen in figure 7 the grain refinement effect is small with a Zr content of 0.17 wt. %. A smaller grain size of ~ 100 µm can be achieved when the Zr content is increased up to 0.25 wt. %. It should be noted that both Zr concentrations are below 0.6 wt. % Zr i.e. the additions are hypoperitectic. The smaller grain size with the higher Zr content suggests a growth restriction effect. After 80 minutes the grain sizes increases slightly. At 120 minutes stirring was stopped for 60 minutes. A further sample was taken after 180 minutes. The grain size increased dramatically and subsequently declined strongly after 5 minutes stirring to approximately the size observed at 120 minutes.

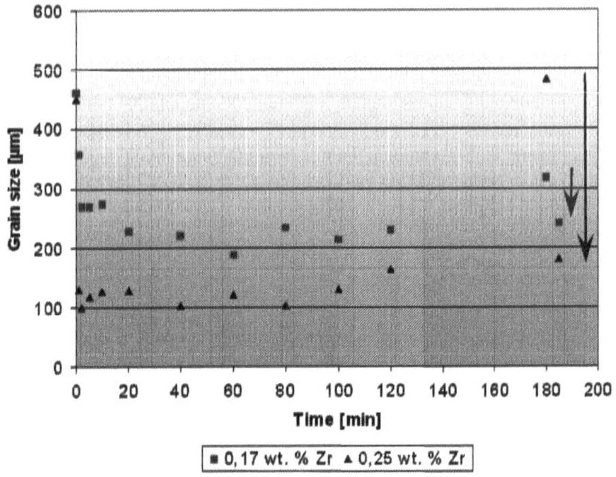

Figure 7: Grain size vs. time.

This implies that the borides settle to the bottom of the steel crucible i.e. behave similar to boride particles in Al melts [5]. Likewise, with stirring it was possible to reactivate the settled particles but not to obtain the ultimate grain size measured after 5 minutes.

Figure 8 shows the non refined state of the Mg showing a columnar structure. The microstructure altered significantly from columnar to equiaxed with grain refinement. Figures 9 and 10 show typical examples of uniformly equiaxed grains in the microstructure 2 minutes after adding the master alloy to the Mg melt.

The transition of a columnar to uniformly equiaxed microstructure can be seen as the minimum requirement for successful grain refinement. It should be noted that some Al_3Zr phases are present in the grain refined microstructure, which may be detrimental to further mechanical processing.

Figure 8: Microstructure of the non refined, as-received pure Mg.

Figure 9: Microstructure at 2 minutes after adding the 0.17 wt. % Zr to the Mg melt.

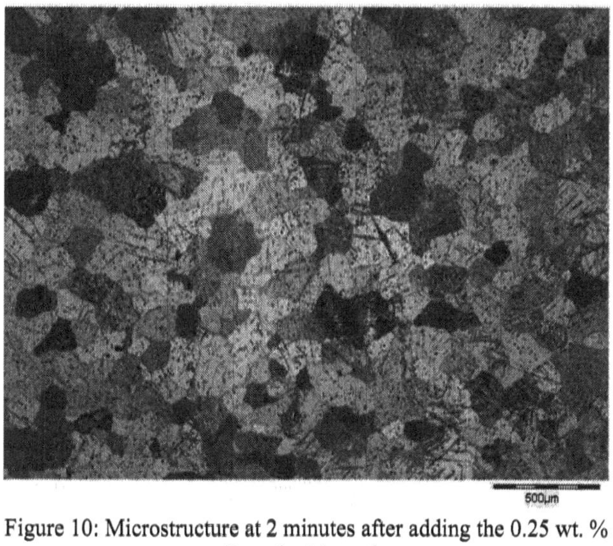

Figure 10: Microstructure at 2 minutes after adding the 0.25 wt. % Zr to the Mg melt.

Grain refinement of Mg using synthetic ZrB_2 particles

Figure 11 shows a scanning electron micrograph of the synthetic ZrB_2 particles used. The ZrB_2 particles displayed a hexagonal platelet morphology. The good lattice match on the ZrB_2 particles (table 1) occurs only on the basal planes. The average ZrB_2 particle size lies in the range of 0.5 to 4 µm and corresponds well with the stated average diameter given by the supplier. However, it is noticeable that clustering can occur and cluster sizes can be in the order of 10 µm. The particle sizes are therefore of the same magnitude as a critical homogeneous nuclei (diameter: d_{crit} ~ 3 µm) with an undercooling of 1°C.

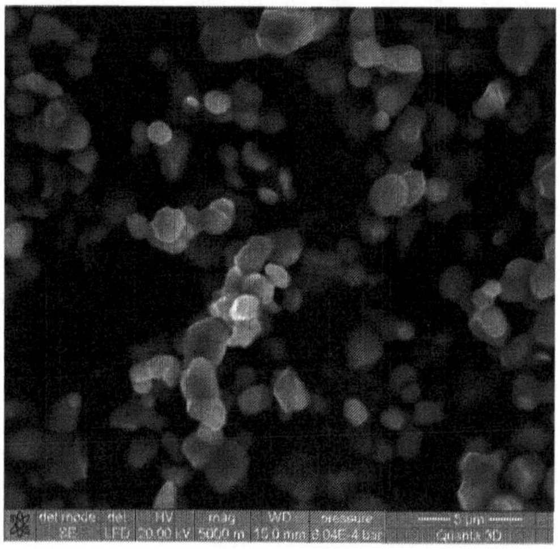

Figure 11: ZrB_2 synthetic particles.

In figure 12 two test series with different contents of ZrB_2 particles resulted in significant grain refinement despite only a moderate growth restriction effect. In the first series with 0.1 wt. % ZrB_2, an ultimate grain size of 100 µm was achieved with a contact time of 5 minutes. After this the grain size increased slightly to 130 µm, within experimental error. Figure 13 presents a typical microstructure after 2 minutes showing a uniform grain size with an equiaxed morphology. Some twinning was visible within the microstructure.

The second series was performed with a ZrB_2 content of 0.5 wt. %. An ultimate grain size of 60 µm was obtained after 5 minutes. Further holding resulted in an increase of grain size up to 120 µm, within experimental error.

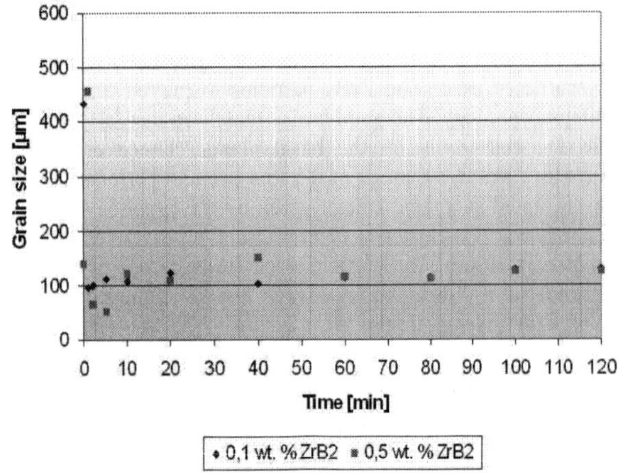

Figure 12: Grain size vs. time.

Figure 14 shows the effect of increasing the addition level to 0.5 wt. % ZrB_2 that, in comparison to figure 13, revealed a much finer grain size. Again an equiaxed uniform microstructure was visible with some twinning. Generally the crystal structure changed in both series from a columnar to an equiaxed structure.

Figure 13: Microstructure at 2 minutes after adding the 0.1 wt. % ZrB_2 particles to the Mg melt.

Figure 14: Microstructure at 2 minutes after adding the 0.5 wt. % ZrB_2 particles to the Mg melt.

Summary and Conclusions

Successful grain refinement of Mg-Al alloys can be obtained by producing in situ ZrB_2 particles by a reaction of conventional Al grain refiner master alloys (Al-Ti-B) with Al-Zr master alloys. In situ ZrB_2 particles are formed in the master alloy and result in the grain refinement of Mg alloys containing Al. Ultimate grain sizes of 100 µm can be obtained. Fading of the grain refining performance is observed after approximately 80 minutes. This appears to be caused by sedimentation of the nucleant particles and could be partly overcome by renewed stirring. Some effect of growth restriction can be observed, the higher Q – value results in a finer grain size. A metallographic examination showed some Al_3Zr phases in the samples, which were refined using the produced master alloy.

In contrast to the simple addition method of the pre-produced master alloys, addition of synthetic ZrB_2 particles offer further advantages. In one respect the addition of the powder is somewhat more complex in its procedure to avoid oxide free nucleation particles but nevertheless a finer grain size and less fading is observed. An ultimate grain size of 60 µm after 2 minutes could be obtained despite only moderate growth restriction by free Al in the alloy. No Al_3Zr phases were present in the microstructure. Grain refinement was mainly obtained by the provision of nucleation sites with a higher number density of heterogeneous particles resulting in a finer grain size. Surprisingly, fading of the grain refining mechanism was less than that for the in situ grain refining particles.

The main results can be summarized as follows:

- Successful grain refinement of Mg-Al alloys was possible with the addition of ZrB_2 particles.
- Significant grain refinement has been obtained with ZrB_2 particles with only moderate growth restriction.
- Fading of the grain refiner performance appears to be associated with settlement of active nucleation sites.
- Synthetic ZrB_2 particles performed better than in situ formed particles from the master alloy route.
- The exact nucleation mechanism of synthetic ZrB_2 particles requires further work.

Acknowledgements

The authors gratefully acknowledge the financial support of the Austrian Federal Ministry of Economics and Labour under the pro-kis[03]-programme. Furthermore we would like to thank TREIBACHER Industrie AG for providing synthetic particles.

References

[1] M.M. Avedesian, H. Baker (Eds.), ASM Specialty Handbook Mg and Mg Alloys, ASM International, Materials Park, OH; 1999, pp. 4 – 6.

[2] K.U. Kainer (Eds.), Mg, DGM, 2000, pp. 11 – 15.

[3] Y.C. Lee, A.K. Dahle, D.H. StJohn, The role of solute in grain refinement of Mg, Metallurgical and Materials Transactions A, Volume 31A, 11/2000, pp. 2895 – 2906.

[4] M. Qian, D.H. StJohn, M.T. Frost, Heterogeneous nuclei size in Mg-Zr alloys, Scripta Materialia 50, 2004, pp. 1115 – 1119.

[5] P. Schumacher, A.L. Greer, J. Worth, P.V. Evans, M.A. Kearns, P. Fisher, A.H. Green, New studies of nucleation mechanisms in aluminium alloys: implications for grain refinement practice, Materials Science and Technology Vol. 14, 1998, pp. 395 - 404.

[6] P. Villars, L.D. Calvert (Eds.), Pearson´s Handbook of Crystallographic Data for Intermetallic Phases, ASM, 1985.

[7] Y.C. Lee, A.K. Dahle, D.H. StJohn, Grain refinement of Mg, Mg Technology, 2000, pp. 211 -218.

[8] A Hellawell, Solidification and casting of metals, The Metals Society, London, 1979, pp. 161 - 168.

[9] E.F. Emily, Principles of Mg Technology, Pergamon Press, Oxford, 1966, pp. 127 – 155.

[10] Y.C. Lee, A.K. Dahle, D.H. StJohn, Grain Refinement of Magnesium, Magnesium Technology 2000,The Minerals, Metals & Materials Society, 2000, pp. 211 – 218.

[11] D. StJohn, M. Easton, The application of a new analytical method for the assessment of the grain refinement of light alloys, Proceedings of the 2nd International Light Metals Technology Conference, 2005, pp. 63 – 68.

[12] E.A. Brandes, G.B. Brook (Eds.), Smithells Metals Reference Book, Butterworth-Heinemann, seventh edition, 1999, pp. 8-23.

[13] Standard test procedure for aluminium alloy grain refiners 1990 TP-1, The Al Association, Inc. Washington, DC, 2006.

[14] G.J. Zhang, M. Ando, J.F. Yang, T. Ohji, S. Kanzaki, Boron carbide and nitride as reactants for in situ synthesis of boride-containing ceramic composites, Journal of the European Ceramic Society 24, 2004, pp. 171 – 178.

[15] Y.X. Chen, D.X. Li, G.D. Zhang, Microstructural studies of in-situ formed MgB_2 phases in a Mg alloy matrix composite, Materials Science and Engineering A337, 2002, pp. 222 – 227.

[16] H.Y Wang, Q.C. Jiang, X.L. Li, J.G. Wang, In situ synthesis of TiC/Mg composites in molten Mg, Scripta Meaterialia 48, 2003, pp. 1349 – 1354.

SOLID SOLUTION EFFECTS ON THE TENSILE BEHAVIOUR OF CONCENTRATED Mg-Zn ALLOYS

A. H. Blake, C. H. Cáceres[1]
Co-operative Research Centre for Cast Metals Manufacturing (CAST)
Materials Engineering, School of Engineering, The University of Queensland
Brisbane, Queensland, 4072, Australia

Keywords: magnesium; magnesium-zinc alloys; solid-solution softening; prismatic slip; pyramidal slip.

Abstract

Solute in solution hardens the basal planes but causes solid solution softening on the prismatic (and possibly pyramidal) planes in dilute (< 0.5 at.% Zn) Mg-Zn alloys. Solid solution softening lowers the strain-hardening rate and increases the ductility of the alloys with respect to the pure Mg metal. In the concentrated (0.5 ~ 2.4 at.% Zn) alloys, solid solution hardening of the basal planes is extensive but it is not clear whether solid solution softening of the secondary slip systems still occurs. Therefore, solid solution effects on the strain hardening rate and ductility of cast polycrystalline Mg-Zn alloys, with Zn contents between 0 and 2.4 at.% have been studied. A constant grain size was obtained in all alloys by adding a small amount of Zr to the melt.

The strain-hardening rate is low for dilute concentrations increasing monotonically above 1 at.% Zn. The ductility goes through a maximum at very low concentrations of Zn, decreasing for higher concentrations. This suggests that the solid solution hardening gradually offsets the solid solution softening effects at high concentrations of Zn.

Introduction

A major drawback of Mg alloys is their relative lack of ductility, which mainly stems from magnesium's plastic anisotropy and limited number of active slip systems. The number of independent slip systems for basal and prism slip is only four [1], one short of satisfying the Von Mises criterion for arbitrary change of shape. It has been proposed [2, 3] that hexagonal close packed (HCP) materials may not need five independent slip systems, as twinning in unfavourably oriented grains can offset the requirement for strain to be equal in each grain, negating the need for slip along the c axis. The prevalent form of twinning in Mg is $\{10\bar{1}2\}$, which can only produce a modest 7% axial strain in any given grain [4], so the maximum ductility afforded by this mechanism is naturally capped. Alloys can be far more ductile than pure Mg, even if their tendency to twin is less marked. Thus, the activation of a pyramidal slip system, $\{hkil\}\langle 11\bar{2}\bar{3}\rangle$, which would provide five independent slip systems by itself [1, 4-7] is often considered. Indeed, second order pyramidal slip, $\{11\bar{2}2\}\langle\bar{1}\bar{1}23\rangle$, has been observed in pure magnesium and Mg-Li alloys [8, 9], and it has been pointed out [10] that the activation of <c+a> dislocations, observed at temperatures above room temperature in pure Mg, results in lower flow stress and strain hardening rate.

The plastic anisotropy of Mg is also reflected by the way the metal behaviour is affected by solute in solution. The effects of solute on basal and prismatic slip in dilute magnesium alloys [11-15] and more recently on pyramidal slip [8, 9] have been well established by studies using single crystals. In Mg-Zn single crystals [12, 13] the critical resolved shear stress for basal slip ($CRSS_{basal}$) in the dilute (c < 0.5 at.%) concentrations increases according to $c^{2/3}$, where c is the atomic content of zinc. In contrast, the addition of Zn causes solid solution softening on the prismatic planes, i.e., the $CRSS_{prism}$ is lowered. The $CRSS_{pyramidal}$ in Mg-Li alloys also decreases with increasing solute concentration [5, 8, 9]. The solid solution softening is often ascribed to a change in the c/a ratio, which affects the core structure of the dislocations [5, 8, 9]. However, Akhtar and Teghtsoonian [12, 13] have shown that solution softening is a general solid solution effect in Mg, irrespective of the solute effect on the c/a ratio, and likely to be connected with a reduction in the stress necessary to overcome the Peierls-Nabarro friction.

The respective softening and hardening effects of zinc on the $CRSS_{prism}$ and $CRSS_{basal}$ in Mg-Zn single crystals of dilute concentrations are shown in Figure 1, together with the tensile ductility of polycrystals. The rapid increase in $CRSS_{prism}$ followed by a decrease at very low concentrations of Zn closely corresponds with the minimum in elongation at about 0.018 at.%. This effect has been ascribed to an initial increase in the $CRSS_{prism}$ at very low Zn contents.

The increase in ductility above 0.018 at.% shown in Figure 1, which corresponds with the lowering of the $CRSS_{prism}$, is of more interest for the present discussion. Akhtar and Teghtsoonian [16] showed that the increased ductility of polycrystalline Mg-Zn alloys corresponds with a decrease in the strain hardening rate for concentrations between 0.018 and 0.42 at.% Zn. The increased ductility was ascribed to stress relaxation by prism slip at the head of pile-ups created by basal slip [16]. Similarly, experiments in Mg-Li polycrystals indicate a decrease in strain hardening rate and increased ductility with Li contents up to 5 at.% [5], also attributed to enhanced secondary slip.

Recent work [17] in the concentrated (c = 0.5 ~ 2.4 at.%) alloys has shown that Zn causes extensive hardening, ascribed to short-

[1] *Corresponding author. Address: Materials Engineering, School of Engineering, The University of Queensland, QLD 4072 Australia.*
E-mail: c.caceres@minmet.uq.edu.au ; Phone 61-7-33654377; Fax: 61-7-33653888.

range order effects on the basal slip system. Short-range order is likely to increase the CRSS in all slip systems, not just in those involving the basal plane, and this might progressively offset the solid solution softening on the $CRSS_{prism}$. A straightforward way of estimating the effects of solid solution softening and hardening on the different slip systems is thus to assess the dependence of the strain-hardening rate and ductility on the concentration of solute. It is anticipated that the strain-hardening rate should be low, and the tensile ductility high, up to some critical concentration of Zn, above which the opposite should be observed.

The experiments described in this work were organised to examine the concentration dependence of the strain-hardening rate and the tensile ductility of binary Mg-Zn alloys, over the range of 0 to 2.4 at.% Zn (0 to 6.2 wt.%). The maximum solid solution concentration of Zn in Mg is 2.4 at.% at 340 °C while at room temperature it is about 0.5 at.%. Alloys with concentrations below and above 0.5 at.% will be referred to as dilute and concentrated, respectively.

A constant grain size was obtained in all alloys by adding a small amount of Zr to the melt. Magnesium can dissolve up to 0.65 wt.% Zr (0.17 at.%) at 500°C, so at low concentrations of Zn the amount of Zr in solution may have had a measurable effect on the solid solution hardening. However, this possibility has been ignored in this work.

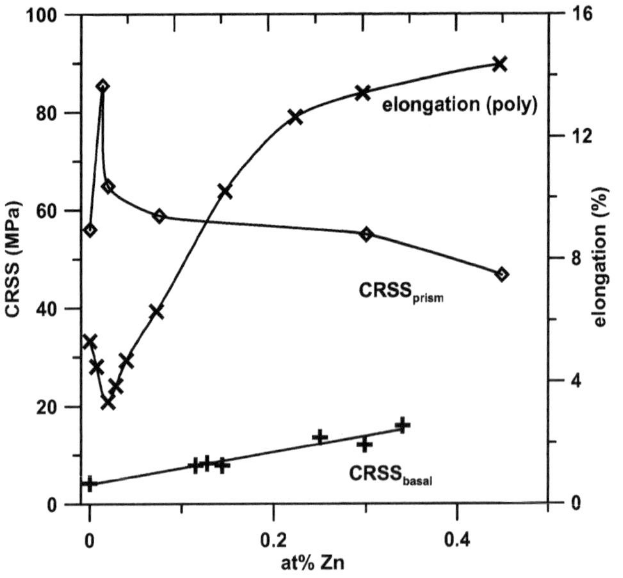

Figure 1. CRSS of the basal and prismatic slip systems and elongation of polycrystalline Mg-Zn alloys, for dilute solute concentrations of Zn. Data points replotted from references [11, 12, 16]. Data for prism slip at 250K.

Experimental

Casting

Magnesium-zinc alloy castings with nominal compositions from 0 to 2.4 at.% Zn were made by melting commercially pure (99.7%) magnesium and zinc in boron nitride coated steel crucibles in an electrical resistance furnace. Grain refining was obtained by the use of zirconium from a commercial [18] master alloy (23:77 Zr:Mg), added to a nominal composition of 1 wt.% Zr. Composition data are detailed in Table I.

Melt protection was provided by the use of a commercial cover gas [19]. Additions were made at 750°C and stirred mechanically for 10 min before pouring under a protective atmosphere into steel moulds preheated to 250°C. The cast cylinders were 90mm × 50mm ∅ and 100mm × 60mm ∅.

Table I: Grain size, hardness and Zn and Zr content of alloys

Nominal Zn (at.%)	Zn (at.%) [wt.%]	Zr (at.%) [wt.%]	Grain Size (μm)
0.2	(0.21) [0.56]	(0.15) [0.57]	47
0.4	(0.40) [1.07]	(0.18) [0.67]	39
1	(0.98) [2.60]	(0.15) [0.55]	34
2	(1.85) [4.83]	(0.12) [0.45]	35
2.4	(2.66) [6.85]	(0.26) [0.98]	41

Heat Treatment

The cast cylinders were solution treated under an argon atmosphere and quenched into water. The times and temperatures used are detailed in Table II, following the schedule proposed by Lagowski and Meier [17, 20].

Table II. Solution heat treatment

Zn content (at.%)	Time (h)	Temperature (°C)
0 - 1	3	520
2	4	470
2.4	10	450

The castings were kept at −5°C between quenching and machining and between machining and tensile testing to reduce possible solute segregation effects, although hardness testing shows that the increase of strength with time at room temperature in the most concentrated alloy is minimal [17] .

Grain Size

For each alloy a section was polished to 1 μm and etched with acetic-picral acid. The grain size was then determined by the linear intercept method in accordance with ASTM E112-88, with no less than 300 intercepts counted for each composition. Table I details the average grain size for each alloy.

Tensile Testing

Cylindrical specimens were machined post heat treatment, with a gauge length of 20 mm and diameter of 6 mm. Testing was performed on a screw driven machine at 1 mm/min with a 12.5 mm extensometer attached to the specimens. Digital files were stored for further analysis.

Solute Dependence of the Tensile Behaviour

Flow Behaviour

Representative true stress-true plastic strain flow curves for each alloy are shown in Figure 2. It is seen that the dilute compositions have quite similar flow curves despite the increase in yield stress with Zn content, while the flow stress and strain hardening rate increases rapidly for the concentrated alloys. In Figure 3 the strain-hardening rates are compared. All of the curves initially exhibit a very high hardening rate, followed by a region of nearly constant strain hardening at strains between 1 and 4%, and a gradual transition to a low hardening rate at strains above 10%. Note the similar strain-hardening rate of the dilute compositions in Figure 3.

Agnew *et al.* [5, 6] used neutron diffraction to show that the elastic plastic transition is very protracted in Mg alloys, with three stages in the strain hardening rate. The first stage exhibits an extremely high strain-hardening rate, while in stage II and III, a decreasing strain-hardening rate is observed. The flow stages are illustrated in Figure 4, using the 0.2 flow curve from Figure 2. In compression, stage I is terminated by a sharp transition to a low strain-hardening rate region which extends for about 7% strain [21], due to profuse tensile $\{10\bar{1}2\}$ twinning. It has been suggested that the lower strain hardening rates of stages II and III stem from [5] the progressive activation of secondary deformation modes.

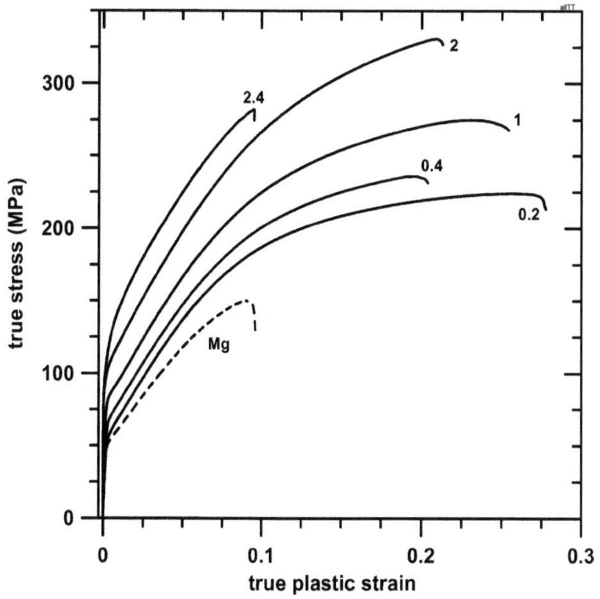

Figure 2. Tensile flow curves of Mg-Zn alloys. Numerals indicate the nominal at.% Zn. The dashed line curve represents a specimen of pure Mg of similar grain size to the alloys.

A possible mechanistic account of the three deformation stages is as follows. Stage I (strains below 1%): activation of basal slip in favourably oriented grains, and profuse $\{10\bar{1}2\}$ twinning in pure Mg metal. Stage II (strains between 1 and 10%): prism slip is now activated in sites of high stress concentration, such as grain boundaries and triple points. The incidence of twinning in the alloys is limited, more so in tension [5], but it is more extensive in the pure Mg. The activation of prism slip in stage II relieves some of the stress concentrations caused by the continued basal slip, particularly in the alloys due to the solution softening effect, and the strain-hardening rate is reduced in comparison with stage I. However, prism slip is only able to accommodate strain about the c-axis, and since many hard-oriented grains remain undeformed, the polycrystal is a mixture of elastic and partly plastic grains [5, 6]. Hence, the strain-hardening rate is still quite high. Stage III (strains beyond 10%): The significant reduction of strain-hardening rate indicates that the internal stresses approach saturation [5, 6]. The latter is naturally accounted for by the activation of pyramidal slip, which enables general plasticity. Hence, in stage III, basal, prism and pyramidal slip are activated. Alternatively, the decreasing strain hardening rate at high stresses has been ascribed to activation of compressive $\{10\bar{1}1\}$ and $\{30\bar{3}2\}$ twinning [22]. The connection between the strain hardening behaviour shown in Figure 3 and the flow stages identified in Figure 4 is evident.

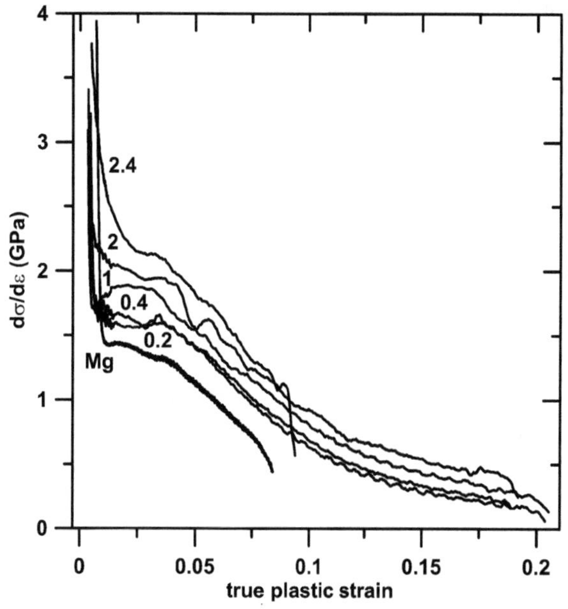

Figure 3. Strain-hardening rate of alloys. Labels represent nominal at.% Zn. The plateau observed at 0.01~0.04 strain is connected to the onset of stage II in Figure 4.

Ductility Behaviour

The strain-hardening rate at 2.5% strain, i.e., at the plateau in Figure 3, is compared in Figure 5 with the yield stress and tensile ductility plotted as functions of the Zn content. The yield stress data points are mean values of not fewer than 4 specimens. Some of the specimens failed at relatively low strains due to casting defects, so the points representing ductility are the maximum values out of 4 to 8 tests. It is seen that the tensile ductility is generally larger in the alloys than in the pure metal, with a maximum at the lowest concentrations. For Zn contents of 1 at.% and above, the hardening rate is higher, and the yield stress increases rapidly with the Zn content while the ductility steadily decreases.

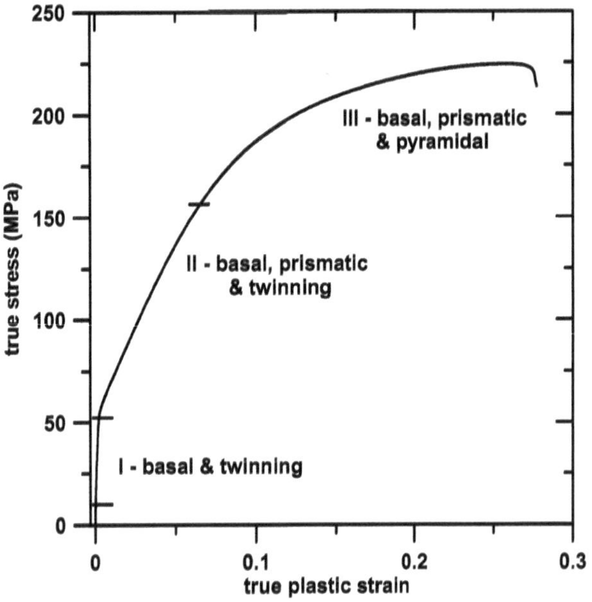

Figure 4. Stages of deformation modes in a Mg-Zn alloy deformed in tension.

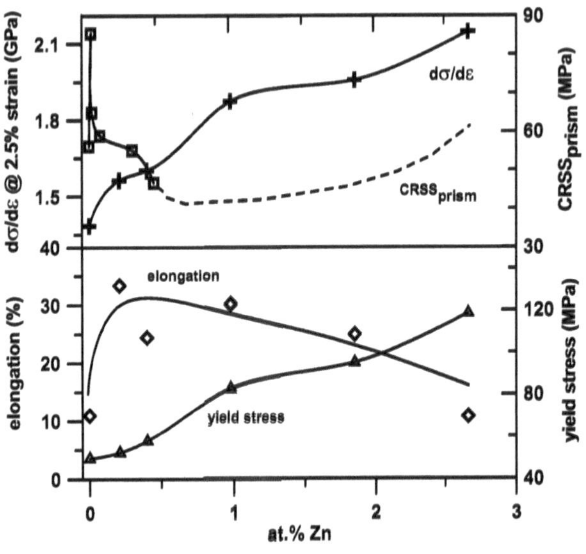

Figure 5. CRSS$_{prism}$, strain-hardening rate, elongation to fracture and yield stress as a function of the concentration of Zn. The CRSS$_{prism}$ data as in Fig. 1. The dashed line suggests the possible solute dependence of the CRSS$_{prism}$ in the concentrated alloys.

Solute Effects on Secondary Slip

The results of Figure 5 generally conform to the arguments presented in the introduction, i.e., the ductility is largest at low and intermediate concentrations of Zn, when the strain-hardening rate is reduced. This behaviour can be taken as due to increased activity of prism slip in the first 10% strain (stage II in Figure 4) and possibly pyramidal slip at larger strains (stage III in Figure 4). On the other hand, the increased strain hardening rate and decreased ductility at concentrations above 1 at.% Zn suggest that, indeed, the solid solution hardening increasingly offsets the solution softening effects. This is schematically indicated in Figure 5 by the dashed line extrapolated from the CRSS$_{prism}$ data.

Summary and Conclusions

Solid solution effects on the strain hardening rate and ductility of polycrystalline Mg-Zn alloys have been studied over the range of 0 to 2.4 at.% Zn.

An increase in the strain-hardening rate above 1 at.% Zn coincides with a decreasing tensile elongation, suggesting that solid solution softening of secondary slip is gradually offset by the solid solution hardening at high solute concentrations.

Acknowledgements

CAST was established and is funded in part by the Australian Government's Co-operative Research Centres Program.

References

1. Yoo, M.H., "Slip, twinning, and fracture in hexagonal close-packed metals," *Metallurgical Transactions,* 12A (1981), 409-418.

2. Hutchinson, J.W., "Creep and plasticity of hexagonal polycrystals as related to single crystal slip," *Metallurgical Transactions,* 8A (1977), 1465-1469.

3. Kocks, U.F. and D.G. Westlake, "The Importance of Twinning for the Ductility of HCP Polycrystals," *Transactions AIME,* 239 (1967), 1107-1109.

4. Partridge, P.G., "The crystallography and deformation modes of hexagonal close-packed metals," *Metallurgical Reviews,* 12 (1967), 169-194.

5. Agnew, S.R., M.H. Yoo, and C.N. Tomé, "Application of texture simulation to understanding mechanical behavior of Mg and solid solutions containing Li or Y," *Acta Materialia,* 49 (2001), 4277-4289.

6. Agnew, S.R., et al., "Study of slip in a magnesium alloy by neutron diffraction and modelling," *Scripta Materialia,* 48 (2003), 1003-1008.

7. Yoo, M.H., et al., "Nonbasal deformation modes of HCP metals and alloys: role of dislocation sources and mobility," *Metallurgical and Materials Transactions A*, 33 (2002), 813-822.

8. Ando, S. and H. Tonda, "Pyramidal Slip in Magnesium Alloy Single Crystals," *Materials Science Forum*, 419-422 (2003), 87-92.

9. Tonda, H. and S. Ando, "Effect of Temperature and Shear Direction on Yield Stress by {1122}<1123> Slip in hcp Metals," *Metallurgical and Materials Transactions A*, 33A (2002), 831-836.

10. Máthis, K., et al., "The evolution of non-basal dislocations as a function of deformation temperature in pure magnesium determined by X-ray diffraction," *Acta Materialia*, 52 (2004), 2889-1894.

11. Akhtar, A. and E. Teghtsoonian, "Solid solution strengthening of magnesium single crystals - I. Alloying behaviour in basal slip," *Acta Metallurgica*, 17 (1969), 1339-1349.

12. Akhtar, A. and E. Teghtsoonian, "Solid solution strengthening of magnesium single crystals - II. The effect of solute on the ease of prismatic slip," *Acta Metallurgica*, 17 (1969), 1351-1356.

13. Akhtar, A. and E. Teghtsoonian, "Substitutional solution hardening of magnesium single crystals," *Philosophical Magazine*, 25 (1972), 897-916.

14. Levine, E.D., W.F. Sheely, and R.R. Nash, "Solid-solution strengthening of Magnesium single crystals at room temperature," *Transactions of the Metallurgical Society of AIME*, 215 (1959), 521-526.

15. Lukác, P., "Solid solution hardening in Mg-Cd single crystals," *physica status solidi (a)*, 131 (1992), 377-390.

16. Akhtar, A. and E. Teghtsoonian, "Solid solution hardening in magnesium alloys," *Transactions of Japan Institute of Metals*, 9, supplement (1968), 692-697.

17. Cáceres, C.H. and A. Blake, "The strength of concentrated Mg-Zn solid solutions," *physica status solidi*, 194 (a) (2002), 147-158.

18. Qian, M. and D.H. St. John, "A New Zirconium-rich Master Alloy for the Grain Refinement of Magnesium Alloys", 6th International Conference Magnesium Alloys and their Applications, Wolfsburg (Germany), 2003.

19. Cashion, S.P. and N.J. Ricketts, "Industrial application of the AMCover cover gas system for magnesium melt protection," 6th International Conference Magnesium Alloys and their Applications, Wolfsburg (Germany). 2003.

20. Lagowski, B. and J.W. Meier, "Characteristics of sand-cast magnesium-zinc alloys," *Transactions of the American Foundrymen's Society*, 72 (1964), 561-574.

21. Tomé, C.N., et al., "The relation between texture, twinning and mechanical properties in hexagonal aggregates," *Materials Science Forum*, 408-412 (2002), 263-268.

22. Koike, J., "Enhanced deformation mechanisnms by anisotropic plasticity in polycrystalline Mg alloys at room temperature", TMS Winter Meeting, Charlotte, North Caroline, 2004.

Age hardening behavior of Mg-1.2Sn-1.7Zn alloy containing Al

T. T. Sasaki, T. Ohkubo, K. Hono
National Institute for Materials Science, 1-2-1 Sengen, Tsukuba, Ibaraki, 305-0047, Japan

Keywords: Mg-Sn, Precipitation hardening, Al

Abstract

We have investigated the age hardening behavior of Mg-1.2at.%Sn (T5) based alloys, which are solution treatable at lower temperature as compared to the previously studied Mg-2.2Sn (T10) based alloys. To enhance the age hardening response with the reduced Sn content, Zn and Al were alloyed with the Mg-1.2at.%Sn alloy. Particularly, Mg-1.7Zn-1.2Sn-2.0Al alloy (TZA543) showed the peak hardness of 79 VHN at 200°C. The peak aged Mg-1.2Sn-1.7Zn-2.0Al alloy consisted of cuboidal β_2' precipitates with the size of less than 50 nm, rod shaped precipitates, and lathplate shaped Mg_2Sn phase.

Introduction

There are strong demands for the development of high strength wrought magnesium alloys applicable to transportation vehicles and aircrafts. Since commercial wrought alloys such as Mg-Zn-Zr (ZK) and Mg-Al-Zn (AZ) show only moderate age hardening responses, not much attention has been paid to the precipitation hardening of wrought alloys. Recent investigations showed yield strength can be readily increased by aging rolled sheets and extruded rods of microalloyed Mg-Zn alloys microalloyed with Ca and Ag **[1]**, demonstrating the potential of heat treatable wrought mangnesium alloys.

Mg-Sn based alloys appear to be promising as heat resistant alloys for both cast and wrought applications because the Mg_2Sn phase has a higher melting temperature than those in Mg-RE (RE: Rare Earth) systems **[2]**. In fact, some of precipitation hardened Mg-Sn based alloys exhibit excellent creep properties **[3, 4]**. For better formability, extrusion and rolling temperature should be selected above the solvus temperature of alloys. However, the solution heat treatment temperature of the previously studied Mg-2.2Sn based alloy (>500 °C) is too high for thermomechanical processing **[5]**. One way to reduce the solution treatment temperature is to decrease the Sn content. However, it also reduces the age hardening responses. Since the typical size of Mg_2Sn precipitates is rather coarse (~1 μm) **[5-8]**, the significant refinement of the precipitates may compensate the degradation of the age hardening response even with the reduced supersaturation. In Mg-Sn-Zn alloys, $MgZn_2$ type phase such as β_1' phase provides the heterogeneous nucleation site for the Mg_2Sn phase **[8, 9]**, thereby increasing the number density of the Mg_2Sn precipitates. In this work, 1.7at.%Zn was alloyed with Mg-1.2at.%Sn so that the alloy can be solution treated at 450 °C . Since the addition of Al is reported to refine the Mg_2Sn phase in the Mg-2.2Sn-0.5at.%Zn alloy **[8, 10]**, we examined the effect of Al alloying expecting the improvement of the age hardening behavior of the Mg-1.2at.%Sn-1.7at.%Zn alloy (TZ54).

Experimental Procedure

Alloy ingots with nominal compositions of Mg–1.2Sn-1.7Zn-*x*Al (*x*=0, 0.5, 1 and 2) in at.% were prepared by induction melting using steel crucibles under an Ar atmosphere and by casting into iron moulds. Table 1 shows the compositions of the alloys in both at.% and wt.%.

Table 1: Chemical composition of alloys

at.%	wt.%
Mg-1.2Sn-1.7Zn	Mg-5.5Sn-4.3Zn
Mg-1.2Sn-1.7Zn-0.5Al	Mg-5.4Sn-4.3Zn-0.5Al
Mg-1.2Sn-1.7Zn-1.0Al	Mg-5.4Sn-4.2Zn-1.0Al
Mg-1.2Sn-1.7Zn-2.0Al	Mg-5.4Sn-4.2Zn-2.1Al

The ingots were subjected to homogenization at 350 °C for 24 hr followed by 450 °C heat treatment for 24 hr in a He-filled Pyrex tube. The small samples that were cut out from the ingots were solution heat treated for 0.5 hr s at 450 °C in the He-filled Pyrex tubes and then water quenched. Single aging was performed on alloys at 200 °C in an oil bath for various aging times. The age hardening responses were measured by the Vickers hardness tester under a load of 500 g. Transmission electron microscope (TEM) specimens were prepared by the twin-jet polishing technique using a solution of 5.3 g LiCl, 11.16 g $Mg(ClO_4)$, 2,500 ml methanol, and 100ml 2-butoxy-ethanol at about −50 °C and 90 V. Some of the specimens were finished for surface cleaning by ion milling using a Gatan Precision Ion Polishing System (PIPS). Transmission Electron Microscope (TEM) observation was carried out using FEI TECNAI T20 TEM operating at 200 kV.

Results and Discussion

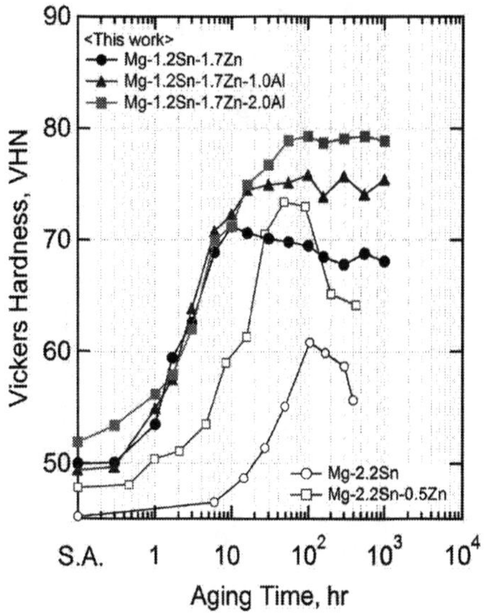

Figure 1: Variation in Vickers Hardness as functions of time during aging at 200°C on Mg-1.2Sn-1.7Zn, Mg-1.2Sn-1.7Zn-1.0Al and Mg-1.2Sn-1.7Zn-2.0Al alloys. Included is the data obtained from Mg-2.2Sn and Mg-2.2Sn-0.5Zn alloy **[5]**.

Figure 1 shows variations in Vickers hardness as functions of time during 200 °C aging for Mg-1.2Sn-1.7Zn (TZ54), Mg-1.2Sn-1.7Zn-1.0Al (TZA541) and Mg-1.2Sn-1.7Zn-2.0Al (TZ542) alloys. The age hardening curves show that the addition of Al substantially increased the peak hardness. The peak hardness increased from 71 to 79 VHN by the addition of 2at.% Al while the time to reach the peak hardness is delayed by the Al addition. Mg-1.2Sn-1.7Zn alloy reached the peak hardness after 10 h of aging at 200 °C, and Mg-1.2Sn-1.7Zn-2.0Al alloy took 100 h to reach the peak hardness. While Harosh et al. reported two hardness peaks appeared in an Mg-5.6Sn-4.4Zn-2.1Al alloy during aging at 225 °C **[11]**, the Mg-1.2Sn-1.7Zn-2.0Al alloy showed only one hardness peak. The Mg-1.2Sn-1.7Zn based alloys do not overage as quickly as Mg-2.2Sn and Mg-2.1Sn-0.5Zn alloys **[5]**. The Mg-1.2Sn-1.7Zn alloy maintained the peak hardness of ~ 70 VHN up to 100 h and the Mg-1.2Sn-1.7Zn-2.0Al alloy maintained the hardness level of ~ 80 VHN up to 1000 h during 200 °C aging.

Figure 2 shows the bright field TEM images obtained from the Mg-1.2Sn-1.7Zn alloy aged at 200 °C. **Figure 2 (a) and (b)** are those after aging at 200°C for 10 h, which corresponds to the peak aged condition, and **Fig. 2 (c) and (d)** are the ones after overaging for 100 h at 200 °C. Note that **Fig. 2 (a) and (c)** are taken from the $[11\bar{2}0]$ zone, and **Fig. 2 (b) and (d)** are from the [0001] zone, respectively. As shown in **Fig. 2 (a) and (b)**, the peak aged Mg-1.2Sn-1.7Zn alloy shows similar microstructure to that seen in the Mg-Zn based alloy subjected to aging **[12, 13]**, i.e., rod shaped β_1' along the [0001] direction of the magnesium matrix, and the other is β_2' plates lying on the (0001) plane of the magnesium matrix. The corresponding SAED pattern in **Fig. 2 (a)** exhibits diffuse streaks along the $[01\bar{1}0]$ directions. This is attributed to the rod shaped β_1'precipitates along the [0001] direction as well as the crystal structure of the precipitate phase.

The overaging for 100 h resulted in the coarsening of the β_1' and β_2' phases as shown in **Fig. 2 (c) and (d)**. The β_1' phase coarsened to ~ 1 μm in length and 40 nm in diameter, and the size of the β_2' phase increased to 200 nm in diameter and 50 nm in thickness. In addition to the coarsened β_1' and β_2' phases, another kind of precipitates is observed as indicated by an arrow. From **Fig. 2 (c) and (d)**, the precipitates have plate like shape forming on the prismatic plane of the magnesium matrix, $(10\bar{1}0)$. **Figure 3 (a)** shows the microbeam diffraction patterns obtained from one of the plate like precipitates. Note that they are recorded from the $[01\bar{1}0]$ zone. The precipitates are identified as the

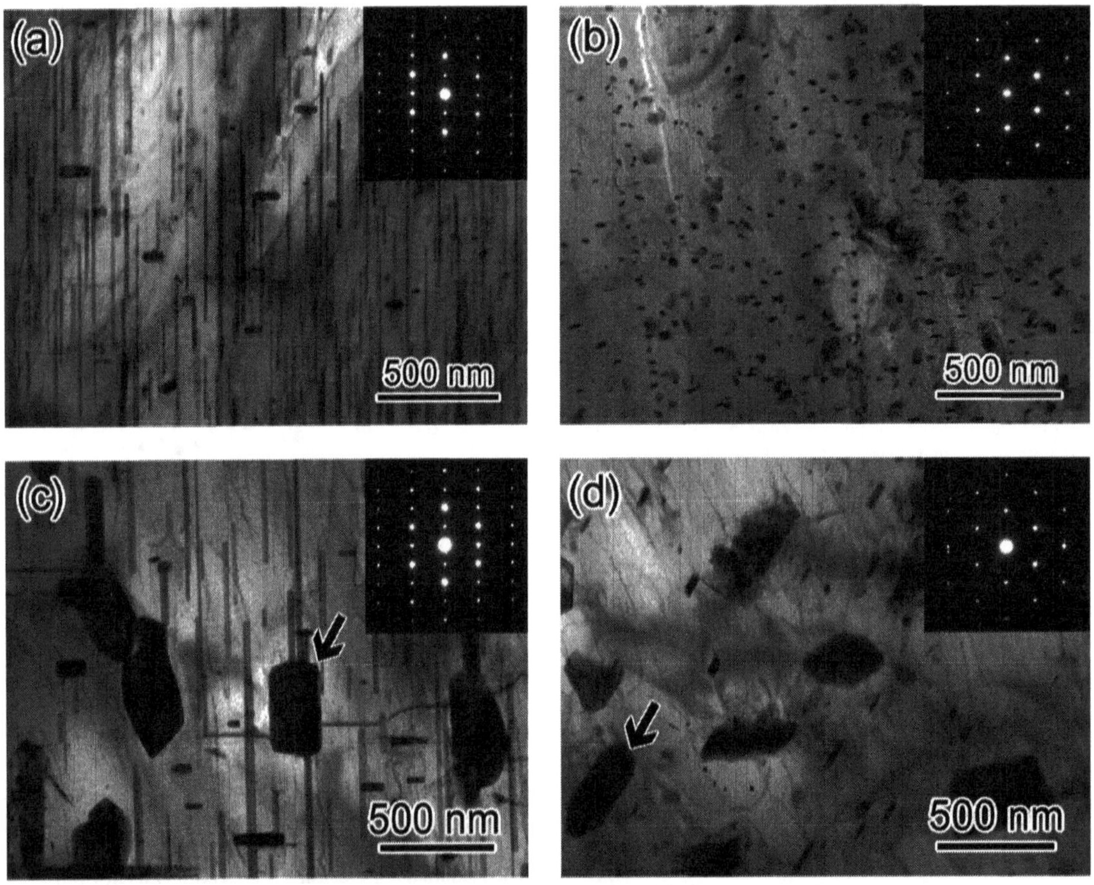

Figure 2: Bright field TEM images of Mg-1.2Sn-1.7Zn alloy subjected to 200 °C aging. (a) and (b) are obtained after 200°C/10hr aging, which is consistent with the peak aging condition, and (c) and (d) are are obtained after aging at 200°C for 100hr. (a) and (c) are taken from the zone axis of $[11\bar{2}0]$, and (b) and (d) are taken from the zone axis of [0001].

Mg_2Sn phase as shown in the calculated diffraction pattern, Fig. 3 (b). The orientation relationship (OR) between the plate like Mg_2Sn phase forming on the prismatic and the magnesium matrix is determined to be $(0001)_{Mg} // (110)_{Mg_2Sn}$, $[11\bar{2}0]_{Mg} // [001]_{Mg_2Sn}$, which is the common OR in the Mg-Sn based alloy **[5-8]**. The late precipitation of Mg_2Sn phase after the peak aged condition is considered to compensate the hardness drop due to the coarsening of the β_1' and β_2' phases.

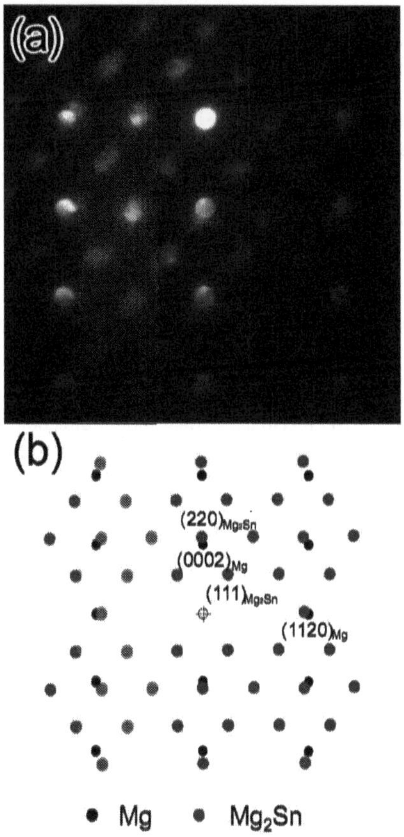

Figure 3: (a) microbeam diffraction patterns obtained from a precipitate indicated by arrow in Fig. 2 (a), and (b) Calculated diffraction patterns. Note that these images are taken from the zone axis of $[01\bar{1}0]$.

Figure 4 shows the bright field TEM images obtained from the Mg-1.2Sn-1.7Zn-2.0Al alloy (TZA542) aged at 200 °C for 100 hr, which corresponds to the peak aged condition. Note that **Fig. 4 (a) and (b)** are taken from the $[11\bar{2}0]$, and [0001] zones, respectively. The peak aged microstructure shows different feature as compared to the peak aged Mg-1.2Sn-1.7Zn alloy. The rod shaped precipitate growing along [0001] direction is refined to ~ 300 nm as indicated by arrowhead 1. In addition, as reported by Oh-ishi et al. **[10]**, the Al addition resulted in the formation of the cuboidal shaped precipitates with the size less than 50 nm as indicated by arrowhead 2. In addition to these precipitates, the lath shaped precipitates forming on the prismatic plane could be also observed as indicated by arrowhead 3. The lath shaped precipitates are the Mg_2Sn phase, and its size was refined to ~200 nm in length and ~75 nm in thickness.

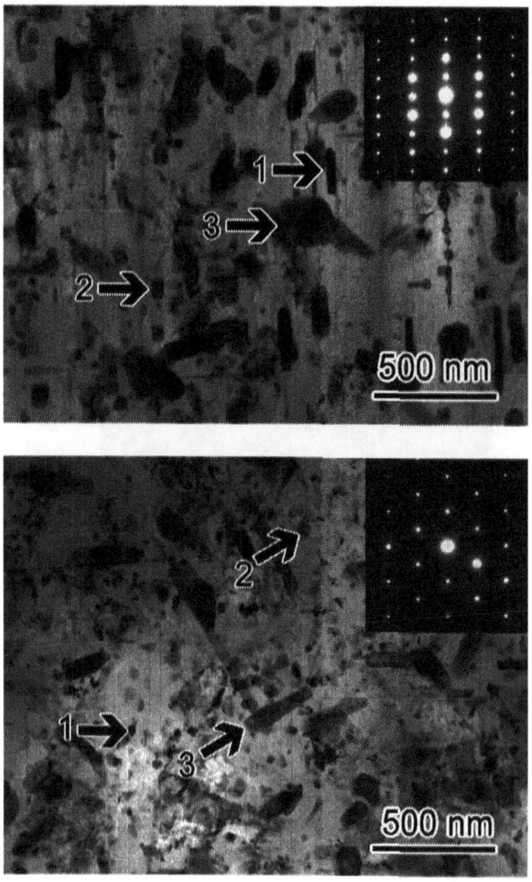

Figure 4: Bright field TEM images of Mg-1.2Sn-1.7Zn-2.0Al alloy aged at 200 °C aging for 100 hr, consistent with the peak aging condition. (a) and (b) are taken from the zone axis of $[11\bar{2}0]$, and [0001], respectively.

Figure 5 (a) shows the micro-beam diffraction pattern obtained from the cuboidal precipitates observed in the 200 °C / 100 hr aged Mg-1.2Sn-1.7Zn-2.0Al alloy taken from the zone axis of [0001]. As shown in **Fig. 5 (b)**, the structure of the cuboidal precipitate is consistent with that of the β_2' phase. However, since the structure of the rod precipitates remains unclear, future work will shed light on the clarification of the phase of the rod shaped precipitates.

In this work, we have investigated the age hardening behavior of Mg-1.2Sn-1.7Zn based alloy containing Al, and found that the peak hardness of the Mg-1.2Sn-1.7Zn based alloy could be substantially increased from 71 to 79 VHN by the addition of 2.0at.% of Al. From the microstructure observation of the Mg-2.1Sn-1.7Zn alloy at peak aged and over aged conditions, **Fig. 2**, the peak hardness was attributed to the precipitation of the $\beta_1'$and β_2' phases, **Fig. 2 (a) and (b)**. Although the hardness is expected to decrease upon the coarsening of the $\beta_1'$and β_2' phases by the over aging, the Mg-1.2Sn-1.7Zn based alloys showed smaller hardness reduction up to 1000 h. This is because Mg_2Sn phase precipitated after the peak aging, and could compensate the hardness decrease due to the coarsening of the β_1' and β_2' phases, **Fig. 2 (c) and (d)**. Since the diffusion of the Sn atoms is much slower as compared to that of the Zn atoms in the magnesium matrix, and Zn and Sn atoms behave repulsively due to the positive enthalpy of mixing as reported previously **[5, 14]**, we can understand the precipitation of the β_1'and β_2' phases prior to the precipitation of the Mg_2Sn phase.

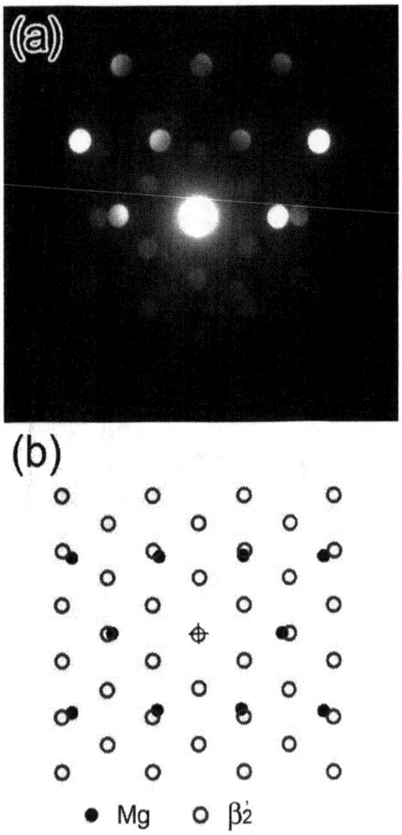

Figure 5: (a) microbeam diffraction patterns obtained from one of cuboidal precipitate seen in 200°C/100hr aged Mg-1.2Sn-1.7Zn-2.0Al alloy, and (b) Calculated diffraction patterns. Note that these images are taken from the zone axis of$[10\bar{1}0]$.

In the Mg-1.2Sn-1.7Zn-2.0Al alloy, on the other hand, the Mg_2Sn phase also contributes to the peak hardness in addition to the rod shaped and cuboidal shaped precipitates as seen in **Fig. 4**. The precipitation of the Mg_2Sn resulted in the increased number density of the precipitates in the peak aged condition, resulting in the enhanced peak hardness in the Mg-1.2Sn-1.7Zn-2.0Al alloy. The time to reach peak hardness was shortened in the Mg-2.2Sn-0.5Zn alloy even by the addition of 1.0at.% Al **[8]**. The presence of the Mg_2Sn phase in the peak aged condition is attributed to the change in the kinetics due to the Al.

The enhanced peak hardness in the Mg-1.2Sn-1.7Zn-2.0Al alloy is also attributed to the increase in the number density of the precipitates, and the change in the morphology of the precipitates. The addition of Al refined the precipitate size as shown in the microstructure of the peak aged samples, **Fig. 2 and 4**. The peak aged Mg-1.2Sn-1.7Zn-2.0Al alloy consisted of fine cuboidal β_2' precipitates with the size of less than 50 nm, rods along [0001], and fine plate shaped Mg_2Sn phase while the Mg-1.2Sn-1.7Zn alloy was mainly strengthened by coarse rod shaped β_1' phase with over 500 nm in length. In the Mg-2.3Zn-0.5Mn alloy, Oh-ishi et al concluded that plate to cuboidal transition of β_2' phase is because of the lattice misfit change between the Mg and precipitate phases by Al addition **[10]**. This may also have caused the shape changes of the precipitates in this system.

Finally, the Mg_2Sn phase observed in the Mg-1.2Sn-1.7Zn had different shape from the ones seen in the previously studied Mg-Sn based alloy **[5-9]**. The plate like precipitates forming on the prismatic plane of the magnesium matrix, $(11\bar{2}0)$ could be observed, **Fig. 2 (c) and (d)**, while the Mg_2Sn phase usually forms on the (0001) plane of the magnesium matrix **[5-9]**. Since the precipitation of the Mg_2Sn on the prismatic plane could be seen even in the Al-free Mg-1.2Sn-1.7Zn alloy, the lattice misfit change due to Al addition might not be the sole reason for the precipitation on the prismatic plane. Therefore, future work will also shed light on the detailed formation mechanism of the Mg_2Sn phase.

Summary

We have investigated the age hardening behavior of the Mg-1.2Sn-1.7Zn-xAl (x=0.5, 1.0 and 2.0). The addition of 2.0at.%Al substantially increased the peak hardness from 71 to 79 VHN while the time to reach peak hardness was delayed. In the Mg-1.2Sn-1.7Zn alloy, the peak hardness is attributed to the precipitation of β_1'and β_2' phases. On the other hand, the refinement of the rod precipitates along [0001] direction and the precipitation of fine cuboidal β_2' phase and Mg_2Sn phase resulted in the increase in the peak hardness in the Mg-1.2Sn-1.7Zn-2.0Al alloy.

Acknowledgement

The authors gratefully acknowledge Dr. C. L. Mendis for her fruitful discussion.

References

[1] C.L. Mendis, K.Oh-ishi, Y. Kawamura, T. Honma, S. Kamado, K. Hono, "Precipitation-hardenable Mg-2.4Zn-0.1Zn-0.1Ca-0.16 Zr (at.%) alloys," *Acta Materialia*, 57 (3) (2009) 749-760.

[2] A.A. Nayeb-Hashemi and J.B. Clark, *Phase Diagrams of Binary Magnesium Alloys*, (Materials Park, OH ASM International, 1988).

[3] M.A. Gibson, X. Fang, C.J. Bettles, C.R. Hutchinson, "The effect of precipitate state on the creep resistance of Mg-Sn alloys," *Scripta Materialia*, 63 (8) (2010) 899-902.

[4] T.T. Sasaki, T. Ohkubo, K. Hono, "Heat-treatable Mg-Sn-Zn wrought alloy," *Scripta Materialia*, 61 (1) (2009) 80-83

[5] T.T. Sasaki, K. Oh-ishi, T. Ohkubo, K. Hono, "Enhanced age hardening response by the addition of Zn in Mg-Sn alloys," *Scripta Materialia*, 55 (3) (2006) 251–254.

[6] C.L. Mendis, C.J. Bettles, M.A. Gibson, C.R. Hutchinson, "An enhanced age hardening response in Mg-Sn based alloy containing Zn" *Materials Science and Engineering A*, 435 (2006) 163-171.

[7] C.L. Mendis, C.J. Bettles, M.A. Gibson, S. Grosse, C.R. Hutchinson, "Refinement of precipitate distributions in and age-hardenable Mg-Sn alloy thorough microalloying" *Philosophical Magazine Letters*, 86 (7) (2006) 443-456.

[8] T.T. Sasaki, K. Oh-ishi, T. Ohkubo, K. Hono, "Effect of double aging and microalloying on the age hardening behavior of

a Mg-Sn-Zn alloy," *Materials Science and Engineering A*, (2011) doi: 10.1016/j.msea.2010.05.010.

[9] A. Gorny, A. Katsman, "Precipitation and stress-influenced coarsening in Mg-based Mg-Zn-Sn-Y and Mg-Zn-Sn-Sb alloys" *Journal of Materials Research*, 23 (2008) 1228-1236

[10] K. Oh-ishi, K. Hono, K.S. Shin, "Effect of pre-aging and Al addition on age hardening and microstructure in Mg-6wt.%Zn alloys," *Materials Science and Engineering A*, 496 (2008) 425-433.

[11] S. Harosh, L. Miller, G. Levi, M. Bamberger, "Microstructure and properties of Mg-5.6%Sn-4.4%Zn-2.1%Al alloy," *Journal of Materials Science*, 42 (2007) 9983-9989

[12] J.B. Clark, "Transmission electron microscopy study of age hardening in a Mg-5 wt.%Zn alloy," *Acta Metallurgica*, 13 (12) (1965) 1281.

[13] G. Mima, Y. Tanaka, "The Aging Characteristics of Magnesium-4 wt% Zinc Alloy" *Transactions of Japan Institute of Metals*, 12 (2) (1971) 71-75.

[14] A.B. Eric and G.B. Brook, *Smithells Metals Reference Book, (7th ed.)*, (Oxford, Butterworth-Heinemann, 1992)

THE RELATIONSHIP BETWEEN MICROSTRUCTURE AND CREEP BEHAVIOR IN AE42 MAGNESIUM DIE CASTING ALLOY

Bob R. Powell, Vadim Rezhets, Michael P. Balogh, and Richard A. Waldo

General Motors Research & Development Center
30500 Mound Road
Warren, Michigan 48090-9055

Abstract

Microstructural analysis of die cast AE42 reveals a correlation between microstructure and creep strength. A lamellar phase $Al_{11}RE_3$ which dominates the interdendritic microstructure of the alloy partly decomposes above 150°C into Al_2RE and Al (forming $Mg_{17}Al_{12}$). The creep strength decreases sharply with these phase changes. A mechanism for the decrease in creep strength of AE42 is proposed whereby reduced presence of lamellar $Al_{11}RE_3$ and/or the presence of $Mg_{17}Al_{12}$ contribute to the observed poor creep strength at the higher temperature. The increased solubility of Al in Mg at higher temperature may also promote the decomposition of $Al_{11}RE_3$.

Introduction

An increased emphasis on weight reduction in the automotive industry is promoting the use of magnesium alloys. Magnesium is one-third less dense than aluminum and four-fifths less dense than iron. Magnesium alloys also have excellent castability that enables thin-wall and complex parts to be cast more easily relative to aluminum.

The Mg alloys currently used in the automobile industry are high-pressure die cast and have either high strength or high ductility at room temperature. The high strength alloy is AZ91D (nominal composition in mass pct, Mg-9Al-1Zn). Its applications include cam covers, baffles, oil adapters, clutch housings, roof frame assemblies, brake pedal assemblies, and steering column brackets. The high ductility alloys are AM60 and AM50 (Mg-6Al-0.2Mn and Mg-5Al-0.2Mn, respectively). Their applications include knee bolsters, instrument panels, braces, steering wheel armatures, and seat risers.

However, there is limited opportunity for the above alloys in powertrain applications such as automatic transmission cases and engine blocks because their mechanical properties decrease rapidly with increasing temperature. There has been successful production implementation of Mg alloys in engines, specifically by VW in the 1970s (1), but there are no Mg engine blocks currently in production due to high operating temperature requirements of the engine and the cost of the alloy.

A die casting alloy that was developed (by Dow) for high temperature applications is AE42 (Mg-4Al-2E where E is rare earths added as misch metal) (2). The alloy was AE42 developed from non-Al Mg alloys in which rare earths were shown to increase creep resistance by forming fine Mg_9RE precipitates along the grain boundaries (3). Aluminum was added to improve castability and room temperature mechanical properties. However, it was discovered that the rare earth reacted preferentially with Al and that slow solidification (sand and permanent mold casting) favored Al_2RE formation and produced no improvement in creep resistance. Under die casting conditions (higher solidification rates), Al_4RE formation was favored and improved creep resistance was achieved. Due to the presence of the misch metal, this alloy carries a significant cost premium over AZ91D, AM50, and AM60. Furthermore, its creep resistance decreases abruptly above 150°C (4).

This paper describes the results of our study of the microstructure of AE42 at temperatures up to 175°C. A mechanism is proposed for the microstructural change that explains the reported decrease in the creep resistance of AE42.

Experimental Procedure

The samples were obtained from Sieracki, who determined their creep behavior (4). They were prepared from Dow Chemical AE42 ingot. With the exception of AE42, the nomenclature RE will be used throughout to refer to all rare earth-containing phases, except where a specific rare earth is intended. The misch metal contained the individual rare earths: 52-55 Ce, 23-25 La, 16-20 Nd, and 5-6 Pr (5).

The samples were prepared by cold-chamber, high-pressure die casting using a steel die with a cored center hole. The center holes were chamfered to clear the passage of the threaded rod. This and degating comprised the only machining that was done to the samples. The samples were individually creep-tested at 22, 100, 125, 150, and 175°C.

Upon receipt of the tested samples (1000 hr), they were subjected to analysis methods including Inductively Coupled Plasma Atomic Emission Spectroscopy (ICP-AES), Electron Probe Microanalysis (EPMA), X-Ray Fluorescence Analysis (XRF), Scanning Electron Microscopy with Energy Dispersive Spectroscopy (SEM/EDS), X-Ray Diffraction (XRD), and Analytical Electron Microscopy with Energy Dispersive Spectroscopy and Electron Diffraction (AEM). AEM samples were prepared either by electropolishing or microtoming.

Results

A cross-section of a typical sample is shown in Figure 1. The fine-grained "skin" that results from die casting is indicated in the figure. The skin extends 0.1 to more than 0.5 mm into the sample. The electron probe microanalysis (EPMA) and XRF results in Table 1 show that the skin is more highly alloyed than the core and that the Al:RE weight ratio is preserved in both regions. The bulk composition of the samples, determined by

ICP-AES is A(3.8)RE(2.5). The bulk Al:RE weight ratio, 1.52, agrees with the EPMA results.

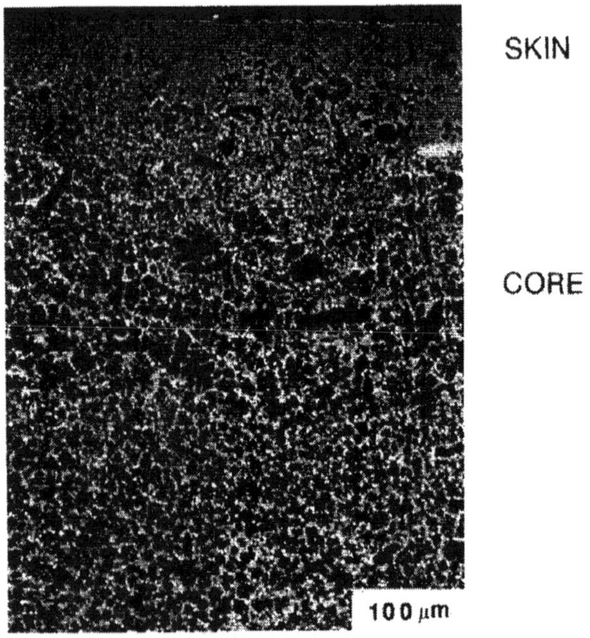

Figure 1. Backscattered electron image of the cross-section of an AE42 specimen showing the skin and core region microstructures.

The microstructures of the skin and the core regions of the tested and untested samples, were characterized. Backscatter electron images of the core regions of three samples are shown in Figure 2. These images were taken about 9 mm from the surface. All of the samples are characterized by primary Mg dendrites (dark gray) and one or more phases in the interdendritic regions. The interdendritic phase morphologies are either lamellar or particulate. The lamellar phase dominates the interdendritic microstructure of the untested sample (Fig. 2A) and the 150°C sample (Fig. 2B), while particulate phase dominates the microstructure of the sample tested at 175°C (Fig. 2C).

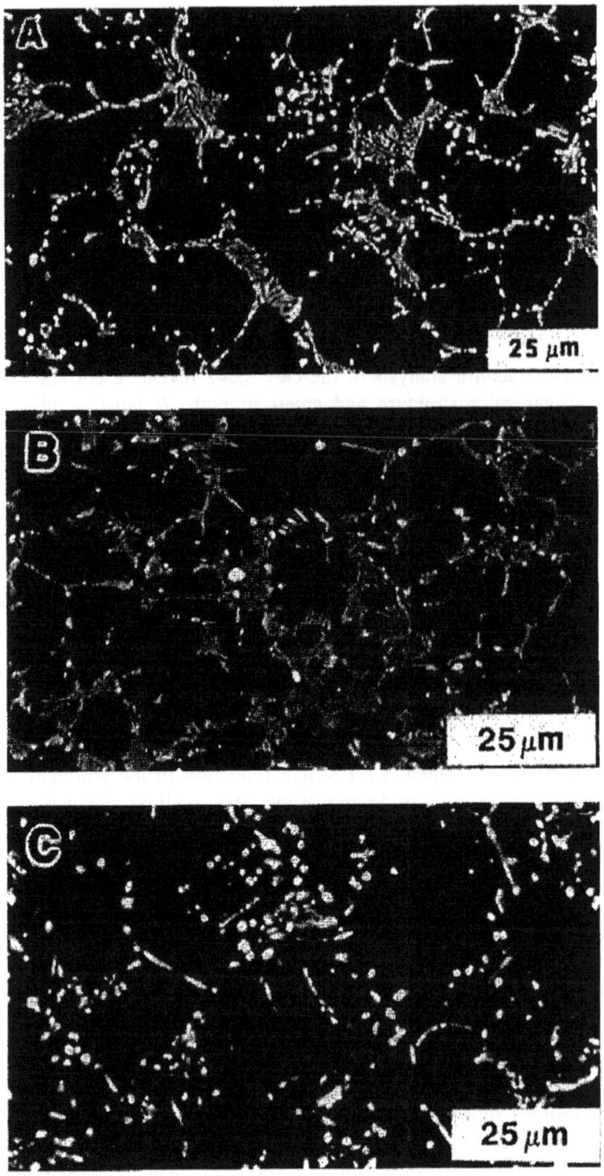

Figure 2. Backscattered electron images of the core regions of AE42 before (A) and after (B) testing at 150°C and 175°C (C).

Table I Chemical Analysis of Skin and Core Areas of Cast Specimens

Analysis Method	Skin Composition	Skin Al:RE Ratio	Core Composition	Core Al:RE Ratio
EPMA	A(3.9)RE(2.5)	1.56	A(3.4)RE(2.1)	1.62
XRF	A(4.1)RE(3.2)	1.28	A(2.9)RE(2.3)	1.26

The microstructures of the skin regions in the same samples are shown in Figure 3. The transition from lamellar to particulate phase in the interdendritic regions with increasing test temperature is less pronounced, but still present. Considering the geometry of the test temperatures, the observed creep behavior during uniaxial compressive loading across the sample (perpendicular to the skins on either side) is probably dominated by the core properties. Thus, the two microstructures; the lamellar phase-dominated structure of the samples that demonstrated good compressive stress retention up to 150°C and the particulate phase-dominated microstructure that developed in the samples during testing 175°C (which exhibited extensive creep) are significant for in depth analysis.

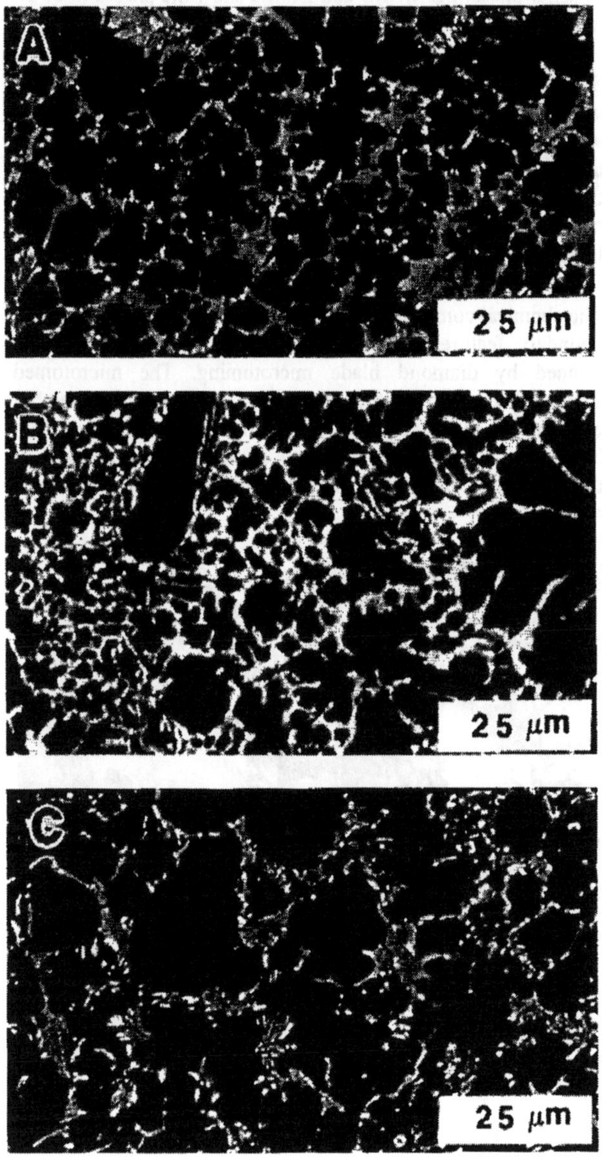

Figure 3. Backscattered electron images of the skin regions of AE42 before (A) and after (B) testing at 150°C and 175°C (C).

Phase Identification

X-ray diffraction patterns of the core regions of samples tested at room temperature, 150 and 175°C are shown in Figure 4. Five distinct phases were identified in the patterns: Mg, $Al_{11}La_3$, Al_4La, Al_2Nd, and $Mg_{17}Al_{12}$. The peak positions for Mg, the major phase, are shifted to higher angles, a shift that is consistent with dissolved Al in the phase (see below). There is some disagreement in the literature as to whether $Al_{11}La_3$ and Al_4La are the same phase (6,7). We shall treat the two phases as synonymous and refer to them as $Al_{11}La_3$.

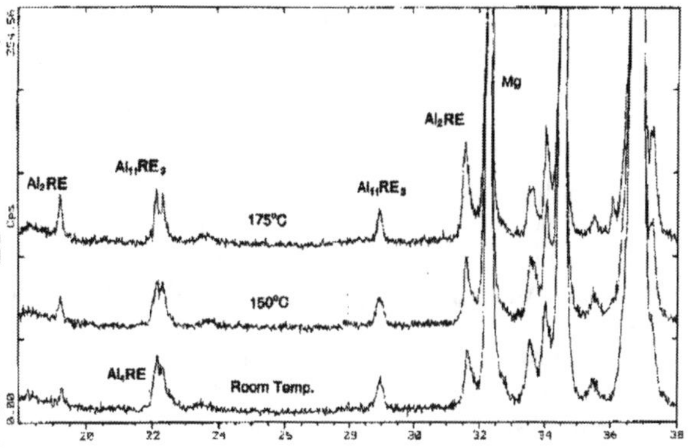

Figure 4. X-ray diffraction patterns (17 to 33° 2θ) of the core regions of samples tested at room temperature, 150 and 175°C.

Neither $Al_{11}La_3$ nor Al_2Nd (which was also identified in our samples) contain only La or Nd. The rare earth in each is a mixture of each of the components of misch metal. These rare earths compounds (La or Nd) and the other rare earths in misch metal (Ce and Pr) are isostructural (8) and will be referred to collectively as RE; e.g., $Al_{11}RE_3$, Al_4RE, and Al_2RE. EPMA and AEM evidence regarding the chemical composition of individual particles will be presented below.

Two changes in the patterns occurred as the test temperature was increased above 150°C (see Table II). The intensity of the Al_2RE peaks increased and at 36°2Θ $Mg_{17}Al_{12}$ appeared. Rietveld analysis of the patterns (9) yielded the following estimates for relative concentrations (in weight percent) of Mg, $Al_{11}RE_3$, Al_2RE, and $Mg_{17}Al_{12}$ in each sample:

Table II Effect of Temperature on Composition by Phase

Test Temp., °C	*Mg*	$Al_{11}RE_3$	Al_2RE	$Mg_{17}Al_{12}$
25	97.5	1.8	0.8	0.0
150	97.7	1.5	0.8	0.0
175	97.0	1.2	1.3	0.6

These changes are consistent with the reaction by which $Al_{11}RE_3$ decomposes to Al_2RE and frees Al which then reacts with Mg to form $Mg_{17}Al_{12}$. The poor creep resistance of AZ and AM alloys is attributed to the presence of the $Mg_{17}Al_{12}$ (10).

Figure 5 shows XRD patterns for the skin and core regions of a sample that was tested at 175°C. Four peaks are labeled. Two of the peaks are $Al_{11}RE_3$ and are less intense in the core pattern than they are in the skin pattern. The other two peaks are the Al_2RE phase and are present only in the core pattern. This data supports the interpretation of the SEM and EPMA micrographs. However, it also suggests that the skin microstructure is more stable than the core (no Al_2RE in the skin diffraction pattern). This observation will be discussed in some detail later.

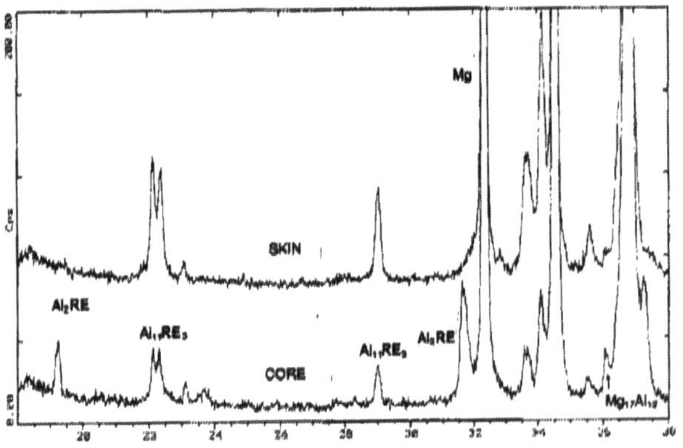

Figure 5. X-ray diffraction patterns (18 to 38° 2θ) of the skin and core regions of samples tested at 175°C.

Primary Mg Dendrites

The chemical composition of the primary Mg dendrites was measured by EPMA and SEM/EDS. Both measurements show that the major element in the primary Mg dendrites is Al (1.0 - 1.5%). Rare earths were not detected in the dendrites of any of the samples, down to the sensitivity of the EPMA, ~ 0.05%. Nor were they detected in analytical electron microscope analysis of microtomed samples, down to its detection limit: ~ 0.2%.

XRD analysis shown in Table III indicates that much of the Al detected in the primary Mg is in solid solution. Measured lattice parameters for Mg were shifted to smaller values, consistent with the reported effect of dissolved Al on the Mg lattice parameters (8). The shifted a_o and c_o values were used to estimate the amount of dissolved Al in the Mg using solubility data from the literature (8). The results are shown below. There is excellent consistency in the amount of dissolved Al calculated from each of the two lattice parameters, a_o and c_o. The solubility limits for Al in Mg (binary solid solution) at each test temperature (8) are included in the table. The data suggest that 1-1.5% Al in Mg in all samples (concentrations obtained from EPMA and SEM/EDS measurements) is completely dissolved during the elevated-temperature tests.

Grain Boundary/Interdendritic Microstructure

Figures 2 and 3 show that the grain boundaries of the AE42 samples contain extensive precipitation. In untested and 150°C samples, the precipitates are predominantly lamellar. Some particulates are present, but the microstructure is mainly lamellar phase. SEM/EDS and EPMA show the phases to be E- and Al-rich.

Samples that were tested at room temperature, 150, and 175°C were analyzed by AEM. The interdendritic regions are filled with the lamellar phase and occasional particulates in the room temperature and 150°C samples (see Figures 6 and 7) similar to the untested sample shown in Figures 2 and 3. Dislocation networks/webs are evident in the primary Mg. Most of the lamellar phase is gone in the sample tested at 175°C (Figure 8).

The chemical composition of the precipitates and the other grain boundary features was determined using samples that were thinned by diamond blade microtoming. The microtomed samples had uniform thickness, about 70 nm, and were thin enough such that precipitates did not overlap under the electron beam. This enabled chemical and crystallographic analysis of the individual precipitate particles and the regions between them.

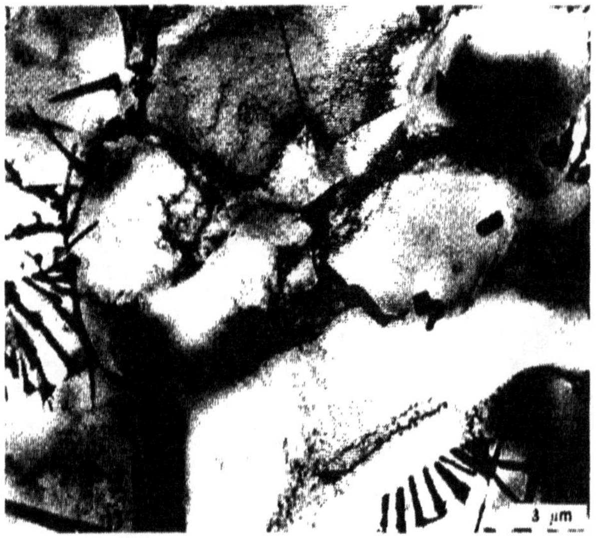

Figure 6. Transmission electron micrograph of the core region of an AE42 sample that was tested at room temperature. Sample prepared by electrochemical thinning.

Table III Effect of Test Temperature on Primary Magnesium Lattice Parameters

Test Temp., °C	a_0 *(nm)*	$\% Al_{est}$	c_0 *(nm)*	$\% Al_{est}$	*Solubility Limit for Al, %*
25°C	0.320644	0.76	0.520573	0.80	0.7
150°C	0.320586	0.95	0.520508	1.01	2.0
175°C	0.320498	1.60	0.520359	1.30	2.5

Figure 7. Transmission electron micrograph of the core region of an AE42 sample that was tested at 150°C. Sample prepared by electrochemical thinning.

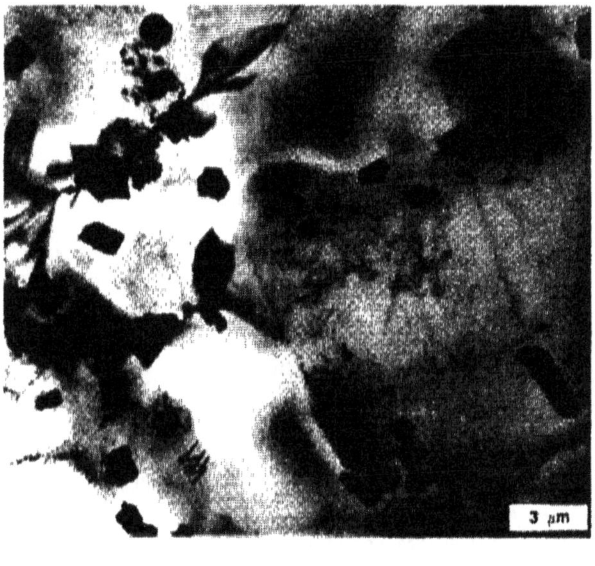

Figure 8. Transmission electron micrograph of the core region of an AE42 sample that was tested at 175°C. Sample prepared by electrochemical thinning.

The interdendritic region comprises generally two phases, the lamellar phase and the particulate phase. The lamellar phase present in samples tested below 150°C is predominantly $Al_{11}RE_3$ and the particulate phase present at 175°C is Al_2RE. Si-rich particulates were observed, but they play a minor role; the overall content of Si in the sample was 0.02%.

There is no evidence of rare earth in the Mg-rich regions between the lamellar and/or particulate phases. The only rare earth observed was in the two phases, $Al_{11}RE_3$ and Al_2RE. There is no evidence of rare earth in solid solution as fine Mg-RE precipitates in the grain boundaries. Both possibilities have been proposed to explain the creep resistance of AE42 (6,11,12).

In summary, the microstructure of the samples that were tested up to 150°C is characterized by:

- primary Mg dendrites with 1-1.5% Al in solution and trace amounts of rare earths
- a lamellar interdendritic phase with the formula $Al_{11}RE_3$ ($\sim Al_4RE$)
- a minor particulate interdendritic phase having the formula Al_2RE
- no Mg-Al-RE phases or $Mg_{17}Al_{12}$
- no rare earth solid solutions or phases (Mg-RE) in the interdendritic regions.

The microstructure of the samples tested at 175°C contains:

- primary Mg dendrites with 1-1.5% Al in solution and trace amounts of rare earths
- an interdendritic region dominated by particulate Al_2RE
- lamellar Al_4RE is the minor phase in the interdendritic region
- $Mg_{17}Al_{12}$ in the interdendritic region
- no other rare earth-containing phases or solid solutions.

These results support the common model of creep resistance that attributes high creep rates of Mg-Al alloys to the presence of $Mg_{17}Al_{12}$ in the microstructure. Below 175°C the $Mg_{17}Al_{12}$ is absent because Al is sequestered in primary Mg solid solution and Al_4RE and creep is slow. At 175°C the Al is released by the decomposition of Al_4RE breaks, $Mg_{17}Al_{12}$ is formed and creep is accelerated.

Discussion

Two models have been reported to explain the improved creep resistance. The first proposes that not all of the rare earth in AE42 forms Al_4RE. The "extra" rare earth remains in solution to provide the creep resistance (13). The other model proposes that by forming Al_4RE the residual aluminum in the alloy is low enough that it can be retained in solution in the primary Mg, the overall result being that there is essentially no $Mg_{17}Al_{12}$ present (14). In our analysis of the tested samples we specifically looked for Mg-RE and Al-RE phases, Al and RE solid solutions in Mg, and $Mg_{17}Al_{12}$.

Proposed Mechanism for Loss of Creep Resistance at 175C

None of the work in the literature suggests a reason for accelerated creep of AE42 at high temperature (T>150C). Our results suggest that $Al_{11}RE$ is unstable and decomposes according to the reaction:

$$Al_{11}RE_3 \Longrightarrow 3Al_2RE + 5Al$$

The released Al subsequently reacts with Mg to form $Mg_{17}Al_{12}$:

$$12Al + 17Mg \Longrightarrow Mg_{17}Al_{12}$$

It is also possible that the increased solubility of Al in the Mg as the temperature is raised also contributes to the decomposition of $Al_{11}RE_3$. Thus, a more general formula (unbalanced) for the reaction that degrades AE42 creep resistance is:

$$Al_{11}RE_3 + Mg ===> Al_2RE + Al \text{ (solid solution in Mg) } + Mg_{17}Al_{12}$$

Accordingly, an AE42 sample that has undergone this reaction has apparently lost its creep resistance because lamellar $Al_{11}RE_3$ is no longer present in the eutectic region to retard creep and/or low-melting $Mg_{17}Al_{12}$ is present which facilitates creep.

Factors Contributing to the Formation and Stability of $Al_{11}RE_3$

In this study, dozens of individual particles were analyzed in the AEM. The objective was to determine whether the particles were $Al_{11}RE_3$ or Al_2RE. Both chemical and crystallographic data were collected. It was discovered that the occurrence of $Al_{11}RE_3$ vs. Al_2RE in the samples correlated with the La:Nd ratio in the particle. Figure 9 shows the RE compositions of about five-dozen particles observed in the core of a sample that was tested at room temperature. Those particles with a low La:Nd ratio (below about 0.7) were Al_2RE, whereas when the La:Nd ratio was high, $Al_{11}RE_3$ formed. This is the first evidence that the formation of either $Al_{11}RE_3$ or Al_2RE is sensitive to individual rare earth elements, an observation that could influence future studies of stabilization of the $Al_{11}RE_3$ phase.

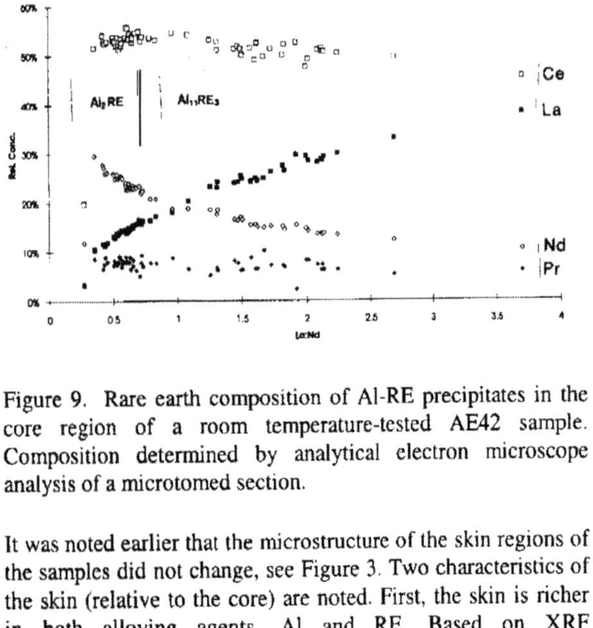

Figure 9. Rare earth composition of Al-RE precipitates in the core region of a room temperature-tested AE42 sample. Composition determined by analytical electron microscope analysis of a microtomed section.

It was noted earlier that the microstructure of the skin regions of the samples did not change, see Figure 3. Two characteristics of the skin (relative to the core) are noted. First, the skin is richer in both alloying agents, Al and RE. Based on XRF compositional analysis, the skin composition is an alloy that could be called ARE43; the core, ARE32. Aune et al. (15) studied the stress relaxation rate of several A-RE alloys. The weight percents of A and RE were varied from 3-4 and 1-4, respectively. They saw no trend in stress relaxation rate for the compositions. Their experiments were done at 150°C and no microstructural analysis was reported. The effect of alloying levels at higher temperature needs further study.

The second difference between the skin and core microstructures is the dendrite size. It is finer in the skin than in the core due to the rapid solidification of the alloy by the die wall during die casting. The finer size may inhibit the proposed decomposition reaction, but the studies to demonstrate this in the A-RE system have yet to be done.

The effect of stress on the decomposition of $Al_{11}RE_3$ was also considered. Samples were exposed to the same thermal history but without the applied stress. The same microstructural changes already described also occurred in the unstressed samples, thereby eliminating stress as a contributing factor to $Al_{11}RE_3$ stability.

Summary and Conclusions

1. The sharp decrease in the creep resistance of AE42 at 175°C reported by Sieracki correlates with a reduction in volume fraction of $Al_{11}RE_3$ in the microstructure of the specimens. The lamellar $Al_{11}RE_3$ phase dominates the interdendritic microstructure of the samples tested at 150°C and below.
2. Above 150°C, $Al_{11}RE_3$ diminishes in volume fraction, Al_2RE increases, and there is concurrent formation of $Mg_{17}Al_{12}$. Creep resistance deteriorates either because of the reduced presence of the lamellar $Al_{11}RE_3$ in the microstructure or because of the appearance of $Mg_{17}Al_{12}$.
3. The existing theory in the literature that AE42 creep resistance is due to rare earth dissolved in Mg is not supported by this work; no evidence of rare earth solid solution was found.

Acknowledgements

The authors would like to thank Edmund G. Sieracki for providing the samples and for helpful technical discussions, Deborah F. Esch and Patrick M. Hanley and Raymond C. Lints for sample analysis, and Alan A. Luo, Paul E. Krajewski, and Anil K. Sachdev for helpful discussions.

References

1. F. Hollrigl-Rosta et al., "Magnesium in Volkswagen," Light Metal Age, (8) (1980), 22-29.
2. K.E. Nelson, "Magnesium Die Casting Alloys," (presented at the 6th SDCE International Die Casting Congress, Cleveland, Ohio, 1970) paper no. 13.
3. W.E. Mercer, "Magnesium Die Cast Alloys for Elevated Temperature Applications," SAE Technical Publication No. 900788, SAE, Warrendale, PA, USA, 1990.
4. E.G. Sieracki, J.J. Velazquez, and K. Kabiri, "Compressive Stress Retention Characteristics of High Pressure Die Cast Magnesium Alloys," SAE Technical Publication No. 960421, SAE, Warrendale, PA, USA, 1996.
5. William.E. Mercer, private communication with author, Dow Chemical Company, 15 March 1998.
6. L.Y. Wei and G.L. Dunlop: "Precipitation Hardening in a Cast Mg-Rare Earth Alloy," in Magnesium Alloys and Their Applications, ed., B.L. Mordike and F. Hehmann, DGM Informationsgesellschaft, Verlag, Germany, (1992), 335-342.
7. A.H. Gomes de Mesquita and K.H.J. Buschow, "The Crystal Structure of So-Called alpha-$LaAl_4$," Acta Cryst., 22 (1967), 497-501.

8. J.L. Murray: Phase Diagrams of Binary Magnesium Alloys, (Metal Park, Ohio: ASM International, 1988) 17-34.
9. H.M. Rietveld, "Line Profiles of Neutron Powder-Diffraction Peaks for Structure Refinement," Acta Cryst., 22 (1967), 151-152.
10. A.A. Luo and M.O. Pekguleryuz, "Cast Magnesium Alloys for Elevated Temperature Applications," J. Mat. Sci., 29 (1994) 5259-5271.
11. L.Y. Wei and G.L. Dunlop, "The Solidification Behavior of Mg-Al-Rare Earth Alloys," J. Alloys and Compounds, 232 (1996) 264-268.
12. E.F. Emley, Principles of Magnesium Technology, (Oxford, England: Pergamon Press, 1966), 267.
13. G. Foerster, "Designing Die Casting Alloys," (paper presented at the 7th SDCE International Die Casting Congress, 1972), No. 9372.
14. D.J. Sakkinen, "Physical Metallurgy of Magnesium Die Castings," SAE Technical Publication No. 940779, SAE, Warrendale, PA, USA, 1994.
15. T.K. Aune, H. Westengen, and T. Ruden, "The Effects of Varying Aluminum and Rare Earth Content on the Mechanical Properties of Die Cast Magnesium Alloys," SAE Technical Publication No. 940777, SAE, Warrendale, PA, USA, 1994.

DEVELOPMENT OF CREEP RESISTANT Mg-Al-Sr ALLOYS

Mihriban O. Pekguleryuz, Eric Baril

Noranda Inc. – Technology
240 Hymus Boulevard
Pointe-Claire, Québec, Canada H9R 1G5

Abstract

There have been attempts since in the 70's to develop creep resistant magnesium diecasting alloys for automotive applications such as automatic-transmission case and engine components. The earliest die casting alloys developed as a result of these activities were the Mg-Al-RE and Mg-Al-Si systems (AE and AS alloys). The shortcomings of these two alloy systems related to high cost or borderline properties have led to renewed activity in the 90's in the development of magnesium alloys with improved elevated-temperature properties. This paper presents the development of a new family of creep-resistant Mg alloys based on the Mg-Al-Sr system. Creep resistance, the tensile yield strength and the bolt-load-retention of these alloys at 150°C and 175°C show improvement over Mg-Al-RE and Mg-Al-Si system. The microstructure of the alloys is characterized by Al-Sr-(Mg) containing intermetallic second phases. The absence of the $Mg_{17}Al_{12}$ phase in the microstructure, either creep-induced or as-cast, is one of the factors that contribute to improved creep-resistance of these alloys over the Mg-Al based diecasting alloys. Furthermore, the alloys exhibit better salt-spray corrosion resistance (0.09-0.15mg/cm2/day) than other commercial magnesium diecasting alloys such as AM60B, AS41, AE42 and the aluminum diecasting alloy A380.

Introduction

The need to reduce weight in transport vehicles has led to growing use of magnesium in automotive applications in the 1990's in applications such as valve covers, instrument panels, seat-frames and steering wheels. Further use of magnesium is targeted for more critical components such as transmission and engine parts that require elevated-temperature performance. Magnesium here faces a challenge in meeting the performance requirements of these components for elevated-temperature (150°C) strength and creep resistance as well as adequate corrosion resistance.

It is known that hexagonal magnesium is prone to creep-deformation even at room temperature at high stresses. Most commercial alloys of magnesium, which are based on the Mg-Al or Mg-Zn system also, suffer from low creep resistance. In the diecast state, the additional complication is the aging of the supersaturated alloys during creep. Discontinuous precipitation from the supersaturated matrix of low-melting point second phases such as $Mg_{17}Al_{12}$ has been concluded recently (1) to contribute to grain boundary migration and creep deformation in most commercial diecasting alloys such as AZ91D (Mg-9%Al-1%Zn) and AM60 (Mg-6%Al).

Rare-earth additions to Mg are known to increase the resistance to creep deformation. Other elements with beneficial effect are thorium, yttrium, scandium, and silver. Most of the commercial magnesium alloy systems that possess creep resistance are based on rare-earth or yttrium containing systems. These alloys, though widely used in aerospace applications, find limited use in the automotive industry due to the cost penalty. Furthermore, automotive applications using magnesium require manufacturing processes that are of high productivity such as diecasting, which may upset the relatively high cost of magnesium metal, when compared to steel or aluminum. The challenge in developing new magnesium-based creep resistant alloys for the automotive industry is to find alloy systems that combine performance, diecastability (or other processes of high productivity) and cost. Mg-Al-Sr system described in this paper offers an optimum combination of these requirements.

Theoretical Background

It has been known since the 50's that rare-earth elements such as Ce improve the creep resistance of magnesium (2). However, the rare-earths are an expensive alloying addition to magnesium, and this has been the main factor why rare-earth containing magnesium alloys have not been widely accepted by a cost-sensitive automotive industry . The R&D activity in the 80's led to the development of Mg-Al-RE alloy system as a diecastable alloy for automotive applications. Designated as AE alloys, this alloy system was promoted by Dow Magnesium in the 90's, but did not find wide use due the high cost of misch-metal sources. AE alloys considered were AE42 (Mg-4%Al-2%RE), AE 21 (Mg-2%Al-1%RE) and AE52 (Mg-5%Al-2%RE). The AE42 diecasting alloy, which exhibits good creep resistance, has the $Al_2(RE)$ and Al_4RE compounds. These intermetallics are thermally and metallurgically stable as seen in the related phase diagrams (Fig.1.a) and would be expected to yield effective grain boundary strengthening and resistance to flow during creep loading (Fig 1.b). The fact that the alloy undergoes no aging and precipitation from the matrix has also been suggested to be the main reason for its good creep resistance at 150°C (1).

The search for alternative creep-resistant Mg alloys of acceptable cost to the automotive industry has led to the development of alkali-earth containing Mg-alloys. The starting point for the selection of alkali-earth elements for creep-resistance was the fact that alkali-earth elements such as Mg, Ca or Sr behave similarly to rare-earth elements (Ce, Nd) in many alloy systems. Examples are inoculation of graphite by either Ce (rare-earth) or Mg (alkali-earth), or the modification of various second phases by Sr, Ca or Ce (4).

Initial work in replacing RE additions with low cost alternatives had led to the development of Mg-Al-Ca alloys (5,6). This system also possesses a structure similar to AE alloys and a similar creep resistance. The structure is characterized by the Al_2Ca second phase (2) at the grain boundaries, which is a stable line compound with a relatively high melting point (Fig. 2.a&b). These alloys ran into diecastability problems when the work was extended to the prototyping stage of automatic transmission cases. (7) The development of Mg-Al-Sr system was a next step in the work aimed at replacing RE additions to magnesium with alkali-earth elements.

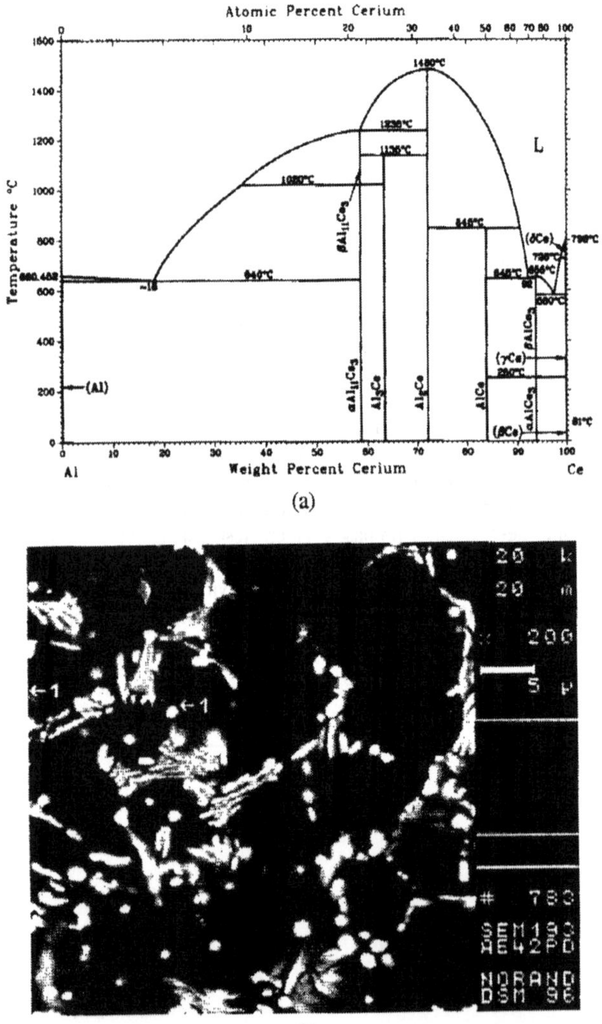

Fig. 1. (a) Al-Ce system (Ref. 3) and (b) SEM micrograph of diecast AE42 alloy showing the Al-RE intermetallics.

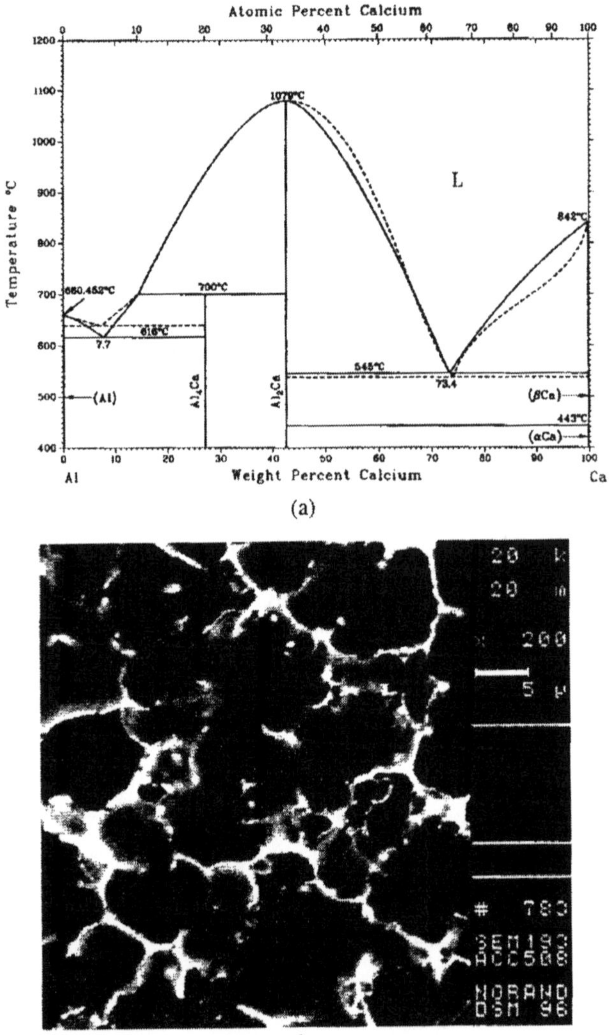

Fig. 2. (a) Al-Ca system (Ref. 3) and (b) SEM micrograph showing the Al_4Ca second phases at grain boundaries of a diecast Mg-5Al-0.8Ca alloy.

Experimental Procedure

Alloy Preparation and Casting

Permanent-Mold Casting - The initial work was carried out on permanent-mold cast samples. Alloys in a wide range of Al (2-12% Al) and Sr (0.5-10%) were investigated. 1-2 kg alloy batches were prepared using commercial Mg-5.0%Al alloy and an Al-Sr master alloy (Timminco Metals) using a Lindberg electric resistance furnace. The alloys were cast into ASTM-standard tensile bars using a steel permanent-mold and superheat of 100ºC. Chemical compositions of the alloys are given in Table I.

Creep testing was performed according to ASTM E139. The total creep extension (% Creep) was measured from each elongation-time curve.

Salt-spray corrosion testing was conducted on diecast corrosion test plates (ASTM standard) of A-alloys as well as on AZ91D, AS41, AE42 magnesium alloys and the A380 aluminum alloy. For each alloy, six (6) plates were subjected to a 200-hour salt spray test in accordance with ASTM B117. An average corrosion rate was then obtained based on the weight loss during the test. Tests were conducted at Intermag technologies and Bodycote Laboratories.

TABLE I. CHEMICAL COMPOSITION OF PERMANENT-MOLD CAST ALLOYS

ALLOY	CHEMICAL COMPOSITION							
	Al (wt%)	Sr (wt%)	Mn (wt%)	Zn (ppm)	Fe (ppm)	Cu (ppm)	Ni (ppm)	Ca (ppm)
Permanent-Mold Cast Alloys								
AC2	4.90	1.59	0.30	43	60	<2	<2	<10
AC4	4.70	1.26	0.33	78	84	122	7	35
AC6	4.89	1.22	0.32	69	41	127	9	40
AC9	4.82	1.07	0.32	42	39	82	3	31
AC10	5.08	1.46	0.29	52	39	150	2	8
Diecast Alloys								
A1	4.90	1.74	0.26	94	23	4	3	18
A2	4.85	1.23	0.29	94	11	2	3	17

Diecasting--Two Sr levels (1.2 and 1.8%Sr) were selected for diecasting trials. A 5% aluminum level was selected for the best compromise between strength and ductility. The alloys were tentatively designated A1 and A2 alloys.

800-kg batches of the A1 and A2, alloys were prepared for diecasting test specimens. A 600-ton Prince cold-chamber diecasting machine of Intermag Technologies (formerly ITM Inc) was used to diecast mechanical (tensile, creep, fatigue) and corrosion test specimens.

Specimens of commercial magnesium alloys, AZ91D (Mg-9%Al-1%Zn), AS41 (Mg-4%Al-1%Si), AE42 (Mg-4%Al-2%RE), and the Al alloy A380 (Al-8%Si-3%-Cu-3%Zn-1.3%Fe) were also diecast to compare with the experimental A-alloys. Chemical analyses (ICP) of alloy samples taken from the melts are shown in Table I.

Mechanical and Corrosion Testing

Tensile and creep tests of the permanent-mold cast and diecast specimens were carried out at room temperature and elevated temperatures at Westmoreland Laboratories. Elevated-temperature tests were performed at 150°C and 175°C with a soaking time of 30 minutes.

Tensile tests were carried out on round and flat tensile specimens. ASTM sub-size flat specimens were used for creep testing at 150°C and 175°C and at 35 MPa and 50 MPa and (tensile stress) for 200 and 500 hours.

Microstructural Characterization

Since the second phase particles in die-cast part are very small (<300 nm), standard cross-sectional techniques used for x-ray analyses in the Scanning Electron Microscope (SEM) are not adequate for obtaining quantitative chemical or morphological information. Accordingly, the Transmission Electron Microscopy (TEM), X-Ray Energy Dispersive Spectroscopy (EDX) and Electron Energy Loss Spectrocopy (EELS) were used to characterize the chemistry and morphology of second phase particles in magnesium die-casting alloys. This allowed the determination of the chemical composition of small areas of the particles with a very high spatial resolution (≈5 nm). Dry ultramicrotomy was used to produce thin sections for STEM imaging work. Two significant advantages of using ultramicrotomy to prepare samples for microanalysis are the following: (1) sectioning with a diamond knife does not change the chemistry of the samples, and (2) the very small thickness obtained (25 to 30 nm) significantly improves the spatial resolution of the analysis. (8)

The ultra-microtomed cross-sections were examined using a *Philips CM30* Scanning Transmission Microscope (STEM) equipped with an *Oxford/Link LZ-4* EDX detector and a *Gatan 666* parallel electron energy loss spectrometer (PEELS). Alternately, the bulk samples polished with the ultramicrotome were examined using a *Hitachi S-4700HQ* Field Emission Gun Scanning Electron Microscope (FEG-SEM). In order to develop a more standard characterization technique, the results from the STEM characterization were matched to more standard optical microscopy and metallographic cross-sectioning techniques.

Results and Discussion

Properties of Permanent-Mold Cast Mg-Al-Sr Alloys

Table II shows the tensile test results at room temperature and 150 °C and creep resistance at 150 °C of permanent-mold cast Mg-Al-Sr alloys and compares them to other magnesium alloys and the aluminum alloy A380, that were also permanent-mold cast using the same mold.

Tensile yield strength of the A-alloys at 150 °C was superior to Mg-Al-RE (AE42), and the aluminum A380 alloy. Creep resistance at 175°C was higher than all other magnesium alloys and the aluminum alloy A380.

Properties of Diecast Mg-Al-Sr Alloys

Tensile test results of the two die cast Mg-Al-Sr alloys (A1 and A2) are shown in Table III. The elevated-temperature tensile strength of the new Mg-Al-Sr alloys were observed to be superior to the creep resistant magnesium alloys AE42 and AS41. At 150-175°C, tensile yield strength (YS) and the ultimate tensile strength (UTS) were higher than AE42 and AS41. Tensile yield and ultimate tensile strength of the alloys at 175 °C also surpassed AZ91D. Ductility of the A1 was comparable to the AZ91D.

Tensile creep performance of Mg-Al-Sr alloys at 150 °C and 175°C and under stress loading of 35-50 MPa is given in Table IV. A1 (Mg-5Al-1.8Sr) and A2 (Mg-5Al-1.2Sr) alloys show higher creep resistance than all magnesium alloys and the A380 aluminum alloy at 150 °C and 175°C under 35 MPa tensile loading. The A1 alloy showed the most resistance. The trend was observed to continue at higher stresses (Table IV & Fig. 3). It can be seen that die cast Mg-Al-Sr alloys offer an optimum combination of creep resistance and tensile strength at 150°C and 175 °C.

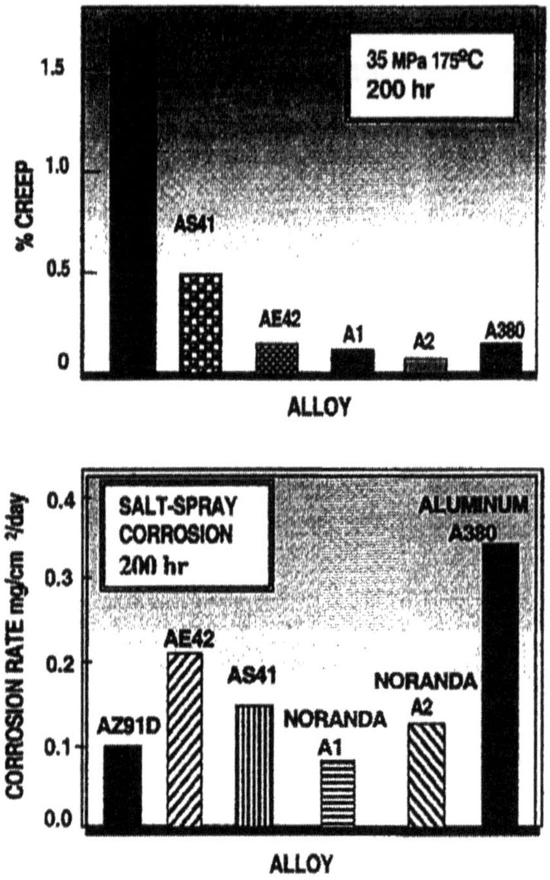

Fig. 3. (a) 175°C creep resistance and (b) salt spray corrosion behavior of diecast magnesium alloys.

TABLE II Properties of Permanent-Mold Cast Magnesium Alloys

ALLOY	PROPERTIES AT 150°C			CREEP EXTENSION, % @ 35 MPa, 150°C, 200 hours
	Ultimate Tensile Strength, MPa	Yield Strength, MPa	Elongation, %	
AZ91	170	83	8.6	0.136 %
AE42	114	45	9.0	0.03 %
AC9	118	58	5.7	0.01 %
AC10	109	62	4.3	0.03 %
AC6,AC4	94	64	3.6	0.01 %
AC2	98	66	2.3	-
A380	175	125	1.1	0.096

* Permanent-mold cast; flat tensile specimens-ASTM standard (average of 3 specimens).

TABLE III. Tensile Properties Diecast Magnesium Alloys

ALLOY	TENSILE STRENGTH (MPa)			YIELD STRENGTH (MPa)			ELONGATION %		
	R.T	150°C	175°C	R.T	150°C	175°C	R.T	150°C	175°C
AZ91	239	170	138	157	105	89	4.7	18.0	20.5
AE42	226	142	121	135	87	81	9.2	22.5	23.1
AS41	249	153	127	132	94	85	8.9	16.8	18.0
Mg-5Al-1.8Sr (A1)	202	164	148	145	108	103	4.0	13.6	14.8
Mg-5Al-1.2Sr (A2)	233	149	133	138	102	97	8.8	16.4	21.4
A380	290	255	248	155	149	154	3.2	6.4	7.1

* High-pressure diecast; round tensile specimens (average of 8 specimens).

TABLE IV Creep and Corrosion Properties of Diecast Magnesium Alloys

ALLOY	35 MPa 200 hr		50 MPa 150 °C		CORROSION RATE (mg/cm²/day)
	150°C	175°C	200hr	500hr	
AZ91D	1.21	1.84	2.7	6.35	0.10
AE42	0.07	0.14	0.06	0.084	0.21
AS41	0.13	0.50	0.45	0.735	0.16
Mg-5Al-1.8Sr (A1)	0.03	0.09	0.03	0.033	0.09
Mg-5Al-1.2Sr (A2)	0.07	0.05	0.07	0.092	0.14
A380	0.18	0.15	0.08	0.099	0.34

* High-pressure diecast; ASTM flat subsize tensile specimens (average of 5 specimens).

Salt-spray corrosion resistance of the Mg-Al-Sr alloys is also comparable to AZ91D and AE42 alloys. The A1 alloy had higher corrosion resistance (lower corrosion rate) to salt-spray than even the AZ91D alloy (Fig. 3)

Microstructure of Mg-Al-Sr Alloys

The resistance to creep deformation of the Mg-Al-Sr alloys seems to be due to dispersoid (second phase) strengthening. SEM EDX analysis shows that strontium in the alloy combines with Al and impedes the formation of the $Mg_{17}Al_{12}$ phase. Since the latter phase is thought to facilitate grain boundary migration thereby decreasing creep resistance (1), its absence in the microstructure is probably one of the main reasons for the superior creep resistance of the Mg-Al-Sr alloys.

As illustrated in Figures 4 and 5, both A alloys show very different microstructure. Indeed, A1 alloy microstructure contains two types of second phase compounds: *A-type* (divorced eutectic type) and *B-type* (non-lamellar type). In contrast, A2 alloy contains only one type of second phase compound, which is a lamellar type (*C-type*). These second-phase compounds are located at the grain boundaries of the •-Mg solid solution. The total volume fraction of second phase particles is approximately the same for both alloys. In A1 alloy, the volume fraction of *B-type* compounds seems to be higher than the fraction of *A-type*.

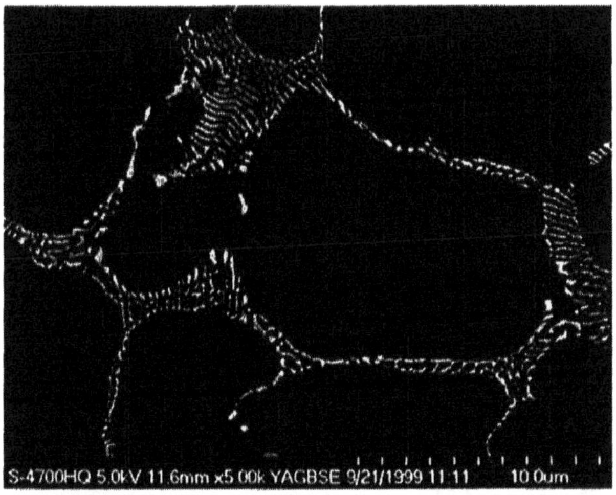

Figure 4: Compositional contrast image, obtained with back scattered electrons imaging in the FEG-SEM, of the diecast structure of Mg-5Al-1.2Sr (A2) alloy showing the second phase surrounding the grain boundaries.

Scanning transmission electron microscopy and x-ray mapping of the ultra-microtomed cross-sections were carried out in order to determine the type of second phases present in the alloys. X-ray mappings of A1 and A2 alloys are shown in figure 6. The presence of three different types of second phase compound was confirmed by the STEM characterization. From the x-ray mappings, no $Mg_{17}Al_{12}$ phases were found. As shown in figure 6, the second phases analysed in the diecast Mg-5Al-1.8%Sr alloy (A1) revealed two types of compounds. One is an Al-Sr intermetallic with a low level of magnesium and the second is a Mg-Sr-Al intermetallic as shown in the x-ray line profile (see figures 7 and 8). Conversely, A2 alloy contains only the eutectic type second-phase compound, which is a Mg-Al-Sr intermetallic.

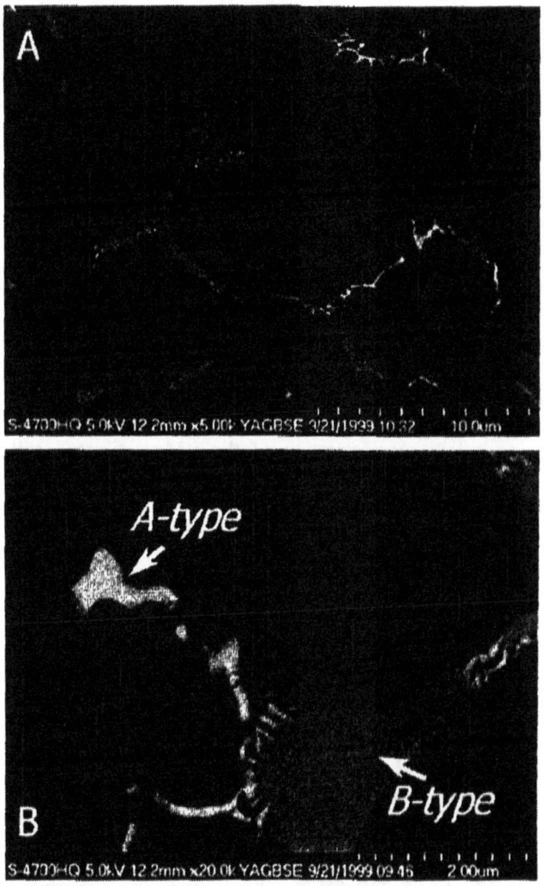

Figure 5: Compositional contrast image (A), obtained with back scattered electrons imaging in the FEG-SEM, of the die cast structure of Mg-5Al-1.8Sr (A1) alloy showing the two types of second phase surrounding the grain boundaries (see zoomed image (B)).

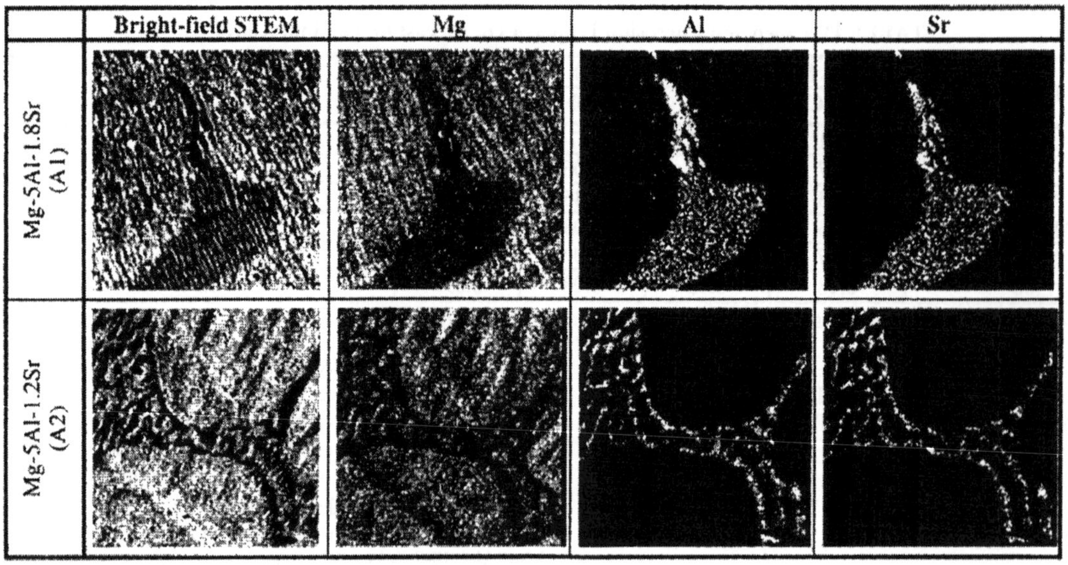

Figure 6: STEM image of the diecast structure of A-alloys and related x-ray intensity maps from Mg, Al and Sr.

The two A-alloys were investigated using STEM and EDX. In order to precisely identify the second phase compounds, the results generated from the x-ray profiles were compiled and the average composition and statistical error were determined (see Table V). From the atomic fraction data, the second phase compounds in the two A-alloys were tentatively designated (see Table VI). Many designations could be attributed to these compounds. As an example, *C-type* compounds in A2 alloy could be a ternary compound $Al_{10}Mg_{13}Sr$ or $Al_{12}Mg_{17}$ binary compound with Sr atoms substituting for Al and Mg atoms. There is no Mg-Al-Sr ternary compounds reported in the literature (9).

These complex compounds could be related to the fast solidification that occurs during die casting. Therefore, non-equilibrium microstructures and metastable compounds could be obtained in the final structure. However, the good creep resistance observed in A-alloys suggests that the microstructure is relatively stable. In addition, the Al content of the •-Mg matrix surrounding the second phase compounds is relatively low suggesting a low supersaturation of the •-Mg. Consequently, these alloys should be less sensitive to discontinuous precipitation ß-$Mg_{17}Al_{12}$.

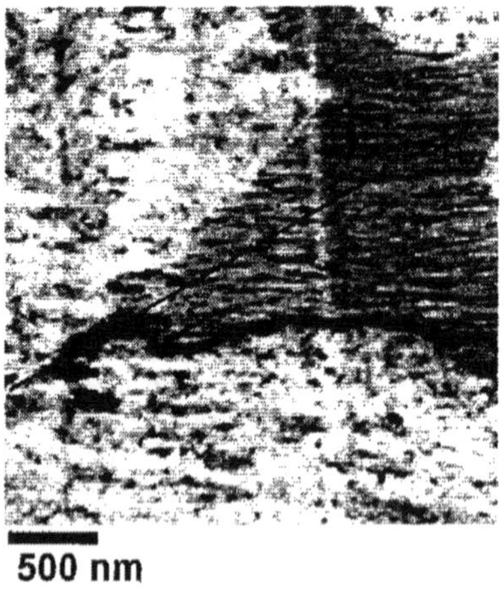

Figure 7: STEM image of the diecast structure of Mg-5Al-1.8Sr alloy showing the location of the x-ray line profile through the two types of second phase (see Figure 5).

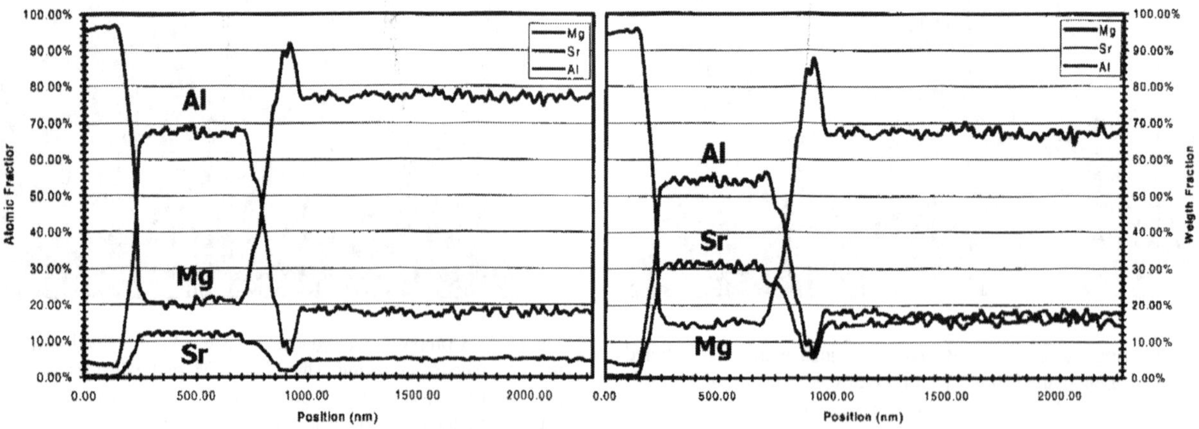

Figure 8: X-ray line profile (see line in figure 5) showing the content of Al, Sr and Mg of the two types second phase particles.

TABLE V Average Composition of the Mg Matrix and the Two Types of Second-Phase Particles

ALLOY	PHASE	WEIGHT FRACTION			ATOM FRACTION		
		Mg	Sr	Al	Mg	Sr	Al
A1	Matrix	95.92±0.58	0.66±0.21	3.42±0.49	96.71±0.47	0.19±0.06	3.11±0.45
	A-type	14.89±0.71	31.01±0.88	54.10±0.92	20.60±0.92	11.91±0.43	67.48±0.93
	B-type	68.62±1.53	14.50±1.06	16.88±0.91	78.10±1.18	4.58±0.37	17.32±0.99
A2	Matrix	97.18±0.77	0.25±0.25	2.57±0.63	97.60±0.60	0.07±0.07	2.33±0.57
	C-type	42.37±5.84	21.30±2.67	36.33±3.41	52.16±5.77	7.32±1.10	40.51±4.76

Table VI-Second Phases in the Mg-Al-Sr Alloys (Tentative)

ALLOY	Second Phases (Tentative designation)
A1 Alloy (Mg-5Al-1.8Sr)	Type A : $Al_{17}Mg_5Sr_3$ or $Al_2Mg_{1-x}Sr_x$
	Type B : $Al_3Mg_{13}Sr$
A2 Alloy (Mg-5Al-1.2Sr)	Type C : $Al_{12-y}Mg_{17-x}$-Sr_{x+y} or $Al_{10}Mg_{13}Sr$

Conclusions

Two Mg-Al-Sr diecast alloy compositions were investigated so as to determine their mechanical properties, creep resistance, corrosion resistance and microstructure. A1 alloy which is of the Mg-5Al-1.8Sr and the A2 alloy of Mg-5Al 1.2Sr composition were presented. The following conclusions can be drawn.

1. Elevated-temperature tensile properties of permanent-mold cast and diecast Mg-Al-Sr alloys were superior to AS41 and AE42 alloys. At 175°C, the ultimate-tensile-strength and yield-strength also surpassed AZ91D and A380.

2. Creep resistance of Mg-Al-Sr alloys at 150°C-175°C were superior to AE42, AS41, AZ91D and aluminum A380. The creep resistance increased with increasing Sr content.

3. Using SEM and STEM, no $Mg_{17}Al_{12}$ second phase particles were observed in the diecast microstructure of Mg-5Al-1.2Sr and Mg-5Al-1.8Sr alloys. The Al supersaturation of •-Mg matrix surrounding the second phase particles was relatively low.

4. Stoichiometry was proposed for the second phases observed in the diecast alloys. Since the crystal structure is not yet known for any of these compounds, the precise identification can not be completed.

The future work in this study will be dedicated to the following :

- An in-depth characterization of the creep behavior will be carried out. Creep mechanisms of the Mg-Al-Sr system are currently under investigation.
- The type of compounds (stoichiometry, coherency with the matrix, melting point, covalency) and whether or not creep-induced precipitation exists will also be explored in detail.
- Diecastability studies are underway and will be addressed in a separate paper.

Acknowledgments

Authors would like to thank Robert Bouchard, Michel Lefebvre and Pierre Vermette of the Noranda Technology Center, Pointe-Claire, Quebec, Canada, for their excellent contribution to the experimental work and to Noranda Magnesium Inc for sponsoring the study.

References

1. M. S. Dargusch, G.L.Dunlop, K. Pettersen, Elevated-TemperATURE Creep and Microstructure of Diecast Mg-Al Alloys", Proc. Conf. Magnesium Alloys and Their Applications, B.L. Mordike, K.U. Kainer (Eds.) Verkstoff-Informationsgesellschaft, Wolsburg, Germany, (1998), 277-282.

2. G.V. Raynor, The Physical Metallurgy of Magnesium and Its Alloys, Pregamon Press, Oxford, UK (1959) 364-390.

3. H. Baker, Alloy Phase Diagrams, ASM, Metals park, Ohio, USA, (1992).

4. M. O. Pekguleryuz, M. M. Avedesian, "Magnesium Alloying—Some Metallurgical Aspects", Proc. Intl. Conf. Magnesium Alloys and Applications, B. L. Mordike, F. Hehman (Eds), Materialkunde, Garmisch-Partenkirchen, Germany, (1992), 213-20

5. M. O. Pekguleryuz and J. Renaud, "Creep Resistance in Mg-Al-ca alloys", 2000 Magnesium Technology, H. Kaplan, J. Hryn and B. Clow (Eds.) TMS, 2000, 279-84

6. M. O. Pekguleryuz and A. Luo, ITM Inc., International Patent Application WO 96/25529, 1996.

7. J.J. Berkmortel, H. Hu, J. Kearns, J. Allison, "Diecastability Assessment of Magnesium Alloys for High Temperature Applications, Part 1 of 2", Publications of the SAE World Congress, Paper 2000-01-1119, Detroit Michigan, (2000),.

8. M.P. Barreto, R. Veillette, and G. L'Espérance, "Development and Application of a Dry Ultramicrotomy Technique for the Preparation of Galvanneal Sheet Coatings" Microscopy Research and Technique, 31 (1995), 293-299.

9. P. Villars, Pearson's Handbook – Desk Edition : Crystallographic Data for Intermetallic Phases, (Materials Park, OH : ASM International, 1997).

PHASE TRANSFORMATION AND CREEP OF Mg-Al-Ca BASED DIE-CAST ALLOYS

Akane Suzuki, Nicholas D. Saddock, J. Wayne Jones, Tresa M. Pollock
University of Michigan, Materials Science and Engineering; 3062 H. H. Dow, 2300 Hayward St.; Ann Arbor, MI 48109, USA

Keywords: transmission electron microscopy, Laves phases, dislocation substructure

Abstract

The microstructure and microstructural stability of die-cast AC53 (Mg-5Al-3Ca) and AXJ530 (Mg-5Al-3Ca-0.15Sr) have been investigated in detail by transmission electron microscopy (TEM). Both alloys have an as-cast microstructure of α-Mg with $(Mg, Al)_2Ca$ (dihexagonal C36) eutectic at grain boundaries. During aging at 573 K, the C36 phase transforms to Al_2Ca (cubic C15) phase. These two phases have a crystallographic orientation relationship of $(0001)_{C36}//\{111\}_{C15}$ and $[2\bar{1}\bar{1}0]_{C36}//[01\bar{1}]_{C15}$, and the transformation from C36 to C15 occurs by a shear-assisted process. Despite this change in the phase constitution, the network structure of the intermetallic compound(s) surrounding α-Mg grains is fairly stable, morphologically, even after prolonged exposure at elevated temperature. In the α-Mg matrix phase, precipitation of Al_2Ca was observed after aging for 360 ks at 573 K. The precipitates are disc-shaped with a habit plane of $\{111\}_{C15}//(0001)_{\alpha}$. AXJ530 shows higher creep resistance than AC53. The dislocation substructure that evolved during creep deformation was investigated in both alloys, and the basal and non-basal slip of ***a***-dislocation and other slip modes of ***a+c***-dislocations were observed. The relationship between creep properties and microstructure is discussed.

Introduction

The development of high temperature structural magnesium alloys for automotive engine and transmission components is driven by the need to achieve high fuel efficiency. To utilize these alloys in commercial components requires good castability for fabrication of large components by high-pressure die-casting, and good creep and corrosion resistance in the temperature range of 150~200°C, at competitive costs [1-2].

For the improvement of high temperature strength and creep resistance of Mg-Al based alloys, the addition of Ca, which is a lower cost and weight element compared to rare-earth elements, has been reported to be effective [3-6]. Furthermore, the AXJ series alloys have been developed by modifying AC53 (Mg-5Al-3Ca, wt%) with small Sr additions [7-10]. The AXJ530 (Mg-5Al-3Ca-0.15Sr, wt%) is reported to have better creep properties than those of the AE42 (Mg-4Al-2RE, wt%), which is the benchmark creep-resistant alloy. However, in Mg-Al-Ca-(Sr) based alloys, a fundamental understanding of the microstructure and the effect of the Sr addition in conjunction with creep properties is lacking, but is important to continued development of this class of alloys.

Die-cast magnesium alloys usually consist of dendritic α-Mg grains with intermetallic phase(s) in the interdendritic region. The dominant creep mechanisms operative in Mg-Al based alloys are considered to be dislocation glide and climb in α-Mg grains at higher stresses and grain boundary sliding at lower stresess [10-11]. The transition point between two regimes is at approximately 70 MPa in AC53 at 175°C [10]. Therefore, grain boundary strengthening and stability, as well as solid-solution and/or precipitation strengthening in α-Mg grains, are the important factors for optimization of the high temperature strength of the fine-grained die-cast alloys [12].

In the grain boundary region of Mg-Al-Ca based alloys, Mg_2Ca and/or Al_2Ca phases are reported to form during final eutectic reaction with the α-Mg phase in preference to the $Mg_{17}Al_{12}$ phase that is observed in the binary system. Mg_2Ca and Al_2Ca are Laves phases with crystal structures of hexagonal C14 and cubic C15, respectively. For an overall alloy composition of Mg-5Al-3Ca, the eutectic compounds have been reported to be Al_2Ca [5], Mg_2Ca [13], $(Mg, Al)_2Ca$ with hexagonal structure [10] and mixtures of these three phases [14]. However, we recently determined the crystal structure of the eutectic compound as dihexagonal C36 in AXJ530 by transmission electron microscopy [15]. The inconsistency in the previous reports is undoubtedly related to the similarity of crystal structures of Laves phases. Since these magnesium alloys are to be used in the as-cast form in automotive components, the stability of microstructure and existing phases at elevated temperatures is important, especially near grain boundaries where higher alloying element concentration arises during solidification.

In this study, the microstructure and existing phases in AC53 and AXJ530 in as-cast form and after annealing at 573 K have been investigated in detail by TEM. The relationship between creep properties and microstructure has been also investigated.

Experimental Procedures

The AC53 and AXJ530 alloys used in this study were supplied by General Motors and had compositions of Mg-4.5Al-3.0Ca and Mg-4.5Al-3.0Ca-0.14Sr (wt%), respectively. The chemical compositions are listed in Table I. To fabricate these alloys, commercial AM50 alloy was used as a base and pure Ca and an Al-90wt%Sr alloy were added to achieve the desired compositions. The alloy was melted at 950 K, and after high speed stirring for 3.6 ks, cold-chamber die-casting with a die temperature of 400 K was utilized to produce cylindrical tensile creep specimens 25.4 mm in gage length and 6.35 mm in diameter.

Metallographic samples were cut from the gage sections of the as-cast specimens, and some pieces were aged at 573 K for 360~3600 ks, followed by a water quench. Each aging treatment was conducted by sealing the sample in a quartz capsule that was

Table I. Chemical Compositions of the Alloys Studied (wt%)

	Mg	Al	Ca	Sr	Mn	Fe	Ni	Cu
AC53	bal.	4.5	3.0	-	0.27	0.003	<0.002	0.003
AXJ530	bal.	4.5	3.0	0.14	0.25	<0.005	<0.002	<0.0005

argon back-filled after evacuation. After holding the capsule in the furnace, the sample was quenched into water quickly by breaking the capsule.

Tensile creep tests were conducted at 448 K in accordance with ASTM E-139. Creep tests were conducted at a stress of 100 MPa, which is in the dislocation creep regime for AC53 [10]. The temperature was maintained within ±1°C, and displacement as a function of time was recorded with a digital data acquisition system using two linear variable differential transformers (LVDTs) having a strain resolution of approximately 10^{-5}. The creep tests were stopped at approximately 1% strain.

The as-cast, aged and post-crept samples were examined by SEM and TEM. For TEM studies, discs 0.12 mm in thickness and 3 mm in diameter were cut from near the center of the die-cast specimen and mechanically polished, followed by twin-jet electron polishing in a solution of methanol with 8 vol.% perchloric acid at 243 K. Subsequently, low-angle ion milling was employed to remove the surface oxide layer introduced during jet polishing. A Phillips CM12 microscope was used in this study.

Results and Discussion

As-cast Microstructure

Figure 1 shows SEM back scattered electron (BSE) images of the as-cast AC53 and AXJ530. The typical microstructure of both alloys contains equiaxed α-Mg grains with a grain size of 5-10 µm. The grain boundary area with bright contrast in BSE images indicates the existence of intermetallic compound(s) with higher Al and Ca content. TEM bright field images (BFIs) of near-grain boundary microstructure are shown in Fig. 2. A eutectic microstructure is observed, especially at the triple points among α-Mg grains. The phases with bright and dark contrast correspond to α-Mg and the intermetallic compound, respectively.

The crystal structure of the eutectic compound was analyzed by electron diffraction. Figure 3 shows BFIs and corresponding selected area diffraction patterns (SADPs) taken from the eutectic regions in AC53 (a~c) and AXJ530 (d~f). Figure 3b is the SADP taken from the eutectic region shown in Fig. 3a, and Fig. 3c was obtained in the same region after tilting the TEM foil by 30° with fixed reciprocal lattice vector, **g**. Both alloys exhibit exactly the same patterns with strong streaks along **g**. These patterns are consistent with the calculated diffraction pattern of the dihexagonal C36 structure (Figs. 3g and h), and the incident beam directions **B** of Figs. 3b and 3e, Figs. 3c and 3f are indexed as $10\bar{1}0_{C36}$ and $2\bar{1}\bar{1}0_{C36}$, respectively. Other SADPs taken with different orientations were also confirmed to be those of the C36 structure, and the eutectic compound was confirmed to be single phase C36, rather than a mixture of C36 and C14 and/or C15 [15]. Thus, the eutectic compound is C36-$(Mg, Al)_2Ca$ phase, regardless of the Sr addition. No specific crystallographic orientation relationship was observed between the eutectic C36 and α-Mg phases within various regions examined.

Crystallographically, the C36 structure is the intermediate structure between hexagonal C14 and cubic C15 structures, and the difference among them is the stacking sequence of the quadruple layers (X, Y and Z) of the close-packed planes [15-16]: XYX'Y... in C36, XY... in C14 and XYZ... in C15. Therefore, the formation of the C36 phase is reasonable in the Mg-Al-Ca ternary system, which has Mg_2Ca (C14) and Al_2Ca (C15) Laves phases. Since there has been no clear experimental evidence for the existence of C36 phase in previous studies of this ternary system, it is not clear whether this phase is metastable or stable. Thus, there are two possibilities for the formation pathway of the C36 phase; one is that the C36 phase is transformed from the original eutectic C14 phase during cooling in the casting process, and the other is that the C36 phase is directly formed by the eutectic transformation. Further ongoing research will determine the transformation pathway during solidification.

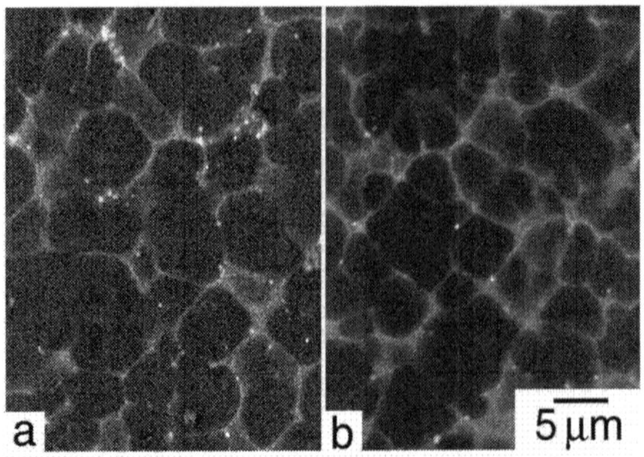

Figure 1. Back scattered electron (BSE) images of as-cast AC53 (a) and AXJ530 (b).

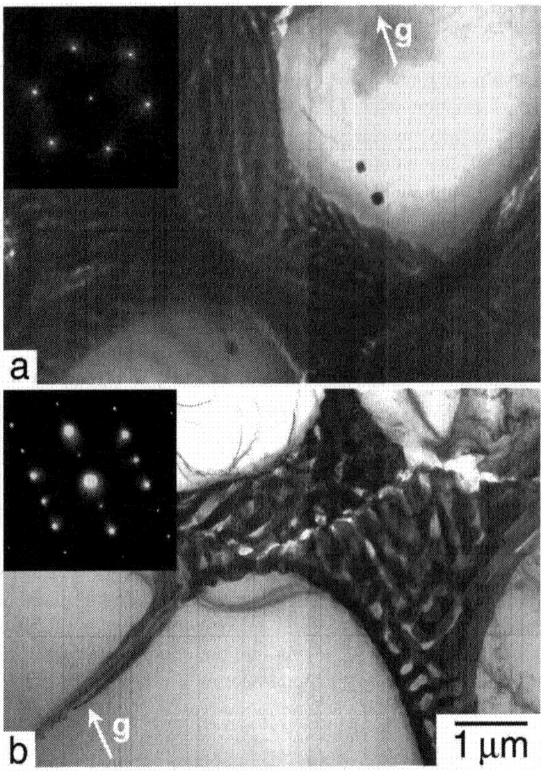

Figure 2. TEM bright field images (BFIs) of as-cast AC53 (a) and AXJ530 (b), taken with **B**=$2\bar{1}\bar{1}3$ and **B**=$2\bar{1}\bar{1}0$, respectively.

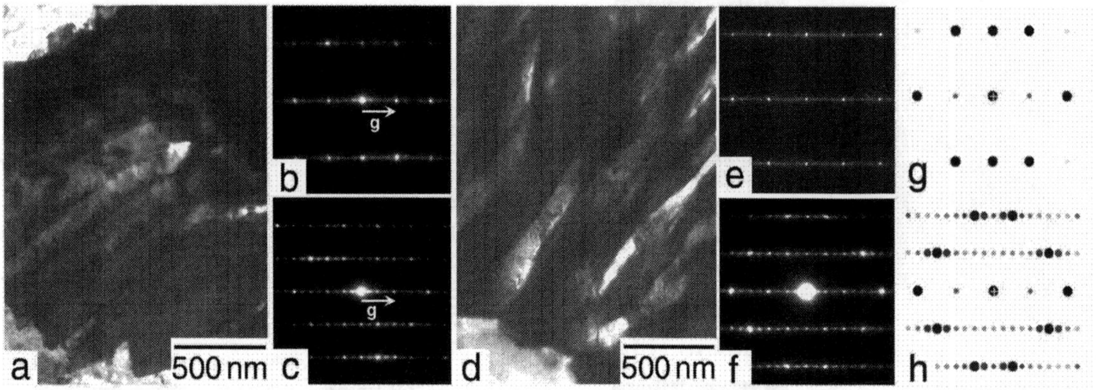

Figure 3. TEM BFIs and selected area diffraction patterns (SADPs) taken from the eutectic regions in AC53 (a~c) and AXJ530 (d~f), together with calculated diffraction patterns of C36 structure (g~h): (a) and (d) BFIs taken with **B**=$10\bar{1}0$, (b), (e), (g) corresponding SADPs and calculated pattern of **B**=$10\bar{1}0$, (c), (f), (h) corresponding SADPs and calculated pattern of **B**=$2\bar{1}\bar{1}0$.

Phase Transformations and Microstructure Change at 573 K

Figure 4 shows BSE images of both alloys after 3600 ks aging at 573 K. The intermetallic phase with bright contrast retains its network morphology surrounding α-Mg grains. Figure 5 presents TEM BFIs of AXJ530 that show microstructural changes after annealing at 573 K. The morphology of the eutectic compound changes to a discrete plate shape after 360 ks (Fig. 5a) and finally becomes spherical after 3600 ks (Fig. 5b). Within the α-Mg grains, small precipitates are observed after 360 ks aging, as indicated by arrowheads. The same change in microstructure were observed in the AC53 alloy.

Transition from C36 to C15 in the Eutectic Compound. Figure 6 shows TEM BFIs and SADPs taken from the eutectic region in the aged AXJ530 samples. Observation along $<2\bar{1}\bar{1}0>_{C36}$ reveals a high density of planar faults parallel to the close-packed planes with a roughly periodic spacing of several nm (Fig. 6a). The corresponding SADP (Fig. 6b) exhibits extra reflections, as schematically shown in Fig. 6c. These reflections are identified as those of C15 structure with orientation of 011_{C15}, and there are two variants of C15 with fcc-type twin relationship. Since the 0004_{C36} reflection coincides with $\bar{1}1\bar{1}_{C15}$ reflection, the two phases have the following crystallographic relationship $(0001)_{C36}//\{111\}_{C15}$, $[2\bar{1}\bar{1}0]_{C36}//[01\bar{1}]_{C15}$.

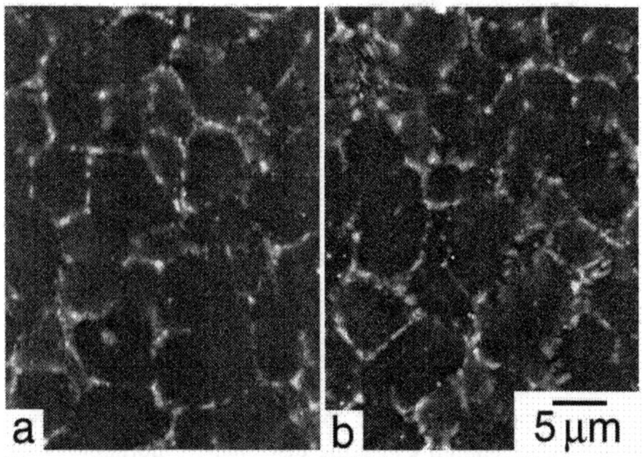

Figure 4. BSE images of AC53(a) and AXJ530(b) after 3600 ks aging at 573 K.

After 3600 ks aging, only the twinned C15 phase is present, as is apparent in BFI (Fig. 6d) and SADP (Figs. 6e and 6f). Similar results were also obtained for AC53. These results demonstrate the C36 phase transforms to C15 phase at 573 K. The formation of twinned C15 could be caused by a shear-assisted transformation on the close-packed planes [16].

Figure 7 shows the change in the lattice parameters of the eutectic compound with aging in AXJ530, together with those of the stoichiometric Mg_2Ca and Al_2Ca [17]. In order to directly compare the lattice parameters of the three Laves structures, they

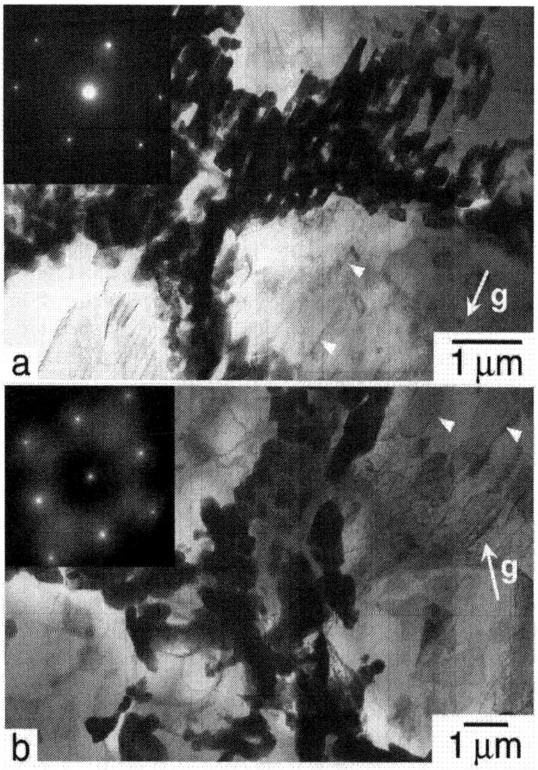

Figure 5. TEM BFIs of AXJ530 aged at 573 K for: (a) 360 ks (**B**=$4\bar{2}\bar{2}3$) and (b) 3600 ks (**B**=$01\bar{1}1$).

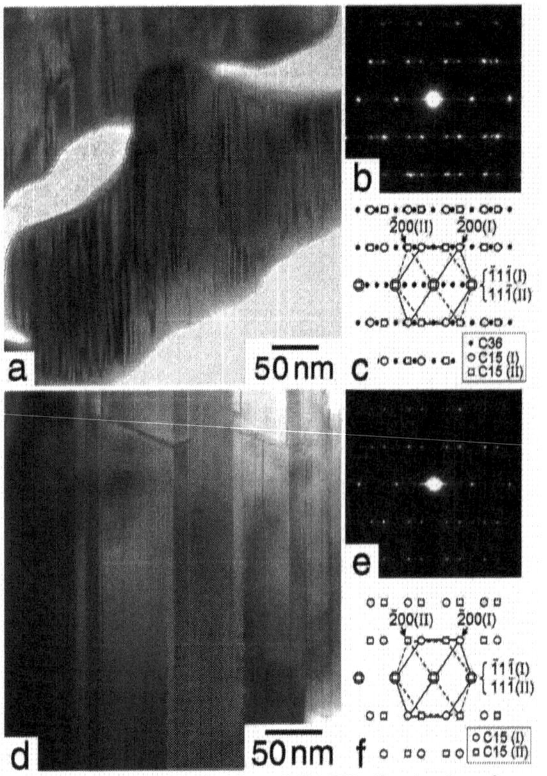

Figure 6. TEM BFIs (a, d) and SADPs (b, e) taken from the eutectic compound of AXJ530 aged at 573 K for: (a~c) 360 ks and (e~f) 3600 ks ($\mathbf{B}=2\bar{1}\bar{1}0_{C36}//011_{C15}$). (c) and (f) are the schematic illustrations of (b) and (e), respectively, showing occurrence of transformation from C36 to C15.

are converted to the nearest-neighbor Ca atom spacing on the close-packed planes, d_a ($d_a = a_{C14} = a_{C36} = a_{C15}/\sqrt{2}$), and the spacing of the quadruple layers, d_c ($d_c = c_{C14}/2 = c_{C36}/4 = a_{C15}/\sqrt{3}$), because they are common parameters among these Laves phases [15]. The larger atomic radius of Mg causes larger d_a and d_c of Mg_2Ca than those of Al_2Ca. The values obtained from the eutectic compound fall between the stoichiometric values, and they decrease with aging. These facts indicate that the composition of the eutectic compound shifts toward Al_2Ca. Therefore, C36→C15 transition is associated with a rejection of supersaturated Al in the eutectic α-Mg phase via diffusion during aging, because the solubility limit of Al in α-Mg at 573 K is lower than at the eutectic transformation temperature.

The C36→C15 transition induced by diffusional redistribution of Mg and Al atoms does not require separate nucleation of the product phase, but proceeds within the matrix C36 phase. This results in the stability of the near-grain boundary microstructural morphology even though a change in the phase constitution occurs.

Precipitation of Al_2Ca. Precipitates observed in α-Mg grains are shown in Fig. 8. They are disc shaped, several hundred nm in diameter, several nm in thickness, and lie on the basal planes. The number density of precipitates is higher in AXJ530 (Fig. 8b) than AC53 (Fig. 8a). Figure 9 shows TEM BFI of AXJ530 aged for 3600 ks, taken with $\mathbf{B}=10\bar{1}0_{\alpha}$. The corresponding SADP indicates the existence of C15 phase with an orientation of $10\bar{1}$. Therefore, these precipitates are Al_2Ca phase which have the following crystallographic relationship with α-Mg matrix phase;

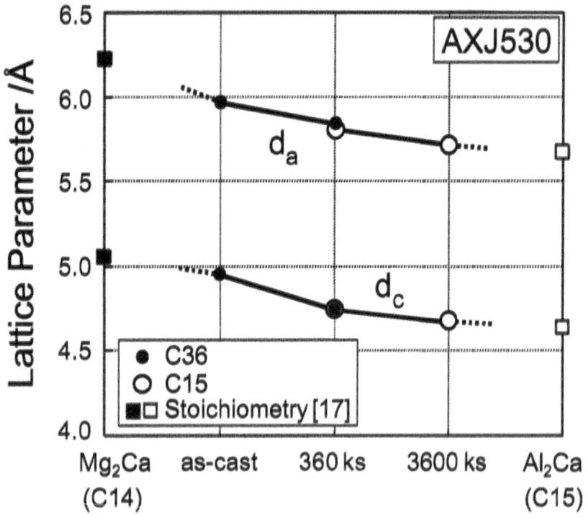

Figure 7. Change in the lattice parameters of the eutectic Laves compound with aging at 573 K, together with the values of stoichiometric Mg_2Ca and Al_2Ca [17].

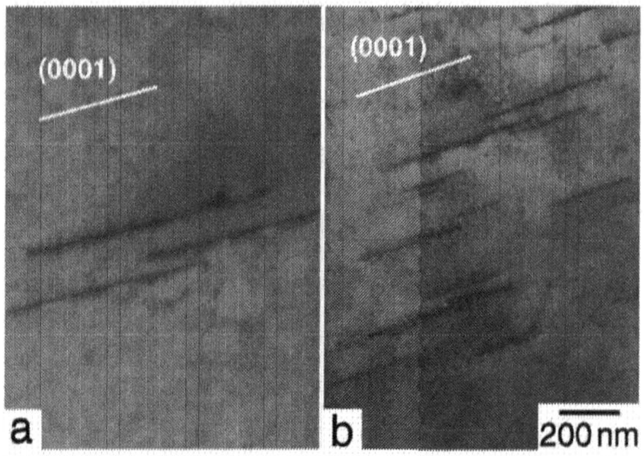

Figure 8. TEM BFIs of AC53 (a) and AXJ530 (b) aged at 573 K for 360 ks, showing precipitates in α-Mg ($\mathbf{B}=2\bar{1}\bar{1}0$).

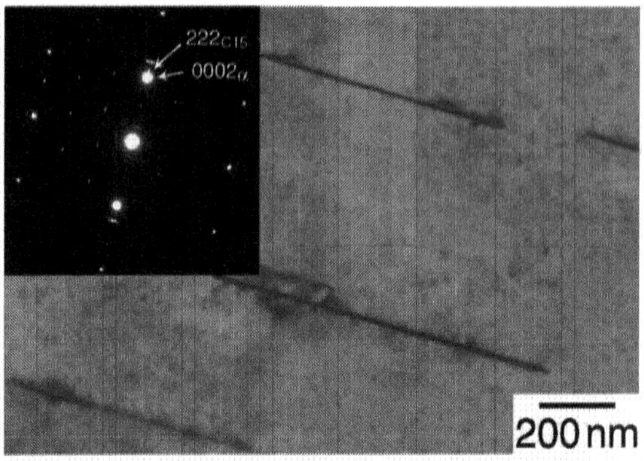

Figure 9. TEM BFI and corresponding SADP of AXJ530 aged at 573 K for 3600 ks ($\mathbf{B}=10\bar{1}0_{\alpha}//10\bar{1}_{C15}$).

$$\{111\}_{C15}//(0001)_{\alpha},\ [10\bar{1}]_{C15}//[10\bar{1}0]_{\alpha}.$$

The lattice misfits measured along $[10\bar{1}0]_{\alpha}$ and $[1\bar{2}10]_{\alpha}$ are +3% and -2%, respectively, while that along $[0001]_{\alpha}$ is -10%. This large difference should cause the disc-shaped morphology of the precipitates with large aspect ratio.

Creep of AC53 and AXJ530

Creep curves of AC53 and AXJ530 at 448 K and 100MPa are shown in Figure 10. The AXJ530 pre-aged at 573 K for 360 ks was also tested in order to investigate the effect of precipitates on the creep resistance. As-cast specimens exhibit primary creep followed by gradually decreasing creep rates. AXJ530 exhibits higher creep resistance than AC53, and the minimum creep rate (when interrupted at 1% strain) is 4×10^{-9} s^{-1}, which is lower by a half order of magnitude than that of AC53. On the other hand, the pre-aged AXJ530 has the lowest creep resistance, and the minimum creep rate is higher than that for as-cast AXJ530 by one order of magnitude.

Figures 11 and 12 show TEM BFIs of AC53 and AXJ530 after 1% creep deformation, respectively. No precipitation within the α grains was observed in these creep interrupted samples. Dislocations were characterized using a series of two-beam conditions. From the calculation of **g**·**b** values for perfect dislocations in hcp structure, under the two-beam condition **g**=$10\bar{1}1$, 2/3 of ***a***-dislocations, 4/6 of ***a***+***c*** and all ***c***-dislocations are visible. On the other hand, under the two-beam condition **g**=0002, pure ***a***-dislocations are invisible, but ***a***+***c*** and ***c***-dislocations are visible. In Fig. 11a taken with **g**=$\bar{1}01\bar{1}$, straight and curved dislocations are observed. The line directions of the straight dislocations are parallel to the trace of the basal planes, and those

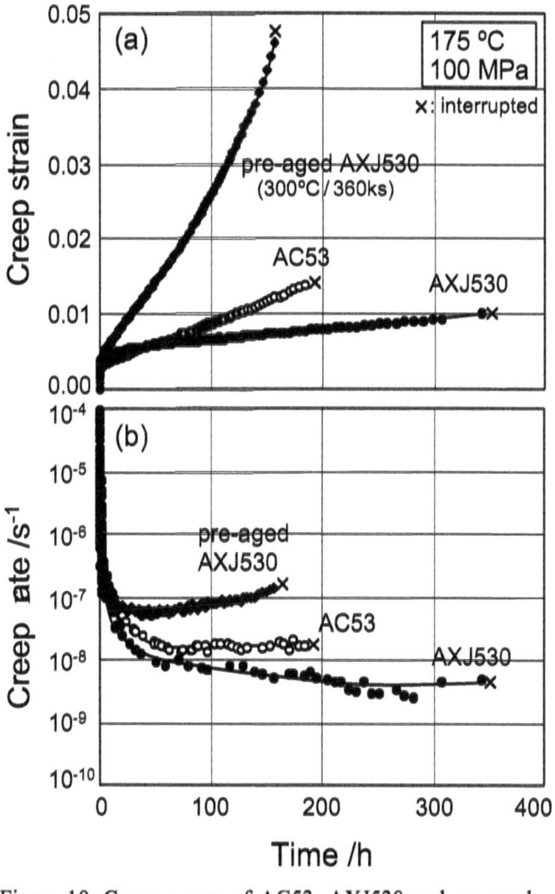

Figure 10. Creep curves of AC53, AXJ530 and pre-aged AXJ530 (573 K / 360 ks) at 448 K, 100 MPa: (a) strain - time curves and (b) strain rate - time curves.

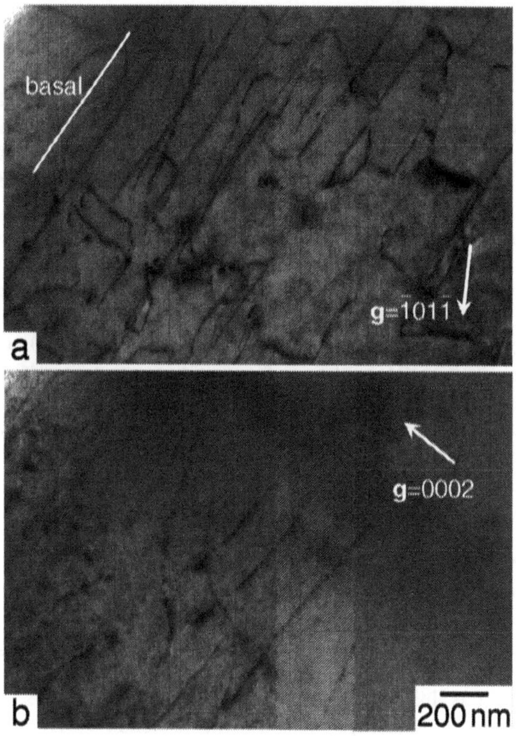

Figure 11. Dislocation substructure of AC53 after 1% deformation at 448 K, 100 MPa (**B**=$1\bar{2}10$) taken with: (a) **g**=$\bar{1}01\bar{1}$ and (b) **g**=0002, showing the ***a***- and ***a***+***c***-dislocations.

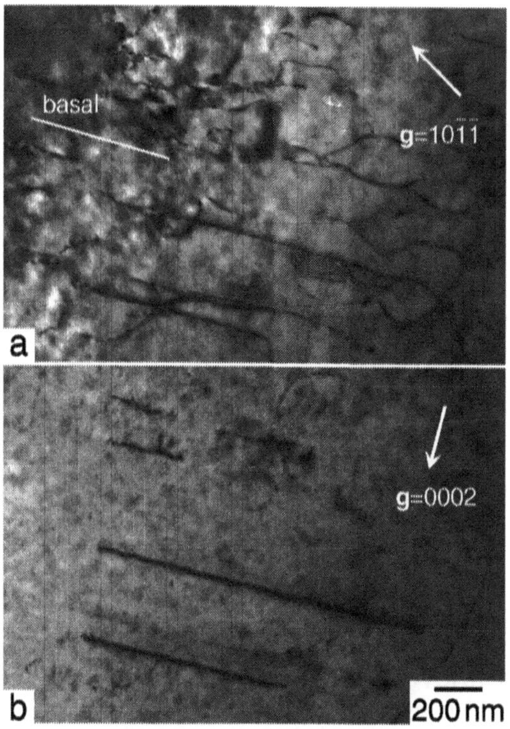

Figure 12. Dislocation substructure of AXJ530 after 1% deformation at 448 K and 100 MPa (**B**=$1\bar{2}10$) taken with: (a) **g**=$10\bar{1}1$ and (b) **g**=0002, showing the ***a***- and ***a***+***c***- dislocations.

of the curved dislocations are extending from the trace. In Fig. 11b taken with **g**=0002, most of the dislocations observed in Fig. 11a are out of contrast, indicating they are the ***a***-dislocations. But a fraction of the straight dislocations are visible, indicating they have ***c***-component, and they are determined to be ***a+c***-dislocations by using another **g**. In the AXJ530 after 1% creep deformation, dislocations with ***c***-component are observed along the trace of the basal plane, as shown in Fig. 12b, in addition to the curved ***a***-dislocations (Fig. 12a). By analyzing with another **g**, the dislocations observed in Fig. 12b are also determined to be ***a+c***-dislocations. Therefore, the basal and non-basal slip of ***a***-dislocations and another slip mode of ***a+c***-dislocations are active in both alloys. There was no significant difference in the dislocation substructure in AC53 compared to AXJ530. Other factors, such as deformation in the vicinity of grain boundaries or solidification induced variations in solid solution strengthening elements, should be examined to understand the higher creep resistance of AXJ530 compared to AC53.

In spite of the formation of Al_2Ca precipitates after aging at 573 K in both alloys, no precipitates were observed in the creep interrupted samples (Figs. 11 and 12). Taking into consideration that the intermetallic compound at grain boundaries was not spheroidized after 360 ks aging, and that the pre-aged sample does not exhibit any break-up of the network of the compound at grain boundaries, the deterioration of creep resistance by the pre-aging treatment is likely caused by a decrease in the solid-solution strengthening in α-Mg grains via the depletion of Al and Ca atoms to form precipitates. Therefore, this result indicates the importance of solid-solution hardening for the creep resistance of these die-cast alloys in the dislocation creep regime.

Conclusions

Microstructure, phase transformations and creep properties of AC53 and AXJ530 were investigated in this study, and the following conclusions were obtained;

1. The crystal structure of the initial eutectic compound was identified as the C36 structure.
2. The eutectic C36 phase transforms into the C15 (Al_2Ca) phase at 573 K.
3. Precipitation of the disc-shaped C15 phase in grain interiors occurs at 573 K, and the density is higher in AXJ530 than in AC53.
4. The minimum creep rate of AXJ530 is lower than that of AC53. In both alloys, ***a***- and ***a+c***-dislocations are observed, and there is no significant difference in the dislocation substructure.
5. Pre-aging at 573 K decreases the creep resistance of AXJ530, indicating the importance of the solid-solution strengthening of α-Mg.

Acknowledgement

The authors acknowledge the support of NSF FRG Grant No. DMR-0309468. The authors also would like to thank C. J. Torbet and C. Henderson of University of Michigan for their assistance and Dr. B. R. Powell of General Motors for useful discussions and for providing the die-cast materials.

References

1. B.R. Powell, L.J. Ouiment, J.E. Allison, J.A. Hines, R.S. Beals, L. Kopka, and P. Ried., "The Magnesium Powertrain Cast Components Project," *Magnesium Technology 2004*, ed. A.A. Luo (Warrendale, PA: TMS, 2004), 3-10.

2. R. Beals, L. Kopka, J.E. Allison, J.A. Hills, and B. McCune, "Fundamental Research Needs for the Magnesium Powertrain Cast Components (MPCC) Project," *Magnesium Technology 2004*, ed. A.A. Luo (Warrendale, PA: TMS, 2004), 11-117.

3. F. Hollrigle-Rosta, E. Just, J. Kohler, and H.-J. Melzer, "Magnesium in Volkswagen," *Light Metal Age*, 1980, August: 22-29.

4. G.V. Raynor., *The Physical Metallurgy of Mg and Its Alloys* (Oxford, UK: Pergamon Press, 1959), 288.

5. R. Ninomiya, T. Ojiro, and K. Kubota, "Improved Heat Resistance of Mg-Al Alloys by the Ca Addition," *Acta Metall. Mater.*, 43 (1995), 669-674.

6. Y. Terada, N. Ishimatsu, R. Sota, T. Sato, and K. Ohori, "Creep Characteristics of Ca-added Die-Cast AM50 Magnesium Alloys", *Materials Science Forum*, 419-422 (2003), 459-464.

7. B.R. Powell, V. Rezhets, A.A. Luo, and B.L. Tiwari, "Low-Cost, Creep-Resistant Magnesium Alloy for Powertrain Components," US Patent No. 6,264,764 dated July 24, 2001.

8. B.R. Powell, A.A. Luo, V. Rezhets, and B.L. Tiwari, "Development of Creep-Resistant Magnesium Alloys for Powertrain Applications: Part 1 of 2," SAE Technical Paper No. 2001-01-0422, (Warrendale, PA: SAE, 2001).

9. A.A. Luo, M.P. Balogh, and B.R. Powell, "Tensile Creep and Microstructure of Magnesium-Aluminium-Calcium-Based Alloys for Powertrain Applications: Part 2 of 2," SAE Technical Paper No. 2001-01-0423, (Warrendale, PA: SAE, 2001).

10. A.A. Luo, M.P. Balogh, and B.R. Powell, "Creep and Microstructure of Magnesium-Aluminum-Calcium Based Alloys," *Metall. Mater. Trans.*, 33A (2002), 567-574.

11. M.S. Dargusch, and G.L. Dunlop, "Elevated Temperature Creep and Microstructure of Die Cast Mg-Al Alloys," *Magnesium Alloys and Their Applications*, ed. B.L. Mordike and K.U. Kainer (Frankfurt, Germany: Werkstoff-Informationsgesellschaft, 1988) 277-282.

12. I.P. Moreno, T.K. Nandy, J.W. Jones, J.E. Allison, and T.M. Pollock, "Microstructural stability and creep of rare-earth containing magnesium alloys," *Scripta Mater.*, 48 (2003), 1029-1034.

13. K. Ozturk, Y. Zhong, A.A. Luo, and Z.-K. Liu, "Creep Resistant Mg-Al-Ca Alloys: Computational Thermodynamics and Experimental Investigation," *JOM*, 2003, November: 40-44.

14. Y. Zhong, A.A. Luo, J.O. Sofo, and Z.-K. Liu, "Laves Phases in Mg-Al-Ca Alloys," *Magnesium Technology 2004*, ed. A.A. Luo (Warrendale, PA: TMS, 2004), 317-323.

15. A. Suzuki, N.D. Saddock, J.W. Jones, and T.M. Pollock, "Structure and transition of eutectic $(Mg, Al)_2Ca$ Laves phase in a die-cast Mg-Al-Ca base alloy," *Scripta Mater.*, 51 (2004), 1005-1010.

16. K.S. Kumar, and P.M. Hazzledine, "Polytypic transformation in Laves phases," *Intermetallics*, 12 (2004), 763-770.

17. W.B. Pearson, *A Handbook of Lattice Spacing and Structure of Metals and Alloys* (Oxford, UK: Pergamon Press, 1958), 136 and 158.

Magnesium Diecasting Alloys for High Temperature Applications

Mihriban O. Pekguleryuz

McGill University, Department of Metals and Materials Engineering
3601 University Street, Montreal, QC, Canada H3A 2B2

A. Arslan Kaya
Tubitak Marmara Research Center, P.O.B.21, Gebze, 41470 Turkey

Abstract

New growth area for automotive use of magnesium is powertrain applications such as the transmission case and engine block. These applications see service conditions in the temperature range of 150-200C under 50-70 MPa of tensile and compressive loads. In addition, metallurgical stability, fatigue resistance, corrosion resistance and castability requirements need to be met. A decade of research and development has resulted in a number of creep-resistant magnesium alloys that are potential candidates for elevated-temperature automotive applications. These alloys are mostly based on rare-earth and alkaline earth element additions to magnesium. This paper gives an overview of the various magnesium alloy systems for use in elevated-temperature applications.

Introduction

Automotive use of magnesium is currently limited to a few select applications such as the instrument panel, steering wheel, and valve cover. The alloys used in these applications utilize Mg-Al based casting alloys such as the AM series, Mg- (2-6)%Al and the AZ91 alloy (Mg-9%Al-1%Zn) which offer good combination of room-temperature strength and ductility, good salt-spray corrosion resistance and excellent diecastability. Major new growth areas in automotive use for Mg alloys are the powertrain applications, namely, the transmission case and the engine block. These applications require good elevated temperature performance at service conditions of 150-200°C and stresses in the range of 50-70 MPa. Commercial AM and AZ type of Mg alloys do not possess this required elevated temperature performance.

In the last decade, concentrated effort has been expanded in developing and optimizing Mg alloys with good creep resistance. Mg alloys for high temperature applications require considerations on several properties in conjunction with the creep strength. Castability, especially diecastability, of a new alloy must be at an acceptable level, as this is the major processing route for production of parts made of Mg alloys. Corrosion and oxidation resistance is important. In addition, to be able to compete with relatively lower cost metals such as aluminum and steel, the alloys should not have additional cost penalty by containing costly alloying elements. Majority of Mg parts used in automotive applications is high-pressure diecast. This is due to the excellent fluidity of Mg alloys providing good die filling, the compatibility of liquid Mg with die materials resulting in long die-life and the high productivity of the die casting process upsetting the material cost factors. In this review, background on creep resistance of Mg alloys is presented. Creep resistant Mg alloys are discussed with respect to alloy systems, microstructure, creep behavior and comparative properties are given whenever available.

Creep Deformation

Creep is time-dependent slow and continuous deformation in time. While in elastic-plastic deformation strain is a function of stress only, in creep strain is a function of stress, time and temperature.

$\varepsilon = f(\sigma)$ elastic/plastic deformation,

$\varepsilon = f(\sigma, T, t)$ creep deformation

Creep deformation is due to the thermally activated (short-range) component of the stress,

$\sigma = \sigma_E + \sigma^*$ (normal stress)

$\tau = \tau_u + \tau^*$ (shear stress)

where σ^*, τ^* are thermal and σ_E, τ_u are athermal components. The athermal component involves movement of dislocations passed stress fields whose origins are distant on an atomic scale, about 300 atomic distances. This long-range component is associated with work hardening. The thermal component is due to atom vibrations. It involves dislocation motion passed obstacles whose stress fields are more localized. Hence, while the long-range component involves the movement of long segments of parallel dislocations, the thermal component can only cause movement of small sections of a dislocation (e.g. dislocation jogs, or the cutting of a dislocation by another). This nevertheless causes slow and permanent deformation called creep. Since atoms vibrate anywhere above absolute zero, thermal component exists at very low temperatures and creep is possible even at very low temperatures, given sufficient time.

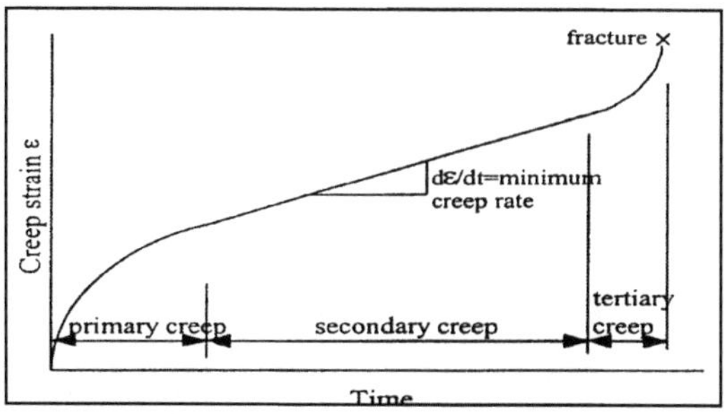

Fig. 1. Typical creep strain vs. time curve showing the three stages of creep.

Fig. 1 shows a typical creep-strain versus time curve for a metal, where three distinct regions can be observed. After an initial and instantaneous restructuring of the metal, primary stage creep occurs with increasing creep rate. Here the work hardening is higher than the stress recovery. In the secondary creep (steady-state) stage the metal experiences a three-way balance between work-hardening, increasing stress and stress recovery resulting in steady-state creep (constant creep-rate). The tertiary creep stage exhibits increasing creep rate due to necking resulting finally in fracture.

Mathematical description of creep that considers a thermally activated process capable of acting in the direction of the stress or opposite to it as it passes over an energy barrier defines creep–rate as shown below (1). Hence, creep rate at a given stress is extremely temperature sensitive.

$$? = A\,e\,(-q/kT).\ e(v\tau^*/kT)$$

There are two distinct creep regions, (1) dislocation creep at high stresses where strain-rate has a power law dependence on stress and (2) diffusional creep at high temperatures with a linear dependence of strain-rate on stress. The mechanisms of creep at micro-level are recovery mechanisms as summarized in Table I. Magnesium, having h.c.p.-crystal structure, possesses fewer slip systems compared to, for example, f.c.c structure, and therefore its ability to deform is limited especially at low temperatures. The slip system at the room temperature is basal {0001} $\langle 11\bar{2}0\rangle$ and above 250°C, $\{10\bar{1}1\}$ and $\{10\bar{1}2\}$ pyramidal and $\{10\bar{1}0\}$ prism slips are activated (3). Room temperature deformation occurs mainly by slip on the basal planes. In addition, twinning on the pyramidal $\{10\bar{1}2\}$ planes provides means of deformation. Mechanisms of creep in pure Mg have been studied (4) and it has been observed that pure Mg exhibits basal slip on the {0001} planes, twinning and sub-grain formation (Fig.2.a&b) in the 90-300°C temperature range at stress levels of 8-70 MPa. These are mechanisms in the primary creep stage. At the higher temperatures, above 250°c, non-basal slip on the pyramidal and prismatic slip planes (Fig. 2.c) and grain boundary deformation and sliding occur (steady-state creep). Creep mechanisms in a metal or alloy can be identified by determining the activation energy for creep from Arrhenius relationship of strain rate and reciprocal temperature. Minimum creep rate in steady-state (secondary) creep is obtained from long- term creep tests. The value of the activation energy and the stress exponent would give clues as to which thermally activated creep mechanisms listed in Table II are predominant in a particular temperature and stress range.

Creep mechanism in AZ91D (Gg-9Al-1Zn) alloy has been studied by Dargusch et al (5). It was found that in the 125-175°c temperature range and with σ=50 MPa, the activation energy, Q in $? = A\sigma^n \exp(-Q/rt)$ was 30-45 kJ/mol and n=2. This activation energy was too small compared to self-diffusion of Mg (135 kJ/mol), diffusion of al in Mg (143 kJ/mol) or grain boundary diffusion (80 kJ/mol). The value did fit with the activation energy Q for discontinuous precipitation of $Mg_{17}Al_{12}$ = 30 kJ/mol. Creep induced precipitation of $Mg_{17}Al_{12}$ and grain boundary migration were observed in the evolved microstructure after creep (Fig. 3). Since such precipitation would aid easy grain boundary sliding and migration, the creep was attributed to creep-induced precipitation of $Mg_{17}Al_{12}$ and the involvement of this phase in the creep deformation of Mg-Al based alloys was established. At stresses higher than 50 MPa, Q was determined to be 95 kJ/mol and n was 5, indicating a change in the creep mechanism. No mechanism was named in the study for the high stress range but looking at table ii this activation energy seems to fall in the range of activated cross slip, as well as grain boundary diffusion for magnesium. Other potential creep mechanisms would be dislocation intersection or overcoming of *Peierls stresses.*

Creep mechanism in AM50 (Mg-5%Al) alloy has been studied by Terada et al (6) At 100 MPa, and in the temperature range of 150-225°C, the activation energy was determined to be 126 kJ/mol, which the authors attributed to *dislocation climb.* It would fit very well with a creep mechanism described as "movement of dislocations with jogs" (Table II).

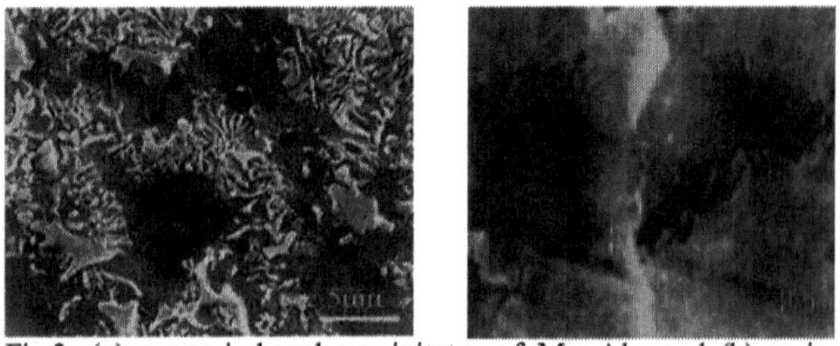

Fig.3. (a) creep induced precipitates of $Mg_{17}Al_{12}$ and (b) grain boundary migration in AZ91D (ref.5).

TABLE I. CREEP MECHANISMS (1)

Low to Moderate Temperatures: Stress-activated Recovery	**High temperatures :** ***Thermally Activated Recovery***	
	Dislocation Creep	**Diffusion creep**
Grain sub structures. In HCP structures like Mg these are elongated and parallel to grain boundaries, in FCC they are equiaxed (Fig. 3.a). In HCP Mg, twinning (Fig. 3b).	Dislocation Intersection, Dislocation climb, Movement of dislocation atmospheres, Cross-slip, Grain-boundary shear	• Fast diffusion paths such as dislocation cores and grain boundaries • Bulk diffusion (vacancy and interstitial • $T>0.3T_M$

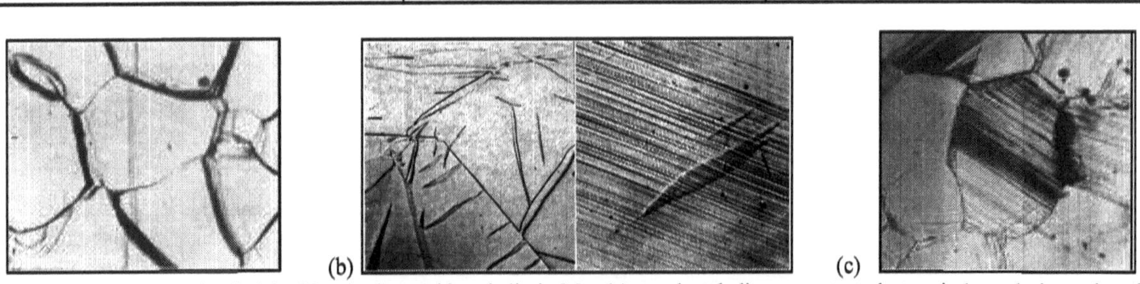

(a) (b) (c)

Fig. 2 (a) Sub-grain formation in Mg (b) twinning and basal slip in Mg. (c) non-basal slip concentrated at grain boundaries rather than grain interiors at high temperatures (Ref. 3).

TABLE II. STRAIN-RATE AND ACTIVATION ENERGY FOR THERMALLY ACTIVATED CREEP MECHANISMS (1)

Creep Mechanism	Creep Rate / Activation Energy
Dislocation Intersection - work required to force a dislocation through the stress field of another can be thermally aided. Mg single crystals	? = $A e [-(q_l - \tau^* b^2 l)/kT]$
Activated Cross Slip - Thermal activation to unite partial dislocations and to breakdown into partials on different slip systems.	q =100 kJ/mol for high stacking fault metals
Movements of Dislocations with Jogs - formation (q_f) and movement (q_m) of vacancies. Self diffusion ($q_d = q_f + q_m$).	? = $A e [-(q_d - \tau^* b^2 x)/kT]$ $q_{d\,Mg}$ 135 kJ/mol, q_{Mg} 125 kJ/mol.
Dislocation Climb - Climb of dislocations over sessile dislocations or precipitates. Vacancy diffusion to and from edge dislocations (q_d) and formation of jogs (q_j). Activation energy, $q = q_d + q_j$	? = $A\sigma^{\alpha} e (- q_d) / kT$? = $Be^{\beta\sigma} e (- q_d) / kT$ q in the range of q_d (135kJ/mol)
Movement of Dislocation Atmospheres - Diffusion of solute atoms and the viscous behavior of the solute atmosphere	? = $A \sigma e (- q_s) / kT$ q_{Al} in Mg 143 kJ/mole
Grain Boundary (GB) Shear - relative movement of grains, deformation in narrow region adjacent to the GB, shear direction lies in the boundary with the max resolved shear stress. Effective at high temps, usually when the recovery temperature is achieved. GBs recover before the grains, softening. Discontinuous. Accounts for 30% of the deformation. q ***Grain boundary diffusion: 80-100 kJ/mol***	

3. Creep Resistant Mg Alloys for High Pressure Diecasting

3.1. Mg-Al-Si Alloys

The first development in creep resistant Mg diecasting alloys for automotive applications has been the development of Mg-Al-Si diecasting alloys by Volkswagen in the 70's. Silicon addition in AS alloys leads to the formation of the Mg_2Si precipitates that has low density, high hardness, low thermal expansion coefficient, and perhaps most importantly, unlike the β-$Mg_{17}Al_{12}$ phase, a high melting temperature (1085°C). Si lacks the ability to tie up Al rendering it ineffective in hindering the formation of β-phase and discontinuous precipitation for a given level of Al. These alloys hence have two types of intermetallics, the β- ($Mg_{17}Al_{12}$), and the Mg_2Si phase with high temperature stability, which is deemed to pin dislocations and the grain boundaries thereby imparting increased creep resistance to the alloy. The Mg_2Si that has a face centered cubic crystal structure (7) and is a metallurgically stable phase that forms in Chinese script morphology under slow cooling conditions, which would exhibit very low ductility. In the diecast alloy, a fine distribution of the strengthening phase, Mg_2Si, is achieved with resultant high temperature resistance and a satisfactory level of toughness (8). The creep resistance achieved is moderate and the alloys are difficult to diecast.

The AS41 (Mg-4Al-1Si) and AS21 (Mg-2Al-1Si) alloys are the two standardized Mg-Al based heat-resistant alloys in use that display some improvement in creep properties in the range 130-150°C. AS41 offers better heat and wear resistance than AZ91, albeit with reduced fluidity of its melt. AS41 has moderate castability, but due to lower Al content, castability in AS21 becomes even a bigger problem. Foerster (9) suggested that silicon improves the fluidity of a Mg alloy when 4%Al is present, making AS41 more castable than AS21. Considerable amount of data on Mg_2Si has been reported by Beer et al. (7). However, silicon addition must be accompanied by a reduction in Al content to lower the amount of $Mg_{17}Al_{12}$ in the microstructure in order to improve the creep resistance. Creep mechanism in AS21 alloy was studied by Dargusch and Dunlop (5, 10). The same activation energy of 30-45 kJ/mol and n=2 were determined and creep induced precipitation of $Mg_{17}Al_{12}$ was observed (Fig. 4). The improved creep resistance of the AS21 over AZ91D was attributed to the smaller amount of the creep-induced precipitate and the grain boundary pinning provided by the Mg_2Si precipitates.

Blum et al. (11) have reported that at high stresses AS21 displays smaller maximum deformation resistance compared to AZ91 whereas under low stress the opposite is true. It is also interesting to note, as reported by Aune and Ruden (12), that up to 200°C, alloys of the AZ, and AS, series all appear to exhibit the same relative decrease in strength as the temperature is increased, and that at all temperatures AZ91 shows superior yield strength. The same authors also revealed that as 200°C is approached except for AS41 these alloys would display similar levels of ductility. It is possible to obtain a compromise between castability and creep resistance in this alloy system by going to an AS31 (Mg-3Al-1Si) composition. Recently, a modified version of AS21x with small additions of rare earth elements has been developed. Rare earth additions at trace levels do not affect the creep resistance but improve the corrosion resistance as well as castability (13).

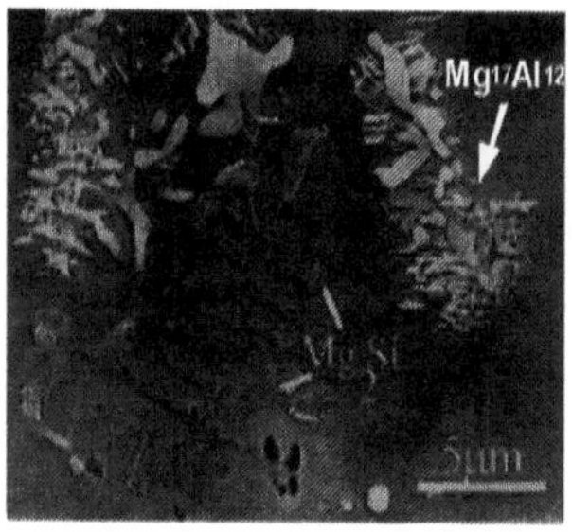

Fig. 4. Microstructure of AS21 after 120 hr creep deformation at 150 C and 50 MPa showing the Mg_2Si and creep induced precipitates of $Mg_{17}Al_{12}$ (Ref. 5).

3.2 Mg-Al-RE Alloys

A major development in creep resistant Mg alloys has been the emergence of rare-earth (RE) containing alloys (AE41, AE42). This group of Mg-Al-RE alloys involves at least one, and in general a mixture, of the REs as precipitate forming alloying addition in their compositions. Rare earth alloying additions are expensive and therefore a cheaper substitute known as mischmetal, a mixture of several rare earth elements that is enriched in one of the constituents is generally used. AE series constitutes the standardized alloys in the group of Mg-Al-RE alloys. This alloy system exhibits major improvement in creep resistance due to the complete suppression of the formation of the β-($Mg_{17}Al_{12}$) phase and the presence, instead, of the Al-RE containing intermetallics (14). In a review by Aune and

Westengen on mechanical properties of various Mg alloys, it has been shown that the ultimate tensile strength increases with the Al and RE content in the alloy, and that tensile yield strength primarily depends on the amount of RE and attains a limiting maximum level (14). For a given RE content, the creep strength decreases with increasing Al content, while for a given Al content it increased with increasing amount of RE addition (15).

There have been a number of studies aimed at identifying the second phases in AE alloy family. Polmear has claimed that Mg-Al alloys containing RE addition are suitable only for diecasting since slow cooling rates lead to the formation of coarse Al_2RE particles (16). $Mg_{12}Ce$ type particles were also detected at the grain boundaries during creep. Both of these precipitates, unless they coarsen, are considered to inhibit grain boundary sliding and thus being beneficial for creep resistance (16). Precipitation of Al_4RE compound in the alloy AE42 has also been reported (5, 17-19).
Depending on the composition, precipitates of other types of compounds have also been mentioned in the literature. Pettersen et al (17) suggested that when RE/Al weight ratio is above 1.4 all of the aluminum will be tied up as $Al_{11}RE_3$ in which case further precipitation of other phases such as another type of Al-Re phase or $Mg_{12}RE$ becomes possible. In samples produced by cold chamber diecasting they observed $Al_{11}RE_3$ and $Al_{10}RE_2Mn_7$ formation at the grain boundaries. Their TEM observation has revealed that the major phase was $Al_{11}RE_3$ (a body-centered orthorombic cell with lattice parameters a=4.5Å b=13.2Å and c=9.9Å). No special orientation relationship was found between this phase and Mg-matrix. EDS measurements gave an atomic ratio Al:RE=3.3±0.6 which is claimed to be in good agreement with the proposed stoichiometric ratio 11:3. The relative amounts of REs were reported to be (at%) $La_{27\pm2}Ce_{51\pm2}Pr_{6\pm1}Nd_{16\pm1}$.

This phase was found to co-exist with a minor phase $Al_{10}RE_2Mn_7$. The diffraction patterns from this phase were indexed according to a hexagonal system with lattice parameters a=9.0Å and c=13.1Å. This structure was found in agreement with the published data for $Al_{10}Ce_2Mn_7$ phase. There was, a problem due to an unexpected reflection, (1011); several pictures indicating –h+k-l=3n reflections, which is not consistent with any space group. The authors explained this as being due to a mirror image of the reciprocal space with the basal plane as the mirror. A possible physical interpretation given is that growth of twins with the basal planes as the twinning-plane may be influenced by the substitution of RE for Ce. EDS analysis gave $Al_{49\pm2}RE_{12\pm2}Mn_{39\pm2}$ which is in reasonable agreement with the proposed 10:2:7 ratio. The relative compositions of the RE components was $La_{10\pm1}Ce_{50\pm3}Pr_{9\pm1}Nd_{31\pm2}$. Lanthanum was said to go into $Al_{11}RE_3$ phase, while Nd into $Al_{10}RE_2Mn_7$ phase. They attributed the creep resistance to a strongly branched morphology precipitate morphology at the grain boundaries contributing to grain boundary pinning.

Dargusch et al have studied creep in AE42 alloy (5) and have observed that just as AZ91 and AS21 alloy, the activation energy was in the range of 30-40 kJ/mol, indicating the same creep mechanism of creep-induced-precipitation resulting in grain-boundary migration. $Mg_{17}Al_{12}$ was not observed but Al-supersaturation was noted. The difference between AZ and As alloys and the AE was that the onset of the second creep mechanism at the high stress region at 150°C was not seen in AE42 up to 100 MPa and that the creep strain was lower than AZ91 and AS21 (Fig. 5).

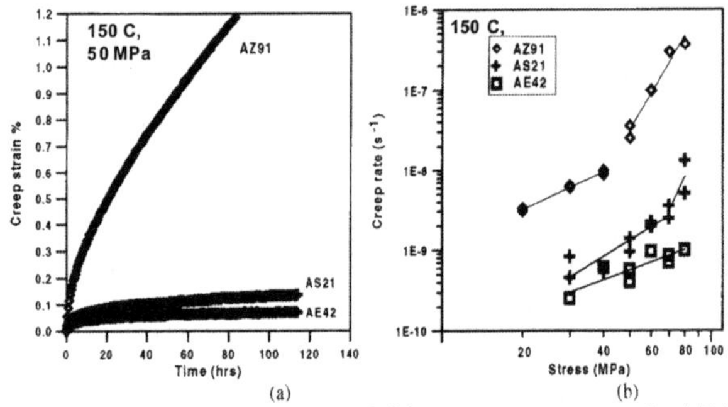

Fig. 5. (a)Creep strain vs time and (b) creep rate vs stress in AE42, AS21 and AZ91 (Ref 5).

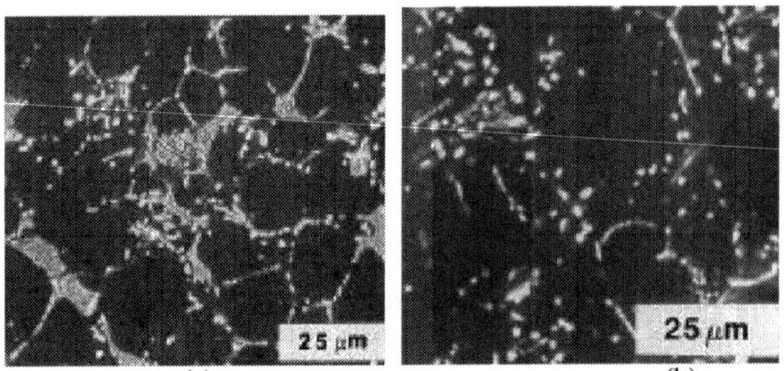

(a) (b)

Fig.6. AE42 microstructure (a) diecast (b) 175C exposure (20).

In a more recent paper by Powell et al (20) on the microstructural evolution of AE42 (Fig. 6), it was reported that two intermetallics $Al_{11}RE$, Al_2RE and practically no $Mg_{17}Al_{12}$ were present in the diecast structure. Creep testing at 150°C & 175°C showed a change in microstructure and in the relative amounts of these phases (Table III). These are consistent with the reaction by which $Al_{11}RE_3$ decomposes to Al_2RE and frees Al, which then reacts with Mg to form $Mg_{17}Al_{12}$, which is accepted as being responsible for poor creep resistance in Mg alloys (21). To see if stress played a role on the decomposition of $Al_{11}RE_3$ samples were exposed to the same thermal history but without the applied stress (20): the same microstructural changes were observed and it was concluded that $Al_{11}RE_3$ suffered merely from thermal instability at 175°C. A relationship was discovered between the relative amounts of $Al_{11}RE_3$ vs. Al_2RE and the La:Nd. (below 0.7 Al_2RE was seen and above 0.7 $Al_{11}RE_3$ seemed to form). Formation of either $Al_{11}RE_3$ or Al_2RE may be sensitive to individual rare earth elements, and that this might influence the stabilization of the $Al_{11}RE_3$ phase in AE42 alloy.

Table III. Change in Amounts of phases (wt AE42 alloy (20).

T (C)	Mg	$Al_{11}RE_3$	Al_2RE	$Mg_{17}Al_{12}$
As-cast 25C	97.5	1.8	0.8	0.0
150	97.7	1.5	0.8	0.0
175	97.0	1.2	1.3	0.6

AE42 alloy was perceived to have issues with cost, castability and fatigue strength and was not considered for widespread automotive use for a long time. In recent studies a fresh look at the AE alloy family is presented (22). Different combinations of Al and RE are investigated namely Mg-3%-5%RE (AE35), and Mg-8Al-2RE (AE82). Future studies on AE42 should address the thermal stability at 175°C and conduct a thorough search in the ternary Mg-Al-RE alloy system to determine compositions that offer both creep resistance, thermal stability and good castability. Phase diagram studies in the ternary system are also important

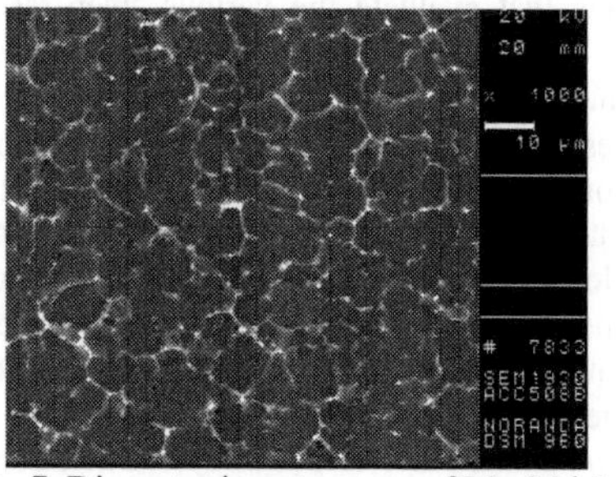

Fig.7. Diecast microstructure of Mg-5Al-0.8Ca (Ref. 28).

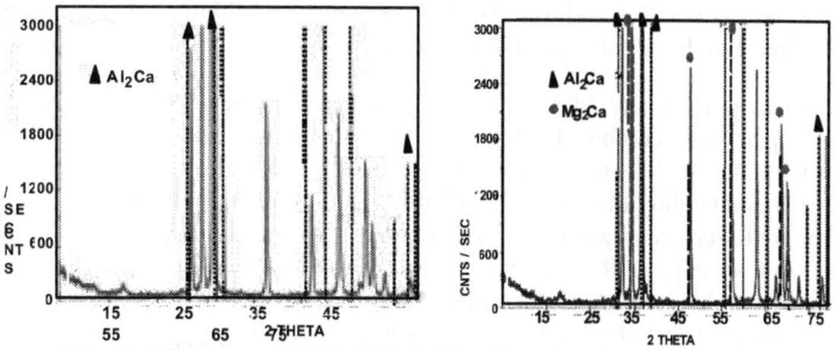

Fig. 8. XRD from (a) Mg-5Al-0.6Ca alloy showing Al_2Ca. (b) Mg-5Al-0.8Ca showing the presence of Al_2Ca and Mg_2Ca.

3.3. Mg-Al-Ca Alloys

Ca addition, as a cheaper and lighter alternative to RE elements, also contributes to high temperature properties (23-28). When Ca is added to Mg-Al binary alloys, the type of precipitating compound is found to depend on the Ca/Al mass ratio. When this ratio is more than ~0.8 presence of both Mg_2Ca and Al_2Ca were detected that resulted in considerable increase in hardness, while below this ratio only Al_2Ca was observed to have formed (Figs. 7 & 8). Both types of precipitates were observed to form along the grain boundaries with an increasing volume fraction as the Al and Ca content increased (27).

Pekguleryuz et al have shown that the creep resistance in Mg-Al-Ca ternary alloys (AX alloys) in the range of 2-6 wt% Al and 0.6-1.0wt% Ca show creep resistance at 150°C similar to the AE42 alloy with the added benefit of good corrosion resistance (25, 28, 29). The Al_2Ca compound has stability at high temperatures and the formation of the β-$Mg_{17}Al_{12}$ phase is suppressed. Beneficial effect of Ca addition was also proven in rapidly solidified AZ91 alloy. Gjestland et al (29) have shown that despite the smaller grain size compared to conventional AZ91, rapidly solidified structure involving fine Al_2Ca particles exhibited drastically smaller secondary creep rate. Although useful, Ca addition is not problem-free. Calcium makes the alloy prone to hot cracking and die-soldering especially when the level is above 1% (23-25, 30). Subsequent addition of 8% Zn to the alloy has improved diecastability but the properties were found to vary over a wide range (31). The second phase in this alloy was a Mg-Al-Ca-Zn containing complex phase. At this time, none of these experimental alloys with Ca addition has yet led to a commercially successful alloy, More work with the base alloy systems containing Ca is needed to optimize these alloys for diecasting, however, work has been carried out to use these alloys for semi-solid applications (32). Mg-Al-Ca system can be further investigated for optimum castability.

Recently, Luo (33) has reported that increasing the ca to above 2% improves castability by eliminating hot-tearing. Terada (6) et al studied creep mechanisms in Mg-5Al-1.7Ca alloy. At 150°c, the change in the stress exponent occurred at around 120 MPa, and in the low stress regime (100 MPa), in the 175-250°C range, the activation energy was determined to be 142 kJ/mol. While the authors attributed this to dislocation climb, the mechanism can be better described by the mechanism of "movement of dislocation atmospheres" (diffusion of solute atoms and the viscous behavior of the solute atmosphere) since Q is identical to Q_{Al} 143 kJ/mole (diffusion of al in Mg) (Table II).

3.4. Mg-Al-Sr System

Mg-Al-Sr (AJ) alloys are a new addition to the creep-resistant magnesium alloys (34-43). Various alloy compositions such as AJ51 (Mg-5Al-1Sr), AJ52 (Mg-5Al-2Sr) and AJ62 (Mg-6Al-2Sr) have been studied. AJ52 shows the highest creep resistance, while AJ62x gives an excellent combination of creep performance and castability.

Pekguleryuz et al report that the alloys show different microstructures based on the Sr/Al ratio (42). For Sr/Al ratio below about 0.3, Al_4Sr intermetallic is the only second phase in the structure (Fig. 9a&b). When the Sr/Al ratio is higher, a second intermetallic phase (a ternary Mg-Al-Sr) is also observed. Sr/Al controls the formation of $Mg_{17}Al_{12}$ as well (Figs.9 c&d). When the Sr/Al ratio is very low, there would be insufficient amount of Sr to bind all Al and the excess Al would form the $Mg_{17}Al_{12}$ phase. It is reported that the ternary Mg-Al-Sr phase is a new compound, which is under study. The superior creep resistance of AJ52 and AJ62 alloys is attributed to the existence of this second phase in addition to Al_4Sr. Al_4Sr is a tetragonal phase and no phase coherency with the primary Mg phase has yet been reported.

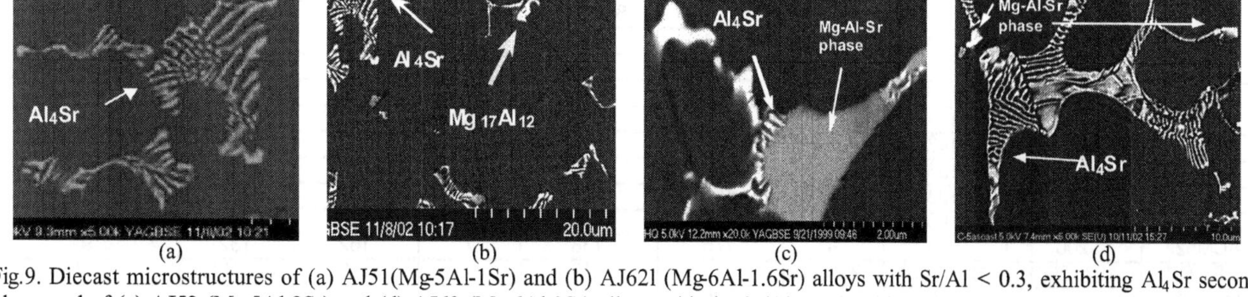

(a) (b) (c) (d)

Fig.9. Diecast microstructures of (a) AJ51(Mg-5Al-1Sr) and (b) AJ621 (Mg-6Al-1.6Sr) alloys with Sr/Al < 0.3, exhibiting Al_4Sr second phase and of (c) AJ52 (Mg-5Al-2Sr) and (d) AJ62 (Mg-6Al-2Sr) alloys with the Sr/Al > 0.3 exhibiting two types of second phases, Al_4Sr and a ternary Mg-Al-Sr compound.

3.5. Mg Diecasting Alloys With Combined Additions of Alkaline Earth and/or Rare Earth Elements

In recent years, many Al-containing Mg alloy systems have been developed with combined additions of rare-earth and alkaline earth elements. Nissan patent (44) on a Mg-Al-Ca-RE alloy and later a Honda alloy (45) ACM522 (Mg-5Al-2Ca-2RE) both claim improved creep resistance over AE42 alloy. ACM522 alloy is based on specific Al/RE/Ca ratios to ensure Al-RE precipitates in addition to Al-Ca intermetallics. This alloy is used by Honda in oil pans for their hybrid car. The microstructures of these alloys exhibit a combination of Al-RE and Al-Ca intermetallic mixtures. The Mg-8Al-2.5RE-1.6Ca-1.3Mn version of the shows Al_2Ca coupled- eutectic phase as well as needle shaped Al-RE (Ca, Mn) intermetallics and Al-Mn-RE intermetallics (46) Another alloy system based on Mg-Al-Ca-RE with optional additions of Sr and Zn is a patent by Dead Sea Magnesium and Volkswagen AG (47). Two alloy formulations code-named MRI 153 and MRI 230 are reported to exhibit good elevated-temperature performance (48) at 150°C and 180°C respectively. Furthermore MRI 153 exhibits good diecastability similar to AZ91 alloy. The alloy system of this patent is quite complex, having a number of elements and would lend itself to further development to find optimum combinations of strength, creep resistance and castability.

Two separate alloy systems with combined additions of Sr and Ca but no rare-earths have been developed by Noranda (49) and General Motors (50). The Noranda alloy is a Mg-Al-Sr-Ca alloy with small amounts of Sr and Ca (AJX Alloy). Code-named as Noranda N alloy, it has creep resistance comparable to AS41 at 150°C but creep resistance improves at 175°C being superior to AE42 alloy (46). The GM version is a Mg-Al-Ca-Sr alloy with substantial Ca and a small amount of Sr (AXJ Alloy) and exhibits superior creep resistance (51). All these complex alloys offer great potential for high creep resistance and the major effort should be invested on improving diecastability of these alloy systems compositions that offer the highest creep resistance. This can be achieved by systematic and fundamental castability research or by trace element effects.

3.6. Comparative Properties of Magnesium Diecasting Alloys for Powertrain Applications

Formal comparative evaluation of the various alloys is in its initial stages. Automotive partnership programs such as USCAR and EUCAR have projects that evaluate the various creep resistant magnesium alloys. Important performance requirements for powertrain components, among others, are tensile and compressive creep resistance, bolt-load-retention, yield strength, fatigue resistance, corrosion resistance and diecastability as well as cost and availability. It is rather difficult to make a rigorous comparison of the alloys from published data due to differences in sample, test conditions and type of test. Creep tests are carried out at different stress-temperature conditions and results are reported at different creep time. An attempt is made in Tables IV & V to summarize some of the data on the various alloys. Other data can be found in a recent review by A. Luo (33). Formal evaluation programs would be of great benefit to end users in identifying appropriate alloys for powertrain application development programs.

Diecastability is a special performance requirement of these alloys. The factors that contribute to good diecastabilty are low surface tension and relatively low liquidus for adequate fluidity, resistance to hot-tearing, optimum degree of wettability to avoid excessive reaction with steel dies, etc. Generally, increasing Al in Mg-Al based alloys improves diecastability by decreasing the liquidus and improving fluidity. The same can be achieved by increasing the diecasting temperature, within reasonable limits so as not to sacrifice die-life. A larger freezing range may also help diecastability by increasing the length of time the metal stays liquid and hence eliminate cold-shuts. Alloying elements that decrease the thermal conductivity or increase heat capacity are also beneficial for diecasting (40). Each alloy would require special set of diecasting parameters to fit the specific characteristics of the alloy.

Mg diecasting alloys for elevated-temperature applications are coming of age. A number of Al-containing alloys with single or combined additions of rare-earth and/or alkaline-earth elements have been developed. Creep resistance in these alloys is attributed to grain boundary strengthening by intermetallic phases. Developing alloy compositions that avoid detrimental creep-induced-precipitation which contributes to grain boundary migration is the key learning of the past decade. These new alloys are providing commercially viable alloys for automotive power-train applications such as oil-pans, transmission-case and engine components.

TABLE IV. TENSILE PROPERTIES (42, 48, 33)

ALLOY	20°C			150°C			175°C		
	UTS, MPa	YS, MPa	%E	UTS, MPa	YS, MPa	%E	UTS, MPa	YS, MPa	%E
AZ91D	204 (240)	139 (160)	3.1 (3)	169 (157)	105 (108)	16 (23)	138	89	21
AS41	197 (215)	136 (140)	4 (6)	153 (155)	94 (95)	17 (24)	127 (127)	85 (85)	19 (25)
AS21x*	210	121	5.5	130	87	20	110	78	23
AE42	220 (230)	128 (145)	10 (11)	140 (157)	88 (100)	23 (20)	121 (130)	81 (91)	23 (23)
AX51	192	128	7	161	102	7	-	-	-
AX52	228	161	13	-	-	-	171	133	23
ZAX850	219	146	5	159	117	11			
AJ52x	212	134	6	163	110	12	141	100	18
AJ62x	240	143	7	166	116	27	143	103	19
AJ62Lx	276	153	12	163	108	19	143	109	27
AJX500	236	133	10	152	101	16	137	98	20
AXJ531	238	190	8	-	-	-	196	146	15
ACM522	200	158	4	175	138	7	152	132	9
MRI153*	197 (250)	157 (165)	2.2 (5)	-	118	-	139	113 (110)	3.4
MRI230D	235	180	5	205	150	16	-	-	-
A380	290 (290)	155 (180)	3.2 (3)	251 (257)	148 (159)	6.4 (7)	248 (228)	154 (148)	7.1 (10)

TABLE V. CREEP OF DIECAST ALLOYS (42, 48, 33)

ALLOY	TENSILE CREEP, %						% Compressive Creep 150°C, 70 MPa, 200 hrs
	@ 50 MPa, 200 hrs		@ 50 MPa, 500 hrs		70 MPa, 100 hrs	70 Mpa, 200 hrs (53)	
	150°C	175°C	150°C	175°C	175°C	175°C	
AZ91D	2.7	*	6.35	-	-	-	21
AS41	0.05	2.48	0.07	-	-	-	6.13
AS21x	0.19	1.27	-	-	-	8.95	3.97
AE42	0.06	0.33	0.08	0.44	-	0.18	2.16
AX52	-	-	-	-	0.06	-	-
AXJ531	-	-	-	-	0.06	-	-
AJ52x	0.04	0.05	0.03	0.09	-	0.14	0.24
AJ62x	0.05	0.05	-	-	-	-	1.73
AJ62Lx	0.13	0.29	-	-	-	-	-
MRI 153	0.18	-	-	-	-	-	-
MRI 230D	-	-	-	-	-	0.24	-
A380	0.08	0.04	0.10	0.05	-	0.22**	0.03

4. References

1. R. E. Reed-Hill, Physical Metallurgy Principles, 2nd Edition, D. Van Nostrand, New York, 1973.
2. M. F. Ashby, D. R.H. Jones, Engineering Materials 1 Pergamon Press, Oxford, 1980.
3. E. F. Emley, Principles of Mg Tech., Pergamon Press, London, 1966.
4. G.V. Raynor, The Physical Metallurgy of Mg and its Alloys Pergamon Press, New York, 1959
5. M.S.Dargusch, G.L. Dunlop, K. Pettersen, Mg Alloys & Their Applications, B.L. Mordike, K.U. Kainer, Eds.,Werkstoff-Informationsgesllschaft, Germany, 1998, pp.277-82.
6. Y. Terada, N.Ishimatsu, R. Sota, T.Sato, K. Ohori, Mg Alloys, Mat. Sci. Forum, TransTech, Switzerland, vols. 419-22, 2003, pp. 459-64.
7. S.Beer, G.Frommeyer, E. Schmid, Mg Alloys & Their Applications DGM, B.L. Mordike, F. Hehmann, Eds., April 1992, pp.317-324.
8. M.Pekguleryuz, A. Luo, C. Aliravci, Light Metals, CIM, Canada, Aug-Sep 1993, pp. 409-416.
9. G.S. Foerster, Designing Diecasting Alloys Light Metal Age, v.30, 9-10, 1972, pp. 11-13.
10. M.S. Dargush, G.L. Dunlop, Materials 98, Inst. Mtls Eng., Australasia Ltd., Ed. Michael Ferry, Wollongong, Australia, 1998, p.579.
11. W. Blum, B. Watzinger, P. Weidinger, Mg Alloys and Their Applications, B.L. Mordike, K.U. Kainer, Eds., Werkstoff-Informationsgesellschaft, Frankfurt, Germany, 1998, p.49-60.
12. T. Kr. Aune, T.J. Ruden, SAE 920070, Detroit, Michigan, Feb, 1992.
13. K. Pettersen, H. Westengen, J.I. Skar, M. Videm, L-Y. Wei, Mg Alloys and Their Applications K.U. Kainer, Ed., DGM, Wiley-VCH, Weinheim, Germany, 2000, pp. 29-34.
14. T.Kr. Aune and H. Westengen, SAE Paper No 950424, Detroit, Michigan, February, 1995.
15. W. Mercer II, SAE No 900788, Detroit, Michigan, February, 1990.
16. I.J. Polmear, Light Alloys, Metallurgy of Light Metals, 3rd Edition, Arnold, London, 1995.
17. O. Holta, M. Videm, H. Westengen and D.L. Albright, Diecasting Innovation, NADCA, Indianapolis T95-053, Oct 1995, p.177-182.
18. G. Pettersen, H. Westengen, R. Hoier, O.Lohne, Mat. Sci. & Eng., A207, 1996 pp.115-120.
19. L.Y. Wei, G.L. Dunlop, J. Alloys & Compounds, 232, 1996, p. 264.
20. B. R. Powell, V. Rezhets, M.P. Balogh, and R.A. Waldo, J. Metals, TMS, August 2002, pp. 34-38.
21. A.A. Luo, M.O. Pekguleryuz, J. Mat. Sci., vol.29, 1994, pp. 5259-71.
22. P.Bakke, K.Pettersen, H.Westengen, 2003 Mg Technology, TMS, San Diego, March 2003, pp. 171-176.
23. F. Hollfigl-Rosta, E. Just, J. Kohler and H.J. Melzer, Light Metals Age, August 1980, p.22-29.
24. H.J. Fuch, British Patent 847, 992,1960.
25. M.O. Pekgüleryüz, A. Luo, Patent PCT/CA96/00091, Aug 22, 1996.
26. M. Pekguleryuz, A. Luo, Patent WO96/25529 (1995)
27. R. Ninomiya, T. Ojiro, K. Kubota, Acta Metall., vol. 43, No.2, 1995, pp.669-674.
28. M. O. Pekguleryuz, J. Renaud, 2000 Mg Techn., H. Kaplan, J. Hryn and B. Clow, Eds. TMS, 2000, pp. 279-84.
29. H. Gjestland, G. Nussbaum and G. Regazzoni, Mat. Sci. And Eng., A134, 1991, pp. 1197-1200.
30. F. Holbrigl-Rosta, E. Just, J. Kohler and H.J. Melzer, World Mg Conf., IMA, New York, 1980, p. 38.
31. A. Luo and T. Shinoda, "US 5,855, 997, January 5, 1999.
32. EP 0799901A1, Mazda 1997.
33. A. A. Luo, Mg Alloys 2003, Mat. Sci.. Forum, Trans Tech, Switzerland, vols. 419-422, 2003, pp. 57-65.
34. M. Pekguleryuz, P. Labelle, US Pat. No. 6,322,644, Nov 27th, 2001 and PCT WO 01/44529
35. M. Pekguleryuz, Mg Alloys, Mat.Sci. Forum, Y. Kojima, T. Aizawa, S. Kamado, Eds., Trans. Tech., Switzerland, 2000, pp. 131-140.
36. M. O. Pekguleryuz, E. Baril, 2001 Mg Tech., J. Hryn, Ed., TMS, New Orleans, March 2001 pp.119-136.
37. P. Labelle, M. Pekguleryuz, M. Lefebvre, R. Bouchard, SAE Paper 02M-161, 2002.
38. D. Argo, M. O. Pekguleryuz, P. Labelle M. Dierks, T. Sparks, T. Waltematte, 2001 Magnesium Techn., J. Hryn, Ed., TMS, New Orleans, March 2001 pp.131-125
39. M.Pekguleryuz, P.Labelle, D.Argo, M.Dierks, T.Sparks, T. Waltematte, World Mg Conf., IMA, Brussels, May 2001, pp. 27-33.
40. D.Argo, M.Pekguleryuz, P.Labelle, P.Vermette, R.Bouchard ,M. Lefebvre, Mg Technology, H. Kaplan, Ed., TMS, Seattle, Feb 2002, pp. 87-93.
41. Y.Fasoyinu, T.Castles, R.Bouchard, M.Sahoo, M.Pekguleryuz, P. Labelle, 2003 Mg Tech., TMS, San Diego, March 2003, p. 207.
42. M. Pekguleryuz, P. Labelle, E.Baril, D. Argo, 2003 Magnesium Tech., TMS, San Diego, March 2003, pp. 201-207.
43. M. Pekguleryuz, P. Labelle, E.Baril, D. Argo, SAE Paper 2003-01-190, Detroit, March 2003.
44. K. Samato, Y. Yamamoto, N. Sakate, and S. Hirabara, EP 0 799 901 A1 August 10 1997.
45. S. Koike, K. Wasizu, S. Tanaka, T. Baba and K. Kikawa, SAE Paper 200-01-1117, Detroit, March, 2000.
46. M. Pekguleryuz, Mat. Trans., Platform Sci. and Techn. for Advanced Mg Alloys, Japan Inst. Metals, 2001, Vol. 42, No. 7, pp. 1258-1267.
47. B. Bronfin, E. Aghion, S. Schuman, P. Bohling, and K.U. Kainer, US Patent 6,139,651, Oct 31, 2000.
48. E.Aghion, B.Bronfin, H.Friedrich, S.Schumann, F. von Buch., 2003 Mg Tech., TMS, San Diego, March 2003, pp. 177-182.
49. M. Lefebvre, M. Pekguleryuz, P. Labelle, US Pat. No 6, 342,180, Jan 29, 2002 and PCT WO 02/099147
50. B.R. Powell, V. Rezhets, A.A.Luo, J.J. Bommarito, and B.L.Tiwari, US Patent 6,264,763, July 24, 2001.
51. A. Luo, M. Balough, B. R. Powell, SAE Paper 2001-01-0423, Detroit, Michigan, 2001.

FIRE-PROOF EVALUATION OF CAO ADDED MG-3AL, MG-6AL, AND MG-9AL MG CAST PRODUCTS

Jin-kyu Lee and Shae K. Kim
Eco Materials & Processing Department, Korea Institute of Industrial Technology,
7-47 Songdo-Dong, Yeonsu-Gu, Incheon 406-840, Korea

Keywords: Mg alloy, Fire-proof, CaO, DTA, Chip test, Product test

Abstract

Mg products are easily oxidized and burned when they are exposed to high temperature or fire by accident. In order to solve these problems, the fire-proof solution has been developed by adding CaO in Mg alloys. The fire-proof was evaluated by three methods: quantitative DTA for small sphere specimen, furnace ignition test for burrs and machined chips, and torch ignition test for products. DTA was carried out for obtaining quantitative ignition temperature data with respect to specimen geometry and test environment; the furnace ignition test for burr and chip ignition temperature data; and the torch test for ignition temperature data for manufactured products. This paper discusses the results of fire-proof properties of 1wt.%, and 1.5wt.% CaO added Mg-3Al, Mg-6Al, and Mg-9Al Mg alloys compared with other high temperature Mg alloys such as AE44, AS21, MRI153, and MRI230. Eco-Mg alloy will be low-cost and fire-proof Mg alloys for airplane and train applications in terms of preventing poisonous gas generation, inhalation burn, and ignition generation.

Introduction

Mg alloys are becoming important light weight structural materials for airplane and train applications [1-4]. However, Mg alloys are easily oxidized and burned when they are exposed to high temperature of fire by accident. Fire-proof solution for Mg product of airplane and train applications has been studied by developing new Mg alloys. New Mg alloy development includes three categories: (1) development of Mg-Al based alloys with Si (forming AS alloys), with RE (forming AE alloys), with Ca (forming AX alloys), with Sr (forming AJ alloys), with Zn and Ca (forming AZX alloys), with Ca and RE (forming MRI alloys), with Ca and Sr (forming AXJ alloys); (2) development of Mg-RE (forming WE alloys); (3) development of Mg-RE-Zn (forming MEZ alloys) [5-12]. These alloys are mostly used by diecasting and sand casting. However, these alloys are very expensive compared with general Mg alloys such as AZ91 and AM60 Mg alloys. More than that, these alloys have process limitation, in turn, shape limitation according to their corresponding manufacturing processes. The research to add metallic Ca into Mg alloy has been carried out for fire-proof solution. It has been revealed to be very useful in terms of fire-proof solution. However, fire-proof solution with Ca added Mg alloys contains generally as much as 1.0wt.%Ca, which in turn causes another problems such as material cost, fluidity decrease, hot tearing susceptibility, die soldering, and Ca subtraction during recycling, etc. Fire-proof Eco-Mg alloys of CaO added Mg-Al based alloy have been researched and developed. Eco-Mg alloys, which could increase fluidity, reduce hot tearing susceptibility, and reduce die soldering tendency, will be low-cost and fire-proof Mg alloys for airplane and train applications in terms of preventing poisonous gas generation, inhalation burn, and ignition generation [13-16].

This paper discusses the results of fire-proof properties of CaO added Mg-3Al, Mg-6Al, and Mg-9Al Mg alloys. The fire-proof was evaluated by three methods: quantitative DTA for small sphere specimen, furnace ignition test for burrs and machined chips, and torch ignition test for products. DTA was carried out for obtaining quantitative ignition temperature data with respect to specimen geometry and test environment; the furnace ignition test for burr and chip ignition temperature data; and the torch test for ignition temperature data for manufactured products. The results are compared with other high temperature Mg alloys such as AS21, AE44, MRI153, and MRI230.

Experimental Procedures

Al containing Mg alloys (3, 5 and 9wt.%) were melted in a steel crucible in an electric resistance furnace under SF_6 and CO_2 gases. The desired faction of CaO was added into molten Mg-Al alloys at 680°C. The CaO added melts were cast into steel mold preheated at 100°C without protective gas. Chemical compositions were determined by ICP-mass spectrometer. In order to observe the change by CaO addition, the cast billets were polished and observed using an Olympus PME3 microscopy.

The fire-proof was evaluated by three methods: quantitative DTA for small sphere specimen, furnace ignition test for burrs and machined chips, and torch ignition test for products. The sample weight for DTA test was about 1mg and the heating rate was 10°C/min. DTA was carried out under dry air atmosphere. The alloy chips for furnace ignition test from the as-cast ingot by drilling without cutting oil. Chips on a copper boat were inserted into an electric resistance furnace at 800°C in an ambient atmosphere. Temperature change was monitored by a thermocouple equipped with the copper boat and the ignition was detected by abrupt temperature rise. The torch ignition test for ignition temperature data for manufactured products was carried out under an ambient atmosphere. The torch ignition test was photographed for ignition time and ignition tendency at real time.

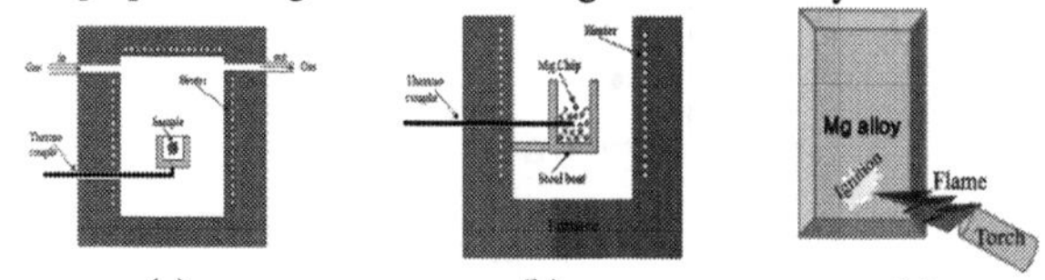

(a) (b) (c)

Figure 1. Schematic illustration for (a) DTA, (b) furnace ignition test, and (c) torch ignition test.

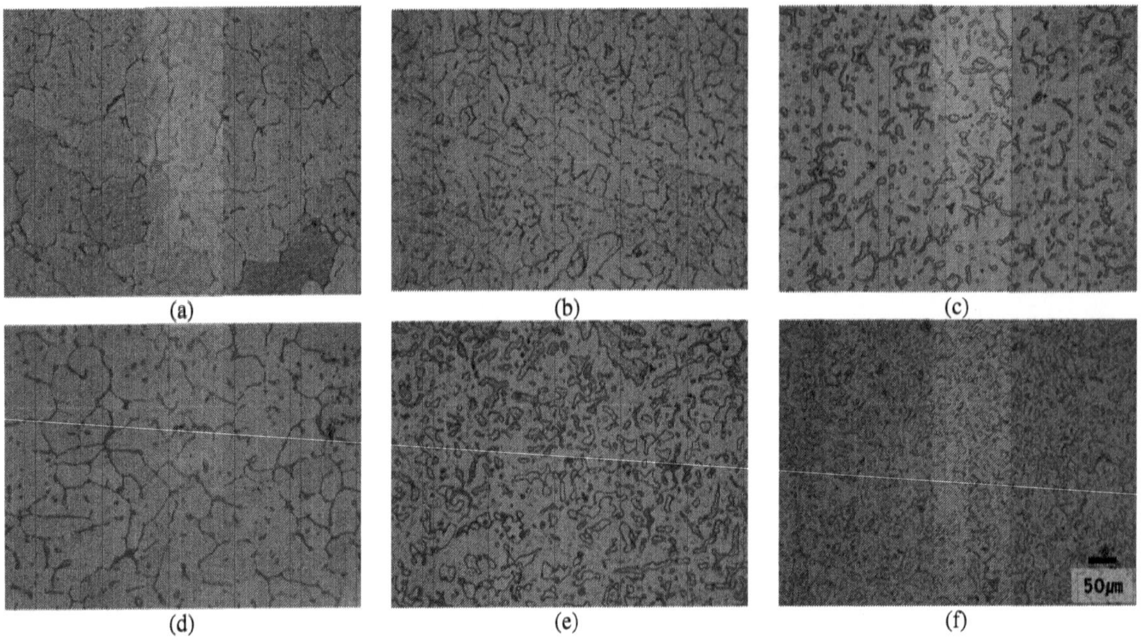

Figure 2. As cast microstructures of (a) Mg-3Al-1.13wt.%CaO, (b) Mg-3Al-1.5wt.%CaO, (c) Mg-6Al-1wt.%CaO, (d) Mg-6Al-1.65wt.%CaO, (e) Mg-9Al-1.02wt.%CaO, and (f) Mg-9Al-1.5wt.%CaO Mg alloys.

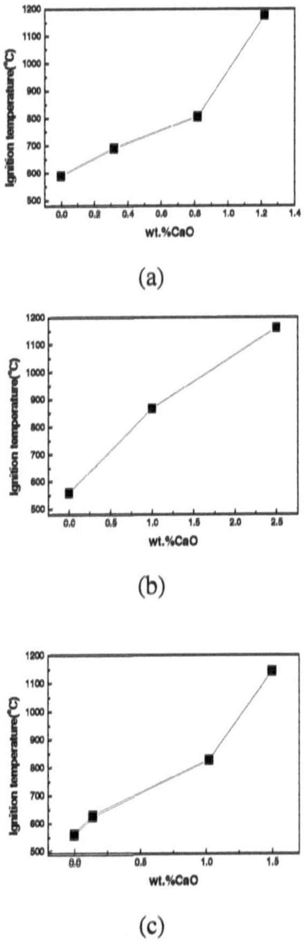

Figure 3. Dependence of ignition temperatures on CaO amount of (a) Mg-3Al, (b) Mg-6Al, (c) Mg-9Al Mg alloys under dry air atmosphere by DTA.

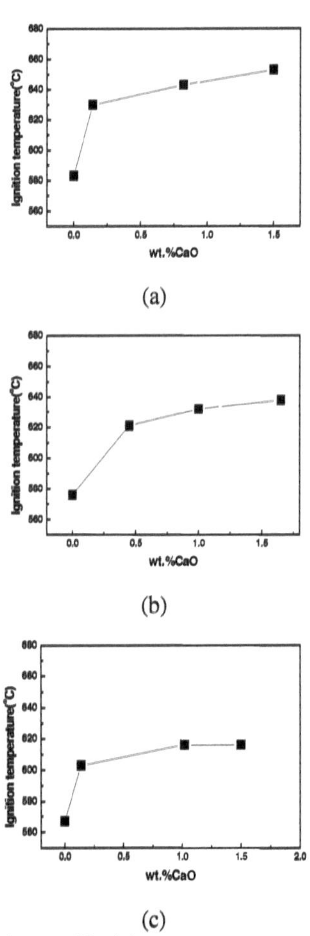

Figure 4. Dependence of ignition temperatures on CaO amount of (a) Mg-3Al, (b) Mg-6Al, and (c) Mg-9Al Mg alloys under an ambient atmosphere by furnace ignition test.

Results and Discussions

Fig. 2 shows the microstructures of as-cast Mg-Al-CaO Mg alloys. The microstructure of as-cast Mg-3Al-1.12wt.%CaO, illustrated in Fig. 2 (a), obviously was equiaxed α-phase and grain boundaries. Grain boundary consisted of β-phase ($Mg_{17}Al_{12}$), Al_2Ca and (Mg, Al)$_2$Ca intermetallic compounds. Lee et al. used EPMA and TEM to study the CaO behavior in AZ91D melt [13-16]. They confirmed that CaO in AZ91D melt was reduced to Ca and then formed Al_2Ca and (Mg, Al)$_2$Ca phases without forming solid solution in Mg matrix. CaO existed after reduction as Al_2Ca and (Mg, Al)$_2$Ca in AZ91D through EPMA and TEM analysis. The reason for the reduction of CaO and formation of Al_2Ca and (Mg, Al)$_2$Ca in AZ91D Mg alloy could not be explained by thermodynamic mechanism. This is simply because CaO is more stable than MgO in terms of thermodynamic. Fig. 2 (b) shows the microstructure of as-cast Mg-3Al-1.5wt.%CaO Mg alloy. The as-cast microstructure of Mg-3Al-1.5wt.%CaO was refined by the additional 0.4wt.%CaO and showed the sub-grain boundary. Fig. 2 (c) shows the microstructure of as-cast Mg-6Al -1wt.%CaO Mg alloy. The as-cast microstructure of Mg-6Al-1wt.%CaO under non-equilibrium solidification consisted of the coarse primary α-phase dendrites and bulk skeletal β-phase divorced eutectic compounds, which was mainly distributed at grain boundaries. Fig. 2 (d) shows the microstructure of Mg-6Al-1.65wt.%CaO alloy. The total volume fraction of the β-phase of Mg-6Al-1.65wt.%CaO alloy was decreased compared with Mg-6Al-1wt.%CaO Mg alloy because of CaO addition. Grain boundary consisted of β-phase, Al_2Ca and (Mg, Al)$_2$Ca intermetallic compounds. Fig. 2 (e) and (f) show the microstructure of Mg-9Al-1.02wt.%CaO and Mg-9Al-1.5wt.%CaO Mg alloys, respectively. The as-cast microstructure of Mg-9Al-1.02wt.%CaO under non-equilibrium solidification consisted of the coarse primary α-phase dendrites and bulk skeletal β-phase divorced eutectic compounds, which was mainly distributed at grain boundaries. The total volume fraction of the β-phase of Mg-9Al-1.02wt.%CaO alloy, as expected, was much than those of Mg-3Al-CaO and Mg-6Al-CaO Mg alloys. Grain size of Mg-9Al-1.02wt.%CaO Mg alloy decreased and β-phase volume increased as increasing Al amount compared with Mg-3Al-CaO and Mg-6Al-CaO Mg alloys. The microstructure of Mg-9Al-1.5wt.%CaO compared with Mg-9Al-1.02wt.%CaO Mg alloy was refined by the additional 0.4wt.%CaO and showed the sub-grain boundary.

Fig. 3 shows the results of DTA for Mg-3Al-xCaO, Mg-6Al-xCaO, and Mg-9Al-xCaO Mg alloys under dry air atmosphere. Considering the excellent fire-proof performance of Mg-3Al-xCaO alloys, the ignition temperature of Mg-3Al-xCaO Mg alloys increased with increasing CaO content as shown in Fig. 3 (a). The ignition temperatures of Mg-3Al, Mg-3Al-0.32wt.%CaO, Mg-3Al-0.82 wt.%CaO and Mg-3Al-1.22wt.%CaO Mg alloys were 590°C, 691°C, 806°C, and 1177°C, respectively. From the results of DTA for Mg-3Al-1.22wt.%CaO, the ignition temperature reached 1177°C. It could be explained by the increased Al_2Ca phase fraction together with decreasing eutectic phase on the grain boundary. Fig. 3 (b) shows the results of DTA for Mg-6Al-xCaO. The ignition temperatures of Mg-6Al, Mg-6Al-1wt.%CaO, and Mg-6Al-2.5wt.%CaO Mg alloys were 558°C, 842°C, and 1161°C, respectively. Fig. (c) shows the results of DTA for Mg-9Al-xCaO Mg alloys. The ignition temperature of Mg-9Al, Mg-9Al-0.14wt.%CaO, Mg-9Al-1.02wt.%CaO, and Mg-9Al-1.5wt.%CaO were 559°C, 626°C, 826°C, and 1142°C, respectively.

Fig. 4 shows the results of furnace chip ignition test under an ambient atmosphere at 800°C. The ignition is the initiation of burning, the process by which Mg alloys are oxidized at a rapid rate. The ignition of Mg-Al-CaO Mg alloys is an exothermic reaction and its heat of burning is as high as about 25,000 kJ/kg. The ignition temperature of Mg-3Al-xCaO Mg alloy was much higher than that of Mg-3Al Mg alloy as shown in Fig. 4 (a). The ignition temperature of Mg-3Al was 583°C. The ignition temperatures of Mg-3Al-0.14wt.%CaO, AZ31-0.82wt.%CaO, and 1.5wt.%CaO were 630°C, 643°C, and 653°C, respectively. With increasing CaO content, the ignition temperature of Mg-3Al-xCaO Mg alloys increased. The ignition of Mg was occurred because of porous MgO surface oxide film at high temperature. The ignition behavior of Mg could be explained though Pilling-Bedworth ratio, R [17]. Pilling-Bedworth ratio means the ratio of oxide volume to metal volume. Because the Pilling-Bedworth ratio for Mg is 0.83, MgO oxide film on the surface of molten Mg cannot act as a protective layer to prevent ignition resistance. On the contrary, CaO added Mg alloy could not be explained by Pilling-Bedworth theory. The ignition resistance of CaOadded Mg was suppressed due to the mixed dense CaO (0.78) and MgO surface films [18]. Fig. 4 (b) shows the result of Mg-6Al-xCaO Mg alloy under an ambient atmosphere. The ignition temperature of Mg-6Al Mg alloy was 576°C. The ignition temperatures of Mg-6Al-0.45wt.%CaO, Mg-6Al-1wt.%CaO, and Mg-6Al-1.65wt.%CaO were 621°C, 632°C, and 638°C, respectively. With increasing CaO amount, the ignition temperature of Mg-6Al-xCaO Mg alloys increased. Fig. 4 (c) shows the result of Mg-9Al-xCaO Mg alloy under an ambient atmosphere. The ignition temperature of Mg-9Al Mg alloy was 567°C. The ignition temperatures of Mg-9Al-0.14wt.%CaO, Mg-9Al-1.02wt.%CaO and Mg-9Al-1.5wt.%CaO were 603°C, 616°C, and 604°C, respectively. With increasing CaO amount, ignition temperature of Mg-9Al-xCaO Mg alloys increased. This result might indicate that small CaO addition was of interest for fire-retardation of process burrs and machined chips of cast and wrought Mg alloys.

Fig. 5 compares the results of DTA for Mg-3Al-xCaO, Mg-6Al-xCaO and Mg-9Al-xCaO as a function of Al content under dry air atmosphere. With increasing Al content, the ignition temperature of Mg-Al-xCaO Mg alloys decreased. Fig. 5 shows clearly the linear relationship between the ignition temperature and Al content. The ignition temperatures of all Mg-xAl-1wt.%CaO Mg alloys were increased at about 300°C compared with Mg-xAl alloys without CaO.

Fig. 6 shows the results of DTA for commercial high temperature Mg alloys under dry air atmosphere. The ignition temperatures of AS21, AE44, MRI153 and MRI230 were 591°C, 602°C, 1057°C and 1165°C, respectively.

Fig. 7 compares the results of DTA for Mg-Al-CaO and MRI Mg alloys under dry air atmosphere. Fig. 7 shows the ignition temperature of Mg-xAl-CaO Mg alloys as an amount of Al in Mg alloys, in comparison with commercial MRI 153 and MRI 230 Mg alloys. The ignition temperatures of Mg-xAl-CaO and MRI230 reached over 1140℃.

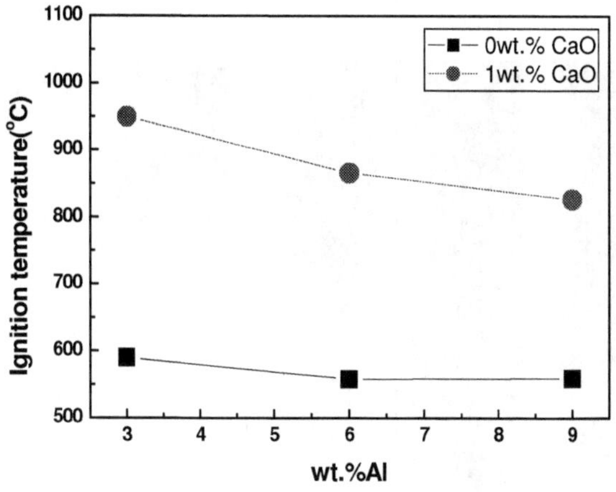

Figure 5. Dependence of ignition temperatures for Mg-3Al-xCaO, Mg-6Al-xCaO and Mg-9Al-xCaO as a function of Al content under dry air atmosphere by DTA.

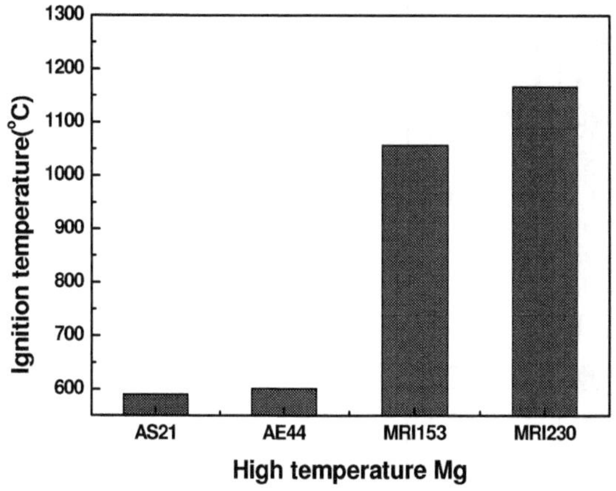

Firgure 6. Dependence of ignition temperatures for AS21, AE44, MRI153, and MRI230 Mg alloys under dry air atmosphere by DTA.

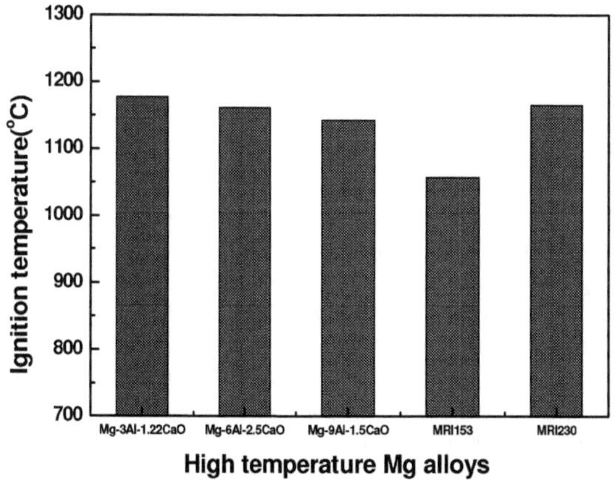

Figure 7. Dependence of ignition temperatures for Mg-3Al-1.5CaO, Mg-6Al-1.65CaO, Mg-9Al-1.5CaO, MRI153, and MRI230 Mg alloys under an ambient atmosphere by DTA.

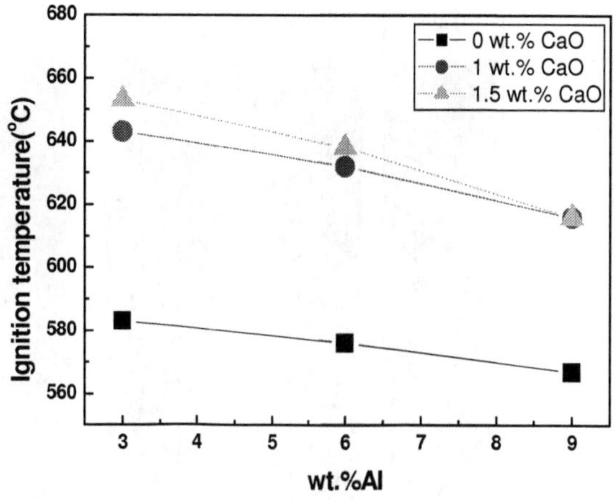

Figure 8. Dependence of ignition temperatures for Mg-3Al-xCaO, Mg-6Al-xCaO and Mg-9Al-xCaO as a function of Al content under an ambient atmosphere by furnace ignition test.

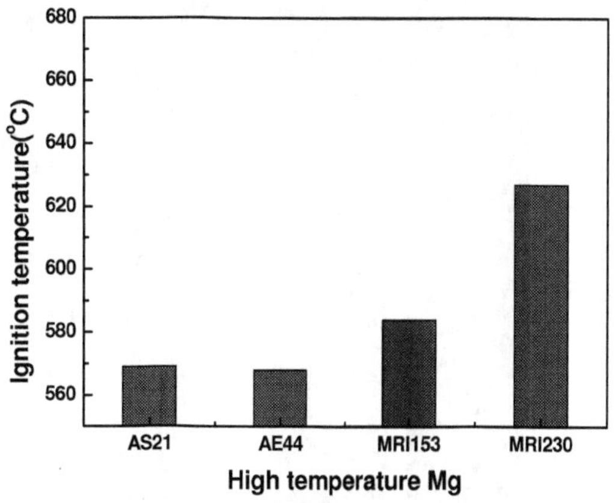

Figure 9. Dependence of ignition temperatures for AS21, AE44, MRI153, and MRI230 Mg alloys under an ambient atmosphere by furnace ignition test.

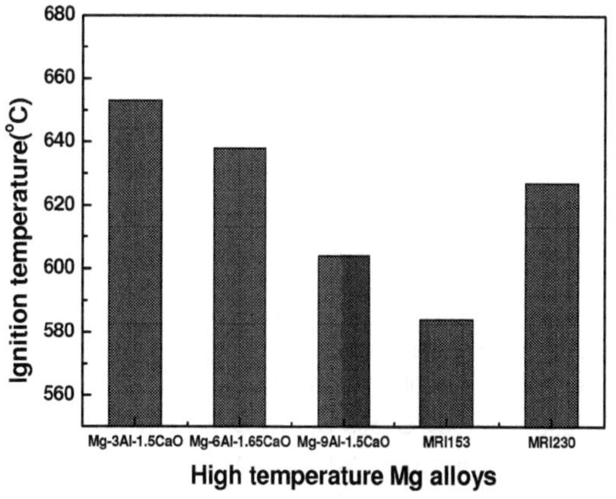

Figure 10. Dependence of ignition temperatures for Mg-3Al-1.5CaO, Mg-6Al-1.65CaO, Mg-9Al-1.5CaO, MRI153, and MRI230 Mg alloys under an ambient atmosphere by furnace ignition test.

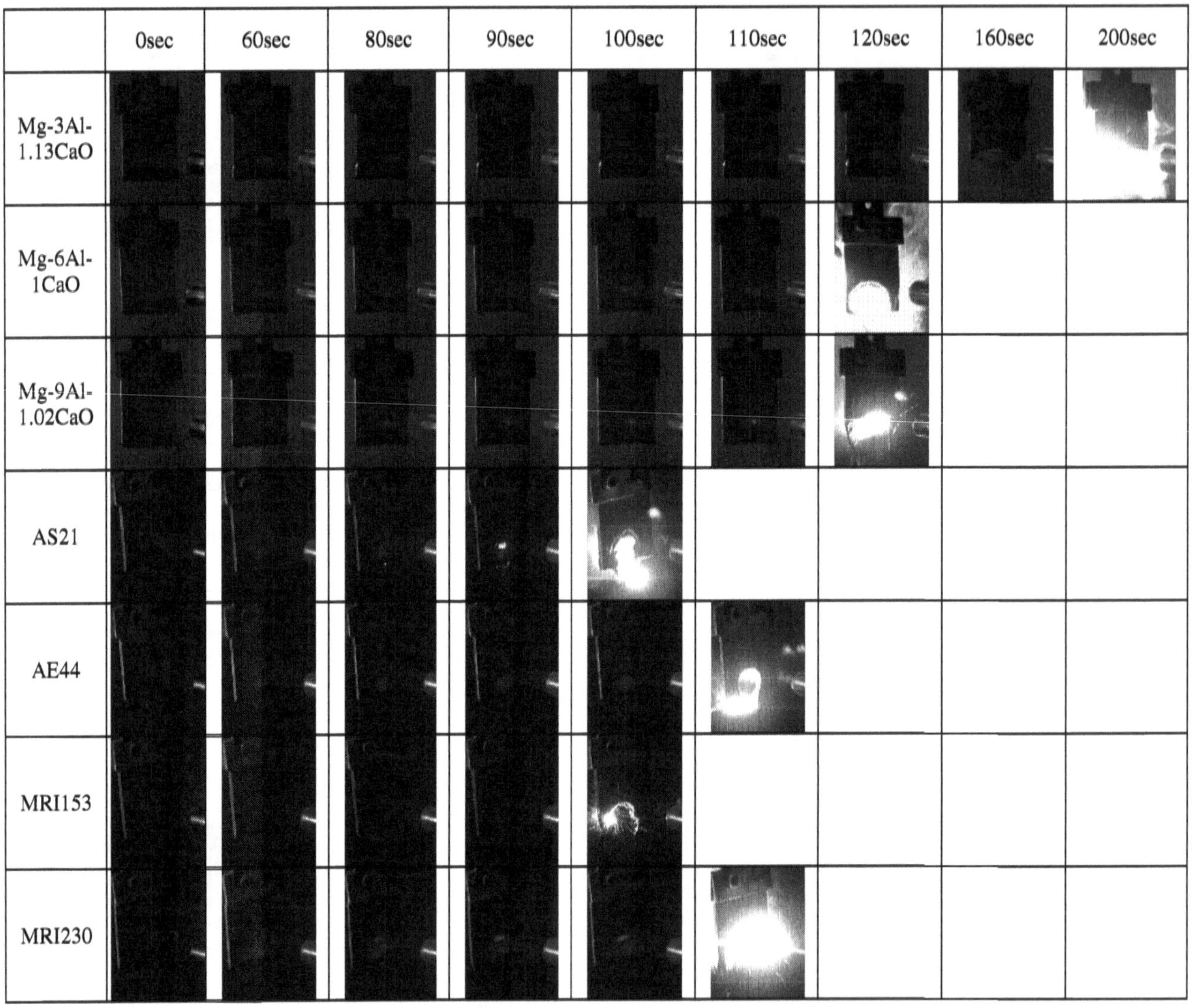

Figure 11. Torch ignition results of Mg-3Al-1.13CaO, Mg-6Al-1CaO, Mg-9Al-1.02CaO, AS21, AE44, MRI153 and MRI230.

Fig. 8 compares the results of furnace chip ignition test for Mg-3Al-xCaO, Mg-6Al-xCaO and Mg-9Al-xCaO as a function of Al content under an ambient atmosphere. With increasing Al content, the ignition temperature of Mg-Al-xCaO Mg alloys decreased. Fig. 6 shows clearly the linear relationship between the ignition temperature and Al content. The ignition temperatures of all Mg-xAl-1wt.%CaO Mg alloys were increased at about 60°C compared with Mg-xAl alloys without CaO.

Fig. 9 shows the results of furnace chip ignition test for commercial high temperature Mg alloys under an ambient atmosphere. The ignition temperatures of AS21, AE44, MRI153 and MRI230 were 569°C, 568°C, 584°C and 627°C, respectively. The ignition temperature of MRI230 Mg alloy was much higher than those of AS21 and AE44 Mg alloys.

Fig. 10 compares the results of furnace chip ignition test for Mg-Al-CaO and MRI Mg alloys under an ambient atmosphere. Fig. 10 shows the ignition temperature of Mg-xAl-1.5CaO Mg alloys as an amount of Al in Mg alloys, in comparison with commercial MRI 153 and MRI 230 Mg alloys. The ignition temperatures of Mg-3Al-1.5wt.%CaO and Mg-6Al-1.65wt.%CaO were much higher than the other high temperature Mg alloys.

Fig. 11 shows the results of torch test for Mg-Al-CaO and high temperature Mg alloys under an ambient atmosphere. The cast product was continuously heated by torch until the ignition occurred. The ignition time was defined at the time when the cast product was ignited. Part of cast product was slowly melted by torch and part of molten Mg alloys was ignited. The ignition time of Mg-3Al-1.13wt.%CaO, Mg-6Al-1wt.%CaO, and Mg-9Al-1.02wt.%CaO were 200 seconds, 120 seconds, and 120 seconds, respectively. This result might indicate that proper amount CaO addition was of interest for fire-proof Mg products when they were exposed to a fire or a spark under an ambient atmosphere. The ignition time of AS21, AE44, MRI153, and MRI230 were 100 seconds, 110 seconds, 100 seconds and 110 seconds, respectively. After the ignition, the fire of Mg-3Al-1.13wt.%CaO, Mg-6Al-1wt.%CaO, Mg-9Al-1.02wt.%-CaO, MRI 153, and MRI230 Mg alloys were extinguished. However, the fire of AS21 and AE44 Mg alloys continuously propagated by self-heating. The behavior of the ignition by torch was discussed from AES result as shown in Fig. 12. Fig. 12 shows the AES depth profile sputtered from the surface of cast product. In cast product of Mg-9Al-1.02CaO Mg alloy, there was the thin oxide layers mixed dense CaO and MgO surface films. Fig.13 shows the AES depth profile sputtered from the surface of Mg-9Al-0.3CaO Mg alloy after oxidation at 450℃ for 7 hours. The surface scale of alloy consisted of the thin oxide layer mixed with MgO and CaO filmes.

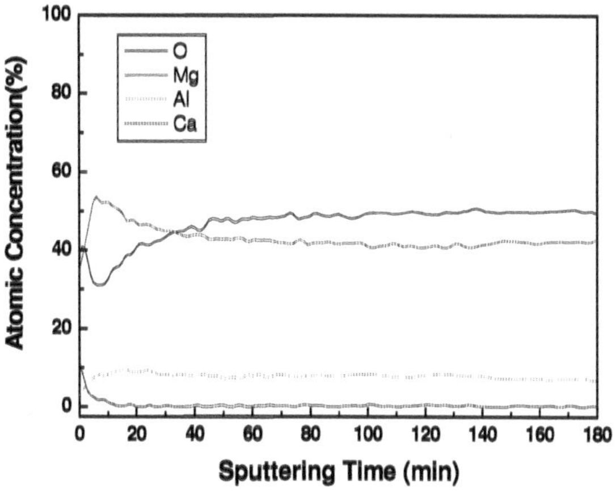

Figure 12. The surface oxide film analysis of as-cast product for Mg-9Al-1.02CaO by AES.

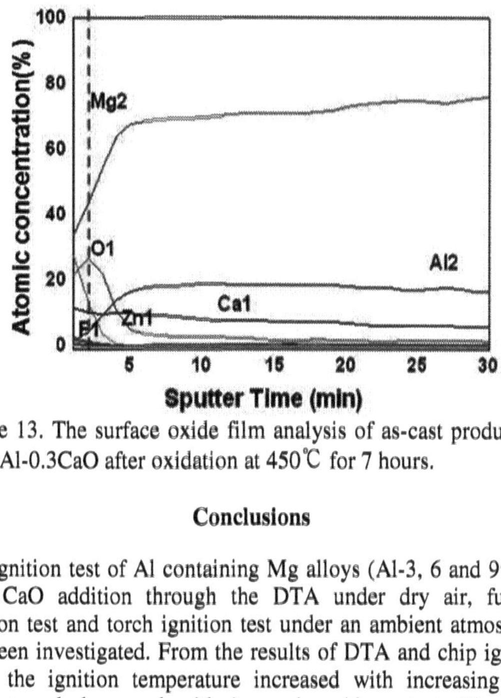

Figure 13. The surface oxide film analysis of as-cast product for Mg-9Al-0.3CaO after oxidation at 450℃ for 7 hours.

Conclusions

The ignition test of Al containing Mg alloys (Al-3, 6 and 9wt.%) with CaO addition through the DTA under dry air, furnace ignition test and torch ignition test under an ambient atmosphere has been investigated. From the results of DTA and chip ignition tests, the ignition temperature increased with increasing CaO amount and decreased with increasing Al content. The DTA ignition temperatures of Mg-3Al-1.22wt.%CaO, Mg-6Al-2.5wt.%CaO, and Mg-9Al-1.5wt.%CaO reached 1177°C, 1161°C and 1142°C, respectively. It could be explained by the increased Al_2Ca phase fraction together with decreasing eutectic β-phase on the grain boundary. This result could be seen that the effect of small CaO additions was of interest for fire-retardation of process burrs and machined chips of cast and wrought Mg alloys. The chip ignition temperatures of Mg-3Al-1.5wt.%CaO and Mg-6Al-1.65wt.%CaO were much higher than those of the other high temperature Mg alloys such as AS21, AE44, MRI153 and MRI230. This result could be seen that the effect of small CaO additions was of intersest for fire-retardation of process burrs and machined chips of cast and wrought Mg alloys. From the torch test results, the ignition resistance by CaO addition in Mg-Al alloys was greatly improved. The ignition of Eco-Mg alloys was suppressed due to the mixed dense CaO and MgO surface films. This result might indicate that proper amount of CaO addition was of interest for fire-proof Mg products when they were exposed to a fire or a spark under an ambient atmosphere. Eco-Mg alloy will be low-cost and fire-proof Mg alloys for airplane and train applications in terms of preventing poisonous gas generation, inhalation burn, and ignition generation.

Reference

1. B. L. Mordike and T. Ebert, "Magnesium Properties-application-potential," *Materials Science and Engineering*, A302 (2001), 37-45.

2. I. J. Polmear, "Magnesium alloys and applications," *Materials Science and Technology*, 10 (1994), 1-16.

3. I. J. Polmear, "Recent Developments in Light Alloys," *Materials Transactions, JIM*, 37 (1996), 12-31.

4. H. Friedrich and S. Schumann, "Research for a new age of magnesium in the automotive industry," *Journal of Materials Processing Technology*, 117 (2001), 276-281.

5. B. Bronfin, E. Aghion, F. von Buch, S. Schumann and M. Katzir, US patent, No.7041179, 2006.

6. B. R. Powell, V. Rezhets and A. A. Luo, US patent, No. 6264763.

7. M.O Pekguleryuz, and J. Renaud, "Creep resistant in Mg-Al-Ca alloys," *Proc. Magnesium Technology* 2000, (2000), 279-284.

8. M.O Pekguleryuz, E. Baril, "Development of creep resistant Magnesium-Aluminum-Strontium Alloys." *Proc. Magnesium Technology* 2001, (2001), 119-136.

9. M. Sakamoto and S. Akiyama, "Suppression of ignition and burning of molten Mg alloys by Ca bearing stable oxide film" Journal of Material Science Letter, 16 (1997) 1048-1050.

10. M.O Pekguleryuz, and E. Baril, " Creep resistant magnesium diecasting alloys based on alkaline earth elements," *Materials Transactions*, 42 (2001), 1258-1267.

11. J. R. TerBush, A. Suzuki, N. D. Saddock, J. W. Jones and T. M. Pollck, "Microstructure and creep of diecast magnesium alloys" *Proc. Magnesium Technology* 2007, (2007).

12. T. K. Aune and T. J. Ruden, " High temperature properties of magnesium Die Casting Alloys" *SAE Technical paper Series*, 920070, (1992), 1-7.

13. J. K. Lee et al., "Effect of CaO addition on ignition behavior in molten AZ31 and AZ91D Magnesium alloys", *Rare Metals*, 25 (2006) 155-159.

14. S. K. Kim et al., "The behavior of CaO in Mg alloys", *Proc. Magnesium Technology* 2005, (2005), 285-289.

15. S. H. Ha et al., "Effect of CaO on oxidation resistance and microstructure of pure Mg", *Materials Transactions*, 49 (2008) 1081-1083.

16. J. K. Lee et al., "Development of Novel Environment-friendly magnesium alloys", *Advanced Materials Research*, 47-50 (2008) 940-943.

17. N. B. Pilling, R. E. Bedworth, "The oxidation of metal in high temperature", Journal International Metals, 29 (1923) 529-582.

18. B. S. You, "The effect of calcium addtions on the oxidation behavior in magnesium alloys", *Scripta materials*, 42 (2000) 1089-1094.

THE ROLE OF RARE EARTH ELEMENTS IN STRUCTURE AND PROPERTY CONTROL OF MAGNESIUM DIE CASTING ALLOYS

Dr. Per Bakke, Dr. Håkon Westengen
Hydro Magnesium Competence Centre, Hydro Aluminium R&D Centre, N-3908 Porsgrunn, Norway

Abstract

The performance of magnesium die cast parts is governed by the microstructure and by the distribution of structural features which occur as a result of the chemical composition and processing history of the alloy. The elevated temperature properties, especially mechanical strength under creep conditions, are primarily determined by the grain structure, the elements in solid solution and the effectiveness of second phase particles in stabilizing the grain boundaries. The current emphases in alloy development focus on the utilization of elements with low solubility in the solid state, leading to the formation of stable precipitates during solidification. Such elements include the rare earths, as well as silicon, strontium and calcium. A detailed analysis of the various microstructural features and attributes is given for a new family of rare earth-containing alloys. The optimization of alloy composition is addressed in terms of blending advantageous microstructural characteristics with phase equilibria considerations.

Introduction

An initial metallurgical screening [1,2] and a second screening based on raw material cost and availability, indicate that basically only silicon (Si), calcium (Ca), strontium (Sr) or rare-earth (RE) (in addition to Al, Zn and Mn) can be considered as alloying elements to magnesium die casting alloys. Possibly tin (Sn) and antimony (Sb) can also be added to this list. With Mg-Al-Si (AS) alloys being limited to service up to 150°C maximum [2], RE additions become the next most economical addition for higher temperature use. Extensive work on the Mg-Al-RE (AE) alloy system over the past several years has identified alloys with unique combinations of properties [1-6].

The beneficial effect of rare earth elements (Ce) to Mg was discovered in the 1930's [7,8]. Sand and gravity casting alloys e.g. EZ33 and ZE41, grain-refined with small additions of Zr and subsequently heat treated, have been in use for some time [9,10]. AE alloys for die casting were extensively examined in the 1970's [11-13], culminating with AE42 in the early 1990's [14-16].

Although AE42 is the current benchmark with respect to elevated temperature properties and corrosion, the die casting alloys of the AE system have not received the attention they probably deserve from the property and metallurgical points of view. One reason for this may be inaccurate rumors that the castability of these alloys is inferior. To the contrary, recent work comparing the performance of various creep resistant alloys rank AE42 as the alloy having the best combination of properties at room temperature [17] and die castability similar to AZ91D for large automotive (power train) components [18]. High cost has also been an argument against AE alloys. This was true in the 1990's when the prices of mischmetal (MM) were very high. New entrants and expansions by existing producers of MM have resulted in additional capacity at the mining, processing and product levels, leading to a significant and sustainable price reduction [19]. Hydro Magnesium has recently patented a new family of AE die casting alloys including AE44, AE35 and AE63.

Principles for alloy development - The toolbox

Alloy development is the art and science of combining an appropriate chemical composition with the possibilities of the processing method to create microstructures that provide the material properties desired. Figure 1 shows schematically the microstructural details that influence the alloy properties.

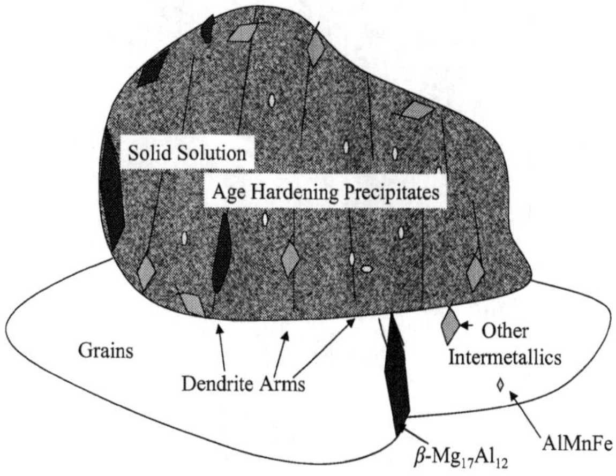

Figure 1. Schematic microstructure showing constituents that influence properties [6].

Solid solution hardening

The α-magnesium phase is strengthened by solid solution hardening. Potent elements include Al, Zn and some RE elements [20]. Since creep occurs preferentially in the grain boundary regions where any eutectic components would be located [21], it is important that the eutectic temperature is not too low. For example the Mg-Zn alloy system has a eutectic temperature of about 340°C, and ternary elements are required to increase the eutectic temperature. Alloying with Al provides a eutectic temperature around 430°C. Additions of elements like RE, Ca and Sr may further increase the eutectic temperature.

Grain boundary hardening

Die casting is characterized by rapid cooling. Depending on the wall thickness, the cooling rate will typically be in the range 10-1000°K/s; the resulting grain size is very fine, typically between 5 and 100 μm. This leads to a significant hardening effect predictable from the Hall-Petch relationship [22]. The fine grain size is harmful to creep properties, but it is known that ultrafine-grained magnesium alloys deform superplastically at elevated temperatures [23].

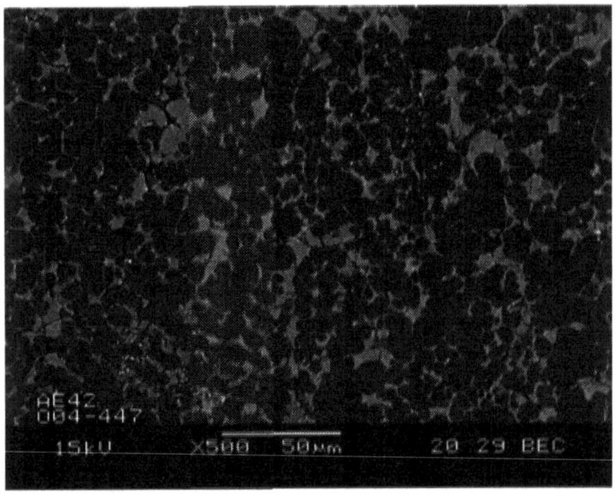

Figure 2. SEM-BEC of typical die cast microstructure of AE42.

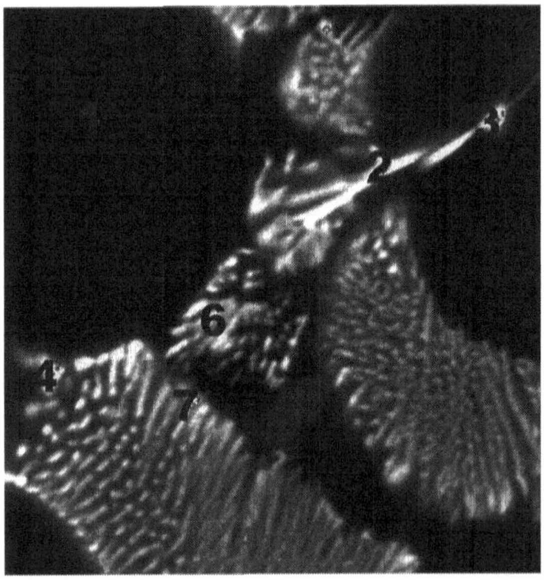

AE42	Mg	Al	Ce	La	Nd	Pr	Mn	ΣRE	RE/ (RE+Al)	Likely phase
1	98.65	1.18	0.00	0.00	0.00	0.00	0.10	0.00	0.00	
2	87.28	8.80	1.81	0.96	0.57	0.33	0.08	3.67	0.29	$Al_{11}RE_3$
3	91.33	5.22	1.66	0.79	0.44	0.23	0.00	3.12	0.37	Al_2RE
4	80.33	14.42	1.68	1.14	0.28	0.24	1.82	3.34	0.19	$Al_{11}RE_3$
6	86.56	10.43	1.41	1.04	0.09	0.04	0.35	2.58	0.20	$Al_{11}RE_3$
7	84.31	12.38	1.66	1.07	0.22	0.18	0.19	3.13	0.20	$Al_{11}RE_3$

Figure 3. SEM-EDS analyses of dispersoids in AE42.

Dispersion hardening

Alloying elements that form a dense distribution of intermetallic particles during solidification plays a key role in providing creep resistance to die casting alloys. Such particle distributions may have a high dispersion level, f/r, where f is the volume fraction of particles and r is the mean particle radius. These particle dispersions become increasingly efficient in reducing the creep rate arising from thermal activation by forming three-dimensional networks along the grain boundaries [24]. An additional positive effect is observed when the dispersoid-forming element combines with the solid solution hardening element to reduce the amount of

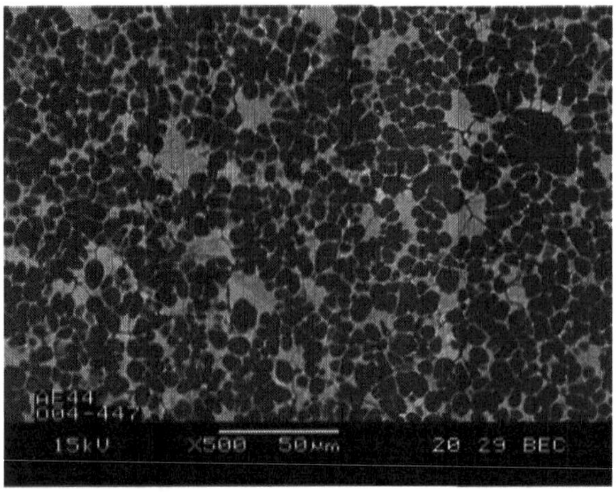

Figure 4. SEM-BEC of typical die cast microstructure of AE44.

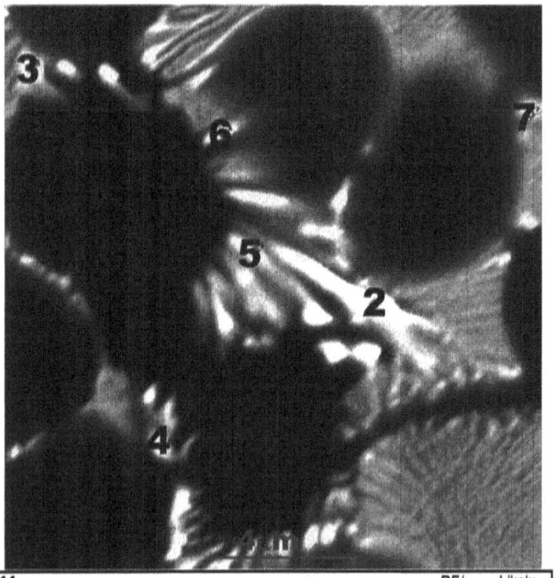

AE44	Mg	Al	Ce	La	Nd	Pr	Mn	ΣRE	RE/ (RE+Al)	Likely phase
1	98.26	1.40	0.03	0.00	0.03	0.07	0.12	0.13	0.08	
2	88.56	7.69	1.68	1.10	0.47	0.33	0.09	3.58	0.32	Al_2RE
3	86.11	10.33	1.71	1.19	0.30	0.24	0.02	3.44	0.25	Al_3RE
4	85.88	10.79	1.38	1.25	0.25	0.20	0.19	3.08	0.22	$Al_{11}RE_3$
5	90.23	6.66	1.39	0.82	0.39	0.22	0.15	2.82	0.30	Al_2RE
6	88.30	8.18	1.60	0.87	0.52	0.25	0.10	3.24	0.28	Al_3RE
7	90.67	6.67	1.18	0.89	0.26	0.25	0.00	2.58	0.28	Al_3RE

Figure 5. SEM-EDS analyses of dispersoids in AE44.

low melting eutectic along the grain boundaries. Rare earth elements, Ca and Sr in Mg-Al alloys are examples of this.

Age hardening

Age hardening in the usual sense of performing a full T6 heat treatment is generally not considered to be compatible with the desired properties and high volume production rates of die cast parts. Among the reasons for this are a) during solution treatment, the loss in strength due to grain coarsening can be larger than the increase caused by subsequent age hardening, and b) the formation of blisters due to expansion of gas porosity upon heating can be a problem unless special precautions are taken.

However, age hardening is commonly used for coarse-grained, sand cast parts; an example is provided by WE43 alloy, where age hardening precipitates are formed in extremely high densities during a final age hardening [25]. These particle dispersions provide both strength and creep resistance, due to the high thermal stability of the particles formed. This explains the paradox that while magnesium die casting alloys commonly show lower creep resistance than aluminium alloys, some magnesium sand casting alloys have superior creep strength to the aluminium casting alloys [26].

Experimental

Ingots of AE42 and AE44 were melted in a 15 kW resistance heated furnace. The special AE alloys, containing Ce, Nd and combinations, rather than MM, were prepared from a basis of molten AM20 to which pure Al, Ce and/or Nd were added. The various AE alloys were die cast on a 420 ton Bühler SC42D Evolution machine, with a five cavity plate/bar die containing two 3 mm test plates and three test bars (tensile bar, creep bar and fatigue bar [2].

Specimens for microstructure studies were cut and prepared from one arbitrary test plate in the as cast condition. The specimens were examined on a JSM5600, Pioneer, Scanning Electron Microscope (SEM) system. Energy Dispersive Spectroscopy (EDS analysis) was performed at 8000X magnification with an accelerating voltage of 15 kEV.

Specimens from a series of alloys were tensile creep tested for minimum 150 h in accordance with ASTM E139, at e.g. 175°C/40 MPa and 175°C/75 MPa. Tensile creep testing was also performed on samples that were aged at 175°C for 500 h. A minimum of two specimens was tested at each condition.

Results and discussion

Dispersoid phases in AE alloys

The microstructure of die cast Mg-Al based alloys generally consists of a grain interior of α- Mg(Al) solid solution, and a grain boundary zone (grain "mantle") which is eutectic Mg-Al (Al-enriched α Mg with β - $Mg_{17}Al_{12}$ embedded).

The rare earth used in the AE alloys is usually mischmetal (MM), a mixture of Ce, La, Nd and Pr, where Ce (approx. 50%) and La (about 30%) are the main constituents. AE42 contains 2-3 wt% Ce-rich MM (RE). Intermetallic phases (dispersoids) are distributed in the grain boundary zone as an almost continuous network, Fig. 2. SEM-EDS analysis indicates that the intermetallic phase formed as die cast mainly is $Al_{11}RE_3$, Fig. 3. (RE/(Al+RE) close to 0.21). This is also reported in the literature [27, 28]. The formation of $Al_{11}RE_3$ reduces the effective amount of Al, which otherwise would lead to Al-enrichment and subsequent formation of $Mg_{17}Al_{12}$ in the grain boundary zone. The effect on creep is primarily to inhibit grain boundary sliding. The grain interior is not significantly influenced by the addition of RE [1]. This is also indicated in Fig. 3, pos. 1. In alloys with a RE/Al ratio corresponding to AE42, the $Al_{11}RE_3$ phase has a lamellar structure with a relatively large extension from the grain boundaries. With increasing Al content the $Al_{11}RE_3$ phases are coarsening, and more concentrated close to the grain boundaries. The positive effect on creep is also reduced since the intermetallic particles then increasingly are embedded in eutectic Mg-Al.

At higher RE/Al ratios the ductility is reduced due to formation of a continuous skeleton of an intermetallic phase decorating the grain boundaries [1,2]. On the other hand, the creep strength will improve.

AE44 contains around 4 wt% RE. As seen from Fig. 4, the continuous network of intermetallic phases on the grain boundaries is even more predominant than in AE42. SEM-EDS analysis indicates that in AE44 the grain boundary phase formed in the die cast specimens is mainly Al_2RE, rather than $Al_{11}RE_3$ (and Al_3RE). (in Al_2RE the RE/(Al+RE) ratio is close to 0.33). This means that with an increasing RE/Al ratio, Al_2RE gradually becomes the more dominant phase relative to $Al_{11}RE_3$. It seems that at a RE/Al ratio (wt% basis) of slightly less than unity, the Al_2RE phase is the more dominant. The phase diagrams of Al-Ce, Al-La, Al-Nd and Al-Pr all indicate stable phases of Al_3RE and Al_2RE [29]. The transition from $Al_{11}RE_3$ to Al_2RE (possibly through Al_3RE) will be investigated in more detail in a future paper [30].

Phases in ACe and ACeNd alloys

In order to further study the dispersoid phases in grater detail, "synthetic" AE alloys were prepared, die cast and examined. In these alloys MM was replaced by pure Ce and Nd, and combinations. From Fig. 6 it is seen that the die cast microstructure of ACe44 is very similar to that of AE44 (Fig. 4). Looking at the phases, Fig. 7, SEM-EDS analysis indicates that in the alloy with 4% Ce mainly $Al_{11}Ce_3$ and/or Al_3Ce is found, and not Al_2Ce as might be expected from AE44.

In a second alloy Ce and Nd were added in equal amounts to produce ACeNd422. The microstructure of this alloy is also very similar to that of AE44 (Fig. 4) and ACe44 (Fig.6). The SEM-EDS analysis indicates that also in the ACeNd422 alloy mainly $Al_{11}(Ce,Nd)_3$ is found, and not $Al_2(Ce,Nd)$ (or $Al_3(Ce,Nd)$) as might be expected from the study of AE44 (Fig.3).

In a third alloy only Nd was added to produce ANd43. Also in this alloy there are strong indications that $Al_{11}Nd_3$ and possibly Al_3Nd are present. (Nd/(Al+Nd) ratios corresponding to 0.21 and 0.25, respectively). One may speculate if lanthanum (La) in the AE alloys is instrumental in the promotion of Al_2RE, or if it is the mixture of the different RE elements (Ce, La, Nd and Pr) in a certain ratio that are important. This is presently subjected to further investigations [30].

Stability of dispersoids in AE alloys

It has been reported that sharply decreasing creep strength of AE42 above 150°C is attributed to a partial decomposition of $Al_{11}RE_3$ into Al_2RE and Al, subsequently forming $Mg_{17}Al_{12}$ [26]. As a practical consequence, prolonged exposure of AE42 to elevated temperatures could have a negative influence on the creep properties. This could be highly relevant for e.g. engine blocks or other high temperature exposed automotive components in service. With a higher RE/Al ratio than that of AE42, possibly producing Al_2RE during solidification, this unwanted phase transformation may no longer occur.

To study this experimentally, die cast specimens of AE44 and AE42 were subjected to a special heat treatment (aging) at 175°C for 500 hours. Tensile creep tests of as cast and aged specimens were undertaken at 40 and 75 MPa at 175°C. The results are shown in Figs. 10 and 11, respectively. From Fig. 10 it is seen that at 175°C/40 MPa the aging of AE42 has a negative effect on the creep properties. On the other hand, AE44 seems to remain unaffected by the aging process, indicating that no phase

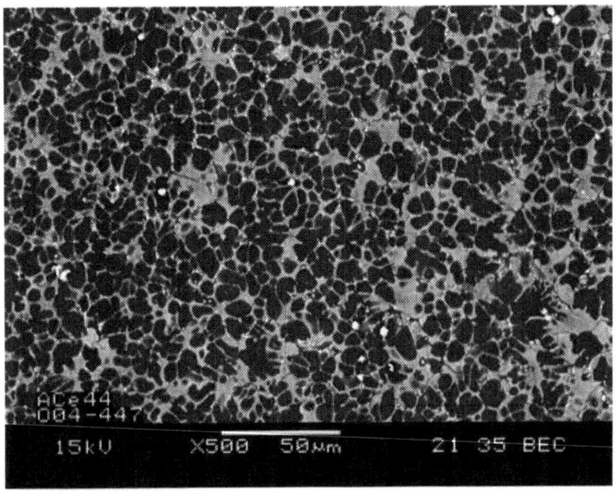

Figure 6. SEM-BEC of die cast microstructure in ACe44.

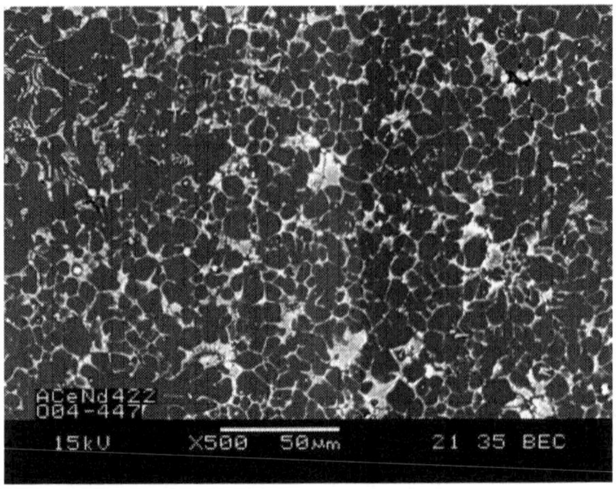

Figure 8. SEM-BEC of die cast microstructure in ACeNd422.

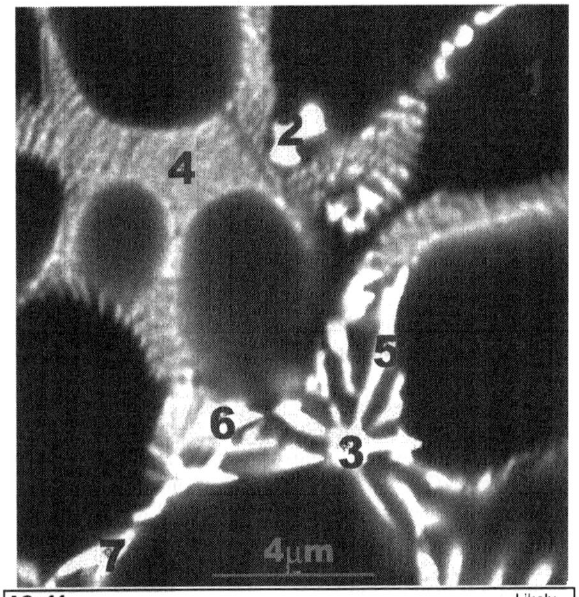

ACe44	Mg	Al	Ce	Mn	Ce/(Ce+Al)	Likely phase
1	98.25	1.45	0.12	0.10	0.08	
2	89.36	8.14	2.26	0.18	0.22	$Al_{11}Ce_3$
3	85.88	10.10	3.75	0.02	0.27	Al_3Ce
4	91.13	6.84	1.81	0.15	0.21	$Al_{11}Ce_3$
5	88.56	8.50	2.71	0.11	0.24	Al_3Ce
6	92.37	5.82	1.62	0.10	0.22	$Al_{11}Ce_3$
7	96.65	2.40	0.81	0.15	0.25	Al_3Ce

Figure 7. SEM-EDS analyses of dispersoids in ACe44.

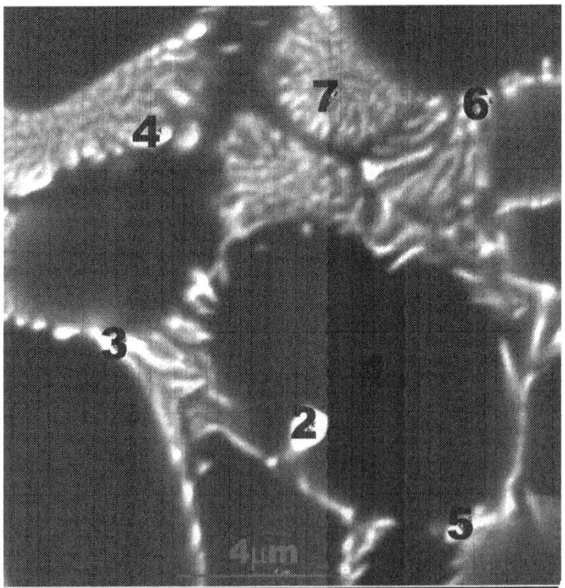

ACeNd422	Mg	Al	Ce	Nd	Mn	(Ce+Nd)/ (Ce+Nd+Al)	Likely phase
1	98.55	1.27	0.00	0.00	0.13	0.00	
2	97.80	1.79	0.15	0.16	0.06	0.15	$Al_{11}(Ce,Nd)_3$
3	88.67	8.99	1.14	0.82	0.28	0.18	$Al_{11}(Ce,Nd)_3$
4	84.83	12.36	1.27	1.09	0.36	0.16	$Al_{11}(Ce,Nd)_3$
5	91.44	6.75	0.86	0.76	0.09	0.19	$Al_{11}(Ce,Nd)_3$
6	93.82	4.67	0.75	0.56	0.11	0.22	$Al_{11}(Ce,Nd)_3$
7	91.39	6.55	0.98	0.80	0.21	0.21	$Al_{11}(Ce,Nd)_3$

Figure 9. SEM-EDS analyses of dispersoids in ACeNd422.

transformation has taken place. It is noticed that at this condition AE42 (as cast) exhibits creep properties at the level of AE44.

From Fig 11 where the load has been increased to 75 MPa, it is seen that AE44 in aged and as cast conditions remains the same. Furthermore, it is noticed that at this condition AE44 exhibits significantly better creep strain and creep rates than AE42. Tensile creep tests at 150°C/50 MPa indicate that AE42 in the as cast and aged condition behaves similarly.

Notice that the frequency of data recording during the start-up of the creep tests has not been sufficient to produce reliable data during the initial stages of the phase I creep. As a consequence, the value of the creep strain may be affected by a constant error throughout the experiments. Alternatively, the creep rates should be used as a measure of creep performance. Creep rates obtained at 100 h are listed in Table 1. These are, however, not the same as the minimum creep rate during phase II. During the lowest stress level (40 MPa) it is not likely that AE44 has fully reached the

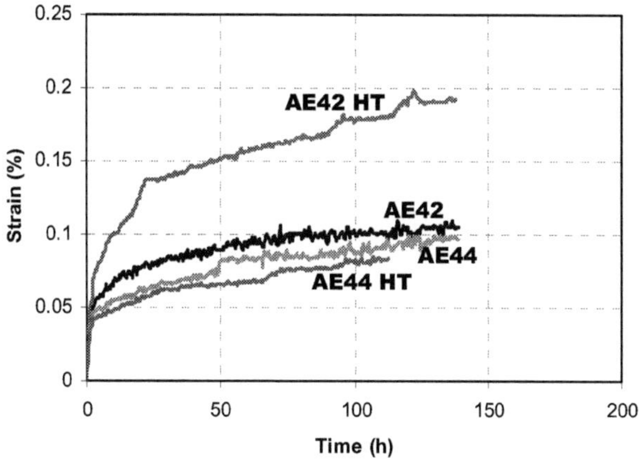

Figure 10. Creep strain as a function of time at 175°C/40 MPa; as cast and aged (HT) specimens of AE42 and AE44. The aging consisted of exposure to 175°C for 500 h.

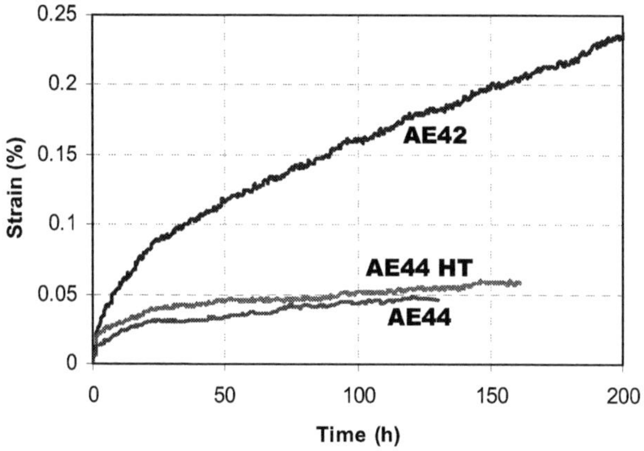

Figure 11. Creep strain as a function of time at 175°C/75 MPa; as cast and aged (HT) specimens of AE42 and AE44. The aging consisted of exposure to 175°C for 500 h.

Table 1 Creep rates at 100 h (h^{-1}).

	175°C	
	40 MPa	75 MPa
AE42	1.1E-06	8.4E-06
AE42 HT	5.8E-06	n/a
AE44	2.5E-06	1.8E-06
AE44 HT	2.3E-06	1.5E-06

steady state phase II creep. This may explain why the creep rates of AE44 at 100 h apparently are higher than that of as cast AE42.

Conclusions

This initial study indicates that at a critical RE/Al ratio the dispersoid phase stabilizing the grain boundaries of AE alloys changes from $Al_{11}RE_3$ (and Al_3RE) to Al_2RE. In the alloy AE44, recently patented by Hydro Magnesium, the Al_2RE phase is dominant. However in the "synthetic" alloys ACe44, ACeNd422 both containing 4wt% rare earth elements the dominating phases still seems to be $Al_{11}Ce_3$ (possibly with some Al_3Ce) and $Al_{11}(Ce,Nd)_3$, respectively. The reason for this is not yet understood.

Furthermore, it is verified experimentally at 175°C/40 MPa that the creep properties of AE42 are significantly better in the as cast condition than in the aged condition. Most probably this is attributed to a phase change in AE42 from the metastable $Al_{11}RE_3$ to Al_2RE with subsequent formation of the brittle $Mg_{17}Al_{12}$. The creep properties of AE44 at 175°C, 40 and 75 MPa, are not affected by aging.

Acknowledgements

The following persons are acknowledged for contributing to this work: Darryl Albright, Kari Straume Andersen, Sten Bjørneboe, Per R. Gjertsen, Haavard Gjestland, Jan Ivar Klovland, Ketil Pettersen, Lisabeth Riopelle, Stian Sannes, Vidar Sjøberg, Jan Ivar Skar, Eva Thorjussen, Jørild M. Svalestuen, Chenguo Tian, Arve Wilhelmsen, Jo Willekens.

References

1. K. Pettersen, P. Bakke, D. Albright, "Magnesium Die Casting Alloy Design", *Magnesium Technology 2002*, Ed.: H.I. Kaplan, TMS-AIME, 2002, pp. 241-246.
2. P. Bakke, H. Westengen, K. Pettersen, D. Albright, "Die cast magnesium alloys for high temperature applications", SAE Technical Paper 2003-01-0189, 2003
3. H. Westengen, P. Bakke, "Magnesium die casting alloys for use in applications exposed to elevated temperatures: Can they compete with aluminium?", *Magnesium Alloys 2003*, Eds: Y. Kojima, T. Aizawa, K. Higashi, S. Kamado, Trans Tech Publications Ltd., Switzerland, 2003, pp.35-40.
4. P. Bakke, H. Westengen, S. Sannes, D. Albright, "Powertrain Components - Opportunities for the Die Cast AE Family of Alloys", SAE Technical Paper 2004-01-0655, 2004
5. P. Bakke, H. Westengen, K. Pettersen, "Enhanced ductility and strength through RE addition to Magnesium die casting alloys", *Magnesium Technology 2003*, Ed.: H.I. Kaplan, TMS-AIME, 2003, pp.171-176
6. H. Westengen, P. Bakke " Advances in Magnesium Alloy Development", *Proc. 61th IMA Conference*, New Orleans, LA, May 2004.
7. A. Beck, *The Technology of Magnesium and its Alloys*, F.A. Hughes & Co. Ltd, London UK, 1940, pp.250-252 *(Translation from German)*
8. A. Beck, Ibid. p.286
9. E.F. Emley, *Principles of Magnesium Technology*, Pergamon Press, London UK, 1966, pp. 308-341
10. W. Unsworth, "The role of rare earth elements in the development of magnesium base alloys", Intl. J. of Materials and Product Technology, Vol.4, no.4, 1989, pp.359-378
11. K.E. Nelson, "Magnesium die casting alloys", *Proc. 6th SDCE*, Paper no. 13, 1971
12. G.S. Foerster, "Designing die casting alloys", *Proc. 7th SDCE*, Paper no. 9372, 1973
13. G.S. Foerster, "Improved Magnesium Die Casting Alloys", *8th SDCE Intl. Die Casting Exposition & Congress*, Paper no G-T75-112, 1975
14. W.E. Mercer II, "Magnesium Die Cast alloys for Elevated temperature applications", SAE Technical Paper No. 900788, 1990
15. J.S. Waltrip, "Fresh look at some old magnesium die casting alloys for elevated temperature applications", *Proc. 47th IMA Conf.*, Cannes, 1990, pp.124-129
16. T.Aune, H.Westengen, and T.Ruden, ''The effects of varying aluminium and rare-earth content on the mechanical properties of die cast magnesium alloys'', SAE Technical Paper No. 94077, 1994

17. A.P. Druschitz, E.R. Showalter, J.B. McNeill, D.L. White, "Evaluation of structural and high-temperature magnesium alloys at room temperature", *Magnesium Technology 2002*, Ed. H.I. Kaplan, TMS-AIME, 2002
18. J. Berkmortel, H. Hu, J.E. Kearns, J.E. Allison, "Die castability assessment of magnesium alloys for high temperature applications", SAE Technical Paper 2000-01-1119, 2000
19. P. Bakke, H. Westengen, " Magnesium Die Casting Alloys for Elevated Temperature Applications", *12th Magnesium Abnehmer Seminar*, Aalen, Germany, Sept. 2004
20. A.A. Nayeb-Hashemi, J.B. Clark, *Phase Diagrams of Binary Magnesium Alloys*, ASM International, Metals Park, OH, USA, 1988
21. M.S. Dargush, "The Role of Microstructure in the Creep of Die Cast Magnesium Alloys", Ph.D. thesis, University of Queensland, Australia, 1998
22. H. Gjestland, H. Westengen, "High pressure die casting – Getting the best out of magnesium", *10th Magnesium Abnehmer Seminar*, Aalen, Germany, Sept. 2002
23. H. Watanabe, H. Tsutsui, T. Mukai, K. Ishikawa, Y. Okanda, M. Kohzu, K. Higashi, "Superplastic Behaviour in Commercial Wrought Magnesium Alloys", *Magnesium Alloys 2000*, Materials Science Forum Vols. 350-351, Trans Tech. Publ., Switzerland 2000, pp.171-176
24. R.W. Evans, B. Wilshire, *Introduction to Creep*, The Institute of Materials, London, UK, 1993, pp.61-63
25. J.F. King, "Development of practical high temperature magnesium casting alloys", *Magnesium Alloys and Their Applications*, Ed.: K.U. Kainer, Wiley-VCH, Weinheim, Germany, 2000, pp. 14-22
26. W. Unsworth, "The role of rare earth elements in the development of magnesium base alloys", Int. J. of Materials and Products, Vol.4, No.4, 1989, pp.359-378
27. B. Powell, V Rezhets, M.P. Balogh, R.A. Waldo, "The relationship between microstructure and creep behavior in AE42 magnesium die casting alloy", *Magnesium Technology 2001*, Ed.: J. Hryn, TMS-AIME, 2001. pp.175-181
28. M. O. Pekguleryuz, A. Arslan Kaya, "Magnesium die casting alloys for high temperature applications", *Magnesium Technology 2004*, Ed.: A. Luo, TMS-AIME, 2004, pp.281-287
29. T.B. Massalski, Ed., *Binary Alloy Phase Diagrams*, ASM, Metals Park, OH, 1986
30. P. Bakke, H. Westengen et al., "Phase formation and microstructures of die cast AE alloys", To be published

From *Magnesium Technology 2009*, E.A. Nyberg, S.R. Agnew, N.R. Neelameggham, and M.O. Pekguleryuz, Editors

Directionally controlled precipitation on twin-boundaries in Mg-Zn-Y alloys

Julian M. Rosalie[1], Hidetoshi Somekawa[1], Alok Singh[1], Toshiji Mukai[1].

[1]NIMS (National Institute for Materials Science); Sengen 1-2-1, Tsukuba, Ibaraki, 305-0047, Japan.

Keywords: magnesium, precipitation, twinning.

Abstract

Precipitation strengthening in Mg-Zn-Y alloys occurs via the formation of a fine dispersion of β_1' (Mg_4Zn_7) rods aligned parallel to the hexagonal axis. In this study, texture and controlled deformation have been used to control the directionality and morphology of the precipitates and their effect on the ageing response. The resulting precipitates were studied via conventional and high-resolution transmission electron microscopy.

Texture was developed through extrusion at 300°C, followed by controlled room-temperature deformation to generate microstructures where yield occurred either with, or without, twinning. The ageing response was monitored using hardness testing. Both compressive and tensile deformation accelerated the ageing response. Compression parallel to the extrusion axis generated a high volume fraction of twins. Precipitates on the twin boundaries assumed a low aspect-ratio morphology different from the usual high-aspect ratio rods.

Introduction

Alloys based on the magnesium-zinc binary system display age-hardening behaviour, due to the precipitation one or more intermetallic phases. Optimal hardness in these alloys is generally associated with a microstructure consisting of a fine dispersion of rod-like β_1' precipitates aligned parallel to the hexagonal axis[1, 2].

The precipitation strengthening response of the alloy was found to be sensitive to pre-ageing deformation of cold-rolled material[3]. Cold-rolling reductions of $\leq 10\%$ accelerated the ageing process and increased the peak hardness, while in more heavily cold-worked cases this effect was balanced by annealing out of the defects, and the hardness decreased from the as-rolled value. It was noted that the β_1' precipitates were formed on both dislocations and twin boundaries and it was suggested that both defects were preferential nucleation sites[3].

Deformation in magnesium alloys is intrinsically more complex than in cubic metals, owing to the lower symmetry of the structure, and the subsequent twinning behaviour[4, 5]. Despite the importance of twinning as a deformation mode, there has been little systematic study on understanding the interaction between pre-ageing deformation, texture and precipitation strengthening in magnesium alloys.

There has been some disagreement over the composition and structure of the β_1' phase, but recent work[6, 7] supports the contention that is monoclinic, with composition Mg_4Zn_7 and lattice parameters; $a = 25.96\,\text{Å}$, $b = 5.24\,\text{Å}$, $c = 14.28\,\text{Å}$, $\beta = 102.5^o$ [8].

The β_1' structure possess a single plane which has a close match in lattice parameters with the matrix. The (010) plane of the precipitate has an interplanar spacing of 5.24Å, close to that of the Mg (0001) basal plane (5.20Å). This gives rise to the elongated rods $\parallel [0001]_{Mg}$ which form with the orientation relationship $[010]_{\beta_1'} \parallel [0001]_{Mg}$. Several variants of this orientation relationship have been reported, and involving matching of either; $(603)_{\beta_1'} \parallel (01\bar{1}0)_{Mg}$ [6], $(\bar{2}01)_{\beta_1'} \parallel (2\bar{1}\bar{1}0)_{Mg}$, $(200)_{\beta_1'} \parallel (10\bar{1}0)_{Mg}$, or $(\bar{2}02)_{\beta_1'} \parallel (\bar{1}100)_{Mg}$ [7] planes. Each of these orientation relationships corresponds to either rotation about, or reflection across, the common axis [7]. A further orientation relationship with $[001]_{\beta_1'} \sim\parallel [10\bar{1}0]_{Mg}$ and $(250)_{\beta_1'} \sim\parallel (0001)_{Mg}$ has also been reported[6].

High-resolution transmission electron microscopy revealed the presence of a complex domain substructure within the rod or lath-shaped particles[6, 7]. It was argued[7] that domains correspond to different variants of the possible orientation relationships.

The addition of yttrium to Mg-Zn alloys provides a number of benefits, including elevating the eutectic temperature and reducing the rate of overageing[9]. The precipitation sequence in Mg-Zn-Y alloys followed that of the binary Mg-Zn alloy, with the formation of a rod-like phase predominating in the intragranular regions. A number of additional intergranular phases have been identified in the ternary system, including a cubic W-phase, a hexagonal phase of composition Mg_3ZnY[10] and the icosahedral (I-phase) with composition $Zn_{50}Mg_{42}Y_8$[11]. It has been proposed that the β_1' phase is structurally related to the icosahedral phase[12].

The present work addresses the relationship between deformation and precipitation strengthening by controlled deformation of a textured Mg-Zn-Y alloy. Manipulation of the direction (tensile vs. compressive) of deformation enables the development of microstructures in which twins are either prevalent or absent. The objective of this study is to examine the role of twin boundaries on the ageing sequence of the alloy.

Experimental details

An alloy with composition Mg-2.8at.%Zn-0.4at.%Y was produced by direct chill casting from high-purity elements. Following homogenisation (15 h) at 350°C the billet was extruded with an extrusion ratio of 12:1 at 300°C. Compression (10 mm height, 5 mm diameter cylinders) and tensile (10 mm gauge length, 3 mm diameter rods) samples were machined from the extruded material. The specimens were encapsulated in argon, solutionised at 400°C for 1 h and quenched with water.

A number of specimens were deformed in either tension or compression using an Instron mechanical test apparatus up to an engineering stain of 3%, at at strain rate of $1 \times 10^{-3}\,s^{-1}$). By extrapolation from the maximum of the stress-strain curve, the overall plastic strain was calculated to be 1.7% for extended samples and 2.2% for compressed samples.

Samples were prepared for optical microscope analysis using standard methods, with acetic-picral (5 mL acetic acid, 6 g picric acid, 10 ml H_2O, 100 mL ethanol) and acetic glycol (20 mL H_2O, 60 mL ethylene glycol, 20 mL glacial acetic acid, 1 mL conc. nitric acid) etchants.

Samples were aged at 150°C in an oil bath for times ranging from 0.5 h to 2048 h. The ageing response was examined by Vickers indentation hardness testing, using a 1 kg load.

Foils for electron microscope analysis were produced by cutting slices from the aged samples, mechanically grinding to $\approx 70\,\mu m$ and thinning to perforation using a Gatan precision ion polishing system.

Diffraction contrast TEM studies were conducted on a Philips CM200 instrument. High resolution TEM was carried out using a JEOL 4000 EX instrument, operating at 400 kV and with a point resolution of 0.18 nm.

Results

Microstructural evolution

The variation of the Vickers indentation hardness with ageing time at 150°C is shown in Fig. 1. The non-deformed sample reached a peak hardness of 70 ± 2 HV after ageing for 256 h. Both compressed and extended sample displayed accelerated ageing, reaching peak hardness values of 73 ± 2 HV and 74 ± 1 HV, respectively after 48 h of ageing. The hardness increment (peak minus minimum hardness), however, was considerably lower for the sample deformed in compression ($\Delta HV = 10$) than in tension ($\Delta HV = 19$). This was also lower than the hardness increment in the non-deformed sample ($\Delta HV = 15$).

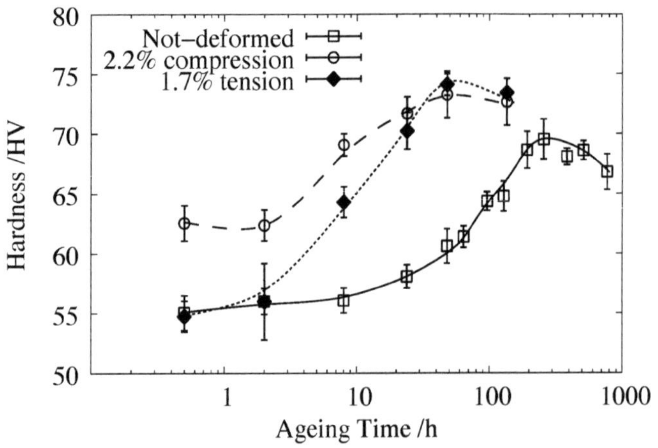

Figure 1: A graph showing ageing curves for the Mg-Zn-Y alloy at 150°C with three deformation conditions; non-deformed, compressed parallel to the extrusion direction and extended parallel to the extrusion direction. Both compressive and tensile deformation accelerated the precipitation hardening process; however the hardness increment for compression was reduced relative to the non-deformed material. Conversely, tension parallel to the extrusion direction resulted in a more extensive response to ageing heat treatment.

Grain structures

In the extruded condition the microstructure consisted of equi-axed grains with an average linear intercept grain size of $7 \pm 0.7\,\mu m$. Coarse intermetallic precipitates were observed in as discrete particles in lines running parallel to the extrusion direction.

Following solution treatment, the average linear intercept grain size had increased to $13 \pm 1\,\mu m$. Similar values were observed following compressive ($13 \pm 1\,\mu m$) and tensile ($13 \pm 2\,\mu m$) deformation.

The principle difference observed between the solution-treated, compressed and extended microstructures was the high volume fraction of twins ($26 \pm 3\%$) present in the compressed samples. Twins were essentially absent in the as-extruded and extended samples (1% volume fraction with an uncertainty of ±1%).

Location and morphology of precipitates

The twin-matrix relationship in the extruded and compressed samples was determined using selected area diffraction. With the matrix viewed along the $[2\bar{1}\bar{1}0]$ direction, the twins were also observed in a $< 2\bar{1}\bar{1}0 >$ orientation, but rotated 86° around the viewing direction. This indicated that materials was twinned across a $\{10\bar{1}2\}$ plane.

The intragranular region of the twinned samples was occupied by a population of fine rod-like particles, like those observed in the non-deformed specimens and reported by previous workers. The twin boundary regions were decorated by a series of larger blocky precipitates precipitates, as shown in Fig. 2.

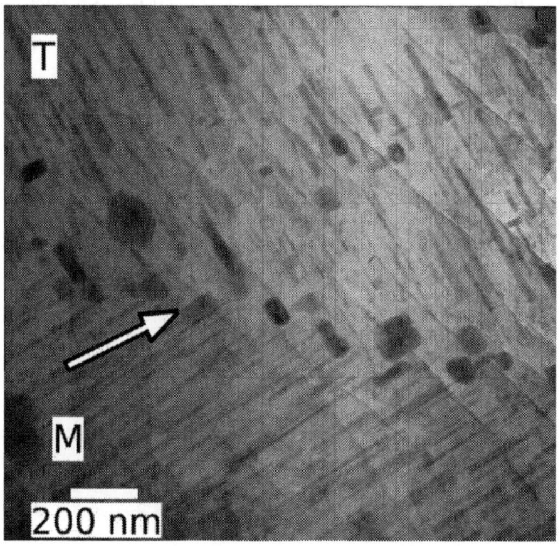

Figure 2: An electron micrograph showing a twin (T) within the Mg matrix (M) in a foil compressed and aged for 48 h. The twin boundary is decorated by larger, blocky particles unlike the rod-shaped β_1' precipitates found within both matrix and twin.

A typical example of such a precipitate can be seen in Fig. 3. In this micrograph the matrix (dark) is close to a [0001] orientation, with the twin (light) approximately 4° from a $[1\bar{1}00]$ direction. Two precipitates are observed on the twin-boundary, with is inclined to the viewing direction. Intragranular β_1' precipitates are viewed in cross-section in the matrix.

Centred dark-field (CDF) imaging was used to clarify whether all of the twin-boundary precipitates belonged to

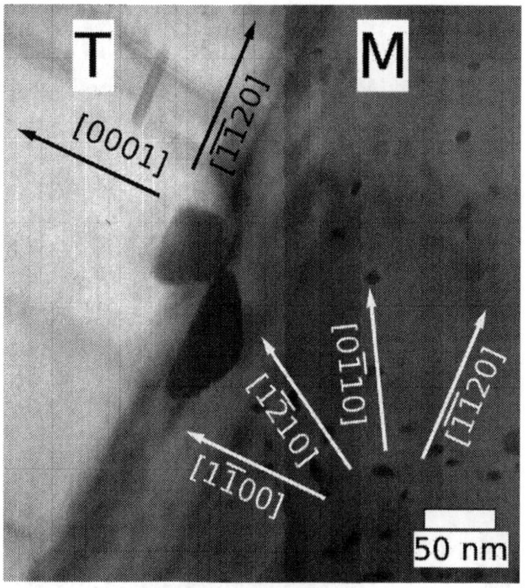

Figure 3: An electron micrograph showing two β_1' precipitates on a twin boundary. The matrix (dark) is close to a [0001] orientation, with the twin (light) approximately 4° from a $[1\bar{1}00]$ orientation. The image was obtained in material compressed and aged for 48 h.

a single subgrain (matrix or twin) or whether a mixture of orientations was present. The micrograph presented in Fig. 4a shows a twin-matrix interface decorated with irregular intragranular precipitates. In Fig. 4b the same region is shown in dark-field mode, using a matrix **g**-vector of $\frac{1}{2}\bar{1}010$. In this orientation the intragranular β_1' phase in the matrix is in a diffracting condition, while the same phase in the twin is not. Of the twin boundary preciptitates, a number are observed to diffract strongly (see points labelled A and D) with others (at B and C) do not.

Structural details of precipitates

High resolution electron microscopy (HRTEM) showed precipitates lying along the boundary between matrix and twin. The precipitate morphology was irregular and differed considerably between instances of this precipitate. A complex substructure was observed within the precipitates as can be seen in Fig 5.

Discussion

The results obtained in this study highlight the sensitivity of the precipitation hardening response of Mg-Zn-Y alloys

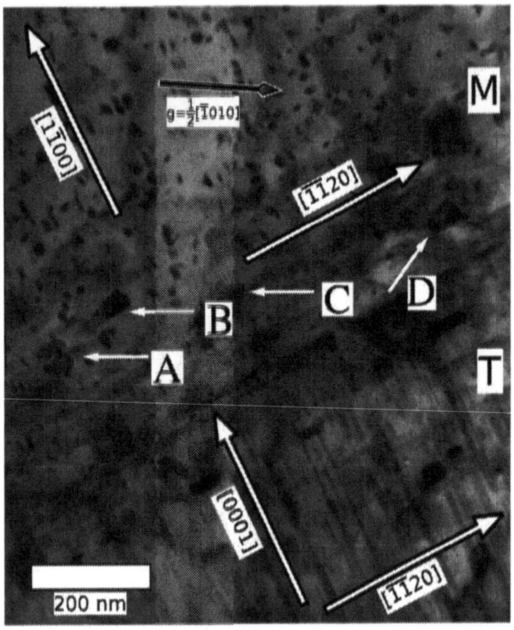

(a)

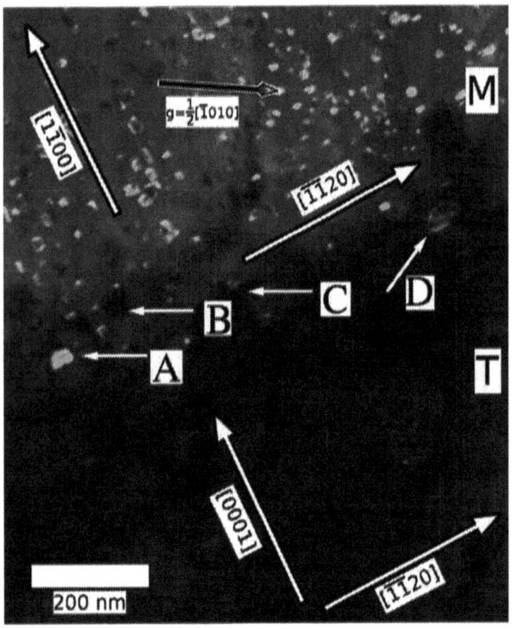

(b)

Figure 4: Bright field (a) and Centred dark-field (b) electron micrographs showing β_1' precipitates on a twin boundary in peak-aged, compressed and aged material. In (b) the precipitates labelled A and D are in a diffraction condition for a matrix **g**-vector of $\frac{1}{2}\,\bar{1}010$, like those of the β_1' phase in the matrix. Precipitates B and C show negligible intensity under these conditions. The matrix orientation was approximately parallel to [0001] with with the twin close to a $[1\bar{1}00]$ orientation,

to the presence of twins. Such twins were only observed where the alloy was compressed parallel to the extrusion direction. It should be noted that extrusion of the alloy rod will generate a tube texture, with the hexagonal axes projecting radially from the central axis of the rod. Compression normal to the extrusion direction results in c-axis *tension* which has been shown to cause extensive $\{10\bar{1}2\}$ twinning at stresses just above the yield point[5, 13]. c-axis compression is not expected to yield twins except under severe deformation[5, 14].

The hardness curves reported herein show the effect of a number of influences. Although the extended and compressed materials reach near-identical hardness, the microstructures developed are considerably different, with the irregular twin-boundary precipitates only observed in the compressed samples. The presence of twins in the compressed material will provide an addition strengthening mechanism[15]. Indeed, calculations using many-body simulations concluded that $\{10\bar{1}2\}$ twins would absorb basal plane dislocations[16].

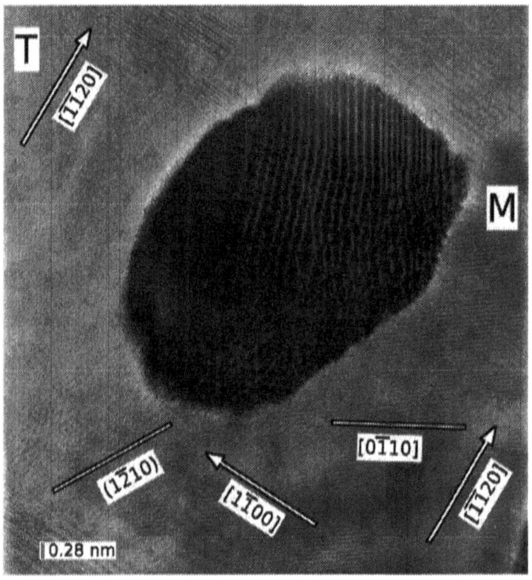

Figure 5: A high resolution transmission electron micrograph showing a single β_1' precipitate situated at a twin boundary. The matrix (M,right) is in a [0001] orientation, with the twin (T) approximately 4° from $[01\bar{1}0]$. A domain substructure is visible within the precipitate, which was larger in size than those observed in the surrounding region and irregular in shape.

The heat-treatment induced increase in the hardness of each sample can be ascribed to precipitation strengthening (with a corresponding loss of solid-solution strengthening). The grain and twin-boundary contribution to the hardness should be constant, since the ageing temperature of 150° is too low for significant grain growth and twin boundary migration.

Tensile and compressive deformation both accelerated the ageing process, with samples reaching maximum hardness after 48 h compared to 256 h for non-deformed material. Tensile deformation (*c*-axis compression) resulted in a more pronounced increase in hardness, whereas compression (*c*-axis tension) reduced the ΔHV value, even compared to non-deformed samples. To explain this observation, it is necessary to compare the microstructures of the twinned and non-twinned specimens following ageing.

Transmission electron microscopy revealed that the twin-boundaries were decorated with blocky, irregular precipitates. If these twin-boundary precipitates are approximated as ellipsoids, then in a peak aged condition, the volume of a single twin-boundary precipitate will be an order of magnitude greater than a rod-like intragranular precipitate. This suggests that the presence of these precipitates will act as a solute sink and impede the formation of the strengthening intragranular rod-like precipitates and reducing the total number of rod-like precipitates.

The twin-boundaries are thought to be resistant to dislocation glide[15] and it appears unlikely that the nucleation of a blocky, low aspect ratio precipitate will necessarily provide further strengthening.

Centred dark-field imaging showed twin-boundary precipitates belonging to either matrix or twin, in approximately equal proportions (See Fig 4). It was thought that this indicates that the twin-boundary precipitates were oriented with their usual *b*-axis aligned with the *c*-axis of either matrix or twin.

Fourier transforms of different regions of the high-resolution TEM images of the precipitates suggested that the orientation of each region of the precipitate differed from the surrounding regions. This shows the presence of the domain structures reported previously[6, 7]. It was noted that the domains substructure was quite regular and individual domains shared sharply defined interfaces with one another. The external morphology of all such precipitates observed in cross-section, however, was irregular and asymmetric, as shown in Fig. 5.

The effect of greater or lesser amounts of deformation on such a textured sample has yet to be established. The amount of deformation applied in this study is low compared to both previous work in magnesium alloys[3]. Although further *c*-axis tension would be expected to increase the number of twin present, the major effect would probably be the broadening of the existing twins. This would be able to occur with relatively little change in the twin-boundary surface area and hence may not result in significantly greater amounts of twin-boundary precipitation.

Increased *c*-axis compression should give rise to high dislocation line densities, with a correspondingly greater impact on the precipitation behaviour. It is therefore expected that the precipitation behaviour will be more sensitive to the level of *c*-axis compression than *c*-axis tension.

The precipitation behaviour in these strongly textured alloys was greatly dependent on the direction in which deformation was applied. In a randomly-oriented sample the potential for twin formation will depend on the orientation of each individual grain with respect to the deformation. It is probable that pre-ageing deformation of such a material would result in a greater degree of variability in mechanical properties, since the precipitation behaviour observed in each grain would depend on the presence of twins, and thus on the orientation of the grain with respect to the applied strain. This highlights the importance of understanding the texture of precipitation strengthened magnesium alloys in order to produce consistent, high-quality products.

Conclusions

The precipitation of the β_1' monoclinic phase has been examined on twin boundaries in a Mg-Zn-Y alloy. The twin boundary precipitates were characterised by an irregular, blocky morphology unlike the rod-like precipitates of the same phase formed within the intragranular regions. The precipitation morphology could be controlled via the direction of pre-ageing deformation applied. The larger, irregular precipitates were only observed for samples deformed in *c*-axis tension. This dependence on the type of pre-ageing deformation applied could give rise to additional variability in the mechanical properties of products produced from randomly oriented, or weakly textured materials.

Acknowledgments JMR gratefully acknowledges the support of the Japan Society for the Promotion of Science (JSPS).

References

[1] G. Mima and Y. Tanaka. The aging characteristics of magnesium-4wt% zinc alloy. *Transactions of the Japan Institute of Metals*, 12(2):71–75, 1971.

[2] L. Sturkey and J. B. Clark. Mechanism of age-hardening in magnesium-zinc alloys. *Journal of the Institute of Materials*, 88:177–181, 1959–60.

[3] G. Mima and Y. Tanaka. The main factors affecting the aging of magnesium-zinc alloys. *Transactions of the Japan Institute of Metals*, 12(2):76–81, 1971.

[4] M. H. Yoo. Slip, twinning and fracture in hexagonal close-packed metals. *Metallurgical and Materials Transactions A*, 12A:409–418, 1981.

[5] E. W. Kelley and W. F. Hosford, Jr. Plane-strain compression of magnesium and magnesium alloy crystals. *Transactions on the American institute of mining and metallurgical engineers*, 242, 1968.

[6] X. Gao and J. F. Nie. Characterization of strengthening precipitate phases in a Mg-Zn alloy. *Scripta Materialia*, 56:645–648, 2007.

[7] A. Singh and A. P. Tsai. Structural characteristics of β_1' precipitates in Mg-Zn-based alloys. *Scripta Materialia*, 57:941–944, 2007.

[8] Ya. P Yarmolyuk, P. I. Kripyakevich, and É. V. Mel'nik. Crystal structure of the comound Mg_4Mn_7. *Krystallografiya*, 20:329–331, 1975.

[9] L. Y. Wei, G. L. Dunlop, and H. Westengen. Precipitation hardening of Mg-Zn and Mg-Zn-RE alloys. *Metallurgical and Materials Transactions A*, 26A(7):1705–1716, 1995.

[10] D. W. Deng, K. H. Kuo, Z. P. Luo, D. J. Miller, M. J. Kramer, and K. W. Dennis. Crystal structure of the hexagonal Zn_3MgY phase. *Journal of alloys and compounds*, 373(1-2):156–160, 2004.

[11] A. P. Tsai, A. Niikura, A. Inoue, T. Masumoto, Y. Nishida, K. Tsuda, and M. Tanaka. Highly ordered structure of icosahedral quasicrystals in Zn-Mg-RE (RE= rare earth metals) systems. *Philosophical Magazine Letters*, 70(3):169–175, 1994.

[12] Q. B. Yang and K. H. Kuo. A new description of pentagonal Frank-Kasper phases and a possible structure model of the icosahedral quasicrystal. *Acta Crystallographica A*, A43:787–795, 1987.

[13] M. R. Barnett. Twinning and the ductility of magnesium alloys Part I: "Tension" twins. *Materials Science and Engineering A*, 464(1-2):1–7, aug 2007.

[14] M. R. Barnett. Twinning and the ductility of magnesium alloys Part II. "Contraction" twins. *Materials Science and Engineering A*, 464(1-2):8–16, aug 2007.

[15] C. H. Cáceres, P. Lukác, and A Blake. Strain hardening due to $\{10\bar{1}2\}$ twinning in pure magnesium. *Philosophical Magazine*, 88:991–1003, 2008.

[16] A. Serra, D. J. Bacon, and R. C. Pond. Twins as barriers to basal slip in hexagonal-clos-packed metals. *Metallurgical and Materials Transactions A*, 33A:809–812, 2002.

APPLICABILITY OF Mg -Zn-(Y, Gd) ALLOYS FOR ENGINE PISTONS

Kazutaka Okamoto[1], Masato Sasaki[2], Norikazu Takahashi[2], Qudong Wang[3], Yan Gao[3], Dongdi Yin[3], Changjiang Chen[3]
[1]Hitachi, Ltd.; Hitachi, Ibaraki 319-1292, Japan
[2]Hitachi Automotive Systems, Ltd.; Atsugi, Kanagawa 243-8510, Japan
[3]Shanghai Jiao Tong Univ.; Shanghai 200240, China

Keywords: Gravity casting, Heat treatment, Tensile strength, Creep strain, Fatigue strength

Abstract

Commercial magnesium alloys have a great potential for structural applications in automotive due to their significant weight saving. However, they have poor creep resistance at temperature over 125°C, thus making them inadequate for power train applications such as engine pistons, which are operated at temperature up to 300°C. Recently, creep resistant magnesium alloys with rare-earth elements and Zn have been developed, hence the applicability of Mg-Zn-(Y, Gd) alloys for engine pistons was investigated in this paper. Gravity casting was performed with Mg-Zn-(Y, Gd)-Zr alloy, followed by T6 treatment. Effects of the amount of alloying elements on the mechanical properties of tensile strength and creep strain were evaluated. Nominal composition of Mg-2Zn-11Y-5Gd-0.5Zr was selected for the actual piston cast trial and its high cycle fatigue test was conducted comparing to the current aluminum cast alloy of A336 (JIS AC8A) for pistons. At room temperature, the fatigue strength is 27% lower than A336, while it is 35% higher at 300°C. It is suggested that Mg-2Zn-11Y-5Gd-0.5Zr alloy shows attractive high temperature mechanical properties higher than A336, hence it is promising as a candidate material for the engine piston application.

Introduction

Magnesium alloys have a great potential for structural applications in automotive and aerospace industries due to their significant weight savings, thus improving fuel economy and lessening environmental impact. The most significant magnesium applications are in castings, such as instrument panel, transfer cases, valve/cam covers, various housings and brackets, and steering components in automobiles, with commercial magnesium alloys of AZ91, AM50 and AM60. These alloys offer an excellent combination of mechanical properties, corrosion resistance and castability.

On the other hand, power train components are made of heavy parts, hence the specific weight of magnesium, which is 2/3 of aluminum, is very attractive for weight reduction of transmission case or engine block. Although aluminum engine piston is not so weighty, generally about 0.3kg, it is reciprocated at higher velocity over 10m/s, so that the reciprocating mass can be reduced. Furthermore, weight reduction of circumference parts such as connecting rod, less NVH (Noise, Vibration and Harshness) issues, higher performance of engine are expected. However, these components are operated at higher temperatures above 175°C, engine block up to 200°C, and engine pistons up to 300°C, thus making commercial magnesium alloys inadequate for major power-train applications because of their poor creep resistance at temperatures above 125°C.

The requirements for high temperature creep resistance and weight savings have made it imperative to develop new magnesium alloys to decrease use of conventional aluminum and iron castings in vehicles. Both diffusion controlled dislocation climb and grain boundary sliding/shearing have been reported as creep mechanisms in magnesium alloys, depending on the alloy system, microstructure, and stress and temperature regimes. In magnesium alloys, aluminum is often added to obtain a good combination of strength, ductility and castabilty. However, the poor thermal stability of $Mg_{17}Al_{12}$ intermetallics (β phase, with a eutectic temperature of 437°C) and its discontinuous precipitation can result in substantial grain boundary sliding/shearing at elevated temperatures. Moreover, the accelerated diffusion of aluminum solute in magnesium matrix and the self-diffusion of magnesium at elevated temperatures contribute to creep deformation in Mg-Al based alloys. Based on the above arguments and alloy strengthening mechanisms, possible approaches to improve creep resistance in magnesium alloys include solution strengthening, precipitation strengthening and dispersoid strengthening. Most of magnesium alloy systems were developed using these approaches to obtain good elevated/high-temperature creep resistance, namely Mg-Al-Sr (AJ), Mg-Al-Ca(-Sr) (AX(J)), Mg-Al-RE (AE), Mg-Y-Nd (WE), etc. Among those alloys, the enhancement of creep resistance by rare earth (RE) elements is particularly common. RE has a quite large solubility, resulting in solid solution strengthening, decreasing stacking fault energy, thus splitting basal dislocation into partials and increasing resistance of dislocation glide and climb [1-3]. Furthermore, Zn addition significantly enhances the age hardening response and the creep resistance with uniform and dense distribution of basal precipitate plates [4] and/or long period stacking order (LPSO) [5-13], namely Mg-Zn-RE alloy.

In this research, high temperature properties of Mg-Zn-(Y, Gd) cast alloy were evaluated so as to investigate the applicability of the cast to engine pistons, considering those of the current aluminum cast alloy of A336 (JIS AC8A). Conventional gravity casting was carried out, followed by T6 heat treatment, and both creep strain and high cycle fatigue tests as well as tensile test at room and elevated temperature were conducted. Our development goal was set as summarized in Table 1.

Table 1 Development goal for the mechanical properties of Mg-Zn-(Y, Gd) cast alloy

Property and Test Conditions	Goal
Ultimate Tensile Strength	
RT	> 230MPa
300°C	> 150MPa
Creep Strain	
250°C , 80MPa for 20hrs	< 0.30%
300°C , 50MPa for 20hrs	< 0.40%
Fatigue Strength	
300°C, 10^7cycles	> 65MPa

Experimental

Materials to be Evaluated

The chemical compositions of the magnesium alloys are summarized in Table 2. The actual chemical composition of the alloys were evaluated by an inductively couples plasma (ICP).

All these alloy master ingots were prepared with high purity Mg (99.95%) and Zn (99.95%), Mg-25mass%Gd, Mg-25mass%Y and Mg-30mass%Zr alloys. After smelting using an electric resistance furnace under the mixed cover gas of CO_2 and SF_6, the molten metal was pored into a preheated (200°C) cast iron block mold with a wall thickness of 20mm, see Fig.1.

Table 2 Nominal and actual chemical composition of Mg-Zn-(Y, Gd)-Zr cast alloy (mass %)

Nominal comp.	Actual comp.				
	Zn	Y	Gd	Zr	Mg
Mg-10Y-5Gd-0.5Zr	-	8.9	5.4	0.4	bal.
Mg-2Zn-**10Y-5Gd**-0.5Zr	2.1	10.5	4.6	0.4	bal.
Mg-2Zn-**5Y-15Gd**-0.5Zr	1.9	3.6	12.5	0.4	bal.
Mg-2Zn-**5Y-5Gd**-0.5Zr	2.3	4.1	5.0	0.4	bal.
Mg-2Zn-**11Y**-5Gd-0.5Zr	2.2	10.7	5.6	0.4	bal.
Mg-2Zn-**12Y**-5Gd-0.5Zr	2.0	11.9	5.6	0.4	bal.
Mg-2Zn-**14Y**-5Gd-0.5Zr	2.2	14.4	4.9	0.5	bal.
Mg-**0.5Zn**-10Y-5Gd-0.5Zr	0.6	7.2	4.1	0.4	bal.
Mg-**2Zn**-10Y-5Gd-0.5Zr	2.1	10.5	4.6	0.4	bal.
Mg-**3Zn**-10Y-5Gd-0.5Zr	3.5	9.1	4.8	0.5	bal

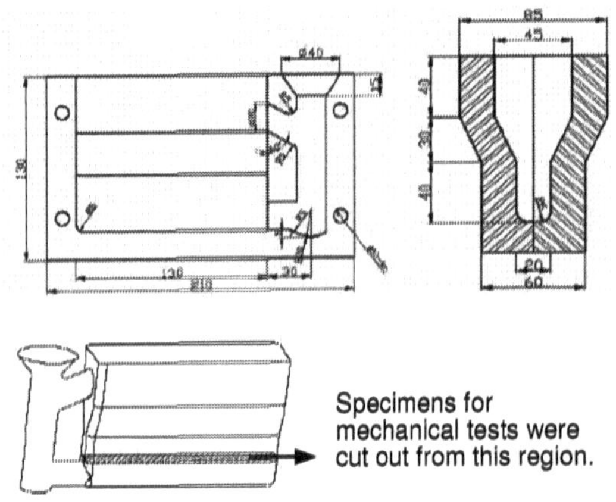

Fig. 1 Schematic drawings of cast iron block mold

Microstructure and Hardness

Microstructure observation was carried out using an optical microscope (OM) and a scanning electron microscope equipped with an energy dispersive X-ray spectrometer (SEM-EDS). Specimens were cut and polished, then etched with 5mass% HNO_3 ethanol. Differential scanning calorimetry (DSC) as well as Vickers hardness tests with an indentation load of 49N were also performed in order to decide the heat treatment condition.

Mechanical Properties

Mechanical properties were evaluated with the specimens shown in Fig. 2, cut from the cast magnesium alloy. Tensile test was performed at both room and elevated temperatures with an initial strain rate of 0.5mm/min. Creep strain was evaluated at 200°C, 250°C and 300°C with an initial load of 120MPa, 80MPa and 50MPa, respectively. Furthermore, high cycle tensile-tensile fatigue test was carried out at room temperature, 250°C and 300°C, up to 10^7 cycles. The stress ratio and frequency were set to R=0 and 30Hz, respectively.

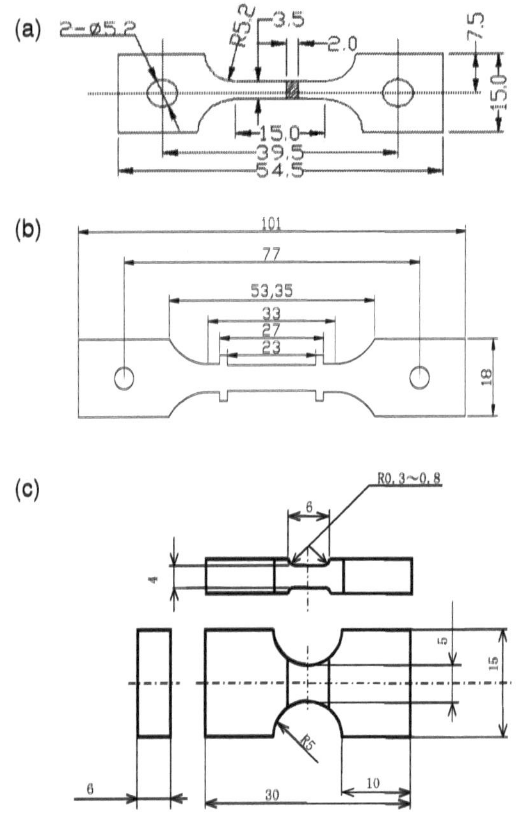

Fig. 2 Schematic drawings of mechanical test specimens for (a) tensile test, (b) creep test and (c) fatigue test

Results and Discussions

Heat Treatment

DSC was conducted for both Mg-10Y-5Gd-0.5Zr and Mg-2Zn-10Y-5Gd-0.5Zr alloys as shown in Fig. 3, and the peak temperatures were measured at 567°C and 540°C, respectively. In order to apply a same heat treatment condition to both alloys, temperature for solution heat treatment was set to 535°C, just below the eutectic temperature of Mg-2Zn-10Y-5Gd-0.5Zr alloy.

Figure 4 shows as-cast and as-T4 microstructure of both alloys. According to the previous research [14], by adding Zn, the resultant microstructure is remarkably changed with long period stacking order formation (seen as a lamellar like morphology in Fig. 4). After 16hrs at 535°C, the eutectic intermetallics of Mg-(Y, Gd) were almost solutionized in α-Mg matrix of Mg-10Y-5Gd-0.5Zr alloy, while the remnant of the eutectic intermetallics was still observed in Mg-2Zn-10Y-5Gd-0.5Zr alloy so that Zn addition decreases the solubility of Y and Gd into α-Mg matrix. The average grain size was approximately 100μm.

After the solution heat treatment, aging treatment was applied at various temperature of 225, 250 and 300°C. Figure 5 shows the age hardening response of Mg-10Y-5Gd-0.5Zr alloy. The peak hardness decreased with increasing aging temperature. At 225°C, the hardness increased rapidly by holding 1hr, and the peak hardness of around135Hv was obtained at about 16-24hrs. After reaching the peak hardness, a plateau region was observed up to 64hrs, then the hardness gradually decreased.

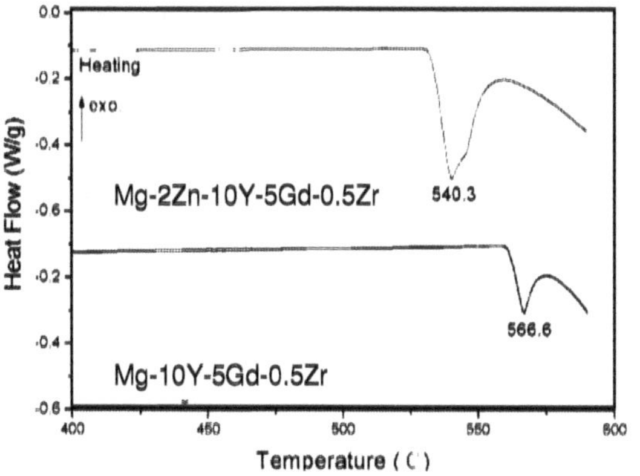

Fig.3 DCS curves for the magnesium alloys

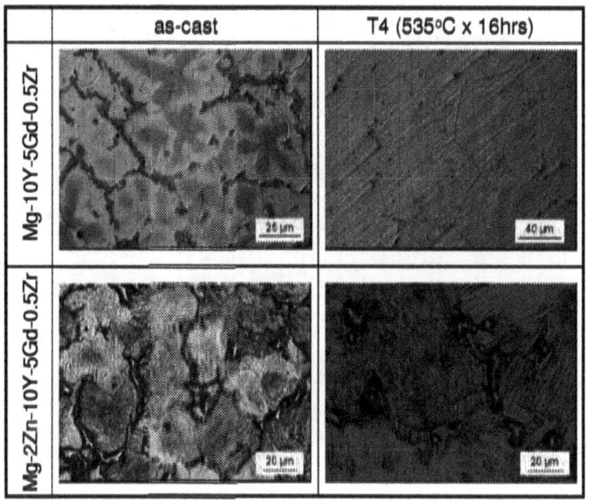

Fig.4 Microstructure of the magnesium alloys

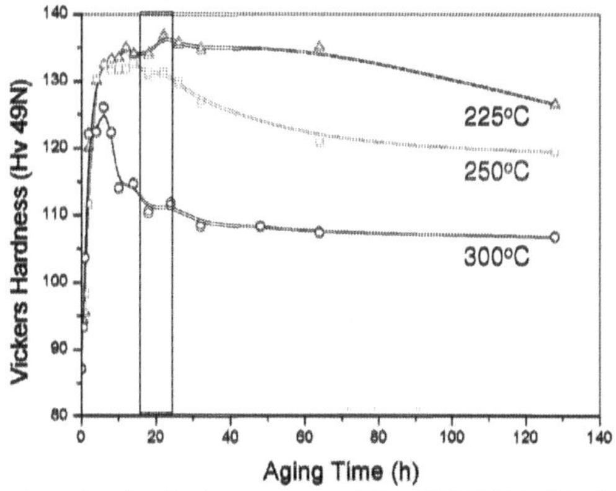

Fig.5 Age hardening response of Mg-10Y-5Gd-0.5Zr alloy

Tensile Strength

In order to investigate the effects of the total amount of RE elements, effects of Y and Gd content on the tensile properties were evaluated. Table 3 summarizes the tensile properties of Mg-2Zn-(Y, Gd)-0.5Zr alloys. It is clearly shown that both 0.2% proof stress ($\sigma_{0.2}$) and ultimate tensile strength (UTS) increased with increasing the total amount of (Y, Gd) while the elongation (ε) remarkably decreased down to less than 1%. However, the melt loss is quite significant for Mg-2Zn-5Y-15Gd-0.5Zr alloy, so that the amount of Gd was selected to 5mass %. Figure 6 shows stress-strain curves of Mg-2Zn-10Y-5Gd-0.5Zr alloy at 300°C and 350°C, respectively. Even at 300°C, UTS is higher than 200MPa, which meets our development goal. However, UTS remarkably decreased over 300°C with significant increase in the elongation up to 28%.

Table 3 Effect of (Y, Gd) on the tensile properties

	$\sigma_{0.2}$ / MPa	UTS / MPa	ε (%)
Mg-2Zn-**10Y**-**5Gd**-0.5Zr	248	273	0.6
Mg-2Zn-**5Y**-**15Gd**-0.5Zr	236	305	0.9
Mg-2Zn-**5Y**-**5Gd**-0.5Zr	126	251	12

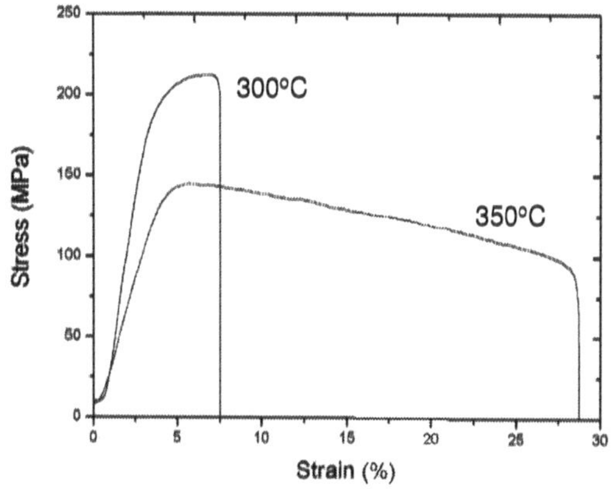

Fig. 6 Stress-strain curves of Mg-2Zn-10Y-5Gd-0.5Zr alloy at elevated temperatures

Creep Strain

In order to determine the composition range, effects of Zn and Y content on the creep properties were investigated.

It was found that the addition of Zn to Mg-10Y-5Gd-0.5Zr alloy remarkably improves the creep resistance, as shown in Table 4, summarizing the creep strain and secondary creep rate at 250°C, 80MPa. Furthermore, Zn content of 2mass % is most affective for the improvement of creep resistance.

Figure 7 shows creep curves of the magnesium alloys at 300°C, 50MPa. Amount of Y, a primary alloying element for these magnesium alloys, has a strong influence on the creep strain. Creep stain is minimized and our goal was achieved for the alloys with Y content of 11-12mass %. However higher Y was demonstrated to be not beneficial for the improvement of creep resistance and the material cost will increase. According to Mg-Y binary diagram, the maximum solubility of Y into α-Mg matrix at the eutectic temperature is about 12.5mass %, hence over high Y addition results in the existence of large amount of eutectic phase, which can not be dissolved into the matrix completely even after the solution heat treatment. These thermal unstable intemetallic compounds may deteriorate the creep resistance at α-Mg grain boundaries.

Figure 8 shows the microstructure evolution of Mg-2Zn-11Y-5Gd-0.5Zr alloy during the creep test at 300°C, 50MPa for 0-100hrs. It can be seen that the lamellar like morphology is developed as increasing the testing time and maybe the structure causes higher creep resistance.

Table 4 Effect of Zn on the creep properties at 250°C, 80MPa

	Creep strain for 20hrs (%)	Secondary creep rate / s^{-1}
Mg-10Y-5Gd-0.5Zr	0.36	44.6×10^{-9}
Mg-**0.5Zn**-10Y-5Gd-0.5Zr	0.16	6.19×10^{-9}
Mg-**2Zn**-10Y-5Gd-0.5Zr	0.12	3.89×10^{-9}
Mg-**3Zn**-10Y-5Gd-0.5Zr	0.28	11.8×10^{-9}

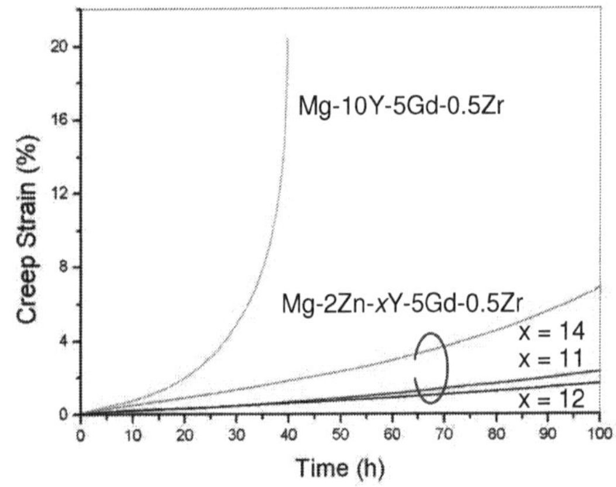

Fig. 7 Creep curves of the magnesium alloys at 300°C, 50MPa

Fig.8 Microstructure evolution in Mg-2Zn-11Y-5Gd-0.5Zr alloy during the creep test at 300°C for (a) 0hr, (b) 20hrs, (c) 50hrs and (d) 100hrs

Fatigue Strength

Based on the mechanical properties of various Mg-Zn-(Y, Gd)-Zr alloys above mentioned, Mg-2Zn-11Y-5Gd-0.5Zr alloy was selected for further investigations. Table 5 summarizes the high cycle tensile-tensile fatigue test at room temperature, 250°C and 300°C, comparing to those of A336-T6. Although the fatigue strength of the magnesium alloy at room temperature and 250°C is 30-40% lower than A336, it is about 85% of A336 at 300°C.

Microstructure evolution was observed after the tests at 250°C and 300°C, within the α-Mg grain interior near the boundaries as described in Fig. 9, which may strengthen the grain boundaries and prevent its deformation.

Figure 10 shows the fracture surface of Mg-2Zn-11Y-5Gd-0.5Zr alloy. It is observed that cleavage like cracks are predominant and facet size of the cleavage plane is almost the same as the grain size. Furthermore, fatigue crack propagates perpendicular to the small steps, so that it is considered that the LPSO along with the basal plane of hexagonal α-Mg results in the cleavage cracks.

Table 5 High cycle fatigue strength of Mg-2Zn-11Y-5Gd-0.5Zr and A336 aluminum alloys

	RT	250°C	300°C
Mg-2Zn-11Y-5Gd-0.5Zr	70MPa	59MPa	53MPa
A336	103MPa	96MPa	63MPa

Fig. 9 Microstructure of Mg-2Zn-11Y-5Gd-0.5Zr alloy after testing for 9.5×10^6 cycles at 55MPa, 300°C

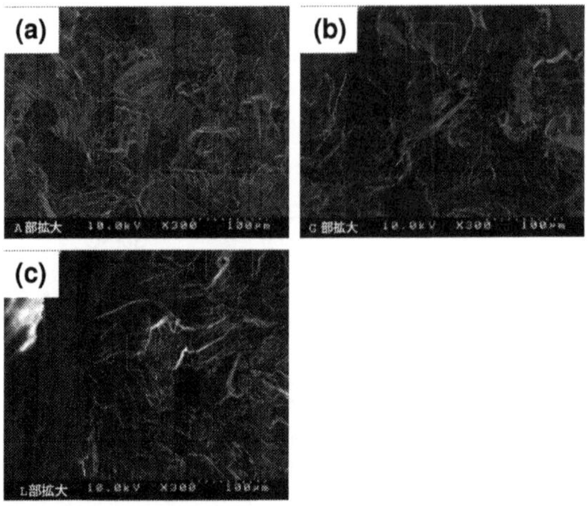

Fig. 10 Fracture surface of fatigue specimens of Mg-2Zn-11Y-5Gd-0.5Zr alloy tested at (a) RT, (b) 250°C and (c) 300°C

Piston Cast Trial

Gravity casting was performed using a piston mold illustrated in Fig. 11. Specimens were cut out from the crown surface for metallurgical and mechanical investigations. The actual chemical composition is analyzed to Mg-2.1Zn-10.9Y-4.6Gd-0.4Zr (mass %). T6 of 535°Cx16-20hrs and 225°Cx24hrs was applied.

Figure 12 show both as-cast and as-T6 microstructures. In case of as-cast, Gd is enriched in α-Mg matrix, whose average grain size is about 100μm, while it is grown to about 150μm via T6 treatment. Zn and Y are detected in the eutectic intermetallics at α-Mg grain boundaries. Furthermore, lamellar like morphology is observed along the α-Mg grain boundaries, which suggests LPSO is formed originated from Y element [6-9]. Yamasaki et al. [9] reported that RE can be categorized into two types; type-I additives such as Y form LPSO during solidification, and type-II additives such as Gd form during high temperature exposure. In case of as-T6, it is suggested that Gd plays an important role forming LPSO. Furthermore, Yamada [11, 12] and Honma [13] reported that 0.3-1.0at. % Zn containing Mg-Gd-Y-Zr alloy shows room temperature tensile strength over 400MPa with 14H-LPSO at α-Mg grain boundaries and metastable β' phase, via solution heat treatment at 500°C followed by aging treatment at 225°C. Therefore, it is considered that the similar type of microstructure evolution can be expected for this alloy.

Figure 13 shows the tensile properties of Mg-2Zn-11Y-5Gd-0.5Zr alloy. It is observed that UTS shows an inverse temperature dependence up to 200°C, and it decreases over the temperature. As mentioned above, the aging treatment at 225°C might form metastable β' phase, so that further investigations of microstructure evolution and phase stability of the resultant microstructure are required. Table 6 summarized the creep properties of Mg-2Zn-11Y-5Gd-0.5Zr alloy at elevated temperatures. It can be said that the creep resistance of the piston specimen is almost comparable to that of block specimen.

Table 6 Creep properties of Mg-2Zn-11Y-5Gd-0.5Zr alloy at elevated temperatures

Temperature / °C	Creep strain for 20hrs (%)	Secondary creep rate / s^{-1}
200	0.17	0.97×10^{-9}
250	0.32	3.65×10^{-9}
300	0.33	4.72×10^{-9}

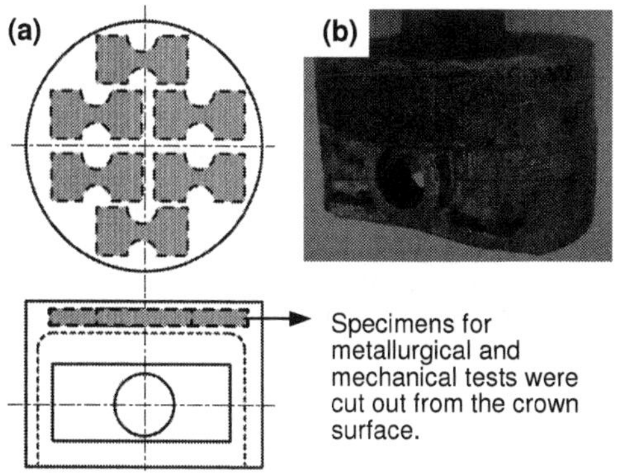

Fig.11 Actual piston cast trial
(a) schematic drawing of piston, (b) as-cast piston

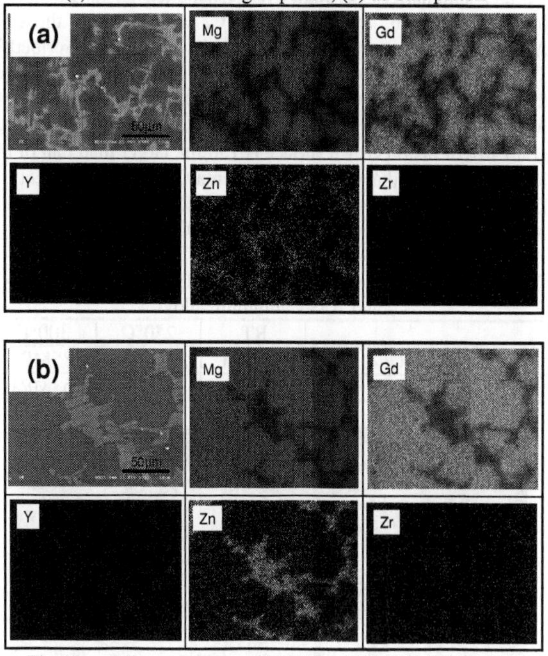

Fig.12 Microstructure of Mg-2Zn-11Y-5Gd-0.5Zr alloy
(a) as-cast and (b) as-T6

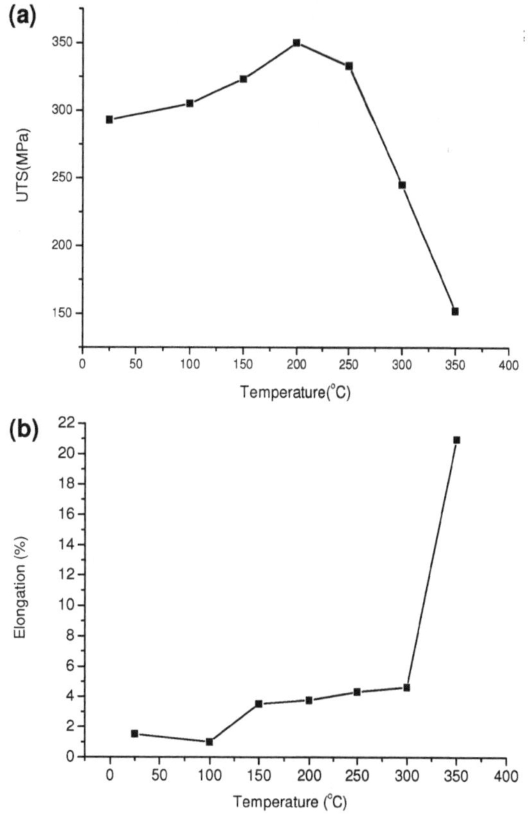

Fig.13 Tensile properties of Mg-2Zn-11Y-5Gd-0.5Zr alloy
(a) UTS and (b) elongation

Table 7 summarizes the high cycle tensile-tensile fatigue test at room temperature, 250°C and 300°C, comparing to those of A336-T6. Similar to the results of the block samples described in Table 5, the fatigue strength at room temperature is 27% lower than A336, while almost the same or even 35% higher strength were obtained at 250°C or 300°C. It is thought that the average grain size of piston crown surface is 150μm, larger than that of the block sample, so that the cooling rate may significantly affect the LPSO formation. Furthermore, microstructure evolution was also observed during the tests at 250°C and 300°C as seen in Fig. 14.

Table 7 High cycle fatigue strength of Mg-2Zn-11Y-5Gd-0.5Zr and A336 aluminum alloys

	RT	250°C	300°C
Mg-2Zn-11Y-5Gd-0.5Zr	75MPa	90MPa	85MPa
A336	103MPa	96MPa	63MPa

Fig. 14 Microstructure of Mg-2Zn-11Y-5Gd-0.5Zr alloy after testing for $3.3x10^6$ cycles at 90MPa, 300°C

Conclusions

Mechanical properties of Mg-Zn-(Y, Gd) alloys were evaluated by using gravity cast and T6 treated specimens and its applicability for engine pistons was investigated. The results obtained are summarized as follows:

- Effect of the amount of the alloying elements, Zn and Y of Mg-*x*Zn-*y*Y-5Gd-0.5Zr (mass %) alloy on the creep property at elevated temperature were studied. It is suggested that the addition of 2mass %Zn and 11mass %Y minimizes the creep stain at elevated temperatures.
- Mg-2Zn-11Y-5Gd-0.5Zr alloy was selected for high cycle fatigue test. Fatigue strength at room temperature and 250°C is 30-40% lower than A336, while it is about 85% of A336 at 300°C.
- Actual pistons of Mg-2.1Zn-10.9Y-4.6Gd-0.4Zr (mass %) were gravity cast. In the comparison to A336, almost the same or even 35% higher fatigue strength were obtained at 250°C or 300°C.

Mg-2Zn-11Y-5Gd-0.5Zr cast alloy exhibits excellent high temperature mechanical properties such as tensile strength, creep resistance and fatigue strength, equivalent or even higher in the comparison with the current aluminum alloy for piston. Therefore, it can be concluded that this magnesium alloy is very promising as a candidate material for piston applications, since specific strength design for this alloy piston is not required. However, there are still manufacturing related issues remained, such as optimization of casting design, heat treatment schedule for the stability of microstructure at elevated temperatures, wear resistant surface treatment strategy, and so on. Therefore further investigations are on going via piston trial manufacturing to pursue technical possibility of Mg-Zn-(Y, Gd) alloys for engine pistons.

References

1. I.A.Anyanwu, S.Kamado and Y.Kojima: "Aging characteristics and high temperature tensile properties of Mg-Gd-Y-Zr alloys", Mater. Trans. 42 (2001), pp. 1206-1211
2. I.A.Anyanwu, S.Kamado and Y.Kojima: "Creep properties of Mg-Gd-Y-Zr alloys", Mater. Trans. 42 (2001), pp. 1212-1218
3. Y.Gao, Q.Wang, J.Gu, Y.Zhao, Y.Tong and J.Kaneda: "Effects of heat treatments on microstructure and mechanical properties of Mg-15Gd-5Y-0.5Zr alloy", J. of Rare Earth 26 (2008), pp. 298-302
4. J.F.Nie, X. Gao and S.M. Zhu: "Enhanced age hardening response and creep resistance of Mg-Gd alloys containing Zn", Scripta Mater. 53 (2005), pp. 1049–1053
5. Y.Kawamura, K. Hayashi, A. Inoue and T. Masumoto: "Rapidly solidified powder metallurgy Mg97Zn1Y2 alloys with excellent tensile yield strength above 600MPa", Mater. Trans. 42 (2001), pp. 1172-1176
6. E.Abe, Y.Kawamura, K.Hayashi and A.Inoue: ""Long-period ordered structure in a high-strenght manocrystalline Mg-1at%Zn-2at%Y alloy studied by atomic-resolution Z-contrast STEM", Acta Mater. 50 (2002), pp. 3845-3857
7. T.Itoi, T.Seimiya, Y.Kawamura and M.Hirohashi: "Long period stacking structures observed in Mg97Zn1Y2 alloy", Scr. Mater. 51 (2004), pp. 107-111
8. M.Yamasaki, T.Anan, S.Yoshimoto and Y.Kawamura: "Mechanical properties of warm-extruded Mg-Zn-Gd alloy with coherent 14H long periodic stacking ordered structure precipitate", Scr. Mater. 53 (2005), pp. 799-803
9. M.Yamasaki, M.Sasaki, M.Nishijima, K.Hiraga and Y.Kawamura: "Formation of 14H long period stacking ordered structure and profuse stacking faults in Mg-Zn-Gd alloys during isothermal aging at high temperature", Acta Mater. 55 (2007), pp. 6798-6805
10. Y.Kawamura and M.Yamasaki: " Formation and mechanical properties of Mg97Zn1RE2 alloys with long-period stacking ordered structure", Mater. Trans. 48 (2007), pp. 2986-2992
11. K.Yamada, Y.Okubo, M.Shiono, H.Watanabe, S.Kamado and Y.Kojima: "Alloy development of high toughness Mg-Gd-Y-Zn Zr alloys", Mater. Trans. 47 (2006), pp. 1066-1070
12. K.Yamada, Y.Okubo, S.Kamado and Y.Kojima: "Precipitate microstructures of high strength Mg-Gd-Y-Zn-Zr alloys", Advanced Materials Research 11-12 (2006), pp. 417-420
13. T.Honma, T.Ohkubo, S.Kamado and K.Hono: "Effect of Zn additions on the age-hardening of Mg-2.0Gd-1.2Y-0.2Zr alloys", Acta Materialia 55 (2007), pp. 4137-4150
14. Y.Gao, Q.Wang, J.Gu, Y.Zhao and D.Yin: "Comparison of microstructure in Mg-10Y-5Gd-0.5Zr and Mg-10Y-5Gd-2Zn-0.5Zr alloys by conventional casting", J of Alloys and Compounds 477 (2009), pp. 374-378

PART 6: WROUGHT PROCESSING

Suveen N. Mathaudhu

While the large percentage of Mg-alloy products are used in the as-cast state, a number of applications require some form of wrought processing to form the geometries (e.g., sheets, beams, tubes) for specific applications. In many cases the mechanical properties can be simultaneously improved during the deformation, however there are still a number of barriers to more widespread usage. The deformation mechanisms of Mg-alloys, owing to their anisotropic plastic response from the hexagonal close-packed atomic structure, are extraordinarily complex and continue to be studied. Furthermore, unlike most Fe- and Al-based alloys, elevated temperatures are typically needed in both the Mg-alloy and tooling to induce enough plasticity for wrought processing. The papers selected for this section report on these major issues for mostly Mg-Al-Zn (AZ) alloys, which have historically been the workhorse alloys for wrought applications. Key themes include quantification of the complex deformation mechanisms, texture formation and resulting plastic anisotropy, and novel low temperature wrought processing methods.

As mentioned, the relevant deformation mechanisms must be known and thoroughly understood to maximize processability of Mg-alloys. Ebeling and colleagues presented a highly detailed study that examines deformation mechanisms and textural formation of AZ alloys during hot-rolling, and in parallel, compared the experimental results with viscoplastic self-consistent (VPSC) models. The complexity of property prediction was clearly demonstrated, and it was shown that a variety of slip and twinning modes are necessary to accurately predict mechanical response during deformation. Sloof and colleagues explored a similar theme in the next paper where they systematically reported the constitutive hot-working response of AZ alloy over a wider range of temperatures and strain rates than previously reported. Importantly, many of the constitutive behaviors were found to be strain dependent, and new constitutive predictive equations were developed to account for these new findings. Chun and Davies presented the development of a novel approach to quantifying the complex slip, twinning, texture banding, and texture formation phenomena using in-grain misorientation axes (IGMA) distribution functions. This method has since proven to be a very useful tool for studying deformation mechanisms.

In a paper that pulls together the themes of the prior three papers, Agnew reported on the plastic anisotropy that results from the complex slip mechanisms and textural formation in AZ alloy sheet through an astute set of experiments and computational simulations. The anisotropic plastic deformation phenomena reported in this paper are the root cause of the poor formability at low temperatures, and thus the historically limited workability into wrought parts. Stutz and colleagues recognized that tailoring textures could be used to improve the low temperature formability, and reported on their results for Mg-Zn-Ce-La-Zr (ZE10) and AZ alloys with different starting textures. The more random basal orientations in the texture of the ZE alloy result in better low temperature formability than the AZ alloy, and show the impact of alloy control on wrought processability. Similarly, the effects of Ce on the texture of Mg was reported by Mishra and colleagues, who showed that even a small percentage (0.2%) of Ce altered the texture in a way that dramatically improved formability. These reports forecast the potential for designing alloys and processes that can be used to manufacture parts at lower temperatures, and thus lower energy premiums. While low temperature formation would be a large step toward enabling economical manufacturing, the knowledge of the resultant microstructures at various temperatures and rates is still tremendously important. Beer and Barnett characterized the flow stress and microstructural formation in AZ31 under torsional and compression deformation at elevated temperatures strains and strain rates. Livescu and colleagues also study deformation of Mg-alloys under high deformation rates and under various orientations. Both papers present important findings at high strains and strain rates.

The knowledge gained on processing mechanisms and thermomechanical conditions has resulted in a number of new tools and complex component geometries. Luo and Sachdev presented the mechanical properties and microstructures of AZ31 alloy tubes formed at various temperatures. While elevated temperature processing was not completely successful (it resulted in dynamic recrystallization-induced instabilities and cavity merging), the results indicated the potential to design dynamically recrystallized grain structures in wrought products. Kurz reported on heated hydro-mechanical deep drawing, wherein similar to the tube formation, high temperatures are needed for plasticity. With proper tooling and die design, and carefully selected temperatures and processing rates, enhanced formability and deep drawing was achieved at temperatures as low as 150°C. In the last paper of the section, Sagapuram and colleagues present a single step extrusion-machining process for fabricating Mg-alloy strips and sheets with a variety of textures based on the processing temperatures. The non-basal textures generated during low temperature processing demonstrate enhanced formability. It is fully expected that continued alloy and process development, paired with intelligent exploitation of deformation mechanisms, will result in processes similar to this one, which will enable future applications of high-strength Mg-alloy wrought products.

Deformation Mechanisms of AZ31 Magnesium alloy

T. Ebeling[1], Ch. Hartig[1], T. Laser[1], M.R. Nürnberg[3], R. Bormann[1,2]
[1] Hamburg University of Technology, Inst. of Materials Science and Technology, 21073 Hamburg, Germany
[2] GKSS Research Centre Geesthacht, Inst. for Materials Research, 21502 Geesthacht, Germany
[3] GKSS Research Centre Geesthacht, Magnesium Innovation Centre MagIC, Max-Planck-Str. 1, D-21502 Geesthacht, Germany

Keywords: Magnesium, Deformation Mechanisms, Modeling, VPSC

Abstract

A detailed investigation of the deformation mechanisms plays an important role for a better understanding of texture evolution and anisotropic behavior of magnesium wrought alloys. Therefore, room temperature deformation tests of AZ31 hot rolled sheets and extruded bars have been performed. The experimental tensile textures of the rolled sheets show a reorientation resulting in a {0001}⟨101;¯0⟩ main component.

The results were compared with simulations using a viscoplastic self-consistent model to gain further information about deformation mode activities, texture evolution and mechanical properties. Thereby three simulation sets were executed: The first set included the results of both the rolled sheets and the extruded bars, while the others considered the results separately.

Based on the simulations, it will be demonstrated that the activity of the basal, prismatic, pyramidal slip and the twinning mode is required to achieve comprehensive results for the mechanical properties as well as for the texture evolution.

1. Introduction

Magnesium wrought alloys have attained an increasing interest in research during the last years for saving weight of structural parts. However, the use of semi-finished components like sheets is still limited due to the insufficient formability at room-temperature, which can be traced back to the lack of adequate activation of slip modes [1]. As a consequence, a high plastic anisotropy appears during deformation of magnesium, and slight asymmetries in the texture of rolled sheets can have significant effects on their mechanical properties. For this reason, it is of fundamental interest for the processing of the magnesium alloys to gain more detailed insight into the deformation behavior of magnesium alloys. In the present work we will use the well-known magnesium wrought alloy AZ31 and study the texture evolution during tensile and compression of rolled and extruded samples.

The predominant slip mode of magnesium is the basal slip (bas) mode which provides only two independent slip systems. Thus a fulfillment of the von Mises criterion of five necessary independent slip systems requires an activation of the hard non basal slip modes [2]. The prismatic slip (pri) offers two additional independent slip systems. The two commonly observed deformation modes that offer a deformation part in direction of the c-axis are the <c+a> pyramidal slip mode (pyr), which provides five independent slip systems, and the predominant {101;¯2}⟨101;¯1⟩ mechanical tensile twinning mode (ttw) [3]. According to [4] the twinning mode occurs soon after the onset of plastic deformation and plays an important role in texture evolution of magnesium. In prior investigations it could be shown that an activation of these four deformation modes leads to good simulated results in respect to the stress-strain behavior and the texture evolution [5].

However, the determination of especially the twinning mode by evaluation of tensile tests of rolled sheets (RS) in rolling and in transverse direction is difficult because the as-rolled texture does not favor twinning activity. The approach to gain more detailed information about the twinning mode is to use additionally results of tensile and compression tests of extruded bars (EB). Due to the texture of extruded samples enabling more frequent twinning than the texture of rolled sheets, the effects of twinning on simulation results are more relevant.

2. Experiment and Modeling

2.1 Experimental Procedures

AZ31 samples (nominal composition of 3 wt.% Al, 1 wt.% Zn, 0.3 wt.% Mn, balance Mg) in solution treated (20h@400°C) stress-relieved condition were hot-rolled in a two-high rolling mill (diameter 200 mm) at 400°C and extruded on an industrial-like extrusion press (8.32 MN) at the Extrusion Research and Development Center (ERDC) of the Technische Universität Berlin.

The hot rolling process was carried out up to a total strain of φ=1.6 achieved in three runs. Each run comprised several reduction steps that were followed by a reheating (3min@450°C). A run was followed by subsequent annealing 30min@400°C. This procedure was chosen in order to obtain a comparable microstructure and dislocation density in comparison to the as-extruded samples. All sheets were tested in as-rolled conditions. Details about the extrusion process can be found in [6].

For tensile-testing in rolling (RD, 0°) and transverse direction (TD, 90°), specimens of the size 25 x 6.5 x 1.5mm (gage length, width, thickness) were cut out of the rolled sheets. Tensile and compression specimens of the extruded bars were cut out in sizes of l_0 = 25mm, d_0 = 5mm and h_0 = d_0 = 10mm respectively.
The tensile and compression tests were carried out at room temperature using a universal ZWICK testing machine with a constant strain rate of $10^{-3}\cdot s^{-1}$. The longitudinal elongation was measured by means of an inductive extensometer. At least three samples of each material and direction were tested.

Metallography was carried out using a picric acid as etchant [7] to reveal grains in prepolished sections. The average grain size was determined from several micrographs using a computer-aided linear intercept measurement.

The texture samples were prepared from the middle sections of the rolled sheets, extruded bars and the middle part of the broken tensile specimens respectively. Six incomplete pole figures were measured up to a pole distance of α = 85° using a D8 X-ray diffractometer (BRUKER AXS Inc.) with a special parallel-beam optics (polycapillary optics) and Cu-K_α radiation.

2.2 Modeling Procedures

To perform the simulations a viscoplastic self consistent model (VPSC) [8] implemented into a genetic optimization algorithm [9] was utilized. The main goal of the simulations was the calculation of best fitting critical resolved shear stresses (CRSS) and hardening parameters for the three simulations sets shown in Table 1:

Table 1: Executed Simulation Sets

Simulation	Samples	Experimental Basis
RS	rolled sheets (RS)	tension RD & TD
EB	extruded bars (EB)	tension & compression ED
RS+EB	rolled sheets and extruded bars	tension RD & TD tension & compression ED

As boundary conditions of the VPSC model, an elongation in load direction, zero shear strains and zero stresses perpendicular to the load direction were prescribed. The rate sensitivity for all slip and twinning modes was set to n = 19. The hardening of the involved individual slip and twinning modes was taken into account using the general voce hardening law (Equation 1) describing the critical resolved shear stress τ^s of the deformation systems as a function of the total shear strain Γ as follows:

$$\tau^s = \tau_0^s + \left(\tau_1^s + \theta_1^s \cdot \Gamma\right) \cdot \left[1 - \exp\left(\frac{\theta_0^s \cdot \Gamma}{\tau_1^s}\right)\right] \qquad \text{Eq. 1}$$

τ_0^s is the CRSS at the onset of plastic deformation, $\tau_0^s + \tau_1^s$ is the saturation shear stress and θ_0^s is the extrapolated work hardening for zero flow stress. θ_1^s is a hardening term which takes contributions from work hardening stage IV into account and was set to zero.

To execute the simulations, the measured orientation distribution functions (ODF's) were converted to discrete orientations and used as input quantity. The total true plastic shear strain was in the observed range between 0.005 and 0.15. Parameter optimization was performed by calculating a theoretical stress-strain curve in the quoted range and minimizing the weighted least squares differences between theoretical and experimental stress. With the obtained CRSS's the simulation was calculated again in order to achieve detailed information of texture evolution.

3. Results and Discussion

3.1 Experiments

3.1.1 Rolled Sheets (RS)

The microstructure of the AZ31 samples in as-rolled conditions is shown in Figure 1. It consists of almost recrystallized grains. An average grain size of 12 µm is revealed.

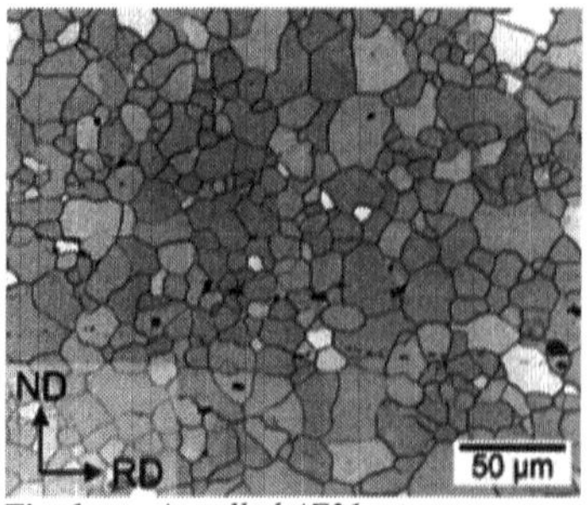

Fig. 1: As-rolled AZ31 microstructures

Mechanical properties were obtained from stress-strain curves and are collected in Table 2. An anisotropy could be observed with regard to the 0.2% offset yield strength which showed higher values in TD than in RD.

Table 2: Strength Values of As-Rolled AZ31

Sample material	Test / Direction	$R_{p0.2}$ [MPa]	A [%]
AZ31, RS	tensile RD	147.2	16.8
	tensile TD	168.4	2.8

The measured as-rolled textures are shown in Figure 2. A basal texture developed which can be characterized as $(\varphi_1, \Phi, \varphi_2) = (\pm 90°; < 20°; \varphi_2)$. At lower strain rates a splitting of the basal pole was observed. This is a hint of the activity of the <c+a> pyramidal slip mode.

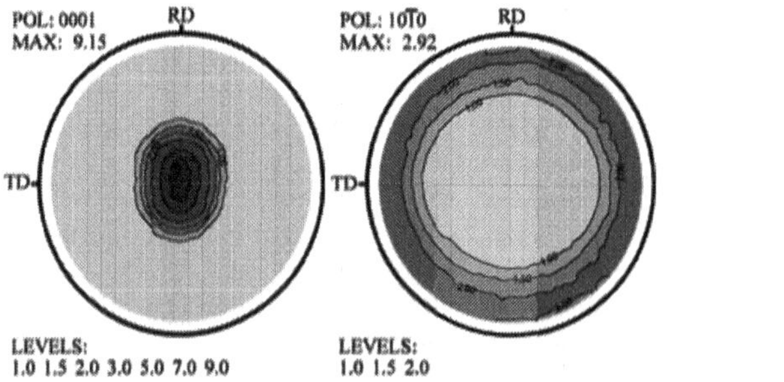

Fig. 2: Experimental bas and pri pole figures of as-rolled AZ31

During the tensile test in RD, the prismatic {101;¯0} planes rotate into a position perpendicular to the sheet plane resulting in a {0001}⟨101;¯0⟩ main texture component (Figure 3). Thus four prismatic {101;¯0} planes were favorably reoriented for prismatic slip during tensile testing.

The experimental textures of the tensile test in TD are represented in Figure 4. The texture development is not in an advanced state in this case, because the sample broke at 2,8% tensile strain. In simulation and other experiments [5] it could be shown, that in this case the prismatic {101;¯0} planes rotate into a favourite position for prismatic slip too resulting in a {0001} ⟨112;¯0⟩ texture.

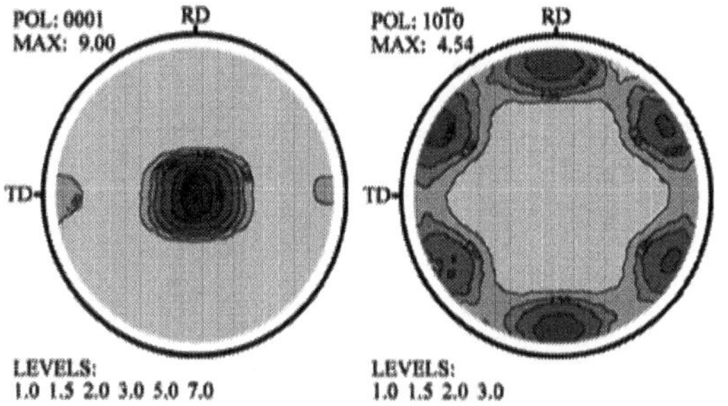

Fig. 3: *Experimental bas and pri pole figures of AZ31, ex post tensile test in RD*

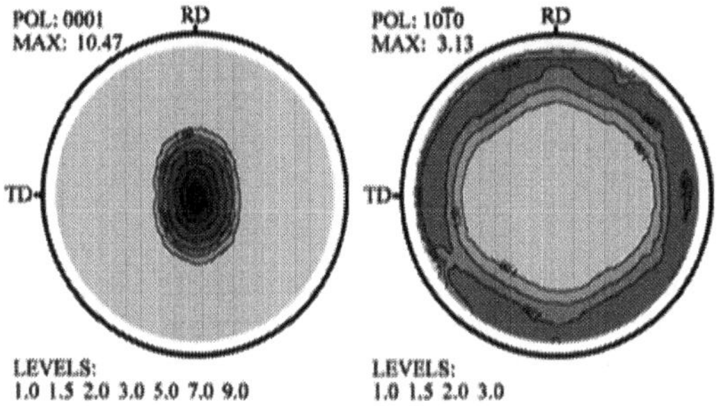

Fig. 4: *Experimental bas and pri pole figures of AZ31, ex post tensile test in TD*

3.1.2 Extruded bars (EB)

The microstructures of the AZ31 samples in as-extruded conditions are shown in Figure 5. They consist of almost recrystallized grains with an average grain size of 12 μm. Few elongated grains in extrusion direction could be found.

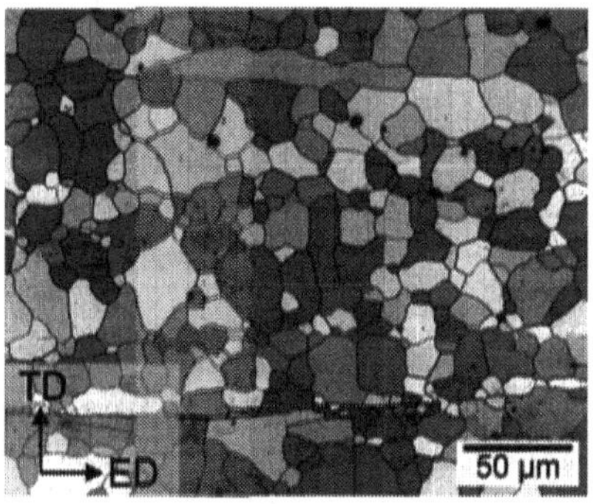

Fig. 5: *As-extruded AZ31 microstructures*

Mechanical properties were obtained from stress-strain curves and are collected in Table 3. A strong tensile-compression strength asymmetry is present in the extruded samples. While the 0.2% offset yield strength is approx. 115 MPa in compression, it reaches a value of approx. 180 MPa in tension.

Table 3: *Strength values of AZ31 as-extruded (EB)*

Sample material	Test / Direction	$R_{p0.2}$ [MPa]	A [%]
AZ31, EB	tensile ED	181.9	21.9
	compression ED	113.5	13.2

In Figure 6 the texture of as-extruded AZ31 is shown. It can be described as $(\varphi_1,\Phi,\varphi_2)=(\varphi_1;>75°;\varphi_2)$, i.e. normal vectors of the basal planes are oriented parallel to the radial direction of the extruded bar. Thereby the prismatic planes are oriented perpendicular to the extrusion direction.

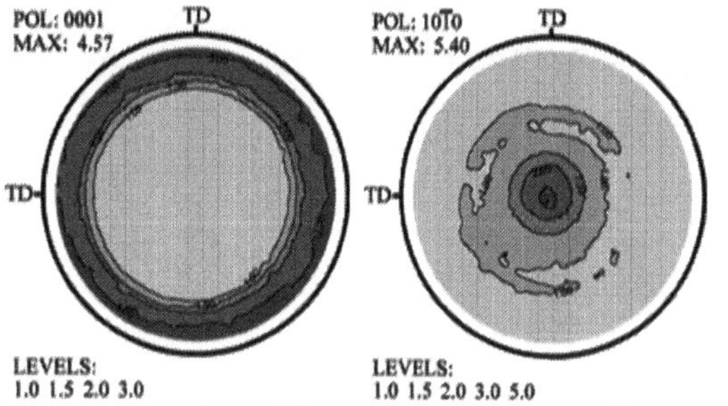

Fig. 6: *Experimental bas and pri pole figures of as-extruded AZ31, inverse Polfigures can be found in [6]*

3.2 Modelling

The combination of slip and twinning modes that were used gives the best fits for the three simulation sets using the voce-parameters represented in Table 4. The CRSS's obtained from the three model calculations are similar except the CRSS of the tensile twinning mode: While the commonly used simulation of RS tests leads to very high τ_0^s, the simulation of the EB tests as well as the simulation of the RS+EB tests are leading to much lower values.

It is evident that changes of the parameters of the tensile twinning mode lead only to slight differences of the simulation of the RS tests results. The texture of the extruded bars favors the activity of the tensile twinning mode during compression. This gives a high sensitivity of the simulation result to the used voce parameters of the tensile twinning mode. In regard to the simulation set of the RS+EB tests, the prismatic slip is five times, the pyramidal slip almost 12.5 times and the tensile twinning two times harder than the basal slip. Additionally, the pyramidal mode shows strong hardening in contrast to the other deformation modes. A high CRSS at the onset of deformation as well as strong hardening of the <c+a> pyramidal slip mode applies analogously to all simulation sets.

Table 4: *CRSS's and hardening parameters according to Voce-Equation (Equation 1) of AZ31 using the experimental data of the rolled sheets (RS) or/and the extruded bars (EB)*

deform. Mode	slip plane/ direction	used exp. data	τ_0^s [MPa]	τ_1^s [MPa]	θ_0^s [MPa]
bas	{0001} <21;¯1;¯0>	RS	28.2	476.4	277.0
		EB	26.2	309.6	122.0
		RS+EB	31.2	33.0	150.0
pri	{101;¯0} <21;¯1;¯0>	RS	149.2	80.3	274.0
		EB	167.7	45.1	227.8
		RS+EB	153.2	58.3	520.0
pyr	{1;¯21;¯2} <1;¯21;¯3>	RS	261.0	500.0	2560.9
		EB	300.9	155.0	1494.0
		RS+EB	391.6	262.7	2245.9
ttw	{101;¯2} <101;¯1>	RS	147.9	1.6	12.7
		EB	66.3	1.9	20.3
		RS+EB	61.5	1.9	24.7

In Table 5 the error values using a least squares method for the three parameter sets are represented. It could be seen, that the highest error value appears, if the voce parameters obtained for the simulation of the RS tests applied on the data of the extruded samples (cf. Figure 11). In contrast to this the data of the simulation of the EB tests seem to be applicable with an acceptable increase of the error value on the data of rolled samples. Finally, the simulation of the RS+EB tests gives good agreement with the experimental data of the extruded and the rolled samples. In the following mainly the results of the simulations of the RS+EB tests are illustrated.

Table 5: *Weighted relative error calculated using the least squares*

Relative error [%]		Parameter set (sim.)		
		RS	EB	RS+EB
Stress-strain curve (exp.)	RS	0,7	8,7	3,3
	EB	82,4	5,2	6,5

3.2.1 Rolled Sheets

In the stress-strain curves of the tensile tests in RD and TD of the rolled sheets (Figure 7), an anisotropic behavior could be observed: The yield in RD is lower than the yield in TD.

The slip mode activities could be evaluated by using the experimental initial textures and the achieved voce parameters in vpsc-simulation. The as-rolled sheet shows alike activities in RD and TD (Fig. 6), means high, decreasing basal activity and moderate, increasing prismatic activity. While the twinning mode shows slight activity at the onset of deformation, the pyramidal mode is almost inactive, especially at the onset of deformation.

The simulated tensile texture evolution of AZ31 represented in Figure 9 including the four described deformation modes is in good accordance with the experimental texture. The simulated texture of the tensile test in TD is represented in Figure 10 and it shows a good agreement to the experimental texture.

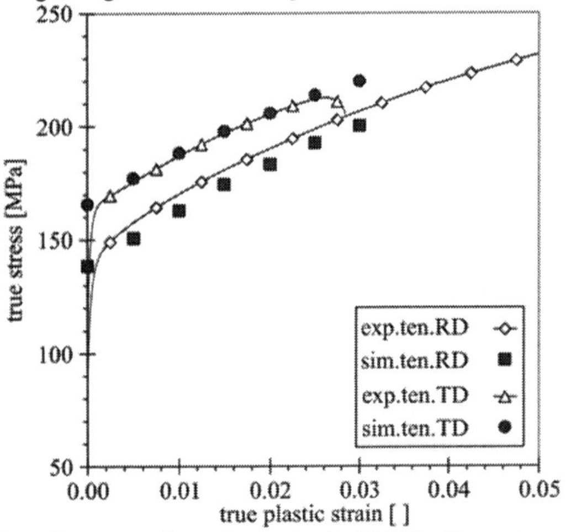

Fig. 7: *Experimental and simulated (RS+EB Parameters used, q.v. Table 3) stress-strain curves of as-rolled AZ31*

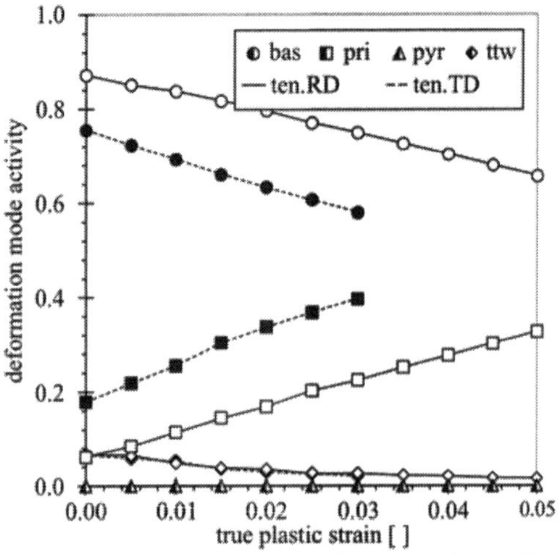

Fig. 8: *Deformation mode activities during tensile in RD and TD according the simulations using the RS+EB Parameters, q.v. Table 3*

It is important to note that the simulations of the texture evolution of the AZ31 sample during tension with an inactive prismatic slip mode do not show the characteristic evolution of the {0001}⟨101;¯0⟩ main texture component. The main component develops only with an active prismatic-slip mode, hence it is a result of the prismatic slip. These results demonstrate that already at room temperature prismatic slip activity is necessary to explain the measured textures. The activity of the prismatic slip at room-temperature deformation of AZ31 was found in prior investigations during tensile tests using the Electron back scattering diffraction method (EBSD) [10] and in situ neutron scattering [11]. Furthermore, a Taylor model calculation reveals high internal incompability stresses with an inactive prismatic slip mode [12]. Finally, model calculations using a prismatic slip mode are leading to Lankford coefficients in the right magnitude (0.6 to 3.7, q.v. experiments in [13]), while the inactivation of the prismatic slip results in values around 0.

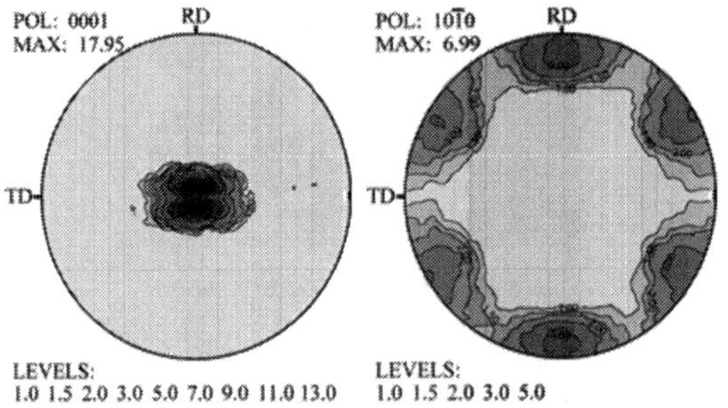

Fig. 9: *Simulated (RS+EB Parameters used, q.v. Table 3) bas and pri pole figures of AZ31, ex post tensile test in RD, q.v. measured texture in Fig. 3*

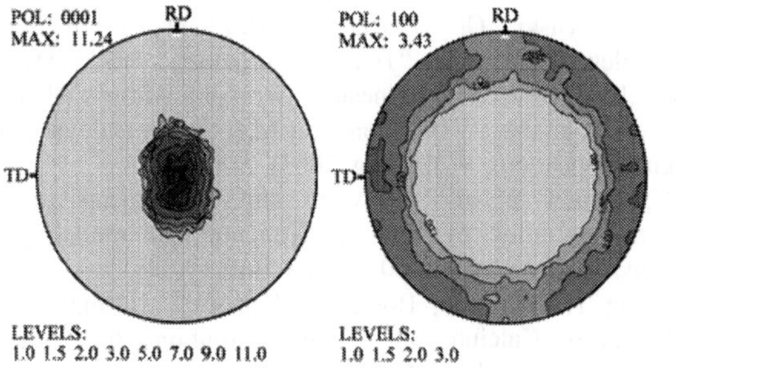

Fig. 10: *Simulated (RS+EB Parameters used, q.v. Table 3) bas and pri pole figures of AZ31, ex post tensile test in TD, q.v. measured texture in Fig. 4*

3.2.2 Extruded bars

In Figure 11 the simulated and experimental tension-compression stress-strain curves are compared. A strong tension-compression strength asymmetry could be observed in the experimental curves. At higher deformations the hardening of the stress-strain curve decreases during tension, but during compression it increases drastically.

A good accordance between the experimental and simulation results could be observed using the voce parameters based on the ES+EB simulation set. In contrast to this it is evident, that the voce parameters based on the RS simulation set do not lead to a good agreement. Especially the tension-compression asymmetry could not be reflected.

The deformation mode activities (Fig. 12) are completely different in tension and compression. In tension the activity of basal slip is predominant at the onset of deformation but prismatic slip becomes predominant by an ongoing strain. Due to the texture the tensile twinning mode is favored in compression. Thereby it is the predominant deformation mode at the onset of deformation. At higher strain rates the activity of the twinning mode decreases and the basal mode becomes predominant. The high activity of the soft tensile-twinning mode is the reason for the strong asymmetric tension-compression behavior that was observed in the stress-strain curves.

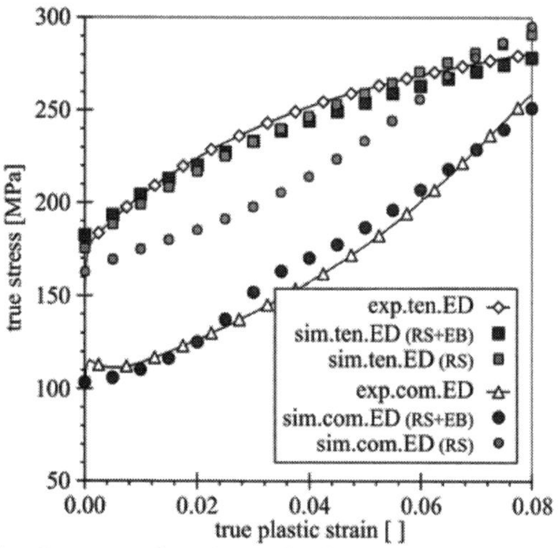

Fig. 11: *Experimental and simulated (RS+EB Parameters used, q.v. Table 3) stress-strain curves of as-extruded AZ31*

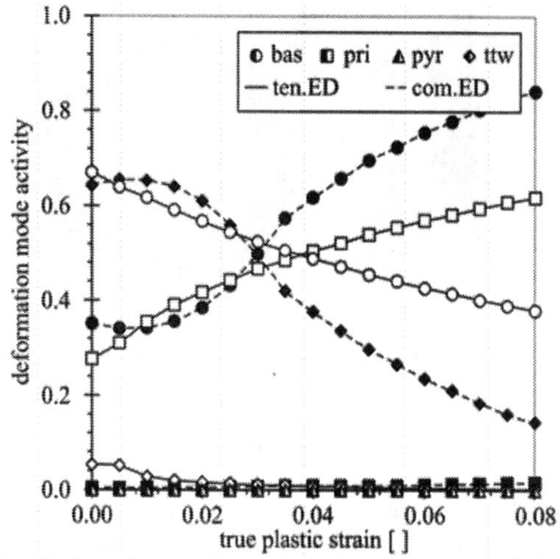

Fig. 12: *Deformation mode activities during tensile in RD and TD according the simulations using the RS+EB Parameters, q.v. Table 3*

The inactivity of the pyramidal <c+a> slip mode for the as-extruded bars (as well as for the as-rolled sheets) gives the idea to perform simulations with a pyramidal slip mode which is set as inactive. But this leads to very high internal incompability stresses, because this slip mode is the only one which is capable to accommodate strains along the c-axis [10,11]. Therefore <c+a> slip has always been included.

The simulated texture of the extruded bar after tension in extrusion direction is shown in Figure 13. An analysis of the orientation distribution function reveals a strong {101;¯0}-fibre texture. Twinning occurs only marginal. By contrast, the twinning mode is predominant at the onset of compression of the extruded bar in extrusion direction. This is due to the twinning favoring texture and leads to a strong basal fibre (Figure 14).

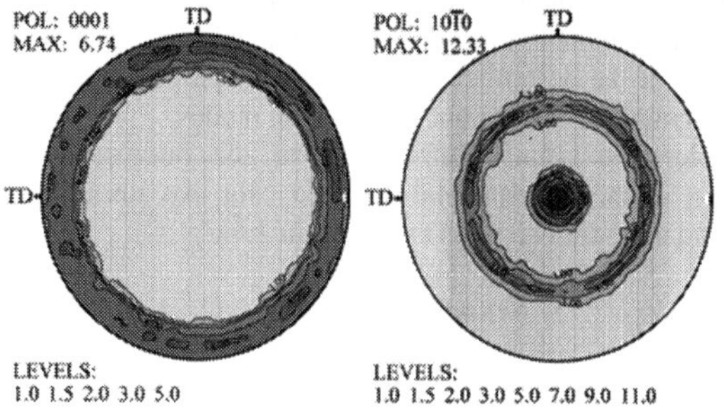

Fig. 13: *Simulated (RS+EB Parameters used, q.v. Table 3) bas and pri pole figures of AZ31, ex post tensile test in ED*

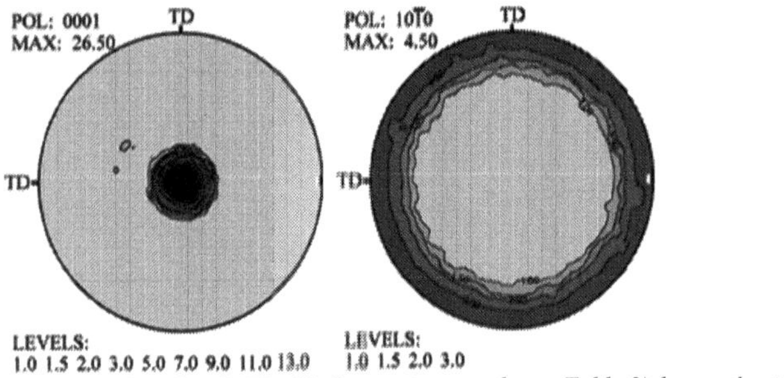

Fig. 14: *Simulated (RS+EB Parameters used, q.v. Table 3) bas and pri pole figures of AZ31, ex post compression test in ED*

4. Conclusions

The best simulation results are obtained by means of the basal, prismatic, <c+a> secondary pyramidal slip modes and the {101;¯2}⟨101;¯1⟩ tensile twinning mode. Even though the simulations show an almost inactive pyramidal slip mode, it was not deactivated because to very high internal incompability stresses and the capability to accommodate strain along the c-axis. The deactivation of the prismatic slip mode leads to wrong texture evolution as well as wrong calculated lankford-coefficients.

At room-temperature the pyramidal slip mode has comparatively high CRRS values (factor 9.0 - 12.5 compared with the basal slip) and strong hardening. The basal slip and the tensile twinning (factor 2.0) modes are soft in regard to the simulations of the EB tests and the simulation of the RS+EB tests. Concerning the simulation of the RS tests the tensile twinning is eminently hard.

The experimental texture evolution and the mechanical response during tensile and compression could be reproduced using the illustrated voce-parameters. In order to obtain universal material parameters by means of deformation simulations, some aspects should be considered:

- The use of several textures to perform deformation simulations to get more detailed information about a material, e.g. voce parameters of the deformation systems and their activities. The textures should favor different deformation modes.
- Nevertheless, it is also important that the microstructural derivation of the experimental obtained data like stress-strain curves and measured textures is almost the same.

5. Acknowledgements

This study was supported by the German Research Foundation (DFG). Thanks are due to the Extrusion Research and Develop-ment Center (ERDC) of the Technische Universität Berlin for careful technical support. The authors appreciate software support (VPSC6) by C.N. Tomé.

6. References

1. S.B. Yi, H.G. Brokmeier, K.U. Kainer, T. Lippmann, "Texture Development of AM20 Tensile Samples Under Load" (Paper presented at the 6th Int. Conf. on Magnesium Alloys and Their Applications, Wolfsburg, 2003), 220-225.
2. S.R. Agnew, M.H. Yoo, C.N. Tomé, "Application of texture simulation to understanding mechanical behavior of Mg and solid solution alloys containing Li or Y". *Acta Materialia*, 49 (2001), 4277-4289.
3. A. Styczynski, Ch. Hartig, R. Bormann, F. Kaiser, J. Bohlen, D. Letzig, "Texture Formation and Texture Modelling of AZ31 Magnesium Wrought Alloy", (Paper presented at the 6th Int. Conf. on Magnesium Alloys and their Applications, Wolfsburg, 2003). 214-219.
4. S.R. Agnew et al, "Study of slip mechanisms in a magnesium alloy by neutron diffraction and modeling", *Scripta Materialia*, 48 (2003), 1003-1008.
5. Ebeling T, Hartig C, Bohlen J, Letzig D, Bormann R, "Effects of Calcium on Texture Evolution and Plastic Anisotropy of the Magnesium Alloy AZ31" (Paper presented at the 7th Int. Conf. on Magnesium Alloys and Their Applications, Dresden, 2006), 158-164.
6. T. Laser, Ch. Hartig, T. Ebeling, M.R. Nürnberg, D. Letzig, R. Bormann, "The Influence of Calcium and Rare Earth Metals on the Microstructure and Mechanical Properties of Mg-3Al-1Zn during Extrusion" (Paper presented at the Int. Conf. "Magnesium Technology 2008", New Orleans, 2008)
7. V. Kree, J. Bohlen, D. Letzig, K.U. Kainer, „The metallographical examination of magnesium alloys", *Praktische Metallographie-Practical Metallography*, 41 (2004), 233-246.
8. C.N. Tomé, R.A. Lebensohn, "Visco-plastic self-consistant (VPSC) V6", 2003
9. M. Eldred, "DAKOTA: A Multilevel Parallel Object-Oriented Framework for Design Optimization, Parameter Estimation, Uncertainty Quantification, and Sensitivity Analysis", http://www.cs.sandia.gov/DAKOTA
10. Z. Keshavarz, M. Barnett, "EBSD analysis of deformation modes in Mg-3Al-1Zn", *Scripta Materialia*, 55 (2006), 915-918
11. S.R. Agnew, D.W. Brown, C.N. Tomé, "Validating a polycrystal model for the elastoplastic response of magnesium alloy AZ31 unsing in situ neutron diffraction", *Acta Materialia*, 54 (2006), 4841-4852.
12. A. Styczynski, Ch. Hartig, J. Bohlen, D. Letzig, "Cold rolling textures in AZ31 wrought magnesium alloy", *Scripta Materialia*, 50 (2004), 943-947.
13. Ch. Hartig, A. Styczynski, F. Kaiser, D. Letzig, "Plastic anisotropy and texture evolution of rolled AZ31 magnesium alloys", *Materials Science Forum*, 495-497 (2005), 1615-1620.

From *Magnesium Technology 2007*, R.S. Beals, A.A. Luo, N.R. Neelameggham, and M.O. Pekguleryuz, Editors

CONSTITUTIVE BEHAVIOR OF WROUGHT MAGNESIUM ALLOY AZ61

F.A. Slooff[1], J. Zhou[1], J. Duszczyk[1], L. Katgerman[1]
[1]Delft University of Technology; Delft, Mekelweg 2 2628 CD The Netherlands

Keywords: Magnesium, Hot workability, Constitutive analysis, Compression testing

Abstract

The data on the hot workability of the magnesium alloy AZ61 available in the open literature are based on the tests over narrow ranges of temperatures and strain rates and thus are of limited use for bulk-forming processes. The objective of the present research was to perform a detailed constitutive analysis of the alloy, on the basis of the results obtained from uniaxial compression tests at temperatures from 250 to 450 °C and strain rates from 0.01 to 100 s^{-1}. The shape of the flow curves suggested the occurrence of twinning followed by dynamic recrystallization. The alloy did not display steady-state flow stress behavior such that the constitutive analysis of the alloy using the hyperbolic equation must take strain into account. It was found that the constitutive constants were indeed strain dependent. A method to apply the hyperbolic constitutive equation at given strains was developed, with which an accurate strain-stress database for the alloy could be established.

Introduction

In order to reduce the emissions from automobiles, the automotive industry is eager to use lightweight materials. Since magnesium is lightweight and has a high specific strength, it is a suitable material for reducing the automobile weight and thus the exhaust. However, the use of wrought magnesium alloys has yet been quite limited, because of their low deformability caused by the restricted number of available slip systems. It is a challenge to define the deformation limits of wrought magnesium alloys and to provide the industry with in-depth knowledge about their deformability.

In most of the preceding research, attention has been paid to the wrought magnesium alloy AZ31 [1-3]. The alloy AZ61 has not received a lot of attention, although it has a higher strength due to the presence of the β-phase [4-5]. The studies on the deformability of the alloy AZ61 conducted so far have covered relatively narrow ranges of temperatures and strain rates [6-8], relative to those encountered in the commercial bulk-forming processes such as hot extrusion. Extrapolations often lack in scientific rationale and lead to a low level of confidence in accuracy. To investigate the deformation behavior of a magnesium alloy, various techniques are available such as uniaxial compression testing and torsion testing. The ability to apply high strain rates to the material makes hot uniaxial compression testing suitable for the constitutive analysis.

In the current work, the magnesium wrought alloy AZ61 was uniaxial hot compressed over a wide range of temperatures at different strain rates. The objectives were to obtain material behavior data for constitutive analysis and to perform this analysis with a strain-dependent method, as the magnesium alloy displayed non-steady-state deformation behavior so that the classic constitutive equations could not be applied regardless of strain.

Experimental Procedures

The as-cast AZ61 alloy having a composition as shown in Table I was extruded into round bars with a diameter of 15.8 mm, at a reduction ratio of 10. These bars were machined into cylindrical specimens, having a diameter of 10 mm and a height of 12 mm. The height of the specimens was parallel to the extrusion direction. During machining the surface of the specimens was kept as smooth as possible, to exclude any surface effects on the hot deformation behavior.

Table I. Composition of the AZ61 alloy used for testing.

Element	Al	Zn	Mn	Si	Cu	Mg
Wt%	5.6	0.81	0.18	0.08	0.003	Bal.

For the hot uniaxial compression tests, a DSI Gleeble 3800 system was used. Wide ranges of strain rates and temperatures were chosen for the hot compression tests. Temperature was varied from 250 °C to 450 °C with an interval of 50 °C. At every temperature tests were performed at strain rates of 0.01, 0.1, 1, 10 and 100 s^{-1}. The specimen was heated at a rate of 10 °C s^{-1} and kept for 60 to 120 seconds at the set temperature in low vacuum to equalize the temperature throughout the specimen. The specimen was then hot compressed at a constant strain rate to a true strain of 1, which resulted in a final height of about 4.4 mm. The Gleeble system used thermocouple feedback to control the A/C resistance heating to reach and maintain the set temperature. A thermocouple was spot welded onto the surface of the specimen at the mid height. To minimize the friction, graphite foils were attached to the water-cooled tungsten carbide anvils using a nickel containing paste. After the uniaxial deformation, air quench was applied to cool the specimen down to room temperature. Three repeat tests under each set of the parameters were performed.

Results

The raw data from the Gleeble 3800 system were not suitable for immediate constitutive analysis; these were processed as described elsewhere [9]. The most important adjustment made was the correction for (adiabatic) deformation heating, which was significant at a strain rate of 10 s^{-1} and higher. At a strain rate of 100 s^{-1}, the mechanical work done on the specimen could cause a temperature increase over 100 °C. Since the current work concerns the influence of deformation temperature on flow stress, the correction for this temperature change is of great importance. Moreover, the constitutive analysis requires isothermal data and the correction for temperature deviations from the set temperature will benefit the final results. To give an impression of the deformation applied on the cylindrical specimens, Figure 1 shows five different specimens, which were obtained after stopping the tests at different strains. The far left specimen is an initial specimen before deformation, followed by the specimens deformed to true strains of 0.075, 0.15, 0.4 and 1.

Figure 1. Example of an initial specimen (left) and the specimens after deformation to strains of 0.075, 0.15, 0.4 and 1.

The corrected true stress - true strain curves from the compression tests at 250, 350 and 450 °C are shown in Figure 2. It is clear that the flow stress varies greatly with strain, especially when strain rate is high.

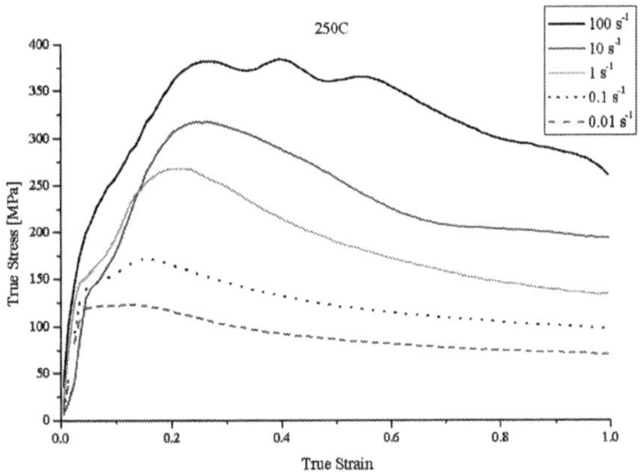

Figure 2a. True stress – true strain curves for AZ61 at 250 °C and various strain rates.

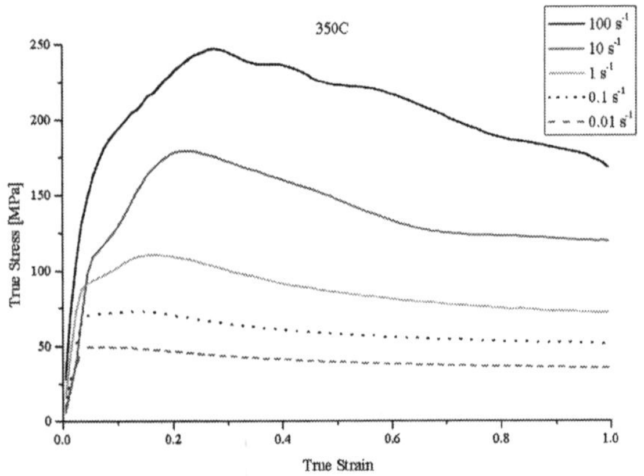

Figure 2b. True stress – true strain curves for AZ61 at 350 °C and various strain rates.

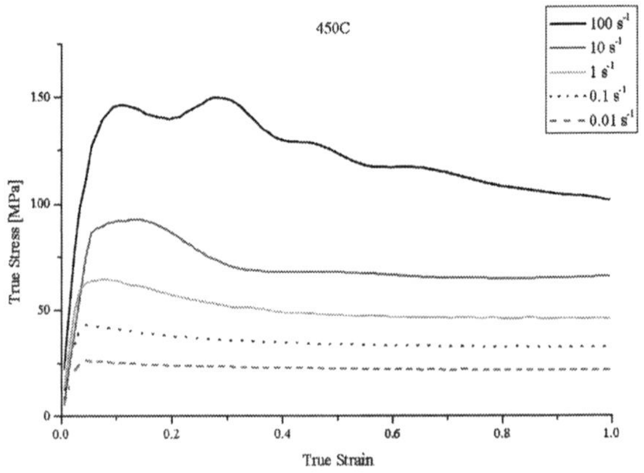

Figure 2c. True stress – true strain curves for AZ61 at 450 °C and various strain rates.

Figure 3 compares of the flow stress curves of the AZ61 alloy with those of the aluminum alloy 7020 at a constant strain rate of 10 s^{-1} and different temperatures. It is obvious that the fluctuations in the flow stress of the magnesium alloy AZ61 during deformation require a different approach to constitutive analysis than the aluminum alloy 7020, since the flow stress of the aluminum alloy tends to be steady throughout deformation and a distinct stress peak is absent.

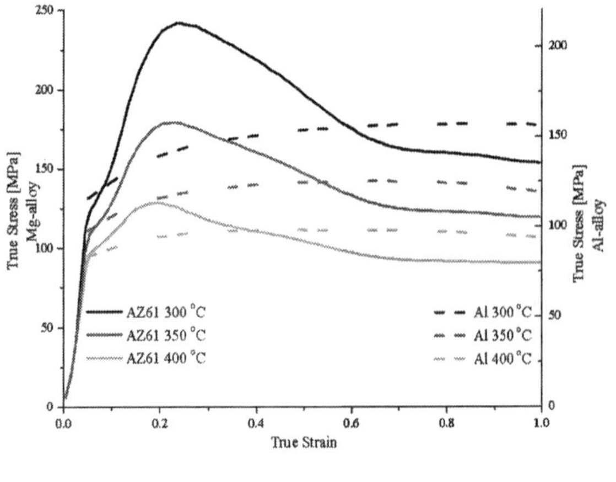

Figure 3. True stress – true strain curves of the AZ61 magnesium alloy in comparison to the aluminum alloy 7020 at a strain rate of 10 s^{-1} and various temperatures.

The flow curves of AZ61 show strain hardening, followed by work softening. This results in a peak in the flow stress, which is less pronounced at a lower strain rate and at high temperatures. The results of the compression tests at a strain rate of 0.1 s^{-1} and higher and at 250 °C, show a nearly linear increase in stress just before the stress peak is reached. It has been reported [1] that this is an indication of twinning. The twinning is followed by dynamic recrystallization (DRX), causing the flow stress to reduce after the stress peak. The shift of the flow stress peak to a higher strain with increasing strain rate is also an indication of DRX [10].

To show the influence of strain rate and temperature on the maximum flow stress, in Figure 4, the peak flow stress is plotted against strain rates for all the tests.

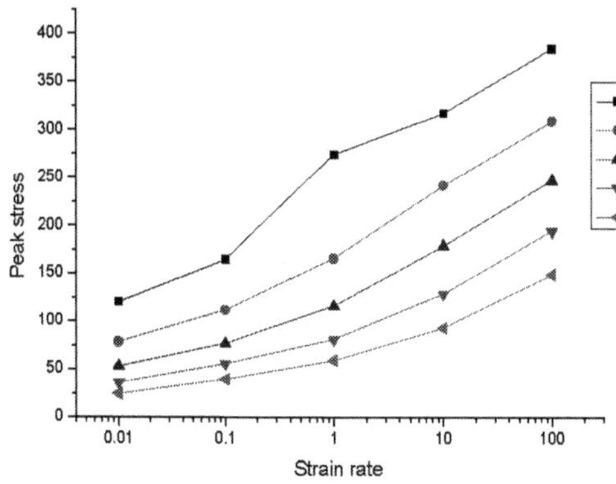

Figure 4. Peak stresses as a function of strain rate.

It is clear from Figure 4 that the peak stress of the material is sensitive to temperature and strain rate. It increases with increasing strain rate or reducing temperature. The lower the deformation speed or the higher the deformation temperature, the less pronounced is the stress peak. Zhou *et al* [7] reported the cracking of the material at 250 °C and a strain rate of 0.01 s^{-1}. However, the tests at 250 °C in the present investigation were all successful. Even at a strain rate of 100 s^{-1}, the specimen after deformation to the final strain of 1 had some sheared shape, but no fracture was observed.

Constitutive equation is customarily used in hot working to calculate the force for deformation at a given temperature. In the case of deformation under low stresses such as creep, the favorite constitutive equation is the power law (1).

$$\dot{\varepsilon} = A'\sigma n' \qquad (1)$$

In which $\dot{\varepsilon}$ is the strain rate, *A'* and *n'* are material constants and σ is the stress. For hot workability analysis the exponential law (2) may be used,

$$\dot{\varepsilon} = A''\exp(\beta\sigma) \qquad (2)$$

where *A"* and β are material constants. Equation (2) breaks down at high temperatures and at strain rates below 1 s^{-1} [11]. These two equations do not take temperature into account, while actually, temperature varies during deformation depending on material properties, strain rate and other factors. Equation (3) that uses the Zener-Hollomon parameter *Z* and includes two variables strain rate and temperature, probably is the most used constitutive equation for hot working processes.

$$A[\sinh(\alpha\sigma)]^n = \dot{\varepsilon}\exp\left(\frac{Q}{RT}\right) = Z \qquad (3)$$

In this equation, α is the stress multiplier, an additional adjustable constant which brings $\alpha\sigma$ into a correct range to make constant *T* curves in the log[sinh($\alpha\sigma$)] versus ln[$\dot{\varepsilon}$] plots linear and parallel [12]. *A* and *n* are constants, *R* is the universal gas constant and *Q* the activation energy for deformation. In the literature the constitutive equations are mostly used for materials having a rather stable flow stress and thus strain is not factored in. It is clear from Figures 2 and 3 that this is not the case for AZ61.

With corrected stress-strain curves, it is possible to further process the data to obtain the material constants in Equation (3) to determinate the Zener-Hollomon parameter at different strains. The stress multiplier α is adjusted to obtain ln[sinh($\alpha\sigma$)] versus ln($\dot{\varepsilon}$) graphs in which the lines are as parallel as possible. When this is accomplished, the average slope of these graphs is $m = 1/n$, the inverse stress exponent. Figure 5 shows an example of such a graph at a true strain of 0.2.

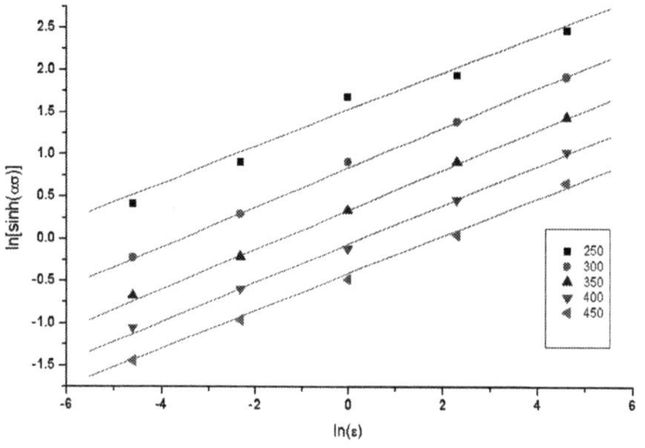

Figure 5. ln[sinh($\alpha\sigma$)] versus ln($\dot{\varepsilon}$) for AZ61 at a strain of 0.2.

Since the flow stress of the magnesium alloy during uniaxial compression varies significantly with strain during deformation (as can be seen in Figure 2), the calculations are performed at fixed strains. At high strain rates and low temperatures the changes of the flow stress with strain become much greater.

With the α and *n* values obtained, the activation energy for the deformation can be determined. Rearranging equation (3) leads to:

$$\ln(\dot{\varepsilon}) = \ln(A) + n\ln[\sinh(\alpha\sigma)] - \frac{Q}{RT} \qquad (4)$$

Rearranging (4) and differentiating with respect to *1/T* gives an expression for the activation energy *Q*, Equation (5) or simply Equation (6).

$$Q = R\left(\frac{\partial \ln \sinh \alpha\sigma}{\partial \frac{1}{T}}\right)_{\dot{\varepsilon}}\left(\frac{\partial \ln \dot{\varepsilon}}{\partial \ln \sinh \alpha\sigma}\right)_T \qquad (5)$$

$$Q = RSn \qquad (6)$$

The term *S* in Equation (6) is the average slope of the lines in the ln[sinh($\alpha\sigma$)] versus *1000/T* plot, as described by Equation (7), and *n* the reciprocal slope of Figure 5 and represented by Equation (8).

$$S = \frac{d\ln[\sinh(\alpha\sigma)]}{d\left(\frac{1}{T}\right)} \qquad (7)$$

$$n = \left(\frac{\partial \ln \dot{\varepsilon}}{\partial \ln \sinh \alpha\sigma} \right) \quad (8)$$

With the calculated values of the activation energy at selected strains, the Z values can be calculated, using Equation (9). Finally, $\ln(A)$ from Equation (4) can be determined, which is the intercept in the $\ln(Z)$ versus $\ln[\sinh(\alpha\sigma)]$ plot, described by Equation (10), a natural logarithm of the right-hand side of Equation (3).

$$Z = \dot{\varepsilon} \exp\left(\frac{Q}{RT} \right) \quad (9)$$

$$\ln(Z) = \ln(A) + n \ln[\sinh(\alpha\sigma)] \quad (10)$$

Figure 6 is an example of the plot of $\ln[Z]$ against $\ln[\sinh(\alpha\sigma)]$ at a strain of 0.2.

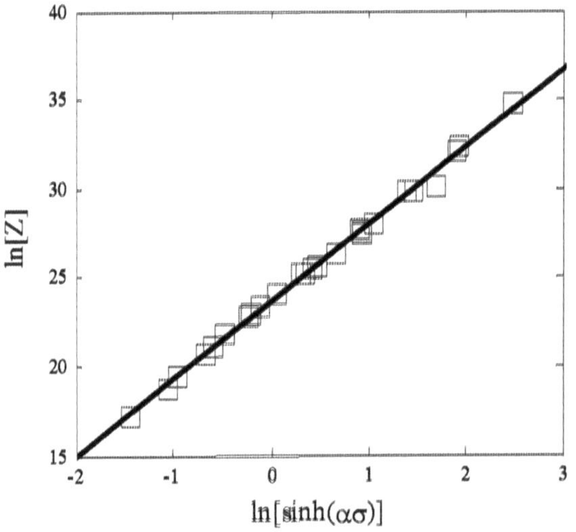

Figure 6. $\ln[Z]$ versus $\ln[\sinh(\alpha\sigma)]$ plot for AZ61 at 0.2 strain.

As can be seen from Figure 6, the data at a strain of 0.2 fit in the hyperbolic sine law very well. Similar graphs were produced at a given strain interval, making it possible to examine the dependence of the constitutive constants on strain.

With all the graphs obtained, it is possible to plot the constitutive parameters such as the stress exponent n, activation energy Q and the stress multiplier α against the true strain.

Figure 7 shows the variation of the activation energy with strain. At a strain of 0.2 the activation energy reaches a value of 131 kJ/mol, which is approximately the value of self-diffusion for magnesium (133-138 kJ/mol) as reported elsewhere [12, 13]. According to [14] the order of magnitude of Q/RT_m for close packed metals such as magnesium should be approximately 18. The value of Q/RT_m at the peak stress obtained is 17.6, being quite close to the theoretical value.

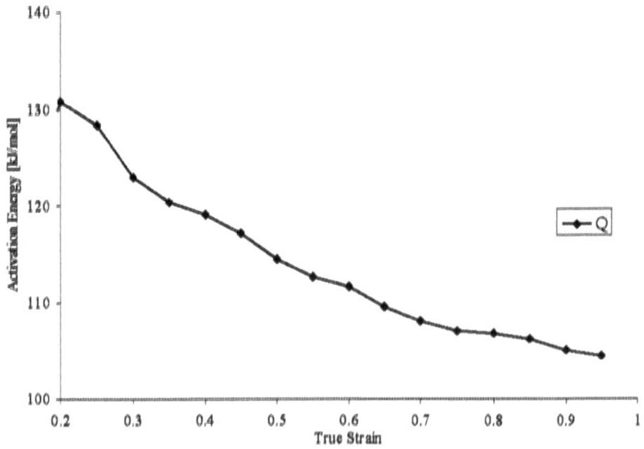

Figure 7. The activation energy versus strain for AZ61.

During hot deformation of magnesium, dynamic recrystallization takes place; while deformation proceeds, new grains are formed. An increased number of grain boundaries and thus enhanced diffusion may explain the reduction of the activation energy during deformation. The increased ease of diffusion may reduce the activation energy required to about 92 kJ/mol [15]. Another explanation for the reducing activation energy with strain is the method used for determining the α values. To make the slopes of the $\ln[\sinh(\alpha\sigma)]$ versus $\ln(\dot{\varepsilon})$ as close to each other as possible, low α values were used with increasing strain. As a result, the slopes of the $\ln[\sinh(\alpha\sigma)]$ versus $1000/T$ were relatively flat. Since the average slope of these plots gives the S value for the calculation of Q, a low Q value can be obtained.

Figure 8 shows the stress exponent n and the stress multiplier α versus strain. The α value is said to influence both the n and Q values. In the literature an α value of 0.052 MPa^{-1} is often used for calculation, also for some aluminum alloys [16, 17, 18, 19]. However, Spigarelli chose three different values of this constant in his creep tests between 120 and 180 °C, in order to find out the impact of fixing the α value. He found that the α value did not affect the Q value too much, but the impact on the n value was much greater: it varied between 2 and 5. In the present research the α value was calculated as a function of strain, and thus a more consistent insight could be obtained into the correlations of α and n with strain. As can be seen from Figure 8, the α value starts at approximately 0.01 MPa^{-1} and decreases to a steady value of 0.004 MPa^{-1}, which is much lower than that used by Spigarelli. The value of the stress exponent n varies inversely with α. The calculated n value is comparable with that found in the literature [12, 16]. However, in contrast to the stress multiplier α, n does not seem to stay at a stable level but increases with increasing strain.

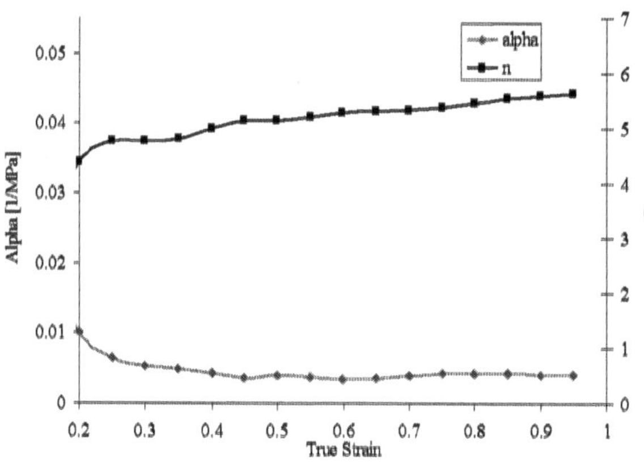

Figure 8. Stress exponent n and stress multiplier α versus strain.

The constitutive constants obtained at selected strains can now be used to recalculate flow stresses using Equation (3) and compare the results with the measured values. Figure 9 clearly shows a reasonable agreement of the calculated stresses with the actually measured ones.

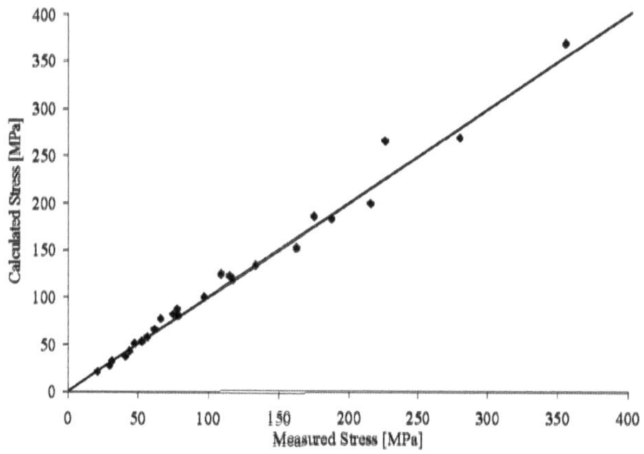

Figure 9. Comparison between calculated and measured stresses at a strain of 0.6.

For an average strain rate sensitivity value v, a plot of the peak stress versus the Zener-Hollomon parameter can be made. For this plot, the flow stress curves displaying a peak may be used. By fitting a power law in the form of Equation (11), the v value can be determined.

$$\sigma = CZ^{v} \tag{11}$$

Such a plot is shown in Figure 10. A v value of 0.13 is in the same order as that found by Beer and Barnett for AZ31 [1]. Adding the data from the experiments without a distinct peak stress led to a similarly shaped curve which deviates itself slightly from Equation 11.

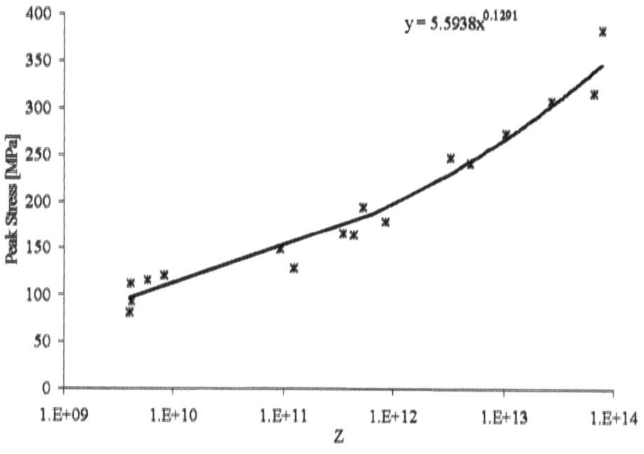

Figure 10. Variation of the peak stress as a function of the Zener-Hollomon parameter.

The results of the constitutive analysis show that applying the sine hyperbolic law at given strains provide strain-dependent constitutive parameters. Comparison between the calculated stresses to measured ones confirms that the method of analysis at constant strain works for the wrought magnesium alloy AZ61. If only the peak stress is used for constitutive analysis, the changing flow stress with strain during deformation will result in inaccurate prediction of deformation resistance during a bulk forming process such as extrusion. With the current approach, an accurate stress-strain database can be established.

Conclusions

1. Wrought magnesium alloy AZ61 can be deformed between 250 °C and 450 °C at strain rates up to 100 s^{-1} without cracking.

2. The strain softening and the shift of the peak stress to higher strain with increasing strain rate suggest the occurrence of dynamic recrystallization.

3. The constitutive parameters were calculated as a function of strain and these indeed displayed strain-dependence.

4. The flow stresses calculated from the hyperbolic sine equation at given strains agrees with the stresses measured during uniaxial compression testing. This constitutive law was proven to fit with the data over the wide ranges of temperatures and strain rates for the AZ61 magnesium alloy.

Acknowledgement

The authors gratefully acknowledge the *Innovatieve Onderzoeksprogramma* (IOP) for funding this research under project code IOT 3003

References

1 A.G Beer, M.R Barnett, "Influence of Initial Microstructure on the Hot Working Flow Stress of Mg-3Al-1Zn", *Materials Science and Engineering*, A423 (2006), 292-299.

2 S.-B Yi et al, "Deformation and Texture Evolution in AZ31 Magnesium Alloy During Uniaxial Loading", *Acta Materialia*, 54, 2006, 549-562.

3 Q. Guo, "Behaviour of AZ31 Magnesium Alloy During Compression at Elevated Temperatures", *Materials Science and Technology*, 21, 2005, 1349-1354.

4 Datasheet AZ31: Magnesium AZ31B, Engineering Alloys Digest Inc. 1962, © ASM international 2002

5 Datasheet AZ61: Magnesium AZ61A, Engineering Alloys Digest Inc. 1959, © ASM international 2002

6 S. Kleiner, P.J. Uggowitzer, "Mechanical Anisotropy of Extruded Mg-6% Al-1% Zn Alloy", *Materials Science and Engineering*, A379 2004, 258-263.

7 H.T. Zhou et al, "Microstructural Evolution of AZ61 Magnesium Alloy During Hot Deformation", *Materials Science and Technology*, 20, 2004, 1397-1402.

8 Y. Yoshida, et al, "Realization of High Strength and High Ductility for AZ61 Magnesium Alloy by Severe Warm Working", *Science and Technology of Advanced Materials*, 6, 2005, 185-194.

9 F.A. Slooff et al, "Data Correction for Constitutive Analysis of Wrought Magnesium Alloys Based on Compression Tests", submitted for publication in the proceedings of the 7th International Conference on Magnesium Alloys and their Applications, 6 - 9 November 2006, Dresden, Germany.

10 M.R. Barnett, "Influence of Deformation Conditions and Texture on the High Temperature Flow Stress of Magnesium AZ31", *Journal of Light Metals*, 1, 2001, 167-177.

11 H.J. McQueen, N.D. Ryan, "Constitutive Analysis in Hot Working", *Materials Science and Engineering*, A32, 2002, 43-63.

12 L. Li, J. Zhou, and J. Duszczyk, "Determination of the Flow Stress of AZ31 B Magnesium Alloy During Hot Deformation", *Journal of Materials Processing Technology*, 172, 2006, 372-380.

13 A. Mwembela, H.J. McQueen, "Hot Workability of Mg Alloys AZ31 and AZ31-Mn", *Recent Developments in Light Metals*, 1994, 365-373.

14 D.A. Porter and K.E. Easterling, *Phase Transformations in Metals and Alloys* (Boca Raton, Florida, CRC Press, 1992, 2004 reprint), 75-81.

15 H.K. Kim, W.J. Kim, "Microstructural Instability and Strength of an AZ31 Mg Alloy After Severe Plastic Deformation", *Materials Science and Engineering*, A385, 2004, 300-308.

16 S. Spigarelli et al, "Analysis of the Creep Behaviour of a Thixoformed AZ91 Magnesium Alloy", *Materials Science and Engineering* , A289, 2000, 172-181.

17 R.V. Ravi Kumar et al, "Grain Refinement in AZ91 Magnesium Alloy During Thermomechanical Processing", *Materials Science and Engineering*, A359, 2003, 150-157.

18 H.J. McQueen, M. Sauerborn, "Hot Workability and Extrusion Modelling of Magnesium Alloys", Carl Hamser Verlag, Munchen, Z. Metallkunde,. 96, 2005, 638-644.

19 E. Cerri et al, "Comparative Hot Workability of 7012 and 7075 Alloys after Different Pretreatments", *Materials Science and Engineering*, A197, 1995, 181-198.

From *Magnesium Technology 2010*, S.R. Agnew, N.R. Neelameggham, E.A. Nyberg, and W.H. Sillekens, Editors

THE EVOLUTION OF IN-GRAIN MISORIENTATION AXES (IGMA) DURING DEFORMATION OF WROUGHT MAGNESIUM ALLOY AZ31

Y.B. Chun[1], C.H.J. Davies[1]
ARC Centre of Excellence for Design in Light Metals, Department of Materials Engineering, Monash University, Victoria 3800, Australia

Magnesium, Electron back-scattering diffraction, In-grain misorientation axes (IGMA), Slip, Deformation twinning, Deformation bands

Abstract

Understanding deformation mechanisms is a prerequisite for the development of more formable magnesium alloys. We have developed a novel approach based on analysis of in-grain misorientation axes which allows identification of the dominant slip system for a large number of grains. We investigated the effects of orientations and temperatures on active deformation mechanisms during the rolling of AZ31, including slip, deformation twinning and deformation banding. The IGMA analysis suggests that increasing rolling temperature promotes activation of prism <a> slip which enhances the rollability of the plate favorably oriented for this slip mode. The approach also reveals an orientation-dependent occurrence of deformation banding and its crystallographic relationship with parent grain. It is concluded that IGMA analysis can be effectively used to study deformation mechanism in hcp metals, and can be used as a criterion for validating some crystal plasticity models.

Introduction

Magnesium and its alloys have attracted great interest for structural applications owing to their low density and hence high specific strength. In this regard, die-cast magnesium alloys have expanded their application in automotive and electronic industries. However, the use of wrought magnesium alloys for structural applications has not developed although they have superior mechanical properties [1] to their cast counterparts. One factor that limits the widespread application of wrought magnesium alloys is their limited low-temperature formability [2], which was attributed to an insufficient number of operable slip modes in magnesium. Although this is not strictly true (deformation including non-basal <a> [3-5] and pyramidal <c+a> slip [6-9] and twinning have all been observed at room temperature), slip on non-basal plane is not readily activated at room temperature because of the high critical resolved shear stress (crss) [7, 10].

At elevated temperatures, magnesium alloys become quite ductile and are readily formable and therefore wrought magnesium alloys are usually hot formed to avoid cracking. Increased activity of pyramidal <c+a> slip has been observed at higher temperatures [6, 7] and is thought to be responsible for the improved formability since this mode alone can provide five independent slip systems and also accommodate strain along the c-axis. A better understanding of the underlying mechanisms responsible for enhanced formability of magnesium alloys at elevated temperatures is important because this knowledge may be reflected in alloy and/or process design strategies to develop more formable wrought magnesium alloys.

The current work aims to enhance the understanding of deformation mechanism of wrought magnesium alloy AZ31. For this purpose, we developed a technique based on analysis of the IGMA distribution and permits investigation of deformation mechanism for a large number of grains. Thus we can identify the active slip mode operating in individual grains. Using this technique, effects of orientation and rolling temperature on deformation mechanism of commercial AZ31 alloy were investigated, with particular attention being focused on activation of prism <a> slip.

Experimental procedure

The material used in the current study is wrought magnesium AZ31 plate. Received in the thermally stabilized condition, the material exhibited strong basal texture (Fig. 1(a)) and consisted of equiaxed grains with an average grain size of 20 µm. The 4 mm-thick plates were machined from the 76mm thick as-received material in different orientations so as to give four different starting textures (Fig. 1).

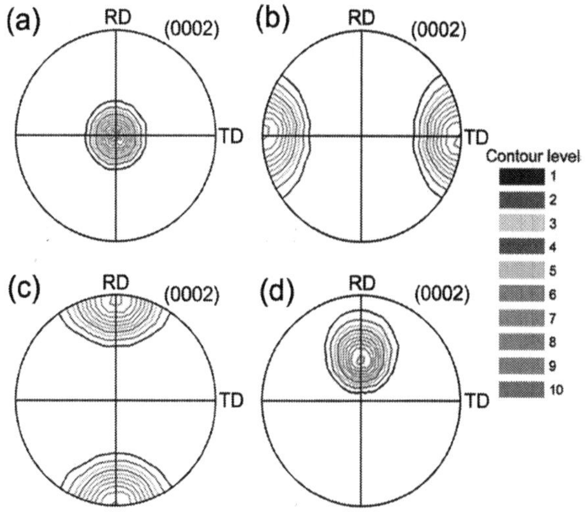

Figure 1. Starting textures, represented by (0002) pole figures, of AZ31 plates used for rolling: (a) the NB, (b) TB, (c) RB and (d) N45RB plates.

The plates thus prepared were rolled to various reductions from 10% to 44% in thickness at 25°C (cold rolling) and 200°C (warm rolling). Microstructures of the rolled plates were characterized by using EBSD for which JEOL 7001F field-emission-gun scanning-electron microscope equipped with Nordlys II detector and HKL Technology Channel 5.0 acquisition system were used. EBSD specimens were prepared by cutting transverse cross sections from the rolled plate which were then electro-polished in a solution consisting of 200 ml nitric acid and 800 ml methanol at 25°C under the applied voltage of 5V, following mechanical grinding. In EBSD analysis, high-angle grain boundaries (HAGB) were assigned to points with a lattice misorientation, θ, exceeding 10°. The total area of EBSD measurement for each material is 90,000 μm^2 (450 μm×200 μm). The microstructures characterized by EBSD were presented as inverse-pole-figure (IPF) maps in which the colors shown correspond to the crystallographic directions parallel to the TD of the rolled plate (The ND, TD and RD have their usual meaning in the rolled plate).

'IGMA analysis' is based on slip-induced lattice rotation and determination of its rotation axis. The principle of this technique was tested for Zr [11] and Be [12,13], using the Laue X-ray technique to identify the axis of lattice rotation in deformed single crystals. Determination of the active slip mode is based on the following assumptions: 1) the Laue asterism arises from a bending of the crystal lattice under the action of slip, 2) this bending takes places about an axis (termed the 'Taylor axis'), and gives rise to a rotation of all crystallographic planes about this axis, and 3) the Taylor axis is at a crystallographic direction of low indices lying in the slip plane and perpendicular to the slip direction. Under the second assumption, the Taylor axis is defined as:

$$T_S = n_S \times d_S \quad (1)$$

where T_S, n_S and d_S are the Taylor axis, slip plane normal and slip direction, respectively, for a given slip mode, S. It is thus possible to determine the dominant slip mode in a deformed single grain simply by matching the Taylor axis for a given slip mode to the experimentally-measured IGMA. The Taylor axes for slip modes available in Mg are summarized in Table 1.

Table 1. Slip modes available in magnesium and corresponding Taylor axes

Deformation mode	Number of slip variants	Taylor axis	Number of variants of the Taylor axis
(0001)<1120>	3	<1010>	3
{1010}<1120>	3	<0001>	1
{1122}<1123>	6	<1010>	3

The application of IGMA analysis has been restricted to single crystals deformed to strains below ~0.1, due to the nature of the Laue X-ray technique employed and the heavy surface upheavals involved at large strains. By employing EBSD technique to determine the IGMA, we have extended its applicability to polycrystals deformed to larger strains. The distribution of IGMA for a given deformed grain is obtained from EBSD data by the following procedure: 1) selection of a deformed grain of interest, (ii) determination of the misorientation axes for all possible neighboring lattice-point pairs in the grain, and (3) plot of the misorientation axes on standard unit triangle.

The use of the EBSD technique to determine IGMA, however, needs to be approached with caution due to the limited angular resolution of the EBSD analysis: the minimum misorientation angle reliable for analysis should be determined beforehand. In preliminary work we found that the angular resolution is 0.5° which is taken as the lower cut-off angle for IGMA analysis. The upper cut-off angle is set to 2.0° since the preliminary work revealed that the qualitative distribution of IGMA is not significantly affected by the upper cut-off angle. More importantly a higher misorientation angles might result from the formation of deformation bands, and the misorientation axes between matrix and the deformation band do not deliver information on active slip modes.

Results and discussion

<u>Orientation and temperature effects on rollability</u>

Rollability of AZ31 was highly dependent on starting texture and was enhanced by increasing the rolling temperature. All four starting textures were cold-rollable without introducing edge cracks up to a modest 10% reduction. However, further increase in cold reduction resulted in initiation of cracks which formed at ±45° to the rolling plane: the cracks began to form at cold reductions of 14%, 19%, 14%, and 25% for the NB, TB, RB and N45RB plates, respectively. Increasing rolling temperature to 200°C improved the rollability of all plates, and the most remarkable enhancement was found for the TB plate: the NB, TB, RB and N45RB plates were deformable up to 21%, 44%, 21% and 40% reductions, respectively.

<u>Microstructure evolution during 10% rolling</u>

The microstructures of the 10% cold and warm rolled plates are presented as IPF maps for the TD in Figures 2 and 3, respectively. All the plates exhibit the twinned microstructures irrespective of the rolling temperatures, although the twinning propensity of the rolled plate varies depending on the starting texture: the RB plates were most twinned while the NB plates showed the least twinned microstructure.

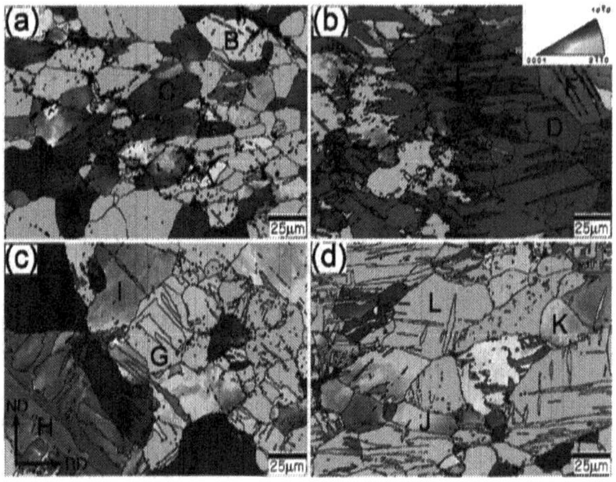

Figure 2. IPF maps of the 10% cold-rolled AZ31 plates: (a) the NB, (b) TB, (c) RB and (d) N45RB plates.

Analyses of the misorientation angle distributions and misorientation axes of the rolled plates revealed that each plate with different starting textures exhibits a distinct distribution pattern while the distribution patterns are not significantly affected by rolling temperature. Active twinning modes were deduced from the misorientation angle/axis pairs of the twin boundaries developed in the 10% cold-rolled plates (Figure 4).

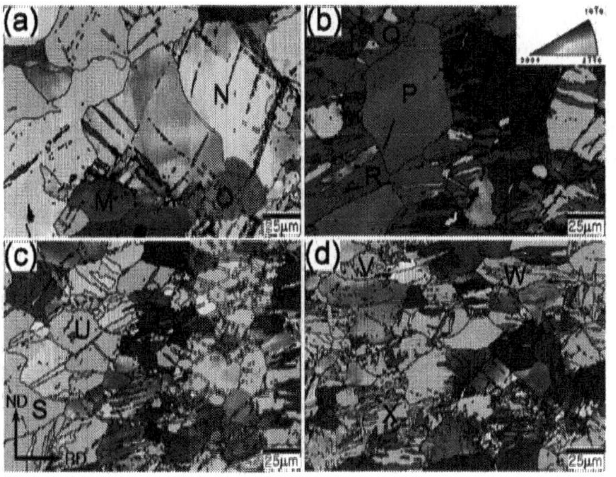

Figure 3. IPF maps of the 10% warm-rolled AZ31 plates: (a) the NB, (b) TB, (c) RB and (d) N45RB plates.

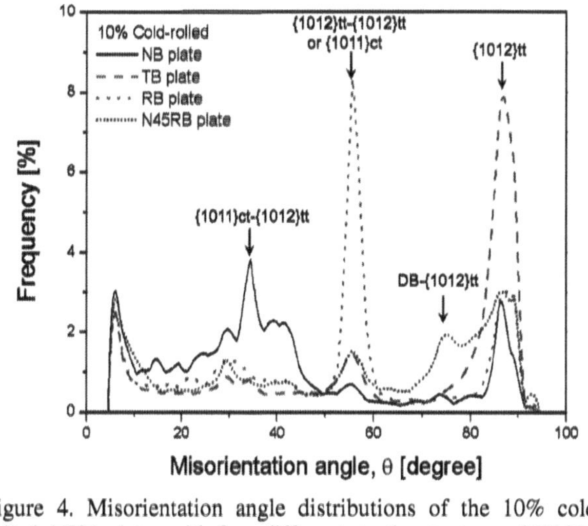

Figure 4. Misorientation angle distributions of the 10% cold-rolled AZ31 plates with four different starting textures. {1012}tt, {1011}ct and DB represent {1012} tension twin, {1011} compression twin and deformation band.

In the case of the NB plate, most of twins existed in the form of {1011}–{1012}double twins (~38°<1120>), while {1012} tension twins (~86°<1120>) dominated in the TB plate. In the RB plates, two distinct peaks at ~60° and ~86° are notable but are found to result from the same source, i.e., {1012} tension twin: the former peak is a result of impingement of two different {1012} twin variants activated within a grain [14]. For the N45RB plate, a peak at 70~80°, which has not been reported yet, was observed and the corresponding boundaries to the peak were found to be misoriented about the <1120> axis. This misorientation relationship is a result of {1012} tension twinning within a deformation band. The mechanism governing this phenomenon will be discussed in detail later in a separate section.

Distribution characteristics of IGMA

From the deformed grains marked as 'A~L' and 'M~X' on the IPF maps in Figures 2 and 3, the distribution patterns of IGMA were analyzed (Figures 5 and 6, respectively). Each deformed grain exhibits a characteristic distribution of IGMA which is quite dependent on its orientation. For the 10% cold-rolled plates, grains with c-axes in a near-ND showed either a uniform distribution (grains A, C and G) or a distribution with moderate intensities around <hki0> (grain B), whereas moderate to strong intensities around the <0001> axis were observed in the IGMA distributions for the grains with c-axes in a near TD (grains D, E and F). For the grains with c-axes in the directions tilted by ~45° from the ND to either the RD (grains J, K and L) or the TD (grains H and I), the IGMA populated the <hki0> axes.

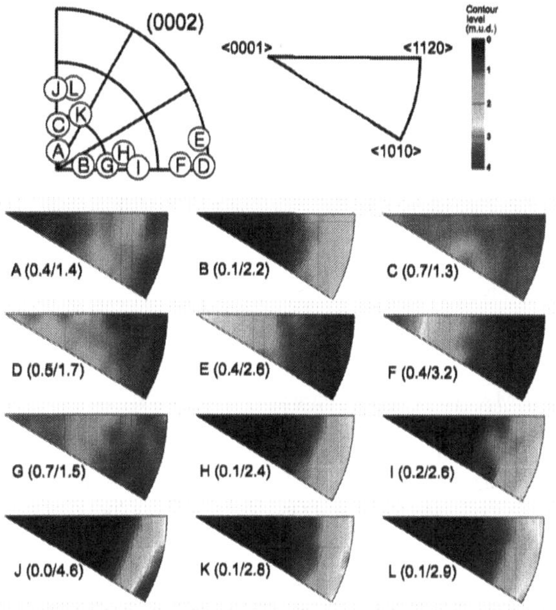

Figure 5. Orientations and IGMA distributions of the deformed grains in the 10% cold-rolled AZ31 plates. 'A~L' correspond to the grains shown in Figure 2. The minimum/maximum intensities (unit: multiple uniform distribution) of the IGMA distribution are given in parentheses. A-C, D-F, G-I and J-L grain are from the NB, TB, RB and N45RB plates, respectively.

The distributions of IGMA for the 10% warm-rolled plates exhibited similar patterns to those for the 10% cold-rolled plates. It is noteworthy, however, that the grains with c-axes in the TD (grains P, Q and R) showed stronger intensities around the <0001> axis in the warm rolled plates than in the cold-rolled counterparts. In addition, the orientation space exhibiting preferred distributions of IGMA around the <0001> axis expanded as the rolling temperature increased: the grain with c-axes in the directions tilted by ~45° from the ND to the TD (grain O) also showed a distribution in which the IGMA populate the <0001> axis (c.f. grains H and I in Figure 5 with grain O in Figure 6).

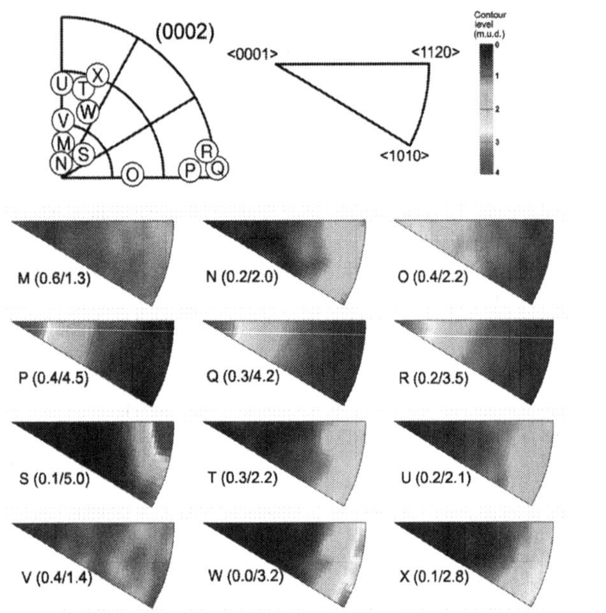

Figure 6. Orientations and IGMA distributions of the deformed grains in the 10% warm-rolled AZ31 plates. 'M~X' correspond to the grains shown in Figure 3. The minimum/maximum intensities (unit: multiple uniform distribution) of the IGMA distribution are given in parentheses. M-O, P-R, S-U and V-X grain are from the NB, TB, RB and N45RB plates, respectively.

The IGMA distributions with strong intensities around the <hki0> axes, which were observed in the grains with c-axes in the directions tilted by ~45° from the ND to the RD, can be attributed to dominant activation of basal <a> slip, since those grains are favorably oriented for this slip mode and because the <1010> axis is the Taylor axis of basal <a> slip (Table 1): the preferred distribution of IGMA around the <hki0> axes can develop as a result of co-activation of variants of basal <a> slip. One may ascribe such a distribution to an activation of {1122}<1123> slip since the <1010> axis is shared as the Taylor axis by the two slip modes (Table 1). However, activation of the <c+a> slip mode can be excluded from consideration because the critical resolved shear stress for this mode is quite high [7, 10] and the grains mentioned above are unfavorably oriented for the slip mode.

On the other hand, grains with c-axes in the TD showed a consistent distribution pattern of IGMA, i.e., preferred distribution of IGMA around the <0001> axis. Since the <0001> axis is the Taylor axis of prism <a> slip and those grains are favorably oriented for this slip mode, such a distribution can be attributed to dominant activation of prism slip. Increasing rolling temperature strengthens the intensity of IGMA around the <0001> axis and expands orientation space in which such distributions can develop. This suggests an enhanced activity of prism <a> slip at elevated temperatures, which can in turn explain the remarkable enhancement of rollability of the TB plate at 200°C. The origin of the enhanced activity of prism <a> slip at elevated temperature is not clear yet. Reduced Peierls stress at elevated temperature may a possible explanation for lowered critical resolved shear stress for prism <a> slip, but it seems so far unclear why the extent of reduction in Peierls stress is greater for prism <a> slip than those of other slip modes.

A corollary of this investigation is that the IGMA distribution may be used as a comparative criterion in validating the results obtained from deformation modeling. Good agreements in the deformation texture and stress-strain response between simulation and experiment have been important criteria for substantiating the validity of modeling and the combination of relative crss values thus obtained. In this regard, the IGMA distribution can be added as another criterion since it also reflects the active slip modes and the interaction between them within the individual grains. It seems to be useful in determining the relative crss values and/or hardening rates of slip modes governing plastic deformation of metals if comparison of the experimentally-observed IGMA distributions with those obtained from modeling is made.

Orientation-dependent deformation banding

Grain boundaries with 70~80°<1120> relation to the matrix were observed in both cold-rolled and warm-rolled N45RB plates (we also found the same feature in the N45RB samples which are deformed in tension along the RD.). These boundaries were found to correspond to those formed between the matrix and the tension twin formed within deformation bands as shown in Figure 7(a). In fact, the nature of the band is not clear at the moment, but we have used the term 'deformation band' since the latter is an umbrella term describing any bands which are induced by deformation.

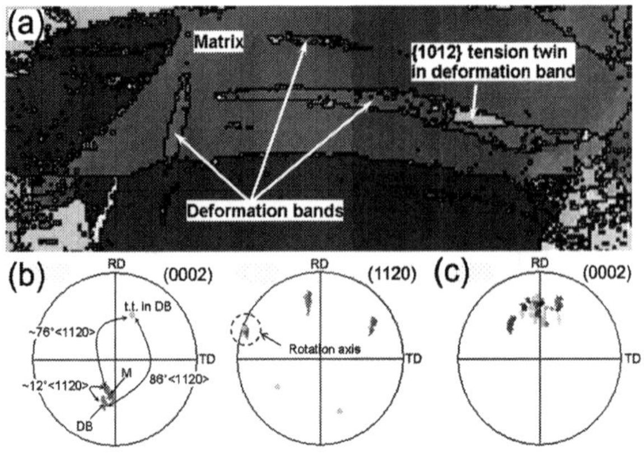

Figure 7. (a) IPF map of the 10% cold-rolled N45RB plate and (b) (0002) and (1120) discrete pole figures of the matrix, deformation bands and {1012} tension twin formed within deformation band. Shown in (c) are orientations of the grains in which deformation bands were observed. M, DB and t.t. on the (0002) discrete pole figure in (b) represent the matrix, deformation band and tension twin, respectively.

Such deformation bands were found to have a special orientation relationship of 5~15°<1120> with the surrounding parent matrix material but their misfit angles varied depending on the amount of deformation. It was also found that the deformation bands are usually formed in regions of the matrix with c-axes tilted by 55~75° from the ND to the RD (Figure 7(c)). Detailed analysis of the deformation band revealed that the formation of the bands results in reorientation of c-axes towards the RD by their misfit angles, i.e., 5~15° (Figure 7(b)). This in turn creates a condition under which the tension twins are readily activated in the deformation bands. This eventually leads to a formation of boundaries with 70~80°<1120> relation between the matrix and the tension twin since the tension twin and deformation band have 86°<1120> and 5~15°<1120> relations, respectively, with their parent grains. Since such bands were found to form only in the grains which are favorably oriented for basal <a> slip, it is assumed that the band is formed as a result of strain localization. The deformation banding seems to act as an additional deformation mechanism other than slip and twinning, and contributes indirectly to the overall strain accommodation by promoting tension twinning.

Conclusions

The present study demonstrates that deformed grains in rolled AZ31 exhibit characteristic IGMA distribution, the patterns of which vary depending on crystallographic orientations. The dominant slip mode activated in individual grains can be deduced from the IGMA analysis, which highlights the usefulness of the technique for studying deformation mechanism of Mg alloys. IGMA analyses for rolled AZ31 plates also provided experimental evidence for the activation of basal <a> slip in the grains favorably oriented for that slip mode, and revealed an increased activity of prism <a> slip at elevated temperature. It is also found that deformation bands, which form in the grains with certain orientations, reorient c-axes towards the RD and thereby promote activation of tension twinning.

Acknowledgements

This work is supported by the Australian Research Council (ARC) through the ARC Centre of Excellence for Design in Light Metals.

References

1. C.J. Bettles and M.A. Gibson, "Current wrought magnesium alloys: strengths and weaknesses," *Journal of Metals*, 57 (2005), 46-49.

2. M.R. Barnett, N. Stanford, P. Cizek, A. Beer, Z. Xuebin, and Z. Keshavarz, "Deformation mechanisms in Mg alloys and the challenge of extending room-temperature plasticity," *Journal of Metals*, 61 (2009), 19-24.

3. P. Ward Flynn, J. Mote, J.E. Dorn, "On the thermally activated mechanism of prismatic slip in magnesium single crystals," *Transactions of the Metallurgical Society of AIME*, 221 (1961), 1148-1154.

4. R.M. Quimby, J.D. Mote, J.E. Dorn, "Yield point phenomena in magnesium-lithium single crystals," *Transactions of the ASM*, 55 (1962), 149-157.

5. J. Koike, R. Ohyama, "Geometrical criterion for the activation of prismatic slip in AZ61 Mg alloy sheet deformed at room temperature," *Acta Materialia*, 53 (2005), 1963-1972.

6. J.F. Stohr, J.P. Poirier, "Electron microscope study of {1122}<1123> pyramidal slip in magnesium," *Philosophical Magazine*, 25 (1972), 1313-1329.

7. T. Obara, H. Yoshinga, S. Morozumi, "{1122}<1123> Slip system in magnesium," *Acta Metallurgica*, 21 (1973), 845-853.

8. S.R. Agnew, M.H. Yoo, C.N. Tome, "Application of texture simulation to understanding mechanical behavior of Mg and solid solution alloys containing Li or Y," *Acta Materialia*, 49 (2001), 4277-4289.

9. M.H. Yoo, J.R. Morris, K.M. Ho, S.R. Agnew, "Nonbasal deformation modes of HCP metals and alloys: Role of dislocation source and mobility," *Metallurgical and Materials Transactions A*, 33 (2002), 813-822.

10. E.W. Kelly, W.F. Hosford Jr., "Plane-strain compression of magnesium and magnesium alloy crystals," *Transactions of the Metallurgical Society of AIME*, 242 (1968), 5-13.

11. E.J. Rapperport, C.S. Hartley, "Deformation modes of zirconium at 77°, 575°, and 1075°K," *Transactions of the Metallurgical Society of AIME*, 21 (1960), 869-877.

12. P. Bastien, P. Pointu, "Determination des elements de glissement des metaux hexagonaux par L'asterisme des taches de Laue et application a L'etude de la deformation du beryllium a haute temperature," *Journal of Nuclear Materials*, 5 (1962), 101-108.

13. W. Taylor, A. Moore, "Tensile deformation modes in polycrystalline beryllium near the ductile-brittle transition," *Journal of Nuclear Materials*, 13 (1964), 23-27.

14. M.D. Nave, M.R. Barnett, "Microstructures and textures of pure magnesium deformed in plane-strain compression," *Scripta Materialia*, 51 (2004), 881-885.

PLASTIC ANISOTROPY OF MAGNESIUM ALLOY AZ31B SHEET

S. R. Agnew

Department of Materials Science and Engineering
University of Virginia
Charlottesville, Virginia 22904-4745

Abstract

The anisotropic plastic responses of commercial magnesium alloy AZ31B sheets in the stress-relieved and soft-annealed conditions are found to be similar. Measurements of the r-value (r = d□$_w$/d□$_t$) for tensile samples oriented 0, 45, and 90° from the rolling direction highlight a strong in-plane anisotropy, with r(0°)~1 and r(90°)~3.5. This in-plane anisotropy is somewhat surprising given the nearly ideal basal texture present in the material. Simulations of the plastic response which incorporate the initial texture and the strain hardening of the individual grain-level deformation modes enable us to conclude that the slight off-basal character of the texture is enough to induce a change in the balance between the activities of basal and prismatic slip. The tendency of the grain's basal poles to tilt from the normal direction towards the rolling direction is enough to enable easy basal slip to accommodate the majority of the strain during tension along the rolling direction (0°). During tension along the transverse direction (90°), however, the majority of the strain is accommodated by the harder prismatic slip mode, which induces little change in the thickness, while producing a strong reduction in the width.

Introduction

Interest in the application of lightweight magnesium alloys has been growing over the past decade[1,2]. One problem facing the application of wrought magnesium alloys, in particular, is their poor cold forming properties. Historically, this has been linked with an inadequate number of independent slip systems[3]. An issue that has received less attention is the impact of the crystallographic texture induced during primary forming operations, such as rolling, upon subsequent forming steps, such as sheet stamping. It is well established that texture induced plastic anisotropy in sheet metal can give rise to a variety of formability problems. The earing behavior of deep drawn aluminum cans is a classical example[4]. Continuum mechanical models for predicting sheet metal formability utilize to a variety of material parameters including the strain hardening rate 'n', rate sensitivity 'm', out-of-plane anisotropy parameter 'r', as well as in-plane anisotropy[5].

During this series of short reports[6,7], the anisotropic plasticity of commercial magnesium alloy AZ31B sheet material has been explored experimentally by measuring the tensile stress-strain response and r-values of samples oriented 0, 45, and 90° from the rolling direction (RD). The r-value is defined as the ratio between the width and thickness strains, $r = \varepsilon_w/\varepsilon_t$, at an elongation of □$_l$ = 20% or at the limit of uniform elongation, for less ductile metals[8]. Further, samples have been compression tested along the sheet normal direction (ND) to determine how the compressive stress-strain response differs from the tensile.

Finally, these properties have been modeled theoretically using the viscoplastic self-consistent approach[9]. Not only are the stress-strain response and the anisotropy modeled, but also the evolution of texture during the mechanical tests, which may be used as another means of validating the simulations. The current article serves to summarize the recent results, provide insights into the underlying physical phenomena that govern the macroscopic behavior of polycrystalline magnesium alloys, and suggest directions for future research in the area of magnesium forming.

Experimental Procedure

Commercial magnesium alloy AZ31B (with nominal composition 3 wt% Al, 1 wt% Zn, balance Mg) sheet material with 0.125" (3.175 mm) thickness was received in the stress-relieved (H24) temper. Some of the sheet material was heated to 340°C followed by furnace cooling to produce the soft-annealed (O) temper. The crystallographic textures were measured at the sheet midsections using X-ray diffraction with reflection geometry and Cu-K□ radiation. The (0002), $(10\bar{1}0)$ and $(10\bar{1}1)$ pole figures were collected to a sample tilt of 80°. Complete pole figures and orientation distribution functions were determined using the WIMV subroutine of the popLA software[10]. Metallography was performed on three perpendicular planes using an acetic-picral etch[11]. Tensile samples with a reduced section 12.7 x 2.8 x 2.4 mm were electro-discharge machined. The tests were conducted at constant crosshead speed, giving an initial strain rate of $4x10^{-3}$ s^{-1}. The objectives were to determine any differences in the r-values of the two different tempers and to test for in-plane anisotropy. Since AZ31B sheet typically only has a fracture ductility of ~20%, tests were interrupted at the limit of uniform elongation. The gage sections were measured after the test with a micrometer to determine the width and thickness strains, and hence the r-values. The r-values were also calculated using the width and length strains, $r = -\varepsilon_w/(\varepsilon_l + \varepsilon_w)$, assuming incompressible plasticity *via* the relationship □V = □$_l$ + □$_w$ + □$_t$ = 0. Compression tests on short cylindrical through-thickness samples (Ø6.3 x 3.3 mm) were performed using Teflon as a lubricant, again with an initial strain rate of $4x10^{-3}$ s^{-1}. The interest in this case was to measure the strain distribution within the rolling-transverse direction (RD-TD) plane by defining an "r-value" type parameter (*r(comp)* = $\square_{RD}/\square_{TD}$).

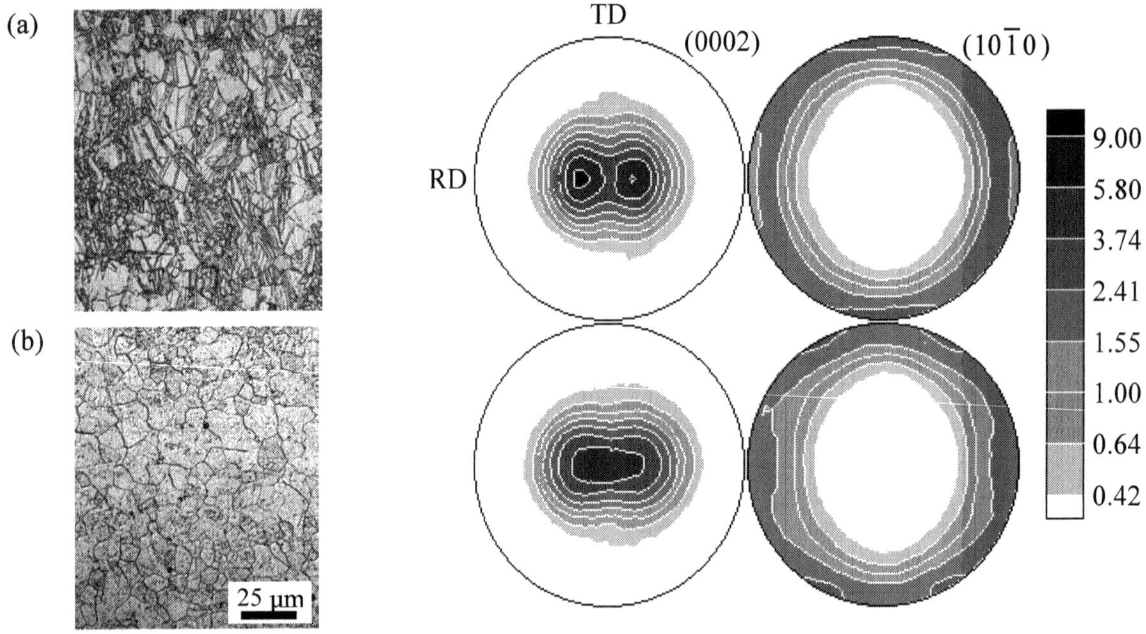

Figure 1: Microstructures and textures of AZ31B sheet material in the (a) stress-relieved (H24) and (b) soft-annealed (O) tempers. (Full pole figures are equal area projections recalculated from the orientation distributions.)

Experimental Results

The microstructure in the H24 temper consisted of pancake shaped grains full of mechanical twins. The soft-annealed O temper had an equiaxed grain structure free of twins (Fig. 1). In general, the textures of the two conditions were similar, although the stress-relieved material did show a slightly stronger, more off-basal character than the annealed (Fig. 1). This that indicates a small rotation of the grains about the transverse direction occurs during recrystallization.

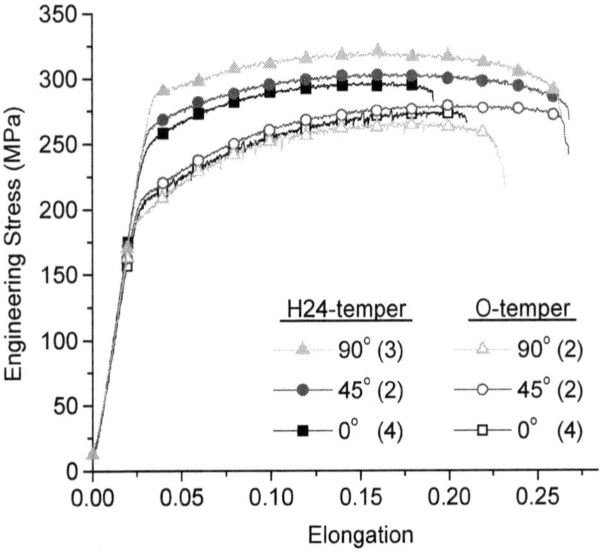

Figure 2. Engineering stress vs. tensile elongation of AZ31B as a function of temper and sample orientation (0, 45, and 90° from the rolling direction.) Note that strain includes machine compliance.

The strain to failure was similar between the two tempers, however, the uniform elongation (the strain at the ultimate strength) was slightly smaller for the hard H24 temper (□~12%) than for the soft O temper (□~14%) (Fig. 2). Although the hardening behaviors of the two tempers were distinct, the different sample orientations of a given temper exhibited very similar hardening behavior. In the case of the H24 temper, the 90° samples are clearly stronger than the 45°, which are stronger than the 0° oriented samples. For the O-temper, sample variations precluded drawing such a conclusion, and it may only be stated that the strengths of the three orientations are similar. Through-thickness compression tests of the annealed material (Fig. 3) emphasize that the compressive yield strength is similar to the tensile, however, the work hardening rate is much higher.

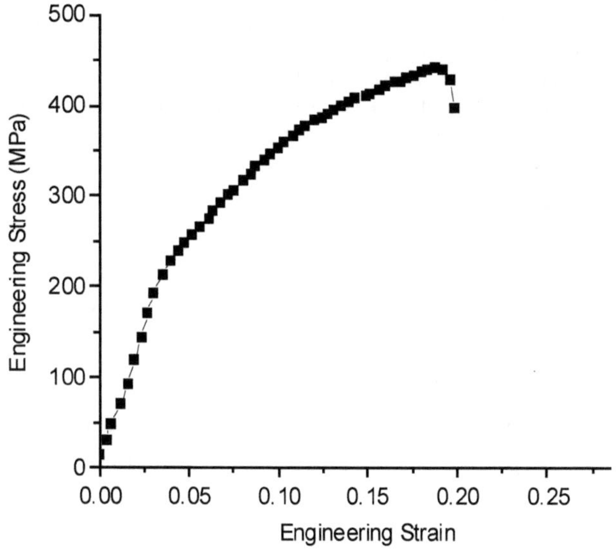

Figure 3: Compression test results from the O-temper material demonstrate that the compressive yield strength is similar to yield in tension. Again, note that strain includes machine compliance.

After the uniform elongation in tension was determined, tests were interrupted at that strain in order to make measurements of the r-value. The r-values show a strong in-plane anisotropy (Fig. 4) that would not be expected from a cursory examination of the texture, which essentially consists of a basal fiber centered on the sheet normal direction. Both tempers exhibit approximately the same anisotropy parameters so all further experiments reported in this study will be for material in the O-temper.

After the strong anisotropy was observed at the limit of uniform elongation, a series of interrupted tests were performed to measure the evolution of the anisotropy with plastic strain (see Fig 4b). During the course of these interrupted tests it was observed that there is a small yield point peak (barely observable in Fig. 2) that returns upon restarting interrupted tests. This suggests a strong interaction between dislocations and the solute atoms in the alloy[12]. In addition to the interrupted tensile tests, three compression tests were performed to different strain levels; and the r-values, all close to *r(comp)* ~2, are also plotted in Fig. 4b.

Modeling Procedure

Simulations of the anisotropic stress-strain response were performed using the viscoplastic self-consistent approach[9], as documented in a previous study[13]. The experimentally measured sheet texture (shown in Fig. 1) was represented by 1000 discrete orientations and used as input. The strain path was introduced as mixed boundary conditions $\dot{\varepsilon}_1 = 1\times10^{-2}$, $\sigma_w = 0$, $\sigma_t = 0$, and off-diagonal strains equal to zero. Other inputs used as fitting parameters were the critical resolved shear stresses (CRSSs) and hardening parameters of the different deformation modes known to be active in magnesium alloys. For this study, the basal *<a>*, prismatic *<a>*, pyramidal *<c+a>* slip, and $\{10\bar{1}2\}$ tensile twinning modes were considered. As a point of emphasis, the dominant slip mode in all hexagonal metals has a slip direction of $<2\bar{1}\bar{1}0>$ or *<a>* contained by the basal plane. For magnesium the primary slip plane is the basal plane, for titanium it is the prismatic plane. Irregardless, strains along the hexagonal c-axis may only be accommodated by hard deformation modes such as pyramidal slip or twinning.

Modeling Results

The texture of the AZ31B sheet under investigation was such that twinning played little role during tensile deformation, so the focus of this discussion will be on slip. Initially, the individual slip systems were assumed to exhibit no hardening in order to determine the approximate ratios between the CRSS values necessary to reproduce the essential aspects of the anisotropy. The anisotropy was characterized by simulating the tensile stress-strain response along the 0° (RD) and 90° (TD) directions. The simulated r-values at each straining step of 1% effective strain were calculated and compared with experimental measurements (Fig. 4b). In the present case, significant *<c+a>* slip was actually shown to suppress the tensile r-value, which was experimentally observed to be r > 1 for nearly all the samples, particularly those of 90° orientation (Fig. 4). Similarly, if basal *<a>* slip completely dominated the straining, low r-values were predicted. Only prismatic *<a>* slip promoted the necessary high r-values. The experimental r-value evolution was fit by adjusting the CRSS ratio between basal and prismatic slip. The best fit (see Fig. 4b) was obtained for a CRSS ratio of 1:3, which indicates that prismatic slip is approximately three times harder than basal slip in magnesium alloy AZ31B.

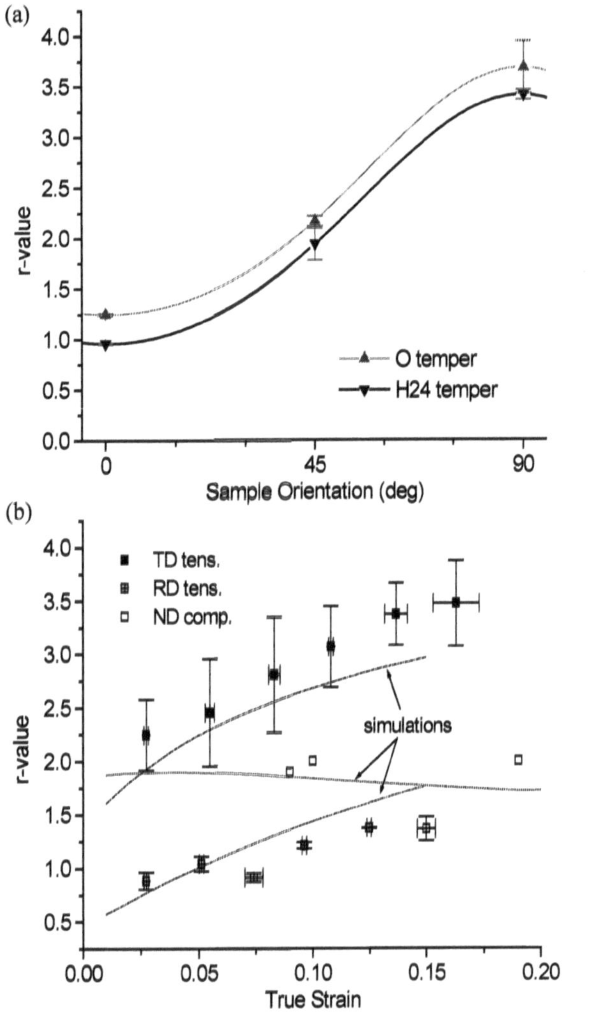

Figure 4: Plots of (a) the r-value of AZ31B as a function of temper and sample orientation (0, 45, and 90° from the rolling direction.) (b) The development of r-value as a function of strain along sheet transverse (TD), rolling (RD) and normal (ND) directions.

The complete hardening response of the annealed material was simulated using an empirical Voce hardening law to describe the hardening of individual slip systems *s*:

$$\tau^s = \tau_0^s + (\tau_1^s + \theta_1^s \Gamma)\{1 - \exp(-\frac{\theta_0^s \Gamma}{\tau_1^s})\}\cdot$$

The □^{s} and □^{s} values are the CRSS and strain hardening parameters of the sth slip system respectively, which were used as fitting parameters. The accumulated strain of the polycrystalline aggregate is □. The parameters determined previously for AZ31B plate[13] were used to formulate an initial guess (see Table 1). The simulations of the r-value evolution and typical stress-strain response are shown in Figures 4 and 5, respectively. Figure 6 shows the fraction of the total strain that is accommodated by each of the slip and twinning modes under consideration averaged over all of the grains. It shows greater prismatic slip activity for the 90° sample orientation, which explains its higher flow strength (Fig. 5) because the prismatic slip mode is harder than the basal slip mode. It also explains the higher r-value of the 90° sample (Fig. 4) because the majority of the grains have their c-axis close to the sheet normal direction and prismatic slip produces neither strain along the c-axis nor rotations of the c-axis during in-plane tension. Another way of conceptualizing this result is to realize that if a hexagonal close packed single crystal were tensile tested perpendicular to its c-axis, prismatic slip would produce only width strain and no thickness strain, giving an infinite r-value.

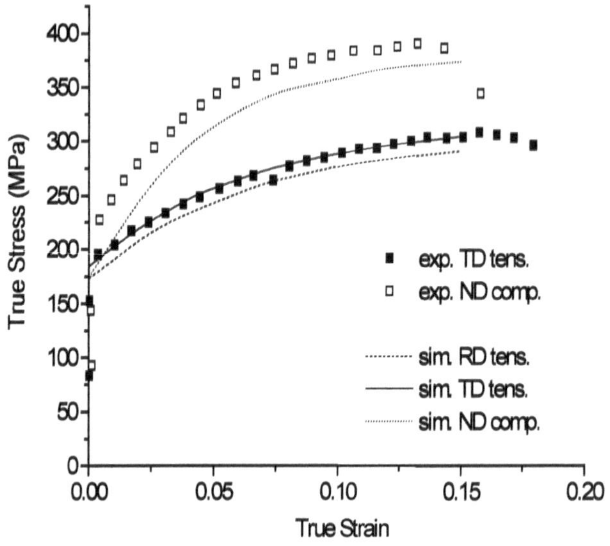

Figure 5: Simulated plastic stress-strain response, fitted to typical flow curves from the annealed material. (Same flow curves as in Figures 2 & 3, with the elastic portion removed and converted to true stress and true strain.)

In addition to simulating the stress-strain response and the anisotropy, the viscoplastic self-consistent modeling approach also provides information about the deformation texture evolution. Figures 7a and 7b show the textures which result from tensile testing along the rolling and transverse directions. Figures 7c and 7d show simulated results, which compare favorably with the experimental measurements. This result gives further confidence that the modeling is indeed capturing the essence of the deformation mechanisms and not simply a result of good "curve-fitting". A closer examination of these textures can provide further insight. For tension along the rolling direction, the fact that the basal planes are slightly inclined towards the rolling direction enables basal slip to accommodate a substantial amount of the strain (Fig 6). During straining, the basal poles rotate toward coincidence with the sheet normal direction (Fig. 7a&c). This rotation will eventually exhaust basal slip and prismatic slip will be forced to compensate (see Fig. 6). For tension along the transverse direction, very few grains are initially favorably oriented for basal slip, whilst prismatic slip is favored (Fig. 6), this does not lead to rotation of the basal poles (Fig. 6b&d).

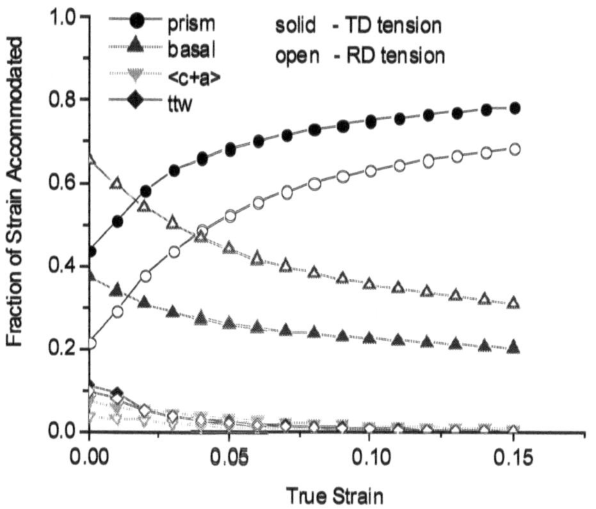

Figure 6: Fraction of strain accommodated by each deformation mode for the different loading paths: in-plane tension along the rolling (RD) and transverse (TD) directions.

Discussion

The current results have stressed the importance of prismatic slip in the general plasticity of polycrystalline magnesium alloys. Initially, this result seems to be in opposition to recently published work by the author, which emphasized the significance of <*c+a*> slip in magnesium[13]. However, the conclusions of that study were based solely upon results obtained from compression tests and compression texture analysis. The results of the two studies may be reconciled as follows. During in-plane tension of basal textured sheet, the Schmid factors for <*c+a*> slip on $\{11\bar{2}2\}$ pyramidal planes and <*a*> slip on prismatic planes are competitive. Therefore, whichever slip mode has a lower CRSS will be active. As can be seen in Table 1 and Figure 6, it is prismatic slip that will be activated during tension. During compression, regardless of the initial texture, basal slip and tensile twinning will cause a basal texture to rapidly develop such that the Schmid factor for prismatic slip is null, while the Schmid factor for <*c+a*> slip is maximized. Thus, compression of a basal slip dominated hexagonal close packed metal, like magnesium, will always call for the activation of the <*c+a*> slip mode. If the alloy in question does not exhibit <*c+a*> slip, it will be likely to behave in a brittle fashion, even during compression. During through-thickness compression, prismatic slip activity is negligible and <*c+a*> slip accommodates a significant fraction of the strain (Fig. 8). The anisotropic stress-strain response during compression was also accurately predicted by the current simulations (Fig. 4b & 5).

Table 1. Strain of Voce strain hardening parameters for the individual deformation modes under consideration

Material	Deformation Mode	$\square_0$	$\square_1$	$\square_0$	$\square_1$
AZ31B plate	basal <*a*>	30	30	300	0
1" (25.40 mm)[13]	pyramidal <*c+a*>	90	80	800	0
	tensile twin	15	0	30	30
AZ31B sheet	basal <*a*>	45	40	300	0
0.125"	prism <*a*>	110	50	300	0
(3.175 mm)[6,7]	pyramidal <*c+a*>	130	170	4000	0
	tensile twin	15	0	30	30

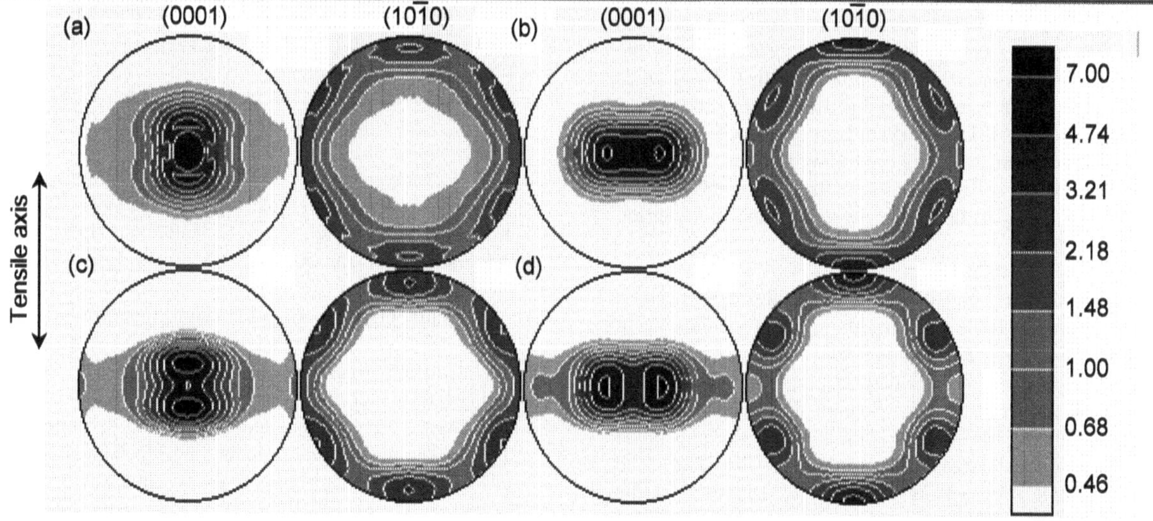

Figure 7: Experimental pole figures collected from tensile samples tested along the (a) rolling direction (RD) and (b) transverse direction (TD) are compared with the simulated pole figures from the (c) RD and (d) TD tests.

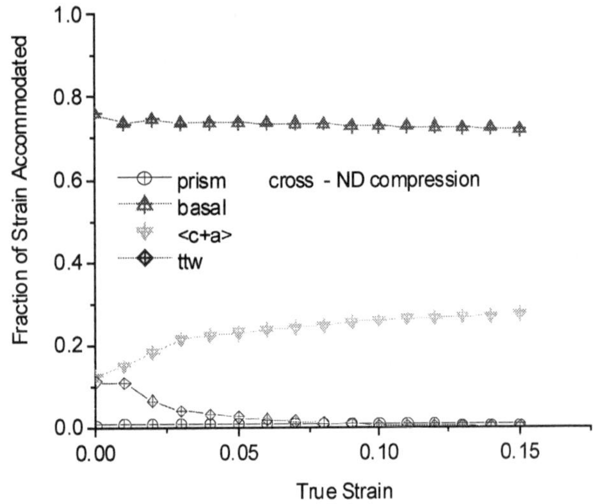

Figure 8: Fraction of strain accommodated by each deformation mode for the different loading paths: (a) in-plane tension along the rolling (RD) and transverse (TD) directions, and (b) through-thickness compression.

Summary

Measurements of the r-value for tensile samples oriented 0, 45, and 90° from the rolling direction highlighted a strong in-plane anisotropy for commercial magnesium alloy AZ31B sheet material, with r(0°)~1 and r(90°)~3.5 (Fig. 4). The anisotropic behaviors of stress-relieved and soft-annealed magnesium alloy AZ31B sheet are similar. In fact, the only reasons the O-temper might be expected to have better forming characteristics are that it has a lower yield strength, a higher initial strain hardening rate, and a slightly higher uniform elongation (Fig. 2). The plastic anisotropy was simulated using a viscoplastic self-consistent model that incorporated the initial texture and the strain hardening behavior of the individual grain-level deformation modes. It was determined that the reason for the in-plane anisotropy was the initial texture. The tendency for grains to have their basal poles rotated slightly from the sheet normal direction towards the rolling direction (Fig. 1) was sufficient to induce a difference in the slip mode activity during tension along the rolling (RD) vs. transverse (TD) directions. Basal <*a*> slip was substantial during RD tension, while prismatic <*a*> slip sustained the majority of the strain during TD tension (Fig. 6).

Finally, through similar testing and modeling of through-thickness compression, it was possible to clarify the roles of secondary slip modes (prismatic <*a*> and pyramidal <*c+a*>) during the deformation of polycrystalline magnesium. During in-plane tension of basal textured sheet the Schmid factors for prismatic <*a*> and pyramidal <*c+a*> slip are similar. Thus the slip mode with the lower critical resolved shear stress (CRSS) will be activated. The simulation results of this study suggest that the CRSS value for prismatic <*a*> slip is lower than that of pyramidal <*c+a*> slip in alloy AZ31B (see Table 1). The net result is that prismatic slip is activated during tension and <*c+a*> slip is only required during compression. It is the relative contribution of the different slip modes that is responsible for the different levels of anisotropy observed along different directions.

The anisotropies highlighted in this study are, at least partially, responsible for the poor cold forming behavior of wrought magnesium alloys. It is also recognized that elevated temperature formability of magnesium can be excellent. Future work will seek to extend the currently developed understanding to deformation of magnesium alloys to elevated temperatures.

Acknowledgements

Research sponsored by the U.S. Dept. of Energy, Assistant Secretary for Energy Efficiency and Renewable Energy, Office of Transportation Technology, Office of Heavy Vehicle Technology, at the Oak Ridge National Laboratory operated by UT-Battelle, LLC, under contract DE-AC05-00OR22725.

References

1. Magnesium Technology 2001, Ed. J. N. Hryn (Warrendale, PA: TMS, 2001).
2. Magnesium Alloys and Their Applications, Ed. K. U. Kainer (Weinheim, Germany: Wiley-VCH, 2000).
3. C. S. Roberts, "The Deformation of Magnesium", in Magnesium and Its Alloys, (New York: John Wiley, 1960), p. 81-107.
4. H. Mecking, "Motivation", in Texture and Anisotropy, Ed. U. F. Kocks, C. N. Tomé, and H.-R. Wenk, (New York: Cambridge, 1998), pp. 2-4.
5. K. S. Chan, "M-K Approach to Calculating Forming Limit Diagrams", in Forming Limit Diagrams: Concepts, Methods, and Applications, Ed. R. H. Wagoner, K. S. Chan, and S. P. Keeler, (Warrendale, PA: TMS, 1989), pp. 73-110.
6. S. R. Agnew, "Measurement and Simulation of Plastic Anisotropy in Magnesium Alloy AZ31B Sheet", in Modeling the Performance of Engineering Structural Materials, Eds. D. R. Lesuer and T.S. Srivatsan, TMS, Warrendale, PA, in press 2001
7. S. R. Agnew, "Plastic Anisotropy in Magnesium Alloy AZ31B Sheet", in Proc. Plasticity 2002, Ed. A. Kahn, in press, 2002.
8. A. K. Ghosh, S. S. Hecker, and S. P. Keeler, in Workability Testing Techniques, Ed. G. E. Dieter, (Metals Park , OH: ASM, 1984), p. 167.
9. R. A. Lebensohn and C. N. Tomé, "A Self-Consistent Anisotropic Approach", Acta Metall., 41 (1993), 2611-2624.
10. J. S. Kallend, U. F. Kocks, A. D. Rollet and H.-R. Wenk, "Operational Texture Analysis", Mat. Sci. Eng., A132 (1991), 1-11.
11. ASM Handbook: Metallography and Microstructures, Vol. 9, Ed. K. Mills et al., (Metals Park, OH: ASM, 1985) p. 426.
12. A.H. Cottrell, "Interaction of dislocations and solute atoms", Acta Metall., 3 (1954) 533.
13. S. R. Agnew, M. H. Yoo, and C. N. Tomé, "Application of Texture Simulation to Understanding Mechanical Behavior of Mg Alloys", Acta Mater., *in press*, 2001.

From *Magnesium Technology 2011*, W.H. Sillekens, S.R. Agnew, N.R. Neelameggham, and S.N. Mathaudhu, Editors

FORMABILITY OF MAGNESIUM SHEET ZE10 AND AZ31 WITH RESPECT TO INITIAL TEXTURE

Lennart Stutz[1], Jan Bohlen[1], Dietmar Letzig[1], Karl Ulrich Kainer[1]
[1]GKSS-Forschungszentrum Geesthacht GmbH; Max-Planck-Straße 1; Geesthacht, D-21502, Germany

Keywords: Formability, magnesium sheet, forming limit, texture, dynamic recrystallisation

Abstract

The commercial application of magnesium alloy sheets is hindered by their low formability and therefore technological and economic constraints. Tailoring the texture has been identified as playing a major role in enhancing the formability of magnesium sheets. In this study, the formability of magnesium sheet ZE10 and AZ31 was investigated. While the texture of AZ31 is of unfavourable basal type, ZE10 shows a significantly different texture with the basal planes being randomly distributed. More basal planes are oriented favourably for basal slip, promising higher formability and thus, lower process temperature. Formability is assessed by means of forming limit diagrams as a representation of material response to various strain paths. Nakajima tests were carried out from room temperature to 250°C. Local strain data is correlated with the microstructural evolution. This study reveals a significant influence of texture on formability with ZE10 sheet showing superior formability compared to AZ31.

Introduction

Applying magnesium sheets in lightweight design bears a significant weight saving potential [1, 2]. Substitution of conventional materials, such as aluminium and steels, can help to reduce vehicle weight and thus, greenhouse gas emission and fuel consumption. Conventional sheets of alloys such as AZ31 show limited formability at low temperature which poses a serious obstacle to commercial applications. Distinct basal textures are formed during the rolling process and hamper the ability to strain hardening which is a prerequisite for improved formability [3]. For deformation in the thickness direction, the majority of basal planes are oriented unfavourably. To realise strain in thickness direction and thus in <c> direction of the majority of grains, twinning and <c+a> pyramidal slip are other options. While the former can only accommodate a low degree of plastic deformation, the latter is thermally not activated below 225°C in pure magnesium. Thus, forming components from magnesium sheets with typical basal type texture requires elevated process temperatures which results in high production costs.

Sheets such as ZE10 or ZW41 with weaker texture were reported to show generally superior formability [4-8] or competitive formability at lower process temperature. This study assesses the formability of two magnesium sheets with significantly varied texture, AZ31 representing a conventional sheet of typical basal type texture and ZE10 with a much more weak texture.

Material and Experimental Procedures

The forming behaviour of magnesium sheets (thickness 1.5mm) of two different alloys was investigated. Conventional AZ31-B (Mg-3Al-1Zn-Mn) sheet and sheet of alloy ZE10 (Mg-1Zn-Ce-La-Zr) were received in annealed condition. The initial microstructure shows a recrystallised microstructure consisting of equiaxed grains with an average grain size of 10 - 11 μm for both cases, see figures 1 and 2. It is notable that ZE10 shows a higher contrast between different grains compared to AZ31. This implies that the orientation of grains differs more in ZE10 than in AZ31.

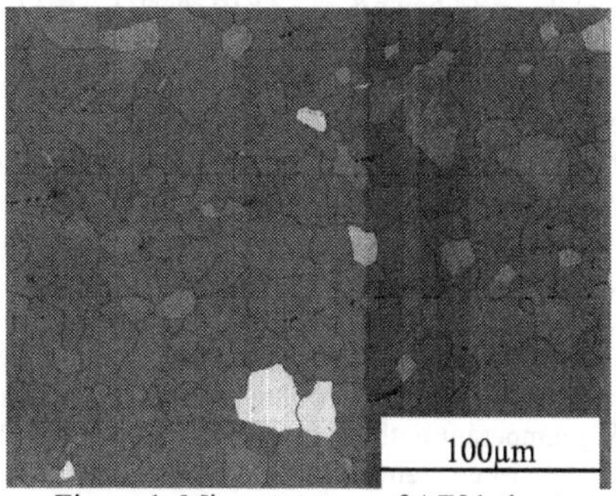

Figure 1: Microstructure of AZ31 sheet

Figure 2: Microstructure of ZE10 sheet

Pole figures of the sheets were measured by a Panalytical X-ray diffractometer setup using Cu K_α radiation (40kV, 40mA). The resulting texture is presented as a recalculated (0002) pole figure from the orientation distribution function in figures 3 and 4. The measured distribution of basal planes differs significantly. AZ31 shows a typical texture of basal type where most grains are aligned parallel to the sheet plane. The angular spread of the orientation distribution is slightly broader to the rolling direction than to the transverse direction. This type of texture is typical for hot rolled magnesium sheets.

In contrary, the basal poles of ZE10 sheet are distributed more randomly. Furthermore, the maximum intensity is found at around 45° towards the transversal direction as a spread double peak. The overall distribution of basal planes is spread towards the transversal direction. Such a texture implies that more basal planes are oriented favourably for basal slip when loaded in sheet plane, promising higher formability.

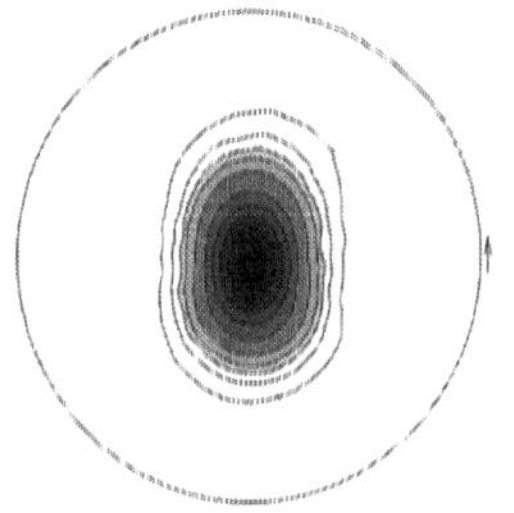

Figure 3: 0002 pole figure of AZ31 sheet, I_{max}= 14.6 m.r.d.

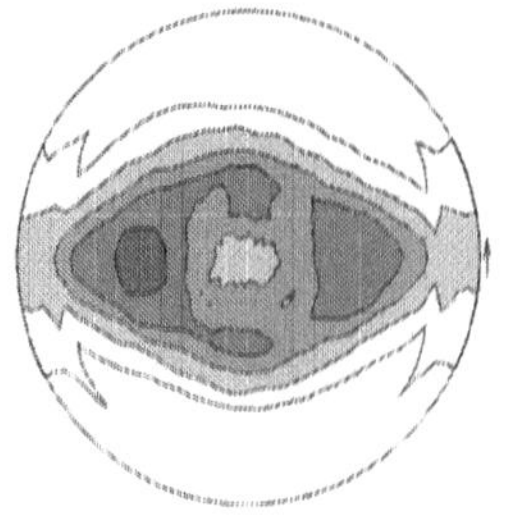

Figure 4: 0002 pole figure of ZE10 sheet, I_{max}= 3.6 m.r.d.

Nakajima type tests were conducted, using an Erichsen 145-60 sheet metal testing machine. A hemispherical punch deformed the sheets at a speed of 1mm/s until fracture. Punch displacement and punch force were recorded digitally at a rate of 5Hz. The blank holder force was 400kN. Oiled PTFE foil of 0.05mm thickness was used as lubricant. Hasek type specimens [9], manufactured by water jet cutting, covered various strain paths from biaxial stretching over plane strain to deep drawing, refer to figure 5. All specimens were cut along the rolling direction. For each geometry and testing temperature, three specimens were tested. Tests were conducted at room temperature, 150°C, 200°C and 250°C respectively. All involved components of the testing device were heated to isothermal condition before testing. The use of an optical deformation measuring system ARAMIS allowed in situ strain measurements. A statistical pattern was applied on the sheet surface and high resolution pictures were recorded digitally during the tests at a rate of 12Hz. From the calculated strain data, forming limit curves according to ISO 12004 as well as strain path plots, were derived in the forming limit diagram (FLD). The FLD, as introduced by Keeler [10] is a powerful tool to assess the formability of sheets. The major strain as highest local strain in sheet plane is plotted versus the perpendicular local minor strain. Figure 5 shows a schematic FLD, specific strain paths are highlighted. At the onset of necking, severe localisation of strain, the so called forming limit curve is plotted.

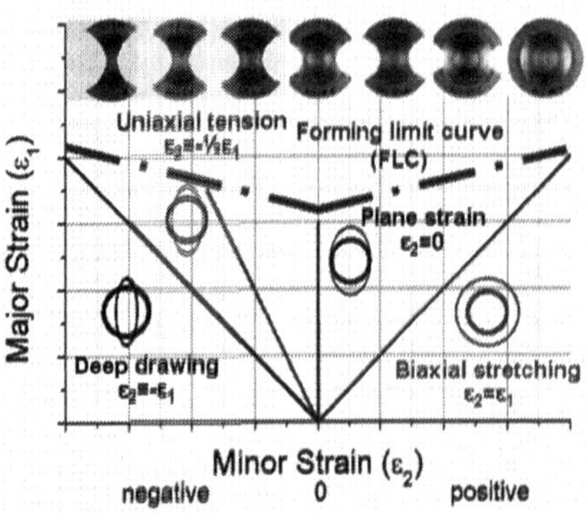

Figure 5: Schematic forming limit diagram. Hasek type specimens cover various strain paths.

Micrographs of the microstructure of deformed sheets were taken, using standard preparation techniques based on picric acid etching [11]. Samples were cut from the deformed specimen close to the crack location. The micro sections were grinded with SiC sandpaper and polished with SiO_2 suspension. Observation direction for all micrographs was the transversal direction (TD).

Results

The mechanical response of the sheets is plotted as punch force – punch displacement curves in figure 6 for AZ31 and 7 for ZE10. For both sheets, specimens are shown that represent the outermost edges of the FLD. While the specimens marked with squares are deformed by biaxial stretching and thus, show a positive minor strain, the specimens marked with triangles represent specimens with negative minor strain. As a general tendency the specimens with positive minor strain require higher punch forces compared to their counterparts with negative minor strain. This can easily be explained by the larger area of contact and thus, amount of deformed material. In the case of AZ31, specimens tested at room temperature and specimens with negative minor strain tested at 150°C show an abrupt drop in punch force upon fracture of the specimens. The specimen tested at room temperature fracture at comparable low forces. All other specimens show a curve that can be separated in four segments: after an initial increase in punch force, a linear relationship between punch force and punch displacement is observed. The slope of this linear relationship decreases with increasing temperature and is generally higher for specimens with positive minor strain (biaxial stretching). Following up, a plateau force is reached and maintained until a rather sudden drop in punch force. The plateau is associated with diffuse necking of the specimen while the latter drop in force is due to crack initiation. It is notable that while the maximum punch displacement, the limiting dome height, increases with temperature in the case of biaxial stretching, the maximum limiting dome height is reached at 200°C for specimens with negative minor strain followed by a decrease of maximum punch displacement at 250°C.

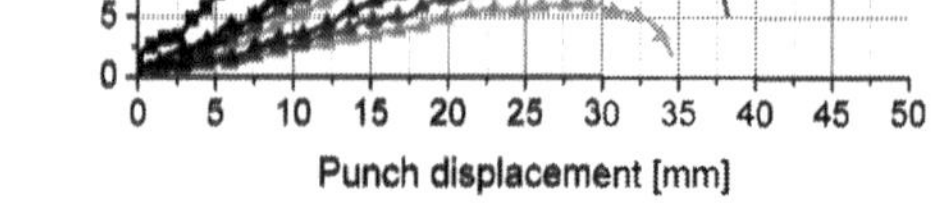

Figure 6: Punch force – punch displacement diagram for AZ31

In comparison, ZE10 shows simpler trends. Only specimen tested at room temperature show a brittle behaviour. With increasing temperature, the necessary punch force decreases and maximum punch displacement – the maximum dome height – increases. For biaxially stretched specimens, only slight differences in maximum dome height are achieved with increasing temperature from 150°C to 250°C. The limiting dome height of specimens with negative minor strain is always below their biaxially stretched counterparts. It is noteworthy that the maximum punch forces of biaxially stretched ZE10 are generally much higher than the punch forces of AZ31 specimens at the same temperature.

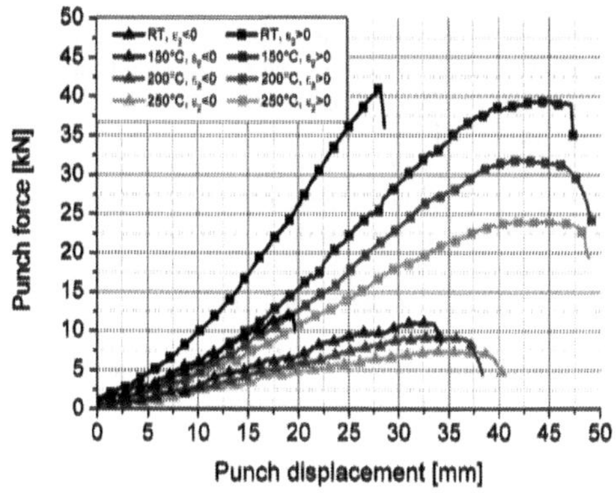

Figure 7: Punch force – punch displacement diagram for ZE10

By evaluation of the strain measurements as obtained by the ARAMIS system, the strain response of the material in terms of major strain vs. minor strain can be plotted in the FLD. Figures 8 and 9 show the strain response of both sheets until fracture of the specimens. For all tested specimens, there is an approximately linear relationship between minor and major strain. Deviations from this linearity are found at high values of major strain, close to fracture of the specimens.

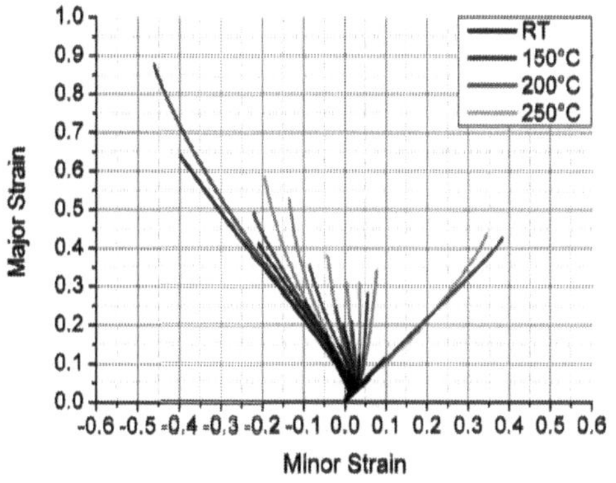

Figure 8: Strain paths of ZE10 sheet, testing temperature RT to 250°C

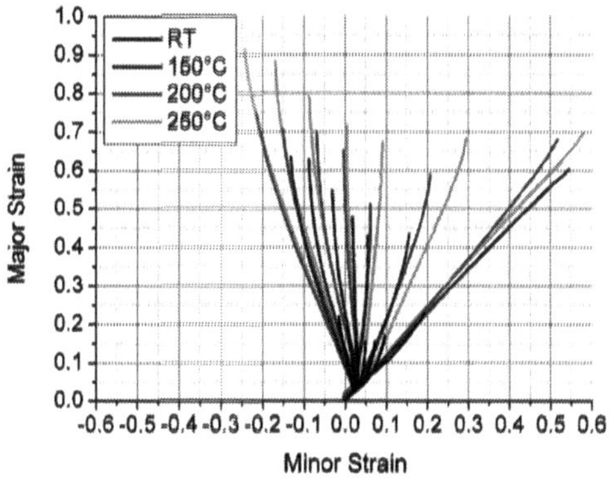

Figure 9: Strain paths of ZE10 sheet, testing temperature RT to 250°C

As a matter of tendency it is found that achieved values of major strain are higher for ZE10 than for AZ31. However, especially on the left hand side of the FLD, the slope of the measured strain paths is generally steeper for ZE10 than for AZ31. While for ZE10 it is found that an increase in temperature results in higher values of maximum major strain, AZ31 shows an exception from the rule at the outermost left side of the FLD. As already reported for the maximum dome height, maximum formability is reached at 200°C here and the strain path of specimens, tested at 250°C, does not even reach the maximum major strain as measured at 150°C.

Forming limit curves of the sheets according to ISO 12004 are shown in figures 10 and 11. The forming limit curve approximates the onset of necking of the material, beyond, strain localisation takes place until fracture. For AZ31, room temperature formability turns out to be poor. An increase of testing temperature to 150°C leads to a significant increase in formability under strain paths of negative minor strain, but gain in the domain of positive minor strain is low. Significant increase in stretch formability is found when increasing the test temperature to 200°C and beyond. The maximum major strain is reached with specimen tested at 200°C under negative minor strain. This result corresponds with the already observed findings in punch force – punch displacement curves and strain paths. The minimum major strain is found for specimen with minor strain close to zero, regardless of the temperature. This is a typical finding for sheet metals, the corresponding major strain value is commonly denoted FLD_0. For AZ31, FLD_0 increases with temperature from 0.10 at room temperature to 0.27 at 250°C.

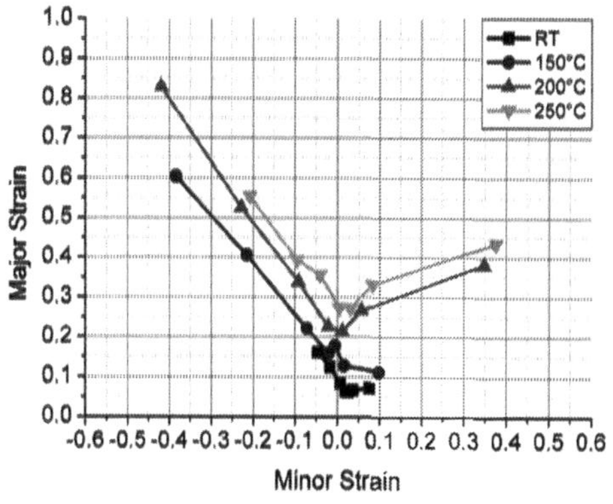

Figure 10: Forming limit curves of AZ31 sheet, testing temperature RT to 250°C

The forming limit curves of ZE10 sheet from room temperature to 250°C, as presented in figure 11, show generally higher values of major strain compared with AZ31. At room temperature formability is already higher than for AZ31 at 150°C under biaxial stretching. FLD_0 depends strongly on temperature, showing an increase from 0.14 to 0.53 from RT to 250°C. ZE10 generally outperforms AZ31 in terms of formability. At 150°C, competitive formability can be achieved, the degree of deformation under plane strain condition and biaxial stretching is unmatched by AZ31 even at 250°C. Figure 11 confirms that formability under positive major strain does not increase significantly from 150°C to 250°C, for negative minor strain and plane strain condition a certain gain in formability is notable.

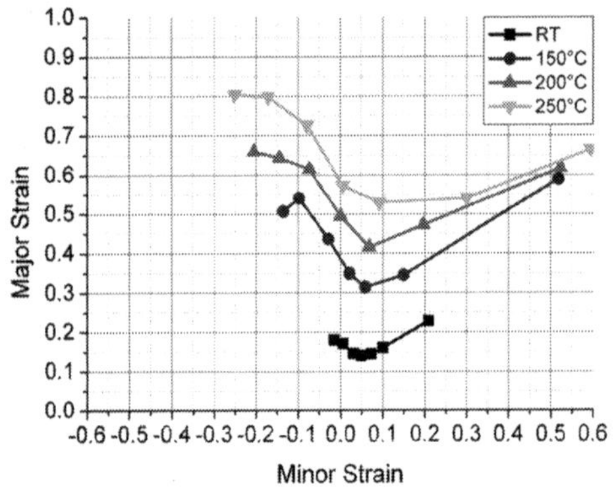

Figure 11: Forming limit curves of ZE10 sheet, testing temperature RT to 250°C

The microstructure of tested specimens is shown in figures 12 and 13. The micrographs show the microstructure close to the crack location. Figure 12a shows clearly that twins are formed during deformation of an AZ31 specimen with negative minor strain at room temperature. Twinning can contribute to the overall deformation of hcp material. Dynamic recrystallisation takes place at elevated temperatures. For the same strain path, figure 12b shows a specimen, deformed at 200°C. A duplex structure of fine, recrystallised grains at the former grain boundaries of large, unrecrystallised grains is found. Figure 12c shows a similar structure, however, the fine grains at the former grain boundaries are larger. Under biaxial stretching, fewer twins are found at room temperature, figure 12d. At elevated temperatures, dynamic recrystallisation is initiated, at 200°C the microstructure is partially recrystallised and voids have formed in the material, figure 12e. Figure 12f shows a fairly recrystallised microstructure of a specimen, tested at 250°C under a strain path with positive minor strain (biaxial stretching). The microstructure consists of small grains, much smaller than the initial grain size.

ZE10 shows a different behaviour. At room temperature, fewer twins are formed as shown in figure 13a. At elevated temperatures, no evidence of dynamic recrystallisation can be found under a strain path with negative minor strain, see figures 13b and 13c. Instead, stretched unrecrystallised grains are found, forming a pan cake like microstructure. Under positive minor strain, severely deformed microstructurse can be observed in the vicinity of the crack location, see figures 13d-f. Twins obviously play no role for the deformation under this strain path at elevated temperatures. Only at room temperature, very few twins can be observed in the microstructure, see figure 13d. With increasing temperature, the degree of elongation of the grains increases. At testing temperature of 250°C, individual grains are hard to identify, the microstructure consists of very flat, stretched grains.

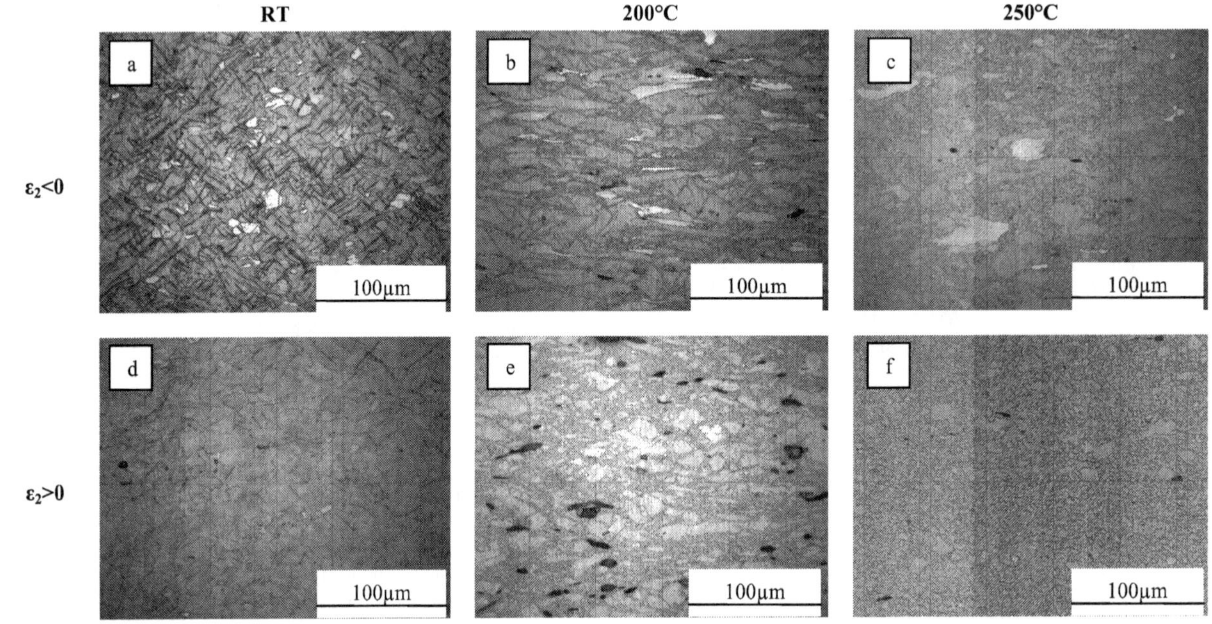

Figure 12: Micrographs of AZ31 sheet, tested at RT, 200°C and 250°C, negative minor strain (a-c) and positive minor strain (d-f)

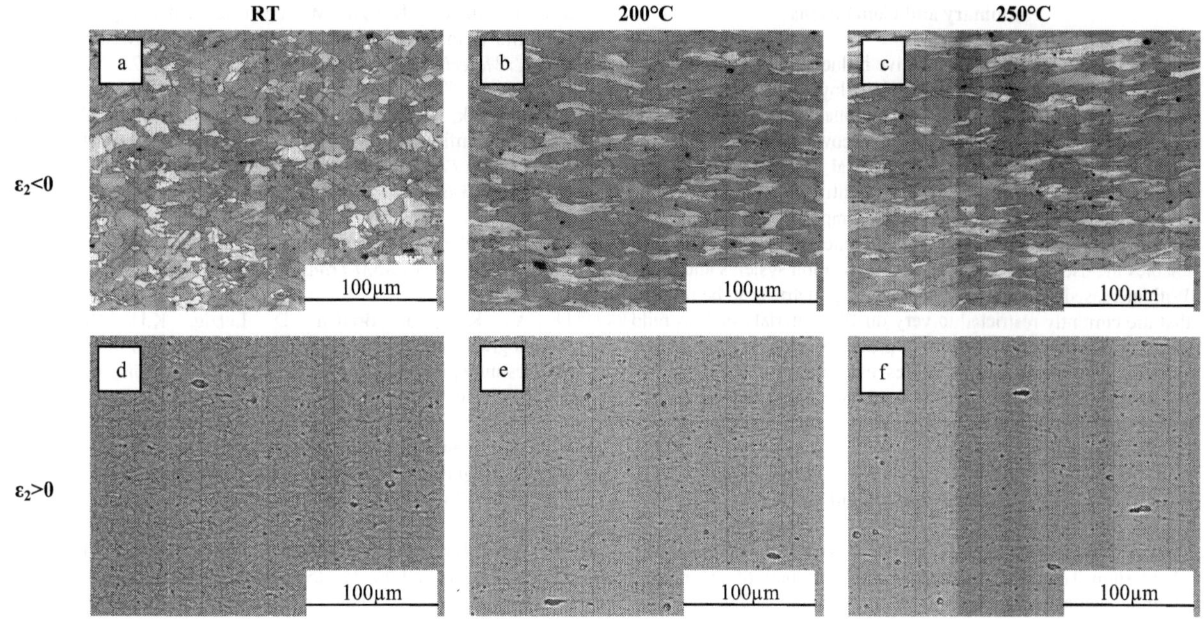

Figure 13: Micrographs of ZE10 sheet, tested at RT, 200°C and 250°C, negative minor strain (a-c) and positive minor strain (d-f)

Discussion

The punch displacement - punch force curves of the both materials provide a first impression on the superior formability of ZE10 compared to conventional AZ31 sheet. It is notable that the limiting dome height and the achieved maximum major strain in the forming limit curves are in good, yet qualitative agreement. High limiting dome height equals good formability in the FLD as reported by Dreyer et al. [4]. The significant differences in strain response of the two sheets are mainly determined by their difference in texture. It is well known that a less pronounced texture can result in an overall higher formability of magnesium sheets [4-8]. The commonly found basal type texture of AZ31 basically does not favour basal slip as a result of the orientation even if a low CRSS favours it in comparison to other slip mechanisms. The deformation via this main slip system is hampered for all strain paths. In case of uni-axial tension tests this was confirmed by Agnew [12]. Strain in thickness direction and thus, perpendicular to the majority of basal planes, is difficult to realise at low temperatures. Twinning can only accommodate low degrees of plastic deformation in <c> direction, formability of AZ31 at room temperature turns out to be poor. At elevated temperatures of 200°C and above, activation of pyramidal <c+a> slip can compensate for the unfavourable orientation of basal planes, with increasing temperature, formability of AZ31 rises significantly especially for strain paths with negative minor strain. Under these strain paths, increasing deformation can be achieved up to 200°C. Interestingly, a further increase of temperature does not lead to a further increase in deformation but to a decrease. This indicates that a consideration of the activity of slip modes does not cover all aspects that are involved. ZE10 shows generally higher formability at any testing condition. The texture is much weaker and many basal planes are oriented favourably for basal slip. <a> slip can be activated much more easily and can realise higher deformation compared to AZ31 under any strain path. In summary, it can be considered that differences in the activation of deformation mechanisms are related to the texture of the material and that the main reason for the higher formability of ZE10 compared to AZ31 is the weaker texture. This is also concluded by Bohlen et al. [6] where the main features of uni-axial tensile tests were modelled based on the different textures rather than based on significant differences in critical stresses required for respective deformation mechanisms. Figures 12a-c reveal one specific feature which changes in the case of AZ31 if the temperature increases. Starting from a deformed microstructure at room temperature there is a considerable formation of necklace type structures of small recrystallised grains. It is noteworthy that this formation, which is observed after fracture of the samples, is visible up to tests at 200°C whereas higher temperatures lead to considerable grain growth. This behaviour was reported before [13, 14] and can explain highly localised deformation in the fine recrystallised grains between the stretched, unrecrystallised grains. At 250°C, the grains highly ductile zones are not observed anymore due to grain growth. In case of ZE10 the finding is different where no hint of ongoing recrystallisation is observed in this sheet. It is described that rare earth containing magnesium alloys generally show hampered dynamic recrystallisation [15, 16]. Furthermore, the texture of this sheet itself leads to changes in the activity of deformation mechanisms including a decrease of <c+a> pyramidal slip activity. This also results in the delay of dynamic recrystallisation [5]. It is noteworthy that ZE10 shows superior formability especially under biaxial and plane strain conditions. The majority of failures in deep drawing processes happen under plane strain conditions. It should be pointed out that ZE10 shows superior formability under plane strain conditions at 150°C which is not surpassed by AZ31 even at 250°C.

Summary and Conclusions

The present study proves the massive influence of texture on sheet formability of magnesium alloys. The alloy ZE10 shows superior formability, compared to the conventional alloy AZ31. This is illustrated by forming limit curves that cover various strain paths from deep drawing processes to biaxial stretching. At testing temperatures as low as 150°C, competitive formability can be achieved. Lowering the process temperatures in forming procedures such as deep drawing saves energy and cost as well as enables the use of less sophisticated lubrication systems and tools. With increased formability, component geometries can be realised that are currently restricted to very ductile materials such as mild steels or require complex forming processes such as super plastic forming. This study clearly proves that changing the texture of magnesium sheets can pave to road to commercial success of magnesium sheets by making forming processes less expensive or offering bigger process windows.

Acknowledgement

The material used in this work is supplied by Salzgitter Magnesium Technology GmbH (SZMT) as part of the project "Mobile with magnesium". The authors appreciate this contribution by Dipl. Ing. Sebastien Wolff and Dr.-Ing. Peter Juchmann from SZMT and the financial support from the WING programme of the German Ministry of Education and Research (BMBF) under contract no. 03X3012H.

References

1. B.L. Mordike, T. Ebert, "Magnesium Properties – applications – potential", *Materials Science and Engineering A*, 302 (2001), 37-45.

2. H. Friedrich, S. Schumann, "Research for a "new age of magnesium" in the automotive industry", *Journal of Materials Processing Technology*, 117 (2001), 276-281.

3. W.F. Hosford, R.M. Caddell, *Metal forming: Mechanics and Metallurgy*, (Upper Saddle River, NJ: PTR Prentice Hall, 1993) 68–79.

4. C.E. Dreyer, W.V. Chiu, R.H. Wagoner, S.R. Agnew, "Formability of a more randomly textured magnesium alloy sheet: Application of an improved warm sheet formability test", *Journal of Materials Processing Technology*, 210 (2010), 37-47.

5. S. Yi, J. Bohlen, F. Heinemann, D. Letzig, "Mechanical anisotropy and deep drawing behaviour of AZ31 and ZE10 magnesium alloy sheets", *Acta Materialia*, 58 (2010), 592-605.

6. J. Bohlen, M.R. Nürnberg, J.W. Senn, D. Letzig, S.R. Agnew, "The texture and anisotropy of magnesium–zinc–rare earth alloy sheets", *Acta Materialia*, 55 (2007), 2101–2112.

7. Y. Chino, K. Sassa, M. Mabuchi, "Texture and stretch formability of a rolled Mg–Zn alloy containing dilute content of Y", *Materials Science and Engineering A*, 513-514 (2009) 394-400.

8. E. Yukutake, J. Kaneko, M. Sugamata, "Anisotropy and Non-Uniformity in Plastic Behavior of AZ31 Magnesium Alloy Plates", *Materials Transactions*, 44 (2003), 452-457.

9. V. Hasek, " Untersuchung und theoretische Beschreibung wichtiger Einflußgrößen auf das Grenzformänderungsschaubild", *Blech Rohre Profile*, 25 (10) (1978), 493-499, 213-220, 285-292, 493-499, 613-627.

10. S.P. Keeler, "Plastic instability and fracture in sheet stretched over rigid punches", *ASM Transaction*, 56 (1964), 25-48.

11. V. Kree, J. Bohlen, D. Letzig, K.U. Kainer, " Metallographische Gefügeuntersuchungen von Magnesiumlegierungen", *Praktische Metallographie*, 41 (2004), no. 5: 233-246.

12. S.R. Agnew, "Plastic Anisotropy of Magnesium Alloy AZ31B Sheet", *Magnesium Technology 2002*, (Warrendale, PA: TMS, 2002), 169-174.

13. S.E. Ion, F.J. Humphreys, S.H. White, "Dynamic recrystallisation and the development of microstructure during the high temperature deformation of magnesium", *Acta Metallurgica*, 30 (10) (1982), 1909-1919.

14. J.A. del Valle, M.T. Pérez-Prado, O.A. Ruano, "Texture evolution during large-strain hot rolling of the Mg AZ61 alloy", *Materials Science and Engineering A*, 355 (2003), 68-78.

15. K. Hantzsche, J. Bohlen, J. Wendt, K.U. Kainer, S.B. Yi, D. Letzig, "Effect of rare earth additions on microstructure and texture development of magnesium alloy sheets", *Scripta Materialia*, 63 (2010), 725–730.

16. J. Bohlen, S. Yi, D. Letzig, K.U. Kainer, "Effect of rare earth elements on the microstructure and texture development in magnesium–manganese alloys during extrusion", *Materials Science and Engineering A*, 527 (2010) 7092–7098.

Influence of Cerium on Texture and Ductility of Magnesium Extrusions

Raja K. Mishra[1], Anil K. Gupta[2], P. Rama Rao[3], Anil K. Sachdev[1], Arun M. Kumar[4] and Alan A. Luo[1]

[1] Materials and Processes Laboratory, General Motors Research & Development Center, Warren, MI, USA
[2] National Physical Laboratory, New Delhi, India
[3] International Advanced Research Centre for Powder Metallurgy and New Materials, Hyderabad, India.
[4] India Science Laboratory, General Motors Research & Development Center, Bangalore, India

Keywords: Magnesium, Texture, Extrusion, Cerium, Mechanical Properties.

Abstract

Unalloyed Mg and Mg-0.2%Ce alloy were extruded as solid rounds at 400°C in a 500 ton vertical extrusion press at 10 mm/sec. Tensile tests revealed a significant increase in elongation with only a small Ce addition of 0.2%. The addition of cerium caused a decrease in yield strength and an increase in work hardening delaying the onset of instability. EBSD analysis shows that the small Ce addition markedly alters the texture of the extruded rods during recrystallization, and orients the c-axes of the grains at an angle to the extrusion axis whereas they are mostly normal to the extrusion axis in Mg. This reorientation favors basal slip activity leading to higher elongation. While Mg deforms mainly by twinning and fracture initiates due to voids nucleating at twin intersections, the Mg-0.2Ce alloy shows void nucleation along nonintersecting shear bands parallel to the tensile axis. The results of this study point to a design approach that combines alloying and processing to alter the slip distribution which can enhance the formability of Mg alloys.

Introduction

Although the cost of magnesium has decreased dramatically in recent years [1], its use in wrought products like sheet and extrusions has been limited due to the poor workability of magnesium alloys in as cast, primary and subsequent fabrication stages [2]. While there are increasing attempts to enhance hot workability during extrusion, there is limited effort to improve the ductility of the primary fabricated material in the as-extruded or in the as-rolled conditions. Earlier efforts have alluded to the fact that alloying additions which can indeed change the crystal structure from HCP to BCC could be a promising approach [3], but these approaches have not found commercial acceptance since either large amounts of alloying are needed (17 a/o in the case of Li), or additions such as Be require special precautions. Recent approaches to increasing the extrudability and ductility of Mg alloys have been to explore the Mg-Al system, leading to the development of AM30 (Mg-3%Al) magnesium alloy [4]. This alloy shows better extrudability compared to AZ31 as well as improved room temperature ductility. However, both extrudability and formability are much lower than those routinely obtained for the workhorse AA6063 aluminum alloy used in the automotive industry. There is therefore a growing need for obtaining high formability wrought magnesium alloys for automotive applications [5]. However, what is not obvious is how to choose the alloying additions which accomplish the desirable effects.

Alloying behavior of metallic materials in regard to phase equilibria has been investigated by systematically determining binary, ternary and, in some cases, multi-component equilibrium diagrams. However, similar systematic studies of such alloying effects on deformation, forming and fracture behavior of Mg are far more demanding and have not been a research tradition. It is only through sporadic investigations that some understanding of alloying effects on Mg deformation has been gained. It is generally recognized that mechanical workability of Mg can be improved by alloying with Zn, Ca and certain RE elements [6-9]. This effect has been ascribed to a lowering of the critical resolved shear stress (CRSS) by alloying [8] resulting in changes to the microtexture during fabrication. The CRSS for prismatic slip in Mg at room temperature has been reported [10] to be about 100 times higher than that for basal slip. In contrast, Koike et. al. [11] have shown that the CRSS for <a + c> pyramidal slip and <a> prism slip at room temperature are merely 2.5 and 2.2 times that of <a> basal slip. If the difference in the CRSS of these slip systems can be further reduced, prismatic slip would offer two independent slip modes beyond the original two basal slip systems for a total of four, and pyramidal slip would offer four independent slip modes on its own [12], fulfilling the Taylor criterion of five independent slip systems to be available for general deformation. It is expected that this would increase the formability of Mg. Furthermore, if slip is easy, the effect of twinning can be minimized, and fracture can be delayed, making it also possible to have high ductility.

In one study, Barnett et. al. have speculated that addition of a small quantity of Ce to Mg reduces rolling forces by distributing deformation away from shear bands [13]. This can occur if Ce addition changes the CRSS values which in turn will alter the microtexture during hot forming. In the present study, we investigate the effect of a small amount of Ce on mechanical properties and microtexture in extruded rods. The microstructure-property relations for room temperature deformation of extruded Mg-0.2Ce are presented here and compared with those of extruded unalloyed Mg rods.

Experimental Procedure

Unalloyed Mg and Mg-0.2%Ce alloy melts of about 100 kg were prepared using commercially pure Mg ingots and Mg-20%Ce master alloy in a steel crucible with SF_6/CO_2 protective cover gas. Billets of 75 mm diameter and 230 mm length were cast at about 700°C into a permanent mold. The billets were preheated to 400°C for two hours and extruded in a Wellman Enefco™ 500-ton multipurpose vertical hydraulic press fitted with a circular die. Solid rods, approximately 15mm in diameter corresponding to an extrusion ratio of 25:1, were extruded using boron nitride lubricant at a billet speed of 10mm/sec and air cooled.

The samples for microstructural evaluation were obtained by scrapping the front 0.15 meters of the extruded rod where it had attained a steady state of extrusion. Metallographic samples were prepared by standard methods, and the polished samples were etched in a solution containing 20 ml glacial acetic acid, 50 ml picric acid, 10 ml methanol and 10 ml de-ionized water. A Nikon™ optical microscope was used to record the microstructure in both longitudinal and transverse directions of the extruded rods. The polished samples were also examined by electron backscattered diffraction (EBSD) in a LEO™ 1450 scanning electron microscope operating at 20kV fitted with a TSL™ EBSD camera [14]. Detailed chemical mapping of Ce, Al, O, Mn and Mg were obtained from polished samples sectioned normal to the extrusion axis using a Cameca™ SX100 electron probe microanalyzer operating at 20 KV.

Tensile specimens with a 25 mm gauge length and 6.25 mm gauge diameter were machined from the extruded rods and pulled to failure in a Universal Testing Machine (Instron™, Capacity: 0-50 kN) at an average strain rate of $0.66x10^{-3}/s$. At least three specimens were taken from different locations of the steady state portion of the extruded rod. Uniaxial compression tests were carried out in a United Testing FRM60™ machine equipped with a 27 ton (approx 250kN) load cell on 12.7 mm compressometer gauge on 37.4mm long samples machined to a diameter of 12.8 mm (length to diameter ratio of 3:1 to confirm to the ASTM E9 test procedure) to a maximum of 10% strain at a strain rate of 0.005/min.

Results

Smooth extrusions were obtained at an extrusion speed of 10 mm/s for both the materials. Fig. 1 shows typical microstructures obtained for Mg and the Mg-0.2%Ce alloy in the transverse direction. The micrographs from longitudinal sections are similar, showing no anisotropy in grain morphology. The morphology indicates a fully recrystallized and nearly equi-axed grain structure. Optical images taken from the mid-section of the deformed sample normal to the fracture surface (Fig. 2) show nearly equal amounts of twinned grains in both Mg and Mg-Ce samples, even though the fracture strain of the latter is over three times the former.

The room temperature tensile properties are listed in Table 1 and the actual tension plots are shown in Fig. 3. The tensile yield strength of Mg-0.2%Ce is about 35% lower than that of Mg. The stress-strain curve exhibits strain hardening behavior characteristic of slip dominated deformation. The most significant difference in the mechanical properties of the two samples is in the tensile ductility, which is substantially higher in Mg-Ce samples compared to that in Mg samples.

Fig. 4 shows electron microprobe maps of the Ce and O distribution taken from a polished surface of the Mg-0.2%Ce sample. Taken together, they show that Ce particles are associated with oxygen in many instances, suggesting that Ce forms a mixture of Mg-Ce intermetallics and oxide particles dispersed in the microstructure. A finite concentration (detection limit of the microprobe is ~0.1 wt%) of Ce was consistently observed in the matrix, away from particles. This could be either due to Ce in solution or particles of a Ce-Mg phase dispersed in the matrix. The larger intermetallic particles are distributed both inside the grains as well as on the grain boundaries, and their relative volume fraction was estimated to be about 1%.

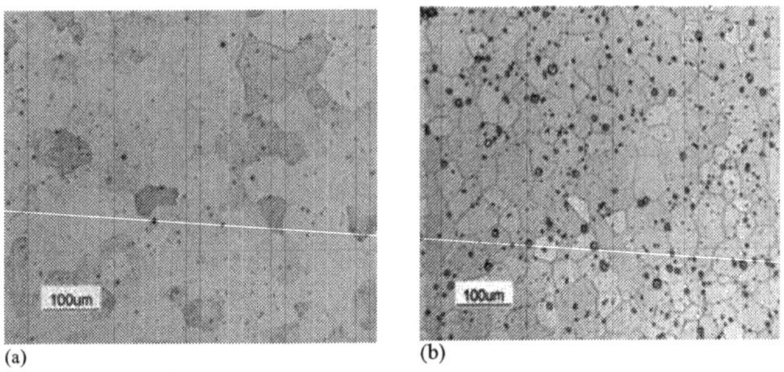

(a) (b)

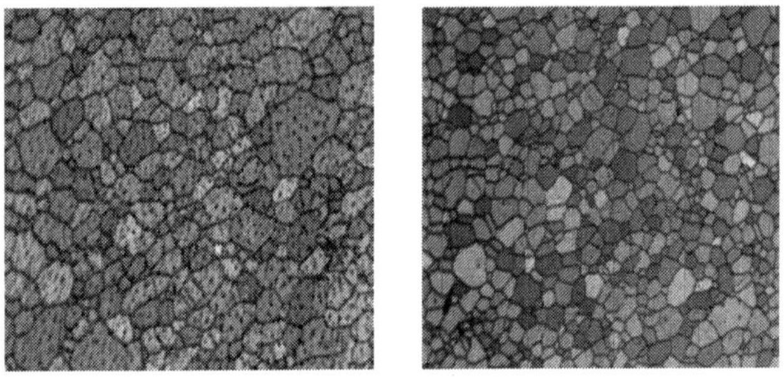

(c) (d)

Fig 1. Optical micrographs and EBSD image quality maps of unalloyed magnesium (a,c) and Mg-0.2%Ce (b,d) following a 25:1 extrusion ratio. Average grain diameters are 34mm for Mg compared to 26mm for the Mg-0.2Ce alloy. Samples are taken normal to the extrusion direction.

The EBSD maps (inverse pole figure – IPF maps) in Figs. 5 (a) and (b) taken from transverse sections show that both materials have grains that have equiaxed shapes with straight boundaries, characteristic of a fully recrystállized extrusion. About 2% of the grains in the Mg-Ce alloy are twinned, but nearly none appear twinned in the unalloyed Mg samples. The average grain diameters measured from EBSD data are ~34 μm and 26 μm, respectively, for Mg and Mg-Ce samples after accounting for the twinned grains. The grain size measured with EBSD vis-à-vis by conventional method is being debated in the literature, and GM is currently part of a round robin measurement to create a new protocol by ASTM.

Table 1: Tensile, compression, and ductility values of Mg and Mg-0.2Ce alloys tested at room temperature

	Yield strength (tension), MPa	Yield strength (compression), MPa	UTS (tension), MPa	Total elongation (%)	n-value in tension based on hardening law of $\sigma = K\varepsilon^n$
Mg	106	53.5	155	9.1	0.24
Mg-0.2Ce	68.6	55.8	170	31	0.34

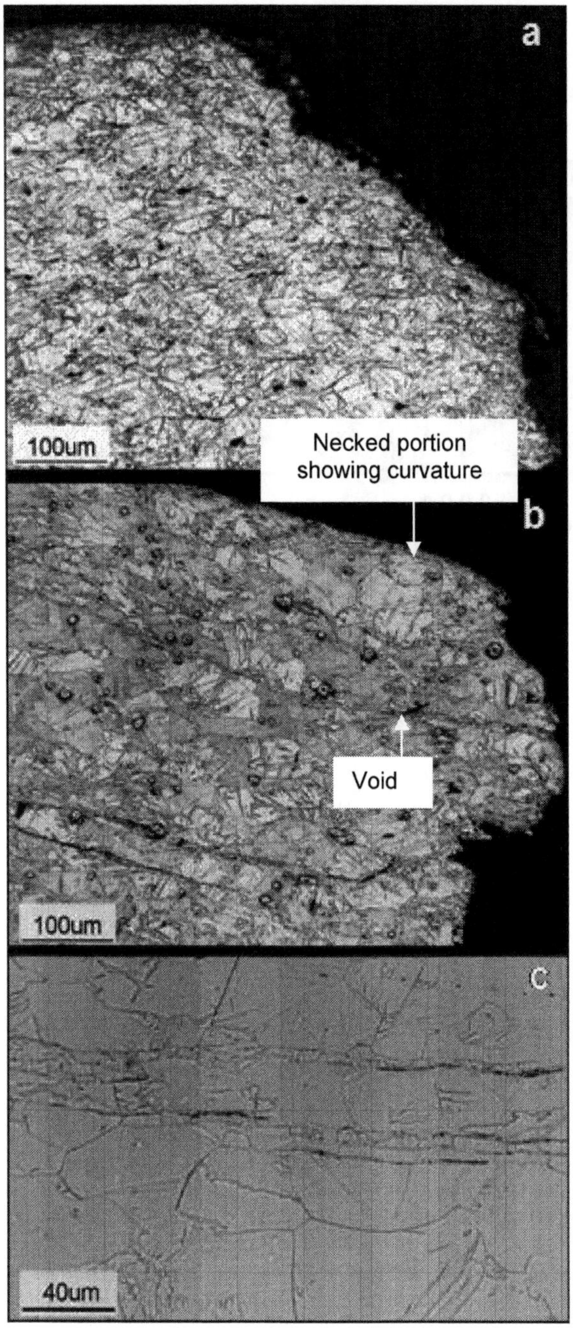

Fig 2. Optical micrographs taken from a cross-section normal and adjacent to the fracture surface. (a) Mg, (b) Mg-0.2Ce and (c) higher magnification on a shear band in (b). Note the curvature in (b) caused by necking of the sample which is absent in (a). Void formation close to a shear band is shown in (b).

The color distribution in the IPF maps (Figs. 5 (a) and (b)) show noticeable differences in the orientation of grains in the unalloyed Mg and Mg-0.2%Ce samples. While nearly all the grains are oriented with their basal planes parallel to the extrusion axis in the unalloyed Mg sample (basal plane normal perpendicular to the extrusion axis), this is not so in the Mg-Ce sample (note the higher proportion of red colored grains in Fig. 5(b) compared to Fig. 5(a)). The texture plots, with the axis of symmetry parallel to the extrusion axis for Mg and Mg-Ce samples in Figs. 5 (c) and (d), show the grains in Mg to have a perfect ring texture around the extrusion axis while the grains in Mg-Ce sample have their basal plane at an angle between 40 and 50 degrees to the extrusion axis. These textures translate to Schmid factors of over 0.4 in 56% of the grains in the Mg-Ce alloy compared to 36% of the grains in Mg, Fig. 6. The Taylor factor maps for the two microstructures show that over 80% of the grains in Mg have a Taylor factor of less than 1 compared to only 30% of such grains in the Mg-Ce sample. In other words, the Mg sample has a majority of grains in hard orientations for basal slip in contrast to the Mg-Ce alloy sample which has a large number of grains in a more formable orientation. An important effect of Ce, therefore, is to change the texture of the material drastically during recrystallization following extrusion.

Previous research has shown that a sharp initial texture and the polarity of deformation twins in magnesium lead to pronounced yield asymmetry associated with the activation of {10-12} twinning during extension (but not contraction) along the c-axis [15-17]. The compressive flow curves in Fig. 7 from Mg and Mg-Ce samples deformed along the extrusion axis show nearly identical compressive yield strength values in pure Mg and Mg-Ce samples. The threshold deformation required to nucleate the {10-12} extension twins in both samples is the same. However, the abrupt transition of pure Mg from elastic to plastic mode in compression is to be contrasted with the gradual transition in the Mg-Ce sample in Fig. 7. This behavior suggests that once twinning starts, the material deforms by twinning exclusively in pure Mg because of its texture, while in Mg-Ce samples, deformation proceeds with a combination of twinning and slip [14].

Fig. 8 shows fracture surfaces of tensile bars of Mg and Mg-0.2%Ce samples. In both cases, the fracture is intragranular. The cleavage planes are prominently observed inside the grains in the Mg samples that also contain evidence of void formation. In the Mg-Ce samples, on the other hand, no cleavage is seen. The dimples on the fracture surface of the Mg-Ce sample are demarcated by intense slip at the boundary region. Small particles inside the voids suggest particle-initiated fracture in the alloy. The difference in the nature of the fracture surfaces correlates well with the microstructure obtained from polished sections normal to the fracture surface in Fig. 2. Very little necking is noted in Mg in these sections compared to the Mg-Ce alloy. This is evident from the curved shape observed in the failed Mg-Ce sample (Fig. 2b). Crack initiation in the Mg sample can occur at twin intersections which can then propagate and expose twin planes successively to create the cleavage fracture surface appearance as in Fig. 8a. Mg-Ce sample in Fig 2b shows nonintersecting shear bands parallel to the tensile axis and void formation confined to the shear bands (Fig. 2b).

Discussion

The solubility of Ce in Mg is about 0.1% at 500°C according to the phase diagram [18]. Any excess Ce will go to form intermetallics and oxide particles as was observed. The fully recrystallized grain structure in both samples suggests grain nucleation and growth to have been completed during cooling of the extruded sample. The small amount of extension twins in the Mg-Ce alloys are most likely a result of compressive thermal stresses during cooling. The smaller grain size in the Mg-Ce alloy is believed to be due to the intermetallic particles that influence nucleation and growth of the recrystallized grains. A smaller grain size should result in an increase in YS due to the Hall-Petch effect [19]. Consequently, the lower YS observed in the Mg-Ce alloy could have been still lower if the alloy would have had a coarser grain size as in the unalloyed Mg sample.

A decrease in yield strength accompanied by an increase in ductile fracture toughness as a result of small solute additions has been noted in a few other instances. A notable example is the addition of Co to Fe where the scavenging of C in solid solution in the iron causes a reduction in the yield strength of the iron [20]. Somewhat similar results on the binary Mg-0.2Ce alloy presently obtained afford an opportunity for understanding their underlying cause.

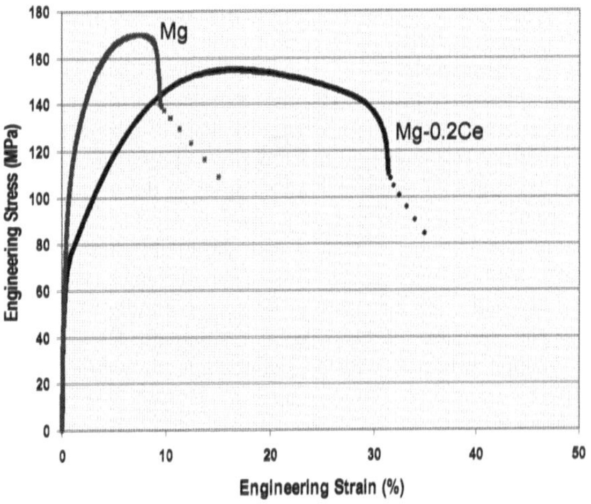

Fig 3. Tensile data for Mg (a) and Mg-0.2Ce (b) after extruding rods with a ratio of 25:1.

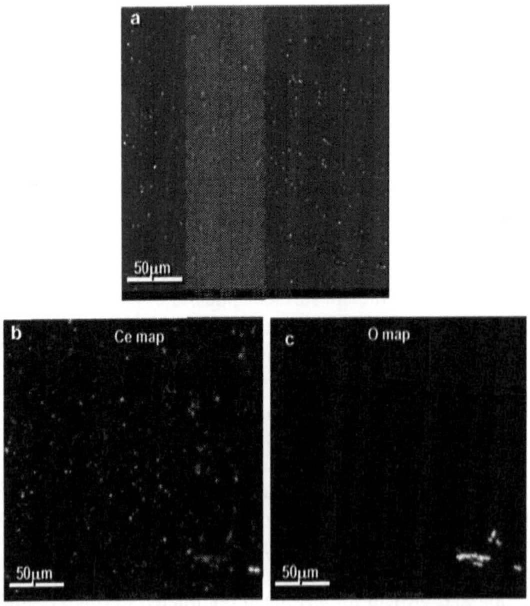

Fig 4. Microprobe analysis showing (a) backscattered image, (b) Ce distribution and (c) oxygen distribution in Mg-0.2%Ce alloy following an extrusion ratio of 25:1.

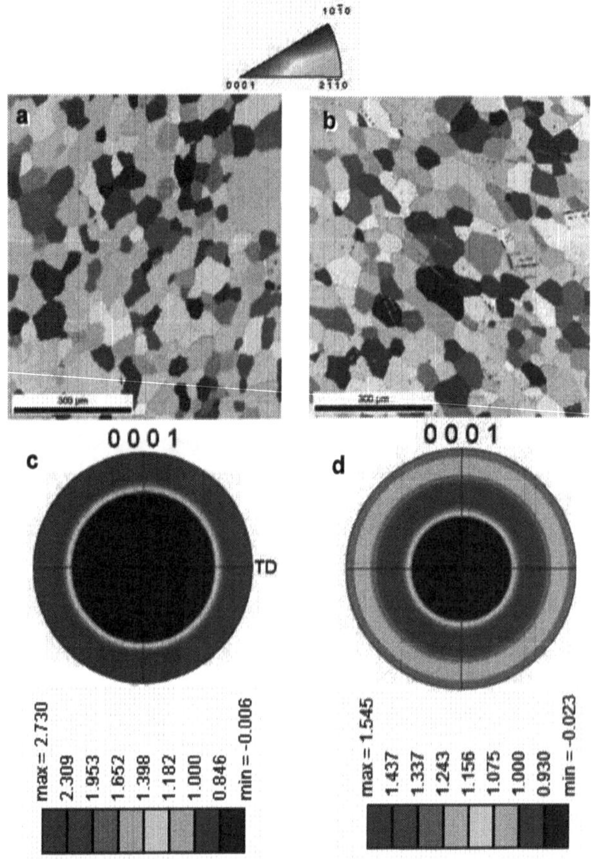

Fig 5. EBSD analysis and texture plots of fully recrystallized Mg (a and c) and Mg-0.2%Ce (b and d) following an extrusion ratio of 25:1.

The present study suggests that the reduction of tensile yield stress (Fig. 3) for Mg-Ce is due to an ease of basal slip because of favorably oriented grains. This feature is effectively precluded in Mg samples because of the unfavorable orientation of the grains for basal slip. Mg can deform by {10-11} twins (contraction) and {10-11} - {10-12} twins (double twins) in tension and it is expected that both samples will have an equivalent amount of these twins when the material has been deformed to high strains. This is reflected in the observations of a similar amount of twinning in both Mg and Mg-0.2Ce samples, Figs. 2 (a) and (b), with the essential difference between the two being the amount of slip in the latter sample, in addition to twinning.

Slip activity in the Mg-Ce sample is enhanced due to favorable grain orientation. The higher strain hardening coefficient of Mg-Ce extruded samples points to the ease of dislocation multiplication due to cross slip. The change in slip behavior due to Ce addition results in an increased capacity of the material for stored work in the shear bands. The suggestion pertaining to texture softening effect and redistribution of slip away from the shear bands during rolling of Mg-Ce alloys [13] needs to be rationalized with respect to the observation of shear bands and void formation. More studies are needed to understand the interactions between Ce, shear band formation, associated texture changes and possible events that lead to localization, voids and failure as deformation proceeds.

It is the combination of lower yield strength and smaller grain size, accompanied by higher work hardening, changed slip distribution, and a texture favoring basal dislocation activity in the Mg-Ce sample, that gives it the high observed ductility. Since particle distribution can be changed by changing the amount of alloying additives as well as extrusion temperature and ratio, it is expected that grain size can be controlled through processing. The effect of Ce on the flow stress of different slip systems is not well known, but the current results suggest that Ce addition can potentially decrease the anisotropy of CRSS for different slip systems and increase the amount of dislocations in the material during deformation. Li additions have been shown to favor <c+a> dislocation activity in Mg that otherwise is predisposed to slip by basal dislocations [21]. A potential topic for future investigation can be to study if Ce addition to the Mg lattice changes the electronic charge distribution in the Mg lattice to alter the CRSS by reducing barriers to dislocation glide on all the slip systems.

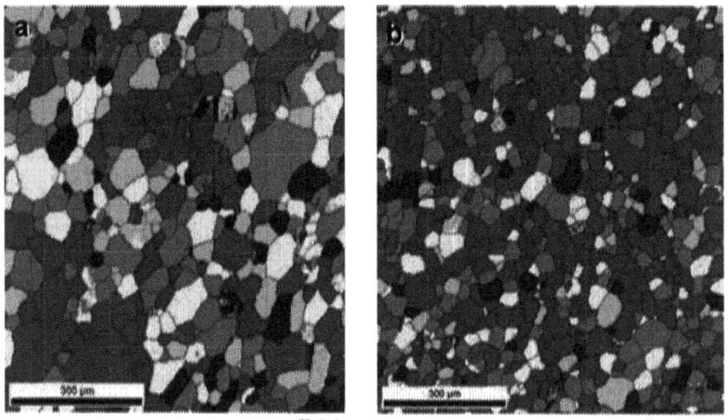

	Min	Max	Total Fraction	Min	Max	Total Fraction
	9.72485e-017	0.0999723	0.070	4.67237e-005	0.100023	0.052
	0.0999723	0.199945	0.128	0.100023	0.2	0.061
	0.199945	0.299917	0.172	0.2	0.299976	0.123
	0.299917	0.399889	0.271	0.299976	0.399952	0.209
	0.399889	0.499861	0.359	0.399952	0.499929	0.555

Fig 6. Schmid factor maps for basal slip for Mg (a) and Mg-0.2%Ce (b) following an extrusion ratio of 25:1.

Summary

The addition of only 0.2% Ce has demonstrated an unusual increase in the ductility of Mg. The Ce addition appears to favor the formation of recrystallized grains with their basal planes oriented at 40-50 degrees to the extrusion axis. The Mg-Ce alloy thus demonstrates an ease of basal slip which is effectively precluded in the extruded rods of Mg because a majority of the grains are oriented with their basal plane parallel to the extrusion axis. While twinning dominates Mg deformation, both twinning and slip activities occur in the Mg-Ce sample. The Ce addition also reduces the yield strength, refines the grain size and leads to a substantial increase in work hardening rate that delays the onset of instability. The change in deformation behavior manifests in different fracture behavior. A cleavage structure exposing twin surfaces is observed in unalloyed Mg due to cracks initiating at intersecting twin boundaries that subsequently interlink, while the fracture surface of the Mg-0.2Ce alloy exhibits a classical dimple structure as a result of slip accumulation and ductile tear.

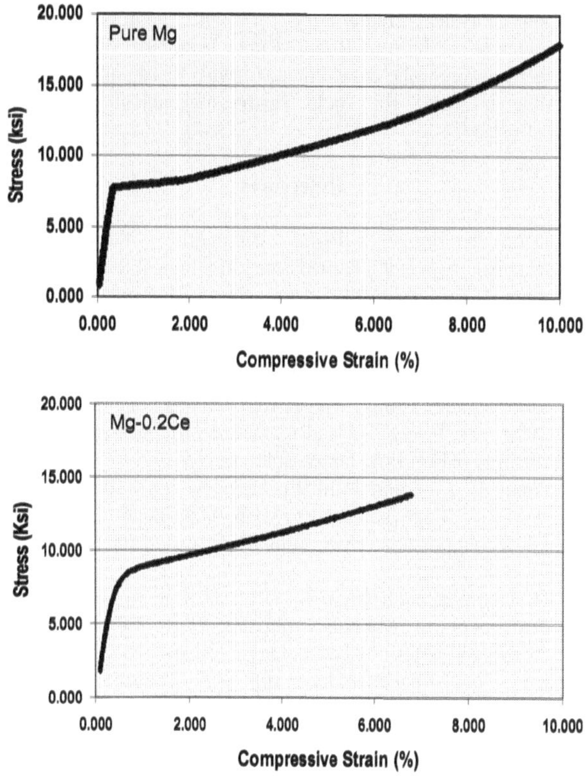

Fig 7. Compression data for pure Mg (a) and Mg-0.2Ce (b) after extruding rods with a ratio of 25:1

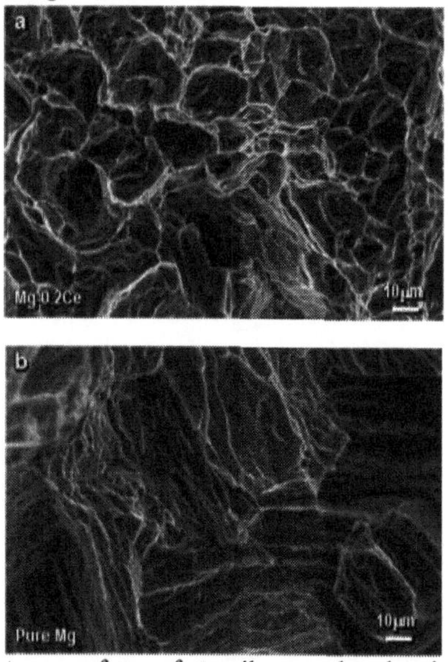

Fig 8. Fracture surface of tensile sample showing dimples characteristic of ductile intragranular fracture in Mg-0.2Ce alloy (a) and a mixture of dimples and cleavage planes in unalloyed Mg (b)

Acknowledgments

Dr. Gupta thanks General Motors for funding this project at the National Physical Laboratory. The authors gratefully acknowledge the help of Robert Kubic Jr. for EBSD data, Rick Waldo for microprobe data, Pinaki Biswas, Shashank Tiwari and S Vijayalakshmi for help with mechanical tests and microstructure.

References

1. D.A. Kramer, "Mineral Commodity Summaries – Magnesium Metal", United States Geological Survey, 2006.
2. Alan A. Luo, SAE 2005 Transactions - Journal of Materials and Manufacturing, SAE, Warrendale, PA, U.S.A., 411-421.
3. D. L. Olson, D. W. Wenman, V. I. Kaydanov, C. Suryanarayana and D. Eliezer, "The Search for Room Temperature Cubic Magnesium Alloys" in **Magnesium 2000**, eds. E. Aghion and D. Eliezer, Magnesium Research Institute (MRI) Ltd., Israel, 2000, pp165-172.
4. Alan A. Luo and Anil K. Sachdev, "AM30 – A New Wrought Magnesium Alloy", in Magnesium Technology 2007 Symposium, TMS Annual Meeting, (2007), Orlando, Florida, USA.
5. H. Dieringa, J. Bohlen, N. Hort, D. Letzig, K.U. Kainer, "Advances in Manufacturing Processes for Magnesium Alloys", Magnesium Technology 2007, eds. R.S. Beals, A.A. Luo, N.R. Neelameggham and M.O. Pekguleryuz, TMS, Warrendale, PA, (2007).
6. P. Menzen, **The Technology of Magnesium and Its Alloys**, A. Beck (ed), F. A. Hughes and Co. Ltd., London, (1940), 34.
7. J. C. McDonald, Trans. Metall. Soc. AIME, 212, (1958), 45
8. S. L. Couling, J. F. Pashak and L. Sturkey, Trans. ASM, 51, (1959), 94
9. B. C Wonsiewicz and W. A. Backofen, Trans. Metall. Soc. AIME, 239, (1967), 1422.
10. S. E. Ion, F. J. Humphreys, S. H. White, Acta Metall. 30, (1982),1909.
11. J. Koike, T. Kobayashi, T. Mukai, H. Watanabe, M. Suzuki, K. Maruyama, K. Higashi, Acta Mater. 51, (2003), 2055.
12. S. R. Agnew, O. Duygulu, Int. J. Plasticity. 21, (2005), 1161.
13. M. R. Barnett, M. D. Nave, and C. J. Bettles, Materials Science and Engineering, A386, (2004), pp205-211.
14. L. Jiang, J. J. Jonas, R. K. Mishra, A. A. Luo, A. K. Sachdev and S. Godet Acta Materialia, 55, (2007), 3899-3910.
15. E.W. Kelley and W.F. Hosford, Trans. TMS-AIME, 242, (1968), 5.
16. E.A. Ball and P.B. Prangnell, Scripta Metall. Mater., 31, (1994), 111.
17. J. Koike, Metall Mater Trans A, 36, (2005), 1689.
18. C. S. Robert, **Magnesium and its Alloys**, Wiley, New York, (1960).
19. R Armstrong, I. Codd, R. M. Douthwaite and N. J. Petch, Phil Mag 7, (1962), 40.
20. M. Srinivas, G. Malkondaiah and P. Rama Rao, Proceedings of the Royal Society, London, (1994), 447, 223.
21. S. Ando, K. Nakamura, K. Takashima and H. Tonda, Mater. Trans., (1992), JIM, 42, 2845.

THE HOT WORKING FLOW STRESS AND MICROSTRUCTURE IN MAGNESIUM AZ31

A.G. Beer and M.R. Barnett

CRC for Cast Metals Manufacturing (CAST)
School of Engineering and Technology, Deakin University, Geelong, VIC, 3217

Abstract

The evolution of flow stress and microstructure for wrought magnesium alloy AZ31 was characterised using torsion and compression testing. Temperatures ranging between 300°C and 450°C and strain rates between $0.001s^{-1}$ and $1s^{-1}$, were employed. Constitutive equations were developed for the flow stress at a strain of 1.0 for torsion, and 0.6 for compression. The flow stress was found to be strongly dependent on deformation mode at low strains. This can be explained in terms of the influence of the deformation accommodating processes of prismatic slip and dynamic recrystallisation (DRX). At higher strains, when the change in flow stress with strain is lower, the flow stress was relatively insensitive to deformation mode. Optical microscopy carried out on torsion samples quenched after twisting to strains between 0.2 and 2 revealed dynamically recrystallised (DRX) grains situated on pre-existing grain boundaries. The average grain size was reduced from 22.5μm down to 7.3 μm after a strain of 2, with the initial grain size being halved after a strain of 0.5.

Introduction

There is increasing pressure on the automotive industry to reduce vehicle emissions due to government regulations, such as the Corporate Average Fuel Economy (CAFE) standards in the U.S.A. Reducing the weight of vehicles, for increased fuel efficiency, is now a high priority and it has spurred the growth of magnesium consumption over the last decade at an annual rate of 15% [1]. Most of the consumption of magnesium alloys by the automotive industry has been in the form of die castings. In 1997, the total magnesium consumption was 333,700 tonnes, 29% of which was in the form of pressure and gravity die castings [2]. Although a relatively small market (1% of 1997 total consumption), wrought magnesium products have the advantages over castings of higher strength and ductility, and so are more suited to structural applications.

Wrought magnesium alloys can be deformed using primary fabrication methods such as rolling, extrusion and forging. For wrought magnesium products to possess a cost advantage over castings, and be a viable alternative to existing materials such as aluminium, optimisation of processing techniques and final properties is essential. An understanding of the deformation behaviour of wrought magnesium alloys is required to achieve this optimisation. This has been the focus for several recent studies [3-5], that have investigated the hot deformation behaviour of magnesium alloy AZ31, using an initial "as-cast" microstructure.

This paper, however, deals with the deformation behaviour of wrought AZ31, which has a finer grain size and stronger texture than as-cast product. The deformation of wrought, or pre-worked, materials occurs in primary fabrication processes such as multi-step forging and rolling operations, and in the extrusion of billets that have been scalped and pre-extruded.

In the present work, torsion and compression tests are employed to determine the influence that deformation conditions and deformation mode have on the flow stress and microstructural evolution of wrought AZ31.

Experimental Procedure

The material employed in the present study was wrought magnesium alloy AZ31 (Mg-3%Al-1%Zn-0.2%Mn). It was supplied in 18mm diameter extruded rod form, with an initial grain size of 22.5μm.

Samples for torsion were machined such that the torsion axis lay parallel to the extrusion direction, having a gauge length (L) of 20mm and radius (r) of 6.7mm. A description of the torsion machine has been published elsewhere [6]. Torsion testing was conducted at temperatures of 300°C, 350°C, 400°C and 450°C and at strain-rates of 0.01, 0.1 and $1s^{-1}$. The samples were heated to the desired temperature and held for approximately 3 minutes to equalise any temperature fluctuation. All torsion tests were carried out at an equivalent true strain of 1.0.

A section of the extruded bar was hot rolled, at 350°C, into 10mm plate and subsequently annealed at 350°C for one hour. Channel die specimens were cut, orientated such that compression occurred in the rolling direction with extension in the transverse direction. Channel die compression testing was conducted at temperatures of 300°C, 350°C and 400°C and at strain-rates of 0.003, 0.03 and $0.3s^{-1}$. The samples were heated to the desired temperature and held for approximately 5 minutes to equalise any temperature fluctuation. All compression tests were stopped at an equivalent true strain of 0.6.

Another series of torsion tests were conducted, at a constant temperature and strain rate of 350°C and $0.01s^{-1}$ respectively. Samples were twisted to strains ranging from 0.2 to 2, and immediately quenched so as to retain the microstructure.

For metallographic examination, the quenched samples were sectioned to allow the observation of the microstructure just below the surface of the gauge length, perpendicular to the radius. The specimens were cold-mounted in epoxy resin and fine ground with SiC paper using 320, 500, 800 and 1200 grit. Polishing involved diamond polishing through 6, 3, and 1 micron, and then an alumina oxide polish. Specimens were immersed in 10%nital (20ml nitric acid, 180ml ethanol) for 30 seconds to chemically polish the surface and remove fine scratches. Specimens were then etched with acetic picral (6g picric acid, 100ml ethanol, 5ml acetic acid and 10ml of water) for 5-10 seconds.

Experimental Results

Torsion and Compression Results

The equivalent true stress-strain behaviour of wrought AZ31, from the torsion and compression experiments, can be seen in Figure 1 (Torsion for strains up to 1.0) and Figure 2 (Compression for strains up to 0.6).

Similar trends can be distinguished for both deformation modes. As strain is increased, the flow stress increases up to a peak stress and then steadily decreases. Also, as the temperature is increased, or the strain rate is reduced, the level of the peak flow stress is reduced. The main difference observed between the two deformation modes is the rate of work hardening and the strain to which the peak flow stress occurs. For torsion, the rate of work hardening is comparably lower and the strain to peak stress higher than that of compression.

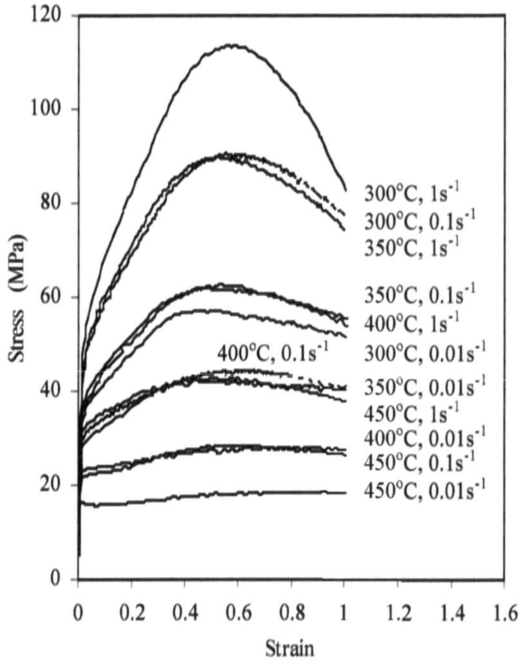

Figure 1. Flow curves obtained by torsion of extruded AZ31.

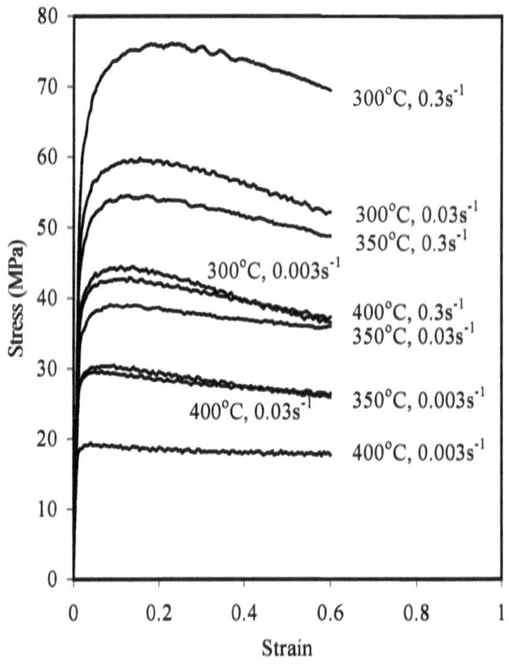

Figure 2. Flow curves obtained by compression of extruded AZ31.

The stress at a strain of 1.0 for torsion, and 0.6 for compression, are analysed below. A plot of these stresses against strain rate is shown in Figures 3(a) and 3(b) respectively. It is apparent that the relationship between stress and strain rate can be described by a power law. This type of relationship is common and is often incorporated in an expression that also includes temperature of the form:

$$\sigma = AZ^m \quad (1)$$

where Z is the Zener-Hollomon parameter ($\dot{\varepsilon}\exp(Q/RT)$), A and m are constant, R is the ideal gas constant (= 8.314 J/g mol K), and Q is the effective activation energy for deformation.

Regression analysis was carried out on the plots of stress versus strain rate, and the power law exponent, m, was determined. The average value of m was found to be 0.14 for torsion and compression.

Transposing Equation 1 results in the expression:

$$\ln(\sigma) = \frac{mQ}{RT} + m\ln(\dot{\varepsilon}) + \ln(A) \quad (2)$$

The plots of ln(σ) versus 1000/RT in Figures 4(a) and 4(b) yield values of mQ of 24 kJ/mol for torsion and 22 kJ/mol for compression. The two corresponding effective activation energies are 166 kJ/mol and 153 kJ/mol respectively.

The constant, A, in Equation 1, was thus found to be 0.72 for torsion and 0.85 for compression. The flow stresses predicted using these values are shown in Figure 5 along with the experimental results.

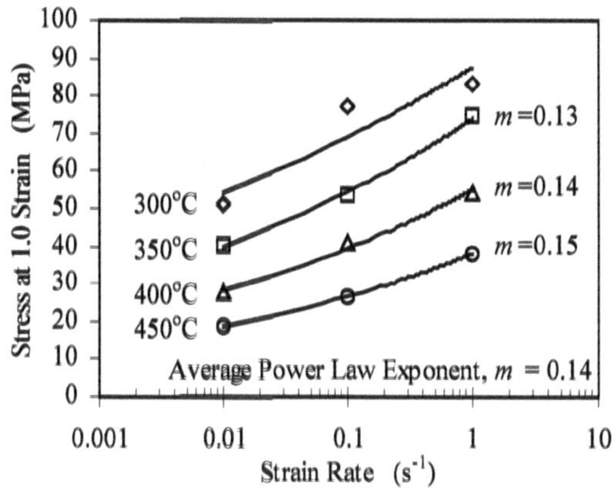

(a)

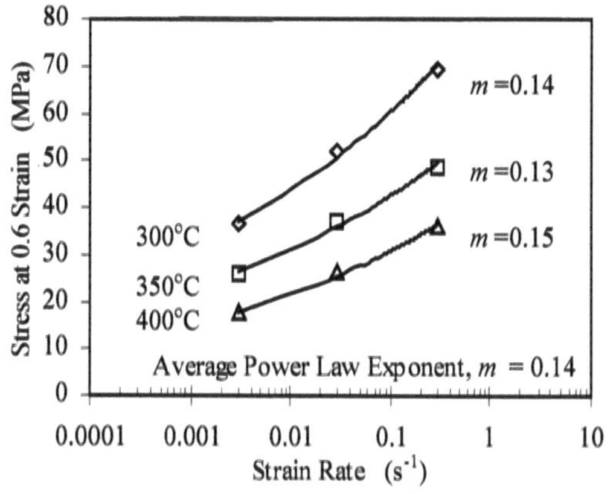

(b)

Figure 3. Stress versus Strain Rate for (a) torsion (stress at 1.0 strain), and (b) compression (stress at 0.6 strain).

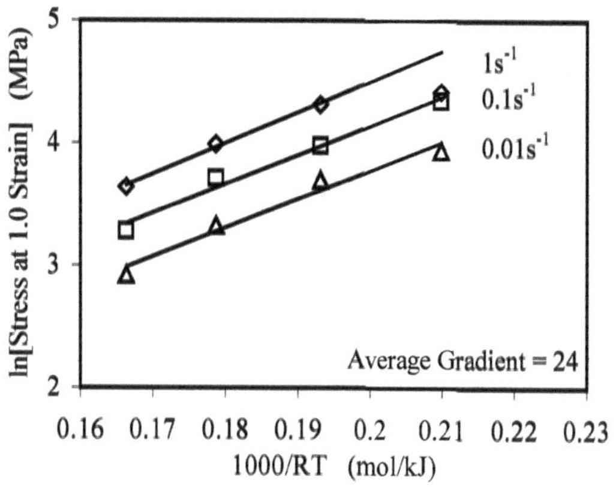

(a)

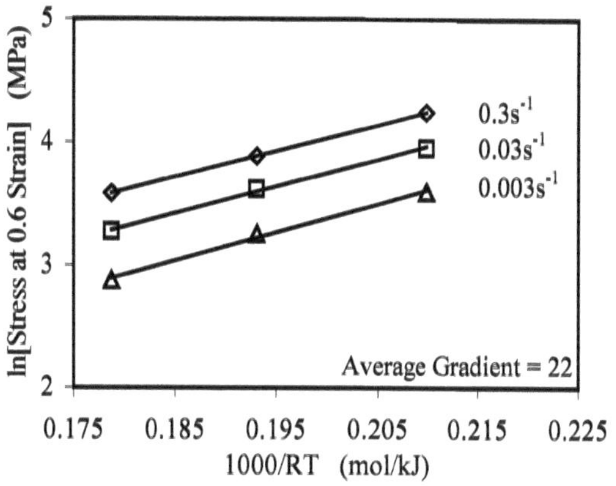

(b)

Figure 4. ln(Stress) versus 1000/RT for (a) torsion (stress at 1.0 strain), and (b) compression (stress at 0.6 strain).

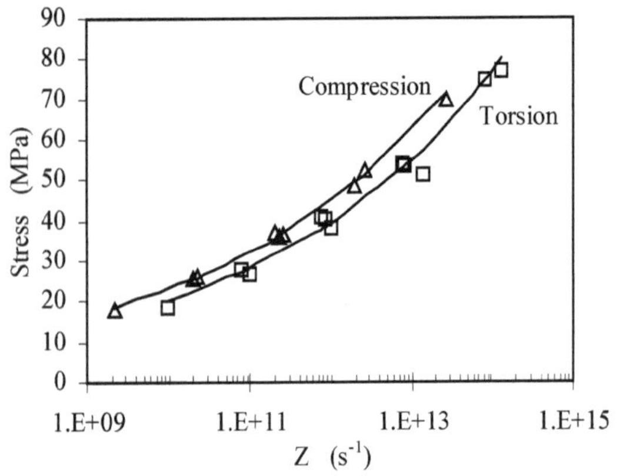

Figure 5. Stress versus the Zener-Hollomon parameter for both torsion and compression.

It is evident that the actual, and predicted, flow stresses for both torsion and compression are fairly similar at the highest strains examined. The compression results are higher than the torsion results by approximately 10%.

Torsion with Increasing Strain Results

The microstructural evolution of AZ31 with increasing strain can be seen in Figure 6., along with the corresponding flow curve for the conditions studied. The initial wrought structure of extruded AZ31 (strain = 0) was found to have an average grain size of 22.5μm. As the strain is increased to 0.2, grain boundaries become serrated. Dynamically recrystallised (DRX) grains are observed at prior grain boundaries at a strain of 0.4. As the strain is further increased to 0.6, DRX grains become more prevalent. At strains of 0.8 and 1.0, DRX grains are more extensive but large

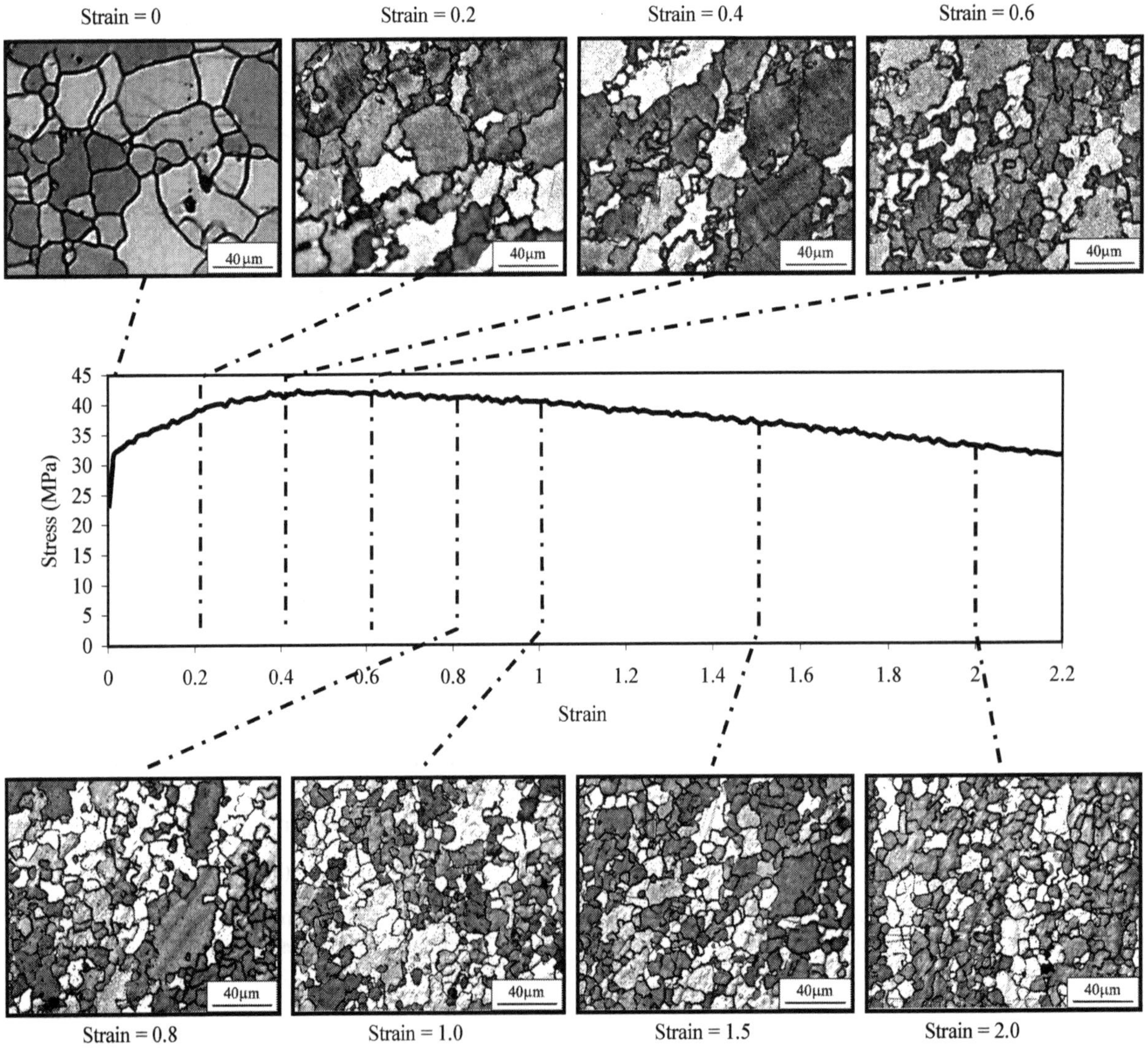

Figure 6. The microstructural evolution of wrought magnesium alloy AZ31 with increasing strain (at a constant temperature and strain rate of 350°C and 0.01s^{-1} respectively).

initial grains still remain. There is a small increase in the extent of DRX as the strain is increased to 1.5 and 2.0 but some large grains still remain (a completely recrystallised structure is not achieved) even after a strain of 2.0.

The average grain sizes were measured for all strains and these are plotted against strain in Figure 7. The average grain size was reduced from 22.5μm down to 7.3 μm after the maximum imposed strain of 2.

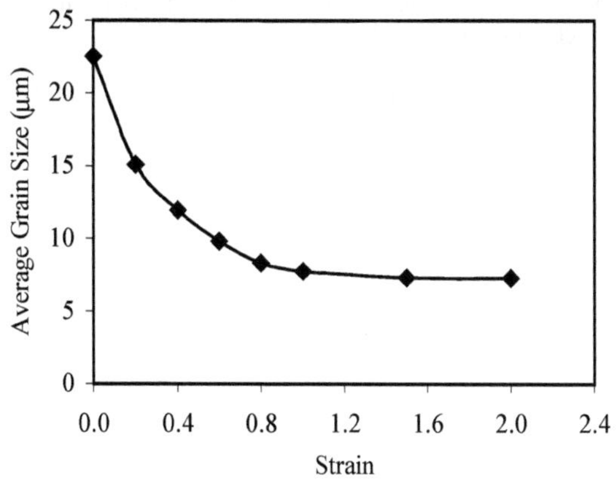

Figure 7. Average grain size versus strain for wrought AZ31 deformed at 350°C and $0.01s^{-1}$.

After a strain of 0.5, which approximately corresponds to the peak flow stress in this situation, the initial grain size was halved. The grain size reduces up to a strain of 1.0, after which an increase in strain does not yield a significant reduction in grain size. It is apparent that a steady state microstructure is established at a strain of 1.0 even though the stress continues to decrease with strain.

Discussion

The main difference observed between the flow behaviours for both deformation modes is in the flow stress at low strains. In compression, work hardening occurs more rapidly and the peak flow stress is reached at lower strains compared to torsion. These differences are not fully understood but it is likely that they are related to the limited slip systems available in the hcp lattice and the strong textures of wrought alloys.

The basal planes in wrought alloys are often found lying parallel to the direction of metal flow [7]. In the extruded bar used for the torsion samples, the basal planes lie predominantly parallel to the extrusion direction, with the *c* axes parallel to the radial direction. Once twisted, the plane of maximum shear stress is perpendicular to many basal planes but parallel to many prismatic planes.

In the case of the rolled bar used for the compression samples, the basal planes lie perpendicular to the normal direction, with the *c* axes parallel to the normal direction. Once subjected to compression along the rolling direction (and extension along the transverse direction), the planes of maximum shear stress coincide with many of the prismatic planes but lie perpendicular to many of the basal planes. In this sense there is some similarity in the relationship between the two different deformation modes and the material texture.

Though basal slip is the predominant slip system in magnesium, the high temperatures and the textures employed in the present work mean that prismatic slip is likely to be playing a significant role in the deformation behaviour. In the torsion samples it is possible that a number of grains deform (in simple shear) by prismatic slip on a single system. However, in the compression samples multiple prismatic slip is far more likely. This difference may account for the higher work hardening behaviour in the compression samples.

We turn now to the difference in the strain to the peak stress between the two deformation modes. This difference is illustrated in Figure 8, which plots the strain to peak stress as a function of the Zener-Hollomon parameter. The strain to the peak stress is often determined by the onset of dynamic (DRX) during deformation. The present results (Figure 6), which show signs of dynamic recrystallisation just prior to the flow stress peak, are consistent with this. It is therefore possible that the differences in strains to peak stresses in the present samples is due to the relative difficulty of initiating dynamic recrystallisation in torsion. This in turn is likely to be related to the high probability of single slip which results in a reduced degree of crystallographic fragmentation and hence a lower likelihood of producing recrystallization nuclei.

Similar arguments for the effect of texture on the DRX kinetics in wrought magnesium alloys have been used by Kaibyshev [8]. Compression of cylindrical samples, with the compression axes forming angles of 0°, 45°, and 90° with the basal plane, revealed that in the sample orientated 45° to the basal plane, deformation took place mostly by basal slip and the kinetics of DRX were sluggish. Conversely, in the samples orientated 0° and 90° to the basal plane, basal slip was impeded, prismatic and pyramidal slip were enabled and DRX was more prevalent.

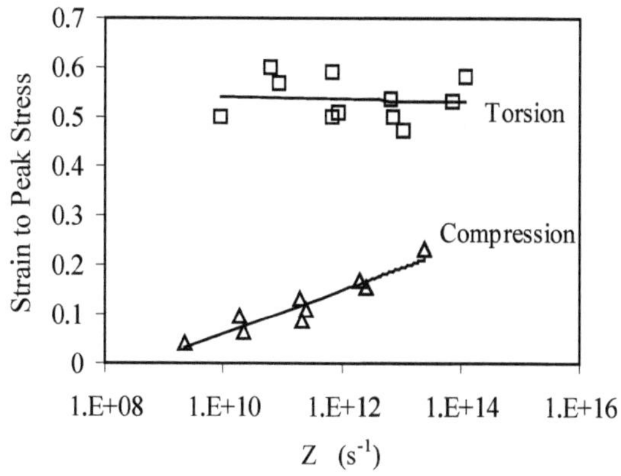

Figure 8. Strain to Peak Stress as a function of the Zener-Hollomon parameter, Z, for both deformation modes.

It is also interesting to note that the strain to the peak stress increases uniformly with *Z* in compression but not in torsion. The behaviour in compression is common to other metals [9] and the reason for the lack of a similar trend in torsion is not clear. It is possible that the inhomogeneous distribution of deformation in torsion plays a role in this phenomenon. This effect requires future investigation and is important for understanding the relationship between microstructure and flow stress.

As the flow stress is sensitive to deformation mode at low strains, constitutive equations used to model the evolution of flow stress will require the prescription of state variables that are sensitive to texture and deformation mode [10]. These developed equations would be more suited to low strain deformation processes such as multiple forging and rolling regimes.

At higher strains, when the change in flow stress with strain is lower, the flow stress is found to be relatively insensitive to deformation mode (Figure 5.). The current power law equation can be used to estimate the flow stress in high strain deformation processes such as extrusion. As some degree of flow softening is observed at higher strains under most of the conditions studied (i.e. a steady state flow stress is not achieved), a more accurate prediction of the flow stress will require that this factor be incorporated into the constitutive equation.

Conclusions

- At low strains, the flow stress, and critical strain for dynamic recrystallisation, is sensitive to deformation mode.
- At high strains, the flow stress of wrought magnesium alloy AZ31 can be described by a power law and is relatively insensitive to deformation mode.

Acknowledgements

CAST was established and is funded in part by the Australian Government's Cooperative Research Centres Program.

References

1. Luo, A. A., (2000), *Materials Comparison and Potential Applications of Magnesium in Automobiles*, Magnesium Technology 2000, Nashville, Tennessee, The Minerals, Metals & Materials Society. pp. 89-98.

2. Avedesian, M. & Baker, H., ed. (1999), *ASM Specialties Handbook*, Magnesium and Magnesium Alloys, ASM International: Materials Park, OH, USA.

3. Kaibyshev, R., & Sitdikov, O., (1996), *Dynamic Recrystallization and Mechanisms of Plastic Deformation*, ReX'96. pp. 203-209.

4. Mwembela, A., Konopleva, E. B. & McQueen, H. J., (1997), *'Microstructural Development in Mg Alloy AZ31 during Hot Working'*, Scripta Materialia, **37**(11), pp. 1789-1795.

5. McQueen, H. J., Myshlaev, M., Sauerborn, M., & Mwembela, A., (2000), *Flow Stress Microstructures and Modeling in Hot Extrusion of Magnesium Alloys*, Magnesium Technology 2000, The Minerals, Metals & Materials Society, Nashville, Tennessee. pp. 355-362.

6. Hodgson, P. D.; Collinson, D. C.; Perrett, B., (1997), *The Use of Hot Torsion to Simulate the Thermomechanical Processing of Steel*, 7th International Symposium on Physical Simulation, Tsukuba, Japan. pp. 219-229.

7. Raynor, G. V., (1959), *The Physical Metallurgy of Magnesium and Its Alloys*. Pergamon Press, New York, pp. 223.

8. Kaibyshev, O. A., (1990), *Relationship between Mechanisms of Deformation and Development of Dynamic Recrystallisation*, Recrystallisation '90, Minerals, Metals & Materials Society, pp. 855-860.

9. Roucoules, C. et al., (1994), Journal of Materials Processing Technology, **45**, pp. 447.

10. Barnett, M. R., (2001), *'Influence of Deformation Conditions and Texture of the High Temperature Flow Stress of Magnesium AZ31'*, Journal of Light Metals, in press.

THE HIGH STRAIN RATE DEFORMATION BEHAVIOR OF HIGH PURITY MAGNESIUM AND AZ31B MAGNESIUM ALLOY

Veronica Livescu[1], Carl M. Cady[1], Ellen K. Cerreta[1], Benjamin L. Henrie[1], George T. Gray III[1]

[1] Los Alamos National Laboratory, MST-8, Mail Stop G755, Los Alamos, NM 87545

Keywords: Magnesium, Mechanical Properties, Deformation Mechanisms, Twinning, Texture, hcp

Abstract

The deformation in compression of pure magnesium and AZ31B magnesium alloy, both with a strong basal pole texture, has been investigated as a function of temperature, strain rate, and specimen orientation. The mechanical response of both metals is highly dependent upon the orientation of loading direction with respect to the basal pole. Specimens compressed along the basal pole direction have a high sensitivity to strain rate and temperature and display a concave down work hardening behavior. Specimens loaded perpendicularly to the basal pole have a yield stress that is relatively insensitive to strain rate and temperature and a work hardening behavior that is parabolic and then linearly upwards. Both specimen orientations display a mechanical response that is sensitive to temperature and strain rate. Post mortem characterization of the pure magnesium was conducted on a subset of specimens to determine the microstructural and textural evolution during deformation and these results are correlated with the observed work hardening behavior and strain rate sensitivities were calculated.

Introduction

Magnesium alloys are extremely attractive structural materials for applications in the automotive and aerospace industries due to their unique combination of light weight, high specific strength and stiffness and high recycling capability. But the use of magnesium alloys has been limited due to their low creep resistance, cold forming capability, and corrosion resistance.

Magnesium exhibits a pronounced anisotropy of mechanical properties during deformation, a typical behavior for metals with hexagonal structure. Hexagonal close packed (hcp) metals have a limited number of active deformation systems and therefore exhibit strong textures upon mechanical processing. It is known that texture plays an important role in the anisotropy of the mechanical properties of hcp metals [1-4]. Many researchers have investigated the deformation mechanisms in hexagonal metals [1, 3, 4]. Since many magnesium products are subject to hot deformation and therefore bear strong textures, it is of practical significance to make use of texture to enhance material properties (i.e. to improve plasticity).

Several experimental and theoretical studies have identified the active deformation mechanisms in magnesium and many of its alloys. Similar to many other hcp metals, deformation is accommodated through both slip and twinning. The activation of these mechanisms is highly dependent upon temperature, strain rate and loading direction with respect to the specimen's texture. Many studies have concentrated on the characterization of texture formation and deformation behavior of polycrystalline wrought alloys [1, 0, 0], but only few researchers have tried to correlate mechanical anisotropy, deformation mechanisms, and texture evolution [0, 0]. In addition, modeling of crystal plasticity and texture [0-0] gained attention in the last decade and is used to determine the physical processes behind the deformation behavior and to interpret experimental investigations.

In spite of these studies, constitutive models which predict the deformation behavior of metals and alloys have had limited success when it comes to the behavior of low symmetry, highly textured materials. These models require mechanical test data over a range of strain rates and temperatures and in the case of magnesium and its alloys there is less mechanical data available over a broad range of strain rates, temperatures and material orientations. While many investigations have considered the effects of texture and have examined the superplastic behavior, only limited data at high strain rates exists for many of these materials.

The aim of this study is a more profound understanding of the influence of strain rate on the mechanical behavior of pure magnesium and AZ31B magnesium alloy. The mechanical response will be discussed in terms of the operative deformation mechanisms and evolution of texture and microstructure during compressive deformation of pure magnesium and AZ31B magnesium alloy over a range of experimental parameters.

Experimental Procedure

The axial compression deformation behavior has been investigated for high-purity magnesium and AZ31B magnesium alloy. The high-purity magnesium is a commercially available hot rolled plate, with chemistry given in Table I. The AZ31B magnesium alloy is a commercial purity alloy plate in the stress-relieved H-24 condition, with chemistry given in Table II.

Electron backscattered diffraction (EBSD) and X-ray diffraction were utilized to determine the initial texture of both alloys. Results indicate that each material has a strong basal texture with the (0001) aligned with the plate normal of the pure material as the extrusion axis of the AZ31B alloy. These results are given in Figure 1.

Cylindrical compression specimens from both materials were machined. Pure magnesium specimens were machined in three orthogonal directions. The first two specimens had a compression axis that was within the plane of the rolled plate: in-plane longitudinal (IP-L) and in-plane transverse (IP-T). The third specimen had its compression axis aligned with the normal of the rolled plate and is referred to as through thickness (TT). The high-purity magnesium specimens were then heat treated at 200°C for 30 minutes in vacuum. This yielded an equiaxed grain size of approximately 65μm. The AZ31B was tested in the as-received condition, also in three orthogonal directions and had a grain size of approximately 50μm.

All mechanical tests were performed in compression at temperatures of -75°C, 20°C and 100°C. Quasistatic compression tests were performed using an MTS-880 universal testing machine at strain rates of 0.001/s, 0.01/s, and 1/s. Initial tests were performed to between 15 and 20% strain in a single loading to reveal the hardening behavior. Subsequently, interrupted strain tests at 0.001/s and temperatures of -75°C, 20°C, and 100°C were conducted on the pure magnesium specimens. The tests were interrupted at strains of 3% and 8%. Dynamic tests were performed also at temperatures of -75°C, 20°C and 100°C using a split Hopkinson pressure bar (SHPB), where strain rates ranging between 2000 and 3500/s were achieved.

Table I: Chemistry of the Pure Magnesium (ppm)

Metal	Al	Ca	Mn	Zr	Zn	Sn	Si	Pb
Pure Mg	30	10	40	10	130	10	40	10

Table II: Chemistry of the AZ31B magnesium alloy (wt %)*

Metal	Al	Mn	Zn	Si	Cu	Fe
AZ31B	3.02	0.30	1.01	0.007	0.003	0.003

*Values are the typical commercial values for the chemistry

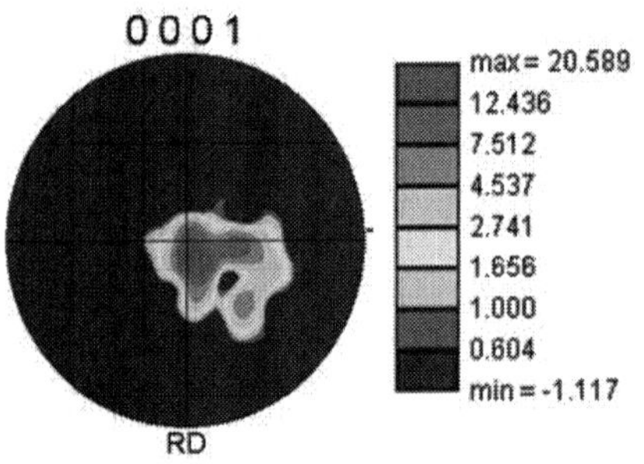

(a)

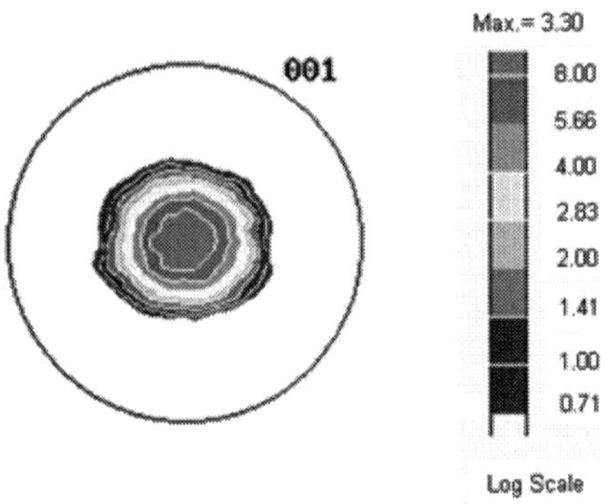

(b)

Figure 1. Initial texture of (a) heat-treated high-purity magnesium and (b) as-received AZ31B alloy.

Quasistatically deformed specimens from the interrupted tests were prepared for a microscopy study of the deformation evolution within the early part of the stress-strain curve. The microstructure of the quasi-static compression specimens was characterized using optical microscopy. Specimens were sectioned perpendicular to the loading axis, mechanically polished and etched and observed on a Zeiss optical microscope at 50 times magnification under polarized light. Sample preparation consisted of grinding on SiC paper with increasingly finer grit, followed by automatic mechanical polishing with 6μm and 1μm oil-based diamond suspension. The grain structure was revealed by etching using a solution of nitric acid (3ml), ethylene glycol (75ml) and water (25ml).

Additionally, the microstructure and texture evolution of this same set of specimens was characterized through electron back-scattered diffraction (EBSD) on a Phillips XL30 electron microscope. These specimens were mechanically ground to a 0.25μm diamond finish, chemically polished in a solution of 1% nitric acid and 99% ethylene glycol and then electro-polished at 0°C and 2V in a solution of 40% phosphoric acid and 60% ethanol.

Results

Effect of Strain Rate and Temperature on the Mechanical Response

Both the high-purity magnesium and AZ31B alloy displayed similar mechanical behavior under identical testing conditions. The stress-strain behavior of both materials is highly dependent upon temperature, strain rate, and orientation of the loading direction with respect to the basal pole. Stress-strain curves as a function of strain rate and temperature are shown in Figures 2 and 3. In the case of the IP specimens the stress-strain data can be separated into two different regions. In the first region, the material deforms elastically up to approximately 20MPa in the case of pure magnesium and 100MPa in the case of AZ31B alloy, and then it yields abruptly. After yielding, the flow curves follow a low hardening rate plateau.

In the case of the pure magnesium and the AZ31B the IP-L and IP-T specimens behaved similarly, with similar yield and hardening behavior, so for clarity only the IP-T data is given in Figures 2 and 3. The IP-L and IP-T specimens have a yield stress that is relatively insensitive to strain rate and temperature as seen in Figures 2(a) and 3(a) and a work hardening behavior that is parabolic, then linearly upwards, followed by a saturation of twinning, a reduction in work hardening, and ultimately leading to failure. For image clarity in Figure 3, we present here only the behavior of the sample tested to a strain rate of 1/s as a function of temperature. Mechanical behavior at all other strain rates is at 20°C. Similar behavior was observed for samples tested at other strain rates.

The TT specimens, compressed along the basal pole direction, are more sensitive to strain rate and temperature and display a concave down work hardening behavior. The effect of strain rate and temperature for the TT specimens are presented in Figures 2(b) and 3(b). Higher yield and flow stresses are observed for specimens tested at higher strain rates and lower temperatures. The yield stress of AZ31B specimens is higher then that of pure magnesium in both IP and TT orientations. Again, for image clarity, we present here only the behavior of the samples tested to strain rates of 1/s as a function of temperature. Similar behavior was observed for samples tested at other strain rates.

Finally, for both materials and in both IP and TT specimens, stresses increased with increasing strain rate and decreasing temperature.

Post Mortem Analysis

After testing specimens were examined optically. All IP pure magnesium and AZ31B specimens had an elliptical cross section after testing. The eccentricity of this ellipse increased with strain for the strains examined in this study. All TT specimens maintained a circular cross section during testing. In both the AZ31B and the pure magnesium failure in compression by shear occurred at high strains. Additionally high rate specimens had been subjected to multiple loadings after the test as the bars were not momentum trapped, so the dynamic specimens were not characterized optically.

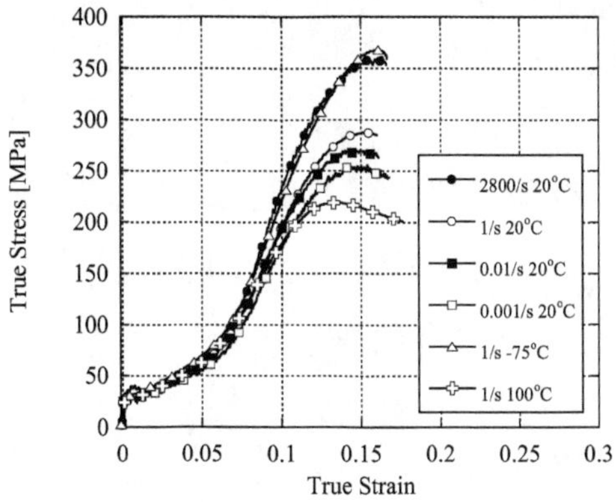

(a)

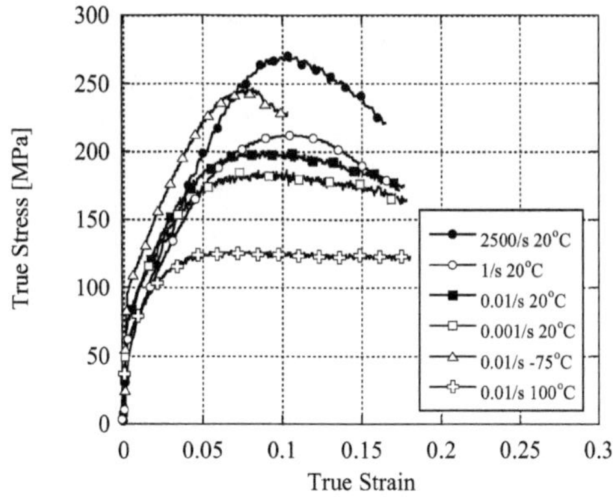

(b)

Figure 2. Stress-strain curves for pure magnesium: (a) IP-T specimens and (b) TT specimens.

Microstructure Evolution

Optical microscopy was performed on an as annealed as well as on several high-purity magnesium specimens subjected to interrupted strain tests. The initial microstructure of the as-annealed material is given in Figure 4. No twins were evident prior to deformation. Figures 5 and 6 detail the microstructure from tests run at 0.001/s strain rate and at three temperatures and two different strains on specimens in the IP and TT orientations, respectively.

The presence of deformation twinning is observed in all IP specimens tested at 3% and 8% strain. Less twinning is observed in the specimen deformed at 100°C at 3% strain than in specimens tested to higher strains or at lower temperatures. Qualitatively, twinning appears to increase with increasing strain and decreasing temperature.

For TT specimens tested in the same conditions, there was no deformation twinning observed in the 3% strain at 20°C and 100°C specimens. Similar to the IP specimens, more twins were seen in the specimens tested at low temperature. Also, twinning increased with increasing strain. However, for all conditions examined the twinning in the TT specimens was less than what was observed in the IP specimens.

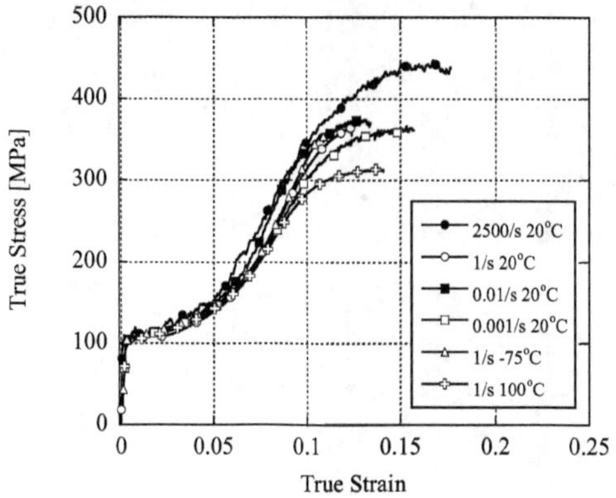

(a)

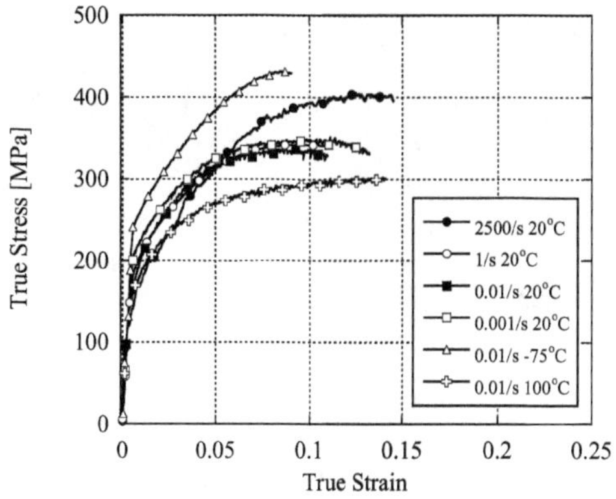

(b)

Figure 3. Stress-strain curves for AZ31B alloy: (a) IP-T specimens and (b) TT specimens.

Figure 4. Microstructure of the as-annealed high-purity magnesium.

Texture Evolution

EBSD analysis of the pure magnesium tested quasi-statically interrupted at 3 and 8% strain was performed on the IP specimens tested at -75, 20, and 100°C. Areas of $1 mm^2$ were scanned to obtain pole figures. While this region is small as compared to the specimen size, the small size of the specimens prevented x-ray analysis and the EBSD information can be used to obtain a qualitative view of the texture evolution. From this analysis 0001 pole figures from specimens with 3% strain tested at 100, 20, -75°C were generated and appear in Figure 7.

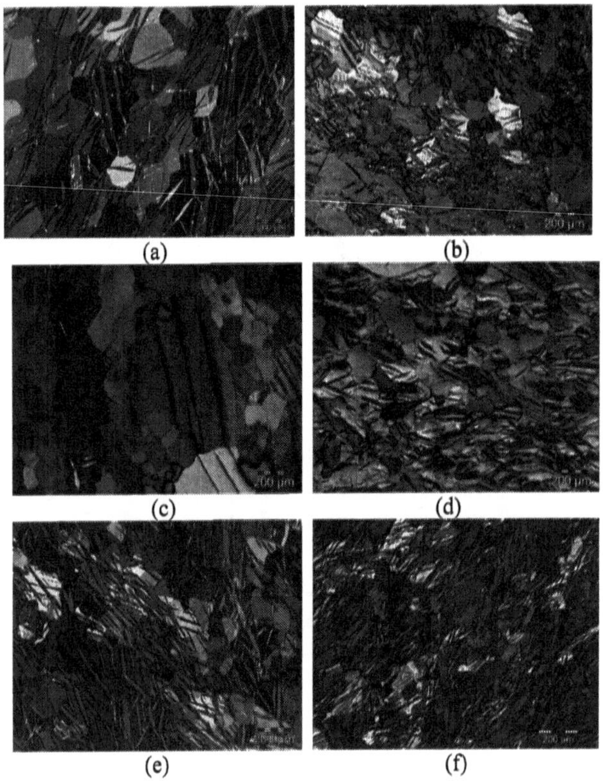

Figure 5. Microstructure evolution in pure magnesium, in-plane orientation, deformed at 3%(left) and 8% strain (right) as a function of temperature: (a) and (b) at 20°C, (c) and (d) at 100°C, (e) and (f) at -75°C.

It can be seen that the texture evolves from a strong basal fiber aligned with the normal of the plate, as observed in Figure 1, to one with the basal fiber rotated almost 90° to a direction within the plane on the plate. In fact this direction is aligned with the compressive axis of the specimen. This is typically achieved by substantial twinning on the $(1\bar{1}02)$ planes [13]. The tendency for this rotation increases with decreasing temperature. In Figure 8 the 0001 pole figure for the IP specimen tested to 3% strain at 100°C is reproduced for comparison with the 0001 pole figure from the specimen tested to 8% strain at the same temperature. Again the reorientation of the specimen increases substantially with strain even at 100°C.

Discussion

The mechanical response of the pure magnesium and the AZ31B alloy is quite similar as is observed from Figures 2 and 3. In both cases the TT specimens display a concave down work hardening behavior where as the IP specimens display a parabolic and then linearly upwards work hardening behavior that does not appear to be influenced significantly by the rolling direction within the plane of the plate. However, while both the AZ31B and the pure magnesium have a yield stress that is relatively insensitive to the temperature and strain rate and a work hardening response that is subtly sensitive to these parameters, the yield and flow stresses are consistently higher in the AZ31B material but the work hardening rates are lower.

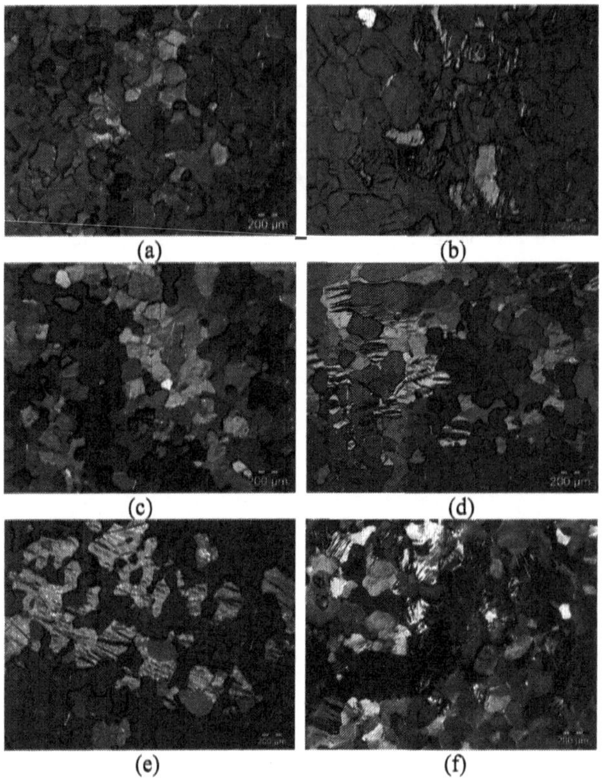

Figure 6. Microstructure evolution in pure magnesium, through-thickness orientation, as a function of temperature: (a) and (b) Deformed at 20°C at 3% and 8% strain respectively, (c) and (d) Deformed at 3% and 8% strain at -75°C respectively, (e) and (f) Deformed at 3% and 8% strain at 100°C respectively.

Agnew et al. [14] discuss the direct influence of the alloying on the mechanical properties of textured materials. Alloying can change the contribution of each deformation mode as well as the strength of that contribution to the overall mechanical response. In this way the subtle difference in chemistry may influence yield behavior of the material. However, the grain size of the AZ31B was smaller than that of the pure magnesium and this too may contribute to the increased yield stress.

In the case of the pure magnesium TT specimens, metallography indicated that even at higher strains and lower temperatures twinning was not a predominant feature of the microstructure (Figure 6). This suggests that slip is the primary deformation mechanism. Agnew et al. [15] and Gehrmann et al. [16] observed similar behavior in hot rolled AZ31B alloy and pure magnesium. A study of the effect of grain size on the mechanical response of an Mg-Zn-Zr alloy indicates that even at high strain rates slip still operates [16]. Here, because the work hardening response, while sensitive to strain rate, maintained similar characteristics of concave down work hardening behavior indicative of deformation controlled by slip at all strain rates, it is likely that the high strain rate deformation is dominated by slip in the TT direction. This is further supported by a similar study where at high strain rates Ishikawa et al. observed a slip dominated behavior in AZ91at temperatures ranging from 298 to 723K [17].

In the case of the IP specimens, deformation is well known to be controlled by slip and twinning [3,4,13-15]. In the current study, the metallography and the texture analysis (Figures 5-8) on the quasi-static specimens indicates that even at higher temperatures (100°C) that twinning is an important deformation mechanism. As observed in numerous other studies of magnesium and other hcp metals, here twinning increased with both increasing strain and decreasing temperature [13-20]. Again, the gross characteristics of the work hardening behavior did not change with strain rate, namely the parabolic behavior that is then followed by a linearly upwards work hardening response indicative of deformation controlled by a combination of slip and twinning. It is, therefore, likely that at high strain rates the deformation behavior is controlled by a combination of slip and twinning as is observed in Hf and Zr [19, 20].

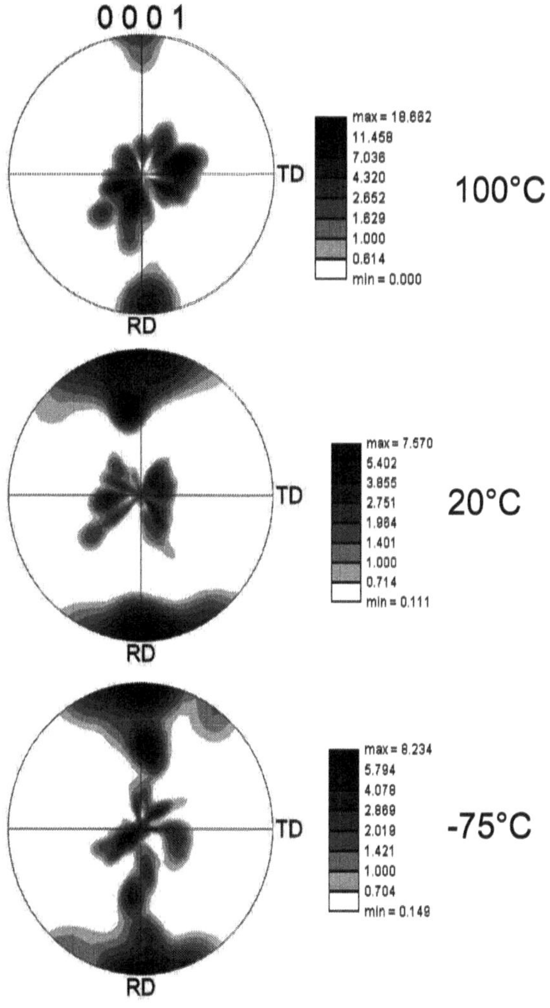

Figure 7. 0001 pole figures of the IP magnesium tested quasi-statically to 3%strain at 100, 20, -75°C from bottom to top.

The strain rate sensitivities (m) for the TT and IP specimens were calculated at three different temperatures and at 5% and 10% strain for the AZ31B and the pure magnesium specimens over strains rates ranging from 10^{-3} to 10^{3}/s. These values appear in Table III.

The value of m is defined as:

$$m = \partial \log \sigma / \partial \log \dot{\varepsilon} \quad (1)$$

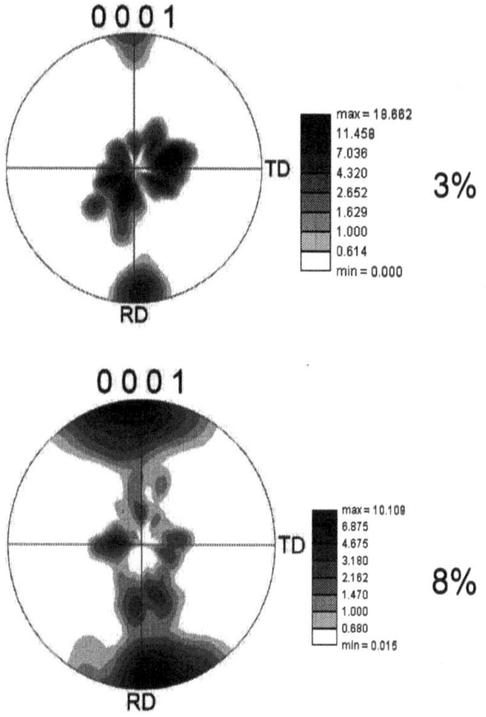

Figure 8. 0001 pole figures of the IP magnesium tested quasi-statically to at 100°C to 3 and 8% strain.

The m-values were calculated at -75, 20, and 100°C and are given in Tables III and IV. In the case of the TT direction, values at 10% strain are not shown because in many of the tests, peak stresses had been achieved between 5 and 10% strain. In these cases the work hardening rates by 10% strain were negative and likely the result of damage and failure mechanisms such as shear banding. All of the calculated values were less than 0.1, which is typical of rate sensitivities for metals deformed at temperatures less than half of the melting point [21]. In the IP direction, the m-value increased with increasing temperature but the trend was less obvious for the TT direction. This indicates that while the strain rate dependence increases with temperature in the IP direction that the dependence is not well correlated with temperature, for this range of temperatures, in the TT direction for both AZ31B and pure magnesium. The strain rate sensitivities were substantially higher for the pure magnesium.

Table III: m-values for the AZ31B

Temperature (°C)	TT, 5% ε	IP-dir, 5% ε	IP-dir, 10% ε
-75	0.01	0.004	0.008
20	0.002	-0.004	0.008
100	0.009	0.005	0.019

Table IV: m-Values for Pure Magnesium

Temperature (°C)	TT, 5% ε	IP-dir, 5% ε	IP-dir, 10% ε
-75	0.046	0.027	0.025
20	0.008	0.034	0.032
100	0.014	0.063	0.061

Conclusions

The following conclusions can be drawn from the results of the current study of the constitutive behavior of pure magnesium and AZ31B alloy:

- The stress-strain behavior of both pure magnesium and AZ31B alloy is highly dependent upon temperature, strain rate and orientation of loading direction with respect to the basal pole.
- In the case of both pure magnesium and AZ31B alloy the IP-L and IP-T specimens behaved similarly, exhibiting a yield stress that is relatively insensitive to strain rate and a work hardening behavior that is parabolic and then linearly upwards. The TT specimens, compressed along the basal pole direction, were sensitive to strain rate and temperature and displayed a concave down work hardening behavior. Higher yield and flow stresses were observed for TT specimens tested at higher strain rates and lower temperatures.
- For both materials and in both IP and TT specimens, flow stresses and rates of work hardening increased with increasing strain rate and decreasing temperature.
- Deformation twinning was present in all IP specimens tested at 3% and 8% strain. Twinning was not seen in TT specimens tested at 3% strain at 20°C and 100°C. In all specimens twinning appeared to increase with increasing strain and decreasing temperature.
- The texture of pure magnesium evolved during deformation from a strong basal fiber aligned with the normal of the plate to one with the basal fiber rotated almost 90° to a direction within the plane on the plate. The evolution of the texture was more significant at higher strains and lower temperatures. Texture evolution correlates well with twinning observed optically and increases with increasing strain and decreasing temperature.
- The strain rate sensitivities of both pure magnesium and AZ31B alloys were calculated for both the TT and IP directions. The values indicate that the work hardening response in the IP direction is more sensitive to strain rate than the TT direction and that the pure magnesium is more sensitive to strain rate than the AZ31B. Finally all m-values were less than 0.1 and usually increased with temperature.

Acknowledgements

The authors wish to thank M. F. Lopez and R.C. Korzekwa for their help with the metallography. This work was performed under the auspices of the US Department of Energy.

References

1. E.W. Kelley and W.F. Hosford, *Trans. AIME,* 242 (1968), 654.
2. D.H. Avery, W.F. Hosford, W.A. Backofen, *Trans. AIME*, 233 (1965), 71.
3. M.H. Yoo and J.K. Lee, *Phil. Mag. A*, 63 (1991), 987.
4. U.F. Kocks and D.G. Westlake, Trans. AIME, 239 (1967), 1107.
5. K.H. Eckelmeyer and R.W. Hertzberg, *Met. Trans.* 1 (1970), 3411.
6. P. Klimanek and A. Pötzsch, *Mater. Sci. Eng. A,* 324 (2002), 145.
7. G.S. Rao and K.L. Murty, *Res. Mech.*, 23 (1988), 363.
8. M.A. Gharghouri et. Al, *Phil. Mag. A*, 79 (1999), 1671.
9. C.N. Tome, R.A. Lebensohn, and U.F. Kocks, Acta Metall. Mater., 39 (1991), 2667.
10. R.A. Lebensohn and C.N. Tome, Acta Metall., 41 (1993), 2611.
11. P.A. Turner and C.N. Tomé, *Acta Mater.*, 42 (1994), 4143.
12. A. Staroselsky and L. Anand, *Int. J. of Plasticity* 19 (2003), 1843-1864.
13. S.R. Agnew, C.N. Tome, D.W. Brown, T.M. Holden, and S.C. Vogel, *Scripta Mat.*, 48 (2003) 1003.
14. S.R. Agnew, M.H. Yoo, and C.N. Tome, Acta Mater. 49 (2001) 4277.
15. S.R. Agnew, D.W. Brown, S.C. Vogel and T.M. Holden, *Mat. Sci. Forum*, 404-407 (2002), 747.
16. R. Gehrmann, M.M. Frommert and G. Gottstein, *mat. Sci. Eng. A*, 395 (2005), 338.
17. T. Mukai, M. Yamanoi, H. Watanabe, K. Ishikawa, and K. Highashi, *Mater. Trans.* 42 (2001), 1177.
18. K. Isjikawa, H. Watanabe, and T. Mukai, *Mat. Lett.*, 59 (2005), 1511.
19. G.C. Kaschner and G.T. Gray, *Mater. Trans A*, 31 (2000) 1997.
20. E. Cerreta and G.T. Gray, *Mater. Trans A*, 35 (2004) 1999.
21. G.E. Dieter, *Mechanical Metallurgy*, New York: McGraw Hill, Inc., 1986.

MECHANICAL PROPERTIES AND MICROSTRUCTURE OF AZ31 MAGNESIUM ALLOY TUBES

Alan A. Luo and Anil K. Sachdev

General Motors Research & Development Center, Warren, MI 48090-9055, USA

Keywords: Magnesium, Mechanical Properties, Deformation, Microstructure, Extrusion, Tube

Abstract

Magnesium alloys are increasingly being used in automotive industry for weight reduction and fuel economy improvement. Extruded tubular sections provide further opportunities in mass-efficient designs of automotive structural and interior applications. In this paper, microstructural evaluation indicates that twinning is the predominant deformation mechanism for magnesium alloys at room and moderate temperatures. Dynamic recrystallization is observed at temperatures as low as 150°C, leading to the formation of fine grains as a "necklace" at prior grain boundaries. These new grains cause strain localization and instability due to a loss in strain hardening, and result in failure by cavitation.

Introduction

Weight reduction for automobile fuel economy has spurred an increase in magnesium consumption over the last decade at an annual rate of 15%. Current automotive magnesium applications include instrument panel beams, front end support, transfer cases, steering components, etc [1, 2]. The die casting process is used to make these magnesium parts due to its high productivity and the excellent castability of magnesium alloys. To maintain the competitiveness of current magnesium components and further expand new applications, wrought magnesium alloys and their manufacturing processes must be developed.

Recently, tubular sections of steel and aluminum alloys are increasingly being used in the automotive industry to replace stamped components, and lead to weight savings, part consolidation and improved vehicle performance. These tubular sections, such as truck frames, engine cradles, roof rails, cross-member and instrumental panel beams, are often bent and hydroformed at room temperature [3, 4].

Magnesium has much less formability than steel and aluminum alloys at room temperature due to its hexagonal (hcp) crystal structure. Table 1 lists the nominal compositions and typical room-temperature tensile properties (yield strength YS, ultimate tensile strength UTS and elongation E_f) of several commercial extruded magnesium alloy tubes [5, 6]. Of these alloys, AZ31 is the most widely used in non-automotive applications. With a higher aluminum content, AZ61A offers slightly higher tensile strength and ductility than AZ31 alloy, but at much lower extrudability. The high-strength Zr-containing ZK60A was designed for applications in racing cars and bicycles, such as wheels and stems [5]. The extrusion speed for making ZK60A tubes is extremely low, rendering it uneconomical for automotive applications. Therefore, AZ31 alloy was selected in this study due to its good combination of mechanical properties and extrudability among commercial magnesium alloys.

This paper investigates the tensile properties of AZ31 alloy tubes at room and elevated temperatures and correlates its behavior to the deformation microstructure that is produced.

Table 1. Nominal compositions and typical room-temperature tensile properties of extruded magnesium alloy tubes [2-4]

Alloy	Temper	Composition, wt.%				Tensile Properties		
		Al	Zn	Mn	Zr	YS, MPa	UTS, MPa	E_f, %
AZ31	F*	3.0	1.0	0.20	-	165	245	12
AZ61A	F	6.5	1.0	0.15	-	165	280	14
ZK60A	F	-	5.5	-	0.45	240	325	13
ZK60A	T5#	-	5.5	-	0.45	268	330	12
AM50	*F*	*5.0*	-	*0.30*	-	*168*	*268*	*18*

* F signifies as extruded.
\# T5 signifies artificially aged after extrusion.

Materials and Experimental Procedures

Magnesium alloy AZ31 tubes of 70 mm nominal outer diameter and 4 mm nominal thickness were obtained from Timminco Metals in Aurora, CO. The chemical composition of the tubes is shown in Table 2. Tensile specimens were machined along the longitudinal direction of the tubes. Only the grip sections of the specimens were flattened and the curved gage sections remained intact. Tensile testing was carried out at room temperature and five elevated temperatures, i.e., 93, 121, 149, 177, and 204°C, according to ASTM E21-92 (1999) procedures. For elevated temperature testing, the specimens were maintained at temperatures for 30 minutes prior to loading. ASTM standard specimens of 2" gage length were used for tests at an initial strain rate of 0.003 s^{-1}. For each condition, at least three specimens were tested and the average properties are reported. The surface, longitudinal, and transverse sections of the tubes were mounted, polished, and etched for microstructural analysis. Optical microscopy was used to examine the microstructure, before and after tensile and testing.

Table 2. Chemical composition of AZ31 alloy tubes (in wt.%).

Alloy	Al	Mn	Zn	Fe	Ni	Cu
AZ31	3.1	0.54	1.05	0.0035	0.0007	0.0008

Results and Discussion

Tensile Properties

Fig. 1 shows a typical tensile curve for AZ31 alloy tube at room temperature. It is evident that AZ31 alloy exhibits serrations in the tensile curve, indicating discontinuous plastic flow during deformation. Serrated flow was also reported in room-temperature tensile curves for Al-Mg based 5xxx sheet materials [7], resulting in Luders bands in parts formed at room temperature [8].

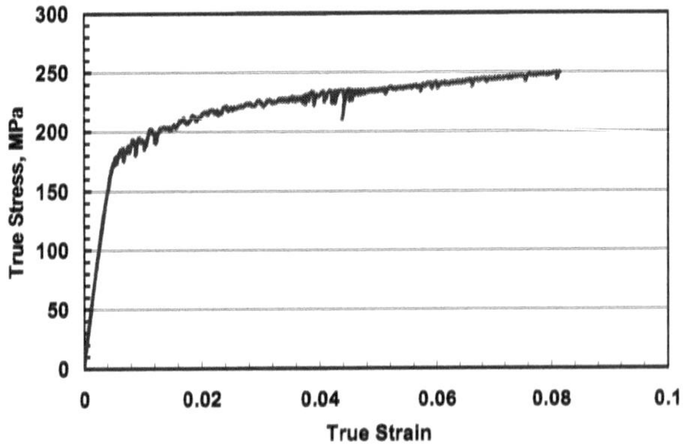

Fig. 1. A typical tensile curve for AZ31 alloy tubes at room temperature.

Fig. 2 shows the typical elevated-temperature tensile curves for AZ31. In general, the yield strength (0.2% YS) and ultimate tensile strength (UTS) decrease, but the percentage elongation increases with temperature. The yield strength data for AZ31 extrusion are in a good agreement with the values reported in the ASM handbook [5]. The changes in YS and UTS with temperature are very significant, but fairly gradual. The temperature effect on the ductility is worthy noting. Compared to the elongation at room temperature, ductility is greatly improved when heated up to 150°C, however, the ductility improvement is less significant between 150 and 200°C.

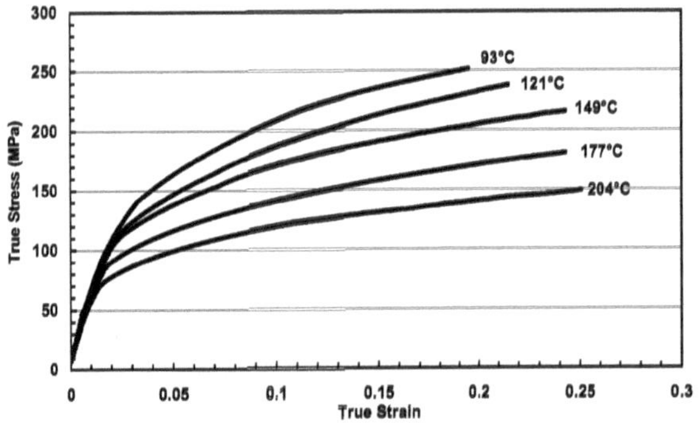

Fig. 2. Tensile curves for AZ31 alloy at elevated temperatures.

Strain-Hardening and Instability

The plastic deformation behavior of magnesium alloys can be explored by the flow equations describing the relationship between true flow stress, σ, and true strain, ε. A power law flow equation, or constitutive equation,

$$\sigma = K\varepsilon^n \quad (1)$$

is frequently used to describe the plastic deformation and strain-hardening of many alloys including magnesium [9], where n is the *strain-hardening exponent* and K is the *strength coefficient.* Eq. (1) was found to fit the σ-ε curve at room temperature and provided an n value of 0.14. In comparison, low-carbon steel has an n value of 0.20-0.25 and aluminum has an n value of 0.20-0.30 [10].

The elevated-temperature flow curves in Fig. 2 can also be fitted by the power-law Eq. (1), and the effect of temperature on strain-hardening exponent is plotted in Fig. 3. It is evident that the n value peaks at about 100°C, above which it gradually decreases with temperature. The tensile behavior of AZ31 sheet specimens was studied by Krajewski [11] at a higher temperature range of 300-500°C. Very limited strain-hardening was observed at 300-400°C and the flow curves were dominated by rapid softening.
To elucidate the strain-hardening behavior and instability in AZ31 alloy, a plot of strain-hardening rate ($d\sigma/d\varepsilon$) vs. true strain (ε) at room temperature is given in Fig. 4. According to Dieter [12], necking or localized deformation begins at maximum load, when the increase in stress due to decrease in the cross-sectional area of the specimen becomes greater than the increase in the load-carrying ability of the material due to strain-hardening. This condition of instability leading to localized deformation (necking) is defined as [12]

$$\sigma = d\sigma/d\varepsilon \quad (2)$$

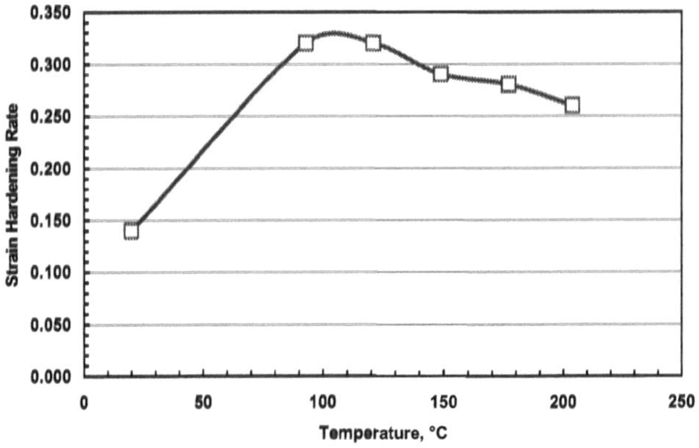

Fig. 3. Strain-hardening exponent vs. temperature for AZ31 alloy.

As shown in Fig. 4, the strain-hardening rate is constantly above the stress level over the entire plastic deformation region, indicating that the necking condition is not reached prior to fracture. The same phenomenon was also found for strain-hardening curves at temperatures up to 177°C, suggesting that strain-hardening can effectively distribute plastic deformation and delay necking until the specimen fails by linking of cavities formed at the grain boundaries, as discussed in the subsequent "Microstructure" section. Fig. 5 shows the strain-hardening rate and stress vs. true strain curves at 204°C. Again, the necking condition was not met despite the very close proximity of the two curves (strain-hardening rate and the flow stress) before the final fracture at plastic strain of about 28%.

Microstructure

Fig. 6 shows the microstructure of AZ31 alloy tubes in three cross sections, tube surface, longitudinal and transverse sections. A certain degree of twinning was observed in the longitudinal and transverse sections. Small micron-sized particles in the microstructure were identified as Al-Mn intermetallics through Energy Dispersive Spectroscopy (EDS).

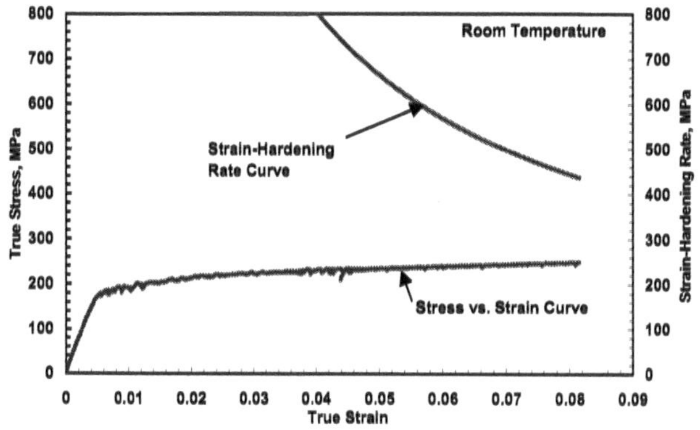

Fig. 4. Strain-hardening rate and stress vs. true strain curves at room temperature.

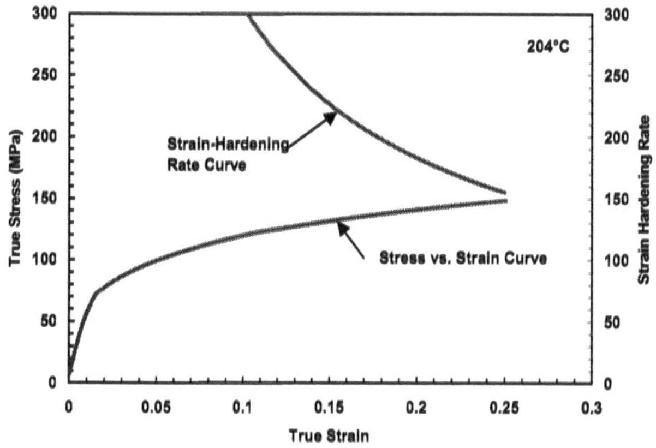

Fig. 5. Strain-hardening rate and stress vs. true strain curves of AZ31 alloy at 204°C.

Fig. 7 shows polished sections near the tensile fracture surfaces. It is obvious that twinning is the predominant deformation mechanism at temperatures up to 149°C. At higher temperatures, dynamic recrystallization (DRX), i.e., the formation of new, small grains (2-3 μm in diameter) during deformation, became evident (Fig. 7(f) and Fig. 8). DRX has been reported in magnesium alloys by a number of authors [13-15]. The morphology of the DRX grain structure in Fig. 8 resembles a necklace structure, which is also observed during hot deformation of other materials such as Ni_3Al intermetallic compounds [16]. Typically, DRX is initiated by strain-induced grain boundary fluctuations, and new grains are nucleated from serrations and bulges of the prior grain boundaries [16]. Additionally, formation of new DRX grains at twin/grain boundary intersections was also observed in AZ31 alloy [13, 15]. In this study, as shown in Fig. 9, the majority of new grains were formed at prior grain boundaries due to deformation-induced serrations and bulging of the grain boundaries (marked "A"). Nucleation of new DRX grains was also observed at intersections of twins and grain boundaries (marked "B" in Fig. 9), indicating twin-assisted DRX mechanism.

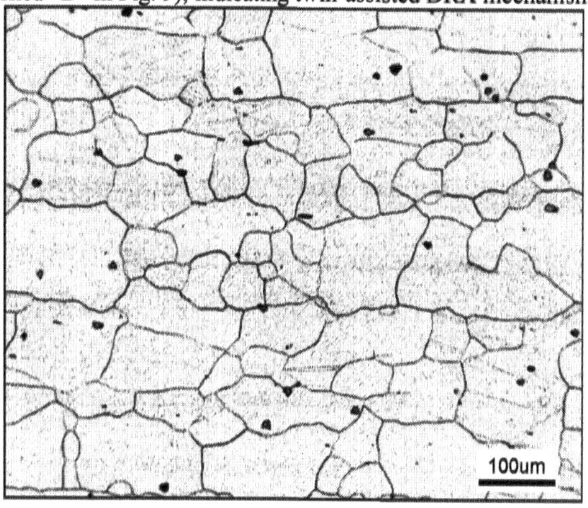

It is very important to discuss the implications of DRX to deformation and formability of magnesium alloys. Enhanced

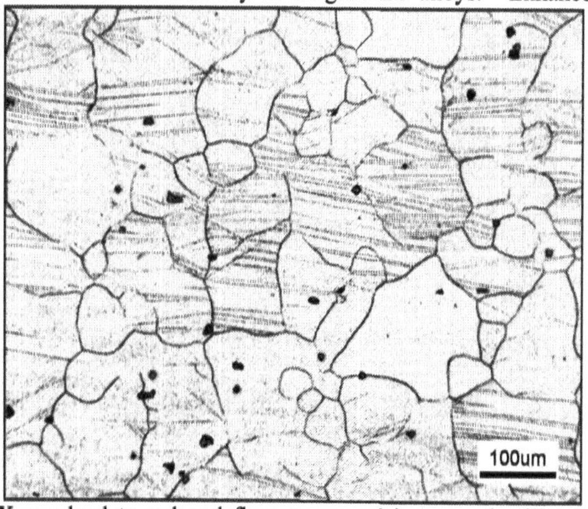

DRX can lead to reduced flow stress and improve ductility for industrial processes of magnesium alloys at elevated temperatures [13]. But, local DRX as observed in this study leads to premature

(a) Tube Surface

(b) Longitudinal Section

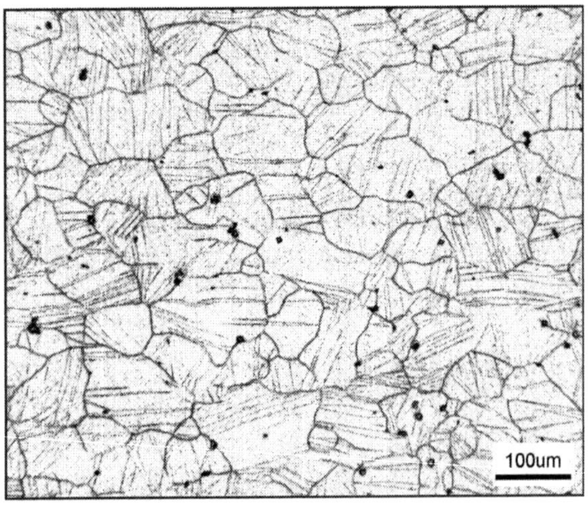

(b) Transverse Section

Fig. 6. Optical micrographs showing the microstructure of AZ31 tubes.

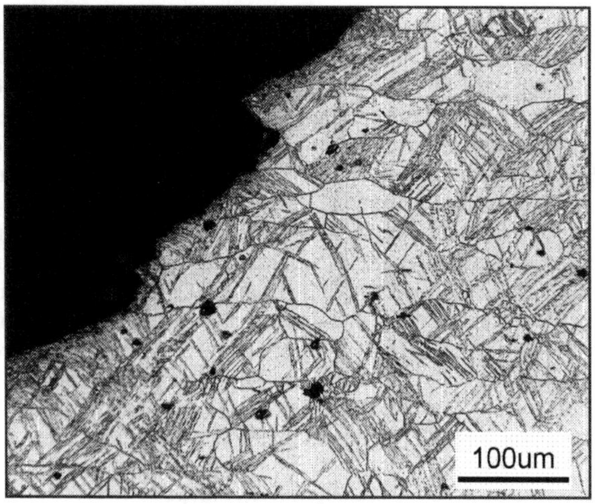

(a) Room Temperature

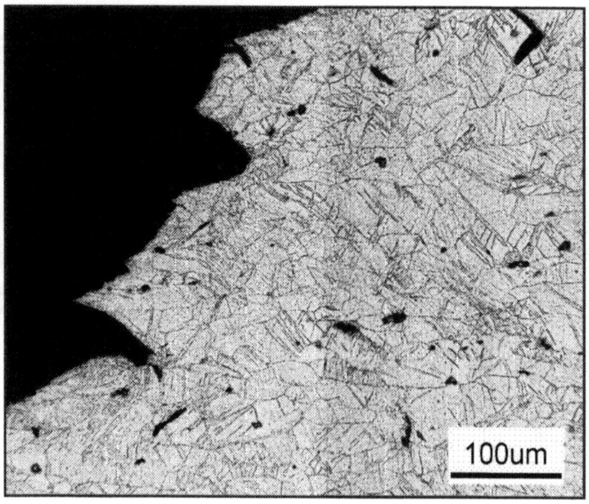

(b) 93°C

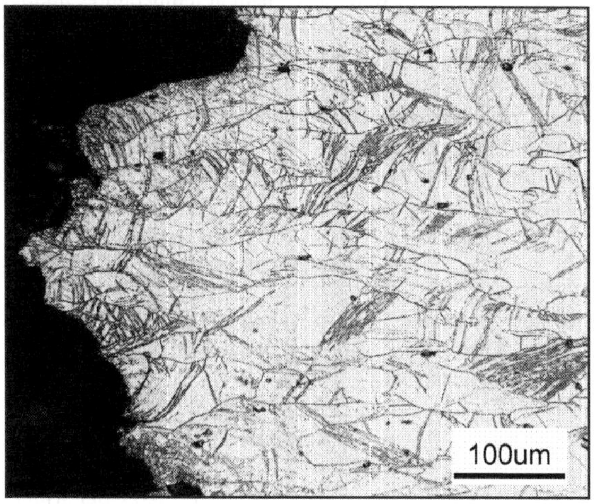

(c) 121°C

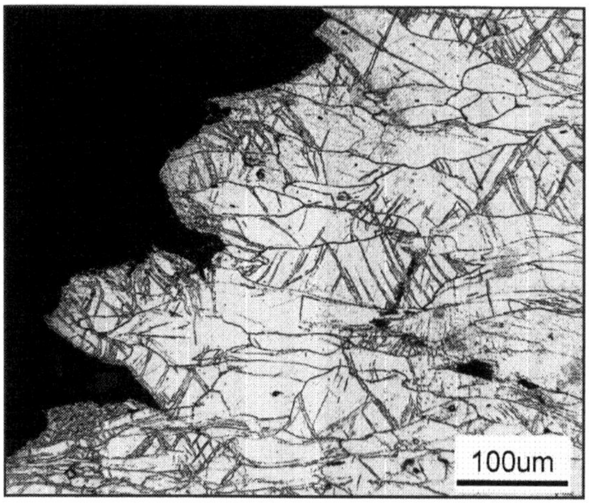

(d) 149°C

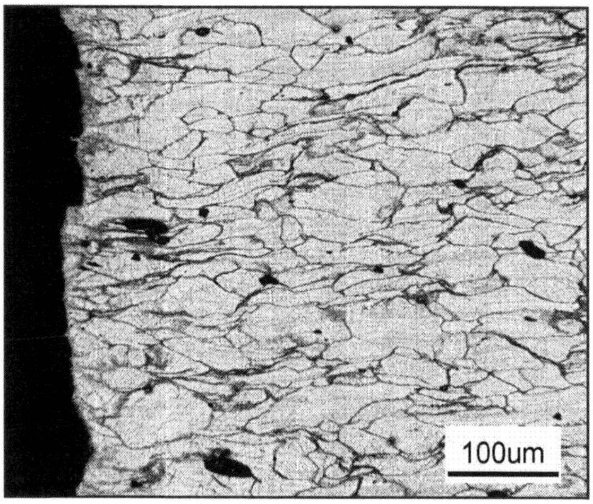

(e) 177°C

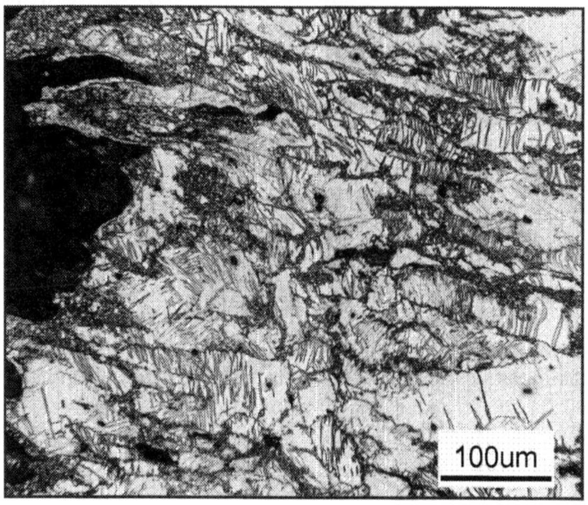

(f) 204°C

Fig. 7. Optical micrographs showing deformation microstructures near fracture surfaces of AZ31 samples tested at various temperatures.

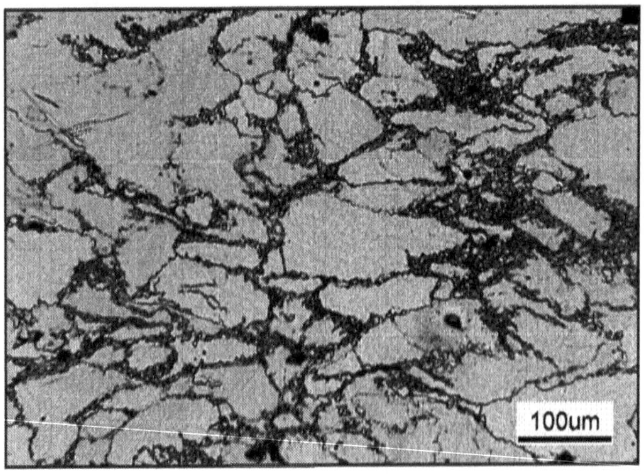

Fig. 8. Optical micrographs showing dynamic recrystallization in an AZ31 specimen tested at 204°C.

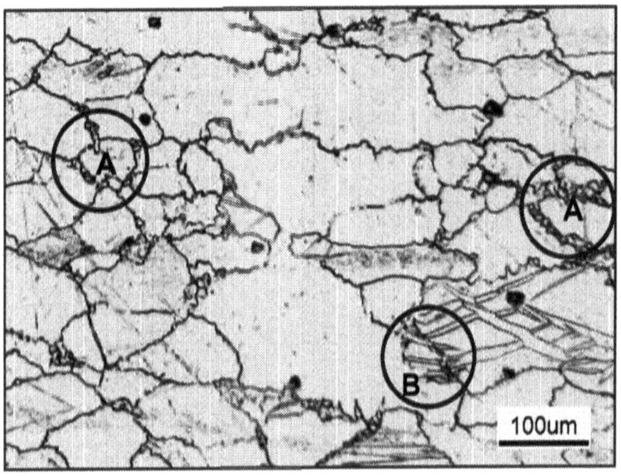

Fig. 9. Optical micrographs showing grain boundary serration and DRX nucleation (marked "A") and twinning-assisted DRX nucleation (marked "B") in an AZ31 specimen tested at 204°C.

flow localization and necking, which results in cavity formation and catastrophic failure during deformation [12]. For room-temperature and moderate temperature forming (below 150°C) where local DRX is absent, strain-hardening is beneficial to formability by delaying deformation instability. Above 150°C, the onset of local DRX results in a decrease in strain-hardening exponent (Fig. 3), leading to softening and extensive void formation, as shown in Fig. 7(f), which prevents the material from achieving its total intrinsic ductility. It is also known that better formability (superplasticity) is observed at higher temperatures (>300°C) due to a total change in uniform deformation and strain localization mechanisms [11] since DRX is now perhaps occurring more uniformly.

Future Opportunities

The most interesting implication of recrystallization (static or dynamic) is the concept of alloy and/or process design to create a completely recrystallized fine starting grain structure for high strength and improved formability. Such attempts have been made by a number of investigators [17-19]. Itoh et al [17] applied thermomechanical processing (15% cold rolling and subsequent annealing at 250°C for 1 hour) to a commercial AZ31 sheet and obtained a fairly uniform grain size of around 7μm through static recrystallization, resulting in 18-20% elongation in the material. In hot-torsion testing of AZ31 samples at 350°C, the average grain size was reduced from 22.5μm to 7.3μm after a strain of 2, due to dynamic recrystallization [18]. Using warming rolling at 350-400°C, Chen et al [19] were able to reduce the grain size of AZ31 sheet specimens from 10μm to 4μm through dynamic recrystallization. This fine-grained material exhibited reasonable stretchability in gas-forming trials performed at intermediate temperatures of 200-350°C. But, how to consistently and economically produce fine grain structure in magnesium sheets and extrusions still remains an important research topic.

Concluding Remarks

Microstructural evaluation indicates that twinning is the predominant deformation mechanism at temperatures up to 150°C for AZ31 alloy. At higher temperatures, local dynamic recrystallization (DRX), leading to the formation of new, small grains (2-3 μm in diameter) during deformation, became evident. These "necklace"-like DRX grains at the prior grain boundaries result in local softening which leads to plastic instability, a reduction in strain-hardening and premature failure by cavity linkage.

Acknowledgments

The authors would like to acknowledge Danielle White and Robert Kubic for carrying out mechanical testing and metallographic analysis.

References

1. A.A. Luo, "Magnesium: Current and Potential Automotive Applications", *JOM Journal of Metals*, 2002, 54, (2), 42-48.
2. J.S. Balzer, P.K. Dellock, M.H. Maj, G.S. Cole, D. Reed, T. Davis, T. Lawson and G. Simonds, "Structural Magnesium Front End Support", SAE Technical Paper 2003-01-0185, SAE International, Warrendale, PA, 2003.
3. C.J. Bruggemann: "Hydroforming of Structure Parts for Personal Cars", presented at International Conference on Hydroforming, Fellbach/Stuttgart, Germany, October 12-13, 1999, pp. 353-371.
4. A.A. Luo and A.K. Sachdev: "Strain Development and Microstructure Evolution in Aluminum Tube Bending and Hydroforming", in *Aluminum 2002*, eds. S. Das and M.H. Skillingbreg, The Minerals, Metals and Materials Society (TMS), Warrendale, PA, 2002, pp. 123-131.
5. ASM Specialty Handbook, *Magnesium and Magnesium Alloys*, eds., M.M. Avedesian and H. Baker, ASM International, Materials Park, OH, 1999.
6. *Timminco Magnesium Wrought Products*, Timminco Corporation Brochure, Aurora, CO.
7. R.A. Ayres and M.L.Wenner, "Strain and Strain Rate Hardening Effects on Punch Stretching of 5182-O Aluminum at Elevated Temperatures", *Metallurgical Transactions*, 10A, 1979, 41-46.

8. The Aluminum Association Publication AT3, "Aluminum for Automotive Body Sheet Panels", The Aluminum Association, Inc., Washington, D.C., 1998.
9. O. Duygulu and S.R. Agnew: "The Effects of Temperature and Strain Rate on the Tensile Properties of Textured Magnesium Alloy AZ31 Sheet", in *Magnesium Technology 2003*, ed., H.I. Kaplan, 237-242; 2003, Warrendale, PA, The Minerals, Metals and Materials Society.
10. W.F. Hosford and R.M. Caddell: "Metal Forming: Mechanics and Metallurgy", Prentice-Hall, Inc., Englewood Cliffs, NJ, 1983.
11. P.E. Krajewski: "Elevated Temperature Behavior of Sheet Magnesium Alloys", in *Magnesium Technology 2002*, ed. H.I. Kaplan, 175-179; 2002, Warrendale, PA, The Minerals, Metals and Materials Society.
12. G.E. Dieter: *Mechanical Metallurgy*, Third Edition, McGraw-Hill, Inc., 1986.
13. M.M. Myshlyaev, H.J. McQueen, A. Mwembela and E. Konopleva: "Twinning, Dynamic Recovery and Recrystallization in Hot Worked Mg-Al-Zn Alloy", *Mater. Sci. & Eng. A*, 2002, Vol. 337, 121-133.
14. X. Yang, H. Miura and T. Sakai: "Dynamic Nucleation of New Grains in Magnesium Alloy during Hot Deformation", *Mater. Sci. Forum*, 2003, Vols. 419-422, 515-520.
15. O. Sitdikov, R. Kaibyshev and T. Sakai: "Dynamic Recrystallization Based on Twinning in Coarse-Grained Mg", Mater. Sci. Forum, 2003, Vols. 419-422, 521-526.
16. D. Ponge and G. Gottstein: "Necklace Formation during Dynamic Recrystallization: Mechanisms and Impact on Flow Behavior", *Acta Mater.*, 1998, Vol. 46, 69-80.
17. G. Itoh, Y. Iseno, and Y. Motohshi, "Recystallized Grain Size in Cold-Rolled and Annealed AZ31 Wrought Magnesium Alloys Affected by Rolling Direction", Magnesium Alloys 2003, *Materials Science Forum*, Trans Tech, Switzerland, vols. 419-422, 2003, pp. 355-358.
18. A.G. Beer and M.R. Barnett, "The Hot Working Flow Stress and Microstructure in Magnesium AZ31, in *Magnesium Technology 2002*, ed., H.I. Kaplan, 193-198, 2002, Warrendale, PA, The Minerals, Metals and Materials Society.
19. Y-H. Chen, Sh. Lee, and J-Y. Wang, "Isothermal Sheet Formability and Microstructure Study of Rolling Processed Magnesium Alloy AZ31," Magnesium Alloys 2003, *Materials Science Forum*, Trans Tech, Switzerland, vols. 419-422, pp. 383-386.

Heated Hydro -Mechanical Deep Drawing of Magnesium Sheet Metal

Gerrit Kurz [2]

Institute for Metal Forming and Metal Forming Machine Tools, University of Hanover, Welfengarten 1A, 30167 Hanover, Germany

Keywords: magnesium, sheet metal, deep drawing, hydro mechanical deep drawing

Abstract

In order to reduce fuel consumption efforts have been made to decrease the weight of automobile constructions by increasing the use of lightweight materials. In this field of application magnesium alloys are important because of their low density. A promising alternative to large surfaced and thin die casting parts has been found in construction parts that are manufactured by sheet metal forming of magnesium. Magnesium alloys show a limited formability at room temperature. A considerable improvement of formability can be achieved by heating the material. Formability increases above a temperature of approximately T = 225 °C.

This paper will give an overview about the heated hydro-mechanical deep drawing process of magnesium sheet metal. It will further show how the process parameters temperature and fluid pressure influence the deep drawing process. The main goal of these investigations was a decrease of the forming temperature by use of the hydro mechanical deep drawing process.

Introduction

An important goal of the automotive industry is to reduce the fuel consumption of cars. The greatest potent ial for savings offered by magnesium alloys can be achieved with components such as sheet metal body shell panels that are exposed to bending stresses [1]. Currently, the weight of the car body is reduced by using high-strength steel with a decreased sheet thickness or conventional lightweight materials, like aluminum. For the future, it can be assumed that aluminum and steel sheets will be replaced by magnesium sheets in some applications. Magnesium possesses particularly favorable material properties if weight-specific values, like the ratio of yield stress $R_{p0,2}$ to density ? (specific strength), is considered.

At present magnesium alloys are mainly processed by die casting in the automobile industry. The die casting technology allows the manufacture of parts with a complex geometry [2]. The mechanical properties of these parts often do not meet the requirements concerning endurance, strength, ductility, etc. A promising alternative to thin, large surfaced parts, like automotive body constructions, are components manufactured by sheet metal forming. Sheet metal formed parts are characterized by advantageous mechanical properties and a sufficient surface without pores in comparison to die casting. Therefore, the substitution of conventional sheet materials by magnesium sheets could lead to essential weight savings.

Influences on the Flow Stress

Conventional magnesium alloys display a hexagonal lattice structure and possess a low formability at room temperature. Investigations of the forming mechanisms of pure magnesium revealed that forming at temperatures up to approximately T = 225 °C is possible mainly by slipping in the basal plane of the hexagonal crystal lattice and by twinning. Temperatures above this level activate further (pyramidal and prismatic) slip planes, causing a strong increase in formability [3], [4].
Evaluations of the deformation properties of magnesium sheets require the determination of the material characteristic mechanical values like anisotropy of flow stresses [5], [6].

The Influence of the Temperature on the Flow Stress

Experiments with magnesium alloy MgAl3Zn1 (AZ31) at room temperature show that the ductility of the material is sufficient to form inner and outer flat vaulted car body parts. A partial heating could be used in areas of higher true strains. At higher drawing ratios, deep drawing with process-annealing or warm deep drawing should be used [7]. For the production of parts at high true strains, it is necessary to heat the deep drawing process.

A key parameter in the forming of magnesium alloys is the forming temperature. Figure 1 displays the flow curves of the magnesium sheet material AZ31 at different temperatures, determined in the uniaxial tensile test according to EN 10002, part V.

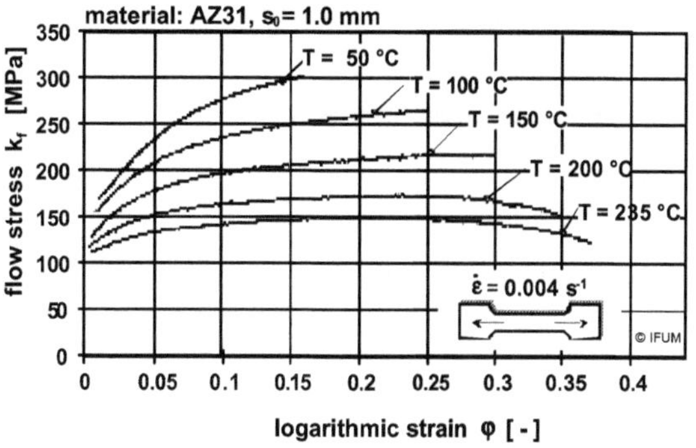

Figure 1: Temperature dependent flow curves of AZ31 (s0 = 1.0 mm) determined in the uniaxial tensile test [8].

It is obvious that the stresses and possible strains largely depend on the forming temperature. The decrease of flow stress in the temperature range above T = 200°C is attributable to temperature dependent relaxation. Furthermore, at higher temperatures the flow stress decreases with increasing straining because of the thermal activated work softening [8].

The Influence of the Strain Rate on the Flow Stress

For the determination of the strain rate sensitivity of magnesium alloys, uniaxial tensile tests at different velocities were carried out. Figure 2 shows the influence of the testing velocity on the flow stress for the sheet material AZ31 at a temperature of T = 200°C. Raising the elongation rate from $\dot{e} = 0.004\ s^{-1}$ (EN 10002) to $\dot{e} = 3.0\ s^{-1}$ leads to a considerable increase in flow stress. This effect is less significant if the experiments are carried out at lower temperatures.
Furthermore, the tensile tests show that increasing the velocity leads to lower maximum elongations of the material at any temperature [8].

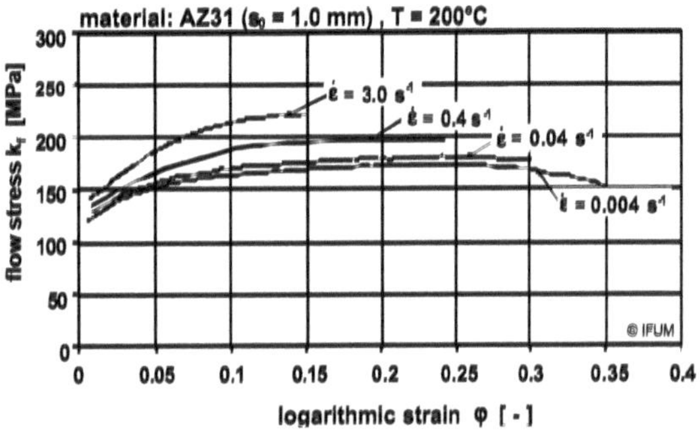

Figure 2: Strain rate dependent flow curves of AZ31B ($s_0 = 1.0$ mm) determined in the uniaxial tensile test at a temperature of T = 200°C [8].

The Influence of the As-Rolled Conditions on the Flow Stress

To investigate the influence of different as-rolled conditions on the flow stress uniaxial tensile tests with sheet materials of different magnesium alloys were carried out at the IFUM. The magnesium sheet material was produced at the Institute for Materials Science (IW) of the University of Hanover, in a joint venture with the IFUM.

The following as-rolled conditions were analysed:

- Rolled.
- Rolled and heat treated.
- Rolled, heat-treated and temper rolled.

In the as-rolled condition, rolled, the sheet material is warm rolled (temperatures: T = 350°C – 500°C). The heat treatment of the as-rolled condition, rolled and heat-treated, means that the sheet material was annealed after rolling for approximately one or two hours at a temperature range from T = 300 °C to T = 400 °C (T8). The sheet material of the third as-rolled condition, rolled, heat-treated and temper rolled, is olled again after rolling and heat treatment at room temperature with a true strain of about 2%.

Figure 3 shows the flow curves of the magnesium sheet material AZ31 in different as-rolled conditions at room temperature and at a temperature of T = 200°C.

The flow curves reveal the dependence of the stresses on the as-rolled conditions, especially at low temperatures. The influence on the flow stress decreases at higher temperatures, because of the thermal activation (above T = 200°C) of the pyramidal and prismatic glide planes. The curves at room temperature demonstrate that the logarithmic strain of the rolled and heat-treated sheet material is higher than the logarithmic strain of the sheet material at other as-rolled conditions. These results make clear that the selection of the right as-rolled condition has an influence on the formability of the magnesium sheet material especially at low temperatures.

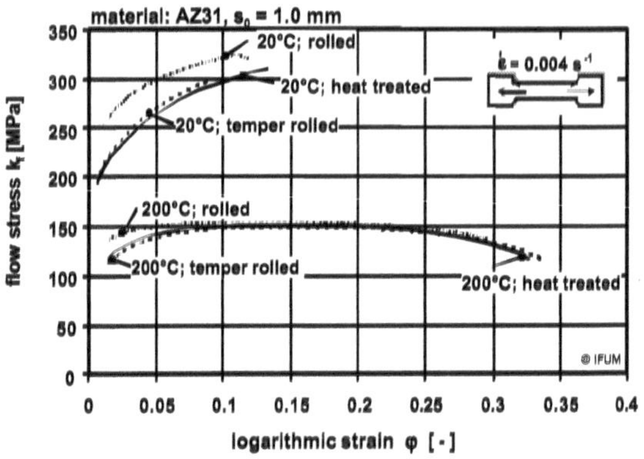

Figure 3: As-rolled conditions dependent flow curves of AZ31 ($s_0 = 1.0$ mm) determined in the uniaxial tensile test.

The Influences on the Deep Drawing Process

This work softening behavior of magnesium at higher temperatures, shown in the uniaxle tensile tests, influences the heated deep drawing process, too. Because of this, the limit drawing ratio in the conventional deep drawing process reaches a maximum close above the temperature (T = 200 °C – T = 250 °C) that activates the further slip planes.

Figure 4 shows the results of the deep drawing tests with MgAl2SE1 (AE21), MgAl3Zn1 (AZ31) and MgZn1SEZr (ZEK100) at temperatures between T = 200 °C and T = 325 °C. The limit drawing ratio $\beta_{0,max}$ can be used to assess formability under deep drawing conditions. It can be seen that at T = 300 °C $\beta_{0,max}$ is optimized ($\beta_{0,max} = 2,2$) for this alloy. The limit drawing ratio for MgZn1SEZr (ZEK100) at T = 225 °C is significantly higher ($\beta_{0,max} = 3,0$) than for MgAl2SE1 (AE21). The results reveal that the limit drawing ratio is largely dependent on the forming temperature and on the alloy composition.

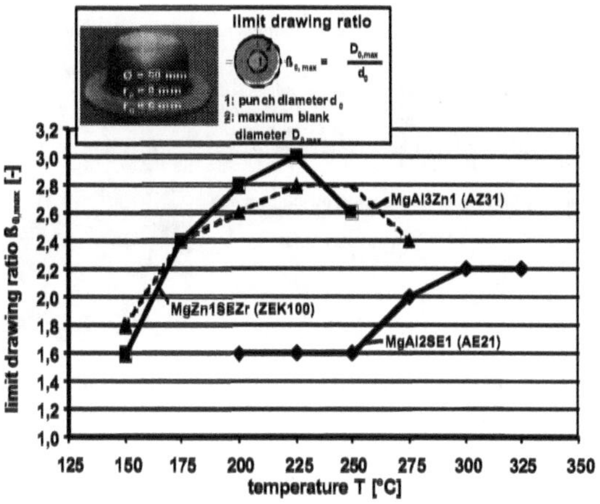

Figure 4: Temperature dependent limit drawing ratio for MgAl3Zn1 (AZ31), MgZn1SEZN1 (ZEK100) and MgAl2SE1 (AE21).

The Hydro-Mechanical Deep Drawing Process

In a conventional deep drawing process, the blank is formed into a hollow part by the punch, the die and the blank holder. The hydro-mechanical deep drawing process is a specific technology of the deep drawing process. In the hydro-mechanical deep drawing process, the die of the tool will be replaced by a fluid. The pressure of the fluid can be controlled. This process is advantageous in the production of complex parts by saving several process steps, because higher drawing depths and drawing ratios can be achieved. Only a blank holder and a punch are needed to produce parts by means of the hydro-mechanical deep drawing process. Thus, the production costs of deep drawing tools of complex part geometries can be decreased in comparison to conventional deep drawing tools [9].

In the hydro-mechanical deep drawing process, the blank will not be drawn into the die as in the conventional deep drawing process, but instead will be pushed in the filled fluid chamber (Figure 5).

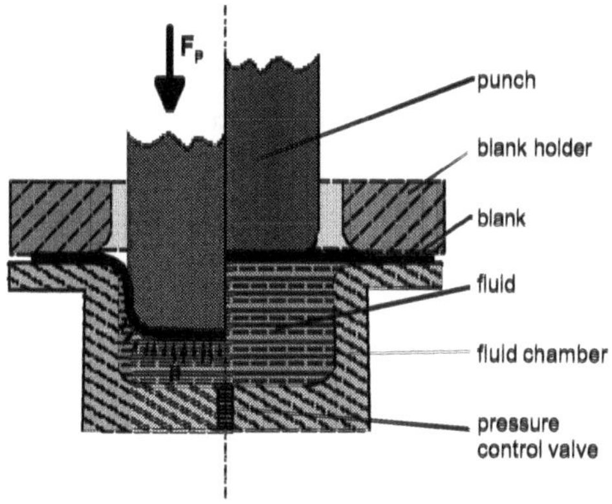

Figure 5: Principle of the hydro mechanical deep drawing process.

The pressure in the fluid chamber increases due to the punch dipping and causes the sheet to move against the punch (Figure 5).

The investigations of Oehler show that for cylindrical deep drawing parts produced by hydro-mechanical deep drawing, the thinning in the wall of the part is lower [10]. This lower wall thinning due to the increased friction between the punch and the sheet is a result of the normal pressure of the fluid. Furthermore, the friction appearing in the conventional deep drawing process between the sheet material and the die edge is reduced. As a consequence of this the radial stretching is lower. This lower radial stretching results in higher limit drawing ratios [11]. Because of this, it is possible to produce at one stroke sharp-edged rectangular components out of thin-walled sheets. Even in areas of the drawn part where exceptionally high strains occur, no cracks appear because of the fluid pressure pushing the sheet against the punch.

Metallographic investigations in steel sheet metal showed a more homogeneous and less deformed microstructure in the critical deformed zones in contrast to the conventional deep drawing process. Furthermore, no difference in the microstructure of the drawn parts can be achieved at different sheet thicknesses as well as different fluid pressures [10].

The Construction of the Applied Heated Hydro-Mechanical Deep Drawing Tool

In order to investigate the influence of the heated hydro-mechanical deep drawing process on the formability of magnesium sheet metal, a heated hydro-mechanical deep drawing tool was developed at the IFUM.

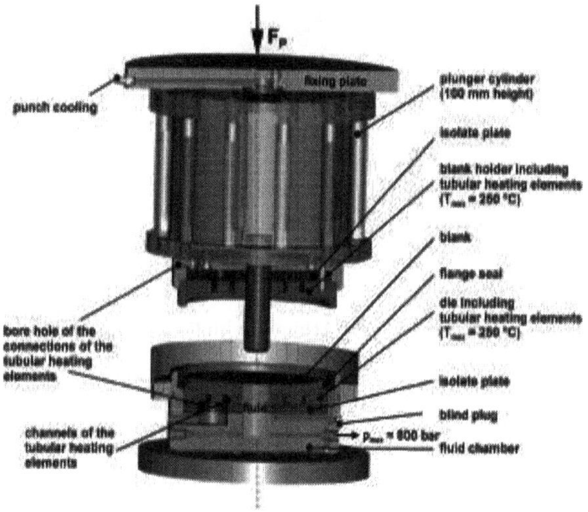

Figure 6: Construction of the heated hydro-mechanical deep drawing tool with radial pressurization

The fluid chamber is fixed on the press table. The pressure in the fluid chamber will be built up by pushing the punch in the fluid and the so effected compression of the fluid. The pressure in the fluid chamber will be controlled by the defined outlet of the fluid. The fluid chamber will be sealed against the atmosphere by the flange of the work piece and the load of blank holder force.

The blank holder force will be realized by a ring-shaped plunger cylinder that is fixed on the ram of the press. The hydraulic circuit of the plunger cylinder is completely separated from the hydraulic circuit of the heated fluid. So that conventional hydraulic oil can be used to run the plunger cylinder. The plunger cylinder will be pushed together during the ram stroke and thus constitutes the blank holder force by hydrostatic pressure. The blank holder force will be controlled by a valve that adjusts the outlet of the hydraulic oil. The hydraulic pressure, that is to extend the plunger cylinder before the stroke, will be realized by an external hydraulic aggregate.

To form the magnesium sheet metal the sheets need to be heated on elevated temperatures. To reach these temperatures, the die and the blank holder will be heated by electric tubular heating elements. The temperature in the die and in the blank holder will be controlled by a temperature control unit type hotset RR508.

A special feature of this heated hydro-mechanical deep drawing tool is the radial pressurization. To realize the radial pressurization, radial and vertical bores are drilled into the die of the tool (Figure 7). This bores can be closed by bind plugs. So that it is possible to work the deep drawing tool both with and without radial pressurization.

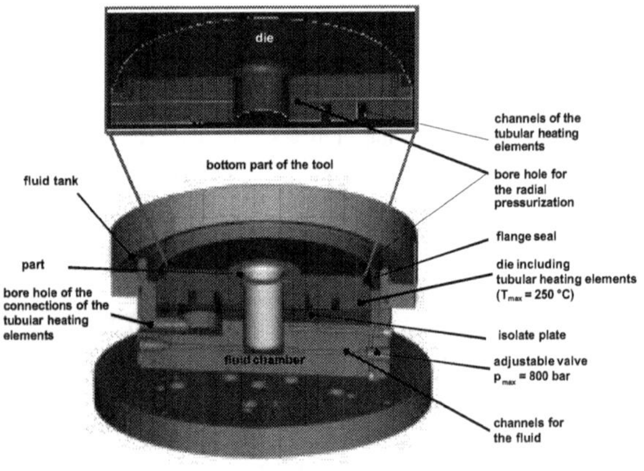

Figure 7: Bottom part of the heated hydro-mechanical deep drawing tool

Deep Drawing Results Achieved with the Heated Hydro-Mechanical Tool

To investigate how the forming limits are affected by the heated hydro mechanical deep drawing process, tests were conducted. The results of these tests were compared with results of conventional deep drawing tests with the same tool geometry. To evaluate the influence of the hydro-mechanical deep drawing process on the forming limits the limit drawing ratio was determined. The results of the deep drawing tests for the magnesium alloy MgAl3Zn1 (AZ31) show that at lower temperatures higher limit drawing ratios can be achieved.

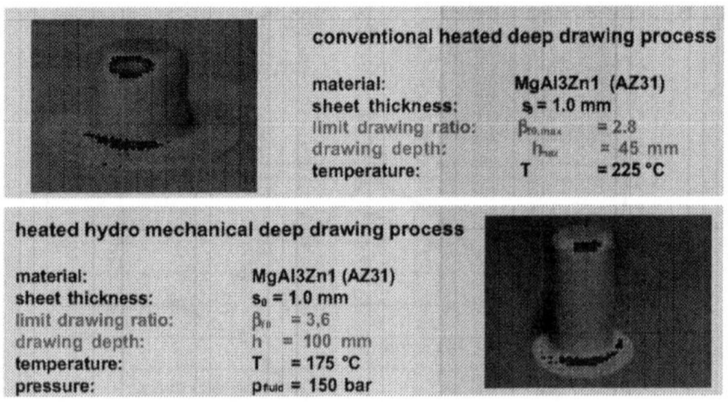

Figure 8: Increased formability by using the heated hydro mechanical deep drawing process (punch geometry Ø 50 mm)

This enhanced formability results from the increased friction between the sheet material and the punch because of the normal pressure of the fluid. Furthermore, the friction in the flange area will be decreased in comparison to the conventional deep drawing process. From these two effect a decreased tensile load results in the work piece in radial direction [10].

The decreased tensile load in the work piece in radial direction influences the formability of magnesium sheet metal very positively. Investigations of Dröder [8] show that the forming limit of the heated deep drawing process depends on the load transmission from the cup bottom via the cup wall in the flange area. For this reason the crack of the magnesium sheet component is placed in the cup wall. The crack in the cup wall results from decreased thermally induced strength in this area.

Deep Drawing Results Achieved with the Heated Hydro-Mechanical Tool and Radial Pressurization

To carry out the heated hydro-mechanical deep drawing tests with a radial pressurization the blind plugs in the die are removed and the pressure, induced by pushing the punch in the fluid, can get into the gap between die and blank holder. Now the fluid pushes against the cross-sectional area of the sheet. This pressure produces a radial shifting of the material in the flange area, which supports the forming process.

The main goal of these tests was the decrease of the forming temperature. Which is why the tests were carried out with a forming temperature of T = 150 °C for the magnesium alloy MgAl2SE1 (AE21).

The results of these experiments shows that the forming limit of this alloy can still be increased. The increase of the formability can be explained by a decrease of the radial strain in the cup wall. The lower radial strain results from the radial shifting in the flange area.

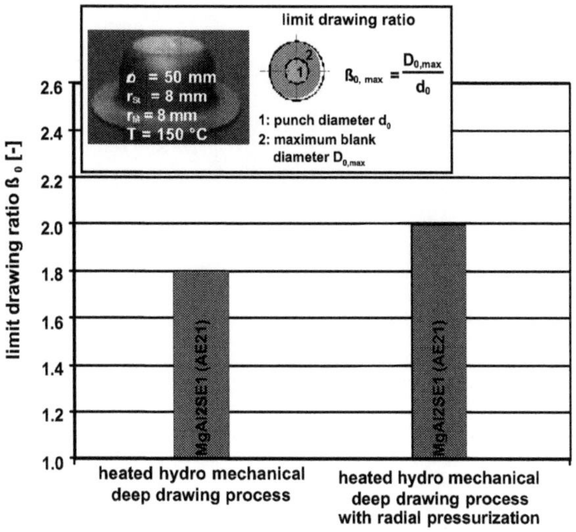

Figure 9: Deep drawing results using the heated hydro mechanical deep drawing tool wit radial pressurization

Acknowledgements

The results shown in this paper are worked out in a joint research project of the Institute of material science and the Institute for Metal Forming and Metal Forming Machine Tools, University of Hanover. The project is financed by the Deutsche Forschungsgemeinschaft (German Research Foundation) (DFG).

Summary

The better properties of wrought magnesium alloys as compared to cast magnesium alloys open a wide spectrum of applications within the automotive engineering sector. Sheet forming with heated hydro mechanical deep drawing tools represents a promising method for the production of large-area and very thin-walled light-weight components, like bodyshell panels. Basic research reveals a considerable potential as to the processes and the material behavior for the production of complex component geometries.

The acceptance of magnesium sheets in automotive engineering will depend on the availability of competitive manufacturing and processing systems for the entire production chain – from the intermediate material to the finished component. Of vital importance in this context is a decrease of the forming temperature realized by the development of new magnesium alloys and roll-optimized sheet materials in the future.

References

[1] Schumann, S.; Friedrich, H., "The use of Magnesium in cars - today and in future" in Magnesium Alloys and their Application, Wolfsburg, April 28.-30., 1998, p. 3-13.

[2] Metz, A., "Karosseriestrukturen mit strangpress-profilen und Mg-Druckgußteilen", Arbeitskreis Au-tomobilbau: "Fortschritte mit Magnesium im Au-tomobilbau", 17./18.02.2000; Bad Nauheim.

[3] Chapman, J.A., Ph.D. Thesis, University of Birmingham, 1963.

[4] Roberts, C. S., "Magnesium and its Alloys", Whiley, New York, 1960.

[5] Doege, E., Dröder, K., „Processing of Magnesium Sheet Metals by Deep Drawing and Stretch Forming", Matériaux & Techniques No. 7-8, 1997.

[6] Doege, E., Dröder, K., Janssen, St., „Umformen von Magnesiumwerkstoffen", DGM-Fortbildungsseminar, Clausthal-Zellerfeld, 28.-30. October 1998.

[7] Wagener, H.-W., Hosse-Hartmann, J.: Zum Tiefziehen von Magnesium-Blech, UTF science IV. Quartal 2001, Meisenbach Verlag, Bamberg. p. 28 – 34.

[8] Dröder, K., "Untersuchungen zum Umformen von Feinblechen aus Magnesiumknetlegierungen", Dr.-Ing. Dissertation, Hanover University, 1999.

[9] Kinneby, B.: „Verfahren der Hydroblechumformung", Internationale Konferenz „Neuere Entwicklungen in der Blechumformung" Fellbach, 23./24. of May 2000, P. 555 – 569.

[10] Oehler, G.: „Gefüge- und Formänderungen; Formgebung beim hydro-mechanischen Tiefziehen", DFBO-Mitteilungen 1977, Bd. 2, H7, S. 126-134.

[11] Danckert, J., Nielsen, K. B.: „Hydromechanical Deep Drawing with Uniform Pressure on the Flange", Annals of the CIRP Vol. 49/1/2000, S. 217 – 220.

NON-BASAL TEXTURES IN MAGNESIUM ALLOY STRIPS PRODUCED BY EXTRUSION-MACHINING

Dinakar Sagapuram[1], Mert Efe[1], Wilfredo Moscoso[3], Srinivasan Chandrasekar[2], Kevin P. Trumble[1]
Center for Materials Processing and Tribology
[1]School of Materials Engineering, [2]School of Industrial Engineering, Purdue University, West Lafayette, IN, USA
[3]Department of Electro-Mechanical and Electronics Engineering, Pontificia Universidad Catolica Madre y Maestra, Santiago, Dominican Republic

Keywords: Magnesium, Non-basal textures, Secondary slip activity, Hall-Petch effect

Abstract

A constrained chip formation technique, large strain extrusion machining (LSEM), was used to produce bulk magnesium alloy (AZ31B) strips with fine grain size (2-6 μm) and non-basal textures. These characteristics are known to enhance the final formability. The deformation temperature during extrusion-machining was varied by preheating the workpiece to a range of temperatures (50°C to 375°C). Microstructural refinement and texture evolution were studied as function of deformation temperature. It was possible to refine the grain size down to ~2 μm by restricting the dynamic grain growth at low to moderate deformation temperatures (below 320°C). LSEM was shown to be capable of resulting in non-basal textures at low deformation temperatures (below 220°C) as well as at elevated deformation temperatures (above 420°C). The influence of active deformation mechanisms and dynamic recrystallization on the texture development is also addressed.

Introduction

As the lightest structural metal, magnesium has attracted much attention due to the significant weight savings it can offer to the discrete product industry. Especially, wrought magnesium in the sheet form is highly desired for the interior automobile body frames. However, wide usage of magnesium sheet is limited by its poor room-temperature formability and high cost. The lack of formability in magnesium and its alloys is a direct result of insufficient number of active independent deformation modes required for the homogeneous deformation. Main deformation modes active at low temperatures (25-200°C) are the basal $\langle a \rangle$ slip and the $\{10\bar{1}2\}$ mechanical twinning [1]. In addition to the limited deformation modes, the plate texture (or basal texture) that develops during conventional sheet rolling also adversely affects the final formability of the sheet. The plate texture is characterized by the (0002) basal poles aligned perpendicular to the plate/sheet surface, and exhibiting fiber symmetry. The near-zero Schmid factor for the basal slip realized during the forming operations, as a result of this plate texture, results in the limited plasticity. Formability is promoted by random texture, or non-basal (or tilted basal) textures, i.e., basal poles tilted at an angle with respect to the sheet normal. However, achieving a random-textured wrought product, especially in hexagonal close packed (HCP) metals, is difficult. Non-basal textures in magnesium achieved by Mukai et al. [2] and Suwas et al. [3] through equal channel angular pressing (ECAP) were previously shown to significantly enhance the ductility/formability. However, ECAP is not suitable for direct sheet making. Besides crystallographic texture, grain size was also known to significantly influence the mechanical properties in Mg. Finer grain size was shown to enhance the strength as well as the ductility [4,5]. Here, large strain extrusion machining (LSEM) is introduced as a deformation processing technique that is capable of producing bulk magnesium strips having non-basal textures, as well as fine grain size (2-6 μm). LSEM offers control over a wide range of shear strains, strain rates, deformation temperatures, etc. In the present study, deformation temperature was varied in order to achieve a variety of textures and microstructures.

Large Strain Extrusion Machining

LSEM is a single-step deformation technique that is capable of producing metallic strip/sheet forms by employing machining and extrusion at the same time. Fig. 1 shows a schematic of the process. The set-up basically consists of a cylindrical workpiece of width, w, rotating on a spindle with a constant surface velocity (also called as deformation speed), V. A strip of final thickness, t_c, having outcoming velocity V_c is obtained by inserting the LSEM tool assembly into the workpiece at a constant feed, t, per revolution. The LSEM tool assembly, consisting of a cutting and a constraining component, results in simultaneous cutting and extrusion. For more detailed description of the process, the reader is suggested to refer Ref. [6].

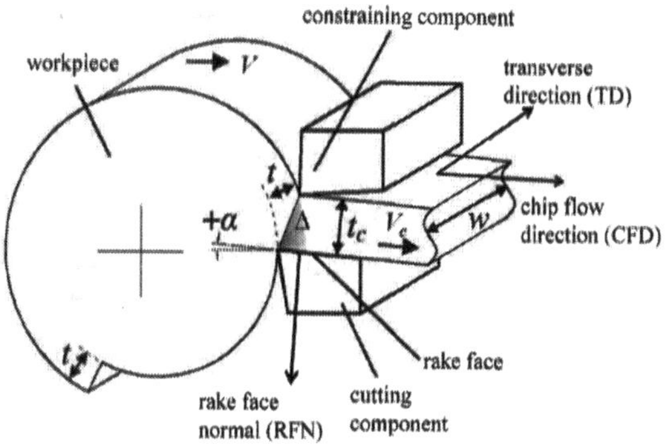

Figure 1. Schematic of LSEM.

The effective strain, $\bar{\varepsilon}$, in LSEM is given by:

$$\bar{\varepsilon} = \frac{1}{\sqrt{3}}\left[\frac{\lambda}{\cos\alpha} + \frac{1}{\lambda\cos\alpha} - 2\tan\alpha\right] \quad (1)$$

where $\lambda = t_c/t$ is the chip compression ratio and α is the rake angle. By controlling λ and α, very high effective strain (up to 3) can be imposed in a material, which helps in refining the microstructure down to the ultrafine grain size regime [6]. The effective strain rate, $\dot{\bar{\varepsilon}}$, in LSEM is given by:

$$\dot{\bar{\varepsilon}} = \frac{\varepsilon V}{\Delta} \qquad (2)$$

where Δ is the thickness of the primary deformation zone (shaded area, Fig. 1), which was experimentally found to be quite narrow (50-150 μm thick). In LSEM, very high strain rates (up to 10^5 s^{-1}) can be obtained by adopting high $\bar{\varepsilon}$ and V. These high strain rates then cause high localized heat generation in the deformation zone. The actual deformation temperature, T_{def}, is therefore given by:

$$T_{def} = T_o + \Delta T \qquad (3)$$

where T_o is the initial workpiece temperature and ΔT is the temperature rise during deformation due to high strain rate effects. Hydrostatic pressure, p, in the deformation zone is given by:

$$\frac{p}{2k} = \frac{1}{2} + 2\tan^{-1}\left(\frac{1}{2\lambda}\right) \qquad (4)$$

where k is the shear strength of the material. Hydrostatic pressures as high as $1.7k$ can be achieved in LSEM. The high heat generation and hydrostatic pressure inherent in the extrusion-machining reduces the need for external heating and aids in smooth extrusion without leading to fracture.

Experimental Procedure

For extrusion-machining, a disc-shaped workpiece (6.4 mm thick) was machined from a magnesium alloy AZ31B (Mg-3 wt.% Al-1 wt.% Zn) tooling plate, supplied by ThyssenKrupp Materials NA, Inc. It should be noted that the thickness of the workpiece, w, corresponds to the final strip width. A constant rake angle, α, of 5° was used. The chip compression ratio, λ, was fixed at ~0.75 by choosing t = 0.25 mm and t_c = 0.19 mm. According to Eq. 1, effective strain, $\bar{\varepsilon}$, was therefore kept constant at 1.1. A constant deformation speed, V, of 0.5 m/s was used. Strain rate, $\dot{\bar{\varepsilon}}$, was therefore fixed at 0.55×10^4 s^{-1}, according to Eq. 2. Deformation temperature, T_{def}, was varied by controlling the initial workpiece temperature (or preheating temperature), T_o. The temperature rise, ΔT of 170°C was estimated from the cutting forces measured using a piezoelectric dynamometer. Details of the heat-partition model used for calculating the temperature rise can be found elsewhere [7]. The preheating temperature was found to have a negligible effect on ΔT. Assuming that the heat-partition model is valid even at higher temperatures, T_{def} can be given by: T_{def} (°C) = T_o (°C) + 170°C. A range of preheating temperatures from 50 to 375°C were used. This accordingly corresponds to a variation of T_{def} from 220 to 545°C.

For optical microscopy, chip flow direction (CFD)-rake face normal (RFN) plane (refer to Fig. 1) of the samples was polished according to standard metallographic procedures and etched with a picric acid solution (100 ml ethanol, 10 ml water, 6 g picric acid, and 5 ml acetic acid). Average grain size was estimated using the linear intercept method. Hardness measurements were carried out on the CFD-RFN plane. For crystallographic texture analysis, an X-ray area detector diffraction system (GADDS, Bruker AXS) was used. A Rietveld refinement program, MAUD [8], was used to analyze the diffraction data and construct the pole figures.

Results

The as-received AZ31B tooling plate exhibited a plate texture (i.e., in-plane basal fiber texture) and an equiaxed microstructure (images not shown here). The grain size was 16 ± 2 μm and the hardness was 58 HV. The effect of T_o on the microstructure development in the LSEM strips is given in Fig. 2. All the strips exhibited equiaxed microstructures. Fig. 3 shows the effect of T_o

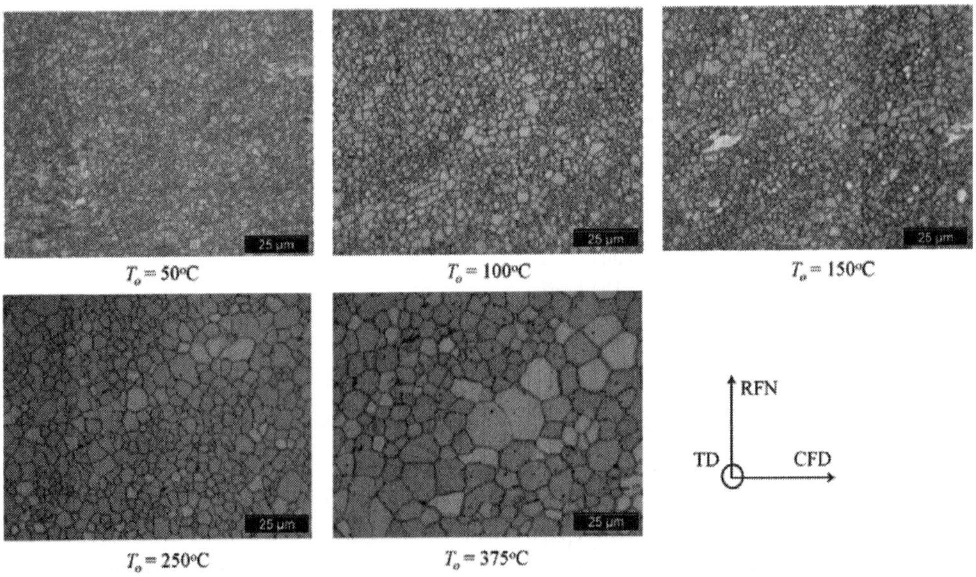

Figure 2. Microstructure development in the LSEM strips as a function of workpiece preheating temperature, T_o. (TD: transverse direction).

on the final grain size, d. Standard deviations are given by the error bars. It can be seen that d remains almost constant initially and increases rapidly when T_o is greater than 150°C. Increase in T_o from 50 to 375°C lead to a corresponding increase in d from 1.7 to 6.4 μm. It is noteworthy that extrusion-machining is capable of refining the microstructure by 2-10 times in a single step, as opposed to the standard multi-step deformation techniques.

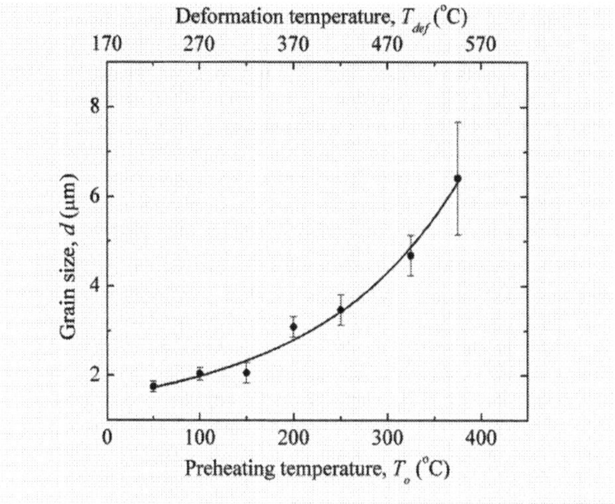

Figure 3. Variation of d as a function of T_o.

The Hall-Petch relationship [9] is typically exhibited by metals and alloys when grain size is in the microcrystalline range. It is given by:

$$H = H_o + K_H\, d^{-1/2} \quad (5)$$

where H_o and K_H are called the Hall-Petch constants. In order to examine the Hall-Petch relationship in the AZ31B strips produced

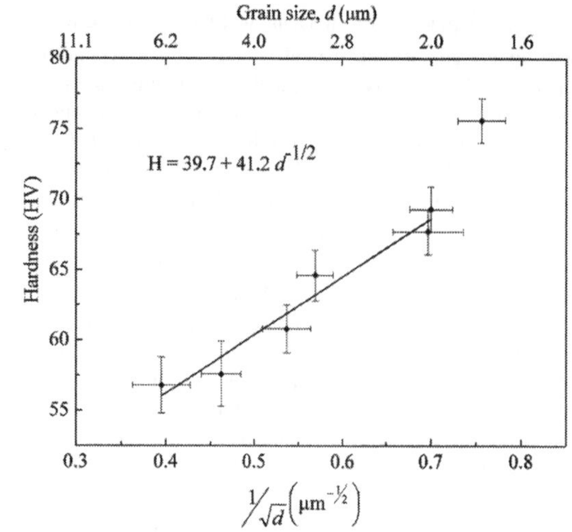

Figure 4. Hall-Petch plot for AZ31B.

by LSEM, hardness is plotted against $d^{-1/2}$ in Fig. 4. The AZ31B strips obeyed the Hall-Petch relationship with H_o = 39.7 and K_H = 41.2.

Evolution of texture with T_o is shown in Fig. 5. The (0002) and $(10\bar{1}0)$ pole figures were measured from the TD-CFD plane (refer to Fig. 1). A mirror symmetry about the CFD axis in the pole figures can be observed. Texture intensity is given in the unit, multiples of a random distribution (mrd). When T_o is low (50°C), extrusion-machining was found to result in two distinct texture components having a similar intensity: a tilted basal texture component, and a basal texture component. The tilted basal texture component is characterized by (0002) basal poles tilted at an angle of ~30° away from RFN toward CFD, while the basal texture component has basal poles aligned parallel to RFN.

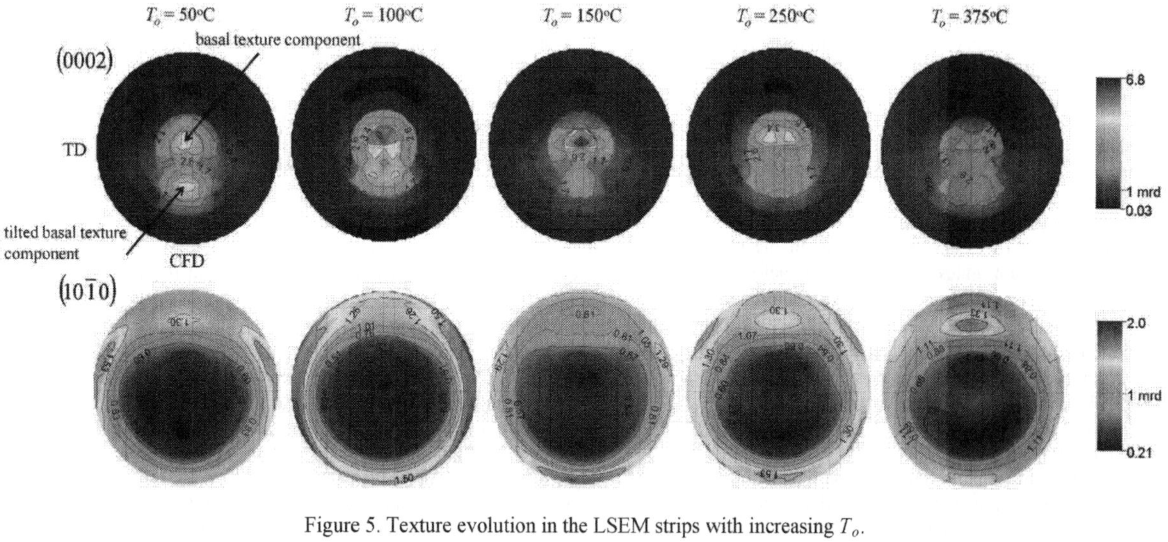

Figure 5. Texture evolution in the LSEM strips with increasing T_o.

From the $(10\bar{1}0)$ pole figures, it seems that both the tilted basal and basal texture components exhibit $<0001>$ fiber symmetry. With increasing T_o, intensity of the tilted basal texture was found to gradually decrease, while that of the basal texture component increased, and reached a maximum value of 6.8 mrd at T_o = 150°C. Correspondingly, the basal texture component no longer exhibits $<0001>$ fiber symmetry. Instead, $(10\bar{1}0)$ poles are aligned parallel to CFD. Further increase in T_o to 250°C resulted in a decrease in texture intensity and splitting of the basal poles about RFN by ± 10-20° along TD. The corresponding $(10\bar{1}0)$ pole figure indicates that the basal poles have regained the fiber symmetry. With further increase in T_o to 375°C, no qualitative changes in the texture were observed. It is interesting to note that, with increasing T_o, the peak intensity value initially increased from 5 mrd at T_o = 50°C, reached a maximum of 6.8 mrd at T_o = 150°C, and then started decreasing, where it finally dropped down to 4 mrd at T_o = 375°C.

Discussion

The fine equiaxed grain structures (Fig. 2) found in the LSEM strips and the grain size variation with deformation temperature (Fig. 3) can be explained by dynamic recrystallization (DRX) followed by dynamic grain growth. Grain size refinement from 16 μm in the as-received material to 1.7 μm in the strip processed at a low temperature of T_o = 50°C is a result of high shear strain in extrusion-machining and restricted grain growth due to warm deformation temperature (T_{def} = 220°C). Grain size remained almost constant up to T_o = 150°C. However, when T_o is greater than 150°C, grain size started to increase rapidly. This suggests the increased activity of dynamic grain growth at moderate temperatures (T_{def} > 320°C ≈ 0.5T_m).

The Hall-Petch effect in AZ31B (Fig. 4) confirms the grain size strengthening widely known in metals. The values of the Hall-Petch constants found in this study (H_o = 39.7, K_H = 41.2) are in fair agreement with the literature, where values of H_o = 40-47 and K_H = 42-50 were reported for the extruded and ECAP'ed AZ31 [10]. It should be noted that the LSEM strips, although having quite different textures (Fig. 5) from one another, still exhibit the Hall-Petch effect. This, indeed, is peculiar since texture is known to have a significant influence on the strength in magnesium alloys.

Deformation texture evolution in a material is primarily influenced by the stress state of deformation, active slip systems, and thermally activated processes like DRX. It is believed that the two texture components observed in the low temperature strip (T_o = 50°C) is a result of extensive basal slip in the primary and secondary deformation zones. The secondary deformation zone arises from the friction between the rake face and the outcoming strip [11], especially under conditions where λ is less than 1. The shear in the primary deformation zone due to the basal slip aligns the basal planes nearly parallel to the shear plane, resulting in the tilted basal texture component. Similarly, the basal texture component occurs due to the shear in the secondary deformation zone, which results in alignment of the basal planes parallel to the rake face. Formation of the basal texture component is typical in rolling/compression type deformation, while the tilted basal texture component is observed in shear-based deformation as in ECAP [3,12]. The increase in T_o to 100°C lead to the strengthening of basal texture component, accompanied by a decrease of tilted basal texture component intensity. This type of textural change with increasing T_{def} in extrusion-machining was previously found to coincide with the beginning of discontinuous dynamic recrystallization (DDRX) [13]. DDRX is a type of DRX, where new grains nucleate near the original grain boundaries, having orientations different from the orientations of deformed grains. The beginning of DDRX was also found to be accompanied by a sharp hardness drop, which seems to be true even in the present study. In Fig. 4, a sharp hardness drop occurs when grain size is increased from 1.7 μm to 2 μm. Hence, the change in texture with increasing T_o from 50 to 100°C is believed to be due to DDRX. However, the reason why DDRX seems to be strengthening particularly the basal texture component is currently unknown, and needs further investigation. With increasing T_o to 150°C, basal texture is further intensified and displays a maximum peak intensity of 6.8 mrd, presumably due to the increased DDRX as a result of higher T_{def}. The reason for the alignment of $(10\bar{1}0)$ poles with CFD for this condition is not clear. Further increase in T_o to 250°C caused splitting of basal poles away from RFN along TD, which is surprising in AZ31B. This kind of basal pole splitting in TD is usually a characteristic of rolling textures of HCP metals like Zr [14], where prismatic $\langle \boldsymbol{a} \rangle$ slip is dominant. ECAP of Mg-4 wt.% Li, which is known to increase prismatic slip activity, was also shown to result in a texture component tilted towards the flow plane normal direction [12]. Since the extrusion-machining geometry can be compared to that of ECAP, it is presumed that increased secondary slip activity (i.e., prismatic $\langle \boldsymbol{a} \rangle$ slip or pyramidal $\langle \boldsymbol{c}+\boldsymbol{a} \rangle$ slip, or both) at elevated temperatures (T_o > 250°C, T_{def} > 420°C) as the reason for the basal pole splitting in TD. The drastic decrease in critical resolved shear stress (CRSS) for secondary slip systems with temperature [15] is also in agreement with this assumption. Further increase in T_o to 375°C lead to no significant changes in texture, except for the decreased peak intensity value. The texture weakening at elevated deformation temperatures is, at present, not clear.

Conclusions

LSEM was shown to be capable of achieving a variety of non-basal textures in fine grained AZ31B strips. This ability of LSEM to control the texture, as well as refine the microstructure, will be very useful in designing the final mechanical properties like strength and formability. Deformation temperature was found to greatly influence the microstructure and texture evolution. Dynamic grain growth was found to be restricted at low to moderate deformation temperatures (≤ 320°C), resulting in fine grain sizes (~2 μm). At elevated deformation temperatures, grain size increased (3-6 μm) due to dynamic grain growth. AZ31B strips were also shown to exhibit the Hall-Petch relationship independent of the texture. Low temperature basal slip resulted in a texture having two components: a tilted basal and a basal texture component. Thermally activated DDRX at moderate temperatures (270-320°C) was found to strengthen the basal texture and weaken the tilted basal texture. At elevated temperatures (≥ 420°C), basal pole splitting in the strip transverse direction was observed. Increased activity of secondary slip systems with temperature is presumed to be the cause for this basal pole splitting. Further increase in temperature seems to weaken the texture without changing the texture character. Finally, formability enhancement in AZ31B as a result of the LSEM non-basal textures (i.e., textures containing the tilted basal texture component or exhibiting the basal pole splitting) and fine grain size looks promising.

Acknowledgements

Support through NSF-CMMI-MPM-0800481 and DOE PNNL-DE-AC06-76RL01830 is gratefully acknowledged.

References

1. I.J. Polmear, "Magnesium Alloys and Applications", *Materials Science and Technology*, 10 (1994), 1-16.

2. T. Mukai, M. Yamanoi, H. Watanabe, and K Higashi, "Ductility Enhancement in AZ31 Magnesium Alloy by Controlling its Grain Structure", *Scripta Materialia*, 45 (2001), 89-94.

3. Satyam Suwas, G. Gottstein, and R. Kumar, "Evolution of Crystallographic Texture during Equal Channel Angular Extrusion (ECAE) and its Effects on Secondary Processing of Magnesium", *Materials Science and Engineering A*, 471 (2007), 1-14.

4. Edward F. Emley, *Principles of Magnesium Technology* (Oxford, New York, Pergamon Press, 1966).

5. L. Jin, D. Lin, D. Mao, X. Zeng, and W. Ding, "Mechanical Properties and Microstructure of AZ31 Mg Alloy Processed by Two-step Equal Angular Channel Extrusion", *Materials Letters*, 59 (2005), 2267-2270.

6. W. Moscoso, M.R. Shankar, J.B. Mann, W.D. Compton, S. Chandrasekar, "Bulk Nanostructured Materials by Large Strain Extrusion Machining", *Journal of Materials Research*, 22 (2007), 201-205.

7. J.H. Wiener, "Shear-plane Temperature Distribution in Orthogonal Cutting", *Transactions of the American Society of Mechanical Engineers*, 77 (1955), 1331-1341.

8. L. Lutterotti, M. Bortolotti, G. Ischia, I. Lonardelli, and H.R. Wenk, "Rietveld Texture analysis from Diffraction Images", *Zeitschrift für Kristallographie Supplements*, 26 (2007), 125-130.

9. E.O. Hall, "The Deformation and Ageing of Mild Steel: II Characteristics of Lüders Bands", *Proceedings of the Physical Society B*, 64 (1951), 742-747.

10. W.J. Kim, Y.K. Sa, "Micro-extrusion of ECAP Processed Magnesium Alloy for Production of High Strength Magnesium Micro-gears", *Scripta Materialia*, 54 (2006) 1391-1395.

11. M. Sevier, H.T.Y. Yang, W. Moscoso, and S. Chandrasekar, "Analysis of Severe Plastic Deformation by Large Strain Extrusion Machining", *Metallurgical and Materials Transactions A*, 39A (2008), 2645-2655.

12. S.R. Agnew, P. Mehrotra, T.M. Lillo, G.M. Stoica, and P.K. Liaw, "Texture Evolution of Five Wrought Magnesium Alloys during Route A Equal Channel Angular Extrusion: Experiments and Simulations", *Acta Materialia*, 53 (2005), 3135-3146.

13. D. Sagapuram, M. Efe, W. Moscoso, S. Chandrasekar, and K.P. Trumble, "Deformation Temperature Effects on Microstructure and Texture Evolution in High Strain Rate Extrusion-Machining of Mg-AZ31B", *Proceedings of the 16th International Conference on Texture of Materials*, Bombay, India, Dec, 2011.

14. Y.N. Wang, and J.C. Huang, "Texture Analysis in Hexagonal Materials", *Materials Chemistry and Physics*, 81 (2003), 11-26.

15. S.E. Ion, F.J. Humphreys, and S.H. White, "Dynamic Recrystallization and Development of Microstructure during the High Temperature Deformation of Magnesium", *Acta Metallurgica*, 30 (1982), 1909-1919.

PART 7: MODELING AND SIMULATION

Alan A. Luo

The materials field has benefited significantly from major advances in the understanding of materials behavior enabled by the development and application of computational methods. Integrated Computational Materials Engineering (ICME) is defined as the integration of materials information, captured in computational tools, with engineering product performance analysis and manufacturing-process simulation. ICME unites manufacturing, materials, and product development into a more holistic computational framework, and offers a new and successful approach for reducing development time/cost of magnesium alloys and products. This part presents 15 papers from computational alloy design to performance prediction.

The first group of papers provide a systemic background of computational thermodynamics and CALPHAD (CALculation of PHAse Diagrams) and their use in magnesium alloy development for various performance requirements. Saunders and colleagues presented models for the calculation of various thermo-physical and physical properties of several multi-component magnesium alloys during solidification for process simulation. Turski and colleagues dealt with the numerical modeling of direct-chill casting of WE43 alloy rolling slabs.

The paper by Abbott and colleagues compared the high-strain deformation behavior of magnesium castings with those of mild steel and aluminum alloys, and suggested some performance benefits of magnesium alloys in terms of energy absorption that actually improved at high strain rates. The next paper by Levesque and colleagues reported numerical modeling of large strain deformation in magnesium using finite element (FE) per grain approach combining conventional FE model with the crystal plasticity theory.

The paper by Solanki and colleagues reported the effect of substitute aluminum in magnesium tension twin, and provided an important understanding of twinning deformation in Mg-Al based alloys. The paper by Kwak and colleagues is one of the first attempts in multi-scale modeling of corrosion in magnesium alloys.

Design Magnesium Alloys: How Computational Thermodynamics Can Help

Zi-Kui Liu

Department of Materials Science and Engineering
The Pennsylvania State University
University Park, PA 16802

Abstract

Thermodynamics has often been viewed applicable to states near equilibrium only although irreversible thermodynamics was already developed in 1950s. The CALPHAD technique of computational thermodynamics developed since early 1970s has helped to change this view. This technique couples the phase diagram and thermochemical properties to explicitly characterize all phases in a system, including stable, metastable, and unstable phases over a wide range of temperature, pressure and composition. The modeling of Gibbs energy of individual phases enables the calculation of driving forces between any intermediate non-equilibrium states for simulating dynamic microstructure evolutions.

As same as the most commercial alloys, magnesium alloys are multi-component in nature with many intermetallic phases. To develop robust alloys that are less sensitive to process variability, phase relations under both equilibrium and non-equilibrium conditions are extremely valuable for the design of alloy compositions and processing procedures. In this presentation, the CALPHAD technique will be discussed. Particular attention will be paid to the phase relations in the Mg-Al-Zn ternary alloys.

Introduction

Alloy development is traditionally conducted by "make-and-see" empirical approach, which is usually costly and time consuming. Alloy theories tend to explain the behavior of materials after their development. The situation is now changing thanks to the accumulated knowledge on alloy theories in the past century and the significant progress in computational techniques in the past decade. One of the emerging fields is the systems materials design[1]. Systems materials design integrates processing, structure and properties through computational thermodynamics, kinetic simulations, and experimental prototype evaluations. The essential feature is to express the design objectives in terms of thermodynamic and kinetic parameters.

The principles of thermodynamics were developed rigorously by Gibbs more than one hundred years ago[2], but he only considered equilibria. Scientists and engineers, generations by generations, have thus been educated and are still taught, in most cases, that thermodynamics is only for equilibria and can not be applied to real processes because of the non-equilibrium nature of most processes. On the other hand, it has been demonstrated, e.g. by Prigogine and Defay[3], that classical thermodynamics can be applied to systems not at equilibrium whereby the affinity (or driving force) for an internal process is evaluated as an ordinary thermodynamic quantity. The driving force for internal processes plays a central role throughout the development and application of thermodynamic principles. In a more recent book on thermodynamics, Hillert[4] retained the entropy production when combining the first and second thermodynamics laws. The driving force thus appears as a natural part of the thermodynamic principles.

In the present paper, some fundamentals of thermodynamics will be reviewed. The system materials design methodology will be briefed. The CALPHAD approach of computational thermodynamics, an acronym for the CALculation of PHAse Diagram, will be discussed and applied to demonstrate the phase relations in the Mg-Al-Zn ternary alloys covering the whole composition ranges. Readers interested in the applications of the CALPHAD approach may find other applications in the literature [5-7].

Fundamentals of Thermodynamics

Common metallurgical experiments are carried out through the control of the temperature and pressure of the surroundings. Gibbs energy is thus the characteristic thermodynamic function to be studied because the Gibbs energy is minimized at equilibrium under constant external temperature and pressure.

Following Hillert's approach[4], the combined first and second thermodynamic laws for the Gibbs energy, G, may be written as

$$dG = -SdT - VdP + \Sigma\mu_i dN_i - Dd\xi \quad (1)$$

where S, T, V, and P are the entropy, temperature, volume and pressure, respectively. μ_i and N_i are the chemical potential and the amount of component i. D is a driving force for an internal process represented by ξ. Under constant temperature and pressure, the driving force is shown below

$$D = \left(\frac{dG}{d\xi}\right)_{P,T} \quad (2)$$

There are many possible internal processes in a system. It is not feasible to construct a single expression of total Gibbs energy for a system as a function of internal processes. On the other hand, if the Gibbs energy of individual phases can be modeled as a function of temperature, pressure and compositions, the total Gibbs energy for any configurations of a system involving those phases can then be calculated, and so can the driving force for any internal processes. The equilibrium states are thus a special case only when the driving forces for all possible internal processes are zero or negative. The CALPHAD approach of computational thermodynamics is based on such philosophy.

System Materials Design

System materials design integrates the three subsystems of a material system, i.e. processing, structure and properties, through computational thermodynamics, kinetic simulations, property modeling, and prototype evaluation. The schematic flowchart of the system materials design is shown in Fig. 1.

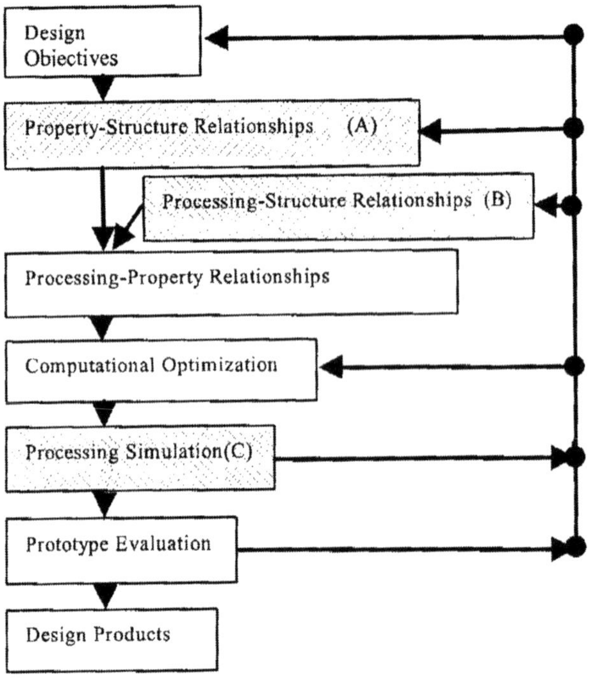

Figure 1: System materials design flowchart

System materials design begins with the identification of design objectives, i.e. the components in the property subsystems. Viewing the structure being dictated by the properties, a property-structure relationship can be constructed. Combining the property-structure relationship with the processing-structure relationships, the processing-property correlation can be developed and processing variables can be optimized, followed by computer simulations of prototype materials. The simulation results are then compared with the designed objectives. If the differences are large, iterations will be needed to modify the previous steps as indicated by the connectors in Fig. 1. With a favorable comparison, those prototypes are experimentally evaluated and the results are compared with the design objectives. Again, iterations may be needed until the agreement is satisfactory. In this way, the amount of experimental work can be considerably reduced, and the design products can be more tolerant to process variability, in comparison with the conventional trial-and-error method.

In Fig.1, three boxes are shaded, representing three databases needed to perform a rigorous system materials design:

- (A) Property database: Contain mechanical property data of individual structure components to correlate the property and structure subsystems,
- (B) Thermodynamic database: Contain the Gibbs energy expressions of individual phases to provide relative phase stability with respect to processing conditions,
- (C) Kinetic database: Contain atomic mobility data of individual species and interfacial properties to provide processing time scale under available driving forces.

As seen in Fig.1, computational thermodynamics is a critical ingredient in reducing the empiricism in developing quantitative processing-structure relationships and providing thermodynamic driving forces for process simulations.

Computational Thermodynamics

CALPHAD has been under development since 1970's[8]. It is capable of predicting the phase behavior in experimentally uninvestigated regions of a multi-component system from the extrapolation of their low-order systems. The development of CALPHAD techniques in the past three decades has been discussed by Saunders and Miodownik[9].

Fig. 2 illustrates the flowchart of the CALPHAD approach. In this approach, the Gibbs energy of individual phases is modeled, and the model parameters are collected in a thermodynamic database. Models for the Gibbs energy are based on the crystal structures of the phases. For pure elements and stoichiometric compounds, the most commonly used model is the one suggested by the Scientific Group Thermodata Europe (SGTE)[10] and has the following form (for simplicity, the pressure dependence and the magnetic contribution are not shown here),

$$G_m - H_m^{SER} = a + bT + cT\ln(T) + \sum d_i T^i \quad (3)$$

The left hand side of the equation is defined as the Gibbs energy relative to a standard element reference state (SER), where H_m^{SER} is the enthalpy of the element in its stable state at 298.15°C. Coefficients, a, b, c and d_i are the model parameters. The SGTE data for pure elements have been compiled by Dinsdale[11].

For multicomponent solution phases, the Gibbs energy has the following general formula,

$$G = G^o + G_{mix}^{ideal} + G_{mix}^{xs} \quad (4)$$

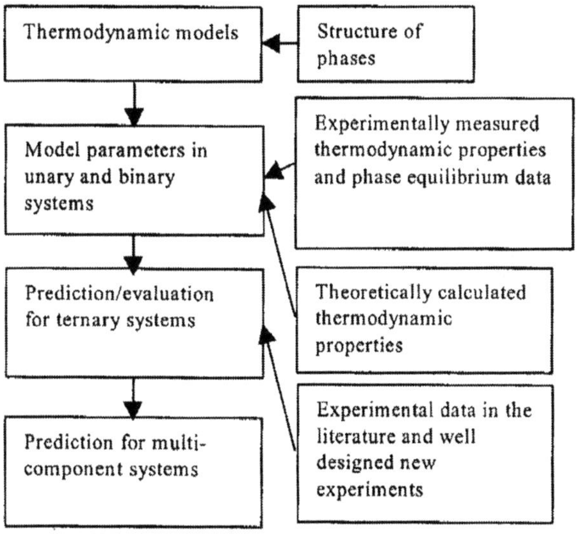

Figure 2: A schematic diagram of the CALPHAD approach

where G^o is the contribution from the mechanical mixing of pure components, G_{mix}^{ideal} the ideal mixing contribution, and G_{mix}^{xs} the excess Gibbs energy of mixing due to non-ideal interactions. Sublattice models have been used to describe solution phases[12, 13]. For example, in an A-B binary system with both components entering two sublattices, the sublattice model is written as $(A,B)_a(A,B)_b$, where subscripts a and b denote the number of sites of each sublattice, respectively. The three terms in eq. 4 are written as,

$$G^o = y_A^I y_A^{II} G_{A:A}^o + y_A^I y_B^{II} G_{A:B}^o + y_B^I y_A^{II} G_{B:A}^o + y_B^I y_B^{II} G_{B:B}^o \quad (5)$$

$$G_{mix}^{ideal} = aRT\left(y_A^I \ln y_A^I + y_B^I \ln y_B^I\right) + bRT\left(y_A^{II} \ln y_A^{II} + y_B^{II} \ln y_B^{II}\right) \quad (6)$$

$$G_{mix}^{xs} = y_A^I y_B^I \left(y_A^{II} \sum_{k=0} L_{A,B:A}^k (y_A^I - y_B^I)^k + y_B^{II} \sum_{k=0} L_{A,B:B}^k (y_A^I - y_B^I)^k \right) + y_A^{II} y_B^{II} \left(y_A^I \sum_{k=0} L_{A:A,B}^k (y_A^{II} - y_B^{II})^k + y_B^I \sum_{k=0} L_{B:A,B}^k (y_A^{II} - y_B^{II})^k \right) \quad (7)$$

where y^I:s and y^{II}:s are the site fractions of A or B in the first and second sublattices, respectively. $G_{I:J}^o$:s are the Gibbs energy of the compound I_aJ_b, expressed by eq.3. $L_{A,B:*}^k$ ($L_{*:A,B}^k$) is the interaction parameter between component A and B in the first (second) sublattice. In this notation, a colon separates components occupying different sublattices, and a comma separates interacting components in the same sublattice. These equations can be generalized for phases with multicomponents and multisublattices[13], and they reduce to a random substitutional model when there is only one sublattice.

Thermodynamic modeling begins with the evalution of thermodynamic descriptions of unary and binary systems. By combining thermodynamic descriptions of consititutive binary systems and ternary experimental data, thermodynamic descriptions of ternary systems are developed. These descriptions cover the whole composition and temperature ranges including the experimentally uninvestigated regions.

Calculations for the Mg-Al-Zn Ternary Alloys

Magnesium is the eighth most abundant element in the Earth's crust and the third most plentiful element dissolved in seawater[14]. It is the lightest structural material with a density of 1.741 g/cm^3, in comparison with the densities of Al (2.70 g/cm^3) and Fe (7.86 g/cm^3). This makes magnesium alloys particularly attractive for transportation for weight reduction and higher fuel efficiency[15-17].

The current principal commercial magnesium alloys are based on the Mg-Al-Zn ternary system[18]. In this section, calculations will be made for the ternary system including liquidus surface, isopleth, and Scheil solidification. The thermodynamic description of the system has been evaluated and is available in the literature[19]. It will be used in the present calculations.

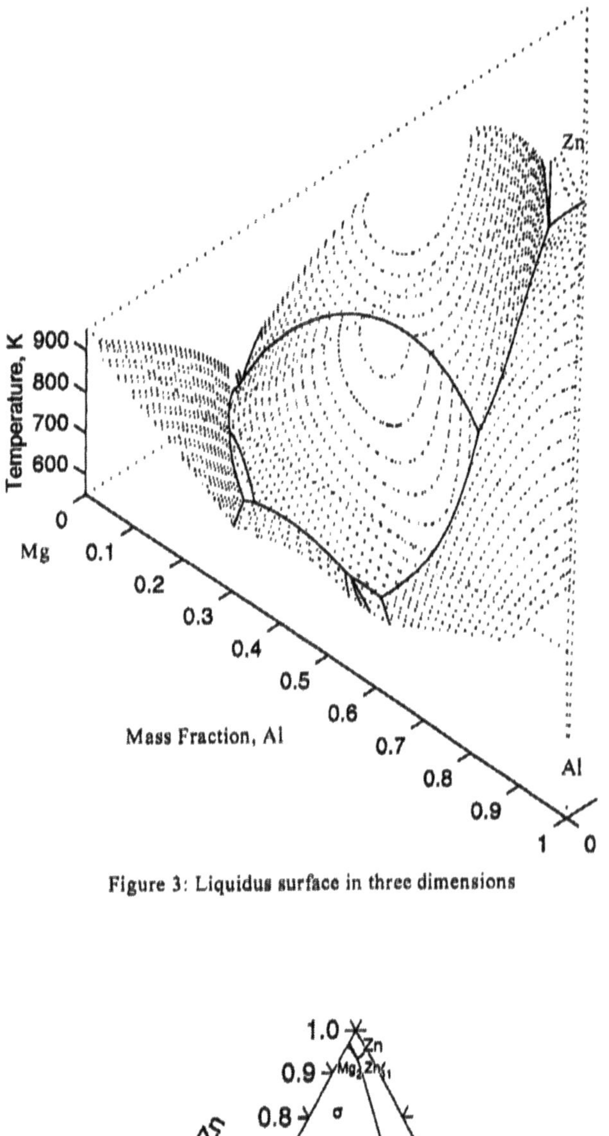

Figure 3: Liquidus surface in three dimensions

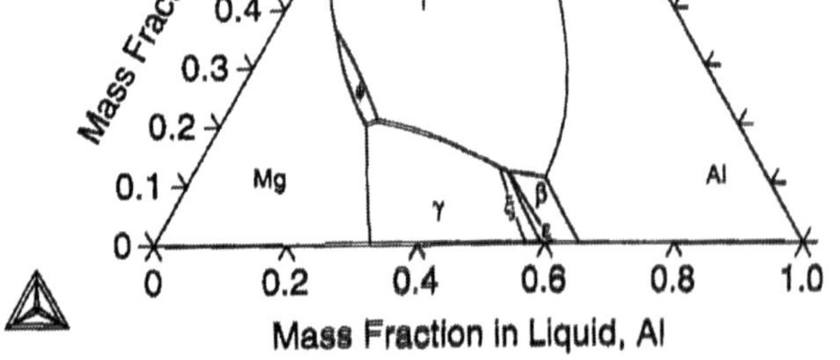

Figure 4a: Liquidus projection on the composition triangle

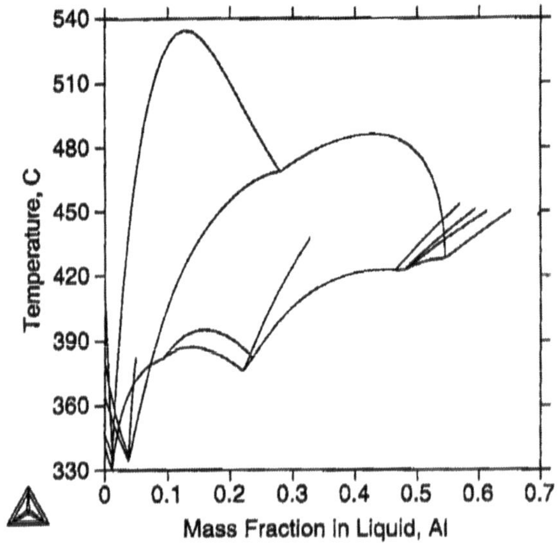

Figure 4b: Liquidus projection on the temperature and the composition of Al

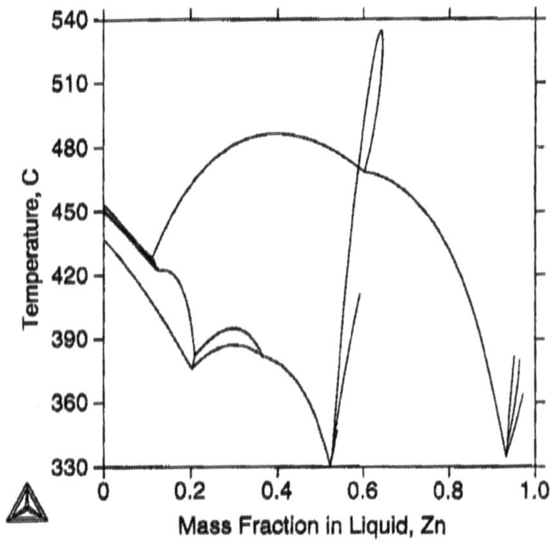

Figure 4c: Liquidus projection on the temperature and the composition of Zn

Liquidus surface

The liquidus surface of the Mg-Al-Zn ternary system is shown in Fig. 3 in three dimensions. Its projections are presented in Fig. 4 in three directions, i.e. the compositions of Al and Zn in Fig. 4a, the temperature and the composition of Al in Fig. 4b, and the temperature and the composition of Zn in Fig. 4c, respectively. The solid lines in Figs. 3 and 4 represent three phase equilibria on the liquidus surface. The dotted lines in Fig.3 are the isotherms. The primary solidification phases from the liquid phase are marked in Fig. 4a. As seen, these diagrams present a clear view of the liquidus surface topology. The liquidus surface provides useful information in understanding solidification process of alloys, especially the change of the liquid composition and phases formed from the liquid.

Isopleth

Isopleth provides useful guidelines for designing alloy composition, and heat treatment parameters. Two isopleth diagrams are shown in Figs. 5 and 6 for $w_{Al}/w_{Zn}=9$ and $w_{Zn}=0.01$, respectively, to examine the phase relations at compositions close to the most commonly used AZ91 alloy. Fig. 5 represents the isopleth starting from pure Mg, while Fig. 6 is the isopleth starting from Mg-1%Zn. Both diagrams show that in order to have a single Mg solid solution phase region for solution treatments, the mass fraction of Al should be less than 0.10. Furthermore, to avoid the direct precipitation of the ϕ phase from the Mg solid solution phase, the mass fraction of Al should be above 0.08. These leave the optimal mass fraction of Al being around 0.09, as indicated by the dotted vertical lines in Figs. 5 and 6. This coincides with Al concentration in the most popular AZ91 alloys, which was established during the trial-and-error development of magnesium alloys. The possible solution treatment temperature in single Mg solid solution phase region can be calculated as a function of alloy concentration.

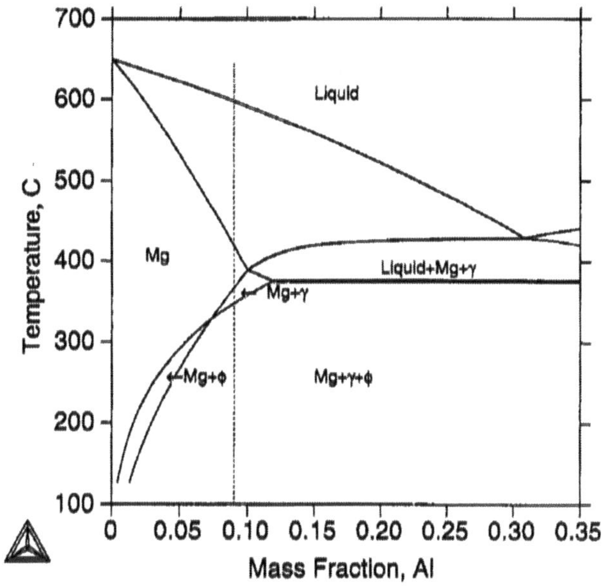

Figure 5: Isopleth of the Mg-Al-Zn ternary system with $w_{Al}=9w_{Zn}$. The dotted line represents AZ91.

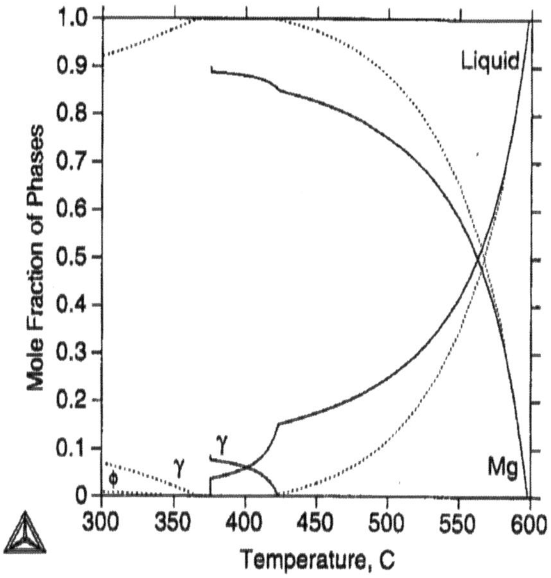

Figure 7: Mole fraction of individual phases as a function of temperature under equilibrium (dotted curves) and Scheil (solid curves) conditions.

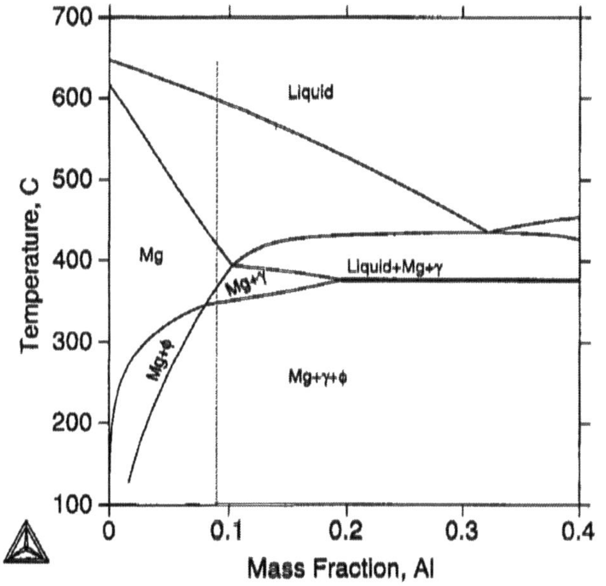

Figure 6: Isopleth of the Mg-Al-Zn ternary system with $w_{Zn}=0.01$. The dotted line represents AZ91.

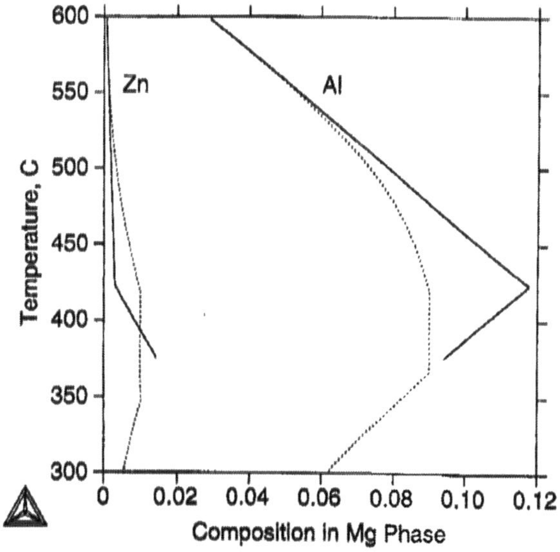

Figure 8: The mass fraction of Al and Zn in the Mg solid solution phase under equilibrium (dotted curves) and Scheil (solid curves) conditions.

Scheil solidification

Scheil solidification is a hypothetical solidification process assuming a homogenous composition in the liquid and no diffusion in the solid phases during solidification. On the other hand, equilibrium calculation is equivalent to assume homogeneous compositions in both liquid and solid phases. In reality, the diffusion in solid phases is neither zero nor infinite as assumed in the Scheil solidification or equilibrium calculations. Nevertheless, the equilibrium calculation and Scheil solidification represent two extreme ends concerning the diffusion in solid phases. Furthermore, both of them can be calculated using the thermodynamic properties of the system alone without involving kinetic properties of the system.

Fig. 7 shows the phase fraction during solidification under equilibrium condition (dashed curve) and Scheil condition (solid curve) of the AZ91 alloy, respectively. With decreasing temperature to the liquidus surface, the Mg solid solution phase starts to form in both cases. Under the equilibrium condition, all liquid will solidify into the Mg solid solution phase, while under the Scheil condition, the γ phase co-precipitates with the Mg solid solution phase from the liquid phase when the liquid composition reaches the three-phase equilibrium of Liquid+Mg+γ (see Fig. 4a). With further decrease of temperature, the liquid composition reaches the invariant four phase equilibrium of Liquid+Mg+γ+ϕ, and all liquid is solidified (see Fig. 4a).

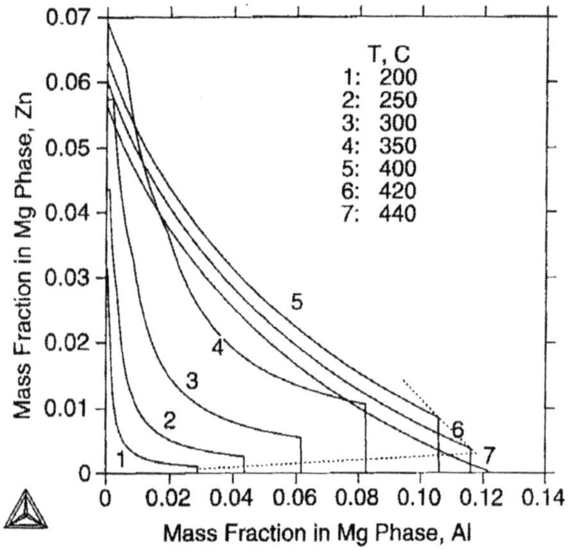

Figure 9: The phase boundary of the single Mg solid solution phase at different temperatures. The dotted curve is from the Scheil calculation.

Solid phase transformation

Typical solid phase transformation in magnesium alloys is the formation of intermetallic phases. The supersaturation of the Mg solid solution phase with respect to the formation of intermetallic compounds is thus an important issue in controlling microstructure evolution at elevated temperatures. Fig. 9 presents the phase boundaries of the single Mg solid solution phase at different temperatures in the Mg-Al-Zn ternary system. As discussed above, during Scheil solidification, it is assumed that there is no diffusion in the solid phase. The concentration in the solidified Mg solid solution is thus not homogenous, but follows the profile shown in the solid curves in Fig. 8 for the AZ91 alloy. This concentration is plotted in the dotted curve in Fig. 9. With the thermodynamic property available, one can calculate the thermodynamic driving force for the formation of intermetallic phases. As an example, the driving force for the

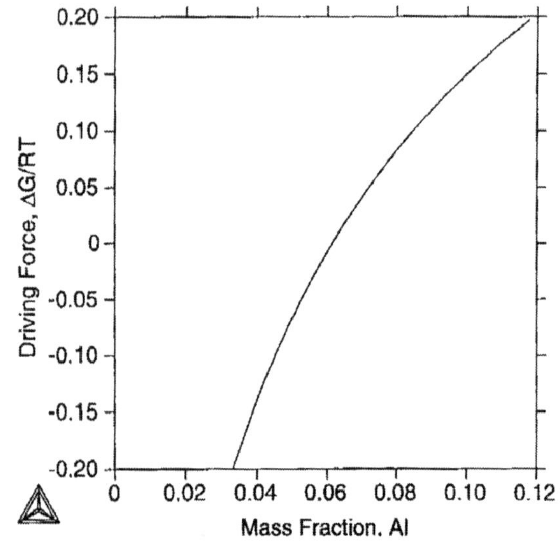

Figure 10: The normalized driving for the formation of the γ phase from the Mg solid solution phase with the comcentration per the dotted curve in Fig. 9.

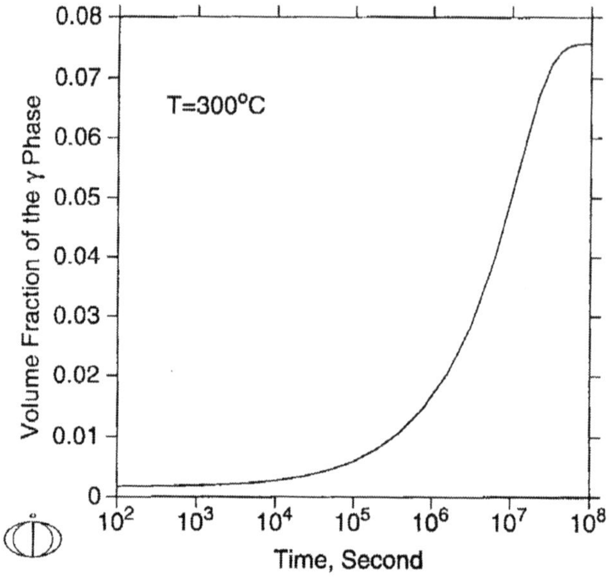

Figure 11: The formation of the γ phase from the supersaturated Mg solid solution phase of the AZ91 alloy at 300°C.

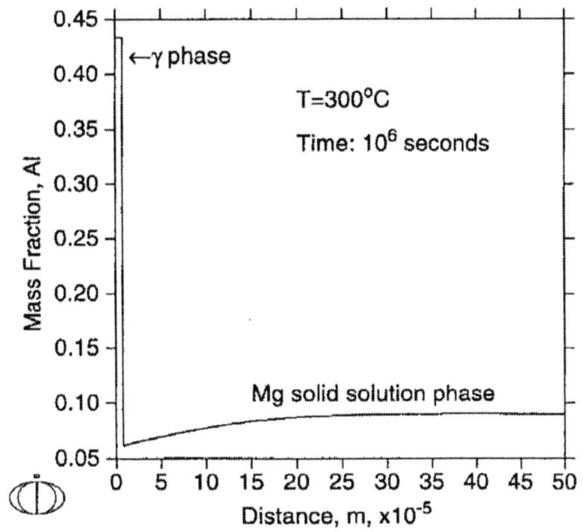

Figure 12: The concentration profile of Al during the formation of the γ phase from the Mg solid solution of the AZ91 alloy at 300°C and 10^6 seconds.

formation of the γ phase from the Mg solid solution phase with the concentration along the dotted line in Fig. 9 is presented in Fig. 10 for the temperature of 300°C. The driving force is normalized to RT with R being the gas constant and T the absolute temperature. As seen, the part of the Mg solid solution phase has a positive driving force and is supersaturated with respect to the formation of the γ phase.

Combining the thermodynamic date with available diffusion data, the process of the formation of intermetallic phases can be simulated. As an example, the precipitation of the γ phase from supersaturated Mg solid solution phase in the AZ91 alloy at 300°C is simulated using a computer program, Dictra[20]. The diffusion data in the Mg solid solution phase are estimated from experimental measurements in the literature[21]. Diffusion in the γ phase is neglected due to the lack of data. The volume fraction of the γ phase is plotted with respect to time in Fig. 11. The concentration profile of Al is plotted in Fig. 12 after 10^6 seconds. These simulation results are useful in helping us understanding microstructure evolution during magnesium alloy processing.

Summary

Computational thermodynamics combines the thermodynamic principles and numerical modeling. It provides quantitative description of thermodynamic properties of multicomponent materials through modeling unary, binary, ternary, and lower order systems and extrapolating them to higher order systems. It is a powerful tool in understanding the complex phase relationships in multicomponent materials under both equilibrium and non-equilibrium conditions.

In addition to equilibrium calculations, the thermodynamic descriptions developed in computational thermodynamics can be used to calculate metastable equilibria by excluding phases that may not form due to slow kinetic processes. Furthermore, combined with diffusion data, thermodynamic descriptions can be used to simulate phase transformations in multicomponent materials.

Through equilibrium calculations, driving force calculations, and phase transformation simulations, computation thermodynamics forms an important component in alloy development with higher robustness and less empiricism. It also helps to reduce the amount of costly experimental work to find optimal alloy composition and processing procesures.

Acknowledgement

The computer programs, Thermo-Calc and Dictra, are licensed from The Foundation of Computational Thermodynamics through ThermCalc AB in Sweden. The author is grateful to Dr. Helen Liang and Prof. Austin Chang for providing a computer file of the Mg-Al-Zn ternary system.

Reference

1. G. B. Olson, "Computational Design of Hierarchically Structured Materials", *Science*, Vol.277, 1997, pp. 1237-1242.
2. J. W. Gibbs, *The Collected Works of J. Willard Gibbs*, Yale Univ. Press, New Haven, 1948.
3. I. Prigogine and R. Dufay, *Chemical Thermodynamics*, Longmans Green, London, 1954.
4. M. Hillert, *Phase Equilibria, Phase Diagrams and Phase Transformations*, Cambridge University Press, Cambridge, 1998.
5. M. Hillert, "Predictions of Alloy Constitution and Phase Transformations by Computer Modelling. An Aid in Alloy Design and Materials Processing", *Scand. J. Metall.*, Vol.26, 1997, pp. 215-223.
6. L. Kaufman, "Calculation of Phase Diagrams for Commercial Materials", *Journal of Phase Equilibria*, Vol.14, 1993, pp. 413-417.
7. K. Hack, Ed., *The SGTE Casebook: Thermodynamics at Work*, Materials Modelling Series, The Institute of Materials, London, 1996.
8. L. Kaufman and H. Bernstein, *Computer Calculation of Phase Diagrams with Special Reference to Refractory Metals*, Academic Press, New York, 1970.
9. N. Saunders and A. P. Miodownik, *CALPHAD (Calculation of Phase Diagrams): A Comprehensive Guide*, Pergamon, Oxford ; New York, 1998.
10. I. Ansara and B. Sundman, "The Scientific Group Thermodata Europe", P. S. Glaeser, Ed., Computer Handling and Dissemination of Data, Elsevier Science Pub. Co, 1987, pp. 154-158.
11. A. T. Dinsdale, "SGTE Data for Pure Elements", *CALPHAD*, Vol.15, 1991, pp. 317-425.
12. B. Sundman and J. Agren, "A Regular Solution Model for Phases with Several Components and Sublattices, Suitable for Computer Applications", *J. Phy. Chem Solids*, Vol.42, 1981, pp. 297-301.
13. J. O. Andersson, A. F. Guillermet, M. Hillert, B. Jansson and B. Sundman, "A Compound-Energy Model of Ordering in a Phase With Sites of Different

Coordination Numbers", *Acta Metall.*, Vol.34, 1986, pp. 437-445.

14. D. Kramer, "Magnesium", U.S. Geological Survey--Minerals Information, 1997.
15. G. Cole, "Magnesium Applications in Ford's PNGV Mondeo", IMA-54, Magnesium Trends, Toronto, Canada, International Magnesium Association, 1997, pp. 24-31.
16. H. Friedrich and S. Schumann, "The Use of Magnesium in Cars - Today and in the Future", IMA-55, A Global Vision for Magnesium, Coronado, CA, USA, International Magnesium Association, 1998, pp. 1-7.
17. G. Craig, "Use of WE-43A-T6 Magnesium in the MD600 Helicopter Drive System", IMA-55, A Global Vision for Magnesium, Coronado, CA, USA, International Magnesium Association, 1998, pp. 19-23.
18. P. Aroule, "Magnesium Demand and Supply", IMA-55, A Global Vision for Magnesium, Coronado, CA, USA, International Magnesium Association, 1998, pp. 36-46.
19. H. Liang, S. L. Chen and Y. A. Chang, "A Thermodynamic Description of the Al-Mg-Zn System", *Metall. Mater. Trans. A*, Vol.28A, 1997, pp. 1725-1734.
20. J.-O. Andersson, L. Hoglund, B. Jonsson and J. Agren, "Computer Simulation of Multicomponent Diffusional Transformations in Steel", G. R. Purdy, Ed., Fundamentals and Applications of Ternary Diffusion, Pergmon Press, 1990, pp. 153-163.
21. C. J. Smithells, E. A. Brandes and G. Brook, *Smithells Metals Reference Book*, Butterworths-Heinemann, Oxford ; Boston, ed. 7th / edited by E.A. Brandes, G.B. Brook., 1992.

A THERMODYNAMIC DATABASE FOR MAGNESIUM ALLOYS

Marie Piché[1], Arthur D. Pelton[1] and Christine Brochu[2]
[1] CRCT, École Polytechnique, C.P. 6079, Succ. 'Centre-Ville' Montréal, Québec,Canada, H3C 3A7
[2] Noranda Technology Center, 240 boul. Hymus, Pointe-Claire, Montréal, Québec, H9R 1G5

Abstract

A complete thermodynamic database for the *Mg-Al-Mn-Fe-Be* system has been developed. The goal of this work is to develop a better understanding of the chemical interactions occurring during the alloying process of magnesium-based alloys. The database consists of evaluated and optimized model parameters, which give all the thermodynamic properties of all phases as functions of temperature and composition.

For the binary systems without beryllium, the parameters were taken from the COST 507 Project (1). The binary systems with beryllium were critically evaluated/optimized in the present work, as were all systems of three or more components. The models are used to predict the properties of the phases in regions of composition and temperature where data are unavailable. The optimized set of model parameters was obtained by considering simultaneously all available literature data for phase equilibria and thermodynamic properties of all phases. Complete details of the optimizations will be presented elsewhere.

The database can be used with the FactSage thermodynamic software (2) to calculate complex chemical equilibria involving the solid, liquid and gaseous phases. Phase diagram sections of any type can be calculated and plotted. The alloying process can be followed for temperatures between 500 and 900°C, with calculation of the equilibrium phase assemblage at any temperature. Non-equilibrium (Scheil-Gulliver) cooling can also be simulated. The software and database can be used to search for improved alloying techniques and increased elemental recovery by calculating elemental distributions between liquid magnesium and sludge, enthalpy balances, phase saturation, intermediate phase formation upon alloying, etc.

Introduction

The main interest in magnesium of the automotive and aeronautic industries is its low density, compared to aluminum and steel, which reduces vehicle weight and fuel consumption. To improve the physical and chemical properties, aluminum, zinc, manganese, and beryllium are added in the preparation of AM50, AM60 and AZ91 magnesium alloys. The alloying recipe is different for each foundry but the form in which the alloying elements are added is essentially the same (3).

Aluminum is added in solid form because its melting temperature (660°C) is near that of magnesium (650°C). To dissolve manganese, which has a melting temperature of 1246°C, it is added in the form of sintered manganese-aluminum powder briquettes or as $MnCl_2$. Manganese is added to increase corrosion resistance and remove iron impurity. Because of the low melting temperature (419°C) of zinc, it is added in solid form (3). An aluminum-beryllium master alloy is added to molten magnesium to improve the dissolution of beryllium that has a melting temperature of 1287°C. Beryllium additions retard oxidation during melting of alloy ingots (4).

A major objective of magnesium alloy producers is to minimize the loss of alloying elements during production. Each company has its own alloying sequence that improves the yield. The use of experimental tests in the laboratory or in production is expensive and time-consuming. To help the magnesium alloy producer with the optimization of the alloying recipe, a thermodynamic magnesium database for *Mg-Al-Mn-Fe-Be* alloys has been developed.

In a thermodynamic "optimization," all available thermodynamic and phase equilibrium data for a system are evaluated simultaneously in order to obtain one set of model equations for the Gibbs energies of all phases as functions of temperature and composition. From these equations, all of the thermodynamic properties and the phase diagram can be back-calculated. In this way, all the data are rendered self-consistent and consistent with thermodynamic principles. Thermodynamic property data, such as activity data, can aid in the evaluation of the phase diagram, and phase diagram measurements can be used to deduce thermodynamic properties. Discrepancies in the available data can often be resolved, and interpolations and extrapolations can be made in a thermodynamically correct manner. A small set of model parameters is obtained. This is ideal for computer storage and calculation of properties and phase diagrams.

When used with the FactSage thermodynamic software (2), the magnesium database allows the magnesium alloy producer to reduce the amount of experimentation and rapidly explore a wide variety of possible conditions without loss of production time. Thermodynamic calculations permit a better understanding of the dissolution of alloying in magnesium and of phase precipitation at different compositions and temperatures. In the future, more elements will be added to the database such as Zn, Si and rare earths.

This article discusses the methodology used to develop the magnesium database and some of its applications. Complete details of the optimizations will be presented elsewhere. See also (5).

Thermodynamic Models

The magnesium database contains Gibbs energy equations for the liquid phase and twenty-five solid phases. From the Gibbs energy equation for a phase, as a function of temperature and composition, one can calculate every thermodynamic property: enthalpy, entropy, heat capacity, activity, etc.

The database contains data for all the stoichiometric compounds of all the binary systems: Al_2Fe, Al_5Fe_2, $Al_{30}Mg_{23}$, $Al_{140}Mg_{89}$, Al_4Mn, Al_6Mn, $Al_{11}Mn_4$, $Al_{12}Mn$, Be_5Fe, $Be_{12}Fe$, $Be_{13}Mg$, Be_5Mn

and $Be_{12}Mn$. For stoichiometric compounds, the molar Gibbs energy is expressed as:

$$g_{A_aB_b} = ag_A^o + bg_B^o + \Delta G_{form}(T) \tag{1}$$

where $\Delta G_{form}(T)$ is the Gibbs energy of formation of the A_aB_b compound expressed as

$$\Delta G_{form}(T) = \Delta H - T\Delta S \tag{2}$$

where ΔH, the enthalpy of formation, and ΔS, the entropy of formation, are constant, T is the absolute temperature and g^o_i is the molar Gibbs energy of the pure element i expressed as:

$$g_i = a + bT + cT\ln T + dT^2 + eT^3 + fT^{-1} + hT^{-9} + g^{mag} \tag{3}$$

where *a, b, c, d, e, f* and *h* are coefficients and g^{mag} is the magnetic contribution to the Gibbs energy (1).

The magnesium database includes data for one liquid solution and sixteen solid solutions. A solution can be a non-stoichiometric intermediate phase like γ-*AlMg* where the Al/Mg ratio can vary or a terminal solution of an element dissolving other elements; for example, solid magnesium can dissolve up to 16 at. % *Al* (figure 1). Furthermore, solutions can also be formed by binary compounds in which other elements are soluble; an example is the solution of manganese in Al_2Fe. The terminal solid solutions are: fcc for *Al* and *Fe*, bcc for *Be, Fe* and *Mn*, ordered bcc (B2) in the system *Al-Fe*, hcp for *Be* and *Mg*, cubic centered (A12) for *Mn* and primitive cubic (A13) for *Mn*.

The models used for all solutions are sublattice models based on the Compound Energy Formalism (6-8). For example, suppose $(i,j)_x(i,j)_y$ is a two-sublattice binary solution with element **i** and **j** on each sublattice and with $x + y = 1$. (For example, for a simple binary substitutional liquid solution with only one sublattice, $(i,j)_x$, x is equal to 1). The molar Gibbs energy of the solution is expressed as:

$$g = g^{ref} + \Delta g^{ideal} + g^E \tag{4}$$

where :

$$g^{ref} = Y_i^I Y_i^{II} g^o_{i:i} + Y_i^I Y_j^{II} g^o_{i:j} + Y_j^I Y_i^{II} g^o_{j:i} + Y_j^I Y_j^{II} g^o_{j:j} \tag{5}$$

$$\Delta g^{ideal} = xRT(Y_i^I \ln Y_i^I + Y_j^I \ln Y_j^I) + yRT(Y_i^{II} \ln Y_i^{II} + Y_j^{II} \ln Y_j^{II}) \tag{6}$$

$$g^E = Y_i^I Y_j^I \left(Y_i^{II} \sum_{v=0} L^{(v)}_{i,j:i}(Y_i^I - Y_j^I)^v + Y_j^{II} \sum_{v=0} L^{(v)}_{i,j:j}(Y_i^I - Y_j^I)^v \right) + Y_i^{II} Y_j^{II} \left(Y_i^I \sum_{v=0} L^{(v)}_{i:i,j}(Y_i^{II} - Y_j^{II})^v + Y_i^I \sum_{v=0} L^{(v)}_{j:i,j}(Y_i^{II} - Y_j^{II})^v \right) \tag{7}$$

where Y^I_i is the molar fraction of i on the sublattice I, $g^o_{i:j}$ is the molar Gibbs energy of the $(i)_x(j)_y$ "compound", R is the gas constant, T is the absolute temperature, $L^{(v)}_{i:i,j}$ is a binary interaction parameter between i and j on sublattice II and the overall mole fraction of the element i is :

$$X_i = xY_i^I + yY_i^{II} \tag{8}$$

The $(i)_x(j)_y$ "compounds" may be either real compounds or hypothetical solution "end members". In the latter case, their Gibbs energies $g^o_{i:j}$ are actually mixing parameters of the solution which are evaluated in the optimization.

The parameters $g^o_{i:j}$ and the binary parameters $L^{(v)}$ are first found by critical evaluation/optimization of the binary systems. The compound Energy Formalism is then used to estimate the properties of ternary and multicomponent solutions from these binary parameters via the Muggianu approximation (1). If experimental data are available in ternary or multicomponent systems, then these can be used in a second optimization step to determine additional ternary interaction parameters, thereby refining the model.

Magnesium Database

In the case of the binary systems *Al-Fe, Al-Mg, Al-Mn, Fe-Mg, Fe-Mn, Mg-Mn*, the evaluations/optimizations have been published as part of the COST 507 project (1). Beginning in January 1990, fifteen countries of the European Union co-operated on this project to evaluate and optimize all available data, as well as to perform measurements, and to develop a light metals thermodynamic database. An important system in this database is the *Mg-Al* system whose phase diagram is shown in figure 1. Magnesium and aluminum are the major components of the *Mg-Al-Mn-Fe-Be* alloys in the present study.

The binary systems *Al-Be* and *Fe-Be* have been fully evaluated/optimized in the present work. *Mg-Be* and *Mn-Be* were evaluated approximately because only a few phase equilibrium data are available. In the case of *Mg-Be*, only two experimental data points on the liquidus and the decomposition temperature of the $MgBe_{13}$ compound are available. For the *Mn-Be* system, no equilibrium data are available and the only information available is the stoichiometry of the intermediate compound. Figure 2 shows the optimized liquidus of the *Al-Be* system.

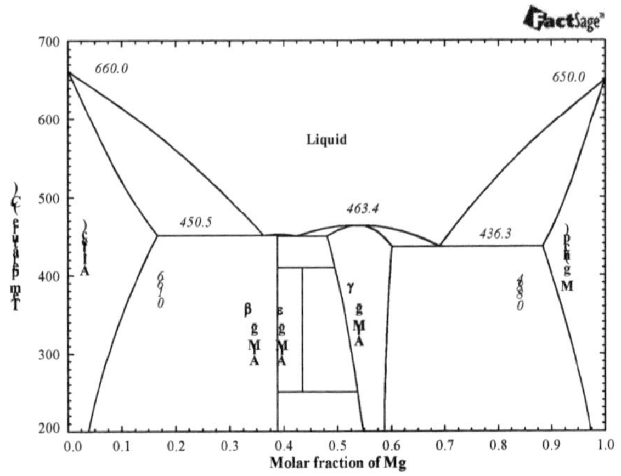

Figure 1 : Optimized *Al-Mg* phase diagram from COST 507 (1).

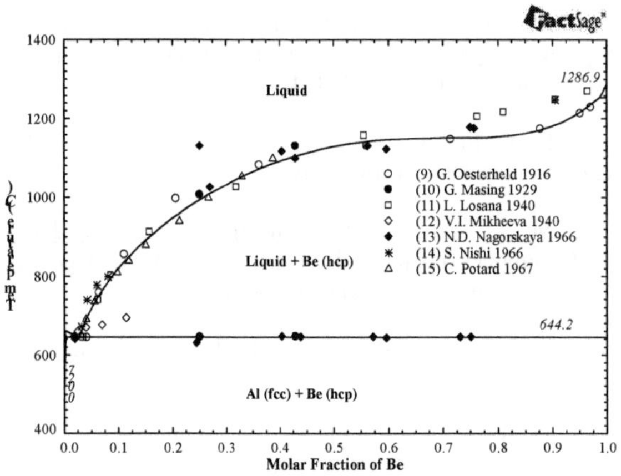

Figure 2 : *Al-Be* phase diagram optimized in the present study.

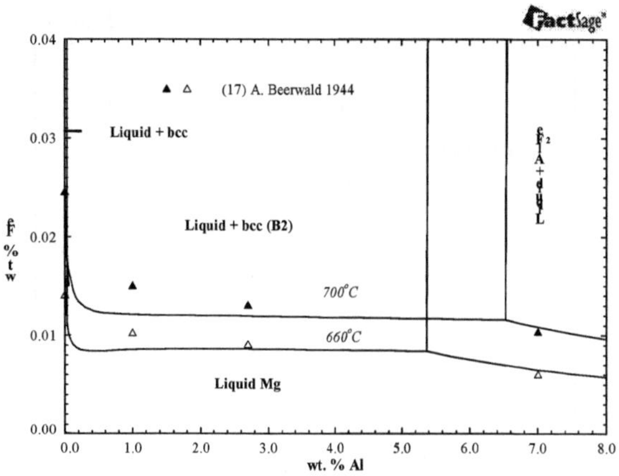

Figure 3 : Magnesium-rich region at 660°C and 700°C of the *Al-Fe-Mg* phase diagram optimized in the present study.

The ternary system *Al-Fe-Mn* has been evaluated/optimized by Jansson (16) and his optimized parameters are used in the present magnesium database. For the *Al-Fe-Mg* system, the optimized diagram is presented in figure 3. A ternary interaction parameter was included.

For the *Al-Fe-Mg-Mn* system, data are available in the Mg-rich region. This system has been optimized in the present study. The data available for the *Al-Fe-Mg-Mn* system at 2.7 wt. % *Al* are presented in figure 4. Figures 5 and 6 show the *Al-Fe-Mg-Mn* phase diagram at 7.0 wt. % *Al* for different temperatures. Most of the experimental data are from Beerwald (17) and the difference between the experimental data points and the optimized phase diagram is less than 50 ppm. No ternary parameters were required for this system.

For all the other multi-component system in the database, especially those with beryllium, no data were available in the Mg-rich region. Hence, as explained above, thermodynamic properties are estimated from the binary parameters by means of the models.

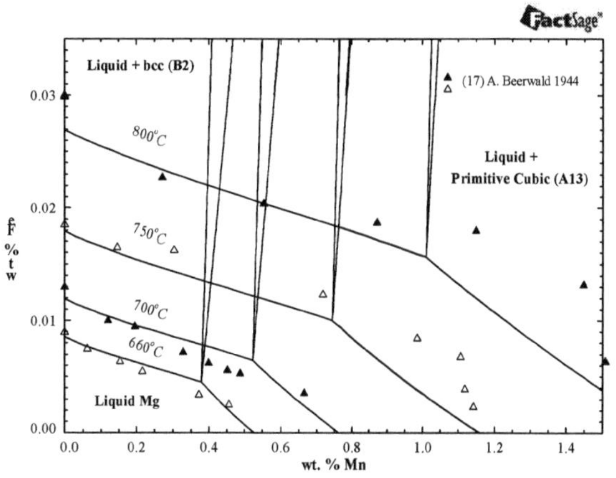

Figure 4 : Magnesium-rich region at 660°C, 700°C, 750°C and 800°C of the *Al-Fe-Mg-Mn* phase diagram at constant 2.7 wt. % Al optimized in the present study.

Applications

The magnesium thermodynamic database, when used with the FactSage thermodynamic software (2), permits the calculation of phase equilibria and thermodynamic properties of *Mg-Al-Mn-Fe-Be* alloys between 500°C and 900°C. Using these tools, alloying processes for magnesium can be simulated. The understanding of the phase behaviour when alloying elements are added at various temperatures and in different sequence can increase recoveries. The FactSage thermodynamic software (2) can also be used with the magnesium database to calculate the course of rapid non-equilibrium (Scheil-Gulliver) cooling in which it is assumed that a solid phase, once precipitated, does not change its composition.

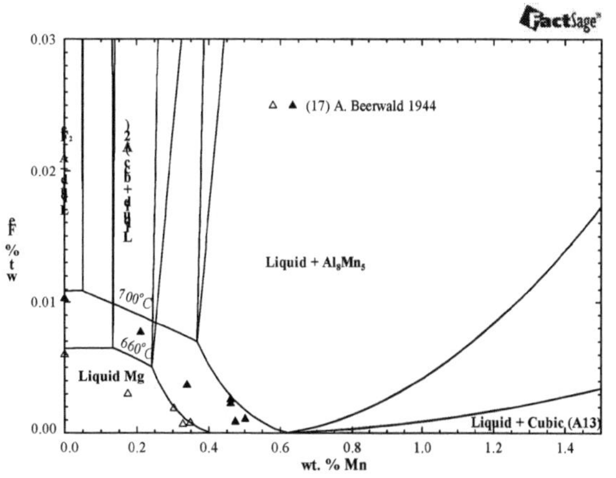

Figure 5 : Magnesium-rich region at 660°C and 700°C of the *Al-Fe-Mg-Mn* phase diagram at constant 7.0 wt. % Al optimized in the present study.

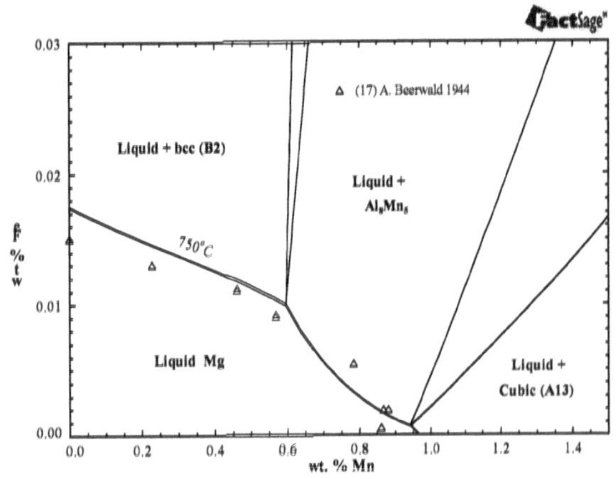

Figure 6 : Magnesium-rich region at 750°C of the *Al-Fe-Mg-Mn* phase diagram at constant 7.0 wt. % Al optimized in the present study.

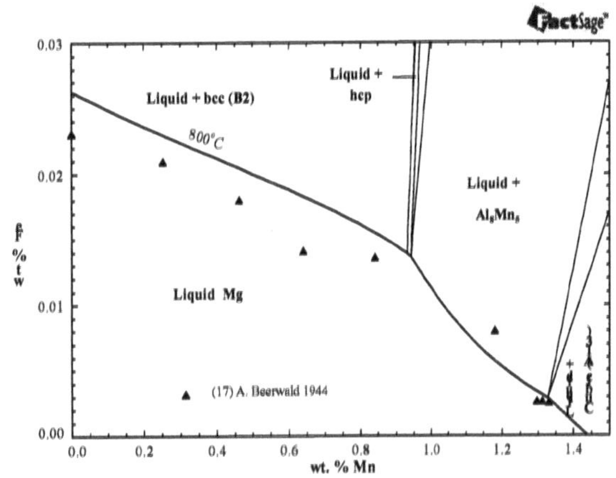

Figure 7 : Magnesium-rich region at 800°C of the *Al-Fe-Mg-Mn* phase diagram at constant 7.0 wt. % Al optimized in the present study.

Iron removal by manganese is illustrated in figures 3 to 7. For example, in figure 6, starting at 0 wt. % Mn at 750°C, the maximum solubility of iron in *Mg-Al* alloy is 0.017 wt. % *Fe*. If 0.4 wt. % *Mn* is added to the melt, a solid phase bcc (B2) is formed in equilibrium with the liquid phase that now has a lower iron concentration. The bcc (B2) phase is an ordered bcc phase in the *Al-Fe* system that dissolves manganese. If more manganese is added, the iron content of the melt will continue to decrease and other solid phases containing iron will be formed: Al_8Mn_5 and the cubic (A13) phase.

REFERENCES

1. I. Ansara, A. T. Dinsdale and M. H. Rand eds., COST 507 : Thermodynamical Database for Light Metal Alloys, vol. 2 (Office for Official Publications of the European Communities, Luxembourg, 1998), 396.

2. Factsage, www.factsage.com, 2001.

3. M. M. Avedesian and H. Baker, Magnesium and Magnesium Alloys (ASM Specialty Handbook, Materials Park, OH: ASM International, 1999), 314.

4. E. F. Emley, Principles of Magnesium Technology, (Pergamon Press, Oxford, 1966), 1013.

5. M. Piché, Détermination des données thermodynamiques de phases de magnésium, (M. Sc. App. thesis, École Polytechtenique de Montréal, 2002).

6. B. Sundman and J. Ågren, J. Phys. Chem. Solids, 42, (1981), 297.

7. M. Hillert, Phase Equilibria, Phase Diagrams and Phase Transformations, (Cambridge: Cambridge University Press, 1998).

8. T.I. Barry et al., J. Phase Equilib., 13, (1992), 459.

9. G. Oesterheld, "Alloys of Beryllium with Aluminum, Copper, Silver and Iron," Z. Anorg. Chem., 97 (1), (1916),6-40.

10. G. Masing and O. Dahl, "Beryllium Containing Aluminium Alloys," Wiss. Veroeff. Siemens-Konsern, 8, (1929), 248-256.

11. L. Losana, "The Beryllium-Aluminum System," Alluminio, 9, (1940), 8-13.

12. V. I. Mikheeva, "The Solubility of Beryllium in Aluminum," Izv. Akad. Nauk SSSR Otd. Khim. Nauk., 5, (1940), 775-782.

13. N. D. Nagorskaya and al., "Phase Diagram of Part of the Al-Be-Mg System," Izv. Akad. Nauk SSSR, Metally, 5, (1966), 137-147.

14. S. Nishi, "Study on Binary Diagram of Al-Be Alloy," Light Metals Jpn., 16, (1966), 5-8.

15. C. Potard, G. Bienvenu and B. Schaud, "Distribution des impuretés fer, cuivre, molybdène entre les phases solide et liquide dans les alliages aluminium-béryllium," (Thermodynamics of Nuclear Materials, Proc. IAEA Conf., Vienna, 1967), 809-825.

16. A. Jansson, Phase Diagram Evaluations and Applications in Light Metal Alloys (Doctoral Thesis, Vetenskap Och Konst, Stockholm, 1997), 24.

17. A. Beerwald, "On the Solubility of Iron and Manganese in Magnesium and Magnesium-Aluminum Alloys," Metallwissenschaft, Metalltecknik, 23, (1944), 404-407.

COMPUTATIONAL THERMODYNAMICS AND EXPERIMENTAL INVESTIGATION OF Mg-Al-Ca ALLOYS

Koray Ozturk, Yu Zhong and Zi-Kui Liu
Department of Materials Science and Engineering
The Pennsylvania State University
University Park, PA 16802, USA

Alan A. Luo
General Motors Research and Development Center
Materials and Processes Lab
30500 Mound Road, Warren, MI 48090-9055, USA

ABSTRACT

The thermodynamic properties of the ternary Mg-Al-Ca system are investigated, based on the three binary systems, i.e. the Al-Mg, Ca-Mg and Al-Ca systems. The ternary system contains six different intermetallic compounds with five of them treated as stoichiometric compounds. The calculated ternary system and Scheil simulations are used to understand the microstructures and phase relationships during solidification. Metallographic samples are prepared for the optical microscopy and SEM observations. The phase amounts were measured and compared with the calculated results. The calculations will be used to direct new experiments in the project for fine-tuning of the phase relationships.

INTRODUCTION

Magnesium is the eighth most abundant element in the Earth's crust and the third most plentiful element dissolved in seawater[1]. It is the lightest structural metal with a density of 1.741 g/cm^3, in comparison with the densities of Al (2.70 g/cm^3) and Fe (7.86 g/cm^3). This makes magnesium alloys particularly attractive for transportation applications such as automobiles and helicopters for weight reduction and higher fuel efficiency [2-4]. The world consumption of magnesium alloys in the automobile industry has experienced an 18% annual increase from 1990 to 1996 and a 32% increase from 1996 to 1997[5]. This rapid growth has highlighted the need for a greater understanding of factors that influence the properties of magnesium alloys and for a wider range of magnesium alloys with lower thermal expansion, higher fatigue strength, higher creep strength, and better corrosion resistance[3, 4]. There is an urgent need for an efficient route of alloy development with reduced cost and time. The emerging concept of system materials design provides such opportunities[6]. In the system materials design, the essential component is the computational thermodynamics, often referred to as CALculation of PHAse Diagrams, or CALPHAD. In the present report, we will report our recent work on computational thermodynamics of the Mg-Al-Ca system along with some experimental observations of Mg-Al-Ca ternary alloys.

Recently, there has been significant experimental research activities on Mg-Al-Ca based alloys with or without additional alloying elements [7-14]. This family of alloys seems promising for automotive powertrain applications due to their excellent creep resistance and low cost. The current project was initiated with the interactions between the Pennsylvania State University and General Motors and further supported by the National Science Foundation. The objective is to study the fundamental thermodynamics and kinetics of Mg-Al-Ca based alloys and to provide understanding and guidance of system design of magnesium alloys through computational modeling and experimental investigations.

METHOD

CALPHAD has been under development since the early 1970's[15]. The development of CALPHAD techniques in the past three decades was discussed by Saunders and Miodownik[16]. This technique was briefly summarized in a previous publication by one of the authors[17] and will not be repeated here. Only the procedure for the development of the thermodynamic description of the Mg-Al-Ca system will be discussed, as the detailed evaluation of the thermodynamic database will be described elsewhere in the near future.

The thermodynamic descriptions of the pure Mg, Al, and Ca elements are taken from the SGTE database[18]. In the Mg-Al-Ca ternary systems, there are three binary systems, i.e. Mg-Al, Mg-Ca, and Al-Ca. The thermodynamic descriptions of the Mg-Al[19, 20] and Mg-Ca[21] are available in the literature. The Al-Ca system has been studied in the present work. The Mg-Al-Ca ternary system is predicted from the combination of the three binary systems. Experimental work is underway to study the phase relationships in the complete Mg-Al-Ca ternary system guided from the predictions.

Two alloys, Mg-4.5%Al-1.9%Ca and Mg-4.5%Al-3.0%Ca, (all in weight percent unless otherwise specified), were supplied by General Motors R&D Center. The alloys were made by die casting at Lexington Die Casting, Lakewood, NY. The melt temperature was controlled at around 677°C with the die surface temperature maintained at 350°C. Chemical analysis using ICP/AES (Inductively-Coupled Plasma/Atomic Emission Spectroscopy) techniques was performed, and the other alloying elements and impurities are typically 0.27%Mn, 0.003%Fe and 0.002%Cu.

Samples were cut from the 6mm diameter, die-casting tensile-test bar and prepared for optical (Nikon Epiphot) and SEM (Philips XL 20) observations under either polished or polished and etched conditions. When etched, the etchant is 5% HNO_3 in methanol. The volume fractions of phases were measured using the Princeton Gamma Tech (PGT) Version 7 image analysis software on photos taken from the optical microscope with at least five areas randomly chosen from each sample.

RESULTS AND DISCUSSIONS

Computational Study

The calculated phase diagrams of the three binary systems are shown in Fig. 1. It is to be noted that there were two new phases recently reported in the Al-Ca system, i.e. AlCa and Al_3Ca_8[22].

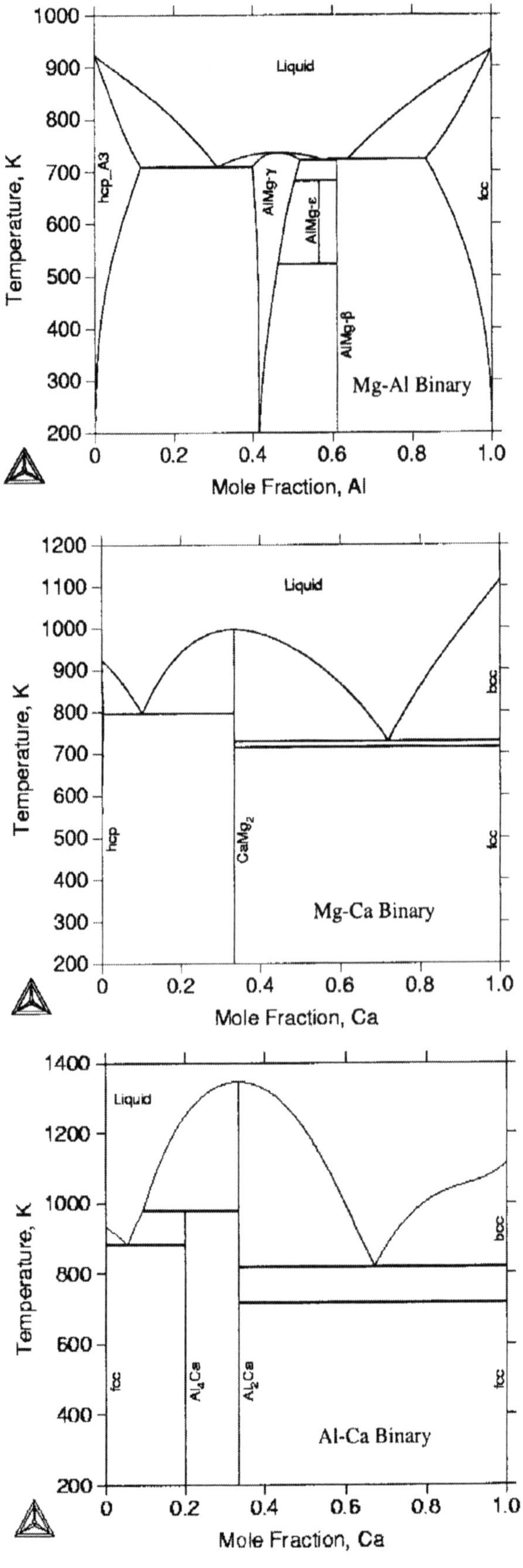

Figure 1: The three calculated binary phase diagrams.

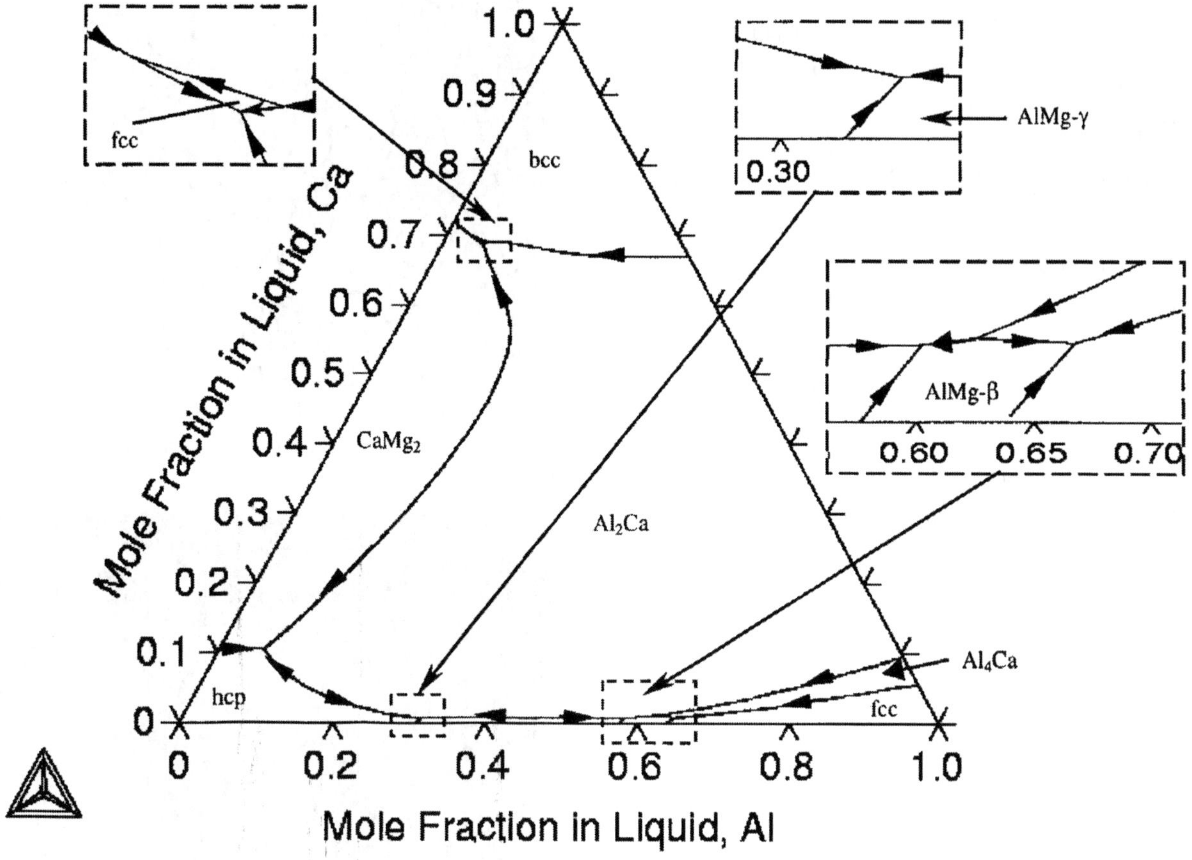

Figure 2: The calculated liquidus projection of the Mg-Al-Ca ternary system.

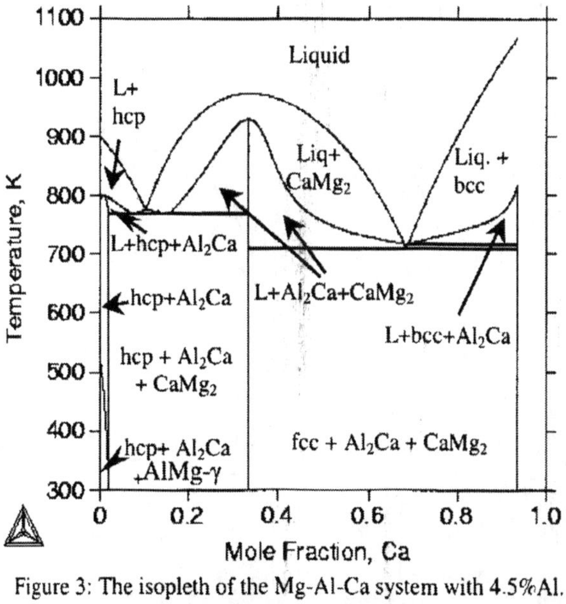

Figure 3: The isopleth of the Mg-Al-Ca system with 4.5%Al.

We are carrying out our own experimental work to verify them. If verified, we will revise the Al-Ca binary system to include these two new phases. However, it should be pointed out that these two phases are at the Ca-rich side of the Al-Ca phase diagram and do not affect the current calculation of magnesium alloys significantly.

The thermodynamic description of the Mg-Al-Ca ternary system was obtained by combining the data of the three binary systems assuming no ternary solubility in the all binary compounds due to the lack of experimental data. No ternary interactions were introduced either. The predicted liquid projections of the Mg-Al-Ca system are presented in Fig. 2, with respect to the composition triangle. Fig. 3 shows the isopleth at 4.5%Al with the temperature and the Ca content plotted, which depicts the effect of the Ca content on phase equilibria and phase relationships at different temperatures.

For the two Mg-Al-Ca alloys, the equilibrium phase fractions (dotted lines) and the phase fractions from Scheil simulations (solid lines) are plotted in Fig. 4 as a function of the temperature. In the Scheil simulations, it is assumed that the liquid phase is homogenous all the time, and there is no diffusion in the solid phases. Therefore, the equilibrium calculations and the Scheil simulations represent two extreme situations. In both alloys, the hcp phase solidifies first from the liquid phase, followed by the hcp+Al_2Ca eutectic solidification. The liquidus temperature of the Mg-4.5%Al-1.9%Ca alloy is about 7 degrees higher than that of the Mg-4.5%Al-3.0%Ca alloy, but the eutectic temperatures are only about 2 degrees apart in the Scheil simulations and about 1 degree in the equilibrium calculations. The equilibrium amounts of the Al_2Ca phase are 4.0% and 6.0% in the Mg-4.5%Al-1.9%Ca and Mg-4.5%Al-3.0%Ca alloys, respectively, while the amount in the Scheil simulations, are 4.3 and 4.5%, respectively. However, in

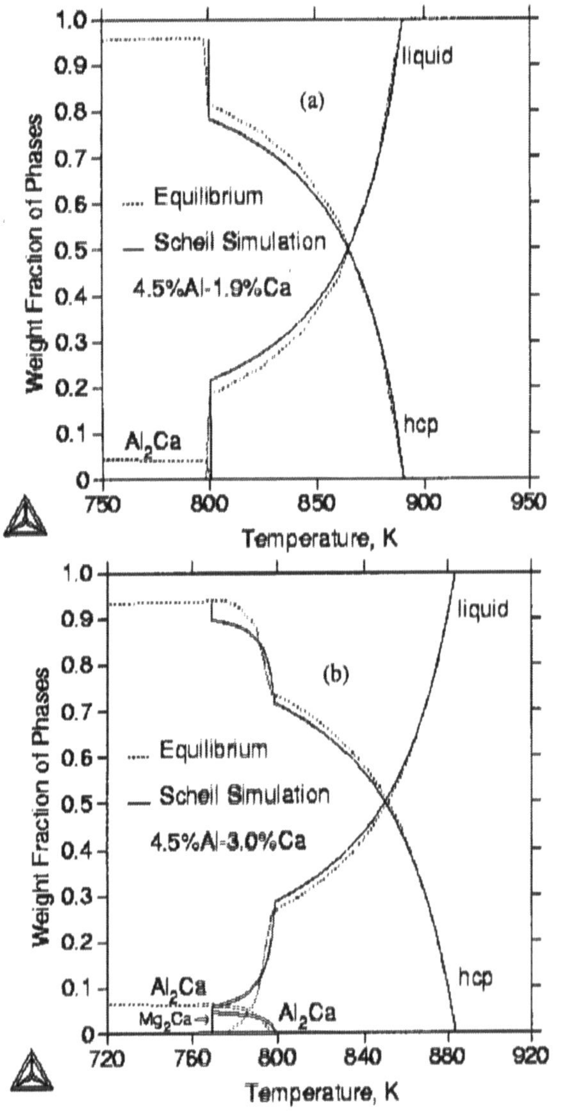

Figure 4: Phase fractions as a function of the temperature for the two alloys: (a) Mg-4.5%Al-1.9%Ca and (b) Mg-4.5%Al-3.0%Ca.

the Mg-4.5%Al-3.0%Ca alloy, there is 1.5% Mg_2Ca phase formed. The amount of the eutectic structures are about 18.5% and 21.5% for the Mg-4.5%Al-1.9%Ca, and 26.2% and 28.5% for the Mg-4.5%Al-3.0%Ca for the equilibrium calculations and Scheil simulations, respectively, i.e. there is 7% more eutectic structure in the Mg-4.5%Al-3.0%Ca alloy.

Experimental Observations

The microstructures of both alloys are very similar to each other under optical microscope and SEM. One optical micrograph of an etched Mg-4.5%Al-1.9%Ca sample is shown in Fig. 5, while backscattered and secondary electron SEM images of unetched samples are shown in Figs. 6 and 7. Die casting porosities were present in both optical and SEM images, and their fractions were subtracted from the calculations during image analysis. The measured volume fractions of the eutectic areas are 18.9±2.3% and 25.1±2.1% for the Mg-4.5%Al-1.9%Ca and Mg-4.5%Al-

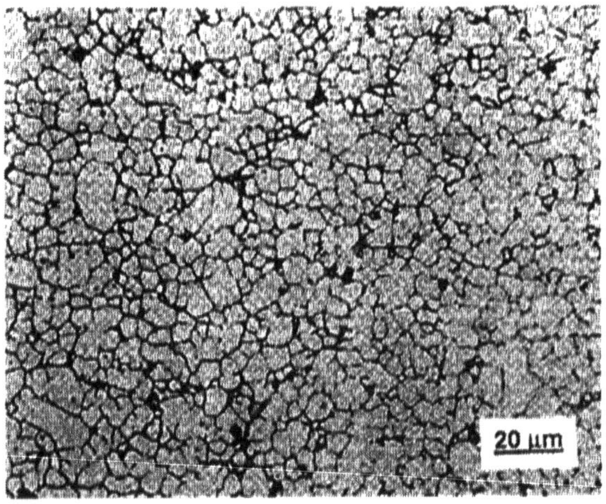

Figure 5: An optical micrograph of the etched Mg-4.5%-1.9%Ca alloy.

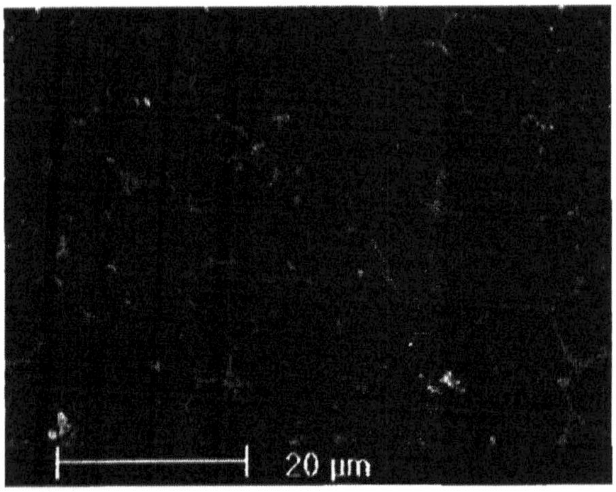

Figure 6: A backscattered electron SEM image of an unetched Mg-4.5%Al-3.0%Ca alloy.

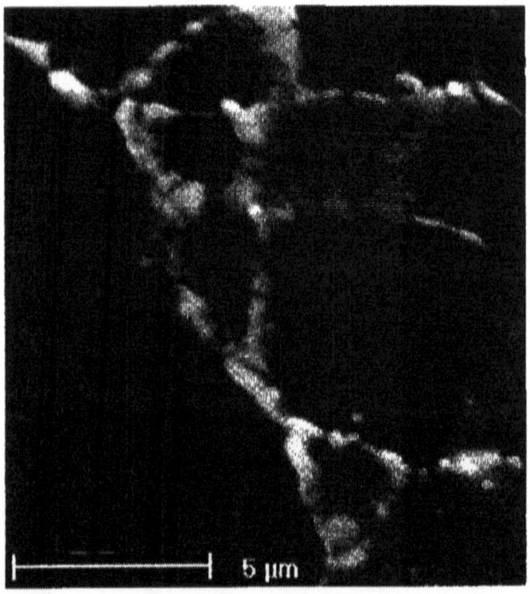

Figure 7: A secondary electron SEM image of an unetched Mg-4.5%Al-1.9%Ca alloy.

3.0%Ca, respectively. With the densities of the hcp-Mg and Al_2Ca being 1.738 and 2.395g/cm^3, the calculated volume fractions of the eutectic areas, using the Scheil simulation results, are 20.6% and 27.3% for the two alloys, respectively, which are in good agreement with the experimental measurements. These areas consist of two types of eutectic structures, the usual lamella eutectic structure as shown by the arrow A, also observed in TEM[13], and the divorced eutectic structure as shown by arrow B, i.e. there is no lamella morphology and the hcp-Mg phase is part of the matrix. This may contribute to the fact that the measured volume fractions are slightly lower than the calculated ones.

It was recently suggested that Mg and Al may have mutual solubility in the intermetallic compound[13]. More detailed analytical TEM study is underway to measure the composition of the intermetallic phase with TEM/EDS. Assuming that Mg and Al have a complete mutual solubility with an ideal mixing, even though Al_2Ca and Mg_2Ca have different crystal structures, i.e. fcc (C15) for Al_2Ca and hexagonal (C14) for Mg_2Ca, the amounts of the intermetallic compound remain almost the same as in the above calculations because they are primarily determined by the Ca contents in the two alloys. The mole fraction ratio of Al and Mg in the hypothetical $(Mg,Al)_2Ca$ compound is found to be close to 6:1 in the Scheil simulation. More experimental investigation on the ternary solubility is currently undertaken in our laboratory.

SUMMARY

The thermodynamic description of the Mg-Al-Ca has been developed with the data of the Al-Ca binary system from the present work. Using this database, various calculations have been made to understand the phase relationships during the die-casting process and the effect of the Ca content on the microstructures. The calculated results and experimental observations carried out in the present work are in good agreement. More work is underway to refine the Ca rich side of the Al-Ca binary system by studying the recently found intermetallic compound using TEM/EDS. This work builds a foundation to study multicomponent Mg-Al-Ca based alloys.

ACKNOWLEDGEMENTS

This work is supported by the NSF CAREER Award under the grant DMR-9983532, the Scholarship from The Ministry of National Education of Turkey for Mr. Ozturk, and the Pennsylvania State University MRI Fellowship for Mr. Zhong. The Thermo-Calc program is licensed from The Foundation for Computational Thermodynamics, Stockholm, Sweden. General Motors Corporation is acknowledged for supplying the alloy specimens and permission of publishing this paper.

REFERENCE

1. D. Kramer, "Magnesium" (U.S. Geological Survey--Minerals Information, 1997).
2. G. Craig, Use of WE-43A-T6 Magnesium in the MD600 Helicopter Drive System, IMA-55, A Global Vision for Magnesium, Coronado, CA, USA (IMA 1998), pp. 19-23.
3. H. Friedrich and S. Schumann, The Use of Magnesium in Cars - Today and in the Future, IMA-55, A Global Vision for Magnesium, Coronado, CA, USA (IMA 1998), pp. 1-7.
4. A. Luo, Materials Comparison and Potential Applications of Magnesium in Automobiles, H. I. Kaplan, J. N. Hryn, B. B. Clow, Eds., Magnesium Technology 2000, Nashville, TN (Minerals, Metals and Materials Society/AIME, PA, USA, 2000), pp. 89-98.
5. P. Aroule, Magnesium Demand and Supply, IMA-55, A Global Vision for Magnesium, Coronado, CA, USA (International Magnesium Association, 1998), pp. 36-46.
6. G. B. Olson, *Science*,**277**, (1997) 1237-1242.
7. F. M. Knoop, K. U. Kainer, B. Smola and B. L. Mordike, Production and Processing of PM Magnesium and Mg-Al-Ca Alloys, Powder Metallurgy World Congress (PM '94), Paris (Editions de Physique (France), 1994), pp. 2237-2240.
8. T. Ebert, K. U. Kainer and B. L. Mordike, Mechanical properties of spray formed magnesium alloys, Magnesium Alloys and their Applications, Wolfsburg, Germany (Werkstoff-Informationsgesellschaft mbH, Hamburger Allee 26, Frankfurt, D-60486, Germany, 1998), pp. 563-568.
9. C. Shaw and H. Jones, *Mater. Sci. Tech.*,**15**, (1999) 78-84.
10. T. Tsukeda, A. Maehara, K. Saito, M. Suzuki, J. Koike, K. Maruyama and H. Kubo, Mechanical properties and microstructure of heat-resistant Mg-Al-Ca alloys formed by thixomolding, H. I. Kaplan, J. N. Hryn, B. B. Clow, Eds., Magnesium Technology 2000, Nashville, TN, USA (TMS/AIME, PA, USA, 2000), pp. 395-402.
11. M. O. Pekguleryuz and J. Renaud, Creep resistance in Mg-Al-Ca casting alloys, H. I. Kaplan, J. N. Hryn, B. B. Clow, Eds., Magnesium Technology 2000, Nashville, TN (TMS/AIME, Warrendale, PA, USA, 2000), pp. 279-284.
12. A. Luo and T. Shinoda, *Magnesium Alloy Having Superior Elevated-Temperature Properties and Die Castability*, Patent 5,855,697, 1999
13. A. Luo, M. P. Balogh and B. R. Powell, "Tensile creep and microstructure of magnesium-aluminum-calcium alloys for powertrain applications: Part 2 of 2" *SAE2001* (Detroit, MI, 2001).
14. B. R. Powell, A. Luo, V. Rezhets, J. J. Bommarito and B. L. Tiwari, "Development of Creep-Resistant Magnesium Alloys for Powertrain Applications: Part 1 of 2" *SAE2001* (Detroit, MI, 2001).
15. L. Kaufman and H. Bernstein, *Computer Calculation of Phase Diagrams with Special Reference to Refractory Metals* (Academic Press, New York, 1970).
16. N. Saunders and A. P. Miodownik, *CALPHAD (Calculation of Phase Diagrams): A Comprehensive Guide* (Pergamon, Oxford ; New York, 1998).
17. Z. K. Liu, Design magnesium alloys: how computational thermodynamics can help, Magnesium Technology 2000 (USA), Nashville, TN, USA (TMS, Warrendale, PA, 2000), pp. 191-198.
18. A. T. Dinsdale, *CALPHAD*,**15**, (1991) 317-425.
19. P. Liang, et al., *Z. Metallkd.*,**89**, (1998) 536-540.
20. I. Ansara, A. T. Dinsdale and M. H. Rand, Eds., *COST 507: Thermochemical Database for Light Metal Alloys*, vol. 2 (European Commission, 1998).
21. R. Agarwal, J. J. Lee, H. L. Lukas and F. Sommer, *Z. Metallkd.*,**86**, (1995) 103-108.
22. A. Pisch, F. Hodaj, C. Colinet, D. Kevorkov, J. G. a. R. and Schmid-Fetzer, Enthalpy of Formation of Some Ca-containing Intermetallic Compounds, L. Kaufman, S. Allen, C. Lupis, Eds., CALPHAD XXIX, M.I.T., Cambridge, MA (CALPHAD Inc., 2000), pp. 83.

Computational Thermodynamics and Experimental Investigation of the Mg-Al-Ca-Sr Alloys

Yu Zhong, Koray Ozturk and Zi-Kui Liu
Department of Materials Science and Engineering
The Pennsylvania State University
University Park, PA 16802

Alan A. Luo
General Motors Research and Development Center
Materials and Processes Lab
30500 Mound Road, Warren, MI 48090-9055 USA

ABSTRACT

The thermodynamic properties of the quaternary Mg-Al-Ca-Sr system are investigated by combining the six binary systems, i.e. the Al-Ca, Al-Mg, Al-Sr, Ca-Mg, Ca-Sr and Mg-Sr. The constructed quaternary database is used to understand the microstructures and phase relationships of the two alloys (Mg-5.7%Al-3.1%Ca-0.15%Sr and Mg-5.0%Al-3.1%Ca-0.07%Sr). Scheil simulations and equilibrium calculations were performed for the solidification process of the alloys and compared with experimental observations.

INTRODUCTION

The light-weight magnesium alloys are particularly attractive for transportation applications such as automobiles and helicopters for weight reduction and higher fuel efficiency (1-3). Mg-Al alloys are one group of these magnesium-based alloys that are under development. Their use, however, is limited to low temperature applications. Therefore, a large amount of effort has been made to increase the service temperature of these alloys. The addition of alkaline earth elements into Mg-Al alloys has been found to be beneficial not only to keep the costs low but also to improve the mechanical properties, especially the creep resistance at elevated temperatures (>150°C)(4). Ca and Sr have been found to be the two alkaline earth elements which affect high temperature performance positively. Parallel to the experimental investigations, there is a great need for an efficient route of alloy development with reduced cost and time. The emerging concept of system materials design provides such opportunities(5). In the system materials design, the essential component is the computational thermodynamics, often referred to as CALculation of PHAse Diagrams, or CALPHAD.

The current project was initiated with the interactions between the Pennsylvania State University and General Motors and supported by the National Science Foundation. In the previous paper, the results for the Mg-Al-Ca ternary system were reported(6). In the present paper, our preliminary results in the Mg-Al-Ca-Sr system are presented and compared with the experimental data. The ultimate objective is to study the fundamental thermodynamics and kinetics of Mg-Al-Ca-Sr based alloys and to provide understanding and guidance of system design of magnesium alloys through computational modeling and experimental investigations.

THEORETICAL APPROACH

CALPHAD has been under development since early 1970's(7). The development of CALPHAD techniques in the past three decades was discussed by Saunders and Miodownik(8). This technique was briefly summarized in a previous publication by one of the authors(9). In this paper, only the general procedure for the development of the thermodynamic description of the Mg-Al-Ca-Sr system will be discussed, as the detailed evaluation of the thermodynamic database is still under development and will be published in the near future.

The thermodynamic descriptions of the pure Mg, Al, Ca and Sr elements are taken from the SGTE database(10). The Mg-Ca(11) and Mg-Al(12) binary systems are obtained from the literature. The Mg-Sr, Al-Ca, Al-Sr and Ca-Sr binary systems are carefully assessed in the present work by using critically reviewed experimental data in the literature. There are 15 binary compounds in these binary systems and they are listed in Table 1 with their crystal structures shown.

Table 1: The binary compounds and their crystal structures in the Mg-Al-Ca-Sr system.

Phase	Space Group	Crystal Structure	Prototype
Al_4Ca	I4/mmm	bct ($D1_3$)	Al_4Ba
Al_2Ca	$Fd\bar{3}m$	fcc (C15)	Cu_2Mg
AlCa	---	---	---
Al_3Ca_8	P1	triclinic	Tl_3Yb_8
$Mg_{17}Sr_2$	$P6_3/mmc$	---	Th_2Ni_{17}
$Mg_{38}Sr_9$	$P6_3/mmc$	---	---
$Mg_{23}Sr_6$	$Fm\bar{3}m$	$D8_a$	$Mn_{23}Th_6$
Mg_2Sr	$P6_3/mmc$	(C14)	$MgZn_2$
Mg_2Ca	$P6_3/mmc$	(C14)	$MgZn_2$
AlMg-□	$R\bar{3}$	---	$Co_5Cr_2Mo_3$
AlMg-□	$Fd\bar{3}m$	---	Cd_2Na
AlMg-□	$I\bar{4}3m$	cbcc (A12)	□-Mn
Al_4Sr	I4/mmm	bct ($D1_3$)	Al_4Ba
Al_2Sr	Imma	---	$CeCu_2$
A_7Sr_8	$P2_13$	---	Al_7Sr_8

There are four ternaries in the system, i.e. Mg-Al-Ca, Mg-Al-Sr, Mg-Ca-Sr and Al-Ca-Sr currently under investigations. As can be seen in Table 1, some of binary compounds have the same crystal structure, which indicates that they are the same phase and will have extensive solubility into each. However, due to the incompleteness of our work in these aspects, in the present paper, the thermodynamic properties of the ternary systems are predicted simply from the combination of those of their constitutive binary systems.

EXPERIMENTAL PROCEDURE

Two alloys, Mg-5.7%Al-3.1%Ca-0.15%Sr and Mg-5.0%Al-3.1%Ca-0.07%Sr, (all in weight percent unless otherwise specified), were supplied by General Motors R&D Center. The alloys were made by die casting at Lexington Die Casting, Lakewood, NY. The melt temperature was controlled at around 677°C with the die surface temperature maintained at 350°C. Chemical analysis using ICP/AES (Inductively-Coupled Plasma/Atomic Emission Spectroscopy) techniques was performed, and the other alloying elements and impurities are typically as 0.27%Mn, 0.003%Fe and 0.002%Cu.

Samples were cut from the 6mm diameter, die-casting tensile-test bar and prepared for optical (Nikon Epiphot) microscopy and SEM (Philips XL 20) observations under either polished only or polished and etched conditions. When etched, the etchant was 5% HNO_3 in methanol. The volume fractions of phases were measured using the Image-Pro Plus Version 4.5 image analysis software on photos taken from the optical microscope with at least ten areas randomly chosen for each sample.

RESULTS AND DISCUSSIONS

To view phase equilibrium in a quaternary system, one typically uses isopleth sections. In Fig. 1, the isopleth sections at 5%Al and 3.1%Ca withrespect to the temperature and the Sr content was calculated, to depict the effect of the Sr content on phase equilibria and phase relationships at different temperatures.

For the two Mg-Al-Ca-Sr alloys, the equilibrium phase fractions (dotted lines) and the phase fractions from Scheil simulations (solid lines) are plotted in Fig. 2 as a function of the temperature. In the Scheil simulations, it is assumed that the liquid phase is homogenous all the time, and there is no diffusion in the solid phases. Therefore, the equilibrium calculations and the Scheil simulations represent two extreme situations.

In both alloys, the hcp phase solidifies first from the liquid phase, followed by the hcp+Al_2Ca eutectic solidification. The liquidus temperature of the Mg-5%Al-3.1%Ca-0.07%Sr alloy is about 3 degree higher than that of the Mg-5.7%Al-3.1%Ca-0.15%Sr alloy, but the eutectic temperatures are only about 0.61 degree apart in the Scheil simulations and about 0.91 degree in the equilibrium calculations.

The equilibrium amounts of the Al_2Ca phase are 5.74% and 5.75% in the Mg-5%Al-3.1%Ca-0.07%Sr and Mg-5.7%Al-3.1%Ca-0.15%Sr alloys, respectively, while the amount in the Scheil simulations, they are 5.96% and 6.75%, respectively. There are 0.99% and 0.23% Mg_2Ca phase formed in the Mg-5%Al-3.1%Ca-0.07%Sr and Mg-5.7%Al-3.1%Ca-0.15%Sr

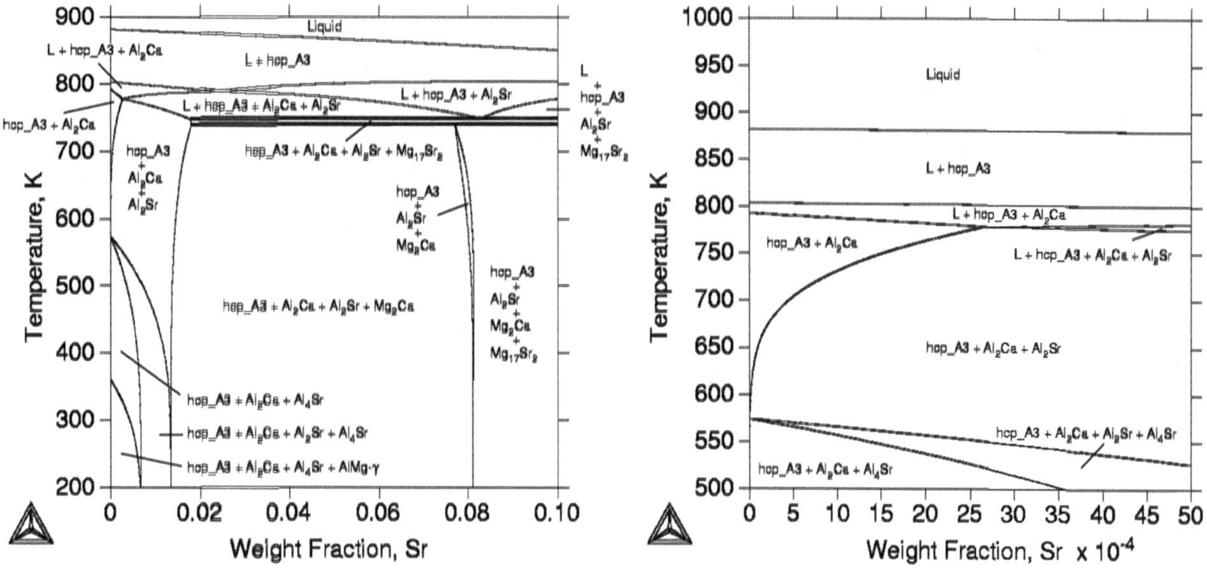

Figure 1: Calculated isopleth sections of the Mg-Al-Ca-Sr system with 5%Al-3.1%Ca. The righ diagram is an enlargement to show the phase diagram with low Sr contents.

alloys in the Scheil simulation, respectively. While in the equilibrium calculations, the Mg_2Ca phase is not stable, but the Al_2Sr phase.

The weight fraction of the eutectic structures are about 24.8% and 26.9% for the Mg-5%Al-3.1%Ca-0.07%Sr and Mg-5.7%Al-3.1%Ca-0.15%Sr for the equilibrium calculations, respectively. While in the Scheil simulations, they are 26.9% and 29.1% for the Mg-5.7%Al-3.1%Ca-0.15%Sr, respectively, i.e. there is about 2% more eutectic structure in the Mg-5.7%Al-3.1%Ca-0.15%Sr alloy.

The microstructures of both alloys are very similar to each other. Optical micrographs from the two alloys are shown in Figure 3. The SEM micrograph and the EDS result of the intermetallic phase in the Mg-5.7%Al-3.1%Ca-0.15%Sr alloy are shown in Fig. 4. The measured volume fractions of the white areas are 17.4±1.5% and 19±1.3% for the Mg-5%Al-3.1%Ca-0.07%Sr and Mg-5.7%Al-3.1%Ca-0.15%Sr, respectively.

With the densities of the hcp and Al_2Ca being 1.738 and 2.395g/cm^3, the calculated volume fractions of the eutectic structure, using the Scheil simulation results, are 21.1% and 23% for the two alloys, respectively.

The comparison between the calculated and experimental results on eutectic fractions in the two die cast alloys is summarized in Table 2. The difference between the experimental and

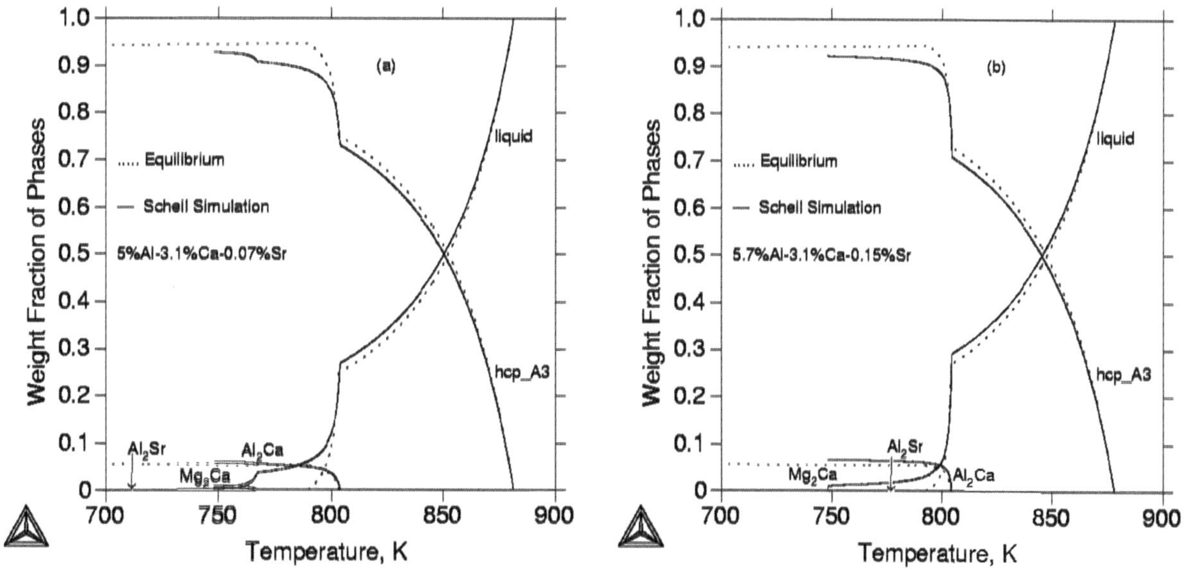

Figure 2: Phase fractions as a function of the temperature for the two alloys.

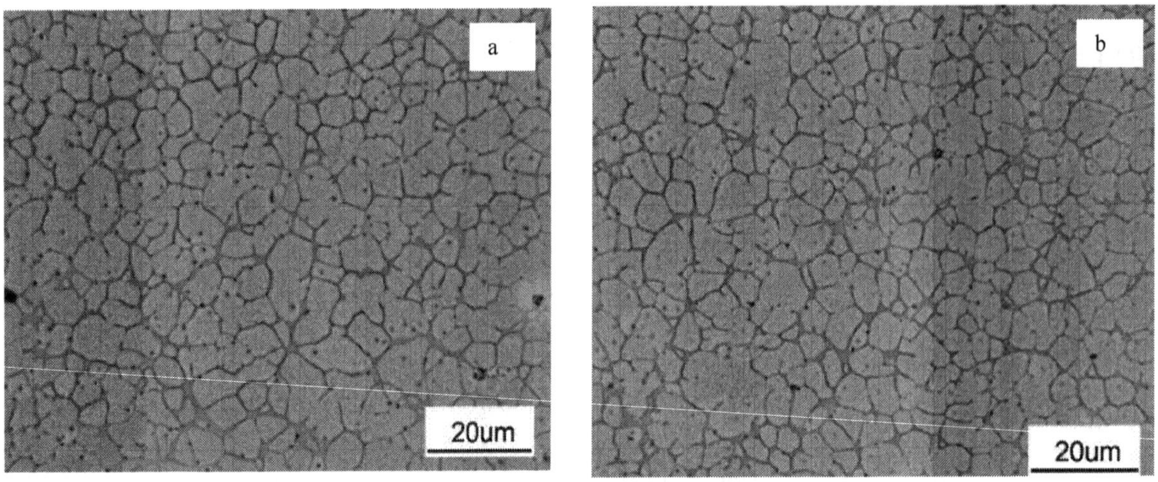

Figure 3: Optical micrographs of the two alloys. (a) Mg-5%Al-3.1%Ca-0.07%Sr and (b) Mg-5.7%Al-3.1%Ca-0.15%Sr.

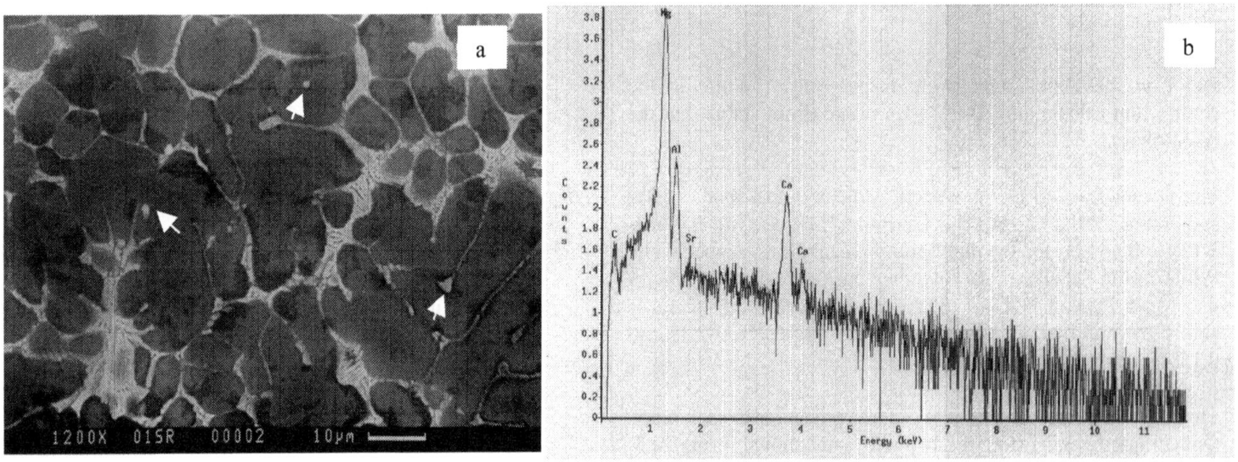

Figure 4: The SEM micrograph (a) and the EDS analysis of the intermetallic phase (b) of the Mg-5.7%Al-3.1%Ca-0.15%Sr alloy.

calculated values is about 4%. This discrepancy can be due to several reasons. First, the thermodynamic database is a simple combination of all binary system as the key ternary systems are still under development. Secondly, there are many degenerated eutectic structure as shown by arrows in Fig. 4 where the hcp phase in the eutectic structure is part of the hcp matrix. For example, assuming 20% of the hcp phase in the eutectic structure becomes part of the hcp matrix, the calculated fraction of the eutectic structure would be reduced by about 4%. Therefore, it is desirable to experimentally measure the amount of the intermetallic phases only so a more direct comparison can be made.

Therefore, the agreement between the calculated and experimental volume fractions of the eutectic structure can be considered satisfactory. Furthermore, the EDS analysis indicates that the intermetallic compound is rich in Al and Ca, indicating the existence of the Al_2Ca phase in accordance with the calculation. The significant Mg peak in the EDS result is probably from the hcp phase in either the matrix or the eutectic structure.

Table 2. Calculated and Experimental Results on Eutectic Fractions in Two Die Cast Alloys

Alloy	Equilibrium Simulation		Scheil Simulation		Experi-mental
	Wt.%	Vol.%	Wt.%	Vol.%	Vol. %
Mg-5%Al-3.1%Ca-0.07%Sr	24.8	19.5	26.9	21.1	17.4±1.5
Mg-5%Al-3.1%Ca-0.07%Sr	26.5	21.2	29.1	23.0	19.0±1.3

In the future work, the details of the structure and composition of the intermetallic compounds in the eutectic structure will be examined by TEM equipped with an EDS system to determine the solubility of Mg in the Al_2Ca phase and Al in the Mg_2Ca phase. With these data, the addition of the third component into the Al_2Ca and Mg_2Ca phases will be considered. Together with available data of various ternary systems in the literature, a more complete thermodynamic database of the Mg-Al-Ca-Sr system will be developed to better describe the phase equilibria in the alloys.

SUMMARY

The thermodynamic description of the Mg-Al-Ca-Sr has been developed with the data of the six binary in the present work. Using this database, various calculations have been made to understand the phase relationships during the die-casting process and the effect of the Sr content on the microstructures. The calculated results and experimental observations carried out in the present work are in reasonable agreement. More work is underway to refine the thermodynamic database and experimentally determine the structure and composition of the intermetallic phases.

ACKNOWLEDGEMENTS

This work is supported by the NSF CAREER Award under the grant DMR-9983532, the Scholarship from The Ministry of National Education of Turkey for Mr. Ozturk. The Thermo-Calc program used for all calculations is licensed from The Foundation for Computational Thermodynamics, Stockholm, Sweden. Mr. Bob Kubic of GM R&D is acknowledged for his assistance in experimental work.

REFERENCES

1. G. Craig, Proceeding of the 55th Conference of the International Magnesium Associate: A Global Vision for Magnesium, (International Magnesium Association, 1998), pp. 19-23.
2. H. Friedrich and S. Schumann, Proceeding of the 55th Conference of the International Magnesium Associate: A Global Vision for Magnesium, (International Magnesium Association, 1998), pp. 1-7.
3. A. Luo, Magnesium Technology 2000, H. I. Kaplan, J. N. Hryn, B. B. Clow, Eds., (TMS, PA, USA, 2000), pp. 89-98.
4. K. E. Mann, Giesserei, Vol.44, 1957, pp. 301.
5. G. B. Olson, Science, Vol.277, 1997, pp. 1237-1242.
6. K. Ozturk, Y. Zhong, Z. K. Liu and A. A. Luo, Magnesium Technology 2001, J. N. Hryn, Eds., (TMS, PA, USA, 2001), pp. 113-117.
7. L. Kaufman and H. Bernstein, Computer Calculation of Phase Diagrams with Special Reference to Refractory Metals, (Academic Press, New York, 1970).
8. N. Saunders and A. P. Miodownik, CALPHAD (Calculation of Phase Diagrams): A Comprehensive Guide, (Pergamon, Oxford ; New York, 1998).
9. Z. K. Liu, Magnesium Technology 2000, H. I. Kaplan, J. N. Hryn, B. B. Clow, Eds., (TMS, PA, USA, 2000), pp. 191-198.
10. A. T. Dinsdale, CALPHAD, Vol.15, 1991, pp. 317-425.
11. R. Agarwal, J. J. Lee, H. L. Lukas and F. Sommer, Z. Metallkd., Vol.86, 1995, pp. 103-108.
12. P. Liang, H. L. Su, P. Donnadieu, M. G. Harmelin, A. Quivy, P. Ochin, G. Effenberg, H. J. Seifert, H. L. Lukas and F. Aldinger, Z. Metallkd., Vol.89, 1998, pp. 536-540.

NEW PHASES IN MG-AL-CA SYSTEM

Yu Zhong[1], Alan Luo[2], J.F. Nie[3], and Jorge O. Sofo[4], Zi-Kui Liu[1]

[1]Department of Materials Science and Engineering, Pennsylvania State University, University Park, PA 16802 USA

[2]General Motors Research and Development Center, Materials and Processes Lab 30500 Mound Road, Warren, MI 48090-9055, USA

[3]School of Physics and Materials Engineering, Monash University, Victoria 3800, Australia

[4]Department of Physics, Pennsylvania State University, University Park, PA 16802 USA

Abstract

Mg-Al-Ca alloys exhibit appealing creep strength at temperatures over 100°C. It is believed that it is due to the substitution of the γ-$Al_{12}Mg_{17}$ phase by laves phases of higher thermal stability. In the literature, the C15-Al_2Ca phase in the Al-Ca binary system and the C14-Mg_2Ca phase in the Ca-Mg binary system have been reported. Their relative stability in the ternary system may play an important role in the development of the Mg-Al-Ca alloys. Furthermore, recent experiments and first-principles calculations in the literature revealed the existence of a C36 laves phase in the Al_2Ca-Mg_2Ca pseudo-binary system in addition to the C14-Mg_2Ca and C15-Al_2Ca laves phases. In the present work, special quasirandom structures (SQS) for all three laves phases were constructed. The structures possess local pair and multisite correlation functions that mimic those of the corresponding random structures. First-principles calculations were carried out based on the SQS developed to predict the enthalpy of formation in the Al_2Ca-Mg_2Ca pseudo-binary system. It was observed that the enthalpy of formation of C36 is very close to that of C14 at the Mg_2Ca end and decreases with the addition of small amount of Al, while the enthalpy of formation of C14 increases with the addition of Al. It is thus energetically plausible that C36 is stable in the Al_2Ca-Mg_2Ca pseudo-binary system. Experimental investigations were done using diffusion couple of Mg-Ca alloy and pure Al. EPMA analysis of the diffusion zone found the existence of two ternary phases $(Mg,Al)_2Ca$ and $Al_2(Mg,Ca)$.

Introduction

The addition of Ca has been found to improve the creep resistance of Mg-Al based alloys [1], owing to the replacement of the $Mg_{17}Al_{12}$ phase by more stable laves phases [2, 3]. Most literature data indicated the existence of two laves phases in the Mg-Al-Ca system[4-6], i.e. C14-Mg_2Ca and C15-Al_2Ca, except a recent report by Amerioun et al. [7] showing the C36 phase in the Mg_2Ca-Al_2Ca pseudo-binary system through experimental study and first-principles calculations.

Amerioun et al. [7] experimentally studied the phase equilibria of $CaAl_{2-x}Mg_x$ Alloys (nominal compositions x=0.33, 0.67, 0.75, 1.0, 1.25, 1.33, 1.5, 1.67) by heating stoichiometric amounts of Al_2Ca and Mg_2Ca to 700°C for 10 days and slowly cooling at a rate of 20°C/hr to room temperature. The alloy compositions were analyzed using EDS (Energy-Disperse Spectroscopy) method. Up to 10 at.% Mg loss were revealed by EDS analyses. The solubility range for the C14 phase was found to be x>1.51. The solubility range of the C36 phase was from x=0.67 to x=1.5. The solubility range of the C15 phase was with x<0.24. However, because their experiments are not done in equilibrium condition, the existence of C36 is still not clear, and the composition ranges of these three laves phases are still in question. They also calculated the stability of the laves phases in the Al_2Ca-Mg_2Ca pseudo-binary system using VCA (Virtual-Crystal Approximation). A total energy difference as a function of composition diagram is drawn to discuss the stability of these three phases at 0 K. Further investigation of the stability of these three phases are still needed, because of the less accurate VCA method used, the limited k-points, and the potential files with only 2 valences considered for Mg and Ca.

More recently, Grobner et al.[8] investigated the Mg-Al-Ca phase equilibria with X-ray powder diffraction analysis, metallographic analysis, SEM/EDX/WDX and DTA. 25 samples with different compositions were prepared. The results show large solubility range for C14 laves phase and C15 laves phase, up to 23 at.%Al and 6 at.%Mg, respectively. However, because only one sample, Al20Ca20Mg60, was sealed in silica tubes and annealed at 573K for 3 week before phase identification, other samples (as cast, from arc-melting, rapidly cooled) cannot provide accurate equilibria information at 573K. The phase identification of the sample Al20Ca20Mg60 shows that it lies in the C14+C15+Mg-hcp three phase region. As X-ray diffraction patterns of C14 and C36 laves phases are very close, it is very difficult to distinguish C14 and C36 laves phases by X-ray refinement. Especially, the author didn't consider the possibility of the existence of C36 in their X-ray refinement; it is thus inconclusive to exclude the existence of C36 in their samples.

In our previous investigation[9], two sublattice model $(Mg, Al)_2(Ca)_1$ is chosen for these three laves phases, C14, C15, and C36. The random mixing of Mg and Al were considered in the first sublattice. It showed the first-principles calculation results for these three laves phase with one unit cell. Supercells were also constructed for the simplest structure C15. Due to the complex crystal structures, C14 and C36 have much more atoms in one unit cell and much more sites in the first sublattice than C15. So there are so many different supercells in the pseudo-binary, which can't be constructed duo to the limit of current computational power.

On the other hand, special quasirandom structures (SQS) proposed by Zunger et al. [10-12] were extensively used to calculate binary random alloys. They are specially designed *small-unit-cell* periodic structures with only a few (2–32) atoms per unit cell, which closely mimic the most relevant, near-neighbor pair and multisite correlation functions of the random substitutional solutions. A single DFT calculation of an SQS can give many important alloy properties [10-13] (e.g., equilibrium bond lengths, charge transfer, formation enthalpies, etc.) which depend on the existence of those distinct local environments. It has been largely used in binary systems.

In the present work, first-principles calculations and experiments were carried out to investigate the phase equilibria in the Mg-Al-Ca system and compared with the thermodynamic calculations based on the CALPHAD approach [14] using Thermo-Calc[15]. SQS for the C15 phase $(Mg_x,Al_{1-x})_2Ca$ at compositions x=0.25, 0.50, and 0.75 and for the C14 and C36 phases at composition x=0.50 were constructed, respectively. The first-principles calculation results using those SQS are compared with the results from one unit cell and supercell in terms of the enthalpy of formation of the three laves phases in the Mg-Al-Ca system. Diffusion couple of a Mg-Ca alloy and pure Al was made to examine the existence of C36 phase, its solubility range and composition of other phases. The current project was initiated with the interactions between the Pennsylvania State University and General Motors and supported by the National Science Foundation.

Special Quasirandom Structures

For a binary $A_{1-x}B_x$ substitutional alloy, the configurationally-dependent properties (such as the energy) can be characterized very efficiently by a "lattice algebra": [10, 11, 16, 17]. Pseudo-spin variables are assigned to each site, S_i=-1(+1) if an A (B) atom sits at site *i*. We further define geometric figures, *f*, symmetry-related groupings of lattice sites, e.g., single site, nearest-neighbor pair, three-body figures, etc. These figures, *f*=(*k*,*m*), can have *k* vertices and span a maximum distance of *m* (*m*=1,2,3..., are the first, second and third-nearest neighbors, etc.). By taking the product of the spin variables over all sites of a figure and averaging over all symmetry-equivalent figures of the lattice, the correlation functions $\overline{\Pi}_{k,m}$ [10, 11] can be obtained. For the perfectly random $A_{1-x}B_x$ alloys, there is no correlation in the occupation between various sites, and therefore $\overline{\Pi}_{k,m}$ simply becomes the product of the lattice-averaged site variable, which is related to the composition by $\langle S_i \rangle = 2x-1$. Thus, for the perfectly random alloy, the pair and multisite correlation functions $\overline{\Pi}_{k,m}$ are given quite simply as $\overline{\Pi}_{k,m} = (2x-1)^k$.

The SQS approach is to find small ordered structures that possess $(\overline{\Pi}_{k,m})_{SQS} \cong \langle \overline{\Pi}_{k,m} \rangle_R$ for as many figures as possible. Admittedly, describing random alloys by small unit-cell periodically-repeated structures will surely introduce erroneous correlations beyond a certain distance. However, since interactions between nearest neighbors are generally more important than interactions between more distant neighbors, we can construct SQS that exactly reproduce the correlation functions of a random alloy between the first few nearest neighbors, deferring errors due to periodicity to more distant neighbors.

Table 1. Pair correlation functions of SQS-*N* structures of laves phases.

Phase	Composition		$\overline{\Pi}_{2,1}$	$\overline{\Pi}_{2,2}$	$\overline{\Pi}_{2,3}$	$\overline{\Pi}_{2,4}$	$\overline{\Pi}_{2,5}$	$\overline{\Pi}_{2,6}$	$\overline{\Pi}_{2,7}$	$\overline{\Pi}_{2,8}$	$\overline{\Pi}_{2,9}$
C15	$(Mg_{0.5},Al_{0.5})_2Ca$	SQS-24	0	0	0	0	0	0	0	-0.1667	0.3333
		Random	0	0	0	0	0	0	0	0	0
	$(Mg_{0.5},Al_{0.5})_2Ca$/ $(Mg_{0.5},Al_{0.5})_2Ca$	SQS-24	0.25	0.25	0.3333	0.25	0.0833	0.25	0.25	0.1667	0.4167
		Random	0.25	0.25	0.25	0.25	0.25	0.25	0.25	0.25	0.25
C14	$(Mg_{0.5},Al_{0.5})_2Ca$	SQS-48	0	0	0	0	0	0	0	0	0.3333
		Random	0	0	0	0	0	0	0	0	0
C36	$(Mg_{0.5},Al_{0.5})_2Ca$	SQS-48	0	0	0	0	0	0	0	0	0
		Random	0	0	0	0	0	0	0	0	0

First-principles calculations

Ultrasoft pseudopotentials were used in the first-principles calculations of total energies[18, 19], as implemented in the Vienna ab initio Simulation Package (VASP) [20-23]. The generalized gradient approximation (GGA) [24, 25] calculations were employed.. Extensive tests of k-point sampling indicated that total energy differences were converged to within ~0.1 kJ/mole.

First-principles calculations were carried out for binary compounds (Al_2Ca and Mg_2Ca) and ternary phases in the Al_2Ca and Mg_2Ca pseudo-binary system, and also for pure elements, Mg, Al and Ca that will be published in somewhere else. C15 has a face centered cubic structure with 6 atoms in a unit cell, and C14 and C36 have hexagonal structures with 12 and 24 atoms in a unit cell, respectively.

Experimental Procedures

Mg-Ca alloy with 30wt.% Ca was purchased from Timminco Limited. The diffusion couple was prepared using Mg-Ca alloy with pure Al (99.99%). Small pieces of each were cut and shaped into discs approximately 0.5 inches in diameter and 0.125-0.25 inches thick. One side of each disc was polished down to 0.05 μm alumina. One aluminum disc and one Mg-Ca alloy disc were placed inside a stainless steel tubular clamp [26], with the polished faces of the discs facing each other. The clamp was sealed in a Pyrex tube filled with helium at one-third atmosphere pressure to prevent oxidation during the heat treatment. The tube was placed in a furnace and isothermally heat treated for 2 weeks at 688K and then quenched in water.

An EPMA analysis was performed on the diffusion couple to determine the composition of phases across the diffusion zone by using Cameca SX50 and a PET crystal and K-alpha X-Ray line for Calcium and a TAP crystal and K-alpha X-Ray lines for both magnesium and aluminum. The beam current was 12nA and the voltage was 15.0 kV.

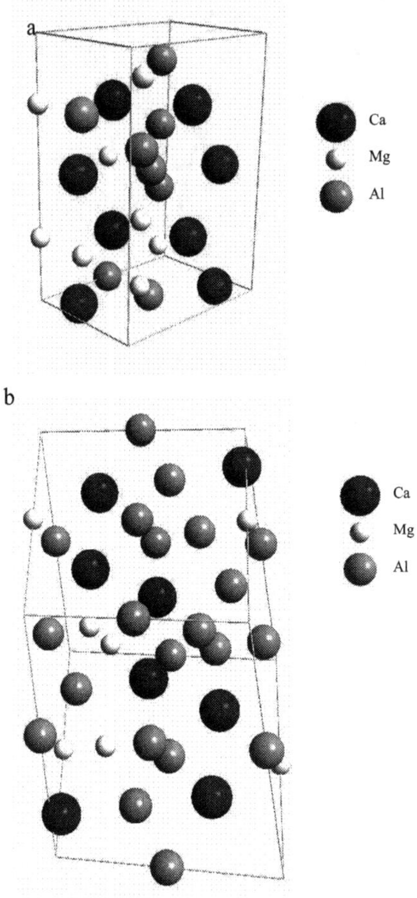

Figure 1: Crystal structure of the SQS-24 structures in their ideal, unrelaxed forms for laves C15 phase (a) $(Mg_{0.5},Al_{0.5})_2Ca$ and (b) $(Mg_{0.25},Al_{0.75})_2Ca$, respectively.

Results and Discussion

Two-sublattice model $(Mg, Al)_2(Ca)_1$ is chosen for these three laves phases, C14, C15, and C36. The random mixing of Mg and Al in the first sublattice is considered and simulated by means of the SQS approach. In the present study, various SQS for C15 at compositions 0.25, 0.50 and 0.75 and C14 and C36 at composition 0.50 were generated using the *gensqs* code in the Alloy-Theoretic Automated Toolkit (ATAT) [27]. The atomic arrangements in the SQS are given in Figure 1 to Figure 3. Similarly to constructing supercells manually, C15 is the simplest structure among the three laves phases and thus easy to find SQS's satisfying our criterion. SQS for these three laves phases at composition x=0.50 were obtained by requiring that its pair correlation functions be identical to those of the random alloy up to the seventh-nearest neighbor. However, for composition 0.75, no SQS satisfy this criterion. Therefore, a SQS-24 structure was chosen which has the best pair correlation functions (correlation functions are identical to random alloys up to the second-nearest neighbor). SQS-48 structures for x=0.75 were constructed to fit our criterion. The pair correlation functions of the SQS-N structures are compared with those of the corresponding random alloys in Table 1.

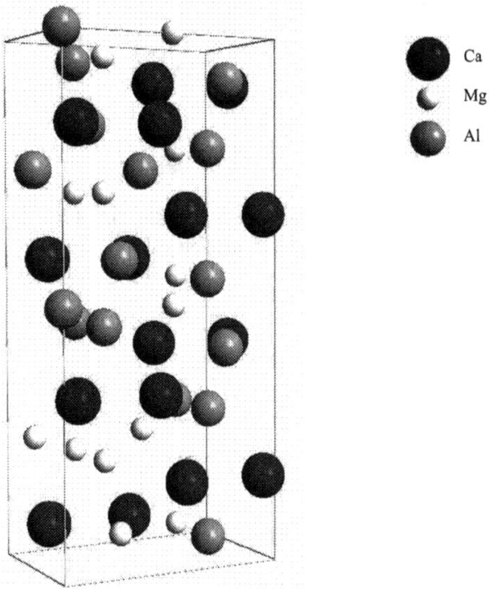

Figure 2: Crystal structure of the SQS-48 structures in their ideal, unrelaxed forms for laves C14 phase $(Mg_{0.5},Al_{0.5})_2Ca$.

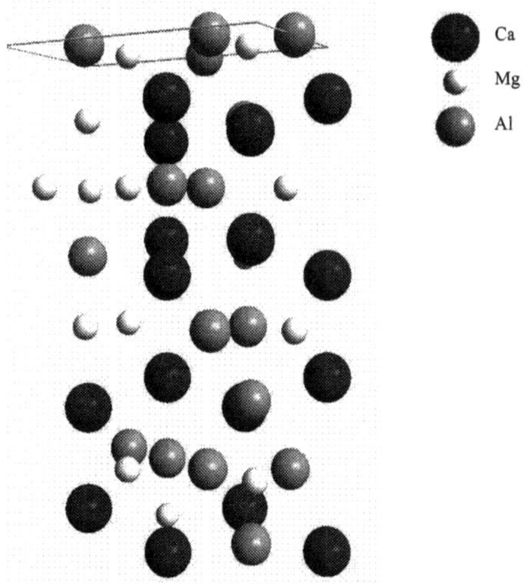

Figure 3: Crystal structure of the SQS-48 structures in their ideal, unrelaxed forms for laves C36 phase $(Mg_{0.5},Al_{0.5})_2Ca$.

First-principles calculation results for C15 with one unit cell, supercell, and SQS are shown in Figure 4. Our SQS calculation results are between those of one unit cell and supercell. As SQS better simulate the random mixing of the Mg and Ca atoms in the first sublattice, it is believed the first-principles calculation results from SQS represent the reality better. Similar calculations were also done for C14 and C36 at the $(Mg_{0.5},Al_{0.5})_2Ca$ composition.

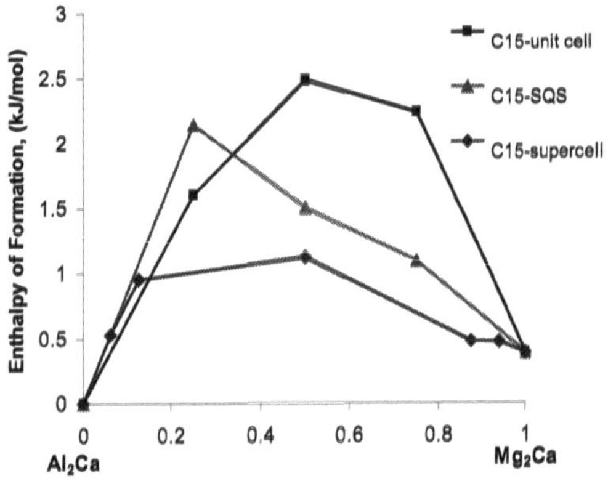

Figure 4: First-principles calculation results for C15 laves phases with one unit cell, supercell, and SQS.

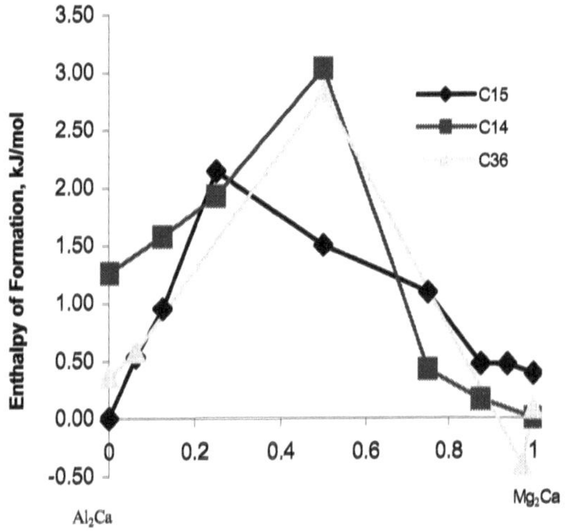

Figure 5: First-principles calculation results from one unit cell, supercell, and coupled with SQS structures for C15, C14, and C36 laves phases.

Figure 5 shows the first-principles calculation results for C15, C14, and C36 from one unit cell, supercell, and coupled with SQS. It can be seen that C15-Al_2Ca and C14-Mg_2Ca are the stable phases, and C36 is a very competitive phase. Especially for the Mg_2Ca end-member, the difference between C36 and C14 is very small (about 80J/mol), within the first-principles error range. It's thus hard to tell which phase is more stable at 0K. Moreover, a very small entropy difference can affect the stability of these two phases at higher temperatures. Nevertheless, it is clear that C36 has a very similar stability to C14 at Mg_2Ca side of the pseudo-binary system at 0K. The results show C15 and C14 are not stable in the middle, and C36 is stable in the Mg_2Ca rich side. More first-principles calculations and experiments are needed to refine the stability range of the C36 phase.

Figure 6 shows backscattering image for the cross section of Mg-Ca alloy and pure Al diffusion couple. According to the Z contrast of this image, the dark region is Mg-hcp phase, as Z of Mg is 12. The lightest region contains highest amount of Ca as Z of Ca is 20. EPMA analyses were done to identify phases involved, such as β phase, γ phase, Al_2Ca phase, Mg_2Ca phase, and Mg-hcp phase.

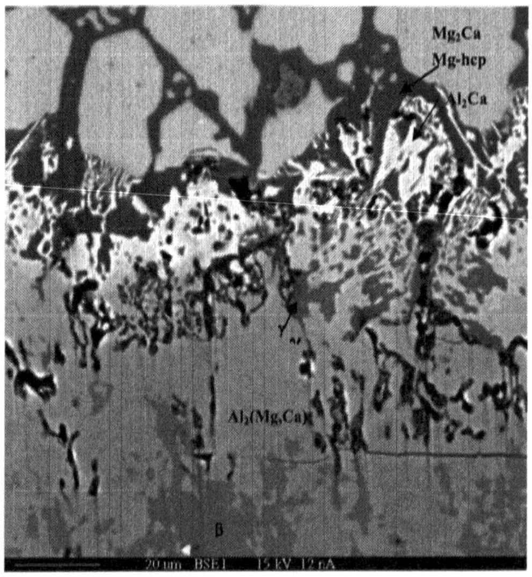

Figure 6. Backscattering Image for the cross section of Mg-Ca alloy and pure Al diffusion couple with the new phase $Al_2(Mg,Ca)$.

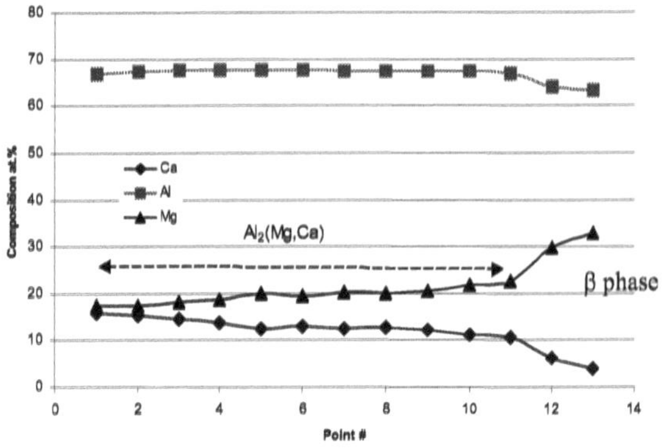

Figure 7. Detail EPMA analysis for the composition profile for $Al_2(Mg,Ca)$.

A broad region above the β phase region was identified as a new phase $Al_2(Mg,Ca)$ as indicated in Figure 6. Detailed EPMA analysis was carried out with points starting from the $Al_2(Mg,Ca)$ phase region and the final point in the β phase region. The composition profile for this phase was shown in Figure 7. It is observed that this phase contains about 66.7at.% Al, while Mg concentration vary from 17.3at.% to 22.5at.%. $Al_2(Mg,Ca)$ phase was labeled in the isothermal section of Mg-Al-Ca ternary system as 688K in Figure 10. The crystal structure of this phase and the stability of this phase will be investigated in the future.

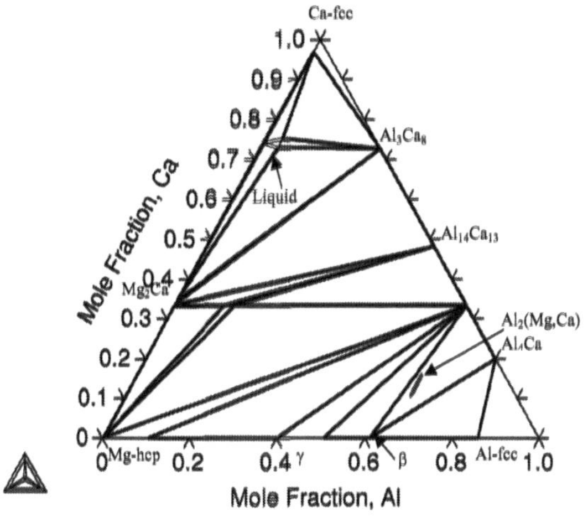

Figure 8. Isothermal section of Mg-Al-Ca ternary system at 688K.

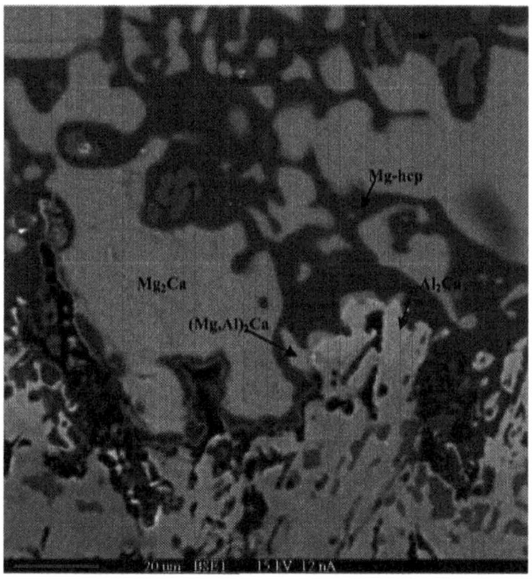

Figure 9. Backscattering Image for the cross section of Mg-Ca alloy and pure Al diffusion couple with the ternary phase $(Mg,Al)_2Ca$.

Small $(Mg,Al)_2Ca$ phase region is found between Mg-hcp and Al_2Ca phase region. By using EPMA, the composition of one point in the $(Mg,Al)_2Ca$ phase region was found to be $Mg_{0.536}Al_{0.161}Ca_{0.303}$ as shown in Figure 9. The composition of another point is $Mg_{0.301}Al_{0.370}Ca_{0.329}$ as shown in Figure 8. This phase can be C14 laves phase as well as C36 laves phase. However, The composition of second point is far from the solubility range of C14 laves phase (23 at.% Al) as claimed in [8] and is in the solubility range of C36 laves phase (from 33 at.% to 67 at.% Al) as claimed in [7]. This phase thus has high possibility to be C36. More investigation is underway to identify the crystal structure of this phase.

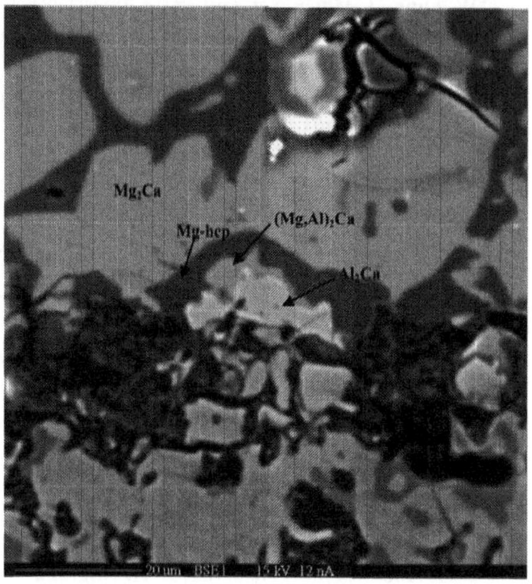

Figure 10. Backscattering Image for the cross section of Mg-Ca alloy and pure Al diffusion couple with the ternary phase $(Mg,Al)_2Ca$.

Summary

The first-principles calculations from SQS coupled with the results from one unit cell and supercell provide us the enthalpy of formation information of the three laves phases: C15, C14, and C36 in the Al_2Ca-Mg_2Ca pseudo-binary system. This is useful for further investigation of the interaction parameters in the first sublattice for these three laves phases in Mg-Al-Ca ternary thermodynamic database.

The EPMA analysis of the diffusion couple shows the possible existence of the C36 laves phase in the Al_2Ca-Mg_2Ca pseudo-binary system, and the existence of a new ternary phase $Al_2(Mg,Ca)$. Further investigation is under way to identify the crystal structure of these two phases.

Acknowledgement

This work is supported by the NSF CAREER Award under the grant DMR-9983532.

Reference:

1. A. A. Luo, M. P. Balogh and B. R. Powell, *Metall. Mater. Trans. A*, 33, (2002) 567-574.
2. M. O. Pekguleryuz and E. Baril, Minerals, Metals and Materials Society/AIME, Magnesium Technology, New Orleans, LA, USA (2001), pp. 119-125.
3. A. Luo, M. P. Balogh and B. R. Powell, "Tensile creep and microstructure of magnesium-aluminum-calcium alloys for powertrain applications: Part 2 of 2" *SAE2001* (Detroit, MI, 2001).
4. J. A. Catterall and R. J. Pleasance, *J. Inst. Metals,* 86, (1957-58) 189-192.

5. R. Ninomiya, T. Ojiro and K. Kubota, *Acta Metall. Mater.*, 43, (1995) 669-674.
6. M. Liu, Q. Wang, Z. Liu, G. Yuan, G. Wu, Y. Zhu and W. Ding, *J. Mater. Sci. Lett.*, 21, (2002) 1281-1283.
7. S. Amerioun, S. I. Simak and U. Haussermann, *Inorganic Chemistry*, 42, (2003) 1467-1474.
8. J. Grobner, D. Kevorkov, I. Chumak and R. Schmid-Fetzer, *Z. Metallkd.*, 94, (2003) 976-982.
9. Y. Zhong, A. Luo, J. Sofo and Z. K. Liu, A. Luo, Ed., Magnesium Technology 2004, Charlotte, NC, USA (2004), pp. 317-323.
10. A. Zunger, S. H. Wei, L. G. Ferreira and J. E. Bernard, *Phys. Rev. Lett.*, 65, (1990) 353-356.
11. S. H. Wei, L. G. Ferreira, J. E. Bernard and A. Zunger, *Phys. Rev. B*, 42, (1990) 9622-9649.
12. K. C. Hass, L. C. Davis and A. Zunger, *Phys. Rev. B*, 42, (1990) 3757-3760.
13. C. Jiang, C. Wolverton, J. Sofo, L.-Q. Chen and Z.-K. Liu, *Phys. Rev. B*, 69, (2004) 214202.
14. N. Saunders and A. P. Miodownik, *CALPHAD (Calculation of Phase Diagrams): A Comprehensive Guide* (Pergamon, Oxford, New York, 1998).
15. J. O. Andersson, T. Helander, L. H. Hoglund, P. F. Shi and B. Sundman, *CALPHAD*, 26, (2002) 273-312.
16. Z. W. Lu, S. H. Wei and A. Zunger, *Phys. Rev. B*, 44, (1991) 3387-3390.
17. Z. W. Lu, S. H. Wei and A. Zunger, *Phys. Rev. B*, 45, (1992) 10314-10330.
18. D. Vanderbilt, *Phys. Rev. B*, 41, (1990) 7892-7895.
19. G. Kresse and J. Hafner, *J. Phys.-Condes. Matter*, 6, (1994) 8245-8257.
20. G. Kresse and J. Hafner, *Phys. Rev. B*, 47, (1993) 558-561.
21. G. Kresse, Thesis, Technische Universitat Wien (1993).
22. G. Kresse and J. Furthmuller, *Phys. Rev. B*, 54, (1996) 11169-11186.
23. G. Kresse and J. Furthmuller, *Comput. Mater. Sci.*, 6, (1996) 15-50.
24. J. P. Perdew, J. A. Chevary, S. H. Vosko, K. A. Jackson, M. R. Pederson, D. J. Singh and C. Fiolhais, *Phys. Rev. B*, 46, (1992) 6671-6687.
25. J. P. Perdew, J. A. Chevary, S. H. Vosko, K. A. Jackson, M. R. Pederson, D. J. Singh and C. Fiolhais, *Phys. Rev. B*, 48, (1993) 4978-4978.
26. S. Kim and Y. A. Chang, *Metall. Mater. Trans. A*, 31, (2000) 1519-1524.
27. A. van de Walle, M. Asta and G. Ceder, *CALPHAD*, 26, (2002) 539-553.

THERMODYNAMIC DATABASE OF Mg-Al-Ca-Sr: A RESOURCE FOR ALLOY DEVELOPMENT AND IMPROVEMENT

Hongbo Cao, Jun Zhu, Chuan Zhang, Y Austin Chang
Department of Materials Science and Engineering, University of Wisconsin-Madison, 1509 University Avenue, Madison, WI 53706, USA

Keywords: Magnesium Alloys, Thermodynamics, Scheil Simulation, Phase Diagram

Abstract

A thermodynamic database of Mg-Al-Ca-Sr, a basic quaternary for a number of commercial magnesium alloys such as AXJ530 (Mg-4.5Al-3.0Ca-0.14Sr, weight percent hereafter, unless noted otherwise) was developed. Calculated isopleth from Mg-5Al-3Ca to Mg-5Al-3Sr agrees with experimental data reported in the literature. Moreover, using this database and a simple kinetic model proposed by Scheil, the calculated solidification paths of several alloys are likewise in accord with experimental data. Thus, the availability of this database enables one to exploit new alloy compositions for promising properties which can be rapidly validated with few experiments.

Introduction

Magnesium alloys with the lowest density among the structural metals has innate advantage for the application in the transportation industry when conservation of the energy consumption and minimization of the environmental degradation become crucial. However, as compared to most aluminum and ferrous alloys, the limited creep resistance hinders the application of Mg alloys in the powertrain components, such as engine blocks and transmission cases, where the service temperature can be as high as 175°C~250°C. Therefore, during the last decade, there has been a great deal of effort to develop creep-resist Mg alloys. It was found that the additions of Ca and Sr or both to Mg-Al based alloys can significantly improve the creep resistance [1-4], such as AJ62 and AXJ530 with a base composition of Mg-6Al-2Sr and Mg-4.5Al-3Ca-0.14Sr respectively. Therein, Ca and Sr are two promising elements not only because of their low density and low cost, but also improving their properties for practical applications. However, the thermodynamic database available in the literature was obtained from an extrapolation from those of the constituent binaries [5]. It is known that such a thermodynamic database is not reliable [6]. For instance the ternary C36 Laves phase is not included in that database for Mg-Al-Ca.

In the present study, a thermodynamic description of the quaternary Mg-Al-Ca-Sr system was developed. This database focuses primarily on the Mg-rich phase equilibria. An isopleth from Mg-5Al-3Ca-0Sr to Mg-5Al-0Ca-3Sr, was calculated, which is in good agreement with the experimental data reported in the literature. Moreover, using this database and a simple kinetic model proposed by Scheil, the calculated solidification paths of several alloys are likewise in accord with experimental data. Thus, the availability of this database enables one to exploit new alloy compositions for promising properties which can be rapidly validated with few experiments.

Thermodynamic Modeling

In Mg-Al-Ca-Sr quaternary system, there are four constituent ternaries. The two key ternaries, Mg-Al-Ca [7] and Mg-Al-Sr [8], containing both Mg and Al, have been experimentally investigated and thermodynamically modeled in our previous work. For Mg-Al-Ca ternary system, we have obtained a thermodynamic description, which is in excellent agreement with our own experimental data, as well as data in the literature [9, 10]. Using this description, coupled with Scheil simulation, we have successfully proved the ternary eutectic reaction in the Mg-rich corner with clear experimental evidences obtained with the help of thermodynamic calculations. [7]. In addition, two saddle points (temperature maximum) were predicted thermodynamically and confirmed experimentally [11]. For the Mg-Al-Sr ternary system, a thermodynamic description was obtained by us based on the experimental data from three designated alloys [8]. It is also in good agreement with the experimental data from a subsequent study [12], though we did not use their data for optimization of model parameters. The thermodynamic description of Mg-Ca-Sr was obtained by extrapolation of the constituent binaries [13]. Zhong et al [14] performed first principles calculations on the end members of the Laves C14 phase, $Mg_2(Ca,Sr)$, by assuming binary solubilities of C14-Mg_2Ca and C14-Mg_2Sr. However there is no experimental data to support their assumption. Therefore, in the present study, C14 was still treated as a line compound in both Mg-Ca and Mg-Sr binaries.

For the non-Mg containing Al-Ca-Sr, since the current alloys development is centered in the phase equilibria of Mg-rich region and experimental data are very sparse in the literature, the thermodynamic description was obtained by extrapolation. Once the thermodynamic descriptions of the four constituent ternary systems are obtained, the thermodynamic description of the Mg-Al-Ca-Sr quaternary can be readily achieved by extrapolation from the descriptions of the four constituent ternaries. In most cases, the extrapolation usually works quite well, and the only exception is that in some rare cases when quaternary phases do exist in the system, where optimization of the model parameters for the quaternary phases in question is necessary [6]. It is worth while to note that since the descriptions of Mg-Ca-Sr and Al-Ca-Sr are obtained by extrapolation, the description of this quaternary may only give the topological features when going to higher Ca and Sr contents. In this study, all the phase diagram calculations were performed using the software Pandat [15].

Results and Discussion

Different from binary and ternary systems, it is difficult to directly visualize the phase equilibria and phase transformation information for a quaternary system. A vertical section of the quaternary system, i.e., an isopleth is usually used to view the phase equilibria relations. Figure 1 shows the calculated isopleth with constant Mg and Al content (92 wt.% Mg, 5 wt.% Al) and Sr contents varying from 0-3 wt.%. Also shown in this diagram, the reported data from [16] on the liquidus and solidus temperatures of seven Mg-Al-Ca-Sr quaternary alloys detected by thermal analysis. The squares denote the liquids temperatures, while the circles correspond to the solidus temperatures. It is clear that there is excellent agreement between our calculation and the experimentally measured temperatures. With the increasing Sr concentration, the liquidus temperatures increase slightly, while the corresponding solidus temperatures decrease initially, reach a minimum and then increase again.

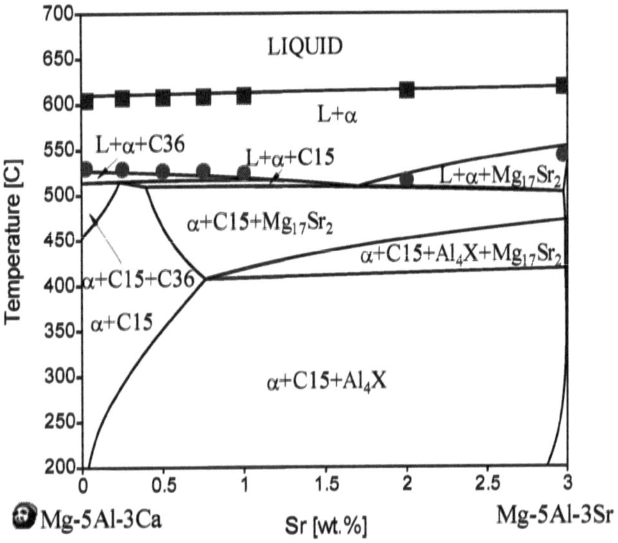

Figure 1. Calculated isopleth (temperature-composition) of the Mg-Al-Ca-Sr quaternary system with 92 wt.% Mg and being 5 wt.% Al, Ca and Sr exchanging with each other. Also shown in this diagram, the reported data from [16] on the liquidus and solidus temperatures of seven Mg-Al-Ca-Sr quaternary alloys detected by thermal analysis during the solidification. The squares denote the liquids temperatures, while the circles correspond to the solidus temperatures.

As one can see that for alloys in this particular composition range, the preferable primary phase of solidification is the α(Mg) phase. This is also consistent with the microstructure observations by Suzuki et al [16], where all the seven cast alloys solidify with the primary phase of α(Mg). According to the calculation, when Sr concentration of an alloy is 0-1.16 wt.%, the second solidified phase is C36-$(Mg,Al)_2Ca$; when Sr content is 1.16-1.70 wt.%, the second phase is C15-Al_2Ca; when Sr content is higher than 1.70 wt.%, the second phase is $Mg_{17}Sr_2$. This is also in line with the identified intermetallic phases in the as-cast microstructure by Suzuki et al. According to [16], the major intermetallic compound observed at the primary α(Mg) dendrite cell boundary region changes from the Al-rich C36-$(Mg,Al)_2Ca$ phase to the Mg-rich $Mg_{17}Sr_2$ phase with increasing Sr concentration. In addition, the calculation also provides that several different solid phases will be stable at lower temperatures, such as α(Mg) + C15, α(Mg) + C15 + Al_4X, etc. Here it should be noted that the Al_4X actually denotes $Al_4(Ca,Sr)$, a combination of Al_4Ca and Al_4Sr. Since these two binary phases have exactly the same crystal structure, i.e. the body center tetragonal structure, it is reasonable and also necessary to combine these two into one phase in the quaternary thermodynamic description.

For all the seven alloys reported in [16], solidification paths have been calculated based on the Scheil model and the thermodynamic description obtained in the current study. The Scheil model, with the assumptions of complete mixing in the liquid but no diffusion in the solid, can be used with reasonable confidence to predict the phase formation sequence for substitutional alloys.

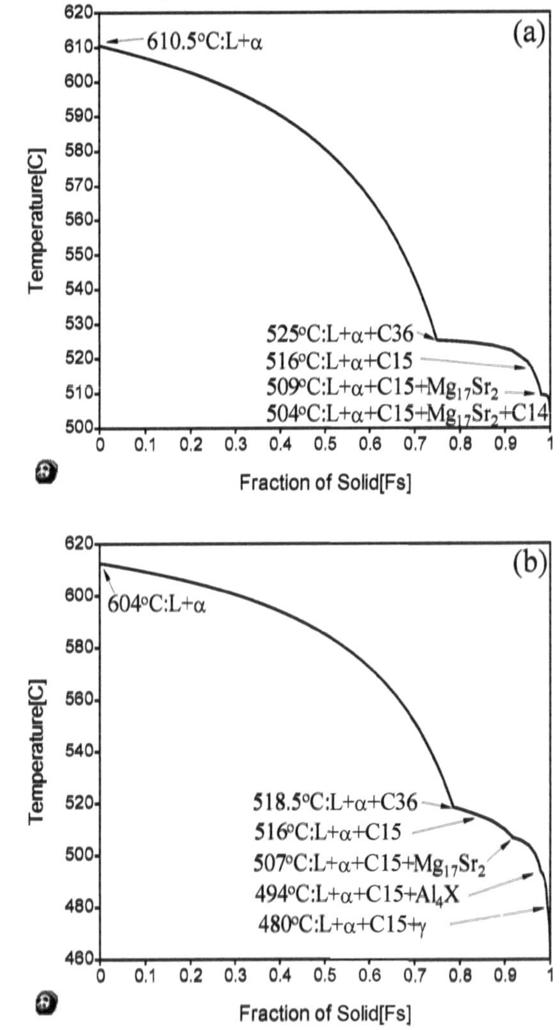

Figure 2. Calculated volume fraction of solid as a function of temperature under Scheil solidification condition for two typical quaternary alloys: (a) Mg-5Al-2.75Ca-0.25Sr, (b) Mg-5Al-2Ca-1Sr.

Figure 2 shows the calculated solidification paths for two typical Mg-Al-Ca-Sr quaternary alloys, (a) Mg-5Al-2.75Ca-0.25Sr, (b) Mg-5Al-2Ca-1Sr. According to the calculation, solidification consists of multiple stages. For both alloys the primary phase, i.e. the first phase to form from the liquid upon solidification, is α(Mg), usually a coarse dendrite. The second and third stages of solidification would be the eutectic-like α+C36 and α+C15-Al_2Ca. For a quaternary system, these phases do not freeze at a fixed temperature and generally are found surrounding the primary phase in the microstructure. The subsequent solidification stage $Mg_{17}Sr_2$ will form. After that, for alloy Mg-5Al-2.75Ca-0.25Sr, the solidification will end at the quaternary eutectic invariant reaction: $L \rightarrow \alpha + C15 + Mg_{17}Sr_2 + C14\text{-}Mg_2Ca$. While for alloy Mg-5Al-2Ca-1Sr, there will be negligible amount of Al_4X (about 0.2%) and γ-$Mg_{17}Al_{12}$ (about 0.4%) phase formed in the final solidification stage. From above discussion, it can be seen that there is a fairly good agreement between the calculated solidification paths and the experimentally observed microstructures from the ac-cast alloys [16]. The only exception is that the calculations predict that there will be small amount of C15 phase formed in both alloys, about 0.4 % and 3 % respectively. However, Suzuki et al did not report the existence of C15 in any of their as-cast microstructures. The possible reason, which might account for this, could be two folds. On the one hand, it is possible that there is indeed very small amount of C15 phase, but it is very difficult to detect experimentally in the as-cast alloys. On the other hand, it is also possible that solidification path calculation is not accurate enough due to the fact that the descriptions of the two constituent ternary systems were obtained by extrapolation, as mentioned early, or the assumption of no diffusion in solid for the Scheil model is not fully valid.

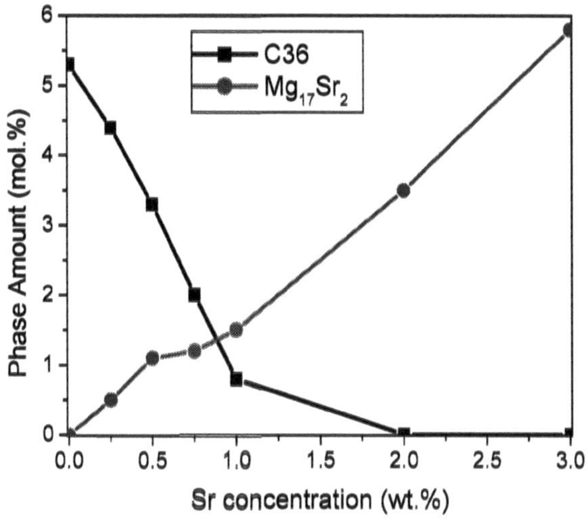

Figure 3. Phase amounts of C36 and $Mg_{17}Sr_2$ as a function of the Sr content from the Scheil calculation for the series of alloys with compositions of Mg-5Al-(3-x)Ca-xSr.

Figure 3 shows the amounts of the C36 and $Mg_{17}Sr_2$ phases as functions of Sr content calculated based on Scheil model and the Mg-Al-Ca-Sr thermodynamic description obtained in the present study. It is clear that in this series of alloys with compositions of Mg-5Al-(3-x)Ca-xSr, the amount of C36 decreases while that of $Mg_{17}Sr_2$ increases when Sr concentration increases. The exact same trend was observed by Suzuki et al [16].

Base on the above discussion, the good agreement between the calculation and the experimental data leads us to conclude that the current Mg-Al-Ca-Sr thermodynamic description can be used as a viable guidance for new alloys development or improvement of the existing alloys. In the future, it would be desirable to have the other two ternary systems, Mg-Ca-Sr and Al-Ca-Sr modeled thermodynamically, when some key thermochemical or phase equilibria data become available, rather than just extrapolation from the corresponding constituent binaries.

Summary

A thermodynamic database of Mg-Al-Ca-Sr was developed based on the descriptions of the constituent binaries and key ternaries. Using this description, the calculated isopleth with 92 wt.% Mg and 5 wt.% Al is in good accordance with the experimental data in the literature. It is very useful to understand the equilibrium phase relationships at low temperatures. The calculated solidification paths based on the Scheil model is also in line with the microstructure observations in the as-cast alloys. The good agreement between the calculation and the experimental data suggests that the present thermodynamic description is reliable for the Mg-rich region of Mg-Al-Ca-Sr system and can be considered as a useful tool for alloys development and improvement.

Acknowledgement

This research was supported by the National Science Foundation (FRG grant no. NSF-DMR-06-05662) and Wisconsin Distinguished Professorship. The authors wish to thank Professor R. Schmid-Fetzer of Clausthal University of Technology, Professor S. Kou of University of Wisconsin, Professor J. W. Jones and Professor T. M Pollock of University of Michigan for their helpful discussions.

References

[1] R. Ninomiya, T. Ojiro, and K. Kubota, "Improved heat resistance of Mg-Al alloys by the Ca addition", *Acta Metall Mater*, 43(2)(1995), 669-674.

[2] B.R. Powell et al., "Development of creep-resistant magnesium alloys for powertrain applications: Part 1 of 2". *Society of Automotive Engineers*, [Special Publication] SP(2001), SP-1622(Light Metal Applications for the Automotive Industry: Aluminum and Magnesium), 29-36.

[3] A.A. Luo, B.R. Powell, M.P. Balogh, "Creep and microstructure of magnesium-aluminum-calcium based alloys", *Metall Mater Trans A* 33(2002), 567-574.

[4] Y. Nakaura, A. Watanabe, K. Ohori, "Effects of Ca,Sr additions on properties of Mg-Al based alloys", *Mater Trans*, 47(4)(2006), 1031-1039.

[5] Y. Zhong et al., "Computational thermodynamics and experimental investigation of Mg-Al-Ca-Sr alloys", in *Magnesium Technology 2002* (Editors: Kaplan H.I.), TMS, 2002, 69-73.

[6] Y.A. Chang et al., "Phase diagram calculation: past, present and future", *Prog Mater Sci*, 49(2004), 313-345.

[7] H. Cao et al., "Experiment coupled with modeling to establish the Mg-rich phase equilibria of Mg-Al-Ca", *Acta Mater*, 2007, accepted.

[8] H. Cao et al., "Experimental investigation and thermodynamic modeling of the Mg-Al-rich region of the Mg-Al-Sr system", *Int J Mat Res (formerly Z Metallkd)*, 97(4)(2006), 422-428.

[9] A. Suzuki et al., "Solidification paths and eutectic intermetallic phases in Mg-Al-Ca ternary alloys", *Acta Mater*, 53(2005), 2823-2834.

[10] Schmid-Fetzer R. private communication, Clausthal University of Technology, Clausthal-Zellerfeld, Germany, 2006.

[11] H. Cao et al., "A computational/directional solidification method to establish saddle points on the Mg-Al-Ca liquidus", *Scripta Mater*, 2007, in press

[12] M.A. Parvez et al., "Experimental study of the ternary magnesium-aluminum-strontium system", *J Alloys Comp*, 402(2007), 170-185.

[13] Y.M. Muggianu, M. Gambino, J.P Bros, "Enthalpies of formation of liquid alloys bismuth-gallium-tin at 723.deg.K. Choice of an analytical representation of integral and partial

excess functions of mixing", *J Chimie Physique*, 72(1)(1975), 83-88.

[14] Y. Zhong et al., "Thermodynamics modeling of the Mg-Sr and Ca-Mg-Sr systems", *J Alloys and Comp*, 421(2006), 172-178.

[15] Pandat 7.0-Phase diagram calculation software for multicomponent systems, Computherm LLC, 437 S. Yellowstone Dr., Suite217, Madison, WI, 53719, USA, 2007.

[16] A. Suzuki et al., "Effect of Sr additions on the microstructure and strength of a Mg-Al-Ca ternary alloy", *Metall Mater Trans A*, 38(2007), 420-427.

From *Magnesium Technology 2008*, M.O. Pekguleryuz, N.R. Neelameggham, R.S. Beals, and E.A. Nyberg, Editors

The Mg-Al-Zn-Mn-Ca-Sr alloy system: Backbone of understanding phase formation in AXJ alloys and modifications of AZ and AM alloys with Ca or Sr

Andreas Janz, Joachim Groebner, Rainer Schmid-Fetzer,
Clausthal University of Technology, Institute of Metallurgy, Robert-Koch-Str. 42, D-38678 Clausthal-Zellerfeld, Germany

Keywords: Mg Alloy, Mg-Al-Zn-Mn-Ca-Sr, Calphad, Phase Diagrams, Computational Thermodynamics

Abstract

Progress is reported on experimental study and thermodynamic modeling of phase formation in key subsystems pertinent to the Mg-Al-Zn-Mn-Ca-Sr alloy system, or abbreviated as (Mg)-AZMXJ. This ongoing work enables thermochemical calculations, which are an important tool for the focused alloy development and process optimization.

As an example, additions of Ca and Sr to alloys of the AZ series may provide improved mechanical and corrosion properties. Understanding the contribution of phase formation to these effects and, based on that, tackling a purposeful optimization is hardly possible without that information. This applies also in the case of Sr-addition to AM, essentially forming AJ alloys or, including the Ca-addition, to the potential offered by the AXJ system.

Introduction

The group of Mg-Al-based alloys forms the current basis of commercially used magnesium alloys. Especially for the AZ and AM series micro-alloying with rare earth elements or Ca and Sr has been extended to increase the mechanical properties especially at high temperatures [1]. New alloy series have been developed on the basis of AM resulting in the groups of AJ and AXJ alloys.

For the ongoing development of these alloy systems and the optimization of their process parameters, thermodynamic modeling has become an increasingly important tool. The Calphad approach [2, 3] has been recognized as a powerful method to calculate the phase diagram and thermodynamic quantities for multi-component systems with quite a high degree of accuracy. Importantly, the reliable thermodynamic calculation provides the basis for understanding multi-component solidification behavior.

We continued our ongoing work on the key subsystems of Al-containing Mg alloys, e.g. Mg-Al-Zn [4], Mg-Mn-Zn [5], Mg-Al-Mn-Zn [6], Mg-Al-Sr [7], Mg-Al-Ca [8], by combining the datasets to develop one reliable database for all alloys within this alloy system.

Key experiments have been performed to evaluate the reliability of the thermodynamic calculations and to refine it if needed.

Results and discussion

The starting basis for the thermodynamic modeling was literature data as far as available. Key samples have been produced to fill gaps of information and later to check the calculations. For the analysis a wide range of equipment was available like DSC, DTA, XRD, SEM and EPMA.

As an example, the assessed transformation temperatures from DSC analysis of two AZ31 alloys with different additions of Ca are superimposed into the calculated phase diagram in Fig. 1, they are in very good agreement with the calculation. Even the solid/solid transformation reaction from the C36-(Al ,Mg)$_2$Ca to the C15-Al$_2$Ca was detected in this experiment.

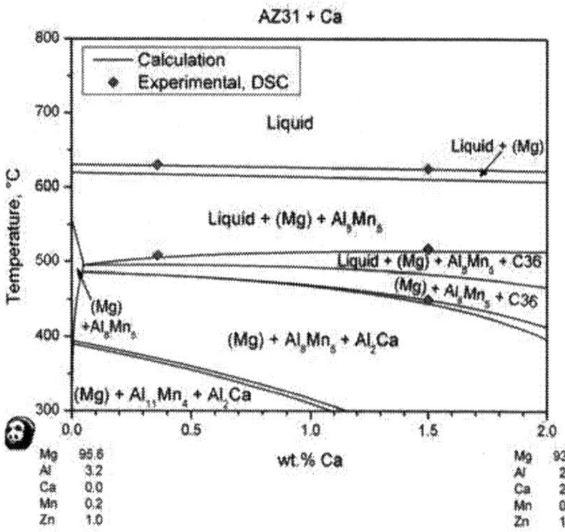

Fig. 1: Calculated vertical phase diagram section of an AZ31-alloy with increasing Ca-content. Superimposed are temperatures assessed from multiple runs of DSC analysis.

The information one gets from the equilibrium phase diagram alone is sometimes misleading in its interpretation for *the real world.* E.g. when the amount of a phase in an equilibrium transformation is very small, as was discussed for the primary precipitation of Al_8Mn_5 in AZ-alloys in [6], data obtained from thermal analysis or DSC may be correlated to the wrong reaction because the heat exchanged within this reaction is too small to be detected.

Therefore calculations of the solidification under both equilibrium and rapid cooling conditions using the Scheil model (assuming no diffusion in the solid state but equilibrium in the liquid phase and at the liquid/solid interface, [9]) have been additionally performed for various alloys in the AJ and AXJ alloy systems. Results are presented in the diagrams displaying phase amount vs. temperature in Fig. 2 for an AJ52x alloy and in Fig. 3 for AXJ530 with two different compositions (1.7 wt.% and 3.3 wt.% Ca as given as the limits of the range of nominal compositions for AXJ530 in [10]). The exact alloy compositions are given on top of each diagram in wt.%. The left part of each diagram shows the frozen-in amounts during the solidification.

For all three alloys Al_8Mn_5 is the primary precipitate, although this phase does not form in an amount much more than 20ppm. The secondary precipitating phase (Mg) will give the first signal in the thermal analysis.

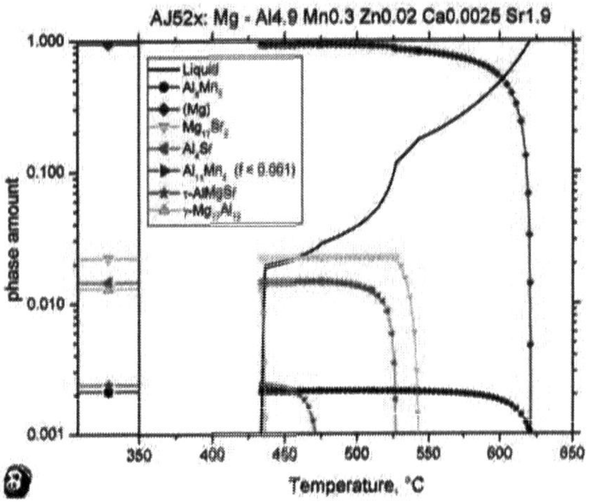

Fig. 2: Evolving phase amounts vs. temperature during the calculated solidification of an AJ52x-Alloy using the Scheil model.

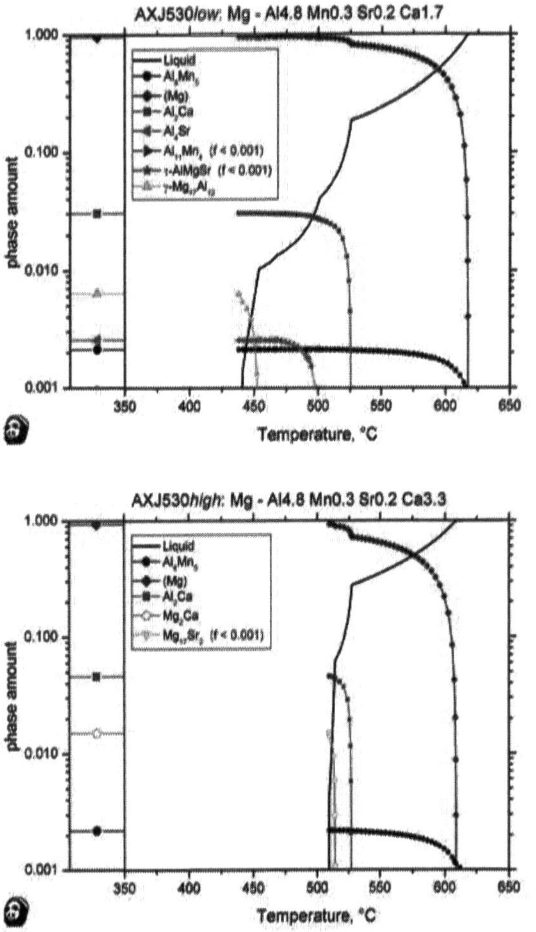

Fig. 3: Evolving phase amounts vs. temperature during the calculated solidification of two AXJ530-alloys with different Ca-content using the Scheil model.

It is noted for the AXJ530 alloys that by increasing the content of Ca by 1.6 wt.% the solidus temperature increases by nearly 70 K, resulting in a much narrower solidification interval of only 135 K.

The phases expected from the calculation to evolve during the solidification of the three samples are much different and especially due to the different chemical composition of the samples, in terms of the Ca and Sr content.

We define the solidus temperature by a liquid phase fraction of 10^{-3}, seen at the lower bound in Figs. 2 and 3. The sensitivity of these alloys on the Ca content regarding castability is emphasized in **[10]** and may be explained with this strong impact on the solidification interval. It also highlights the need to define exact alloy compositions within a "nominal composition range".

It is most interesting, that the higher alloyed sample, AXJ530*high*, exhibits a smaller number of precipitated phases compared to AXJ530 *low*.

Conclusion

By proper thermodynamic modeling of the Mg-Al-Zn-Mn-Ca-Sr alloy system we get a reliable data basis for multicomponent calculations in any of the already commercially used alloy series like AZ, AM and AXJ and furthermore for new developments.

It is beneficial to combine this thermodynamic description with a software requiring no initial knowledge of the phase diagram from the end user or, even worse, has limitations for the calculations or needs starting values for a proper calculation. We have been using the Pandat software **[11]** successfully for all our calculations.

With these tools, it is possible to calculate the presented information and more for any specific real composition of an alloy as well as - for complete alloy series - with equilibrium phase diagrams.

Acknowledgement

This work is supported by the German Research Foundation (DFG) in the Priority Programme "DFG-SPP 1168: InnoMagTec".

References

[1] A.A. Luo, International Materials Reviews 49 (2004), 13-30

[2] L. Kaufman and H. Bernstein, Computer Calculation of Phase Diagrams with Special Reference to Refractory Materials (Academic Press, New York), 1970

[3] Y.A. Chang, S.L. Chen, F. Zhang, X. Yan, F. Xie, R. Schmid-Fetzer, W.A. Oates, Progr. Mater. Sci., 49 (2004), 313

[4] M. Ohno, D. Mirkovic, R. Schmid-Fetzer, Mater. Sci. Eng. A 421 (2006), 328-337

[5] M. Ohno, R. Schmid-Fetzer, Z. Metallkunde 97 (2006), 526-532

[6] M. Ohno, D. Mirkovic, R. Schmid-Fetzer, Acta Materialia 54, (2006), 3883-3891

[7] A. Janz, J. Gröbner, D. Mirkovic, M. Medraj, Jun Zhu, Y. A. Chang, R. Schmid-Fetzer, Intermetallics 14 (2007), 506-519

[8] H. Cao, C. Zhang, J. Zhu, G. Cao, S. Kou, R. Schmid-Fetzer, Y. A. Chang, Acta Mat (accepted)

[9] Scheil E., Z Metallkd 34 (1942), 70

[10] Magnesium Alloys. Science, Technology and Applications. Edited by E.Aghion and D.Eliezer http://www.magnesium.org.il/muaheb/Templates/showpage.asp?TMID=84&FID=913 , 141

[11] http://www.computherm.com/

From *Magnesium Technology 2010*, S.R. Agnew, N.R. Neelameggham, E.A. Nyberg, and W.H. Sillekens, Editors

Thermodynamics and Constitution of Mg-Zn-Ce alloys

Chen-nan Chiu[1,2], Artem Kozlov[1], Joachim Gröbner[1], and Rainer Schmid-Fetzer[1]
[1]Institute of Metallurgy, Clausthal University of Technology, Robert-Koch-Str. 42, D-38678 Clausthal-Zellerfeld, Germany
[2]Department of Chemical Engineering, National Tsing Hua University, 101, Sec. 2, Kuang Fu Rd., Hsin-Chu 300, Taiwan, ROC

Keywords: Calphad modeling, ternary phase diagram, Mg-Zn-Ce alloys

Abstract

Ternary Mg-Zn-Ce phase equilibria were investigated by key samples to determine the solubilities, primary crystallizing phases and invariant reactions. Alloys were prepared from pure elements and investigated by DTA/DSC and SEM/EDS. A consistent thermodynamic Calphad-type model of the ternary Mg-Zn-Ce system is developed. Phase diagram sections at constant 300°C, at constant 85 at.% Mg and the liquidus projection of the Mg-Zn-Ce ternary system were calculated and they compare reasonably well with all available experimental data, especially in the Mg-rich region. This work may, thus, form a reliable basis for the application of computational thermodynamics in Mg-Zn-Ce-X multicomponent alloys.

Introduction

Mg-Zn alloys are considered to have great potential, compared to conventional Mg-Al alloys, for improved mechanical properties and creep resistance at elevated temperature, as reviewed by Bamberger [1]. In order to attain improved creep resistance and strength at elevated temperatures new magnesium alloys with Zn are being developed by adding rare earth (RE) elements (Ce, Gd, and also Y) to form precipitates in these alloys by age hardening [2, 3]. Cerium is a major component of misch-metal and thus is considered as the most important part in these industrially used rare earth elements. It has been reported that addition of Ce effects excellent age hardening behavior for the Mg-1.5Zn-0.2Ce (wt.%) alloy [4]. Therefore, the Mg-Zn-Ce ternary alloy system plays a prominent role among the Mg-Zn-RE systems.

Precise knowledge of the phase diagram and thermodynamic properties of the Mg-Zn-Ce system is necessary for a better understanding of alloy design. It is also a key system for the construction of a thermodynamic database for multicomponent Mg-based alloys. A thermodynamic description of the Mg-Zn-Ce ternary system has not been proposed in the literature. In order to exploit the application potential of computational thermodynamics it is essential to not only know the phase diagram but also the corresponding thermodynamic description in a consistent manner. This also leads to the most effective way to study the ternary phase diagram by a combination of selected key experimental work jointly with thermodynamic modeling using the calculation of phase diagrams (CALPHAD) method. In the present work on the Mg-Zn-Ce alloy system we report on our approach to develop such a consistent ternary thermodynamic description for the first time..

Experimental data in the literature

No complete phase diagram exists in the literature for this system. Experimental phase diagram data in the Mg-Zn-Ce system is limited to the isothermal section at 300°C, at 350°C and some works in the Mg-rich corner. The ternary phase relations were first investigated by Korolkov and Saldau [5]. The liquidus projection in the Mg-rich corner was drawn based on thermal analysis of alloys at six constant mass ratios (Ce:Zn = 1:10, 1:4.5, 1:2, 1:1, 2:1, and 5:1). A ternary eutectic was reported with liquid composition Mg-47.5wt.%Zn-2.5wt.%Ce (Mg-25.9at.%Zn-0.6at.%Ce) at 341 to 343°C without providing explanation [5]. Solvus lines at 20, 200, 300, and 335°C were also given. It is noted that in their study low purity Ce (96 %) was used and the adopted Ce-Mg binary phase diagram differs from the currently accepted one. Therefore, the combined solubility of Ce and Zn in Mg is considered to be an estimated value. Even though, the results indicate that the solubility of Zn in (Mg) is lowered by the addition of Ce.

An isothermal section at 300°C in the Mg-$MgZn_2$-CeMg-CeZn region was given by Melnik et al. [6]. They inspected this region from 150 alloys annealed for 240 hours by using X-ray diffraction analysis. The result shows that most of Ce-Mg binary phases extend with a solid solubility into the ternary system. CeMg and CeZn form a continuous solid solution, and both crystallize in the same CsCl structure. Large solubilities of Zn in $CeMg_3$ and $CeMg_{12}$ at 300°C were found, 29 at.% and 8 at.% Zn, respectively. Four ternary compounds, $CeMg_7Zn_{12}$, $Ce(Mg_{0.5-0.85}Zn_{0.5-0.15})_9$, $CeMg_3Zn_5$, and $Ce_2Mg_3Zn_3$, were discovered at 300°C, designated as τ_1 to τ_4 in the review of data up to the year 2000 by Kolitsch et al. [7]. The homogeneity range of the τ_2 phase extends from 9.1 to 45.5 at.% Zn (81.9 to 45.5 at.% Mg) at 9 at.% Ce at 300°C. The τ_4 phase is homogeneous from 35 to 45 at.% Zn and from 40 to 30 at.% Mg for 25 at.% Ce at 300°C. The crystal structures of τ_2 is reported to be Th_2Ni_{17} structure [8], while that of τ_4 is given as $MgLi_2Ag$ or $AlMnCu_2$ structure by Melnik et al. [9]. The structures of τ_1 and τ_3 are still unknown. Due to the same Th_2Ni_{17} structure of τ_2 and $CeMg_{10.3}$, which corresponds to Ce_2Mg_{17} used in this work, it was deduced that Ce_2Mg_{17} transforms into τ_2 ternary phase gradually after zinc is added into the binary compound Ce_2Mg_{17} [8]. It is noted that Ce_2Mg_{17} only exists in a narrow high temperature range between 608 and 616°C in the Ce-Mg binary system [10]. Melnik et al. [6] reported the τ_1 phase to be in equilibrium with Mg_7Zn_3 at 300°C. However, the result is contradictory to the binary Mg-Zn phase diagram [11] in which the Mg_7Zn_3 phase (or $Mg_{51}Zn_{20}$) is not stable at 300°C. The reported isotherm [6] was thus modified and re-plotted in the critical review by Kolitsch et al. [7].

Two isoplethal sections at 24 wt.% Zn and 34 wt.% Zn in the Mg-rich region were established by Drits et al. [8]. They prepared the alloys in corundum crucibles under VI-2 flux in an electric furnace and examined the alloys by differential thermal analysis (DTA), microstructural, electron-microprobe, and x-ray diffraction analyses. The ternary phases τ_1 and τ_2 are also found in their study, and they labeled τ_2 as $Ce(Mg_{0.5-0.9}Zn_{0.5-0.1})_{10.1}$.

Kevorkov and Pekguleryuz [12] recently studied the Mg-rich corner of the Mg-Zn-Ce phase diagram at 350°C by using a diffusion couple technique. They examined the solid solubilities of binary and ternary solutions by EDS and EPMA. In addition to the four ternary phases reported previously [6], two new ternary phases, $Ce_2Mg_{53}Zn_{45}$ and $Ce_6Mg_{(8-15)}Zn_{(86-79)}$, were found in their work. The decagonal quasicrystal reported in many of the Mg-Zn-RE systems was not observed in the Mg-Zn-Ce system [13]. Pavlyuk et al. reported a Heusler-type alloy with the composition $CeMgZn_2$ [14] and a cubic phase $Ce_{20}Mg_{19}Zn_{81}$ [15]. In this work only three ternary phases T1 $CeMg_7Zn_{12}$, T2 $Ce_2Mg_{53}Zn_{45}$ and T3 $CeMg_3Zn_5$, relevant for Mg alloys, were considered.

Experimental procedures

The Mg-Zn-Ce ternary alloy samples were prepared from pure elements. The weighted elements were pressed carefully to pellets which were sealed under argon atmosphere by welding in thin-walled tantalum crucibles to avoid evaporation and oxidation. No reactions between the crucibles and the samples were observed.

After testing the tightness of the tantalum crucibles in a separate furnace, the samples were measured by DTA in a Netzsch DTA 404S apparatus or by differential scanning calorimetry (DSC) in a heat-flux cylindrical Calvet-type calorimetric system Multi HTC 96 (Setaram, Caluire, France). The heating/cooling rates applied were 1 and 5 K/min.

After DTA and DSC measurements, tantalum crucibles were cut to take out the samples, and the alloys were ground and polished immediately under ethanol to avoid oxidation. Then, the microstructures of the investigated alloys were observed by scanning electron microscopy (SEM), and the composition of the phases was determined by energy dispersive X-ray spectroscopy (EDS) analysis. Ten different compositions of Mg-Zn-Ce ternary alloys were investigated in this work, given in Table 1.

Table 1. Investigated sample compositions

Nr.	Sample composition in at.%
1	Ce13-Mg85-Zn2
2	Ce10-Mg85-Zn5
3	Ce5-Mg85-Zn10
4	Ce9-Mg82-Zn9
5	Ce42-Mg56-Zn2
6	Ce9-Mg45-Zn46
7	Ce25-Mg35-Zn40
8	Ce14-Mg35-Zn51
9	Ce5-Mg35-Zn60
10	Ce11-Mg34-Zn55

Development of the Mg-Zn-Ce ternary thermodynamic description

The thermodynamic parameters of the constituent binary systems are taken from the previous works of Cacciamani et al. [10] (Ce-Mg), Wang et al. [16] (Ce-Zn), and Agarwal et al. [11] (Mg-Zn) and these thermodynamic data sets are considered to be fairly satisfactory.

The ternary system was modeled using the Calphad method. Phases with ternary solid solutions, Ce(Mg,Zn), $Ce(Mg,Zn)_3$ and $Ce(Mg,Zn)_{12}$ were modeled by the Compound Energy Formalism (CEF). The ternary phases T1, T2 and T3 were modeled as stoichiometric compounds. The software package "Pandat" [17] is utilized for all calculations in this work.

The phase T4 reported by [12] is not modeled in this work, since no formation temperature is available for this phase. The composition lies far outside the Mg-rich area which was focused on in this work. Also the phases $CeMgZn_2$ (T5) and $Ce_{20}Mg_{19}Zn_{81}$ reported by [14] and [15] were not considered in the present thermodynamic modeling.

Experimental results

Temperatures obtained by thermal analysis as extracted from the DTA/DSC signals of some samples are shown below in the calculated phase diagram. Each heating/cooling signal is based on at least 2 repeated cycles of a single sample. Invariant reactions were recognized from the peak shape. The interpretations of the experimentally obtained thermal signals are related to the thermodynamic calculation and the corresponding microstructures of slowly solidified samples

For the Zn-rich samples 6, 7, 9 and 10 the interpretation of DTA signals proved to be difficult. The microstructures of the alloys solidified slowly under DTA condition show local inhomogeneities of the formed phases. During polishing a powdering of Zn-rich samples was observed. Therefore no artifact-free photos of the microstructure could be obtained.

Thermodynamic calculation of phase equilibria

Ternary parameters for the solid phases were adjusted on the one hand to meet the measured ternary invariant and liquidus temperatures. Additionally, the assessed data from literature for the homogeneity ranges of $CeMg_{12}$ and $CeMg_3$ and tie-line triangulation at 300°C and own EDS measurements were used. The phases $CeMg_{12}$ and τ_2 from the assessment of Kolitsch et al. [7] were treated as one phase starting from the binary $CeMg_{12}$ phase and dissolving up to 40 at.% Zn. The $CeMg_3$ and τ_4 phases from that assessment [7] were also unified into a single phase starting from the binary $CeMg_3$ phase and dissolving as a line-compound up to 48 at.% Zn by substituting Mg. Although Melnik et al. [6] described different crystal structures for the corresponding phases, we treat them as one phase. This not only simplifies the calculation, moreover [12] did not find any two-phase field between these phases and plotted the $CeMg_{12}$ - τ_2-phase as one line and the $CeMg_3$ - τ_4- phase with dashed triangle in their phase diagram. The maximum Zn-solubility of $CeMg_3$ is set to the value reported by [12] (about 48 at.% Zn at 350°C) and reproduced by the calculation as 48.1 at.% Zn (350°C) and 47.9 at.% Zn (300°C), respectively . The maximum Zn-solubility of $CeMg_{12}$ is set to 40 at.% Zn at 300°C as measured in our own samples. At 350°C a value of 41.5 at.% Zn is calculated, which is lower than the value reported by [12] (about 49 at.% Zn).

The liquidus temperatures of these phases from the thermal analysis (DTA/DSC) are also used to determine the thermodynamic parameters. The Zn-solubilities of $CeMg_2$ phase is measured and modeled at 1 at.% Zn, which gives an excellent fit to the measured liquidus temperature of sample #5, which is also very close to the transition reaction U1. The compositions of the ternary phase reported by [6] and [12] in Table 1 are confirmed for the phases T1, T2 and T3 by own EDS measurements. Their parameters are selected to reproduce both their solid state equilibria and the measured formation temperatures in Table 2: 735°C for T1 in sample #9, 562°C for T2 in sample #6, and 680°C (P2) for T3 in sample #10. The invariant reaction at about 680°C measured in samples #8 and #10 can be allocated to two different reactions, E2 and P2. These reactions are calculated closely spaced at 681 and 680°C, respectively. Depending on the solidification path the alloy will pursue, P2 is passed under equilibrium condition but E2 under Scheil condition.

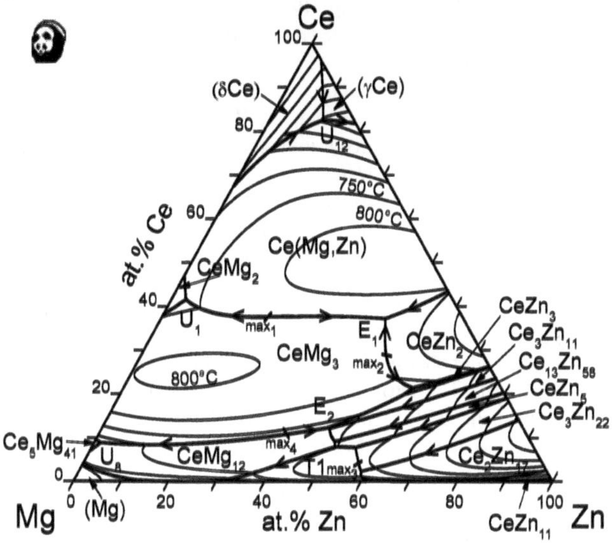

Figure 1. Calculated liquidus projection of the Mg-Zn-Ce system.

Fig. 1 shows the calculated liquidus projection of the Mg-Zn-Ce system based on the proposed thermodynamic description. It is dominated by the three binary Ce-Mg phase with large Zn solubility: Ce(Mg,Zn), $CeMg_3$ and $CeMg_{12}$. Beside the numerous Ce-Zn phases in the Zn-rich part, larger primary fields are only visible for the (Mg) and the ternary T1 phase. The Ce_5Mg_{41} and especially Ce_2Mg_{17} phases show only small primary fields, which is also due to their small solid solubility of Zn. The calculated invariant reactions are given in Table 2.

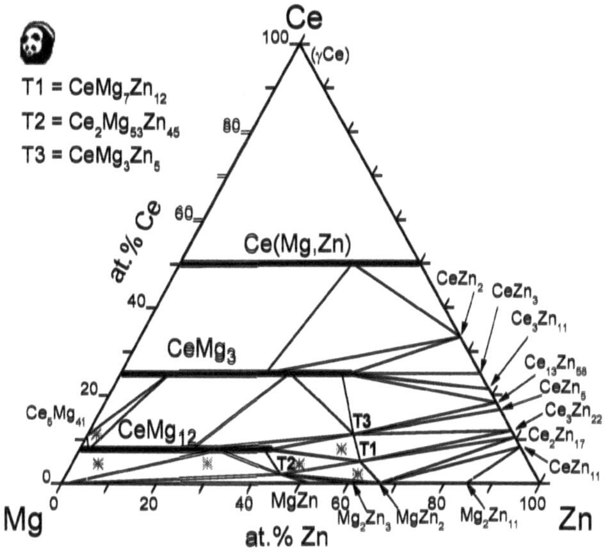

Figure 2. Calculated isothermal section at 300°C. Tie-triangles marked with * are validated by experimental data.

The calculated isothermal section at 300°C is given in Fig. 2. Three-phase equilibria proven by experimental data are marked with an asterisk. The Ce-solubility in (Mg), which is only $8\cdot10^{-8}$ at.% Ce in the binary Mg-Ce, is even further reduced by addition of Zn to $0.02\cdot10^{-8}$ at.% Ce at the three-phase point (Mg) + $CeMg_{12}$ + T2. This edge is at 1.3 at.% Zn (3.4 wt.% Zn), which means that (Mg) with less Zn is saturated with $CeMg_{12}$ while beyond that (Mg) + T2 prevails up to the limit of 2.4 at.% Zn in (Mg) at the (Mg) + T2 + MgZn equilibrium.

Table 2. Calculated invariant reactions compared with measured temperatures of this work.

Typ	Reaction	Temperature [°C]	
		Calc.	Exp.
max1	L → $CeMg_3$ + Ce(Mg,Zn)	770.3	
max2	L → $CeMg_3$ + $CeZn_2$	766.2	
E1	L → $CeMg_3$ + $CeZn_2$ + Ce(Mg,Zn)	739.1	
max3	L + Ce_3Zn_{22} → T1	735.2	735
U1	L + $CeMg_3$ → Ce(Mg,Zn) + $CeMg_2$	729.7	
U2	L + $CeZn_2$ → $CeMg_3$ + $CeZn_3$	726.7	
U3	L + Ce_3Zn_{22} → T1 + Ce_2Zn_{17}	724.9	
U4	L + $CeZn_3$ → Ce_3Zn_{11} + $CeMg_3$	720.6	
U5	L + Ce_3Zn_{22} → T1 + $CeZn_5$	715.3	
P1	L + T1 + $CeZn_5$ → $CeMg_{12}$	697.1	
U6	L + $CeZn_5$ → $CeMg_{12}$ + $Ce_{13}Zn_{58}$	694.9	
max4	L → $CeMg_{12}$ + $CeMg_3$	682.6	
U7	L + $Ce_{13}Zn_{58}$ → $CeMg_{12}$ + Ce_3Zn_{11}	681.9	
E2	L → $CeMg_{12}$ + Ce_3Zn_{11} + $CeMg_3$	681.0	678
P2	$Ce_{13}Zn_{58}$+ $CeMg_{12}$ + Ce_3Zn_{11}→ T3	680.2	678
U8	L + $CeMg_3$ → $CeMg_{12}$ + Ce_5Mg_{41}	616.7	615
U9	L + Ce_5Mg_{41} → $CeMg_{12}$ + Ce_2Mg_{17}	613.2	
max5	L → $MgZn_2$ + Ce_2Zn_{17}	589.8	
U10	L + Ce_2Zn_{17} → T1 + $MgZn_2$	588.1	
U11	L + Ce_2Zn_{17} → $MgZn_2$ + $CeZn_{11}$	581.3	
U12	L + δCe → Ce(Mg,Zn) + γCe	542.3	
max6	L + T1 → T2	489.5	
U13	L + T1 → $CeMg_{12}$ + T2	480.4	
U14	L + T1 → T2 + $MgZn_2$	479.1	
U15	L + $CeMg_{12}$ → (Mg) + T2	420.3	
U16	L + $MgZn_2$ → T2 + Mg_2Zn_3	415.8	
U17	L + $MgZn_2$ → Mg_2Zn_{11} + $CeZn_{11}$	381.2	
E3	L → Mg_2Zn_{11} + (Zn) + $CeZn_{11}$	367.2	
U18	L + Mg_2Zn_3 → MgZn + T2	347.0	
U19	L + (Mg) → Mg_7Zn_3 + T2	341.02	
E4	L → MgZn + Mg_7Zn_3 + T2	340.99	

The phase diagram section calculated at constant 85 at.% Mg, including the DSC signals from samples #1 to #3, is shown in Fig. 3. The strong heating and cooling signal shown by sample #1 at the invariant reaction U8 is most decisive. Also the liquidus temperatures of samples #1 and #2 are well reproduced by the calculation while the liquidus temperature of sample #3, calculated at 639°C, could not be detected.

Solidification paths and microstructures of Mg-rich alloys are given in a more comprehensive publication [18]. They also support the developed thermodynamic description.

Conclusion

On the experimental side our work focuses on the Mg-rich part of the ternary system. Sample compositions were chosen from 34 to 85 at.% Mg from preliminary calculations indicating these to be most decisive for determination of thermodynamic parameters and verification of invariant liquidus reactions and solid state equilibria up to pure magnesium. The final thermodynamic modeling, however, covers the entire range of the Mg-Zn-Ce system. The Zn-Ce-rich region of the Mg-Zn-Ce system is calculated based on the extrapolation from the thermodynamic descriptions of the edge binary phases. Additional experimental data particularly in the Zn-rich corner are required to establish the phase equilibria of this ternary system completely.

All solid solubilities of Zn in Mg-Ce intermetallic phases are due to substitutional exchange of Mg by Zn. The large Zn-solubilities of the $CeMg_{12}$ and $CeMg_3$ phases are treated as two single phases ($CeMg_{12}$ up to 40 at.% Zn and $CeMg_3$ up to 48 at.% Zn at 300°C). This might be considered as simplification since [6] described different crystal structures and two-phase field between these phases. However, these observations could not be confirmed by [12] and in this work.

It is remarkable that extremely small solubilities of Ce in Mg-Zn liquid alloys are observed beyond which Ce-containing intermetallic phases precipitate, namely $CeMg_{12}$, $Ce_2Mg_{53}Zn_{45}$ (T2), and $CeMg_7Zn_{12}$ (T1). The (Mg) phase crystallizes jointly only with $CeMg_{12}$ and T2, the transition between these two secondary phases occurs at the invariant reaction U15 at 420°C with a liquid composition of Mg76.6-Zn23.4-Ce0.003 (at.%). The transition in the solid state equilibria of (Mg) at 300°C occurs at 1.3 at.% Zn (3.4 wt.% Zn), which means that (Mg) with less Zn is saturated with $CeMg_{12}$; in the (Mg) solid solution from 1.3 to 2.4 at.% Zn the secondary phase is T2.

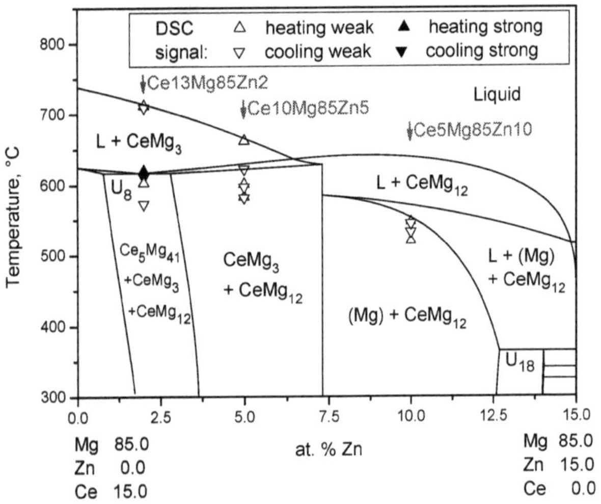

Figure 3. Calculated vertical phase diagram section at constant 85 at.% Mg including the DSC signals from three samples, #1 to #3, detailed in Table 1.

Acknowledgement

This study is supported by the German Research Foundation (DFG) in the Priority Programme "DFG-SPP 1168: InnoMagTec" under grant no. Schm 588/27.29.

References

[1] M. Bamberger, "Structural refinement of cast magnesium alloys," *Mater. Sci. Technol.* 17 (2001) 15-24.

[2] Y.M. Zhu, A.J. Morton, J.F. Nie, "Improvement in the age-hardening response of Mg–Y–Zn alloys by Ag additions," *Scripta Mater.* 58 (2008) 525-528.

[3] J.F. Nie, K. Oh-ishi, X. Gao, K. Hono, "Solute segregation and precipitation in a creep - resistant Mg–Gd–Zn alloy," *Acta Mater.* 56 (2008) 6061-6076.

[4] Y. Chino, K. Sassa, M. Mabuchi, "Tensile properties and stretch formability of Mg–1.5 mass% Zn–0.2 mass% Ce sheet rolled at 723 K," *Mater. Trans.* 49 (2008) 1710-1712.

[5] A.M. Korol'kov, Ya.P. Sal'dau, "Solubility of Zn and Ce in Mg in the Solid State," *Izv. Sekt. Fiziko-Khimich Analiza, Akad. Nauk SSSR* 16 (1946) 295-306.

[6] E.V. Mel'nik, M.F. Kostina, Ya.P. Yarmolyuk, O.F. Zmii, "Study of magnesium-zinc-cerium and magnesium-zinc-calcium ternary systems," in: *Mater. Vses. Soveshch.Issled. Razrab, Primen, Magnievyhk Splavov* (1978) 95–99.

[7] U. Kolitsch, P. Bellen, S. Kaesche, D. Macciò, N. Bochvar, Y. Liberov, P. Rogl, "Cerium-Magnesium-Zinc," in: G. Petzow, G. Effenberg (Eds.), *Ternary Alloys: A Comprehensive Compendium of Evaluated Constitutional Data and Phase Diagrams*, vol. 17, Weinheim: VCH Verlagsgesellschaft, MSI GmbH, Stuttgart, Germany, 2000, pp. 168-176.

[8] M.E. Drits, E.I. Drozdova, I.G. Korol'kova, V.V. Kinzhibalo, A.T. Tyvanchuk, "Investigation of polythermal sections of the Mg–Zn–Ce system in the Mg-rich region," *Russ. Metall.* 2 (1989) 195-197.

[9] E.V. Mel'nik, O.F. Zmii, E.E. Cherkasim, "On the Structure of the $Ce_2Mg_3Zn_3$ Compound," *Vestn. L'viv. Univ. Ser. Khim.* 19 (1977) 34-36.

[10] G. Cacciamani, G. Borzone, R. Ferro, "System Ce–Mg", in: I. Ansara, A.T. Dinsdale, M.H. Rand (Eds.), COST 507—thermochemical database for light metal alloys, July 1998. European Commission EUR 18499 EN, 1998, pp. 137–140.

[11] R. Agarwal, S.G. Fries, H.L. Lukas, G. Petzow, F. Sommer, T.G. Chart, G. Effenberg, "Assessment of the Mg-Zn system," *Z. Metallkd.* 83 (1992) 216-223.

[12] D. Kevorkov, M. Pekguleryuz, "Experimental study of the Ce–Mg–Zn phase diagram at 350ºC via diffusion couple techniques," *J. Alloys Compd.* 478 (2009) 427-436.

[13] T.J. Sato, E. Abe, A.P. Tsai, "Composition and Stability of Decagonal Quasicrysatls in the Zn-Mg-rare-earth Systems," *Philos. Mag. Lett.* 77 (1998) 213-219.

[14] V. Pavlyuk, P. Solokha, G. Dmytriv, B. Marciniak, V. Paul-Boncour, "The Heusler-type alloy $MgZn_2Ce$," *Acta Crystallogr.*, Sect. E: Struct. Rep. Online E63 (2007) i161.

[15] V. Pavlyuk, P. Solokha, O. Zelinska, V. Paul-Boncour, A. Nowik-Zajac, "$Ce_{20}Mg_{19}Zn_{81}$: a new structure type with a giant cubic cell," *Acta Crystallogr.*, Sect. C: Cryst. Struct. Comm. C64 (2008) i50-i52.

[16] C.P. Wang, X. Chen, X.J. Liu, F.S. Pan, K. Ishida, "Thermodynamic modeling of the Ce-Zn and Pr-Zn systems," *J. Alloys Compd.* 458 (2008) 166-173.

[17] S.L. Chen, S. Daniel, F. Zhang, Y.A. Chang, X.Y. Yan, F.Y. Xie, R. Schmid-Fetzer, W.A. Oates, "The PANDAT software package and its applications," *CALPHAD* 26 (2002) 175-188.

[18] C. Chiu, J. Gröbner, A. Kozlov and R. Schmid-Fetzer, "Experimental study and thermodynamic assessment of ternary Mg-Zn-Ce phase relations focused on Mg-rich alloys," *Intermetallics* (2009) accepted.

MODELLING OF THE THERMO-PHYSICAL AND PHYSICAL PROPERTIES FOR SOLIDIFICATION OF MG-ALLOYS

N.Saunders[1], X.Li[2], A.P.Miodownik[1] and J.-P.Schille[2]

[1]Thermotech Ltd., Surrey Technology Centre, The Surrey Research Park, Guildford GU2 7YG, U.K.
[2]Sente Software Ltd., Surrey Technology Centre, The Surrey Research Park, Guildford GU2 7YG, U.K.

Abstract

The thermo-physical and physical properties of the liquid and solid phases are critical components in casting simulations. Such properties include the fraction solid transformed, enthalpy release, thermal conductivity, volume and density all as a function of temperature. Due to the difficulty in experimentally determining such properties at solidification temperatures, little information exists for multi-component alloys. As part of the development of a new computer program for modeling of materials properties (JMatPro) extensive work has been carried out on the development of sound, physically based models for these properties. Wide ranging results will presented for Mg-based alloys including more detailed information concerning the density change of the liquid that intrinsically occurs during solidification due to its change in composition.

Introduction

The increasing use of casting software makes it ever more urgent that the thermo-physical and physical properties of Mg alloys are well characterized as they are critical input for almost all types of process modeling. Obtaining these properties at low temperatures can be time-consuming and expensive procedure if all relevant properties are considered. Experimental measurement becomes far more problematical at high temperature, especially if the liquid phase is involved. To this end it becomes highly desirable to calculate thermo-physical and physical properties over the whole solidification range for as wide a range of alloys as possible.

The present paper describes a general methodology to calculate properties such as density, thermal conductivity, specific heat (C_p), solidification shrinkage etc. for multi-component alloys of all types. The property models that are described in the present paper have also been linked to the simulation of non-equilibrium solidification based on the Scheil-Gulliver (SG) model. Hence it is possible to directly input calculated values into casting simulation packages.

The current work forms part of the development of a more generalized software package (JMatPro) for the calculation of wide ranging of materials properties [1]. A feature of the new program is that great store has been placed on using models that, as far as possible, are based on sound physical principles rather than purely statistical methods. Thus many of the shortcomings of methods such as regression analysis can be overcome. For example, the same model and model parameters are used for density calculations all alloy types, whether it be for a commercially pure Al-alloy or a complex Ni-based superalloy.

The paper will describe the SG solidification model that directly calculates fraction solid, Cp, enthalpy and latent heat of solidification. Detail concerning the creation of a molar volume database that enables a variety of properties to be calculated, such as solidification shrinkage, density, thermal expansion coefficient, will then be presented. The calculation of thermal conductivity will also be discussed. Examples of the linking of the solidification models with the physical property calculations will be made and properties during solidification will be calculated for a series of Mg-alloys and the results presented. A significant advantage of the current method is that properties for each phase are calculated so fine detail can be obtained; for example the density change of the liquid during the solidification, which is governed both by an intrinsic change as the temperature is changed and also by the composition changes that accompany solidification.

Background

Solidification modeling.

Recently the application of a so-called 'Scheil-Gulliver' modelling via a thermodynamic modeling route has led to the ability to predict a number of critical thermo-physical properties for alloys such as Ni-based superalloys [2,3,4], Al-alloys [5,6] and cast irons [7]. Such calculations can be computationally very fast and are currently used within solidification packages such as ProCAST [2].

For equilibrium solidification described by the lever rule and with linear liquidus and solidus lines, the composition of the solid (C_s) as a function of the fraction solid transformed (f_S) is given by

$$C_s = \frac{kC_o}{f_s(k-1)+1} \quad .....(1)$$

where k is the partition coefficient and C_o is the composition of the original liquid alloy. This can be re-arranged to give

$$f_s = \left(\frac{1}{1-k}\right)\left(\frac{T_L - T}{T_f - T}\right) \quad .....(2)$$

where T_L and T_f are the equilibrium liquidus and solidus temperatures. A complimentary limiting case to equilibrium solidification is to assume that solute diffusion in the solid phase is small enough to be considered negligible and that diffusion in the liquid is extremely fast, fast enough to assume that diffusion is complete. In this case eq.1 can be re-written as

$$C_s = kC_o(1-f_s)^{k-1} \quad ....(3)$$

and eq.2 as

$$f_s = 1 - \left(\frac{T_f - T}{T_f - T_L} \right)^{\left[\frac{1}{k-1}\right]} \quad .....(4)$$

The treatment above is the traditional derivation of the Scheil equation but it has quite severe restrictions when applied to multi-component alloys. It is not possible to derive this equation, using the same mathematical method, if the partition coefficient, k, is dependent on temperature and/or composition. The Scheil equation is applicable only to dendritic solidification and cannot, therefore, be applied to eutectic alloys that are common type for Mg-alloys. Further it cannot be used to predict the formation of intermetallics during solidification.

Using thermodynamic modelling all of the above disadvantages can be overcome. The process that physically occurs during 'Scheil' solidification can be envisaged as follows (Fig.1). A liquid of composition C_0 is cooled to a small amount below its liquidus. It precipitates out solid with a composition $C_{S,1}$ and the liquid changes its composition to $C_{L,1}$. However, on further cooling to the initial solid cannot change its composition due to lack of back diffusion and it is effectively 'isolated'. A local equilibrium is then set up where the liquid of composition $C_{L,1}$ transforms to a liquid of composition $C_{L,2}$ and a solid with composition $C_{S,2}$, which is precipitated onto the original solid with composition $C_{S,1}$. This process occurs again on cooling where the liquid of composition $C_{L,2}$ transforms to a liquid of composition $C_{L,3}$, and a solid with composition $C_{S,3}$ grows on the existing solid. This process occurs continuously during cooling and when k<1 leads to the solid phase becoming lean in solute in the center of the dendrite and the liquid becoming more and more enriched in solute as solidification proceeds. Eventually, the composition of the liquid will reach the eutectic composition and final solidification will occur via this reaction.

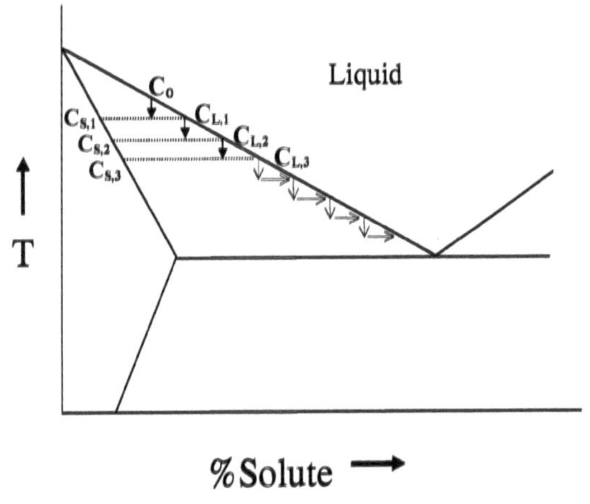

Figure 1. Schematic representation of solidification occurring under Scheil-Gulliver conditions

Any appearance of secondary phases can be easily taken into account in this approach with the assumption that no back diffusion occurs in them. Therefore, all transformations can be accounted for including the final eutectic solidification. The approach described here is based on an isothermal step process but as the temperature step size becomes small it provides results that are almost completely equivalent to that which would be obtained from continuous cooling. A further and very significant advantage of using a thermodynamic approach is that the heat evolution during solidification is a straightforward product of the calculation. The limit to the SG simulation is that some back diffusion will take place. However, if the degree is small, good results will still be obtained. The application of the SG model to Al-alloys provides excellent results [6,8] and it would be expected that similarly good results would be found for Mg-alloys.

Recently a thermodynamic database for Mg-alloys has been developed that can be used in the type of calculation just described [9]. As well, a new software package JMatPro [1] can be utilised so that physical property models can be linked to SG solidification calculations.

The Calculation of C_p, Enthalpy and Latent Heat of Solidification.

As mentioned above, an advantage of using the CALPHAD methodology for solidification modeling is that as well as calculating the fraction solid transformed, the phases formed themselves and their amounts, heat evolution is calculated. This is critical input to casting simulation software. Figures 2 & 3 show respectively the fraction solid transformed and the enthalpy of a AE42 alloy during casting.

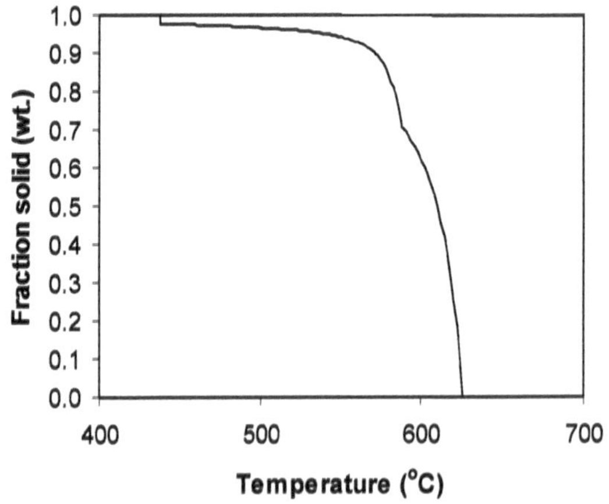

Figure 2. Fraction solid vs. temperature for a AE42 Mg-alloy during solidification

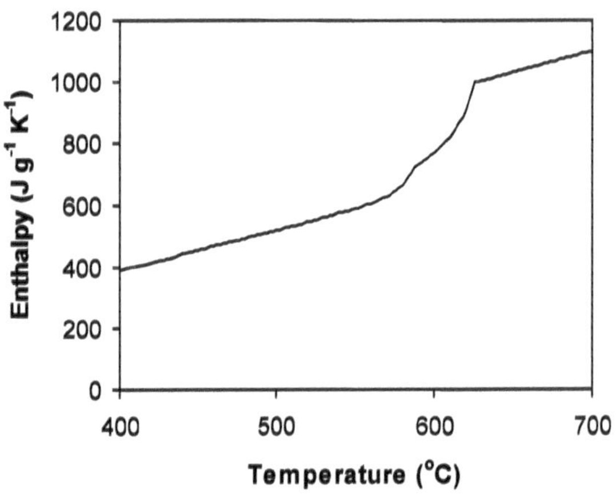

Figure 3. Calculated enthalpy vs. temperature for a AE42 alloy during solidification

At the completion of solidification, information concerning the solid phases formed during solidification is retained and their properties extrapolated below the solidus. This is particularly useful as within a finite element mesh, the temperature can range from well below the solidus near the mould wall to above the liquidus in the feeders. A further utilization is possible by combination with modulus calculations, also calculated by JMatPro, to calculate residual stresses in castings.

Molar Volume Calculations

A major achievement of the JMatPro development project has been the development of an extensive molar volume database that can be linked to its thermodynamic calculation capability and hence provide volume data for the phases involved in the calculation. Presently, an extensive database of parameters exists for most of the major phases in Al-, Fe-, Mg-, Ni- and Ti-alloys which has been tested extensively in the solid state against lattice parameter measurements (both at room temperature and where available at high temperatures) and experimentally reported linear expansion data. Volume calculations are linked to the thermodynamic models such that, once a thermodynamic calculation is made, volume can be directly calculated.

For pure magnesium both the behaviour of the liquid and solid phases are quite well characterized. Isotropic linear expansion measurements combined with a molar volume calculated from lattice parameter or density measurements provide a sound basis for pure Mg. The volume difference on melting is characterized as is the density of the liquid with temperature [10]. Figure 4 shows the calculated molar volume of the HCP and liquid phases as a function of temperature, with experimental data for the liquid phase calculated from density measurements.

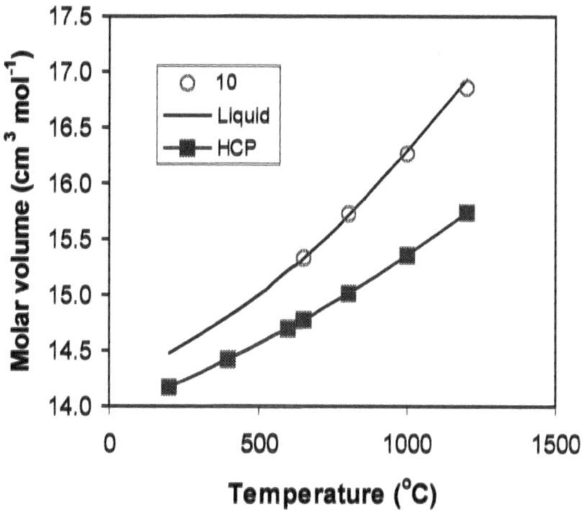

Figure 4. Calculated volumes of solid and liquid phases of Mg as a function temperature.

The liquid phase has been extrapolated below the melting point based on an extrapolation method based on the principle that the transition from the solid to liquid/amorphous state will be accompanied by an increase in volume associated with the larger volume of the liquid/amorphous phase in comparison to that of crystalline structure [11]. This means that the volume of the liquid should never reach that of the solid. We have also assumed that at a temperature close to the glass transition temperature, T_g, (assumed ~$0.3T_m$ for metallic elements) the difference in volume will tend to reach a minimum and below T_g the volume difference will be fairly constant. We have found that quite simple temperature functions give both a good fit to the temperature dependence of the density at high temperatures and a reasonable low temperature extrapolation based on the above principle. While not so important for Mg, as the properties of the liquid phase are extrapolated over a relatively small range of temperature, this principle becomes more important when considering high melting point element additions such as Zr, Fe, Mn etc..

The volume of the liquid and solid phases is represented by simple mixture models, similar to those used for modeling of thermodynamic excess functions in multi-component alloys [12]. In some alloys the stable crystal structure of the element in solid solution may be different to that of the solid solution itself. In this case we have assumed the molar volume of the metastable state to be the same as that for the stable structure. We have invariably found that if there is a deviation from "ideal mixing" a volume contraction is involved and that the magnitude of contraction is strongly governed by the magnitude of the thermodynamic interaction.

It is possible to consider what interaction effect may be present in some Mg solid solutions, for example with Al additions. For other elements this has not been possible, as their solid solubility is so small. In these cases an "ideal" extrapolation is used. It should be noted that the magnitude of likely interaction term is unlikely to be large enough to cause significant error for volume calculations. This is because the level of element addition to Mg is small and the major alloying elements such as Zn and Al have small interactions, which should not lead to significant deviations from ideality.

Linking the volume database to the SG calculations now means that a number of solidification properties can be calculated. Fig.5 shows the calculated shrinkage during casting of a ZMC711 alloy.

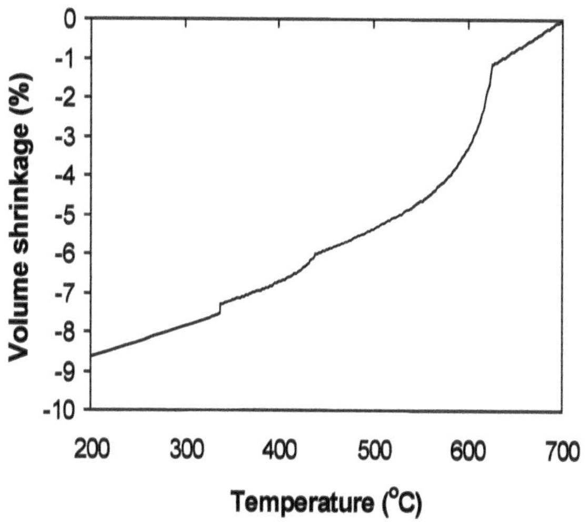

Figure 5. Calculated volume shrinkage for a ZMC711 during solidification

While Fig.5 shows the overall change in volume during casting, it is straightforward to gain more detail by looking at the density change and, in particular, the density change of the liquid. In this case we have plotted the density of the liquid in the mushy zone for an AS41 alloy (Fig.6)

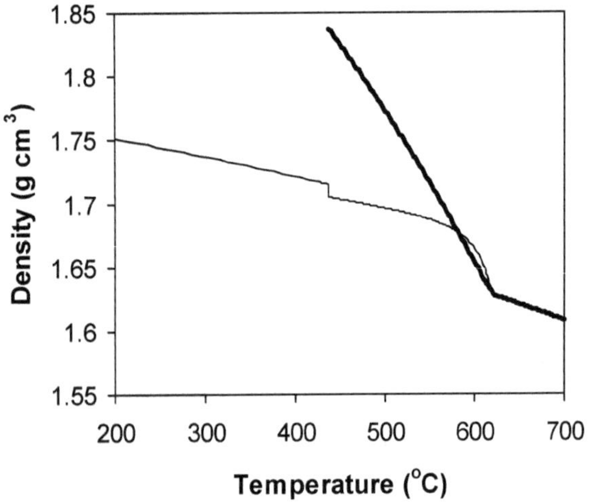

Figure 6. Calculated density of a AS41 alloy during solidification. Bold line shows density of the liquid

What is apparent is that the liquid density is changing much more rapidly than the total density in the mushy zone. This can be understood by analyzing the composition changes in the liquid and Figure 7 shows the Al level in the liquid.

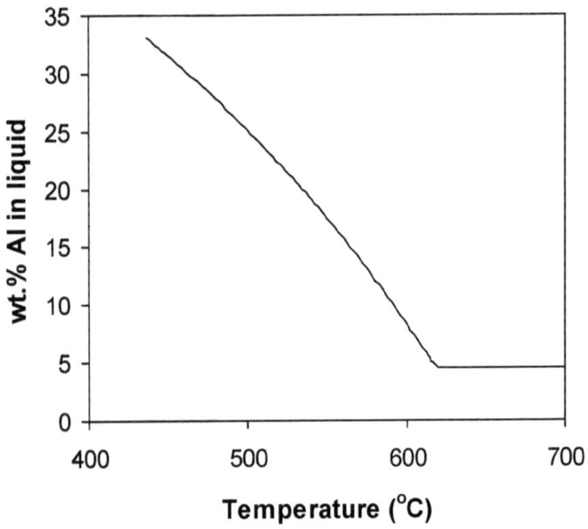

Figure 7. Al concentration of the liquid phase during solidification of an AS41 alloy.

The partition coefficient is such that the Al level increases quite sharply, which directly causes the density to increase. The total density is governed both by the density of the liquid and solid state and in the latter part of solidification is controlled by the lower density Mg solid solution which predominates. A corollary calculation made from volume calculations is to obtain the thermal coefficient of expansion and this is shown for an AZ91 alloy in Figure 8.

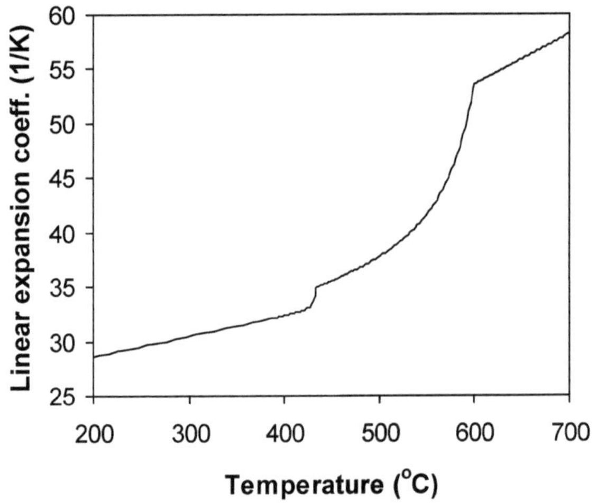

Figure 8. Calculated linear expansion coefficient density of a AZ91 alloy during solidification.

<u>Thermal Conductivity Calculations</u>

The calculation of thermal conductivity is a complex physical phenomenon. There are sharp changes on alloying in the solid state and the behaviour, in terms of mixing models, is more difficult to match. On alloying, a "bath tub" shape is often seen, where the thermal conductivity falls sharply in the dilute range and then forms a fairly flat plateau in the concentrated region [13]. Because of the lack of information concerning the thermal conductivity of binary liquid Mg-alloys alloys we cannot directly assess coefficients for alloying effects in the liquid. However, information does exist for thermal conductivity in the liquid state in pure elements and we have evaluated parameters using this information [14,15] as a first basis.

Fig.9 shows the calculated thermal conductivity in Al-Mg solid alloys with measured [16] thermal conductivities for the Al and

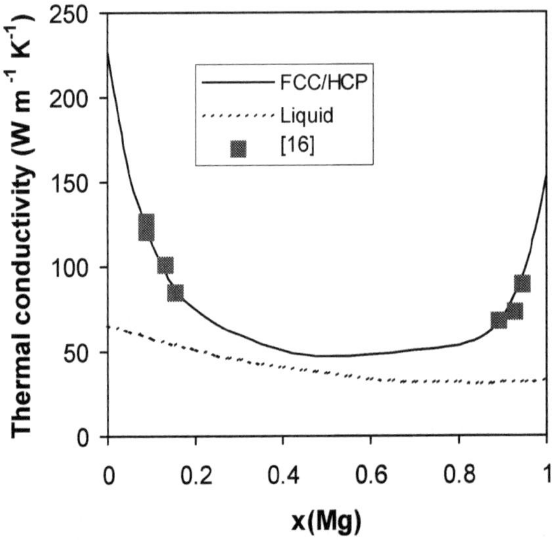

Figure 9. Calculated and experimental thermal conductivities of Al-Mg alloys at 100°C.

Mg solid solutions shown for comparison. In this case we have assumed that the thermal conductivity of the FCC and HCP forms is the same and can be represented using one curve. A "bath tub" curve is clearly seen, but of interest is that an "ideal" extrapolation of the liquid thermal conductivity at this temperature provides thermal conductivities rather close to those in the plateau region. To obtain the usually observed result for metals, in that the thermal conductivity of the liquid should be less than that of the solid, only a small negative interaction coefficient is needed. We have invariably found this result in our assessment work and we have evaluated interaction terms based on this principle.

Extensive validation of solid state multi-component alloys of many types shows good results [1] and where results are available for multi-component liquid alloys good agreement is usually found [17]. Figure 10 shows calculated thermal conductivities for an AZ91 alloy during casting.

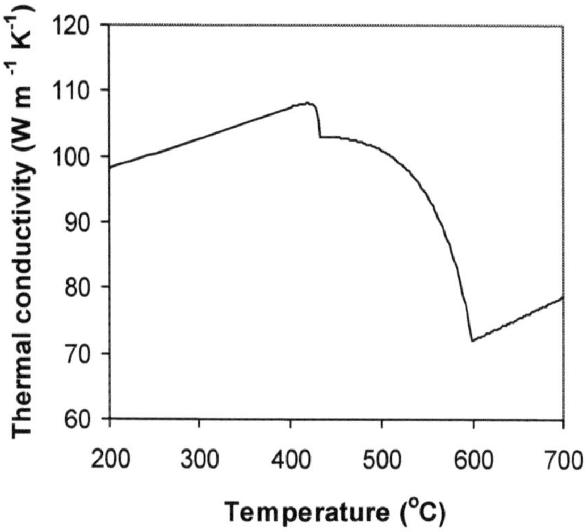

Figure 10. Calculated thermal conductivity of an AZ91 Mg-alloy during solidification

Summary and Conclusions

Models have been developed for the calculation of the various thermo-physical and physical properties with the aim of providing thermo-physical and physical properties for various types of multi-component-alloys during solidification. The present paper describes calculations particularly for Mg-alloys and presents results for fraction solid, enthalpy, Cp, thermal conductivity, density, linear expansion coefficient and volume shrinkage during solidification.

Using known relationships between various properties it is also possible, from the results shown here, to calculate electrical conductivity, electrical resistivity and thermal diffusivity. Work is currently undergoing to model viscosity that will complete a very comprehensive set of properties of value to process modelers of all types. A further, significant advantage of the approach described here, is that it is possible to obtain important properties for each phase individually. For example, the density of the liquid phase during the solidification process is automatically calculated.

References

1. N. Saunders et al., Materials Design Approaches and Experiences, eds. J.-C.Shao et al. (Warrendale, PA: TMS: 2001), 185
2. W.J. Boettinger et al., Modelling of Casting, Welding and Advanced Solidification Processes, VII, eds. M.Cross et al., (Warrendale, PA: TMS, 1995), 649
3. N. Saunders, Superalloys 1996, eds. R.Kissinger et al. (Warrendale, PA: TMS, 1996) 115
4. B.A. Boutwell et al., Superalloys 718, 625, 706 and Various Derivatives, ed. E.A. Loria, (Warrendale, PA: TMS, 1996), 99
5. N.Saunders, Materials Science Forum, 217-222 (1996), 667
6. N.Saunders, Light Metals 1997, ed. R. Huglen (Warrendale, PA: TMS, 1997), 911
7. R.A. Harding and N. Saunders, Trans.American Foundryman's Society, 105, (1997), 451
8. N. Saunders, Solidification Processing 1997, eds. J. Beech and H.Jones (Sheffield: Univ.Sheffield, 1997), 362
9. Mg-DATA, a database for thermodynamic calculations of multi-component Mg-alloys, Thermotech Ltd., Surrey Technology Centre, The Surrey Research Park, Guildford GU2 7YG, U.K.
10. G.V. Raynor, The Physical Metallurgy of Magnesium and its Alloys, (London: Pergamon Press, 1959)
11. J.L. Finney, The Structure of Non-Crystalline Materials, ed. P.H. Gaskell (London: Taylor & Francis, 1977), 35
12. N. Saunders and A.P. Miodownik, CALPHAD – Calculation of Phase Diagrams, Pergamon Materials Series vol.1, ed. R.W. Cahn (Oxford: Elsevier Science,1998)
13. J.B.Austin The Flow of Heat in Metals (Cleveland, OH: ASM, 1942)
14. T. Iida and R.L. Guthrie "The Physical Properties of Liquid Metals", (Oxford: Clarendon Press, 1988)
15. K.C.Mills, B.J.Monaghan and B.J.Keene, Int.Met.Rev., **41**, (1996), p209
16. Y.S. Touloukian, R.W. Powell, C.Y. Ho and P.G. Kliemens, "Thermo-physical Properties of Matter: Vol.1" (IFI Plenum Press, New York, 1970)
17. N. Saunders et al., to be presented at Modeling of Casting, Welding and Advanced Solidification Processes, Sandestin, Florida, May 25-30, 2003

From *Magnesium Technology 2007*, R.S. Beals, A.A. Luo, N.R. Neelameggham, and M.O. Pekguleryuz, Editors

Constitution of Magnesium Alloys

R. Schmid-Fetzer, J. Gröbner, D. Mirković, A. Janz, A. Kozlov
Clausthal University of Technology, Institute of Metallurgy, Robert-Koch-Str. 42, D-38678 Clausthal-Zellerfeld, Germany

Keywords: Mg alloy, thermodynamic database, Calphad, phase diagrams, computational thermodynamics, applications

Abstract

Multi-component magnesium alloys exhibit a complex constitution, requiring computational thermodynamics for a quantitative treatment that goes beyond "just" phase diagrams. The basis for this approach is a thermodynamic Mg alloy database, which is developed in an ongoing long-term project in our group since many years. Three distinctive features of this database are highlighted in this report: (i) Combination of own key experimental work with theoretical modeling to generate consistent data, (ii) Systematic quality control of the database using a variety of elaborated cross-checks, and (iii) Complete and entire composition range descriptions for all pertinent binary or ternary subsystems whenever possible. The latter point is decisive for the capability to use this tool for new alloys, far beyond the composition limits of conventional Mg alloys. It is demonstrated by correctly predicting the phase formation in aluminum-rich six component alloys. Also, new Mg-solder alloys can be tackled that are essentially Zn-rich.

Introduction

Quantitative access to all available thermodynamic and phase diagram data in a consistent, numerical and easily readable form is – at least for multicomponent alloys – virtually impossible without the use of computational thermochemistry. The fundamental idea goes back to the introduction of the Calphad method by Larry Kaufman in 1970 [1]. In a nutshell it is to calculate all these data, ranging from stable and metastable phase equilibria down to the chemical potentials, from a unique set of thermodynamic model parameters of the alloy system. A review of the Calphad method [2] and a recent summary on assessment techniques and database design [3] are given elsewhere.

Major thermodynamic databases are mainly available from developers of commercial thermodynamic software packages such as Pandat (www.computherm.com), Thermocalc (www.thermocalc.com), Factsage (www.factsage.com), MTDATA (www.npl.co.uk/mtdata), more sources are given in [3]. A specific aspect in developing a magnesium alloy database, in comparison to steel and Al alloys, is the substantially smaller pool of experimental data in the published literature. In addition to thermodynamic assessment work it is thus mandatory to perform original experimental work in a combined approach in order to generate a trustworthy Mg alloy database. This is the typical approach in our ongoing work in this field, some results of which will be shown below. Sample compositions for key experiments are selected by preliminary thermodynamic calculations. Such samples are prepared in high purity and the workhorse methods to analyze phase equilibria or transitions are X-ray diffraction (XRD), metallographic analysis, scanning electron microscopy with energy dispersive X-ray microanalysis (SEM/EDX), differential thermal analysis (DTA) and differential scanning calorimetry (DSC). It is the purpose of this study to highlight some real world applications of this approach and also to outline the current scope and structure of the database and the important aspect of its validation.

Selected applications of the Mg alloy database

Phase equilibria in Mg-Al-Zn-Mn system provide crucial information for the advancement of AZ and AM alloy series. Specifically, the liquidus and solidus data reported for commercial AZ and AM alloys are generally based on thermal analysis. The following two points have been clarified by performing own experiments combined with computational thermodynamics [4]: (i) The measured "liquidus" temperature generally does not represent the actual equilibrium one, in other words, the primary precipitate for these Mg-alloys cannot be detected in the thermal analysis. (ii) The measured "solidus" does not correspond to the equilibrium solidus and not even to the end of non-equilibrium solidification process. The measured "solidus" is often associated with the *starting* precipitation of $Mg_{17}Al_{12}$ phase and, importantly, the solidification process of these Mg-alloys ends at much lower temperature. This example shows that it is possible to predict even non-equilibrium effects [4].

In specific studies on AZ31 based alloys it could be shown that the current thermodynamic description can be successfully combined with a phase field simulation approach aiming at a microstructure control [5]. It was also shown helpful in better understanding the influence on manganese on the microstructure of AZ31 and a possible relation with grain refining [6].

Al-rich alloys in the system Al-Zn-Mg-Cu-Sc-Zr were also studied by thermodynamic calculations [7]. Phase formation was compared with experimental data obtained by DTA and microstructural analysis. Calculated phase diagrams, such as shown in Fig. 1, phase amount charts and enthalpy charts together with non-equilibrium calculations under Scheil conditions reveal significant details of the complex phase formation. All calculated data are predictions from the current thermodynamic database developed for Mg-alloys and not from a specialized Al-alloy database. Our basic concept is to assess the entire phase equilibria in the complete composition range of each subsystem as far as possible.

As a result, this database is also quite useful for Al-alloys with adequate alloying elements. That, in turn, provides support for a reasonable application of this database for advanced Mg-alloys beyond the conventional composition ranges.

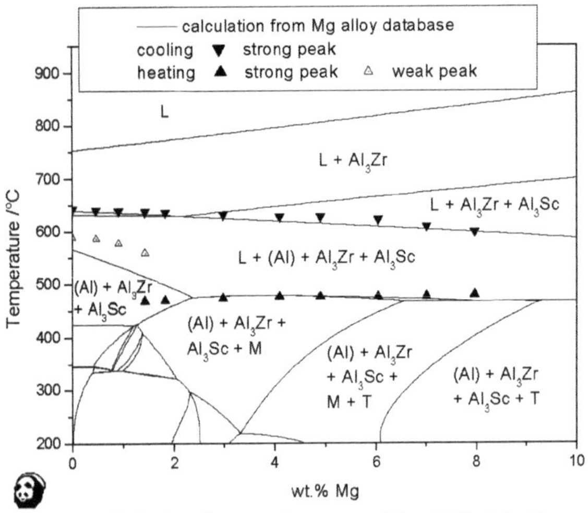

Figure1: Calculated vertical section of the Al-Zn-Mg-Cu-Sc-Zr phase diagram for constant 8 wt.% Zn, 2 wt.% Cu, 0.3 wt.% Sc, and 0.3 wt.% Zr in the Al-rich region [7].

Current scope of the Mg alloy database

The thermodynamic Mg alloy database is developed in an ongoing long-term project in our group since many years. This database currently includes most of the elements used in Mg-alloys including some micro-alloying elements and impurities, and covers most of the phases appearing in commercial magnesium alloys. 17 components (Ag, Al, Ca, Ce, Cu, Fe, Gd, Li, Mg, Mn, Nd, Sc, Si, Sr, Y, Zn and Zr) and a total of 285 phases are included in the current database. 80 of them are modeled as solution phases; the additional 205 phases are stoichiometric. Complete or partial thermodynamic descriptions are developed for many binary and ternary systems as listed in Tables I and II. This means the current database works in a much wider composition range in many sub-systems, as demonstrated above in Fig. 1. The ongoing effort is to complete the description of important interactions among these elements, as well as the extension to new elements.

Table I lists all the binaries in the 17-component system. Thermodynamic descriptions are fully developed for the binaries in green color, which means that there is no composition limits if calculations are carried out for these binary systems. Binaries in yellow color are also developed in the entire composition range, but need further improvements. For these binaries, phase relationships are calculated with larger deviation. No model parameters are developed for those binaries shown white.

Table I. Current Status of Binary Systems

	Al	Ca	Ce	Cu	Fe	Gd	Li	Mg	Mn	Nd	Sc	Si	Sr	Y	Zn	Zr
Ag	Ag-Al	Ag-Ca	Ag-Ce	Ag-Cu	Ag-Fe	Ag-Gd	Ag-Li	Ag-Mg	Ag-Mn	Ag-Nd	Ag-Sc	Ag-Si	Ag-Sr	Ag-Y	Ag-Zn	Ag-Zr
Al		Al-Ca	Al-Ce	Al-Cu	Al-Fe	Al-Gd	Al-Li	Al-Mg	Al-Mn	Al-Nd	Al-Sc	Al-Si	Al-Sr	Al-Y	Al-Zn	Al-Zr
Ca			Ca-Ce	Ca-Cu	Ca-Fe	Ca-Gd	Ca-Li	Ca-Mg	Ca-Mn	Ca-Nd	Ca-Sc	Ca-Si	Ca-Sr	Ca-Y	Ca-Zn	Ca-Zr
Ce				Ce-Cu	Ce-Fe	Ce-Gd	Ce-Li	Ce-Mg	Ce-Mn	Ce-Nd	Ce-Sc	Ce-Si	Ce-Sr	Ce-Y	Ce-Zn	Ce-Zr
Cu					Cu-Fe	Cu-Gd	Cu-Li	Cu-Mg	Cu-Mn	Cu-Nd	Cu-Sc	Cu-Si	Cu-Sr	Cu-Y	Cu-Zn	Cu-Zr
Fe						Fe-Gd	Fe-Li	Fe-Mg	Fe-Mn	Fe-Nd	Fe-Sc	Fe-Si	Fe-Sr	Fe-Y	Fe-Zn	Fe-Zr
Gd							Gd-Li	Gd-Mg	Gd-Mn	Gd-Nd	Gd-Sc	Gd-Si	Gd-Sr	Gd-Y	Gd-Zn	Gd-Zr
Li								Li-Mg	Li-Mn	Li-Nd	Li-Sc	Li-Si	Li-Sr	Li-Y	Li-Zn	Li-Zr
Mg									Mg-Mn	Mg-Nd	Mg-Sc	Mg-Si	Mg-Sr	Mg-Y	Mg-Zn	Mg-Zr
Mn										Mn-Nd	Mn-Sc	Mn-Si	Mn-Sr	Mn-Y	Mn-Zn	Mn-Zr
Nd											Nd-Sc	Nd-Si	Nd-Sr	Nd-Y	Nd-Zn	Nd-Zr
Sc												Sc-Si	Sc-Sr	Sc-Y	Sc-Zn	Sc-Zr
Si													Si-Sr	Si-Y	Si-Zn	Si-Zr
Sr														Sr-Y	Sr-Zn	Sr-Zr
Y															Y-Zn	Y-Zr
Zn																Zn-Zr

: Full description : Full description for major phases : Extrapolation

In addition to the binaries, thermodynamic descriptions for several key ternaries are also developed. Ternary Mg-systems containing the most important alloying elements Al, Ca, Ce, Li, Mn, Si and Zn are listed in Table II. Thermodynamic descriptions for the ternaries in green color are fully developed; the current database therefore works in the whole composition and temperature ranges in these ternary systems. Thermodynamic descriptions for those in the yellow color are also developed, but need further validation by experimental data. Thermodynamic descriptions for the remaining two ternaries (shown white) are obtained by extrapolations of the constituent binaries. Table II is comprehensive and shows all possible ternary systems for this 8 component subset of the Mg-database.

Table II: Key Ternary Systems for the Mg-Al-Ca-Ce-Li-Mn-Si-Zn subset of the Mg-database

Mg-Al-Ca	Mg-Al-Ce	Mg-Al-Li	Mg-Al-Mn	Mg-Al-Si	Mg-Al-Zn
	Mg-Ca-Ce	Mg-Ca-Li	Mg-Ca-Mn	Mg-Ca-Si	Mg-Ca-Zn
		Mg-Ce-Li	Mg-Ce-Mn	Mg-Ce-Si	Mg-Ce-Zn
			Mg-Li-Mn	Mg-Li-Si	Mg-Li-Zn
				Mg-Mn-Si	Mg-Mn-Zn
					Mg-Si-Zn

Additionally, other thermodynamic descriptions for ternary system are included in the Mg-database The Mg-containing systems are listed in Table III. The non-Mg-containing systems are listed in Table IV. The color code is the same as in Table II.

Table III: Additional selected ternary Mg-Systems included in the Mg-database

Mg-Ag-Al	Mg-Ag-Cu	Mg-Al-Cu	Mg-Al-Gd	Mg-Al-Sc	Mg-Al-Sr	Mg-Al-Y
Mg-Cu-Li	Mg-Cu-Si	Mg-Cu-Y	Mg-Cu-Zn	Mg-Li-Gd	Mg-Mn-Gd	Mg-Mn-Sc
Mg-Mn-Y	Mg-Mn-Zr	Mg-Y-Zr				

Table IV: Additional selected ternary non-Mg-Systems included in the Mg-database

Ag-Al-Cu	Al-Ca-Ce	Al-Ca-Fe	Al-Ca-Li	Al-Ca-Si	Al-Ce-Gd	Al-Ce-Nd
Al-Ce-Si	Al-Ce-Y	Al-Cu-Li	Al-Cu-Mn	Al-Cu-Nd	Al-Cu-Si	Al-Cu-Zn
Al-Fe-Mn	Al-Fe-Si	Al-Gd-Nd	Al-Gd-Y	Al-Li-Mn	Al-Li-Si	Al-Mn-Sc
Al-Mn-Si	Al-Nd-Y	Al-Si-Y	Al-Si-Zn	Al-Sn-Zn	Ca-Fe-Si	Ca-Li-Si
Ca-Sr-Zn	Cu-Fe-Si	Cu-Sn-Zn	Fe-Mn-Si	Mn-Y-Zr		

Database validation

Some general aspects concerning the development of the Mg-database with a larger number of additional references validating its content were given earlier [8]. The most important aspects of quality assurance were given specifically in [9] and are further developed in more general terms [3]. These important aspects, concerning artifact-free and thermodynamically reasonable construction, have been used to scrutinize the Mg alloy database and will not be discussed further at this point.

The current thermodynamic database for magnesium alloys has furthermore been extensively tested and validated using the published experimental data [10, 11, 12, 13, 14, 15, 16, 17 18, 19]. Some sub-quaternary systems of this database have been critically assessed: Mg-Al-Ca-Ce, Mg-Al-Ca-Li, Mg-Al-Ce-Li, Mg-Al-Cu-Zn, Mg-Mn-Y-Zr, but not published yet. The quaternary systems Mg-Al-Li-Si [20] and Mg-Ce-Mn-Sc, Mg-Gd-Mn-Sc, Mg-Mn-Sc-Y [21, 22] were thermodynamically modeled and used for technical applications. These comparisons are condensed into a small number of diagrams, highlighting important data, as shown below.

For the reliability of the calculated phase diagrams the fitting of the invariant temperatures are of paramount importance. Since the measured temperatures of the invariant reactions are not affected by super-cooling related problems, these nonvariant data are perfect criteria for comparison of experimental with calculated data as shown in Figure 2.

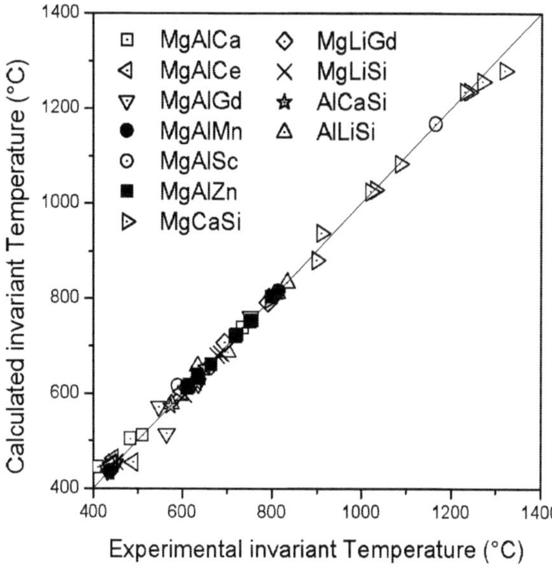

Figure 2: Invariant temperatures for various ternary alloys included in PanMagnesium 5.0: Comparison between calculated and experimental data.

For the Mg-Al-Mn system the reliability is checked in detail [23]. The experimental data are plotted in Figure 3a and 3b as comparison between calculated results and experimental data. The same was done for the system Mg-Mn-Zn [24]. The same experimental data are shown on the liquidus surface of the Mg-rich corner in Figure 4, also indicating the primary crystallizing phase.

From *Magnesium Technology 2007*, R.S. Beals, A.A. Luo, N.R. Neelameggham, and M.O. Pekguleryuz, Editors

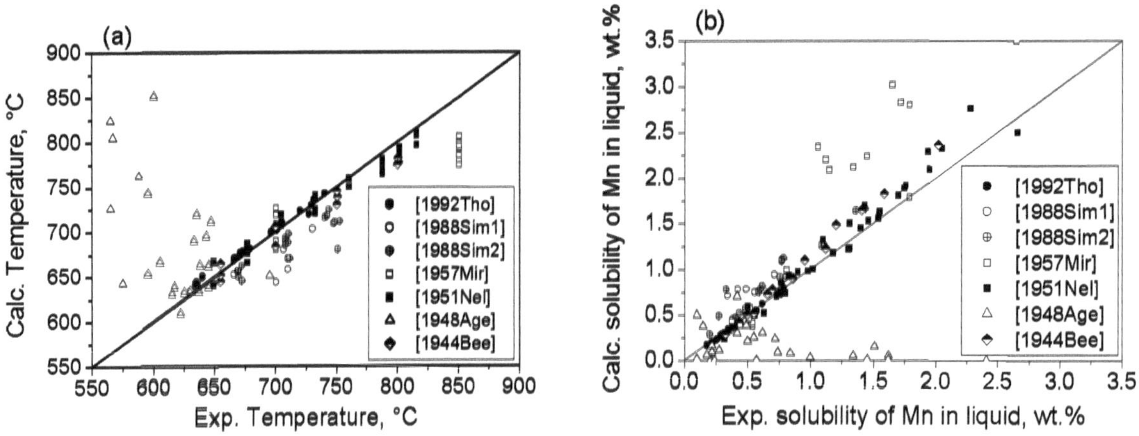

Figure 3: Comparison between calculated results and experimental data for all alloy samples in the Mg-Al-Mn system. (a) Liquidus temperature at a given composition, (b) Solubility of Mn in liquid at a given temperature and Al composition. The straight line in Figs. (a) and (b) is a visual aid corresponding to perfect agreement between experimental values and the calculated results from the present thermodynamic model [23].

This comparison in Figure 3 enables easy identification of those groups of experimental data that are not consistent with the bulk of experimental work. There is a reasonable agreement with this bulk of experimental data and the calculated values. Moreover, there is a reasonable agreement with the primary solidifying phases as shown in Figure 4.

The experimental data for the commercially very important Mg-Al-Zn alloys are shown in Figures 5a and 5b. The liquidus surface of the Mg-rich corner is given in Figure 5a. The same experimental data is plotted in Figure 5b as comparison between calculated results and experimental data [25].

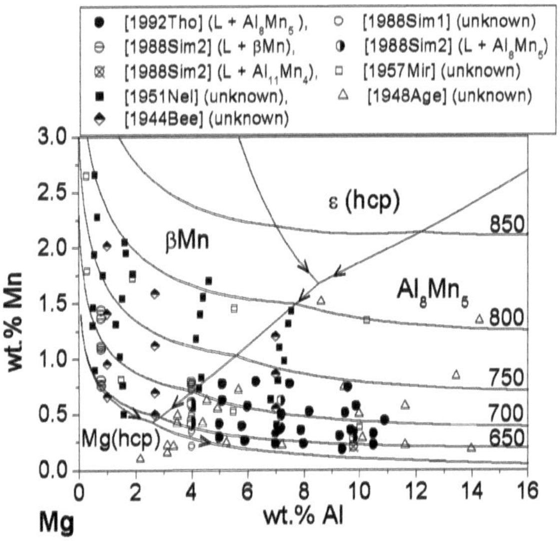

Figure 4: Calculated partial liquidus surface, the thick lines indicate monovariant reaction lines and the thin lines represents the isotherms. Superimposed are the compositions of experimental alloys further compared to calculated data of the liquidus surface in Figure F3. The primary solid phase is specified in some experimental data.

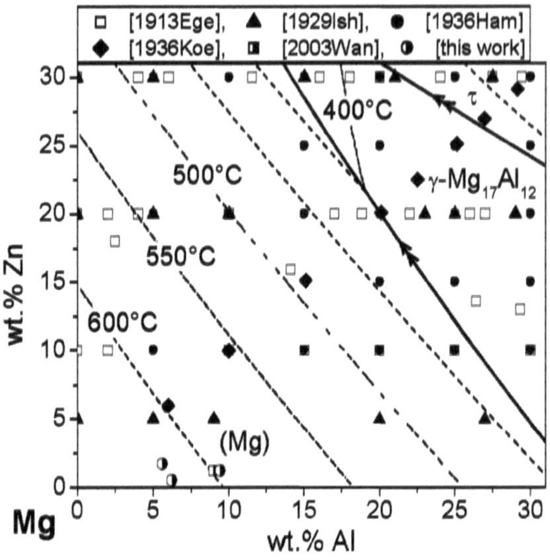

Figure 5a: Calculated partial Mg–Al–Zn liquidus surface and experimental alloy compositions. The thick lines indicate monovariant reaction lines and the dashed lines represent isotherms at an interval of 50 °C [25].

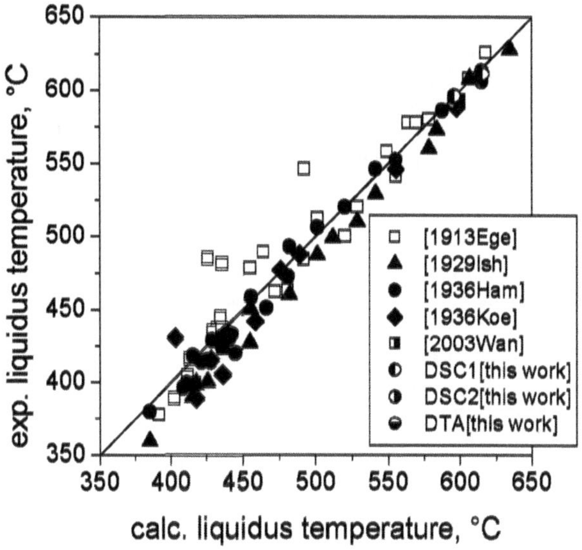

Figure 5b: Comparison between calculated and experimental liquidus temperature for all alloy samples in the Mg-Al-Zn system. The straight line is a visual aid corresponding to perfect agreement between experimental and calculated results[25].

Measured liquidus temperatures for other ternary Mg-alloy systems are compared in Figure 6 with calculations from the magnesium database. These miscellaneous Mg-X-Y systems include the alloying elements Al, Ca, Ce, Gd, Li, Sc, and Si.

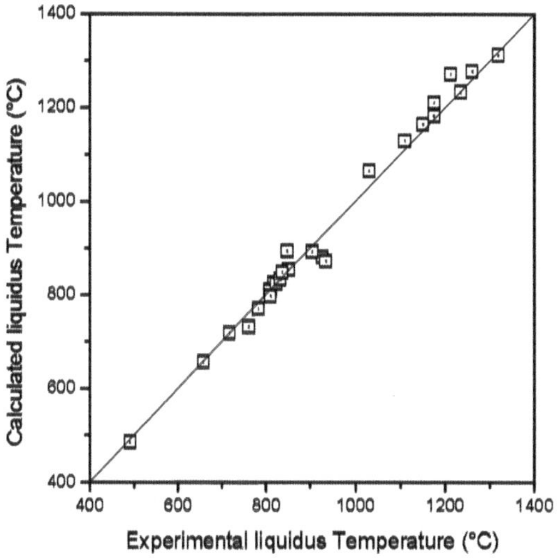

Figure 6: Liquidus temperatures for various ternary magnesium alloys outside the Mg-Al-Mn-Zn system, with alloying elements Al, Ca, Ce, Gd, Li, Sc, Si: Comparison between calculated and experimental data.

Additionally a selected comparison of experimental data and calculated phase equilibria is given below. This demonstrates the feasibility to perform reasonable calculations with the Mg database even in some very high alloyed regions with vanishing Mg-content. For the Al-Li-Si system a comparison between calculated and experimentally measured DTA data is given in Figure 7 [26]. Figure 8 shows a calculated vertical section in the Al-Ce-Si system at constant 90 at.% Al including the DSC/DTA signals measured [16].

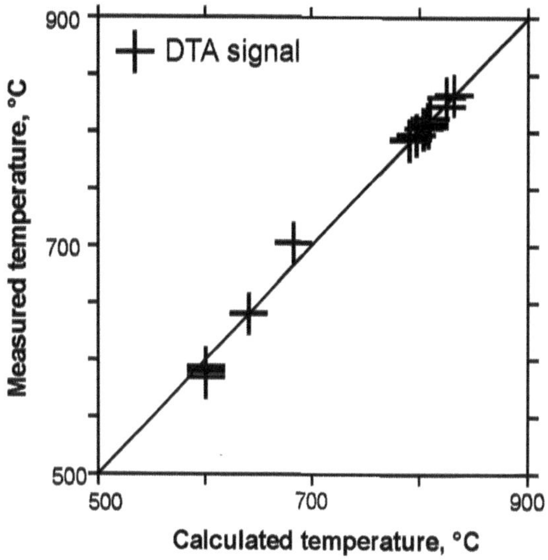

Figure 7: Comparison between calculated and experimentally measured DTA data for the Al-Li-Si system [6].

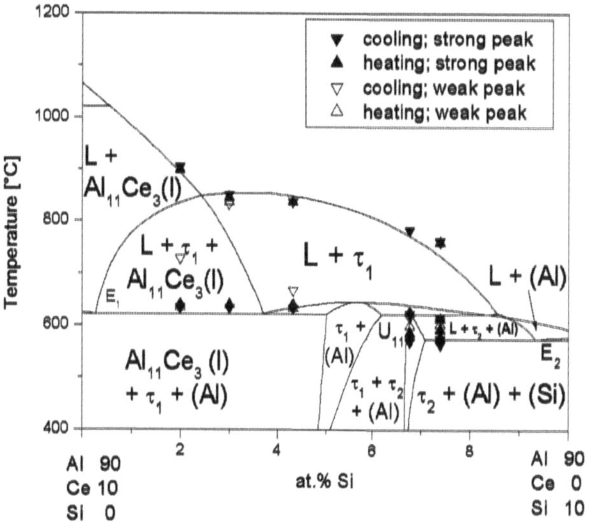

Figure 8: Calculated vertical section $Al_{90}Ce_{10}$ - $Al_{90}Si_{10}$ at constant 90 at.% Al including the DSC/DTA signals measured in [16].

Outlook

The effort is ongoing in our group to complete the description of important interactions among the included elements, as well as the extension of the Mg database to new elements. This has to be done by performing original key experiments combined with Calphad assessments subject to quality assurance and also incorporating first-principle calculations where appropriate, as depicted in Ref. [19]. Another goal is to demonstrate additional useful applications of this database within the tool of computational thermodynamics to Mg alloy development and the advancement of processing of Mg alloys.

This study is supported in part by the German Research Foundation (DFG) in the Priority Programme "DFG-SPP 1168: InnoMagTec"

References

1. L. Kaufman, H. Bernstein, *Computer Calculation of Phase Diagrams with Special Reference to Refractory Materials* (Academic Press, New York, 1970)
2. Y. Austin Chang, Shuanglin Chen, Fan Zhang, Xinyan Yan, Fanyou Xie , Rainer Schmid-Fetzer,W. Alan Oates: Phase Diagram Calculation: Past, Present and Future. *Progress Materials Science*, 49 (2004) 313-345
3. R. Schmid-Fetzer, D. Andersson, P. Y. Chevalier, L. Eleno, O. Fabrichnaya, U.R. Kattner, B. Sundman, C. Wang, A. Watson, L. Zabdyr, M. Zinkevich: "Assessment techniques, database design and software facilities for thermodynamics and diffusion: Group Report from the Ringberg Workshop on Thermodynamic Modeling and First Principles Calculations". *Calphad* (2006) in print.
4. M. Ohno, D. Mirkovic and R. Schmid-Fetzer, "Liquidus and Solidus Temperatures of Mg-rich Mg-Al-Mn-Zn Alloys.," *Acta Materialia,* 54 (2006), 3883–3891.
5. B. Böttger, J. Eiken, M. Ohno, G. Klaus, M. Fehlbier, R. Schmid-Fetzer, I. Steinbach and A. Bürig-Polazek, "Controlling microstructure in magnesium alloys: a combined thermodynamic, experimental and simulation approach," *Adv. Eng. Mat.* 8 (2006), 241-247.
6. T. Laser, M. R. Nürnberg, A. Janz, Ch. Hartig, D. Letzig, R. Schmid-Fetzer and R. Bormann, „The influence of manganese on the microstructure and mechanical properties of AZ31 gravity die cast alloys," *Acta Materialia* 54 (2006), 3033-3041.
7. J. Gröbner, L.L. Rokhlin, T.V. Dobatkina, R. Schmid-Fetzer, "Predictive calculation of phase formation in Al-rich Al-Zn-Mg-Cu-Sc-Zr alloys using a thermodynamic Mg-alloy database," *J. Alloys Compounds*, (2006) accepted.
8. R. Schmid-Fetzer and J. Gröbner, "Focused Development of Magnesium Alloys Using the Calphad Approach," *Adv. Engineer. Mat.* 3 (12) (2001), 947-961.
9. R. Schmid-Fetzer, A. Janz, J. Gröbner and M. Ohno, "Aspects of quality assurance in a thermodynamic Mg alloy database," *Adv. Engineer. Mat.* 7 (12) (2005), 1142-1148.
10. P. Liang, T. Tarfa, J.A. Robinson, S. Wagner, P. Ochin, M.G. Harmelin, H.J. Seifert, H.L. Lukas and F. Aldinger, "Experimental investigation and thermodynamic calculation of the Al-Mg-Zn system," *Thermochim. Acta* 314 (1998), 87-110.
11. J. Gröbner, R. Schmid-Fetzer, A. Pisch, C. Colinet, V.V. Pavlyuk, G.S. Dmytriv, D.G. Kevorkov and O.I Bodak, "Phase equilibria, calorimetric study and thermodynamic modelling of Mg-Li-Ca alloys," *Thermochim. Acta* 389 (2002), 85-94.
12. J. Gröbner and R. Schmid-Fetzer, "Thermodynamic Modeling of Al-Ce-Mg Phase Equilibria Coupled with Key Experiments," *Intermetallics* 10 (2002), 415-422.
13. J. Gröbner, D. Kevorkov, I. Chumak, and R. Schmid-Fetzer, "Experimental Investigation and Thermodynamic Calculation of Ternary Al-Ca-Mg Phase Equilibria," *Z. Metallkd.* 94 (2003), 976-982.
14. J. Gröbner, D. Kevorkov, and R. Schmid-Fetzer, "A new Thermodynamic data set for the binary System Ca-Si and Experimental Investigation in the ternary System Ca-Mg-Si," *Intermetallics* 11 (2003), 1065-1074.
15. D. Kevorkov, R.Schmid-Fetzer and F. Zhang, "Phase Equilibria and Thermodynamics of the Mg-Si-Li System and Remodeling of the Mg-Si System," *J. Phase Equil. Diff.* 25 (2004), 140-151.
16. J. Gröbner, D. Mirkovic, Rainer Schmid-Fetzer, "Thermodynamic Aspects of Constitution, Grain Refining and Solidification Enthalpies of Al-Ce-Si Alloys," *Metall. Mater. Trans. A*, 35A (2004), 3349-3362.
17. H. Cao, J. Zhu, C. Zhang, K. Wu, N. D. Saddock, J. W. Jones, T. M. Pollock, R. Schmid-Fetzer and Y. A. Chang, "Experimental Investigation and Thermodynamic modelling of the Mg-Al-rich region of Mg-Al-Sr System," *Z. Metallkd.* 97 (2006), 422-428.
18. A. Janz, J. Gröbner, D. Mirković, M. Medraj, J. Zhu, Y. A. Chang and R. Schmid-Fetzer, "Experimental study and thermodynamic calculation of Al-Mg-Sr phase equilibria," *Intermetallics* 14 (2006) (accepted).
19. M. Ohno, A. Kozlov, R. Arroyave, Z.K. Liu, and R. Schmid-Fetzer, "Thermodynamic modeling of the Ca-Sn system based on finite temperature quantities from first-principles and experiment," *Acta Materialia* (2006) accepted.
20. D. Kevorkov, "Thermodynamics and Phase Equilibria of the Mg-Al-Li-Si System" (Ph. D. thesis, TU Clausthal, 2001).
21. A. Pisch, J. Gröbner and R. Schmid-Fetzer, "Application of Computational Thermochemistry to Al- and Mg-alloy processing with Sc additions," *Mat. Sci. Eng. A* 289, (2000), 123-129.
22. J. Gröbner, A. Pisch and R Schmid-Fetzer, "Selection of Promising Quaternary Candidates from Mg-Mn-(Sc, Gd, Y, Zr) for Development of Creep-resistant Magnesium Alloys," *J. Alloys Comp.* 320/2, (2001), 296-301.
23. M. Ohno, R. Schmid-Fetzer, "Thermodynamic assessment of Mg-Al-Mn phase equilibria, focusing Mg-rich alloys," *Z. Metallkd.* 96 (2005), 857-869.
24. M. Ohno and R. Schmid-Fetzer, "Mg-rich phase equilibria of Mg-Mn-Zn alloys analyzed by computational thermochemistry," *Int. J. Materials Research (Z. Metallkd,)* 97 (2006), 526-532.
25. M. Ohno, D. Mirkovic and R. Schmid-Fetzer, "Phase equilibria and solidification of Mg-rich Mg-Al-Zn alloys," *Materials Science & Engineering A*, (2006) 421, 328-337.
26. J. Gröbner, D. Kevorkov and R. Schmid-Fetzer, "The Al-Li-Si System, Part 2: Experimental study and Thermodynamic Calculation of the polythermal equilibria," *J. Solid State Chem.* 156 (2001), 506-511.

From *Magnesium Technology 2010*, S.R. Agnew, N.R. Neelameggham, E.A. Nyberg, and W.H. Sillekens, Editors

COMPUTER MODELING OF DC CASTING MAGNESIUM ALLOY WE43 ROLLING SLABS

M. Turski[1], J. F. Grandfield[2], T. Wilks[1], B. Davis[3], R. DeLorme[3], K. Cho[4]
[1] Magnesium Elektron, Rake Lane, Swinton, Manchester, United Kingdom
[2] Grandfield Technology Pty Ltd, Brunswick, Victoria, Australia
[3] Magnesium Elektron North America, Madison, Illinois, USA
[4] U.S. Army Research Laboratory, APG, Maryland, USA

Keywords: DC casting, Elektron WE43, magnesium, computer modeling, cracking

Abstract

During direct chill (DC) casting, significant stresses can develop within the material leading to cracking within the cast slab. The situation is made worse for higher strength magnesium alloys, such as Elektron™ WE43, which exhibits high strength at elevated temperatures. Consequently, the temperature and stress field must be well understood during the casting process to avoid failure during casting.

ALSIM is a fully coupled thermomechanical finite element modeling code capable of simulating the DC casting process. ALSIM was used to simulate vertical DC casting of Elektron™ WE43 slab. Validation of the model was carried out by the insertion of thermocouples during casting and post-mortem sectioning of cast slabs containing cracks. Agreement has been found between the modeling predictions and thermocouple measurements and post-mortem examination of cast material. The use of this validated model has allowed optimized casting parameters to be developed to produce crack-free Elektron™ WE43 slab.

Introduction

Magnesium is the lightest of all engineering materials, being over 30% lighter than aluminum and 4 times lighter than steel. For this reason, the demand for magnesium alloys is increasing where weight-saving engineering solutions are required. From the range of wrought magnesium alloy systems available, magnesium-aluminum-zinc (AZ) alloys are most commonly used in the form of sheet and plate; however, such alloy systems exhibit a significant loss of strength at temperatures above 100 °C. In applications where higher operating temperatures are required, magnesium alloys containing rare earths are typically used. One example of such an alloy is Elektron™ WE43, which is based on the magnesium-yttrium-neodymium-zirconium alloy system. Elektron™ WE43 alloys offer high strength at ambient temperature as well good creep resistance at temperatures up to 300 °C. This alloy also offers good corrosion resistance. Elektron ™ WE43 has also been shown to provide superior ignition resistance [1]. This has been attributed to the protective nature of the oxide layer produced above its melting point.

Elektron™ WE43 was originally developed as a sand casting alloy for aerospace and motor sport applications; however, this alloy has also been used to produce extruded and forged parts. The work presented in this paper is a part of the collaborative project with U.S. Army Research Laboratory (ARL) to establish a commercial production route of wrought magnesium alloy plates and to develop high strength ultra lightweight magnesium plates [2] for military vehicle applications [3]. The first step of such a production route has been to develop a process for DC casting Elektron™ WE43 slabs.

Early attempts to DC cast smaller slab resulted in significant cracking of the alloy (Figure 1). This was attributed to residual stresses that developed in the casting during cooling from the molten state. In the case of Elektron™ WE43 alloy the situation is exacerbated by the low thermal conductivity and high strength of the alloy. As a result of these early trials, computer modeling was employed to determine the optimum parameters required to cast Elektron™ WE43 slab. This paper will outline the modeling process along with preliminary results from casting trials for this alloy.

Modeling of the DC casting process has previously been applied to solve a range of casting problems such as hot tearing [4], porosity [5], cold cracking [6] and macro-segregation [7]. The majority of these models has been applied to aluminum billets and slabs, due to the larger scale of this industry compared to the magnesium industry; however, modeling of magnesium DC casting has been, of late, used as a tool to analyze and improve the process [8].

Figure 1. A typical trouser leg crack formed during DC casting of Elektron™ WE43 alloy slab.

Computer Modeling of the DC Casting Process

Computer simulation of the DC casting process was carried out using the finite element modeling (FEM) package, ALSIM [9]. ALSIM was developed by the Norwegian Institute for Energy Technology (IFE) specifically to model fluid flow, heat transfer and stresses in DC casting. The model is a fully coupled thermal, mechanical and fluid flow model with additional modules available to predict macro-segregation and porosity.

For this application, a quarter symmetry 3D mesh was used (Figure 2). The main elements of the model are the mold, cast distributor, cast material and the dummy block. Given the complex nature of the DC casting process, the cast material is divided into 3 sections, the Eulerian domain, an expansion layer and the Lagrangian domain. The Eulerian domain is fixed in space and size during casting, and only represents the molten metal, so only a pure Eulerian formulation is used for this region. During the start up, the computational nodes follow the downward movement of the bottom block. For the bottom block and the solidified material the material is simulated using the Lagrange formulation, while the nodes in the fluid material are simulated using Eulerian-Langrangian calculations. In this formulation, the convective terms in the transport equations for turbulence parameters (describing molten metal fluid flow) are modified analogous to the momentum equations. In this way, the starting block and solidified parts of the cast slab can be reduced to a pure Lagrangian formulation. Growth of the mesh during casting takes place in the expansion layer, shown in green. During casting, this layer of elements grows vertically until a maximum thickness is reached (in this case ~ 15 mm). At the critical thickness the elements divide, thus creating a new layer.

Figure 2. Mesh and geometry used for computer modeling of the DC casting process.

Thermal properties

Thermophysical measurements were made on as-cast Elektron™ WE43 alloy to provide the model with accurate values of thermal conductivity, density, thermal expansion and the latent heat of fusion as a function of temperature. The thermophysical properties for this alloy were measured due to the lack of published data in the literature. Thermal conductivity was determined through separate measurements of the specific heat capacity, density and thermal diffusivity, at 100K steps from room temperature to the melting point of the alloy. The specific heat capacity and latent heat of fusion were measured using differential scanning calorimetry (DSC). Thermal diffusivity measurements were made using the laser flash technique, and density measurements were made using a pushrod dilatometer.

Mechanical properties

Mechanical properties were measured under compressive loading using a Gleeble 3500 thermomechanical testing machine located at University of British Columbia. Compression samples, 15 mm long and 10 mm in diameter were machined from as-cast Elektron™ WE43 material. Tests were conducted with strain rates of 0.1/s, 0.01/s and 0.001/s at temperatures of 25, 200, 300, 400 and 500 °C, producing 15 tests in total. During testing, each sample was brought up to the respective test temperature by resistive heating at 5°C/s, and held for 60 seconds. True strain was measured directly via a dilatometer through conservation of volume with the relation $\varepsilon = -2\ln(D/D_0)$, with D being the diameter during the test, and D_0 being the diameter before the start of the test including thermal expansion. True stress was calculated directly from the force measured by the load cell divided by the cross section area of the sample during the test. The results from these tests at the strain rate of 0.1 /s are shown in Figure 3. These results show the alloy maintained a significant level of strength up to 300°C. After 300°C the strength of the alloy reduced rapidly.

The strain-hardening response of the alloy within the model is described by the Ludwig equation, given below:

$$\sigma = K\left(\dot{\epsilon} + \dot{\epsilon}_o\right)^m \left(\epsilon + \epsilon_o\right)^n$$

where σ is the stress in MPa, $\dot{\varepsilon}$ is the strain rate in s^{-1} and ε is the strain. The strength coefficient, K, the strain sensitivity coefficient, m, and the work hardening coefficient, n, are parameters fit typically as a function of temperature. The onset of plasticity was determined using the Von Mises yield criteria, and the material was assumed to strain-harden isotropically.

Mechanical properties were also measured under tensile loading, in order to assess the ductility of the material as a function of temperature. Tests were carried out at 25, 200 and 300 °C. Uniaxial loading was carried out on specimens with a 5.6 mm diameter x 28 mm gauge length, at a crosshead displacement of 0.5 mm/min. Figure 4 shows the results from these tests. The ductility is at its maximum at >30% at 300°C, reducing to ~6 % at room temperature. The yield stress values are similar to the compression tests; however, the UTS values are lower. This is due to the higher plastic strains achieved during compressive loading (up to 35 %), resulting in the material strain-hardening to a higher value of stress. The similarity between the tensile and compressive yield stress values suggests that the material is isotropic for monotonic loading.

Initial Simulations and Casts

Simulations

Once the mold geometry had been defined, a range of casting parameters was simulated as part of a sensitivity study. The aim of this study was to determine a set of casting parameters suitable for initial casting trials. The simulation software allowed the whole casting process to be simulated, including initial filling of the mold and DC cast start up and stopping practices. This allowed the effect of a range of parameters to be assessed including:

- Fill time of the mold.
- Hold time.
- Melt level in the mold.
- Melt distribution.
- Ramp time taken from start to full casting speed.
- Casting temperature.
- Casting speed.

- Water flow rate.
- Ramp down time.

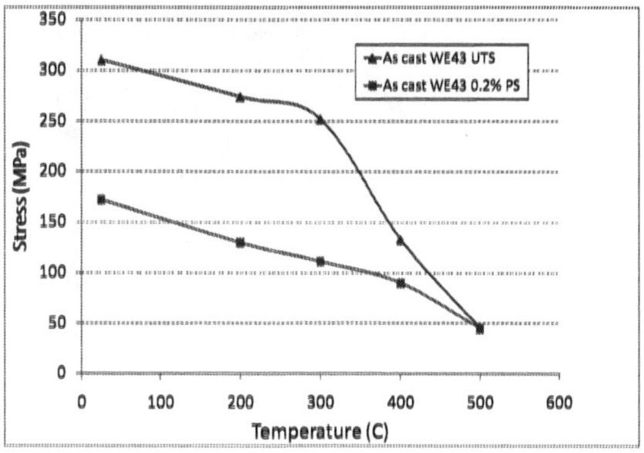

Figure 3. Change in proof stress and UTS of as-cast WE43 as a function of temperature as measured during compression testing

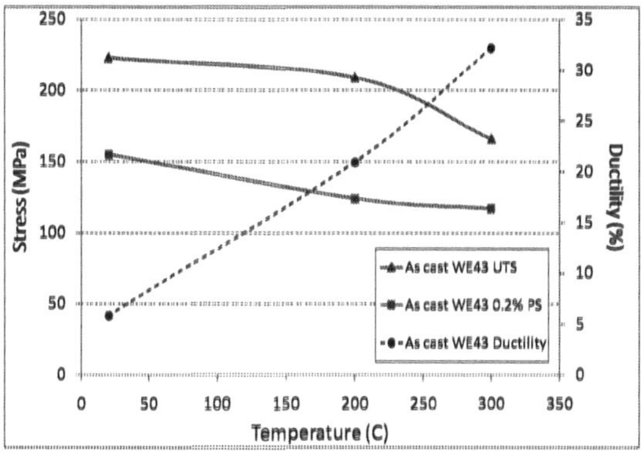

Figure 4. Change in proof stress, UTS and ductility of as–cast WE43 as a function of temperature as measured during tensile loading.

The results from these simulated campaigns were graded in terms of severity of predicted butt curl, and magnitude of plastic strain and tensile stress within the slab. The aim of this exercise was to minimize the accumulation of plastic strain, stress and butt curl. Based on the results from these simulations, an initial set of parameters were determined for the initial casting campaign.

Initial DC casts

Three small scale trials were carried out using an experimental DC casting facility at Magnesium Elektron. The aim of this first campaign was to measure the temperature within the slab during casting and to determine the level of cracking and distortion. Thermal measurements were taken to validate the thermal boundary conditions applied to the model and the subsequent predicted thermal fields, while regions containing distortion and cracking were measured to validate the mechanical model.

For the initial trials the cast weight was restricted to ~500 kg per cast. A range of casting parameters was trialed based on the initial modeling results. Temperatures were measured using a thermocouple tree that was frozen into the melt during casting. Five thermocouples were attached to the tree, the exact positions of which were subsequently determined by sectioning the cast slab. The locations of the 5 thermocouples are shown schematically in Figure 5. Three 500 kg slabs were successfully cast. Although no catastrophic cracking was observed, a number of surface cracks were found at the slab base and at the corners (Figure 6).

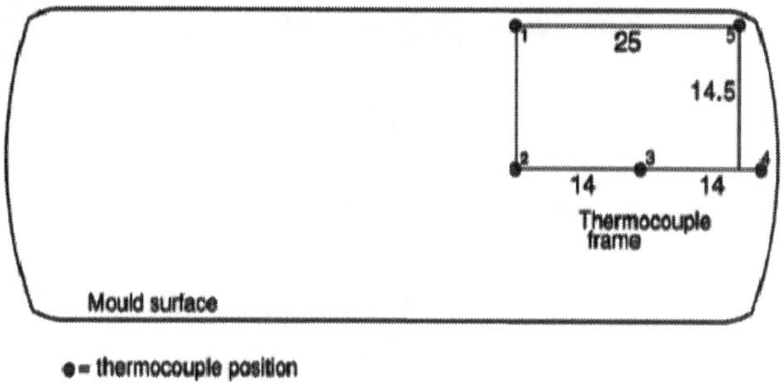

Figure 5. Locations of the thermocouple measurement points with respect to the DC mold surface.

A larger, 1,500 kg scale DC casting trial was carried out in the production foundry, based on the experience gained from the first batch of 500 kg slabs casts. The aim of this larger scale trial was to assess the effect of a longer slab, where steady state casting conditions would be attained. Based on the experience from earlier casts, the casting conditions were optimized to limit the level of cracking within the slab.

Thermal Validation of DC Casting Model

Initial comparisons between thermocouple measurements and temperature predictions extracted from the model showed good agreement between the locations close to the center of the casting, but poor agreement between measurements and predictions at locations close to the mold wall. At locations close to the mold wall the initial cooling rate predicted by the model was higher than the measurements. This suggests that the heat transfer coefficient used for the model through the mold wall was too high. Consequently, changes to the heat transfer coefficient were made to improve agreement between the model predictions and measurements. Figure 7 shows a comparison between the thermocouple measurements and model predictions as a function of distance below the metal level, after a reduction in the heat transfer coefficient. This graph shows agreement between the model predictions and the experimental measurements. One unexpected feature of the graph was the crossover between the short face (TC4) and rolling face (TC1) measurements (i.e., if the mold provides constant cooling around the perimeter, then the short faces of the as-cast slab should always be cooler than the long faces). This led to the investigation of the mold water spray alignment: cooling water distribution was found to be uneven and was subsequently rectified.

Validation of Mechanical Model

Although only mild surface cracking was observed on each of the cast slabs (Figure 6), significant internal cracking was observed using ultrasonic examination and sectioning of the slabs (Figure 8). Figure 8a shows a section taken through the slab thickness, this figure shows large cracks ~20 cm in length, running parallel to the casting direction and orientated ~45° to the slab faces. The cracks are located close to each of the corners. Figure 8b shows regions of cracking identified using ultrasonic inspection. It is clear that cracking has occurred along the length of slab, close to

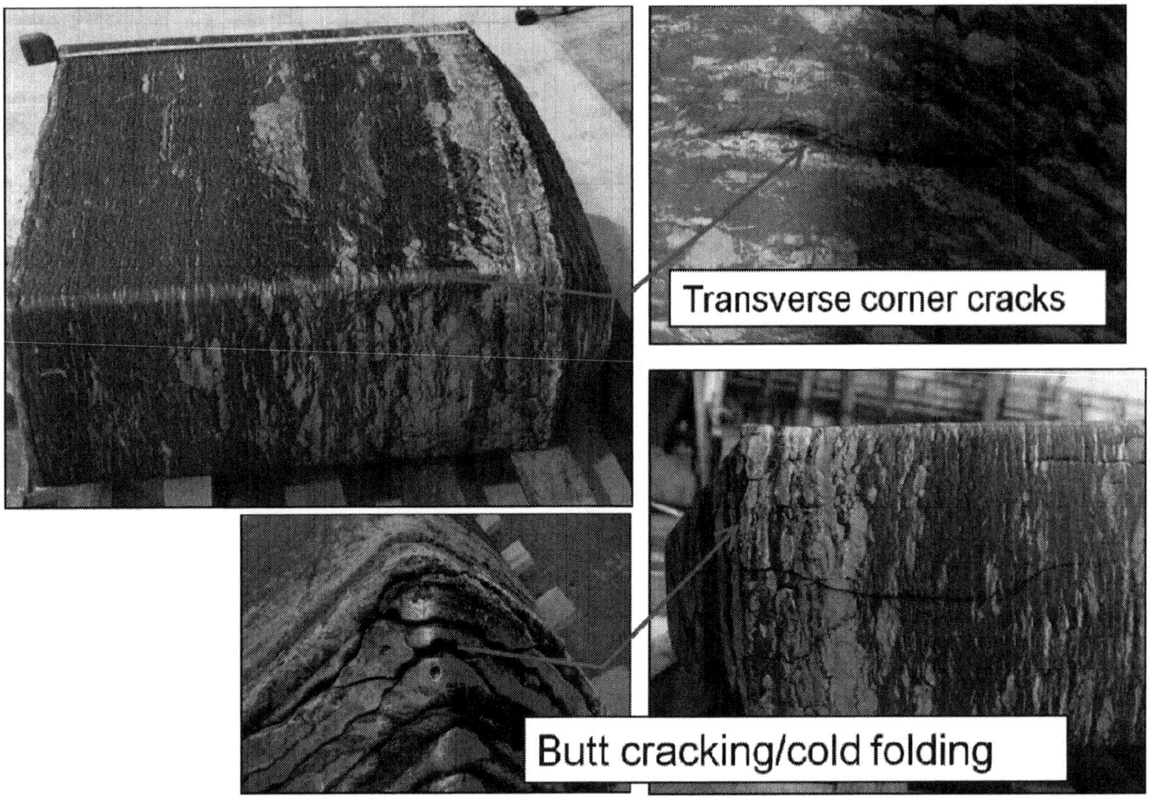

Figure 6. Photographs showing the surface condition of DC cast Elektron™ WE43 alloy

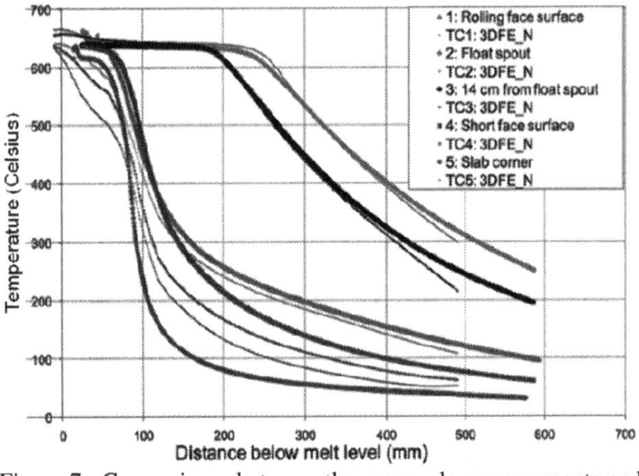

Figure 7. Comparisons between thermocouple measurements and predicted temperatures as a function of distance below the melt level.

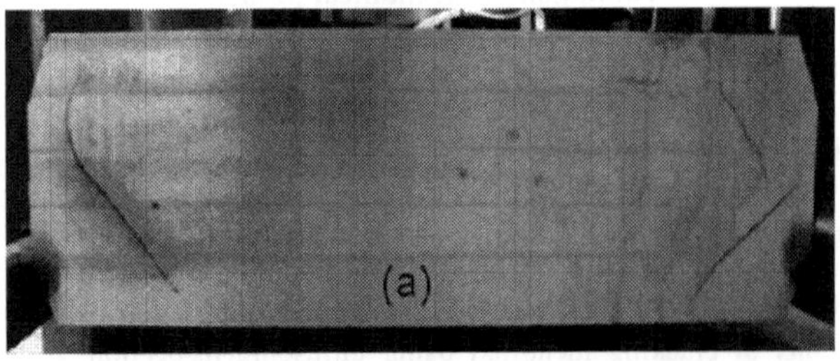

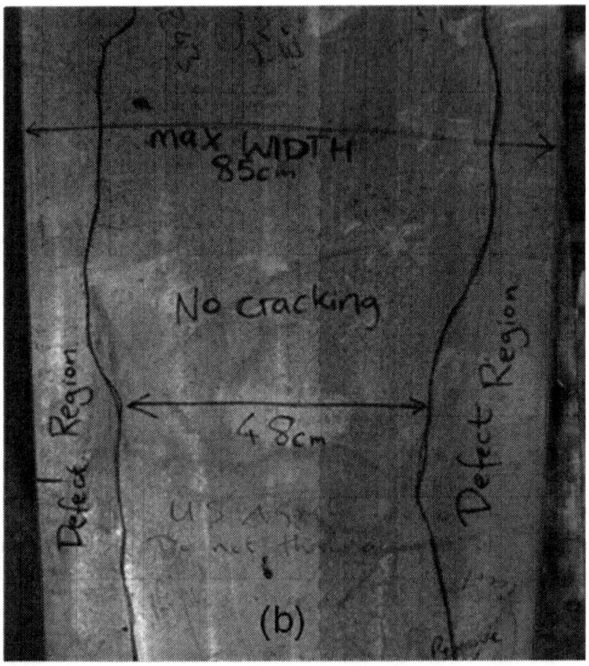

Figure 8. (a) Internal cracking regions as identified using ultrasonic inspection (b) Through thickness internal cracking observed in a section taken from (a).

each of the short faces. Figure 9 shows the predicted X, Y and Z components of stress within the slab after casting. These stress fields show a typical quenching stress distribution, with high compressive stresses (up to -200 MPa) at the surface, balanced by highly tensile stresses (~160 MPa) in the mid thickness of the slab. However, while such a stress field would be expected, it does not correlate to regions of cracking found within the slab.

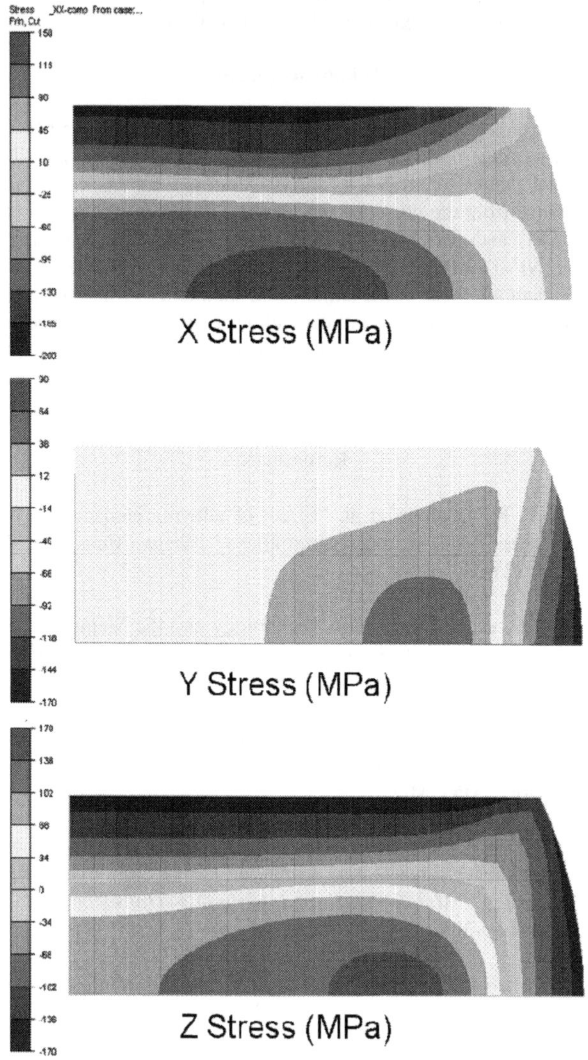

Figure 9. Predicted residual stress fields within slab cross section during DC casting.

Figure 10 shows the XY component of stress through the slab thickness, the stress field has been mirrored for clarity. This figure shows a high concentration of tensile residual stress within the region where cracking had been located. Not only does the location of the tensile stress field correspond to the location of the crack, but the component of stress is also acting perpendicular to the crack face (components x,y,z and xy are shown in Figure 10). This strongly suggests that this component of stress is related to the cracking observed. This region of tensile stress would be expected to act along the whole slab, as the computer model predicted the same stress distribution along the slab length.

One aspect of the internal cracking observed that is not clear is the mechanism of cracking. The magnitude of peak stress shown in Figure 10b is 60 MPa. This is significantly lower than the yield

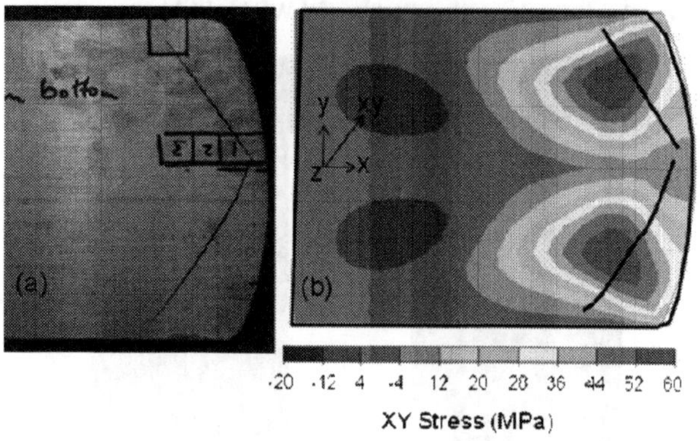

Figure 10. (a) Internal cracking observed in a cast slab. (b) Overlay of internal cracks found in slab with predicted concentration of XY component of stress.

stress of the material (Figures 3 and 4). The peak plastic strain (< 1%) predicted within the vicinity of the crack was also much lower than the ductility of the alloy in the as-cast state (Figure 4). It is possible that the cracks initiated close to the base of the slab where significant cold folding occurred during cast startup. Figure 6 shows an example of a crack that initiated at a cold fold and propagated along the length of the slab. This suggests that cracking may have initiated at startup and propagated along the slab length as the XY component of stress further up the slab started to accumulate. Oxide particles could also play a role in initiating cracks.

Further computer modeling after validation

A new program of simulations was carried out, after validation of the model had taken place. The severity and location of the cracking within the cast slabs helped to focus the direction of the modeling to reduce the residual stresses within the slab. In essence, an extended 'virtual casting campaign' was carried out, allowing different casting parameters to be adjusted systematically using the computer model. This helped to develop optimum casting parameters, without the cost of many 'trial and error' DC casting trials in the foundry.

At the end of the process, a casting route was identified allowing the level of stress within the slab to be reduced dramatically. The resultant stress fields at the end of casting are shown in Figure 11. Figure 11a shows the original principal stress field as predicted for the 1st casting campaign. Figure 11b shows the principal stress distribution at the end of casting, using the parameters determined from the second round of modeling. Both figures are plotted using the same color scaling.

A full scale casting trial was carried out using these casting parameters determined from the second round of modeling. Two slabs, each over 2 m in length, were successfully cast (Figure 12). These were inspected ultrasonically and found to be free from internal cracking. These slabs were machined and successfully rolled to produce Elektron™ WE43 plate.

Summary

This paper has provided an overview of the developmental process applied to determining the casting parameters of the high strength magnesium alloy Elektron™ WE43. FEM computer modeling was an essential part of this process. In order to ensure the accuracy of the model, thermophysical and mechanical properties were gathered for this alloy. Initial trials showed that this alloy is prone to a significant level of cracking during casting, due to the relatively low thermal conductivity and high mechanical strength of the alloy. These initial trials were useful in validating the model and determining the mechanism of cracking. Subsequent modeling work on the validated model allowed optimized casting parameters to be determined, significantly reducing the level of internal stress during casting. Application of these parameters resulted in Elektron™ WE43 slabs being successfully cast. This has helped Magnesium Elektron to develop a large scale commercial process for the production of high strength Elektron™ WE43 magnesium alloy plate. It is hoped that this progress will help to further develop the market for high strength wrought magnesium alloy plate and sheet products.

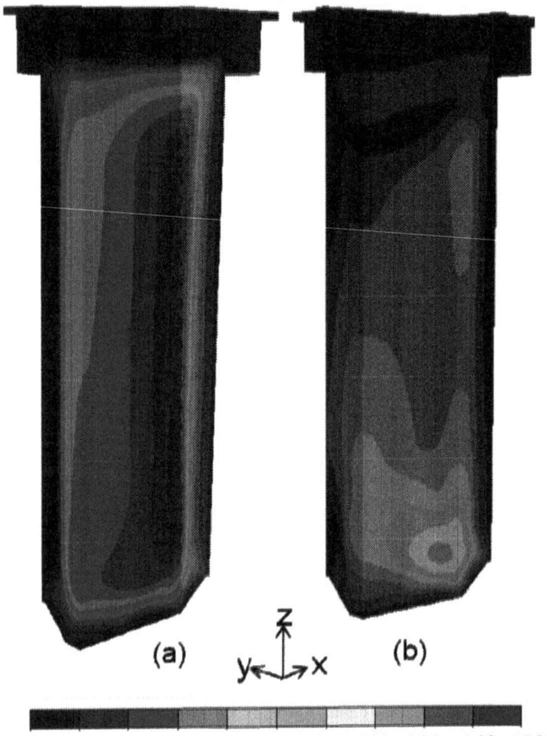

Figure 11. Left: principal stress field in cast slab after initial optimization study. Right: principal stress field in cast slab after 2nd round of modeling activity. All units are in MPa.

Figure 12. Successfully cast Elektron™ WE43 slab

Figure 11b shows a dramatic reduction in the overall stress field, with reductions in some areas of the slab from 160 MPa to ~30 MPa.

Acknowledgments

The authors gratefully acknowledge Drs. D. Mortensen and H. G. Fjaer of IFE for advice on computational modeling with ALSIM. Special thanks to MENA's Martyn Alderman for his vision and guidance from the outset of this project. Research was sponsored by ARL and was accomplished under Cooperative Agreement W911NF-07-2-0073. The views and conclusions made in this document are those of the authors and should not be interpreted as representing the official policies, either expressed or implied, of ARL or the US Government. The US Government is authorized to reproduce and distribute reprints for Government purposes notwithstanding any copyright notation hereon.

References

1. N. V. Ravi Kumar et al. "Effect of alloying elements on the ignition resistance of magnesium alloys," *Scripta Materialia*, 49 (2003), 225-23.

2. K. Cho et al "Magnesium Technology and the Manufacturing for Ultra Lightweight Armoured Ground Vehicles," *Proceedings of the 2008 Army Science Conference* (2008).

3. MIL-DTL-32333 (MR), "Detail Specification Armor Plate, Magnesium Alloy, Az31b, Applique," Army Research Laboratory, APG, MD.

4. M. M'Hamdi, A. Mo and H.G. Fjær, "TearSim: A two-phase model addressing hot tearing formation during aluminium direct chill casting", *Metallurgical and Materials Transactions*, 37A (2006), 3069-3083.

5. P. D. Lee et al, "Modeling of porosity formation in direct chill cast aluminum-magnesium alloys" *Materials Science and Engineering A* 328 (2002), 213-222.

6. W. Boender et al, "Numerical Simulation of DC casting; Interpreting the results of a thermo-mechanical model" Light Metals 2004, 679-684.

7. D. G. Eskin et al, "Structure Formation and the macrosegregation under different process conditions during DC casting," *A* 384 (2004), 232-244.

8. J.F. Grandfield, V. Nguyen and I.F. Bainbridge. "Stresses and Cracking During Direct Chill Casting of AZ31 Alloy Billet" Magnesium Technology 2009, 129-133.

9. D. Mortensen, "A mathematical model of the Heat and Fluid Flows in Direct-Chill Casting of Aluminum Sheet Ingots and Billets" Metallurgical and Materials Transactions B, 30B (1999), 119-133.

MAGNESIUM FOR CRASHWORTHY COMPONENTS

T. Abbott[1], M. Easton[1] and R. Schmidt[2]
CRC for Cast Metals Manufacturing (CAST), PO Box 69M, Monash University 3800, Australia
[1]Monash University, School of Physics & Materials Engineering, Melbourne Vic 3168, Australia
[2]Industrial Research Institute Swinburne, Melbourne Vic 3122, Australia

Abstract

Most applications of magnesium in automobiles are for non-structural components. However, the light weight properties of magnesium make it attractive in structural applications where energy absorption in a crash is critical. Because most deformation in a crash occurs as bending rather than simple tension or compression, the advantages of magnesium are greater than anticipated simply from tensile strength to weight ratios. The increased thickness possible with magnesium strongly influences bending behavior and theoretical calculations suggest almost an order of magnitude greater energy absorption with magnesium compared to the same weight of steel. The strain rate sensitivity of steel is of concern for energy absorption. Mild steels exhibit a distinct yield point which increases with strain rate. At strain rates typical of vehicle impact, this can result in strain localization and poor energy absorption. Magnesium alloys with relatively low aluminum contents exhibit strain rate sensitivity, however, this is manifest as an increase in work hardening and tensile / yield ratio. This behavior suggests that the performance of magnesium alloys in terms of energy absorption actually improves at high strain rates.

Introduction

With the emergence of new materials in automotive applications there is a tendency among design engineers to be cautious until the properties of the new materials are fully characterized. Often the properties which receive the greatest attention are those where the current materials have some deficiencies. These deficiencies are not necessarily characteristic of the new material.

A case in point is strain rate sensitivity for crash worthiness of steel and magnesium. Low carbon steels exhibit a distinct yield point during deformation, usually followed by a drop in load followed by work hardening. As the strain rate is increased, the yield point tends to increase such that it exceeds the post yielding ultimate tensile strength. In this situation there is no effective work hardening, resulting in the localization of deformation so that only a small part of the component participates in energy absorption. This characteristic tensile property and strain rate sensitivity of plain carbon steel is shown in Figure 1 [1].

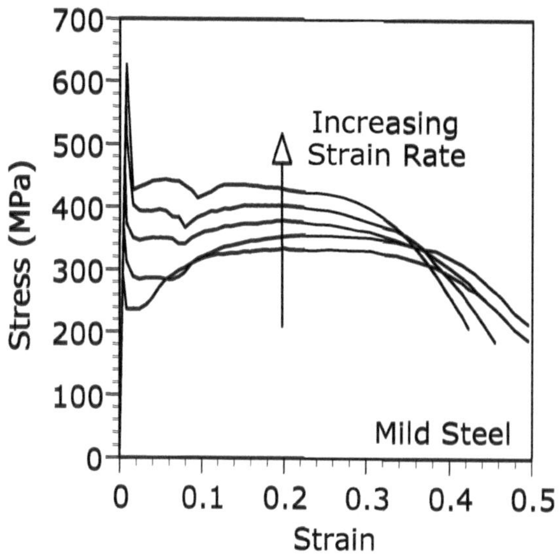

Figure 1. Tensile properties of mild steel tested at a range of strain rates ($0.001s^{-1}$ - $106s^{-1}$). Adapted from [1].

In the case of aluminum alloys, for example 6061, their behavior is essentially strain rate insensitive [1].

There appears to be a strong strain rate sensitivity in pure magnesium [2], although literature data is very limited. The absence of a distinct yield point in magnesium suggests that the issues of strain localization will not be the same as for steel. However, the significance of strain rate sensitivity for structural magnesium components is unclear.

In this paper the natural advantages of using magnesium for energy absorption components are discussed. This is followed by discussion of an experimental investigation into strain rate sensitivity in common magnesium alloys.

Magnesium for Energy Absorption

The advantages of light metals, and in particular magnesium, are often highlighted in terms of strength to weight ratios for tensile properties. In the case of magnesium, the strength relative to steel, on a weight basis, is more than four times the value on a volume basis. While this appears attractive it actually understates the advantages of magnesium in many structural applications.

In an automobile crash situation much of the deformation of structural components occurs in bending modes rather than simple tension or compression. If steel components are replaced with magnesium and the component thickness is increased such that the weight is kept constant then the difference in bending properties and energy absorption is dramatic.

Using a methodology developed by Grzebieta and Murray [3,4] an assessment can be made of the energy absorption of steel and magnesium. This analysis considers a vertical strut subjected to impact loading. The strut has an initial kink or imperfection and undergoes buckling. As the strut is compressed, the load increases to a peak and can be described by an elastic analysis. After reaching a peak load denoted the squash load (yield strength by the cross-sectional area) the load decreases rapidly and is described by rigid plastic behavior (Figure 2). The energy absorbed by the strut corresponds to the area under the curve and is therefore mostly dependent upon the rigid plastic curve and the deflection before fracture.

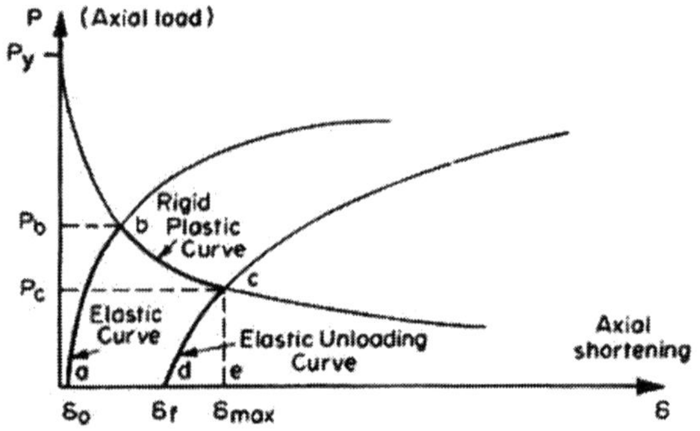

Figure 2. Behaviour of struts according to Grzebieta and Murray [3].

The optimum material behavior is to have a low squash load and a high value for the rigid plastic deformation. This is to minimize the initial jolt but absorb as much energy as possible. The point at which fracture of the strut occurs is also of great importance, that is, the more the strut can be squashed without breaking the better.

The equation developed for rigid plastic behavior is:

$$\delta^P = \frac{M_P^2}{DP^2}\left[1-\left(\frac{P}{P_y}\right)^2\right]^2$$

where δ^P is the plastic vertical deflection, M_p is the plastic bending moment, D is the span length, P is the load and P_y is the squash load.

This equation indicates that the maximum load is related to the yield strength of the material (as well as the size of the kink) and the amount of load that the plastic hinge can absorb is related to the yield strength and the thickness of the strut.

Using the properties of mild steel and magnesium alloy AM60, this analysis can illustrate the relative energy absorption properties of the two materials. Figure 3 shows a comparison of steel plates 2.5 and 10mm thick and AM60 plates 5 and 10mm thick.

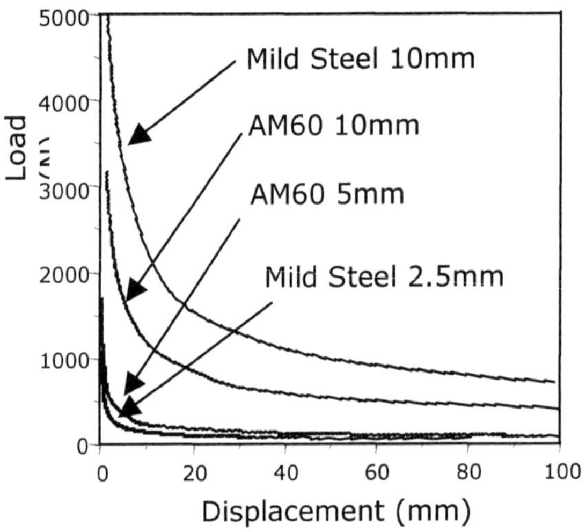

Figure 3. A comparison of the rigid plastic behaviour of mild steel and AM60. In the theoretical calculations the mild steel was assumed to have a yield stress of 280MPa whilst AM60 was assumed to have a yield strength of 150MPa. 2.5mm thick mild steel has the same weight as 10mm thick AM60.

Comparison of the area under the curves for the 10mm AM60 and 2.5mm steel plates (Figure 3) suggests that the energy absorbed by magnesium is close to an order of magnitude higher than for the same weight of steel. Even at half the weight of steel (5mm AM60 compared to 2.5mm steel) the energy absorbed by magnesium is greater.

These comparisons are for idealised models and the actual relative behaviour of magnesium and steel needs to be measured experimentally. However, this analysis suggests a significant benefit can be obtained by applying magnesium to structural applications. The model's validity has been tested previously for steel and fits well [3]. The model's assumption of no work hardening during plastic deformation would produce greater inaccuracies for magnesium. That is, the model should underestimate the performance of magnesium components.

Strain Rate Sensitivity

An experimental program was undertaken to examine mechanical properties and strain rate sensitivity of magnesium alloys produced by high pressure die casting. As-cast tensile bars were produced using a Toshiba 250 tonne Cold Chamber die casting machine. Tests on other samples from this facility have been reported in a number of papers [5,6].

Alloys AZ91, AM60 and AS21 were cast as flat tensile specimens with a 50mm long and 10mm wide gauge length and thicknesses of 1mm, 2mm, 5mm and 10mm. The samples were tested using a high strain rate hydraulic (MTS) mechanical testing machine. A 50mm extensometer was used and the nominal test speeds were 1, 10 and 100mm/s. The strain rates were measured from data logged by the extensometer and were determined to average 0.012, 0.12 and 1.2 /s respectively. All combinations of the above strain rates and section thicknesses were tested for AZ91 and AM60 alloys. For AS21 alloys 1mm thick samples and the intermediate strain rate were excluded due to limited sample availability. A limited number of tests at a strain rate of 0.00012/s were also conducted.

In Figure 4 the original flow curves are shown for 5mm thick AM60 and AS21 samples. The relatively high strain rate sensitivity of AS21 is clearly evident.

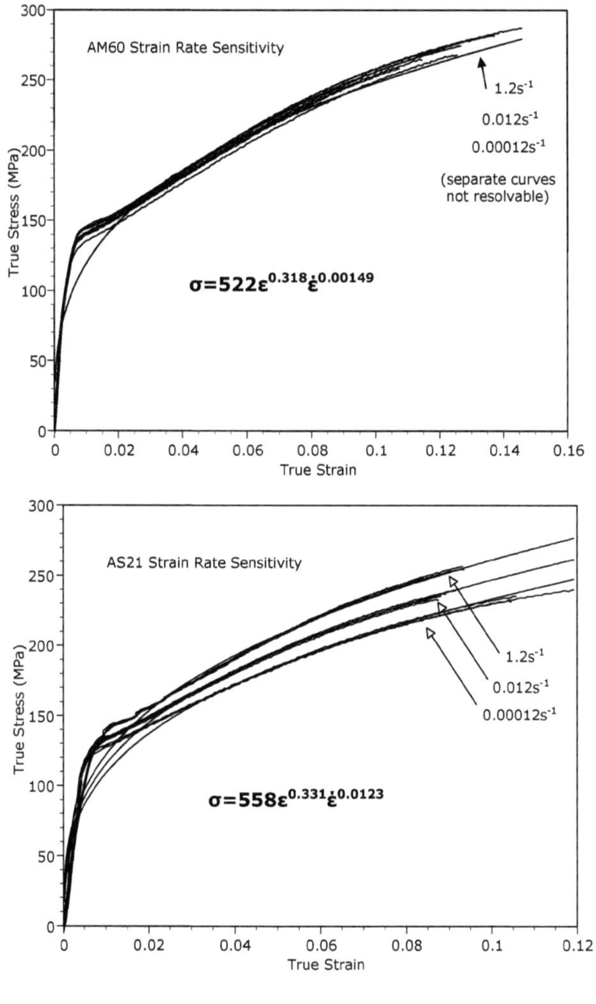

Figure 4. Stress strain curves of 5mm thick as-cast test bars of alloy AM60 (top) and AS21(bottom). A best fit model of stress as a function of strain and strain rate is also shown.

In Figure 5 the fracture stress is plotted against the elongation to failure for each alloy, section thickness and strain rate. This representation shows an effect of section thickness in AZ91 and section thickness and strain rate in AS21. For AM60 a small effect of strain rate is discernable.

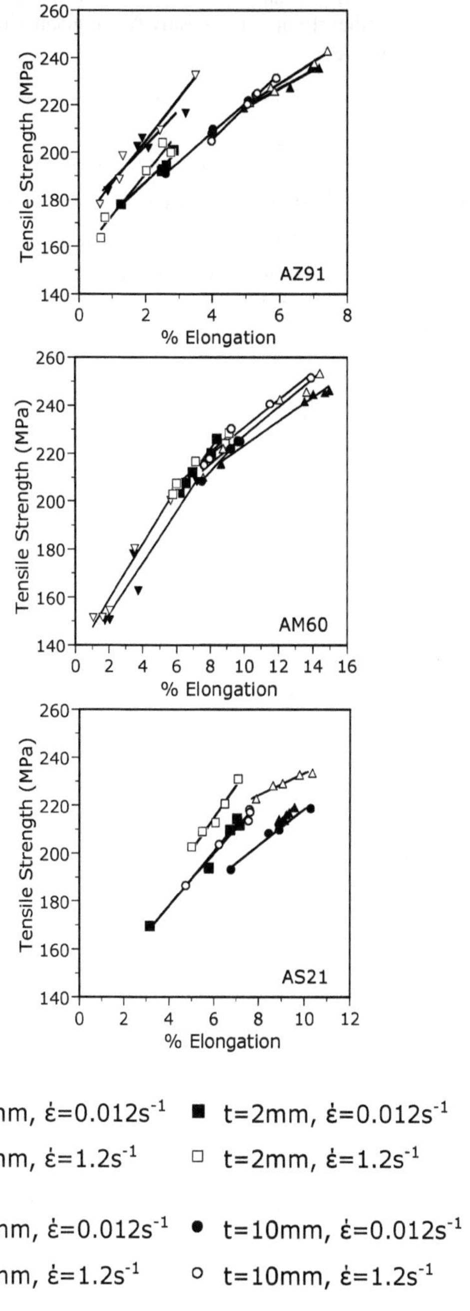

Figure. 5. Effects of section thickness (t) and strain rate ($\dot{\varepsilon}$) on the tensile strength and total elongation of AZ91, AM60 and AS21 alloys.

Discussion

The experimental results indicate a significant strain rate sensitivity for AS21 but not for AM60 or AZ91. The main difference between these alloys is their aluminum content. A previous report [2] has shown this to be a significant factor influencing the strain rate sensitivity of magnesium alloys, as shown in Figure 6.

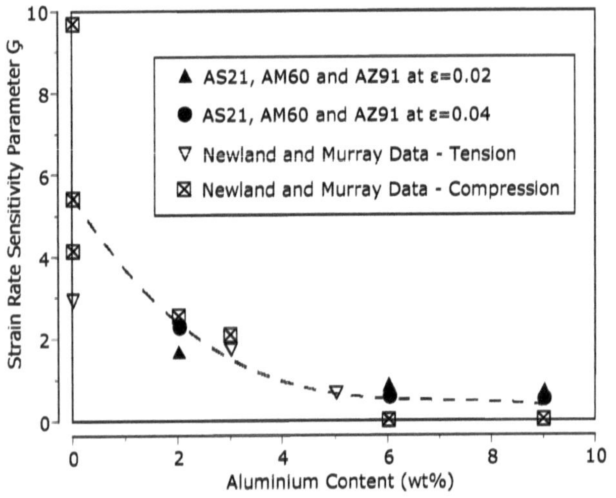

Figure 6. Comparison of strain rate sensitivity data to literature reports. In this case the strain rate sensitivity parameter (Ģ) is equal to ln(dε/dt)/(σ-K) where K is a constant. Newland and Murray data from [2].

In Figure 6 the strain rate sensitivity results for the present work are plotted with previous data measured by Newland and Murray [2]. The pronounced decrease in strain rate sensitivity with aluminum content is clearly evident.

The measured strain rates fall short of typical rates encountered in vehicle impacts by a factor of 100-200 times [7]. However, if the observed behavior extends to the higher strain rates then magnesium appears to have a significant additional advantage over and above that from being light weight.

The form of strain rate sensitivity is different to that associated with the yield point in steel. While high strain rates in mild steel reduces the tensile to yield ratio, in magnesium the increased work hardening rate increases this ratio. This suggests that the most rapidly deforming regions of a magnesium component will exhibit the highest strength, encouraging the spread of deformation to other parts of the component and therefore maximizing energy absorption.

Summary

While magnesium has in the past been largely overlooked for structural applications, there are significant benefits obtainable from its light weight in terms of energy absorption. The potential benefits are greater than the ratios of densities would suggest because deformation in a vehicle impact involves bending as well as tensile and compressive modes.

An additional benefit may be obtained if low aluminum content alloys can be utilized. Alloys with 2% aluminum exhibit an increased work hardening rate at high strain rates which raises the tensile to yield ratio and is likely to benefit energy absorption.

Acknowledgements

CAST was established under the Australian Government's Cooperative Research Centres Scheme. The support of Australian Magnesium Corporation is also appreciated – both in terms of financial support of CAST and constructive discussions with several of their staff.

References

1. N. Jones "Structural Impact" Cambridge University Press, Cambridge, (1989), 337-339.

2. C.A. Newland and M.T. Murray "Strain Rate Dependent Behaviour of Magnesium-Based Alloys", First Australian Congress on Applied Mechanics, The Institute of Engineers, Australia, (1996) 73.

3. R.H. Grzebieta and N.W. Murray, "The Static Behavior of Struts with Initial Kinks at Their Centre Point", Int. J. Impact Engineering, 3(3), (1985), 155-165.

4. R.H. Grzebieta and N.W. Murray, "Energy Absorption of an Initially Imperfect Strut Subjected to an Impact Load", Int. J. Impact Engineering, 4(3), (1986), 147-159.

5. P.D.D. Rodrigo and V. Ahuja "Effect of Casting Parameters on the formation of Pore/Segregation Bands in Magnesium Die Castings", Magnesium 2000 - Proceedings of the Second Israeli Conference on Magnesium Science and Technology (2000) 97-104.

6. A.L. Bowles, J.R. Griffiths and C.J. Davidson, "Ductility and Skin Effect in High Pressure Die-Cast Mg-Al Alloys", Magnesium Technology 2001, The Minerals, Metals and Materials Society, (2001), 161-168.

7. R.H. Grzebieta private communication.

NUMERICAL MODELLING OF LARGE STRAIN DEFORMATION IN MAGNESIUM

Julie Lévesque [1], Kaan Inal [1], Kenneth W. Neale [1], , Raja K. Mishra [2] Alan A. Luo [2]
[1]Department of Civil Engineering, University of Sherbrooke, 2500, boul. de l'Université, Sherbrooke, Québec, J1K 2R1, Canada
[2]General Motors Research and Development Center, 30500 Mound Road, Warren, MI 48090-9055, U.S.A

Keywords: Crystal Plasticity Modelling, Magnesium Alloys, Twinning

Abstract

In this paper, a new constitutive framework based on a rate-dependent crystal plasticity theory is presented to simulate large strain deformation phenomena in HCP metals such as magnesium. In this new model the principal deformation mechanisms considered are crystallographic slip and deformation twinning. The new framework is incorporated into in-house finite element (FE) codes. Simulations of uniaxial tension and compression for the magnesium alloy AM30 are performed and the results are compared with experimental observations of the specimens deformed at 200°C. Limitations of the current modelling approaches are also discussed.

Introduction

Numerical models based on crystal plasticity theories (e.g. Asaro and Needleman [1]) have been successfully used to model macroscopic (stress-strain curves) and microscopic (texture evolution) quantities during finite plastic straining. However, most of these models consider only crystallographic slip as the major deformation mechanism. Thus these models cannot be employed to simulate the behavior of certain metals such as hcp magnesium, where one of the principal deformation mechanisms is twinning.

Recently, Kalidindi [2] has presented a formulation to incorporate twinning in crystal plasticity models. This formulation has the advantage of keeping a constant number of orientations during the simulations.

In this work, a constitutive model has been implemented into an in-house code and calibrated using the stress-strain curves for AM30 alloy. The final texture has been compared with experimental results to determine the validity of the model. The relative accumulated slip on the various slip systems during deformation has also been calculated.

Constitutive model

The plasticity model for polycrystals presented by Asaro and Needleman [1] and modified by Kalidindi [2] is employed in this work. Accordingly, the total deformation of a crystallite is taken to be the result of three distinct physical mechanisms: crystallographic slip, twinning and elastic lattice distortion. The slip systems that are known to be active in magnesium alloys and that are being used in this work are the summarized in Table I.

Extension twinning is considered to be potentially active in a crystal (in compression) when the basal planes are parallel (or with an angle up to 45°) to the major deformation axis. The same condition applies to contraction twinning, except that it is active (in tension) when the basal planes are parallel to the major deformation axis. Based on measured texture data for extruded tubes [3], extension twinning is assumed to be active in 84% of the grains under a compressive load, while contraction twinning is assumed to be active in 16% of the grains under a tensile load.

Table I: Slip and Twinning Systems Used in the Analyses

Slip systes	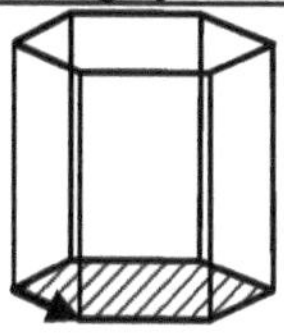 $\{0001\}\langle 1\bar{2}10\rangle$ Basal	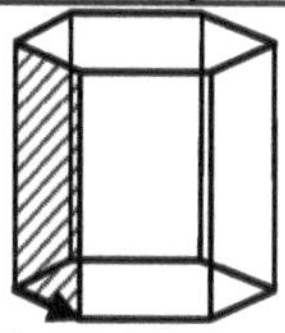 $\{10\bar{1}0\}\langle 1\bar{2}10\rangle$ Prismatic
	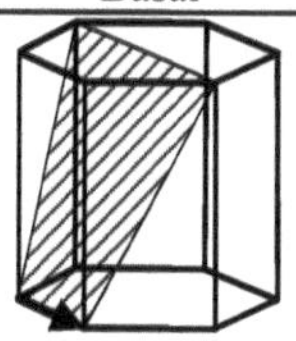 $\{1\bar{1}01\}\langle 11\bar{2}0\rangle$ Pyramidal	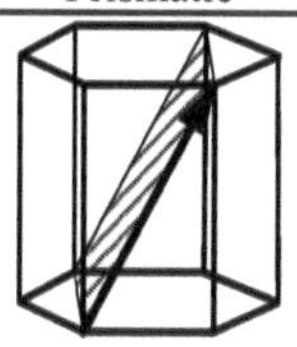 $\{1\bar{2}12\}\langle 1\bar{2}1\bar{3}\rangle$ <c+a>
Twinning systems	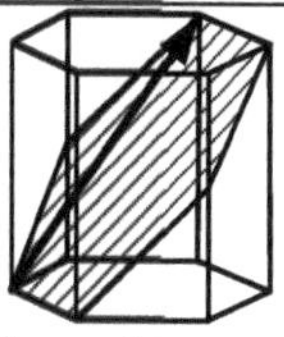 $\{10\bar{1}2\}\langle \bar{1}011\rangle$ Extension	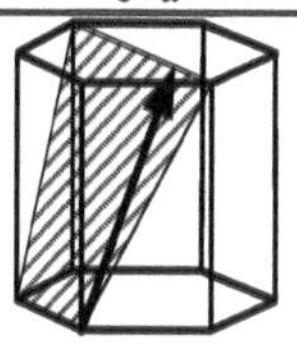 $\{1\bar{1}01\}\langle \bar{1}102\rangle$ Contraction

Plastic deformation is envisaged to occur as a set of simple shears along the various slip and twinning systems, leaving the lattice and the slip system vectors ($s_{(\alpha)}, m_{(\alpha)}$) not only essentially undistorted, but also unrotated. Next, the material and lattice are considered to deform elastically and rotate rigidly from the plastically deformed state to the current configuration. The deformation gradient $\boldsymbol{F}$ is written as

$$\boldsymbol{F} = \boldsymbol{F}^{*}\boldsymbol{F}^{P} \quad (1)$$

where F^P consists of crystallographic slip and twinning, while the elastic deformation and any rigid body rotation are embodied in F^*. From (1), the spatial gradient of velocity can be written as

$$L = \dot{F}F^{-1} = L^* + L^P \tag{2}$$

where

$$L^* = \dot{F}^* F^{*-1},\ L^P = F^*\left(\dot{F}^P F^{P^{-1}}\right)F^{*-1} \tag{3}$$

Taking symmetric and antisymmetric parts of the above relations leads to the elastic and plastic strain-rates $\mathbf{D}^*$ and $\mathbf{D}^\mathbf{P}$, the so called plastic spin $\mathbf{W}^\mathbf{P}$, and the spin $\mathbf{W}^*$ associated with the rigid lattice rotation (Eq. 4).

$$D = D^* + D^P,\ W = W^* + W^P \tag{4}$$

The vectors $s^{(\alpha)}$ and $m^{(\alpha)}$ are regarded as lattice vectors so that they stretch and rotate by

$$s^*_{(\alpha)} = F^* s_{(\alpha)};\ m^*_{(\alpha)} = m_{(\alpha)} F^{*-1} \tag{5}$$

By introducing the following symmetric and skew-symmetric tensors for each slip system α,

$$P_{(\alpha)} = \frac{1}{2}\left[s^*_{(\alpha)} \otimes m^*_{(\alpha)} + m^*_{(\alpha)} \otimes s^*_{(\alpha)}\right] \tag{6}$$

$$W_{(\alpha)} = \frac{1}{2}\left[s^*_{(\alpha)} \otimes m^*_{(\alpha)} - m^*_{(\alpha)} \otimes s^*_{(\alpha)}\right] \tag{7}$$

the plastic strain-rate and spin for the crystal can be, respectively, written as

$$D^P = \left(\sum_{\beta=1}^{N^T} f_{(\beta)}\right)\left[\sum_{\alpha=1}^{N^S} P_{(\alpha)}\dot{\gamma}_{(\alpha)} + \sum_{\beta=1}^{N^T} P_{(\beta)}\dot{f}_{(\beta)}\gamma^{tw}\right] + \sum_{\beta=1}^{N^T}\left[f_{(\beta)}\sum_{\alpha=1}^{N^S}\left(P^{tw\beta}_{(\alpha)}\dot{\gamma}^{tw\beta}_{(\alpha)}\right)\right] \tag{8}$$

and

$$\Omega^P = \left(\sum_{\beta=1}^{N^T} f_{(\beta)}\right)\left[\sum_{\alpha=1}^{N^S} W_{(\alpha)}\dot{\gamma}_{(\alpha)} + \sum_{\beta=1}^{N^T} W_{(\beta)}\dot{f}_{(\beta)}\gamma^{tw}\right] + \sum_{\beta=1}^{N^T}\left[f_{(\beta)}\sum_{\alpha=1}^{N^S}\left(W^{tw\beta}_{(\alpha)}\dot{\gamma}^{tw\beta}_{(\alpha)}\right)\right] \tag{9}$$

where $\dot{\gamma}_{(\alpha)}$ is the shear rate on the slip system α, $f_{(b)}$ is the volume fraction of the grains that have been twinned to the β system, γ^{tw} is the amount of shear associated with twinning and $\dot{f}_{(\beta)}$ is the rate of twinning of the system β. The superscript twβ indicates that the parameter considered is calculated for a region twinned according to the system β. $\mathbf{N}^\mathbf{S}$ and $\mathbf{N}^\mathbf{T}$ represent the number of slip systems and the number of twinning systems, respectively.

The slip and twinning rates to be substituted into Eq. (8) and (9) are assumed to be governed by the following power-law expressions:

$$\dot{\gamma}_{(\alpha)} = \dot{\gamma}_{(0)}\, sgn\, \tau_{(\alpha)}\left|\frac{\tau_{(\alpha)}}{g_{(\alpha)}}\right|^{1/m} \tag{10}$$

$$\dot{f}_{(\alpha)} = \frac{\dot{f}_{(0)}}{\gamma^{tw}}\, sgn\, \tau_{(\alpha)}\left|\frac{\tau_{(\alpha)}}{g_{(\alpha)}}\right|^{1/m} \tag{11}$$

where $\dot{\gamma}_{(\alpha)}$ and $\dot{f}_{(\alpha)}$ are reference shear and twinning rates taken to be the same for all the slip systems and twinning systems, $\tau_{(\alpha)} = P_{(\alpha)} : \sigma$ is the resolved shear stress on the slip system α, $g_{(\alpha)}$ is its hardness and m is the strain-rate sensitivity index. The functions $g_{(\alpha)}$ characterize the current strain-hardened state of all slip and twinning systems.

A simple power-law hardening is used to calculate the hardness of the various slip and twinning systems during deformation. An expression accounting for the Hall-Petch effect is also incorporated in the formulation (this expression, presented in Salem et al. [4], represents the increase in the hardness due to twin boundaries restricting the dislocation displacement).

$$h_{(\alpha)} = h_{(0)}\left[\frac{h_{(0)}\gamma_a}{\tau_{(\alpha)}n}\right]^{n-1} + h_{HP}\left(\sum_{\beta=1}^{N^T} f_{(\beta)}\right)^{0.5} \tag{12}$$

where $h_{(0)}$ is the initial hardness of the slip systems, n is the hardening exponent, h_{HP} is a coefficient for the Hall-Petch effect and γ_a is the sum of the accumulated slip on all the slip systems; i.e.

$$\gamma_a = \int_0^t \sum_{\alpha=1}^{N^S}\left|\dot{\gamma}_{(\alpha)}\right| dt \tag{13}$$

The new model presented in this paper is based on the Taylor assumptions [5], thus the deformation gradient is uniform over the twinned and untwinned regions. However, twinning causes a reorientation of the lattice in the twinned parts of a grain. This is taken into account when calculating, for example, the elasticity tensor. The tensor for the twinned regions $E^{tw\beta}$, is obtained by the following transformation law

$$E^{tw\beta}_{ijkl} = E^{mt}_{mnop} Q_{im} Q_{jn} Q_{ko} Q_{lp} \tag{13}$$

In Eq.13, Q is a transformation matrix defined as [5]

$$Q_{ij} = 2m_i m_j - \delta_{ij} \tag{14}$$

where $\boldsymbol{m}$ is the twin plane normal.

The Cauchy stress, σ, over a grain can be obtained by averaging the stresses in the twinned and un-twinned parts of the grain;

$$\sigma = \left(1 - \sum_{\beta=1}^{N^T} f_{(\beta)}\right)\sigma^{mt} + \sum_{\beta=1}^{N^T} f_{(\beta)}\sigma^{tw\beta} \qquad (15)$$

where σ^{mt} is the Cauchy stress in the matrix (un-twinned regions of the grain) and $\sigma^{tw\beta}$ is the stress in β-twinned region.

Results and Discussion

The material investigated in this study is extruded tubes of the new wrought alloy AM30 (Mg-3%Al-0.3%Mn), developed at General Motors Research and Development Center [6]. Stress-strain curves in tension and compression along the extrusion direction, as well as texture data, have been provided by McGill University [3]. As can be observed in Figure 1, basal planes in the material are strongly aligned parallel to the extrusion direction.

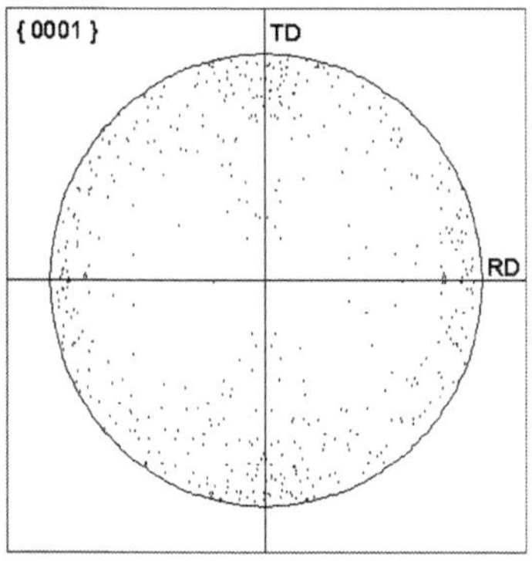

Figure 1. Initial texture of the AM30 extruded tubes

At a moderate temperature of 200°C and a strain rate of 0.1 s^{-1}, four different families of slip systems are believed to be activated [7,8], for a total of 18 slip systems. However, because of the strong texture of the extruded tubes, slip on the basal, prismatic and pyramidal planes is restricted. A great amount of twins present after 14% deformation in compression, but very little twinning is observed in tension test specimens after 14% deformation [3].

Crystallographic slip occurs when the resolved shear stress acting on a slip system is higher than its critical resolved shear stress (CRSS). For the simulations presented in this paper, relative CRSS of the basal, prismatic, pyramidal and <c+a> systems are taken to be 1, 4, 5 and 6 respectively.

The stress-strain curves simulated for tension and compression, as well as the evolution of the volumetric fraction of twins during compression are presented in Figures 2-3. Note that the predicted macroscopic and microscopic quantities are obtained by integrating the single crystal constitutive law at an integration point. Thus the simulations presented in this paper do not account for possible geometric imperfections in a specimen. Figures 4-5 present the relative accumulated slip on the different slip systems during deformation. The amount of twinning is very small in tension at 200°C, probably because the resolved shear stress of the contraction twinning systems is much higher than that of the extension twinning. During simulations of tension, following experimental observations, no twinning was allowed.

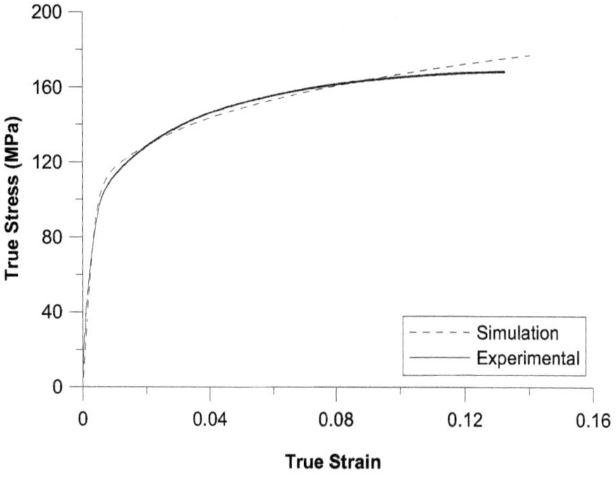

Figure 2. Uniaxial tension stress-strain curves at 200°C

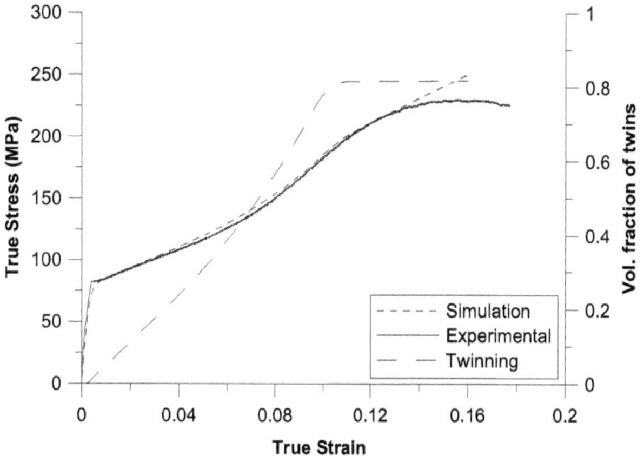

Figure 3. Uniaxial compression stress-strain curves at 200°C

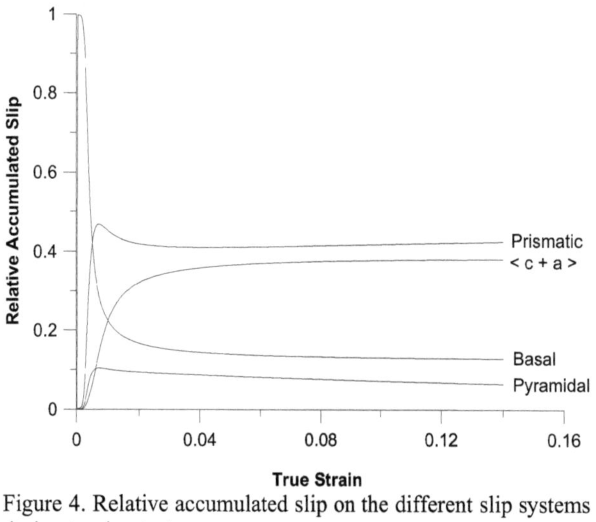

Figure 4. Relative accumulated slip on the different slip systems during tension test

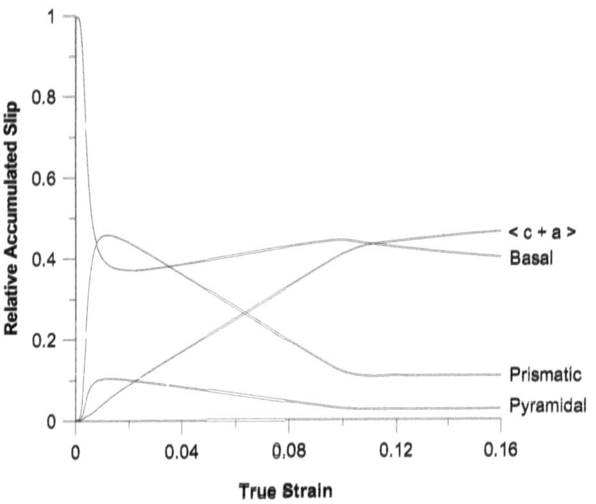

Figure 5. Relative accumulated slip on the different slip systems during compression test

Simulations of tension have shown that crystallographic slip occurs first on the basal systems, since their critical resolved shear stress (CRSS) is lower than that of the other slip systems. Prismatic slip is then activated and becomes the major deformation mode during the deformation process. This can be explained by the fact that some of the prismatic planes are well aligned for their activation. Finally, the <c+a> slip is activated (which is also an important deformation mechanism at this temperature).

Simulations of compression have shown that twinning is the major deformation mode up to around 10% deformation. Furthermore, basal slip is much more active in compression than in tension, which is expected since twinning reorients the lattice, favouring basal slip.

The experimental and simulated evolved textures during tension at 13% strain and compression at 20% strain are presented in Figures 6-7. It can be seen that the experimental results of the measured texture evolution [3] are in good agreement with the simulations.

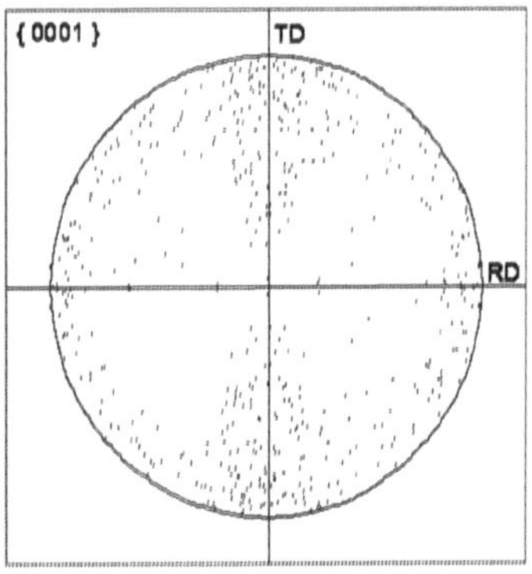

(a)

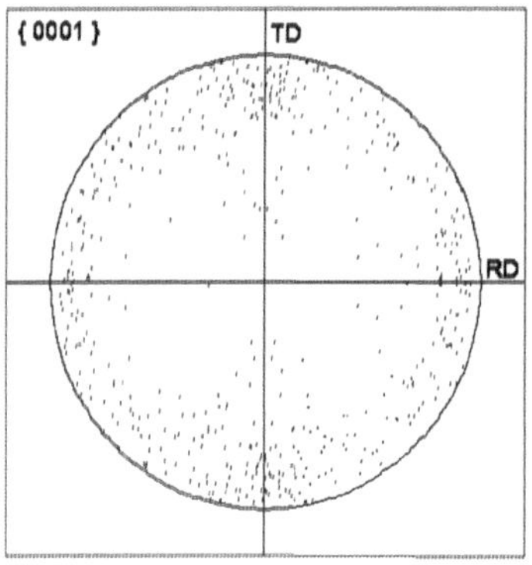

(b)

Figure 6. (a) Experimental [3] and (b) simulated texture after 13% tension

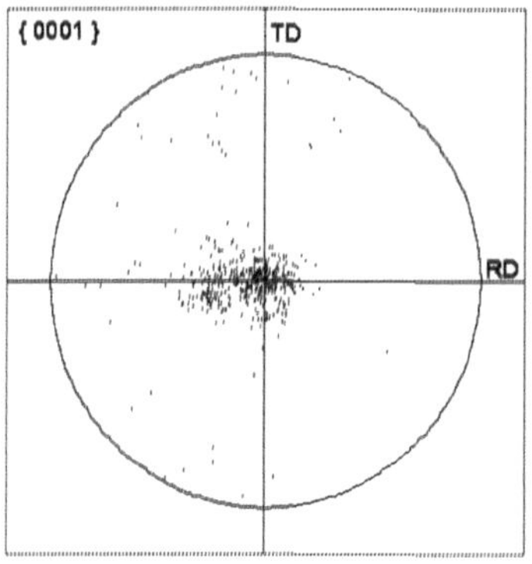

(a)

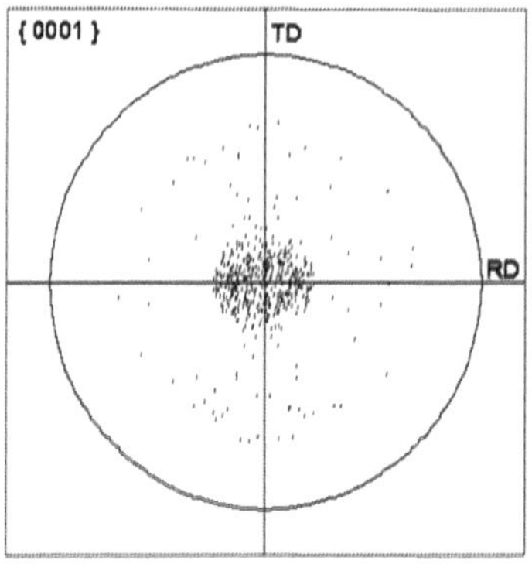

(b)

Figure 7. (a) Experimental [3] and (b) simulated texture after 20% compression

CONCLUSION

In this paper, a new model that can account for both crystallographic slip and deformation twinning has been presented. This model can accurately simulate macroscopic and microscopic features of HCP crystals during various deformations modes. However, the predictions of the rate of strain hardening due to twinning during compression can still be improved. Furthermore, due to the lack of experimental measurements, the predicted relative activities of the slip systems are yet to be validated.

It should be mentioned that a relatively simple hardening model was employed in the simulations, and it is obvious that this model needs to be improved. However, experimental data on the activities of the various slip systems in magnesium alloys at different deformation modes and temperatures is very scarce, if not totally inexistent, for AM30 alloys. In order to develop an accurate hardening model, more experimental data such as the activities of the various slip systems at different levels of deformation and as the evolution of the volumetric fractions of twins are necessary.

Future work will concentrate on the improvement of the hardening model. A new model that can account for the interactions between deformation twinning and crystallographic slipping will be developed with the aid of experimental observations (McGill University). The new model will be validated by comparing simulations of uniaxial tension, compression and ring hoop tension with the corresponding experiments.

ACKNOWLEDGEMENTS

This work was supported by the Natural Sciences and Engineering Research Council of Canada (NSERC) and General Motors of Canada. K. W. Neale holds the Canada Research Chair in Advanced Engineered Material Systems, and the support of this program is gratefully acknowledged. The authors acknowledge L. Jiang of McGill University for providing the experimental results. Prof. John Jonas of McGill University and Dr. Anil Sachdev of General Motors provided many helpful suggestions.

REFERENCES

1. R.J. Asaro and A. Needleman, "Texture Development and Strain Hardening in Rate Dependant Polycrystals", *Acta Metallurgica*, 33 (1985), 923-953.

2. S.R. Kalidindi, "Incorporation of Deformation Twinning in Crystal Plasticity Models", *Journal of the Mechanics and Physics of Solids*, 46 (1998), 267-290.

3. L. Jiang, J.J. Jonas, A.A. Luo, A.K. Sachdev and S. Godet, "Twinning-induced Softening in Polycrystalline AM30 Mg Alloy at Moderate Temperatures", *Scripta Materialia* (2006), in press.

4. A. A. Salem, S. R. Kalidindi and S. L. Semiatin, "Strain Hardening Due to Deformation Twinning in α-Titanium: Constitutive Relations and Crystal-Plasticity Modelling", *Acta Materialia*, 53 (2005), 3495-3502.

5. P. Van Houtte, "Simulation of the rolling and shear texture of brass by the Taylor theory adapted for mechanical twinning", *Acta Metallurgica*, 26 (1978), 591-604.

6. A.A. Luo and A.K. Sachdev, "Magnesium Wrought Alloy Having Improved Extrudability and Formability", US Patent Application, March 4, 2004.

7. T. Obara, H. Yoshinga and S. Morozumi, "{1 1 -2 2} <-1 -1 2 3> Slip System in Magnesium", *Acta Metallurgica*, 21 (1973), 845-852.

8. K. Máthis, K. Nyilas, A. Axt, I. Dragomir-Cernatescu, T. Ungár and P. Lukáč, "The Evolution of Non-Basal Dislocations as a Function of Deformation Temperature in Pure Magnesium Determined by X-ray Diffraction", *Acta Materialia*, 52 (2004), 2889-2894.

From *Magnesium Technology 2012*, S.N. Mathaudhu, W.H. Sillekens, N.R. Neelameggham, and N. Hort, Editors

ATOMS-TO-GRAINS CORROSION MODELING FOR MAGNESIUM ALLOYS

Hyunwook Kwak[1], Jie Xiao, and Santanu Chaudhuri[1*]

[1]ISP/Applied Sciences Laboratory, Washington State University, WA 99210-1495

Keywords: Corrosion Protection, First-Principles Calculations, MgO, Rare-Earth Doping

Abstract

Development of new Mg-alloys needs greater efforts in systematic evaluation of corrosion protection methods under service environments. A deeper understanding of microscale and nanoscale metallurgy is needed to slow down corrosion. Results from our multiscale modeling effort will be presented to demonstrate the power of first-principle theories in predicting the composition-dependent kinetics of corrosion reactions. Our recent results show that the corrosion prevention properties of Mg-alloys can be significantly enhanced by doping rare-earth elements such as Cerium that protects the oxide layer from rapid hydrolysis. Furthermore, we are developing kinetic Monte Carlo and finite-element analysis (FEA) based models to extend the predictions from the atomic length scales to nano- and microscale models that will include grains and alloy microstructure. The potential for incorporating mechanical and corrosion performance in continuum length scale combined with the insights from first-principles based surface chemical models can have powerful impact on the development of Mg-alloys.

Introduction

Our initial multiscale simulation results on Mg alloys indicate that corrosion mechanisms in Mg alloys are unique.[1] Therefore it is important that corrosion prevention strategies need to include consideration for specific mechanistic pathways to prevent failures and find mature alloys for structural applications. From optimizing composition of promising alloys, increasing the stability of the oxide layer, and obtaining a uniform microstructure through carefully controlled processing conditions, all play a part in reducing corrosion vulnerabilities. Our simulation effort is geared towards understanding different galvanic and dissolution processes in hot and humid environments. Some theoretical development will be needed to address the role of electrochemical field inside the pits at the intermetallic particle boundaries. In alloys such as AZ31, AZ80 and AZ91D, there are different phases of intermetallic precipitates that are formed in the alloy matrix. Intermetallic impurities or inclusions such as Al_2Mn, $Mg_{17}Al_{12}$ etc. have the potential to act as a cathode and expedite the oxide film and alloy etching where the Mg-alloy matrix acts as anode. In this current work, we will discuss stability of MgO film without intermetallic impurities across different length scales and identify the links between elemental doping of the MgO surface and the correlation with metal loss from atomistic to microscale.

Figure 1 illustrates the framework we are developing for understanding and prescribing predictive pathways to slow-down corrosion. It consists of four layers of modeling blocks, which are bi-directionally coupled and closely connected with a block of experimental data. At the bottom layer, first-principles-based calculations are responsible for generating thermodynamic information, such as Gibbs free energy of activation for a chemical or electrochemical reaction. This information is critical for kinetic Monte Carlo and molecular dynamics (kMC/MD) simulation of corrosion phenomena on and nearby the metal-electrolyte interface at nanoscale, e.g., oxide layer formation, ion adsorption, desorption, and diffusion through oxide/hydroxide layer, and metal dissolution, etc. On the other hand, reaction kinetics derived from first-principles calculation and kMC/MD simulation can be fed into an upper layer. At this layer, finite element models describe the growth of a single pit, where metal microstructure, pit surface morphology, and corrosion environment are systematically taken into account. Detailed investigation on the growth of a single pit will eventually lead to a successful study of the evolution of an ensemble of pits at macroscopic time and length scales by using the damage function analysis (DFA) method. Finally, corrosion performance evaluation (such as life time quantification) will be conducted at the top layer.

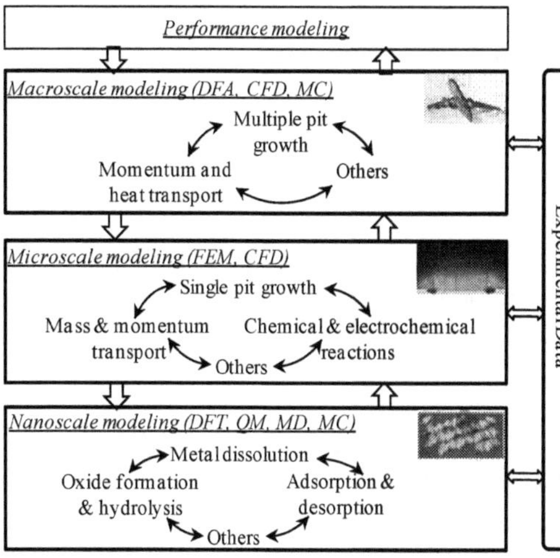

Figure 1. Multiscale modeling framework of corrosion

Oxide film on top of Mg alloys plays an important role on the corrosion protection of pure magnesium and Mg alloys. Presence of alloying compounds in the alloy is also important for the oxide films because the alloying species can transport to the metal surface and alter the composition of the oxide film. This observation is backed by numerous previous work where traces of excessive alloying compounds were detected at the protective films on top of Y-enriched and Ce-enriched Mg alloy systems. The primary emphasis of first-principles based *corrosion modeling effort* is to develop understanding of interfacial reactions as a function of surface composition of alloys under humid and corrosive conditions. Mg-alloys are particularly vulnerable in humid environment due to hydrolysis of MgO layers to Mg()H)$_2$ as shown in figure 2 . Once this barrier layer is damaged, galvanic processes leads to rapid loss of Mg due to its position in the electrochemical series.

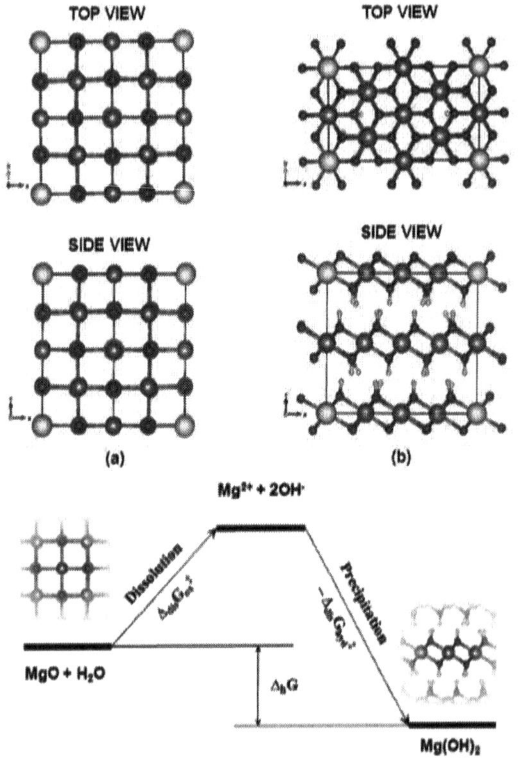

(c)

Figure 2 (a) The unit cell of MgO, and (b) Mg(OH)$_2$ from top and side-view of atomic arrangements. (c) We calculate the thermodynamics from a two-step reaction model for the hydroxylation of the oxide films on Mg alloy.

Methods

<u>Thermochemical properties from first principles</u>

The key to our thermodynamic description of the hydroxylation process for the doped oxides is the use of first-principles methods to calculate the overall change in thermochemical potential during the following hydroxylation process,

$$\mathrm{MgO(periclase)} + \mathrm{H_2O_{(l)}} \longrightarrow \mathrm{Mg(OH)_2\,(brucite)} \quad (1)$$

For oxide films in contact with aqueous solution at room temperature, we consider a system that is exposed to a solution reservoir in which temperature, pressure, and activities of solute species are kept constant at standard condition. The appropriate thermodynamic potential to describe such a system is Gibbs free energy. Based on the definition of Gibbs free energy (G = U – TS + pV), the temperature-dependent free energy of a solid can be approximated as follows,

$$G \simeq U_{\mathrm{elec}} + F_{\mathrm{vib}}(T) + pV \quad (2)$$

The first term indicates the electronic total energy that corresponds to the ground state energy of a system at 0 K. The second term corresponds to the free energy contribution from the vibrational excitation of solids $(F_{\mathrm{vib}}(T) = U_{\mathrm{vib}}(T) - TS_{\mathrm{vib}}(T))$. For gas phase substances, similar expression can be given,

$$G \simeq U_{\mathrm{elec}} + F_{\mathrm{trans}}(T) + F_{\mathrm{rot}}(T) + F_{\mathrm{vib}}(T) + pV \quad (3)$$

where the translational and the rotational free energy of a gas molecule ($F_{\mathrm{trans}} + F_{\mathrm{rot}}$) is also considered. From our methodology, the first three terms of Eqn. (2) are calculated from electronic structure and phonon frequency calculations based on DFT. Change in the pV term is neglected for solids since they are at least several orders of magnitude smaller than the change in the internal energy at standard temperature and pressure condition.

From the phonon frequency calculations of solids, vibrational contribution of the thermodynamic function ($F_{\mathrm{vib}}(T)$) can be calculated using a direct method. Direct method uses first-principles-based force constants to form up a dynamical matrix. The force constants are calculated from the Hellmann-Feynman forces induced by the displacement of individual atoms. In this formalism, dynamical matrix of a crystal is defined as,

$$D(k;\mu\nu) = \frac{1}{\sqrt{M_\mu M_\nu}} \sum_{\mathbf{m}} \Phi(0,\nu;\mathbf{m},\nu)\, e^{-2\pi i \mathbf{k}\cdot[\mathbf{R}(0,\mu)-\mathbf{R}(\mathbf{m},\nu)]} \quad (4)$$

where M_μ, M_ν and $\mathbf{R}(0,\mu)$, $\mathbf{R}(\mathbf{m},\nu)$ are masses and positions of atoms μ and ν, respectively. The summation index **m** runs over all atoms of the crystal. **k** is the wave vector in the first Brillouin zone and $\Phi(0,\nu;\mathbf{m},\nu)$ is the force constant matrix which can be constructed from the induced Hellman-Feynman forces by atomic displacements. Phonon frequencies $\omega(\mathbf{k},j)$ and the polarization vector $\mathbf{e}(\mathbf{k},j)$ are given from the eigenvalues and the eigenvectors of the dynamical matrix:

$$\omega^2(\mathbf{k},j)\mathbf{e}(\mathbf{k},j) = D(\mathbf{k})\mathbf{e}(\mathbf{k},j) \quad (5)$$

with the index j for the normal modes. With this, total phonon density of states can be evaluated from $\omega(\mathbf{k},j)$ and $\mathbf{e}(\mathbf{k},j)$ as,

$$g(\omega) = \frac{1}{nd\Delta\omega} \sum_{\mathbf{k},j} \delta_{\Delta\omega}\left[\omega - \omega(\mathbf{k},j)\right] \quad (6)$$

where n is the number of sampling wave vectors, d is the dimension of the dynamical matrix, $\Delta\omega$ is the frequency interval, and the delta function $\delta_{\Delta\omega}$ is defined as

$$\delta_{\Delta\omega}(x) = \begin{cases} 1 & \text{if } -\frac{\Delta\omega}{2} < x < +\frac{\Delta\omega}{2} \\ 0 & \text{otherwise.} \end{cases} \quad (7)$$

For a given phonon density of states, Sternik and Parlinski showed that the vibrational contribution of the free energy, $F_{\mathrm{vib}}(T)$, can be calculated within the harmonic approximation as,

$$F_{vib}(T) = rk_B T \int_0^{\infty} d\omega\, g(\omega) \ln\left[2\sinh\left(\frac{\hbar\omega}{2k_B T}\right)\right] \quad (8)$$

Using Eqns (2) and (8), free energies of solid MgO and $Mg(OH)_2$ crystals at constant temperature can be calculated solely based on first-principles methods.

Effect of doping on MgO hydroxylation

The fundamental question is, how can we improve the stability of the MgO layer? In this work, we applied the cationic doping concentration of 1 out of 16 Mg sites (6.25 at%Mg) for both MgO and $Mg(OH)_2$. For the doped $Mg(OH)_2$ crystals, 4×2×2 orthorhombic supercell with $16(Mg(OH)_2)$ was built with one cationic doping site. Electronic energy, thermal energy, and entropy values for MgO and $Mg(OH)_2$ are all in close agreement with previous first-principles study by Duan and Sorescu. The calculated Gibbs free energy of hydroxylation (-26.5 kJ / mol) also agrees very well with the prediction from experimental database (-27.1 kJ / mol). The negative Gibbs free energy indicates that the oxide film degradation via hydroxylation occurs spontaneously. The results of the dissolution thermodynamics is summarized in Tables I and II.

Table I First-principles-calculated thermochemical properties of reactants and products from the overall hydroxide reaction shown in Eqn. (1). All units are in kJ / mol.

$T = 298$ K, $p = 1$ bar	U_{elec}	$F_{vib}(T)$	pV	$E_{rot}(T)$	$G(T,p)$
MgO (s)	-1156.1	10.82	-	-	-1145.2
H_2O (g)	-1370.0	17.90	2.48	7.44	-1352.1
$Mg(OH)_2$ (s)	-2601.2	68.90	-	-	-2532.3
$\Delta_h G = G(Mg(OH)_2) - \left[G(MgO) + G(H_2O_{(g)}) + \delta G^{g\to l}_{H_2O}\right] = -26.5$ (kJ/mol)					

In Table I, the first-principles calculated energy and entropy values for the oxide and the hydroxide crystals are shown for standard temperature and pressure condition (298 K, 1 bar). Electronic energy, thermal energy, and entropy values for MgO and $Mg(OH)_2$ are all in close agreement with previous first-principles studies. The calculated Gibbs free energy of hydroxylation at standard condition is -26.5 kJ / mol. This agrees very well with the prediction from the experimental database (-27.1 kJ / mol). The negative Gibbs free energy indicates that the oxide film degradation via hydroxylation occurs spontaneously. We also found that the thermodynamic driving force for the hydroxylation mostly attributes to the decrease in electronic energy ($\Delta_h U_{elec} < 0$) which compensates the increase in thermal energy ($\Delta_h F_{vib} > 0$).

Subsequently, we calculated hydroxylation free energies for the doped oxide and hydroxide crystals with six different types of cationic doping (Zn^{2+}, Al^{3+}, Y^{3+}, Ce^{3+}, Ce^{4+}, and Zr^{4+}). The results are summarized in Table II **Table** along with electronic ($\Delta_h U_{elec}$) and thermal ($\Delta_h F_{vib}$) contributions to $\Delta_h G$. When doped with cationic impurities, a clear trend was found for $\Delta_h G$ with respect to the charge state of the impurity atoms. While doping with isovalent Zn^{2+} shows minimal impact, doping with higher charge state (3+ and 4+) elements leads to a significant increase in the hydroxylation free energy. For Ce^{4+} and Zr^{4+} doped oxides, the doping effect is significant enough to change the sign of $\Delta_h G$ from negative to positive. As a result, hydroxylation reaction becomes no longer spontaneous with the doping. The similar trend found in $\Delta_h U_{elec}$ suggests that such doping effect has the electronic origin. The trend is most apparent from a comparison between Ce^{3+} doping and Ce^{4+} doping, where significant difference between hydroxylation free energies (26.8 kJ / mol) can only attribute to the difference in their impurity charge state. The change in sign from a highly favorable reaction to unfavorable one is a potentially powerful information on use of doping in slowing down failure of MgO grains.

Table II First-principles calculated electronic, vibrational, and overall free energy changes during the hydroxylation of the doped MgO crystals.

(298 K, 1 bar)	$\Delta_h U_{elec}$ (kJ/mol)	$\Delta_h F_{vib}$ (kJ/mol)	$\Delta_h G$ (kJ/mol)
Undoped	-75.2	+40.2	**-26.4**
Zn^{2+}-doped	-73.1	+40.5	**-24.0**
Al^{3+}-doped	-53.8	+40.8	**-4.5**
Y^{3+}-doped	-59.1	+40.3	**-10.2**
Ce^{3+}-doped	-61.1	+41.0	**-11.5**
Ce^{4+}-doped	-33.4	+40.2	**+15.3**
Zr^{4+}-doped	-31.1	+40.6	**+18.1**

First-principles-assisted kinetic Monte Carlo (kMC) model

Let us examine first how similar information can be useful in understanding nanometer scale evolution of surface with doping elements such as Ce. The first model we are developing is for the alloy exposed to electrolyte without the MgO layer on the top. We will eventually describe the MgO grains and their hydrolysis. To analyze nanometer-scale corrosion of Mg surfaces, we employ lattice-based kinetic Monte Carlo (kMC) technique. While the dissolution parameters for Mg atoms on the surfaces that are in contact with aqueous solution will be set from experimental polarization curve of pure Mg, other parameters for atomistic diffusion process will be directly extracted from first-principles DFT calculations. For a given surface morphology and the alloying composition, the outcome of the simulation will provide the corrosion current density (*e.g.* dissolution rate) and the surface morphology of the alloy surface as a function of time. The model will also provide selective dissolution mechanism during the corrosion process and will give insights on the corrosion performance of a given alloying composition.

Simulation setup

In KMC simulation, atoms and molecules are modeled in a coarse-grained way which would represent averages of behaviors that arises from the transitions between different atomistic-scale configurations. We use lattice-based kMC module from SPPARKS code from Sandia National Laboratory to describe dissolution and diffusion of metal atoms at alloy surfaces. The same code will be used in future to describe grains.

For the lattice-based kMC simulation, we employ (100)-facing fcc structured slab surface model with lattice parameter of 4 Å. The 2D periodic surface unit cell is 8 nm × 8 nm wide (800 lattice sites per layer) and 34 atomic-layer thick. The rest of the 6 layers of atomic lattices are initially assigned as solution sites, *i.e.* non-occupied sites. The resulting slab is 6 nm thick with 27200 lattice sites that are initially occupied with Mg or RE atoms. At the beginning of simulation, metal sites are assigned by randomly distributing Mg and RE atoms with a given alloying composition, *i.e.* the doping concentration. Both upward and downward facing surfaces in the slab are exposed to the solution. During the simulation, only diffusion and dissolution steps are considered.

Dissolution model

Selective dissolution of Mg alloy surface is simulated by allowing only Mg atoms to dissolve from the surface while more cathodic RE element remains at the surface. For each Mg atom, propensity of its dissolution is determined by Arrhenius-type formula:

$$r_{\text{dissolution}} = \nu_{\text{diss}} \exp\left[-\left(E_{\text{diss}}(n) - \phi\right)/\text{kT}\right] \quad (9)$$

where $r_{\text{dissolution}}$ is dissolution rate, ν_{diss} is frequency for the dissolution process, ϕ is overpotential, and E_{diss} is dissolution barrier for each Mg atom as function of number of nearest occupied sites n. We employed the frequency value of $\nu_{\text{diss}} = 10^4$ sec^{-1} from the similar kMC studies.

Diffusion model

Diffusion of Mg and RE atoms at the surface is modeled with hopping mechanism between the nearest-neighbor sites. Instead of using the conventional bond-breaking scheme for diffusion model, we use site-energy model in which final state atomic configuration as well as initial state atomic configuration is taken into account. In site-energy model, diffusion barrier is the sum of site-dependent energy $E_{\text{site}}(n)$ and the constant diffusion barrier E^0_{diff}. The interaction energy between Mg and RE atoms ($E^{\text{Mg-RE}}$) could also affect the diffusion barrier. The resulting expression for the rate of diffusion is as follows:

$$r_{\text{diffusion}} = \nu_{\text{diff}} \exp\left[-\left(E_{\text{site}}(n) + E^0_{\text{diff}} + E^{\text{Mg-RE}}\right)/\text{kT}\right] \quad (10)$$

We set $\nu_{\text{diff}} = 10^{13}$ sec^{-1} based on the Debye frequency of solid magnesium at room temperature (~ 1.2×10^{13} sec^{-1}). The site-dependent energy and the diffusion barrier will be determined by performing first-principles DFT calculations of Mg structures embedded with vacancies, surfaces, islands, and clusters.

Selective dissolution of RE-doped Mg surface

As preliminary analysis, we performed KMC simulation on the selective dissolution of 10 at% Ce-doped Mg surface. Based on first-principles transition-state calculation of surface vacancy diffusion and surface ad-atom diffusion on Mg (100) surface, we obtained the average diffusion barrier of 0.40 eV for hopping Mg atoms between the same atomic configurations. For Ce atoms, slightly higher diffusion barrier of 0.50 eV was applied. For both types of atoms, 0.25 eV of cohesive energy per bonding was applied for each occupied nearest neighbor site to assign the site-dependent energies. Diffusion barrier for Mg atoms was assumed to be the same as the diffusion barrier at the surface. To reflect the segregation of RE elements within the alloy matrix, a repulsive interaction potential of 0.05 eV / site was set between Mg and Ce atoms.

Figure illustrates the doping effect on the dissolution of alloy surfaces by plotting the surface current density and the surface composition for the first 50 *ms* of selective dissolution with a given overpotential of -0.3 *V*. First, increase of the dissolution rate becomes noticeably slower with the Ce doping when compared with pure Mg dissolution. It is also found that the surface composition of Mg gradually decreases from its bulk composition (90 *at%*) and converges to a much lower concentration (~75 *at%*). This indicates that the selective dissolution of Mg atoms leads to a surface segregation of Ce atoms near surface so that it would slow down the Mg dissolution. Nanometer-scale evolution of 10 *at%* Ce-doped Mg surface depicted in Figure confirms this argument. The figure clearly shows that the selective dissolution of Mg leads to higher concentration of segregated Ce clusters at the surface.

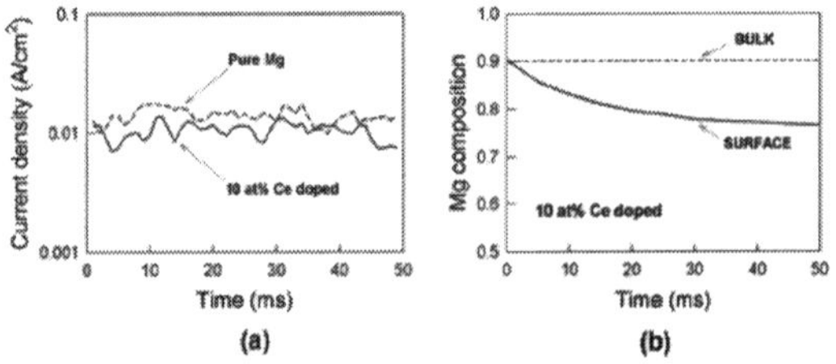

Figure 3. (a) Dissolution rate (in the unit of current density) of pure and 10 at% Ce-doped Mg surfaces, and (b) surface Mg composition of the Ce-doped surface. Dashed line in (b) indicates the bulk Mg composition of the alloy (10 at%). Both graphs are plotted for the first 50 ms of KMC simulation.

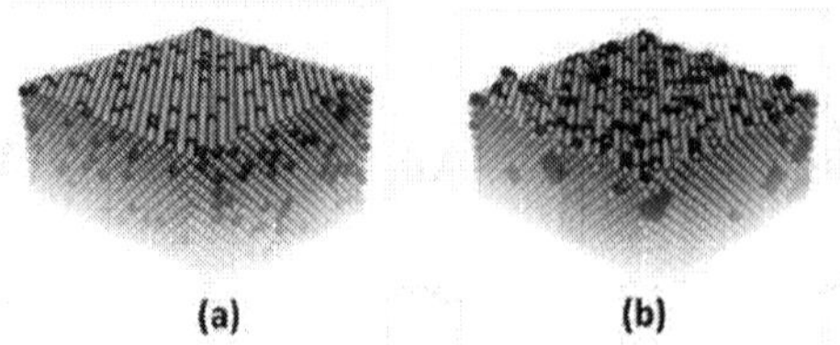

Figure 4. Snapshot from the dissolution of 10 at% Ce-doped Mg surface (8nm × 8nm) at (a) t = 0 ms and (b) t = 50 ms. Grey spheres represent Mg atoms and black spheres represent Ce atoms.

Before we describe the larger length-scale beyond kMC, it shows that elemental design of surfaces and understanding the fundamental science aspects of failure of MgO surface layer is critical for better design of alloy phase. In addition, we have results from reactive surfaces that are treated using coupled electrochemical models using computational fluid dynamics (CFD), both negative differential effect and effects of RE doping is currently being implemented in the microscale models. Long-term goal of our research is to improve the description of failure of grains and dissolution mechanics by coupling DFT-kMC and CFD methods.

Fluid-dynamics Simulation of Corrosion at a Pit

The calculated thermodynamic quantities from DFT can be fed into FEM model by resorting to an information integration method described in this section. Rate constants for chemical and electrochemical reactions can be calculated from Gibbs free energy of activation by:

$$k = \frac{k_B T}{h} \exp\left(-\frac{\Delta G^{\neq}}{RT}\right) \tag{11}$$

where k_B is Boltzmann's constant and h is Planck's constant. The surface coverage of reactive species (θ_j) in electrochemical reactions can be related to the solution chemistry using a Langmuir model [4]:

$$\theta_j = \frac{r_j a_j}{1 + \sum_k r_k a_k} \tag{12}$$

$$r_j = \frac{k_B T}{h} \exp\left(-\frac{\Delta G_{ads,j}}{RT}\right) \tag{13}$$

where a_j is the activity of the j-th reactive species at the metal surface and the coefficient r_j is a scaled adsorption constant. Conventionally, both anodic metal dissolution and cathodic reduction reactions are assumed to follow an empirical Tafel's law and the parameters in Tafel's equation are fitted from experimental data or from first-principles calculated kinetics with corrections for potential. In this work, the general form of current density is:

$$i = i_a + i_c \tag{14}$$

$$i_a = \sum_k \sum_j n_m^k F k_{a,j}^k \left(\theta_{j,s}^k\right)^{\omega_1} \exp\left(\frac{\alpha_{a,j}^k n_m^k F(\psi - \phi)}{RT}\right) \tag{15}$$

$$i_c = -\sum_j n_j F k_{c,j} \left(\theta_{o,s}\right)^{\omega_2} \exp\left(-\frac{\alpha_{c,j} n_j F(\psi - \phi)}{RT}\right) \tag{16}$$

In this way, anodic and cathodic current densities are quantitatively correlated to the applied potential (ψ), electrostatic potential in the solution (ϕ), temperature (T), and most importantly, the material and solution chemistry, which can be reflected by the reaction rate constant (k) and surface coverage of reactants (θ). Note that the subscript j represents the j-th aggressive ion in Eq. (15) and the j-th reduction reaction in Eq. (16). It is widely recognized that the negative difference effect (NDE) is a critical phenomenon in magnesium corrosion. Evolution of gas bubbles from a pit can perturb ion diffusion, and hence the growth of a pit. In order to quantitatively characterize such phenomenon, H_2 gas bubble generation as well as bubble redistribution due to multi-phase flow need to be fully taken into consideration. Figure 5 gives H_2 bubble distribution in systems with different electrolyte flow rates. Electrochemical and chemical reactions involved in this system were assumed to be the same as those in pure Mg corrosion.

$$Mg \rightarrow \equiv Mg^+ + e^- \quad \text{and } Mg^+ \rightarrow Mg^{2+} + e^-$$

$$Mg^+ + H_2O \rightarrow Mg^{2+} + OH^- + 0.5H_2 \text{ and } 2H^+ + 2e^- \rightarrow H_2$$

$$Mg^{2+} + 2OH^- \rightarrow Mg(OH)_2 \text{ and } MgO + H_2O \rightarrow Mg(OH)_2$$

In a stagnant electrolyte system (Figure 5a), gas bubbles arise along the peripheral lip of the pit, which was confirmed in experiments. As we increased the flow speed, the volume fraction of gas bubbles was reduced and the most prominent volume fraction appeared on one side of the pit surface facing towards the flow direction (see figure 5b). The NDE phenomenon was clearly identified (see figure 5d), where the hydrogen evolution rate increased rather than decreased with increasing potential. We have also developed a detailed pit growth and propagation model in the CFD scale using surface dissolution equations for intermetallic particles around Al-2024 alloy using a similar approach. [2, 5]

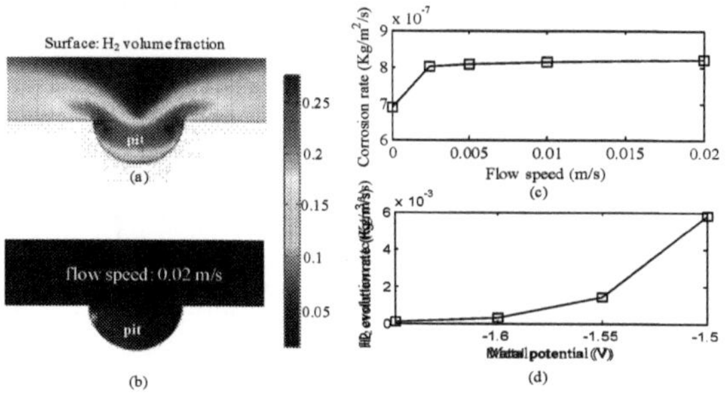

Figure 5. The effect of multiphase flow on corrosion: (a) H_2 bubble volume fraction in a stagnant system, (b) H_2 bubble volume fraction in a flowing system, (c) corrosion rate as a function of flow speed, and (d) H_2 evolution rate

In conclusion, robust mesoscale model of grain structure and incorporation of intermetallic phase across the length scale will be required for more realistic simulations that allow us to explore the effects of doping and alloy composition in a more comprehensive fashion. Such model will therefore be used to quantify the effect of microstructure and chemical environment in determining failure of surface film and the subsequent metal loss from damaged surface or pits.

Acknowledgments

Research was sponsored in part by the Army Research Laboratory and was accomplished under Cooperative Agreement Number W911NF-10-2-0112. The views and conclusions contained in this document are those of the authors and should not be interpreted as representing the official policies, either expressed or implied, of the Army Research Laboratory or the U.S. Government. The U.S. Government is authorized to reproduce and distribute reprints for Government purposes notwithstanding any copyright notation herein. The rest of the funding is from the Office of Naval Research grant #s N00014-04-1-0688 and N00014-06-1-0315.

References

1. H. Kwak and S Chaudhuri; "Cationic doping of MgO surfaces to build corrosion protection in Mg alloys." *Journal of Alloys and Compounds* 509, 32 (2011) 8189-8198.
2. J. Xiao, H. Kwak, S. Chaudhuri, "Understanding the Role of Intermetallic Particles in Localized Corrosion" *Proceedings of MS&T 2011, Columbus, OH*; 2011; p1374-1385.
3. H. Kwak, J. Xiao, S. Chaudhuri, "First-Principles-Based Analysis on the Role of Rare-earth Doping on Mg Alloy Corrosion" *Proceedings of MS&T 2011, Columbus, OH*; (2011) 1386-1396
4. A. Anderko, N. Sridhar, D.S. Dunn, A General Model for the Repassivation Potential as a Function of Multiple Aqueous Solution Species, *Corrosion Science*, 46(2004) 1583-1612
5. J. Xiao, S. Chaudhuri; "Predictive modeling of localized corrosion: An application to aluminum alloys"; *Electrochim Acta* 56, 16(2011), 5630-5641

EFFECT OF SUBSTITUTED ALUMINUM IN MAGNESIUM TENSION TWIN

K.N. Solanki[1], A. Moitra[1], M. Bhatia[1]
[1]Center for Advanced Vehicular Systems, Mississippi State University, Mississippi State, MS, USA

Keywords: Magnesium, Aluminum, Twin Nucleation, Molecular Dynamics, Single crystal

Abstract

Atomistic simulations are performed in order to study the Aluminum substitution effect on Magnesium twinning mechanism. Multiple twin boundaries are found in pure Mg single crystal under tensile loading condition along $\langle\bar{1}012\rangle$ crystallographic direction. However, no twinning has occurred under compression loading. Al substitution has been done for 2, 5, 7, and 10% doping. For 2 and 5% Al substitution, number of twins increase when the system is monitored under tensile loading. On the other hand, for 7 and 10% Al doping under tensile loading, no twin has been found. We found that dislocation-twin and dislocation-dislocation interaction are majorly responsible for this behavior and it is important that which one is prevalent.

Introduction

Alloyed magnesium is the lightest metal with specific strength in practical use today. There are many practical application products such as vehicular industries. However its wide-spread use is greatly hampered by its anisotropic plastic response which limits its formability at room temperature [1]. This comes partly from the hexagonal-closed packed crystal structure, where slip in the basal plane is nearly one-hundred times easier than in perpendicular planes. Slip system of Mg is composed of three slips, which are parallel to basal, prismatic and pyramidal slips. Since the (critically resolved shear stress) CRSS of pyramidal slip is much larger than that of the other slips, it is difficult to activate pyramidal slip at room temperature. Although several slip systems have been reported, the slip is commonly in the closed packing basal direction, and does not produce strains parallel to the c axis. It is known that hcp Mg shows much poor ductility compared with fcc Al due to its crystal anisotropy. Magnesium also twins (a deformation mechanism involving extended stacking changes) at fairly low strains. However, because the wrought magnesium alloy is expensive and is poor in formability, the reduction of manufacturing costs and the improvement of formability are required.

Mechanistic explanation of the plastic behavior of material is based upon the dislocation and twin nucleation. This plastic behavior of the material influences on the formability of the material. Formability of elements and their alloys are affected by strain hardening rate, strain rate sensitivity, degree of anisotropy, stress-state at fracture etc. It is well known that traditional magnesium sheet alloys exhibit poor formability via conventional forming operations at ambient temperatures. However, it is also known that at high temperatures and moderate strain rates, magnesium alloys can be extremely ductile and readily formable. Traditional explanations for these phenomena relate to the behavior of the hexagonal close packed (hcp) magnesium single crystals. The plastic properties of magnesium single crystals were studied in detail during the 1950s and 60s with extensive application of slip trace analyses and constrained deformation experiments [2, 3]. A major conclusion of these studies was that magnesium crystals only possess two independent easy slip systems within the (0001) or basal plane. Thus, it appeared that magnesium, with its hcp crystal structure fails to satisfy the Taylor criterion [4] requiring five independent easy slip systems for homogeneous, generalized ductility of a polycrystalline aggregate. Other slip mechanisms observed [5, 6, 7, 8] during the early studies include the slips in prismatic and pyramidal planes. With reference to the Taylor criterion, it is noted that prismatic slip would offer two independent slip modes beyond the original two of basal slip for a total of four (still short of the required five for uniform plasticity). Pyramidal slip would offer four independent slip modes on its own (the same as basal and prismatic together). Although many investigators [9] performed c-axis compression experiments on single crystals designed to activate dislocations with non-basal Burgers vectors, no evidence of their activity was observed through slip trace analysis. Instead, some twinning mechanism was observed to barely precede fracture [8, 9]. Only after transmission electron microscopy (TEM) was applied to the problem during the early 1970s [10], non-basal dislocations were confirmed to be active within magnesium. In addition to dislocation slip, magnesium exhibits a strong propensity for mechanical twinning, especially the $\{10\bar{1}2\}\langle10\bar{1}\bar{1}\rangle$ tension twin [2]. Twinning may help a material to satisfy the Taylor criterion [11], however, it must be recognized that twinning is a polar mechanism (only allowing simple shear in one direction, rather than both forward and backward directions like dislocation slip). Furthermore, the amount of strain that twinning can accommodate is directly proportional to the volume fraction of crystal which has twinned. Thus there is a finite amount of strain that may be accommodated by twinning, even under ideal circumstances. Deformation twinning results in a sudden reorientation of the volume fraction of the crystal that is being twinned rather than the gradual reorientation that characterizes slip. Finally, although twinning mechanisms offer more options for plastic accommodation, they generally require an accommodation mechanism of their own in order to avoid large internal stresses leading to failure. Wang et al. recently studied [12] in detail the mechanism of $(10\bar{1}2)$ twin nucleation, however the postulated mechanisms for twin nucleation in models have not been validated so far.

Molecular dynamics (MD) simulations have been successful and powerful tool in revealing the deformation mechanisms in materials. Studies related to the deformation process speculate the major role of deformation twinning for magnesium. New alloy development involves new solute additions and deformation to affect mechanical properties. In present paper, we observe the Al substitution effect on the Mg twinning for four different amount of Al substitution.

Method

For the purpose of present study, single crystal magnesium samples were created of 20 × 20 × 5.5 nm^3 dimension. The samples were first relaxed using an NVT ensemble for 10 nS, and then they were relaxed independently in three directions with NPT ensemble for another 10 nS. Atomistic simulations were carried out using an MD time step of 1 fS. A periodic boundary condition was adopted in all directions. A tensile deformation along the Y $(10\bar{1}2)$ direction is performed at a strain rate of 10^8 S^{-1}, whereas deformation in X and Z directions that are on the $(10\bar{1}2)$ plane, are carried out by maintaining a zero pressure by Parrinello-Rahman method [13]. Throughout our simulations the temperature was maintained at 300 K. At a particular interval of time we collect the snapshots of the atomic structure to analyze the microstructure.

It has been already found [14] that the interstitial Al atoms in Mg crystal raise the internal energy prohibitively large via density functional theory calculations. However it is easier for the Al atoms to substitute the Mg atoms, and occupy their positions. Accordingly, we randomly doped different amount of Al atoms in Mg single crystal by substitution. The size of the simulation cell, strain rate, and temperature were kept same as for the case of single crystal Mg simulations and Mg with Al substitution. The molecular dynamics simulations in this work employed the embedded atom method (EAM) potential of Sun [15]. This potential has a better agreement on vacancy formation energy than the Liu's [16] potential, and apart from that, the potential is developed by fitting also the vacancy migration energy. It is important to note here that a recent study [17] while investigating the effect of precipitate in aluminum alloys, proposed that thermally activated vacancy movement plays a central role in grain growth mechanism.

Results and Discussion

Stress strain curves for tensile loading with several % of Al substitution are shown in Figure 1. For pure Mg sample, two distinctive peak stress can be observed, where the first peak represent the dislocation nucleation, the second peak correspond to the twin nucleation. A substitutional atom solution relieves hydrostatic stresses and interacts with nearby dislocations. Dislocations surrounded by atmospheres can produce plastic flow in two ways: at low stress dislocations cannot escape from the atmosphere, and solute atoms migrate with the dislocations, however at large stress those dislocations can tear apart and become highly mobile and able to produce rapid flow under small stress. Interstitial Al atoms locally distort the Mg lattice and produce a strain field locally. For 2% Al substitution, the dislocation nucleation peak is increased, however the twin nucleation peak is decreased compared to the pure Mg sample. With increased Al substitution (more than 2%) these two peaks converged to one peak, such that it is indistinguishable to identify the dislocation or twin nucleation from the stress-strain curve. Applied stress generates the dislocations (and hence the yielding) easily for the samples where Al exists more as substitutional atoms.

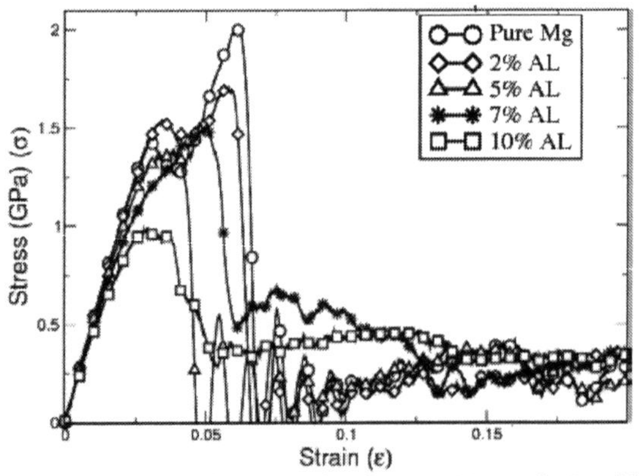

Figure 1: Stress strain curves for the pure Mg, and Mg with several % of Al substituted, under tensile loading.

We found that under tensile loading a twinning occurs, however even at quite larger strain no twinning occurs for the same direction if compressive load is provided. That confirms the fact that twinning is a polar mechanism, unlike dislocations that occurs for applied load in both directions. Figure 2 shows the structures of the sample with pure Mg for compressive loading in a) and tensile loading in b). We found that the twin formation occurs at 6.8 % of strain for pure Mg in tension. Also, we found that the initial twin volume grows with strain and secondary twins nucleate within the primary twin.

Figure 3 shows the microstructure of the samples with different amount of Al substitution. We found that in Figure 3a at 2% of Al substitution number of twins are four at 6.8% of strain. At Figure 3b with 5% Al substitution number of twins are four at 5.8% strain. In Figures 3c and 3d at 7% and 10% Al substitution, respectively we found no twinning occurs even at large strain. It has been also found that strain at which twin nucleates, decreases with increasing amount of Al doping up to 5% of substitution.

Figure 4 shows the evolution of twin-volume with strain for our models. It can clearly be seen that the twin nucleation occurs at higher strain for pure Mg model compared to the models where up to 5% Al is substituted. We can also find that the twin volume increases faster for the cases where Al is substituted compared to the pure Mg case. In our calculations, we have found that the angular rotation of twinned volume have increased up to 87° that in a very good agreement with the previous results. Since for 7 and 10% Al substitution, we do not find any twinning formation, the figure does not contain any data for that higher % s of Al substitution.

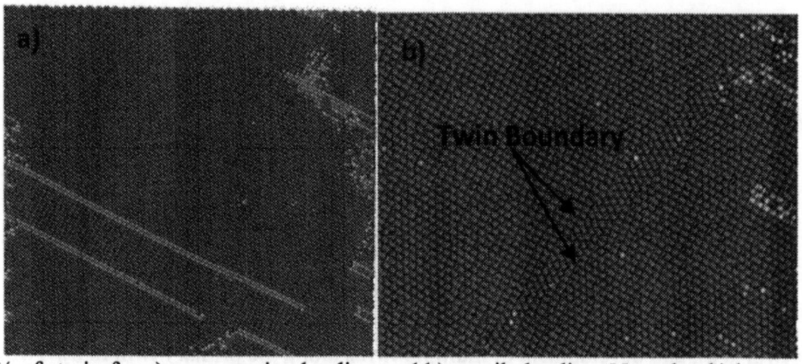

Figure 2: Sample at 6.8% of strain for a) compressive loading and b) tensile loading. Note that b) has a twin formation, while a) does not.

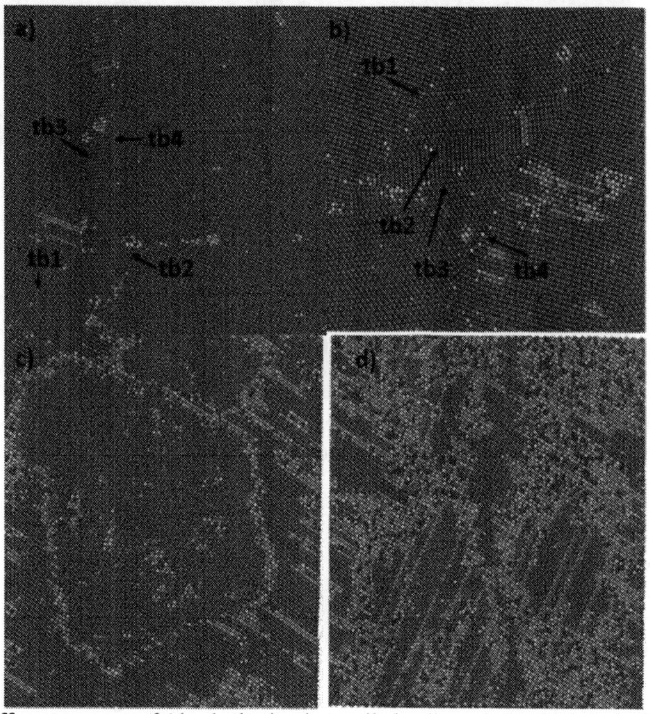

Figure 3: Mg samples with different amount of Al substitution in tensile loading. a): At 6.8% strain for 2% Al substitution, b): At 5.8% strain with 5% Al substitution, c) At 7% strain with 7% Al, and d) At 10% strain with 10% Al substitution. Note that you may find 2 more twins in b), but due to periodic boundary condition, they represent the same twins. Also note that c) and d) has no twins.

We further analyze the microstructure and revealed an interesting interaction of twin boundary with dislocation loop. Figure 5 shows a twin emission and the tip of twin is marked as 'Tip'. Figures 5a and 5b are for a pure Mg tensile deformation at 4.6% and at 9% strain. Figure 5b shows that at larger strain how the number of twins increases and how the secondary twin forms from the primary twin boundary. We have observed that secondary twin always forms from the primary twin boundary plane. Twin boundaries are marked by numbers in Figure 5b.

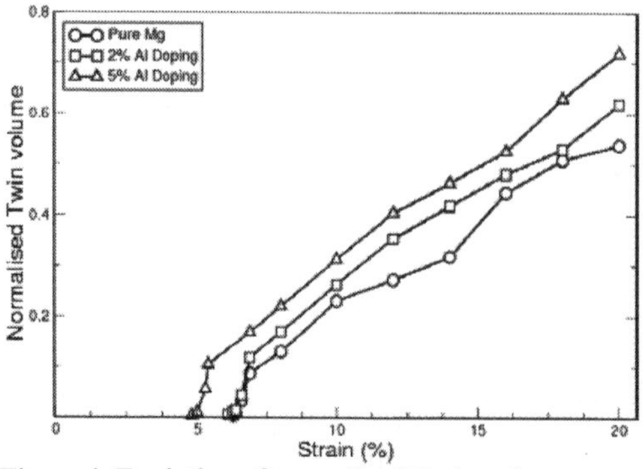

Figure 4: Evolution of normalized Twin volume.

Figure 5: Emission of a twin at a), and development of multiple twins along the primary twin in b). Twin tip is marked by tip.

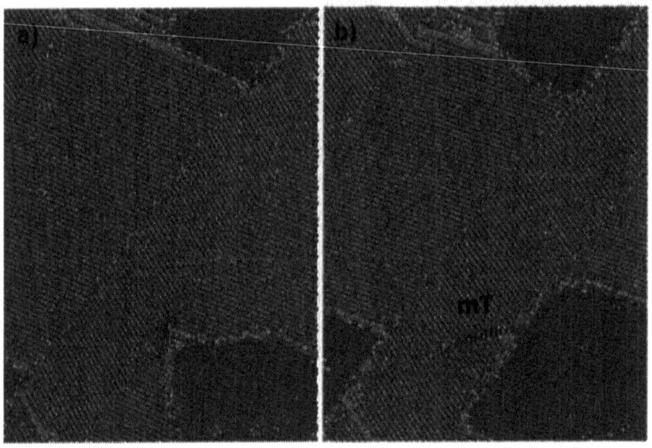

Figure 6: Interaction of dislocation with twin. In a) a dislocation loop emission toward the twin boundary is shown. In b) dislocation growth along the twin boundary is shown. Note a secondary twin formation in b) marked by mT.

In Figures 6a and 6b, we show that the twin boundary interaction with a dislocation loop. It is found that the twin boundary is very rigid and the dislocation loop cannot pass through the twin boundary. In Figure 6a, we show the initial emission of dislocation towards an existing twin boundary at 5% strain. Figure 6b shows the same dislocation-twin interaction at 9.6% strain. We found that even at large strain the dislocation loop grows along the twin boundary plane. This behavior of dislocation-twin interaction certainly due to the fact that twin boundary is very rigid, such that a dislocation cannot penetrate through the twin boundary; instead the dislocation grows along the twin boundary plane. In Figure 6b, however a micro-twin is observed inside the primary twin, marked as mT in the figure. At a later strain (not shown here) this micro-twin also grows. Thus, the interaction compels us in believing the fact that twin- dislocation interaction gives rise to more twins. When there are substituted Al atoms in the Mg system, dislocations occur at less stress state (See Figure 1); these dislocations enhance the twin-dislocation interaction that finally produces more twins. However, when the amount of Al substitution reaches to 7% atomic percentage, enormous number of dislocations are generated. These dislocations then interact mutually. As such, the twin-dislocation interaction seizes and dislocation-dislocation interaction becomes prevalent. When these dislocation-dislocation interactions become predominant, twin formation becomes sluggish. The interaction between two dislocations or a dislocation with a twin is inherently a dynamic nature of interaction of two defected lines. The atoms inside those defected-lines although possess a symmetrical nature of crystal structure (hcp for this case). These atoms being intelligent being at a given time knowing all the forces that causes them to move, try to dissipate the stress in such a fashion that the interaction energy of those two defected-lines minimized. Typically for two dislocation loops interaction, the atoms inside them are trying to reorient themselves to minimize the interaction energy, and hence no effort is being taken to penetrate the twin boundary. Naturally, when the dislocation-dislocation interaction becomes prevalent than the dislocations-twin interaction (which is 7% or higher for the present study), new twin formation does not take place.

In order to increase the formability of Mg we need to get the homogeneous and generalized ductile behavior for Mg. Since the slip systems are limited, and Mg fails to satisfy Taylor criterion, twinning is the alternative to increase the formability of Mg. We found in the present paper that how the evolution of twinning changes with the Al substitution. It should be kept in mind that the present calculation deals a tensile loading condition only for a particular orientation of single crystal Mg. In reality, there are several grains oriented in different directions with multiple grain boundary and different amount of multi-dimensional defects in it. Nevertheless, the nature of the dislocation-dislocation interaction or the dislocation-twin interactions discussed in the paper remains the same for polycrystalline aggregate. Perhaps it could be a nice future study to see the dislocation-twin interactions for a polycrystalline Mg sample with low angular misorientation.

Conclusion

Atomistic simulations are performed in order to study the Aluminum substitution effect on Magnesium twinning mechanism. Deformation twinning are found in pure Mg under tensile loading condition, however no twinning has occurred under compressive loading. It has been found that for 2 and 5% of Al substitution, twinning occurs at low strain compared to pure Mg case, but for 7 and 10% Al substitution, no twinning has been found. The interaction of dislocations with other dislocations and twin are held responsible for this behavior particularly for this orientation. For small (2 and 5%) amount of Al substitution the dislocation interacts with twin and their interaction gives rise to more twins, however for 7% or more Al substitution, dislocation-dislocation interaction becomes prevalent and no twins occur.

Acknowledgments

This material is based upon work supported by the Department of Energy and the National Energy Technology Laboratory under Award Number DE-FC26-02OR22910 and the Office of Naval Research under contract No. N00014-09-1-0661. This report was prepared as an account of work sponsored by an agency of the United States Government. Neither the United States Government nor any agency thereof, nor any of their employees, makes any warranty, express or implied, or assumes any legal liability or responsibility for the accuracy, completeness, or usefulness of any information, apparatus, product, or process disclosed, or represents that its use would not infringe privately owned rights. Reference herein to any specific commercial product, process, or service by trade name, trademark, manufacturer, or otherwise does not necessarily constitute or imply its endorsement, recommendation, or favoring by the United States Government or any agency thereof. The views and opinions of authors expressed herein do not necessarily state or reflect those of the United States Government or any agency thereof. Such support does not constitute an endorsement by the Department of Energy of the work or the views expressed herein. This work was in part supported by the Center for Advanced Vehicular Systems (CAVS) at the Mississippi State University. Computer time allocation has been provided by the High Performance Computing Collaboratory (HPC2) at Mississippi State University.

References

[1] M. Avedesian, H. Baker, ASM Specialty Handbook (1999).
[2] C. Roberts (1964).
[3] B.Wonsiewicz, W. Backofen, Trans. TMS-AIME 239 (1967) 1422–1431.
[4] G. Taylor, Journal of Inst. Metal. 62 (1938) 307–338.
[5] P. Ward Flynn, J. Mote, J. Dorn, TMS-AIME 221 (1961) 1148–1154.
[6] R. Quimby, J. Mote, J. Dorn, Trans. ASM 55 (1962) 149–157.
[7] R. Reed-Hill, W. Robertson, Trans. ASM 220 (1957) 496–502.
[8] R. Reed-Hill, W. Robertson, Trans. ASM 221 (1958) 256–259.
[9] F. Yoshinaga, R. Horiuchi, Trans. JIM 4 (1963) 1–8.
[10] T. Obara, H. Yoshinga, S. Morozumi, Acta Metall. 21 (1973) 845–853.
[11] U. Kocks, D. Westlake, Trans. ASM 239 (1967) 1107–1109.
[12] J. Wang, J. P. Hirth, C. N. Tom´e, Acta Materialia 57 (2009) 5521–5530.
[13] M. Parrinello, A. Rhaman, J. Appl. Phys. 52 (1982) 7182.
[14] B. Jelinek, J. Houze, S. Kim,M. F. Horstemeyer, M. I. Baskes, S.-G. Kim, Phys. Rev. B 75 (2007) 054106.
[15] D. Y. Sun, M. I. Mendelev, C. A. Becker, K. Kudin, T. Haxhimali, M. Asta, J. J. Hoyt, A. Karma, D. J. Srolovitz, Phys. Rev. B 73 (2006) 024116.
[16] X.-Y. Liu, J. B. Adams, F. Ercolessi, J. A. Moriarty, Modelling Simul. Mater. Sci. Eng. 4 (1996) 293–303.
[17] P. S. De, J. Q. Su, R. S. Mishra, Scripta Materialia 64, 1 (2010) 57-60.

PART 8: JOINING

Eric A. Nyberg

Effectively joining magnesium has been a challenge since its inception as an industrial and structural metal. Research over the last 15 years has resulted in methods that now make magnesium joining viable for not only Mg-to-Mg but Mg-to-dissimilar metal joints. Industry has capitalized on this progress as we now see more complex use of magnesium in automotive, aerospace, tools, and electronic applications. Joint strength and performance, mitigation of galvanic corrosion, and cost-effective joining methods have been primary areas of research that have made outstanding progress. This section highlights 11 key papers that address development and performance of mechanical fusion and solid-state joining.

The first five papers are founding articles on various methods of joining magnesium alloys. The first paper was written by Johnson and Threadgill of The Welding Institute (TWI), where "Friction Stir Welding" (FSW) was invented. This article demonstrated that FSW will be an important solid-state joining method for a variety of Mg alloys. The following paper by Lathabai and colleagues uses molten fusion methods, namely Gas Tungsten Arc (GTA) and Laser Beam (LB) welding. It highlights the point that the relatively more rapid cooling rates associated with LB welding produces a finer microstructure, leading to better joint performance. The paper by Pinheiro and colleagues demonstrates a novel approach to Mg joining, using a "drill-and-fill" method referred to as "Friction Hydro Pillar Processing" (FHPP). This process is essentially friction stirring a creep resistant Mg alloy stud into a pre-drilled hole in a magnesium base metal plate. The consumable stud is upset into the baseplate of magnesium creating a solid state fused joint. Mini-tensile results suggested post-joint performance in the stud matching the base metal strength. An excellent review by Jana and colleagues summarizes results for both FSW and Ultrasonic Welding. A variety of tools, process parameters, and surface conditions are used and demonstrate their impact on mechanical, corrosion and general weld properties. The paper by Anopuo and colleagues compares the creep performance of a mechanical bolted joint with a time-dependent stress-strain response modeled using a finite element program based on a power-law creep relationship.

Friction stir welding has demonstrated significant advantages in magnesium research and development. As a result, the remaining six papers all relate to FSW or Friction Stir Spot Welding (FSSW). As early as 2000 we received the first FSW of Mg paper presented by Nagasaw and colleagues. This was an important paper as it set forth the robust potential and positive results associated with joining AZ31 using FSW. The paper by Skar and colleagues demonstrates how FSW tooling plays an important role in the performance of Mg FSW. Evidence of diminished mechanical and corrosion properties, due to iron contamination from tool degradation is motivation for subsequent FSW tool development. As example, tool design and process parameters are the basis for the paper by Zettler and colleagues.

As the FSW methods became more refined we begin to see the process applied to actual components. In 2007, the first paper applying FSW to Mg alloys used in an oil pan assembly was presented by Hunt and colleagues as part of the USCAR Magnesium Powertrain Cast Component (MPCC) Project.

Finally, two papers on the fatigue performance of FSSW magnesium joints are given. The paper by Jordon and colleagues on the experimental fatigue results of AZ31 FSSW joints was the basis that has subsequently led to important modeling efforts by Jordon. The paper by Badarinarayan and colleagues is one of many important papers resulting from the multi-national Canada-China-US Magnesium Front End Research and Development

(MFERD) project. Badarinarayan compares montonic and cyclic fatigue performance of FSSW to resistant spot welds of similar nugget size.

FRICTION STIR WELDING OF MAGNESIUM ALLOYS

Dr Richard Johnson, [1]Dr Philip Threadgill [1]
[1]TWI Ltd, Granta Park, Great Abington, Cambridge CB1 6AL, UK

Abstract

An investigation has been carried out on the friction stir welding of four magnesium alloys. These consisted of one wrought and three die-cast magnesium alloys, including those containing manganese such as AM50 and AM60, and also zinc such as AZ91 and AZ31 (wrought material). All of the alloys have each been successfully welded to themselves and also to each other, without any problems from the trapped gases in the cast materials, but the tolerance box of processing parameters to ensure that sound welds are produced has been found to be more restrictive than those seen in friction stir welding aluminium alloys.

Introduction

Magnesium is one of the most abundant elements in the earth's surface, with virtually inexhaustible supplies in the oceans. Over recent years the industrial output of magnesium alloys has been rising by almost 20% per annum, which is faster than that of any other metal. The increased use of aluminium and magnesium alloys is of great interest to the automotive industry, with the goal of reducing the weight of road vehicles to make them more fuel-efficient or to increase the vehicle specification without adversely affecting its fuel efficiency.

In recent years there has been a renewed interest in the use of magnesium parts for body components, many of which have made by pressure die casting [1,2,3]. These have limited ductility, contain gas occlusions, and are frequently difficult to weld satisfactorily by fusion welding techniques. With the major proportion of magnesium alloys being made by casting there has not previously been an extensive need for improved weldability to be developed. The solid state joining technique of friction stir welding (FSW) Fig. 1, was patented in 1991 [4], and was initially used to extend the weldability of aluminium alloys, some of which were difficult to join by fusion welding techniques because of cracking and porosity problems. The scope of this new welding process has since been extended to the welding of lead, zinc, copper, titanium and ferrous alloys, with some considerable success.

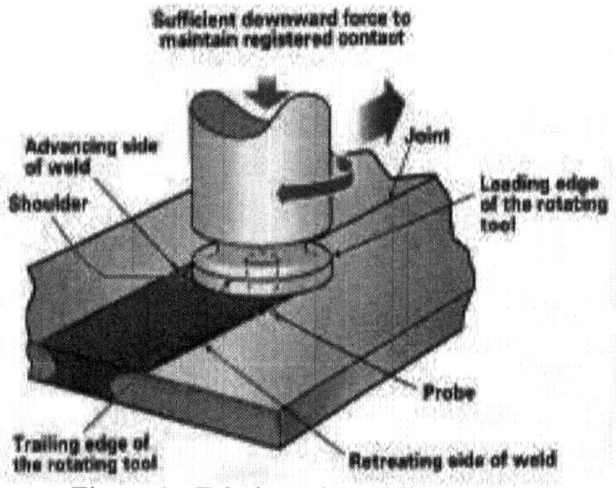

Figure 1: Friction stir welding process.

Of particular interest to the joining of cast magnesium parts is the success of FSW cast and wrought aluminium materials together. The cast material contained significant porosity, indicating trapped gases to be present, but whereas fusion welding methods would have encountered problems from their presence, FSW created a sound weld with no porosity in the weld bead or the immediate HAZ [5,6], Fig. 2. TWI has been studying the FSW of magnesium alloys, and is currently participating in a European collaborative project, MagJoin, to assess the weldability of magnesium alloys to themselves, to each other, and to aluminium alloys.

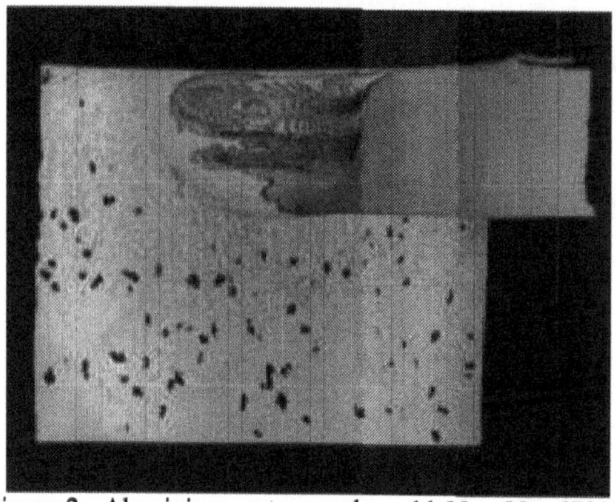

Figure 2: Aluminium cast-wrought weld. Neg. No. 67006.

Experimental Results of Friction Stir Welding Trials on Magnesium Alloys

The work undertaken at TWI has concentrated on four magnesium alloys: three cast alloys, AM50, AM60 and AZ91; and one wrought alloy, AZ31. The AM alloys contain aluminium and manganese as the major alloying elements, while the AZ alloys have aluminium and zinc as the major alloying elements, and all of these alloys are of interest for applications in the automotive industry.

The trials were performed on small plates, 140 x 100 x 6mm in size, except that the AZ31 plates were slightly thicker at 6.4mm. The latter plates were not machined down to match the other plate thicknesses, but the FSW tool was set for a full penetration weld in the 6mm materials, and was allowed to plough slightly into the thicker AZ31. The FSW tools used were of the plain threaded pin or the MX Triflute™ design, and a range of rotation and traverse speeds were investigated.

The first trials were performed on the AM50 alloy plates, and the rotation speed was varied from 250-500rpm, and the traverse speed varied from 160-450mm/min. It was determined that 355rpm and 160-224mm/min resulted in sound welds, and 355rpm was adopted as the standard rotation speed for the other magnesium alloys to be welded, so that a direct comparison of the weld quality could be established. In these other alloys the traverse speed was varied in the range 160-315mm/min, and 160mm/min selected as the best with which to compare full penetration welds.

It was found relatively easy to weld the AZ91 and AZ31 alloys to themselves, and smooth weld surfaces were achievable, particularly with the AZ31 alloy. However, with the AM50 and AM60 alloys, although some smooth weld beads were achievable, under certain conditions these materials were found to adhere to the FSW tool, and would initially cause a slightly torn appearance to the weld surface, and then the FSW tool began to gouge material out of the plates to leave a surface void, Fig. 3. It was also found that the wrought alloy AZ31 and the softest cast alloy AM50 could be processed at slightly higher traverse speeds than the other two alloys.

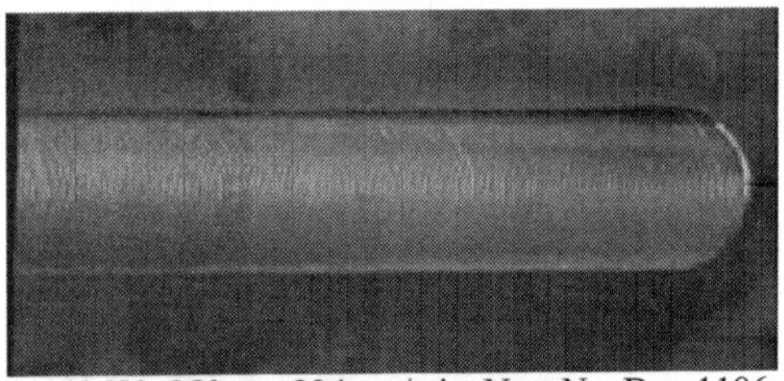

(a) AM50, 250rpm-224mm/min. Neg. No. Dscn1186.

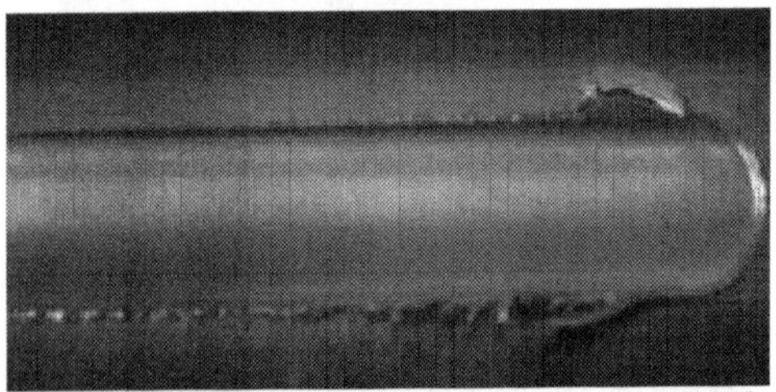

(b) AZ31, 355rpm-160mm/min. Neg. No. Dscn1149.

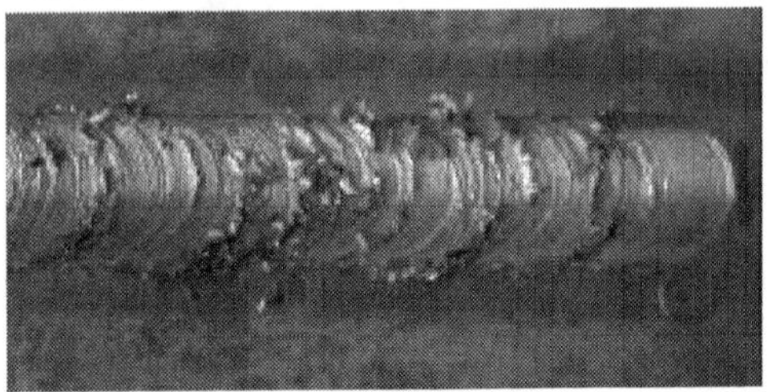

(c) AM50, 355rpm-224mm/min. Neg. No. Dscn1181.

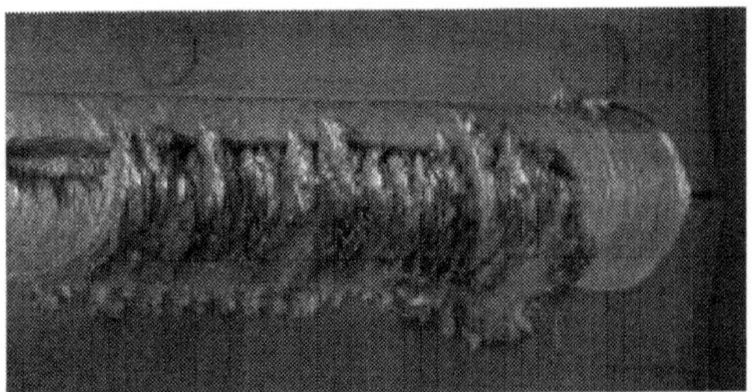

(d) AM60, 355rpm-315mm/min. Neg. No. Dscn1182.

Figure 3: Weld surfaces of magnesium alloys welded to themselves.

The weld macrosections revealed that at a traverse speed of 160mm/min the AZ91 weld nugget was quite parallel-sided, whereas the other three alloy weld nuggets were broader in shape. At 224mm/min the AZ31 weld nugget remained broad whereas the AZ91 and AM60 weld nuggets were now more parallel-sided and possibly beginning to leave a potentially unwelded root at the bottom of the weld. The AM50 weld nuggets also became more parallel-sided at the higher traversing speeds, and at 315mm/min the FSW tool penetration was quite evidently reduced and leaving a potentially unwelded root, Fig. 4.

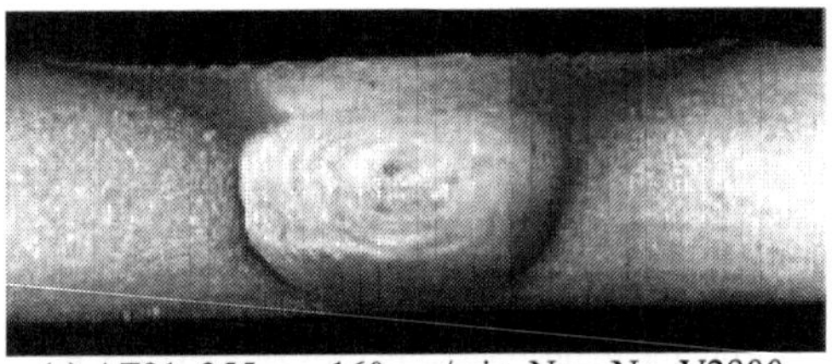

(a) AZ91, 355rpm-160mm/min. Neg. No. V2900.

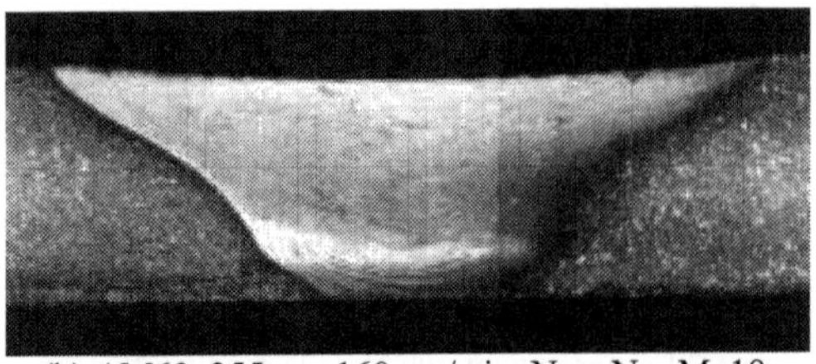

(b) AM60, 355rpm-160mm/min. Neg. No. Mg19.

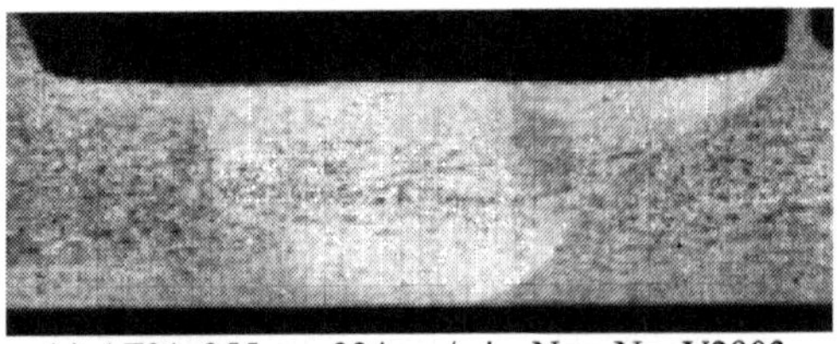

(c) AZ31, 355rpm-224mm/min. Neg. No. V2903.

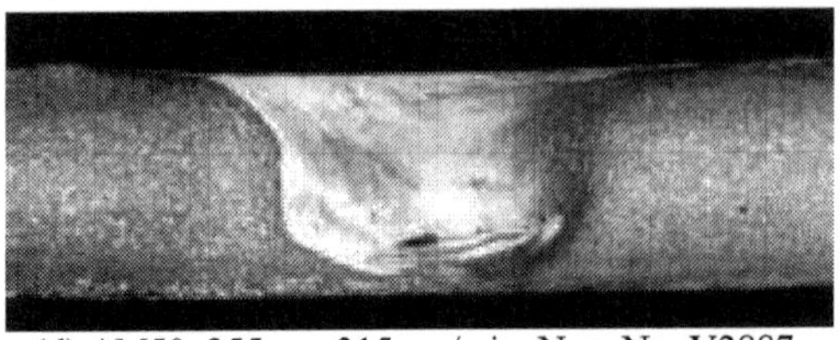

(d) AM50, 355rpm-315mm/min. Neg. No. V2887.

Figure 4: Macrosections of 6mm plate magnesium alloys welded to themselves.

The welding of the dissimilar magnesium alloys to each other were initially performed at a traverse speed of 160mm/min, the speed at which the full penetration of the FSW tool had been verified in all of the alloys. In general the weld surfaces were quite smooth, except that the AZ91-AZ31 weld appeared somewhat rougher. Selected macrosections of these welds, as seen in Fig. 5, show that the AM50-AM60 weld is quite parallel-sided, as are the welds between these alloys and AZ91, but all of the welds with the AZ31 revealed a significantly broader weld nugget.

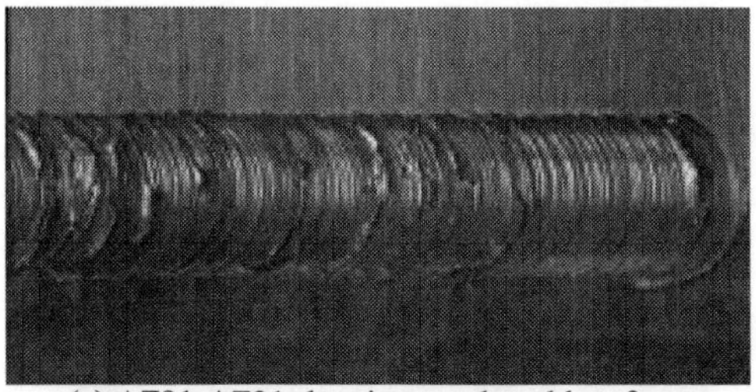

(a) AZ91-AZ31 showing rough weld surface.

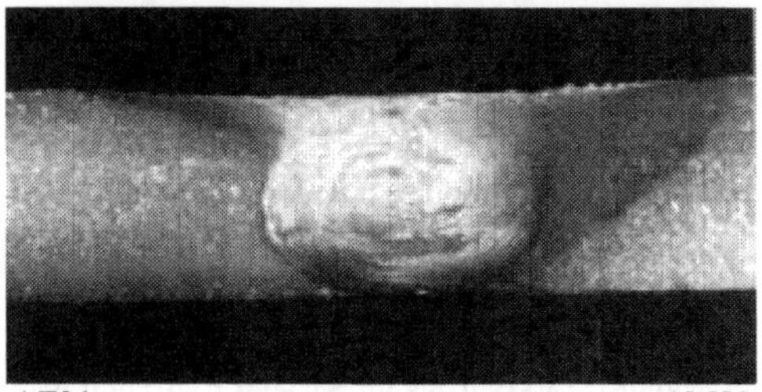

AZ91..AM50

(b) AZ91-AM50 weld.

Figure 5a-b: Dissimilar alloy 6mm plate weld macrosections, 355rpm-160mm/min.

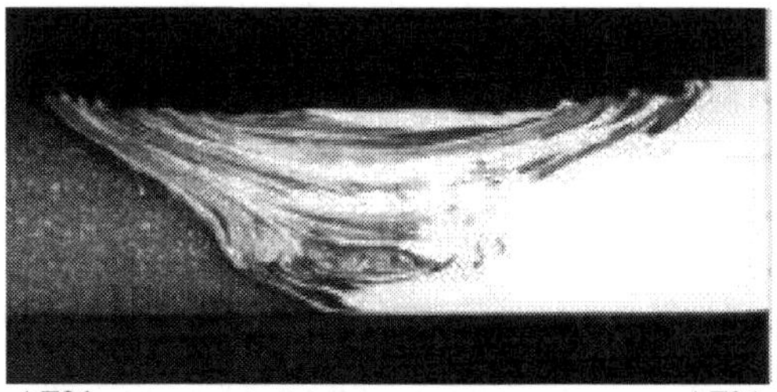

AZ91..AZ31

(c) AZ91-AZ31weld

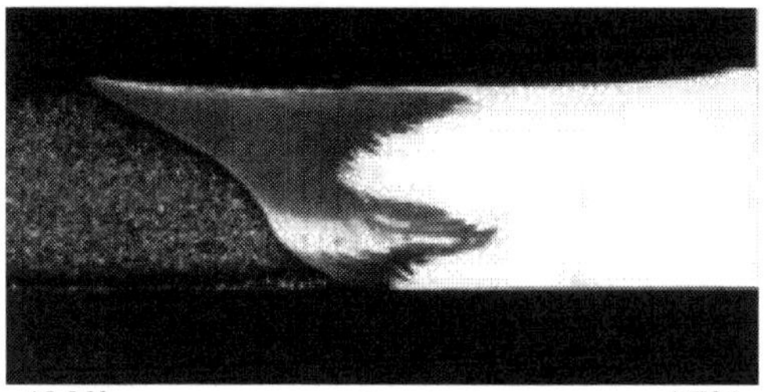

AM60..AZ31

(d) AM60-AZ31 weld

Figure 5c-d: Dissimilar alloy 6mm plate weld macrosections, 355rpm-160mm/min

When welding the dissimilar magnesium alloys together, it was found that in some pairings the alloys could be stirred into each other with equal ease, but in others there was a marked difference in the appearance of the weld surfaces depending on the which alloy was stirred into the other. In order to emphasise this effect, a series of welds was performed at a higher traverse speed than the initial dissimilar alloy welding trials, this time at 250mm/min, and the results can be seen in Fig. 6. In particular it was found preferable to stir the other alloy into AZ31 rather than the opposite way round, while this orientation effect was not seen to be as critical with the welds made without AZ31. It should be noted that a similar effect has been noted in the FSW processing of aluminium alloys [7], and is probably caused by the asymmetric nature of the FSW process and the different strengths and flow stresses of the two materials being welded at the elevated temperature.

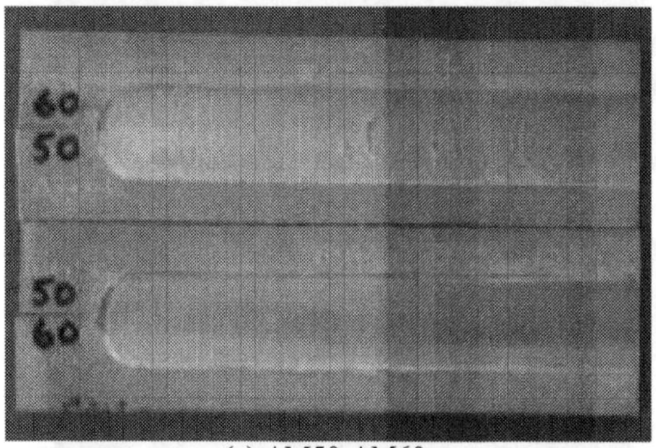

(a) AM50-AM60.

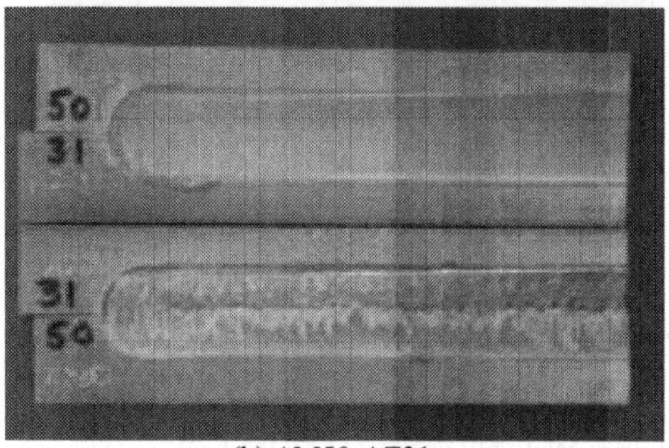

(b) AM50-AZ31.

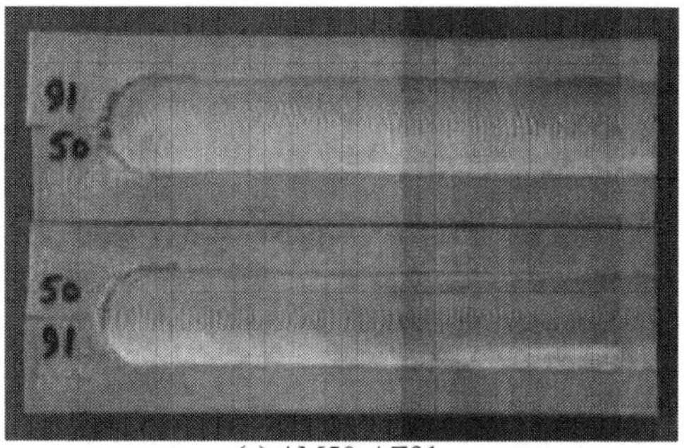

(c) AM50-AZ91.

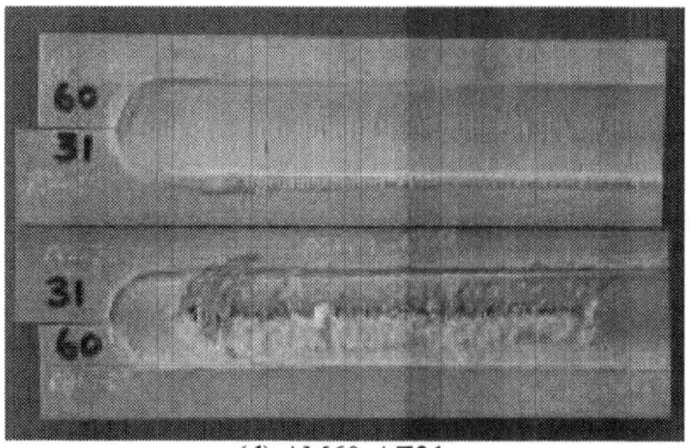

(d) AM60-AZ31.

Figure 6a-d: Weld surfaces from dissimilar magnesium alloy welding trials.

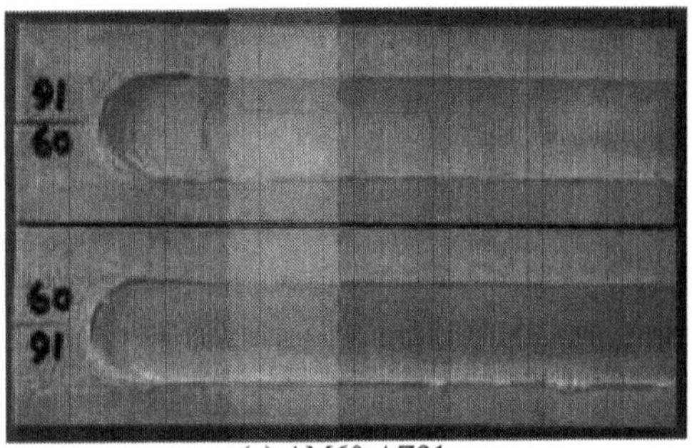

(e) AM60-AZ91.

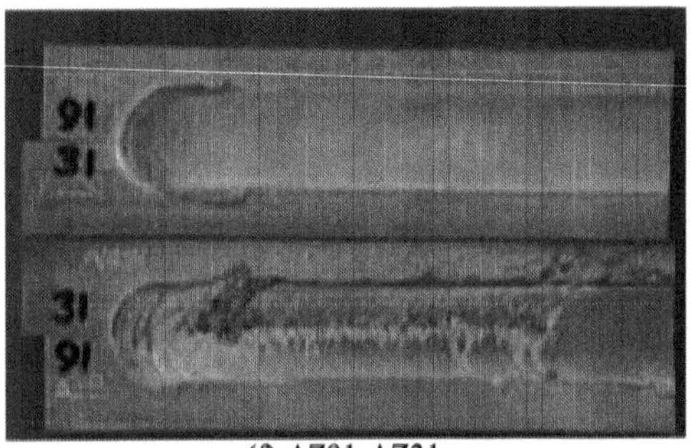

(f) AZ91-AZ31.

Figure 6e-f: Weld surfaces from dissimilar magnesium alloy welding trials.

The macrosections of the dissimilar alloy welded joints showed a fine intermixing of the two alloys. In particular some aluminium-manganese intermetallic particles were clearly visible in the AZ31 alloy, and were relatively unaffected by the FSW process. These particles served to show clearly that they remained in the AZ31 material even when the intermixing was extremely fine, such as at the bottom of the weld nuggets, although occasionally they were found to be situated along the interfaces between the alloys, Fig. 7.

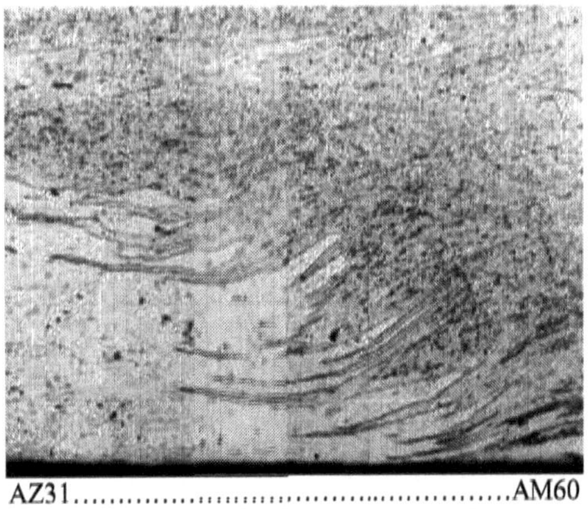

AZ31...AM60

(a) Bottom of weld nugget, Mag. X?.

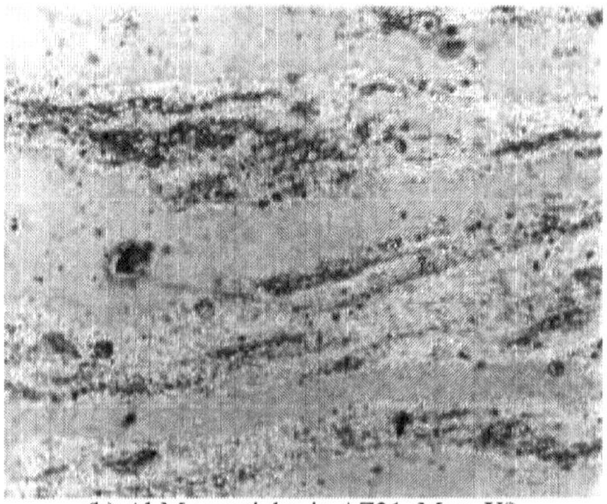

(b) Al-Mn particles in AZ31, Mag. X?.

Figure 7: AM60-AZ31, (a) intermixing of alloys, (b) intermetallic particles in AZ31 alloy.

Hardness measurements were made on selected welds at a load of 2.5kgf, and traverses made at both 2mm and 4mm below the top surfaces of the welds. This was to check if there were significant differences between the top of each weld nugget, where the tool shoulder affects the weld nugget width, and the lower region where the pin alone dictates the nugget width. On the single alloys welded to themselves the plots showed some degree of scatter, with in general a little hardening discernible in the weld nugget region, presumably due to the fine grain size there, but no great differences between the 2mm and 4mm readings. The exception, as expected, was AZ31, which derives much of its strength from being rolled as a wrought product, and here the weld nugget showed a reduced hardness with respect to the parent material. Hardness measurements were also made on the dissimilar alloy joints, where there was again a degree of scatter seen in the plots. In general there did not appear to be any significant hardening other than by the grain size refining of the weld nugget, and only slightly lower values in the HAZ regions, which can be seen in Fig. 8.

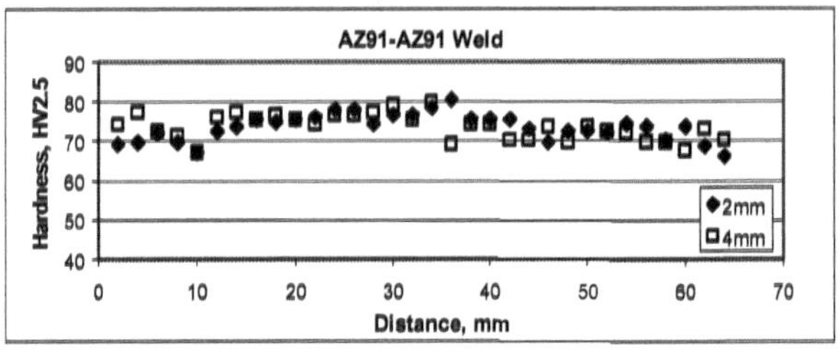

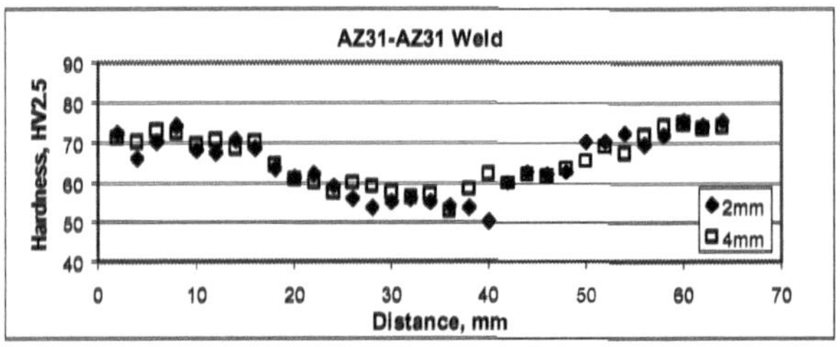

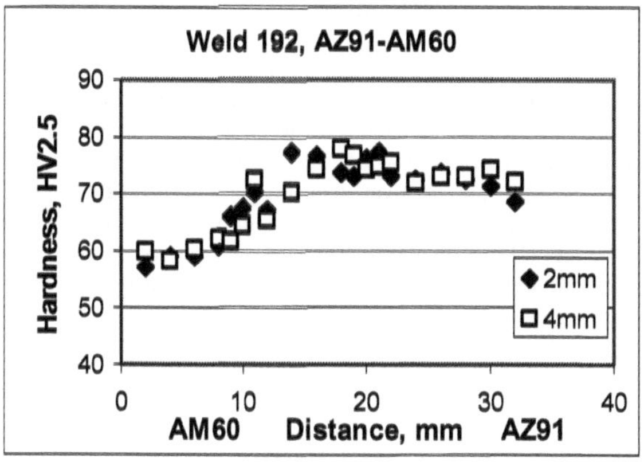

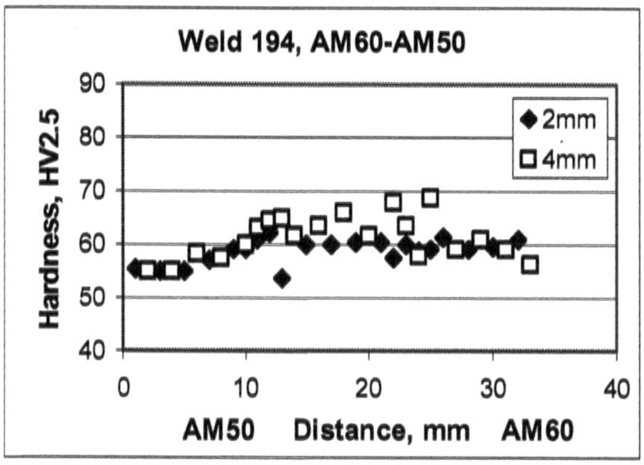

Figure 8: Hardness traverses of like-like and dissimilar alloy welds.

Additionally some tensile tests were made on the magnesium alloys welded to themselves, and these showed that for AM50, AM60 and AZ91 the proof stresses of the welded test specimens were quite similar to those of the parent plates, and that the ultimate tensile stresses were about 10-15% lower for AM50 and AZ91, but slightly higher for AM60. With the AZ31 welds, because much of the parent strength is derived from work hardening during fabrication, it was not surprising to see that both proof and ultimate tensile stresses in the welded specimens were appreciably lower. In all cases the elongations were lower in the welds, as expected when introducing a slightly harder weld nugget and two HAZ regions.

Discussion

The preliminary FSW processing trials on the four selected magnesium alloys have shown that, although three of them are made by high pressure die casting, they can be welded together satisfactorily and without any problems from the trapped gases contained in the materials. Welding each alloy to itself could be readily performed, although the processing parameters were somewhat lower than would be the case in aluminium alloys of the same thickness. The AM alloys were found to tend to adhere to the FSW tool during welding, and this could lead to surface defects on the resultant weld surfaces, and this was seen with plain tool steel FSW tools and also with those coated with titanium nitride. It is intended to assess whether or not other tool coatings might reduce the effect.

When welding dissimilar magnesium alloys together, again there were some problems with the AM alloys adhering to the FSW tool. Additionally, it was found that placing one alloy on the advancing side of the weld was not always as successful as placing it on the retreating side of the weld. This was especially found to be the case with the wrought AZ31 alloy, where better quality weld surfaces resulted from that alloy being placed on the retreating side of the weld. The FSW process is asymmetrical, because of the tool rotation, and the heating effects on the advancing and retreating sides of the joint will necessarily be different. It is interesting that the next phase of the MagJoin programme is to try to weld aluminium and magnesium alloys together, and the first trials have shown the same effect – that placing the aluminium on one side of the joint can result in a different weld quality from placing it on the other side. However, a more systematic study of the FSW processing parameters is required before definite conclusions can be drawn as to how these dissimilar materials may best be welded together.

This work has demonstrated the suitability of friction stir welding for joining magnesium, and it is expected that interest in the method will increase. The process is suitable for adaptation to robotic welding, and there is an increasing volume of literature that has demonstrated this, for a variety of robot designs. Similarly, 5 or 6 axis CNC systems can also be used, and at least one such system for complex three dimensional welds has been delivered into the aerospace industry. The use of such flexible systems will be of value to the automotive industry, as few joints are one-dimensional, as used here.

This work has only described butt welds, although work on aluminium alloys has demonstrated very well that the process can be adapted for use with lap welds, although greater care is required in specifying procedures and developing parameters and tool designs to avoid various defects which can form at the interface. Preliminary studies on lap welds in magnesium have proved encouraging, and there is no reason to assume that this should not be reduced to commercial practice

One drawback of friction stir welding is the need for some backing system to react the downforce. This is simple in straight line one-dimensional welds, but clearly more complex in three dimensional welds. However, with high part numbers expected in automotive applications, the cost of this is likely to be acceptable. For lap welds, which are never fully penetrating through all the layers, the components themselves may contribute to the rigidity, thus reducing, although not eliminating, this requirement.

At present, travel speeds which can be achieved with magnesium alloys are very much less than obtainable with aluminium alloys. This is believed to be a consequence of the close packed hexagonal crystal structure, and limited slip systems compared to aluminium alloys. However, the speeds currently obtainable are economically viable, particularly when assessed against the high quality obtainable in the welds. Efforts are in place to improve these speeds, and hence the productivity which can be obtained. A further minor contribution to process economics is the tool life. There is no critical evaluation available to determine expected tool life, but experience gained at TWI has indicated that this is not likely to be a problem. However, the need for periodic tool cleaning in some alloys, as mentioned earlier, is noted.

Concluding Remarks

A study of friction stir welding four common magnesium alloys has been conducted, in which alloys have been welded to themselves and to each other. It has been possible to develop procedures giving sound welds for all combinations, and initial indications are that mechanical properties will meet expectations.

Acknowledgements

The author acknowledges that this welding research has been possible through the internal TWI Corporate Research Programme. This has also enabled TWI to be a partner in a three-year European collaborative project, MagJoin (GROWTH Project No. GRD1-1999-10918), to investigate the joining of magnesium alloys to themselves and to aluminium alloys, particularly for potential automotive applications. TWI has worked solely on the development of FSW in this project, while other partners have investigated different welding techniques.

Table 1: *Tensile test results of parent and welded magnesium alloy plates.*

Sample	Proof Stress [MPa]	Ultimate Stress [MPa]	Elongation [%]	Fracture location of welded specimens
AM50, typical [8]	125	230	15	
AM50, parent	104	218	10.0	
AM50, parent	117	215	10.5	
AM50, FSW	115	180	4.5	Parent/HAZ
AM50, FSW	110	164	4.0	Weld
AM60, typical [8]	130	240	13	
AM60, parent	121	173	*	
AM60, parent	77	103	1.0	
AM60, FSW	118	190	4.5	Parent/HAZ
AM60, FSW	110	198	6.5	Parent/HAZ
AZ91, typical [8]	160	250	7	
AZ91, parent	144	203	2.0	
AZ91, parent	150	201	2.0	
AZ91, FSW	157	177	2.0	Parent/HAZ
AZ91, FSW	153	183	1.5	Parent/HAZ
AZ31, typical [8]	200	255	12	
AZ31-H24, typical [8]	220	290	15	
AZ31, parent	219	288	6.5	
AZ31, parent	205	292	10.0	
AZ31, FSW	127	201	3.5	Weld
AZ31, FSW	124	201	4.0	Weld

* fractured outside gauge length
-H24 indicates the material has been strain-hardened and partially annealed

References

1. S. Juttner: Return of the Light Alloy Brigade – Welding of Magnesium Alloys, Welding & Metal Fabrication, Vol 66, No 1 (1998).

2. G. S. Cole: The Potential for Magnesium to Reduce Vehicle Mass by 100kg, Automotive Light Metals, Vol 1, Issue 1 (2001).

3. H. Friedrich and S. Schumann: Turning Vision to Reality for the Second Age of Magnesium, Automotive Light Metals, Vol 1, Issue 1 (2001).

4. W. M. Thomas et al: Friction Stir Butt Welding, GB Patent Application 9125978.8, December 1991, (1991).

5. R. Johnson: Friction Stir Welding for Castings, Aluminium Castings Conference, Northampton (1998).

6. S. W. Kallee and A. Mistry, Friction Stir Welding in the Automotive Body-in-White Production, 1st International Symposium on Friction Stir Welding, Thousand Oaks (1999).

7. C. J. Dawes et al: Development of the New Friction Stir Technique for Welding Aluminium, TWI GSP 5651 Project, Phases I, II and III (1994-97).

8. M. M. Avedesian and H. Baker: Magnesium and Magnesium Alloys, ASM Speciality Handbook, ASM (1999).

WELDING AND WELDABILITY OF AZ31B BY GAS TUNGSTEN ARC AND LASER BEAM WELDING PROCESSES

S. Lathabai[1], K.J. Barton[1], D. Harris[1], P.G. Lloyd[1], D.M. Viano[1] and A. McLean[2]
[1]CSIRO Manufacturing and Infrastructure Technology, P.O. Box 4, Woodville, SA 5011, Australia
Department of Mechanical Engineering, University of Adelaide, Australia

Abstract

Welding will play an important role in the fabrication of modular lightweight structures based on magnesium alloy die castings, extrusion profiles and wrought products. Minimisation of rejection rates during fabrication requires that satisfactory weldability be established for a particular combination of materials and welding procedures. In this paper, we present the results of a study to quantify the weldability of wrought alloy AZ31B by gas tungsten arc (GTA) and laser beam (LB) welding processes. The susceptibility to weld metal solidification cracking was evaluated using the Circular Patch weldability test. Operating windows of welding parameters for crack-free and porosity-free GTA and LB welding were identified, based on which welding procedures were developed for sheet and plate AZ31B. The microstructure and mechanical properties of welded test plates were assessed, leading to a better understanding of microstructural development and structure-property relationships in GTA and LB weldments in AZ31B.

Introduction

Magnesium alloys with their characteristic low relative density are used in a wide variety of applications where light weight is important [1]. Other attractive properties include higher strength, elastic modulus and elevated temperature properties relative to polymers and polymer-based composites, fatigue resistance, damping capacity, dimensional stability, machinability, and low resistance to X-rays and thermal neutrons. In recent years, magnesium alloys are increasingly used in automotive applications in a constant effort to reduce the overall weight of vehicles, leading to higher fuel efficiency and lower emissions [2]. Due to their excellent castability, complex magnesium die castings can consolidate several components and eliminate assembly steps [2].

Welding is an important step in the fabrication of modular lightweight structures consisting of magnesium alloy die castings, extrusion profiles and wrought products. Procedures have been developed about forty years ago for the arc welding of magnesium alloys and more recently, for the laser welding of these materials [3-5]. However, information on the science and technology of magnesium welding is limited when compared with the more widely used materials such as steels and aluminium alloys. In order to facilitate the fabrication of components with minimal rejection rates, it is essential that satisfactory weldability be established with the particular combination of materials and welding procedures. Due consideration needs to be given to the unique properties of magnesium alloys when developing procedures for welding them. Under ambient conditions, magnesium melts at 649°C and boils at 1090°C and is easily oxidised. As in the case of aluminium, the solubility of hydrogen in magnesium decreases significantly when it solidifies. The rejection of the dissolved gas during cooling can lead to porosity formation in the weld [3]. On the other hand, the combination of low heat capacity and low heat of fusion of magnesium means that low energy inputs are required for welding and high welding speeds are possible [3].

The primary objective of this investigation was to study and quantify the weldability of the most widely used wrought magnesium alloy, AZ31B, by the gas tungsten arc (GTA) and the laser beam (LB) welding processes with particular attention to its hot cracking behaviour. The operating window of parameters for welding using each process resulting in crack-free and porosity free welds were identified. Procedures for GTA and LB welding of AZ31B sheet and plate materials were developed using these parameters and the quality of test plates welded using these procedures were assessed. This led to a better understanding of the microstructural development and structure-property relationships in GTA and LB weldments in AZ31B.

Experimental Program

Materials

The wrought alloy AZ31B was procured in the form of a sheet, 3.2 mm thick, and plate 12.7 mm thick. The sheet conformed to ASTM AZ31B-H24. The plate was a tooling plate material, an alloy conforming to AZ31B specification, which is produced to very close flatness tolerances to make it suitable for the construction of tooling and jigs [2]. The chemical compositions of the two materials are presented Table I. It is seen that whereas the sheet material contains 0.41 wt.% Mn, usually added for improved corrosion resistance, the plate materials does not contain any Mn.

The AZ31B-H24 sheet had a partially recrystallised microstructure and fine grain size of about 8 μm. Two distinctly different second phase particles were observed in this material: fine, dark etching particles, aligned in the rolling direction and larger cuboid particles. Energy dispersive spectroscopic (EDS) analysis in the scanning electron microscope (SEM) indicated that whereas the fine particles contained mainly Mg and Al with some Zn, the coarse, angular particles contained mainly Mn, Al and Mg. The fine particles are fragmented $Mg_{17}Al_{12}$ (γ phase); the coarse particles, Mn-Al intermetallic compounds. The 12.7 mm thick plate had a much coarser, fully recrystallised, equiaxed grain structure with an average grains size of about 90 μm and displayed only the fine $Mg_{17}Al_{12}$ particles.

The mechanical properties of the two materials in the longitudinal (rolling) direction are presented in Table II. The higher strength of the sheet material is related to its significantly finer grain size and partially recrystallised microstructure.

Table I. Chemical composition (in wt.%)

Material	Mg	Al	Zn	Mn	Fe	Cu	Ni
AZ31B-H24 (3.2 mm)	Bal.	3.01	0.91	0.41	<0.01	0.004	<0.002
AZ31B (12.7 mm)	Bal.	3.19	1.07	<0.005	<0.01	0.008	<0.002

Table II. Mechanical Properties

Material	0.2% PS (MPa)	UTS (MPa)	Elongation (%)	Reduction in area (%)	Microhardness ($HV_{0.05}$)
AZ31B-H24 (3.2mm)	223	286	16.0	18.7	58-70
AZ31B (12.7 mm)	135	233	20.8	21.7	50-55

Welding Equipment and Consumables

GTA welding was carried out with an inverter power source (Thermalarc GTSW 250) using alternating current (AC), which provides good arc cleaning of the surface oxides [3]. The same power source was used for welding in the direct current, electrode negative (DCEN) mode. AWS ER AZ61A filler wire, 1.6 mm in diameter, available in spool form, was used for carrying out automatic GTA welding.

LB welding was carried out using a Lumonics 3.5 kW continuous wave Nd:YAG laser. The laser beam was delivered to the work piece via a 600 µm diameter single core optical fibre and manipulated by robotic control. The robot used was a six-axis Motoman SK16X capable of carrying a payload of 16 kg. The welding head chosen contained a 200 mm focal length lens which was focussed onto the material surface to provide a 600 µm spot size.

Weldability Testing

One of the factors affecting the weldability of Mg alloys containing Al and Zn is the susceptibility to hot cracking [3]. Hot cracking can be in the form of either fusion zone solidification cracking or liquation cracking in the weld heat affected zone (HAZ) when low melting second phase constituents present at the grain boundaries undergo localized liquation. In this research program, the Circular Patch (CP) test was used to study and quantify the hot cracking behaviour of AZ31B.

The CP test is a self restraint test in which a simple circular weld bead is deposited on the test material which is restrained in a fixture. The unique design of the CP test subjects both the weld fusion zone and the HAZ to high stress, the magnitude of which increases with decreasing bead diameter. The details of the CP test sample, fixture design and the testing procedure can be found in Ref. [7]. The CP test was carried out exclusively on the AZ31B-H24 sheet material.

The CP fixture consisted of a solid copper chill, 200x200x20 mm^3, fastened to a steel plate, 200x200x25 mm^3, which provided support to the fixture. Specimens with dimensions 100x100 mm^2 were cut from the 3.2 mm thick sheet and holes were drilled at the four corners and at the centre.

After through cleaning with acetone, the specimen was clamped between the copper chill and a top restraining plate via five restraining bolts, one at the centre and four at the corners. The restraining plate was also made of solid copper, 150x150x19 mm^3 in dimension and had a 75 mm diameter hole to allow access for the welding torch. The five bolts were torqued to 50 Nm to ensure consistent restraint condition and to maintain contact between the copper chill and sample during testing.

A typical testing procedure consisted of producing a full penetration autogenous circular weld, continuous over 360° [6]. As the welding progressed around the diameter of the circular track, at some point around the circumference of the weld, sufficient strain was accumulated to induce cracking in the sample. Once initiated, this centreline crack usually propagated to the end of the weld. The cracked portion of the weld length, Θ_c, obtained by measuring the angle between the point where centreline cracking began and the weld crater, provided an index of the cracking susceptibility of the material.

Tests were carried out using both GTA and LB welding. In the former case, the fixture was securely attached to a turntable positioned under the welding torch. In the latter case, the laser beam was manipulated by the robot, which can be programmed to weld along a circular path. A limited number of tests carried out using GTAW-AC showed no cracking. Hence, the remaining tests were all carried out in the DCEN mode. The main variables were the welding current and travel speed. In the case of LB welding, autogenous weld beads were deposited using a spot size of 600 µm and surface focus. Here, the main variables were the laser beam power and the travel speed.

Quantitative crack data was obtained by taking digital photographs of the cracked specimens with a low power stereomicroscope and using an image analysis program to measure the length (in degrees) of the uncracked portion of the weld. The results were evaluated as a function of the welding variables. Polished cross-sections of cracked specimens were examined by SEM. X-ray elemental mapping using both an energy dispersive and a wavlength dispersive spectrometer (EDS and WDS) provided further insight into the hot cracking mechanism in this Mg alloy.

Welding Procedure Development

The weldability test results provided guidelines for selection of welding parameters for GTA and LB welding which should result in crack-free welded joints. Based on these, procedures were developed for welding test plates using the two materials. In each case, two plates, each 200x200 mm^2 in dimension, were welded together to form a test plate. The welding direction was parallel to the rolling direction. Prior to welding, the joint area was thoroughly cleaned by brushing with stainless steel wool to remove the surface oxide, followed by degreasing with acetone.

The welded test plates were subjected to radiography, microstructural characterisation, mechanical testing and chemical analysis. The microstructures of transverse sections of the weld test plates were examined using light optical microscopy and SEM. Microchemical information was obtained by generating X-ray elemental maps of selected regions of the fusion zone and the HAZ. Transverse tensile tests were carried out on all the test plates. All weld tensile testing could be carried out only in the case of the 12.7 mm thick material. Samples for chemical analysis were carefully extracted from the weld metal in each test plate, ensuring that no parent material was included in the sample.

Results and Discussion

Weldability Testing

GTA welding CP test results. A limited number of tests carried out using GTA welding with AC showed no cracking. The remaining tests were all carried out in the DCEN mode. The welding parameters used and the crack length are shown in Table III. The main variables were the welding current and travel speed. In one series of tests, the travel speed was held constant at 200 mm/min and the arc voltage at ~12.2 V while the welding current was varied from 115 A to 155 A. In another series of tests, different combinations of welding current and travel speed corresponding to a constant weld heat input of 439 kJ/m were used to deposit the test welds.

The cracking data in Table III shows that in DCEN tests carried out at a constant travel speed of 0.2 m/min, as the welding current is increased the crack length increases, indicating a higher susceptibility to cracking at higher currents. It is also evident that relatively minor changes in the welding current result in large increases in cracking susceptibility. The tests carried out at a constant heat input while varying the welding current and travel speed appropriately, also show that when welding with GTA-DCEN, small changes in welding current profoundly affect the hot cracking susceptibility. Nelson and co-workers have proposed that the welding current could affect cracking susceptibility through its influence on the weld bead geometry [7]. These investigators found that at a constant travel speed, as the welding current increased, the bead cross-sectional area increased, and this correlated with increased cracking in the test.

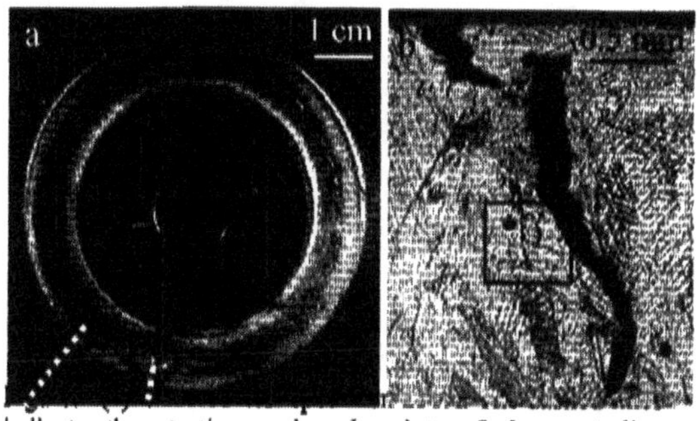

indicate the starting and end points of the centreline solidification crack (b) Cross-sectional view of the crack.

On the other hand, no fusion zone cracking was detected in tests carried out with GTA-AC welding. Clearly, this is again related to the influence of the welding parameters on the bead geometry. It is well known that for the same welding current, the DCEN mode results in higher weld bead penetration and than AC welding [3]. This suggests that besides good arc cleaning action of surface oxides [3], AC welding could provide some resistance to weld metal solidification cracking.

Figure 1(a) is a macrograph of the CP test specimen 250A welding using GTA-DCEN. The fusion zone centreline cracking is clearly visible. Figure 1(b) shows a cross-sectional view of this crack, which is seen to extend to a considerable depth. The characteristic microstructure resulting from cellular-dendritic solidification is clearly evident. Some fine, crack-like features were present within the fusion zone, highlighted by the frame in Figure 1(b). Examination at higher magnification showed that these fine cracks were backfilled with a grey constituent, similar in appearance to the fine, second phase particles present at the cell boundaries.

Table III. Circular Patch Test: GTA Welding

Electrode material	: Tungsten-ceria, 3.2 mm diameter
Electrode geometry	: Conical tip, 90° included angle
Electrode-work distance	: 1.6 mm
Shielding gas	:75%Ar-25%He
Gas flow rate	: 10 L/min

Specime	Current (A)	Voltage (V)	Speed (m/min)	Heat Input (kJ/m)	Θ_c (°)
AC Tests					
AC250A	145	12.5	0.250	435	0.0
AC300A	175	12.2	0.300	512	0.0
AC300B	175	12.2	0.300	512	0.0
DCEN, Constant travel speed tests					
200D	115	11.6	0.200	400	9.4
200E	120	12.2	0.200	439	0.0
200A	125	12.2	0.200	458	182.6
200B	125	12.2	0.200	458	221.7
200C	125	12.2	0.200	458	186.0
200F	135	12.2	0.200	494	162.9
200G	135	12.2	0.200	494	183.9
200H	135	12.2	0.200	494	198.1
200I	145	12.2	0.200	531	223.9
200J	145	12.2	0.200	531	166.9
200K	155	12.2	0.200	572	171.4
DCEN, Constant heat input tests					
200E	120	12.2	0.200	439	0.0
225A	135	12.2	0.225	439	87.1
225B	135	12.2	0.225	439	166.1
225C	135	12.2	0.225	439	42.8
250A	145	12.5	0.250	435	319.0
250B	145	12.5	0.250	435	225.3
250C	145	12.5	0.250	435	189.4

Figure 2 presents higher magnification Mg, Al, Zn and Mn X-ray maps of this region. The X-ray maps show that the fusion zone microstructure of this autogenous weld in AZ31B-H24 consists of primary dendrites, rich in Mg, with a fine cellular structure in the interdendritic regions. The second phase particles at the cell boundaries contain Al, Zn and Mg; a few particles also contain Mn, as seen from the Mn map. The two fine cracks appear to be filled with a substance containing Al, Zn and Mg. Preliminary semi-quantitative analysis indicates this material is similar in composition to the Al-Zn-Mg particles at the cell boundaries. Reference to the Mg-Al phase diagram provides some information about the phases present in fusion zone, although AZ31B is complex alloy with many alloying elements, [7].

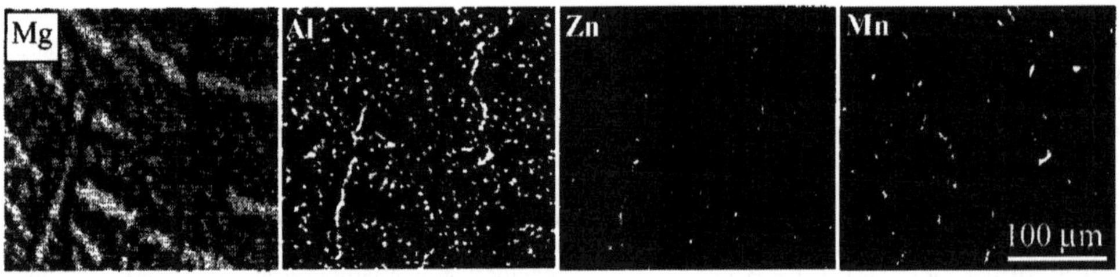

Figure 2. Higher magnification Mg, Al, Zn and Mn X-ray maps of the area outlined in Figure 1(b), showing details of the weld metal microstructure and the material within the backfilled cracks.

The Mg-Al phase diagram shows that the major phase present in Mg-rich alloys is α-Mg, a solid solution containing Al [7]. Under equilibrium conditions a eutectic reaction occurs at 437°C at which liquid decomposes into a two phase structure consisting of α-Mg and an intermetallic compound, γ, $Mg_{17}Al_{12}$. In systems containing eutectic reactions, because of the fast, non-equilibrium cooling rates in welds and low solid state diffusivites, solidifiying weld metal increases continually in solute content with respect to the material previously solidified, resulting in microsegregation [8]. The consequence of this microsegregation is that the last liquid to solidify has a eutectic composition and solidifies as two phases. The particles at the cellular boundaries are $Mg_{17}Al_{12}$ containing some Zn, present as a divorced eutectic. The fact that material of a similar composition is present in the cracks indicates that the solut-rich, low melting liquid backfilled the cracks, resulting in healing. Such healing of cracks by backfilling with a low melting eutectic liquid has been observed in association with solidification cracking in other systems [9].

LB Welding CP Test Results. Table IV presents the LB welding parameters used in the CP tests and the results obtained. The evaluation of these tests was complicated by the fact that it was difficult to ascertain if cracking had indeed occurred in any of the

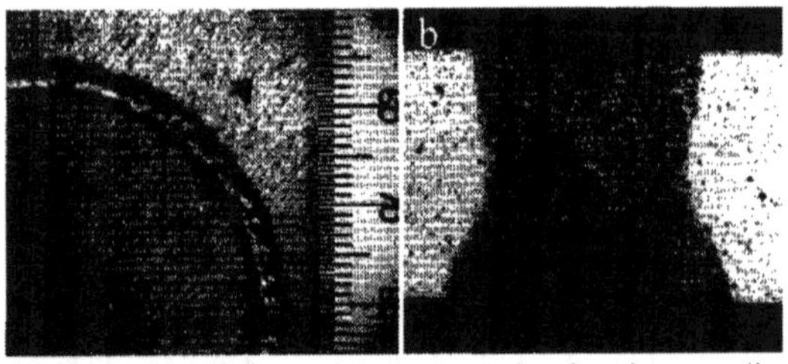

Figure 3. (a) CP test specimen no. 4 showing the centreline crack-like feature. Cross-sectional view of specimen 4.

specimens. Visual examination of the weld beads, particularly those deposited at the slower travel speeds, revealed that a crack-like feature was present along the weld centreline. Examination of these specimens under a low power stereomicroscope indicated that these features were blunter than cracks and had the appearance of minute canyons, rather than cracks. Figure 3(a) shows a CP test specimen with a centreline feature of the type just described. The cross-sectional view of the weld in Figure 3(b) shows that in fact no cracking occurred and that the sagging of the surface of the full penetration weld towards the weld centreline resulted in the canyon-like feature observed.

Careful examination of the cross-sections of all the CP test welds listed in Table IV confirmed that no solidification cracking occurred for any of the welding parameters tested. The combination of low energy input, the associated fast cooling rates and the resulting small weld puddle contributed to prevent solidification cracking in the LB welds.

Table IV Circular Patch Test: Laser Beam Welding

Laser	: 3.5 kW continuous wave Nd-YAG
Spot size	: 600 μm
Focus	: Surface focus
Backing gas	: Ar
Gas flow rate	: 1 L/min

Sample No	Beam Power* (kW)	Travel Speed (m/min)	Heat input (kJ/m)	Total "Crack Length"# (Θ_c)
1	1.8	2.25	48	0.0
2	1.8	2.25	48	0.0
3	1.8	1.75	62	0.0
4	1.8	1.25	86	319.6
5	1.8	1.25	86	258.7
6	1.8	2.25	48	163.4
7	1.8	2.25	48	34.3
8	1.8	1.75	62	281.4
9	1.8	2.75	39	0.0
10	1.8	2.75	39	0.0
11	2.2	2.75	48	44.7
12	2.2	2.75	48	70.9
13	2.4	3.00	48	0.0
14	2.4	3.00	48	0.0
15	2.4	2.25	64	211.7
16	2.4	2.25	64	302.9
17	2.4	1.75	82	241.4
18	2.4	1.75	82	312.2
19	2.4	1.25	115	204.6
20	2.4	1.25	115	309.5
21	1.8	1.25	86	342.2
22	2.4	1.25	115	288.5
23	2.4	1.75	82	279.3

*Power measured at work piece # canyon-like features

A theoretical model by Jarvis of the failure of keyhole welds in thin plates by lengthwise splitting of the weld puddle indicates that failure is dependent on the ratio of the width of the weld bead to its thickness; if this ratio exceeds a certain value, the weld will rupture[10]. For successful keyhole welding of thin plates, the root bead should be narrower than the surface bead [10]. The formation of centreline canyon canyon-like feature in slow travel speed LB test welds that are significantly wider at the root than at the top is probably a precursor to the failure mechanism proposed in the model. The results of the CP test thus served to identify parameters that would cause no centreline sagging in LB welds.

These parameters were used for LB welding procedure development.

Welding Procedure Development

We have seen that while GTA welding in the DCEN mode had a narrow window of welding parameters that resulted in welds free from solidification cracking, welding in the AC mode resulted in crack-free welds over a wide range of welding parameters. In the case of LB welding, relatively slow welding speeds resulted in an undesirable weld bead geometry.

Based on these observations, welding procedures were developed to produce butt joint test plates using semiautomatic multipass GTA-AC welding of the 12.7 mm thick plate material and LB welding of the 3.2 mm thick sheet material. In the former case, a double V edge preparation, ER AZ 61 filler wire and 75%Ar-25%He shielding gas were used. The welding sequence involved three passes on one side, back gouging and five passes on the other side. The welding current was 145 A, the voltage in the range 12.5-16 V, the travel speed 0.18 m/min. The wire feed rate was in the range 0.13-0.615 m/min. Radiography confirmed that the joint was free from porosity or other gross defects.

LB welding was carried out with a machined square edge preparation and no filler metal addition. The plates to be joined were clamped to a welding fixture with a backing groove through which argon was passed at 7 L/min to shield the weld root. In addition, an Ar trailing shield with a flow rate of 40 L/min provided protection of the solidifying weld metal. The butt joint was produced with a beam power of 2.4 kW (measured at the work piece) and a travel speed of 3 m/min. Radiography revealed that there were a few distributed pores in the fusion zone, which were within acceptable limits.

Chemical Analysis and Mechanical Properties. The results of the chemical analysis of the weld metals of the GTA and LB welded test plates are presented in Table V. Due care was taken to ensure that the sample for chemical analysis came entirely from the fusion zone, and that no base material was included. In both cases, the Fe, Cu and Ni contents were less than 0.005 wt.%.

Table V. Weld Metal Chemical Composition (wt.%)

Test Plate	Mg	Al	Zn	Mn
GTA	Bal.	5.61	0.79	0.24
LB	Bal.	2.95	0.87	0.45

It is evident from Table V that the addition of the ER AZ61A filler metal with a nominal Al content of 6 wt.% has resulted in a higher Al content in the GTA weld metal and also contributed some Mn, which was not present in the 12.7 mm thick base material. Although the Zn level is within the specification for AZ31B (0.7-1 wt.%), clearly some loss of this element with a low boiling point (907°C) and a high vapour pressure has occurred. On the other hand, in the case of the LB weld metal, the much lower energy input resulting in a small fusion zone and the inert gas trailing and back shielding provided during welding means that the losses of Zn and Mg were negligible, even without the addition of filler metal.

Table IV presents the results of the tensile tests on the two welded test plates. First let us consider the results of the tensile tests carried out on the GTA welded test plate. The all weld tensile test reveals a high tensile strength for the weld metal, which is consistent with its higher Al content relative to the base material. The transverse tensile test in which the specimen is a composite of the fusion zone, HAZ and base material, resulted in failure at the fusion line or the HAZ, indicating that this is the weakest zone. Nevertheless, the joint efficiency of the test plate, defined as the ratio of the transverse tensile strength to the UTS of the base material (Table II) is about 96%. In the case of the LB welded test plate, the joint efficiency is slightly higher, about 98%.

Table VI. Welded Test Plate Tensile Test Results

Test Plate	Type of Test	UTS (MPa)	0.2% PS (MPa)	Elong-ation (%)	Failure Location
GTA	Trans.	225	---	---	weld/HAZ
	All weld	269	127	20	weld
LB	Trans.	281	---	---	weld
	Trans	279	---	---	weld

Microstructual Characterisation. Figure 4(a) presents a cross-sectional view of the GTA welded test plate. The acetic-picral etchant clearly reveals the various weld passes. Figures 4(b) and 4(c) are higher magnification views of the areas outlined by frames in the final weld pass in Figure 4(a), showing as-deposited weld metal and the fusion line. The fine, cellular, two phase structure of the fusion zone is clearly evident. We have noted above (Figure 2) that the two phases are α-Mg and $Mg_{17}Al_{12}$ containing some Zn.

Figure 4(c) shows the microstructure of the HAZ. Over a distance of ~ 300 μm from the fusion line, the original grain structure present in the base material is still discernible, but within the grains, dark-etching particles, similar in appearance to the second phase particles in the fusion zone, but much finer, are apparent.

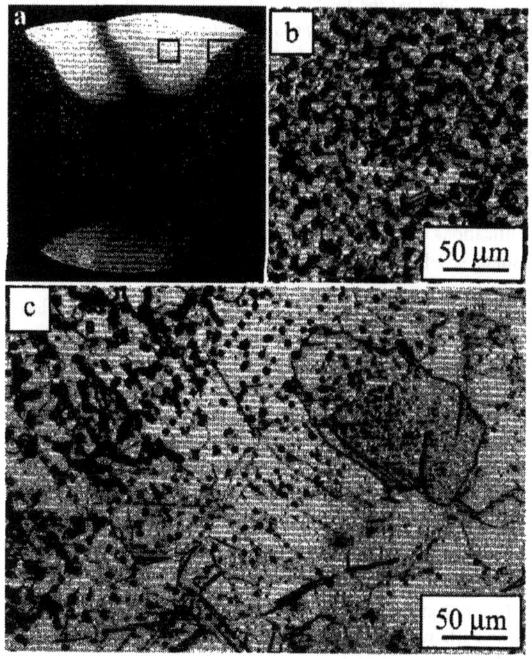

Figure 4. (a) Macrograph of the GTA weld joint. (b) and (c) Optical micrographs of the fusion zone and HAZ. The dashed line in (c) indicates the fusion line. Etchant: acetic-picral

Some of these particles appear to have formed on twin boundaries. Microchemical analysis in the SEM confirmed that these particles contain Al, Zn and Mg, as do the particles in the fusion zone. It is concluded that these particles formed upon solidification of locally melted material adjacent to the fusion line. Localized variation in solute content caused by segregation invariably present in a commercial alloy causes a point to point variation in effective melting point [8]. When the instantaneous temperature adjacent to the fusion line exceeds the effective melting point, localized melting occurs [8].

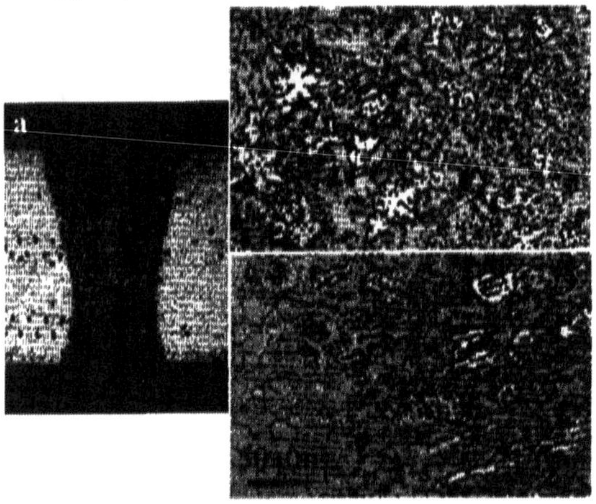

Figure 5. (a) Macrograph of the LB welded test plate. Optical micrographs of (a) the centre of the fusion zone and (b) near the fusion boundary. Etchant: Acetic-picral.

Figure 5(a) depicts the cross-sectional view of the laser welded butt joint. Note that the four rows of evenly spaced dark spots in Figure 5(a) are microhardness traverses. The fusion zone microstructure is significantly finer than that observed in the GTA weld, clearly due to the more rapid solidification in this case. Microchemical analysis confirmed that the two phases present are α-Mg and $Mg_{17}Al_{12}$ containing some Zn. As seen in Figures 5(b) and (c), at the centre of the fusion zone, fine, randomly distributed equiaxed dendrites are present; close to the fusion line, the solidification structure appears cellular. Examination at high magnifications confirmed the presence of a partially melted zone extending ~50-100 μm from the fusion line.

Similar observations have been made in full penetration laser welds in AZ91B [5]. The influence of growth rate, thermal gradient and the solute content on the extent of constitutional supercooling and on the weld metal solidification mode is well recognized [8]. The two dimensional heat flow associated with the keyhole weld resulting in a smaller thermal gradient at the weld centre compared to the fusion boundary, together with relatively higher growth rate and strong convection, caused the formation of the observed equiaxed dendritic structure.

Conclusions

Based on this study of the weldability and welding of AZ31B by GTA and LB welding processes, the following conclusions may be made:

- AZ31B is weldable by GTA welding in both the DCEN and AC modes. In DCEN mode, welding current and travel speed strongly influence solidification cracking behaviour. GTA-AC welding is more tolerant with regard to hot cracking and confers the added benefit of arc cleaning action.
- Hot cracking does not appear to be a problem during LB welding over a range of welding parameters. Careful control of beam power and travel speed is required to avoid surface sagging at the weld centre or even rupture of the weld puddle.
- A multipass weld joint with 96% joint efficiency has been fabricated using 12.7 mm thick AZ31B plate, GTA-AC welding and ER AZ61A filler metal.
- A 3.2 mm thick AZ31B sheet has been LB welded in one pass, at relatively higher welding speeds with no filler metal addition to produce a joint with a joint efficiency of 98%.
- Although some losses of Mg and Zn may occur from weld metals deposited by GTA welding, by appropriate selection of an overmatching filler metal, it is possible to compensate for any loss of weld strength. The losses from LB welding are negligible, provided adequate trailing and back shielding by Ar is available.
- Fusion zone microstructures of both GTA and LB welds consist of α-Mg and $Mg_{17}Al_{12}$ containing some Zn. The more rapid solidification of the LB weld resulted in a finer fusion zone structure with equiaxed dendrites forming at the weld centre.

References

1. M.M.Avedesian and H. Baker, ed., Magnesium and Magnesium Alloys-ASM Specialty Handbook (Materials Park, OH: ASM International, 1999).
2. R. VanFleteren, "Magnesium for Automotive Applications," Advanced Materials and Processes, 5 (1996) 33-36.
3. A.T. D'Annessa et al., "Magnesium and Magnesium Alloys," AWS Welding Handbook, Vol.3 Edition 8, ed. W.R. Oates., (Miami, FL: American Welding Society, 1998).
4. P.G. Sanders et al., "High Power Nd:YAG and CO_2 Laser Welding of Magnesium," Journal of Laser Applications, 11(2), (1999), 96-103.
5. M. Marya and G.R. Edwards, "The Laser Welding of Magnesium Alloy AZ91", IIS/IIW -1470-99 in Welding in the World 44(2), (2000).
6. T.W. Nelson, et al., "Evaluation of the Circular Patch Test for Assessing Weld Metal Solidification Cracking, Part I - Development of a Test Method," Welding Journal, 76(3) (1997), 110s-119s.
7. J.L. Murray, "The Al-Mg (Aluminium-Magnesium) System, Bulletin of Alloy Phase Diagrams, 3(1), (1982), 60-74.
8. W.F. Savage, 1980 Houdremont Lecture, "Solidification, Segregation and Weld Imperfections, Welding in the World, 18 (5/6), (1980), 89-114.
9. J.C. Lippold, et al., "An Investigation of Hot Cracking in 5083-O Aluminium Alloy Weldments," Welding Journal, 56(6) (1977), 171s-178s.
10. B.L. Jarvis, "Keyhole Gas Tungsten Arc Welding: A New Process Variant," (PhD dissertation, University of Wollongog, NSW, Australia, 2001).

From *Magnesium Technology 2009*, E.A. Nyberg, S.R. Agnew, N.R. Neelameggham, and M.O. Pekguleryuz, Editors

MICROSTRUCTURAL AND MECHANICAL ASPECTS OF REINFORCEMENT WELDS FOR LIGHTWEIGHT COMPONENTS PRODUCED BY FRICTION HYDRO PILLAR PROCESSING

Gustavo Pinheiro[1], Jorge dos Santos[1], Norbert Hort[2] and Karl Ulrich Kainer[2]

[1] GKSS-Forschungszentrum; Max-Planck Str. 1; Geesthacht, 21502, Germany
[2] MagIC-Magnesium Innovation Center; Max-Planck Str. 1; Geesthacht, 21502, Germany

Keywords: Magnesium, Friction Welding, High temperature, Microstructure

Abstract

The development of new creep resistant and cost effective die casting magnesium alloys such as AE, MRI, MEZ, ACM, AXJ, AJ, WE have emerged as an alternative to fulfil the actual demands in structural relevant applications as engines blocks, gear and converter boxes. However, magnesium components are in most of the cases screwed with aluminium and steel bolts, which lead the screwed joint to lose the preload force due to relaxation. This barrier limits thus the broad use of magnesium within this segment and should somehow find an adequate solution to be implemented and to help overcoming this limitation. In this context Friction Welding (FW) and particularly Friction Hydro Pillar Processing (FHPP), which can be described as a drill and fill process, appears as an alternative to widespread the use of magnesium. In this context, FHPP is intended to be used to locally reinforce mechanical fastened magnesium components.

In the present work a preliminary experimental matrix was defined and used to determine optimal welding conditions. Furthermore elaborate experimental techniques have been used to describe the process parameters-microstructure-properties relationships and the consequent mechanisms leading to bonding in FHPP welds in dissimilar configurations. The welds have been performed using a hydraulic powered friction welding machine, originally designed and built as a portable stud welding unit, delivering up to 40 kN welding force and 8000 rpm. All welds were monitored, analysed and evaluated using a purpose built data recording system. AZ91 and AXJ magnesium cast ingots have been used in the experimental programme. The results obtained in the course of this study have shown the feasibility of FHPP to produce high strength welds with mechanical properties comparable to those from base material. Defects, like porosity or lack of bonding, were not observed. It could be demonstrated that for dissimilar AXJ to AZ91D welds the consumable member is fully plasticized across the bore of the hole and through the thickness of the workpiece. Mechanical properties of the welded joints have shown values similar to those from AZ91 base material. An increased upsetting indicates no clear variation of tensile strength, with values, in both cases, significantly superior to those from AXJ base material due to the formation of a completely different microstructure in the extruded zone after welding. Hardness values achieve in some points values up to 80HV, which means that in the extruded AXJ material an overmatching condition was created.

Introduction

After the Second World War, when its consumption slumped from 228000 to 10000t per annum, magnesium appears to be again strongly placed among new innovative materials [1]. In the last years magnesium alloys have gained increasing interest from different industry branches. This interest can be attributed to the fact that the material has been overcoming some old barriers as the fairly high base material prices somehow limited to a few technical alloys associated with some costly recycling possibilities [2,3]. Over recent years the industrial output of magnesium alloys has been rising almost 20% per year [4]. This increasing application of lightweight alloys is matching particularly the interests of the automotive industry, with the goal of reducing weight of road vehicles to make them more fuel efficient [5]. In this context several new Mg alloys for different applications were recently developed with the intention of obtaining an adequate combination of different properties. Evidently a combination of creep resistance, castability, mechanical properties, corrosion resistance and affordable cost is hardly achieved and most of the alloys can only address part of the required properties and performance.

Furthermore magnesium alloys need to be in most of the cases mechanical fastened not only in a similar but sometimes in a dissimilar configuration by means of different material screws. Considering the elevated operating temperature experienced by engine blocks or powertrain components, creep resistance of magnesium alloys remains a major issue for such applications. In the literature it has been reported that the use of aluminium and steel screws in magnesium components lead the screwed joint to lose the preload force due to relaxation [6,7]. This barrier limits thus the broad use of magnesium within this segment and should somehow find an adequate solution to be implemented and to help overcoming this limitation. On the other hand, together with alloy development and addition of reinforcement (MMCs), local material engineering processes have been also conceived and considered a method to improve properties and therefore the number of applications for magnesium alloys. In this context Friction Welding (FW) and particularly Friction Hydro Pillar Processing (FHPP), which can be described as a drill and fill process, appears as an alternative to widespread the use of magnesium. Although specialized literature involving FW on magnesium alloys has not expressively contributed to the development of this work and no literature on FHPP of lightweight alloys has been found, FHPP is intended to be used to locally reinforce mechanical fastened magnesium components. Within this approach, regions submitted to stresses imposed by the tightening forces can be compensated by the use of a material with superior properties. In one hand one does not need to fabricate the whole structure from an expensive material, saving costs and on the other the economic pressures of an increasingly competitive global market can be matched.

Based on the current problems involving Mg bolted joints there is an evident need of research on alternative solutions. Hence the global objective is to locally reinforce Mg-components through the welding of a high creep resistance insert (stud) to be used in automotive applications. Within this global objective, some specific points can be also defined as follows: a) To describe the bonding formation mechanisms in FHPP of Mg alloys, leading to the understanding of the microstructural changes during the welding; b) To determine the local mechanical properties in terms of microhardness and transversal tensile tests to provide an analyse of the joint performance based on the microstructural development and joint formation mechanism.

Experimental

AZ91D and an AXJ Mg alloy were selected for the present study. Both materials were delivered in cast ingots and, since it has superior high-temperature properties, AXJ alloy was used to produce the welding consumables. Cylindrical studs with a tapered geometry were chosen to produce the welds in this work. The rotating consumable stud was tightened to the weld head while the AZ91D base plate was held stationary, clamped to the table. A portable hydraulic-powered friction welding system, HSM 3000, was utilized to perform the welds using the continuous drive method. The upsetting, which is considered a process parameter within this study, was varied in 10 and 20 mm, keeping all the other parameter constant. In this case upsetting (also called friction upset distance) is defined as the shortening of the consumable between the beginning of the weld (the instant when both faying surfaces touch each other) and the moment the motor enters the braking phase.

Certain procedures were followed before welding, in order to ensure reproducibility of the welding program and to avoid external influences as much as possible. The studs were cleaned with acetone, to eliminate all possible forms of contamination that could influence the results. The temperature of the hydraulic oil was kept at 20±2°C prior to starting welding, in order to avoid significant changes in the power delivered by the hydraulic machine. The welded specimens were longitudinally cut (in the center of the studs) and etched with a picric acid-based etchant, to allow a complete examination of the joint. Mechanical testing in terms of hardness and transversal tensile tests were performed to evaluate mechanical properties of the joints. Joint performance was also investigated by determining the creep properties of the welded joints.

Results and Discussions

Metallurgical Characterisation: As demonstrated in Figure 1, defect-free welds were produced without visible discontinuities along the weld, independent of the set upsetting. The bonding line could be most of the times easily identified not only because of the different base material microstructures but also due to the fine, dynamically recrystallized microstructure in the AXJ consumable resultant from mechanical working imposed by the process. Figure 2(a) details the transition between the AZ91 base material, where a narrow region with grain growth can be observed, and the AXJ consumable with its above mentioned refined structure. Bottom macrographs, i.e. Figure 2(b), (c) and (d), show a severely deformed area in the vicinity of the joint on the AZ91 side of the weld, where dynamically recrystallized grains appear in an extension from 100 to 400µm from the weld line without however the presence of the eutectic structure. The absence of the eutectic structure in those regions can be explained as follows: The Al-Mg phase diagram shows that a heating temperature higher than 390°C causes the dissolution of the $Mg_{17}Al_{12}$ into the Mg-matrix. Sufficient frictional heating introduced during the process leads to microstructures not containing the eutectic structure in the transition region (RZ) being extended in upper joint regions to the HAZ, i.e. these regions are probably heated at temperatures higher than 390°C during the process. Figure 2(e) presents a micrograph taken over the upper joint parts, where a smooth transition can be observed, maybe due to the role played by the forging force. In those areas the exactly determination of the bonding line was also not possible. Furthermore, in regions close to the side walls (see details in Figure 2(d)), detailed metallographic/SEM investigations elucidate the presence of grouped particles located along the joint line. Such elongated particles are formed by the grouping of smaller second phase particles, apparently fragmented during the welding process, which could not be pushed out of the interface. Such Mg(Al,Ca) based particles have been reported to have a superior strength than that of the matrix being in an intermediate stage of the process anchored and promoting subsequently a kind of stacking process, which evolutes and forms clusters of second phase particles in particular regions along the joint line.

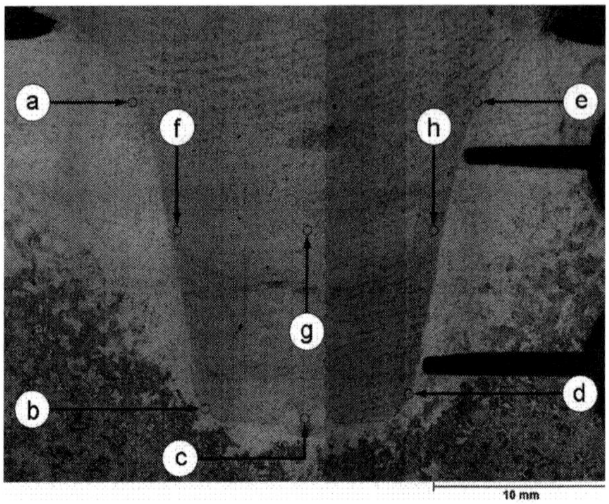

Figure 1: Transversal macrograph of the dissimilar AXJ/AZ welded joint. Details from (a) to (h) are further discussed along the metallurgical characterisation.

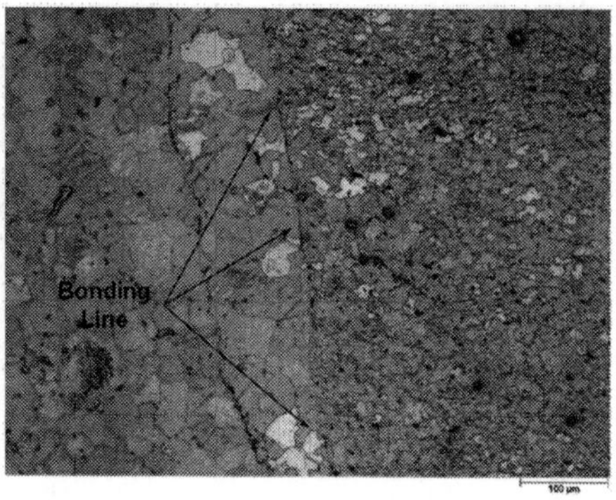

(a)

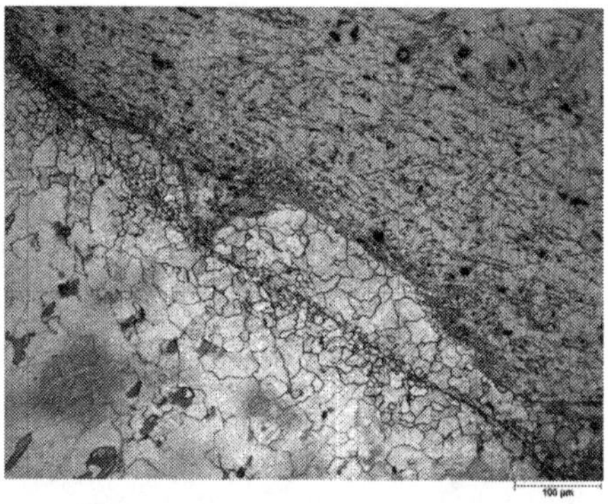

(b)

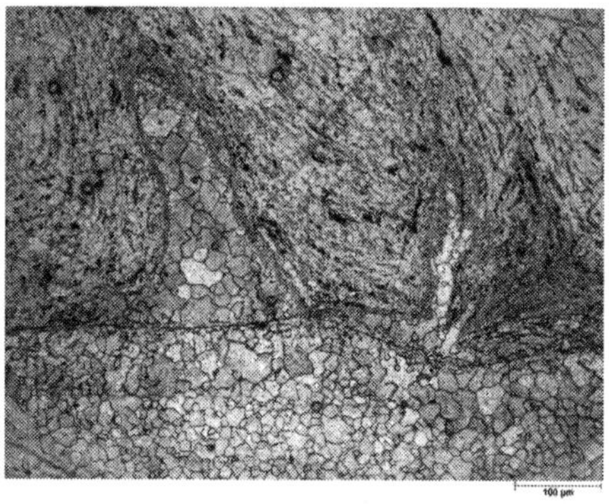

(c)

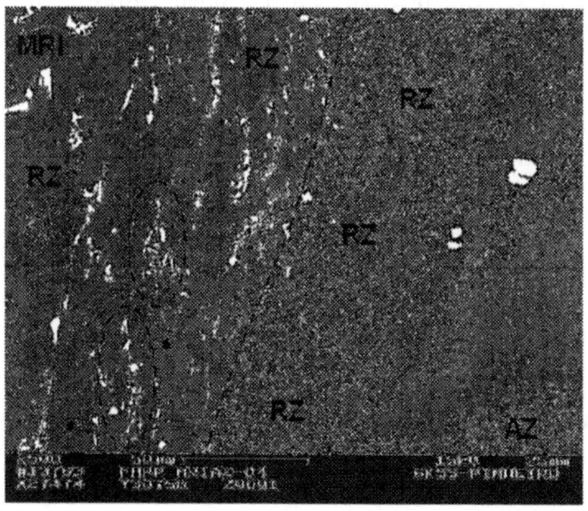

(d)

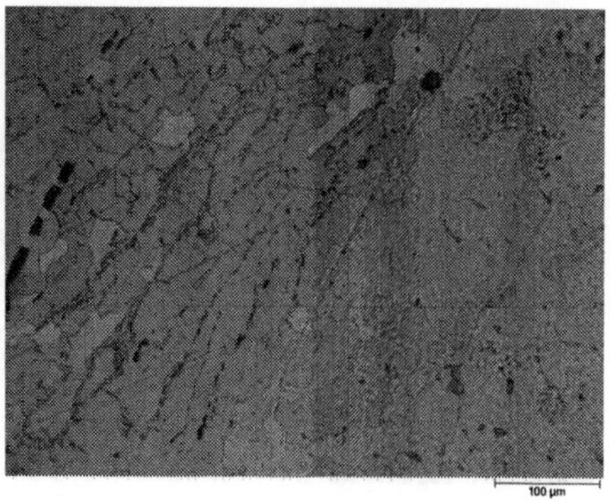

(e)

Figure 2: In (a), (b), (c) and (e), microstructural details along the bonding line of the dissimilar AXJ/AZ91D-T6 welds. In (d) a SEM micrograph showing clusters of particles in regions adjacent to the bonding line evidencing different weld zones.

The "as cast" AXJ ingot material is generally characterized by an α-Mg matrix, some needlelike and polygonal type precipitate, described in the literature as being Al_8Mn_5, Al_2Ca and Al_4Sr, dispersed within the matrix and by the Mg(Al,Ca) eutectic type composition located preferentially along the grain boundaries [8,9]. On the other hand, the microstructure on the AXJ side of the weld is dominated by equiaxed dynamically recrystallized grains formed by the pair mechanical working/temperature imposed during the process. In particular regions of both the vicinity of the joint and the middle of the consumable, a complete degradation of the typical base material second phase particles Mg(Al,Ca) was observed. Figure 3 highlights, under the same magnification, the common base material rounded-shape Mg(Al,Ca) precipitate morphology and the resultant AXJ microstructure after the welding. Fine Al, Ca, Sr and others round and polygonal precipitates dispersed into the α-matrix seems not to be into a large scale strongly influenced by the process, since they are present in the deposit approximately with the same form and size as commonly found in the matrix. Additionally it should be mentioned that the morphology, i.e. size and form of the final precipitates within the welded area, can be roughly controlled by the welding parameters.

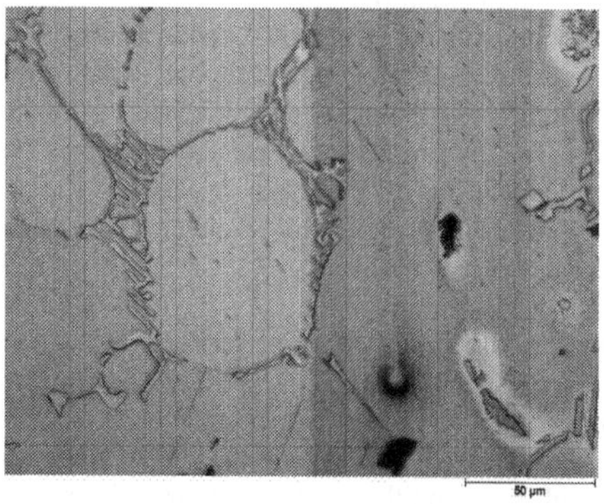

(a)

(b)

Figure 3: Microstructural changes in the consumable in dissimilar MRI230D/AZ91D-T6 welds. In (a) the typical Mg(Al,Ca) eutectics along the grain boundaries and in (b) the same material after welding at mid-thickness at the rotational centre (see position (g) in Figure 1).

As observed during the grain size assessments, the extruded region presents for 10 and 20 mm upsetting welds an inhomogeneous microstructure along the deposit height. In comparison with AXJ base material, where grain size analyses indicate values going from 80 to 120µm, in the deposit grain size varies generally from a finer structure at the bottom, with an average grain diameter ranging from 5 to 13µm, to a coarser structure at the top, where values varying from 18 to 22µm were identified. Generally welds performed with 20mm upsetting presented a more expressive grain growth at the upper regions of the extruded area probably as a consequence of the slightly higher temperatures and also of the longer welding times, which allows not only recrystallization but also grain growth processes subsequently to take place. Furthermore, Figure 4 evidences grain size differences in regions diametrically opposed (see Figure 1, indications "f" and "h"), where theoretically imposed welding conditions were similar and therefore similar final structures should be obtained. Facing with similar inhomogenities, NORTH *et al.*[10] has reported that transient local melting occurs sufficiently often during Al7075-T6 spot welding and directly affects the stir zone formation. Bearing that in mind, the high strain rates imposed by the process associated with local liquid phase formation provide in particular regions the conditions for grain boundary sliding, which influence the RZ formation by reducing or increasing the local heat generation. The above mentioned mechanism varies from region to region dependent of the base material homogeneity and maybe dictates and generates the resultant microstructure inside the extruded area.

According to the evidences presented in the metallographic analysis (see Figures 1 to 4), the joint formation mechanisms within this combination are similar to those for similar AZ91 FHPP welds [11]. Due to the matching correlation material/welding conditions, the plasticized layer is generated through the consumable diameter, arises along the hole leaving beneath a fine dynamically recrystallized deposited material.

(a)

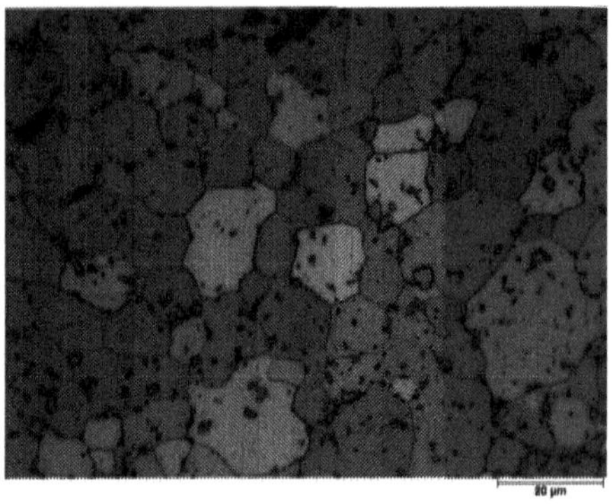

(b)

Figure 4: Grain size assessment within the extruded material. In (a) and (b), respectively, the microstructure in "f" (see Figure 1 - average diameter: 7.3µm) and in "h" (see Figure 1 - average diameter: 11.8µm).

Mechanical Characterization: Hardness surveys along cross sections of the joints indicate that material hardening, in comparison to base material, is easily observed from start to finish of the extruded zone, resulting from the weld thermal cycle. Hardness of AXJ base material was shown to range between 46 and 54 HV, whereas in the extruded area values varied generally between 60 and 70 HV in upper and 60 and 80 HV in lower weld areas (see Figure 5(a) and (b). Scans performed near the top surface are very smooth and present a significant lower scattering even in regions of the HAZ far away from the joint line on the AZ side of the weld. However, in lower stud areas, 20 mm upsetting weld presented slightly superior hardness values.

Similarly to the observed in rotational friction welded magnesium alloys [12], the welding process and the consequent production of a very refined microstructure seems to influence the hardness values in the recrystallized zone of the AXJAZ welds. According to Figure 5(a), scans performed near the top surface are very smooth and present a lower scattering probably due to the higher heat generated on those locations. Since the formation of these areas and the resultant microstructure in 10 and 20 mm-upsetting welds are to some extent similar, a hardening mechanism probably involving fine precipitate dissolution or re-precipitation during cooling took place under these conditions.

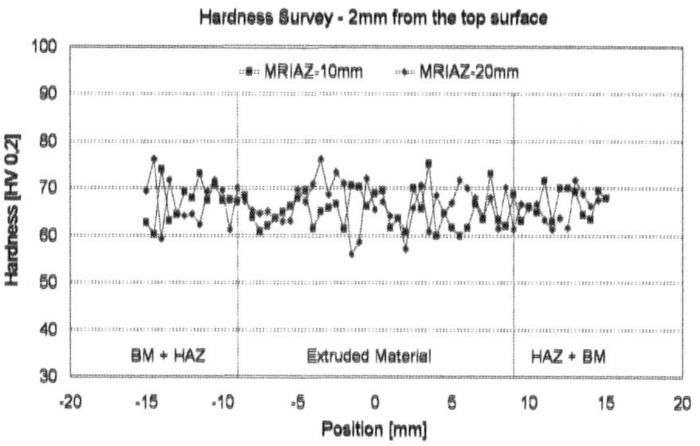

(a)

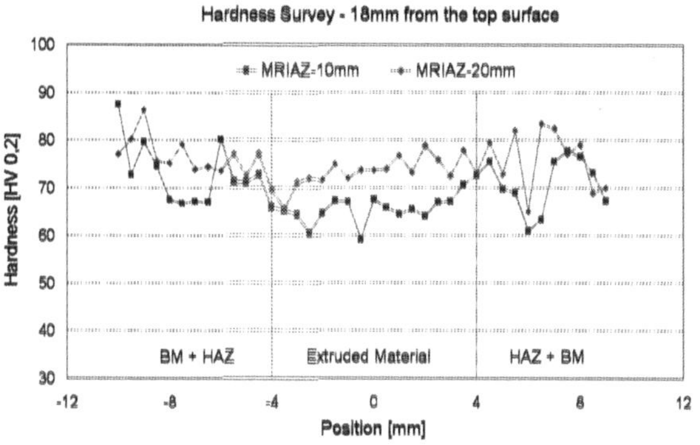

(b)

Figure 5: Horizontal hardness profiles along the joints at 2 and 18mm from the top surface.

The lower scattering of top surface scans can be attributed to the partial dissolution/fragmentation of the eutectic phases within the extruded material and to the dissolution of the lamellar β-phase in the HAZ on the AZ91 side of the weld. The high temperatures achieved on those locations have been considered the reason for the observed differences. Within the extruded area (recrystallized zone - RZ) the scatter is reduced obviously due to a much more homogenised microstructure generated by the hot working imposed by the welding process. Since such high hardness products are dissolved the scatter is reduced. On the other hand, in scans performed across the bottom weld area, the decrease in dispersion is easily observed as soon as the extruded area is reached. In fact, as observed in the metallurgical characterization (see overview in Figure 1), in bottom areas on the AZ91 side of the weld the BM is not in to a deep extension modified in regions very close to the BL. Since the work pieces are at the beginning of the process at room temperature the generated heat is almost instantaneously conducted out of the welding region. Hence the achieved temperatures as well as the dwell times in those locations are not high enough to partially homogenise the microstructure, causing the profiles to show a significantly increased scattering.

In terms of tensile properties, AXJAZ welds presented also satisfactory strengths. Two specimens were machined from each joint and tested in laboratory air at room temperature following the DIN EN 10002 standard [13]. Figure 6 presents schematically the position of "upper" and "lower" tensile samples in relation to the welded area.

Figure 6: Position of specimens relative to the joint (b).

As demonstrated in Table 1, average values in the upper and lower weld areas respectively of 184 and 160% in comparison to the AXJ cast base material were obtained for 10 mm upsetting welds, whereas slightly lower values of 170 and 156% were found for 20mm upsetting joints. Elongations were observed also to match base material values and varied between 1.0 and 1.6% and 0.8 and 2.0% for respectively 10 and 20 mm upsetting welds in lower and upper weld areas.

Table 1: MRIAZ - Tensile testing results.

Sample		Rm [%]	$Rp_{0.2}$ [%]	A [%]
MRIAZ 10mm	Upper	184 ± 8	137 ± 5	1.6 ± 0.1
	Lower	160 ± 2	132 ± 2	1.0 ± 0.1
MRIAZ 20mm	Upper	170 ± 4	110 ± 6	2.0 ± 0.3
	Lower	156 ± 9	130 ± 4	0.8 ± 0.1

Also in this analysis the refined resultant microstructure seems to collaborate to produce high performance welds in terms of tensile testing. In this case the forging force seems to play an important role since upper weld area specimens presented clearly higher strengths if compared with lower specimens. As discussed by Meyer [14], during the final stages of the FHPP welding process the microstructure at the bottom area is to some extent already consolidated and therefore suffers a marginal influence of the forging force. Furthermore, welds with 20 mm upsetting presented at upper weld regions a more expressive grain growth, which probably has contributed to generate in those locations the observed inferior tensile strengths. Failure was located in the 10mm upsetting welds mostly in the base material in a region outside the RZ. On the other hand failure location has moved towards the RZ for higher upsetting welds, which can be to some extent attributed to the inhomogeneities observed on the base material probably associated to the casting process.

Conclusions

Mechanisms of joint formation as well as metallurgical and mechanical properties of dissimilar AXJAZ welds produced by

Friction Hydro Pillar Processing (FHPP) were investigated, determined and assessed. The results and analyses obtained from these investigations lead to the following conclusions:

1- In spite of being a new welding technique, FHPP demonstrated its suitability for joining different magnesium (Mg) materials. The selected combination of base materials was successfully joined using two different upsettings. In both cases defect-free welds were obtained, which suggests a wide operating window for production of satisfactory joints;

2- Within this combination joint formation follows the same sequence presented in previous studies for similar AZ welds. Due to the matching correlation material/welding conditions, the plasticized layer is generated through the consumable diameter, arises along the hole leaving beneath a dynamically recrystallized deposited material. No bonding formation changes were observed with increasing upsetting. The microstructure on the AXJ side of the weld is dominated by equiaxed dynamically recrystallized grains. In particular regions of both the vicinity of the joint and the middle of the consumable, a complete degradation of the typical base material second phase particles Mg(Al,Ca) was observed. Additionally the welds have shown a severely deformed area in the vicinity of the joint on the AZ91 side of the weld, where dynamically recrystallized grains appeared in an extension from 100 to 400μm from the weld line without the presence of the eutectic structure. In comparison with BM, where grain size analyses indicate values going from 80 to 120μm, in the deposit grain size varies generally from a finer structure at the bottom, with an average grain diameter ranging from 5 to 13μm, to a coarser structure at the top, where values varying from 18 to 22μm were identified.

3- Mechanical properties of the welded joints have shown values similar to those from AZ91 base material. An increased upsetting indicates no clear variation of tensile strength, with values, in both cases, significantly superior to those from AXJ base material due to the formation of a completely different microstructure in the extruded zone after welding. Hardness values achieve in some points values up to 80HV, which means that in the extruded AXJ material an overmatching condition was created.

References

[1] MORDIKE B.L. and Ebert T., 2001, "Magnesium: Properties, Applications and Potential", Materials Science and Engineering A302, pp.: 37-45.

[2] KAINER K.U. and von Buch F., 2003, "The Current State of Technology and Potential for Further Development of Magnesium Applications", in Magnesium alloys and Technologies, WILEY-VCH Verlag GmbH & CO. KG aA, Weinheim, pp. 1-22.

[3] PARK S.H.C., Sato Y.S. and Kokawa H., 2003, "Effect of Microtexture on Fracture Location in Friction Stir Welding of Mg alloy AZ61 During Tensile Test", in Scripta Materialia 49, pp.:161-166.

[4] EDGAR R.L., 2006, "Global Overview on Demand and Applications for Magnesium Alloys", published in Proceedings of the 7th International Conference on Magnesium Alloys and Their Application, pp.: 3-8, 06-09 November, Dresden, Germany.

[5] JOHNSON R. and Threadgill P., 2003, "Friction Stir Welding of Magnesium Alloys", TMS, Magnesium Technology 2003 – Edited by Howard I. Kaplan.

[6] PFISTER U., 2005, „Schraubverbindungen und Schraubtechnologie in der Powertrain-Serienmontage", Bad Nauheim / Frankfurt, 7. bis 9. März.

[7] LEITNER H., Godor I., Eichlseder W. und Hinteregger, 2005, „Betriebsfeste Auslegung von Schraubverbindung im Leichtmetallguss", Bad Nauheim 17. 18. November.

[8] OKECHUKWU, A., 2006, "Investigation of bolt load retention behaviour of Mg-alloys", Masters Thesis - Technische Universität Hamburg Harburg (TUHH), Hamburg.

[9] http://www.lam.mw.tu-muenchen.de/Lehre/Download/WfMuA/5_Magnesium.pdf.

[10] NORTH, T.H., Bendzsak, G.J., Gerlich, A., Su, P. and Cingara, G., 2007, "Transient Local Melting in AL7075-T6 Friction Stir Spot Welds", in Materials Science Forum, Vols. 539-543, p. 3826-3831.

[11] PINHEIRO, G.A., 2007, „Ein neues Fügeverfahren zur Erhöhung der Gewindetragfähigkeit im modernen Automobilleichtbau", InnoCarBody, Bad Nauheim / Frankfurt - 2007.

[12] PINHEIRO, G.A., Pankiewicz, C.G., dos Santos, J.F. and Kainer K.U., 2006, "Rotational Friction Welding of a high Creep Resistance Magnesium Alloy", in Proceedings of the 7th International Conference on Magnesium Alloys and Their Applications, Dresden, November.

[13] DIN EN 10002, 1991, "Tensile testing of metallic materials", DIN - Deutsches Institut für Normung e.V., Berlin.

[14] MEYER A., "Friction Hydro Pillar Processing - Bonding Mechanism and Properties", Fakultät für Maschinenbau und Eletrotechnik der Technischen Universität Carolo-Wilhelmina zu Braunschweig, 2003.

SOLID STATE JOINING OF MAGNESIUM TO STEEL

Saumyadeep Jana[1], *Yuri Hovanski*[1], Siva P Pilli[1], David P Field[2], Hao Yu[2], Tsung-Yu Pan[3], M L Santella[3]
[1]Pacific Northwest National laboratory, Richland, WA, 99354, USA
[2]Washington State University, Pullman, WA, 99164-2920, USA
[3]Oak Ridge National Laboratory, Oak Ridge, TN 37381-6096, USA

Keywords: Magnesium, Steel, Friction stir welding, Ultrasonic welding

Abstract

Friction stir welding and ultrasonic welding techniques were applied to join automotive magnesium alloys to steel sheet. The effect of tooling and process parameters on the post-weld microstructure, texture and mechanical properties was investigated. Static and dynamic loading were utilized to investigate the joint strength of both cast and wrought magnesium alloys including their susceptibility and degradation under corrosive media. The conditions required to produce joint strengths in excess of 75% of the base metal strength were determined, and the effects of surface coatings, tooling and weld parameters on weld properties are presented.

Introduction

The U.S. Department of Energy's (DOE) Vehicle Technologies Program (VTP) is leading the development of high-performance, cost-effective materials-and the processes needed to manufacture them-to make advanced vehicles more fuel efficient and affordable. One key goal is to validate a cost-effective 50% weight reduction in passenger-vehicle body and chassis system by 2015 [1]. Recently, a joint project between VTP and U.S. automakers overcame several hurdles to the introduction of lightweight magnesium components, resulting in a magnesium engine cradle for the Chevy Corvette Z06 that is nearly 60% lighter than a steel cradle [1]. A key combination of low density, and high specific strength makes magnesium alloys an ideal choice for automotive applications because of the associated effect of weight reduction on better fuel economy. Though currently there are no adequate wrought magnesium alloys available to meet the requirements of automotive body applications, extensive research is taking place to enhance the properties of magnesium sheet products [2]. Steel, on the other hand, is currently the automaker's material of choice. This is because cold rolled mild steel offers excellent ductility, consistent properties and low overall component production costs. However, in a modern multi-material vehicle, lightweight materials such as aluminum and magnesium alloys can be a challenge to join and attach to the underlying sub-structure, usually composed of steel. Joining methodologies available in the cost environment relevant to automotive manufacturing include resistance spot welding, adhesives, linear fusion welding, hemming, clinching, bolting, and riveting [3].

However, joining Magnesium to steel poses a specific challenge. The maximum solid solubility of Fe in (Mg) is 0.00043 at%, and solid solubility of Mg in (Fe) is nil [3]. Melting points of Mg and Fe are 922 K (649 °C) and 1812 K (1539 °C), respectively. This huge difference in melting points makes it very difficult to melt both metals at the same time as might be required for fusion welding processes. Moreover, both metals are immiscible in a liquid state, and they do not react to form any congruent melting phase(s). Therefore, joining Mg alloys to steels through conventional fusion welding is problematic at best. Recently, laser-GTA hybrid welding technique was tried to join AZ31B alloy to 304 steel in lap configuration with AZ31B on top [4]. Results indicate poor joining strength and Mg-Fe interfacial fracture. The authors attributed such low mechanical strength to severe oxidation at Mg/Fe interface. Joining methods that require a large amount of plastic strain in the magnesium component suffer from magnesium's poor ductility at room temperature.

A recent DOE funded project looked at the feasibility of joining magnesium to steel through solid state route. Two specific joining technologies, e.g., friction stir welding and ultrasonic welding, were examined in this program. Pacific Northwest National Laboratory (PNNL) played the lead role in developing the friction stir welding route. A few recent studies indicate that friction stir welding (FSW) has the ability to join Mg alloys to steel [5,6]. FSW, which is a solid state welding technique invented by Thomas et al. at TWI, UK [7], has the added advantage of minimal associated oxidation because of the solid state nature of the process. A short preliminary effort at PNNL indicated higher joint strength in Mg to steel lap welds when the bottom steel sheet was deformed by the FSW tool [8].

Ultrasonic welding (USW) of Mg alloys to steel sheet was developed at the Oak Ridge National Laboratory (ORNL). The work focused on identifying the factors critical to increasing joint strength while understanding the influence of sonotrode tip design, material stack-up, coatings, and process time.

Summary of the FSW Initiative

Following our initial trial, an effort was made to optimize the FSW tool design in order to increase the Mg/steel joint strength in lap configurations. Two tool designs were tested in this study. The first tool (Tool 1) had a convex scrolled shoulder and a stepped spiral pin. The second tool (Tool 2) had identical features as the first tool plus a short hard WC insert on the tool pin bottom. Both the tools were made of H13 tool steel. A nominally 2.32 mm thick AZ31 magnesium alloy (O temper) was used as the top sheet. Two different steels were used, e.g., (i) HDG 1.5 mm thick HSLA steel and (ii) EG 0.8 mm thick mild steel (MS), as the bottom sheet material. Mechanical performance of the joints was determined through quasi-static unguided lap-shear tests, and fatigue tests. Further, joint interface microstructure was characterized in detail using optical microscope/SEM. Additionally, EBSD technique was employed to learn about the crystallographic information at the joint interface.

The welding was carried out in position controlled mode in a high stiffness, precision FSW machine manufactured by TTI, Inc. and located at PNNL. Complete details about the welding procedure has been published elsewhere [3,8]. Subsequently, three to five lap-shear specimens of width ~30 mm were obtained for each weld covering the entire weld length. Tensile tests were carried out at room temperature at a crosshead speed of 1 mm/min. Both ends of the lap-shear specimens were appropriately shimmed in order to minimize misalignment. Average failure load of a standard 30 mm wide lap-shear AZ31/MS weld coupon varied between 4.6 to 6.8 kN, as a function of process parameters. In the case of AZ31/HSLA weld coupons, the average failure load varied between 3.7 to 9.7 kN. Comparison of the lap-shear data revealed that the failure load was generally higher in the case of AZ31/HSLA welds, reflecting the higher strength of the bottom steel sheet used therein. Moreover, use of Tool 2 helped in lowering the scatter.

For graphical comparison purpose, weld efficiency of all the joints were calculated by comparing the failure load of the joint, with the maximum load carrying capacity of the weaker parent sheet section. Whichever parent sheet section shows a lower total load carrying capacity, is noted to be the weaker one. 0.8 mm steel is the weaker one for AZ31/MS welds, and AZ31 sheet is the weaker one for AZ31/HSLA welds. For calculating joint efficiency, the following equation was used:

$$\text{Joint efficiency} = (\text{Max. load of joint} / (A_w \times TS_{\text{parent sheet}})) \times 100$$

A_w is the area of the weaker parent sheet section, and TS is the tensile strength. Weld efficiency data for all the welds are shown in Figure 1a and 1b. The average minimum joint efficiency is

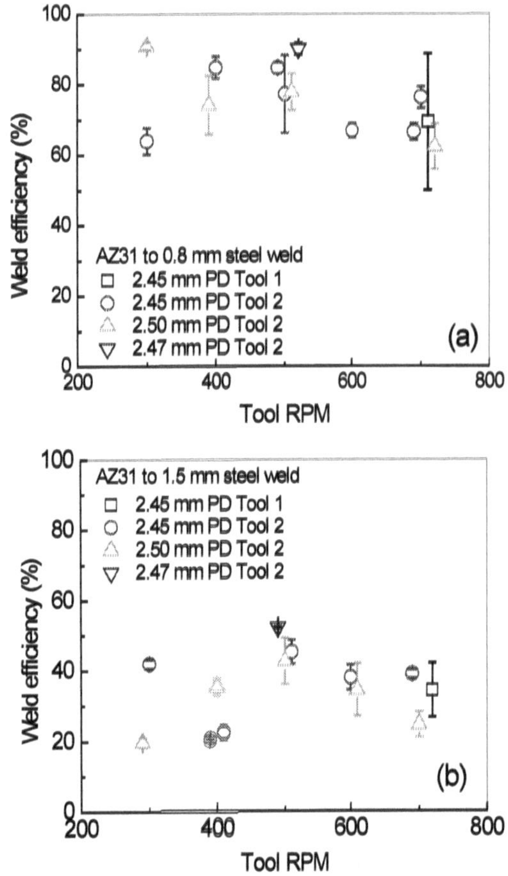

Figure 1. Weld efficiency plots: (a) AZ31/MS, (b) AZ31/HSLA.

>60 % in the case of AZ31/MS welds, and ~40% for AZ31/HSLA welds. Joint efficiency varied as a function of process parameters. Further, the benefit of using Tool 2 in order to obtain uniform strength is evident from the weld efficiency plots. Maximum joint efficiency is achieved at ~500 tool RPM for the AZ31/MS welds, with a tool plunge depth of 2.45 mm (Fig. 1a). For AZ31/HSLA welds, the highest joint efficiency is noted at 500 tool rotation rate, with the tool plunge depth being 2.50 mm (Fig. 1b). Based on the lap-shear test data, a single tool rotation rate of 500 rpm at an intermediate tool plunge depth of 2.47 mm was selected to carry out further welds. It should be noted that the tool travel speed was fixed at 100 mm/min for the entire study. The combination of 500 RPM, 100 mm/min travel speed, and 2.47 mm plunge depth resulted in >90% weld efficiency in the case of AZ31/MS welds. At same process parameters, weld efficiency of AZ31/HSLA weld was noted to ~52%, which was the highest in its class. Additionally, one of the AZ31/MS weld specimens at this particular combination of process parameters showed failure of the steel sheet, leaving the weld intact. The example of such a lap-shear specimen is shown in Figure 2.

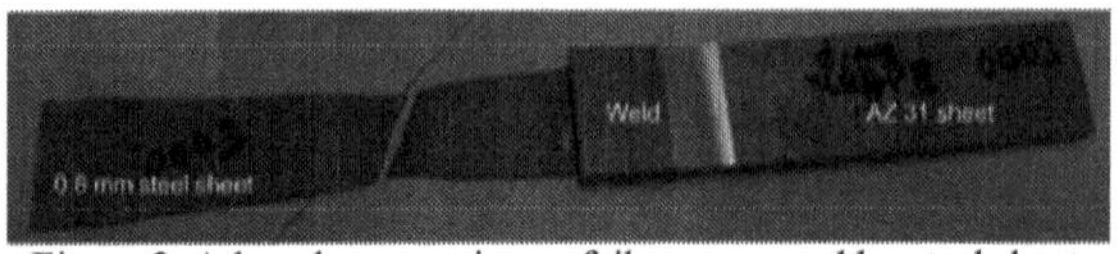

Figure 2. A lap-shear specimen: failure occurred by steel sheet fracture, leaving the weld intact.

Following quasi-static testing, performance of the lap joints at the best process parameters was determined under dynamic loading. A number of 30 mm wide lap-shear weld coupons were tested in a servo-hydraulic fatigue testing machine at a load ratio R of 0.1. The test frequency was 5 Hz. Fatigue tests were conducted in the load-control mode. S-N plot of the lap-shear fatigue specimens is represented in Fig. 3. At similar fatigue lives, the maximum

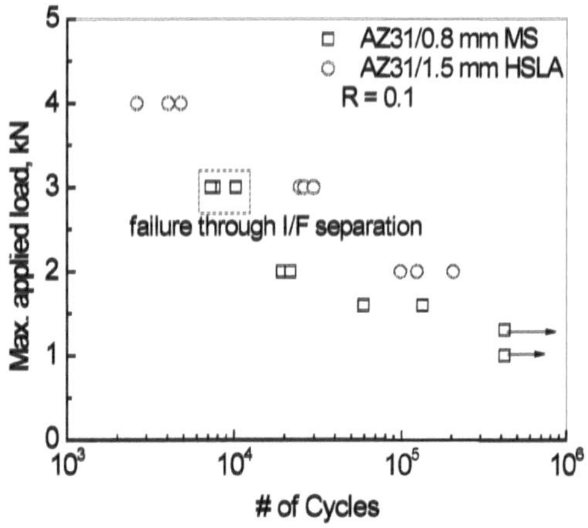

Figure 3. S-N plot for the Mg/steel friction stir lap welded specimens.

applied loads are found to be higher for the AZ31/HSLA welds than the AZ31/MS welds, reflecting the higher failure load of the AZ31/HSLA welds. However, when the maximum applied load is normalized against the respective failure load, as shown in Fig. 4, specimens from both material combinations are noted to have similar fatigue lives at any given applied load to failure load ratio. The major mode of failure was found to be top Mg sheet fracture.

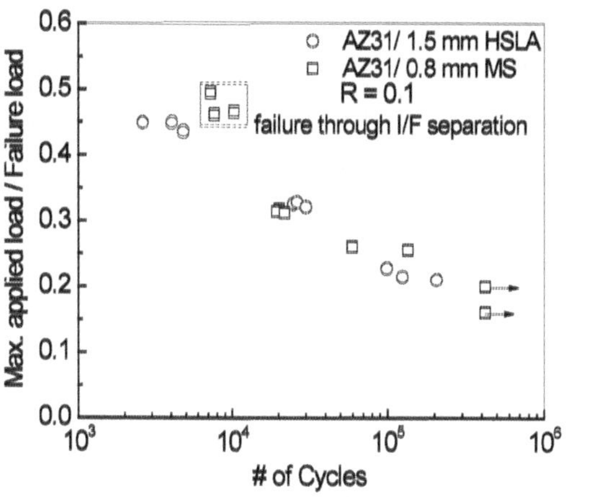

Figure 4. Normalized S-N plot for the Mg/steel friction stir lap welded specimens.

However, three AZ31/MS weld specimens that were tested at 3 kN, showed relatively longer life. Interfacial separation of the top and bottom sheet was the mode of failure for these three specimens. It can be further noted from Fig. 4 that two of the AZ31/MS weld specimens did not fail when applied load to failure load ratio was less than 0.2. Fatigue tests for these two specimens were stopped at ~$5x10^5$ cycles.

A typical cross-section image of a joint fabricated by Tool 2 is shown in Figure 5a. The effect of the WC insert on the tool pin bottom is clearly evident on this image. Presence of the insert resulted in machining of the bottom steel sheet. Further, it led to the formation of two "dove-tail" like features on both ends, as noted by the arrows. Some of the higher magnification SEM images obtained from various locations along the joint interface are shown in Figure 5b and 5c. Microstructural features observed in these images, indicate formation of solidification products through eutectic reaction, due to the melting of the Zn coating present on the steel sheet. However, no such solidification product was found along the region where the steel sheet showed signs of deformation due to the penetrating action of the tool. As the FSW tool plunges through the top Mg sheet, the Zn coating on the bottom steel sheet melts. The liquid Zn layer dissolves Mg from the top Mg sheet, and, hence forms a liquid Zn-Mg alloy. However, this newly formed liquid alloy layer gets squeezed out laterally due to the forging action of the tool.

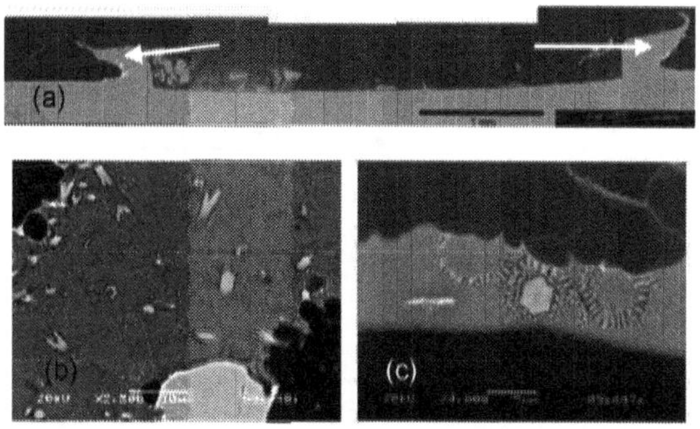

Figure 5. (a) Cross-section of a joint fabricated by Tool 2, (b),(c) Eutectic microstructure at Mg/steel interface.

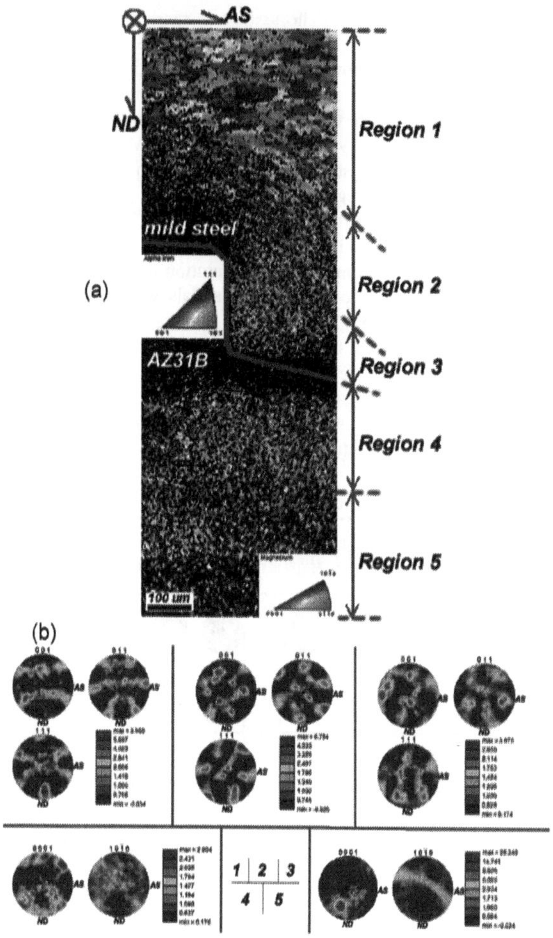

Figure 6. (a) Orientation map of the interface region at the advancing hook of FSW of AZ31B to mild steel, (b) Pole figures of the different regions in shown in (a).

A detailed EBSD study was carried out to characterize the weld nugget and the Mg/steel interface. As shown in Fig 6a, the area around the advancing hook in the FSW of AZ31B to mild steel could be divided into five regions. Region 1 is the base metal of the mild steel and there is no influence of the shear that happens in this region. Region 2 is the region near the interface which lies between the hook area and the base metal. The grains in this region are elongated in the shear direction. Due to the influence of the shear, the grains' thickness decreases as shear increases. Several shear textures form in this region. Region 3 is the region of hook near the weld nugget. The strain is heavy in this region. Grains in this region show a different orientation to the other region of the steel. Region 4 is the interface region of the Mg side. The texture differs to that of the steel side. Region 5 is the region near the weld nugget. The pole figures of each region are shown in Fig.6b.

Fig.6b-1 shows the texture of the base metal of the mild steel. The maximum intensity of the pole figure is 8.069. Fig.6b-2 shows the texture of the interface region of the steel side. The grains in that region rotated in favor of the shear and the strength of the texture decreased to 5.794. The texture of the interface near the weld nugget was influenced by the lower surface of the pin and the side face of the tip. The texture is randomized in this region and the maximum intensity decreases to 3.075. The interface of the Mg side shows a texture much like that of normal direction (ND)-advancing side. The two components of the texture are perpendicular to each other. The maximum strength of the texture is only about 2.904. The texture of the Mg side away from the interface shows a strong texture. The maximum intensity of the pole figures of that region is 25.249.

The pole figures of the weld nugget region are shown in Fig. 7. The texture changes little in the ND in the region far away from the shoulder and the interface. The rotation of shoulder and the pin forms a truncated cone and the crystals within the cone have their basal planes parallel to the cone surface. The direction of the crystalline alignment is due to the local change of the shear direction. The texture under the shoulder is influenced by the lower surface of the shoulder, and the texture changes little in this nugget region. The texture near the interface differs from that of the weld nugget, but the texture strength in this region is low.

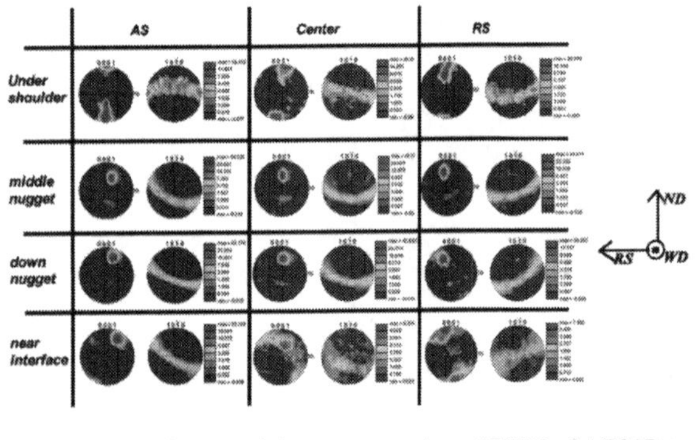

Figure 7. Pole figures of the nugget region of FSW of AZ31B to HSLA

Summary of the USW Initiative

The ultrasonic welds for this study were made with a Sonobond CLF2500 pedestal welding station. The rated conditions at the welding tip were 20 kHz frequency with 25 micrometers (μm) amplitude. The sonotrode tips were made from T1 steel. They typically have flat, rectangular faces with dimensions of either 5 mm x 7 mm or 7 mm x 7 mm. The line pressure to the tip clamping mechanism was adjusted to make the welds under constant nominal pressure of 39 MPa.

The materials used for the current experiments were sheets of 0.8-mm-thick hot-dip-galvanized (HDG) mild steel and 1.6-mm-thick AZ31B-H24. The zinc coating on the steel was about a 9 μm-thick. Prior to welding the surfaces of the AZ31B sheet were buffed with non-metallic abrasive pads to remove surface oxides and produce shiny surfaces. Both metals were also degreased with acetone followed by isopropyl alcohol to remove lubricants and surface debris.

Coupons of AZ31 nominally 30 mm wide x 100 mm long were welded to mild steel coupons of the same size to produce specimens for mechanical testing and metallographic analysis. A 25 mm overlap was used for making lap-welded coupons with spot welds centered in the overlap regions. Specimens were positioned for welding so that the primary vibration direction of the sonotrode was perpendicular to their long axis. Spot welding was typically performed using power of 1500-2500 W, and welding times ranged from 0.2-1.2 s.

The results of the tensile lap-shear testing are presented in Figure 8 where the variations of failure load with welding time are plotted. With both tips there was an initial rapid rise in lap-shear failure load from around 0.5 kN to near 2.9 kN as the welding times increased from 0.2-0.4 s. After the noted increase, failure loads continued increasing with welding times but at a much lower rate. From 0.4-1.2 s, failure loads increased to a maximum average value of 3.7 kN for the 5-mm x 7-mm tip and 4.2 kN for the 7-mm x 7-mm tip. Welds of the AZ31 to bare steel were unsuccessful. Weak bonding occurred in this case, but joints could be easily broken by hand.

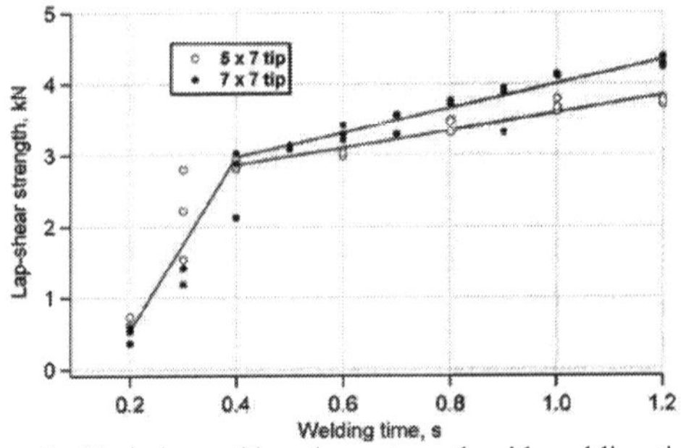

Figure 8. Variations of lap-shear strength with welding time and tip size

The lap-shear strength results can be placed in perspective by comparison with data from resistance spot welds and friction stir spot welds of AZ31. Published work [9] confirmed that AZ31 could be resistance spot welded with lap-shear strengths ranging to 5.7 kN. Lap-shear strengths increased with spot diameters and were about 4.0 kN for diameters with areas equivalent to the 7-mm x 7-mm USW tip. More recently, Lang and co-workers [10] measured lap-shear strengths up to 3 kN for spot welds of AZ31 with nugget diameters of about 7 mm. Another published study showed that AZ31 sheet could be friction stir welded and some joints had lap-shear strengths as high as 4.75 kN. No comparable published data could be found for magnesium-steel spot-welded joints made by these two processes. Nevertheless, strengths for ultrasonically-welded joints at times greater that about 0.5 s consistently exceeded 3 kN and so they are generally consistent with values found for magnesium-magnesium spot welds.

The results of the fatigue testing are shown as Figure 9 where the fatigue life expressed as cycles to failure is plotted against the load range for testing. The USW spot welds were tested at constant values of R = 0.1 and R = 0.2, where R = minimum load/maximum load.

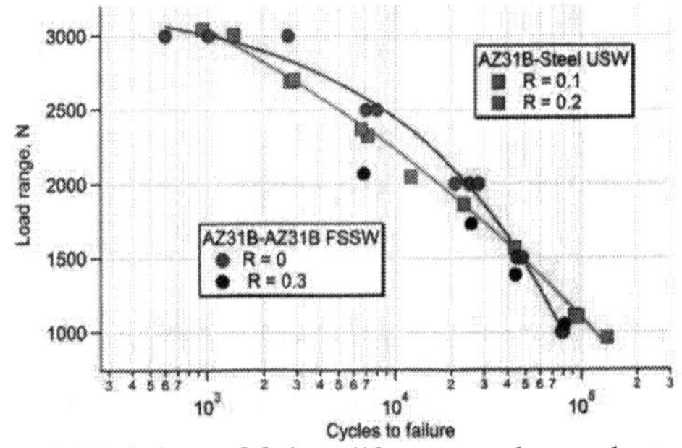

Figure 9. Variations of fatigue life expressed as cycles to failure with loading range

There are few published data for similar or dissimilar spot welds involving magnesium alloys, and Figure 9 compares the USW data to similar data from friction stir spot welds of AZ31-to-AZ31 lap welds. Regression analysis was used for the lines drawn through the USW data at R = 0.1 and the FSSW data of R = 0.0. However, a single line could have been fit through the entire combined data set. This suggests the fatigue properties of the AZ31-steel USW spots welds will not differ significantly from other types of spot welds made using AZ31. A more detailed examination of the fatigue performance for this work was previously documented [11, 12].

Additional results from tensile lap-shear testing are presented in Figure 10 where the variations of failure load with welding time are plotted for two arrangements: Mg-steel, where the sonotrode engaged the AZ31, and Steel-Mg, where the same engaged the hot-dip- galvanized (HDG) steel. The maximum lap-shear failure loads for Mg-steel spot welds are near 4 kN for 1.0 s of welding time. Those lap-shear strengths are in the range of the levels found for resistance spot welding and friction stir spot welding AZ31 to itself [9,10]. Given the disparities of properties between AZ31 and the mild steel, the lap-shear results are remarkable. Figure 10 indicates that similar maximum failure loads can be achieved with the either arrangement. This important result indicates that the USW has desirable flexibility in how it is presented to components in the manufacturing environment.

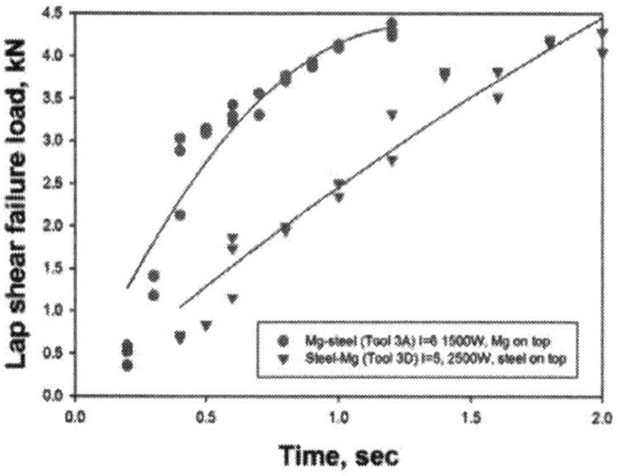

Figure 10. Variations of lap-shear strength with welding time and metal arrangement using a 7-mm x 7-mm sonotrode tip

One additional point of interest was the evaluation of hybid joining. Initial results from this evaluation are presented in Figure 11 where the Mg-steel results are replotted for comparison. Results are shown for three adhesive (Dow Betamate 73305) situations. First, USW spot welds were made through uncured adhesive as is commonly done with resistance spot welding. This approach caused a significant decrease of lap-shear strength. Subsequently, some spot welded coupons were cured. Curing produced lap-shear strengths in the range of 6 kN. As a reference, coupons that were only adhesive bonded and cured were also tested, and these had lap-shear strength of 4-5 kN. These results suggest that USW spot welding through adhesive is not advisable. Presumably, this is at least partial due to the adhesive modifying friction properties at the welded interface. Additional studies will be required to determine how much strength can be preserved in USW spot welds made through uncured adhesives. Once the adhesive is cured, lap-shear strengths can exceed those made by either processes alone. In all of the experiments, the area covered by adhesive was approximately 25 mm x 25 mm, a significantly larger area than that of the USW spot welds, approximately 7 mm x 7 mm.

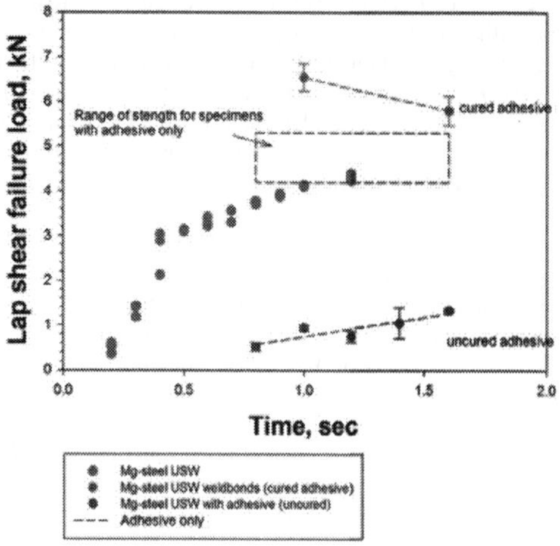

Figure 11. Variations of lap-shear strength with welding time and adhesive cure condition

Finally, results from an evaluation of the effect of corrosion on lap-shear strength are shown in Figure 12. A total of 20 Mg-steel specimens were made for exposure using the Ford Arizona Proving Ground Equivalent Corrosion Cycle (APGE) test, a widely accepted screening test for automotive industry. The exposure consisted of immersion in 5% NaCl bath at room temperature for 15 minutes followed by air-drying for 3 hours. Following this, the dried specimens were placed in a chamber where relative humidity was controlled to 80% and temperature was maintained at 50°C. Each cycle of exposure in the chamber lasted 20.75 hours. Starting at the fifth cycle of exposure, one specimen was tested for lap-shear strength after each cycle. Figure 3 shows that strength decreased linearly with each cycle of exposure up to 12 cycles. From that point on, strengths became very erratic, and many specimens could be broken by routine handling. A compound identified as $Mg(OH)_2$ was deposited in the crevice between the AZ31 and the steel. The deposits were observed after even one exposure, but they continued to thicken with each exposure cycle contributing to a separation force between the sheet specimens. Examination of fracture surfaces found that welded areas also decreased with exposure cycle, but even in the weakest joints there was evidence of metal-metal bonding.

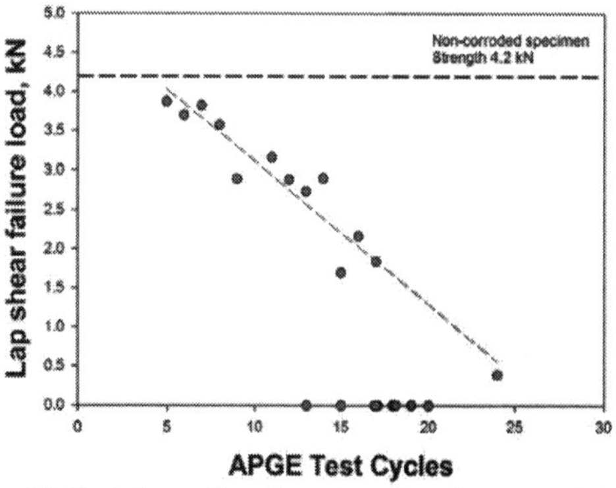

Figure 12. Variations of lap-shear strength and exposure times

Conclusions

A comparative study has been conducted to evaluate the joining potential of Mg/steel couples through two solid state welding routes, e.g., friction stir welding and ultrasonic welding. Initial results in terms of both static and dynamic strengths are very promising. The observed lap-shear strength is higher for FSW route as the bonded area is larger than USW. However, FSW suffers from a restricted configuration of using only Mg as top sheet material. This is not the case for USW method.

The authors would like to acknowledge the support from Dr. Carol Schutte and Mr. William Joost at the US Dept. of Energy, Vehicle Technology, Light Weight Materials Program for this work. This manuscript has been authored by Battelle Memorial Institute, Pacific Northwest Division, under Contract No. DE-AC05-76RL01830 with the U.S. Department of Energy. The United States Government retains and the publisher, by accepting the article for publication, acknowledges that the United States Government retains a non-exclusive, paid-up, irrevocable, world-wide license to publish or reproduce the published form of this manuscript, or allow others to do so, for United States Government purposes.

References

1. www.eere.energy.gov, Vehicle technologies program factsheet, Dec 2010.
2. R.S. Beals, C. Tissington, X. Zhang, K. Kainer, J. Petrillo, M. Verbrugge and M. Pekguleryuz, "Magnesium Global Development: Outcomes from the TMS 2007 Annual Meeting", *JOM*, August 2007, pp. 39-42.
3. S. Jana, Y. Hovanski, G. J. Grant, and K. Mattlin, "Effect of Tool Feature on the Joint Strength of Dissimilar Friction Stir Lap Welds", *Friction Stir Welding and Processing VI*, TMS 2011, pp. 205-211.
4. L.M. Liu, X. Zhao, "Study on the Weld Joint of Mg Alloy and Steel by Laser-GTA Hybrid Welding", *Material Characterization*, 2008, 59, pp. 1279-84.
5. T. Watanabe, K. Kagiya, A. Yanagisawa, and H. Tanabe, "Solid state welding of steel and magnesium alloy using a rotating pin – solid state welding of dissimilar metals using a rotating pin", *Quart J Japan welding soc.*, 2006, 24, pp. 108-15.
6. Y.C. Chen, K. Nakata, "Effect of tool geometry on microstructure and mechanical properties of friction stir lap welded magnesium alloy and steel", *Mat. design*, 2009, 30, pp. 3913-19.
7. W.M. Thomas, E.D. Nicholas, J.C. Needham, M.G. Murch, P. Templesmith, C.J. Dawes, *GB Patent application no. 9125978.8*, December 1991.
8. S. Jana, Y. Hovanski, and G. J. Grant, "Friction Stir Lap Welding of Magnesium Alloy to Steel: A Preliminary Investigation", *Metallurgical and Mat. Trans. A*, 2010, pp. 3173-3182.
9. Klain, P., Knight, D. L., and Thorne, J. P., "Spot Welding of Magnesium with Three-Phase Low Frequency Equipment," *Welding Journal* 32:7-18, 1953
10. Lang: B., Sun, D. Q., Li, G. Z., and Qin, X. F., "Effects of welding parameters on micro-structure and mechanical properties of resistance spot welded magnesium alloy joints *Science and Technology of Welding and Joining* 13:698-704, 2008
11. Teresa J. Franklin, Jwo Pan, Michael Santella, and Tsung-Yu Pan, "Fatigue Behavior of Dissimilar Ultrasonic Spot Welds in Lap Shear Specimens of Magnesium and Steel Sheets," SAE Paper No. 2011-01-0475.
12. Santella, M., T. Franklin, J. Pan, T. Pan, and E. Brown. 2010. Ultrasonic Spot Welding of AZ31B to Galvanized Mild Steel. SAE Paper 2010-01-0975.

From *Magnesium Technology 2009*, E.A. Nyberg, S.R. Agnew, N.R. Neelameggham, and M.O. Pekguleryuz, Editors

ELEVATED TEMPERATURE AND VARIED LOAD RESPONSE OF AS41 AT BOLTED JOINT

Okechukwu Anopuo[1], Guowu Shen[2], Su Xu[2], Norbert Hort[1], Karl Ulrich Kainer[1]

[1]GKSS Research Centre Geesthacht GmbH; Max-Planck Strasse 1, Geesthacht 21502, Germany
[2]CANMET Materials Technology Laboratory, 568 Booth Street, Ottawa, Ontario K1A 0G1

Keywords: Magnesium alloy, Bolt Load Retention, Creep

Abstract

The effective application of Mg alloys as automotive power train components is continuously challenged by the ability of magnesium to withstand fastener clamp load under service condition. The stiffness of a joint is strongly dependent on the elastic moduli of the members of the bolted joint. As deflections on loaded bolted steel components could be ignored at low and elevated temperature condition that of magnesium alloys cannot be overlooked. In this work bolt load retention experiments are carried out on AS41 between stresses of 40 MPa to 70 MPa and temperature of 125 °C to 150 °C. A power law creep relationship coded in finite elemental program is used to describe the time dependent stress-strain response of AS41. The parameters in this relationship are obtained by fitting typical compressive creep test results. A comparison of the model and bolt load retention experiments using load cell measuring techniques shows good agreement.

Introduction

Magnesium alloy offers great deal of advantages as structural material in automotive industry application. There are several component types produced or being develop from magnesium alloys for use in automobile ranging from steering wheel to power train components [1-3]. However, its ability to retain fastener clamp load at elevated temperatures and to what degree, has been an issue of concern at different level of materials and product development. Magnesium alloys are often bolted together with either steel or aluminum alloys when used as power train components. Under the thermo-cyclic operational temperature condition such as that witnessed in power train components, magnesium alloys tend to deform by creep and as a result lose some portion of the initial torque necessary to maintain the bolted parts firmly in position during service. This loss of fastener clamp load at bolted joint due to creep of the component parts are often referred to as bolt load retention (BLR) behavior of the material. Bolt load retention test is therefore an experimental method to quantify the bolt load loss with respect to time for engineering purposes. Several attempts has been made in the past to try to model bolt load retention behavior using stress relaxation or tensile creep test results. However, recent work [4] shows that the BLR is strongly dependent on the configuration of the joint and that the stress distribution within the joint members varies greatly, as such retained clamp load cannot be directly predicted from stress relaxation test. A review on stress relaxation and bolt load retention of magnesium alloys for automotive application reveals that bolt load retention behavior is more related to compressive creep than stress relaxation [5].

Apart from different forms of experimental set ups [6-8] that have been designed in the past to try to predict BLR behavior of materials; efforts have also been made to model this important engineering parameter numerically [9, 10]. In this work, the FEA modeling capability first adopted by the authors of reference [11] in the prediction of BLR behavior of a high pressure die cast (HPDC) AM50 on an instrumented bolt measuring technique is being applied in the prediction of the BLR behavior of a permanent mold AS41 carried out on a high precision load cell measuring system. The time dependent stress-strain response of magnesium alloy AS41 at bolted joint was modeled using compressive creep results obtained from typical cylindrical bolt load retention samples. A power law creep relationship coded in finite elemental program is used to describe this time dependent stress-strain behaviour. The total uniaxial stress-strain relationship of joint material was computed as the sum of elastic, plastic and creep strains. The results of the model and experiment show good agreement.

Experimental Procedures

Permanent mold AS41 Mg alloy was used in this investigation. The AS41 alloy was cast into a detachable permanent steel mold of total height 500 mm with top and bottom diameter of 210 mm and 105 mm respectively. The wall thickness of the steel mold is 22.5 mm. The mold produces a cast billet of 400 mm height and 100 mm bottom diameter with average weight of 8 kg. The melt and mold temperature were 700 °C and 220 °C respectively. The casting was carried out under 0.2 % SF_6 – argon mix cover gas and allowed to cool in air. The chemical composition of the AS41 alloy used in this work are 4.14 wt. % of Al, 1.03 wt. % of Si, 0.242 wt. % of Mn and the balance being Mg. The cast billets were carefully machined to the desired configuration, taking care that less internal stresses or deformation are induced on the samples.

Bolt Load Retention Test

High precision cylindrical load cells were connected to an amplifier which works on the principle of a Wheatstone bridge. The variation of the load is amplified to an output value of 0 to 10 Volts which are digitized via a 24 bit sigma-delta ADC (Analog to digital converter) with serial interface. The ADC readings can then be directly read out through a computer program. In this program an average value is stored over an adjustable time slot. The acquired data can then be preferentially selected and directly plotted for further analysis.
The test samples used in this work measuring 20 mm x 10 mm x 15 mm, that is outside diameter, inside diameter and height respectively were loaded with the M10 steel bolt and load cell as

shown in figure 1. The oil bath with the silicone oil was heated to the test temperature, in this case 125 °C, and 150 °C. The bolt couple was then torqued to the desired load; here 10 kN and 17 kN were used which correspond to 40 MPa and 70 MPa respectively. The bolted couple was usually torqued to about 5 % above the desired preload and was held for 1 hour at ambient temperature before inserting into the test temperature. This was done to reduce the effect of short term relaxation which occurs after torquing the test sample at room temperature. The computer program is set to record the data every 10 seconds for the first 20 minutes after placing the bolted couple in the oil bath. The data taking rate was manually adjusted afterwards to record every 15 minutes. The initial recording mode is very important so as to be able to monitor the increase in load as a result of thermal expansion mismatch as the bolted couple is heated to the test temperature. The duration of the test was restricted to 72 hours, after which the bolted couple was allowed to cool down to room temperature in the oil bath.

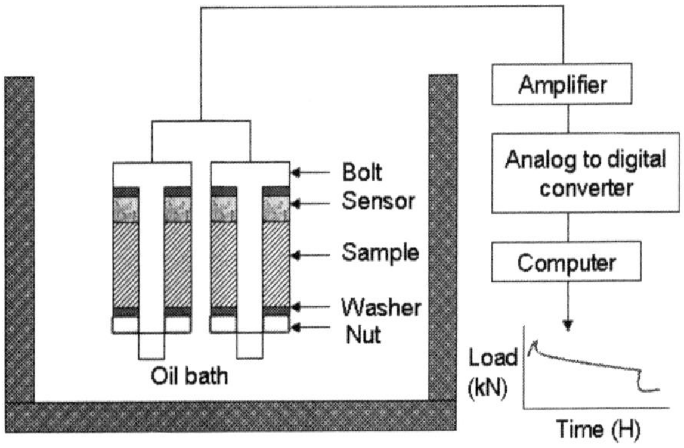

Figure 1. Bolt load retention testing equipment

Creep Investigation

Uniaxial compressive creep test were carried out on typical bolt load retention test specimen on an Applied Testing System (ATS) machine. The tests were performed under constant stress condition of 40 MPa, 70 MPa and 100 MPa at temperatures of 125 °C and 150 °C. The initial loading rate was 0.5 of the desired load per minute. At the appropriate stress level, a soaking time of about 1 hour 30 minutes was observed after which the creep test starts automatically. The creep deformation was constantly measured via an extensometer attached to the sample.

Other information of material properties which were used for the finite element modeling of the joint behavior were obtained by typical experiments that was carried out in this work. These include typical compression test of AS41 magnesium alloys at all the investigated temperatures from which information like yield strength, compressive strength, elastic moduli etc were obtained. Tests were also carried out to obtain the thermal expansion coefficient of the investigated alloy at the appropriate temperature. Information concerning the steel bolts, steel nuts, washers and so on were sourced from open literature.

Results and Discussions

The laboratory simulation of magnesium alloy behavior at bolted joint tries to reflect real service situation as closely as possible. The application of torque in a bolt generates a clamping load on the members of the bolted joint. The tightening of the bolt causes it to be strained in tension while the washers and the magnesium sample are strained in compression. At room temperature, the initial preload is approximately equal to the fastener clamp load with time varying only by few percents. At elevated temperature, during the inception of the bolt load retention test, increase in clamp load is experienced which is a combination of thermal expansion mismatch within the bolted couple and compressive yield of the magnesium component. This is subsequently followed by time dependent deformation due to creep of the magnesium alloy. It is assumed that the M10 steel bolt and washers used in this work expanded within the elastic regime only, considering the applied temperature. This is approximately about 0.1 of the melting temperature of steel bolt. The plastically deformed magnesium alloy causes a reduction in the original specimen height since the sample is deformed under compressive mode. The result of specimen height reduction after deformation brings about a decrease in the combined overall clamp length of the bolted couple and this subsequently leads to loss in clamp load. The degree of loss of clamp load depends among other things, on the operating temperature, stress involved and the compressive creep behavior of the magnesium alloy.

Application of Finite Element Analysis on Bolted Joint

The tightening of the bolted couple was such that during the application of torque on the bolt head, the end part of the M10 steel bolt was assumed to undergo uniform displacement. The bolt itself is strained in tension concurrently during tightening of the bolt. By surface contact, the head of the bolt exerts compressive stress on the AS41 magnesium alloy through the washer and the high precision steel load cell which is between the bolt head and the magnesium alloy. Figure 2 shows the schematic representation of a cylindrical through bolt permanent mold AS41 magnesium alloy used in this work.

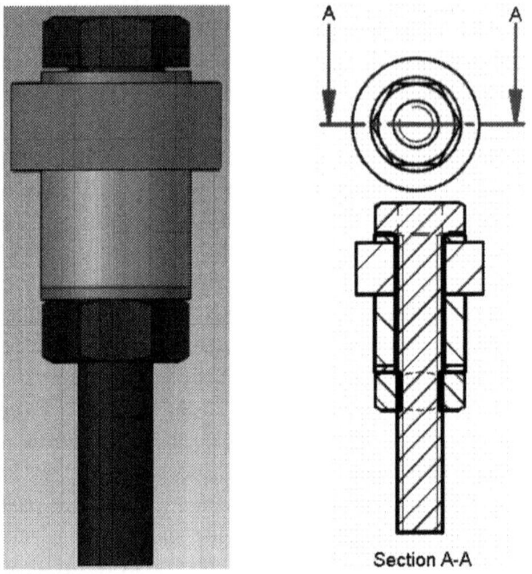

Figure 2. Bolted couple used for modeling bolted joint

When the desired preload is reached, in this case 10 kN and 17 kN which correspond to 40 MPa and 70 MPa respectively. The end of the bolt was locked with a steel nut assuming no further displacement of the steel bolt occurred or any sliding between the steel bolt and the nut. This also allows for the assumption that the grip length, which is the combined length of all the members of the bolted joint clamped by the bolt head and the nut is constant during simulation. It is important to mention that since the stress distribution between the bolt head, members of the clamped couple and the nut was studied in this model, the details of the bolt and nut thread was not modeled in this work. A finite element program ADINA [12] was applied in the simulation of this bolt joint. The configuration of the bolt joint consists of two steel washers of 2 mm thickness each with 20 mm external diameter and a through hole of 11 mm. The AS41 permanent mold magnesium sample measures 15 mm thickness, 20 mm external diameter and 10 mm internal diameter. A steel load cell measuring 11 mm thickness, 30 mm external diameter and 11 mm through hole for passing the M10 steel bolt. The load sensing part of the load cell being equal to the external diameter of the sample. The combined thickness clamped by the bolt head and the nut is 30 mm. The bolt joint was modeled axisymmetrically and was mirrored to get a complete bolt joint mesh. Figure 3 shows finite element mesh of bolted joint used in this work.

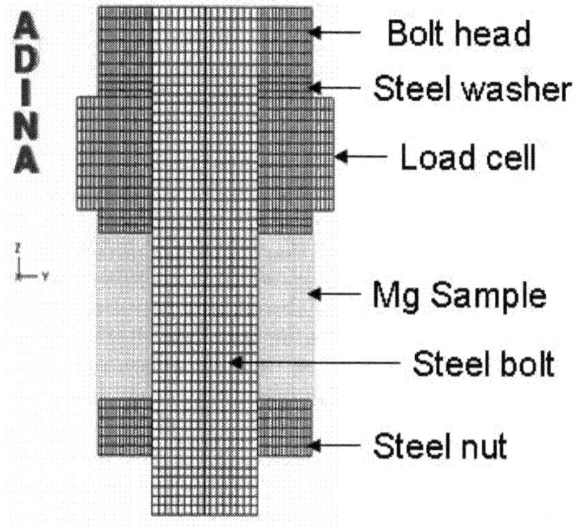

Figure 3. Finite element mesh for the bolt joint

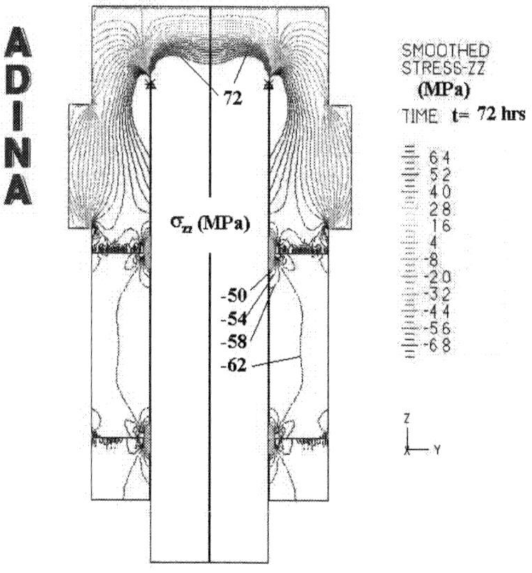

Figure 4. Stress distribution in a bolt joint torqued with a 70 MPa stress

The stress distribution in a typical bolt joint with a load cell is shown in figure 4. From the model, it could be observed that the stresses on the bolted joint are not uniformly distributed within its members. It has been observed that the non-uniformity of the stress within the joint members affects the analysis of the effective stress area and as such application of simple model becomes increasingly difficult [11]. This invariably influences the overall deformation response of the joint. This however demands application of models that makes accommodation for such behavior at bolted joint.

Creep Evaluation Using a Three-Parameter Creep Equation

The knowledge of creep behavior of AS41 magnesium alloy in compressive mode is needed in order to adequately model the bolt load retention response of AS41. To do this, a "three-parameter creep model" coded in finite element program was used to evaluate the compressive creep of AS41. Equation 1 shows the power-law creep model that was used in this work.

$$\frac{d\varepsilon_c}{dt} = C\sigma^m t^n \qquad (1)$$

where ε_c is the time dependent creep deformation, σ is the equivalent stress and t is time. The other parameters (C,m,n) are temperature dependent material constants which are evaluated using regression process. Hence equation 1 can be further expressed as equation 2 and the values of the constants could be determined.

$$\ln d\frac{\varepsilon_c}{dt} = \ln C + m\ln\sigma + n\ln t \qquad (2)$$

Compressive creep test were carried out at various stress levels at temperature of 125 °C and 150 °C in other to evaluate the three parameters used in this work. Since a through hole cylindrical AS41 magnesium alloy used for BLR was also employed in the creep test, this produces a gauge diameter of 17 mm, height of 15 mm and stressed area of 227 mm^2. The creep deformation with respect to time is shown in figure 5 and 6 at different stress levels for temperature of 125 °C and 150 °C respectively.

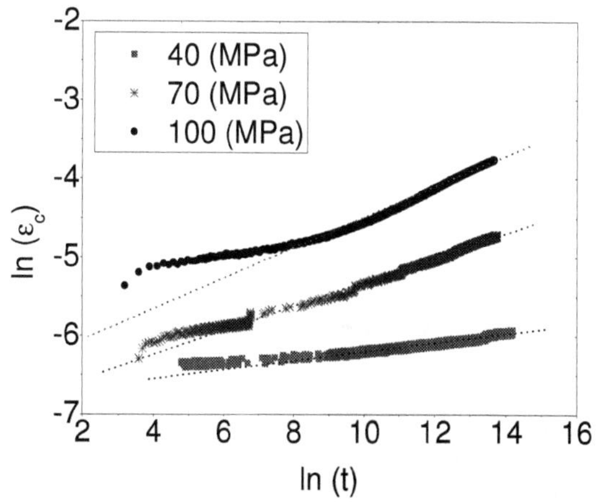

Figure 5. Plot of the natural logarithm of creep strain against time at various stress levels for 125 °C.

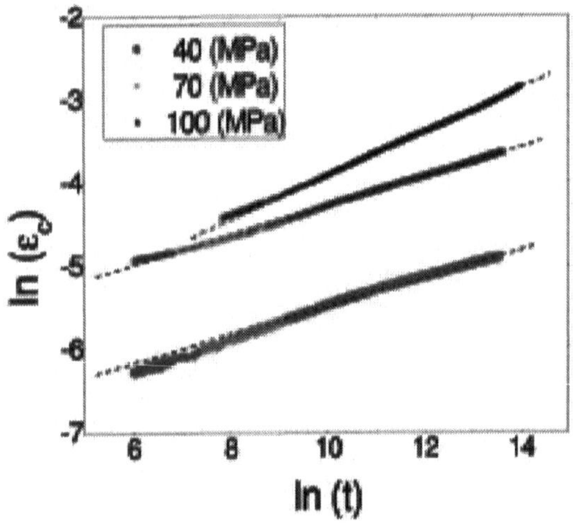

Figure 6. Plot of the natural logarithm of creep strain against time at various stress levels for 150 °C.

The slope of the plot of the natural logarithm of creep against that of time yields the value of n at a particular stress as shown in equation 2. The parameter n used was therefore obtained by taking the average of n values at different stress levels. For the computation of the parameter m, constant time condition was applied so that a plot of the natural logarithm of creep strain against various stress (σ) conditions yields the slope m as shown in equation 2. At constant time ln(t) = 10 and ln (t) = 12.5, figure 7 shows the plot of creep strain against stresses (σ). The stresses were in Pascal where as the times were in seconds. The fitted lines in figure 7 at various determined time scale are approximately parallel to each other indicating the creep strain dependence on the mth power of the applied stress (σ).

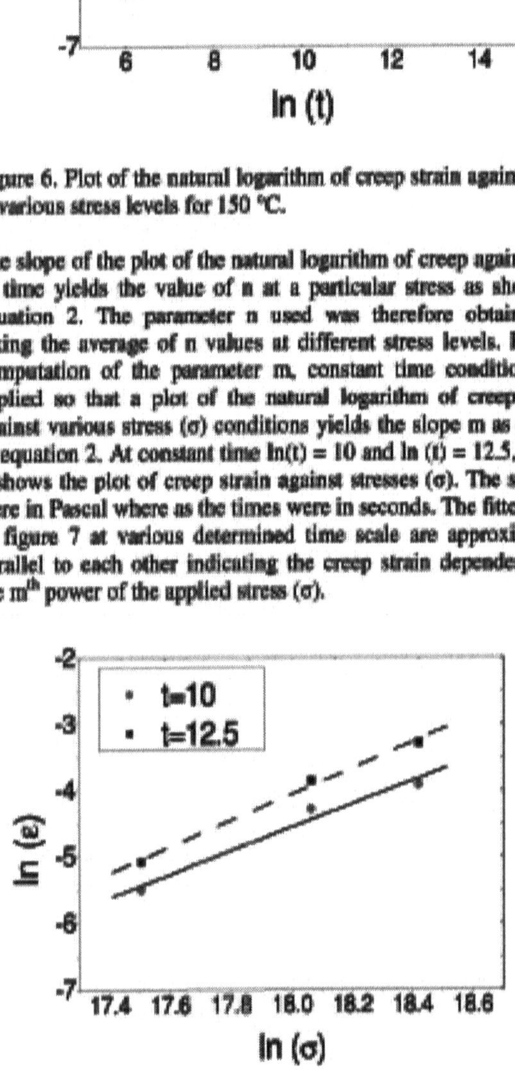

Figure 7. Plot of creep strain against stress at different times for 150 °C.

The creep material properties used for permanent mold AS41 magnesium alloy in FEA for 125 °C were: C = 4.38078 x 10^{-22}, m = 2.20 and n = 0.1956. At 150 °C, the values were 5.0276 x 10^{-18} for C, 0.1692 for n and 1.8681 for m. The results of the typical compressive creep experiments are compared with that of the three-parameter creep model as represented in equation 2. As could be seen in figure 8, the experiment and the model show reasonable agreement.

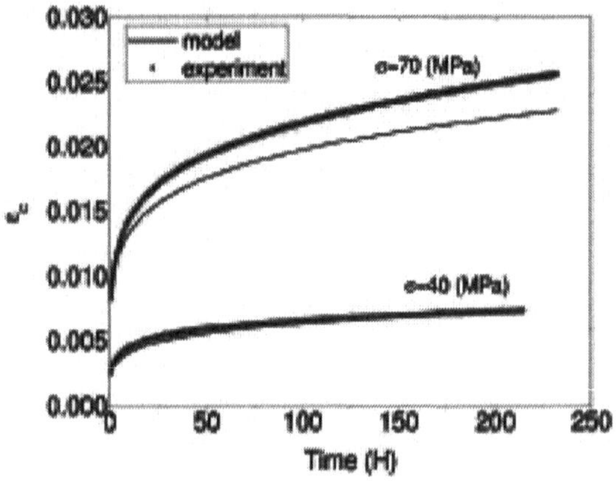

Figure 8. Comparative creep strain of AS41 as a function of time at different stress levels for 150 °C. Solid lines are model predictions while symbols are experimental data

Retained Clamp Load at Bolted Joint

Since the elevated temperature behavior of magnesium alloy at bolted joint is influenced by the alloys compressive yield and creep properties. It follows that the overall bolt load retention behavior of AS41 magnesium alloy, which is the time-dependent stress retained at the bolted joint should be modeled taking into consideration the stress-strain relationship of the alloy at relevant temperature. The comparison of the three-parameter creep equation with respect to experimental creep results in figure 8 which shows reasonable agreement indicates that this creep model can be used in the material model for bolt load retention simulation. The uniaxial stress-strain behaviors of joint material as expressed in reference [11] are given as equation 3 and 4.

$$\varepsilon_T = \varepsilon_e + \varepsilon_p + \varepsilon_c \quad (3)$$

$$\varepsilon_T = \frac{\sigma}{E(T)} + A(T)\sigma^N + C\sigma^m t^n \quad (4)$$

The application of this model takes into account the stress-strain response of AS41 within the elastic regime just before the yield deformation was attained. The plastic and creep deformation at the investigated temperature was also included. Hence in equation 3 and 4, ε_T represents the overall strain, ε_e the elastic strain, ε_p the plastic strain and ε_c the creep strain. E is the elastic modulus, T the temperature, A and N are material constants.

In order to take into account the metal temperature change with respect to time, transient thermal analysis was conducted for the simulation and subsequently structural analysis was performed to evaluate the time dependent bolt load loss as a result of variation in temperature gradient during FE simulation. In the case of AS41 magnesium alloy at 125 °C, based on the fact that not all the thermal plastic properties of AS41 alloy was available at the temperatures from 20 °C to 125 °C, the material model to simulate the period of tightening the bolt and putting it into the oil bath was changed from thermal plastic to thermal elastic. Hence a bi-linear relationship was used in the thermal-plastic-creep material model for simulation of the creep regime. The time dependent retained clamp load measured experimentally and that of the model prediction for AS41 magnesium alloy at 125 °C and 150 °C are shown in figures 9 and 10. The varied load levels presented in this work were reasonably predicted by the model considering the application of load cell measuring technique used.

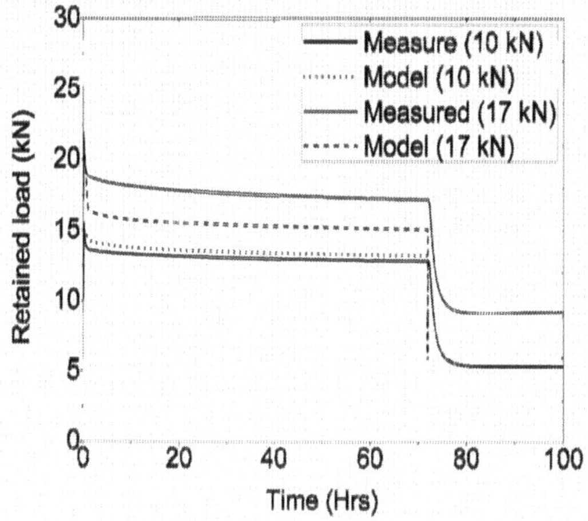

Figure 9. Time dependent retained clamp load at bolted joint for 125 °C. Broken lines are model while solid lines are measured values

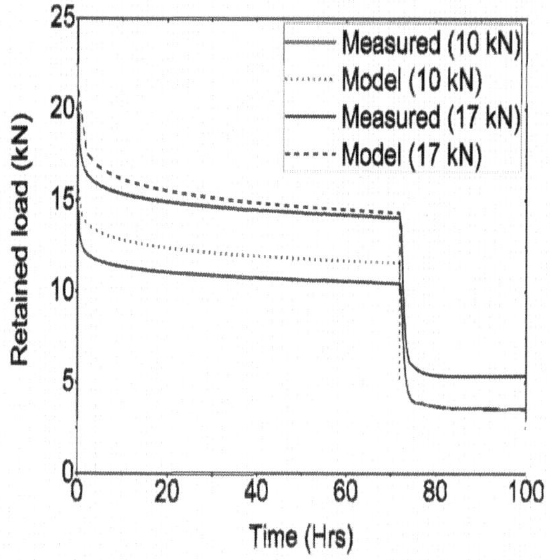

Figure 10. Time dependent retained clamp load at bolted joint for 150 °C. Broken lines are model while solid lines are measured values

Different phases of the BLR experiments which include load increase as a result of thermal expansion difference of the couple during BLR measurements at the inception of the test were correctly predicted. The first phase of the slopes which were characterized by prominent bolt load loss was also evident by the model which was consequently followed by asymptotic decrease in retained clamp load. In general, it could be seen that considering the large scatter of the creep and BLR response of permanent mold AS41 magnesium alloy, experiment and model show good agreement.

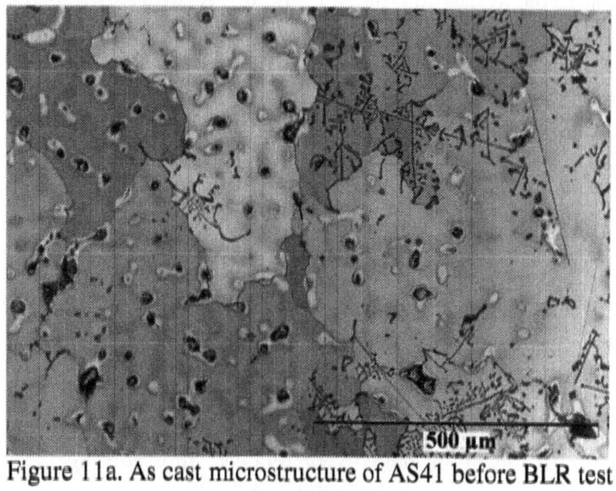

Figure 11a. As cast microstructure of AS41 before BLR test

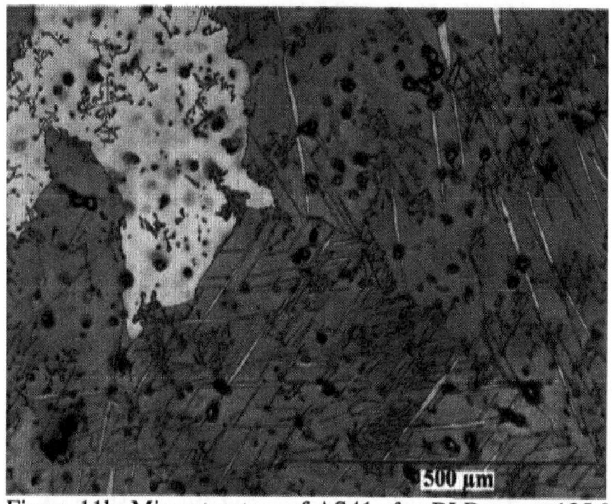

Figure 11b. Microstructure of AS41 after BLR test at 125 °C with 17 kN load

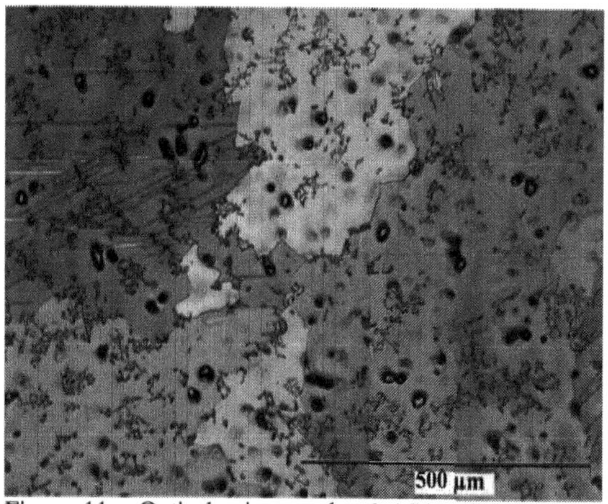

Figure 11c. Optical micrograph of AS41 after BLR test at 10 kN/150 °C

For comparison the optical microstructures of permanent mold AS41 magnesium alloy are show in figure (11a-c). The micrographs were taken from the surface of AS41 magnesium alloy that was in direct contact with the load cell from where compressive load was transferred to the sample. Figure 11a shows the as cast micrograph of AS41 before BLR test. Figure 11b shows the AS41 after BLR test at 125 °C and loaded at 17 kN. The micrograph is characterized with high density of twins in comparison to the as cast micrograph. Figure 11c with lower load at 150 °C shows lesser amount of twins. The twins are expected to have developed from the stress induced by the steel bolt on the magnesium alloy.

Summary

The application of "three parameter creep model" coded in finite element program was used in the prediction of retained clamp load at bolted joint using a load cell testing technique. The model and experiment shows reasonable agreement with respect to varied load and temperature conditions studied in this work. For the BLR behavior of permanent mold AS41 magnesium alloy investigated in this work, 83 % and 80 % of the initial load were retained at 125 °C for 10 kN and 17 kN load respectively after 72 hours. 53 % of the initial loads were retained after cooling the bolted couple from test temperature at 125 °C to room temperature. At 150 °C, 70 % of the investigated load condition were retained after 72 hours where as 33 % of the initial load were retained after cooling down to room temperature. The experimental measurement of BLR using load cell technique can be used as a first approximation in the study of bolt load retention behavior of magnesium alloys at elevated temperature during alloy development.

References

1. S. Schumann and F. Friedrich, "The use of magnesium in cars-today and in Future", *Proc. Conf. on Magnesium Alloys and their Applications*, ed. B.L Mordike and K. U. Kainer. (Wolfsburg, Germany: Werkstoffinformationsgesellschaft 1998) 3-13

2 Joseph C. Benedyk, "Magnesium: Where Have We Been, Where Are We Going" *Light Metal Age*, 2008.

3 F. Hollrigl-Rosta et al., "Magnesium in the Volkswagen", *Light Metal Age*, 1980, 22-29.

4 S. Xu, et al., "Bolt-load Retention Testing of Magnesium Alloys for Automotive Applications", (Detroit, MI: SAE International Congress and Exposition 2006) SAE Technical Paper Series 2006-01-0072

5 S. Xu, J. P. Thomson and M. Sahoo, "A Review on Stress Relaxation and Bolt-load Retention of Magnesium Alloys For Automotive Applications", *Can. Metall. Quart.*, 43 (4) (2004), 489-505.

6. T. K. Aune and T. J. Ruden, "High Temperature Properties of Magnesium Die Casting Alloys", (Detroit, MI: SAE International Congress and Exposition 1992), SAE Technical Paper Series 920070.

7. E. G. Sieracki, J. J. Velazquez and K. Kabiri, "Compressive Stress Retention Characteristics of High Pressure Die Casting Magnesium Alloys", (Detroit, MI: SAE International Congress and Exposition 1996), SAE Technical Paper Series 960421.

8. K. Pettersen and S. Fairchild, "Stress Relaxation in Bolted Joint of Die Cast Magnesium Components", (Detroit, MI: SAE International Congress and Exposition, 1997), SAE Technical Paper Series 970326.

9. J. Arimond, "Bolt Load Retention Modeling from Creep Performance Data", (Detroit, MI: SAE International Congress and Exposition 1995) SAE Technical Paper Series 950485.

10. C. J. Bettles and M. S. Dargusch, "Creep and Bolt Load Retention of Sand Cast Elektron MEZ", *Proc. Conf. on Magnesium Alloys and their Applications*, ed. K. U. Kainer, (Munich, Germany: Werkstoffinformationsgesellschaft 2000) 705-710.

11. G. Shen, S. Xu, "Finite Element Simulation of Bolted Joints and Magnesium Bolt-Load Retention Behavior", (Detroit, MI: SAE International Congress and Exposition 2007), SAE Technical Paper Series 2007-01-1032.

12. ADINA, *Theory and Modeling Guide*, Volume 1, Report ARD 01-7, ADINA Research and Development Inc., (Watertown, USA, 2001).

STRUCTURE AND MECHANICAL PROPERTIES OF FRICTION STIR WELD JOINTS OF MAGNESIUM ALLOY AZ31

T. Nagasawa[1], M. Otsuka[1], T. Yokota[1] and T. Ueki[2]

[1] Department of Materials Science and Engineering
[2] Department of Mechanical Engineering
Shibaura Institute of Technology
Shibaura 3-9, Minato-ku, Tokyo 1088548, Japan

Abstract

The applicability of friction stir welding to hot rolled sheet of commercial magnesium alloy AZ31 plates has been investigated. Friction stir weld joint showed mechanical strength comparable to that of base material, though the ductility remained at one half of that of the latter. The results are consistent with the microstructure which is characterized by a fine grained bond layer bounded by intermediate grained base metals. It is found that both anodizing treatment and insertion of aluminum foil between batting faces do not degrade the joint properties at all. The results suggest that friction stir welding can be potentially applied to magnesium alloy.

Introduction

A revolutionary method of solid phase welding, which permits a wide range of parts and geometries to be welded, and called friction stir welding (FSW), was invented by W. Thomas and his colleagues of The Welding Institute (TWI), UK, in 1991(1).

Friction stir welding is a relatively simple process(2) (see Fig.1); a specially shaped cylindrical tool with a profiled probe, made from a hard and wear resistant material relative to the material being welded, is rotated and plunged into the abutting edges of the parts to be joined. After entry of the profiled probe to almost the thickness of the material and to allow the tool shoulder to just penetrate into the plate, the rotating tool is transitioned along the

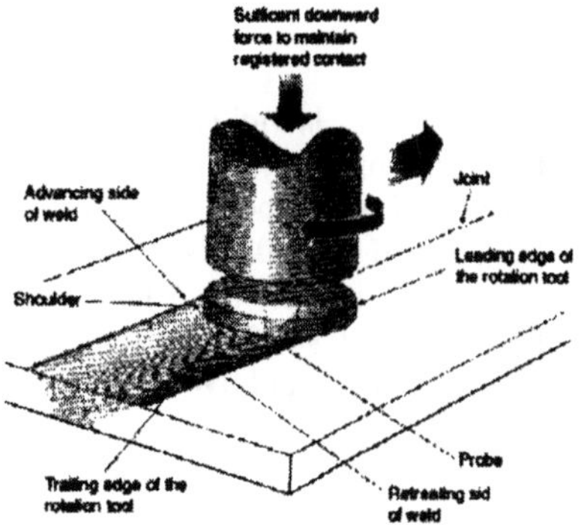

Figure 1: A schematic illustration of friction stir welding(2).

joint line. The rotating tool develops frictional heating of the material, causing it to plasticize and flow from the front of the tool to the back where it cools and consolidates to produce a high integrity weld, in the solid phase.

Subsequent improvement mainly accomplished in the tool design brought the technology forward for application especially to commercial aluminum alloys such as 2219, 2014, 6083, 6082 and 7075(2)-(9).

The process proves predominance for welding non-heat-treatable or powder metallurgy aluminum alloys, to which the fusion welding can not be applied. Thus fundamental studies both on the weld mechanism and on the relation all arrow ship between microstructure, properties and process parameters, have recently been started[(10)-(21)].

On the other hand, very limited information has been obtained about the feasibility of the process for welding commercial magnesium alloys, though they have the potential ability to substitute conventional light weight materials such as aluminum alloys or engineering polymers.

Thus the aim of this study is to examine experimentally the applicability of friction stir welding to commercial magnesium alloy AZ31. The results obtained here are compared with those of other researchers.

Procedure

Materials

The materials tested are hot-rolled plates of commercial magnesium alloy AZ31 of 6mm thick, supplied by Osaka Fuji Kougyo Inc., the nominal composition of which is 3.0%Al, 0.9%Zn, 0.3%Mn, 0.003%Fe, 0.002%Ni, 0.002%Cu, 0.2%Si and balanced Mg (dimension in mass%). A number of rectangular workpieces of 100mm width and 200mm length were machined out of the plates and submitted for FSW.

Welding Equipment

FSW was carried out according to the following sequence. A pair of workpieces free from oil films were abutted along a longitudinal section and fastened rigidly on the cast iron backing bar, which was mechanically fixed on the bed of a vertical type milling machine. A specially designed rotating tool, the details of which are shown in Fig.1, was vertically inserted along the abutted joint until the shoulder face came in touch with the top face of the workpiece. In this case, the height and diameter of pin was 5.8mm and 5.0mm, respectively.

In this paper, tool rotation speed and transition speed was held constant at 1750rpm and 88mm/min, respectively, though it is confirmed by the preliminary tests that the process can be successfully applied at the rotation speed ranging from 44 to 246mm/min and at the translating speed ranging from 980 to 1750rpm.

Evaluation of the Performance of Joints

In order to evaluate the performance of friction stir weld joint of AZ31 alloy plate, several methods were employed.

Thermal History: The change in temperature with time at three points on the lower surface of workpiece, the one (point A) just on the joint face and the other two (points B and C) 5mm distant from the joint, was monitored with a themocouple during the welding.

Metallography: Microstructure observation of weld joint was made on the two planes, one parallel to and the other perpendicular to the tool transitioning direction. Prior to the observation, the mechanically polished surface was chemically etched with an agent, the standard composition of which is 0.4g of picric acid, 13ml of ethanol, 3ml of glacier acetic acid and 3ml of boiled water.

Mechanical Tests: A series of 2mm thick sheets were cut out of the welded joint so that the longitudinal rolling direction of the sheet was parallel to the tool transitioning direction. Tensile test pieces, the gauge section of which was located within the welded zone and had a size of 30mm long, 4mm wide and 2mm thick, were cut out of the sheets. In order to discuss the fracture mechanism, local strain profile was obtained by measuring the change in length between preliminarily stamped points. Sharpy impact tests were carried out on the specimen having a cross section of 6mm x 6mm in which U-notch of 2mm depth was cut through the weld so that its edge laid perpendicular to the plate.

Hardness: The microstructure of welded joints was evaluated by measuring hardness distribution along a central line drawn on a plane perpendicular to the tool transitioning direction.

Effect of Surface Finish:

FSW should in principle be insensitive to the morphology of the surface to be welded. To confirm this prediction, three kinds of weld specimens were prepared: the first was anodized before abutting and then welded, and had oxide films of 12μm thickness. The second was welded after aluminum foil of 15μm thickness was inserted into a butt plane. The third was a single plate which was friction stirred under the condition described in the previous section. Thereafter, the mechanical properties were measured by the conventional tensile tests.

Results and Discussion

Thermal History

Figure 2 shows the change in temperature with time at three points just beneath the lower surface of abutted plate, i.e., point A just upon the abutted plane, points B and C 5mm distant from the

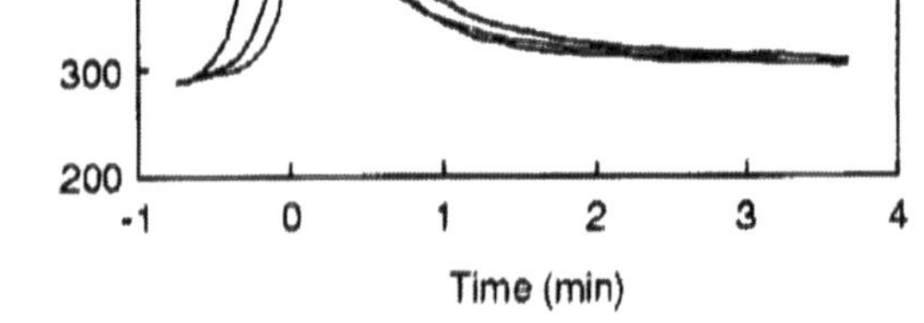

Figure 2: Change in temperature with time, at three points just beneath the lower surface of abutted plate. Point A: Upon the abutted plane, points B and C: 5mm distant from the plane.

plane. The fact that the material is heated up to 733K demonstrates that that FSW proceeds in the solid state. The fact that the material is held above 573K for 30s at point A suggests that the recrystallization occurs during friction stir welding. On the other hand, the peak temperature at points B and C are about 573K . There exists between points A and B (or C) a temperature gradient as large as 30Kmm^{-1}, meaning that only the material within a narrow range adjacent to the abutted plane is heated and softened.

Microstructure

In Fig. 3 are shown both macroscopic and microscopic (c, d, e) structures observed on the plane perpendicular to the tool translation direction. It is obvious that the nugget is well developed at the center of the weld (a, b). This contains an annual ring type inner structure, consisting of concentric ovals (a, d). The characteristic structure demonstrates that the materials plastcized by frictional heat flow not only horizontally but also vertically describing spiral loci. A similar structure has often been observed in some aluminum alloys[2-6]. Appendages to the nugget are seen at the upper surface of the weld and extend to the edges of the tool shoulder.

Considerably uniform and recrystallized grains of about 16μm in diameter are seen in the weld (Fig. 3e), in contrast to the coarse grains of about 56μm in the base metal (Fig. 3c). The heat-affected zone often observed in the friction stir weld joint of aluminum alloy plate is not observed in the present case. This is probably because the grains in the base metal have been recrystallized during hot rolling and hence insensitive to the thermal disturbance.

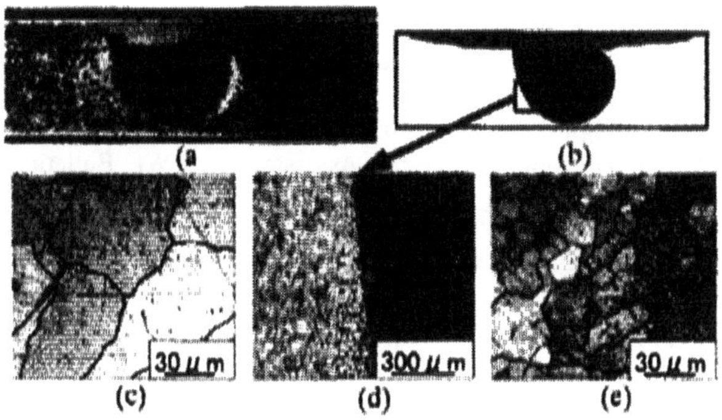

Figure 3: Macroscopic (a, b) and microscopic (c, d, e) structures observed along the same longitudinal rolling direction of the sheet as the tool transition direction.

Hardness Distribution

Figure 4 shows the hardness profile measured along a central line on the transverse plane perpendicular to the welding direction. No difference in the hardness is observed between the base metal and the weld. The result is in contrast to that in the welds of some wrought aluminum alloys that the hardness decreases drastically in the stirred region, and may be explained by the above mentioned fact that both the base metal and the weld consist of recrystallized grains[5].

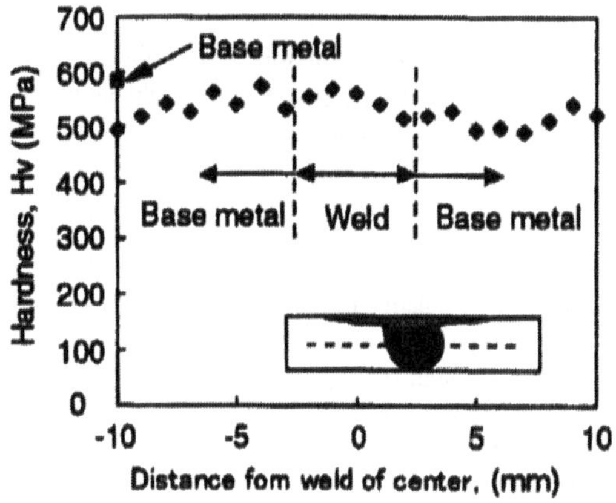

Figure 4: Hardness profile measured along a central line on the transverse plane perpendicular to the welding direction.

Mechanical Properties of FSW Joint

Figure 5 shows nominal tensile stress-nominal strain curves of the weld joint compared to that of the base metal. Both yield stress and ultimate tensile strength of the joint is almost at the same level as that for the base metal. The weld joint has an elongation of about 12%, which is sometimes sufficient for many practical applications, though the base metal exhibits ductility much higher than that of the joint. A remarkable dimple structure is observed on the fracture surface in each case, as shown in Fig. 6.

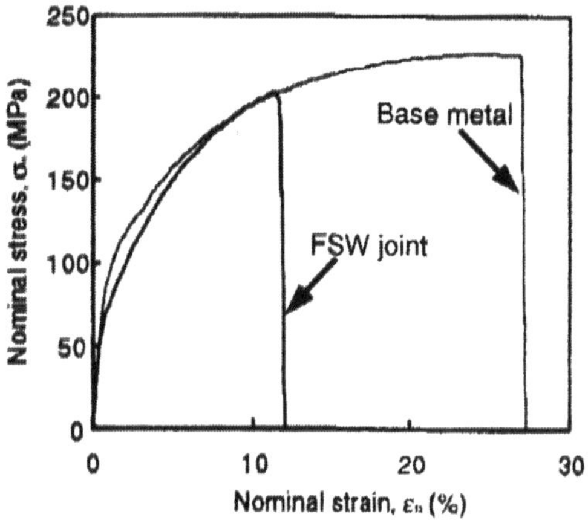

Figure 5: Nominal stress vs. nominal strain curve of the weld joint compared to that of base metal.

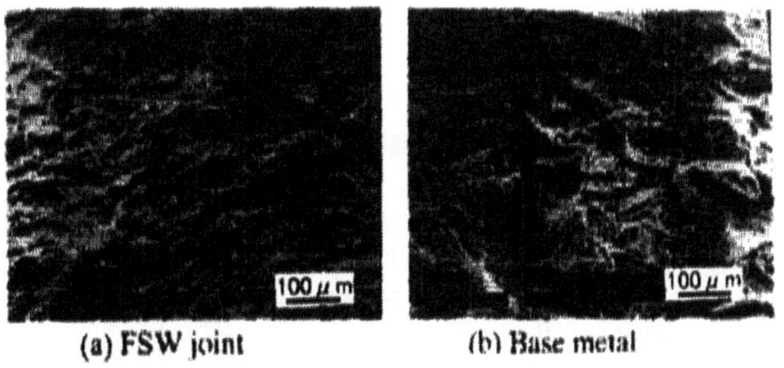

Figure 6: SEM micrographs showing fracture surface of the weld and the base metal.

The Sharpy impact value of the weld was 7.7J, while the base metal showed 6.1J. The superior toughness of the weld should be attributed to the fine and uniform grain structure brought about by FSW.

Effect of Surface Finish

Figure 7 shows nominal tensile stress-nominal strain curves for the specimens prepared by various preliminary treatments described earlier and through subsequent welding. It is seen that both strength and ductility of the welds are affected neither by the surface oxide films, nor by the absence of a butt plane. A separate experiment proved that the surface roughness of the plate to be welded has no effect on the mechanical properties of joint. In summary, FSW is a process quite insensitive to the extent of surface finish.

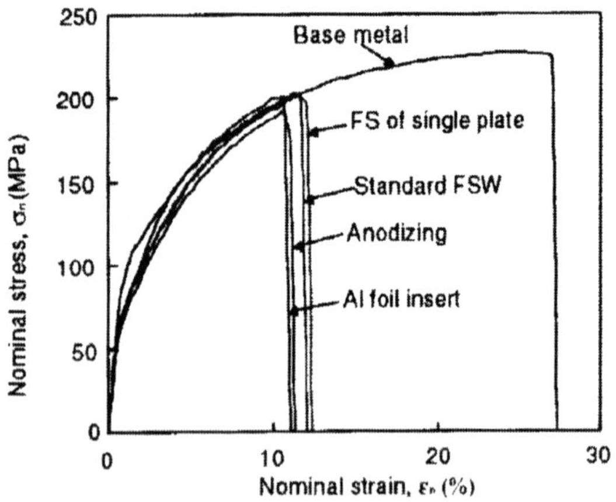

Figure 7: Nominal tensile stress-nominal strain curves of the specimens prepared by various preliminary treatments. (See text.)

Comparison with the Previous Results

To the best of our knowledge, no other comprehensive paper has been published regarding the FSW of magnesium alloys except one by Shimizu and his colleague[22]. As in this case, they also had observed fine and recrystallized grain structure and as well as the flat profile of hardness. However, both the strength and the ductility of FSW joint are much lower than those of the base metal in comparison to our results. This is probably because of their peculiar probe with a V-profile which would be inadequate to stir the lower half material of the plate to be welded, and would hence be apt to cause crack-like weld defects called "kissing bond"[2].

Conclusions

The applicability of friction stir welding in air to hot-rolled plate of magnesium alloy AZ31 has been investigated. The results obtained at a fixed set of rotation speeds and tool transition speed are summarized as follows:

(1) The friction stir weld joint of hot rolled plate of magnesium alloy AZ31 has the strength comparable to that of the base metal. This is because fine and recrystallized grain structure is established during the welding.

(2) An annual ring structure observed on a plane in line with to the welding direction strongly suggests that the flow lines of the material, plasticized by the frictional heat, have vertical components as well as horizontal ones, resulting in the effective mixing of materials.

(3) The mechanical performances of the welds are fairly insensitive to the morphology of the base metal sheet.

Acknowledegments

This study was supported in part by the Ministry of Education, Science, Culture and Sports of Japan for Grant-in-Aid for Scientific Research through Contract No.800190 and in part by the Light Metals Educational Foundation Inc., Osaka through Contract No.LM25-46.

References

1. W. M. Thomas E. D. Nicholas, J. C. Needham, M. G. Murch, P. Templesmith and C. J. Dawes: "Friction Stir Butt Welding", International Patent Application PCT/GB92/02203, GB Patent Application 9125978.8, 6 Dec. 1991 and US Patent 5,460,317.

2. E. D. Nicholas: "Developments in the Friction Stir Welding of Metals", Proc. of ICAA-6 (6th Int. Conf. Aluminum Alloys), Toyohashi, Japan, (1998), 139-151.

3. C. J. Dawes and W. M. Thomas: "Friction Stir Joining of Aluminum Alloys", TWI Bulletin, **36**(1995), 124-127.

4. C. J. Dawes and W. M. Thomas: "Friction Stir Process Welds Aluminum Alloys", Welding J., March, (1996), 41-45.

5. P. Threadgill: "Friction Stir Welds in Aluminium Alloys - Preliminary microstructural Assessment", TWI Bulletin, **38**(1997), March/April.

6. W. M. Thomas, C. J. Dawes, M. Gittos and D. Andrews: "Friction Stir – Where We Are, and Where We Are Going...", TWI Bulletin, **39**(1998), 44-50.

7. C. J. Dawes: "An Introduction to Friction Stir Welding and Its Development", Welding and Metal Fabrication, January, (1995), 14-16.

8. M. Enomoto: "Application of Friction Stir Welding to Aluminium Alloys", J. Light Metal Welding, **36**(1998), 75-79.

9. T. Shinoda: "Recent Developments in New Joining Process, Friction Stir Welding", J. Japan Welding Society, **67**(1998), 60-63.

10. T. Shinoda and Y. Kondo: "Friction Stir Welding of Aluminum Plate", J. Japan Friction Welding Assoc., **2**(1995), 3-8.

11. H. R. Shercliff and M. J. Russell: "Analytical Modelling of Friction Stir Welding" Preprints of INALCO 98 (7th Int. Conf. Joints in Aluminium), Cambridge, UK, (1998), 185-195.

12. O. Frigaad, O. Grong and O. T. Midling: "Modelling of Heat Flow Phenomena in Friction Stir Welding of Aluminium Alloys" Preprints of INALCO 98 (7th Int. Conf. Joints in Aluminium), Cambridge, UK, (1998), 197-207.

13. L. E. Svensson, J. Karlsson, B. Karlsson, H. Larsson and L. Karlsson: "Microstructure and Properties of Friction Stir Welded Aluminium Alloys" Preprints of INALCO 98 (7th Int. Conf. Joints in Aluminium), Cambridge, UK, (1998), 221-230.

14. M. W. Mahoney, W. Harrigan and J. A. Wert: Friction Stir Welding SiC Discontinuously Reinforced Aluminuium" Preprints of INALCO 98 (7th Int. Conf. Joints in Aluminium), Cambridge, UK, (1998), 231-236.

15. T. Hashimoto, N. Nishikawa, S. Tazaki and M. Enomoto: "Mechanical properties of Joints for Aluminum Alloys with Friction Stir Welding Process" Preprints of INALCO 98 (7th Int. Conf. Joints in Aluminium), Cambridge, UK, (1998), 237-247.

16. G. Rhodes, M. W. Mahoney, W. H. Bingel, R. A. Spurling and C. C. Bampton: "Effects of Friction Stir Welding on Microstructure of 7075 Aluminum", Scripta Mater., 36(1997), 69-75.

17. M. W. Mahoney, C. G. Rhodes, J. G. Flintoff, R. A. Spurling and W. H. Bingel: "Properties of Friction Stir welded 7075 -T651 Aluminum" Metall. Mater. Trans., 29A(1998), 1955.

18. J. Hagstrom and R. Sandstrom: "Static and Dynamic Properties of Joints in Thin-Walled Aluminium Extrusions, Welded with Different Methods" Proc. of ICAA-6 (6th Int. Conf. Aluminum Alloys), Toyohashi, Japan, (1998), 1447-1452.

19. H. Larsson, L. Karsson and L.-E. Svensson: "Characteristics of Friction Stir Welds in AA5083 and AA6082 Aluminium" Proc. of ICAA-6 (6th Int. Conf. Aluminum Alloys), Toyohashi, Japan, (1998), 1471-1476.

20. O. Frigaad, B. Bjorneklett, O. Grong and O. T. Midling: "Process Modelling Applied to Friction Stir Welding of Al-Mg-Si Alloys" Proc. of ICAA-6 (6th Int. Conf. Aluminum Alloys), Toyohashi, Japan, (1998), 1477-1482.

21. H. S. Yang: "Microstructural Development in Friction Stir Welding of Aluminum Alloys" Proc. of ICAA-6 (6th Int. Conf. Aluminum Alloys), Toyohashi, Japan, (1998), 1483-1488.

22. B. Shimizu, K. Kato, H. Tokisue: "Mechanical Properties of AZ31 Magnesium Alloy Joints by Friction Stir Welding" (Paper presented at the 95th Fall Meeting of Japan Institute of Light Metals, 31 October 1998), 253-254.

Friction Stir Welding of Magnesium Die Castings

Jan Ivar Skar[1], Haavard Gjestland[2], Ljiljana Djapic Oosterkamp[3], Darryl L. Albright[4]
[1]Norsk Hydro ASA, Corporate Research Centre P.O. Box 2560, N-3907 Porsgrunn, Norway
[2]Hydro Aluminum, Competence Centre Magnesium, P.O. Box 2560, N-3907 Porsgrunn, Norway
[3]Hydro Aluminum, R&D Materials Technology Karmøy, N-4264 Håvik, Norway
[4]Hydro Light Metals, Magnesium Market Development, 39209 Six Mile Road, Suite 200, Livonia, MI48152, USA.

Keywords: Magnesium, Die Castings, Welding, Friction Stir

Abstract

Friction stir welding (FSW), being a solid-state process, is an attractive method for joining magnesium die castings. In this study, FSW of AZ91D and AM50A plates was performed both on the individual alloys and to join them together. The welds were sound and free from defects, except for small surface cracks in AM50A; a fine microstructure characterized the weld zones. The tensile properties of specimens transverse to the weld zone were measured, as were the corrosion properties. The tensile properties were somewhat lower than the base metal, with the largest percentage decrease found in the elongation of AM50A, perhaps due to the surface cracking. The corrosion resistance of the weld zone was relatively poor, most likely due to iron contamination from wearing of the tool. Further optimization of the FSW tool design and process parameters must take place to improve the reliability of FSW for magnesium die castings.

Introduction

Due to the increased use of magnesium components in the automotive industry, there is increasing interest in the development of supporting technologies. Most of these components are die castings, since the excellent castability of magnesium makes it possible to produce large, near net shape components with a relatively small number of locations for joining. However, magnesium components are joined to other materials and to each other; examples are inner door frame structures joined to aluminum outer sheet. Bolting is the main joining technique, followed by adhesive bonding. Welding has not yet found major use for joining magnesium die castings in automotive applications. One reason for this, as mentioned earlier, is that joining is not needed to the same extent as for other metals; another factor is the relatively poor quality of welds. Porosity in the casting itself has given weld porosity, and there is a tendency for formation of brittle intermetallic phases (β-phase-$Al_{17}Mg_{12}$) in the weld [1,2]. Methods using more focused energy input like Electron Beam Welding (EBW) and Laser Beam Welding (LBW) have provided better welds [1,3-6]. The investment cost of these methods is rather high, and thus the methods have not yet been found attractive for commercial use on magnesium components.

An alternative to EBW and LBW is Friction Stir Welding (FSW). FSW is a solid-state welding method. The welding is done using a rotating cylindrical tool, which consists of the larger diameter shoulder and the smaller diameter profiled pin. In the simplest butt geometry, the shoulder is in intimate contact with the material to be welded, and during rotation, frictional heat plasticizes the material. The pin, which is inserted into the joint, shears the surfaces together. As the tool progresses, the material behind the tool is cooled down to form a solid-state weld. The method was developed for aluminum and is patented [7]. A number of publications about FSW of magnesium have been published in recent years [8-13]. In general, the findings are that sound and pore-free welds are prepared on all alloys and there is a fine, recrystallised microstructure in the weld zone. Different alloys have successfully been welded together. The tensile properties approach the properties of the base metal, except that the elongation seems to be reduced. Only in AZ31, a wrought alloy, has a clear reduction of the tensile properties been found [12]. Recently, it has been reported that magnesium has been FSW to aluminum alloys, but the welds were unfortunately brittle due to formation of Al-Mg intermetallic phases [14]. Information about the corrosion properties of welds made by FSW has not been found in the literature.

In the present study, FSW of AZ91D and AM50A has been performed both in individual alloys and in welding the two alloys together. The tensile properties as well as the corrosion properties have been tested.

Experimental

Die cast test plates with the dimensions 150x100x3 mm were used. The composition of the plates is given in Table I. Two and two plates were welded together longitudinally to the melt flow in the die.

A similar tool as used for FSW of aluminum alloys was used. It was comprised of a cylindrical shoulder of 15 mm and a 5 mm pin with three circumferential fins. Welding parameters, the force, rotation and welding speed were varied in order make sound welds. When acceptable welds were obtained, a number of plates were welded together for tensile and corrosion testing. Table 2 shows the parameters used for the different alloys and alloy combination.

Table I. Composition of die cast plates used in the trials

Alloy	Element (wt%)						
	Al	Zn	Mn	Si	Fe	Cu	Ni
AZ91D	8.9	0.79	0.24	0.0093	0.0017	0.0013	0.0010
AM50A	4.8	0.001	0.30	0.0053	0.0011	0.0003	0.0009

Table II. Welding parameters (The grey cells give the final parameters for the samples tested)

Alloy	Force (N)	Rotation (rpm)	Speed (mm/s)	Tilt (°)
AZ91D	9000	1100	2	4
AM50A	8000	1500	4	4
	8000	1100	2.5	4
	8000	1200	3	4
AZ91D to AM50A	8500	1100	2.5	4
	8500	1200	2.5	4
	8500	1100	2	4

Tensile test specimens according to ISO 6892 were prepared transverse to the weld so that the weld was placed in the middle of the gauge length. The gauge dimensions were 65x10x3 mm. Similar tensile specimens were cut longitudinally from un-welded plates of the same batch for determination of the tensile properties of the base metal. The tensile testing was done using an Instron Model 4500 machine, and the testing was done according to ISO 6892.

Welded specimens were corrosion tested in ASTM B117 salt spray for 240 hours. The specimens were tested in the as received condition, except that the specimens were degreased prior to testing. The rear side of the weld was facing the salt spray. Corrosion rate was determined by weight loss measurements. Local the pit depth was measured using a Uniscan 100A profile scanner, which utilizes the reflection of a laser beam to determine the depth profile.

Results

Effect of welding parameters

It was observed that relatively small changes in process parameters resulted in large variations in weld quality. It seems therefore that the process window for FSW of magnesium is narrower than what is found for aluminum.

It was experienced that magnesium had a tendency to adhere to the tool, and that the tool sometimes was stuck into the material.

Weld quality and microstructure

When the proper process parameters were established, sound welds were made. Figure 1 gives an overview of the weld. Figures 2a, 3a and 4a give images of the welded cross-section of AZ91D, AM50A and AZ91D welded to AM50A. The weld nugget is somewhat larger than what is normally observed in friction stir welding of aluminum alloys.

There is no porosity or defects in the welds, but some surface irregularities are found. On AM50A surface cracks are found in the weld zone, as seen from Figure 3e.

The base material, Figures 2b and 3b, reveals a typical high-pressure die cast structure with α-magnesium grains with eutectic (mixture of α- and β-phase) on the grain boundaries. In the transition zone, Figures 2c and 3c, a change in the microstructure from the base structure in the upper right corner of the micrographs to a more deformed structure in the lower left corner can be seen. The amount of eutectic on the grain boundaries seems to gradually diminish toward the weld zone. The weld zone seems to consist of recrystallized equiaxed α-magnesium grains, Figures 2d and e and 3d and e. The structure in the weld zone reminds of a solution heat-treated structure, i.e., the β-phase is dissolved into the α-matrix

Figure 1. Friction stir weld on AZ91D

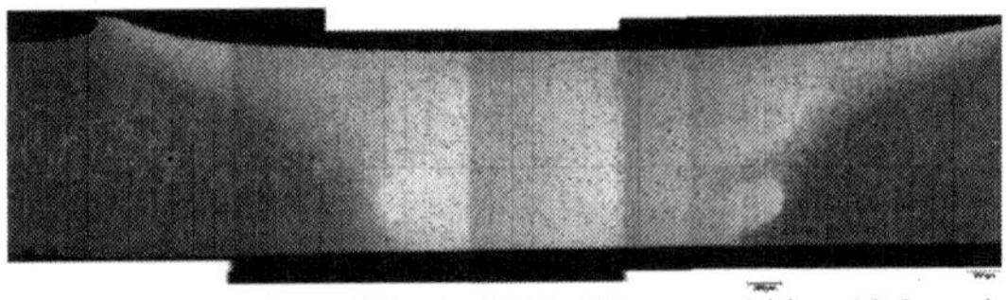

a) Overview of the weld (Full image width = 13.9 mm)

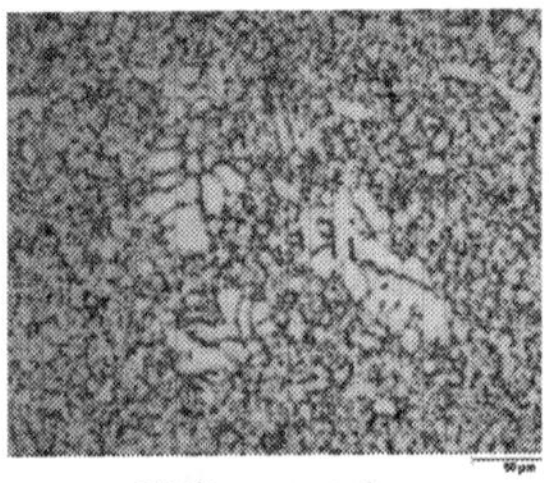

b) Base metal

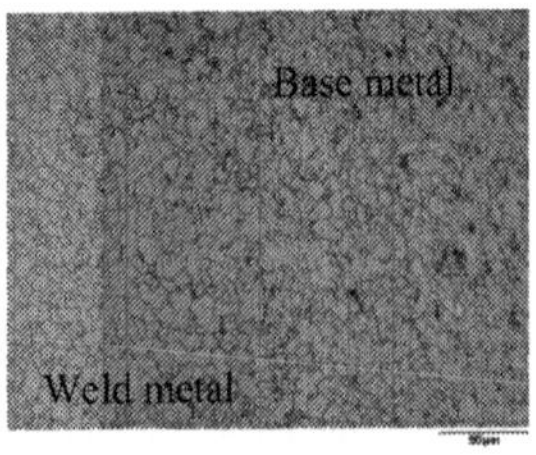

c) Transition zone

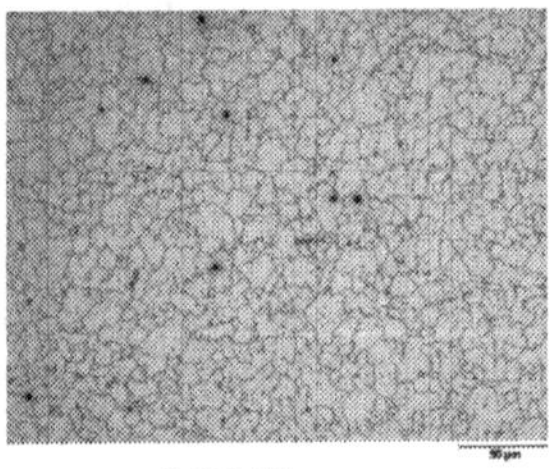

d) Weld zone

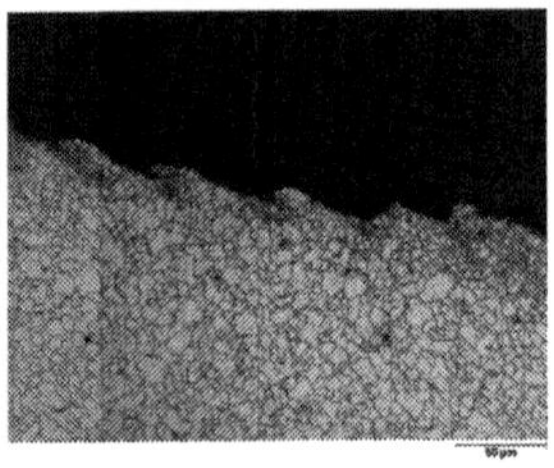

e) Surface

Figure 2. Microstructure of friction stir welded AZ91D (Full image width for b) – e) = 390 μm)

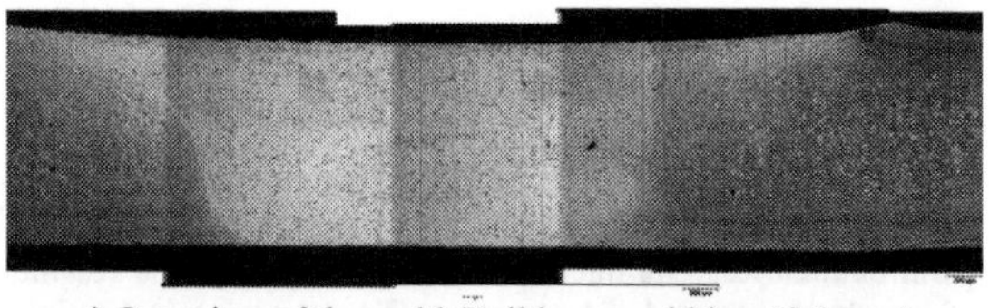

a) Overview of the weld (Full image width = 13.1 mm)

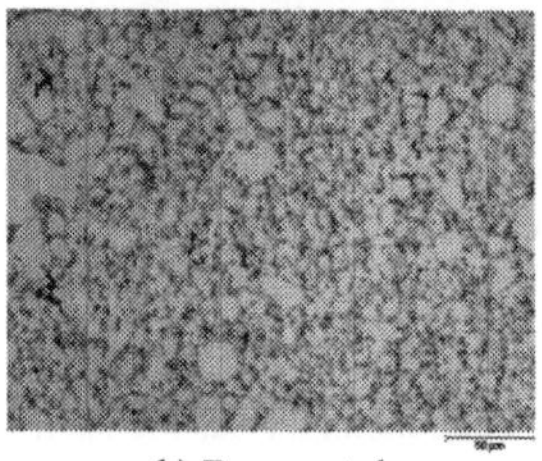

b) Base metal

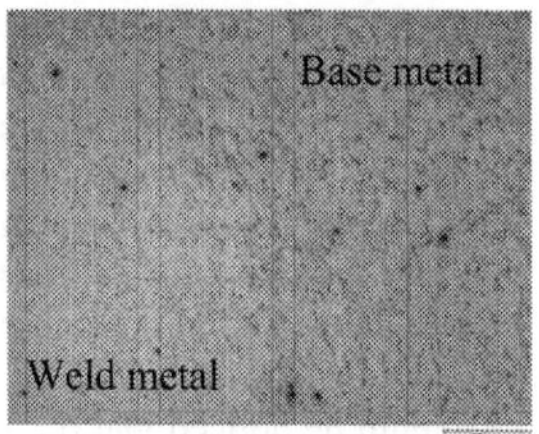

c) Transition zone

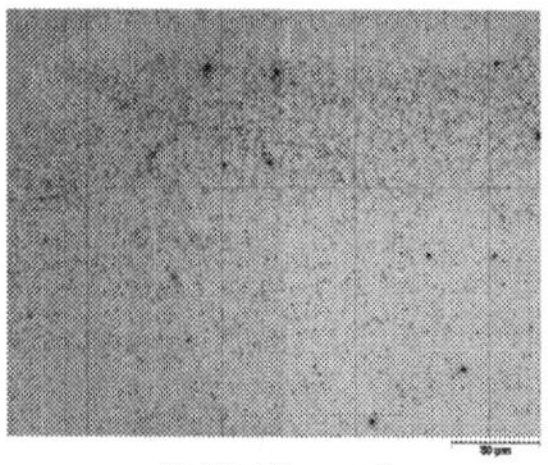

d) Weld metal

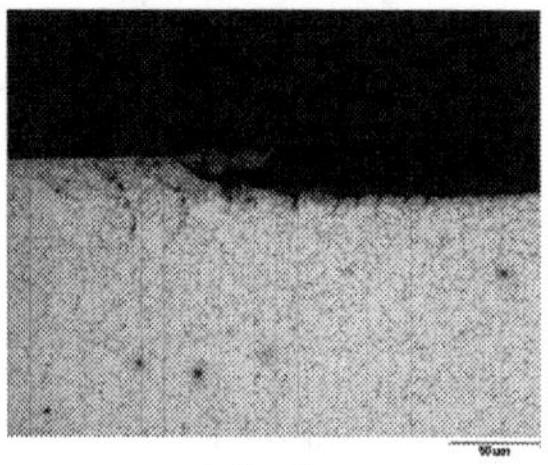

e) Surface

Figure 2. Microstructure of friction stir welded AM50A (Full image width for b) – e) = 278 μm)

For AZ91D welded to AM50A a heavily deformed zone with layers of metal being stirred into each other is seen from Figure 4a. The separate zones of AZ91D and AM50A in the weld exhibit the same microstructure as two alloys welded individually.

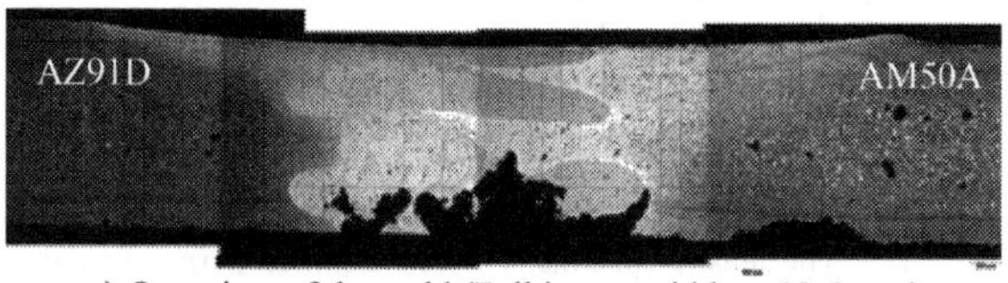

a) Overview of the weld (Full image width = 13.6 mm)

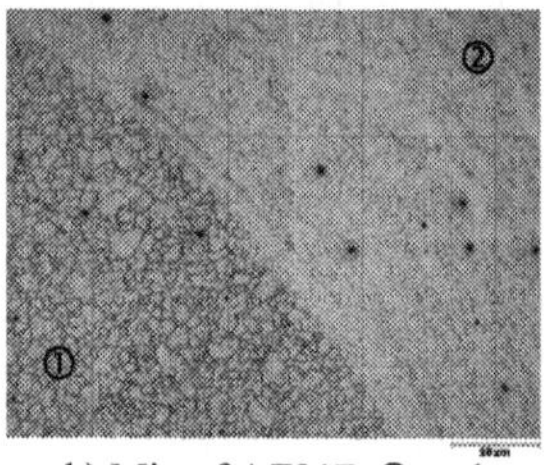

b) Mix of AZ91D ① and AM50A ② weld metal

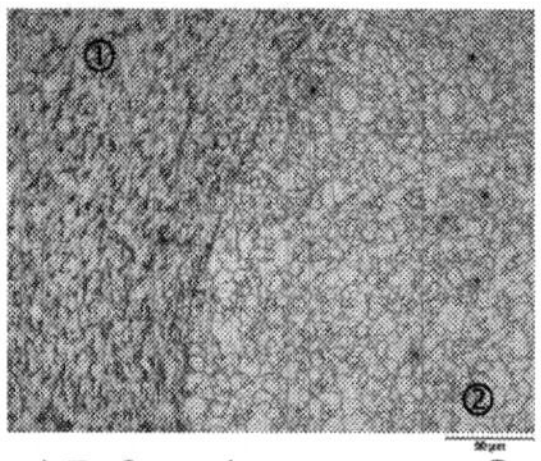

c) Deformed cast structure ① and AZ91D weld metal ②.

Figure 4. Microstructure of AZ91D welded to AM50A. The bottom of figure a) shows the corrosion attack after testing in ASTM B117 salt spray (Full image width for b) and c) = 278 μm)

Tensile properties

The results from the tensile testing are given in Figure 5. The scatter reported as one standard deviation is shown in the plot.

For AZ91D there is no significant difference in tensile yield strength (TYS) between weld and base metal. The ultimate tensile strength (UTS) is slightly reduced in the weld metal compared to the base metal. The elongation is reduced to about half in the weld metal compared to base metal. The failures were found in the base metal, except for one test where the failure occurred in the weld.

For AM50A the TYS is the practically the same for the weld and the base metal, while UTS is reduced in the weld metal to about 70% of the base metal. The most dramatic drop is seen in the elongation, from 9.3% to 2.4%. The scatter in elongation both in the weld and the base metal was rather large. The failures were found in the weld zone.

When AZ91D is welded to AM50A, the tensile properties are in-between the properties found for welded AZ91D and AM50A, except that the elongation is somewhat higher for the mixed weld compared to the other welds. The failures were found in the base metal of AM50A.

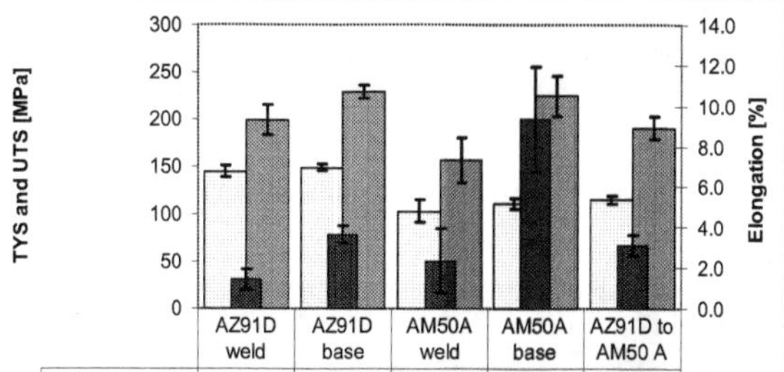

	AZ91D weld	AZ91D base	AM50A weld	AM50A base	AZ91D to AM50 A
□ TYS [MPa]	145	149	103	111	115
▨ UTS [Mpa]	199	229	157	225	191
■ Elongation [%]	1.4	3.6	2.4	9.3	3.1

Figure 5. Tensile properties of base metals and welds of friction stir welded specimens

Corrosion properties

The corrosion rates, given in Table III, show that the welded plates are more corroded than the reference samples. Furthermore, corrosion rates determined by weight loss do not give a satisfactory description. Figures 6 – 8 show the corrosion attack from one end of the weld to the midpoint of its length. As seen from Figures 6 – 8, the corrosion attacks are localized in the weld. In some regions there is perforation of the samples, showing very poor corrosion resistance. The weld zone of AM50A shows significantly more corrosion attack than the weld zone of AZ91D. It was noted that the corrosion attack in all of the welds was greater at the ends than at the midpoint.

For AZ91D welded to AM50A there is preferential a attack of AM50A as can be seen from Figure 4a.

Table III. Corrosion rates of welded AM50A, AZ91D and AM50A welded to AZ91D, ASTM B117

Alloy/sample	Corrosion rate (mg/cm^2·day)
AM50A reference	0.07 ± 0.02
AM50A welded	0.30 ± 0.06
AZ91D reference	0.040 ± 0.007
AZ91D welded	0.077 ± 0.006
AZ91D welded to AM50A	0.28

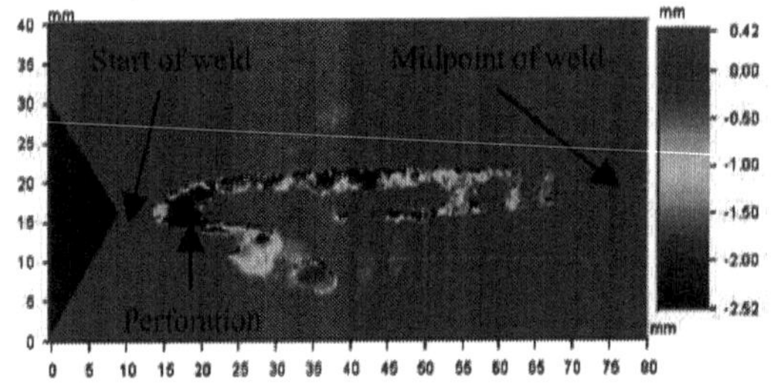

Figure 6. Corrosion attack in the weld zone of AM50A

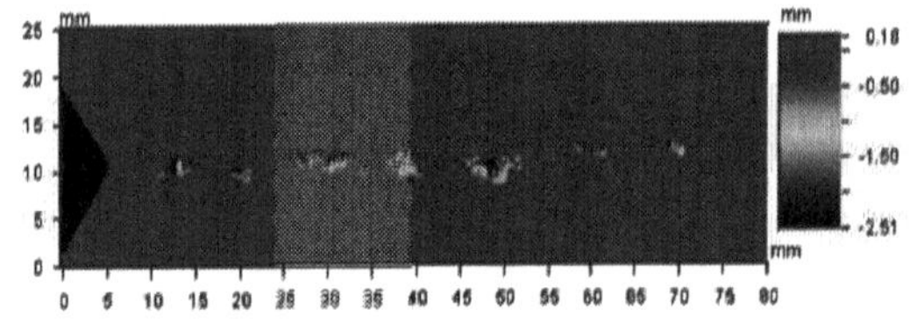

Figure 7. Corrosion attack of weld zone of AZ91D

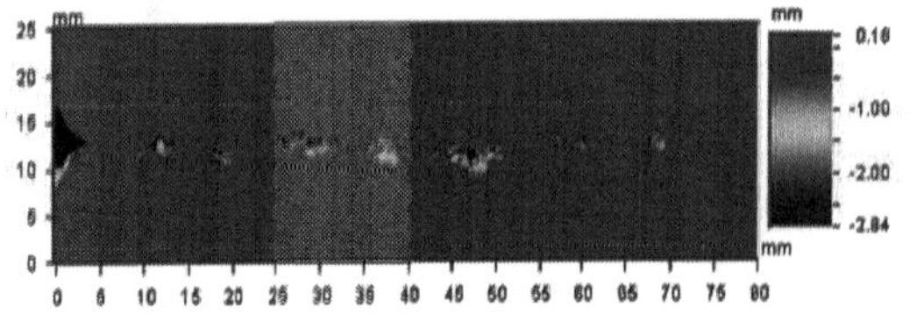

Figure 8. Corrosion attack of weld zone of AZ91D welded to AM50A

Discussion

The present results confirm that sound welds can be made on die cast magnesium using friction stir welding. The welding parameters used to produce these welds ended up at a rotation of 1100 rpm and a speed of 2 mm/s for AZ91D and 1200rpm and 3 mm/s for AM50A. When welding AZ91D to AM50A the parameters were 1200 rpm and 2.5mm/s. With respect to rotation, the present work used significantly higher rotation speed than, for example, Lockyer et al for plates of similar thickness. Their rotation speed was around 500 rpm; the weld speed, however, was similar to what was used herein; around 2 mm/s [9]. Johnson et al found that the best welds were achieved with 355 rpm and speeds varying from 2.6 – 3.7 mm/s for 6 mm thick plates. They ended up with the same parameters for all alloys, including AZ31 (355 rpm and 2.7 mm/s) [12].

FSW produces somewhat smoother welds on AZ91D than on AM50A. On AM50A there is a tendency for small surface cracks. There was also a tendency for the magnesium alloys to adhere to the tool, with AM50A being worse than AZ91D. Johnson et al reported similar findings [12].

In both AZ91D and AM50A welds, the microstructure shows equiaxed recrystallized grains almost free from precipitates at the grain boundaries. The microstructure of the weld zone resembles of a solution treated T4 structure (T4 treatment is heating to 420°C for 24 hours). Combination of the heat generated during welding and the forces used to forge the plasticized material together has contributed to the formation of the T4-like structure, even though the metal has been at elevated temperature only for some seconds. The structure is similar to what is reported by others; additionally a grain refinement in the weld was reported [9]. Examination of the specimens shown in Figures 2 and 3 did not show further grain refinement in the weld.

It should be noticed that porosity found the in die casting itself is not found in the welds. This is evident from Figure 4a, showing porosity in AM50A base metal but no porosity in the weld.

The weld metal when welding AZ91D to AM50A consists of discrete zones. The microstructure in the AZ91D zone is the same as the structure when AZ91D was welded to itself. The same is the case for AM50A.

The tensile yield strength of the welded specimens was similar to the base metal for AZ91D and AM50A. The ultimate tensile strength was reduced for both alloys, and the reduction was about 10% for AZ91D and as much as 30% for AM50A. The elongation for AM50A was significantly lower for the welded specimens; it was reduced from 9.3% to 2.4%. AZ91D being an alloy characterized with high strength and low ductility shows the same trend but to less extent. The fractures in the specimens were located in different zones. For AM50A, most of the fractures were in the weld zone, while all the other but for 1 specimen in the AZ91 samples were in the base material. For AZ91D these findings confirm those of Lockyer and Johnson [9,12]. On AM50A the elongation and UTS reported herein are lower than reported by Lockyer and Johnson on the same alloys [9,12].

Regarding the tensile properties, two preliminary comments are important:

1. The tensile specimens were machined perpendicular to the welding direction. This means that the gauge length covered the base material, the transition zone and the weld zone. As the microstructure in these three zones is different, the deformation during tensile testing is not homogenous over the extensometer length. A calculation of the elongation based on a 50 mm extensometer length will thus give a mean elongation, and not the elongation in the different zones.
2. The properties of the reference material (base material) in this investigation have been tested in a direction parallel to the weld. The properties in the plates may differ depending on testing direction, as there are some porosity bands in the castings due to the metal flow during filling. These bands are parallel to the weld, and consequently tensile measurements in this directions (but outside these bands) should be better than perpendicular to the flow direction.

Consequently, the elongation determined on the welded specimens is probably not the true elongation of the weld, and the welds may have better elongation than reported. To determine the true properties of the weld, tensile specimens must either be placed along the weld, or hourglass specimens with the weld placed in the thinnest region should be used.

In principal the microstructure found in the weld metal and the lack of porosity in the welds should improve the elongation of the welded material. The opposite is found in the present work as well as reported by other investigators [9,12]. This may partially be explained by the previous comments; the reported elongation is an average over 50mm, and the weld metal is tested transversely and the base metal longitudinally to the metal flow during die casting. For AZ91D these factors may explain the difference found between weld and base metal with respect to elongation. For AM50A, however, another important factor is the surface defects in the weld seen in Figure 3e. Most likely these surface cracks act as notches during tensile testing and cause the reduced elongation and thus the fracture mainly occurs in the weld metal of AM50A.

The AZ91D to AM50A welded specimens show tensile properties that are about the average of welds made of the individual alloys, except that the elongation is slightly higher. The fracture is in AM50A, which is natural since this is the weakest alloy. The elongation of AM50A base metal may therefore also govern the elongation on the welded specimens, giving slightly higher elongation of the specimens taken from the welds of dissimilar alloys compared to those taken from welds on individual alloys.

The corrosion properties of the weld material were poor, and higher corrosion rates were found for AM50A than for AZ91D. The AZ91D to AM50A welds showed rates comparable to AM50A alone. The welds were most likely contaminated with iron from wearing of the tool. Figure 9 shows the tool before and after welding. Attempts were made to analyze the composition of the weld with Emission Spectroscopy in order to check the iron content. These analyses did not detect significantly higher iron content in the weld than in the base metal. Thus, it has not been confirmed that the welds are contaminated with iron. The analyses are local, and it is difficult to hit the spots that eventually are contaminated with iron. Therefore the analyses are not conclusive. However, the tool was worn and the wearing can be due to the fact that the temperature locally has been above 430°C. At this temperature the β-phase of the magnesium alloy starts to melt, i.e., the metal is in the two-phase region of the phase diagram. The β-phase is aggressive to steel and iron is easily dissolved. Thus, it is very likely that the iron from the tool enters the weld zone of magnesium, and that this is the primary reason for the poor corrosion resistance.

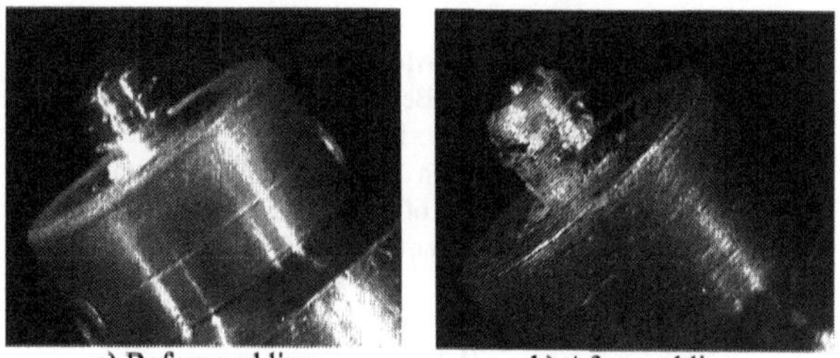

a) Before welding b) After welding

Figure 9. The tool used for welding

A large amount of material was deposited on the tool after it was retracted from the joint line. The wearing of the tool is seen only after the tool is cleaned. Other authors have not reported wearing of the tool, but since wearing is only observed after cleaning, it cannot be excluded that tool wearing also happened in other investigation. The wearing may be a consequence of the high rotation speed used in this work. Most likely this generates more heat and increases the temperature locally. For friction stir welding of magnesium alloys, the temperature should be kept below 430°C to avoid partially melting the β-phase. If this is not practical, other tool material than steel or coated steel tool is recommended. This may also solve the problem with magnesium alloys sticking to the tool.

For commercial use of FSW on magnesium there is a need for improvement both with respect to tensile properties and corrosion properties. The AM-alloys are usually selected because of their high ductility. The ductility of AM50A is reduced due to FSW, most likely caused by defects in the weld surface. Hence, more careful selection of welding parameters or improved tool design is needed in order to maintain the ductility of AM-alloys after welding. Otherwise, welds must be located in positions with low stresses. AZ91D is for applications requiring high strength, and the ductility is of less importance. For components exposed to a non-corrosive environment, the properties of welded AZ91D reported herein may be acceptable. The corrosion properties of the welds must be improved for friction stir welded magnesium to be used in a corrosive environment. The corrosion issue is independent of the magnesium alloys to be welded.

Summary and conclusions

Friction stir welding has successfully been done on AZ91D and AM50A, as well as welding the two alloys together. The welds have a recrystalized equiaxed structure with the β-phase dissolved into the α-grains. This structure should have a potential of giving improved ductility but possibly reduced strength. This is not found, most likely due to surface defects of the weld in the case of AM50A and defects in the base metal in the case of AZ91D. By optimizing the weld parameters and tool design, surface defects may be reduced or eliminated. This would be important for welding the ductile AM-alloys. Another drawback is the poor corrosion properties of the welds. Most likely this is due to wearing of the tool, which gives iron contamination of the welds locally. Partially melting of the β-phase that easily dissolves iron, may cause the wearing. Again optimized welding parameters and tool design may help, but selecting either another tool material than steel or coated steel tool may be necessary to give corrosion resistant weld material.

Acknowledgment

The authors want to thank Jan Bersaas and Svein Arve Halvorsen at R&D Materials Technology, Karmøy for doing the welding. We acknowledge Per Gjertsen and Arve Wilhelmsen for the tensile testing and preparation of the samples for metallographic studies. Eva Thorjussen is acknowledged for doing the corrosion testing.

References

1. C. Lehner, G. Reinhart, L. Schaller, "Welding of die casted magnesium alloys for production Machines", *J. of Laser Application,*11, (5), 1999, 206-210.
2. A. Munitz, C. Cottler, A. Stern, G. Kohn, "Mechanical properties and microstructure of gas tungsten arc welded magnesium AZ91D plates", *Materials Sci.&Eng* A302 (2001), 68-73.
3. A. Munitz, C. Cottler, H. Shaham, G. Kohn, "Electron Beam Welding of Magnesium AZ91D Plates", *Welding Journal,* 79 (7), 202-208.
4. G. Aichele, "New developments in welding technology for light metals, part II", *Aluminium,* 77, (7/8), 2001, 575-584.
5. H. Zhao, T. Debroy, "Pore Formation during Laser Beam Welding of Die-cast Magnesium Alloys AM60B – Mechanism and Remedy", *Welding Journal,* 2001, August, 204-s – 210-s.
6. K. G. Watkins, "Laser Welding of Magnesium Alloys", *Magnesium Technology 2003,* Ed. H. Kaplan, (TMS Annual Meeting, San Diego, CA, USA, March 2-6, 2003), 153-156.
7. W. M. Thomas et al "Friction Stir Butt Welding", GB Patent Application 9125978.8, December 1991, (1991).
8. T. Nagasawa, M. Otsuka, T. Yokota, T. Ueki, "Structure and mechanical properties of friction stir weld joints of magnesium alloy AZ31", *Magnesium Technology 2000*, Ed. H. I. Kaplan, J. Hryn, B. Clow. (TMS Annual Meeting, Nashville, Tn, USA, March 12-16), 383-387.
9. S. A. Lockyer, M. J. Russel, "Mechanical Properties of Friction Stir Welding in Magnesium Alloys", *Magnesium Alloys and Their Applications.* Ed. K. Kainer.
10. S. W. Kallee, W. M. Thomas, E. D. Nicholas, "Friction Stir Welding of Lightweight Materials", Proceedings *Magnesium Alloys and Their Applications.* Ed. K. Kainer, Wiley-VCH Verlag GmbH, Weinheim, Germany.
11. A. Schram, C. Kettler, K. Mazac, C. Goldstein, "Riebschweissen von Magnesiumlegierungen", *Schweissen & Schneiden*, 52, (2000), Heft 6, 349-353.
12. R. Johnson, P. Threadgill, "Friction Stir Welding of Magnesium Alloys", *Magnesium Technology 2003,* Ed. H. Kaplan, (TMS Annual Meeting, San Diego, CA, USA, March 2-6, 2003), 147-152.
13. A. Stern, A. Munitz, G. Kohn, "Application of welding technologies for joining of Mg alloys: Microstructure and mechanical properties", *Magnesium Technology 2003,* Ed. H. Kaplan, (TMS Annual Meeting, San Diego, CA, USA, March 2-6, 2003), 163-167.
14. R. Johnson, "Magnesium alloys get the friction stir treatment". *The bulletin – the technical journal for industrial members of twi.* July/August 2003. TWI, Granta Park, Great Abbington, Cambridge, UK, page 16.

THE EFFECT OF PROCESS PARAMETERS AND TOOL GEOMETRY ON THERMAL FIELD DEVELOPMENT AND WELD FORMATION IN FRICTION STIR WELDING OF THE ALLOYS AZ31 AND AZ61

R. Zettler[1], A.C. Blanco[1], J.F. dos Santos[1] and S. Marya[2]

[1]GKSS Forschungszentrum, Institute for Materials Research, Max-Planck-Str. 1, Joining Technology, Geesthacht D-21502 Germany
[2]Ecole Central Nantes, Welding Joining and Forming Processes, Nantes, France

Keywords: AZ31, AZ61, FSW, Thermal Cycle, Tool Design

Abstract

An increase in the use of magnesium (Mg) in the car manufacturing industry has raised questions concerning its weldability. Friction Stir Welding (FSW) has the advantage of achieving metallic bonding below that of the melting point of the base material thus avoiding many of the metallurgical problems associated with the solidification process. The present study presents the results of a development program carried out to investigate the response of Mg alloys AZ31 and AZ61 to different FSW tool geometries and process parameters. Temperature development across the weld zone was monitored and the produced welds have been subjected to microstructural analysis and mechanical testing. Defect free welds have been produced with optimised FSW-tool and parameters. The microstructure of the welded joint resulted in similar ductility and hardness levels as compared to that of the base material. The results also demonstrated that tool geometry plays a fundamental role in the response of the investigated alloys to the FSW process.

Introduction

Friction Stir Welding (FSW) is a solid state joining technique patented in 1991 by The Welding Institute, TWI, England [1]. The process typically employs a cylindrical and non-consumable tool consisting of a shoulder and smaller diameter pin. The tool is rotated and plunged into the join line between clamped and abutting workpieces so that the shoulder is in intimate contact with the workpiece surface and the pin is buried within the through thickness of the workpieces. Friction between the tool and the workpieces generates heat causing a plastified zone to form under the tool shoulder and around the pin. The rotating tool is then traversed along the join line. The pin then stirs the locally plastified material forcing it to flow in the direction of tool rotation [2] where it is then deposited behind the pin to cool and consolidate.

FSW was initially developed to join aluminium (Al) and its alloys. The process has proven to be particularly successful, especially when it comes to the joining of Al 2xxx and 7xxx series alloys previously considered difficult if not impossible to weld by conventional fusion welding processes. Furthermore, FSW has already been applied to and proven capable of joining a number of different materials; these include magnesium, copper, titanium and steel.

An increasing interest by the transportation industry for lightweight design as well as weight reduction to gain greater fuel efficiency has driven research to develop new high strength and lightweight materials, particularly Mg and its alloys. With the further development of Mg by the automobile industry as an alternative to both steel and aluminium the search is now on to find cost efficient and reliable ways of joining this material. Previous studies [3-5] have demonstrated that re-solidification of both Mg cast and wrought alloys after Gas-Tungsten Arc Welding (GTAW), Laser Beam Welding (LBW) and Electron Beam Welding (EBW) produce a microstructure formed of two phases: α-Mg and β-$Mg_{17}Al_{12}$. In the Mg wrought alloy AZ31, β compounds resolidify at the grain boundaries forming a continuous brittle network, which reduce considerably the mechanical properties of this alloy. High heat inputs (HI) characteristic of the GTAW process can also lead to expansion of pre-existing pores in the base material during welding further diminishing the strength of the alloy. Lower HI and higher cooling rates in LBW and EBW processes give rise to smaller heat affected zones (HAZ), finer grain structures and less porosity; however, solidification cracking is only prevented when small weld puddles can be produced or as in some instances with the addition of specific filler metals.

FSW has the advantage over conventional fusion welding techniques in that it is a solid state welding process whereby no bulk melting of the parent material is shown to occur. In addition, the joining process requires no additional filler material or shielding gas when it comes to the joining of Al and Mg alloys. Due to its lower heat input, evaluated in the present study in the form of temperature measurements in relation to a "Weld Pitch" (*WP = Rotational Speed (RS) / Transverse Speed (TS)*) ratio, FSW of Mg alloy AZ31 typically eliminates problems associated with trapped gases [6]. The lower heat input also contributes to a reduction in the formation of the β phase ($Mg_{17}Al_{12}$) [7]; this intermetallic compound being reported to be one of the key reasons for the brittleness observed in Mg welds [8].

Under the influence of the FSW tool and processing parameters, a fine recrystallized microstructure is formed in the weld zone [9]. This microstructure consisting typically of much finer and eqiaxed grains compared to that of the base material promotes strength levels in the stir zone comparable to those of the base material. Microstructural observations report more uniform distributions of grain size and dislocation density in the region of the weld zone; such microstructures also contribute to a more homogeneous hardness profile observed when FSW Mg alloys [8,10].

As it is the action of the FSW tool in conjunction with welding parameters that plays the fundamental role in the production of the thermal input and plastic deformation it is of key importance to better understand how the tool contributes to the eventual microstructure and weld properties generated during processing. This study forms part of a much larger work involving the FSW of cast and wrought Mg alloys. As such it presents preliminary results concerning the FSW of two wrought Mg alloys AZ31 and AZ61 as welded using two different tool profiles.

Experimental Procedure

Friction stir butt welds were produced in extruded magnesium alloys AZ31 and AZ61 (**Table 1**) having a plate thickness of 2.5mm. Welding was performed at the GKSS-Forschungszentrum Geesthacht, Germany using a *Tricept®* 805 Robot. Further details concerning the robot and FSW unit can be found in the literature [10,11]. All workpiece dimensions measured 340mm in length and 100mm in width. Prior to FSW each workpiece was lightly sanded then cleaned with acetone in the region of the join.

Friction stir welds were produced using 2 shoulder profiles in conjunction with one pin profile, **Table 2**. The tool pin, classified as pin "019", measured 5mm in diameter and contained a thread for the length of pin plunged into the workpieces. The two welding tool shoulders classified as "017" and "Sc" were made of the same material as that used to produce the tool pin. Both shoulders measured 13mm in diameter. Shoulder "017" possessed a concave profile and was tilted away from the direction of welding at an angle of 1.5 degrees. Shoulder "Sc" possessed a spiral or scroll profile and was tilted between 0 and up to 1 degree away from the direction of welding.

All workpieces were clamped against the work-table of the robot onto a steel backing bar. The clamping system further utilised a vacuum table. The generated vacuum avoided the necessity of engaging mechanical fixtures although side clamps were used during FSW to avoid workpiece separation. All friction stir welds were performed in the direction of plate extrusion.

Three energy or heat inputs were investigated, **Table 2.** The energy inputs hereafter classified in terms of a weld pitch (WP) ratio (revs/mm) for both AZ31 and AZ61 alloys were: hot (WP = 7.33), cold (WP = 4.66) and intermediate (WP = 6). Axial force was held constant at 5.5 kN for all welds.

FSW temperatures were measured in response to the HI. Measurements were conducted using 0.5mm diameter k-type thermocouples, which were mounted in the workpieces at either side of the join (**Figure 1a**). Furthermore, additional thermocouples were placed across the width of the backing bar 0.6mm below the surface (i.e. interface between the workpieces and backing bar), **Figure 1b**. All thermocouples were inserted and secured to the bottom of each hole by taping them to the workpieces and or backing bar. Thermal conducting paste was packed into all holes to fix the thermocouples. Weld temperatures were then recorded using a National Instruments SCXI-1000 amplifier and Labview software. Temperatures were sampled at 20 Hz.

Macro and micrographs were produced of the base materials and of each weld transverse to the weld travel direction using standard metallographic techniques. Welds were examined using a Leica QW550 Optical Light Microscope equipped with Qwin software.

Mechanical testing of the friction stir welds consisted of bend tests for both the face and root of each weld. Microhardness tests were performed using a Shimadzu HMV-200 machine where a load of 1.96 N over five seconds was applied for each indentation. Indentations were conducted at workpiece mid-thickness at a spacing of 0.5mm over a length of approximately 15mm either side of the weld join line.

Results and Discussion

The surface quality resulting from the different tool shoulders and process parameters is presented in **Figure 2**. It was observed that WP=6 initially produced surface defects in AZ31 alloy (for both tools) indicating some thermal instability at the beginning of the weld. Weld flash formation was in general not significant and mainly formed when approaching the end of the weld. An inability at the end of the weld to dissipate heat formed during welding typically led to a temperature increase at the end of each weld and thus contributed to the formation of the weld flash. In general tool Sc-019 produced smoother weld surfaces when compared to tool 017-019.

Macrographs of the welds produced in the Mg alloys AZ31 and AZ61 can be found in **Figure 3a**. The effect of the different shoulders (concave and scroll) on the thermal input and hence on the weld formation is clearly evident. The welds produced with the concave shoulder (017-019) are characterised by a deep flow arm running directly into the stir zone. Counterpart welds produced using the scroll shoulder have a much reduced shoulder flow zone as witnessed by the thin flow arm above the stir zone. Some welds demonstrated running voids (defects in and around the stir zone) and/or lack of penetration (LOP) defects, the latter caused in general by an incorrect adjustment of the pin length before welding. LOP defects however can also be attributed to a lack of material flow and/or due to the relatively low processing temperatures developed during these welds. For both alloys most of the defective welds were produced for WP≥6, regardless of the tool used. No conclusive indication could be found when relating the results from the temperature measurements and the presence of defects. If anything it could be argued that the macrographs of well formed nuggets, which in the same way present better surface quality (see **Figure 2**) were performed under tool/WP combinations allowing for higher temperatures. Nevertheless, it seems that further investigation into material flow as a function of tool geometry and process parameters are required in order to fully understand weld formation when FSW Mg and its alloys.

Figure 3b and **Figure 3c** present micrographs of the stir zone and thermo-mechanically affected zone in both alloys where the effects of thermal history and plastic deformation on the resulting microstructure were recorded.

Although some fluctuation in temperature was observed to occur along the length of the workpieces, measurements revealed a quasi-steady state thermal cycle for all specimens welded in this study. **Figures 4** illustrates the variation in peak temperature as measured along the weld length (measured in the workpiece surface, 10mm away form the join line) for both tools and when FSW the AZ31 and AZ61 alloys, respectively. Workpiece temperature profiles for the AZ31 alloy indicated that thermal stability could be enhanced when WP≥6, i.e. increasing revs/mm of weld length above 6.

Figure 5 presents a typical temperature record obtained during the FSW of AZ31 for WP=7.33 with tool "017-019". Included here are the measurements as recorded in the workpiece surface and backing bar. Temperatures as measured in the workpiece and backing bar revealed that a much more accurate assessment of peak processing temperature could be obtained from thermocouples embedded in the latter, (see **Figure 5**). The variation in peak temperature recorded at the backing bar is due to the different measuring sites in relation to the weld centerline (see **Figure 1b**). Thermal cycles recorded in the workpieces also indicated that temperatures in the retreating side are slightly higher than in the advancing side.

Both Mg alloys demonstrated that for the same WP ratios the scroll shoulder (indicated as "Sc") consistently produced the lower welding temperatures, **Figures 6 and 7**, respectively. In the case of the alloy AZ31 (**Figure 6**), both tools demonstrated similar temperature trends (this was more accentuated for tool "017-019"), however both indicated a lower peak temperature for WP=6. A similar pattern (i.e. lower peak temperature for WP=6) was also observed from the temperature measurements made in the backing bar. This decrease in processing temperature as observed for the AZ31 alloy with regard to WP=6 was not evident for the AZ61 alloy. Temperatures recorded for the AZ61 alloys when considering all investigated tools and WP ratios (**Figure 7**), remained within a band of approximately 50° C. Temperatures could be seen to decrease for tool Sc-019 with increasing WP while temperatures increased for tool 017-019 with increasing WP ratio. In aluminium alloys, especially those of the 2xxx and 6xxx families, it has been demonstrated that regardless of the tool geometry increasing WP (i.e. reducing travel speed in relation to rotational speed thus increasing heat input) should result in higher processing temperatures [13]. It is possible that changes in the slip systems of Mg alloys as a function of temperature and hence flow behaviour at temperature during FSW might be an explanation for this phenomenon. A detailed investigation is currently in progress to better understand the observed behaviour.

A comparison of thermal histories as measured under identical tool and WP ratios for both the AZ31 and AZ61 further revealed very large temperature differences when FSW each alloy,

Figure 8. Typically processing temperatures were 40°C to 80°C lower for the AZ61 as compared to the AZ31 alloy. A possible explanation may be due to the slightly higher thermal conductivity of the AZ61 alloy in comparison to the AZ31 alloy (see **Table 1**).

Microhardness measurements conducted at mid-thickness for both the AZ31 and AZ61 friction stir welds revealed that no significant thermal softening had occurred, **Figures 9 and 10**, respectively. In the case of the alloy AZ61 the lower processing temperature and mechanical working produced when using the "Sc" shoulder contributed to weld overmatch i.e. higher hardness in the weld stir zone as compared to that of the base material hardness.

Bend test results are presented in **Figure 11**. In all cases, the bending angles achieved at the surface of the workpiece were higher than those observed in the root region. In some instances this behaviour can be explained by the presence of a LOP defect. However, low bending angles have also been observed in defect free welds, possibly indicating poor ductility in this region of the weld. The results obtained in the course of this study do not allow any conclusive explanation for this local loss in ductility. It is however possible that due to the low notch sensitivity characteristic of Mg and its alloys, any minor joint line remnant in the root region should act as a crack initiation site. Improved material flow and or proper design of backing elements would possibly alleviate this problem.

Summary

The results obtained in the course of this study allow for the following conclusions:

- Friction stir welds in Mg alloys AZ31 and AZ61 investigated in the present study achieved a quasi-steady heat transfer state.
- Defect free welds could be produced for both the AZ31 and AZ61 alloys with weld pitch lower than 6 rev/mm.
- Greater accuracy in obtaining peak processing temperatures values at the stir zone were achieved when mounting thermocouples in the backing bar directly below the FSW tool.
- The scroll shoulder (i.e. Tool "Sc-019") produced lower processing temperatures than its concave counterpart for the alloys AZ31 and AZ61 over all weld pitch ratios investigated in this study.
- In general lower peak temperatures were observed when FSW the AZ61 alloy in comparison to AZ31.
- A relationship between heat input and weld pitch ratio could not be established for the alloy AZ31. An increase in weld pitch ratio from 4.6 to 7.33 rev/mm resulted in lowest processing temperatures for the intermediate weld pitch ratio 6 rev/mm.
- Microhardness profiles showed "even-matching" welds with little thermal softening evident over all weld pitch ratios for both alloys. Overmatching occurred for the alloy AZ61 at the highest weld pitch ratio.
- Bend tests results showed consistently lower ductility at the root possibly due to minor joint line remnant in the this region.

References

1. Thomas W. M., Nicholas E. D., Needham J. C., Church M. G., Templesmith P., Dawes C. J.: International Patent Application N° PCT/GB92/02203 and GB Patent Application N° 9125978.9 (1991). U. S. Patent N° 5,460,317 (October-1995).
2. London B, Mahoney M, Bingel W, Calabrese M, Bossi RH and Waldron D, "Material Flow In Friction Stir Welding Monitoredwith Al-SiC and Al-W Composite Marker". Friction Stir Welding And Processing II, Proceedings of Symposia 2003 TMS Annual Meeting San Diego, California March 2-6, 2003.
3. Stern A, Munitz A and Kohn G, "Application Of Welding Technologies For Joining Of Mg Alloys: Microstructure And Mechanical Properties". Magnesium Technology 2003, Proceedings of Symposium 2003 TMS Annual Meeting San Diego, California March 2-6, 2003.
4. Lathabai S, Barton KJ, Harris D, Lloyd PD, Viano DM and McLean A, "Welding And Weldability Of AZ31B By Gas Tungsten Arc And Laser Beam Welding Processes". Proceedings of Symposia 2003 TMS Annual Meeting San Diego, California March 2-6, 2003.
5. Zhao H and DeBroy T, "Pore Formation During Laser Beam Welding Of Die-Cast Magnesium Alloy AM60B-Mechanism And Remedy". www.aws.org.
6. Johnson R and Threadgill P, "Friction stir welding of magnesium alloys": TWI Ltd, Cambridge, UK. Proceedings of Symposia 2003 TMS Annual Meeting San Diego, California March 2-6, 2003.
7. Hiroshi Hayashi, "Microstructures And Mechanical Properties Of Friction Stir Weld Joint Of Magnesium Alloys". www.mc.mat.shibaura-it.ac.jp
8. Nagasawa T, Otsuka M, Yokota T and Ueki T "Structure And Mechanical Properties Of Friction Stir Weld Joints Of Magnesium Alloy AZ31. Proceedings of Symposium 2000 TMS Annual Meeting Nashville, Tennessee, March 12-15, 2000.
9. Hwan S, Park C, Sato YS and Kokawa H ,"Effect of micro-texture on fracture location in friction stir weld of magnesium alloy AZ61 during tensile test". Scripta Materialia April 2003.
10. "Propiétés du magnésium et de ses alliages", par FranÇois Bazile. www.techniques-ingenieur.fr
11. von Strombeck A, Schilling C and dos Santos JF, "Robotic friction Stir welding – Tool technology and Applications". Proceedings of the 2nd International Symposium on Friction Stir Welding, June 26-28, 2000, Gothenburg, Sweden.
12. von Strombeck A, Loitz H and dos Santos JF, "Monitoring and Control in FSW". Procedings of the International "QUALISTIR" Seminar. Paper 9, Ocotber 16-17, 2003, Geesthacht, Germany.
13. Zettler R, Lomolino S, dos Santos JF, Donath T, Beckmann F, Lippman T, and Lohwasser D, "A Study on Material Flow in FSW of AA 2024-T351 and AA 6056-T4 Alloys". Proceedings of the 5th International Symposium on Friction Stir Welding, September14-16, 2005, Metz, France.

BASE MATERIAL	
AZ31	AZ61
G (ASTM Grain Size Number): 7.03 Average Grain Diameter (μm): 31.8	G (ASTM Grain Size Number): 6.41 Average Grain Diameter (μm): 38.96

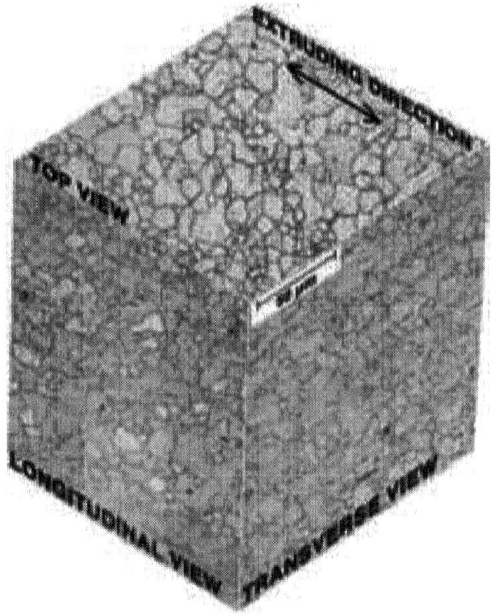

Nominal Chemical Composition (% wt.)

Alloy	Al	Zn	Mn	Si	Cu	Fe	Ni
AZ31	2.5 to3.5	0.5 to1.5	>=0.2	0.1	0.05	0.005	0.005
AZ61	5.5 to 6.5	0.5 to 1.5	>=0.15	0.1	0.05	0.005	0.005

Physical Characteristics

Alloy	Density (gr/cm^3)	Solidus Temperature (°C)	Liquidus Temperature (°C)	Thermal Conductivity at room temperature $(W/(m \cdot °K))$
AZ31	1.78	575	630	83.7
AZ61	1.8	510	610	79.5

Mechanical Properties (at room temperature)

Alloy	$Rp_{0.2}$ (Mpa)	Rm (Mpa)	Elongation (%)	Hardness (HV 0.2)	Bending (°)	Plate Thickness (mm)
AZ31	220	290	15	59.2	165	2.5
AZ61	205	305	12	69.57	161	2.5

Table 1. AZ31 and AZ61 Mg alloys base material data. The values presented for chemical composition and physical characteristics have been taken from reference [9]; the remaining data is based on experiments performed at the GKSS-Forschungszentrum.

Weld Combination AZ31-AZ31

Welding Tools		Shoulder ∅ (mm)	Pin		Tool Tilt (°)
Shoulder	Pin		∅ (mm)	Length (mm)	
017	019	13	5	2.4	1 ÷ 3
Sc	019	13	5	2.4	0

Welding Parameters	Weld Pitch (rev/mm)	Axial Force (kN)
Hot Condition	7.33	5.5
Cold Condition	4.66	5.5
Middle State	6	5.5

Weld Combination AZ61-AZ61

Welding Tools		Shoulder ∅ (mm)	Pin		Tool Tilt (°)
Shoulder	Pin		∅ (mm)	Length (mm)	
017	019	13	5	2.4	1 ÷ 3
Sc	019	13	5	2.4	0
Welding Parameters	Same as above				

Table 2. Welding tool and weld parameters data used for the FSW of Mg AZ31 and AZ61 alloys.

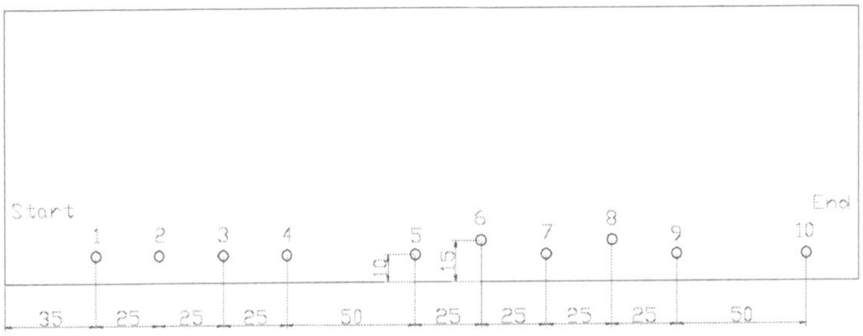

Figure 1(a). Location of the thermocouples over the surface of each workpiece.

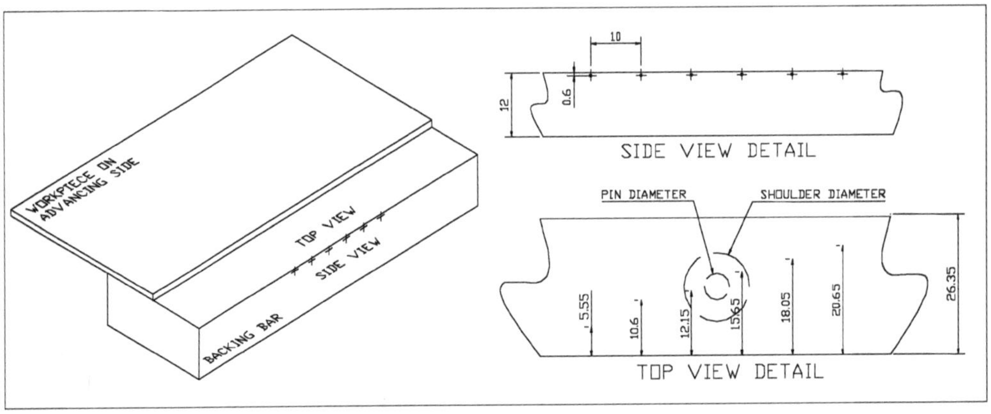

Figure 1(b). Location of the thermocouples in the backing bar.

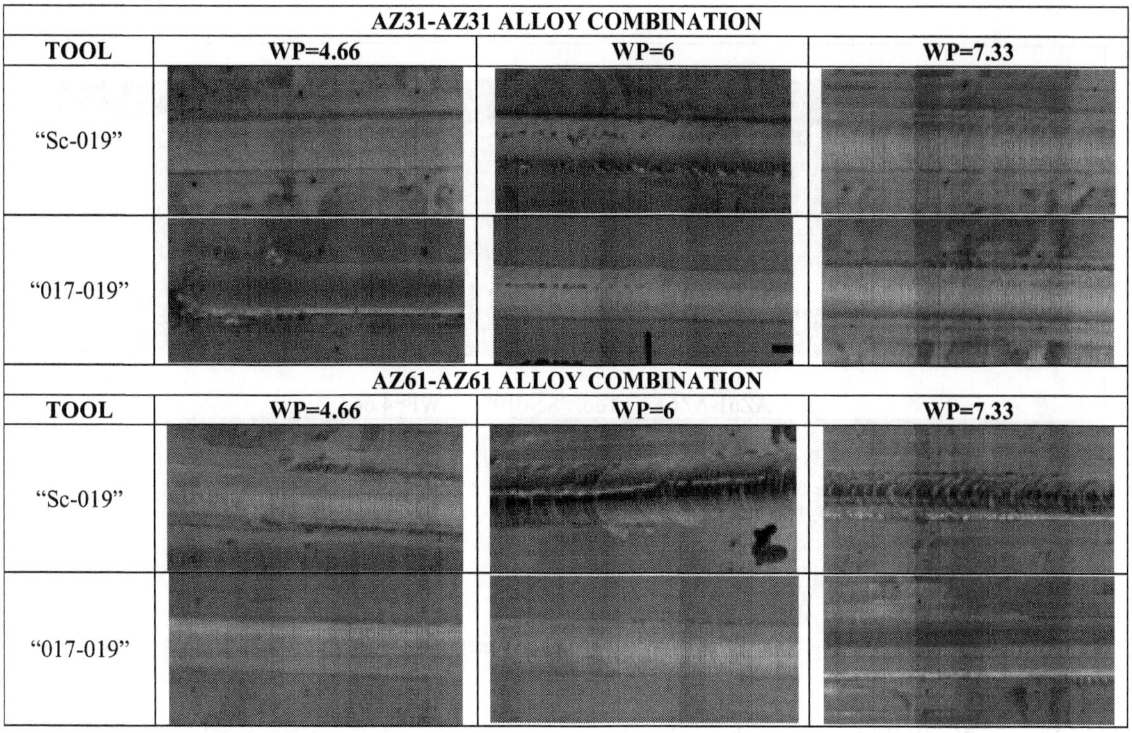

Figure 2 . Surface appearance of the all investigated welds. Flash formation mainly restricted to region near end of the weld. Some welds also suffered from surface breaking voids at the start of the weld.

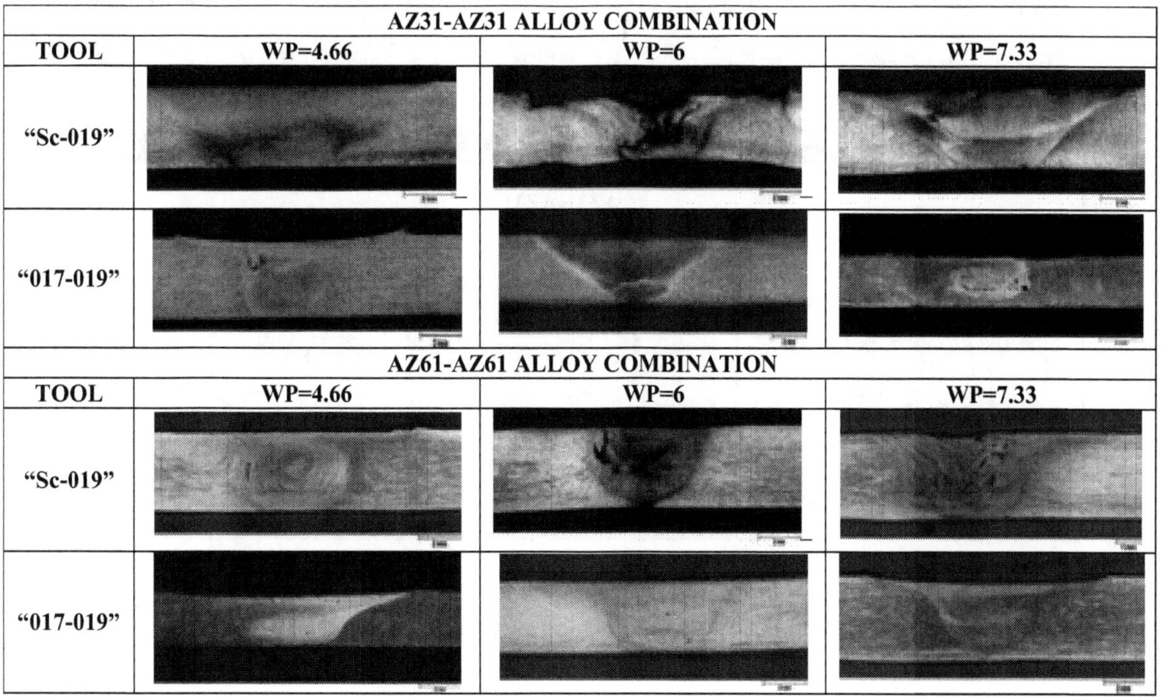

Figure 3a. Macrographs of welds produced during the experimental investigation. Note defects have been included and are characteristic of weld start only.

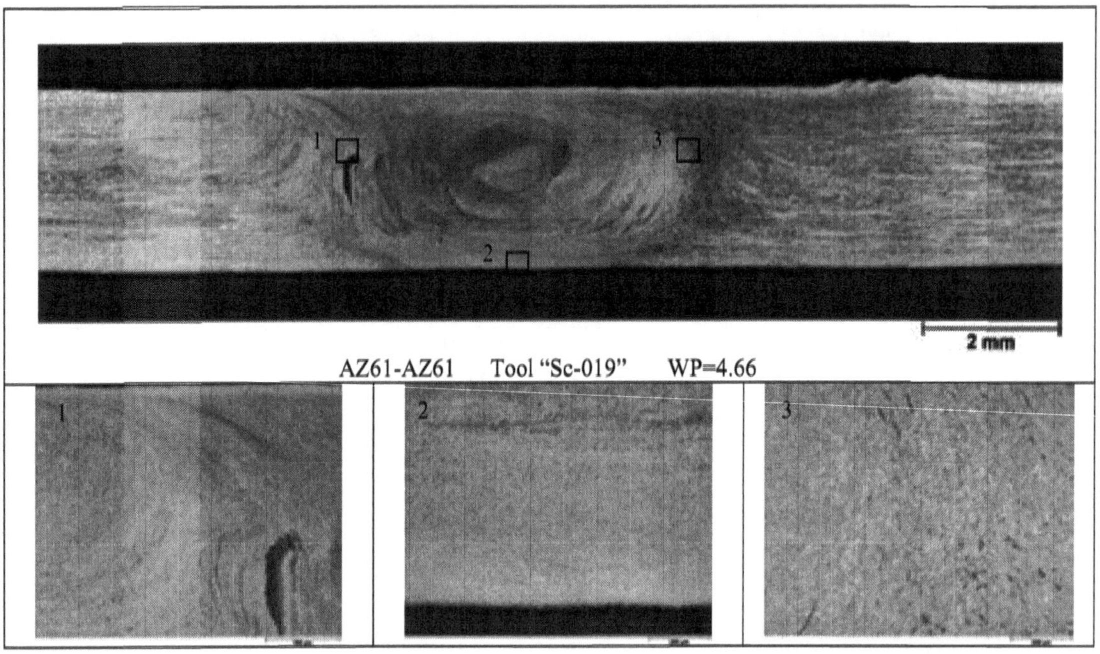

Figure 3b. Weld macrographs and micrographs for AZ61-AZ61 alloy combination. The indication near point "1" is etching artifact and not a defect.

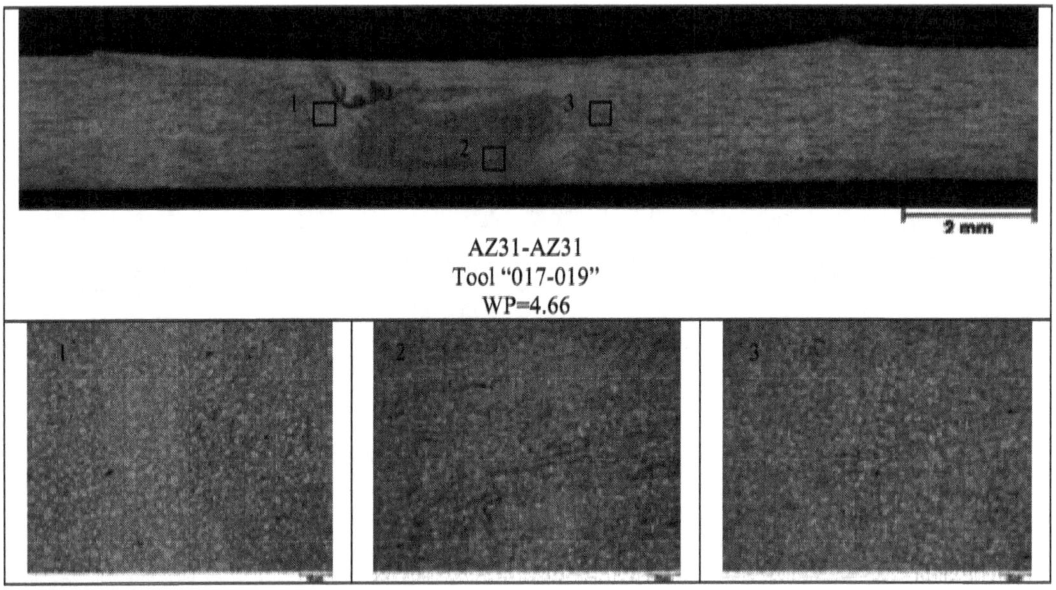

Figure 3c. Weld macrographs and micrographs for AZ61-AZ61 alloy combination. The indication near point "1" is etching artifact and not a defect.

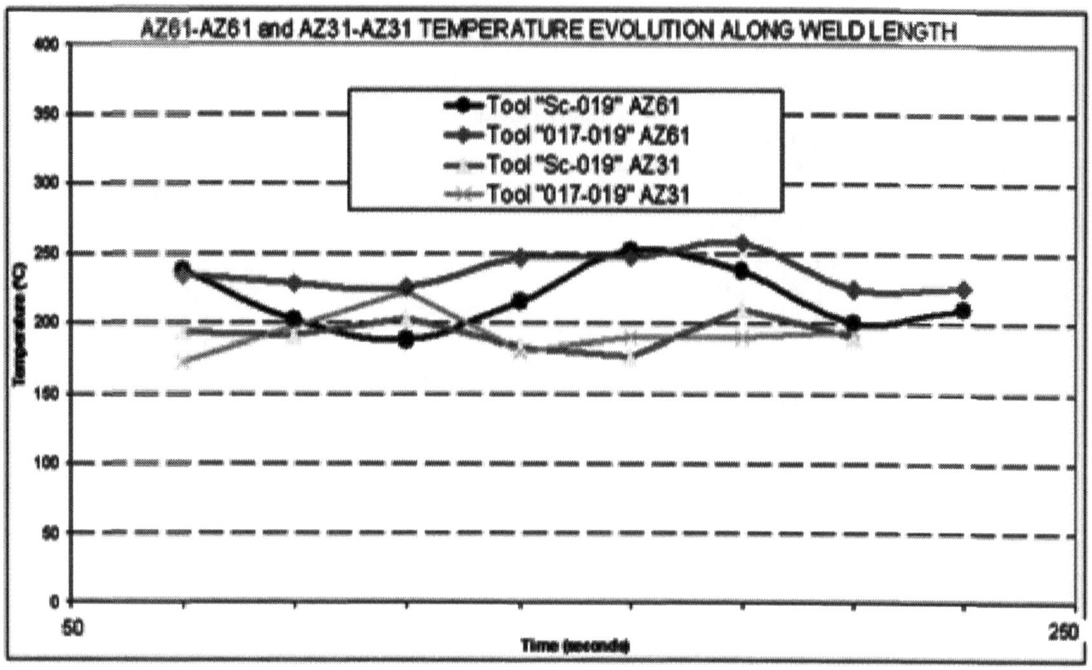

Figure 4. Peak temperatures as measured for WP=6 in workpiece and along length of weld for AZ61-AZ61 and AZ31-AZ31 alloy combination.

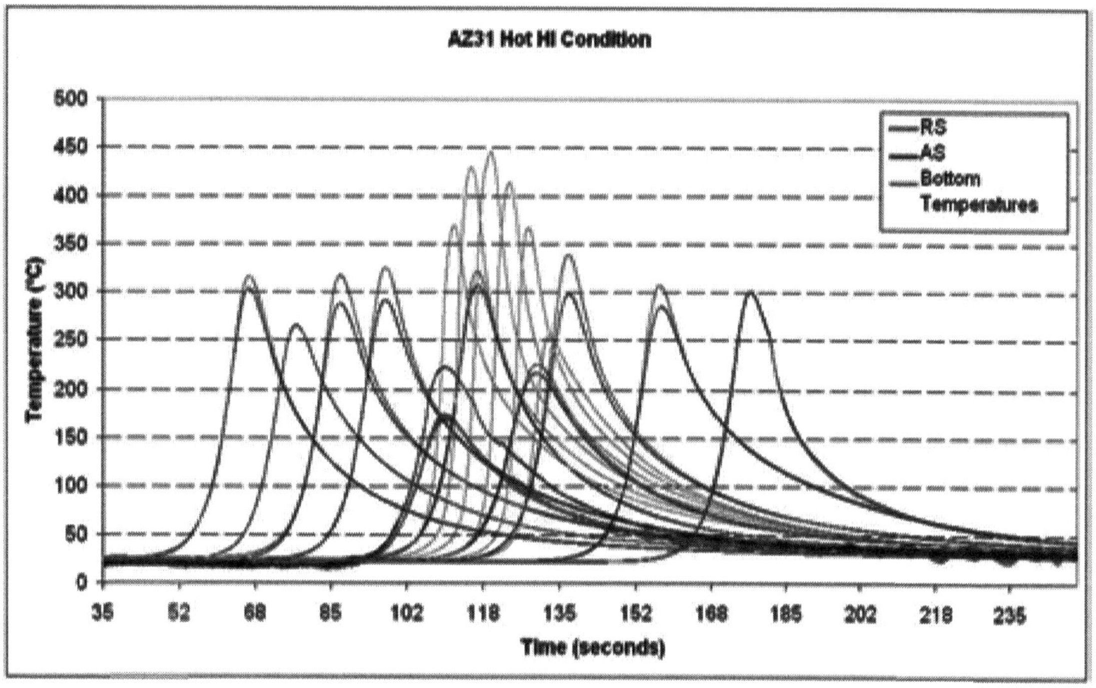

Figure 5. Comparison between temperatures as measured in the workpieces (Advancing Side and Retreating Side) and temperature registered in the backing bar.

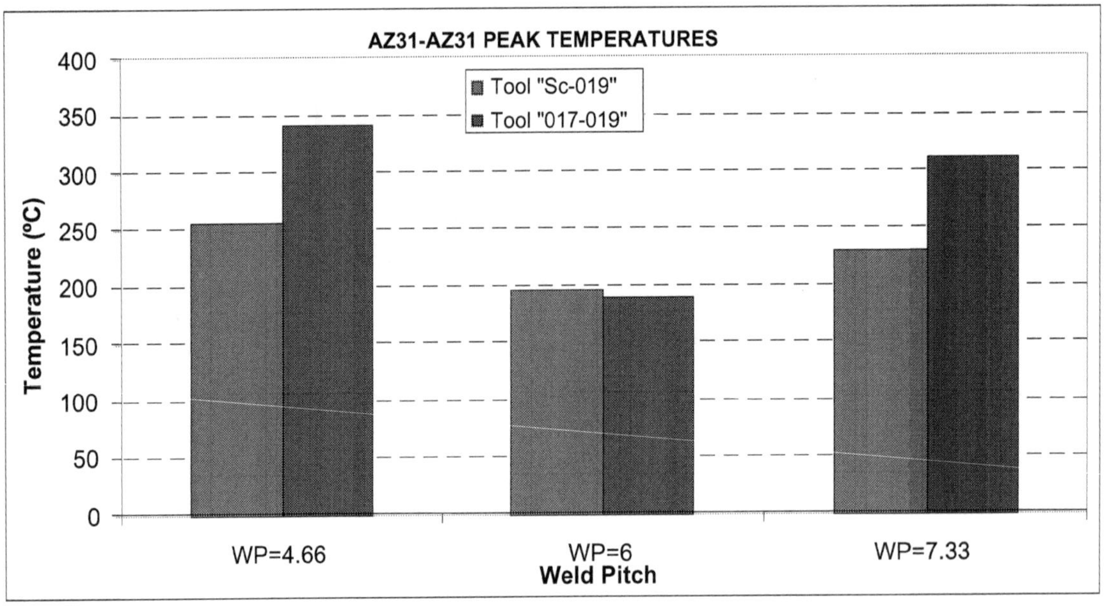

Figure 6. Peak temperatures as measured in both sides of the workpieces under the various WP ratios for the AZ31-AZ31 alloy combination.

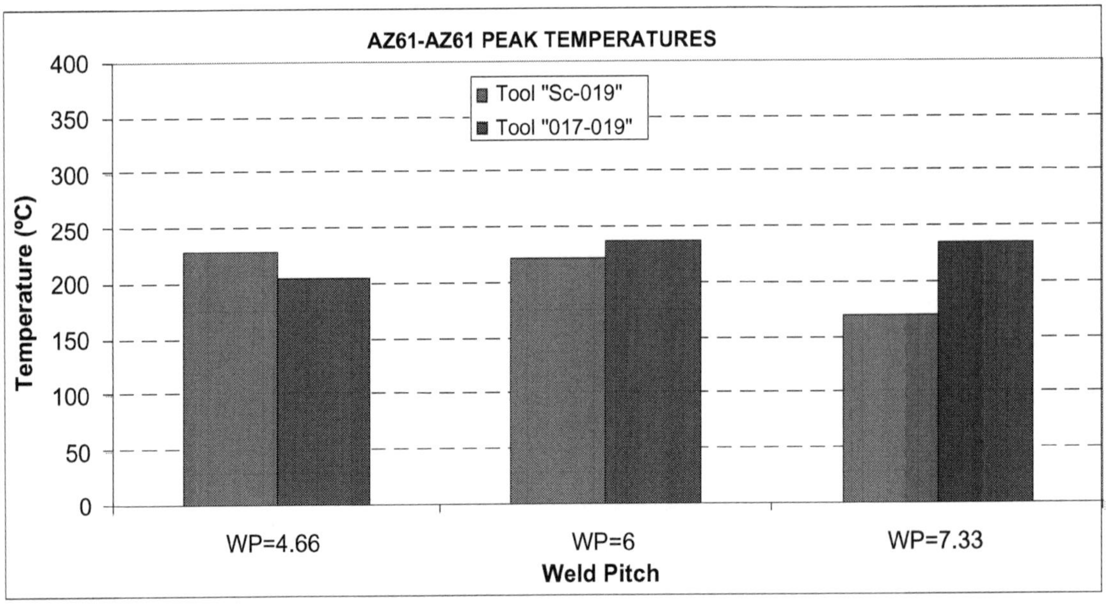

Figure 7. Peak temperatures as measured in both sides of the workpieces under the various weld pitch ratios for the AZ61-AZ61 alloy combination.

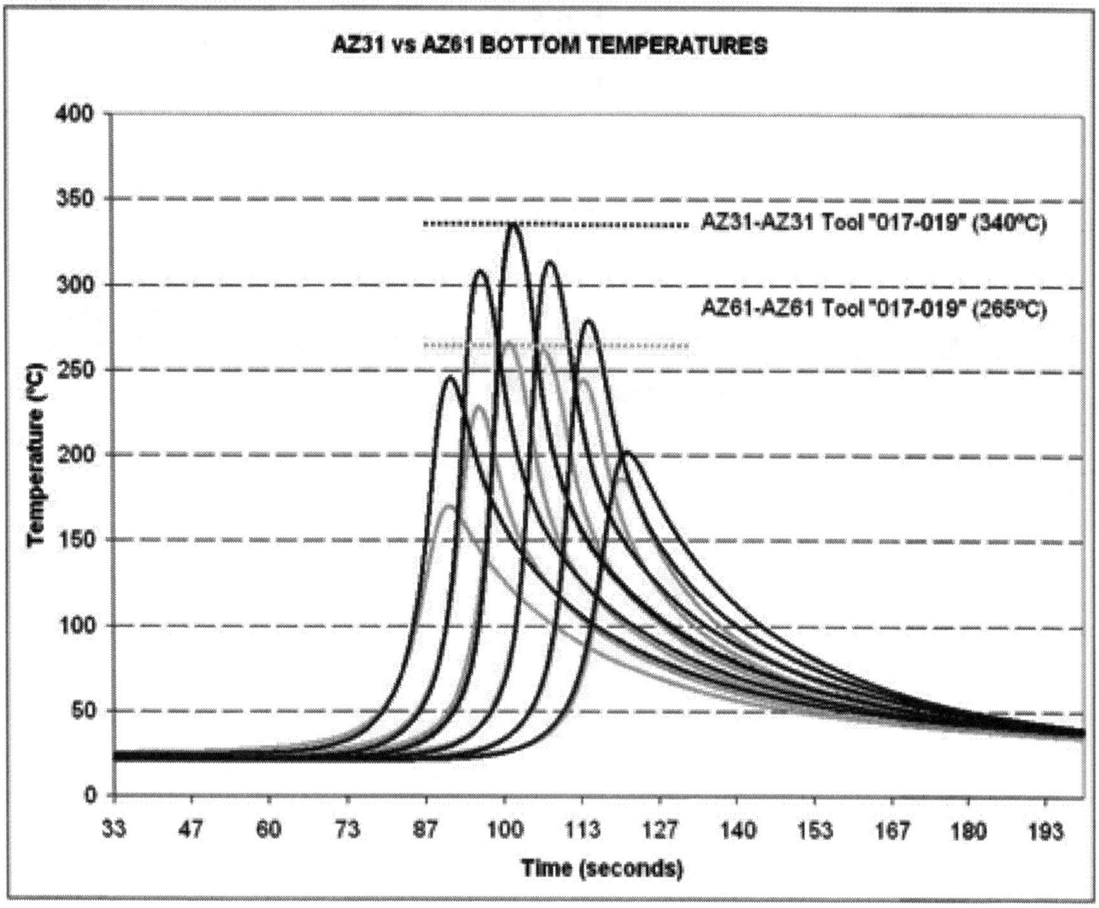

Figure 8. Temperatures as measured by thermocouples placed in the backing bar comparing the thermal response between both AZ31 and AZ61 for WP=6 when FSW using tool 017-019.

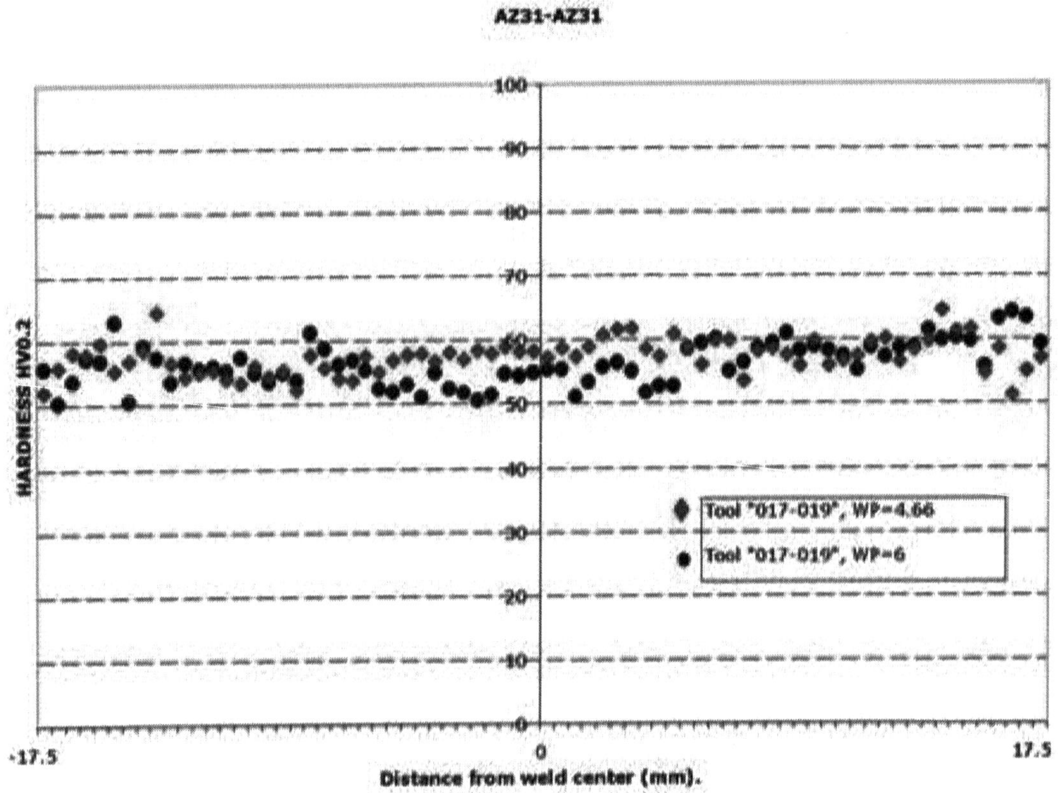

Figure 9. Microhardness tests carried out on the friction stir welded AZ31-AZ31 alloy combinations.

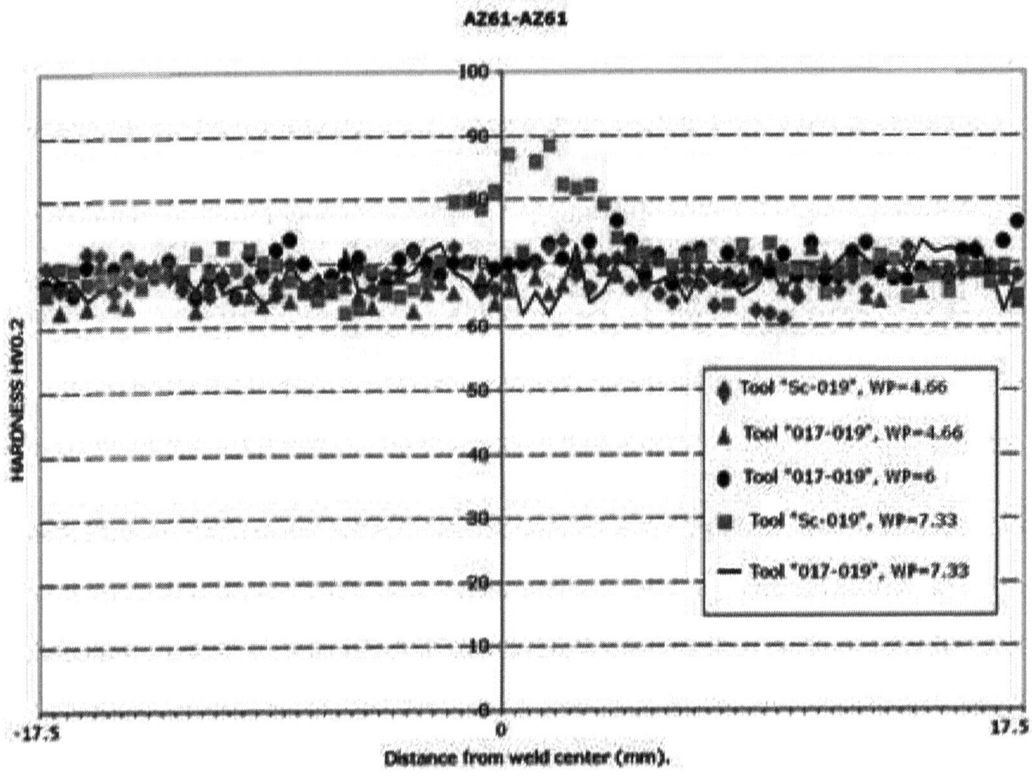

Figure 10. Microhardness tests carried out on the friction stir welded AZ61-AZ61 alloy combinations.

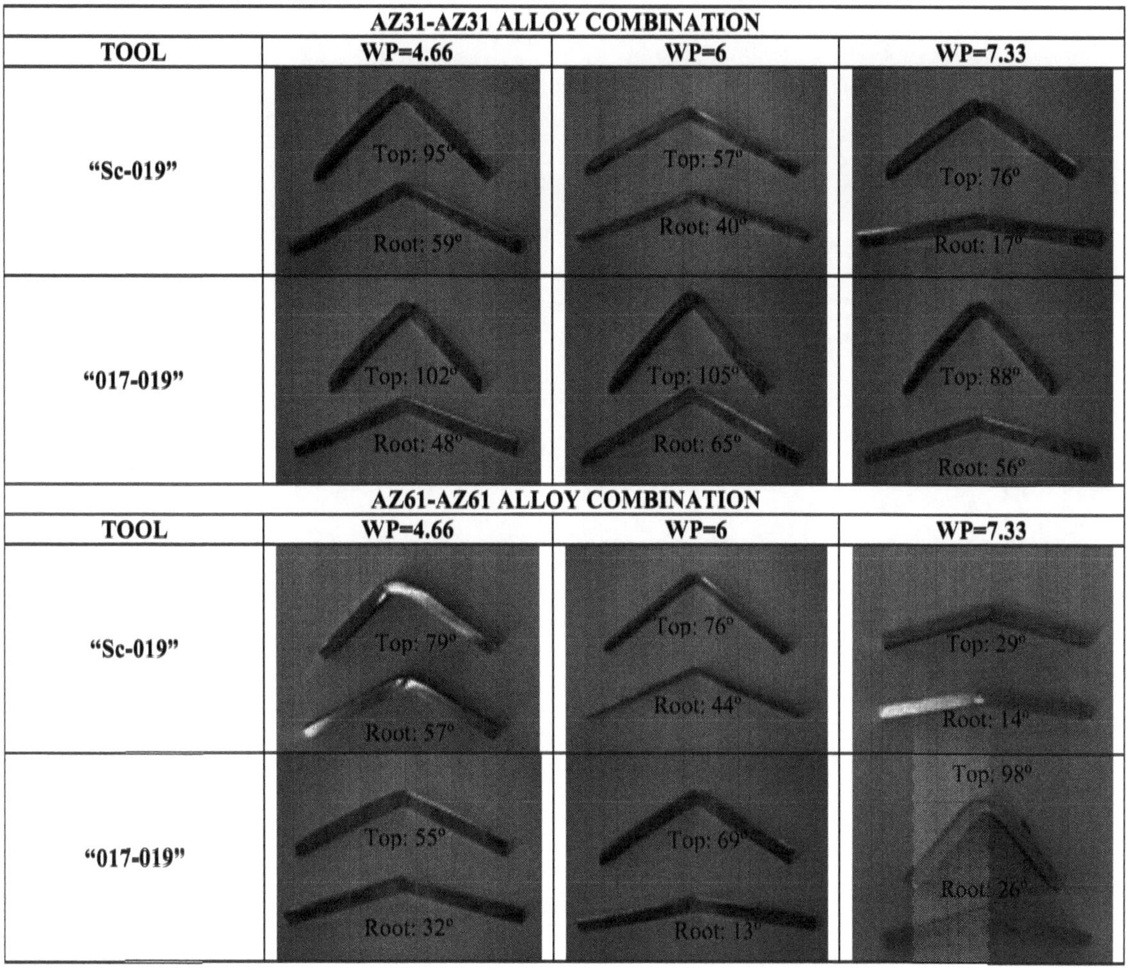

Figure 11. Results of bend tests for all alloy and WP ratios.

From *Magnesium Technology 2007*, R.S. Beals, A.A. Luo, N.R. Neelameggham, and M.O. Pekguleryuz, Editors

FRICTION STIR WELDING OF MAGNESIUM OIL PAN

F. Hunt[1], Q. Yang[1], H. Badarinarayan[1], K. Okamoto[1] and D. Platt[2]
[1] Hitachi America, Ltd. Automotive Products Research Lab.
34500 Grand River Avenue, Farmington Hills, MI 48335, USA
[2] The University of Michigan, Ann Arbor, MI USA.

Keywords: Friction Stir Welding, Magnesium, Automotive Components, Oil Pan

Abstract

In this research, the feasibility of FSW dissimilar magnesium alloys was investigated. Specifically, die cast magnesium MRI-153M and AZ31 (plate) were butt-welded using the FSW technique. The effects of weld parameters (tool rotation (rpm) and weld speed) on the weld quality have been quantified in terms of hardness profile across the weld as well as tensile strength. Weld surface and cross section were also analyzed. Tensile testing showed that the welded material failed in the base material (MRI-153M) and not in the weld region. Weld cross section analysis showed void free welds were achieved at weld speeds up to 500 mm/min. Furthermore, a smooth and glossy surface ripple appeared on the weld beads and only a small amount of burr was observed. The Mg alloys were then welded together in an experimental oil pan cast in the USCAR-Magnesium Powertrain Cast Component Project. Finally, the oil pan assembly was successfully water bath-leak tested in water bath at 82.7KPa (12 psi) pressure for 3 minutes to validate the applicability of FSW.

Introduction

Magnesium is a light metal with a density 2/3rds that of aluminum or about 1.8g/cm^3. Even though Mg is a light-weight material it's specific strength is the highest of any structural metal. In recent years, magnesium usage in automotive industry has increased because of the benefit of vehicle weight reduction on improved fuel economy. However, magnesium is an active metal that oxidizes relatively easily at high temperature. As a result, inert gas shielding is required to prevent oxidation during fusion welding.

Friction stir welding (FSW) is a solid state joining process done at an operating temperature below the melting temperature of the base material. FSW was invented and patented by TWI in 1991[1] An illustration of this joining technique is shown in **Figure 1**. For this process a cylindrical tool with a threaded pin and shoulder is rotated and plunged into metallic material and fed to form a linear weld. The material is heated by friction from the rotating shoulder. The materials on each side of the weld are mixed together by the stirring action of the tool pin. The forward rotating side of the tool is called the advancing side (AS) and the material on the opposite side is called the retreating side (RS) of the weld. As will be explained later, AS and RS materials can be like (or similar) or dissimilar.

FSW has many advantages including: very low distortion, no fumes, porosity or spatter, no consumables (no filler wire), no special surface treatment and no shielding gas requirements. FSW joints have improved mechanical properties and are free from porosity or blowholes compared to conventionally welded materials.

While FSW has been applied to a wide variety of applications[2] for aluminum alloys, fewer coupon welds have been attempted on magnesium alloys[3-5]. Magnesium alloys (like aluminum) have a relatively low melting point (about 650°C) that makes it an ideal material for FSW. In this report, die cast magnesium MRI-153M and AZ31 were FSW. MRI-153M, developed by Dead Sea Magnesium Ltd, is a creep- and corrosion-resistant Mg alloy for die-casting applications. Joint properties such as: hardness, tensile strength, macrostructure and microstructure were investigated. Finally, FSW was applied to a prototype oil pan in support of the USCAR-MPCC project[6]. The goal of this research was to optimize the welding conditions to obtain strong, leak-free welds with minimal burr formation.

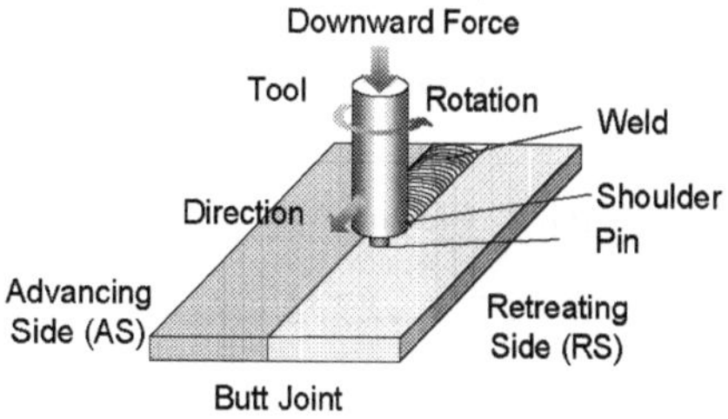

Figure 1. FSW Butt Joint

Experiment

Materials and Welding

AZ31 sheet metal and die-cast MRI-153M were FS welded to each other to evaluate the quality of the joint. All of the welding trials were carried out on a CNC-controlled, 3-dimensional FSW machine (HITACHI GR-3DC5T) with 11kW motor shown in **Figure 2**. The tool, made of tool steel, was tilted 3deg. from vertical to facilitate welding. The tool used in this part of the study had an 8mm shoulder and a 3.0mm diameter pin that was 2.7mm long. The weld plates were tightly clamped during the welding process. The spindle speed was kept constant during welding and the rotating tool pin was plunged into the plate until the tool shoulder came into contact with the surface of the material. This generated the frictional heat to locally soften the material around the tool pin. Then the tool was traversed the desired distance and then retracted to complete the welds.

Figure 2. 3D FSW Machine with 5 Axes

Coupon Welding

Initial FSW was conducted at a tool feed rate ranging from 100-500mm/min with two different tool rotating speeds of 2,000 and 3,000rpms. The plates of AZ31 sheet metal and die cast MRI-153M were butt welded after the edges were machined down to 2.6 mm thick.

The objective of the coupon testing was to determine the optimal FSW conditions, feed rate and rpm, for butt-welding these dissimilar magnesium alloys. In addition too feed rate and spindle speed, the material orientation with respect to the rotating axis of the tool was also varied. In other words, welding was carried out by changing the material on the advancing and retreating sides of the weld joint. This provided a total of twenty combinations of weld parameters to be tested, and resulted in the test matrix shown in **Table 1**.

Table 1. Coupon Weld Matrix

Weld #	Feed Rate (mm/min)	Spindle Speed (rpm)	Advancing Side Material
1	100	3000	MRI 153
2	200		
3	300		
4	400		
5	500		
6	100	2000	MRI 153
7	200		
8	300		
9	400		
10	500		
11	100	3000	AZ 31
12	200		
13	300		
14	400		
15	500		
16	100	2000	AZ 31
17	200		
18	300		
19	400		
20	500		

Results and Discussion

Weld Appearance

Photos of the welds were taken in order to record surface finish and overall weld appearance. **Figure 3** shows the welding conditions that produced some of the best weld surfaces. A smooth and glossy surface ripple appears on the weld beads and only a small amount of burr is observed on the weld.

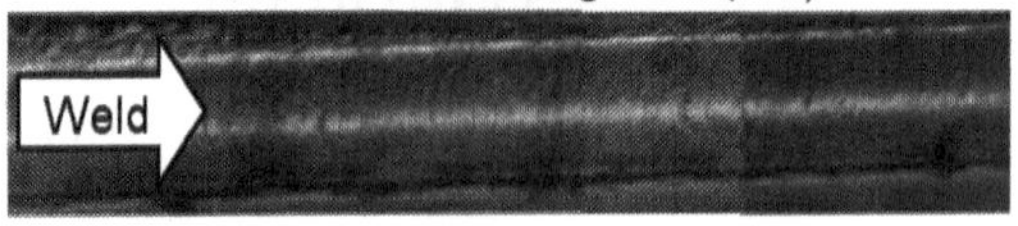

Figure 3. Appearance of Mg FSW Surface

It was sometimes observed that Mg material stuck to the thread pin and tore some material out of the base material, thereby causing a surface defect (indicated by arrows) shown in **Figure 4**. This phenomenon is unique to magnesium alloys and has been previously reported[5] and further investigation is recommended.

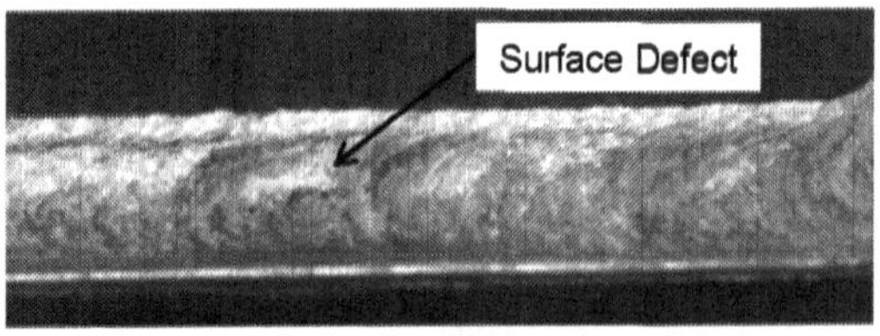

Figure 4 Mg Surface Defect from Stiction

Weld Cross Section

Each weld was cross-sectioned using an abrasive cutter and mounted for grinding and polishing. After each cross-section was mounted and then polished using a liquid suspension with 0.1 microns. Generally, samples are polished using a water-based solution, but that type of solution left oxide residue on the surface of the welds, which made it difficult to view under a high-powered microscope. **Figure 5** shows a welded cross section for weld condition #3 or 3,000 rpm @ 300 mm/min, with MRI-153M on the advancing side (AS) of the weld. As shown there are no voids or holes in the weld region. The pitting that is shown is from the chemical etching solution used to take the photograph.

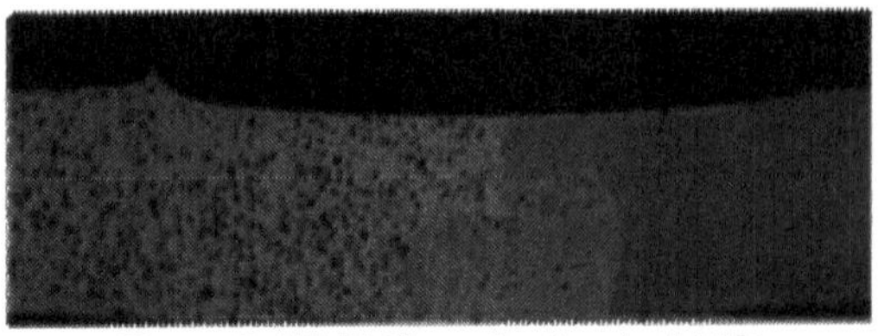

Figure 5. Weld Cross Section (weld #3)

On the other hand **Figure 6** shows a welded cross section for weld condition #10 for the weld conditions at 2,000 rpm @ 500 mm/min, with an advancing side of MRI-153M. These conditions produced an undesirable void in the weld zone.

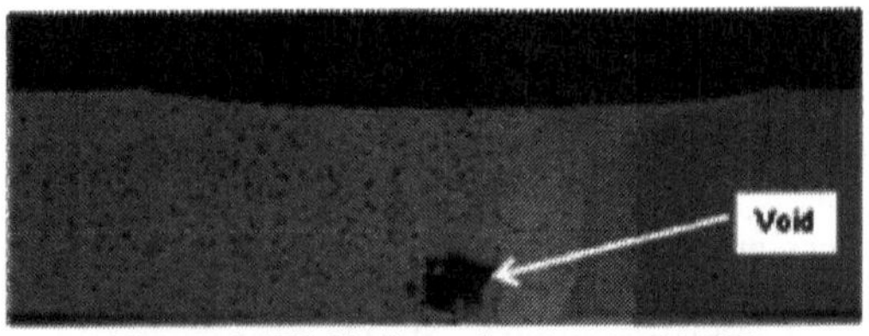

Figure 6. Weld Cross Section (weld #10)

Figure 7A & B show the welding conditions that produced good welds and those that contained voids for the welds with AZ31 and MRI-153M on the advancing side of the weld. The data show that welding with low heat-input from high feed rates and/or low spindle speeds, yielded welds that contained defects (or voids). The cross-sections that had the most mixing and were free of voids, were then chosen for more in depth exploration in the parameter study.

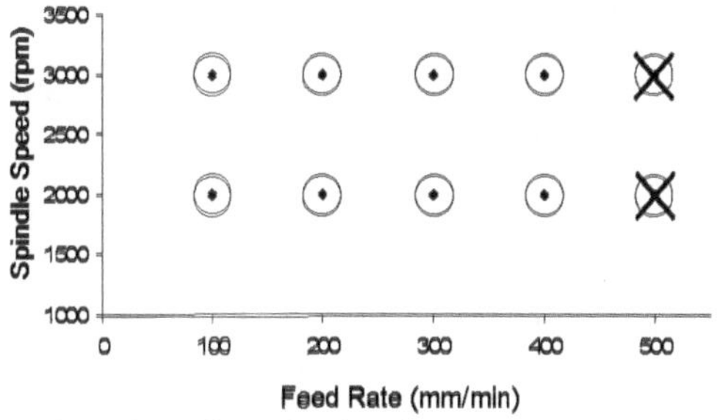

Figure 7A. Weld Cross Section Summary with AZ31 on the Advancing Side (AS) of the Weld

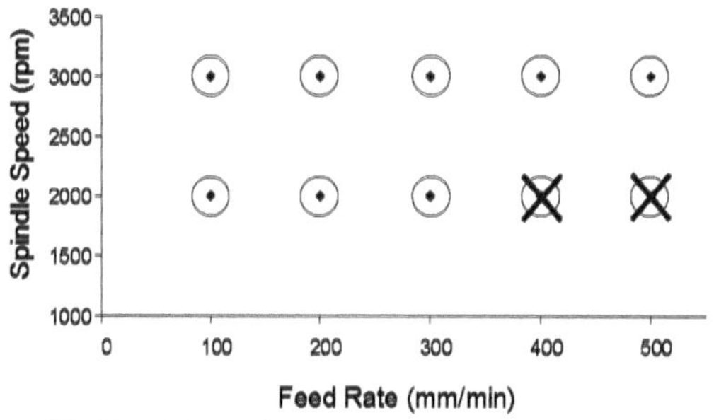

Figure 7B. Weld Cross Section Summary with MRI-153M on the Advancing Side (AS) of the Weld

Hardness Tests

Material hardness was measured using a Vickers micro hardness tester. The optimal welding parameters that resulted from the cross-section study were tested for hardness. For each cross-section, an average of 33 indentations, 0.5 mm apart, were made on the weld surface. This provided data from outside the stir zone, on each edge of the weld, in order to determine how the welding parameters affected both the weld zone and the heat-affected zone. The hardness profiles that resulted from welding conditions weld #3 at 3,000 rpm @ 300 mm/min, with MRI-153M on advancing side (AS) of the weld, weld #13 at 3,000 rpm @ 300 mm/min, with AZ31 on the advancing side (AS) of the weld and weld #18 done at 2,000 rpm @ 300 mm/min with AZ31 on the advancing side (AS) of the weld are plotted in **Figure 8**.

In the graph, the open circles represent AZ31 material. It can be seen from the plot that in the stir zone the hardness of MRI-153M increases and AZ31 decreases in comparison to the hardness of their base metal. Hardness of the base material is between 65-70Hv while that of the stir zone measured around 60Hv. The base metal of AZ31 is work hardened so that the dislocation density of the base metal may be higher than the stir zone. The difference of the dislocation density may cause a hardness decrease in the stir zone[3]. The hardness increase of the MRI-153M in the weld region may be due to the removal of the porosity in the casting resulting from the FSW process.

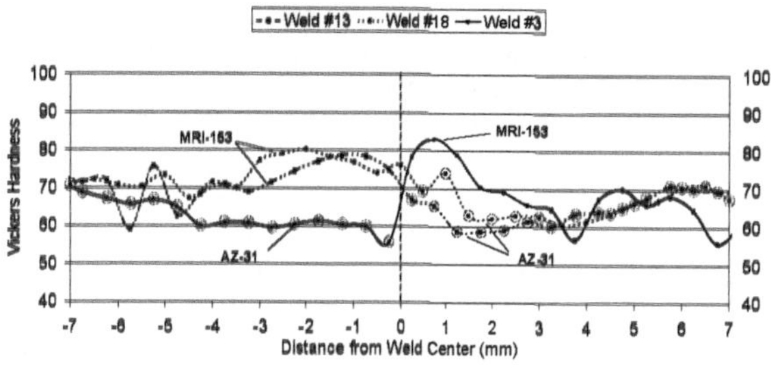

Figure 8. Vickers Hardness Profile

Tensile Tests

Transverse tensile properties were evaluated using an Instron tensile testing machine. The specimens were machined 110 mm long and 16mm wide on the ends with a center section 8mm wide and 40mm long. Tensile tests were performed on the three best sets of welds: those with the parameters that resulted from the Hardness test: #3, 13, and 18. The tensile data for all 3 coupons are plotted in **Figure 9**. The plot shows that there are no big differences in the tensile data for the 3 weld conditions; they all failed at about 200MPa. **Figure 10** shows the samples after testing and as shown #3 sample failed outside the weld region, however, samples #13 and #18 failed in the weld region. Therefore, weld parameters for #3 were chosen for the starting point for oil pan welding.

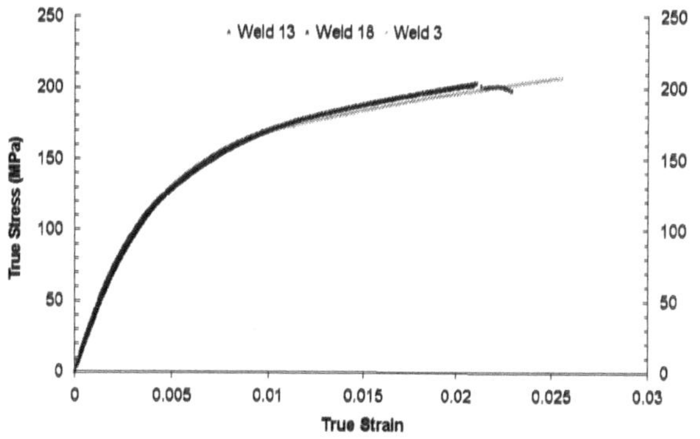

Figure 9. Tensile Test Results

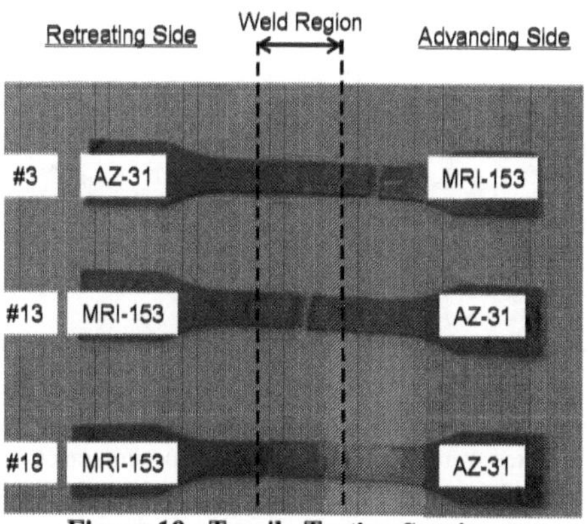

Figure 10. Tensile Testing Specimen

FSW Application: Mg Oil Pan

After coupon testing was completed, the experience was applied to FSW a die cast Mg oil pan. This oil pan is an experimental part that was designed and cast for the USAMP-Magnesium Powertrain Cast Components Project (MPCC). The original oil pan was a 2 piece design with an end plate bolted to the casting to seal its oil sump. The disadvantage of this design is that the eleven bolts required to hold the 2 pieces together add weight and casting complexity. Furthermore, the bolted design has the further disadvantage of the possibility of oil leakage over time.

Therefore, FSW was investigated to weld and seal the two pieces together to reduce the weight and improve the durability of the design.

Figure 11. Bolted Oil Pan Design (Experimental Part for USAMP-MPCC Project)

First the oil pan end plate design was changed to facilitate FSW. The plate was changed to 6.35mm thick sheet stock and the bolt holes were removed. The end plate was then machined with a 2.6 mm thick flange as shown in **Figure 12**. The oil pan casting was changed to facilitate a butt joint with a 2.6 mm thick flange and a run on/off tab was added for FSW.

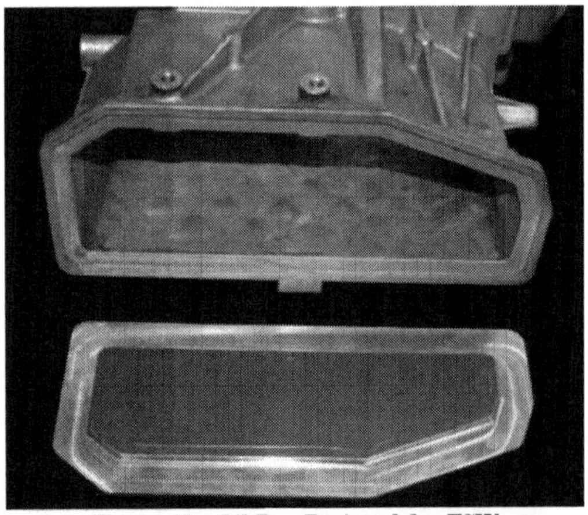

Figure 12. Oil Pan Designed for FSW (Experimental Part for USAMP-MPCC Project)

In order to weld the oil pan a fixture was designed as shown in **Figure 13**. The fixture was made using aluminum and consisted of 1" square aluminum extrusions in the four corners of the fixture. 1" Al top plates were designed to hold the perimeter of the oil pan to facilitate FSW. Also, a wedge support was attached underneath the top plate to prevent movement in the Z direction. A total of five clamps were positioned around the perimeter of the AZ31 plate to firmly hold it to the oil pan during FSW. During welding, the clamps were removed manually one at a time.

Figure 13. Jig and Clamps to Hold Oil Pan and End Plate during FSW

First oil pan welding attempts were made with the 8mm shoulder diameter tool described above. These welds resulted in surface defects. For some oil pans the gap between the oil pan and the plate was 0.5mm which made it difficult to achieve consolidated welds using the small diameter tool. Therefore, a 10 mm diameter tool was chosen because it is more tolerant to larger gaps.

Figure 14 shows the prototype oil pan welded using the 10mm diameter tool at 1,000 rpm and 400mm/min feed rate. As shown there is some burr on the corner of the oil pan. This burr is the result of the AZ31 plate rising up during welding. The surface temperature of the plate was measured during welding.

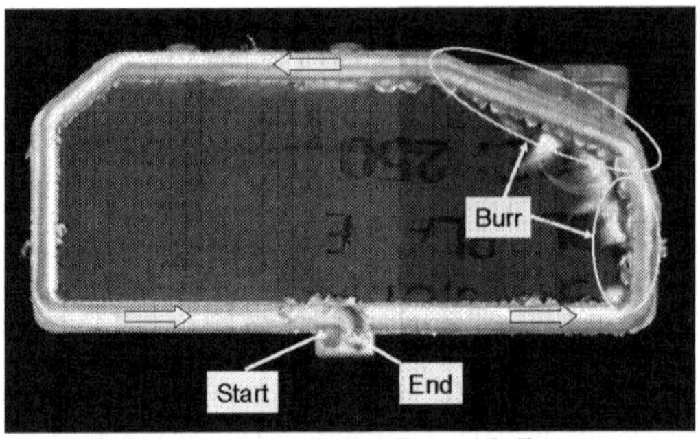

Figure 14. FSW Oil Pan with Burr

Figure 15 shows the temperature measurement locations. **The** AZ31 plate surface temperature was measured at four locations around the perimeter of the plate 7 mm from the center of the weld. K-type thermocouples were surface-mounted using aluminum adhesive tape. **Figure 16** shows the temperature measured on the surface of the AZ31 plate during FSW. The temperature measured at location 3 is lower than locations 1, 2 and 4 because location 3 is in a linear region of the weld and the other spots are in a corner. The corner temperatures are higher than those in the linear regions because of temperature buildup in these confined areas; e.g., the 2-D corner. Furthermore, the temperature at location 1 is higher than locations 2 and 4 because of the tighter radius at that location. It seems also that this temperature buildup significantly expanded the plate at the corners and resulted in joint mismatch.

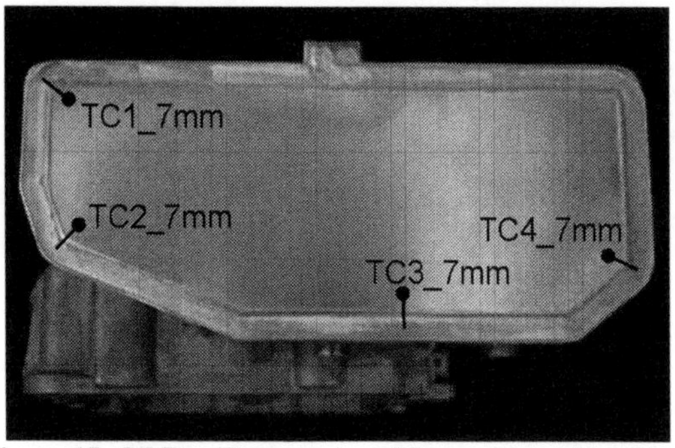

Figure 15. Oil Pan Temperature Measurement Locations

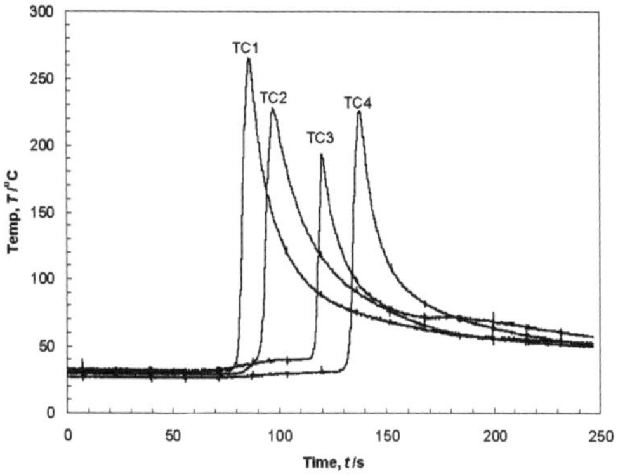

Figure 16. Oil Pan Temperature Measurement

A successful FSW and de-burred oil pan is shown in **Figure 17**. The oil pan demonstrates the capability of welding die cast magnesium MRI-153M to AZ31 plate using the FSW technique. FSW oil pans were successfully leak tested in water bath at 82.7KPa (12psi) for 3 minutes. In the near future, the oil pan will be tested on an engine dynamometer as part of USCAR-Magnesium Powertrain Cast Components Project.

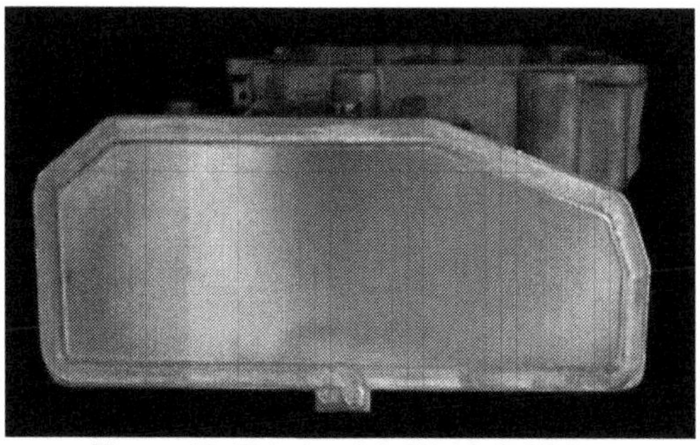

Figure 17. Friction Stir Welded Oil Pan (Experimental Part for USAMP-MPCC Project)

An unsuccessfully FSW oil pan is shown in **Figure 18**. This oil pan was welded with a large gap between the casting and the AZ31 plate of about 0.5 mm. The presence of a large gap necessitates the use of extra plunge depth that can increase burr and form a tunnel in the butt weld seam. Ideally, the gap should be about 0.25 mm between the plate and oil pan. This phenomenon didn't occur too often during this research. Nevertheless, it is important to note for manufacturing tolerance specifications.

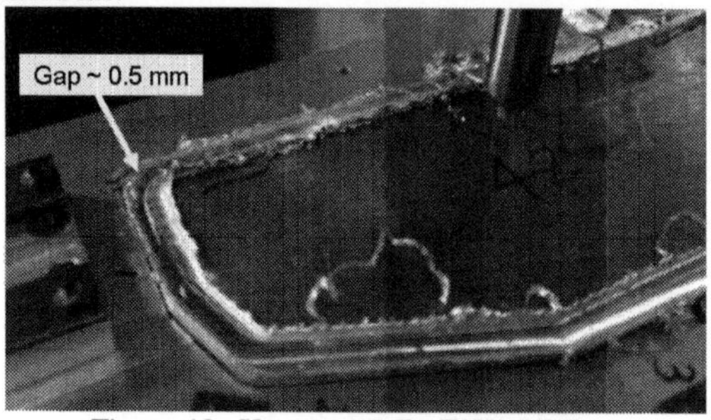

Figure 18. Unsuccessfully FSW Oil Pan.

Conclusions

FSW has been applied to successfully join dissimilar magnesium alloys of AZ31 (plate) to die cast MRI-153M. The results obtained are summarized as follows:

1. FSW of dissimilar Mg alloys resulted in welds with good appearance and joint integrity.
2. FSW process was used to successfully weld an experimental automotive part made of die cast MRI-153M and AZ31 plate for USCAR-Magnesium Powertrain Cast Components Project. The oil pan successfully passed water leakage testing at 82.7kPa (12psi) pressure for 3 minutes.
3. Thermal expansion of the AZ31 top plate during FSW was significant and necessitated proper clamping, gap tolerance and tool plunge depth control.

Acknowledgments

The authors acknowledge that this research was supported, in whole or in part, by the United States Department of Energy National Energy Technology Laboratory under Award Nos.DE-FC05-95OR22363, DE-FC05-02OR22910, and DE-FC26-02OR22910. Reference herein to any specific commercial product, process, or service by trade name, trademark, manufacturer, or otherwise does not necessarily constitute or imply its endorsement, recommendation, or favoring by the United States Government or any agency thereof. The authors further acknowledge Spartan Light Metal Products for the oil pans supplied for this project and Line Precision for machining and testing services.

References

1. W.M.Thomas *et al.*: "Friction stir butt welding", International patent application PCT/GB92/02203.
2. J.G.Wylde: "Friction Stir Welding Creating Opportunities for the Future", J. JILM, 50-5(2000), 189-197.
3. R.Johnson *et al.*, "Friction stir welding of magnesium alloys", Proc. of 4th Int. symposium on FSW, Utah, USA, 2003.
4. Naiyi Li, *et al.* "Friction Stir Welding Magnesium AM60 Alloy", Magnesium Technology, TMS 2004.
5. K. Okamoto, *et al.*, "Friction Stir Welding of Magnesium for Automotive Applications". SAE 2005- 01- 0730.
6. B. Powell *et al.*, "Progress Toward a Magnesium-Intensive Engine: the USAMP Magnesium Powertrain Cast Components Project". SAE 2004-01-0654.

Fatigue Evaluation of Friction Stir Spot Welds in Magnesium Sheets

J.B. Jordon[1], M.F. Horstemeyer[1,2], J. Grantham[1,] H. Badarinarayan[3]
[1]Center for Advanced Vehicular Systems (CAVS), Mississippi State University, Mississippi State, MS 39762
[2] Department of Mechanical Engineering, Mississippi State University, Mississippi State, MS 39762
[3] Automotive Products Research Laboratory, Hitachi America Limited, Farmington Hills, MI 48335, USA

Keywords: Fatigue, Friction Stir; Crack Growth, Magnesium

Abstract

In this study, the fatigue behavior of a magnesium AZ31 alloy in friction stir spot welds is experimentally characterized. Load control cyclic tests were conducted on single weld lap-shear coupons to determine fatigue crack characteristics. Cyclic failure modes included weld nugget pullout and full coupon separation. Good correlation was obtained between fatigue life and linear elastic fracture mechanics. In addition, in-situ fatigue tests were conducted to determine crack growth rates of the friction stir spot welds via an automated digital microscope. The experimentally measured crack growth rate suggests that the assumption of constant crack growth employed in literature is reasonable.

Introduction

In order to achieve the lightest vehicle designs, research on magnesium alloys has produced promising alloys for structural components. Welded joints are one of the prominent issues concerning magnesium alloys in proposed manufactured components. Since so few products are fabricated from magnesium, determining the most appropriate joining technique is the focus of ongoing research. Of the several different types of joining approaches under consideration, friction stir spot welds (FSSW) show promise as a potential joining technique for widespread use. While limited research on FSSW of aluminum alloys is available [1-5], the fatigue behavior of FSSW in magnesium alloys is non-existent.

Linear elastic fracture mechanics has been almost predominantly employed in predicting the fatigue behavior of FSSW in aluminum alloys. The main issue in this approach is that the stress intensity factor (SIF) solutions for FSSW are not readily available. Recently, Li et al [1] adapted a resistance spot weld SIF solution by employing a correction factor based on three dimensional finite element results. The SIF employed in that study was based on the weld radius and was not a function of the crack length. Thus, the main assumption commonly used by Li et al to predict the fatigue life was to assume constant crack growth. While, this assumption has been verified on resistant spot welds, it has not been confirmed for FSSW.

The purpose of this paper is to evaluate the fatigue behavior of magnesium AZ31 in friction stir spot welds. Additionally, linear elastic fracture mechanics are employed to correlate the experimental fatigue-life results. Lastly, a modified Paris law is employed to estimate the crack growth rates determined from in-situ experiments and verify the constant crack assumption regarding cyclic lap-shear testing of FSSW.

Material

Magnesium AZ31 alloy sheets of 2.0mm thickness were chosen for the present study. FSSW coupons were welded in a lap configuration and the individual sheet dimensions were: length 100mm, width 35mm and are welded on an overlap area of 35 x 35 mm. The FSSW tool was made from standard tool steel (H13) material, having a shoulder with diameter 12 mm, pin length of 3.2 mm and left hand threads (M5). The shoulder had a concave profile with the angle of concavity of 10 deg. The welding process parameters were: tool rotation speed 1000 RPM, tool plunge speed of 20mm/min, shoulder plunge depth of 0.5mm and a dwell time of 2.5 sec. During the welding process, the interface between the upper and lower sheet is formed into a hook like shape due to the penetration of the FSSW tool into the bottom sheet [6]. A scanning electron microscope (SEM) image of the interfacial hook of an untested magnesium AZ31 FSSW coupon is shown in Figure 1.

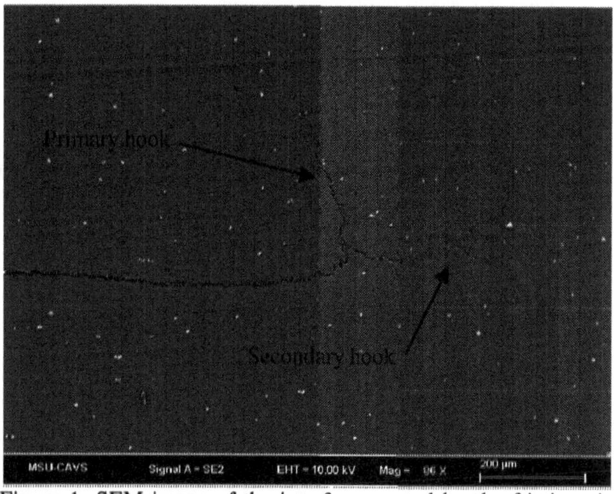

Figure 1. SEM image of the interface created by the friction sir process.

Experimental Procedure

Tensile Tests

Lap-shear tensile tests were conducted in displacement control at a rate of 1 mm/minute at a temperature of 23°C and relative humidity of 50%. In place to doublers typically employed in lap-shear testing, a 1.96 mm thick shim was placed behind one side of the lower hydraulic wedge grips. An identical sized shim was placed in the opposite side of the upper grip. To help insure consistency in the test results, the specimens were gripped 55 millimeters from the center of the weld. Thus, the grip-to-grip distance employed for each coupon was 110 millimeters. An example of the FSSW lap-shear coupon employed in this study is shown in Figure 2.

Figure 2. Geometry of magnesium AZ31 alloy friction stir spot weld lap-shear coupon. Dimensions are in cm.

Fatigue Tests

Lap-shear fatigue tests were cyclically tested under load control, at a temperature of 23°C, relative humidity of 50%, and at three different load ratios (R=0, 0.3, 0.7). The grips employed for the cyclic tests were identical to monotonic tests. The weld zone of the coupons were optically observed for the onset of through-crack initiation using automated digital imaging equipment. For specimens where the onset of a through-crack was not monitored, a change in displacement method based on the digital imaging results was employed to estimate the number of cycles pertaining to initiation of a through-crack.

Crack Growth Tests

For further analysis of the fatigue behavior of FSSW, fatigue tests were conducted on coupons that were sectioned through the center of the coupon. A constraint bracket, as shown was employed to simulate the constraint of the full width coupon. Throughout the life of the coupon, the interfacial hook area was monitored with a long focal microscope that recorded one image every fifth cycle. An automated digital imaging equipment was employed to record the images.

Results and Discussion

Tensile and Fatigue-Life Tests

The results of the three monotonic tests are shown in Figure 3. The load versus displacement results for the monotonic lap-shear tests reveal that the static mechanical behavior of the FSSW is fairly consistent. Furthermore, the ultimate strength of the spot weld is similar to friction stir spot welds in aluminum alloys [1]. Figure 4 presents the experimental results of the fatigue tests conducted at R-ratios of 0, 0.3, and 0.7. The load range is plotted versus the number of cycles required to achieve through-crack initiation as well as final coupon separation. The fatigue behavior of the FSSW in magnesium AZ31 alloy was similar to results reported in the literature on FSSW of aluminum alloys [1-5]. The main fatigue failure modes for this type of weld and alloy are grouped primarily into two types: nugget pullout and full separation. The nugget pull-out failure was observed for the two maximum load levels tested (2.5 and 3.0kN, R=0) and full separation was observed for load levels below 2.5kN. The cause of the two different types of failure modes is related to the shear/tensile stress in the remaining net area of the nugget as the crack propagates around the nugget diameter. When the cyclic load was high (above 2.5kN), the remaining cross section was unable to carry the remaining load and nugget pullout occurred. For lower cyclic loads (below 2.5kN), crack propagation across the full width of the coupon occurred which resulted in complete separation. From Figure 4, it is clear that the differences between the through-crack initiation life and full separation life of the FSSW coupons is significant. Thus, the differences between the two failure definitions are due to the amount of time required for the crack to propagate around the circumference of the weld versus through crack initiation. For comparison purposes the fatigue-life of the half-width coupon versus cycles for through-crack initiation is shown in Figure 4. The fatigue results show that the half-width coupon had similar fatigue lives compared to the full width coupon, which suggest that the constraint employed in the testing had minimal effect on the stress field of the weld zone. A sampling of the digital imaging results from the half-width fatigue tests are shown in Figure 5.

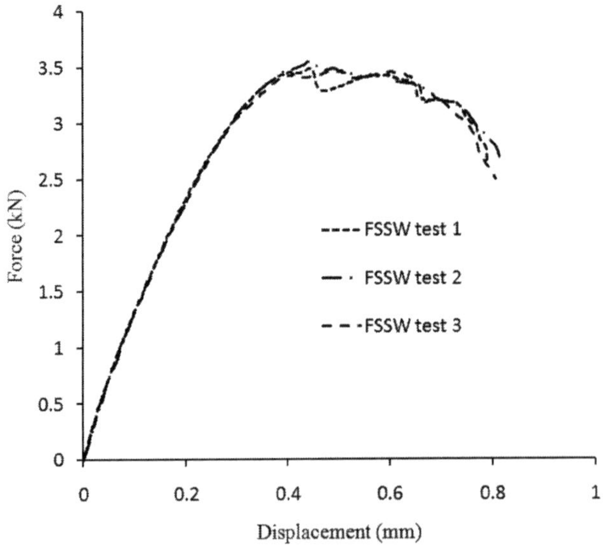

Figure 3. Monotonic lap-shear tension test results of the friction stir spot weld coupons tested to failure.

Fractography Analysis

Post-mortem fractography of the full width coupons reveal that the fatigue crack always grew from the primary hook, and no crack growth was observed from the secondary hook. For 3.0 and 2.5kN fatigue load range tested, the weld nugget pulled out of the top sheet. Below 2.5kN the crack propagated around the circumference of the weld nugget and then propagated through the base material. Figure 6a shows a typical fracture surface of a FSSW tested at a load range of 3kN at R=0. Fractography via the scanning electron microscope (SEM) revealed that the fatigue crack propagated in a semi-circular path around the circumference of the weld until the coupon failed in a tension/shear overload or the crack propagated through the base material. A typical fracture surface image from a coupon fatigued at a load range of 3kN (R=0) is shown in Figure 6a. A higher magnification image of the area indicated in Figure 6b. The fractographs shown in Figure 6 come from a location just above the tip of the interfacial hook in the stir zone.

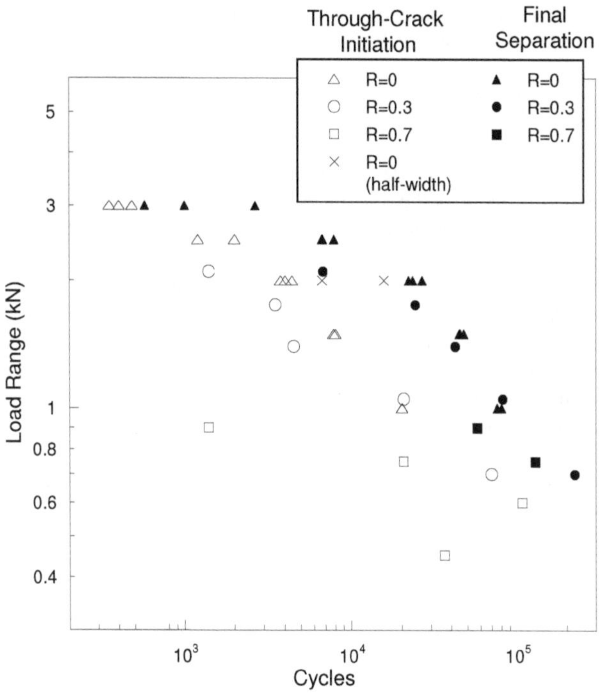

Figure 4. Friction stir spot weld through-crack initiation and full separation life plotted against applied load range. Additionally shown is the through-crack initiation and full separation life for the half-width coupons.

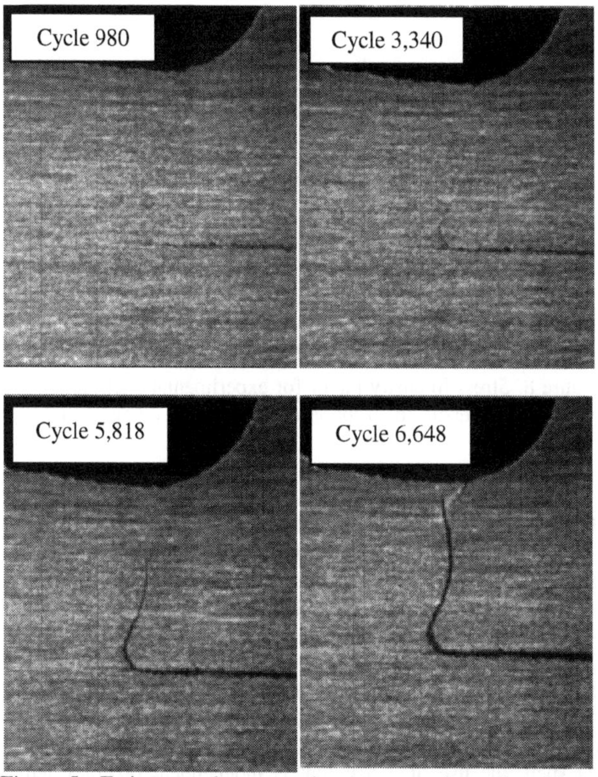

Figure 5. Fatigue crack propagation in magnesium alloy friction stir spot weld for a half-width coupon tested at a load range equivalent to a full width coupon tested at a load range of 2kN and at R=0.

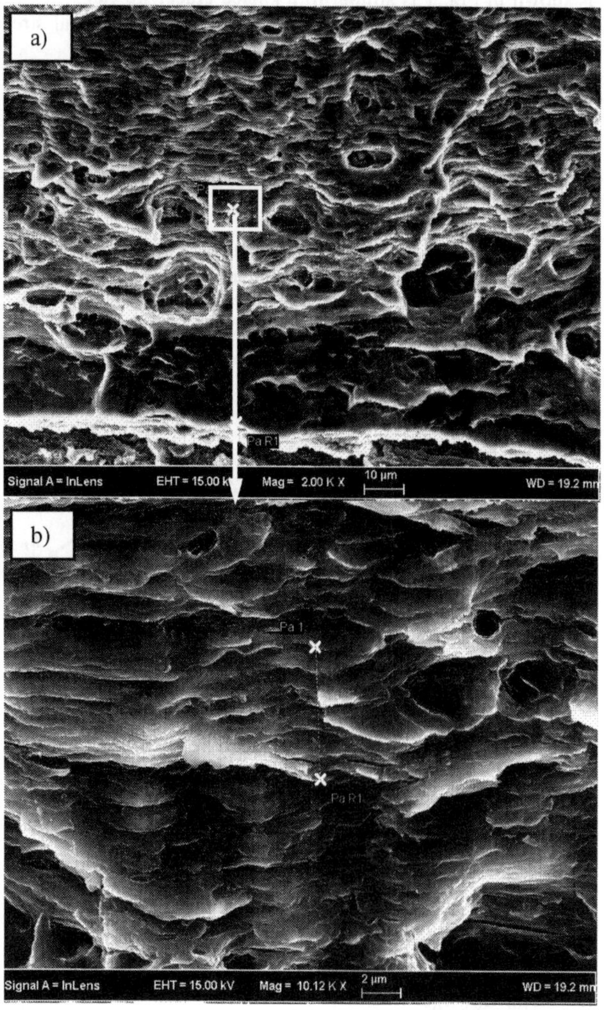

Figure 6. a) Typical scanning electron microscope image of the fracture surface just above the crack tip for a coupon tested at load range of 3kN at R=0. b) A higher magnification of the fracture surface of the area indicated on b).

Linear Elastic Fracture Mechanics

For analysis purposes, stress intensity factors (SIFs) for the FSSW were calculated. However, since no complete well defined solution exists for a FSSW, a SIF for resistant spot welds is employed similar to Lin et al [6]. The method employed by Lin et al. for FSSW is based on a kinked crack modification of the SIF proposed by Cottrell and Rice [7] for an infinitesimally small crack. Equation (1) and (2) are the mode I and mode II global stress intensity solutions for a crack in RSW under tensile loading [13]

$$K_I = \frac{P}{r^{3/2}}\left[0.341\left(\frac{2r}{t}\right)^{0.397}\right] \quad (1)$$

$$K_{II} = \frac{P}{r^{3/2}}\left[0.282 + 0.162\left(\frac{2r}{t}\right)^{0.710}\right] \quad (2)$$

where P is the load, r is the weld radius, and t is the sheet thickness. Since, the interfacial hook between the two materials as shown in Fig. 1 is similar to a kinked crack, we can relate the mode I and mode II global stress intensity values to local stress intensity values as a function of kinked crack angle, θ [7].

$$k_I = \frac{K_I}{4}\left[3\cos\left(\frac{\theta}{2}\right) + \cos\left(\frac{3\theta}{2}\right)\right] - \frac{K_{II}}{4}\left[3\sin\left(\frac{\theta}{2}\right) + 3\sin\left(\frac{3\theta}{2}\right)\right] \quad (3)$$

$$k_{II} = \frac{K_I}{4}\left[\sin\left(\frac{\theta}{2}\right) + \sin\left(\frac{3\theta}{2}\right)\right] + \frac{K_{II}}{4}\left[\cos\left(\frac{\theta}{2}\right) + 3\cos\left(\frac{3\theta}{2}\right)\right] \quad (4)$$

Equations (3) and (4) are the local mode I and mode II stress intensity solutions. Figure 7 is a pictorial representation of the mode I and II stress intensity factors associated with a kinked crack [7].

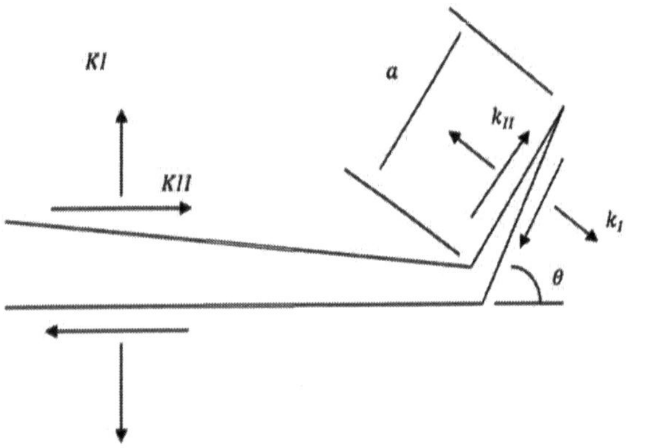

Fig. 7. Schematic of a kinked crack stress intensity factor for a resistant spot weld.

As an alternative to a crack growth approach, the FSSW fatigue life results conducted in this study were plotted against the calculated SIF range and are shown in Figure 8. A model fit of the SIF range to fatigue life correlation is shown in Equation (5).

$$N_f = 10^6 e^{-0.85*\Delta K} \quad (5)$$

where N_f is the number of cycles for full separation life and ΔK is the stress intensity factor range.

While entirely empirical, the linear elastic fracture mechanics (LEFM) approach of correlating the cycles to failure to the SIF range collapses the three R-ratios with good results. Also, the good correlation of the LEFM approach can be viewed as indirectly validating the use of a SIF in crack growth approaches used to predict the fatigue life of FSSW.

Crack Growth Tests

Crack growth rate versus crack length were tabulated for the magnesium AZ31 alloy in FSSW. The crack growth rate (da/dN) was calculated based on in-situ images taken from the half-width coupons tests. In order to predict the crack growth rate and validate the constant growth assumption, a Paris Law approach was used. Due to mode I and mode II loading of the lap-shear coupon, an equivalent mode I stress intensity [9] is employed.

$$k_{eq} = \sqrt{{k_I}^2 + {k_{II}}^2} \quad (5)$$

The Paris law, modified to account for the R-ratio effect [10], was employed as shown in Equation (6)

$$\frac{da}{dN} = C\left(\frac{\Delta k_{eq}}{(1-R)^{(1-\gamma)}}\right)^m \quad (6)$$

where C and m are constants derived from crack growth experiments, R is the loading ratio, g is a R-ratio parameter, and Δk_{eq} is the equivalent stress intensity range. The Paris law constants C=4.2 and m=3.7x10^{-7} were chosen based on crack growth experiments conducted by Tokija et al [11]. The angle of crack propagation, α, from the tip of the interfacial hook was estimated to be approximately 80°. To account for the differences between FSW and FSSW (i.e., the center bore of the FSSW), an equivalent weld radius equal to 2.5mm was chosen. While the authors' acknowledge that this approach is problematic, this is likely the most pragmatic solution in the absence of three dimensional finite element results. Future work will involve three dimensional finite element calculations to fully develop a SIF solution for FSSW.

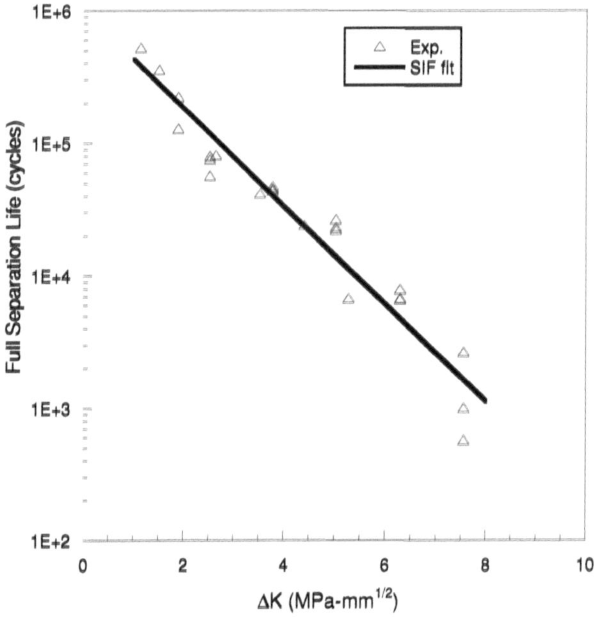

Figure 8. Stress intensity factor for experimental friction stir spot weld results versus full separation fatigue life along with model fit.

The comparison of the estimated crack growth rate to the experimental results are shown in Figure 9. The calculated crack growth rate was found to be higher than the measured rate. The non-conservative estimation is likely the result of using a SIF for RSW and the use of Paris law constants for the base material. While experimental growth rate was observed to exhibit a large degree of scatter, the general trend of the measured growth rate of the FSSW suggests that the rate is constant. This observation was found to be consistent with work done by Cooper and Smith [12]. As such, the use of a SIF that is independent of crack length seems plausible based on the result presented in Figure 9. However, additional experiments are needed to fully characterize the crack growth rate in FSSW.

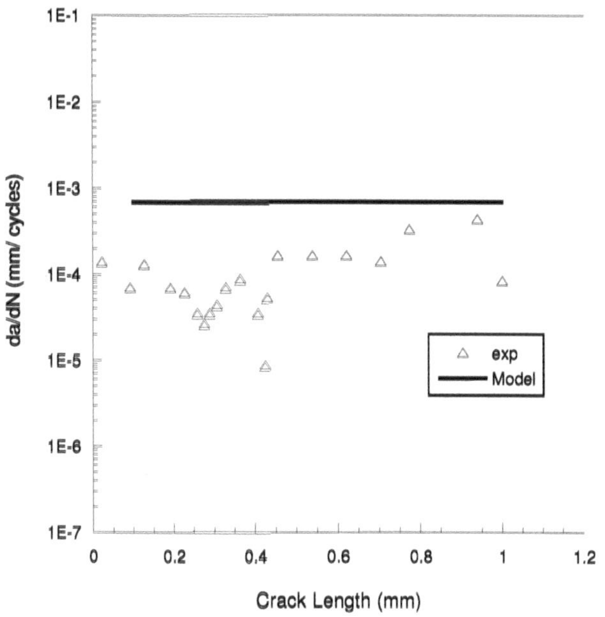

Figure 9. Crack growth rate versus crack length for the friction stir spot weld coupon tested at R=0 and the load range was 1kN. Measurements were taken from half-width coupon tests via in-situ imaging.

Conclusions

A summary of the main conclusions of this work are as follows:

1. The fatigue crack initiated at the interfacial hook located in the stir zone of the spot weld coupon. The crack exhibited two types of Failure modes: nugget pullout and full separation.
2. The experimental fatigue-life results showed considerable differences between full separation life and through-crack initiation life.
3. Good correlation for the three R-ratios (0, 0.3, and 0.7) was obtained between the stress intensity factor range and the number of cycles to full separation.
4. The measured crack growth rate via in-situ tests suggests that the constant crack growth assumption employed here and elsewhere is a reasonable assumption.

Acknowledgments - The authors would like to recognize Richard Osborne, James Quinn, Xuming Su, Alan Luo, John Allison, Robert McCune, Don Penrod, Matthew Castanier, and David Lamb for their encouragement of this study. This material is based upon work supported by the Department of Energy and the National Energy Technology Laboratory under Award Number No. DE-FC26-02OR22910, and the U.S. Army TACOM Life Cycle Command under Contract No. W56HZV-08-C-0236 through a subcontract with Mississippi State University, and was performed for the Simulation Based Reliability and Safety (SimBRS) research program. This report was prepared as an account of work sponsored by an agency of the United States Government. Neither the United States Government nor any agency thereof, nor any of their employees, makes any warranty, express or implied, or assumes any legal liability or responsibility for the accuracy, completeness, or usefulness of any information, apparatus, product, or process disclosed, or represents that its use would not infringe privately owned rights. Reference herein to any specific commercial product, process, or service by trade name, trademark, manufacturer, or otherwise does not necessarily constitute or imply its endorsement, recommendation, or favoring by the United States Government or any agency thereof. The views and opinions of authors expressed herein do not necessarily state or reflect those of the United States Government or any agency thereof. Such support does not constitute an endorsement by the Department of Energy of the work or the views expressed herein.

References

[1] P.-C. Lin, J. Pan, T. Pan, "Failure modes and fatigue life estimations of spot friction welds in lap-shear specimens of aluminum 6111-T4 sheets. Part 1: welds made by a concave tool," *Int. J. Fatigue*, 30 (2008), 74–89.

[2] V -X. Tran, J. Pan, T. Pan, "Fatigue behavior of aluminum 5754-O and 6111-T4 spot friction welds in lap-shear specimens," *Int. J. Fatigue*, 30 (2008), 2175-2190.

[3] P.-C. Lin, J. Pan, T. Pan. "Failure modes and fatigue life estimations of spot friction welds in lap-shear specimens of aluminum 6111-T4 sheets. Part 2: welds made by a flat tool," *Int. J. Fatigue* 30 (2008), 90–105.

[4] D.-A. Wang and C.-H. Chen, "Fatigue live of friction stir spot welds in aluminum 6061-T6 sheets," *J. Mater. Process Technol.*, (2008), 367-375.

[5] Y. Uematsu, K. Tokaji, Y. Tozaki, T. Kurita, S. Murata, "Effect of re-filling probe hole on tensile failure and fatigue behaviour of friction stir spot welded joints in Al-Mg-Si alloy," *Int. J. Fatigue*, 30 (2008), 1956-1966.

[6] H. Badarinarayan, Y. Shi, X. Li, K. Okamoto, "Effect of tool geometry on hook formation and static strength of friction stir spot welded aluminum 5754-O sheets," Int. J. of Machine Tools and Manufacture, 49 (2009), 814-823.

[7] B. Cotterell, J.R. Rice, "Slightly curved or kinked cracks," *Int. J. Fract*, 16 (1980), 155–69.

[8] L.P. Pook, "Fracture mechanics analysis of the fatigue behavior of spot welds." *Int. J. Fracture*, 11 (1975), 173–176.

[9] S.B. Biner, "Fatigue crack growth studies under mixed-mode loading," *Int. J. Fatigue*, 23 (2001), 259-263.

[10] K. Walker, "The effect of stress ratio during crack propagation and fatigue for 2024-T3 and 7075-T6 aluminum. In: *Effects* of Environment and Complex Load History on Fatigue Life," ASTM STP 462, ASTM, Philadelphia, PA, (1970), 1–14.

[11] K. Tokaji, M. Kamakura, Y. Ishiizumi, N. Hasegawa, "Fatigue behaviour and fracture mechanism of a rolled AZ31magnesium alloy," Int. J. Fatigue 26 (2004) 1217–1224

[12] J. F. Cooper and R. A. Smith, "The measurement of fatigue cracks at spot-welds," Int. J. Fatigue 7 (1985) 137–140.

Monotonic and Fatigue Behavior of Mg Alloy Friction Stir Spot Welds: An International Benchmark Test in the "Magnesium Front End Research and Development" Project

H. Badarinarayan[1], S.B. Behravesh[2], S.D. Bhole[3], D.L. Chen[3], J. Grantham[4], M.F. Horstemeyer[4], H. Jahed[2], J.B. Jordon[5], S. Lambert[2], H.A. Patel[3], X. Su[6], and Y. Yang[7]

[1] Automotive Products Research Laboratory, Hitachi America Limited, USA
[2] Mechanical and Mechatronics Engineering Department, University of Waterloo, Canada
[3]Department of Mechanical and Industrial Engineering, Ryerson University, Canada
[4]Center for Advanced Vehicular Systems (CAVS), Mississippi State University, USA
[5]Department of Mechanical Engineering, The University of Alabama, USA
[6]Ford Motor Company, USA
[7]Institute for Metals Research, China

Keywords: Fatigue, Friction stir spot welds; Magnesium alloys; Resistance spot weld

Abstract

This paper presents the experimental results of benchmark coupon testing of monotonic and cyclic conditions on friction stir spot welded coupons of Mg AZ31alloy. The results presented here are a product of a collaborative multinational research effort involving research teams from Canada, China, and the United States. Fatigue tests were conducted in load control at R=0.1 at two different maximum loads: 1kN and 3kN. Good agreement was found between the participating labs regarding the number of cycles to failure. Differences in the failure modes were observed for the two different loading conditions tested. At the higher load, fatigue failure was caused by interfacial fracture. However, at the lower load, fatigue cracks formed perpendicular to the loading direction, which led to full width separation. For additional comparison, the monotonic and cyclic results of the friction stir spot welds are compared to resistance spot welded coupons of similar nugget size.

Introduction

The continued push for more fuel efficient automobiles designs is motivation for ongoing research in lightweight metals. Through lightweight designs comes the need to explore alternatives to traditional metals like aluminum and steel currently used in the manufacturing of automobiles. As such, a collaborative multinational research effort involving researchers from Canada, China, and the United States and joined by Chrysler, Ford, and General Motors, is underway with the goal of developing the ability to build a front end of an automobile constructed of magnesium alloys. Magnesium alloys, with a density of 1.74g/cm^3, weigh less than a quarter of steel and two-third of aluminum, and with abundant reserves on the earth, are attractive substitutes for steel and aluminum for vehicle body structures [1–4]. Replacing the largely steel structure of a vehicle's front end with magnesium can also move the vehicle center of gravity away from the front, improving vehicle drivability.

This research project, called the Magnesium Front End Research and Development program, or MFERD, has already yielded several fatigue characterizations of wrought magnesium alloys [cf. 5-7]. In order to fully meet the project's objectives, characterization of mechanical properties of potential joining techniques is also needed. As such, friction stir spot welding is targeted as a possible joining technique in the fabrication of a front end of an automobile using magnesium alloys.

Friction stir welding has steadily been gaining more wide spread use when high integrity and strength are required. Friction stir spot weld is a recent variant of friction stir welding and is an attractive welding technique due to the solid state nature of the process and the lack of stress relieving that is typically needed. A recent literature review of the friction stir spot friction process can be found in [8]. The solid state nature and the ability of joining dissimilar metals have made friction stir spot welding an attractive welding process.

The fatigue behavior of a friction stir spot weld (FSSW) is highly dependent on process parameters employed to create the weld [9,10]. These parameters include speed, depth of plunge, dwell time, and tool configuration. Up to this point, a majority of studies of the nature of fatigue of FSSW's have been almost exclusively focused on aluminum alloys [9-12]. With regards to FSSW's made of aluminum alloys, the failure modes of quasi-static and cyclic loads of FSSW coupons vary based on the load level [9]. Differences have been observed in the fracture path for quasi-static loads compared to fatigue loads for aluminum alloys [9]. In this study[9-10], FSSW coupons failed by interfacial fracture under quasi-static loading, whereas, under cyclic loading, fatigue cracks initiated and grew from several locations including the interfacial tip and outside the weld zone. In addition, fatigue failure modes of aluminum FSSW made using various tooling were also observed to differ based on the shape of the tooling used to make the weld [10]. Also, fatigue of FSSW's of dissimilar aluminum alloys were observed to have different failure modes compared to FSSW's with identical alloys for the top and bottom sheets [11]. The fatigue failure modes of dissimilar metals (aluminum and steel) were also observed to vary compared with joints made of all aluminum alloys [12]. While the FSSW joining technique of magnesium alloys is documented [13-16], limited published literature exists on the fatigue properties of FSSW coupons of magnesium alloys. Mallick and Agarwal [17] were the first to quantify the fatigue behavior of FSSW's made of a magnesium alloy. However, their characterization of the failure mechanisms under cyclic loading was limited in its presentation.

Resistance Spot Weld (RSW), on the other hand, is currently the most common joining process in the automotive industry and hence is the manufacturers' preferred joining technique. Thus, as have been performed for steel [18-22] and aluminum alloys [23-27], characterization of FSSW and RSW techniques and comparing their monotonic and cyclic behavior, provide the motivation for these types of studies. As such, the purpose of this research is the characterization of FSSW'ed sheets of AZ31B magnesium alloy through a round-robin experimental program and also evaluating its mechanical behavior by a comparative analysis to RSW of the same alloy.

Materials

Magnesium AZ31B alloy sheets of 2.0mm thickness are chosen for the present study. Coupons were welded in lap configuration. The individual sheet dimensions were: length 100mm, width 38mm and were welded on an overlap area of 38 × 38 mm. The FSSW tool was made from standard tool steel (H13) material, having a shoulder with diameter 12 mm, pin length of 3.2 mm and left hand threads (M5). The shoulder was a concave profile with the angle of concavity of 10 deg. The welding process parameters were: tool rotation speed 750 RPM, tool plunge speed of 20mm/min, shoulder plunge depth of 0.1mm and a dwell time of 2.5 sec. In regards to the bonded area in FSSW, the welding process produces an annulus shape, and the average inner and outer diameters in this study were found to be 6.5mm and 9.7mm, respectively. During the welding process, the interface between the upper and lower sheet is formed into a hook like shape due to the penetration of the FSSW tool into the bottom sheet.

For comparison purposes, resistant spot weld (RSW) coupons were made of the same alloy as for the FSSW. The thickness of the sheet as well as the specimen configuration and dimensions were similar to the FSSW. Additionally, the weld nugget was approximately the same. The RSW parameters were: welding current of 34 kA, welding time of 8 cycles (8/60 sec), electrode force of 4 kN, and holding time of 30 cycles (0.5 sec). Figure 1 shows general dimensions for FSSW and RSW coupons employed in this study.

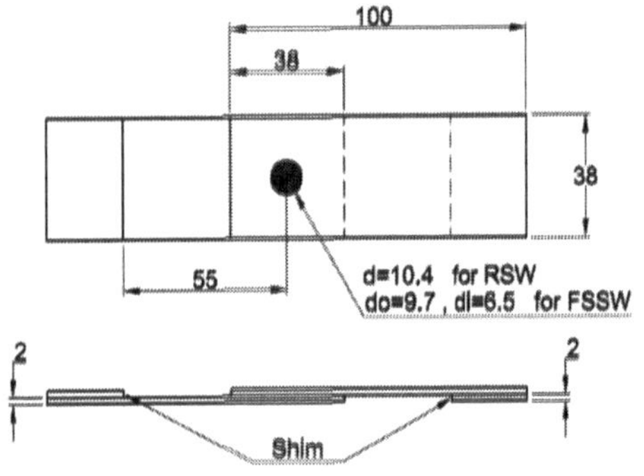

Figure 1. Configuration of friction stir and resistant spot welds single-weld lap-shear coupons. Dimensions are in mm.

Experiments

Mississippi State University (MSU), Ryerson University (RU), University of Waterloo (UW), and the Institute of Metal Research (IMR), all participated in round-robin fatigue testing of lap-shear FSSW coupons. Each institution conducted six (6) load control tests consisting of three (3) specimens conducted at each of the two (2) P_{max} load levels: 3kN and 1kN. The tests were run at $R = \frac{P_{min}}{P_{max}} = 0.1$, and a frequency of 5 Hz. All tests were conducted at ambient temperature and humidity. The test set-up also included certain specifications for mounting including careful attention to grip alignment, grip distance, and use of shims or offsets. The grip-to-grip distance was maintained at 110 mm at each of the labs. In addition, UW also tested RSW coupons under the same conditions as stated for the FSSW coupons.

Results and Discussion

Monotonic Tests

Monotonic tests were performed on a servo-hydraulic load frame under uniaxial displacement rate of 1 mm/min. The test results show that the FSSW coupons exhibited an average ultimate tensile-shear load (UTSL) of 4650±10N. All specimens tested failed in a consistent manner with a partially interfacial failure mode, as shown in Figure 2.

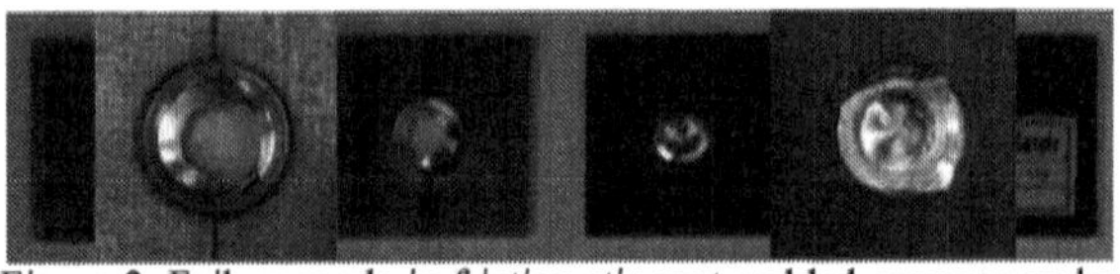

Figure 2. Failure mode in friction stir spot welded coupons under monotonic tensile-shear loading.

Similar to studies on steel FSSW [22] and aluminum FSSW [27], comparing the static and cyclic behavior of emerging FSSW joining technique and commonly used RSW is of interest. Recent research [28] has studied the monotonic and fatigue behavior of RSWs of AZ31B-H24 Mg alloy, and showed that the highest UTSL is achieved using the welding current of 34 kA, and the welding time of 8 cycles. The same welding parameters were utilized in this study, and a solid circular nugget with an average diameter of 10.4 mm was obtained which is close to outer diameter of FSSW (9.7 mm). It should be noted that although FSSW and RSW are of a similar outer diameter, the bonded area are very different (40 mm^2 in FSSW and 85 mm^2 in RSW), due to the different shapes of bonded region. However, comparing the mechanical behavior of these specimens is sound from the application perspective, as the area of coupons contributing in the joints is almost the same in FSSW and RSW specimens. Monotonic testing of tensile-shear RSW specimens yielded an average UTSL of 7620±48N with interfacial failure mode, as shown in Figure 3.

Figure 3. Failure mode in resistant spot welded coupons under monotonic tensile-shear loading.

Load-displacement curves, shown in 4, illustrate that the RSW coupons have higher UTSL compared to FSSW coupons. Higher UTSL of RSW specimens is mainly attributed to the smaller bonded area in FSSW specimens. As mentioned before, RSWs have a solid circular, and FSSWs have an annulus-shaped weld region. Therefore, even for the same outer diameter, a higher UTSL is expected for RSWs.

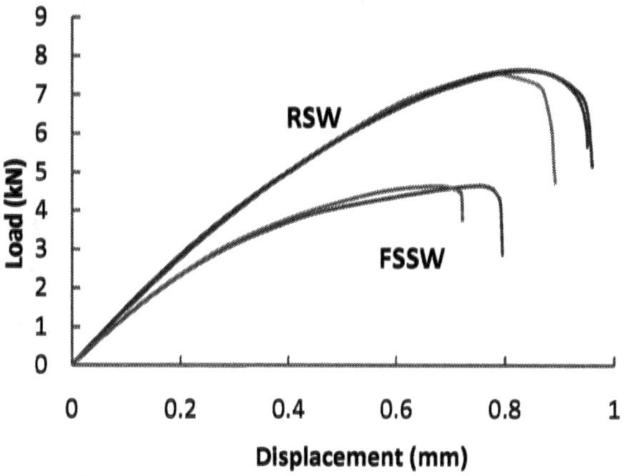

Figure 4. Load-displacement curves for friction stir spot weld and resistance spot weld tensile-shear specimens under quasi-static loading.

Fatigue Tests

Figure 5 displays the round-robin fatigue results of the FSSW from the four institutions. The first observation based on the fatigue life results was the overall good correlation between all of the laboratories. Fatigue failure for this particular study was defined as complete separation of the lap-joint. However, there were some outliers in both sets of load levels tested. One of the tests MSU conducted at the 3kN load level failed much earlier than the rest of the tests compared to the other institutions. At the 1kN level, several specimens tested at IMR failed much later compared to the rest of the group of coupons. However, the variation in the fatigue results overall is fairly consistent compared to other published round-robin fatigue testing programs. It is important to point out that at the higher fatigue load level (3kN), the load is above the elastic limit based on the monotonic load-displacement curve shown in Figure 4. At the 1kN load level, the load is within the elastic range. The number of cycles to failure for the FSSW coupons presented here are in the same order of magnitude of similar FSSW [9-12,17]

For further comparisons, the fatigue life of the FSSW coupons are compared to the fatigue life of the RSW. Fatigue results show that the RSW exhibited better fatigue life at the lower load level (1kN) compared to the FSSW coupons. However at the higher load level (3kN), the FSSW coupons exhibited better fatigue resistance compared to the RSW coupons. For the lower load level (1kN), where we have the same failure modes in FSSW and RSW specimens (coupon width separation, Fig. 7.a), higher fatigue life of RSW could be attributed to larger nugget size which causes lower stress concentration and hence smaller hot spot stress and retardation of crack initiation. However, at the higher load level (3kN), the better fatigue performance of FSSW is due to the different failure modes in FSSW and RSW specimens. FSSW specimens failed in interfacial mode (Fig. 7.b), while coupon failure perpendicular to the loading direction (Fig. 7.a) was observed in RSW specimens. The reason why the fatigue strength in interfacial mode is higher than coupon failure is, in both cases, the mode I stress intensity factor (K_I) is the main factor for fatigue crack propagation. The load component contributing to K_I in interfacial failure mode is normal to the coupon surface at the joint, and the load component contributing to K_I in coupon separation is parallel to the coupon surface or normal to the coupon thickness at the joint. And because the load component parallel to the coupon is larger than the one normal to the coupon surface (as bending rotation in coupon at the joint is less than 45 deg), the coupon separation mode is more critical than interfacial failure. However, this still needs more investigations.

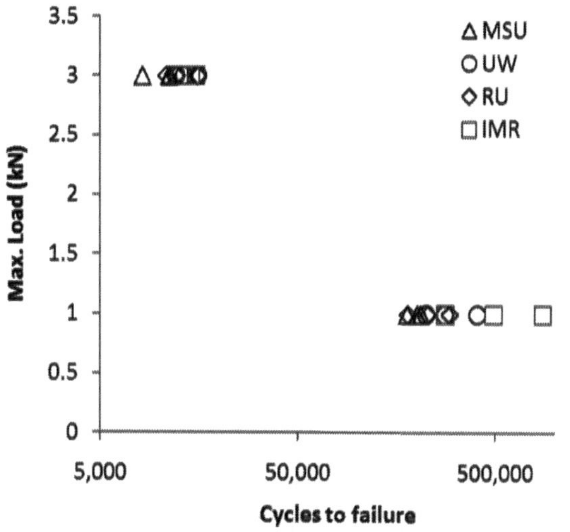

Figure 5. Comparison of Mg AZ31 friction stir spot welds fatigue results from the four universities: Mississippi State University (MSU), Ryerson University (RU), University of Waterloo (UW), and Institute for Metal Research (IMR). Fatigue tests were conducted in load control at R=0.1, at a frequency of 5 Hz and at room temperature.

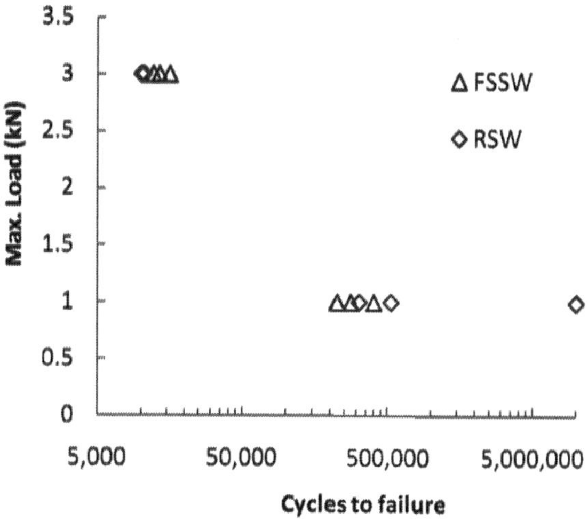

Figure 6. Comparison of fatigue results of the magnesium AZ31 alloy friction stir spot welds to the resistance spot welds tested at University of Waterloo (UW). Fatigue tests were conducted in load control at R=0.1, at a frequency of 5 Hz and at room temperature.

Fractography

The failure modes under cyclic loading were observed to vary for the different load levels tested in the round-robin testing program. All four laboratories reported that the failure mode at the lower cyclic load level of 1 kN was different compared to the higher cyclic load level of 3 kN. That is, the failure at the lower load level occurred perpendicular to the loading direction, while the failure at the higher load level exhibited interfacial fracture, as shown in Figure 7(a) and (b), respectively. The interfacial failure observed for the 3 kN load level likely occurred because as the crack propagated circumferentially around the nugget, the shear/tensile stress in the remaining net area of the nugget increased with each advancement of the crack front. Once the crack had propagated around approximately half of the nugget diameter, the shear/tensile stresses acting on the net area were such that the remaining cross section failed under shear/tensile overload. The other type of fatigue failure occurred at the load level of 1kN. Once the crack had propagated circumferentially around the nugget, the crack then propagated outward through the sheet material.

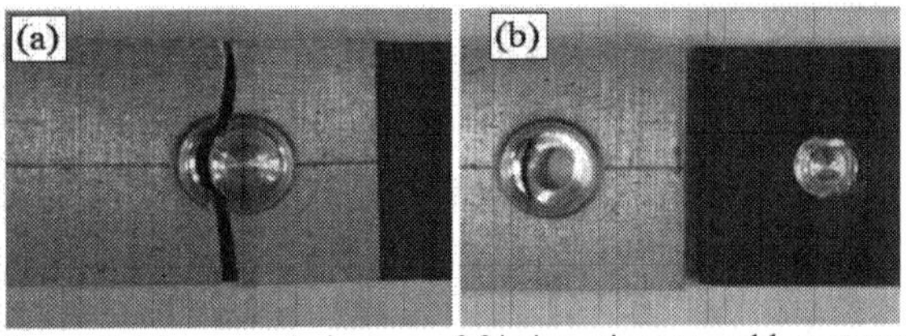

Figure 7. Macroscopic images of friction stir spot weld coupons of magnesium AZ31 alloy samples fatigued at a load level of (a) P_{max}=1 kN, and (b) P_{max}=3 kN (R=0.1, 5 Hz, sine waveform, room temperature).

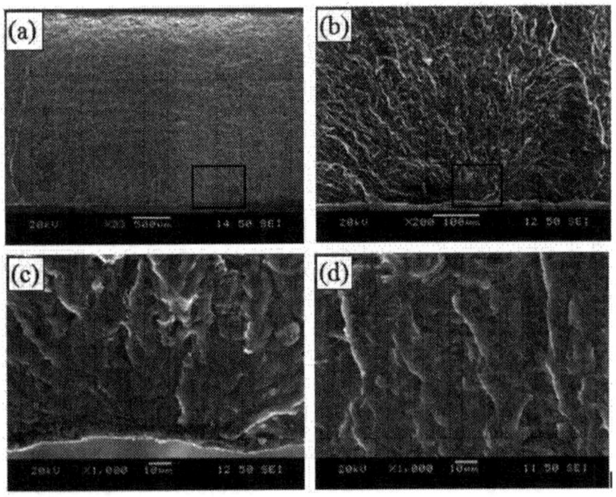

Figure 8. Scanning electron microscope images of fracture surfaces of friction stir spot weld coupons of magnesium AZ31 alloy fatigued at a load level of P_{max}=1 kN, (a) low magnification image near the center of the sample, (b) initiation site in the boxed region in (a), (c) magnified view of the boxed region in (b), and (d) crack propagation area near the center of the sample thickness.

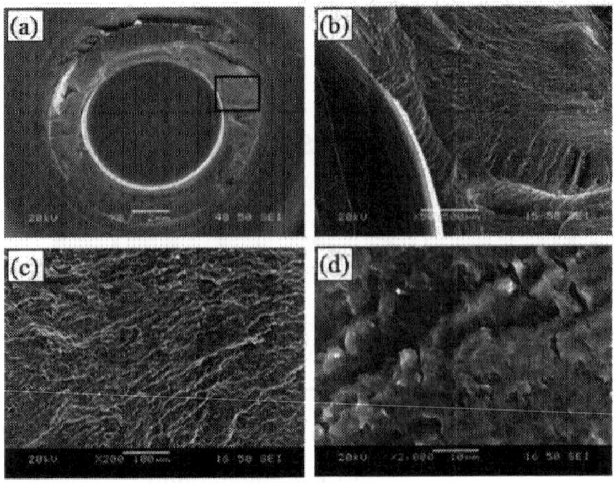

Figure 9. Scanning electron microscope images of fracture surfaces of friction stir spot weld coupons of magnesium AZ31 alloy fatigued at a load level of P_{max}=3 kN, (a) overall view of interfacial fracture of the sample, (b) initiation site in the boxed region in (a), (c) magnified view near the initiation site, and (d) crack propagation area at a higher magnification.

Fracture surfaces of the fatigued specimens were examined under scanning electron microscope (SEM). Figure 8(a-d) and Figure 9(a-d) show the typical SEM images of the coupons tested at P_{max}=1 kN and P_{max}=3 kN, respectively. The low magnification image shown in Figure 8(a) was taken near the center of the sample fractured at P_{max}=1 kN, while 9(a) showed an overall view of interfacial fracture of the sample tested at P_{max}=3 kN. Figure 8(b) and Figure 9(b) showed the boxed region in Figure 8(a) and 9(a), respectively, indicating the crack initiation site. While the failure mode was different (see Figure 7), the fatigue crack initiation at both load levels occurred from the surface. Figure 8(c) and 9(c) showed the higher magnification images near the fatigue crack initiation sites for both load levels. Figure 9(d) shows the fatigue crack propagation area with some striation-like features perpendicular to the crack propagation direction.

Conclusions

A summary of the main conclusions of this work are as follows:

1. The resistant spot weld lap-shear coupons exhibited better monotonic strength compared to the friction stir spot weld coupons with similar specimen dimensions and outer nugget diameter.
2. Friction stir spot weld and resistant spot weld coupons both failed in the interfacial mode under monotonic loading.
3. The fatigue results from the four different testing labs demonstrated consistent fatigue results on the friction stir spot weld lap-shear coupons.
4. Different failure modes were observed for the friction stir spot weld coupons for the two cyclic load levels tested. At the high cyclic load level (3kN), the coupons failed by interfacial fracture. At the lower cyclic load level (1kN), the coupons failed by full width separation.
5. Striations-like features were observed on fracture surfaces of specimens tested at 3kN.

6. The fatigue life the friction stir spot weld coupons were observed to compare closely to the fatigue life of the resistant spot weld coupons. At the higher cyclic load (3kN), the friction stir spot weld coupons exhibited better fatigue life, compared to the resistant spot weld coupons. However, at the lower cyclic load (1kN), the resistant spot weld coupons exhibited better fatigue resistance compared the friction stir spot weld coupons.

Acknowledgments

The authors would like to recognize Richard Osborne, James Quinn, Alan Luo, John Allison, and Robert McCune for their encouragement of this study. This material is based upon work supported by the Department of Energy and the National Energy Technology Laboratory under Award Number No. DE-FC26-02OR22910, AUTO21 Network of Centers of Excellence, Natural Sciences and Engineering Research Council of Canada (NSERC), Premier's Research Excellence Award (PREA). Such support does not constitute an endorsement by the Department of Energy of the views expressed herein. This paper was prepared as an account of work sponsored by an agency of the United States Government. Neither the United States Government nor any agency thereof, nor any of their employees, makes any warranty, express or implied, or assumes any legal liability or responsibility for the accuracy, completeness, or usefulness of any information, apparatus, product, or process disclosed, or represents that its use would not infringe privately owned rights. Reference herein to any specific commercial product, process, or service by trade name, trademark, manufacturer, or otherwise does not necessarily constitute or imply its endorsement, recommendation, or favoring by the United States Government or any agency thereof. The views and opinions of authors expressed herein do not necessarily state or reflect those of the United States Government or any agency thereof.

References

1. K.U. Kainer, Magnesium Alloys and Technology, Wiley-VCH, Cambridge, United Kingdom, 2003.
2. Luo, A.A., "Magnesium: Current and potential automotive applications" JOM. 54 (2): 42–48, 2002.
3. Luo, A.A., Journal of Materials and Manufacturing. SAE Transactions, Warrendale, PA, 411–21, 2005.
4. Friedrich, H.E., and Mordike, B.L., Magnesium Technology—Metallurgy, Design Data, Applications, Springer-Verlag, Berlin, Germany, 2006.
5. S. Begum, D.L. Chen, S. Xu, Alan A. Luo, Met. Mater. Trans. A 39A (2008) 3014
6 C.L. Fan, D.L. Chen, A. A. Luo, Mater Sci. Eng. A 519 (2009) 38
7. J.D. Bernard, J.B. Jordon, M.F. Horstemeyer, H. El Kadiri, J. Baird, David Lamb, Alan A. Luo, "Structure-property relations of cyclic damage in a wrought magnesium alloy," Scripta Materialia, 63 (2010) Viewpoint set no. 47, 751-756.
8. T. Pan, 2007, "Friction stir spot welding (FSSW) – a literature review," Society of Automotive Engineers, Warrendale (PA), SAE Technical Paper No. 2007-01-1702.
9. Lin, P.-C., Pan, J, Pan T., 2008, "Failure Modes and Fatigue Life Estimations of Spot Friction Welds in Lap-Shear Specimens of Aluminum 6111-T4 Sheets, Part 1: Welds Made By a Concave Tool," Int. J. Fatigue, 30(1) pp.74–89.
10. Lin, P.-C., Pan, J., Pan, T., 2008, "Failure Modes and Fatigue Life Estimations of Spot Friction Welds in Lap-Shear Specimens of Aluminum 6111-T4 Sheets. Part 2: Welds Made By a Flat Tool," Int. J. Fatigue, 30(1), pp. 90–105.
11. Tran, V.-X., Pan, J., Pan, T., 2010, "Fatigue behavior of spot friction welds in lap-shear and cross-tension specimens of dissimilar aluminum sheets," Int. J. Fatigue, 32(7), pp. 1022-1041
12. Tran, V.-X., Pan, J., 2010, "Fatigue behavior of dissimilar spot friction welds in lap-shear and cross-tension specimens of aluminum and steel sheets," Int. J. Fatigue, 32(7), pp. 1167-1179
13. Su, P., Gerlich A., and North, T.H., 2005,"Friction stir spot welding of aluminum and magnesium alloy sheets, Society of Automotive Engineers," Warrendale (PA) (2005) [SAE Technical Paper No. 2005-01-1255.
14. Gerlich, A., Su, P., and North, T.H., 2005, "Tool penetration during friction stir spot welding of Al and Mg alloys," J Mater Sci 40 pp. 6473–6481.
15. Pan, T.-Y., Santella, M., Mallick, P.K., Frederick, A., Schwartz, W.J., 2006, "A feasibility study on spot friction welding of magnesium alloy AZ31," In: Proceedings of 63rd annual world magnesium conference, Beijing, China, May 21–24; pp. 179–86.
16. Agarwal, L., Mallick, P.K., and Kang, H.T., 2008, "Spot friction welding of Mg-Mg, Al-Al and Mg-Al alloys", Society of Automotive Engineers, Warrendale (PA), SAE Technical Paper No. 2008-01-0144.
17. Mallick, P.K. and Agarwal, L., 2009, "Fatigue of spot friction welded joints of Mg–Mg, Al–Al and Al–Mg alloys," Society of Automotive Engineers, Warrendale (PA) SAE Technical Paper No. 2009-01-0024.
18. Aota K., Ikeuchi K., "Development of friction stir spot welding using rotating tool without probe and its application to low-carbon steel plates", Welding International, Vol. 23, No. 8, August 2009, pp. 572–580.
19. Ohashi R., Fujimoto M., Mironov S., Sato Y.S., Kokawa H., "Effect of contamination on microstructure in friction stir spot welded DP590 steel", Science and Technology of Welding and Joining, 2009, Vol. 14, No. 3, pp. 221-227.
20. Person N.L., "Tensile-Shear and Fatigue Properties of Resistance and MIG Spot Welds of Some Al Auto Body Sheet Alloys", SAE Paper No. 750463, 1975.
21. Chao Y.J., "Ultimate Strength and Failure Mechanism of Resistance Spot Weld Subjected to Tensile, Shear, or Combined Tensile/Shear Loads", Journal of Engineering Materials and Technology, Vol. 125, APRIL 2003, pp. 125-132.
22. Khan M.I., Kuntz M.L., Su P., Gerlich A., North T. and Zhou Y., "Resistance and friction stir spot welding of DP600: a comparative study", Science and Technology of Welding and Joining, 2007, Vol. 12, No. 2, pp. 175-182.
23. Karthikeyan R., Balasubramanian V., "Predictions of the optimized friction stir spot welding process parameters for joining AA2024 aluminum alloy using RSM", The International Journal of Advanced Manufacturing Technology, April 2010.
24. Thoppul S.D., Gibson R.F., "Mechanical characterization of spot friction stir welded joints in aluminum alloys by combined experimental/numerical approaches", Materials Characterization, November 2009, Vol. 60, No. 11, pp. 1342-1351.
25. Gean A., Westgate S.A., Kucza J.C., Ehrstrom J.C., "Static and Fatigue Behavior of Spot-Welded 5182-0 Aluminum Alloy Sheet", Welding Journal, March 1999, pp. 80s-86s.
26. Hassanifard S., Zehsaz M., Tohgo K., "The Effects of Electrode Force on the Mechanical Behaviour of Resistance Spot-Welded 5083-O Aluminium Alloy Joints", Strain, 2009.
27. Uematsu Y., Tokaji K., "Comparison of fatigue behaviour between resistance spot and friction stir spot welded aluminium

alloy sheets", Science and Technology of Welding and Joining, 2009, Vol. 14, No. 1, pp. 62-71.

28. Behravesh B., Liu L., Jahed H., Lambert S., Glinka G., Zhou Y. , "Effect of Nugget Size on Tensile and Fatigue Strength of Spot Welded AZ31 Magnesium Alloy", SAE Technical Paper, 2010, SAE No. 2010-01-0411.

PART 9: CORROSION, SURFACE TREATMENT, AND COATING

Wim H. Sillekens

Developments in the field...

The unique properties of magnesium and its alloys have made them desirable in a variety of applications and even irreplaceable in some; however, due to their highly active nature, actual use does depend critically on the successful control of corrosion. For example, in the car industry, this corrosion issue has been termed a "show stopper" as it severely limits many groundbreaking applications.

Efforts to combat this corrosion problem were prominent in the past two decades. Particularly since the year 2000, increasing resources have been directed to this area of research, with many impressive achievements being reported including new alloys and applications, but most notably relating to the following aspects.

- The understanding of electrochemical corrosion mechanisms has been deepened. This includes the proposal of the electrochemical reactions involved in the anodic magnesium dissolution and cathodic hydrogen evolution, the explanation of the Negative Difference Effect phenomenon, the development of corrosion simulation and acceleration methods, and the establishment of corrosion-rate estimation and monitoring technologies.
- More comprehensive knowledge of the metallurgical influences on the corrosion behavior has been generated, including the roles of alloying elements in the matrix and of secondary phases. Also, the study of the relationship between microstructure and corrosion has been extended from cast to wrought alloys, as well as to some novel alloys and amorphous materials.
- Corrosion behavior in many solutions representing service environments has been explored. These typical environmental media include lab chloride containing electrolytes, ASTM and industrial standard solutions, coolants, ionic liquids, and simulated body fluids. Other metals in contact with magnesium (galvanic corrosion) as well as applied stress (stress-corrosion cracking) have also been addressed in studies during recent times. Moreover, biocompatibility has now become one of the most challenging research areas, and a considerably increased number of papers have been published on this topic as of late.

In the end, the purpose of corrosion research is to arrive at suitable engineering solutions to current or future corrosion problems. Corrosion prevention and protection strategies for magnesium alloys have therefore gained much attention in the past decade, and the development of surface-treatment and coating techniques has strikingly intensified. This has led to advances in the following main fields.

- **Surface treatment**. Surfaces are modified or changed in state, chemical composition and microstructure by means of cleaning, machining, alloying, beam (electron, ion or laser) processing, anodizing (sparking), and conversion (including many chemical treatments), with the treated surface layer being more stable and corrosion-resistant than the original substrate. These technologies have interested many engineers in various industries; in fact, this field can be considered to represent the most exciting development in the corrosion prevention of magnesium alloys.
- **Coatings**. Being the most practical achievement, there have been metallic and organic coatings developed such as normal electro and electro-less plating, innovative ionic liquid deposition, metallic powder cold spraying, traditional organic painting, electro-less E-coating and novel sol-gel film forming. With the interaction of coatings with magnesium being confined to the coating/substrate interface, tailoring these coatings to acquire some

special physical and chemical properties (particularly a high corrosion resistance) has been the main concern.

- **Inhibition techniques**. This has recently started to attract research interest as an additional corrosion mitigation approach, based on the knowledge regarding the electrochemistry of magnesium and the understanding of the environmental influence on corrosion performance.

As the use of magnesium alloys is expected to further increase, more and more efforts will be committed to the understanding of corrosion and the development of corrosion protection techniques. With this sustained input, future research and development is to target the following needs.

- **Fundamental understanding**. Still limited as yet, the underlying knowledge of corrosion mechanisms is key in developing cost-effective corrosion mitigation techniques. As novel alloys emerge for new applications, basic research on this will remain one of the long-term interests.
- **Corrosion data**. The application in industry requires practically reliable corrosion results. Collection of exposure results will continue to be one of the areas of interest to industrial engineers. The first step will be simulation of the corrosion behavior under a few typical corrosion environments in the laboratory; predicting the corrosion behavior in their service environments will be a long-term goal.
- **Surface engineering**. Being currently still the most cost-effective approach, surface-treatment and coating technologies are highly desired and their further development will remain to be a major focal area for the time to come.
- **Corrosion-resistant alloys**. Although it is extremely challenging to change the nature of magnesium from active to passive, modification of the chemical composition (through alloying with small amounts of some "miracle" elements) and microstructure to improve the corrosion performance will continue to be a focused research topic. It will be a long journey to develop and commercialize such alloys though.

... and their relationship with this book

The papers in this part represent a fair yet exemplary selection of the research topics on the subject as outlined above. Being more or less thematically grouped, they include recent and less recent reports on measuring and evaluating corrosion behavior (by Song *et al.*, Tiwari *et al.*, and Takei *et al.*), corrosion protection methods (by Skar and Albright, and Blanchard *et al.*), coating technologies (by Nibhanupudi and Manavbasi, and Shrestha *et al.*), corrosion mitigation in actual application (by Bretz *et al.*), coatings for enhancing wear resistance (by Mandagie *et al.*), and magnesium alloys and surface modifications for biomedical applications (by Virtanen and Fabry).

Acknowledgment: Useful discussions with Dr. Guangling Song of Oak Ridge National Laboratory are gratefully recognized.

AN HYDROGEN EVOLUTION METHOD FOR THE ESTIMATION OF THE CORROSION RATE OF MAGNESIUM ALLOYS

Guangling Song, Andrej Atrens, David StJohn

CRC for Cast Metals Manufacturing (CAST)
Department of Mining, Minerals and Materials Engineering,
The University of Queensland,
Brisbane, Qld 4072, Australia

Abstract

Due to the negative difference effect, a special electrochemical phenomenon for magnesium and its alloys, significant errors can be introduced into corrosion rate measurements when using traditional electrochemical techniques. Also, the classical weight-loss method only provides final corrosion information (or an integrated corrosion rate), and moreover errors may easily be introduced into the final result during the removal of corrosion products and in the calculation of the corrosion rate.

This paper experimentally and theoretically demonstrates the use of a hydrogen evolution method for measuring the corrosion rate of magnesium and its alloys. The following are shown. 1) The amount of hydrogen collected is equal to the amount of magnesium dissolved. 2) The collection of evolved hydrogen can reveal the instantaneous corrosion rate, and the changing corrosion behaviour of magnesium and its alloys. 3) Smaller theoretical and experimental errors are introduced by the hydrogen evolution collection method. Most importantly, this method is easy to set up and operate, and is suitable for a quick evaluation of corrosion performance and monitoring of the corrosion behaviour of magnesium and its alloys.

Introduction

The uses of magnesium alloys in various industries have been increasing in the past decade, and it is expected that this increasing trend will continue in the coming years. The increasing demand of high quality magnesium alloys has greatly stimulated the research and development of new alloys. Obtaining good mechanical properties to meet the design requirements of magnesium alloy components is always one of the main goals of developing new magnesium alloys. Recently, a number of investigations [1, 2, 3, 4, 5] on some new alloys with excellent mechanical properties have been reported, and the use of these magnesium alloys for industrial purposes appear to be promising.

However, satisfactory mechanical properties are not enough to enable a magnesium alloy to serve in a practical environment, in which the alloy is exposed to environmental corrosion attack. In fact, magnesium is intrinsically highly reactive and its alloys usually have relatively low corrosion resistance, which is actually one of the main obstacles that impedes the use of magnesium alloys in practical environments. Therefore, corrosion performance is a major issue for research and development of magnesium alloys.

Corrosion rate is the most important and basic parameter in terms of the corrosion performance of an alloy. Therefore, measuring the corrosion rate is an essential step in evaluating corrosion performance. The measurement of the corrosion rate provides not only a base for a corrosion investigation, but also an important parameter for industrial designers in the design of components.

Normal methods of measuring the corrosion rate of magnesium alloys mainly include measuring weight-loss (after salt spray or immersion) and estimating the corrosion current density (by means of Tafel extrapolation on polarisation curves, etc.). These methods have been successful in some cases, but their disadvantages are also significant.

The weight-loss rate method has been regarded as the most basic and reliable method. Usually, it is used as a standard, against which the accuracy of corrosion rates obtained by other methods are determined. The most attractive feature of this method is its easily understood definition that directly expresses the corrosion concerns of material producers and users. However, this classical method has some shortcomings. Firstly, the weight-loss rate is actually an average result of a corroding specimen over a certain period of time. How the corrosion rate changes in this period is unknown. Some important corrosion details can therefore be missed from this measurement. Secondly, to accurately measure the weigh-loss caused by corrosion, all of the corrosion products stuck on the specimen surface should be completely removed without removing any un-corroded metal. Experimentally, this is difficult. The effort to keep the un-corroded metal untouched during the corrosion product removal usually leads to incomplete removal of the corrosion products. Sometimes, the un-corroded areas or the metallic substrate underneath the corrosion products are also removed when complete removal of the corrosion products is achieved. Therefore, some experimental errors can be introduced into the final weight-loss rate.

The corrosion current density obtained by electrochemical techniques has been widely and successfully used in corrosion studies of other metals. The advantage of this method over the weight-loss method is its ease of obtaining instantaneous corrosion rates. It can be used to monitor the change of corrosion rate with time, and the measured result provides real time information, rather than an average rate. The corrosion current density has also been employed to investigate and evaluate the corrosion performance of magnesium and it alloys [6, 7, 8, 9, 10]. However, the corrosion of magnesium alloys is a special electrochemical dissolution process, which is always accompanied by a special electrochemical phenomenon, the negative difference effect. It has an abnormal polarisation behaviour that the hydrogen evolution rate increases during anodic polarisation [11, 12]. Due to the negative difference effect on magnesium and its alloys, the estimation of the corrosion current density simply based on a polarisation curve becomes theoretically unfeasible, because the negative difference effect greatly distorts the normal polarisation curve. Even though in some cases, the polarisation curve may exhibit Tafel linear regions, the corrosion current can not be extrapolated from this linear region. The reason is that the reaction mechanism around the corrosion potential is different from that in the Tafel region due to the negative difference effect. Hence the polarisation curve is not a reliable method of estimating the corrosion rate of magnesium and its alloys. Significant theoretical errors result if the estimation of the corrosion rate is simply based on the polarisation curve, which is experimentally demonstrated latter in this paper.

Therefore, the measurement of corrosion rates of magnesium alloys is still an issue that has not been well addressed. Even though the weight-loss method has been widely used, in spite of the possible error caused by the corrosion product removal, the measured result is not very informative. Therefore, it is essential to develop a technique that can not only measure the average corrosion rate like the weight-loss method, but can also monitor the changing corrosion rate with time.

Principle

Pure magnesium

It is well known that the overall corrosion reaction of magnesium in aqueous solutions at its corrosion potential can be expressed as follows [11, 13, 14]:

$$Mg + 2H_2O = Mg^{2+} + 2OH^- + H_2 \quad (1)$$

This means that the dissolution of one magnesium atom generates one hydrogen gas molecule. In other words, the evolution of one mole of hydrogen gas corresponds to the dissolution of one mole of magnesium. Therefore, in theory, measuring the volume of hydrogen evolved is equivalent to measuring the weight-loss of magnesium dissolved, and the measured hydrogen evolution rate is equal to the weight-loss rate if both have been converted into the same units.

It must be stressed that, in the overall corrosion reaction (1), the anodic reaction is dissolution of magnesium and the cathodic reaction is hydrogen evolution. The contribution of oxygen reduction to the cathodic process can be neglected. This is actually a prerequisite for using hydrogen evolution to estimate the corrosion rates of magnesium. Fortunately, in most cases, particularly in aggressive solutions, such as NaCl, the cathodic reaction is mainly hydrogen evolution and the contribution of oxygen reduction is negligible, and thus the prerequisite is automatically satisfied.

From reaction (1), it can be seen that hydrogen evolution is determined by the dissolution of magnesium. The corrosion products do not affect the relationship between the hydrogen evolution and magnesium dissolution. The hydrogen evolution rate directly reflects the corrosion rate of magnesium. Therefore, theoretically the corrosion rate of magnesium measured by hydrogen evolution should be very reliable.

An argument may exist that sometimes the falling of some small metallic particles from the magnesium surface during corrosion due to the undermining effect [11, 15, 16] would cause errors to hydrogen evolution measurement. However, if the experimental set-up is designed in such a way that hydrogen evolution from the undermined particles can also be collected together with the hydrogen from the magnesium specimen, then the total amount of hydrogen collected should still be equal to the total amount of dissolved magnesium.

Magnesium alloys

For magnesium alloys, the corrosion mechanism is more complicated than for pure magnesium. Since most alloys have a multi-phase microstructure, all the microstructural constituents can affect the corrosion rates of the alloys. However, no matter how complicated the interaction among those microstructural constituents in an alloy, the matrix α phase always corrodes preferentially [17, 18, 19, 20]. The other constituents usually act as cathodes to the α phase or act as corrosion barriers that are inert to the corrosion process. The α matrix is a magnesium-based solid solution. It has been shown that the α phase of AZ alloys has the same corrosion reaction mechanism as pure magnesium [18]. This indicates that the overall corrosion rate can be expressed by an equation similar to reaction (1). At least, the dissolution of elemental Mg in the α matrix should follow equation (1).

Not only the elemental Mg is involved in the corrosion of the α matrix, but also other alloying elements in the solid solution as solutes would be dissolved during corrosion. Therefore, simply using reaction (1) to estimate the corrosion rate of a magnesium alloy could result in an error. How significant the error depends on the amount of the other alloying elements involved in the corrosion of the α matrix. A reasonable estimation of the maximum theoretical error can be made based on the solid solubility of the alloying elements in the α matrix. For example, for AZ91 alloys, the theoretical maximum error could be <10% (a certain amount of Al and Zn becomes very stable in the βb phase, so the concentrations of Al and Zn in the α phase are much lower than 9% and 1%). Also, considering the non-uniform corrosion on the α phase, the low alloying areas in the α phase are preferentially corroded while the highly alloying zones remain un-corroded [19, 20]. Thus, the error in practice would be expected to be much lower than the theoretical error. Further more, the dissolution of some alloying elements can also give out hydrogen, which to some extent also reduces the error.

To confirm this, an AZ91 alloy was made in the laboratory whose composition is 8.9%Al and 0.70%Zn. After several hours of immersion in 1N NaCl (pH11), the ratio of Mg:Al:Zn in the test solution was found to be 14.2 : 1: 0.017. The actual dissolved Al was less than 9%, and the dissolution of Zn was negligible.

Therefore, for a magnesium alloy, even though reaction (1) only describes the hydrogen evolution reaction resulting from the dissolution of elemental Mg in the α matrix of the alloy, the theoretical error of using hydrogen evolution to estimate the corrosion rate of the alloy is insignificant. In other words, the hydrogen evolution rate should be a reliable indication of the corrosion of a magnesium alloy.

Experimental Method

An ideal device for hydrogen evolution collection is easy to set up and operate. Figure 1 schematically illustrates a simple set-up that was used for the hydrogen evolution rate measurements in this study. This is actually a classic set-up for detection of the negative difference effect.

The magnesium specimen was put in a beaker containing the test solution. A funnel was placed over the specimen, which ensured the collection of all the hydrogen from the specimen surface as well as from any undermined metal particles at the bottom of the beaker. A burette was mounted over the funnel, and was initially full of the test solution. The hydrogen collected by the funnel went into the burette and gradually displaced the test solution in the burette. In this way the volume of the evolved hydrogen was easily measured by reading the position of the test solution level in the burette.

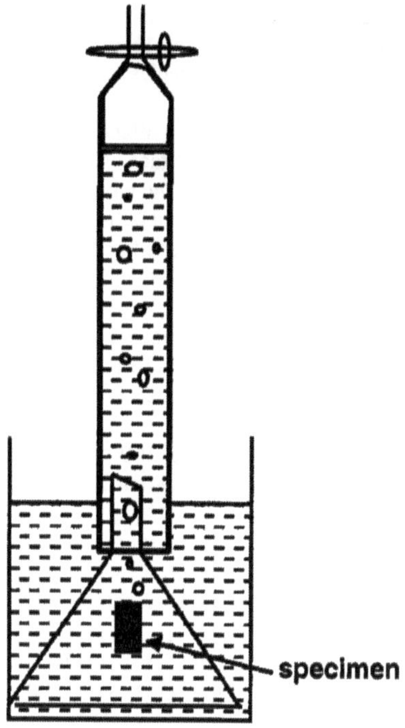

Figure1: Schematic illustration of the set-up for measurement of the volume of hydrogen evolved.

The measured hydrogen volume can be easily converted into the number of gram moles using the ideal gas laws. Simply assuming the number of moles of the evolved hydrogen is equal to the number of moles of the dissolved magnesium, the weight-loss or weight-loss rate of the specimen can be calculated.

The above set-up is very simple. It can be modified into more complicated forms with other functions of obtaining other results during the measurement of hydrogen evolution. For example, the specimen can be connected to a potentiostat as a working electrode, so the corrosion potential or AC impedance of the specimen can also be monitored during hydrogen collection. After that, the specimen can still be weighed to obtain the weight loss and weight-loss rate, and the test solution analysed for the amounts of all elements dissolved from the specimen.

In this study, hydrogen evolved from various specimens was collected in the manner described above. About 450 ml of the test solution prepared with distilled water and analytical grade chemicals in a 500~600ml beaker was used. The size of the burette used was 50 or 100ml.

Slightly different techniques were also employed to measure the weight loss for different groups of specimens.

For specimens 1~4 and 6~8 (see Table I), their weight-losses were measured through ICPAES analysis of the test solution after hydrogen collection. This was performed as follows. After the specimen was removed from the beaker, the solution left in the burette was poured back into the beaker, and approximately 100 ml of 5N HCl was added to the solution to completely dissolve the corrosion products precipitated at the bottom of the beaker. The solution was then made up to 1 litre with distilled water. This solution was analysed by the ICPAES technique for the concentrations of magnesium and other main alloying elements, such as Al, Zn, etc. The concentrations of these metallic ions in the solution were converted into the total amount of dissolved specimen in either number of gram moles or weight loss.

The accuracy of this method could be affected by some corrosion products stuck on the specimen surface, so it is best used for short-term immersion experiments. The advantage of this method is that the un-corroded areas and the metallic substrate underneath the corrosion products remain untouched and no other dissolution occurs.

The weight loss of specimens 9~15 (see Table I) was measured through dissolving corrosion products in a hot chromic solution. After hydrogen collection, the specimen was taken out from the test solution, washed with distilled water, and dried quickly in hot flowing air. Then it was put into a 100 g/L CrO_3 solution at about 90~95°C for about 5~10 min. After that, the specimen was washed with distilled water and dried again, then weighed. The weight was compared with the original weight before the specimen was immersed in the test solution for hydrogen collection to determine the corrosion weight loss.

The weight loss obtained in this way could sometimes be overestimated because the un-corroded areas and the metallic substrate underneath the corrosion products could also be dissolved to some extent. Figure 2 shows the range of weight loss obtained due to dissolution of several un-corroded alloys in chromic solution.

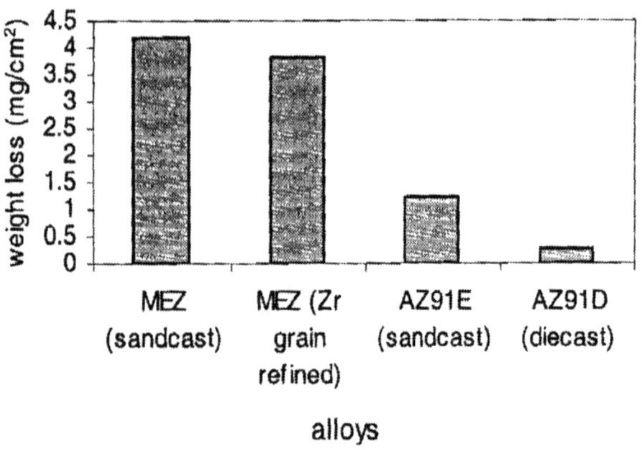

Figure 2: Weight loss of different un-corroded specimens immersed in hot chromic solution 100g/L CrO_3 for 8min.

The dissolution of the un-corroded metal for some alloys in the hot chromic acid is sometimes significant. To reduce the error caused by the dissolution of the un-corroded areas and substrate under the corrosion products during removal of the corrosion products, an un-corroded specimen is usually used as a reference in the chromic cleaning process. It is expected that the weight loss caused by the chromic cleaning on the un-corroded reference specimen is approximately equal to the dissolution of the un-corroded areas and the metallic substrate under the corrosion products. Certainly, this is not an accurate compensation for the overestimated weight loss, as the corroded specimen has an unknown real surface area, so it is impossible to convert the dissolution of the reference specimen into a corrosion factor for the corroded specimen during the chromic acid cleaning.

Similar weight-loss measurement procedures were also applied to specimen 5, but instead of hot chromic acid, a 200g/L CrO_3 + 10g/L $AgNO_3$ solution was used to remove the corrosion products at room temperature. As $AgNO_3$ can precipitate chloride, so this solution can be reused. However, the problem of dissolution of un-corroded areas and the metallic substrate under the corrosion products for some alloys still exits.

In addition to the different weight-loss measurement methods, various test solutions were used in the experiments. Table 1 lists the details of the various specimens and experimental conditions.

Table I Specimens and testing conditions

No.	specimens	testing conditions
1	Mg (99.96%)	immersion in 1N NaCl (pH6), ICPAES analysis for weight loss
2	Mg (99.96%)	immersion in 1N NaCl (pH11), ICPAES analysis for weight loss
3	Mg (99.96%)	immersion in 1N Na(OH), ICPAES analysis for weight loss
4	Mg (99.96%)	immersion in 1N HCl, ICPAES analysis for weight loss
5	Mg (99.6%)	immersion in 5% NaCl, 200g/L CrO_3 +1g/L $AgNO_3$ cleaning corrosion products at room temperature
6	A6	immersion in 1N NaCl (pH11), ICPAES analysis for weight loss
7	AZ21	immersion in 1N NaCl (pH11), ICPAES analysis for weight loss
8	AZ91D diecast	immersion in 1N NaCl (pH11), ICPAES analysis for weight loss
9	AZ91D-1*	immersion in 5% NaCl, hot chromic solution cleaning corrosion products
10	AZ91D-2*	immersion in 5% NaCl, hot chromic solution cleaning corrosion products
11	AZ91D-3*	immersion in 5% NaCl, hot chromic solution cleaning corrosion products
12	AZ91D-4*	immersion in 5% NaCl, hot chromic solution cleaning corrosion products
13	AZ91D-5*	immersion in 5% NaCl, hot chromic solution cleaning corrosion products
14	AZ91D-6*	immersion in 5% NaCl, hot chromic solution cleaning corrosion products
15	AZ91D-7*	immersion in 5% NaCl, hot chromic solution cleaning corrosion products

* same alloy from different castings

Results and Discussion

Accuracy and Reliability

It is essential to determine whether or not the method of estimating corrosion rate through measuring hydrogen evolution is accurate in experiments. It is impossible to obtain the theoretical corrosion rate of a specimen in a certain test solution. Therefore, experimentally measured weight loss has to act as a standard to calibrate the accuracy of the new method.

Comparison of corrosion rates of various specimens measured by weight loss and the hydrogen evolution method is shown in figure 3.

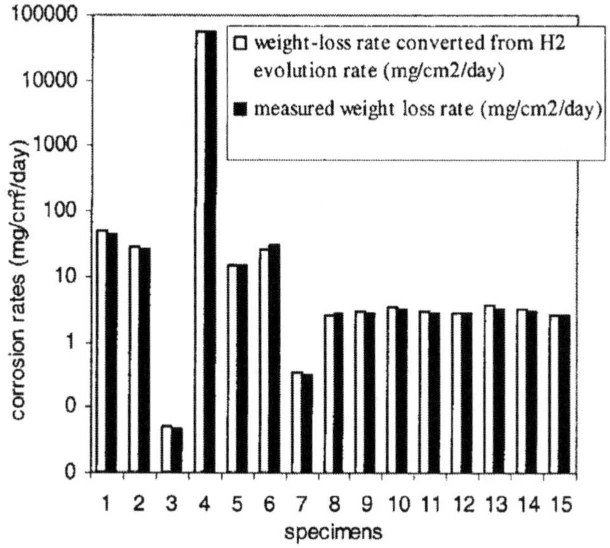

Figure 3: Weight-loss rates and hydrogen evolution rates of various specimens; the hydrogen evolution rates have been converted into weight-loss rates of magnesium.

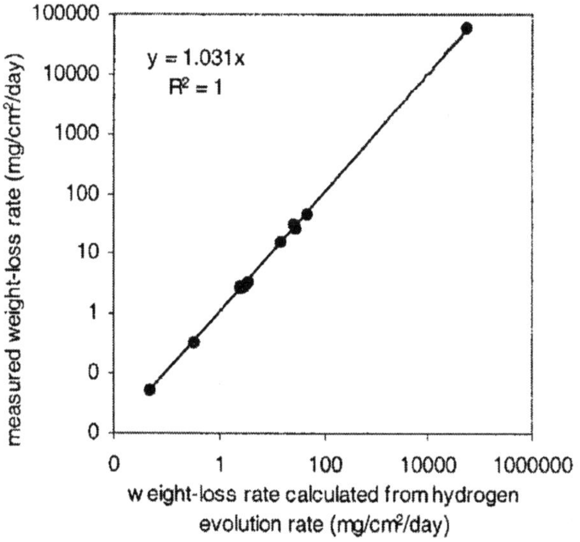

Figure 4: Correlation between the corrosion rates of specimens 1~15 measured by weight loss and hydrogen evolution collection.

It can be seen that the differences between the rates obtained by the two different methods for all the specimens are insignificant. To quantify the relationship between the two measures of corrosion rate, the relationship is shown in Fig.4 along with the correlation coefficient.

The regression line has a slope value of 1.03. It is so close to the ideal value 1 that it can be concluded without doubt that the corrosion rate measured by hydrogen evolution is exactly equal to the weight-loss rate. Also, the regression coefficient R=1 means that the above results are very reliable. This suggests that the hydrogen evolution measurement is a reliable and accurate method for estimating the corrosion rate of magnesium and its alloys.

It is difficult to determine whether the hydrogen evolution rate is more accurate than the weight loss measurement. However, compared with the estimation of corrosion rate based on polarisation curves, hydrogen evolution collection is undoubtedly very reliable. This can be demonstrated by the following results.

The polarisation curves of sand cast MEZ and diecast AZ91D are shown in Figure 5.

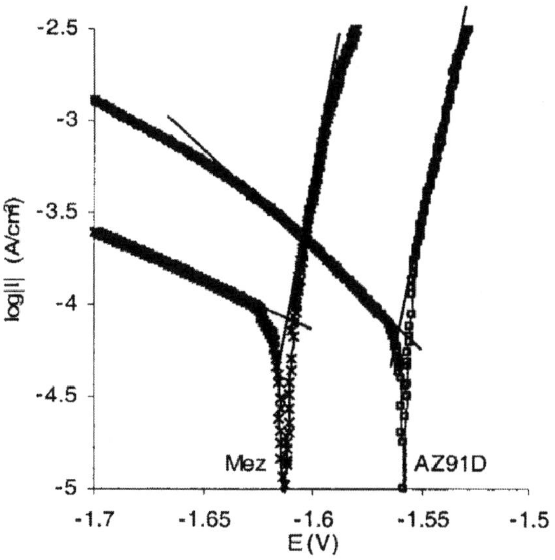

Figure 5: Polarisation curves for sand cast MEZ and diecast AZ91D in 5% NaCl (pH11)

The corrosion rates of these two alloys estimated from their polarisation curves (through Tafel slope extrapolation) are almost the same, i.e. MEZ has a corrosion rate similar to AZ91D if based on the polarisation curves. However, the hydrogen evolution from MEZ was measured to be about 40 times higher than that of diecast AZ91D. This is completely different from the result from the polarisation curves. The weight-loss measurements from a simple half immersion test of these two alloys in 5% NaCl for one week followed by the hot chromic acid cleaning of corrosion products gave a ratio of weight-losses of these two alloys, MEZ:AZ91D=25:1, which is consistent with the results obtain from the hydrogen evolution method. This means that the hydrogen evolution method more reliably estimated the corrosion rates of these two alloys than the polarisation curves. Therefore, the estimation based on the polarisation curves is incorrect.

The incorrect estimation based on the polarisation curves alone results from the negative difference effect. In this case, the Tafel extrapolation is not reliable. In fact, if the theoretical expression of a polarisation curve is unknown, it is impossible to calculate the corrosion rate from the experimental polarisation curve.

Monitoring of corrosion process

In one measurement, the increasing volume of collected hydrogen indicates the increasing amount of specimen corroded. Therefore, by monitoring the rate of volume change during measurement of hydrogen evolution, the corrosion process of the specimen can be recorded.

Figure 6 displays hydrogen evolution rates of various magnesium-rare earth-zinc alloys immersed in 5% NaCl in combination with the weight loss rates of these alloys in the test solution after different periods of time.

It should be stressed that the results obtained by the hydrogen evolution measurements in figure 6 are instantaneous rates whereas the weight-loss results are average rates. They could not be exactly the same. The figure shows a good consistency in the corrosion rate changing trends between the results obtained by weight-loss and hydrogen evolution.

Figure 6 also demonstrates a very attractive feature of the hydrogen evolution method, i.e., the hydrogen evolution curve exhibits more detailed information than the weight-loss

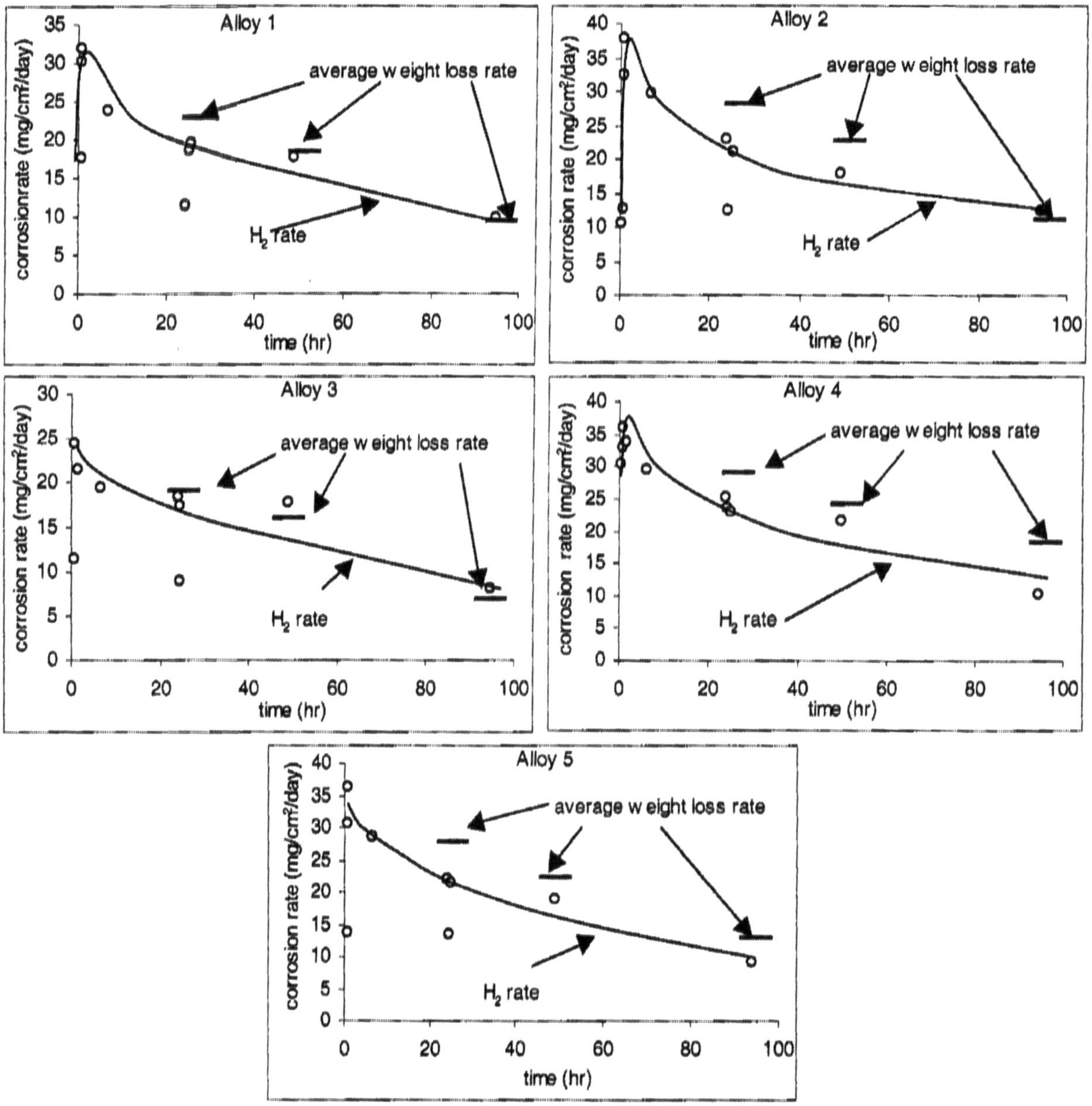

Figure 6: The hydrogen evolution rates (that have been converted into the equivalent weight-loss rates of Mg) and the measured weight-loss rates of magnesium-rare earth-zinc alloys immersed in 5% NaCl at different exposure times.

measurement. For example, the hydrogen evolution rates of most magnesium-rare earth-zinc alloys increased at the start of the test and then gradually decreased, becoming slower and slower with time. This phenomenon is completely missed in the weight loss measurements. The latter only provided the final average information.

Another attractive feature of the hydrogen evolution method is its high experimental efficiency. The results from a sequence of weight-loss measurements can be obtained in one simple experiment (one curve) continuously measuring hydrogen evolution.

Also, the hydrogen evolution method is easier to set up and operate than the weight-loss method. There is no need to remove the corrosion products or analyse the test solution.

Conclusions

The hydrogen evolution method has been shown to be an easy, accurate and reliable method for measuring the corrosion rate of magnesium and its alloys.

The features of this method include:

(1) the amount of the collected hydrogen is equal to the amount of the dissolved magnesium; and
(2) the hydrogen evolution rate reveals the instantaneous corrosion rate.

The advantages of this method are:

(1) smaller theoretical and experimental errors are introduced into the final estimated corrosion rate by the hydrogen evolution collection method;
(2) the experiment is easy to set up and operate; and
(3) the method is suitable for corrosion monitoring of magnesium and its alloys.

Acknowledgment: This work was supported by a grant from the University of Queensland. CAST is funded under the Australian Government's Cooperative Research Centre Program. The authors would also like to thank Mr. Zhiming Shi and Ms Li Zheng for their assistance in some experiments.

References

1. J.K.King, "Development of Practical High Temperature Magnesium Casting Alloys", K.U.Kainer, ed., Magnesium Alloys and Their Applications (Munich, Germany, DGM, 2000), 14-22.

2. K.Pettersen, et al., J, "Creep Resistant Mg Alloy Development", K.U.Kainer, ed., Magnesium Alloys and Their Applications (Munich, Germany, DGM, 2000), 29-34.

3. C.Jaschik, H.Haferkamp, M.Niemeyer, "New Magnesium Wrought Alloys" K.U.Kainer, ed., Magnesium Alloys and Their applications (Munich, Germany, DGM, 2000), 41-46.

4. T.Horie, et al., "Development of High Creep-Resistant Magnesium Alloy Strengthened by Ca Addition", H.I.Kaplan, J.N.Hryn, and B.B.Clow, ed., Magnesium Technology 2000 (TMS and IMA, Nashville, Tennesssee, 2000) 261-269

5. R.D.Carnahan, et al., "Development of Semi-solid Molded Magnesium Components From Alloys with Improved High Temperature Creep Properties", H.I.Kaplan, J.N.Hryn, and B.B.Clow, ed., Magnesium Technology 2000 (TMS and IMA, Nashville, Tennesssee, 2000) 403-409

6. S.Krishnamurthy, E.Robertson and F.H.Froes, "Rapidly Solidified Magnesium Alloys Containing Rare Earth Additions", Henry G.Paris and W.H.Hunt, ed., Advances in Magnesium Alloys and Composites, (International Magnesium Association and the Non-Ferrous Metals Committee, The Minerals, Metals, and Materials Society, Phoenix, Arizona, January 26, 1988) 77-89

7. X.Hallopeau, T.Beldjoudi, C.Fiaud, L.Robbiola, "Electrochemical Behaviour and Surface Characterisation of Mg and AZ91E Alloy in Aqueous Electrolyte Solutions Containing XO_y^{n-} Inhibiting Ions", G.W.Lorimer, ed, Proceedings of the Third International Magnesium Conference, (The Institute of Materials, London, 1997), 713-724.

8. P.L.Bonora, M.Andrei, A.Eliezer, E.Gutman, "Corrosion Behaviour of Stressed Magnesium Alloys", E.Aghion and D.Eliezer, ed., Magnesium 2000, Proceedings of the Second Israeli International Conference on Magnesium Science and Technology, (Dead Sea, Israel, 2000), 410-416

9. A.Eliezer, et al., "Dynamic and Static Corrosion Fatigue of Magnesium Alloys in Electrolytic Environment", E.Aghion and D.Eliezer, ed., Magnesium 2000, Proceedings of the Second Israeli International Conference on Magnesium Science and Technology, (Dead Sea, Israel, 2000), 356-362

10. S.Mathieu, J.Hazan, C.Rapin, P. Steinmetz, "Corrosion Behaviour of Die Cast and Thixocast AZ91D Alloys - The Use of Sodium Linear Carboxylates as Corrosion Inibithors", E.Aghion and D.Eliezer, ed., Magnesium 2000, Proceedings of the Second Israeli International Conference on Magnesium Science and Technology, (Dead Sea, Israel, 2000), 339-346

11. G.G.Perrault, in Encyclopedia of Electrochemistry of the Elements, Vol.VIII, ed. A.J.Bard, (Marcel Dekker, New York, 1978), p262

12. Guangling Song et al., "The Electrochemical Corrosion of Pure Magnesium in 1N NaCl", Corrosion Science, 39 (1997) 855-875

13. M.R.Bothwell, in The Corrosion of Light Metals, ed. H.P.Godard, (John Wiley and Sons, New York, 1967), 257

14 . G.L.Makar and J.Kruger, "Corrosion of Magnesium", International Materials Reviews, 38(1993), 138-153

15. G.L.Makar and K,Kruger, "Corrosion Studies of Rapidly Solidified Magnesium Alloys", J.Electrochem.Soc. 137 (1990), 414-421

16. G.L.Makar, J.Kruger and A.Joshi, "The Effect of Alloying Elements on the Corrosion Resistance of Rapidly Solidified Magnesium Alloys", ed., Henry G.Paris and W.H.Hunt, Advances in Magnesium Alloys and Composites, (International Magnesium Association and the Non-Ferrous Metals Committee, The Minerals, Metals, and Materials Society, Phoenix, Arizona, 1988) 105-121

17. Guangling Song, A. Atrens, M. Dargusch, "Influence of Microstructure on the Corrosion of Diecast AZ91D", Corrosion Science, 41 (1999) 249-273

18. Guangling Song et.al., "Corrosion Behaviour of AZ21, AZ501 and AZ91 in Sodium Chloride", Corrosion Science, 40 (1998), 1769-1791

19. O.Lunder, et al., "The Role of Mg17Al12 phase in the Corrosion of Mg Alloy AZ91", Corrosion, 45(1989), 741-748

20. R.Ambat, et al., "Evaluation of Microstructural Effects on Corrosion Behaviour of AZ91D Magnesium Alloy", Corrosion Science, 42 (2000) 1433-1455

A NOVEL TECHNIQUE TO EVALUATE THE CORROSION BEHAVIOR OF MAGNESIUM ALLOYS

Basant L. Tiwari and John J Bommarito

General Motors R&D Center
30500 Mound Road
Warren, Michigan 48090

Abstract

A novel technique for measuring the intrinsic corrosion rate of magnesium alloys has been developed. As a magnesium alloy sample dissolves in a 5% NaCl solution, the dissolution rate is determined by measuring the amount of HCl added to the NaCl solution to control the pH between 5 and 7. The corrosion rate is determined from the slope of the curve where the dissolution rate reaches a steady state. The technique was used to compare the corrosion behavior of newly developed creep resistant Mg-Al-Ca (AC) alloys with several known magnesium alloys. It was shown that the corrosion rate of AC52 (Mg-5%Al-2%Ca) is comparable to AZ91, meeting one of the several design criteria for new, creep-resistant alloys. This technique provides a fast way to screen the effect of small composition changes on corrosion behavior of Mg alloys. Because of the short test time, this technique provides a powerful tool to accelerate alloy and product development with Mg alloys.

Introduction

A research project was carried out to develop creep-resistant magnesium alloys for powertrain applications [1-2]. One of the requirements of the new alloys was corrosion resistance comparable to AZ91. The salt spray corrosion test normally used to determine the corrosion behavior is based on measuring the weight difference of the samples before and after the tests [3]. It is slow (3 to 10 days), labor intensive, and is error-prone due to the cleaning procedure used. The corrosion samples also lose weight due to crumbling during the tests. Because of these drawbacks, the salt spray method is not suitable for rapidly measuring the intrinsic corrosion rate of the alloys but is very useful in estimating their overall corrosion behavior. Since the test has many drawbacks, a program was initiated to develop a fast corrosion-screening test, which could quickly determine the effect of small composition changes on the corrosion rate of magnesium alloys. This report presents such a method. The novel method will accelerate alloy and product development and could ultimately complement the salt spray test.

Theoretical Consideration

The corrosion of magnesium in salt water is galvanic in nature (Fig. 1). For corrosion to occur, there must be an anodic site on a microscopic scale where magnesium is selectively oxidized according to reaction (1), producing Mg ions in the solution.

$$Mg = Mg^{2+} + 2e^- \quad (1)$$

There must be a cathodic site where hydrogen ions are reduced consuming the electrons produced at the anode and producing hydrogen gas according to reaction (2).

$$2H^+ + 2e^- = H_2 \quad (2)$$

The cathodic and anodic sites must be connected by a water film with appreciable ionic conductivity (such as salt water) and both sites must be electronically connected through the substrate. The combination of reactions (1) and (2) gives the corrosion reaction (3), according to which magnesium simply displaces hydrogen out of the water, producing hydrogen gas.

$$Mg + 2H^+ = Mg^{2+} + H_2 \quad (3)$$

Cathodic reduction of hydrogen described by reaction (2) proceeds at sites which have lower oxidation potential than magnesium (Table I) [4]. These sites consist mostly of impurities such as Fe, Ni, and Cu, which have much lower oxidation potential than magnesium. Intermetallic precipitates such as $Mg_{17}Al_{12}$ and MgZn also provide cathodic sites. The more impurities and precipitates in the magnesium alloy, the more there are cathodic sites for hydrogen evolution to take place and, consequently, more magnesium corrodes. This is the reason why Fe, Ni and Cu are considered detrimental to magnesium corrosion. Therefore, their concentration levels must be kept below the levels specified by ASTM B94 (0.004% Fe, 0.002% Ni, 0.015% Cu), to evaluate the effect of alloying elements such as Ca, Si and Sr on corrosion rates of magnesium alloys.

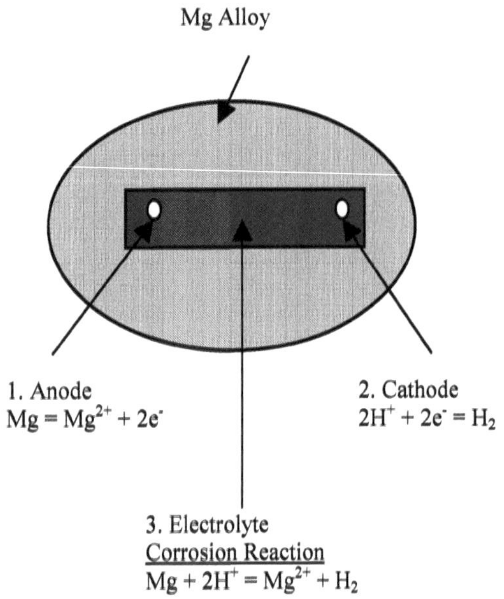

Fig. 1. Schematic presentation of galvanic corrosion concept.

Table I Standard oxidation potential of metals in acid solution [4]

Reaction	Potential, V
$Sr = Sr^{2+} + 2e^-$	2.89
$Ca = Ca^{2+} + 2e^-$	2.87
$Mg = Mg^{2+} + 2e^-$	2.73
$Al = Al^{3+} + 3e^-$	1.66
$Si + 2H_2O = SiO_2 + 4H^+ + 4e^-$	0.86
$Zn = Zn^{2+} + 2e^-$	0.76
$Fe = Fe^{2+} + 2e^-$	0.44
$Ni = Ni^{2+} + 2e^-$	0.25
$H_2 = 2H^+ + 2e^-$	0.00
$Cu = Cu^{2+} + 2e^-$	-0.34

Hydrogen overpotential is the minimum amount of potential, versus a reversible hydrogen electrode, needed to evolve hydrogen on a given substrate. Table II shows that Fe has the lowest value of overpotential [5], and, therefore, it provides the most effective site for hydrogen evolution. In addition, Fe has almost no solubility in magnesium and it does not form any intermetallic compound with magnesium [6]. Therefore, Fe exists in magnesium in elemental form, making it most detrimental to magnesium corrosion. In Mg alloys, however, iron forms intermetallic compounds with the alloying elements such as Mn, which are somewhat less detrimental.

Table II Hydrogen overpotential of various metals [5]

Metals	Overpotential, V
Iron	0.08
Nickel	0.21
Copper	0.23
Zinc	0.70

As pointed above, corrosion of Mg in salt water proceeds according to reaction (3), and the rate of corrosion is normally measured by the weight loss of the magnesium sample over a long testing period. In a short period, however, the weight loss would be negligible. As the corrosion reaction proceeds, the pH of the solution increases and $Mg(OH)_2$ precipitates on the surface of the sample often causing a weight gain. Thus the weight-loss method is not capable of measuring corrosion rates over short times.

Other methods to rapidly determine the corrosion rate are 1) measuring the amount of hydrogen gas evolved [7] and 2) measuring the pH of the salt solution [8] with time. In both methods, however, the hydrogen concentration of the salt solution decreases significantly with time, causing precipitation of $Mg(OH)_2$ on the surface of the samples. This affects the intrinsic corrosion rate. Therefore, both methods are not capable of measuring the effect of a small change in composition on the intrinsic rate of magnesium corrosion.

A novel and probably the only chemical means to rapidly determine the intrinsic rate of magnesium corrosion is to measure the amount of hydrogen ion consumed by reaction (3). This can be done by periodically adding HCl to the salt solution to keep the pH constant over a given period of time. The pH of the solution in this study was kept between 5 and 7, which closely corresponds to the initial pH of 5% NaCl solution and neutral water, respectively. The pH range closely approximates the pH of water accumulated on salted roads. The variation in pH between 5 and 7 does not affect the overall corrosion rate significantly and it also prevents any precipitation of $Mg(OH)_2$ on the magnesium samples.

Measuring the rate of HCl consumption by reaction (4) provides an effective way to determine the corrosion rate. Reaction (4) was obtained by replacing hydrogen ions in reaction (3) with hydrochloric acid.

$$Mg + 2HCl = MgCl_2 + H_2 \qquad (4)$$

Sulfuric acid cannot be used to adjust the pH because it may react with Ca in the alloy forming $CaSO_4$, which has a very low solubility and may precipitate on the surface of the samples.

The rate of corrosion can be determined by Equation (5), which was deduced by performing a mass balance on reaction (4):

$$R_{corr} = [(V_{HCl}/h) \cdot M_{HCl}] \cdot [12.1525/A], \text{ mg/h/mm}^2 \qquad (5)$$

where, (V_{HCl}/h) is the volume (mL) of HCl used per hour, M_{HCl} is the concentration of HCl solution in mol/L, A is the area of a Mg sample in mm^2, and R_{corr} is the corrosion rate in $mg/h/mm^2$.

Experiment

Magnesium Samples

Figure 2 shows the photograph of a magnesium sample used in the corrosion test. The sample size was 39.7 x 39.7 x 10.3 mm with an 11.1 mm diameter hole in the center. The sample was suspended in the salt solution with a nylon string. Samples of AZ91, AM50, ZC63, and several types of AC alloys were tested. All the alloys except ZC63 were die cast. The ZC63 sample was machined from a slab, which was cast in a permanent mold. The ZC63 sample was also slightly larger than the samples of the other alloys with surface area of 5345 mm^2 exposed to the salt solution, compared to 4953 mm^2 for the other samples. Magnesium alloy AM50 was used as the feed stock for preparing the AC alloys. The chemical compositions of the alloys are given in Table III.

Fig. 2. Photograph of Mg alloy corrosion samples.

Table III Chemical compositions of magnesium alloys

Alloys	Chemical Compositions, wt%					
	Al	Ca	Si	Sr	Mn	Zn
AZ91	8.9	-	-	-	0.20	0.7
ZC63	-	-	-	-	0.05	6.2
AM50	4.7	-	-	-	0.27	-
AC52	4.5	1.9	-	-	0.27	-
AC53	4.5	3.0	-	-	0.27	-
AC53+0.3Si	4.5	2.9	0.26	-	0.28	-
AC53+0.3Si+0.1Sr	5.4	2.9	0.27	0.11	0.26	-
AC53+0.3Si+0.15Sr	5.7	3.0	0.28	0.15	0.26	-
AC53+0.03Sr	4.7	3.1	-	0.03	0.30	-
AC53+0.07Sr	5.0	3.1	-	0.07	0.28	-
AC53+0.15Sr	4.7	3.1	-	0.15	0.27	-
AC62+0.1Sr	5.4	1.9	-	0.10	0.30	-
AC62+0.2Sr	5.9	1.8	-	0.17	0.29	-

Reagents

High purity HCl, NaCl, and 1M ± 0.005M NaOH solution were used in the study. The 1M NaOH solution was used to precisely determine the concentration of HCl solution used to adjust the pH of the salt solution by titration.

Experimental Setup and Test Procedure

Figure 3 shows the schematic diagram of the experimental setup used to conduct the corrosion screening tests. It consists of a 1000 mL beaker containing 750 mL of 5% NaCl solution. A magnesium specimen was freely suspended in the salt solution and the solution was gently stirred at a constant speed with a magnetic stirrer. A pH meter was used to monitor the pH of the solution continuously. The initial pH of 5% NaCl solution was close to 5 and it increased as the Mg dissolved in the solution. When the pH reached 7, it was brought down to 5 by adding a few drops of HCl solution of known concentration. The HCl solution was held in a burette mounted over the beaker. Both the cumulative amount of HCl and the dissolution time were recorded. At the end of a test, the leach solution was acidified and made up to 1000 mL. The solution was analyzed for magnesium and other alloying elements. All the tests were conducted at room temperature.

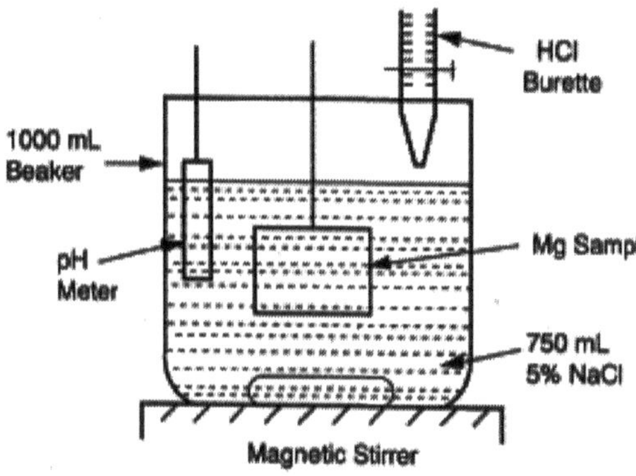

Fig. 3. Schematic of the experimental setup for corrosion tests.

Results and Discussion

Figure 4 shows a typical corrosion curve of AC53 wherein the cumulative amount of HCl added into the salt solution is plotted against time. This curve can be divided into three parts. The corrosion rate is initially fast, but slows down and reaches a steady state in the middle, and then starts increasing again at the end. Initially, the corrosion rate depends on the surface condition of the sample; cast versus machined surface, and the amount of magnesium oxide on the surface. The corrosion rate is mostly controlled by reactions 3 and 6. In reaction 3, magnesium reacts with hydrogen ions, producing hydrogen gas and magnesium ions in the solution.

$$MgO + 2H^+ = H_2O + Mg^{2+} \quad (6)$$

In reaction 6, magnesium oxide reacts with hydrogen ions producing water and magnesium ions. In the middle portion of the corrosion curve, the rate is exclusively controlled by reaction 3, representing the intrinsic corrosion of Mg. The rate depends on the composition and also on the temper of the alloys. In the third portion of the corrosion curve, disintegration of Mg caused by the H_2 bubbles in form of fine particles was observed. This increases the effective surface area and, therefore, the corrosion rate. The slope of the curve in the middle where the rate is constant represents the intrinsic corrosion rate of the alloy. Therefore, corrosion tests were stopped before the samples started to erode and the slope of the curve where corrosion reaches a steady state was used to determine the corrosion rate using Equation 5.

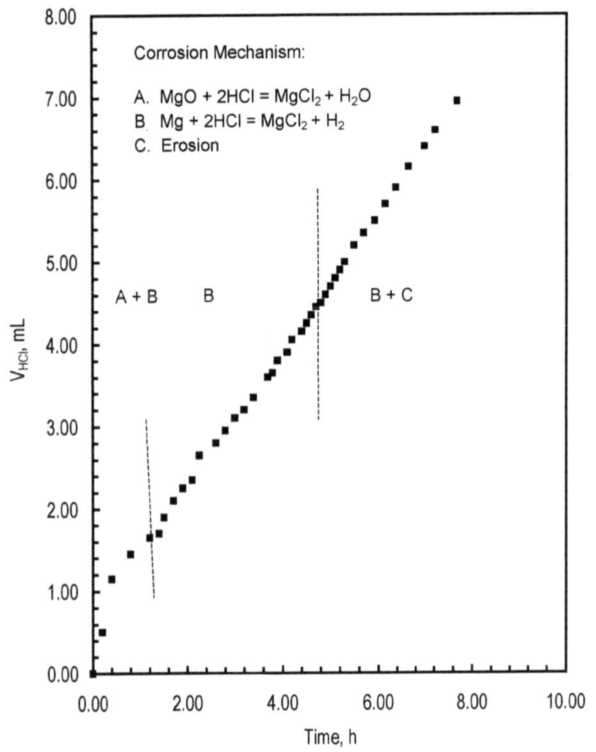

Fig. 4. A typical corrosion curve generated with AC53 Mg alloy and 0.922 M HCl to maintain the pH between 5 and 7.

Figure 5 shows the data points of two corrosion curves produced separately using AC53 alloy. The scatter in the data points of both curves is negligible, suggesting that the corrosion tests are reasonably reproducible.

Figure 6 shows the corrosion curves for ZC63 and AC53. The sample of ZC63 had a machined surface and the corrosion rate is constant from the beginning due to the absence of an excessive oxide layer on the surface. On the other hand, AC53 had a cast surface with oxide on it and, therefore, the initial corrosion rate is higher than ZC63.

Figure 7 shows the corrosion curves for several Mg alloys with straight lines fitted using the least square analysis to the last three hours of the data points where the corrosion rates depend mostly on the compositions of the alloys. The corrosion rates of the alloys were determined by substituting the slopes of the straight lines in Equation 5. The slopes and the corrosion rates for all the alloys are reported in Table IV. The corrosion rates of several known and two new alloys are compared in Figure 8. It is quite obvious from the figure that AZ91 has the lowest rate of corrosion followed by AC52, AC53, ZC63 and AM50.

Since both AC52 and AC53 were prepared by adding Ca to AM50, the addition of Ca first decreases the corrosion rate and then increases it. Calcium initially dissolves in Mg and improves its corrosion properties by lowering its activity. On further increasing the concentration of Ca in Mg, Ca forms an intermetallic compound $(Mg,Al)_2Ca$ with Al and Mg [9], which is probably more cathodic in nature than Mg and therefore increases the corrosion rate. The overall effect of an alloying element on corrosion rate depends on where and in what form and amount it is present in the alloy.

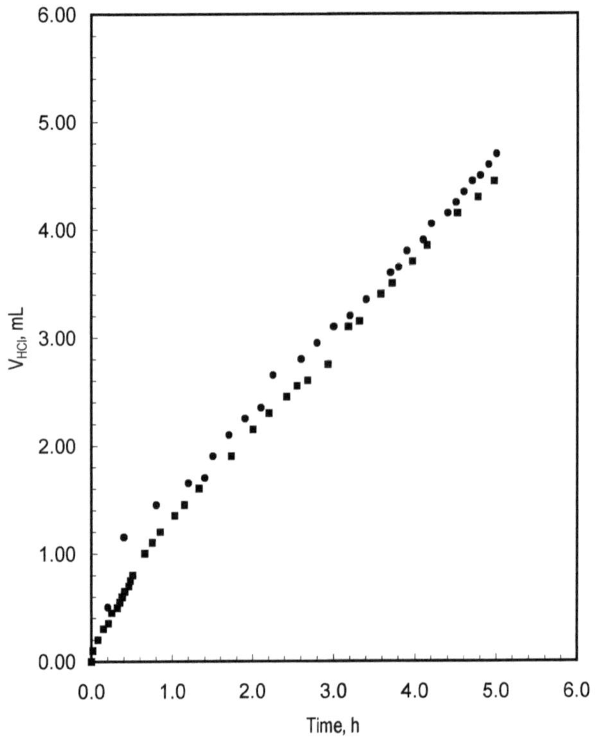

Fig. 5. Comparison of two corrosion curves for AC53 showing good reproducibility of the corrosion technique.

Table IV Summary of the corrosion tests results

Alloy	HCl Conc. (mol/L)	Slope (mL/h)	R_{corr} x10^{-3} (mg/h/mm^2)
AZ91	0.922	0.441	0.997
AM50	0.922	1.191	2.694
ZC63	0.922	1.111	2.328
AC52	0.922	0.511	1.155
AC53	0.922	0.796	1.801
AC53+0.03%Sr	0.461	1.841	2.082
AC53+0.07%Sr+	0.922	0.765	1.730
AC53+0.15%Sr	0.461	1.161	1.313
AC53+0.3%Si	0.461	2.175	2.457
AC53+0.3%Si+0.1%Sr	0.461	1.767	1.999
AC53+0.3%Si+0.15%Sr	0.461	1.320	1.493
AC62+0.1%Sr	0.486	1.566	1.865
AC62+0.2%Sr	0.486	1.499	1.785

Figure 9 shows the corrosion rates for the alloys in %wt loss as determined by the salt spray method. AZ91 again, as compared to the results shown in Figure 8, had the lowest rate of corrosion followed by AM50, AC52, AC53, and ZC63. Unlike the spray method, our method predicts that AM50 has a higher rate of

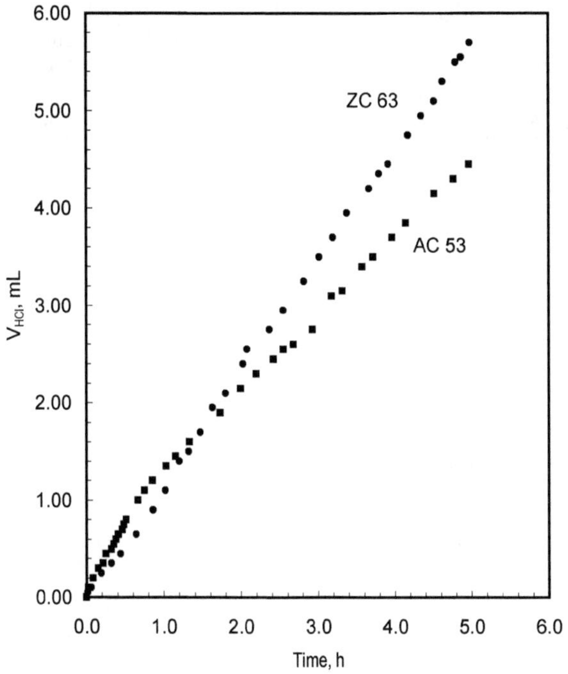

Fig. 6. Comparison of corrosion curves for magnesium alloys ZC63 and AC53.

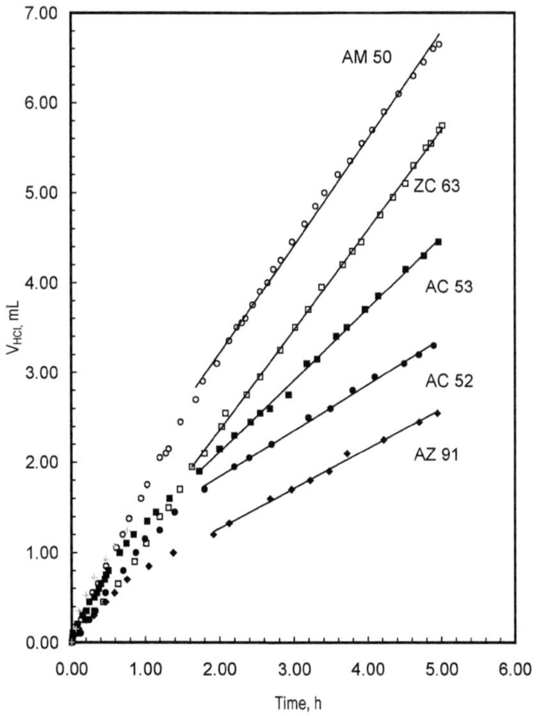

Fig. 7. Corrosion curves for several magnesium alloys showing steady- state slopes.

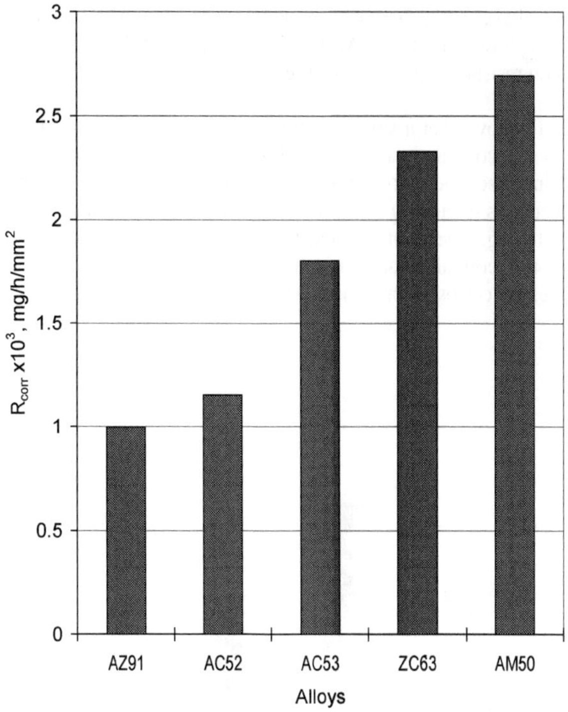

Fig. 8. Corrosion rates of Mg alloys in 5% NaCl at pH 5 - 7.

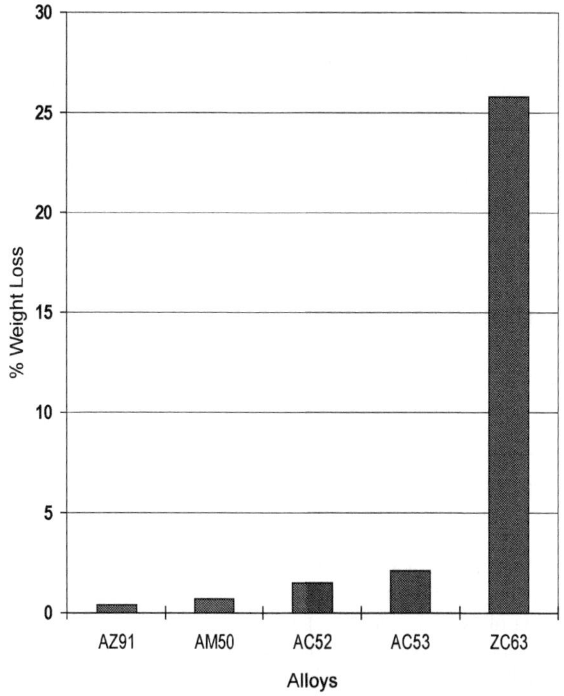

Fig. 9. Corrosion behavior of Mg alloys determined by salt spray method.

corrosion than ZC63. It should be kept in mind that the salt spray method determines the overall corrosion rate that also includes weight loss due to disintegration, and this could be the cause for ZC63 to have a higher overall corrosion rate than AM50. Both methods, however, rank AZ91, AC52, and AC53, for their corrosion resistance in the same order.

Figure 10 shows that a small addition (0.03-0.15%) of Sr to AC53 reduces the corrosion rate of the alloy. Addition of 0.3% Si in AC53 increases the corrosion rate of the alloy (Figure 11), but the rate decreases on adding Sr to AC53+0.3%Si. Both Ca and Sr are anodic to Mg, (see Table I), and therefore both in small amounts reduce the corrosion of magnesium. On other hand, silicon increases the corrosion because it is cathodic to magnesium.

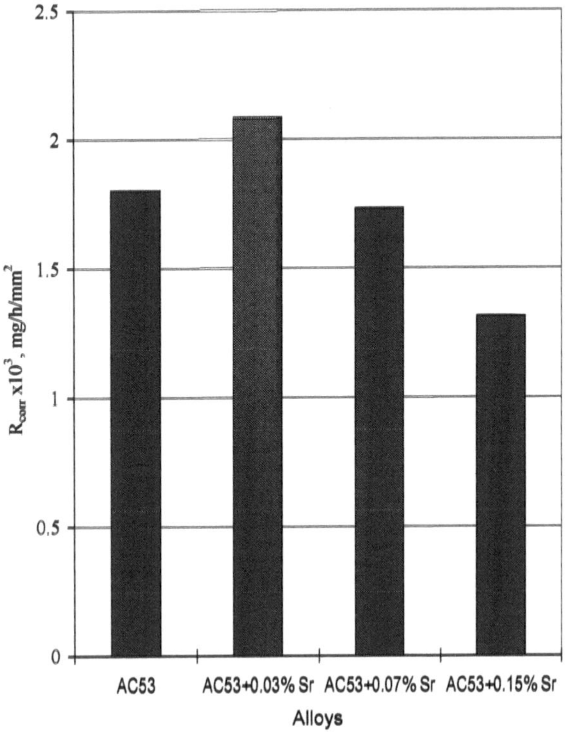

Fig. 10. Effect of Sr on corrosion rate of AC53.

The corrosion rates for all alloys tested are compared in Figure 12. AC52 has a corrosion rate comparable to AZ91, meeting one of the criteria for developing new creep resistant alloys for powertrain applications. The addition of a small amount of Sr, ~0.2%, should further improve the corrosion resistance of AC52 alloy.

The cumulative amount of HCl consumed during a corrosion test was calculated using the concentrations of Mg and Ca in the leach solution of 1000mL, and compared with the measured amount for certain alloys in Table V. The table shows that magnesium is predominantly present in the leach solution. The calculated values are well in agreement with the measured values, confirming that the corrosion of magnesium alloys occur according to reaction 4.

Table V Comparison of measured and calculated HCl (0.922M) usage

Alloy	Mg, mg/L	Ca, mg/L	HCl, mL	
			Measured	Calculated
AZ91	27	<1	2.50	2.45
AC52	36	1.4	3.30	3.31
ZC63	53	-	5.75	5.30

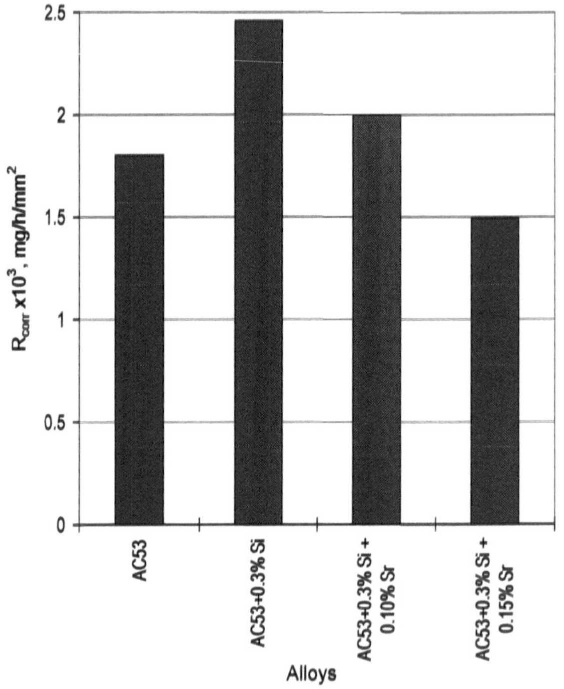

Fig. 11. Effects of Si and Sr on corrosion rate of AC53.

Conclusions

The dissolution of Mg samples immersed in 5% NaCl solution whose pH is controlled between 5 and 7 provides a fast way to screen the effect of composition change on the corrosion behavior of Mg alloys. The intrinsic corrosion rate is determined from the slope of the curve where the rate of dissolution reaches steady state.

The ranking of the Mg alloys for their corrosion behavior by this method is comparable to the ranking determined by the salt spray corrosion test, suggesting that the new screening method is very reliable.

This method is capable of discerning the effect of small changes in alloy composition on corrosion rates. Addition of small amounts of Si (0.3%) to AC alloys increases the corrosion rate, while Sr (0.07%) reduces it.

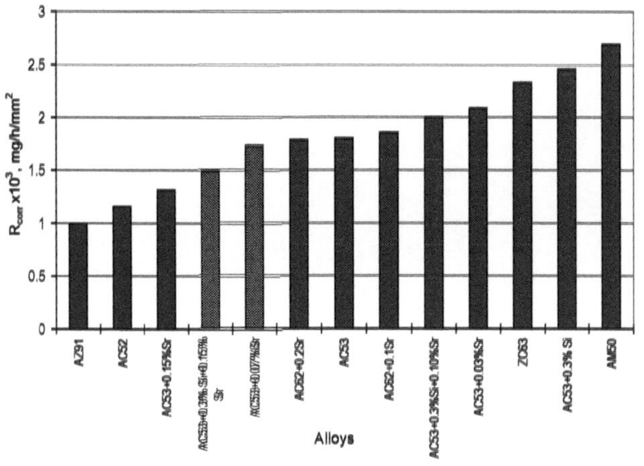

Fig. 12. Corrosion rates of magnesium alloys.

The corrosion rate of new, AC52 (Mg-5%Al-2%Ca), creep-resistant Mg alloy is comparable to AZ91, meeting one of the several design criteria for the new, creep-resistant alloys.

Summary

The salt spray corrosion test is based on the weight difference of the sample before after the test. This method is not only slow, but it does not measure the intrinsic corrosion rate of the alloys. Other methods based on measuring the amount of H_2 evolved or changes in pH also have drawbacks. The novel method based on measuring the dissolution rate of magnesium alloy samples in 5% NaCl solution under a controlled pH as presented in this study overcomes the drawbacks of tests available currently for determining the corrosion behavior of Mg alloys. This method is fast (<5 hours) and capable of discerning the effect of small compositional changes on the corrosion behavior of several magnesium alloys.

Acknowledgments

The authors wish to acknowledge the contributions of the following individuals: Larry S. Thompson for conducting the salt spray corrosion tests; Deborah Esch and Noel Potter for conducting the chemical analysis; Bob Powell, Anil Sachdev and Yar-Ming Wang for editing the manuscript and providing useful suggestions.

References

1. B. R. Powell, V. Rezhets, M. P. Balogh and R. A. Waldo, The Relationship between Microstructure and Creep Behavior in AE42 Magnesium Die-Casting Alloy, Magnesium Technology 2001, The Metallurgical Society Annual Meeting, New Orleans, LA, Feb. 11, 2001.

2. B. R. Powell, A. A. Luo, V. Rezhets, J. J. Bommarito and B. L. Tiwari, Development of Creep-Resistant Magnesium Alloys for Powertrain Applications: Part 1 of 2, SAE Technical Paper 2001-01-0422, Detroit, MI, 2001.

3. "Standard Method for Salt Spray (Fog) Testing, B 117," Annual Book of ASTM Standards, American Society for Testing and Materials.

4. W. M. Latimer, Oxidation Potentials, Second Edition, Prentice-Hall, Inc., 1952.

5. C. L. Mantell, Electrochemical Engineering, Fourth Edition, McGraw-Hill, 1960, p 456.

6. Phase Diagrams of Binary Magnesium Alloys, A. A. Nayeb-Hashemi and J. B. Clark, Editors, ASM Publication, 1998

7. O. Lunder and A. S. Langseth, Measurement of Hydrogen evolution from corroding Mg, SINTEF Metallurgy Corrosion Center (Internal Report), 1992, Norway.

8. D. Weiss, et al., Corrosion Resistance Evaluation of Magnesium and Magnesium Alloys by an Ion Selective Electrode, Proceedings of the First Israeli International Conference on Magnesium Science and Technology, November 10-12, 1997.

9. A. Luo, M. P. Balogh and B. R. Powell, Tensile Creep and Microstructure of Mg-Al-Ca Based Alloys for Powertrain Applications: Part 2 of 2, SAE Technical Paper 2001-01-0423, Detroit, MI, 2001.9.

Corrosion Phenomenon Evaluation of Mg Alloys using Surface Potential Difference Measured by SKPFM

Rei TAKEI[1], Hiroyuki FUKUDA[1], Hisashi IMAI[2], Junko UMEDA[2], Katsuyoshi KONDOH[2]
[1]Department of Mechanical Engineering, Osaka University, 2-1 Yamadaoka, Suita, Osaka 565-0871, Japan
[2]Joining and Welding Research Institute, Osaka University, 11-1 Mihogaoka, Ibaragi, Osaka 567-0047, Japan

Keywords: magnesium, galvanic corrosion, surface potential, SKPFM

Abstract

Magnesium is easy to be corroded because it has small standard electrode potential (SEP). In magnesium alloys, the corrosion resistance of dispersoids hasn't yet evaluated quantitatively as the SEP of dispersoids wasn't almost investigated. This study surveyed the surface potential difference (SPD) between α-Mg in AZ91D magnesium alloy and the dispersoids in that alloy by using Scanning Kelvin Probe Force Microscope (SKPFM). The corrosion resistance was evaluated and correlated with SPD, where this measured SPD represents the difference of work function between α-Mg and the dispersoid. At this experiment, it was obtained that the difference of SEP generally increases as the SPD increases. The good correlation between difference of SEP and SPD was obtained. The SPD between α-Mg and $Mg_{17}Al_{12}$ or Al_6Mn in AZ91D Mg alloy is measured. In addition, in order to obtain the reference of SPD, the area of α-Mg only and the interface area between Fe and α-Mg were respectively scanned. The result shows that $Mg_{17}Al_{12}$ has small SPD to α-Mg and that the SPD between $Mg_{17}Al_{12}$ and α-Mg is mostly the same as that in α-Mg. Therefore, $Mg_{17}Al_{12}$ is considered of no effect on corrosion. On the other hand, Al_6Mn degrades corrosion resistance because the SPD between Al_6Mn and α-Mg is as large as that between Fe and α-Mg.

1. Introduction

Magnesium alloys have high specific strength and stiffness, and are applied to industrial components such as transport products and mobile electronic devices. They are expected to improve fuel efficiency and to reduce CO_2 gas emission in the automobile. Other characteristic properties include the capability of vibrational absorption, the dimensional stability and electromagnetic wave shielding [1]. However, one of the disadvantages of Mg alloys is poor corrosion resistance [2]. Mg itself isn't severely corroded in air because of the formation of a passive film on the surface. When Mg is in contact with other metals which are higher in standard electrode potential (SEP), Mg is easily corroded first. The SEP of Mg compared with other structural materials is low, and so Mg is easy to be corroded [3]. The corrosion behavior of Mg when it is in contact with other pure metals can be evaluated with their values of SEP. In addition, the greater the difference of SEP is, the more corrosion proceeds. On the other hand, in Mg alloys, the local corrosion phenomenon between dispersoids and α-Mg phase hasn't been surveyed quantitatively as SEP of dispersoids wasn't investigated enough yet. When the difference of SEP between those phases are large, corrosion phenomenon generally arises more easily. This indicates that corrosion phenomenon can be evaluated with the value of the difference of SEP. In this study, the surface potential difference (SPD) at the interface between α-Mg and dispersoids could be measured by SKPFM and corrosion resistance of Mg alloys was evaluated. Therefore, the relationship between SPD and difference of SEP was investigated to estimate corrosion resistance. In fact, good relationship between SPD and difference of SEP was obtained. This shows that corrosion resistance in Mg alloy can be evaluated by the SPD at the interface. This paper reports the corrosion phenomenon between α-Mg and Mg alloy dispersoids using the SPD.

2. Experimental procedure

2.1 Principle of SKPFM and surface potential

In the Scanning Kelvin Probe Force Microscope (SKPFM), the measured surface potential is the contact potential difference (CPD) between the measurement point and the needle at the end of cantilever. This needle is coated with PtIr5. This CPD is led by subtracting the work function of sample metal from that of PtIr5 and dividing by elementary electric charge, i.e. $CPD = (\Phi_{PtIr5} - \Phi_{sample})/e$. The variation tendency of this surface potential is different from that of SEP. The work function is the minimum energy needed to remove an electron from a solid into vacuum. It is suggested that the work function is evaluated by the correlating equation using Pauling's electronegativity [4]. This equation is expressed as $\Phi = 2.27\chi + 0.34$ where χ is the Pauling's electronegativity, and the correlation between the work function obtained by this equation and SEP is shown at Figure 1 [5,6].

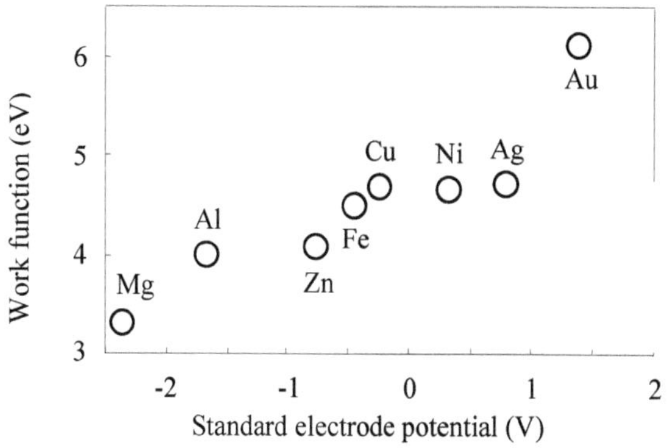

Fig.1 Relationship between standard electrode potential and work function

This correlation indicates that the work function approximately linearly increases with increasing SEP. In addition, the metals which have small SEP are small in work function and are easy to emit electron. Therefore, the surface potential of base metals is high as it is derived by CPD equation. The SPD between phase A and phase B is evaluated as following equation under the condition that the surface potential of phase A, V_A is larger than that of phase B, V_B, where V_A is defined as criterion.

$$\begin{aligned} SPD &= V_A - V_B \\ &= (\Phi_{PtIr5} - \Phi_A)/e - (\Phi_{PtIr5} - \Phi_B)/e \\ &= (\Phi_B - \Phi_A)/e \qquad (1) \end{aligned}$$

SPD is proportional to difference of work functions. In addition, from Fig.1, the difference of work function between one metal and Mg generally increases as the SEP between that metal and Mg increases. From these relationships, SPD is generally proportional to the difference of SEP. Therefore, the local difference of surface potential induces galvanic corrosion. This study evaluates the corrosion resistance quantitatively by measuring SPD.

2.2 Material under test

In order to evaluate the corrosion behavior by SPD in this experimental apparatus, the relationship between surface potential and SEP was first surveyed by using pure metals. These metals are magnesium, aluminum, zinc, iron, nickel, copper, silver and gold. In addition, the SPD between pure Fe and α-Mg of AZ91D Mg alloy is measured. That pure Fe as fine particles was cast into AZ91D Mg alloy. This pair generally shows bad corrosion resistance. So it is used as the reference of SPD which indicates bad corrosion resistance and as the SPD for evaluating whether the pair of metal is good or not for corrosion resistance. Based on these results, the SPD between α-Mg and $Mg_{17}Al_{12}$ or Al_6Mn in AZ91D Mg alloy was measured. Corrosion resistances of these pairs were evaluated by comparing with the pair of α-Mg and Fe, and SPD in α-Mg of AZ91D.

2.3 Way to identify dispersoid

When some phase of pure metals or the interfacial phase of Mg and Fe are scanned, one can identify the scanning point. This is because those phases can be identified with the optical microscope equipped with this SKPFM. The magnification of this optical microscope is 84 to 525 times. But the dispersoids of Mg alloy can't be identified by this optical microscope. One way for identifying dispersoids is the use of indentations of Vickers hardness. This way proposes that identifying the dispersoids is led from relative positions with the indentations. First, one indentation is produced at some position. Then, second and third indentations are produced at 160μm apart from first indentation for horizontal axis direction and at 120μm apart from first indentation for vertical axis direction, as shown at Figure 2. After that, the relative position of the each dispersoid against the three indentations is captured on SEM (scanning electron microscope) image (Figure 3). That dispersoid is identified by EDS (energy-dispersive X-ray spectroscopy) analysis and XRD (X-ray diffraction) analysis. Therefore, one can fix the relative position of dispersoid in the image (Figure 4) obtained by the optical microscope equipped with SKPFM based on the three indentations and identify the dispersoid. The cantilever is moved to this relative point of dispersoid and begins to scan the square area at that point. The square area includes dispersoid and α-Mg, so that the interfacial SPD between dispersoid and α-Mg is obtained.

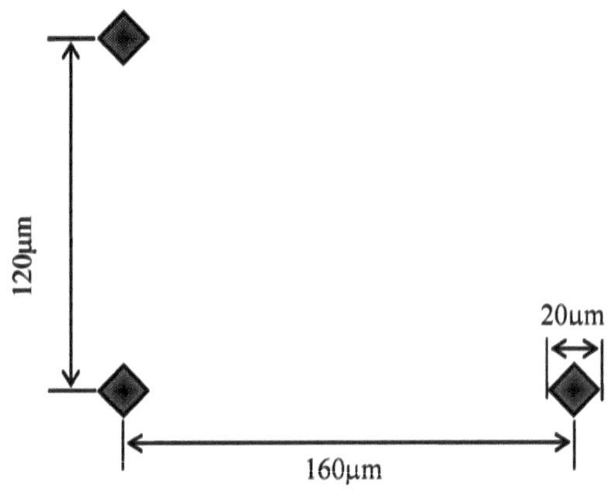

Fig.2 Arrangement of indentations

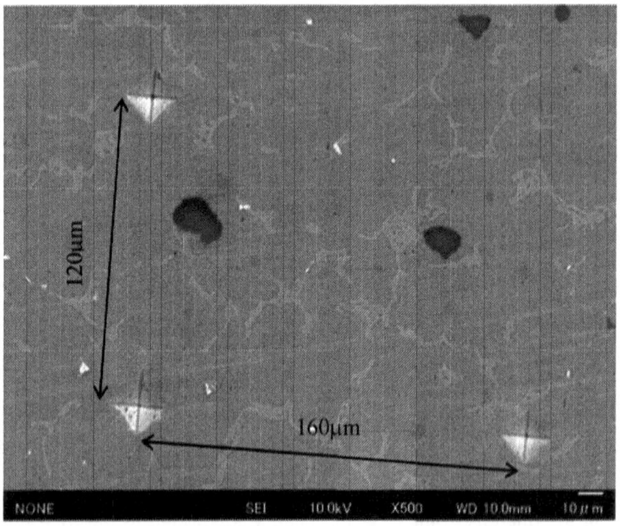

Fig.3 SEM observation of the indentations' arrangement

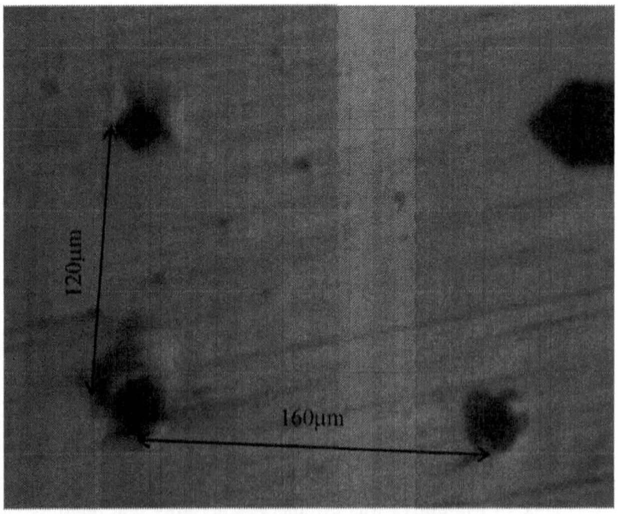

Fig.4 Image obtained by optical microscope equipped with SKPFM

3. Results and discussion

3.1 Relationship between surface potential and standard electrode potential

The measured surface potential of pure metals is indicated in Table 1. Before scanning, the surface polishing is conducted with waterproof abrasive paper. Then, sample surface is finally given the mirror finish by buffing using diamond paste. Figure 5 shows the comparison between measured surface potential and SEP. It can be shown that the surface potential decreases generally as the SEP increases. This result of surface potential corresponds approximately to the theory led by CPD equation and the work function obtained by using Pauling's electronegativity. As this result of surface potential and CPD equation show, the work function of metal decreases as SEP decreases. It generally corresponds well to Fig.1's tendency. From Fig.5 the relationship between SPD based on Mg, i.e. $SPD_{Mg} = V_{Mg} - V$, and difference of SEP based on Mg, i.e. $\Delta SEP_{Mg} = SEP - SEP_{Mg}$, is shown in Figure 6, where SPD based on Mg is calculated by subtracting V from V_{Mg} as the surface potential of base metal is high. The good correlation between ΔSEP_{Mg} and SPD_{Mg} is obtained. From Fig.6, as ΔSEP_{Mg} increases, SPD_{Mg} generally increases. So, corrosion resistance can be evaluated by SPD_{Mg} and in this experiment, SPD is obtained at interface between dispersoid and α-Mg of AZ91D.

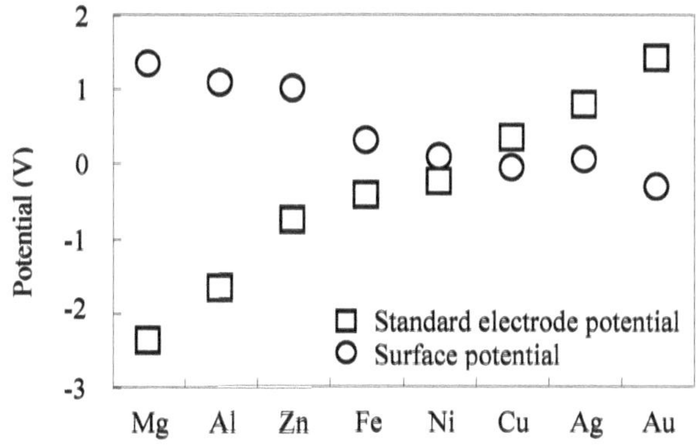

Fig.5 Relationship between standard electrode potential and surface potential

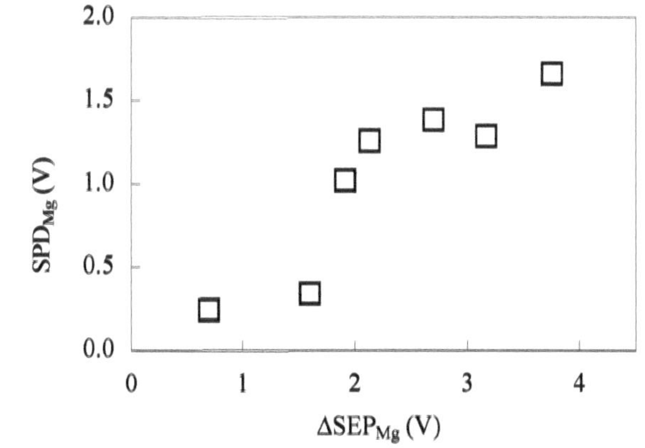

Fig.6 Relationship between ΔSEP_{Mg} and ΔSPD_{Mg}

3.2 Effect of strain induced by indentation on surface potential

The position of dispersoid is identified by relative position from three arranged indentations. The surface potential is very sensitive against surface condition, and so the strain induced by indentation may affect the surface potential around each indentation. Thus, the surface potential around indentation and surface potential far from indentation is measured, and those potentials are compared. The test metal needs to have uniform surface potential so that the variation of surface potential due to the strain is observed. Moreover, the metal should have about the same hardness as Mg alloy so that about the same amount of strain is induced. So, pure Mg is chosen for the test metal. The measurement point is shown in Figure 7 and the scanning area at one time is a square, 3μm on a side. This result is shown in Table 2 and the difference between the surface potential near indentations and that far from indentations is within 0.1 (V), which is the approximately same to SPD in surface potential of pure metal: 0.0~0.2 (V). Therefore, it is estimated that the effect of strain induced by indentation on surface potential can be neglected.

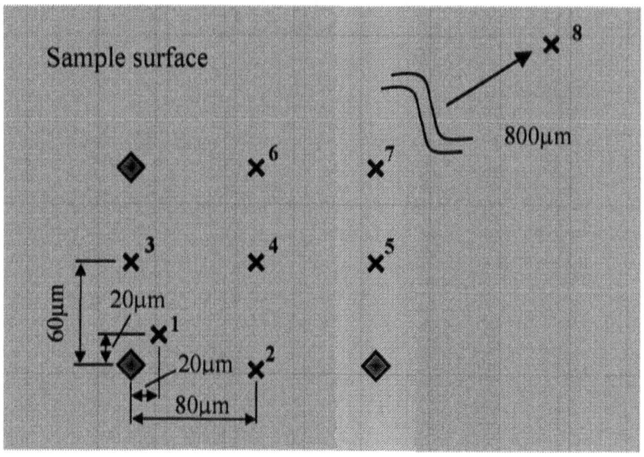

Fig.7 The indentation points and the measurement points around the indentations

3.3 Effect of the dispersoid on corrosion resistance

This study investigates the difference of surface potential between dispersoid and α-Mg which is the criterion of surface potential in this experiment, where the dispersoids are $Mg_{17}Al_{12}$ (β phase) and Al_6Mn. The SPD between those dispersoids and α-Mg is shown in Table 3. The SPD between Fe particle and α-Mg is added to this table to evaluate the level of corrosion resistance. It shows that the SPD between $Mg_{17}Al_{12}$ and α-Mg is as small as that between two points of α-Mg, and is considerably smaller than that between Fe particle and α-Mg. Here, the SPD in α-Mg is larger than expected, because that SPD is measured with macroscopic scale, and that SPD is calculated by the maximum and minimum values in the square. From this result, it is obtained that the dispersoid $Mg_{17}Al_{12}$ scarcely induces corrosion to α-Mg and is a safe dispersoid. In addition, the surface potential of $Mg_{17}Al_{12}$ is higher than that of α-Mg and it indicates that $Mg_{17}Al_{12}$ is more stable than Mg. Therefore, $Mg_{17}Al_{12}$ is corroded before α-Mg. On the other hand, the SPD between Al_6Mn and α-Mg is as large as that between Fe particle and α-Mg. The surface potential of Al_6Mn is lower than that of α-Mg, and so α-Mg is first affected by

From *Magnesium Technology 2010*, S.R. Agnew, N.R. Neelameggham, E.A. Nyberg, and W.H. Sillekens, Editors

Table1. Surface potential of pure metal

Pure Metal	Mg	Al	Zn	Fe	Ni	Cu	Ag	Au
Surface Potential (V)	1.32	1.04	1.00	0.30	0.07	-0.05	0.02	-0.33

Table.2 Surface potential near and far from indentations

Measurement point	1	2	3	4	5	6	7	8
Maximum (V)	1.38	1.37	1.36	1.39	1.40	1.39	1.37	1.25
Minimum (V)	1.31	1.25	1.29	1.31	1.27	1.29	1.28	1.40
Average (V)	1.35	1.31	1.33	1.35	1.34	1.34	1.33	1.33
Difference (V)	0.07	0.12	0.07	0.08	0.13	0.10	0.09	0.15

Table.3 SPD between different phases

Measurement Point	α-Mg – α-Mg	$Mg_{17}Al_{12}$ – α-Mg	Al_6Mn – α-Mg	Fe – α-Mg
SPD based on α-Mg (V)	0.0 ~ 0.2	0.0 ~ +0.1	-0.7 ~ -0.6	-0.7 ~ -0.5

corrosion. This result shows that the dispersoid Al_6Mn reduces the corrosion resistance of AZ91D Mg alloy.

4. Conclusion

Magnesium is low at SEP, and so it is easy to corrode. It is highly affected for that property when it is in contact with other metal. In Mg alloys, the problem is that α-Mg is in contact with dispersoids, but this corrosion resistance has been generally investigated at a macroscopic area. This paper evaluated corrosion resistance at local area, i.e. the interface between α-Mg and dispersoids. Then, this study surveyed corrosion resistance against dispersoids in AZ91 by SPD measured by SKPFM. The results in this study are summarized as follows.

(1)The measured surface potential of pure metal generally decreases, as SEP increases. This surface potential correlates well to SEP, and in addition, SPD also has good correlation with the difference of SEP. In theory, SPD is calculated by the work function of interfacial two metals. This theoretically-derived SPD and the SPD obtained from experimental results have generally similar tendency for difference of SEP.
(2)The SPD between α-Mg and $Mg_{17}Al_{12}$ is as small as the SPD within α-Mg. Besides, the surface potential of $Mg_{17}Al_{12}$ is higher than that of α-Mg. This indicates that $Mg_{17}Al_{12}$ results in more stable than α-Mg and α-Mg becomes safer at that interface. However, the SPD between α-Mg and Al_6Mn is as large as that between Fe and α-Mg, and the surface potential of Al_6Mn against α-Mg is low. So α-Mg first undergoes corrosion. This result shows that the dispersoid Al_6Mn degrades corrosion resistance of Mg alloys.

References

[1] D.F. Miner, J.B. Seastone, Handbook of Engineering Materials, Wiley, New York, 1955, pp.2-273 –2-292
[2] G. Song, B. Johannesson, S. Hapugoda, D. StJohn, Galvanic corrosion of magnesium alloy AZ91D in contact with an aluminium alloy, steel and zinc, Corros. Sci. 46 (2004) 955
[3] M.M. Avedesian, H. Baker, Magnesium and Magnesium Alloys, ASM International, OH, 1999, pp.194
[4] M. Tsukada, Work function, Kyouritsu Publishing Co., LTD., Tokyo, 1983, pp.89-90 (in Japanese)
[5] A.L. Allred, Electronegativity values from thermochemical data, J. Inorg. Nucl. Chem. 17 (1961) 219
[6] P.W. Atkins, Physical Chemistry Sixth Edition, Oxford University Press, Oxford, 1998, pp.934-936

EMERGING TRENDS IN CORROSION PROTECTION OF MAGNESIUM DIE-CASTINGS

Jan Ivar Skar[1)], and Darryl Albright[2)]

[1)]Norsk Hydro Research Centre, P.O. Box 2560, N-3907 Porsgrunn, Norway.
[2)]Hydro Magnesium, 39209 Six Mile Road, Suite 200, Livonia, Michigan 48152, USA

Abstract

Due to the improved corrosion resistance of magnesium alloys brought forward by the development of the high purity grades 20 years ago, the use of die cast magnesium components has moved towards more and more demanding applications. The corrosion resistance is sufficient so that magnesium is used in the under-body environment of vehicles without additional general corrosion protection, although special attention must be given to galvanic corrosion protection. There has recently been an increasing interest in both semi-exterior and exterior automotive parts. Since magnesium alloys are not compatible with standard automotive phosphate treatments, the parts need to be pre-coated prior to assembly. For such components, the challenges are to find efficient, environmentally friendly pre-treatments and robust coating systems that can maintain their integrity through the assembly process. The present paper reviews some of the established and newly developed methods for corrosion protection and finishing of die cast magnesium automotive parts.

Introduction

In the recent years, the annual growth of magnesium alloys has been 10 - 15%. The growth is found in the die casting industry, and a major part of the growth is for automotive applications. The same growth is expected in the next 5 years. The main reasons for the growth is the low weight and excellent castability of magnesium. This allows casting of large, thin wall multi-functional components. An instrument panel structure is an excellent example of such a casting. Most of the present magnesium components are located in the interior in the car, but components are also exposed to the rather corrosive environment found under the car body. If the expected growth becomes a reality, many of the new products will be exposed to rather corrosive environment.

Corrosion Mechanisms of Magnesium.

The alloying elements in magnesium are aluminium, manganese (AM alloys) and sometimes zinc (AZ alloys). Other elements used to alloy magnesium are silicon (AS alloys) and Rare Earth metals (AE alloys). In additions, there are a number of impurity elements like iron, copper and nickel. Together with aluminium, magnesium forms two phases, the α-matrix and the β-phase ($Mg_{17}Al_{12}$). In die-castings the matrix is normally surrounded by the β-phase. Some of the alloying elements and the impurity elements form intermetallic particles and the most important are those of the form $Al_xMn_yFe_z$. X, Y and Z vary with the composition of the alloy. Both the β-phase and these inter-metallic particles are more noble than the matrix. They are also more or less good catalysts for the cathode reaction shown in equation 1. The anode reaction, which occurs mainly in the matrix, is initiated in the vicinity of β-phase and the inter-metallic particles. The anode reaction is described by equation 2.

Cathode reaction: $2H_2O + 2e = 2OH^- + H_2(g)$ (1)

Anode reaction: $Mg = Mg^{2+} + 2e$ (2)

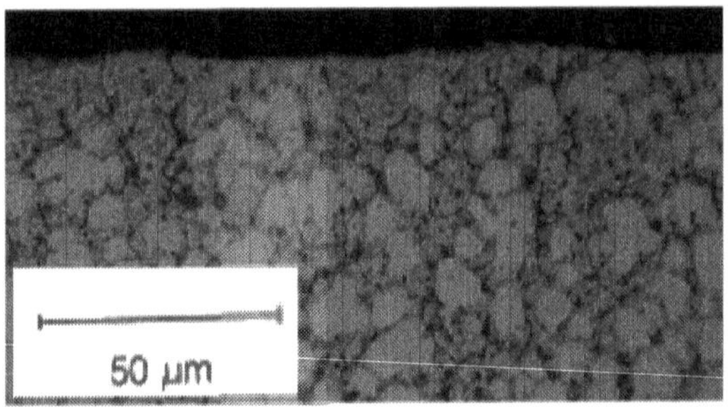

Figure 1: Microstructure of a good quality magnesium die casting. Alloy: AM50A.

The β-phase has a dual effect on the corrosion reaction. In addition, to being an initiator, it also acts as a barrier to corrosion growth in the matrix (1,2,3,4). Hence, the microstructure of the metal is important. The best corrosion resistance is obtained when there is a fine grain structure as shown in Figure 1. The β-phase forms a network around the α-matrix with many corrosion barriers. This is a typical microstructure for a good quality die-casting.

The corrosion resistance of magnesium alloys decreases with decreasing aluminium content as illustrated in Figure 2. The reason is that when the concentration of aluminium decreases, the amount of β-phase decreases. In addition, the intermetallic particles contain less aluminium relative to iron, and thus make these particles more aggressive. They become more noble compared to the matrix and their cathode efficiency increases (5). As a consequence, there are still many sites for corrosion initiation, but less corrosion barriers due to the lower content of β-phase.

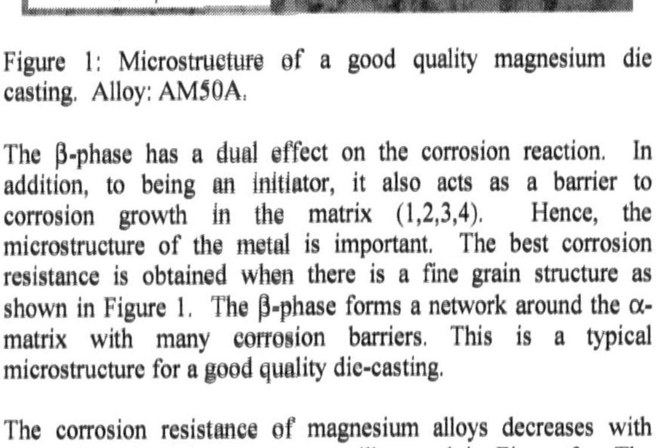

Figure 2. Corrosion rates as a function of aluminium content in high purity AZ and AM alloys. Test conditions: ASTM B117 salt spray for 240 hours. Typical corrosion rates of AE42 and AS21X are also shown.

In Figure 2, the typical corrosion rates of AE42 and AS21X are also indicated. AE42 and AS21X show good corrosion resistance despite the low content of aluminium. The reasons are that these alloys contain Rare Earth metals. The effects of Rare Earth metals are not fully understood, but probably they make the intermetallic particles less noble compared to the matrix. They can also contribute by strengthening the oxide film (4). In AS21X there is also silicon and silicon reduces the solubility of manganese and thereby contribute to better corrosion resistance.

Galvanic Corrosion of Magnesium

The general corrosion resistance of magnesium is good. Transfer cases made of die cast AZ91D, are used under the car body without any protection against general corrosion. However, magnesium is the least noble structural metal used and when connected to any other metal, galvanic corrosion is always a concern. The anode and cathode reactions shown in Equations 1 and 2 also apply for galvanic corrosion. In an assembly between magnesium and another metal, the anode reaction occurs mainly on magnesium and the cathode reaction on the other metal. These two reactions must balance each other, and when the cathode reaction is limited, the anode reaction is limited accordingly. Galvanic corrosion can only occur in the presence of an electrolyte.

The current, I, flowing between the anode and cathode in the assembly can be expressed by the simplified Equation 3.

$$I = (E_k - E_a)/(R_k+R_a+R_e+R_m) = (E_k-E_a)/(R_k+R_e) \quad (3)$$

E_k and E_a = cathode and anode open circuit potentials, respectively,
R_a = anode polarisation resistance,
R_k = cathode polarisation resistance,
R_e = electrolyte resistance,
R_m = metal resistance.
R_e and $R_k >> R_a$ and R_m

The current is proportional to the corrosion rate of magnesium, and by minimising the current, the galvanic corrosion on magnesium is also minimised. This can be achieved by I) proper materials selection, II) selective use of coatings and insulation materials and III) proper design of the assembly (6,7).

I) Proper Material Selection.

The most compatible materials with magnesium are those that provide the lowest potential difference, $(E_k - E_a)$, and the largest polarisation resistance, R_k.

The latter means that the resistance to the cathode reaction is high. The effect of high cathode polarisation resistance is illustrated in Figure 3. The plot shows cathode polarisation curves of Sn/Zn plating (80% Sn and 20%Zn) and aluminium AA6013 and anode polarisation curve of magnesium (AM60B). For the assembly Mg to Al, the potential difference, (E_k-E_a), is about 0.04 V and the current is about 0.001 A. Despite the fact that the potential difference between Mg and the Sn/Zn plating is more than 0.5 V, the current is the same as for the Mg to Al assembly. The reason is the high polarisation resistance (especially of tin) for the cathode reaction. Consequently, the galvanic corrosion on magnesium is the same for the two assemblies, even though one assembly has a driving force of 0.5 V and the other only 0.04 V.

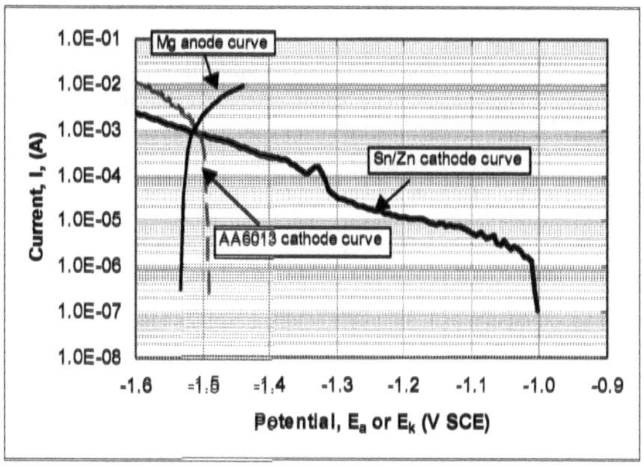

Figure 3: Cathode polarisation curves for Sn/Zn plating and Al and anode polarisation curve for Mg. The assemblies are immersed in 3% NaCl saturated with $Mg(OH)_2$. The polarisation scans started at the open circuit potentials of each metal. The scan rates were 0.05 V/min. The cathode and anode open circuit potentials are found by extrapolating the curves to the x-axis, and the galvanic current is found at the intersection between the anode and the cathode polarisation curves.

The most compatible materials are aluminium of the 5000 and 6000 series (6,8). Figure 4 shows a laser surface scan of a die cast magnesium plate with an AA6013 bolt after exposure in GM9540P for 41 days. The maximum pit depth is less than 0.1 mm, and the corrosion rate only 0.008 mg/cm^2·day. A steel bolt with a poor coating system in the same test environment gives pits deeper than 3.2 mm, and the corrosion rate is 40 times higher (see Figure 5).

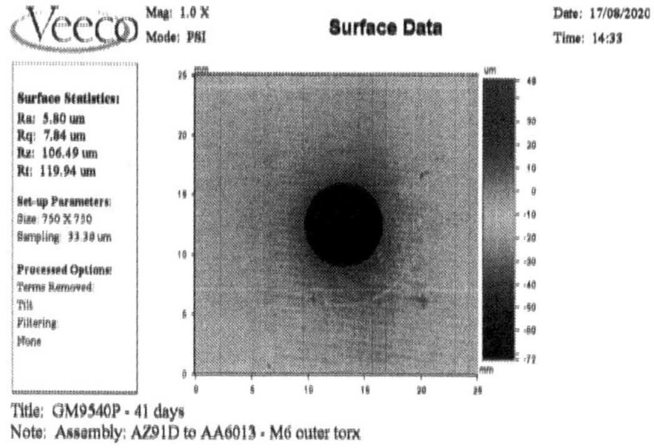

Figure 4: Surface scan of magnesium (AZ91D) with AA6013 aluminium fastener after 41 days in GM9540P test environment. Pit depth < 0.1 mm. Corrosion rate = 0.008 mcd[1]. The technique is based on reflection of a laser beam from the surface. Surface irregularities, such as corrosion attacks, cause changes in the reflection of the laser beam. Equipment: Uniscan OSP-100 from AG Electro Optics.

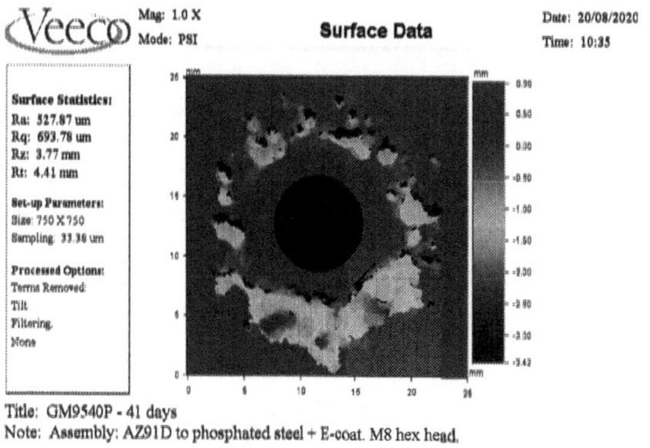

Figure 5: Surface scan of magnesium (AZ91D) with phosphated steel fastener plus E-coat after 41 days in GM9540P test environment. The E-coat peeled off during testing. Pit depth > 3.2 mm. Corrosion rate = 0.28 mcd.

Plastics are inert to galvanic corrosion, and recently fibre reinforced plastic fasteners have been developed. The fasteners are made of PEEK (PolyEtherEtherKetone) and the reinforcement is either of carbon fibres or polyamide fibres. With carbon fibres the galvanic corrosion would not be totally eliminated, since carbon is a very reactive cathode material. It is claimed that these reinforced PEEK fasteners are temperature resistant up to 170°C (9).

Materials that are not compatible with magnesium include steel, stainless steel, copper, nickel, and certain aluminium alloys. One of the latter is A380, which is a common alloy used in automotive engines. Some materials are between these extremes, e.g., tin and to a certain degree zinc.

II) Selective Use of Coatings and Insulation Materials (10,11).

It is impossible to avoid using steel in contact with magnesium. Coatings are then used to improve the compatibility to magnesium. Figure 6 shows corrosion rates and pitting depths on magnesium connected to steel fasteners with various plating systems. Sealed zinc plating and tin/zinc alloy plating reduce both the corrosion rates and pitting depths significantly compared to ordinary zinc plating. The corrosion rates of the sealed zinc plating were only marginally higher than the general corrosion rate of AZ91D in the same test environment. These platings work by passivating the steel surface as shown in Figure 3, i.e., the reaction rate of the cathode reaction is reduced.

Another metal plating under development is aluminium-magnesium alloy plating. The test result of steel fasteners with this type of plating is very promising (12).

Other options are polymer coating or caps covering the bolt head. The polymer insulates the steel from the electrolyte and thus breaks the electric circuit in the assembly (11).

[1] mcd = mg/cm^2·day

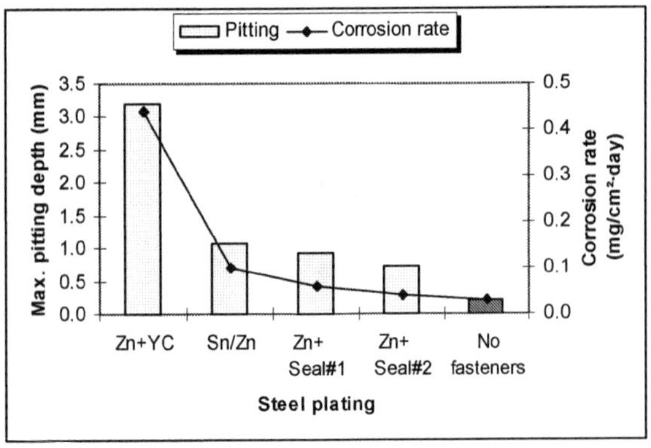

Figure 6: Corrosion rates and max. pitting depths of selected steel fasteners with different types of plating. The fasteners were M10, flanged head fasteners. The magnesium alloy was AZ91D. Test conditions: GM9540P – 80 days. Zn+YC = zinc plating with yellow chromate. Sn/Zn = tin – zinc alloy plating. Zn + Seal #1 and #2 = zinc plating plus yellow chromate and a sealer coat. No fasteners = AZ91D without any fasteners.

III) Proper Design of Assembly.

The design is important for the electrolyte resistance. By providing good drainage of the assembly, the electrolyte film becomes thinner. A thinner electrolyte film increases the resistance of the electrolyte, R_e, and decreases the galvanic current in equation 2. Washers and spacers are also efficient. They give protection by increasing the electrolyte path length and again the electrolyte resistance is increased. The effect of aluminium washers is displayed in Figure 7. In the case of zinc plated steel fasteners with an aluminium washer (AA6082) between the magnesium and the bolt, the washer reduced the pitting depth from more than 3.2 mm down to less than 0.6 mm. The corrosion rate was reduced from 0.16 to 0.02 mg/cm²·day. For steel with sealed Zn plating, the corresponding reductions in pit depths and corrosion rates were from 1.8 to 0.5 mm and 0.08 to 0.02 mg/cm²·day, respectively. Hence, with the aluminium washer, the difference caused by the fastener plating was eliminated.

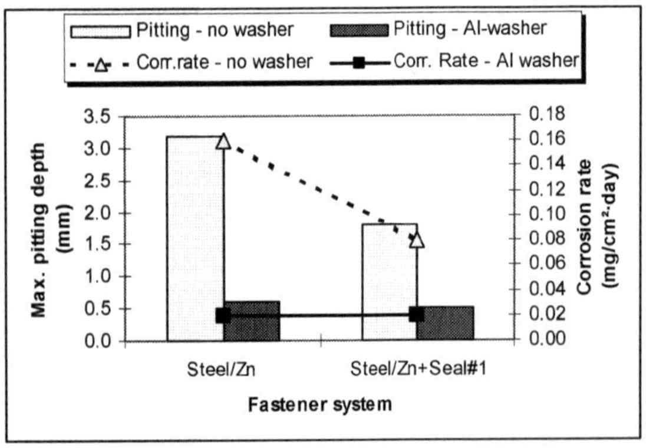

Figure 7: Max. pit depths and galvanic corrosion rates and of magnesium assembly to various types of fasteners (M8 hex head). The magnesium alloy was AZ91D. The washers used were aluminium disc washers (AA6082) with OD = 20 mm, thickness = 3 mm.

Surface Finishing of Magnesium.

Surface finishing of magnesium has mainly been used for decorative purposes, but by the increased use of magnesium in automotive applications as well as the electronic industry, surface finishing for corrosion protection is becoming increasingly important. The finishing of magnesium does not differ much from finishing of other metals, but special products designed for magnesium must sometimes be used (13).
Strong alkaline cleaners nowadays do the cleaning process. This step removes residues of lubricants used in the casting process as well as other organic contamination. Such contamination hinders the effectiveness of chemicals in subsequent treatment steps. Consequently, the quality of the conversion coating is reduced.

Various types of conversion coatings can be used. The most common conversion coating has been a chromate treatment. In the chromate bath, which contains hexavalent chromium and a number of additives, the magnesium is oxidised to magnesium ions, and the chromium is partially reduced to its trivalent state (see Equation 4). These ions react further with the chromate and form a film on the magnesium surface (Equations 5 and 6). The film composition is a mixture of hydrated chromium chromate and hydrous oxides of chromium and magnesium (14).

$$3Mg + Cr_2O_7^{2-} + 14H^+ = 3Mg^{2+} + 2Cr^{3+} + 7H_2O \quad (4)$$

$$2Mg^{2+} + Cr_2O_7^{2-} + 2H_2O = 2MgCrO_4 + 2H^+ \quad (5)$$

$$2Cr^{3+} + Cr_2O_7^{2-} + 3H_2O = 2Cr(OH)(CrO_4) + 4H^+ \quad (6)$$

This chromate film, conversion coating, provides corrosion protection through the corrosion inhibiting properties of the hexavalent chromium present in the film and by the physical barrier of the film itself. The corrosion protection is, however, only a temporary protection, since the hexavalent chromium in the film is consumed over time. More important than the corrosion protection itself is that the chromate promotes the adhesion between paint, lacquers and adhesives. Furthermore, the inhibiting effect of hexavalent chromium in the conversion layer prevents subsequent adhesion loss due to underfilm corrosion (15). As a result, the chromate conversion coating forms an excellent basis for painting, shown in Figure 8. Even after 480 hours in ASTM B117 salt spray, only minor underfilm corrosion is found on chromated magnesium with an E-coat. In contrast, alkaline cleaned magnesium alone gave more than 8 mm underfilm corrosion after 240 hours.

Due to the environmental issues of chromate treatments, the use of these treatments will be strongly limited in the future. This limitation will be introduced from 2003 in Europe (13). Thus, there is a strong need for chrome-free alternatives. The most promising options seem to be complex flouro-compounds, and phosphates, but other special coatings also show good performance (13,16).

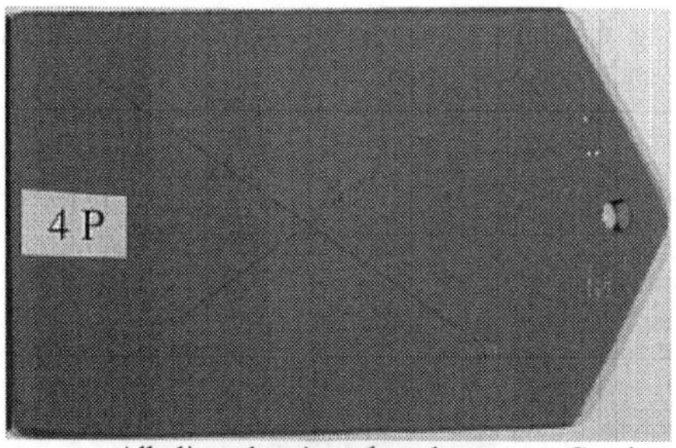

Pre-treatment: Alkaline cleaning plus chromate. Coating 20 μm E-coat. Test conditions: 480 hours ASTM B117 salt spray.

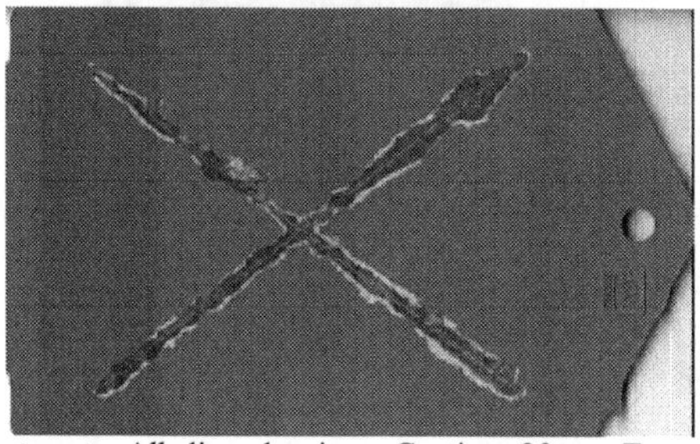

Pre-treatment: Alkaline cleaning. Coating: 20 μm E-coat. Test conditions: 240 hours ASTM B117 salt spray.

Figure 8: Underfilm corrosion from scribe of coated magnesium with different pre-treatments.

The film formation of the complex fluoro-compounds and phosphate is somewhat different than chromate. The fluoro-compound reacts and forms a complex with the magnesium hydroxide (See equation 8) (14). Magnesium fluoride is also most likely incorporated into the conversion coating. The fluoro-complex may be based on zirconium or titanium or a mix of those (14).

$$ZrF_6^{2-} + 4Mg(OH)_2 = ZrF_2(MgO_2H)_4^{2-} + 4H^+ \quad (8)$$

In phosphate treatments, the magnesium is normally oxidised according to equation 1 by oxidising agents in the bath. The phosphate in the bath and magnesium ions forms a magnesium phosphate film on the surface (17).

$$3Mg^{2+} + 2PO_4^{3-} = Mg_3[(PO)_4]_2 \quad (9)$$

Both types of films promote the adhesion of polymer coatings, but they do not have the self healing and inhibiting effect as chromate films have. Despite this lack of extra corrosion protection, these treatments show performance similar to chromate treatments in accelerated corrosion tests. Test results of the chrome free treatments, indicate that these treatments are more sensitive to the cleaning and activation steps, prior to the application of the coating. This is illustrated in Figure 9 showing the underfilm corrosion from a scribe for chromate and three chrome free treatments. When acid pickling was used for activation, the performance of the chrome free treatments approached the performance of chromate, but without this acid pickling the chromate was far better.

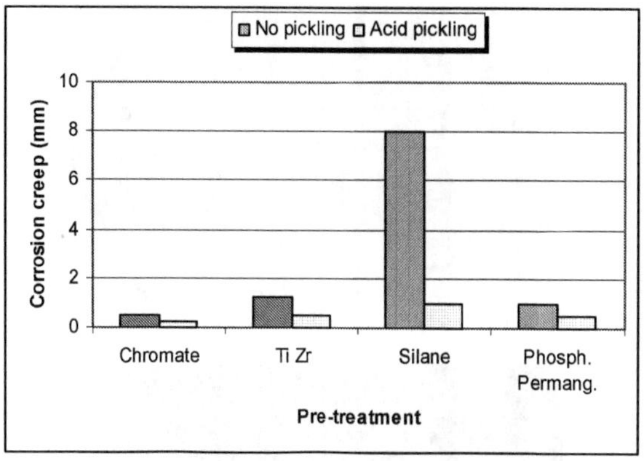

Figure 9: Underfilm corrosion of chromate and three chrome free conversion coating. Magnesium alloy: AZ91D. Coating: E-coat approx. 20 μm. Test conditions: VDA 621-415 3 cycles of one week each. TiZr = Complex flouro-titanate/zirconate, Silane = silane based treatment, Phosph. Permang. = phosphate permanganate.

Anodising is an alternative treatment that can be used as a final finish as well as base for subsequent painting (13,18). In anodising, magnesium is the anode in an electrochemical cell. High voltage is applied and magnesium is dissolved according to reaction shown in Equation 1. The magnesium ions produced at the anode combine with ions in the electrolyte (bath), and thus enhance the natural oxide film either by forming $Mg(OH)_2$ or MgO. Other constituents in the electrolyte can be incorporated into the film (for example chromate, phosphate, silicate and fluoride). Some processes utilise very high voltage, resulting in spark discharging taking place on the magnesium surface. A ceramic-like film is produced (19). Anodising gives excellent corrosion resistance as well as improved resistance to wear. Figure 10 shows that after 80 days in GM9540P standard, there is no corrosion attack at the scribe and only minor attacks at the rest of the surface.

New Challenges for Magnesium Finishing

Recently, magnesium car body components have been launched, and others are planned. These products face special challenges, since they are coated in the automotive finishing lines. Magnesium cannot be treated in the standard phosphate baths used in the automotive industry due to three reasons (20):

- These phosphates do not provide sufficient paint adhesion for magnesium.
- Magnesium ions are dissolved and may contaminate the bath so the bath efficiency for other metals is reduced.
- Metal ions from treatment of other metals can deposit and destroy the corrosion resistance of high purity magnesium.

Magoxide® – 11µm. Less than 1% of surface corroded.

Anomag® – 25µm. No corrosion.

Figure 10: Performance of anodising coatings. The test condition was GM9540P for 80 days. Alloy: AZ91D. Magoxide®: AHC Oberflächentechnik GmbH. Anomag®: Magnesium Technology Licensing Inc.

Therefore, magnesium must be coated prior to assembly on the car body. During assembly, there is a risk of damaging the coating of magnesium, and these damages may be sites for galvanic corrosion initiation. The corrosion rate is proportional to the cathode/anode area ratio. When the coating is damaged, a tiny anode area is created, and it is very small relative to the cathode area. Consequently, the corrosion attack grows rapidly into the metal rather than being more distributed and shallow as when the magnesium is not coated.

Therefore the coating on magnesium must be robust and remain intact during assembly. The coating must also withstand the dynamic loads as components move relative to each other during service. Figure 11 shows a surface scan of AM60B with an E-coat, which has been connected to a zinc plated steel fastener. As seen from the plot, there are several local pit attacks, and these may be as deep as 0.9 mm. Figure 12 shows a similar plot of a specimen with E-coat plus powder. Due to the increased coating thickness, there are no defects in the coating, and consequently there is no corrosion attack. The coating thickness of the latter is about 80 µm.

Other published results show similar trends (20,21,22). A powder coating with a thickness of 60 to 80 µm is able to prevent galvanic corrosion attacks. In order to further reduce the risk for corrosion initiation, plastic or aluminium spacers/washers between the other metal and the coated magnesium have shown to be efficient. (20,22)

Instead of powder coating, anodising is obviously an option. Test results indicate that anodising give similar protection as powder coating. The mechanical strength of anodising is better than standard pre-treatment plus powder coating. Thus, anodising may give better protection than powder coating in service under dynamic load (22).

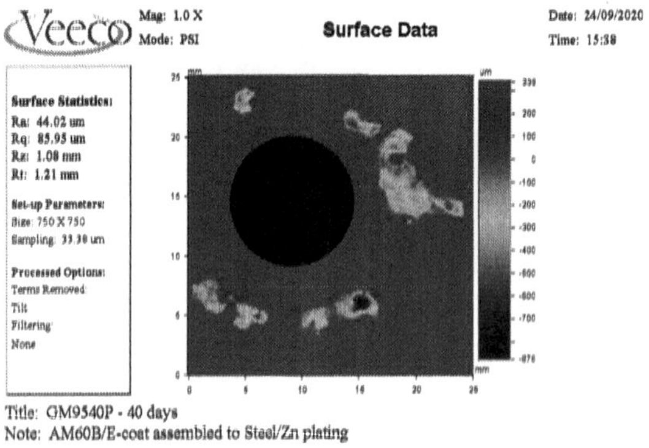

Figure 11: Galvanic corrosion attack on coated magnesium (AM60B) connected to zinc plated steel fastener. Test condition: GM9540P for 40 days. The coating system was a single layer of E-coat. Total coating thickness is 21 µm. Pit depths are down to 0.9 mm.

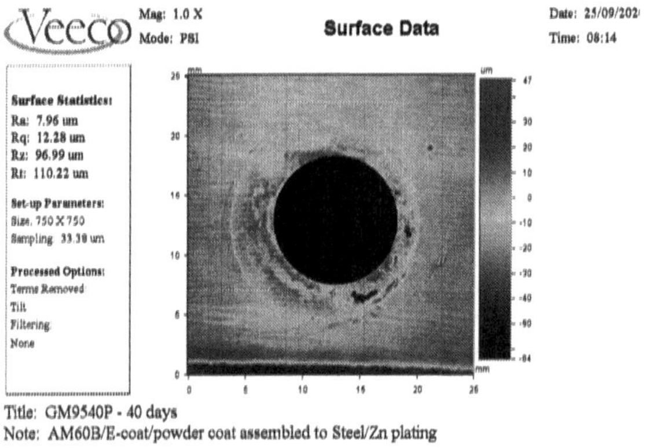

Figure 12: Galvanic corrosion attack on coated magnesium (AM60B) connected to zinc plated steel fasteners. Test condition: GM9540P for 40 days. The coating system was E-coat plus powder. Total coating thicknes was 77 µm. No bare metal exposed. The dark areas are surface of E-coat.

Conclusions

- The corrosion resistance of magnesium alloys is determined by the alloy composition and the microstructure.
- Galvanic corrosion protection is provided by proper material selection, selecetive use of coating and insulation materials, and proper design.
- Chromate pre-treatment gives excellent paint adhesion and corrosion protection. Chrome-free alternatives are complex fluoro-compounds, phosphate and other processes.
- Anodising methods give excellent corrosion protection and are also a good basis for subsequent painting.
- Exterior or semi-exterior magnesium components must be pre-coated prior to assembly and painting of the car. A robust coating system that can withstand the assembly forces and maintain its integrity during the assembly, is required.

References

1. K. Nisancioglu, O. Lunder, T. Kr. Aune, "Corrosion Mechanism of AZ91 Magnesium Alloy", (Paper at IMA, Cannes 1990).

2. T. Kr. Aune, O. Lunder, K. Nisancioglu, "In-situ Microscopic Investigations of Corrosion on Magnesium Alloy", Microstuctual Science, 17, (1989), 231-241.

3. G. Song, A. Atrens, X.Wu, B. Zhang, "Corrosion Behaviour of AZ21, AZ501, and AZ91 in Sodium Chloride", Corrosion Science, 40, No.10, 1998, 1769-1791.

4. K. Nisancioglu, J.H. Nordlien, "Understanding Corrosion Mechanisms of Commercial Alumnium and Magnesium Materials", Jour. of Japan Institute og Light Metals, 50, No.9, 2000, 417-430.

5. O. Holta, H. Westengen, J. Røen, "High Purity Magnesium Die Casting Alloys: Impact of Metallurgical Principles on Industrial Pratice", (Proceedings 3rd Int. Magnesium Conference, Manchester, 1996).

6. J.I.Skar, "Corrosion and Corrosion Prevention of Magnesium Alloys", Materials and Corrosion, 50, 1999,2-6.

7. B. Isecke, "Einflussgrössen auf die Kontaktkorrosion", (Paper at DVM-Tag 1998, Bauteil '98, Berlin, Germany, 6th-8th May, 1998).

8. C. Friedrich, "Fastening of Light Weigth Components with Elements Made of Alumnium", (SAE 2000 World Congress, Detroit, Michigan, 6th-9th March, 2000). Paper 2000-01-1118

9. G. Ahlers-Hestermann, "Verbinden und Befistigen von und an Mg-Bauteile", Fortschritte mit Magnesium im Automobilbau, ed. F.Ebert, Dr. M. Woydt, Bad Nauheim, Germany,18th-19th Sept. 2001.

10. B.Reinholdt, S.G. Klose, Ch. Strobel, "Die silikatische Versieglung verzinkter Oberflächen – ein Schutzsystem für Bauteile in Kontakt mit Magnesium," Materials and Corrosion, 50, 1999, 517-522.

11. D. Hawke, T. Ruden, "Galvanic Compatiblity of Coated Steel Fasteners with Magnesium", (, SAE 1995 World Congress, Detroit, Michigan, ,1995), Paper 950439.

12. H. Lehmkuhl, K.Mehler, B. Reinhold, H. Bongard, B. Tesche, "Elektrolytische Abscheidung von korrosionsstabilen Aluminium-Magnesium Legierungen", Published by Max-Plank-Institut für Kohlenforschung, Mülheim a.d.Ruhr, Germany, Oct. 2000.

13. J.I. Skar, J. Willekens, "Perspective in Magnesium Applications", (Paper at Leichtmetall-Anwendungen. Neue Entwicklung in der Oberflächentechnik, Münster, Germany, 13th-14th of March, 2001). 100-111.

14. A. Nowak, R. Seidel, "Verbesserung der Lackhaftung und des Korrosionsschutzes von Magnesium durch moderne Konversionverfahren", (Paper at Leichtmetall-Anwendungen. Neue Entwicklung in der Oberflächentechnik, Münster, Germany, 13th-14th of March, 2001). 122-131.

15. F. W. Eppensteiner, M. R. Jenkins, "Chromate Conversion Coatings", Metal Finishing, 94 (1A Guidebook&Dir.), 1996, 458-470.

16. J.I.Skar, "Surface Finishing of Magnesium", Fortschritte mit Magnesium im Automobilbau, ed. F.Ebert, Dr. M. Woydt, Bad Nauheim, Germany,18th-19th Sept. 2001.

17. D. Hawke, D. Albrigth, "A Phosphate-Permangnate Conversion Coating for Magnesium", Metal Finishing, Oct. 1995, 34-38.

18. I. Nakatsugawa, "Anodizing Treatments for Magnesium" (Report PC305-I, Institute of Magnesium Technology, Oct. 1993).

19. P. Kurze, D. Banerjee, "Eine neue anodische Beshichtung zur Verbesserung der Korrosionsbeständigkeit von Magnesiumwerkstoffen," Giesserei-Praxis, Nr.11/12, 1996, 211-217.

20. G. G. Wang, K. Stewart, R. Berkmortel, J.I.Skar, "Corrosion Prevention for External Automotive Components", (SAE 2001 World Congress, Detroit, Michigan, ,5th-8th March 2001) Paper 2001-010421.

21. H. Schreckenberger, M. Papke, S. Eisenberg, "The Magnesium Hatchback of the 3-liter Car: Processing and Corrosion Portection", (SAE 2000 World Congress, Detroit, Michigan, 6th-9th March, 2000). Paper 2000-01-1123

22. J.I.Skar, G. Wang, G.Guerci, "Corrosion Protection of Magnesium – When, Where, How", (Paper at 9th Magnesium Automotive and End User Seminar, 27th-28th Sept. 2001, Aalen, Germany).

Acknowledgement

Acknowledgment is given to the following persons for their contribution to the work persented herein: Ms. E. Thorjussen, Norsk Hydro Research Center, Mr. M. Isacsson, Volco Car Corporation, Mr. T. Wendel, Chemetall GmbH, Dr. J. Kresse and Dr. U. Jueptner, Henkel Surface Technologies, Prof. P. Kurze, AHC Oberflächentechnik, Mr. J. MacCulloch, Magnesium Technology Licensing LTD., Mr. Eiji Suzuki, Honda R&D CO., Mr. J. Sibson and Mr. T. Maeda, Honda R&D Europe (UK).

Evaluation of Corrosion Protection Methods for Magnesium Alloys in Automotive Applications

Blanchard, P.J., Hill, D.J., Bretz, G.T., & McCune, R.C.
Ford Motor Co., Research & Advanced Engineering
2101 Village Road, Dearborn, MI 48121

Keywords: Magnesium, Conversion Coatings, Anodized Coatings and Corrosion

Abstract

Magnesium alloys are susceptible to galvanic corrosion. Consequently, it is often necessary to apply coatings to magnesium components for isolation purposes. However, previous publications suggest the effectiveness of commercial coatings can vary widely. Therefore, an extensive corrosion screening study was performed to evaluate pre-treatment and coating systems currently available for use within the automotive industry. This paper focuses on a selection of conversion and anodized coatings. In many instances, these coatings were used in conjunction with either powder coat or an electro-coat to assess the additional protection offered by a supplemental barrier. Scribe test results and corroded area determination after accelerated corrosion testing are presented and used to quantify the pre-treatment performance. These results are supplemented by DC polarization measurements to determine the level of passivation. Finally, SEM micrographs were used to determine coating thickness variability and morphology. The overall performance of each pre-treatment and coating is then assessed with respect to corrosion protection and robustness.

Introduction

In response to continued pressure to reduce vehicle mass, a wide variety of weight reduction opportunities have been explored by automotive OEMs. With this objective, non-ferrous alloys, such as those based upon magnesium, have been examined and applied to a wide variety of body and power-train applications [1]. Unfortunately, a critical drawback of using magnesium is the potential for galvanic corrosion when used in the presence of other metals. This often necessitates the adoption of suitable protection strategies to provide dielectric isolation between components and sub-assemblies. Also, from a vehicle assembly viewpoint, magnesium is susceptible to dissolution in automotive phosphate systems. This introduces the potential for bath contamination, thus requiring some form of protective measure when processing a vehicle through a conventional automotive phosphate and cathodic epoxy electro-coat line.

To overcome these concerns, an extensive screening study was conducted to determine the corrosion performance and robustness of commercial pre-treatment and coating systems. In line with the European End-Of-Life vehicle directive (2000/53/EC), an emphasis was placed on the evaluation of chrome free systems as opposed to those using hexavalent chrome conversion coatings. To this end, results were gathered for a selection of conversion and anodized coatings applied to create either a single layer of protection, or alternatively as a composite system with the addition of powder coating or electro-coating.

A final evaluation of coating performance was based upon results from accelerated cyclic corrosion testing, and electro-chemical methods. The sections that follow detail each of the test methods employed and highlight the key observations and suitability of the candidate systems for future implementation.

Material Selection and Experimental Methods

Experimental Test Plan

Table I contains a summary of the coating combinations. In total, seven primary coatings or pre-treatments were evaluated as part of this study. However, since previous studies [2] recommended the use of supplemental barrier layers, in most cases, either an epoxy electro-coat or powder coat was also applied to the primary coating to determine the incremental benefit of these additional layers.

Table I. Sample identification and coating combinations.

I.D.	Primary Coating	Secondary Coating
A	Hexavalent Cr 6 Conversion Coating	Epoxy-Polyester Powder Coat I
B	Cr Free Conversion Coating I	None
BC	Cr Free Conversion Coating I	Epoxy E-Coat II
C	Cr Free Conversion Coating I	Epoxy E-Coat I
D	Cr Free Conversion Coating I	Epoxy-Polyester Powder Coat I
E	Cr Free Conversion Coating I	Organic Sealer
F	Anodized I - 5μm	None
G	Anodized I - 5μm	Epoxy E-Coat I
H	Anodized I - 5μm	Epoxy-Polyester Powder Coat I
I	Anodized I - 10μm	None
J	Anodized I - 5μm	Polymer Seal
K	Cr Free Conversion Coating II	None
L	Cr Free Conversion Coating II	Epoxy-Polyester Powder Coat I
M	Cr Free Conversion Coating II	Epoxy E-Coat I
N	Cr Free Conversion Coating III	None
O	Cr Free Conversion Coating III	Epoxy E-Coat I
P	Cr Free Conversion Coating III	Epoxy-Polyester Powder Coat I
Q	Cr Free Conversion Coating III	Epoxy Powder Coat I
R	Anodized II - 5μm	None
S	Anodized II - 5μm	Epoxy E-Coat I
T	Anodized II - 5μm	Epoxy-Polyester Powder Coat I
U	Anodized III- 5μm	None
V	Anodized III- 5μm	Epoxy E-Coat I
W	Anodized III- 5μm	Epoxy-Polyester Powder Coat I
X	Anodized III- 10μm	None
Y	Anodized III- 10μm	None

Coating application

The link between inconsistent corrosion performance and sample cleaning has been highlighted in previous publications [3]. Consequently, to eliminate any concerns related to cleaning or pre-treatment procedures, responsibility for application of the primary coatings to the magnesium coupons was given to the individual suppliers. Subsequent electro-coating was performed at Ford Research Laboratory and powder coating by the powder coating supplier. An additional electro-coat sample (BC) and an alternative powder coat (Q) were also included in the test matrix. Sample BC was processed through a standard automotive electro-coat line including phosphating.

Materials
High pressure die cast AM50 magnesium coupons, measuring 127mm x 25.4mm x 2.0mm, were prepared for corrosion testing using the procedures adopted in an earlier study [2]. The magnesium coupons were adhesively bonded to an untreated coupon of aluminum alloy AA-6111-T4 to create a lap shear joint. The aluminum served as a potential cathode within the corrosion test chamber, promoting galvanic corrosion between the two metals if the protective coating failed. To assess the susceptibility to corrosion undercutting, a 25mm long scribe was cut along the center line of the magnesium coupon, ensuring exposure of the base alloy.

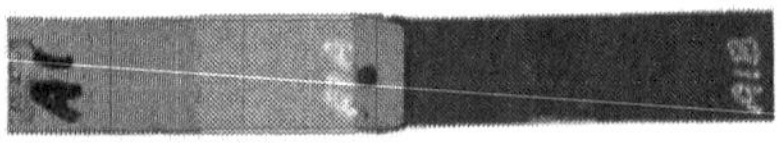

Figure 1. Mg/Al bonded assembly.

Corrosion Testing
Accelerated cyclic corrosion testing was performed according to the Ford Arizona Proving Ground Equivalent (APGE) laboratory test method [4]. A single 24-hour corrosion cycle is shown in Table II. While each 24-hour period is considered a full cycle, phases 1 and 2 were omitted on weekends as the samples remained in the environmental chamber. To determine the relative corrosion protection offered by the coating combination listed in Table I, samples from each batch were subjected to 30 APGE cycles before further examination.

Table II. Ford Motor Company APGE Cyclic Corrosion Test.

Ford APGE accelerated corrosion test cycle	
Phase 1: 15 minutes	Immersion in 5% NaCl sol'n at R.T.
Phase 2: 75 minutes	Air dry at ambient temperature/humidity
Phase 3: 22.5 hrs	Environ. chamber, 50°C, 85% R.H.
Phase 4:	Application of kaolin powder (weekly)

Corrosion Performance
After exposure to 30 APGE cycles, the extent of corrosion was determined using two methods for quantitative analysis. Initially, filiform corrosion measurements were taken from the scribe mark using an optical microscope and graduated scale to determine the resistance to undercutting of the top coat. The second method used to quantify the extent of corrosion was to measure the corroded area as a percentage of the total. This technique not only captured corrosion from the scribe, but also initiation from the coupon edges. For the area determination, the worst sample from each batch was selected and all loose corrosion products and flaking powder coat was removed. The coupon was then placed directly onto an Epson Perfection 3170 scanner to obtain a digital image. An examination area measuring 2149 x 628 pixels (91.0mm x 25.4mm) was then mapped onto the scanned image, as shown in Figure 2.

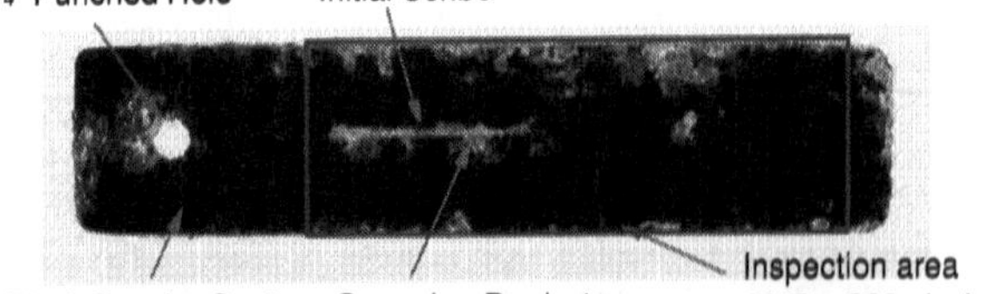

Figure 2. Designated area for corrosion measurement.

Figure 3. Post processed image of a corroded magnesium coupon.

Using Adobe Photoshop Elements 2.0 this image was post processed to improve the contrast and converted from color into greyscale. Certain coupons required a considerable amount of post processing to enhance the contrast between the grey powder coat and grey magnesium corrosion products. The image was then imported into ImageJ [5] and converted into a binary black and white negative image, as shown in Figure 3. A pixel count was then performed using the ImageJ software to determine the extent of corrosion for that single coupon. In each case, the initial scribe line equated to approximately 1% of the area examined.

Coating Thickness Measurement
Measurement of coating thickness, prior to corrosion exposure, was achieved by sectioning a single coupon from each batch. Sheared samples were secured in a colorless epoxy resin mount, ground and then polished prior to examination using an optical microscope and a JEOL JSM-6300F scanning electron microscope.

Electro-Chemical Measurements on Coated AM50 Magnesium
A classical electrochemical technique was employed to assess the protective capability of both primary conversion and secondary coatings of selective samples. For the case of assessing the protective capability of conversion coatings only, DC polarization methods were employed, as reported elsewhere by Rajagopalan et al [6]. In these measurements, the open-circuit or corrosion potential of a particular coated system was measured relative to a Ag/AgCl (saturated KCl) electrode. The corrosion current was assessed from either Tafel extrapolation or polarization resistance using the CorrWare/CorrView software package supplied by Scribner Associates (Southern Pines, North Carolina). There was no attempt to compensate for negative difference effect (NDE) as described by Song et al [7]. Instead the object of the exercise was to generate a comparative measurement of different coatings, and not an absolute measurement of corrosion rate for the alloy. It had been found previously [6] that a pH6, buffered electrolyte containing 1000 wt ppm NaCl in de-ionized water was a suitable medium for making comparative measurements of surface treatment and alloy characteristics of magnesium.

Results and Discussion

Following 30 cycles of accelerated corrosion testing, the AM50/AA-6111 bonded coupons were inspected to identify any clear distinction between the performance of candidate corrosion protection systems. Figure 4 shows the range of performance offered by the coupons examined that employed either an electro-coat or powder coat. In practice, it was anticipated that coupons relying upon a conversion coating alone, would not offer the same degree of protection as the composite systems. However, these coupons remained part of the study to identify any significant differences in the performance of the chrome free conversion coatings.

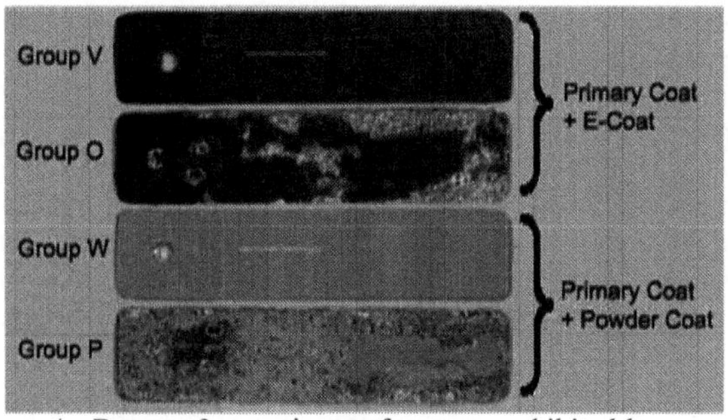

Figure 4. Range of corrosion performance exhibited by coupons using an electro-coat or powder coat.

A visual examination of the coupons concluded that significant performance advantages were offered by some systems over others. This is highlighted in Figure 4 where Group O and V represent systems using the same electro-coat but different pre-treatments. Similarly, W and P were coated with the same powder coat prior to testing.

Figure 5 shows an example of the filiform corrosion observed on the samples examined. For each coupon, the maximum distance of corrosion propagation from the scribe line was measured. An average was then calculated for each group as shown in Figure 6.

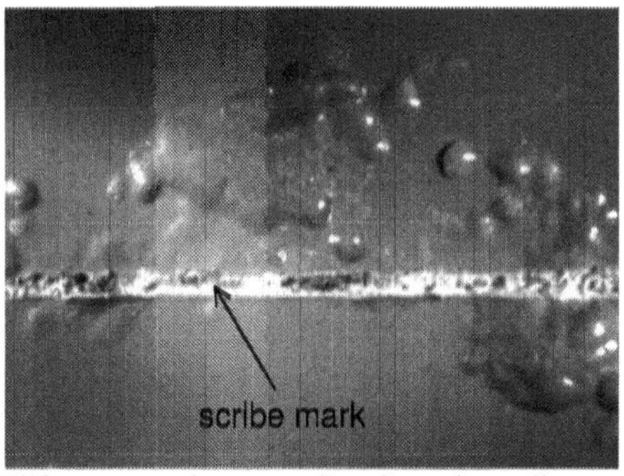

Figure 5. Filform corrosion extending from the scribe mark.

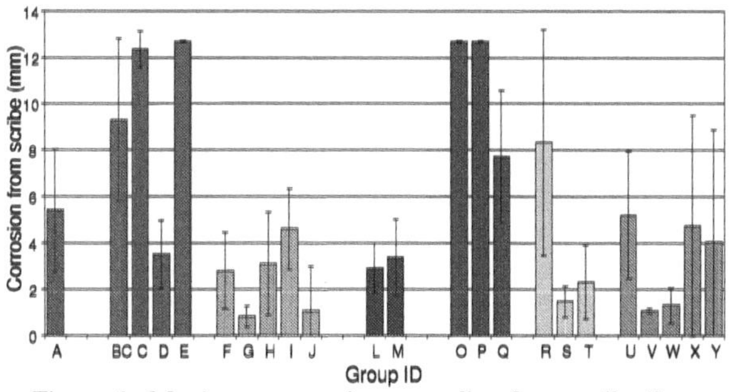

Figure 6. Maximum corrosion extending from scribe line.

Figure 6 shows that samples could be separated into three broad categories: samples that suffered from extensive corrosion undercutting (C,E,O,P), samples showing almost no corrosion propagation (G,J,V,W), and the remaining groups within this range. Further analysis of the corrosion measurements indicated that protection systems offering the most resistance to corrosion undercutting used an anodized coating as a foundation in conjunction with either a powder coat or electro-coat. However, anodized samples without any supplementary barrier were still prone to corrosion attack. In contrast, samples based upon a combination of chrome free conversion coatings and either powder coat or electro-coat exhibited a wide range of performance. Groups D, L and M (all chrome-free) exhibited a similar level of protection to the anodized groups. However, results from the other chrome free conversion coatings indicated extensive undercutting. Interestingly, a comparison between P and Q show a remarkable improvement in performance by switching to the alternative powder coat.

Figure 7 shows the results of the corrosion area measurements. It can be seen that the extent of corrosion varied considerably between the different pre-treatments and secondary coatings. The best samples (D,G,S,T,V & W) showed less than 1% corrosion which represents the initial scribe, the worst samples (K & N) showed in excess of 98% corrosion. The powder coated chrome VI sample (A) that was included as a control resulted in only 6% corrosion. It is interesting to observe that nine other coating systems (D,F,G, J,M,S,T,V & W) had a performance better that this control sample. Of these nine, six (D,G,S,T,V & W) were classified as perfect (<1% corrosion) showing no visible sign of corrosion after 30 APGE cycles. Five of these samples were anodized and included a secondary coating of either electro-coat or powder coat. The only samples that did not follow this trend were samples D and H that used a chrome free conversion coating and powder coat. Considering the anodized samples alone, the only one not to show a near perfect score when used with a secondary coat was sample H that was powder coated. This particular sample showed 8% corrosion due to undercutting from the scribe initiation point. In general terms the anodized coupons exhibited considerably less corrosion than the chrome free conversion coatings (50% Vs 8%) for all coupons within each set. However, it can also be seen that the worst performing samples in each family were those without a secondary coating. When these poor performers were excluded, the split between anodized samples and conversion coated samples becomes even more distinct at 39% and 2%. Unfortunately the improved performance of anodized coating has historically been associated with increased cost, making batch anodizing of automotive components uncompetitive. If, as shown within Figure 7, a secondary coating is required such as electro-coat or powder coat then the cost increases still further.

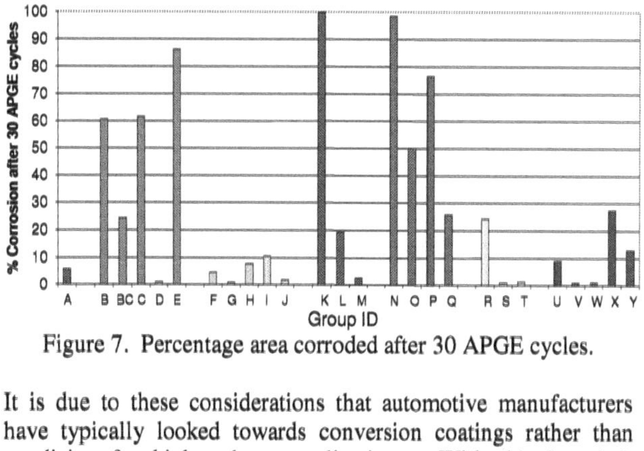

Figure 7. Percentage area corroded after 30 APGE cycles.

It is due to these considerations that automotive manufacturers have typically looked towards conversion coatings rather than anodizing for high volume applications. With this in mind, probably the most interesting coupons tested were samples D and M (conversion coat I with powder coat and conversion coat II with electro-coat). Both of these samples have a thin chrome free conversion coating with a secondary coat, but they both out performed the powder coated chrome VI control sample (A). Sample D exhibited only 1% corrosion and sample M recorded 2%. Considering the spread of results for the conversion coat I and conversion coat II families it can be seen that the secondary coating is the key to its success. For example, sample B which is the conversion coat I without any secondary coating performed extremely poorly, similarly for the conversion coat II pre-treatment without any coating (sample K) exhibited the worst performance of all coupons tested. This observation may suggest that such conversion and secondary coat combinations may not be as robust as the coated anodized samples. In general terms, as with the filiform corrosion measurements, the anodized families generally showed a better average than the non-anodized samples.

Figure 8 shows a scatter plot of the mean undercut corrosion length against the percentage corrosion area. As would be expected, a positive correlation between these two measurements techniques can be seen. It is also possible to group the samples into five distinct groups. Group 1 is populated entirely by conversion coated samples and exhibited very poor corrosion performance.

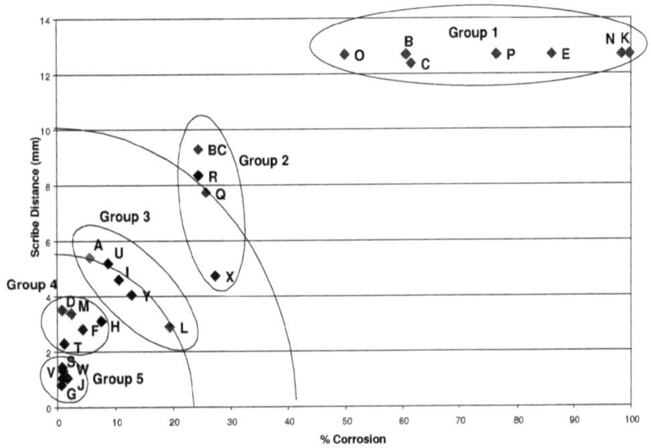

Figure 8. Scatter plot of filiform corrosion vs. area corrosion.

Group 2 showed significantly improved corrosion performance and contained coated conversion samples as well as uncoated anodized samples. Group 3 contains predominately anodized samples as well as the chrome VI control. Group 4 exhibited only minor corrosion based upon area measurement, but still some undercutting from the scribe mark. Finally Group 5 showed little evidence of corrosion after 30 cycles of APGE. The scatter plot shows much more clearly the difference between the anodized and conversion coated coupons. The anodized samples tend to be located towards the bottom left (minimal corrosion) and the conversion coated towards the upper right (extensive corrosion). Using only corroded area as the selection criteria, there were nine pre-treatments better than the powder coated chrome VI control. Using corrosion undercut distance as the selection criteria resulted in 16 pre-treatments being better than the chrome control. Constructing a radial arc from the origin set to the radius of the chrome VI sample has the effect of averaging the two corrosion assessment methods. As can be seen from Figure 8, fourteen pre-treatments are included. In a similar manner, a second arc has been drawn on Figure 8 that corresponds to sample Q (shown in Figure 2). This pre-treatment combination has been tested extensively using a variety of accelerated corrosion methods, including salt spray and full vehicle testing. It has been determined that this pre-treatment is capable of providing adequate corrosion protection and functional requirement for the entire life of an under-hood magnesium component. Using this sample as a benchmark, there are sixteen alternative pre-treatments that provide either the same or better corrosion performance. Again, the most interesting, from an automotive manufacturers perspective, are the lower cost conversion coated samples. Using cost as a selection criterion, samples D, M and L all performed better than the benchmark sample Q. It should be noted, however, that an under-hood magnesium component has primarily a functional rather than visual corrosion requirements. The visual corrosion requirements of a door inner panel, for example, would be more demanding than those for an under-hood application. However, the effect of paint and additional clear-coat layers should also be considered when selecting a pre-treatment for a visible magnesium part such as a door inner. The authors have reported in earlier work that the addition of a paint layer and clear-coat improved the visual corrosion performance of a door hem joint significantly over the unpainted test coupons [2].

Figure 8 can also be used to see dramatic differences with what would seem only subtle changes in pre-treatments. For example, samples P and Q were pre-treated with the same chrome free conversion coating but with different powder coats. Sample Q showed a dramatic improvement in corrosion performance over sample P. It should be noted that sample Q used a unique powder coat to all of the other samples processed. If the trend exhibited between samples P and Q were applicable to all of the primary pre-treatments (conversion coated and anodized) then there would be a similar improvement in corrosion performance for all of the powder coated coupons. Another interesting comparison is between samples L and P. Both samples used the same powder coating and a chrome free conversion coating. The only difference being the selection of the chrome free conversion coating. The same is also true for samples M and O that were electro-coated rather than powder coated. Once again, the chrome free pre-treatment from one supplier outperformed the other. The electro-coated samples again outperformed the powder coated samples. Finally, comparing samples BC and C shows the detrimental effect that electro-coat parameters can have on overall corrosion performance. Sample C was processed using a conventional electro-coat line including the cleaning and phosphating. In contrast, samples BC were electro-coated offline without the cleaning or phosphating. It is likely that either the cleaning or phosphating effectively stripped the conversion coating from the magnesium. This demonstrates that off line electro-coat is the preferred processing route. The same guideline would apply to the anodized samples.

Electro-Chemical Measurements of Conversion Coatings

To provide further rationale behind the performance of the conversion coatings, the open circuit potential was measured on selected samples to determine the level of passivation offered by the different coatings. As shown in Table I, these coatings studied were generally intended to be the pre-treatment layer for a polymer topcoat. Figure 9 shows the variation of the open-circuit potential with material and pre-treatment applied. In general, alloying pure magnesium with aluminum raises the open circuit potential in this electrolyte, and this is also evident for the AM50 alloy considered here relative to a baseline 99.98% magnesium (Timminco). Chromate is the classic magnesium pre-treatment and had the largest influence on open circuit potential, producing an open circuit value approximately 500 mV more noble than the base AM50 alloy. In contrast, chrome free conversion coatings B, K and N had less of an impact, with B offering the largest shift in open circuit potential of 160mV. In practice the large experimental scatter makes any clear distinction between the chrome free conversion coatings difficult. However, a comparison with the chrome based control sample shows none of these conversion coatings offer the same level of performance.

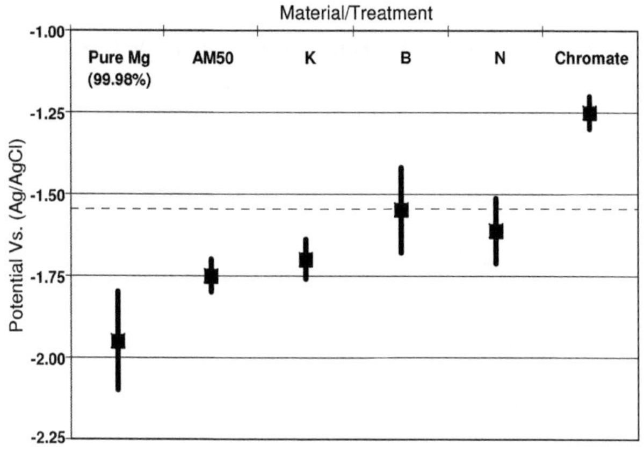

Figure 9 Change in open-circuit potential for chrome based and chrome free conversion coatings.

Film Thickness and Coating Morphology
An examination of coating thickness and morphology was performed in an attempt to identify any correlation between the physical protection offered by the various coating systems and ultimate corrosion performance. Of particular interest was the capability of the coating processes to create a uniform film build over the coupon. Edges of the samples were examined for evidence of thinning as sharp features on parts have historically been sites for corrosion initiation [2]. Figure 10 shows a breakdown of coating thickness of samples coated with the epoxy-polyester powder coat. A nominal coating thickness on the flat areas of the samples ranged from 52μm to 114μm. Hence, the powder coating process is obviously prone to large variations in material deposition within a batch. However, this is not unexpected and typical applications specify an acceptable range of coverage.

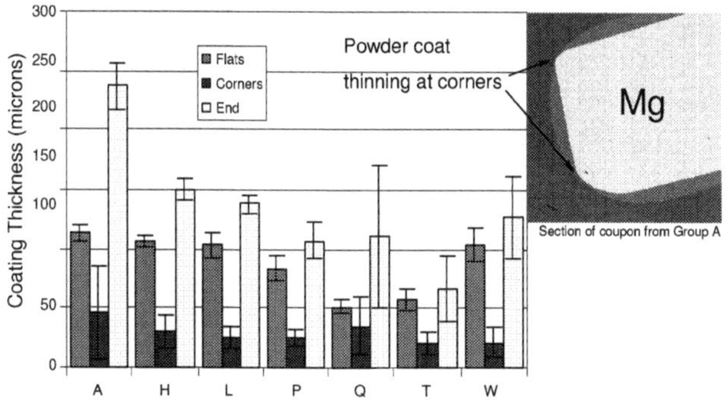

Figure 10. Powder coat thickness measurements.

Of greater concern was the extent of power coat thinning observed at the corners of the samples. Even after deburring and dressing the magnesium coupons prior to coating, the mean coverage at the corners declined to between 21μm to 46μm. Also as the standard deviation indicates, individual samples showed extreme thinning down to less than 5μm. This suggests that pre-treatments susceptible to corrosion undercutting may not have adequate protection along sharp radii and other design features if a powder coat is chosen as the secondary barrier layer. Of the pre-treatments tested, this could explain the poor undercutting performance of sample P. In contrast, the drop off in film thickness, measured on the powder coated chrome VI control sample (A), appears to have more of a secondary effect on the overall corrosion performance.

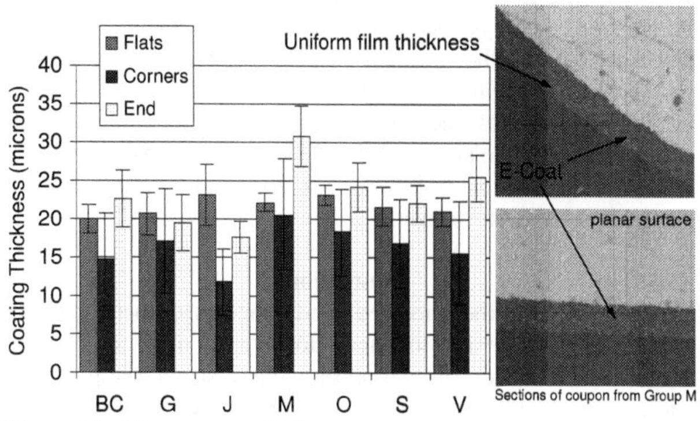

Figure 11. Electro-coat thickness measurements.

Relative to the powder coat measurements, samples subjected to electro-coat exhibited a far more uniform coating coverage. The inserts in Figure 11, from Group M, show the coverage typical of electro-coat around the sample corners. Hence, while some thinning is observed between the samples subjected to electro-coat, the absolute departure from nominal is far less substantial. This suggests that even with a nominal coating thickness of ~20μm vs. ~100μm for powder coating, the electro-coat option may be more robust in practice due to the superior edge coverage. However, a comparison of anodized samples such as V and W, and S and T show no discernable difference between powder coat and electro-coat corrosion performance. More appreciable differences in performance between the two top coat options were observed following analysis of samples applied with the chrome free conversion coatings. However, no consistent trend emerged. For example, while Group M (chrome free and electro-cost) offers superior performance relative to the comparator group L (powder coat), the converse is true for Groups BC (electro-coat) and D (power coat).

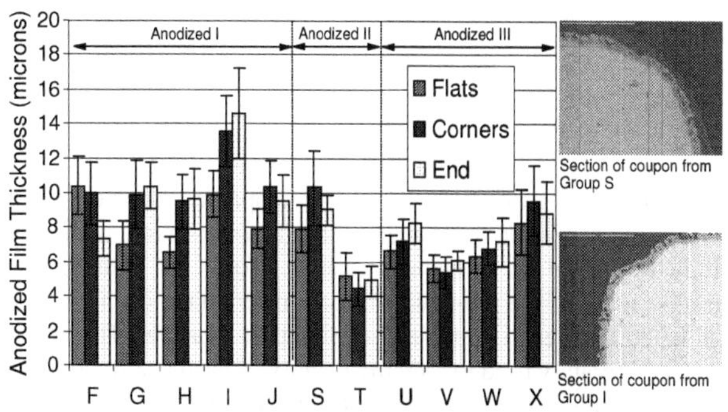

Figure 12. Anodized film thickness measurements.

Of the primary coatings, only the anodizing process produced a continuous film thickness exceeding 5μm. In contrast conversion coatings examined using scanning electron microscopy were typically of thicknesses less than 1μm. Figure 12 shows the film thickness produced by the three anodizing processes under evaluation. In accordance with the process specifications, the anodized films measured between 5μm and 10μm. However, a minimum thickness was specified where possible to reduce processing cost and reduce the risk of creating a brittle oxide layer. Examination of the images in Figure 12 indicates that, as with the electro-coat process, a uniform film was established. In fact, in many cases the test coupons showed a slight increase in film thickness at the corners and ends. This suggests that the anodized pre-treatments may be a more robust alternative to conversion coatings due to more uniform coating coverage. Porosity, on the other hand, is clearly visible within the coatings as shown in Figure 12. This may explain the significant performance improvement exhibited by using an additional sealer or top coat over the anodized film. This can also explain why the choice of electro-coat or powder coat appears not to influence the final corrosion performance.

Conclusions

In general terms the anodized pre-treatments outperformed the chrome free conversion coatings. However, in order to obtain the best out of anodized films it appears that a secondary polymeric top coat or "seal" is necessary. The choice of top coat does not appear to be important as electro-coat, powder coat and sealers seems to offer similarly excellent corrosion performance. Of course financial considerations need to be addressed due to the increased processing costs associated with batch anodizing. However, in this study a deliberate move was made to minimize anodized coating thickness, hence processing cost. For the most part the coating thickness specification was 5μm which was considered the operating limit for the processes investigated. Therefore, even with a minimum coating coverage, the anodized samples still offered the most protection of the samples tested.

Placing all of the results into context, 14 of the groups tested exhibited equivalent or superior performance to the powder coated chrome VI control. Combined with field test data of the control sample, this suggests that chrome free conversion coatings are still worthy of consideration. In fact, certain non-chrome conversion coatings in combination with a polymeric top coat provided corrosion resistance better than the hexavalent chromium chromate control sample. However, the choice of top coat layer appears to be selective. For some of the conversion coatings, electro-coat produced the best results, whereas for others, it was the powder coat that produced the best corrosion performance. This variability was not evident for any of the anodized samples, which suggests that the conversion coating process is less robust than anodizing.

Acknowledgement

Our gratitude is extended to all of the companies that have helped prepare samples for this paper. The data presented within this study is part of an on-going research effort. In light of these results, some of the pre-treatment parameters have been modified by the companies to improve corrosion performance. The authors have deliberately tried to hide the identity of each pre-treatment as not to prejudge or comment on specific companies. It is hoped that the generic results and conclusions drawn from this paper are valid to the community at large without the need to reveal their identity. Special thanks are also extended to Mike Kelly and Chris Borsellino at Ford Central Laboratory for performing all of the corrosion testing. Bill Donlon is also thanked for his support in performing the surface analysis work.

References

1. **Mordike, B.L. and T. Ebert**, "Magnesium Properties - applications – potential," Materials Science and Engineering, pp 37-45, 2001

2. **Bretz, G., Lazarz, K., Hill, J. & Blanchard, P.** "Adhesive Bonding and Corrosion Protection of a Die Cast Magnesium Automotive Door", TMS 2004.

3. **Skar, J.I, Sivertsen, L.K & Oster, J.M**. "Chrome-free Conversion Coatings for Magnesium Die Casting – a Review", ICEPAM – International Conference on Environmental friendly pre-treatment for Aluminum and other Metals, Oslo, Norway, 2004.

4. **Ford Laboratory Test Method BI 123-01.** "Painted Sheet Metal Corrosion Test (Laboratory Simulated Arizona Proving Ground Test)", 2001.

5. ImageJ, National Institute of Health, http://rsb.info.nih.gov/ij/

6. **Rajagopalan, B., Mohanty, P.S., Mallick, P.K., McCune, R.C & Ricketts, M.S.** "Corrosion Performance Evaluation of Magnesium Alloys by Electrochemical Polarization Technique," Magnesium Technology 2004, as held at the 2004 TMS Annual Meeting, Charlotte, N.C., USA, 14-18 Mar. 2004. pp 131-136, 2004.

7. **Song, G., Atrens, A., Stjohn, D., Nairn, J. & Li, Y.** "The Electrochemical Corrosion of Pure Magnesium in 1N NaCl," Corrosion Science 39 [5], pp 855-875, 1997.

Advanced Conversion Coatings for Magnesium alloys

Syam Nibhanupudi, Alp Manavbasi

Metalast International, Minden, NV

Keywords: conversion coating, chromium, hexavalent, trivalent, magnesium, corrosion.

Abstract

Magnesium and its alloys have excellent physical and mechanical properties due to their high strength-to-weight ratio and are ideal for various applications in automotive, aerospace and defense sectors. However, Mg alloys are also highly susceptible to corrosion under harsh environments. Owing to this carcinogenicity as well as environmental impact of hexavalent chromium fueled by stringent environmental regulations, an environmentally green alternative to the carcinogenic hexavalent chromium coatings on magnesium is due.

In this work, a novel trivalent chromium based conversion coating has been developed to improve the corrosion resistance and paint adhesion properties of Mg alloys. Coating performance characterization has been investigated via hydrogen evolution, weight loss measurement and electrochemical corrosion analysis techniques. Results have shown that the novel environmentally green trivalent chromium based coating on magnesium has indeed performed comparable to hexavalent chromium and thus establishing a viable alternative.

Introduction

Magnesium, the eighth most abundant metal on earth, has seen an increase in applications in a variety of sectors ranging from aerospace, military, defense, automobile to commercial mobile phones, sporting goods and handheld tools owing to its high strength-to-weight ratio, machinability, thermal conductivity and weldability [1-4]. Although magnesium alloys have the distinct advantage of high strength-to-weight ratio and high impact resistance, their extremely poor corrosion resistance in aggressive environments especially in assemblies with multiple galvanic couples, limits their usage because of the premature component degradation and replacement. [4-7].

Improving corrosion performance involves a variety of finishing processes including oil applications, wax coating, anodizing, electroplating, conversion coating, painting etc. Metal finishers are often faced with challenges to tailor a better combination of these processes to enhance the product corrosion performance considering various factors ranging from specific applications to environmental conditions. Even though development of new and high purity magnesium alloys and novel surface modification techniques such as ion implantation and laser annealing have mitigated the corrosion problem, magnesium alloys still suffer from performance issues especially under aggressive corrosive environments. High chemical reactivity, complex alloy microstructures, hazardous pretreatment procedures and recycling problems to an extent have discouraged many applications of magnesium [8-10].

Among the typical surface modification techniques, conversion coating, electroplating with alloys like Zn, Ni and Cr and anodizing processes such as DOW 17, HAE, and Tagnite have become widely popular for the surface treatment of magnesium alloys.

Majority of the magnesium alloys are either sand-cast or die-cast and hence are prone to microstructural inhomegeniety, surface imperfections such as porosity and impurities. Magnesium and its alloys are also prone to galvanic corrosion and alloy dissolution owing to lesser nobility of magnesium. Design of a process routine that results in an effective barrier to corrosion as well as wear would be ideal. Hexavalent chromium (Cr^{6+}) based conversion coatings have been used for a long time as a coating material on aluminum, magnesium and other metals in order to increase the corrosion protection, paint adhesion, and adhesive bonding characteristics. However, solutions containing Cr6+ are highly toxic and adversely affect the environment and human health.

A recent Under Secretary of Defense memo, dated April 2009, regarding the elimination and restriction of the Cr6+ from Department of Defense (DoD) weapon systems and platforms shows the importance of novel non-hexvalent technologies for the U.S. military departments. The U.S. and International market needs for a Trivalent Chromium Conversion Coating is predicated on three severe and stringent European Union (EU) Directives: Restriction of Hazardous Substances (RoHS), Waste Electrical & Electronic Equipment (WEEE), and End of Life Vehicle (ELV). Global manufacturing demands compliance even though these directives are implemented in the EU. EU member nation states have established regulations and enforcement of these Directives. In the U.S., heavy regulations from both EPA and OSHA are continuing to be issued and implemented. OSHA has mandated a Permissible Exposure Level (PEL) of 5 ppm of Hexavalent Chromium, an unrealistic level for the vast majority of metal finishers and manufacturers to attain. Compliance to the Directives and OSHA requirements are not an option, they are mandatory. In addition, the EPA Executive Order 12856 showed the need to reduce or eliminate the release of chromates during aircraft coating applications.

There is, therefore, a need for environmentally green conversion coatings that can provide high corrosion resistance and increase the adhesive bonding strength characteristics of the metal surface. Although there are other conversion coatings, which do not contain Cr6+, their corrosion performance and paint adhesion characteristics are not as effective as the Cr6+ based conversion coatings.

Our aim is to at develop an environmentally friendly conversion treatment for magnesium alloys. In this present study, trivalent chromium based environmentally conversion coating for AZ92A magnesium alloy was developed. Furthermore, the performance was gauged against hexavalent chromium coatings using hydrogen evolution test, weight loss measurement techniques and potentidynamic studies.

Experimental set up

Experimental process routine for the study was conducted in accordance with AMS -3171-C, [11]. Sand cast magnesium Alloy AZ92A discs (1.5 in. dia and 0.5 in. thickness) conforming to AMS 4434 [12], have been purchased from Metalmart Inc., with the following compositions.

Al	Zn	Si	Mn	Cu
9.0	2.0	0.3 max	0.10 min	0.25 max

Table 1 : Elemental compositions of AZ92A sand casting in percentages.

Conversion coating samples underwent the following identical process routine for neutral coating-performance comparisons.

Mechanical Grinding: Samples have been ground to 1200p using SiC wafers, rinsed in D.I water and dried.

Pretreatment Process: Ground samples are then wiped with acetone and rubbing alcohol. Samples are then subjected to the pretreatment routines for effective coating depositions as shown below.

Alkaline Cleaning: Sodium Hydroxide based cleaner has been chosen for alkaline cleaning of magnesium alloys for 5 minutes at 140F and a pH> 12.0. Magnesium is very resistant to corrosion by alkalis if the pH exceeds 10.5 which corresponds to the pH of a saturated $Mg(OH)_2$[13]; a $Mg(OH)_2$film is formed on the magnesium surface [3,13]. Dilute alkali solutions show negligible attack at temperatures up to the boiling point [14]. Consequently, a 10% caustic solution is commonly used for cleaning at temperatures up to the boiling point [15].

Surface activation via Acid Pickling: Chromic-Nitric-HF acid pickle in compositions recommended in the AMS-3171-C - specification [11], has been observed to be beneficial in contrast to phosphoric and sulfuric acid etches on AZ92A and hence has been adopted. A 2 minute immersion in an agitated mixture of Ammonium Bifluoride + Nitric acid has been observed to effectively remove the acid smut adhering to the surface in the trial studies.

Conversion Coating: Hexavalent chromium based conversion coatings were applied in accordance with the Type VIII treatment in the AMS-3171-C specification [11]. Iridite #15 was used for hexavalent chromium conversion coating of AZ92A.

Once conversion coated samples are stored in a dehumidifier chamber for a minimum of 24 hour curing period prior testing.

Performance characterization

Hydrogen Evolution and Weight loss measurement:

Hydrogen evolution studies have been conducted on bare AZ92A discs ground to 1200p grit and hexavalent chromium conversion coated samples for base line test criteria. Trivalent chromium conversion coated samples processed and cured were tested for amount of hydrogen evolved under 3.5% NaCl solution and was compared to the baseline criteria for rate of hydrogen evolution per surface area. Hot chromic acid 24 oz/gal at 190F, 8 minutes was used to remove any corrosive products on the alloy surface followed by a rinse in D.I. ,dried and weighed to evaluate the amount of weight loss per surface area due to salt immersion.

DC Polarization studies

Potentiodynamic measurements were performed on the conversion coated and bare AZ92A discs. The experiments were carried out in an aggressive 3.5% NaCl electrolyte. A three-electrode configuration was employed: conversion coated magnesium substrate as working electrode, a carbon rod as a counter electrode, and a SCE (Standard Calomel Electrode) as reference electrode. The amplitude of the perturbation was 10mV and the OCP (Open Circuit Potential) was chosen as bias potential. The examined frequency range was from 10mHz to 100kHz. Results obtained from these tests are used to determine the corrosion current, potential and corrosion rate.

Results and discussion

Hydrogen evolution experiment:

Magnesium dissolution in aqueous environments are often influenced by an electrochemical reaction with water that produces magnesium hydroxide and hydrogen gas. Under film-breakdown agents such as chlorides the stable oxide/hydroxide layers experience surface film break-down followed by the evolution of hydrogen. A stable coating negates this surface film limits this surface film break down vis-à-vis lesser hydrogen evolution. After conversion coated and cured, substrates are submerged in a bath of 3.5% NaCl solution to measure the amount of hydrogen released and the amount of weight loss per surface area. The volume of hydrogen displaced in the inverted burette is used to calculate the amount of hydrogen released per surface area.
Results obtained from the above measurements were used to evaluate the corrosion rate of the alloy (mm/y) with respect to their coating configurations using the following formulae.

Corrosion rate [mm/y] = 2.2785 (hydrogen evolution rate per Surface area [ml/cm^2/d]).

Corrosion rate [mm/y] = 2.10 (weight loss rate [mg/cm^2/d]);

Figure 1 and 2 details the results of the hydrogen evolution test conducted on various conversion coated and bare magnesium samples as well as the calculated corrosion rates. It is clear from these values that the novel trivalent coating enhanced the corrosion protection compared to the uncoated bare substrate and performed close to the hexavalent coating. Figure 3 details the calculated corrosion rates via weight loss measurement tests and it can be observed that in both cases Trivalent chromium conversion coating processed samples performed on par with hexavalent IRIDITE coating by offering comparably effective film breakdown protection.

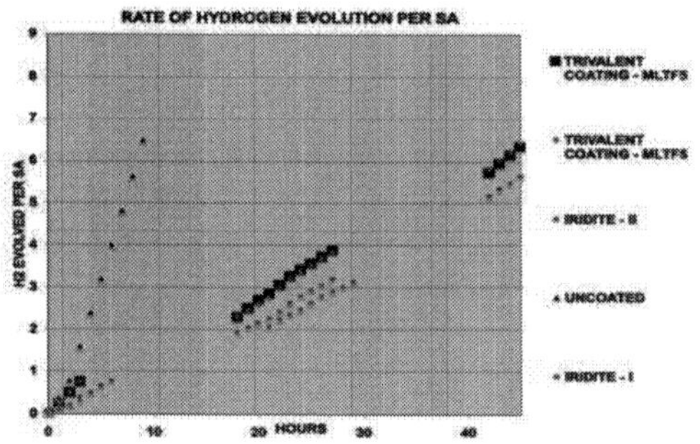

Figure 1: Rate of hydrogen evolution per surface area for various coated surfaces.

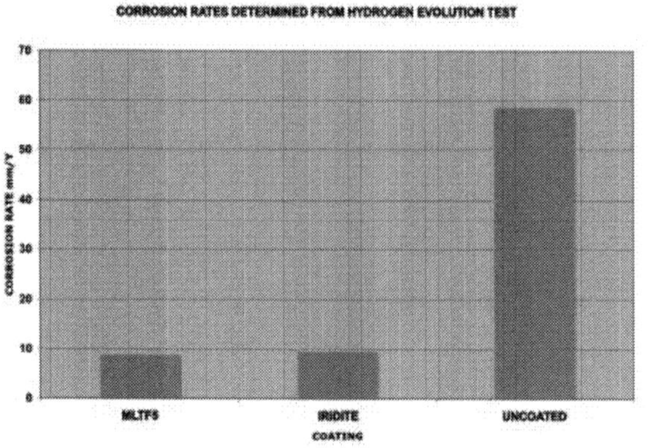

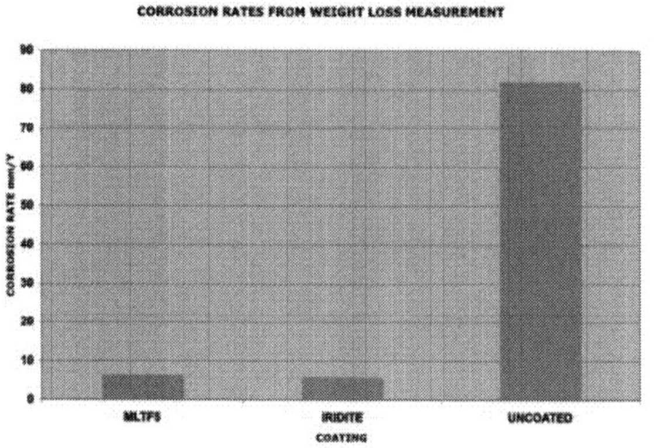

Figure 2 & 3: Corrosion rates calculated from the hydrogen evolution experiment and weight loss measurement tests on various coated surfaces.

DC Polarization studies:

DC polarization tests were conducted on bare as well as conversion coated substrates. Potentiodynamic studies revealed that the trivalent based conversion coating offered much nobler corrosion potential as shown in Figure 4 and Table 2 displays the corrosion current, corrosion potential and corrosion rates from Tafel plots generated using Gamry DC105 corrosion techniques software. It can be observed from these values that the corrosion rate for trivalent chromium based conversion coating on AZ92A has been reduced than for the hexavalent chromium based conversion coating as wells as the bare substrate.

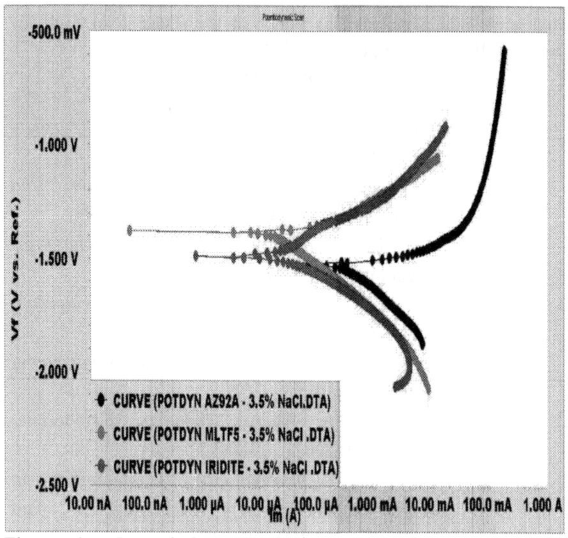

Figure 4: Potentiodynamic study results of various coating surfaces under 3.5% NaCl electrolyte.

Coating	Icorr-Amperes	Ecorr - Volts	Corrosion Rate (MPY)
Uncoated	4.48 mA	-1.53	6.82E+02
Iridite – Cr^{+6}	24.6 μA	-1.49	11.23
MLTF5 – Cr^{+3}	18.1 μA	-1.38	8.26

Table 2: Corrosion current, voltage and corrosion rates obtained from Potentiodynamic tests on AZ92A.

Conclusions

The novel trivalent chromium based conversion coating developed for magnesium alloys has in fact enhanced the corrosion resistance comparable to hexavalent chromium. Based on the results from various tests it can be concluded that;

- Rate of hydrogen evolution per surface area for magnesium substrates with trivalent chromium based conversion coatings is less than bare substrates and comparably same as hexavalent chromium coated substrates.
- Corrosion rates obtained from hydrogen evolution tests and weight loss measurement tests have shown that the trivalent chromium based coatings have indeed enhanced the corrosion resistance of the magnesium alloys.
- DC polarization studies have shown that the trivalent chromium based conversion coatings have offered significant amount of resistance when exposed to harsh chloride environment.

Acknowledgment

This work was performed as a part of the Phase I research for DoD - SBIR Grant proposal, F083-226-0271.

References

1. Natarajan.S, *Corrosion Prevention and Control*, v 51, n 4, p 142-163, December 2004
2. Avedesian.M.M, Baker.H, *Magnesium and magnesium alloys*, ASM International, 1999.
3. Emley.E.F, *Principles of Magnesium Technology*, Pergamon Press New York 1966.
4. G. Song.G, A. Atrens, *Advanced Engineering Materials* 5 (2003) 837–858.
5. G. Song, A. Atrens, *Advanced Engineering Materials* 1 (1999) 11–33.
6. G. Song, A. Atrens, *Advanced Engineering Materials* 9 (2007) 177-183.
7. G. Song, *International Conference on Magnesium*—Science Technology and Applications (2004) E01.
8. S.S. Cho, B.S. Chun, C.W. Won, S.D. Kim, B.S. Lee, H. Baek, *J. Mater. Sci.* 34 (1999) 4311.
9. Y. Li, J. Lin, F.C. Loh, K.L. Tan, *J. Mater. Sci.* 31 (1996) 4017.
10. S. Ono, N. Masuko, *Materials Science Forum* 419 (4) (2003) 897–902.
11. *"Magnesium Alloy, Proceses for Pretreatment and Prevention of Corrosion on"*, Society of Automotive Engineers, Warrendale PA.
12. "*Magnesium Alloy Castings, Sand 9.0Al – 2.0 Zn (AZ92-T6) Solution and Precipitation Heat Treated", specification*, Society of Automotive Engineers, Warrendale PA.
13. G. L. Makar, J. Kruger, *Int. Mater. Rev.* 1993, 38(3), 138.
14. W. S. Loose, *Corrosion and Protection of Magnesium*, ASM Int., Materials Park, OH 1946, pp. 173±260.
15. Froats, T. K. Aune, D. Hawke, W. Unsworth, J. Hillis, *Metals Handbook*, 9th ed., Vol. 13, ASM Int., Materials Park, OH 1987, pp. 740±754.

IMPROVED CORROSION PERFORMANCE OF AZ91D MAGNESIUM ALLOY COATED WITH THE KERONITE™ PROCESS

S. Shrestha [a], A. Sturgeon [a], P. Shashkov [b] and A. Shatrov [b]

[a] TWI Ltd, Granta Park, Great Abington, Cambridge CB1 6AL, UK
[b] Keronite Ltd, Granta Park, Great Abington, Cambridge CB1 6GX, UK

Abstract

This paper describes the results of a study on the corrosion behaviour of die-cast magnesium alloy type AZ91D coated using the KERONITE™ process. This process is a commercially available and environmentally friendly electrolytic coating method applicable to all types of magnesium alloys. The process involves creation of a hard ceramic oxide layer on the surface of a light substrate alloy by plasma electrolytic oxidation in a low concentration alkaline solution.

Corrosion performance was assessed in a 3.5% NaCl solution using an accelerated electrochemical potentiodynamic polarisation method, and a salt spray exposure. Parallel tests were also carried out on an uncoated AZ91D alloy. The study demonstrated an improvement in the corrosion resistance of Keronite coated AZ91D magnesium alloy.

Introduction

Magnesium-based alloys are of growing interest for many industrial applications, which include components in automotive [1], aerospace [2, 3], electronics [2], textile machinery and printing machines due to their favourable strength to weight ratio. Thin walls, high casting rates and long die life are typical advantages that can be achieved by using die-cast magnesium. The most commonly used alloy is AZ91D (9%Al, 1%Zn), which exhibits good mechanical and physical properties with excellent castability [4]. But the relative poor corrosion and wear resistance limits its industrial use to just a few applications [5]. An example of a potential application of the Keronite coating to improve the corrosion resistance of automotive pump casing made of AZ91D alloy is shown in Fig.1a.

Various chemical conversion surface treatments such as chrome pickling or anodising, in combination with alkali resistant organic coating systems, have been used for magnesium alloys to give limited corrosion protection in mild corrosive environments. Due to environmental concerns these traditional processes are being replaced with more environmentally safe surface treatment processes. Techniques for improving the wear resistance of magnesium alloys such as laser surface alloying [6] and thermal spraying [7] have been reported. But these applications also have limitations and can not be used for components with restricted access areas and smaller dimensions.

Thus, there exists potential for new electrolytic processes such as KERONITE™ licensed by Keronite Ltd. (UK), Anomag offered by Magnesium Technology Licensing Ltd. (New Zealand) and Magoxid-Coat licensed by AHC Oberflachentechnic GmbH & Co. OHG (Germany). Unlike traditional acidic chromate and/or fluoride-based processes, these new processes employ alkaline-based chromate-free solutions.

This report presents results obtained from an evaluation of the corrosion performance of die-cast magnesium AZ91D alloy without any coating and with a coating deposited using the KERONITE™ process.

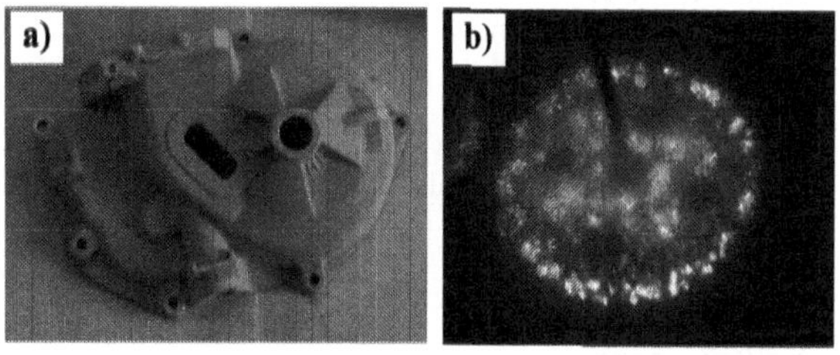

Figure 1: Pump case made of AZ91D and coated with Keronite (a); Plasma discharge surrounding the surface of a component immersed in an electrolytic bath (b).

KERONITE™ Process

Coating of magnesium alloy using the KERONITE™ process involves creation of a plasma discharge around a component immersed in an electrolyte. A modulated AC voltage is applied to the component, creating intense plasma due to micro-arc (Fig.1b) generation at the component surface. This results in oxidation of the component surface (plasma electrolytic oxidation) as well as elemental co-deposition from the electrolyte solution. The electrolyte is a low concentration alkaline solution of proprietary composition and can be disposed of without further treatment. A bath with a capacity of 200 litres and a power rating of 80kW can coat a substrate up to 50dm^2 surface area at a rate of 1-2μm per minute. Substrate preparation requires only degreasing of the component and the process is operated between 20-50°C.

Experimental

Material and Test Specimens

The magnesium alloy considered in this work was a high purity die cast type known as AZ91D (Bal Mg, 9Al, 1Zn, 0.2Mn, max. 0.001Fe, max. 0.001Ni and max. 0.015Cu). A Keronite coating to a thickness of about 35μm was deposited on as-cast alloy samples of size 150x100x1.5mm. For comparison purposes in terms of performance, parallel tests were carried out on uncoated AZ91D alloy.

Coating Characterisation

A coated sample was carefully sectioned to a size of 10x20mm using an abrasive cutting wheel and the cut specimen was mounted in araldite in cross section. Standard metallographic preparation techniques of grinding on abrasive papers and polishing to 1μm diamond finish were used to prepare the cross section. The obtained cross section was gold sputtered in an evaporative coater and observed under a scanning electron microscope (SEM).

Determination of the various crystalline phases present in the coating was undertaken using the X-ray diffraction (XRD) technique. The XRD technique comprised of collecting spectrum using CuK_α radiation from the coating material ground to a fine powder.

Coating hardness was measured on the polished cross section by Vickers micro-indentation with a load of 25g using the procedure described in ASTM E384-99.

Free Corrosion Test

Specimens of size 35x35mm were cut from the larger samples and painted with lacquer 'Electromask' so that the coating/substrate interface was sealed during the exposure period. The specimen was immersed in a 3.5% NaCl solution. The temperature was maintained at 25±1°C and the conductivity of the solution was maintained at 53mS.cm^{-1}. Observations were made on a daily basis over a one-month period and any change on the coating surface was noted.

Salt Spray Test

Specimen size for the salt spray test was 150x100x2mm. The salt spray exposure was carried out as described in the ASTM standard B117-97 'Standard practice for operating salt spray (fog) apparatus'. This test comprised of exposing the coating surface to a 5wt% NaCl solution atomised to create a fog within an enclosed chamber. The coating was supported 15-30° from the vertical. The exposure duration was 1000 hours and temperature was maintained at 35°C. Changes to the coating surfaces were recorded following periodic observations every 24 hours. On completion of the test after 1000-hour exposure, the surface quality was given a ranking number in accordance with ASTM D1654-92 'Evaluation of painted or coated specimens subjected to corrosion environments' using the rating for unscribed areas – procedure B.

Electrochemical Polarisation Test

Specimens were cut to a size of 35x35mm for the electrochemical corrosion tests. A small area at the rear face of the coated specimen was scratched and micro-welded with titanium rod to provide electrical connection for the electrochemical polarisation test. The coated area on the front face exposed to the electrolyte was 1cm^2 in size. Testing of the coating was performed in the as-received condition after degreasing with alcohol. Testing of the uncoated specimen was performed on a surface abraded with 1200-grit sandpaper.

Electrochemical corrosion tests were performed using a three-electrode Avesta cell arrangement attached to a computer controlled potentiostat as shown in Fig.2. Experiments were carried out according to the guidelines described in the ASTM standard G61. The electrolyte was a 3.5% NaCl solution (made from distilled water) with a pH value of about 8.1. The temperature was maintained at 25±1°C. Electrical conductivity was measured at about 53mS.cm^{-1} at 25°C and the solution was deaerated by continuous purging with nitrogen.

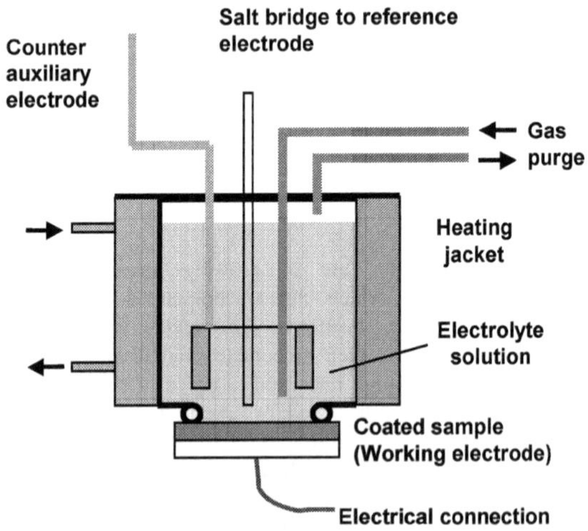

Figure 2: Schematic diagram of the Avesta cell.

The specimens were electrically connected to the Avesta cell. The coated surface was exposed to the electrolyte and the rest potential 'E_{corr}' (also known as free corrosion potential) was stabilised for one hour prior to the anodic polarisation. This potential was measured using a reference saturated calomel electrode (SCE). The area exposed to the electrolyte was then anodically polarised from its rest potential to more positive potential at a rate of 10mV.min^{-1}. This was continued until a current density of 5mA.cm^{-2} was measured in the external circuit between the working electrode (specimen) and a platinum auxiliary electrode. On reaching a corrosion current density of 5mA.cm^{-2} the scan was reversed and the potential reduced in the negative direction back to E_{corr}. A plot of the corrosion current density as a function to the polarisation potential was recorded.

The collected polarisation plots were used to compare the corrosion behaviour of the magnesium alloy in the uncoated condition with that having the Keronite coating. The following features were used to compare the corrosion behaviour:

- Rest potential 'E_{corr}' – a more negative E_{corr} representing greater susceptibility to corrosion attack.
- Breakdown potential 'E_b' – At potentials above E_b, the current rapidly increases with further increase in potential. The constant current region below E_b is referred to as a passive region where little or no corrosion occurs. At potentials above E_b, rapid corrosion is occurring. The presence of a passive region and a more positive value for E_b indicates better resistance to corrosion attack.

Results

Coating Characterisation

A back-scattered SEM image of a typical 35μm Keronite coating is shown in Fig.3, which displays a relatively low porosity coating. A few light grey and dark contrast features appeared in the back-scattered SEM image due to atomic number difference.

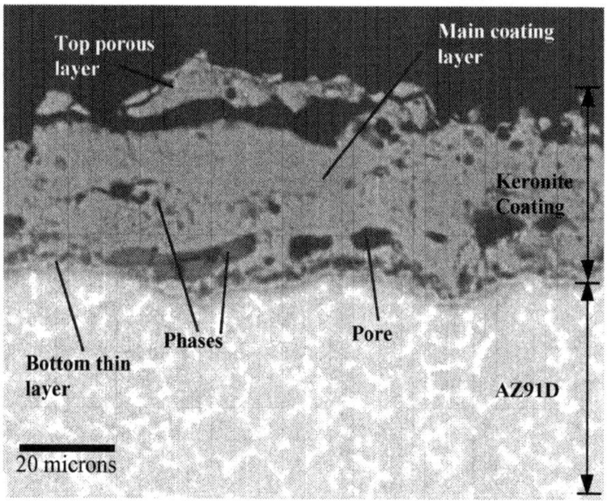

Figure 3: Cross-section SEM image of as-deposited 35μm Keronite coating on die cast AZ91D.

Figure 3 shows a multi layer coating structure, which comprised a top porous layer, a middle main coating layer and a very thin interfacial layer adjacent to the substrate. The existence of similar features has been reported for the Keronite coatings on aluminium alloys [8].

An SEM image of the deposited Keronite coating surface displays an abundance of agglomerated particles formed during the micro-arc oxidation process (Fig.4). X-ray diffraction of the coating layer confirmed spinel ($MgAl_2O_4$) as the major constituent together with existence of stishovite (SiO_2) and silicon phosphide (SiP_2). This is illustrated by the XRD pattern shown in Fig.5.

Micro-indentation hardness values measured on the polished cross section ranged between 345 to 446 HV for the 35μm Keronite coating and about 95 for the AZ91D alloy. Average hardness values of the AZ91D alloy and the 35μm Keronite coating are compared in Fig.6.

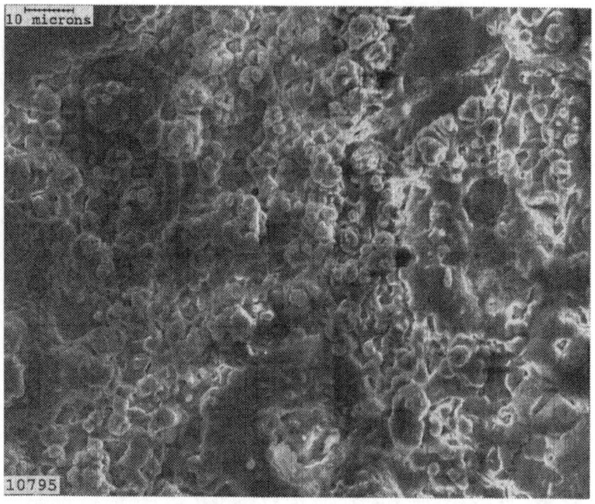

Figure 4: SEM image of the as deposited surface of the Keronite coating on AZ91D.

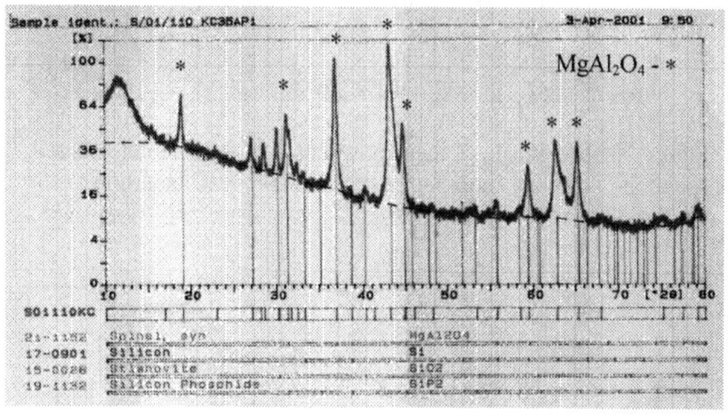

Figure 5: X-ray diffraction patterns collected from the Keronite coating.

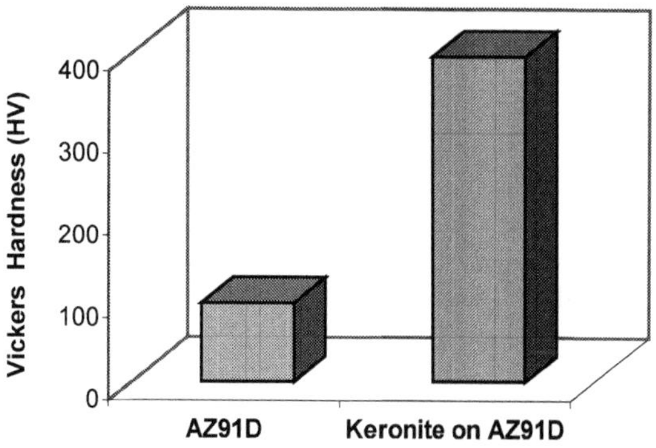

Figure 6: Vickers microhardness values measured on die cast AZ91D and 35μm Keronite coating under a load of 25g.

Free Corrosion

The magnesium alloy coated with Keronite did not show any sign of significant corrosion during the one-month immersion. A small amount of white corrosion product began to appear on the surface after two weeks of immersion. However, there was no significant further increase of this corrosion product during the one-month test period. The uncoated magnesium alloy showed evidence of corrosion after 24 hours of immersion and severe pitting was observed after one month of immersion. The SEM image in Fig.7 shows the surface of the Keronite coating after one-month immersion in salt water.

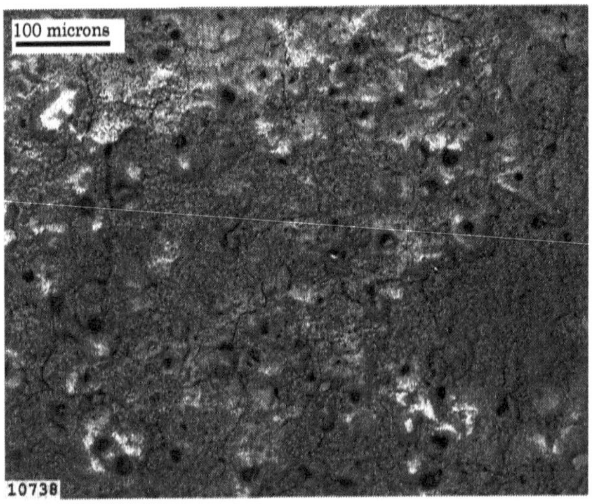

Figure 7: SEM image of the exposed area of the Keronite coated specimen after one-month exposure to 3.5% NaCl solution.

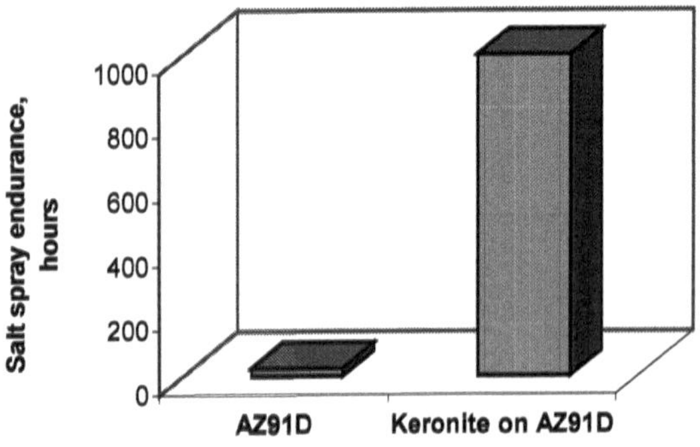

Figure 8: Salt spray (ASTM B117) endurance of the uncoated AZ91D and AZ91D with the 35μm Keronite coating.

Salt Spray Test

Salt spray corrosion test showed little visual evidence of corrosion attack on the surface of AZ91D coated with a 35μm Keronite coating after 1000 hours of exposure. In contrast, uncoated AZ91D displayed significant corrosion of the exposed surface after 24 hours of exposure. Figure 8 shows endurance of the coated and uncoated test coupons to the rating of 9 in accordance with ASTM D1654-92, procedure B. Surface appearance of the uncoated and Keronite coated test coupons after 1000 hours of salt spray exposure are shown in Fig.9.

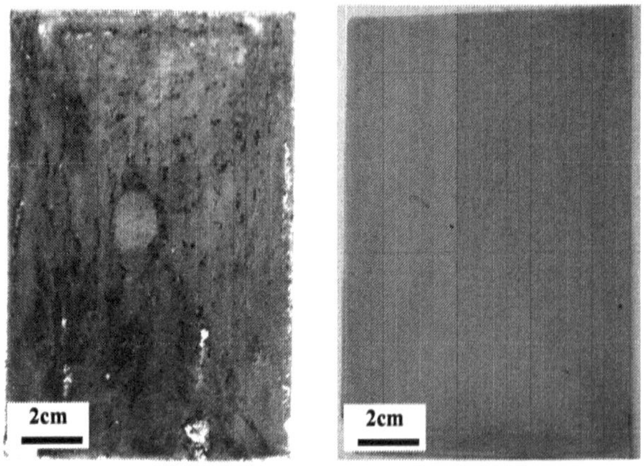

Figure 9: Optical image of the exposed surface of AZ91D (left) and AZ91D coated with 35μm Keronite coating (right) after 1000 hours of salt spray (ASTM B117) test.

Electrochemical Polarisation Test

Accelerated electrochemical test (anodic polarisation) results of AZ91D uncoated and coated with 35μm Keronite are shown in Fig.10. The results demonstrate contrasting behaviour of the coated specimens compared to uncoated alloy. The coated specimen displayed a more noble rest potential 'E_{corr}' compared to that shown by the uncoated specimen. A more noble rest potential in deaerated electrolyte generally signifies less susceptibility to corrosion attack.

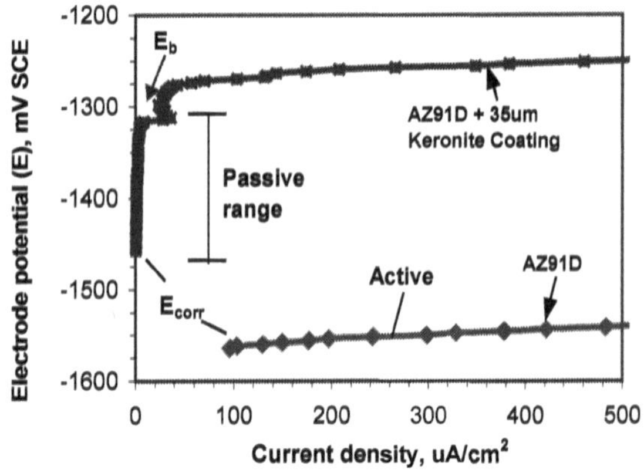

Figure 10: Anodic polarisation (ASTM G61) curves of uncoated AZ91D and AZ91D coated with Keronite in deaerated 3.5% NaCl solution at 25°C.

Moreover, the coated specimen displayed greater resistance to initiation of corrosion attack shown by the presence of a breakdown potential 'E_b' and a passive potential range 'E_b-E_{corr}'. A metal surface, which is not protective and corrodes actively, displays a curve similar to uncoated AZ91D with significant current within a small range of positive potential from its rest potential value.

Discussion

This work shows that the coating formed on the magnesium alloy mainly consists a ceramic layer of magnesium-aluminium oxide compound (Spinel). The coating microstructure in cross section also shows that the main layer of Keronite coating possesses little porosity. The bottom layer is relatively thin (1-2μm) but higher magnification SEM revealed that it is dense. Existence of multiple layers has been reported also for Keronite coatings on aluminium alloy substrates [8] and also on coatings [9] deposited with similar processes (micro-arc oxidation) on aluminium alloys. The presence of a dense ceramic layer on magnesium alloy would be expected to improve its resistance to corrosion attack. This was confirmed by the electrochemical test result with a presence of a passive region over a range of potential more positive to 'E_{corr}'. This was further supported by the resistance to corrosion attack during 1000 hours of salt spray exposure and one-month immersion in salt water. These results indicate the ability of the Keronite coating to act as a barrier coating to stop penetration of corrosive species. The surface of uncoated magnesium displayed both general type of corrosion as well as localised pitting.

Further work is now underway to look at the corrosion behaviour of Keronite coated magnesium alloys in a galvanic couple with steel.

Conclusions

- Microscopic observation showed that the Keronite coating has a relatively homogeneous coating structure with little porosity and multiple layers.
- The coating formed on a soft magnesium alloy is primarily a ceramic layer of Spinel ($MgAl_2O_4$)
- Electrochemical results suggest that the Keronite coating can improve the corrosion resistance of magnesium alloy AZ91D.
- The Keronite coating can survive one-month immersion in a saline solution and 1000 hours in a salt spray environment without significant visual evidence of corrosion attack.

References

1. M. Brettmann, B. Reinhold, and H. Schnattinger, "The use of magnesium in cars – Aspects of corrosion," Magnesium Alloys and Their Applications, ed. K. U. Kainer, (Wiley-VCH, Verlag GmbH, 2000), 271-273
2. B. Landkof, "Magnesium application in the electronic industries," Proc. 2nd Israeli Intl Conf. on Magnesium Science & Technology, ed. E. Aghion and D. Eliezer, (Magnesium Research Institute Ltd, Israel, 2000), 50-56.
3. G. B. Evans, "Applications of magnesium in aerospace," Magnesium Technology, Proc. London Conf., November 1986, (Institute of Metals, London, 1987), 103-109.
4. R. L. Edgar, "Global overview on demand and applications for magnesium alloys," Magnesium Alloys and Their Applications, ed. K. U. Kainer, (Wiley-VCH, Verlag GmbH, 2000), 3-8.
5. G. Govender and J. H. Ferreira, "Erosion corrosion of magnesium AZ91D in the as-cast condition," Proc. 2nd Israeli Intl. Conf. on Magnesium Science & Technology, ed. E. Aghion and D. Eliezer, (Magnesium Research Institute Ltd, Israel, 2000), 371-376.
6. R. Galun, A. Weisheit, and B. L. Mordike, "Properties of laser alloyed surface layers on magnesium base alloys," Magnesium Alloys and Their Applications, ed. B. L. Mordike, and K. U. Kainer, (Werkstoff-Informationsgesellsschaft mbH, Germany 1998), 439-444.
7. U. Lenz, A Weisheit, and B. L. Mordike, "Thermal spraying on the magnesium alloys AZ91D," Magnesium Alloys and Their Applications, ed. B. L. Mordike and K. U. Kainer, (Werkstoff-Informationsgesellsschaft mbH, Germany 1998), 445-450.
8. P. Shashkov, "Environment-friendly ceramic surface for aluminium," Aluminium Today, June-July 2000, 44-45.
9. J. Patel, "Micro-arc Oxidation Process Protects Aluminium Alloys," Materials Progress, (August, 2000).

Adhesive Bonding and Corrosion Protection of a Die Cast Magnesium Automotive Door

G. T. Bretz, K.A. Lazarz, D.J. Hill, & P.J. Blanchard

Ford Motor Co., Research & Advanced Engineering, 2101 Village Rd., Dearborn, MI 48121

Keywords: Magnesium, Conversion Coatings, Adhesive Bond, and Corrosion.

Abstract

It is well known that magnesium alloys, in close proximity to other alloys, are susceptible to galvanic corrosion. Combined with this fact, in automotive applications, it is rare that magnesium will be present in the absence of other alloys such as steel or aluminum. Therefore, in wet applications, where the galvanic cell is completed, it is necessary to isolate the magnesium in order to prevent accelerated corrosion. There are numerous commercial pre-treatments available for magnesium, however this paper focuses on conversion coatings in conjunction with a spray powder coat. By means of example, results for a hem flange joint on an AM50 die cast magnesium door structure will be presented. The outer door skin is an aluminum alloy hemmed around a cast magnesium flange. An adhesive is used between the inner and outer to help with stiffness and NVH (Noise, Vibration and Harshness). Results from bonded lap-shear coupon tests that have been exposed to accelerated corrosion cycles are presented. A second phase of this work considered a surrogate hem flange coupon, which was similarly exposed to the same accelerated corrosion cycle. Results from both of these tests are presented within this paper along with a discussion as to their suitability for use within automotive applications.

Introduction

Magnesium's lower density and high strength-to-weight ratio are appealing to automotive manufacturers since they are able to save weight while maintaining performance. This has resulted in a steady increase in applications over the last decade. There is also a clear trend towards application of magnesium within vehicle closure systems. Examples of concept and low volume closures include the VW Lupo liftgate, the "MAGDOOR" and VW Polo side door [1,2,3]. Perhaps the most prominent magnesium door structures have been those implemented by Mercedes for their CL and SL models.

While the use of magnesium provides an opportunity to reduce weight, new technical challenges are also introduced as the sub-assemblies often include metals far removed from magnesium on the electrochemical series. This potential for galvanic corrosion between magnesium and adjacent metals requires the adoption of protection strategies to provide dielectric insulation between components, thus ensuring long-term corrosion durability and joint integrity. From a body assembly viewpoint, magnesium is also susceptible to dissolution in automotive phosphate systems. This introduces the potential for bath contamination, thus requiring some form of protective measure when processing a door assembly through a conventional automotive phosphate and cathodic epoxy electrocoat (E-coat) system [4].

To identify suitable corrosion protection systems for the closure application shown in Figure 1, a series of experimental studies were performed.

Figure 1. Prototype magnesium door inner panel.

This paper focuses on an evaluation of corrosion performance for a series of commercially available magnesium pre-treatment and powder coat systems. Results from lap-shear adhesive bond tests and accelerated corrosion trials of surrogate hem flanges were used to identify suitable pre-treatments, powder coats and adhesives for future implementation.

Literature Review

Conversion Coatings for Magnesium

Conversion coatings are generally broken into two categories: chrome and non-chrome. Chrome conversion coatings have long been known to provide excellent corrosion protection. The chrome forms a hexavalent complex on the surface that acts as a 'self-healing' protective layer [5]. Small scratches to bare magnesium metal have been shown not to suffer from excessive corrosion undercutting when exposed to a corrosive environment [1]. However, chrome systems are detrimental to the environment and are subject to the European End-Of-Life vehicle directive (2000/53/EC), which is driving for their reduction and eventual elimination. Non-chrome treatments include; permanganate, fluorozirconate conversion coatings and organometallic titanium-zirconium coatings (Ti/Zr) [5]. Although all of these coating are conversion coatings, some researchers report that the pre-treatment process appears to be less robust than the more widespread hexavalent chromium systems [6]. A typical magnesium pre-treatment process is shown in Figure 2. While Step 5 provides the surface chemistry to promote adhesive bonding, the caustic and acid cleaning steps are critical to provide a stable surface condition [7]. Neglecting this fact is often manifested by excellent laboratory results not being demonstrated in a production environment.

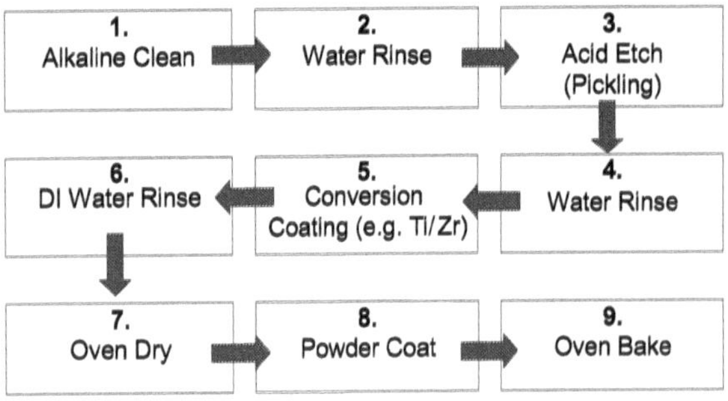

Figure 2. Typical pre-treatment and powder coating process.

Adhesive Systems for Magnesium

Chrome based pre-treatments have shown excellent corrosion performance and a robust foundation for adhesive bonding [8]. Pre-treating a magnesium substrate with a chromic acid etch followed by anodization in a chromate phosphoric acid bath (Dow 17 treatment) not only provides a robust solution to long-term corrosion durability, but also provides for excellent adhesive durability. It has been demonstrated that when used with the Dow 17 pre-treatment, a single-component (1K) nitrile rubber and phenolic blend adhesive film produced excellent long-term corrosion durability results. It was observed that adhesives with a lower tensile modulus produced better results than those with a higher modulus. Eickner proposed that the higher modulus adhesives resulted in higher stresses and hence premature failure of the brittle anodized treatments [8].

In contrast, there is very little published data on the performance of non-chrome based pre-treatments for adhesive bonding. Data exists on bonding aluminum using non-chrome pre-treatments, however, etch and desmutt procedures are suitably different to prevent any direct correlation between different metals. As part of a Brite-Euram EUCAR consortium project, Volvo investigated various pre-treatments for magnesium in conjunction with adhesive bonding for a magnesium door concept [9]. The work, once again demonstrated the robustness of chrome (NH35) pre-treatments as a preparation for adhesive bonding and their long-term corrosion durability through a variety of accelerated corrosion tests. The AM60 magnesium test coupons were bonded to an AC120 (AA6016) aluminum sample using a 1K toughened epoxy adhesive. The aluminum was pre-treated with a Ti/Zr pre-treatment prior to bonding. Nilsson reported that chrome (NH35), HAE anodized and phosphate permanganate pre-treatment on the magnesium produced excellent long-term corrosion results. Unlike the work performed by Schreckenberger et al [1], no powder coat was used on the magnesium. Instead, after bonding the test coupons were phosphated and E-coated.

Nilsson also identified the importance of mold release on the die cast AM60 magnesium coupons. Early experiments showed poor performance with all of the pre-treatments, which was attributed to poor cleaning of mold release agents prior to pre-treating. The only pre-treatment that did not appear to suffer from the presence of mold release was the HAE anodized samples. In a separate study, Volvo continued their work on evaluating different magnesium pre-treatments for adhesive bonding by investigating two-component (2K) adhesives [10]. No discernible differences were noted between the 1K adhesive results of the initial work and for the later phase evaluating 2K adhesives. Once again, the magnesium samples were not pre-treated with a powder coat on top of the conversion coating. Again the NH35 chrome and phosphate permanganate gave excellent long-term corrosion results with almost negligible reduction in the residual lap-shear strength. Interestingly, an observation that was made in the initial study and repeated in the second study was the poor long-term corrosion performance of the Ti/Zr (fluorozirconate) pre-treatment.

Skar and Albright [6] also noted that the chrome-free pre-treatments are more sensitive to the cleaning and oxidation steps prior to the application of the coating. When acid pickling was used for oxidation, the performance of the chrome-free pre-treatments approached the corrosion performance of chrome, but with alkaline cleaning, the chrome was far superior. Similarly, Balzer et al [11] recorded that long-term corrosion durability could be obtained on an automotive magnesium part using a Ti/Zr based conversion coating. In addition to the Ti/Zr conversion coating, Balzer used an epoxy powder coat to provide added protection as well as preventing dissolution within the phosphate bath. Balzer states that special care must be taken to remove all sources of surface contamination from the magnesium part prior to coating.

The durability of a bonded joint is a function of how the adhesive behaves and also the adhered and interfaces between them. After long-term exposure to a corrosive environment, failure usually occurs at one of the interfaces. Failure will occur within the weakest link in the system. For untreated bonded magnesium the weakest link, with respect to long-term corrosion durability is usually in the magnesium oxide, which forms on the surface. Moisture hydrates this oxide causing it to fail when exposed to stressed conditions. It is therefore standard practice to stabilize the oxide by a conversion coating or by anodizing prior to adhesive bonding. Bond failure usually occurs by corrosion undercutting of the adhesive, a suitable corrosion prevention system would arrest this failure mode. A dual protection system consisting of a conversion coating plus an organic powder coat has therefore been the focus of this study.

Material Selection and Experimental Methods

Two parallel test programs provided an opportunity to determine the performance of different coating systems and adhesives when subjected to accelerated corrosion testing. Adhesively bonded lap-shear samples provided a source of bond strength before and after corrosion cycling, while hemmed samples were examined by visual inspection. The 2mm thick pre-treated magnesium coupons were bonded to a 0.9mm thick 6111 aluminum coupon to replicate the material combination used on the prototype door. The intentional inclusion of untreated 6111 aluminum within the corrosion test chamber, gave rise to the formation of a galvanic corrosion cell between the two alloys if the pre-treatment failed. In total, three different conversion coatings were evaluated as candidate pre-treatments. In addition, two different powder coats, and finally, two different adhesives were used. A full test matrix showing the material combinations is contained in Table I. A final list identifying each batch of lap-shear coupons or hem samples is contained in Table II.

	Levels			DOE 1 Lap Shear Coupons	DOE 2 Hem Flange Coupons
Factors	1	2	3	Measurable	Measurable
Conversion Coating	Chrome	Organometallic Ti/Zr	Ti/Zr fluoronate	Initial LSS LSS after APGE Visual Inspection	Visual Inspection at 20, 40, 60 and 100 APGE cycles.
Powder Coat	Epoxy	Epoxy/Polyester Primer			
Adhesive	1K epoxy	2K epoxy			

Table I. Experimental test matrix

Group ID	Conversion Coating	Powder Coat	Adhesive
A	Chromium VI	Epoxy/Polyester Primer	1K epoxy
B	Chromium VI	Epoxy/Polyester Primer	2K epoxy
C	Organometallic Ti/Zr	Epoxy/Polyester Primer	1K epoxy
D	Organometallic Ti/Zr	Epoxy/Polyester Primer	2K epoxy
E	Ti/Zr fluoronate	Epoxy/Polyester Primer	1K epoxy
F	Ti/Zr fluoronate	Epoxy/Polyester Primer	2K epoxy
G	Ti/Zr fluoronate	Epoxy	1K epoxy
H	Ti/Zr fluoronate	Epoxy	2K epoxy

Table II. Group pretreatment/coating/adhesive combinations.

Materials

The AM50 magnesium test samples were high pressure die cast into 2.0mm X 25.4mm X 100mm coupons. Flash was removed from the coupons using a belt sander. All edges were dressed to remove sharp corners. A 6.35mm hole punched through the magnesium at one end (Figure 3) served as a point to hang the coupons while applying pre-treatments and coatings.

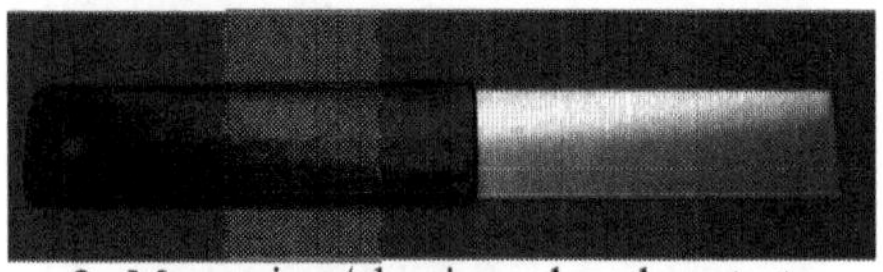

Figure 3. Magnesium/aluminum lap-shear test sample.

The aluminum was supplied by Alcan Products in sheet thickness of 0.9mm, AA6111-T4. Samples of the sheet were cut into 25.4mm X 100mm to match the size of the magnesium coupons. The samples were bonded in the "as received" condition that included a coating of the MP404 stamping oil. After baking of the bonded samples, the AA6111 aluminum was in a T6 temper.

Lap-Shear Adhesive Bond Test Samples

Adhesive was applied to the lap-shear joint, maintaining a bond area of 25.4mm x 12.7mm. Glass beads (0.25mm diameter) were used to control the bond gap. The aluminum coupons were prepared with the material rolling direction parallel to the tensile loading direction. Due to the high viscosity of the 1K-epoxy adhesive, the material was heated by placing the tube in an oven at 47°C until warm. A heat lamp was also used to warm the test coupons during sample preparation to prevent the adhesive from cooling. All adhesives were cured in a preheated laboratory oven at 180°C for 30 minutes. To replicate the assembly sequence, both the 1K and 2K bonded coupons were left for 24 hours at room temperature prior to baking. A typical lap-shear test sample is shown in Figure 3.

Lap-shear tests were carried out according to SAE J1523 [12]. Six bonded lap-shear coupons were tested from each group using an Instron 1127 universal test machine with a 10,000 kg load cell at a rate of 13mm/min. Samples were shimmed within the jaws to reduce peel loads. Following lap-shear testing the mean failure stress and standard deviation were recorded. The coupons were also examined visually to note the extent of corrosion.

Hem Flange Test Samples

Figure 4 shows a schematic of a hem flange section. The geometry is the reverse of a conventional door structure. Typically, the outer aluminum skin would be relatively flat and the die cast magnesium inner would be stepped to form the hem flange. This reversed geometry was chosen as it made use of the same simple (100mm x 25.4mm) flat magnesium coupons used for the lap-shear strength experiments. It was important to replicate the door geometry, so a trough was created for collection of electrolyte during testing. The punched hole in the end of the magnesium coupon simulated a drain hole in an actual door and enabled its effect on corrosion to be studied.

Adhesive was applied to the pre-bent flanged aluminum blank (Figure 5). After placement of the magnesium coupon, a table-top hemmer was used to form the completed hem.

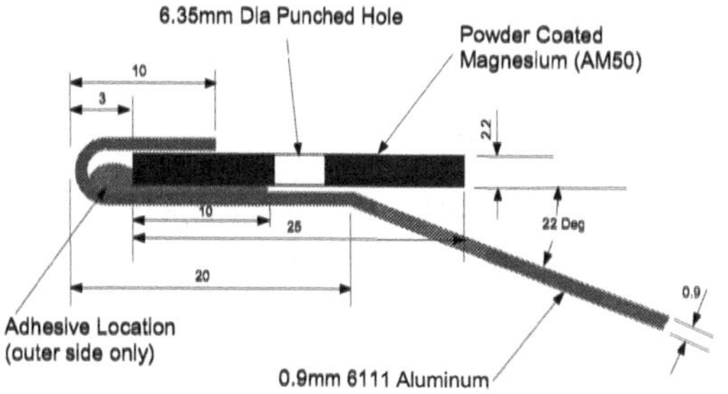

Figure 4. Schematic of a hem flange test coupon.

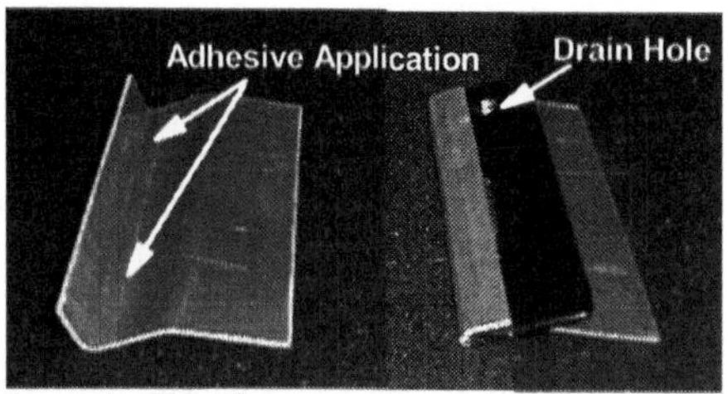

Figure 5. Hem flange test coupon (prior to paint).

After assembly and cure in a preheated laboratory oven at 180°C for 30 minutes, the hem flanges were passed through a regular phosphate and E-coat process (180°C for 20 mins), then primed, base coated and clear coated. Finally, samples were subjected to a low temperature paint bake cycle of 100°C for 45 mins. In the fully painted condition, the hemmed coupons were representative of a finished door hem excluding any over-hem sealer.

Evaluation of the hemmed samples was limited to a visual examination and subsequent ranking of the extent of corrosion. In addition to visual examination of the hemmed assembly, sections were taken from certain samples to determine adhesive distribution, powder coat deposition thickness and corrosion failure mechanisms. For this purpose, a JEOL JSM-840A Scanning Electron Microscope (SEM) with Energy Dispersive X-ray (EDX) was used to analyze the sectioned hems after corrosion exposure.

Corrosion Testing

Following sample preparation, corrosion cycling was performed according to the Arizona Proving Ground Equivalent (APGE) standard [13]. A single 24-hour corrosion cycle is shown in Table III. While each 24-hour period was considered a full cycle, phases 1 and 2 were omitted on weekends with samples remaining in the environmental chamber. A total of 5 complete cycles were executed each working week.

FLTM BI 123-01 APGE 24 hour Test Cycle	
Phase 1: 15 minutes	Immersion in 5% NaCl sol'n at R.T.
Phase 2: 75 minutes	Air dry at ambient temperature/humidity
Phase 3: 22.5 hrs	Environ. chamber, 50°C, 85% R.H.
Phase 4:	Application of kaolin powder (weekly)

Table III. Ford Motor Company APGE cyclic corrosion test.

The lap-shear specimens were subjected to 10, 20 and 30 cycles prior to testing. In contrast, the hem samples that received the additional paint layers were subjected to 100 cycles. Corrosion propagation was captured photographically every 20 cycles through to 100 cycles. Both sample groups were positioned in a horizontal orientation throughout all four phases of the corrosion test. For the hemmed test samples, the orientation simulated actual vehicle positioning, thus forming a trough for the water to drain through the punched hole.

Results

Lap-shear strength results

Figure 6 shows the initial failure stress of each system against the residual stress after 10, 20 and 30 APGE corrosion cycles. From the chart, it is clear that the performance of chromium containing systems with the 1K-epoxy adhesive (Group A) was superior to all other system combinations. The mean lap-shear peak stress and variability reduced only slightly during the corrosion exposure. The second chrome containing system, Group B, also performed well. However, bonded samples using the 2K-epoxy adhesive experienced a more significant drop of 40% in peak stress after exposure to 30 APGE cycles. This trend is also observed when comparing peak stress values for data sets, C/D, E/F and G/H. Therefore, based purely upon the lap-shear results, the 1K toughened epoxy adhesive appears to offer superior APGE exposure performance.

Examination of the lap-shear data for the Ti/Zr conversion coatings shows samples can be grouped into two broad performance bands. Groups C, D and E exhibited residual bond strengths approaching the value of the chrome conversion coating in Group B. However, the performance of the remaining samples in Groups F, G and H, deteriorated rapidly following APGE exposure. A visual inspection of the lap-shear coupons showed the most prominent failure mode within all of the chrome free pre-treatments tested was failure of the powder coat from the magnesium substrate. Undercutting corrosion of the powder coating was so severe in certain cases that no powder coat remained. Note that samples from Group F were so heavily corroded that no lap-shear data was taken for 30 APGE cycles (Figure 7).

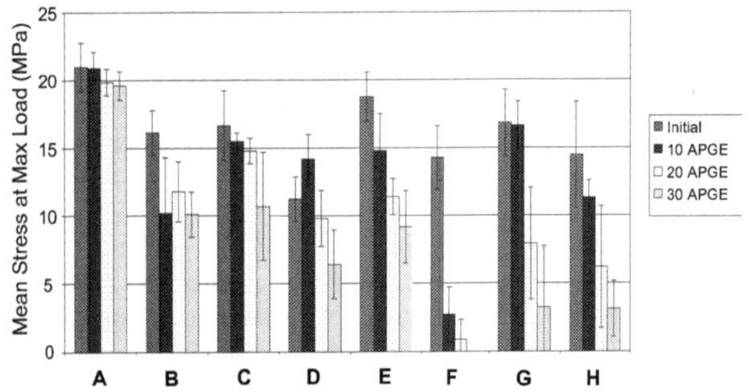

Figure 6. Mean lap-shear peak stress before and after corrosion testing.

In contrast, samples using chrome based conversion coatings (Groups A and B) exhibited limited corrosion over the bulk of the samples. Evidence of corrosion was present on sharp corners and other areas with potentially thin powder coat deposition (Figure 7), but propagation along the magnesium and power coat interface had not occurred. The effect of conversion coating on bond strength can be demonstrated by comparing systems A and E (Figures 6 and 7). Here the same adhesive and powder coat were used. The chrome treated coupons produced a mean lap-shear strength of 19.6MPa after 30 APGE cycles, compared to only 9.2MPa for the Ti/Zr fluoronate conversion coating (50% reduction). Once again, this shows a clear advantage for the chrome conversion coating over the chrome-free alternative.

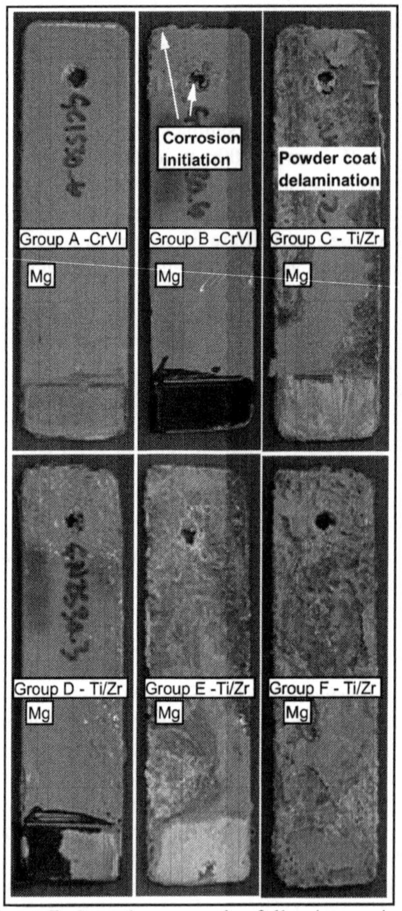

Figure 7. Lap-shear samples following testing.

Within the groups of chrome-free conversion coatings, a comparison of residual bond strength would suggest Group C offers the best performing system. However, Figure 7 places the bond strength data into the context of corrosion performance. As stated previously, the sample from Group B using the chrome conversion coating shows signs of corrosion, but this is limited to the sample edges and the punched hole. On the contrary, the sample from group D (using the same adhesive) shows signs of corrosion damage over the full width of the magnesium coupon. Group C, which offers the best residual lap-shear strength, shows signs of extensive corrosion propagation similar to Group F. In each case, the propagation of corrosion over the sample would appear to originate from the punched hole and sample edge, working along the sample toward the lap-shear joint. Therefore, in practice, the final lap-shear joint strength was influenced largely by the proximity of the corrosion initiation site to the bond area and the propagation rate. This leads to an argument suggesting that any discussions regarding selection of chrome-free conversion coatings should not be limited to an examination of the lap-shear data alone. Specifically, a more robust selection criteria would also consider the overall corrosion performance of the system.

Hem Flange Corrosion Test Results

As the lap-shear sample study indicated, the selection procedure for the pre-treatment and powder coat needs to consider the level of corrosion protection offered by the system. Cognizant of this, the hem flange corrosion study was conducted to highlight the robustness of the different corrosion protection systems to features typically found in a vehicle closure design. Initiation and propagation of corrosion within sample batches was tracked at 20 cycle increments until the study was halted at 100 cycles. Figure 8 shows a comparison of the hem samples A, B, E and F after 100 APGE cycles.

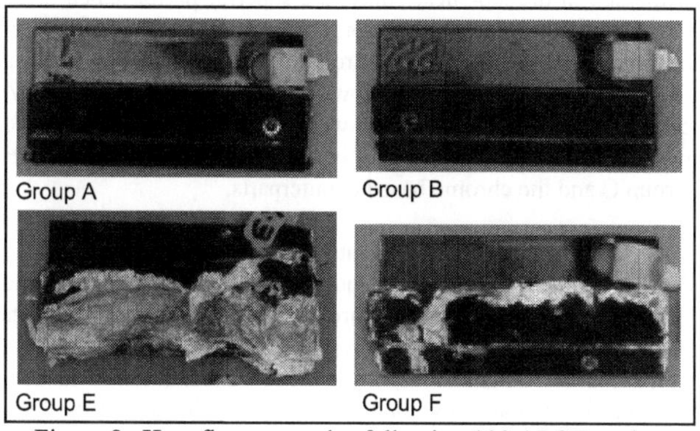

Figure 8. Hem flange samples following 100 APGE cycles.

Group A and B represent samples subjected to the hexavalent chromium pre-treatment and bonded with either 1K or 2K adhesives. As with the lap-shear samples, regardless of adhesive type, there is a clear distinction between the performance of the chrome and non-chrome conversion coatings tested. In the case of the Groups A & B, minimal or no corrosion product was visible on samples following the full 100 cycles. In contrast, the corrosion protection offered by non-chrome alternatives, E and F, was ineffective in preventing complete failure of the hem assembly. The organometallic Ti/Zr pre-treatment applied to groups C and D had mixed results. In some case, samples offered the same corrosion performance to the chrome pre-treatments. However, this was not consistent within the group, with some samples exhibiting severe corrosion damage while others remained relatively intact. A typical failure progression for group F is portrayed in Figure 9.

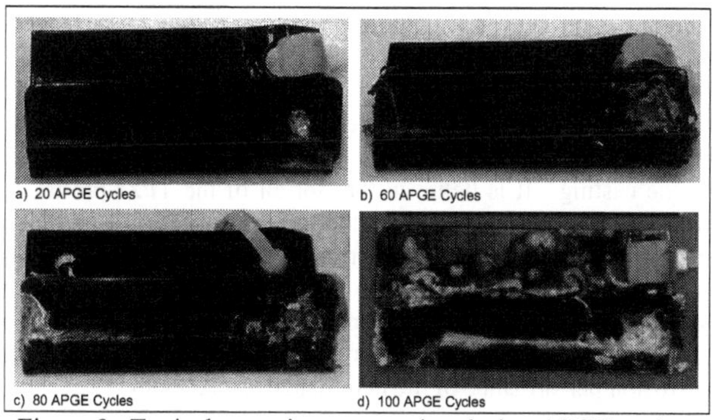

Figure 9. Typical corrosion progression during APGE testing.

The primary initiation site was within the 6.35mm drain hole (Figure 9a). This is consistent with the lap-shear coupons. Groups E and F showed evidence of corrosion at this location as early as 20 cycles into the test. The next stage of corrosion was perforation of the powder coat at the extreme ends of the sample, as shown in Figure 9b after 60 APGE cycles. Once the corrosion damage had reached this extent, severe undercutting of the powder coat continued along the sample compromising any protection that remained. No self-healing capability was exhibited by any of the chrome-free conversion coatings. On the contrary, once undercutting of the powder coat occurred, propagation along the sample appeared to accelerate non-linearly with each corrosion cycle.

The corrosion performance of the hem flange coupons were ranked by visual inspection and compared to lap-shear samples. This visual comparison is shown in Table IV. A clear correlation exists between the lap-shear coupon ranking and corrosion protection offered by the chrome based pre-treatments. In addition samples from Group F are clearly observed as underperforming the others. Comparison of ranking position for the other groups did highlight some inconsistencies. It is interesting to note the corrosion protection offered by pre-treatment H does not translate into equal ranking for lap-shear corrosion. In addition, when comparing Groups E and B, hem samples from Group E showed more severe signs of corrosion relative to other samples than in the lap-shear study.

Following a visual examination of the hemmed samples, sections were taken to identify the root cause for failure. Locations of interest were mid-way along the sample (minimum corrosion), and at the location of the drain hole (maximum corrosion). Figure 10 shows a section through a hem sample from Group B after 100 APGE cycles.

Ranking	Lap-Shear Coupons (After 30 APGE Cycles)	Hemmed Samples (After 100 APGE Cycles)
1	A, B	A, B
	C, D, E	H, D
		C, G
8	F,G, H	E, F

Table IV. Relative ranking of pretreatment and powder coat corrosion performance.

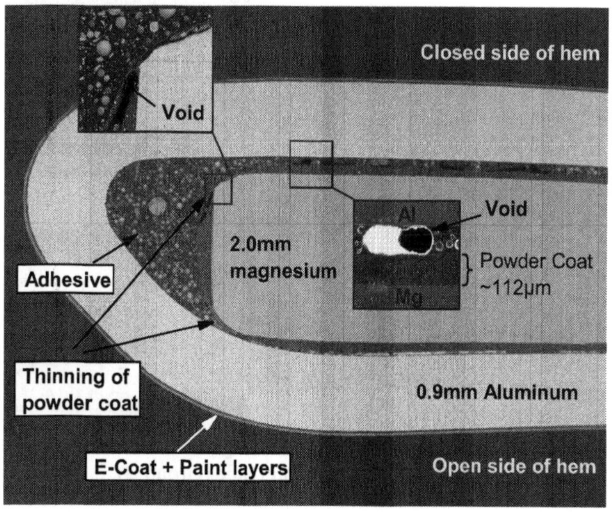

Figure 10. Section of hem sample from Group B taken at 15X.

Examination of the bond gap between the magnesium and aluminum shows the presence of adhesive throughout the hem to the open side of the flange. Thickness measurements show a nominal powder coat thickness of approximately 112µm. However, severe thinning is shown to occur at the edge of the magnesium. Measurements taken, indicate the thickness falls below 17µm regardless of the removal of sharp edges designed to assist the powder coat deposition. Of equal concern is the presence of voids in the powder coat layer, which are shown to reduce the effective coating thickness to less than 5µm in places. It is worth noting that fewer defects were observed along the planar surfaces of the magnesium.

Under 2000X magnification the chrome based conversion coating can be detected. The elemental map shown in Figure 11 shows a continuous layer of chrome, less than 1µm in thickness, at the interface between the magnesium and powder coat.

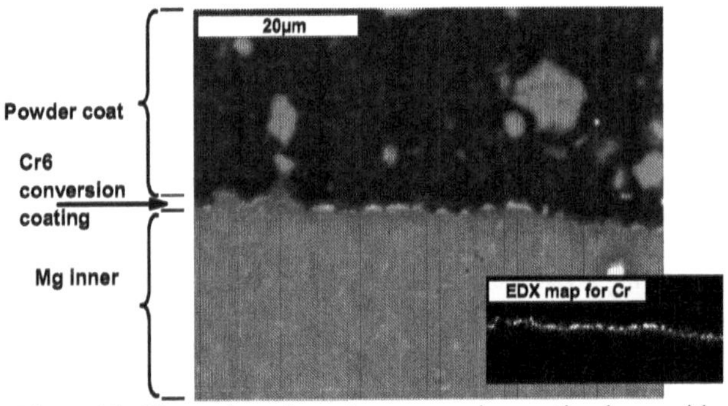

Figure 11. Hexavalent chromium conversion coating layer with X-Ray map (Group B), 2000X., SEI.

Figure 12 shows a low magnification secondary electron image taken through the drain hole in the hem flange sample. The paint layers applied following powder coat are clearly visible from the top left to bottom right of the inset image. The presence of the adhesive in the drain hole is a by-product of excess squeeze out during the hemming process. In practice this would need to be minimized to prevent blockage of the drain hole.

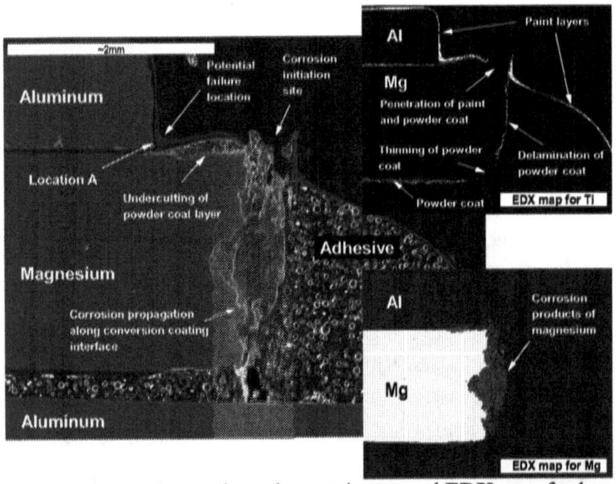

Figure 12. SEM secondary electron image and EDX map for hem section through the drain hole (Group B).

Consistent with the hem section shown in Figure 10 is the apparent thinning of the powder coat layer along the corners of the magnesium. This is observed by referring to the EDX elemental map for titanium. The presence of titanium is clearly visible at the locations of the paint and powder coat. Also marked is the initial point of coating penetration. Once the coating was breached at the corner, corrosion propagation was observed along the interface of the conversion coating. Undercutting of the powder coat and resultant delamination is shown clearly in the X-ray analysis for Ti in Figure 12. However, the extent of corrosion is still limited considering the sample was subjected to 100 cycles of corrosion exposure.

Since all hem flanges showed the onset of corrosion at the drain hole location, it can be proposed that the rate of corrosion propagation is strongly dependent on the choice of pre-treatment and the choice of powder coat is only a secondary effect. As witnessed in the lap-shear study, the superior performance of the hexavalent chromium conversion coatings is attributed to the self healing capabilities of the chrome based conversion coating. Group C samples, applied with the organometallic Ti/Zr conversion coating, performed well compared to other chrome-free alternatives, although there is still a clear distinction between Group C and the chrome based counterparts.

Discussion

Results from both lap-shear and hem flange studies show that Groups A and B, using the chrome based conversion coatings, offer the best protection of the systems tested. Switching between a 2K to a 1K toughened epoxy adhesive caused a shift in bond strength, but the adhesive had no apparent influence on the long-term durability of the joint. The choice of powder coat would also appear to have a negligible influence on the results.

The same generalization can be applied when considering the chrome-free conversion coatings. In all testing, the choice of conversion coating appeared to be the main factor influencing overall performance. A key observation made during testing of the Groups C to H was the broad range of results. While each of these groups included samples exhibiting extensive corrosion, in certain cases, particularly within groups using the organometallic Ti/Zr coating (C and D), the performance of the samples equaled that of the comparator Groups A and B. This is consistent with data reported by Skar and Albright [6] and Balzer et al [11]. Both publications indicate excellent corrosion performance can be achieved using chrome-free conversion coatings. However, Balzer emphasizes the need for special attention to cleaning and etching prior to pre-treatment. Bearing this in mind, discussions with the supplier of the Ti/Zr pre-treatment used within this study, suggests that one possible explanation for the poor performance of samples E through H was the etching of the magnesium prior to coating. As shown in Figure 2, an acid etch is typically used to remove the native magnesium oxide that is formed, prior to applying the Ti/Zr conversion coating. The acid concentration and residence time is dependent upon the oxide thickness or age of the casting. It is believed that for all of the Ti/Zr pre-treated samples tested, this oxide had not been sufficiently removed during the etch process. Unfortunately, being as both the lap-shear coupons and hem flange coupons were processed at the same time, the corrosion performance was similar. Therefore, the addition of the paint and clear coat, did little to improve the corrosion performance of the hem flange test coupons.

With the End-Of-Life European Directive looming, there is an urgent need for chrome-free pre-treatments for magnesium. The Ti/Zr conversion coatings appear to be those most widely promoted. However, the data presented within this paper shows that such systems can be less robust than traditional chrome based solutions. This suggests that further research is required to ensure that a robust system for adhesively bonded joints is available for widespread implementation, and therefore meet the objective of the European Directive.

Conclusions

- The most significant conclusion is the inconsistent performance of the Ti/Zr conversion coatings compared to the hexavalent chromium systems evaluated.

- Differences between the adhesive type and powder coat appear to be only secondary effects compared to conversion coat selection. While the addition of supplementary paint layers extended the life of samples, the ranking performance did not improve for the Ti/Zr converson coatings compared to the hexavalanet chromium.

- Initial lap-shear strength values cannot be used as a differentiator between different pre-treatments. Virtually no difference is seen in initial lap-shear strength between the treated samples and there is no correlation with corrosion performance.

- The lap-shear data after corrosion exposure only provides a limited indication of durability performance. A visual inspection of the magnesium substrate is also required. Corrosion undercutting of the powder coating occurs from specific locations that may not be adjacent to the bonded region and therefore have negligible effect on the lap-shear strength value.

- Powder coat performance is limited by the thinning that occurs at edges or features. It appears that even gently rounded corners suffer from powder coat thinning (up to 85%). Sharp edges produced as a result of punched holes within the magnesium are a cause for concern. Therefore, additional protection may be required to mitigate corrosion around design features such as drain holes and fastener attachments.

- Once corrosion initiation through the powder coat had occurred, the Ti/Zr conversion coatings suffered far more undercutting than the hexavalent chromium pretreatments. The performance and integrity of the powder coating is therefore more critical for the non-chrome based conversion coatings.

- It appears that the non-chrome conversion coatings are less robust, from a manufacturing perspective, than the more traditional hexavalent chromium pre-treatments.

Acknowledgements

Special thanks to Mike Kelly and Swati Saini at Ford Central Laboratories for carrying out the corrosion and lap-shear strength testing. Finally, Bill Donlon, Pete Friedman and Craig Miller all from Ford Research Laboratory are thanked for SEM analysis and assistance in manufacturing the hem flange test coupons.

References

1. **Schreckenberger, H., Papke, M. and Eisenberg, S.**, "The Magnesium Hatchback of the 3-liter Car: Processing and Corrosion Protection", Society of Automotive Engineers Technical Paper Series, 2000-01-1123.
2. **Jonason, P. et al**, "MAGDOOR – Magnesium in Structural Applications", Society of Automotive Engineers Technical Paper Series 1999-01-3198.
3. **Friedrich, H. and Schumann, S.** "Research for a new age of magnesium in the automotive industry", J. Materials Processing Technology, 117, 2001, pp. 276-281.
4. **Gao, G. and Ricketts, M.**, "Evaluation of Protective Coatings on Magnesium for Phosphate Process: Compatibility and Galvanic Corrosion Prevention". Society of Automotive Engineers Technical Paper Series, 2002-01-0081, .
5. **Gray, J.E and Luan, B.**, "Protective Coatings on Magnesium and its Alloys – A Critical Review", Journal of Alloys and Compounds, 336, 2002, pp. 88-113.
6. **Skar, J.L and Albright, D.**, "Emerging Trends in Corrosion Protection of Magnesium Die Castings", Magnesium Technology 2002. TMS 2002.
7. **Wernick, S., Pinner, R. and Sheasby, P.G.**, "The Surface Treatment and Finishing of Aluminum and Its Alloys", Fifth Ed., Vol. 1, ASM International.
8. **Eickner, H.W.**, "Effect of Surface Treatment on the Adhesive Bonding Properties of Magnesium", Forest Products Lab, Report 1865, 1958.
9. **Nilsson, P.** "Pretreatments of magnesium for bonding and painting in a conventional production process", Volvo Internal Report, ER-511441, 1999.
10. **Fröjd, A.**, "Adhesives for Magnesium, Results from the One Year Scab Test", Volvo Internal Report, LM-510994, 1998.
11. **Balzer J.S. et al**, "Structural Magnesium Front End Support Assembly", Society of Automotive Engineers Technical Paper Series, 2003-01-0186.
12. **SAE J1523** "Metal to Metal Overlap Shear Strength Test for Automotive-Type Adhesives", 1993.
13. **Ford Laboratory Test Method BI123-01**. "Painted Sheet Metal Corrosion Test (Laboratory Simulated Arizona Proving Ground Test)", 2001.

PARAMETRIC STUDY OF LASER CLADDING OF AS 21 MAGNESIUM ALLOY WITH ALUMINIUM SILICON/ TUNGSTEN CARBIDE POWDER

M Mandagie [1,2] M Brandt [1,2] Y Durandet[1,2] M Jahedi [1,3]

[1] CRC CAST Metals Manufacturing, [2] IRIS Swinburne University of Technology

[3] CSIRO Manufacturing and Infrastructure Technology

Abstract

Magnesium alloys are of growing interest for a number of applications in a range of industries because of their low densities and high specific strength. An impediment to their increased application is their low resistance to both wear and corrosion compared to that of steel or aluminium alloys. This is due to their relatively low surface hardness and high chemical affinity for numerous elements respectively.

This paper describes the results of experiments investigating Nd:YAG laser cladding of AS21 magnesium alloy with a mixture of aluminium silicon (40 % wt) and tungsten carbide (60 % wt) powder to improve its wear properties. The effects of laser power, scan speed, powder feed rate, and laser spot size on the clad layer thickness were examined. The results indicate that crack and porosity free clad layers up to 2 mm thick can be produced under optimum laser and powder mass flow rate conditions

.

1. Introduction

The light-weight properties and high strength-to-weight ratio of magnesium alloys compared to those of steel or aluminium have made them the material of interest for a number of applications in the automotive and aerospace industries (Zeumer, 1998). In the automotive industry, the drive to lower fuel consumption and emission of harmful pollutants has seen the range of automotive components made from magnesium alloys expand. The automotive components now made from magnesium alloys include the steering wheel, transmission housing, gear box housings, wheel rim and door inner (Zeumer, 1998). However, despite their attractive range of bulk material properties magnesium alloys have relatively low resistance to wear and corrosion which is a serious impediment to their wider application.

Wear and corrosion are essentially surface related phenomena that can be reduced or minimised by suitable tailoring of the surface microstructure and or composition without affecting the bulk. Laser surfacing is one technique that can be used to improve the surface properties of materials (Steen, 1998). Galun et al (1999) performed laser surface alloying of commercially pure magnesium, cast WE 54 T6 and extruded Al 80 and AZ 61 alloys using powders of aluminium, copper, nickel, or silicon with particle sizes of 10 to 45μm. Alloyed layers were produced with thicknesses of 500 to 1200 μm and were generally free of cracks and porosity. The highest wear resistance for laser alloying with single element was obtained with layers containing 30 to 33 % at of copper. With Al and Ni alloying elements, the wear depth produced using the multiple scratch test method was reduced by 90 % compared to untreated pure magnesium.

Hiraga et al (2000) used a CO_2 laser to clad AZ91E alloy substrates with powders of TiC, SiC and Al-30% Si aluminium alloy at varying powder feed rates. Their results show that the wear resistance of the coating, as measured by the pin-on-disk method, was significantly improved compared to the bulk and that further improvement is possible.

This paper reports on the results of laser cladding experiments investigating the use of WC particles in Al/Si matrix for improving the wear resistance of magnesium alloys. The main objective of the study was to determine the operating condition for producing crack and porosity free clad layer. The study has investigated the effects of laser power, scan speed, powder feed rate and laser spot diameter on the formation of the coatings.

2. Experimental Details

Experiments were performed using a mixture of Al/Si (40 % wt) and WC (60 % wt) powder with AS21 magnesium alloy substrate. The Al/Si powder consisted of 88 % Al and 12 % Si with an approximate powder particle size of 45 μm. The composition of the substrate is given in Table 1. The substrate was cut from cast ingots into 10 mm thick slices and its surface machined. Before cladding, the substrate surface was cleaned with alcohol.

Table 1. Composition of AS21 ingot cast substrate

Composition	Al	Zn	Mn	Si	Cu	Ni
AS 21	1.9 –2.5	0.15 0.20	0.35	0.70 – 1.2	0.04	0.002

The experiments were carried out using a 2.5 kW Nd:YAG laser equipped with a 10 m length of 0.6 mm diameter step-index glass fiber, terminated with a 200 mm focal length processing head. Figure 1 shows the Nd:YAG laser cladding set up. Argon was used as the shrouding gas at a flow rate of 2.75 l/min. To minimise the reaction of the substrate with air, a steel shielding box continuously purged with argon was also used in the cladding process. Two series of experiments were conducted, one involving single tracks of length 32 mm and the other overlapped tracks in order to produced a continuous clad layer. For the single track series, the main process parameters were laser power, spot size, powder feed rate and scan speed, as summarized in Table 2. The scan speed was increased in increments of 100 mm/min. For the overlapped track series, the main process parameters were laser power and scan speed as shown in Table 3.

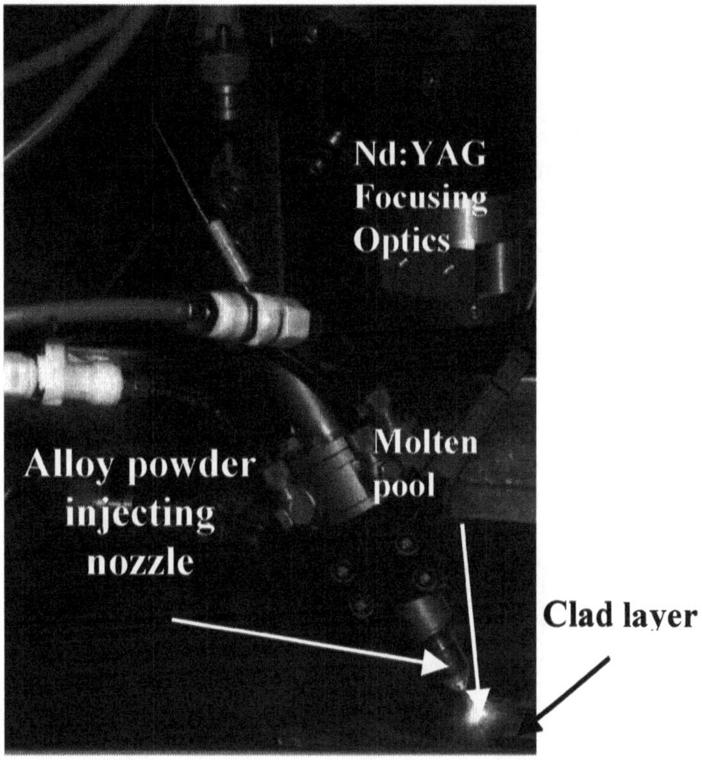

Figure 1. Nd:YAG laser cladding set up

Table 2. Processing parameters used to produce single track clad layer.

Clad sample	Laser Spot diameter (mm)	Laser Power (kW)	Powder Feed Rate (g/min)	Scan Speed (mm/min)
A	3.75	1.4	5.13	100 - 300
B	3.75	1.4	7.50	100 - 300
C	3.75	1.4	10.13	100 - 400
D	4.6	1.8	5.13	100 - 400
E	4.6	1.8	7.50	100 - 500
F	4.6	1.6	5.13	100 - 300

Table 3. Processing conditions used to produce clad layers with overlapped tracks.

Clad sample	Laser Power (kW)	Powder Feed Rate (g/min)	Scan Speed (mm/min)	Track increment pattern (mm)
A1, B1, C1	1.6	10.13	300 - 500	25 mm long x 1.5 increment x 14
D1	1.4	10.13	300	

After cladding, the samples were cut, cleaned with alcohol, cold mounted and polished to 1 µm for measurement and analysis.

3. Results and Discussion

3.1. Single Track Clad Layer

Figure 2 shows the appearance of some single clad tracks produced using processing parameters given in Table 2. The surface of each clad track is relatively rough and its width, when using the laser power of 1.8 kW, increases from the start to finish as the temperature of the substrate increases due to greater coupling of the beam. It is also observed that the formation of a clad layer is a function of scan speed, starting later at higher scan speeds. This again reflects on the coupling of laser light into the powder and substrate.

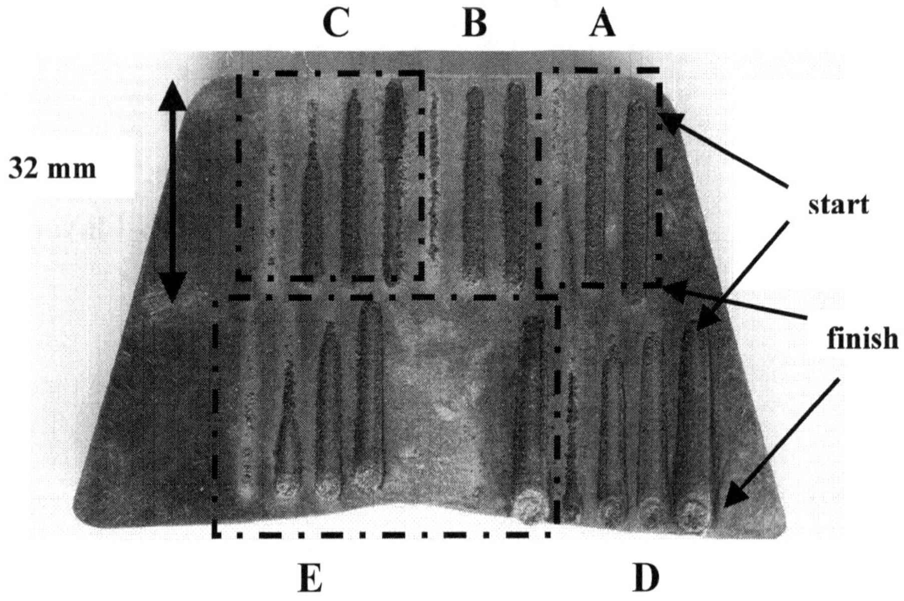

Figure 2. Surface appearance of single clad tracks produced under different processing parameters. For each sample group the scan speed increases from right to left.

It is evident that for each sample group, the faster the scan speed the less of the powder alloy is deposited onto the substrate. For example, for tracks in group A, produced using the laser beam spot size of 3.75 mm, laser power of 1.4 kW, and powder feed rate of 5.13 g/min, sufficient laser power and powder are available to form a clad layer at 100 mm/min, while at 300 mm/min the laser power and powder per unit length are insufficient to form a clad layer. The trend of decreasing layer thickness with increasing scan speed holds true for different laser power, powder feed rate, and laser spot size.

A typical cross section of a clad layer is shown in Figure 3 where it can be seen that significant melting of the substrate occurs. The amount of melting is a function of laser power, scan speed and powder mass flow rate, increasing with increasing laser power, decreasing scan speed and mass flow rate. These parameters also affect the distribution of WC particles in the clad layer. At lower scan speed and/or higher mass flow rate (Figure 3a) more WC particles are found in the clad layer. Also, these particles are located at the bottom of the clad layer. At higher scan speed and/or lower mass flow rate (Figure 3b) less WC particles are deposited per unit length and these particles are distributed more uniformly through out the clad layer.

The distribution of WC particles is governed by melt pool convection currents and its solidification time. Solidification time can be estimated from the dwell time, ratio of spot diameter to scan speed. Dwell time in Figure 3a is 27 ms, while in Figure 3b is 2.5 ms. As the density of WC particles at 15.7 g/cm^3 is about seven times greater that that of Al and Mg the WC particles will sink to the bottom of a melt pool at large dwell times as shown in Figure 3a. Due to the rapid stirring in the melt pool, at short dwell times the particles become trapped throughout the solidified clad layer depth.

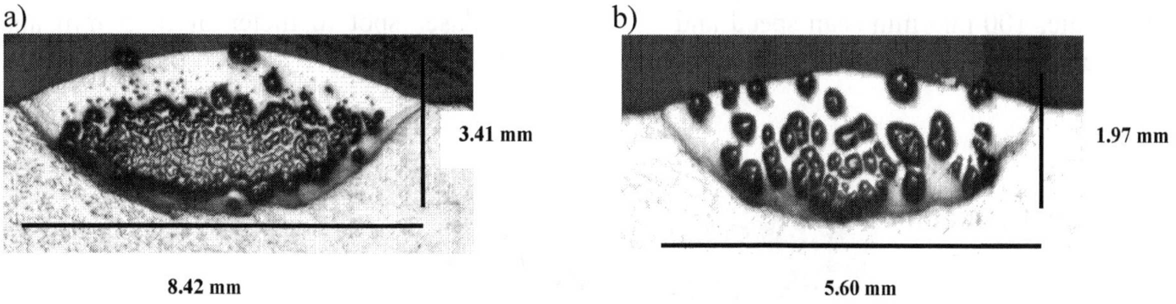

Figure 3. Effect of scan speed on clad layer formation for a) 100 mm/min and b) 300 mm/min for laser power of 1.8 kW, spot size of 4.6 mm and 5.13 g/min powder feed rate

Illustrated in Figure 4 is the dependence of clad thickness and width on scan speed for different laser powers. It can be observed that the layer thickness decreases from about 3.4 mm at 100 mm/min and 1.8 kW to 2 mm at 400 mm and 1.8 kW. The width (measured at half length) decreases from 6.1 mm at 100 mm/min and 1.8 kW to 4.7 mm at 400 mm/min and 1.8 kW. The trend of decreasing thickness with increasing scan speed is also observed for laser powers of 1.6 kW and 1.4 kW.

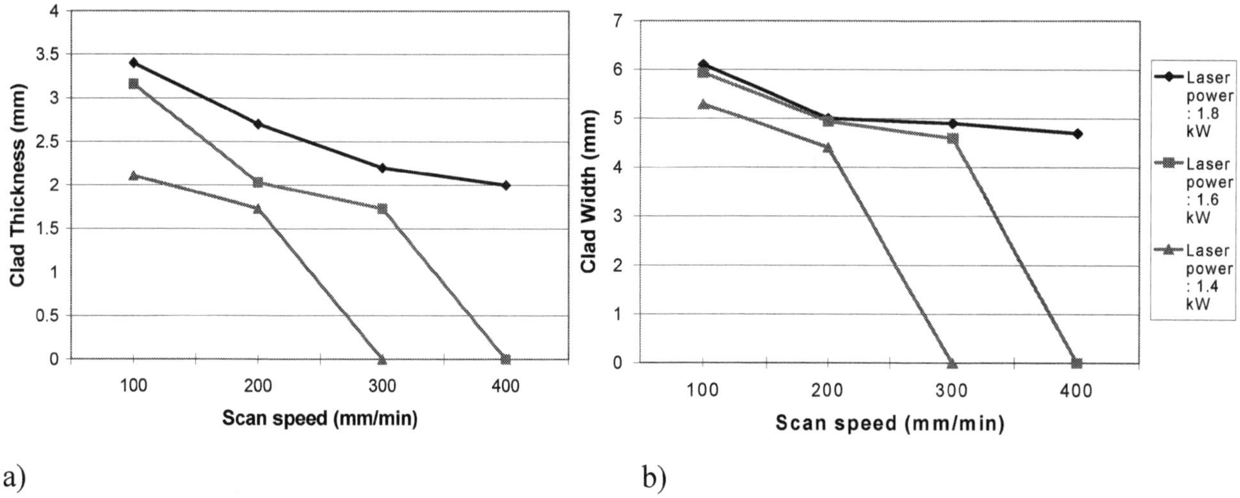

Figure 4. Effect of scan speed on a) layerl thickness and b)layer width for laser spot size of 3.75 mm and powder feed rate 5.13 g/min.

The increase in clad layer thickness with increasing laser power is due to the increase in the amount of substrate that is melted at higher laser power. The decrease in layer thickness with increasing scan speed is associated with decreasing line energy and powder per unit length available for fusion with the substrate. In addition, experiments indicate that the width of clad layer is also dependent on the laser spot size and hence power density. A clad track produced with 1.4 kW laser power, 5.13 g/min powder feed rate, 100 mm/min scan speed and laser spot diameter of 3.75 mm has a width of 5.29 mm. A track produced under the same laser processing parameter except for the laser spot size of 4.6 mm, has a width of 6.36 mm.

3.2. Multiple Track Cladding

Figure 5 shows the appearance of overlapped clad layers under different operating conditions using a laser spot diameter of 3.75 mm and increment between tracks of 1.5 mm.

A1 (400 mm/min)

B1 (500 mm/min)

C1 (300 mm/min)

D1 (300 mm/min)

Figure 5. Surface appearance of overlapped clad layers. Each clad layer is formed from right to left.

For the same laser power and powder feed rate, faster scan speed produced a smoother surface (Figure 5 B1) than the slower one (Figure 5 C1). However, at the faster scan speed more time is needed for the clad layer to be formed similar to single tracks. It is not until the end of the first track that the coating started to be formed.

The effect of laser power on clad layer structure for the constant operating condition is shown in Figure 6. It can be observed that laser power not only affects the clad layer thickness but also the distribution of WC particles within it.

In the case of higher power (Figure 6a), significant melting of the surface occurs and the WC particles are predominantly present at the bottom of the clad layer. Whereas, at the lower laser power, a thicker layer was produced with different distribution of the WC particles. In this case the layer is formed on the surface of the substrate and the WC particles are distributed more uniformly throughout the layer.

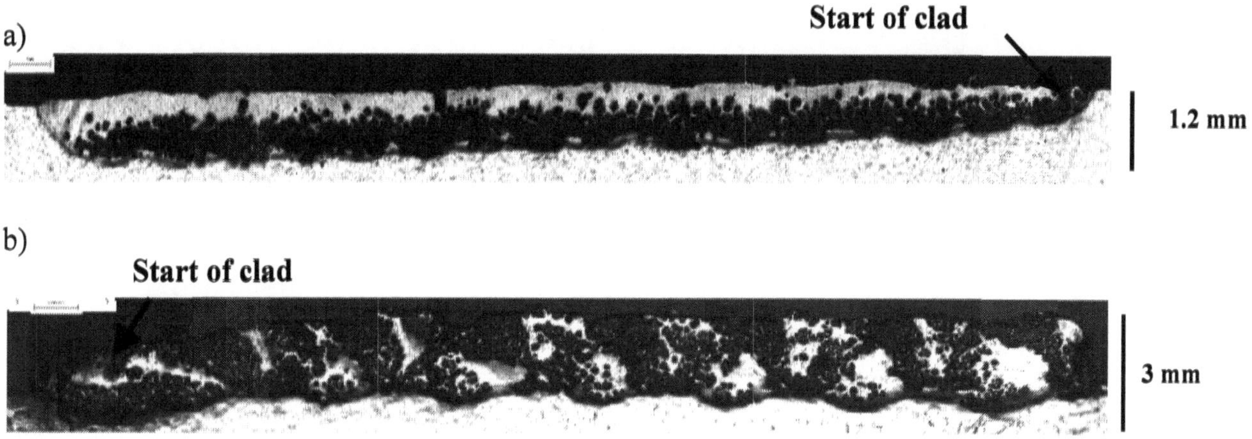

Figure 6. Effect of different laser power on clad layer formation for a) 1.6 kW and b) 1.4 kW for mass flow rate of 7.50 g/min and scan speed of 200 mm/min.

4. Conclusion

A series of experiments was conducted to investigate laser cladding of AS21 magnesium alloy substrate with a mixture of WC/AlSi powder. The results indicate that it is possible to deposit a crack and pore free clad layer. The result also indicates that there are relationships between laser processing parameters and the formation of the clad layer. The slower the scan speed, the thicker the total clad layer produced due principally to more energy deposited into the substrate. The distribution of tungsten carbide particles in the clad layer was shown to be effected by the duration the melt pool stays molten. Further work is in progress to produce samples for wear testing.

5. Acknowledgment

The CRC for CAST Metals Manufacturing (CAST) was established under the Australian Government's Cooperative Research Centres Scheme.

6. References

Galun, R., Weisheit, A., Mordike, B., *"Surface Treatment of Magnesium Alloys with Laser Melting and Laser Alloying"*, 5th European Conference on Laser Treatment of Material, Bremen-Vegesack, 1994.

Hiraga, H., Inuoe, T., Kojima, Y., Kamado, S., Watanabe, S., *"Surface Modification by Dispersion of Hard Particles on Magnesium Alloy with Laser*", 'Magnesium Alloy 2000': 2000, pp: 253-260.

Schuman, S., Friedrich, F., *" The Use of Magnesium in Cars-Today and in Future"*, Magnesium Alloy and Their Application Conference Proceeding, Wolsburg, Germany, 1998.

Subramanian, R., Sircar, S, Mazumder, J, *"Laser Cladding of Zr on Mg for Improved Corrosion Properties"*. Environmental Degradation of Ion and Laser Beam Treated Surfaces, The Minerals, Metals & Material Society, 1989.

Wang A.H and Yue, T.M, *" YAG Laser cladding of Al-Si Alloy on Mg/ SiC composite for the Improvement of Corrosion Resistance"*. Composite Science and Technology, 2001, vol 61, no 11 pp 1549 - 1554.

Zeumer, N., Honsel A.G, Arudy, "*Magnesium Alloy in the New Aeronautic Equipment"*, Magnesium Alloys and Their Application Conference, Germany, 1998.

CORROSION, SURFACE MODIFICATION, AND BIOCOMPATIBILITY OF MG AND MG ALLOYS

Sannakaisa Virtanen[1], Ben Fabry[2]
[1]University of Erlangen-Nuremberg, Institute for Surface Science and Corrosion, Department of Materials Science (WW-4), Martensstrasse 7, 91058 Erlangen, Germany
[2]University of Erlangen-Nuremberg, Center for Medical Physics and Technology, Department of Physics, Henkestraße 91, 91052 Erlangen, Germany

Keywords: Magnesium, corrosion, biodegradation, biocompatibility

Abstract

This manuscript summarizes recent studies on the corrosion behavior of Mg and Mg alloys in simulated biological environments, as well as interactions between corroding Mg (alloy) surfaces and cells. The influence of different types of simulated body solutions on the corrosion behavior of Mg is discussed. The effects of different types of chemical surface treatments on cell adhesion and spreading is presented. Moreover, possible routes to further optimize the corrosion and the biological performance of Mg alloys by surface modification are discussed.

Introduction

Mg-based materials corrode in aqueous biological environments and are of growing interest for use as biodegradable implants (see e.g. reviews [1,2]). For a successful and biologically safe application, a thorough understanding of the corrosion behavior in (simulated) physiological environments is required. In spite of the increasing number of publications in this field, many open questions still exist. In particular, detailed mechanisms of the effects of proteins and cells on Mg corrosion are unknown. Concerning the effects of proteins on the dissolution rate of Mg alloys, contradicting effects have been reported. For the Mg alloy AZ31, albumin addition has been found to show relatively little effect on the electrochemical behavior, but for pure Mg and the alloy LAE442 albumin addition led to strongly accelerated anodic dissolution [3]. On the other hand, corrosion inhibition by albumin addition has been reported for the AZ91 alloy in SBF [4], as well as for Mg-Ca alloy in water and in NaCl solutions [5]. These seemingly contradicting findings could be due to different experimental approaches, but may also indicate that albumin effects on corrosion are alloy-dependent.

Moreover, the effect of Mg dissolution on the biological environment needs to be characterized, as the corrosion reaction is coupled with H_2 gas liberation and surface alkalization. An increasing number of *in vitro* cell culture studies on corroding Mg alloy surfaces have been recently reported (see e.g., [6-11]. However, as different alloys and different cell lines have been used in these studies, it is not easy to compare the results from different studies or to draw general conclusions on the critical Mg alloy corrosion induced effects on the biological performance.

The aim of our studies is to further elucidate the interactions between corroding Mg (alloy) surfaces and cells, by an interdisciplinary approach of electrochemistry, surface modification and analysis, and cell culture testing. Of special interest is the development of novel chemical and biological surface functionalization methods, for tailoring the corrosion rate and biocompatibility.

Experimental

Experiments were carried out on cp-Mg (99.9%, Chempur) as well as on commercial WE43 and AZ91 Mg alloys. Corrosion behavior of the materials was studied in simulated body fluids (SBF 5[12]) and in cell culture medium [Dulbecco's Modified Eagle's cell culture medium (DMEM) with addition of 10% fetal bovine serum (FBS)].

Corrosion rates and modes were studied by Mg ion release measurements (atomic absorption spectroscopy), H_2 gas collection, pH measurements of the medium, electrochemical impedance spectroscopy, and various surface characterization methods.

Different types of surface treatments, including simple chemical passivation in 1 M NaOH or soaking in simulated body fluids, were explored regarding their ability to optimize the corrosion behavior and the biocompatibility. Moreover, functionalization of Mg surface by protein layers was explored. Surface coating with bovine serum albumin (BSA) was achieved by silanization of the surface with 3-aminopropyltriethoxysilane (APTES) and using an ascorbic acid (vitamin C) linker for attachment of BSA. For comparison, magnesium discs were directly incubated in aqueous BSA solution for 24 h at room temperature under gentle stirring of the solution. Details to the experimental procedures for preparation of the protein layers can be found in Ref. [13].

For the cell culture experiments, Mg samples were sterilized under UV irradiation with a wavelength of 260 nm. Cells from a human cervical cancer cell line (HeLa) were cultured in an incubator (5% CO_2, 37°C, 95% relative humidity) for 24 h in 24-well plates containing the differently treated Mg samples. DMEM with addition of 10% FBS and 100 U/ml Penicillin/Streptomycin was used as a cell culture medium. After 24 h in the incubator, the cells were fixed in 2% paraformaldehyde and stained with Alexa red phalloidin to visualize the actin cytoskeleton of the cells, and with Hoechst 33342 to visualize the cell nucleus. Fluorescence microscopy of the stained cells was carried out with a Leica DMI 6000B microscope.

Results and Discussion

Time-dependent corrosion behavior

Corrosion behavior of cp-Mg as well as Mg alloys was studied in SBF solutions as well as in cell culture medium. In agreement with findings from other laboratories [14,15], significantly lower corrosion rates were observed in cell culture medium than in SBF. The good corrosion resistance in cell culture medium has been brought into context of surface passivation by Mg-carbonates

[14,15]; however, detailed mechanisms on the reactions taking place on Mg surface in SBF vs. cell culture medium still need to be elucidated.

In previous studies, different types of simulated physiological solutions have been explored regarding their effect on the corrosion behavior of Mg alloys with an emphasis on the biodegradation behavior. A critical summary of the influence of the experimental parameters, including the electrolyte used, on the corrosion behavior of Mg alloys has been recently presented in [16]. A large number of investigations on Mg corrosion has been carried out in non-buffered NaCl solutions, however, these data have only limited value for understanding the corrosion behavior of Mg alloys in biological environments because body fluids are buffered. For Mg alloys, solution buffering is more critical than for many other metals, as a very strong alkaline pH shift takes place in the vicinity of corroding Mg surfaces. In non-buffered solutions, the pH can reach values >10-11 (starting from a neutral solution). In this case, the pH value is approaching the region of passivity of Mg [17]. Even though the presence of chloride ions in the environment hinders the formation of a stable passive surface, the high pH value suppresses the active dissolution rate. In simulated body fluids, buffered to pH 7.4, less strong pH effects occur. Therefore, different type of surface reactions take place on Mg in buffered or in non-buffered solutions.

Of course, not only buffering is important for the type of surface reactions taking place, but also the exact solution composition, as already discussed above in the case of SBF solutions versus cell culture medium. The nature of the corrosion layers changes from crystalline $Mg(OH)_2$ to an amorphous hydrated, carbonated (Mg,Ca)-phosphate in SBF [18]. The corrosion product layers from Mg and Mg alloys in SBF solutions are not highly protective, and active dissolution of Mg takes place even though the corrosion rate is gradually decreasing with time, with increasing coverage of the surface by the corrosion product layer [18,19].

Bovine serum albumin (BSA) addition showed a surprisingly complex time-dependent influence on the corrosion rate: in the first 6 h of immersion, a higher polarization resistance value was measured in presence of BSA in the solution, but during longer immersion times the polarization resistance again decreased, and for longer exposure times remained even lower than the polarization resistance in SBF in absence of albumin (Fig. 1). The initial increase of the polarization resistance could indicate adsorption of BSA on the surface, which partially blocks the dissolution reaction. The non-steady time-dependence suggests that the albumin adsorption layer changes with time, which is not surprising considering that the Mg alloy surface dissolves with a relatively fast rate during the experiment. This could for instance lead to removal of albumin from the surface. However, it is not clear why in the longer-term experiment the R_p-values remain significantly lower in the presence of albumin in the solution than in albumin-free solution. Clearly, a detailed surface characterization after different exposure times is required to understand the effects of albumin on the electrochemical behavior of Mg.

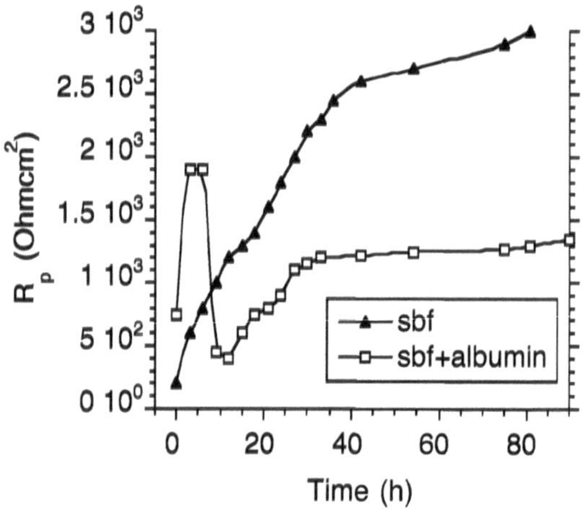

Fig. 1 Polarization resistance (R_p) values as a function of time in solution for the Mg alloy WE43. The R_p-values were determined from electrochemical impedance spectroscopy measurements in sbf or in sbf + 40 g/l bovine serum albumin (BSA) at 37°C.

Cell culturing on corroding Mg surfaces

For conventional cell culture experiments with direct adhesion of the cells on the sample surface, the surface of cp-Mg, without any surface modification was far too reactive for cell adhesion and survival. Even though strong retardation of the corrosion rate takes place for longer immersion times of Mg in cell culture medium, the initial corrosion rate was very high. Immediate hydrogen gas liberation and alkalinization of the culture medium could be observed. Cytotoxicity testing indicated that high Mg-concentrations - corresponding to the dissolution of the samples in the cell culture medium - did not reduce cell survival [20]. Rather, the single most important cause for reduced cell adhesion and survival was the increased pH in the cell culture medium due to Mg corrosion [21].

A viable strategy to achieve cell adhesion on corroding Mg (alloy) surfaces under *in vitro* cell culture conditions is to reduce the initial surface reactivity [22]. For instance, Figure 2 shows the effect of simple Mg surface passivation by soaking in 1 M NaOH. On the NaOH passivated surface, good spreading of the cells can be seen (2a). This in contrast to the polished Mg sample, in this case only very few non-spread cells were seen on the surface (2b). Although the $Mg(OH)_2$ passive film formed on Mg in NaOH is not stable in cell culture medium, it provides sufficient short-term protection to enable initial cell adhesion.

In general, all surface treatments that reduced the initial reactivity of the surface also increased the cell survival rate in our studies, but distinct differences were observed between different types of surface treatments as they altered not only surface reactivity but also chemistry, roughness, and wettability. For example, Mg samples incubated in SBF showed a biphasic behavior. During incubation in SBF, the Mg surface was covered by an amorphous hydrated, carbonated (Mg,Ca)-phosphate layer. Monitoring the cell behavior on such a surface as a function of time showed that initially strong cell spreading took place, but over time cell death occurred. These findings indicate that the surface chemistry and roughness of coatings formed by soaking in SBF are initially beneficial for cell adhesion and spreading, but after longer times the high reactivity of the surface induces cell death, presumably due to the local surface pH increase as these coatings are porous and hence not highly protective. The best coating, until now, in view of an efficient reduction of the corrosion rate and a good biocompatibility was the layer formed by incubating the Mg sample in cell culture medium [20].

Electrochemical experiments (open-circuit potential measurements and impedance spectroscopy EIS) of Mg surfaces were carried out in the absence and presence of a cell adhesion layer on the surface [21]. The electrochemical response correlated with the spreading of the cell layer on the surface. The formation of a cell layer on the Mg alloy surface significantly reduced the corrosion rate, as indicated by a reduced pH increase with time for Mg samples covered with a cell layer.

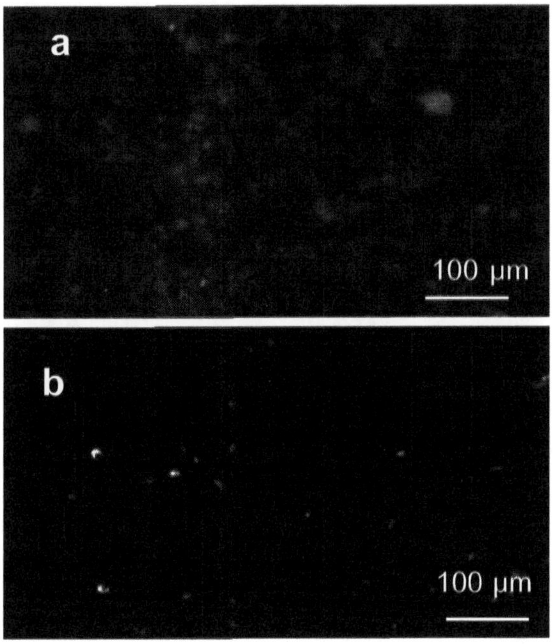

Fig. 2 Actin staining of HeLa cells on cp-Mg surfaces after 24 h culture. a: Mg surface passivated for 24 h in 1 M NaOH prior to cell culture testing; b: polished Mg surface.

Functionalization of Mg surface by proteins

As shown in Fig. 1, protein adsorption on Mg surfaces can have a significant effect on the corrosion behavior. We further studied protein adsorption on Mg surfaces, and also explored the possibility to use protein adsorption layers as a pre-treatment for tailoring the corrosion behavior and biocompatibility. First trials to simply coat Mg by directly soaking in BSA containing aqueous solutions were only partially successful; even though BSA adsorption took place, this was accompanied by corrosion (and H_2 liberation), and hence no homogeneous albumin layer could be prepared on the surface. To achieve a homogeneous protein coating on Mg surfaces, we therefore exploited silane coupling chemistry of a OH^- -terminated Mg surface. To enhance the OH^- termination, pre-passivation in NaOH is feasible, but the method works also for samples simply passivated in air (native oxide layers formed in air after surface polishing). The success of albumin-coating was verified by XPS (X-ray photoelectron spectroscopy) and ToF-SIMS (Time-of-flight Secondary Ion Mass Spectroscopy) measurements [13].

One advantage of such selective protein adsorption is to prevent non-selective protein adsorption from body fluids, and hence to condition the surface to the desired cell reaction. In our work, we are also interested in the effect of such protein coatings on the corrosion behavior. All experiments until now indicate that homogeneous protein coatings can be highly efficient in decreasing the corrosion rate. An example is shown in Fig. 3, illustrating the cumulative H_2 gas liberation as a function of time for albumin-coated Mg, in comparison to ground Mg surface. An efficient protection of the Mg surface by the albumin layer is observed, and the protective effect remains clearly detectable for a period of 5 days at least (longer-term experiments are running, results not shown). The exact amount of H_2 liberated from protein-covered samples varies, depending on the details of the surface pre-treatments. But it is also evident from Fig. 3 that no complete protection of the surface can be achieved, i.e., slow dissolution still takes place, which is desired for the material to be biodegradable.

This approach is currently being further explored on different proteins and different Mg alloys. Moreover, optimization of the protein coatings by a variation of surface pre-treatments, linker molecules, and soaking time is being carried out. Especially in the case of Mg alloys, specific pre-treatments of the heterogeneous multiphase alloy surface may be required, to ensure a homogeneous reactivity towards the linker molecules in the first coating step. For the optimized (homogeneous and well-structured) protein coatings, *in vitro* cell culture studies with suitable cell lines will be carried out in order to study the biological performance of the layers.

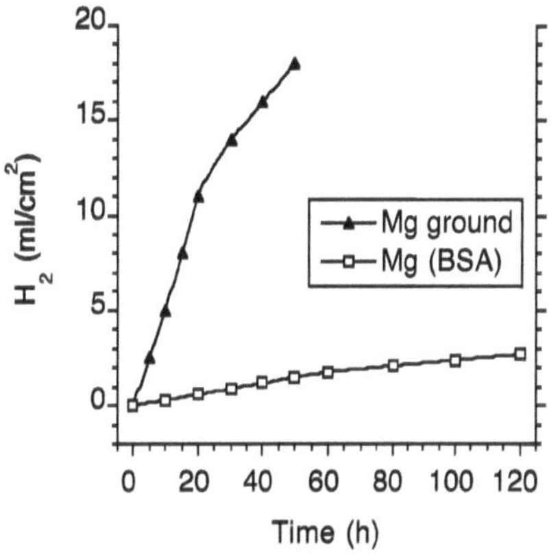

Fig.3 H_2 gas volume as a function of time for ground Mg sample and an albumin-coated Mg sample exposed to SBF.

References

1. M.P. Staiger, A.M. Pietak, J. Huadmai, G. Dias, Magnesium and its alloys as orthopedic biomaterials: a review, *Biomaterials* 27 (2006)1728-1734

2. F. Witte, N. Hort, C. Vogt, S. Cohen, K.U. Kainer, R. Willumeit, F. Feyerabend, Degradable biomaterials based on magnesium corrosion, *Current Opinion in Solid State and Materials Science* 12 (2008) 68-72.

3. W.-F. Mueller, M.F. L. de Mele, M. L. Nascimento, M. Zeddier, Degradation of magnesium and its alloys: Dependence on the composition of the synthetic biological media, *J. Biomed. Mater. Res.* 90A (2009) 487-495.

4. C. Liu, Y. Xin, X. Tian, P.K. Chu, Degradation susceptibility of surgical magnesium alloy in biological fluid containing albumin, *J. Mater. Res.* 22 (2007) 1806-1814.

5. C.L. Liu, Y.J. Wang, R.C. Zheng, X.M. Zhang, W.J. Huang, P.K. Chu, In vitro corrosion degradation behaviour of Mg-Ca alloy in the presence of albumin. *Corrosion Science* 52 (2010) 3341-3347.

6. L. Li, J. Gao, Y. Wang, Evaluation of cyto-toxicity and corrosion behavior of alkali-heat-treated magnesium in simulated body fluid. *Surface & Coating Technology* 185 (2004) 92-98.

7. Y. Zhang, H.-R. Tao, Y.-H. He, G.-Y. Zou, Y. Jian, S.-X. Zhang, B.-L. Zhang, J.-N. Li, C.-L. Zhao, X.-N. Zhang, Cytotoxicity and hemolytic properties of biodegradable Mg-Zn alloy, *Journal of Clinical Rehabilitative Tissue Engineering* 12 (2008) 8162 – 8166.

8. X. Gu, Y. Zheng, Y. Cheng, S. Zhong, T. Xi, In vitro corrosion and biocompatibility of binary magnesium alloys, *Biomaterials* 30 (2009) 484-498.

9. Y.H. Yun, Z. Dong, D. Yang, M.J. Schulz, V.N. Shanov, S. Yarmalenko, Z. Xu, P. Kumta, C. Sfeir, Biodegradable Mg corrosion and osteoblast cell culture studies, *Materials Science and Engineering C* 29 (2009) 1814-1821.

10. E. Zhang, D. Yin, L. Xu, L. Yang, K. Yang, Microstructure, mechanical and corrosion properties and biocompatibility of Mg-Zn-Mn alloys for biomedical application, *Materials Science and Enigineering C* 29 (2009) 987-993.

11. S. Zhang, J. Li, Y. Song, C. Zhao, X. Zhang, C. Xie, Y. Zhang, H. Tao, Y. He, Y. Jiang, Y. Bian, In vitro degradation, hemolysis and MC3T3-E1 cell adhesion of biodegradable Mg-Zn alloy, *Materials Science and Engineering C* 29 (2009) 1907-1912.

12. L. Müller, F.A. Müller, Preparation of SBF with different HCO_3^- content and its influence on the composition of biomimetic apatites. *Acta Biomaterialia* 2 (2006) 181-189.

13. M. Killian, V. Wagener, P. Schmuki, S. Virtanen: Covalent functionalization of metallic magnesium with protein layers, *Langmuir* 26 (2010) 12044-12048.

14. X.N. Gu, Y.F. Zheng, L.J. Chen, Influence of artificial biological fluid composition on the biocorrosion of potential orthopedic Mg-Ca, AZ31, AZ91 alloys, *Biomedical Materials* 4 (2009) 1-8.

15. A. Yamamoto, S. Hiromoto, Effect of inorganic salts, amino acids and proteins on the degradation of pure magnesium in vitro, *Materials Science and Engineering C* 29 (2009) 1559-1568.

16. W.-D. Mueller, M. Lucia Nascimento, M. Fernández Lorenzo de Mele, Critical discussion of the results from different corrosion studies of Mg and Mg alloys for biomaterial applications, *Acta Biomaterialia* 6 (2010) 1749–1755.

17. M. Pourbaix (1974) *Atlas of Electrochemical Equilibria in Aqueous Solutions, 2nd ed.* NACE, Houston.

18. R. Rettig, S. Virtanen, Composition of corrosion layers on a magnesium rare-earth alloy in simulated body fluids, *J. Biomed. Mater. Res. 88A* (2009) 359-369.

19. R. Rettig, S. Virtanen, Time-dependent electrochemical characterization of the corrosion of the magnesium alloy WE43 in simulated body fluids, *J. Biomedical Materials Research A* 85A (2008) 167-175.

20. S. Keim, J.G. Brunner, B. Fabry, S. Virtanen, Control of magnesium corrosion and biocompatibility with biomimetic coatings, *J. Biomedical Materials Research B* (in press, 2010).

21. F. Seuss, S. Keim, M. Can Turhan, B. Fabry, S. Virtanen (to be submitted)

22. C. Lorenz, J. Brunner, P. Kollmannsberger, L. Jaafar, B. Fabry, S. Virtanen, Acta Biomaterialia 5 (2009) 2783-2789.

AUTHOR INDEX
Essential Readings in Magnesium Technology

A

B

C

D

E

F

G

H

I

T

U

V

W

X

Y

Z

SUBJECT INDEX
Essential Readings in Magnesium Technology

A

B

C

D

E

F

G

H

I

J

L

M

N

O

P

R

S

T

U

V

W

MIX
Papier aus verantwortungsvollen Quellen
Paper from responsible sources
FSC® C105338

If you have any concerns about our products,
you can contact us on
ProductSafety@springernature.com

In case Publisher is established outside the EU,
the EU authorized representative is:
Springer Nature Customer Service Center GmbH
Europaplatz 3, 69115 Heidelberg, Germany

Printed by Libri Plureos GmbH
in Hamburg, Germany